Integrated SCIENCE

Sixth Edition

Bill W. Tillery
Arizona State University

Eldon D. Enger
Delta College

Frederick C. Ross
Delta College

McGraw Hill

Connect
Learn
Succeed™

INTEGRATED SCIENCE, SIXTH EDITION

Published by McGraw-Hill, a business unit of The McGraw-Hill Companies, Inc., 1221 Avenue of the Americas, New York, NY 10020. Copyright © 2013 by The McGraw-Hill Companies, Inc. All rights reserved. Previous editions © 2011, 2008, 2007, and 2004. Printed in the United States of America. No part of this publication may be reproduced or distributed in any form or by any means, or stored in a database or retrieval system, without the prior written consent of The McGraw-Hill Companies, Inc., including, but not limited to, in any network or other electronic storage or transmission, or broadcast for distance learning.

Some ancillaries, including electronic and print components, may not be available to customers outside the United States.

This book is printed on acid-free paper containing 10% postconsumer waste.

1 2 3 4 5 6 7 8 9 0 DOW/DOW 1 0 9 8 7 6 5 4 3 2

ISBN 978-0-07-351225-9
MHID 0-07-351225-7

Vice President & Editor-in-Chief: *Marty Lange*
Vice President & Director Specialized Publishing: *Janice M. Roerig-Blong*
Publisher: *Ryan Blankenship*
Sponsoring Editor: *Todd Turner*
Director of Developmental: *Kristine Tibbets*
Senior Developmental Editor: *Mary E. Hurley*
Senior Marketing Manager: *Lisa Nicks*
Senior Project Manager: *Joyce Watters*
Buyer: *Laura Fuller*
Lead Media Project Manager: *Christina Nelson/Judi David*
Design Coordinator: *Brenda A. Rolwes*
Cover Designer: *Studio Montage, St. Louis, Missouri*
(USE) Cover Image: *Person Trekking Uphill in Snow:* © Brand X Pictures/Superstock; *Hot Spring at Yellowstone:* © Royalty-Free/CORBIS; *Girl with Bag of Recyclable Plastic Bottles:* © Cultura/Getty Images RF; *Scientists Looking at DNA Model:* © OJO Images/Getty Images RF.
Photo Research: *David Tietz/Editorial Image, LLC*
Compositor: *Lachina Publishing Services*
Typeface: *10/12 Minion*
Printer: *R. R. Donnelley*

All credits appearing on page or at the end of the book are considered to be an extension of the copyright page.

Library of Congress Cataloging-in-Publication Data

Tillery, Bill W.
   Integrated science / Bill W. Tillery, Eldon D. Enger, Frederick C. Ross.—6th ed.
      p. cm.
   Includes index.
   ISBN 978-0-07-351225-9 (alk. paper)
      1. Science—Textbooks. I. Enger, Eldon D. II. Ross, Frederick C. III. Title.
Q161.2.T54 2013
500—dc23

                              2011045420

www.mhhe.com

Contents

18

Earth's Waters   413

19

Organic and Biochemistry   434

20

The Nature of Living Things   464

21

The Origin and Evolution of Life   496

CHAPTER 22

The History of Life On Earth 525

CHAPTER 23

Ecology and Environment 562

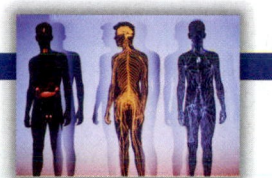

CHAPTER 24

Human Biology: Materials Exchange and Control Mechanisms 597

CHAPTER 25

Human Biology: Reproduction 635

CHAPTER 26

Mendelian and Molecular Genetics 665

Preface

WHAT SETS THIS BOOK APART?

Creating Informed Citizens

Integrated Science is a straightforward, easy-to-read, but substantial introduction to the fundamental behavior of matter and energy in living and nonliving systems. It is intended to serve the needs of nonscience majors who must complete one or more science courses as part of a general or basic studies requirement.

Integrated Science provides an introduction to a scientific way of thinking as it introduces fundamental scientific concepts, often in historical context. Several features of the text provide opportunities for students to experience the methods of science by evaluating situations from a scientific point of view. While technical language and mathematics are important in developing an understanding of science, only the language and mathematics needed to develop central concepts are used. No prior work in science is assumed.

Many features, such as Science and Society readings, as well as basic discussions of the different branches of science help students understand how the branches relate. This allows students to develop an appreciation of the major developments in science and an ability to act as informed citizens on matters that involve science and public policy.

"I especially like the application of the concepts and the connections in this text. We try very hard to show that science has connections to the everyday world and why it's important to see those connections. I don't think science can be taught to nonscience majors unless this type of approach is taken."

—Richard L. Kopec, St. Edward's University

Flexible Organization

The *Integrated Science* sequence of chapters is flexible, and the instructor can determine topic sequence and depth of coverage as needed. The materials are also designed to support a conceptual approach or a combined conceptual and problem-solving approach. The *Integrated Science* ARIS Instructor's Resources offer suggestions for integrating the text's topics around theme options. With laboratory studies, the text contains enough material for the instructor to select a sequence for a one- or two-semester course.

THE GOALS OF *INTEGRATED SCIENCE*

1. **Create an introductory science course aimed at the nonscience major.** The origin of this book is rooted in our concern for the education of introductory-level students in the field of science. Historically, nonscience majors had to enroll in courses intended for science or science-related majors such as premeds, architects, or engineers. Such courses are important for these majors but are mostly inappropriate for introductory-level nonscience students. To put a nonscience student in such a course is a mistake. Few students will have the time or background to move through the facts, equations, and specialized language to gain any significant insights into the logic or fundamental understandings; instead, they will leave the course with a distaste for science. Today, society has a great need for a few technically trained people but a much larger need for individuals who understand the process of science and its core concepts.

2. **Introduce a course that presents a coherent and clear picture of all science disciplines through an interdisciplinary approach.** Recent studies and position papers have called for an interdisciplinary approach to teaching science to nonmajors. For example, the need is discussed in *Science for All Americans—Project 2061* (American Association for the Advancement of Science), *National Science Education Standards* (National Research Council, 1994), and *Science in the National Interest* (White House, 1994). Interdisciplinary science is an attempt to broaden and humanize science education by reducing and breaking down the barriers that enclose traditional science disciplines as distinct subjects.

"The authors obviously feel that emphasizing the interconnectedness of nature should be studied by integrating all of the sciences into a coherent, understandable network of facts, concepts, and interpretations that lead the student to view the universe in a new and enlightened perspective. This philosophy is particularly important in the education of nonscience majors who may never again formally study science."

—Jay R. Yett, Orange Coast College

3. **Help instructors build their own mix of descriptive and analytical aspects of science, arousing student interest and feelings as they help students reach the educational goals of their particular course.** The spirit of interdisciplinary science is sometimes found in courses called "General Science," "Combined Science," or "Integrated Science." These courses draw concepts from a wide range of the traditional fields of science but are not concentrated around certain problems or questions. For example, rather than just dealing with the physics of energy, an interdisciplinary approach might consider broad aspects of energy—dealing with potential problems of an energy crisis—including social and ethical issues. A number of approaches can be used in interdisciplinary science, including the teaching of science in a *social, historical, philosophical,* or *problem-solving* context, but there is

no single best approach. One of the characteristics of interdisciplinary science is that it is not constrained by the necessity of teaching certain facts or by traditions. It likewise cannot be imposed as a formal discipline, with certain facts to be learned. It is justified by its success in attracting and holding the attention and interest of students, making them a little wiser as they make their way toward various careers and callings.

4. **Humanize science for nonscience majors.** Each chapter presents historical background where appropriate, uses everyday examples in developing concepts, and follows a logical flow of presentation. A discussion of the people and events involved in the development of scientific concepts puts a human face on the process of science. The use of everyday examples appeals to the nonscience major, typically accustomed to reading narration, not scientific technical writing, and also tends to bring relevancy to the material being presented. The logical flow of presentation is helpful to students not accustomed to thinking about relationships between what is being read and previous knowledge learned, a useful skill in understanding the sciences.

VALUED INPUT WENT INTO STRIVING TO MEET YOUR NEEDS

Text development today involves a team that includes authors and publishers and valuable input from instructors who share their knowledge and experience with publishers and authors through reviews and focus groups. Such feedback has shaped this edition, resulting in reorganization of existing content and expanded coverage in key areas. This text has continued to evolve as a result of feedback from instructors actually teaching integrated science courses in the classroom. Reviewers point out that current and accurate content, a clear writing style with concise explanations, quality illustrations, and dynamic presentation materials are important factors considered when evaluating textbooks. Those criteria have guided the revision of the *Integrated Science* text and the development of its ancillary resources.

New to This Edition

Chapter 1: A discussion of "Scientific Communication" was added in order to add detail to the discussion of the scientific method, making the topic more appropriate for nonscience majors.

Chapter 3: A discussion of "Simple Machines" and also a new Myths, Mistakes, and Misunderstandings on recycling were added.

Chapter 4: A discussion of "efficiency" was included at a level of depth and detail appropriate for nonscience majors.

Chapter 8: Discussions of potential energy of electrons and uses for semiconductors as well as an Example on frequency and energy of electrons were added.

Chapter 11: The discussion of high-level nuclear waste was updated, and a discussion of what happened at Fukushima I was added.

Chapter 19: A new People Behind the Science biography on polymer chemist, Roy J. Plunkett (inventor of Teflon), was included. Section 19.2, Extraterrestrial Origin for Life on Earth, was rewritten. Also, the Closer Look discussion on enzymes was moved into the main text, while new information on ways to increase the level of 'good' cholesterol was added. These changes improved the relevance of this material for nonscience majors.

Chapter 21: New information on "Goldilocks planets" was added. Also, the material on selection and herbicides was heavily revised, with new material also added.

Chapter 22: This chapter was heavily revised: references to Usher and "theist" were removed; the section on Paleontology and Archaeology was revised, with more emphasis on definite statements and findings; the section on Genus Homo was revised; discussion of the Multiregional Hypothesis was removed; and a cladogram and expanded sense of history were added to the section on Hominin Origins.

Chapter 23: The nitrogen cycle description and diagram were revised. A new People Behind the Science on Jane Lubchenco was also added.

Chapter 24: This chapter was revised to make it more relevant to the nonscience major: medical-related information on and more discussion of eating disorders was added; the use of technical terms in the introduction to the nervous system were eliminated; Concepts Applied on Check Out the Nutrition Labels, Taste versus Smell, and Antagonistic Muscles were added; Science and Society, What Happens When You Drink Alcohol, was added; a new People Behind the Science on Henry Molaison and William Beecher Scoville was added; new information of tanning, gastric reflux, and probiotics was added; and the section on Guidelines for Obtaining Adequate Nutrients and the information on the new MyPlate food guide from the USDA were updated.

Chapter 25: The coverage on sexually transmitted diseases was expanded; a new Myths, Mistakes, and Misunderstandings, Is It Sex?, was added; and the sections on Hormonal Control Methods, Changes in Sexual Function with Age, and fraternal twins were all rewritten in chapter 25.

Chapter 26: Information on stem cells was moved into the main text of the chapter in order to improve the relevancy of this material for nonscience majors.

Appendices: The appendices have been revised and reorganized to provide improved problem-solving assistance for students. The tips and formatting for problem solving have been moved prior to the solutions in order to provide this material to students prior to their viewing of the solutions. A discussion of the methodology for solving multiple-choice type problems was also added to the problem-solving appendix. The answers to the end-of-chapter Applying the Concepts questions were also moved to the appendix.

Questions for Thought: The number of Questions for Thought was increased in all chapters without Parallel

Exercises in order to increase the number of practice questions for students and assignable homework questions for instructors.

THE LEARNING SYSTEM

To achieve the goals stated, this text includes a variety of features that should make student's study of *Integrated Science* more effective and enjoyable. These aids are included to help you clearly understand the concepts and principles that serve as the foundation of the integrated sciences.

OVERVIEW TO INTEGRATED SCIENCE

Chapter 1 provides an overview or orientation to integrated science in general and this text in particular. It also describes the fundamental methods and techniques used by scientists to study and understand the world around us.

MULTIDISCIPLINARY APPROACH

Chapter Opening Tools

Core Concept and Supporting Concepts

Core and Supporting Concepts integrate the chapter concepts and the chapter outline. The Core and Supporting Concepts outline and emphasize the concepts at a chapter level. The supporting concepts list is designed to help studets focus their studies by identifying the most important topics in the chapter outline.

Connections

The relationship of other science disciplines throughout the text are related to the chapter's contents. The core concept map, integrated with the chapter outline and supporting concepts list, the connections list, and overview, help students to see the big picture of the chapter content and the even bigger picture of how that content relates to other science discipline areas.

Chapter Overviews

Each chapter begins with an introductory overview. The overview previews the chapter's contents and what students can expect to learn from reading the chapter. It adds to the general outline of the chapter by introducing students to the concepts to be covered. It also expands upon the core concept map, facilitating in the integration of topics. Finally, the overview will help students to stay focused and organized while reading the chapter for the first time. After reading this introduction, students should browse through the chapter, paying particular attention to the topic headings and illustrations so that they get a feel for the kinds of ideas included within the chapter.

APPLYING SCIENCE TO THE REAL WORLD

Concepts Applied

As students look through each chapter, they will find one or more Concepts Applied boxes. These activities are simple exercises that students can perform at home or in the classroom to demonstrate important concepts and reinforce their understanding of them. This feature also describes the application of those concepts to their everyday lives.

Examples

Many of the more computational topics discussed within the chapters contain one or more concrete, worked **Examples** of a problem and its solution as it applies to the topic at hand. Through careful study of these Examples, students can better appreciate the many uses of problem solving in the sciences.

Follow-up Examples (with their solutions found in Appendix D) allow students to practice their problem-solving skills. The Examples have been marked as "optional" to allow instructors to place as much emphasis (or not) on problem solving as deemed necessary for their courses.

Science and Society

These readings relate the chapter's content to current societal issues. Many of these boxes also include Questions to Discuss that provide students an opportunity to discuss issues with their peers.

Myths, Mistakes, and Misunderstandings

These brief boxes provide short, scientific explanations to dispel a societal myth or a home experiment or project that enables students to dispel the myth on their own.

People Behind the Science

Many chapters also have one or two fascinating biographies that spotlight well-known scientists, past and present. From these People Behind the Science biographies, students learn about the human side of science: science is indeed relevant, and real people do the research and make the discoveries. These readings present the sciences in real-life terms that students can identify with and understand.

Closer Look and Connections

Each chapter of *Integrated Science* also includes one or more **Closer Look** readings that discuss topics of special human or environmental concern, topics concerning interesting technological applications, or topics on the cutting edge of scientific research. These readings enhance the learning experience by taking a more detailed look at related topics and adding concrete examples to help students better appreciate the real-world applications of science.

In addition to the **Closer Look** readings, each chapter contains concrete interdisciplinary **Connections** that are highlighted. **Connections** will help students better appreciate the interdisciplinary nature of the sciences. The **Closer Look** and **Connections** readings are informative materials that are supplementary in nature. These boxed features highlight valuable information beyond the scope of the text and relate intrinsic concepts discussed to real-world issues, underscoring the relevance of integrated science in confronting the many issues we face in our day-to-day lives. They are identified with the following icons:

"A Closer Look: The Compact Disc was, again, an excellent application of optics to everyday life and to something modern students thrive on—CDs and DVDs."

—Treasure Brasher, West Texas A&M University

"Connections—wonderful!!!. . . . . . A Closer Look . . . excellent. Clear, interesting, good figures. You have presented crucial information in a straightforward and uncompromising way."

—Megan M. Hoffman, Berea College

 *General:* This icon identifies interdisciplinary topics that cross over several categories; for example, life sciences and technology.

 *Life:* This icon identifies interdisciplinary life science topics, meaning connections concerning all living organisms collectively: plant life, animal life, marine life, and any other classification of life.

 *Technology:* This icon identifies interdisciplinary technology topics, that is, connections concerned with the application of science for the comfort and well being of people, especially through industrial and commercial means.

 *Measurement, Thinking, Scientific Methods:* This icon identifies interdisciplinary concepts and understandings concerned with people trying to make sense out of their surroundings by making observations, measuring, thinking, developing explanations for what is observed, and experimenting to test those explanations.

 *Environmental Science:* This icon identifies interdisciplinary concepts and understandings about the problems caused by human use of the natural world and remedies for those problems.

END-OF-CHAPTER FEATURES

At the end of each chapter are the following materials:

- *Summary:* highlights the key elements of the chapter
- *Summary of Equations* (chapters 1–9, 11): highlights the key equations to reinforce retention of them
- *Key Terms:* page-referenced where students will find the terms defined in context
- *Applying the Concepts:* a multiple choice quiz to test students' comprehension of the material covered (Answers are included in appendix F.)
- *Questions for Thought:* designed to challenge students to demonstrate their understandings of the topic
- *For Further Analysis:* exercises include analysis or discussion questions, independent investigations, and activities intended to emphasize critical thinking skills and societal issues, and develop a deeper understanding of the chapter content

- *Invitation to Inquiry:* exercises that consist of short, open-ended activities that allow students to apply investigative skills to the material in the chapter
- *Parallel Exercises* (chapters 1–9, 11): There are two groups of parallel exercises, Group A and Group B. The Group A parallel exercises have complete solutions worked out, along with useful comments in appendix G. The Group B parallel exercises are similar to those in Group A but do not contain answers in the text. By working through the Group A parallel exercises and checking the solution in appendix G, students will gain confidence in tackling the parallel exercises in Group B and thus reinforce their problem-solving skills.

"I like the key terms with the page numbers with each one. I always like to see more conceptual- and synthesis-type questions, which is why I like the 'Questions for Thought' and 'For Further Analysis' parts. . . . Exercises such as 'Questions for Thought' number 7, having students think about why oxygen is in Earth's atmosphere but not in Venus or Mars' atmosphere, is a valuable sort of question, because it requires students to know something and apply it."

—Jim Hamm, Big Bend Community College

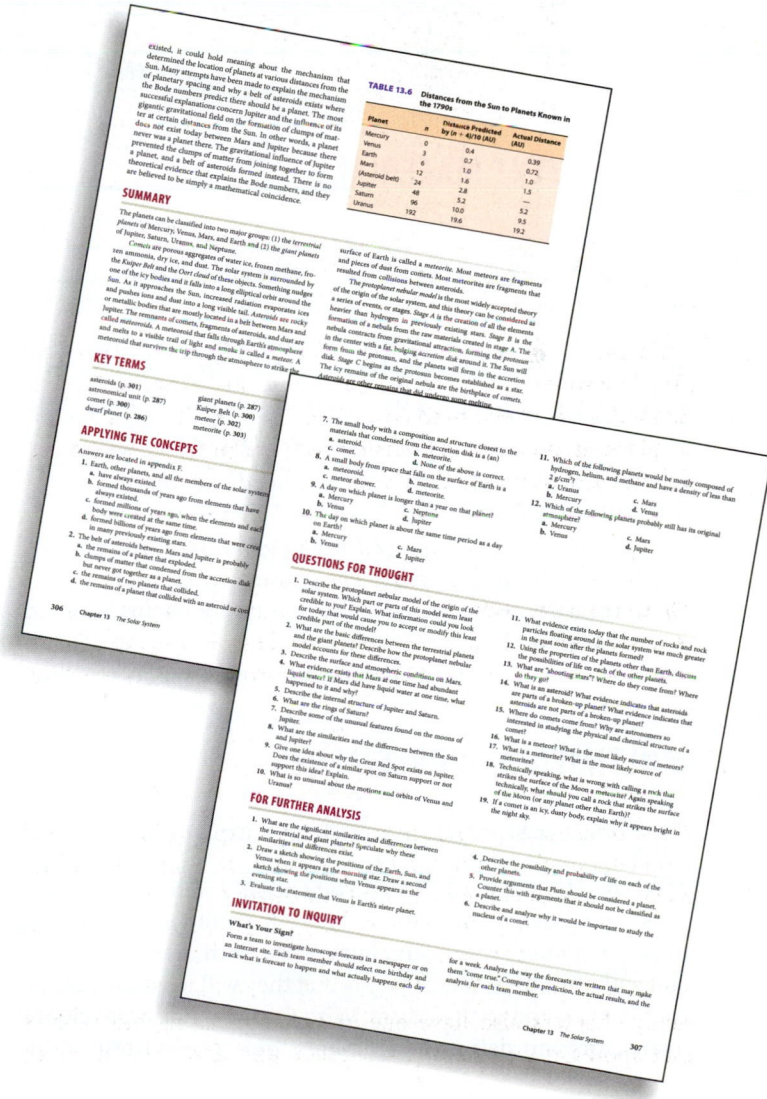

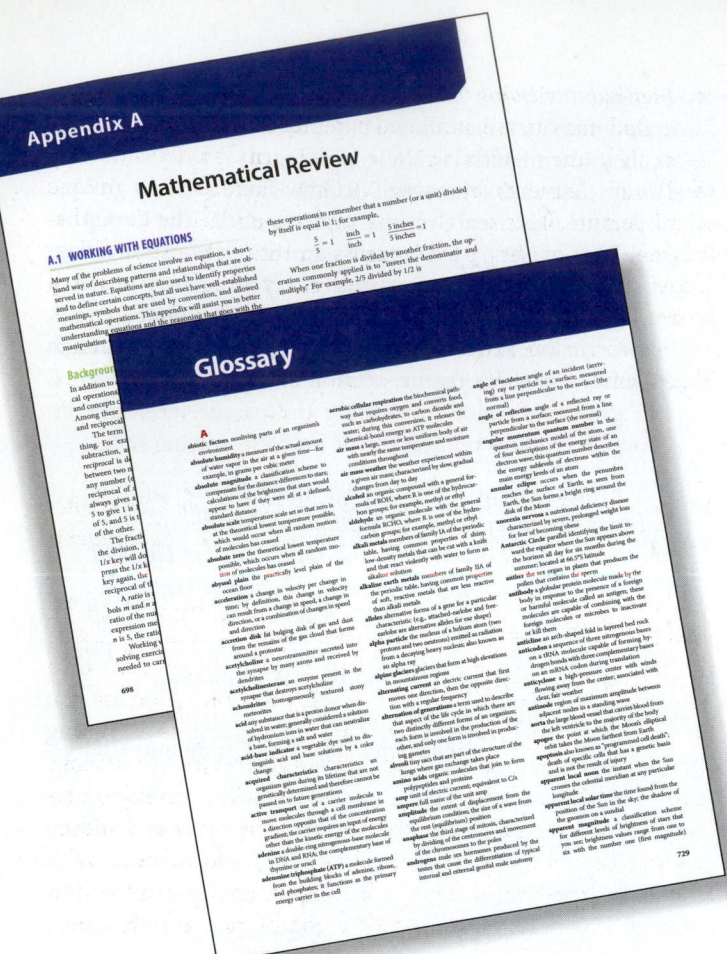

END-OF-TEXT MATERIAL

At the back of the text are appendices that give additional background details, charts, and answers to chapter exercises. Appendix E provides solutions for each chapter's follow-up Example exercises. There are also a glossary of all key terms, an index organized alphabetically by subject matter, and special tables printed on the inside covers for reference use.

> "... many books addressing similar disciplines have a tendency to talk over a student's head, making a student frustrated further in a class they do not want to be attending. ... Personally, I would admit that Integrated Science has a slight edge. The glossary seems up-to-date and centers in on words many nonscience majors may not understand."
>
> —David J. DiMattio, St. Bonaventure University

MULTIMEDIA SUPPLEMENTS

McGraw-Hill Connect Physical Science

McGraw-Hill Connect Physical Science is a web-based assignment and assessment platform that gives students the means to better connect with their coursework, with their instructors, and with the important concepts that they will need to know for success now and in the future. With Connect Physical Science, instructors can deliver assignments, quizzes, and tests easily online. Students can practice important skills at their own pace and on their own schedule. Connect Physical Science includes all Parallel Exercises from the text (in open-ended, numerical entry, and algorithmic format whenever possible) as well as all Questions for Thought in multiple-choice format (written by the authors of the text).

With the *Integrated Science* companion site, instructors also have access to PowerPoint lecture outlines, an Instructor's Manual, an Instructor's Edition Lab Manual, electronic images from the text, clicker questions, quizzes, animations, and many other resources directly tied to text-specific materials in *Integrated Science*. Students have access to self-quizzing, animations, and more.

Presentation Tools

Presentation Tools offer the ultimate multimedia resource center for the instructor. Graphics from the textbook are available in electronic format to create customized classroom presentations, visually based tests and quizzes, dynamic course website content, or attractive printed support materials.

The following assets are available in digital formats, for full-text download and also grouped by chapter:

- **Art and Photo Library:** Full-color digital files of all of the illustrations and many of the photos in the text can be readily incorporated into lecture presentations, exams, or custom-made classroom materials.
- **Worked Example Library and Table Library:** Access the worked examples and tables from the text in electronic format for inclusion in your classroom resources.
- **Animations Library:** Files of animations and videos covering various topics are included so that you can easily make use of these animations in a lecture or classroom setting.
- **Lecture Outlines:** Lecture notes, incorporating illustrations and animated images, have been written to the sixth edition text. They are provided in PowerPoint format so that you may use these lectures as written or customize them to fit your lecture.

Personal Response Systems

Personal response systems bring interactivity into the classroom or lecture hall. A wireless response system gives the instructor and students immediate feedback from the entire class. The wireless response pads are essentially easy-to-use remotes that engage students, allowing instructors to motivate student preparation, interactivity, and active learning. Instructors receive immediate feedback to gauge which concepts students understand. Questions covering the content of the *Integrated Science* text and formatted in PowerPoint are available on the *Integrated Science* companion site.

Computerized Test Bank Online

A comprehensive bank of over nine hundred test questions written by the authors of the text, in multiple-choice format at a variety of difficulty levels is provided within a computerized test bank powered by McGraw-Hill's flexible electronic testing program—EZ Test Online (www.eztestonline.com). EZ Test Online allows you to create the paper and online tests or quizzes in this easy-to-use program!

Imagine being able to create and access your test or quiz anywhere, at any time without installing the testing software. Now, with EZ Test Online, instructors can select questions from multiple McGraw-Hill test banks or author their own, and then either print the test for paper distribution or give it online.

McGraw-Hill Higher Education and Blackboard®

McGraw-Hill Higher Education and Blackboard have teamed up! What does this mean for you?

Do More

1. **Your life, simplified.** Now you and your students can access McGraw-Hill's Connect™ and Create™ right from within your Blackboard course—all with one single sign-on. Say goodbye to the days of logging in to multiple applications.

2. **Deep integration of content and tools.** Not only do you get single sign-on with Connect™ and Create™, you also get deep integration of McGraw-Hill content and content engines right in Blackboard. Whether you're choosing a book for your course or building Connect™ assignments, all the tools you need are right where you want them—inside of Blackboard.

3. **Seamless Gradebooks.** Are you tired of keeping multiple gradebooks and manually synchronizing grades into Blackboard? We thought so. When a student completes an integrated Connect™ assignment, the grade for that assignment automatically (and instantly) feeds your Blackboard grade center.

4. **A solution for everyone.** Whether your institution is already using Blackboard or you just want to try Blackboard on your own, we have a solution for you. McGraw-Hill and Blackboard can now offer you easy access to industry leading technology and content, whether your campus hosts it, or we do. Be sure to ask your local McGraw-Hill representative for details.

McGraw-Hill Tegrity®

McGraw-Hill Tegrity is a service that makes class time available all the time by automatically capturing every lecture in a searchable format for students to review when they study and complete assignments. With a simple one-click start-and-stop process, instructors capture all computer screens and corresponding audio. Students replay any part of any class with easy-to-use browser-based viewing on a PC or Mac. Educators know that the more students can see, hear, and experience class resources, the better they learn. With McGraw-Hill Tegrity, students quickly recall key moments by using McGraw-Hill Tegrity's unique search feature. This search helps students efficiently find what they need, when they need it across an entire semester of class recordings. Help turn all students' study time into learning moments immediately supported by the class lecture.

To learn more about McGraw-Hill Tegrity, watch a 2-minute Flash demo at http://tegritycampus.mhhe.com.

Create ™

Visit www.mcgrawhillcreate.com today to register and experience how McGraw-Hill Create™ empowers you to teach your students your way.

With McGraw-Hill Create™, www.mcgrawhillcreate.com, instructors can easily rearrange text chapters, combine material from other content sources, and quickly upload their own content, such as course syllabus or teaching notes. Content can be found in Create by searching through thousands of leading McGraw-Hill textbooks. Create allows instructors to arrange texts to fit their teaching style. Create also allows users to personalize a book's appearance by selecting the cover and adding the instructor's name, school, and course information. With Create, instructors can receive a complimentary print review copy in 3 to 5 business days or a complimentary electronic review copy (eComp) via e-mail in minutes.

Disclaimer

McGraw-Hill offers various tools and technology products to support the *Integrated Science* textbook. Students can order supplemental study materials by contacting their campus bookstore, calling 1-800-262-4729, or online at www.shopmcgraw-hill.com. Instructors can obtain teaching aides by calling the McGraw-Hill Customer Service Department at 1-800-338-3987, visiting our online catalog at www.mhhe.com, or contacting their local McGraw-Hill sales representative.

As a full-service publisher of quality educational products, McGraw-Hill does much more than just sell textbooks. We create and publish an extensive array of print, video, and digital supplements to support instruction. Orders of new (versus used) textbooks help us to defray the cost of developing such supplements, which is substantial. Local McGraw-Hill representatives can be consulted to learn about the availability of the supplements that accompany *Integrated Science*. McGraw-Hill representatives can be found by using the tab labeled "My Sales Rep" at www.mhhe.com.

PRINTED SUPPLEMENTARY MATERIALS

Laboratory Manual

The laboratory manual, written and classroom-tested by the authors, presents a selection of laboratory exercises specifically written for the interest and abilities of nonscience majors. Each lab begins with an open-ended *Invitations to Inquiry,* designed to pique student interest in the lab concept. This is followed by laboratory exercises that require measurement and data analysis for work in a more structured learning environment. When the laboratory manual is used with *Integrated Science,* students will have an opportunity to master basic scientific principles and concepts, learn new problem-solving and thinking skills, and understand the nature of scientific inquiry from the perspective of hands-on experiences. There is also an **instructor's edition lab manual** available on the *Integrated Science* companion site.

ACKNOWLEDGMENTS

This revision of *Integrated Science* has been made possible by the many users and reviewers of its fifth edition. The authors are indebted to the fifth edition reviewers for their critical reviews, comments, and suggestions. The reviewers were:

Loren Byrne, *Roger Williams University*
Timothy Champion, *Johnson C. Smith University*
Gregory S. Farley, *Chesapeake College*
Laura Frost, *Point Park University*
Eugene Grimley, *Elon University*
Sumitra Himangshu, *Macon State College*
David T. King Jr., *Auburn University*
Steven Levsen, *Mount Mary College*
Gail L. Miller, *York College*
Brie Paddock, *Arcadia University*
Frank Palaia, *Edison State College*
Mark Pilgrim, *Lander University*
Laura Racine, *Norwalk Community College*
David Rosengrant, *Kennesaw State University*
Paramasivam Sivapatham, *Savannah State University*
Alexander Williams, *York College*
Andrew J. Wood, *Southern Illinois University, Carbondale*

The authors would also like to thank the theme integration authors for their contributions to the Instructor's Resources on the companion site. Those contributors include:

Mary Brown, *Lansing Community College*
David J. DiMattio, *St. Bonaventure University*
Tasneem F. Khaleel, *Montana State University–Billings*
G. A. Nixon and Mary Ellen Teasdale, *Texas A&M–Commerce*
Thad Zaleskiewicz and Jennifer Siegert, *University of Pittsburgh–Greensburg*

The authors would also like to thank the following media ancillary authors for their contributions of the PowerPoint Lecture Outlines and the clicker questions, respectively:

Christine McCreary, *University of Pittsburgh–Greensburg*
Jeffrey J. Miller, *Metropolitan State College of Denver*

MEET THE AUTHORS

Bill W. Tillery

Bill W. Tillery is professor emeritus of Physics at Arizona State University. He earned a bachelor's degree at Northeastern State University (1960) and master's and doctorate degrees from the University of Northern Colorado (1967). Before moving to Arizona State University, he served as director of the Science and Mathematics Teaching Center at the University of Wyoming (1969–1973) and as an assistant professor at Florida State University (1967–1969). Bill has served on numerous councils, boards, and committees and was honored as the "Outstanding University Educator" at the University of Wyoming in 1972. He was elected the "Outstanding Teacher" in the Department of Physics and Astronomy at Arizona State University in 1995.

During his time at Arizona State, Bill has taught a variety of courses, including general education courses in science and society, physical science, and introduction to physics. He has received more than forty grants from the National Science Foundation, the U. S. Office of Education, private industry (Arizona Public Service), and private foundations (Flinn Foundation) for science curriculum development and science teacher inservice training. In addition to teaching and grant work, Bill has authored or co-authored more than sixty textbooks and many monographs, and has served as editor of three newsletters and journals between 1977 and 1996.

Eldon D. Enger

Eldon D. Enger is professor emeritus of biology at Delta College, a community college near Saginaw, Michigan. He received his B.A. and M.S. degrees from the University of Michigan. Professor Enger has over thirty years of teaching experience, during which he has taught biology, zoology, environmental science, and several other courses. He has been very active in curriculum and course development.

Professor Enger is an advocate for variety in teaching methodology. He feels that if students are provided with varied experiences, they are more likely to learn. In addition to the standard textbook assignments, lectures, and laboratory activities, his classes are likely to include writing assignments, student presentation of lecture material, debates by students on controversial issues, field experiences, individual student projects, and discussions of local examples and relevant current events. Textbooks are very valuable for presenting content, especially if they contain accurate, informative drawings and visual examples. Lectures are best used to help students see themes and make connections, and laboratory activities provide important hands-on activities.

Professor Enger has been a Fulbright Exchange Teacher to Australia and Scotland, received the Bergstein Award for Teaching Excellence and the Scholarly Achievement Award from Delta College, and participated as a volunteer in Earthwatch Research Programs in Costa Rica, the Virgin Islands and Australia. During 2001, he was a member of a People to People delegation to South Africa.

Professor Enger is married, has two adult sons, and enjoys a variety of outdoor pursuits such as cross-country skiing, hiking, hunting, kayaking, fishing, camping, and gardening. Other interests include reading a wide variety of periodicals, beekeeping, singing in a church choir, and preserving garden produce.

Frederick C. Ross

Fred Ross is professor emeritus of biology at Delta College, a community college near Saginaw, Michigan. He received his B.S. and M.S. from Wayne State University, Detroit, Michigan, and has attended several other universities and institutions. Professor Ross has thirty years' teaching experience, including junior and senior high school, during which he has taught biology, cell biology and biological chemistry, microbiology, environmental science, and zoology. He has been very active in curriculum and course development. These activities included the development of courses in infection control and microbiology, and AIDS and infectious diseases, and a PBS ScienceLine course for elementary and secondary education majors in cooperation with Central Michigan University. In addition, he was involved in the development of the wastewater microbiology technician curriculum offered by Delta College.

He was also actively involved in the National Task Force of Two Year College Biologists (American Institute of Biological Sciences) and in the National Science Foundation College Science Improvement Program, and has been an evaluator for science and engineering fairs, Michigan Community College Biologists, a judge for the Michigan Science Olympiad and the Science Bowl, a member of a committee to develop and update blood-borne pathogen standards protocol, and a member of Topic Outlines in Introductory Microbiology Study Group of the American Society for Microbiology.

Professor Ross involves his students in a variety of learning techniques and has been a prime advocate of the writing-to-learn approach. Besides writing, his students are typically engaged in active learning techniques including use of inquiry-based learning, the Internet, e-mail communications, field experiences, classroom presentation, as well as lab work. The goal of his classroom presentations and teaching is to actively engage the minds of his students in understanding the material, not just memorization of "scientific facts." Professor Ross is married and recently a grandfather. He enjoys sailing, horseback riding, and cross-country skiing.

1

What Is Science?

Science is concerned with your surroundings and your concepts and understanding of these surroundings.

CORE **CONCEPT**

Science is a way of thinking about and understanding your surroundings.

OUTLINE

Measurement is used to accurately describe properties and events (p. 4).

An equation is a statement of a relationship between variables (p. 10).

Science investigations include collecting observations, developing explanations, and testing explanations (p. 12).

Scientific laws describe relationships between events that happen time after time (p. 14).

Physics

▷ Energy flows in and out of your surroundings (Ch. 2–7).

Chemistry

▷ Matter is composed of atoms that interact on several different levels (Ch. 8–11).

Earth Science

▷ Earth is matter and energy that interact through cycles of change (Ch. 14–18).

Astronomy

▷ The stars and solar system are matter and energy that interact through cycles of change (Ch. 12–13).

OVERVIEW

Have you ever thought about your thinking and what you know? On a very simplified level, you could say that everything you know came to you through your senses. You see, hear, and touch things of your choosing, and you can smell and taste things in your surroundings. Information is gathered and sent to your brain by your sense organs. Somehow, your brain processes all this information in an attempt to find order and make sense of it all. Finding order helps you understand the world and what may be happening at a particular place and time. Finding order also helps you predict what may happen next, which can be very important in a lot of situations.

This is a book on thinking about and understanding your surroundings. These surroundings range from the obvious, such as the landscape and the day-to-day weather, to the not so obvious, such as how atoms are put together. Your surroundings include natural things as well as things that people have made and used (figure 1.1). You will learn how to think about your surroundings, whatever your previous experience with thought-demanding situations. This first chapter is about "tools and rules" that you will use in the thinking process.

1.1 OBJECTS AND PROPERTIES

Science is concerned with making sense out of the environment. The early stages of this "search for sense" usually involve *objects* in the environment, things that can be seen or touched. These could be objects you see every day, such as a glass of water, a moving automobile, or a running dog. They could be quite large, such as the Sun, the Moon, or even the solar system, or invisible to the unaided human eye. Objects can be any size, but people are usually concerned with objects that are larger than a pinhead and smaller than a house. Outside these limits, the actual size of an object is difficult for most people to comprehend.

As you were growing up, you learned to form a generalized mental image of objects called a *concept*. Your concept of an object is an idea of what it is, in general, or what it should be according to your idea (figure 1.2). You usually have a word stored away in your mind that represents a concept. The word *chair*, for example, probably evokes an idea of "something to sit on." Your generalized mental image for the concept that goes with the word *chair* probably includes a four-legged object with a backrest. Upon close inspection, most of your (and everyone else's) concepts are found to be somewhat vague. For example, if the word *chair* brings forth a mental image of something with

four legs and a backrest (the concept), what is the difference between a "high chair" and a "bar stool"? When is a chair a chair and not a stool? These kinds of questions can be troublesome for many people.

Not all of your concepts are about material objects. You also have concepts about intangibles such as time, motion, and relationships between events. As was the case with concepts of material objects, words represent the existence of intangible concepts. For example, the words *second, hour, day,* and *month* represent concepts of time. A concept of the pushes and pulls that come with changes of motion during an airplane flight might be represented with such words as *accelerate* and *falling*. Intangible concepts might seem to be more abstract since they do not represent material objects.

By the time you reach adulthood, you have literally thousands of words to represent thousands of concepts. But most, you would find on inspection, are somewhat ambiguous and not at all clear-cut. That is why you find it necessary to talk about certain concepts for a minute or two to see if the other person has the same "concept" for words as you do. That is why when one person says, "Boy, was it hot!" the other person may respond, "How hot was it?" The meaning of *hot* can be quite different for two people, especially if one is from Arizona and the other from Alaska!

FIGURE 1.1 Your surroundings include naturally occurring objects and manufactured objects such as sidewalks and walls.

FIGURE 1.2 What is your concept of a chair? Is this a picture of a row of chairs, or are they something else? Most people have concepts—or ideas of what things in general should be—that are loosely defined. The concept of a chair is one example, and this is a picture of a row of deck chairs.

FIGURE 1.3 Could you describe this rock to another person over the telephone so that the other person would know *exactly* what you see? This is not likely with everyday language, which is full of implied comparisons, assumptions, and inaccurate descriptions.

The problem with words, concepts, and mental images can be illustrated by imagining a situation involving you and another person. Suppose that you have found a rock that you believe would make a great bookend. Suppose further that you are talking to the other person on the telephone, and you want to discuss the suitability of the rock as a bookend, but you do not know the name of the rock. If you knew the name, you would simply state that you found a "_____." Then you would probably discuss the rock for a minute or so to see if the other person really understood what you were talking about. But not knowing the name of the rock and wanting to communicate about the suitability of the object as a book-end, what would you do? You would probably describe the characteristics, or **properties,** of the rock. Properties are the qualities or attributes that, taken together, are usually peculiar to an object. Since you commonly determine properties with your senses (smell, sight, hearing, touch, and taste), you could say that the properties of an object are the effect the object has on your senses. For example, you might say that the rock in figure 1.3 is "big, yellow, and smooth, with shiny gold cubes on one side." But consider the mental image that the other person on the telephone forms when you describe these properties. It is entirely possible that the other person is thinking of something very different from what you are describing!

As you can see, the example of describing a proposed bookend by listing its properties in everyday language leaves much to be desired. The description does not really help the other person form an accurate mental image of the rock. One problem with the attempted communication is that the description of any property implies some kind of *referent*. The word **referent** means that you *refer to,* or think of, a given property in terms of another, more familiar object. Colors, for example, are sometimes stated with a referent. Examples are "sky blue," "grass green," or "lemon yellow." The referents for the colors blue, green, and yellow are, respectively, the sky, living grass, and a ripe lemon.

Referents for properties are not always as explicit as they are with colors, but a comparison is always implied. Since the comparison is implied, it often goes unspoken and leads to assumptions in communications. For example, when you stated that the rock was "big," you assumed that the other person knew that you did not mean as big as a house or even as big as a bicycle. You assumed that the other person knew that you meant that the rock was about as large as a book, perhaps a bit larger.

Another problem with the listed properties of the rock is the use of the word *smooth.* The other person would not know if you meant that the rock *looked* smooth or *felt* smooth. After all, some objects can look smooth and feel rough. Other objects can look rough and feel smooth. Thus, here is another assumption, and probably all of the properties lead to implied comparisons, assumptions, and a not very accurate communication. This is the nature of your everyday language and the nature of most attempts at communication.

1.2 QUANTIFYING PROPERTIES

Typical day-to-day communications are often vague and leave much to be assumed. A communication between two people, for example, could involve one person describing some person, object, or event to a second person. The description is made by using referents and comparisons that the second person may or may not have in mind. Thus, such attributes as "long"

fingernails or "short" hair may have entirely different meanings to different people involved in a conversation. Assumptions and vagueness can be avoided by using **measurement** in a description. Measurement is a process of comparing a property to a well-defined and agreed-upon referent. The well-defined and agreed-upon referent is used as a standard called a **unit.** The measurement process involves three steps: (1) *comparing* the referent unit to the property being described, (2) following a *procedure,* or operation, which specifies how the comparison is made, and (3) *counting* how many standard units describe the property being considered.

The measurement process thus uses a defined referent unit, which is compared to a property being measured. The *value* of the property is determined by counting the number of referent units. The name of the unit implies the procedure that results in the number. A measurement statement always contains a *number* and *name* for the referent unit. The number answers the question "How much?" and the name answers the question "Of what?" Thus a measurement always tells you "how much of what." You will find that using measurements will sharpen your communications. You will also find that using measurements is one of the first steps in understanding your physical environment.

1.3 MEASUREMENT SYSTEMS

Measurement is a process that brings precision to a description by specifying the "how much" and "of what" of a property in a particular situation. A number expresses the value of the property, and the name of a unit tells you what the referent is as well as implying the procedure for obtaining the number. Referent units must be defined and established, however, if others are to understand and reproduce a measurement. It would be meaningless, for example, for you to talk about a length in "clips" if other people did not know what you meant by a "clip" unit. When standards are established, the referent unit is called a **standard unit** (figure 1.4). The use of standard units makes it possible to communicate and duplicate measurements. Standard units are usually defined and established by governments and their agencies that are created for that purpose. In the United States, the agency concerned with measurement standards is the National Institute of Standards and Technology. In Canada, the Standards Council of Canada oversees the National Standard System.

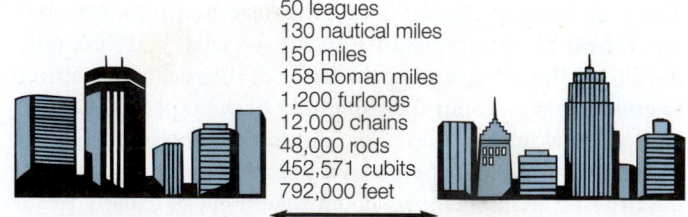

50 leagues
130 nautical miles
150 miles
158 Roman miles
1,200 furlongs
12,000 chains
48,000 rods
452,571 cubits
792,000 feet

FIGURE 1.4 Any of these units and values could have been used at some time or another to describe the same distance between these hypothetical towns. Any unit could be used for this purpose, but when one particular unit is officially adopted, it becomes known as the *standard unit.*

There are two major *systems* of standard units in use today, the English system and the metric system. The metric system is used in all industrialized countries except the United States, where both systems are in use. The continued use of the English system in the United States presents problems in international trade, so there is pressure for a complete conversion to the metric system. More and more metric units are being used in everyday measurements, but a complete conversion will involve an enormous cost. Appendix A contains a method for converting from one system to the other easily. Consult this section if you need to convert from one metric unit to another metric unit or to convert from English to metric units or vice versa. Conversion factors are listed inside the front cover.

People have used referents to communicate about properties of things throughout human history. The ancient Greek civilization, for example, used units of *stadia* to communicate about distances and elevations. The "stadium" was a unit of length of the racetrack at the local stadium (*stadia* is the plural of stadium), based on a length of 125 paces. Later civilizations, such as the ancient Romans, adopted the stadia and other referent units from the ancient Greeks. Some of these same referent units were later adopted by the early English civilization, which eventually led to the *English system* of measurement. Some adopted units of the English system were originally based on parts of the human body, presumably because you always had these referents with you (figure 1.5). The inch, for example, used the end joint of the thumb for a referent. A foot, naturally, was the length of a foot, and a yard was the distance from the tip of the nose to

TABLE 1.1   **The SI Standard Units**

Property	Unit	Symbol
Length	meter	m
Mass	kilogram	kg
Time	second	s
Electric current	ampere	A
Temperature	kelvin	K
Amount of substance	mole	mol
Luminous intensity	candela	cd

the end of the fingers on an arm held straight out. A cubit was the distance from the end of an elbow to the fingertip, and a fathom was the distance between the finger-tips of two arms held straight out. As you can imagine, there were problems with these early units because everyone was not the same size. Beginning in the 1300s, the sizes of the units were gradually standardized by various English kings. In 1879, the United States, along with sixteen other countries, signed the *Treaty of the Meter,* defining the English units in terms of the metric system. The United States thus became officially metric but not entirely metric in everyday practice.

The *metric system* was established by the French Academy of Sciences in 1791. The academy created a measurement system that was based on invariable referents in nature, not human body parts. These referents have been redefined over time to make the standard units more reproducible. In 1960, six standard metric units were established by international agreement. The *International System of Units,* abbreviated *SI,* is a modernized version of the metric system. Today, the SI system has seven units that define standards for the properties of length, mass, time, electric current, temperature, amount of substance, and light intensity (table 1.1). The standard units for the properties of length, mass, and time are introduced in this chapter. The remaining units will be introduced in later chapters as the properties they measure are discussed.

1.4   STANDARD UNITS FOR THE METRIC SYSTEM

If you consider all the properties of all the objects and events in your surroundings, the number seems overwhelming. Yet, close inspection of how properties are measured reveals that some properties are combinations of other properties (figure 1.6). Volume, for example, is described by the three length measurements of length, width, and height. Area, on the other hand, is described by just the two length measurements of length and width. Length, however, cannot be defined in simpler terms of any other property. There are four properties that cannot be described in simpler terms, and all other properties are combinations of these four. For this reason they are called the *fundamental properties.* A fundamental property cannot be defined in simpler terms other than to describe how it is measured.

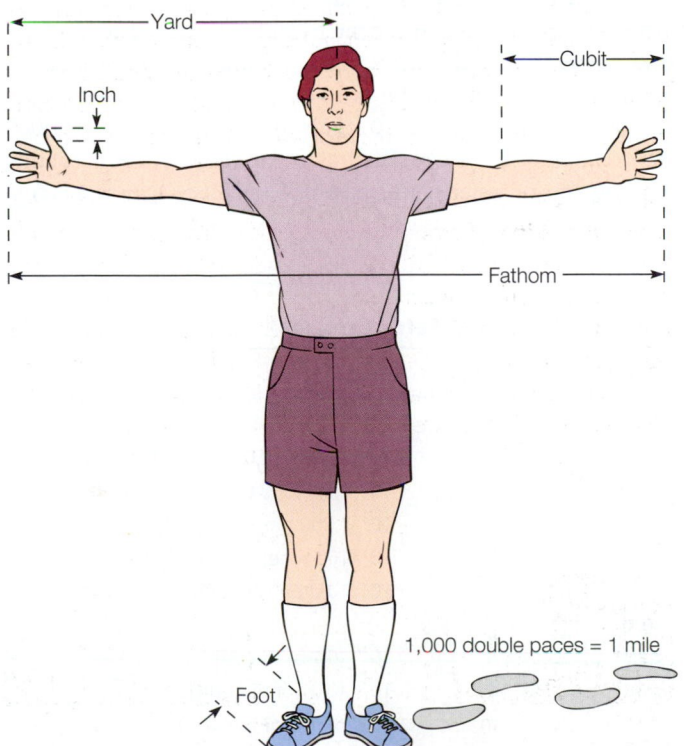

FIGURE 1.5   Many early units for measurement were originally based on the human body. Some of the units were later standardized by governments to become the basis of the English system of measurement.

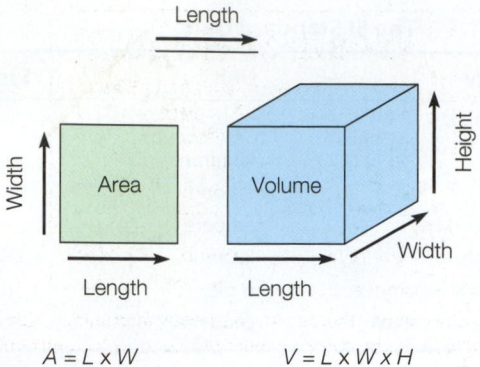

$$A = L \times W \qquad V = L \times W \times H$$

FIGURE 1.6 Area, or the extent of a surface, can be described by two length measurements. Volume, or the space that an object occupies, can be described by three length measurements. Length, however, can be described only in terms of how it is measured, so it is called a *fundamental property*.

These four fundamental properties are (1) *length*, (2) *mass*, (3) *time*, and (4) *charge*. Used individually or in combinations, these four properties will describe or measure what you observe in nature. Metric units for measuring the fundamental properties of length, mass, and time will be described next. The fourth fundamental property, charge, is associated with electricity, and a unit for this property will be discussed in chapter 6.

Length

The standard unit for length in the metric system is the *meter* (the symbol or abbreviation is m). A meter is defined in terms of the distance that light travels in a vacuum during a certain time period, 1/299,792,458 second. The important thing to remember, however, is that the meter is the metric *standard unit* for length. A meter is slightly longer than a yard, 39.3 inches. It is approximately the distance from your left shoulder to the tip of your right hand when your arm is held straight out. Many doorknobs are about 1 meter above the floor. Think about these distances when you are trying to visualize a meter length.

Mass

The standard unit for mass in the metric system is the *kilogram* (kg). The kilogram is defined as the mass of a certain metal cylinder kept by the International Bureau of Weights and Measures in France. This is the only standard unit that is still defined in terms of an object. The property of mass is sometimes confused with the property of weight since they are directly proportional to each other at a given location on the surface of the Earth. They are, however, two completely different properties and are measured with different units. All objects tend to maintain their state of rest or straight-line motion, and this property is called "inertia." The *mass* of an object is a measure of the inertia of an object. The *weight* of the object is a measure of the force of gravity on it. This distinction between weight and mass will be discussed in detail in chapter 2. For now, remember that weight and mass are not the same property.

Time

The standard unit for time is the *second* (s). The second was originally defined as 1/86,400 of a solar day (1/60 × 1/60 × 1/24). Earth's spin was found not to be as constant as thought, so the second was redefined to be the duration required for a certain number of vibrations of a certain cesium atom. A special spectrometer called an "atomic clock" measures these vibrations and keeps time with an accuracy of several millionths of a second per year.

1.5 METRIC PREFIXES

The metric system uses prefixes to represent larger or smaller amounts by factors of 10. Some of the more commonly used prefixes, their abbreviations, and their meanings are listed in table 1.2. Suppose you wish to measure something smaller than the standard unit of length, the meter. The meter is subdivided into ten equal-sized subunits called *decimeters*. The prefix *deci-* has a meaning of "one-tenth of," and it takes 10 decimeters to equal the length of 1 meter. For even smaller measurements, each decimeter is divided into ten equal-sized subunits called *centimeters*. It takes 10 centimeters to equal 1 decimeter and 100 to equal 1 meter. In a similar fashion, each prefix up or down the metric ladder represents a simple increase or decrease by a factor of 10 (figure 1.7).

When the metric system was established in 1791, the standard unit of mass was defined in terms of the mass of a certain volume of water. A cubic decimeter (dm^3) of pure water at 4°C was *defined* to have a mass of 1 kilogram (kg). This definition was convenient because it created a relationship between length, mass, and volume. As illustrated in figure 1.8, a cubic decimeter is 10 cm on each side. The volume of this cube is therefore 10 cm × 10 cm × 10 cm, or 1,000 cubic centimeters (abbreviated as cc or cm^3). Thus, a volume of 1,000 cm^3 of water has a mass of 1 kg. Since 1 kg is 1,000 g, 1 cm^3 of water has a mass of 1 g.

TABLE 1.2   Some Metric Prefixes

Prefix	Symbol	Meaning
tera-	T	10^{12} (1,000,000,000,000 times the unit)
giga-	G	10^{9} (1,000,000,000 times the unit)
mega-	M	10^{6} (1,000,000 times the unit)
kilo-	k	10^{3} (1,000 times the unit)
hecto-	h	10^{2} (100 times the unit)
deka-	da	10^{1} (10 times the unit)
Unit		
deci-	d	10^{-1} (0.1 of the unit)
centi-	c	10^{-2} (0.01 of the unit)
milli-	m	10^{-3} (0.001 of the unit)
micro-	μ	10^{-6} (0.000001 of the unit)
nano-	n	10^{-9} (0.000000001 of the unit)
pico-	p	10^{-12} (0.000000000001 of the unit)

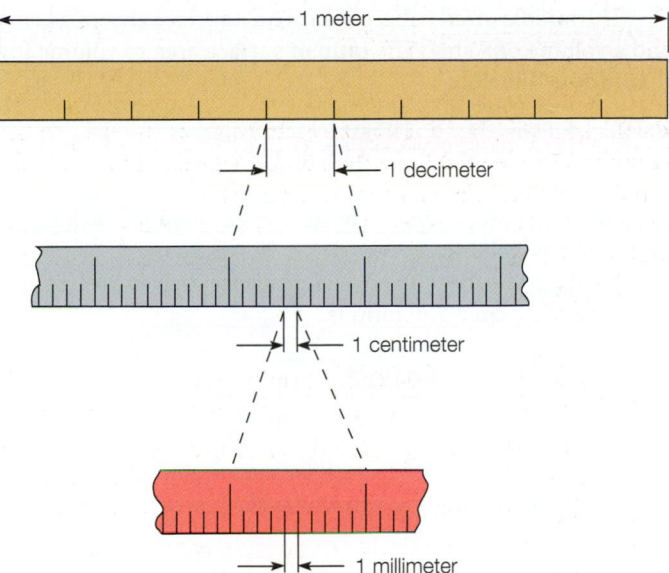

FIGURE 1.7 Compare the units shown above. How many millimeters fit into the space occupied by 1 centimeter? How many millimeters fit into the space of 1 decimeter? Can you express this as multiples of ten?

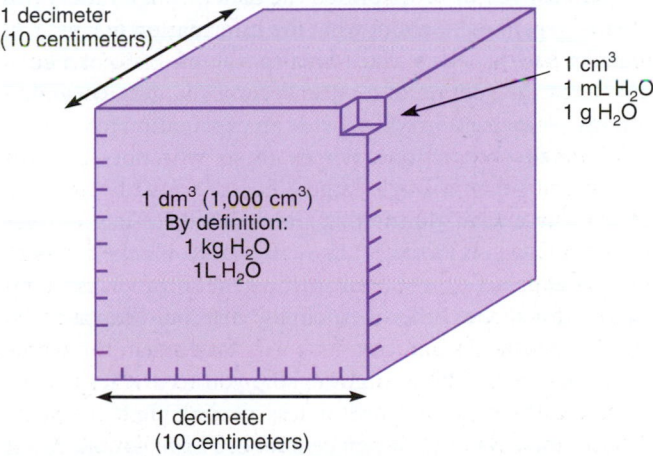

FIGURE 1.8 A cubic decimeter of water (1,000 cm^3) has a liquid volume of 1 L (1,000 mL) and a mass of 1 kg (1,000 g). Therefore, 1 cm^3 of water has a liquid volume of 1 mL and a mass of 1 g.

The volume of 1,000 cm^3 also defines a metric unit that is commonly used to measure liquid volume, the *liter* (L). For smaller amounts of liquid volume, the *milliliter* (mL) is used. The relationship between liquid volume, volume, and mass of water is therefore

$$1.0 \text{ L} \rightarrow 1.0 \text{ dm}^3 \text{ and has a mass of } 1.0 \text{ kg}$$

or, for smaller amounts,

$$1.0 \text{ mL} \rightarrow 1.0 \text{ cm}^3 \text{ and has a mass of } 1.0 \text{ g}$$

1.6 UNDERSTANDINGS FROM MEASUREMENTS

One of the more basic uses of measurement is to *describe* something in an exact way that everyone can understand. For example, if a friend in another city tells you that the weather has been "warm," you might not understand what temperature is being described. A statement that the air temperature is 70°F carries more exact information than a statement about "warm weather." The statement that the air temperature is 70°F contains two important concepts: (1) the numerical value of 70 and (2) the referent unit of degrees Fahrenheit. Note that both a numerical value and a unit are necessary to communicate a measurement correctly. Thus, weather reports describe weather conditions with numerically specified units; for example, 70° Fahrenheit for air temperature, 5 miles per hour for wind speed, and 0.5 inch for rainfall (figure 1.9). When such numerically specified units are used in a description, or a weather report, everyone understands *exactly* the condition being described.

Weather Report

Friday (24 hours ended at 5 P.M.)
Highs—airport 73°F, downtown 76°F
Lows—airport 68°F, downtown 70°F
Rainfall .... 0.26 in
Average wind speed .... 5.2 mph
Relative humidity .... High 85%
Low 75%
Rainfall ± normal to date.....+0.94 in

FIGURE 1.9 A weather report gives exact information, data that describe the weather by reporting numerically specified units for each condition.

Data

Measurement information used to describe something is called **data.** Data can be used to describe objects, conditions, events, or changes that might be occurring. You really do not know if the weather is changing much from year to year until you compare the yearly weather data. The data will tell you, for example, if the weather is becoming hotter or dryer or is staying about the same from year to year.

Let's see how data can be used to describe something and how the data can be analyzed for further understanding. The cubes illustrated in figure 1.10 will serve as an example. Each cube can be described by measuring the properties of size and surface area.

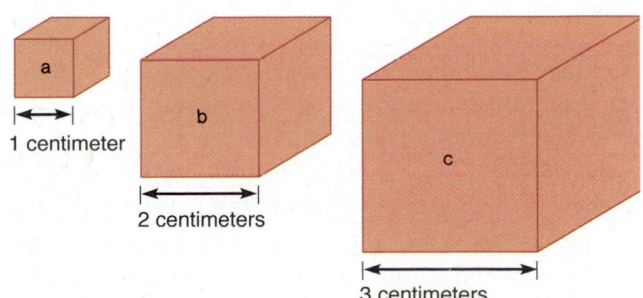

FIGURE 1.10 Cube *a* is 1 centimeter on each side, cube *b* is 2 centimeters on each side, and cube *c* is 3 centimeters on each side. These three cubes can be described and compared with data, or measurement information, but some form of analysis is needed to find patterns or meaning in the data.

First, consider the size of each cube. Size can be described by *volume*, which means *how much space something occupies.* The volume of a cube can be obtained by measuring and multiplying the length, width, and height. The data is

volume of cube *a*	1 cm^3
volume of cube *b*	8 cm^3
volume of cube *c*	27 cm^3

Now consider the surface area of each cube. *Area* means *the extent of a surface,* and each cube has six surfaces, or faces (top, bottom, and four sides). The area of any face can be obtained by measuring and multiplying length and width. The data for the three cubes thus describes them as follows:

	Volume	Surface Area
cube *a*	1 cm^3	6 cm^2
cube *b*	8 cm^3	24 cm^2
cube *c*	27 cm^3	54 cm^2

Ratios and Generalizations

Data on the volume and surface area of the three cubes in figure 1.10 describes the cubes, but whether it says anything about a relationship between the volume and surface area of a cube is difficult to tell. Nature seems to have a tendency to camouflage relationships, making it difficult to extract meaning from raw data. Seeing through the camouflage requires the use of mathematical techniques to expose patterns. Let's see how such techniques can be applied to the data on the three cubes and what the pattern means.

One mathematical technique for reducing data to a more manageable form is to expose patterns through a *ratio.* A ratio is a relationship between two numbers obtained when one number is divided by another number. Suppose, for example, that an instructor has 50 sheets of graph paper for a laboratory group of 25 students. The relationship, or ratio, between the number of sheets and the number of students is 50 papers to 25 students, and this can be written as 50 papers/25 students. This ratio is *simplified* by dividing 25 into 50, and the ratio becomes 2 papers/1 student. The 1 is usually understood (not stated), and the ratio is written as simply 2 papers/student. It is read as 2 papers "for each" student, or 2 papers "per" student. The concept of simplifying with a ratio is an important one, and you will see it time and time again throughout science. It is important that you understand the meaning of *per* and *for each* when used with numbers and units.

Applying the ratio concept to the three cubes in figure 1.10, the ratio of surface area to volume for the smallest cube, cube *a*, is 6 cm^2 to 1 cm^3, or

$$\frac{6 \text{ cm}^2}{1 \text{ cm}^3} = 6 \frac{\text{cm}^2}{\text{cm}^3}$$

meaning there are 6 square centimeters of area *for each* cubic centimeter of volume.

The middle-sized cube, cube *b*, had a surface area of 24 cm^2 and a volume of 8 cm^3. The ratio of surface area to volume for this cube is therefore

$$\frac{24 \text{ cm}^2}{8 \text{ cm}^3} = 3 \frac{\text{cm}^2}{\text{cm}^3}$$

meaning there are 3 square centimeters of area *for each* cubic centimeter of volume.

The largest cube, cube *c*, had a surface area of 54 cm^2 and a volume of 27 cm^3. The ratio is

$$\frac{54 \text{ cm}^2}{27 \text{ cm}^3} = 2 \frac{\text{cm}^2}{\text{cm}^3}$$

or 2 square centimeters of area *for each* cubic centimeter of volume. Summarizing the ratio of surface area to volume for all three cubes, you have

small cube	*a*	6:1
middle cube	*b*	3:1
large cube	*c*	2:1

Now that you have simplified the data through ratios, you are ready to generalize about what the information means. You can generalize that the surface-area-to-volume ratio of a cube *decreases* as the volume of a cube becomes larger. Reasoning from this generalization will provide an explanation for a number of related observations. For example, why does crushed ice melt faster than a single large block of ice with the same volume? The explanation is that the crushed ice has a larger surface-area-to-volume ratio than the large block, so more surface is exposed to warm air. If the generalization is found to be true for shapes other than cubes, you could explain why a log chopped into small chunks burns faster than the whole log. Further generalizing might enable you to predict if large potatoes would require more or less peeling than the same weight of small potatoes. When generalized explanations result in predictions that can be verified by experience, you gain confidence in the explanation. Finding patterns of relationships is a satisfying intellectual adventure that leads to understanding and generalizations that are frequently practical.

The Density Ratio

The power of using a ratio to simplify things, making explanations more accessible, is evident when you compare the simplified ratio 6 to 3 to 2 with the hodgepodge of numbers that you would have to consider without using ratios. The power of using the ratio technique is also evident when considering other properties of matter. Volume is a property that is sometimes confused with mass. Larger objects do not necessarily contain more matter than smaller objects. A large balloon, for example, is much larger than this book, but the book is much more massive than the balloon. The simplified way of comparing the mass of a particular volume is to find the ratio of mass to volume. This ratio is called **density,** which is defined as *mass per unit volume.* The *per* means "for each," as previously

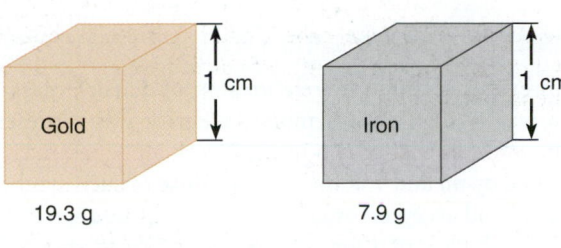

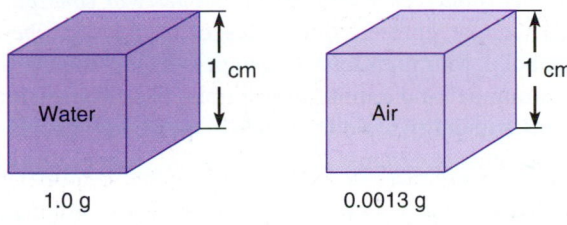

FIGURE 1.11 Equal volumes of different substances do not have the same mass, as these cube units show. Calculate the densities in g/m³. Do equal volumes of different substances have the same density? Explain.

discussed, and *unit* means "one," or "each." Thus "mass per unit volume" literally means the "mass of one volume" (figure 1.11). The relationship can be written as

$$\text{density} = \frac{\text{mass}}{\text{volume}}$$

or

$$\rho = \frac{m}{V} \qquad \text{equation 1.1}$$

(ρ is the symbol for the Greek letter *rho*.)

As with other ratios, density is obtained by dividing one number and unit by another number and unit. Thus, the density of an object with a volume of 5 cm³ and a mass of 10 g is

$$\text{density} = \frac{10 \text{ g}}{5 \text{ cm}^3} = 2 \frac{\text{g}}{\text{cm}^3}$$

The density in this example is the ratio of 10 g to 5 cm³, or 10 g/5 cm³, or 2 g to 1 cm³. Thus, the density of the example object is the mass of *one* volume (a unit volume), or 2 g *for each cm³*.

TABLE 1.3 **Densities of Some Common Substances**

Substance	Density (ρ) (g/cm³)
Aluminum	2.70
Copper	8.96
Iron	7.87
Lead	11.4
Water	1.00
Seawater	1.03
Mercury	13.6
Gasoline	0.680

Fish Density

Sharks and rays are marine animals that have an internal skeleton made entirely of cartilage. These animals have no swim bladder to adjust their body density in order to maintain their position in the water; therefore, they must constantly swim or they will sink. The bony fish, on the other hand, have a skeleton composed of bone and most also have a swim bladder. These fish can regulate the amount of gas in the bladder to control their density. Thus, the fish can remain at a given level in the water without expending large amounts of energy.

Any unit of mass and any unit of volume may be used to express density. The densities of solids, liquids, and gases are usually expressed in grams per cubic centimeter (g/cm³), but the densities of liquids are sometimes expressed in grams per milliliter (g/mL). Using SI standard units, densities are expressed as kg/m³. Densities of some common substances are shown in table 1.3.

CONCEPTS APPLIED

Density Examples

1. What is the density of this book? Measure the length, width, and height of this book in cm, then multiply to find the volume in cm³. Use a balance to find the mass of this book in grams. Compute the density of the book by dividing the mass by the volume. Compare the density in g/cm³ with other substances listed in table 1.3.

2. Compare the densities of some common liquids. Pour a cup of vinegar in a large bottle. Carefully add a cup of corn syrup, then a cup of cooking oil. Drop a coin, tightly folded pieces of aluminum foil, and toothpicks into the bottle. Explain what you observe in terms of density. Take care not to confuse the property of *density*, which describes the compactness of matter, with *viscosity*, which describes how much fluid resists flowing under normal conditions. (Corn syrup has a greater viscosity than water—is this true of density, too?)

Myths, Mistakes, and Misunderstandings

Tap a Can?

Some people believe that tapping on the side of a can of carbonated beverage will prevent it from foaming over when the can is opened. Is this true or a myth? Set up a controlled experiment to compare opening cold cans of a carbonated beverage that have been tapped with cans that have not been tapped. Are you sure you have controlled all the other variables?

If matter is distributed the same throughout a volume, the *ratio* of mass to volume will remain the same no matter what mass and volume are being measured. Thus, a teaspoonful, a cup, or a lake full of freshwater at the same temperature will all have a density of about 1 g/cm³ or 1 kg/L.

EXAMPLE 1.1   *(Optional)*

Two blocks are on a table. Block A has a volume of 30.0 cm³ and a mass of 81.0 g. Block B has a volume of 50.0 cm³ and a mass of 135 g. Which block has the greater density? If the two blocks have the same density, what material are they? (See table 1.3.)

SOLUTION

Density is defined as the ratio of the mass of a substance per unit volume. Assuming the mass is distributed equally throughout the volume, you could assume that the ratio of mass to volume is the same no matter what quantity of mass and volume are measured. If you can accept this assumption, you can use equation 1.1 to determine the density:

Block A

$$\text{mass } (m) = 81.0 \text{ g}$$
$$\text{volume } (V) = 30.0 \text{ cm}^3 \qquad \rho = \frac{m}{V}$$
$$\text{density } (\rho) = ?$$
$$= \frac{81.0 \text{ g}}{30.0 \text{ cm}^3}$$
$$= \frac{81.0}{30.0} \frac{\text{g}}{\text{cm}^3}$$
$$= 2.70 \frac{\text{g}}{\text{cm}^3}$$

Block B

$$\text{mass } (m) = 135 \text{ g}$$
$$\text{volume } (V) = 50.0 \text{ cm}^3 \qquad \rho = \frac{m}{V}$$
$$\text{density } (\rho) = ?$$
$$= \frac{135 \text{ g}}{50.0 \text{ cm}^3}$$
$$= \frac{135}{50.0} \frac{\text{g}}{\text{cm}^3}$$
$$= 2.70 \frac{\text{g}}{\text{cm}^3}$$

As you can see, both blocks have the same density. Inspecting table 1.3, you can see that aluminum has a density of 2.70 g/cm³, so both blocks must be aluminum.

EXAMPLE 1.2   *(Optional)*

A rock with a volume of 4.50 cm³ has a mass of 15.0 g. What is the density of the rock? (Answer: 3.33 g/cm³)

Symbols and Equations

In the previous section, the relationship of density, mass, and volume was written with symbols. Density was represented by ρ, the lowercase letter rho in the Greek alphabet, mass was represented by m, and volume by V. The use of such symbols is established and accepted by convention, and these symbols are like the vocabulary of a foreign language. You learn what the symbols mean by use and practice, with the understanding that *each symbol stands for a very specific property or concept.* The symbols actually represent **quantities,** or *measured properties.* The symbol m thus represents a quantity of mass that is specified by a number and a unit, for example, 16 g. The symbol V represents a quantity of volume that is specified by a number and a unit, such as 17 cm³.

Symbols

Symbols usually provide a clue about which quantity they represent, such as m for mass and V for volume. However, in some cases, two quantities start with the same letter, such as volume and velocity, so the uppercase letter is used for one (V for volume) and the lowercase letter is used for the other (v for velocity). There are more quantities than upper- and lowercase letters, however, so letters from the Greek alphabet are also used, for example, ρ for mass density. Sometimes a subscript is used to identify a quantity in a particular situation, such as v_i for initial, or beginning, velocity and v_f for final velocity. Some symbols are also used to carry messages; for example, the Greek letter delta (Δ) is a message that means "the change in" a value. Other message symbols are the symbol $\therefore$, which means "therefore," and the symbol $\propto$, which means "is proportional to."

Equations

Symbols are used in an **equation,** a statement that describes a relationship where *the quantities on one side of the equal sign are identical to the quantities on the other side. Identical* refers to both the numbers and the units. Thus, in the equation describing the property of density, $\rho = m/V$, the numbers on both sides of the equal sign are identical (e.g., $5 = 10/2$). The units on both sides of the equal sign are also identical (e.g., g/cm³ = g/cm³).

Equations are used to (1) *describe a property,* (2) *define a concept,* or (3) *describe how quantities change relative to each other.* Understanding how equations are used in these three classes is basic to comprehension of physical science. Each class of uses is considered separately in the following discussion.

Describing a property. You have already learned that the compactness of matter is described by the property called density. Density is a ratio of mass to a unit volume, or $\rho = m/V$. The key to understanding this property is to understand the meaning of a ratio and what *per* or *for each* means. Other examples of properties that will be defined by ratios are how fast something is moving (speed) and how rapidly a speed is changing (acceleration).

Defining a concept. A physical science concept is sometimes defined by specifying a measurement procedure. This is called an *operational definition* because a procedure is established that defines a concept as well as telling you how to measure it. Concepts of what is meant by force, mechanical work, and mechanical power and concepts involved in electrical and magnetic interactions can be defined by measurement procedures.

Describing how quantities change relative to each other. Nature is full of situations where one or more quantities change in value, or vary in size, in response to changes in other quantities. Changing quantities are called **variables.** Your weight, for example, is a variable that changes in size in response to changes in another variable; for example, the amount of food you eat. You already know about the pattern, or relationship, between these two variables. With all other factors being equal, an increase in the amount of food you eat results in an increase in your weight. When two variables increase (or decrease) together in the same ratio, they are said to be in *direct proportion.* When two variables are in direct proportion, *an increase or decrease in one variable results in the same relative increase or decrease in a second variable.*

Variables do not always increase or decrease in direct proportion. Sometimes one variable *increases* while a second variable *decreases* in the same ratio. This is an *inverse proportion* relationship. Other common relationships include one variable increasing in proportion to the *square* or to the *inverse square* of a second variable. Here are the forms of these four different types of proportional relationships:

Direct	$a \propto b$
Inverse	$a \propto 1/b$
Square	$a \propto b^2$
Inverse square	$a \propto 1/b^2$

FIGURE 1.12 The volume of fuel you have added to the fuel tank is directly proportional to the amount of time that the fuel pump has been running. This relationship can be described with an equation by using a proportionality constant.

Inverse Square Relationship

An inverse square relationship between energy and distance is found in light, sound, gravitational force, electric fields, nuclear radiation, and any other phenomena that spread equally in all directions from a source.

Box figure 1.1 could represent any of the phenomena that have an inverse square relationship, but let us assume it is showing a light source and how the light spreads at a certain distance (*d*), at twice that distance (2*d*), and at three times that distance (3*d*). As you can see, light twice as far from the source is spread over four times the area and will therefore have one-fourth the intensity. This is the same as $\frac{1}{2^2}$, or $\frac{1}{4}$.

Light three times as far from the source is spread over nine times the area and will therefore have one-ninth the intensity. This is the same as $\frac{1}{3^2}$, or $\frac{1}{9}$, again showing an inverse square relationship.

You can measure the inverse square relationship by moving an overhead projector so its light is shining on a wall (see distance *d* in box figure 1.1). Use a light meter or some other way of measuring the intensity of light. Now move the projector to double the distance from the wall. Measure the increased area of the projected light on the wall, and again measure the intensity of the light. What relationship did you find between the light intensity and distance?

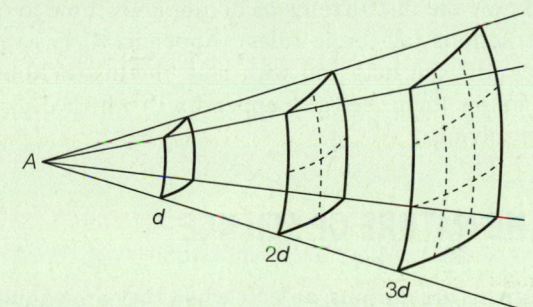

Box Figure 1.1 How much would light moving from point *A* spread out at twice the distance (2*d*) and three times the distance (3*d*)? What would this do to the brightness of the light?

Proportionality Statements

Proportionality statements describe in general how two variables change relative to each other, but a proportionality statement is *not* an equation. For example, consider the last time you filled your fuel tank at a service station (figure 1.12). You could say that the volume of gasoline in an empty tank you are filling is directly proportional to the amount of time that the fuel pump was running, or

$$\text{volume} \propto \text{time}$$

This is not an equation because the numbers and units are not identical on both sides. Considering the units, for example, it should be clear that minutes do not equal gallons; they are two different quantities. To make a statement of proportionality into an equation, you need to apply a *proportionality constant,* which is sometimes given the symbol k. For the fuel pump example, the equation is

$$\text{volume} = (\text{time})(\text{constant})$$

or

$$V = tk$$

In this example, the constant is the flow of gasoline from the pump in L/min (a ratio). Assume the rate of flow is 10 L/min. In units, you can see why the statement is now an equality.

$$L = (\text{min})\left(\frac{L}{\text{min}}\right)$$

$$L = \frac{\text{min} \times L}{\text{min}}$$

$$L = L$$

Here are some tips to consider if your instructor uses a problem-solving approach. You could think of an equation as a *set of instructions.* The density equation, for example, is $\rho = m/V$. The equation tells you that mass density is a ratio of mass to volume, and you can find the density by dividing the mass by the volume. If you have difficulty, you either do not know the instructions or do not know how to follow the instructions (algebraic rules). Appendix G has worked examples that can help you with both the instructions and how to follow them. See also appendix D, which deals with problem solving.

1.7 THE NATURE OF SCIENCE

Most humans are curious, at least when they are young, and are motivated to understand their surroundings. These traits have existed since antiquity and have proven to be a powerful motivation. In recent times, the need to find out has motivated the launching of space probes to learn what is "out there," and humans have visited the Moon to satisfy their curiosity. Curiosity and the motivation to understand nature were no less powerful in the past than today. Over two thousand years ago, the ancient Greeks lacked the tools and technology of today and could only make conjectures about the workings of nature. These early seekers of understanding are known as *natural philosophers,* and they observed, thought, and wrote about the workings of all of nature. They are called philosophers because their understandings came from reasoning only, without experimental evidence. Nonetheless, some of their ideas were essentially correct and are still in use today. For example, the idea of matter being composed of *atoms* was first reasoned by certain ancient Greeks in the fifth century B.C. The idea of *elements,* basic components that make up matter, was developed much earlier but refined by the ancient Greeks in the fourth century B.C. The concept of what the elements are and the concept of the nature of atoms have changed over time, but the ideas first came from ancient natural philosophers.

The Scientific Method

Some historians identify the time of Galileo and Newton, approximately three hundred years ago, as the beginning of modern science. Like the ancient Greeks, Galileo and Newton were interested in studying all of nature. Since the time of Galileo and Newton, the content of physical science has increased in scope and specialization, but the basic means of acquiring understanding, the scientific investigation, has changed little. A *scientific investigation* provides understanding through *experimental evidence,* as opposed to the conjectures based on thinking only of the ancient natural philosophers. In chapter 2, for example, you will learn how certain ancient Greeks described how objects fall toward Earth with a thought-out, or reasoned, explanation. Galileo, on the other hand, changed how people thought of falling objects by developing explanations from both creative thinking and precise measurement of physical quantities, providing experimental evidence for his explanations. Experimental evidence provides explanations today, much as it did for Galileo, as relationships are found from precise measurements of physical quantities. Thus, scientific knowledge about nature has grown as measurements and investigations have led to understandings that lead to further measurements and investigations.

What is a scientific investigation and what methods are used to conduct one? Attempts have been made to describe scientific methods in a series of steps (define problem, gather data, make hypothesis, test, make conclusion), but no single description has ever been satisfactory to all concerned. Scientists do similar things in investigations, but there are different approaches and different ways to evaluate what they find. Overall, the similar things might look like this:

1. Observe some aspect of nature.
2. Propose an explanation for something observed.
3. Use the explanation to make predictions.
4. Test predictions by doing an experiment or by making more observations.
5. Modify explanation as needed.
6. Return to step 3.

The exact approach a scientist uses depends on the individual doing the investigation as well as the particular field of science being studied.

Another way to describe what goes on during a scientific investigation is to consider what can be generalized. At least three separate activities seem to be common to scientists in different fields as they conduct scientific investigations, and these generalized activities are:

- Collecting observations.
- Developing explanations.
- Testing explanations.

No particular order or routine can be generalized about these common elements. In fact, individual scientists might not even be involved in all three activities. Some, for example, might spend all of their time out in nature, "in the field" collecting data and generalizing about their findings. This is an acceptable means of scientific investigation in some fields of science. Yet, other scientists might spend all of their time indoors, at computer terminals, developing theoretical equations that offer explanations for generalizations made by others. Again, the work at a computer terminal is an acceptable means of scientific investigation. Thus, there is not an order of five steps that are followed, particularly by today's specialized scientists. This is one reason why many philosophers of science argue there is no such thing as *the* scientific method. There are common activities of observing, explaining, and testing in scientific investigations in different fields, and these activities will be discussed next.

Explanations and Investigations

Explanations in the natural sciences are concerned with things or events observed, and there can be several different means of developing or creating explanations. In general, explanations can come from the results of experiments, from an educated guess, or just from imaginative thinking. In fact, there are several examples in the history of science of valid explanations being developed even from dreams.

Explanations go by various names, each depending on intended use or stage of development. For example, an explanation in an early stage of development is sometimes called a *hypothesis*. A **hypothesis** is a tentative thought- or experiment-derived explanation. It must be compatible with observations and provide understanding of some aspect of nature, but the key word here is *tentative*. A hypothesis is tested by experiment and is rejected, or modified, if a single observation or test does not fit.

The successful testing of a hypothesis may lead to the design of experiments, or it could lead to the development of another hypothesis, which could, in turn, lead to the design of yet more experiments, which could lead to. . . . As you can see, this is a branching, ongoing process that is very difficult to describe in specific terms. In addition, it can be difficult to identify a conclusion, an endpoint in the process. The search for new concepts to explain experimental evidence may lead from a hypothesis to a new theory, which results in more new hypotheses. This is why one of the best ways to understand scientific methods is to study the history of science. Or you can conduct a scientific investigation yourself.

Testing a Hypothesis

In some cases, a hypothesis may be tested by simply making additional observations. For example, if you hypothesize that a certain species of bird uses cavities in trees as places to build nests, you could observe several birds of the species and record the kinds of nests they build and where they are built.

Another common method for testing a hypothesis involves devising an experiment. An experiment is a re-creation of an event or occurrence in a way that enables a scientist to support or disprove a hypothesis. This can be difficult since a particular event may be influenced by a great many separate things. For example, the growth of a crop, such as corn, is influenced by a variety of factors including sunlight, soil moisture, air and soil temperature, soil nutrients, competition with other plants, and damage caused by insects and diseases. It might seem that developing an understanding of the factors that influence the growth of corn would be an impossible task. To help unclutter such situations, scientists have devised what is known as a *controlled experiment*. A **controlled experiment** compares two situations that have all the influencing factors identical except one. The situation used as the basis of comparison is called the *control group* and the other is called the *experimental group*. The single influencing factor that is allowed to be different in the experimental group is called the *experimental variable*.

The importance of various factors in corn growth would have to be broken down into a large number of simple questions, as previously mentioned. Each question would provide the basis on which experimentation would occur. Each experiment would provide information about one factor involved in corn growth. For example, to test the hypothesis that nitrogen is an important soil nutrient for corn growth, an experiment could be performed in which two identical fields would be prepared and planted with corn. The two fields would need to have the same type of soil, exposure to sunlight, moisture, pest-control treatment, and history of previous crops. In practice in agricultural experiments, the two fields are typically side by side. One corn field would be planted with a known quantity nitrogen-containing fertilizer added to the soil (the experimental group), while the other field (control group) would be grown without any nitrogen-containing fertilizer. Data on growth would need to be gathered in a standard way. One way to do this would be to harvest all the corn plants and weigh them. Another way could be to determine the average height of the plants. After the experiment, the new data (facts) gathered would be analyzed. If there were no differences between the two groups, scientists could conclude that the variable (amount of nitrogen in the soil) evidently did not influence corn growth. However, if there was a difference, it would be likely that the variable was responsible for the difference between the control and experimental groups. In the case of corn growth in this kind of experiment, the presence of nitrogen fertilizer enhances growth.

Accept Results?

Scientists are not likely to accept the results of a single experiment, since it is possible that a variable was overlooked in the process of designing and carrying out the experiment. For example, it is possible that a fungus pest could have affected one field more than the other, or the machinery used to spray for pest control may have malfunctioned on one field, or a local rain shower could have provided more rain to one of the fields. Therefore, scientists repeat experiments with exactly the same conditions and compare the results of several replicates (copies) of the experiment before arriving at a conclusion.

Basic and Applied Research

Science is the process of understanding your environment. It begins with making observations, creating explanations, and conducting research experiments. New information and conclusions are based on the results of the research.

There are two types of scientific research: basic and applied. *Basic research* is driven by a search for understanding and may or may not have practical applications. Examples of basic research include seeking understandings about how the solar system was created, finding new information about matter by creating a new element in a research lab, or mapping temperature variations on the bottom of the Chesapeake Bay. Such basic research expands our knowledge but will not lead to practical results.

Applied research has a goal of solving some practical problem rather than just looking for answers. Examples of applied research include the creation and testing of a highly efficient fuel cell to run cars on hydrogen fuel, improving the energy efficiency of the refrigerator, or creating a faster computer chip from new materials.

Whether research is basic or applied depends somewhat on the time frame. If a practical use cannot be envisioned in the future, then it is definitely basic research. If a practical use is immediate, then the work is definitely applied research. If a practical use is developed some time in the future, then the research is partly basic and partly practical. For example, when the laser was invented, there was no practical use for it. It was called "an answer waiting for a question." Today, the laser has many, many practical applications.

Knowledge gained by basic research has sometimes resulted in the development of technological breakthroughs. On the other hand, other basic research—such as learning how the solar system formed—has no practical value other than satisfying our curiosity.

Questions to Discuss

1. Should funding priorities go to basic research, applied research, or both?
2. Should universities concentrate on basic research and industries concentrate on applied research, or should both do both types of research?
3. Should research-funding organizations specify which types of research should be funded?

The results of an experiment are considered convincing only when there is just one variable, many replicates of the same experiment have been conducted, and the results are consistent. Furthermore, scientists often apply statistical tests to the results to help decide in an impartial manner if the results obtained are *valid* (meaningful; fit with other knowledge), *reliable* (give the same results repeatedly), and show cause-and-effect, or if they are just the result of random events.

During experimentation, scientists learn new information and formulate new questions that can lead to yet more experiments. One good experiment can result in a hundred new questions and experiments. The discovery of the structure of the DNA molecule by Watson and Crick resulted in thousands of experiments and stimulated the development of the entire field of molecular biology. Similarly, the discovery of molecules that regulate the growth of plants resulted in much research about how the molecules work and which molecules might be used for agricultural purposes.

If the processes of questioning and experimentation continue, and evidence continually and consistently supports the original hypothesis and other closely related hypotheses, the scientific community will begin to see how these hypotheses and facts fit together into a broad pattern.

Patterns and experimental results are shared through *scientific communication*. This can be as simple as scientists sharing experimental findings by e-mail. Results are also checked and confirmed by publishing articles in journals. Such articles enable scientists to know what other scientists have done, but also communicate ideas as well as the thinking process. Scientific communication ensures that results and thinking processes are confirmed by other scientists. It can also lead to new discoveries based on the work of others.

Scientific Laws

Sometimes you can observe a series of relationships that seem to happen over and over again. There is a popular saying, for example, that "if anything can go wrong, it will." This is called Murphy's law. It is called a *law* because it describes a relationship between events that seems to happen time after time. If you drop a slice of buttered bread, for example, it can land two ways, butter side up or butter side down. According to Murphy's law, it will land butter side down. With this example, you know at least one way of testing the validity of Murphy's law.

Another "popular saying" type of relationship seems to exist between the cost of a houseplant and how long it lives. You could call it the "law of houseplant longevity." The relationship is that the life of a houseplant is inversely proportional to its purchase price. This "law" predicts that a $10 houseplant will wilt and die within a month, but a 50 cent houseplant will live for years. The inverse relationship is between the variables of (1) cost and (2) life span, meaning the more you pay for a plant, the shorter the time it will live. This would also mean that inexpensive plants will live for a long time. Since the relationship seems to occur time after time, it is called a law.

A **scientific law** describes an important relationship that is observed in nature to occur consistently time after time. Basically, scientific laws describe *what* happens in nature. The law is often identified with the name of a person associated with the formulation of the law. For example, with all other factors being equal, an increase in the temperature of the air in a balloon results in an increase in its volume. Likewise, a decrease in the temperature results in a decrease in the total volume of the balloon.

The volume of the balloon varies directly with the temperature of the air in the balloon, and this can be observed to occur consistently time after time. This relationship was first discovered in the latter part of the eighteenth century by two French scientists, A. C. Charles and Joseph Gay-Lussac. Today, the relationship is sometimes called *Charles' law.* When you read about a scientific *law,* you should remember that a law is a statement that means something about a relationship that you can observe time after time in nature.

Have you ever heard someone state that something behaved a certain way *because* of a scientific law? For example, a big truck accelerated slowly *because* of Newton's laws of motion. Perhaps this person misunderstands the nature of scientific laws. Scientific laws do not dictate the behavior of objects; they simply describe it. They do not say how things ought to act but rather how things *do* act. A scientific law is *descriptive;* it describes how things act.

Models and Theories

Often the part of nature being considered is too small or too large to be visible to the human eye and the use of a *model* is needed. A **model** (figure 1.13) is a description of a theory or idea that accounts for all known properties. The description can come in many different forms, such as an actual physical model, a computer model, a sketch, an analogy, or an equation. No one has ever seen the whole solar system, for example, and all you can see in the real world is the movement of the Sun, Moon, and planets against a background of stars. A physical model or sketch of the solar system, however, will give you a pretty good idea of what the solar system might look like. The physical model and the sketch are both models since they give you a mental picture of the solar system.

At the other end of the size scale, models of atoms and molecules are often used to help us understand what is happening in this otherwise invisible world. Also, a container of small, bouncing rubber balls can be used as a model to explain the relationships of Charles' law. This model helps you see what happens to invisible particles of air as the temperature, volume, and pressure of the gas change. Some models are better than others, and models constantly change along with our understanding about nature. Early twentieth-century models of atoms, for example, were based on a "planetary model," which had electrons in the role of planets moving around the nucleus, which played the role of the Sun. Today, the model has changed as our understandings about the nature of the atom have changed. Electrons are now pictured as vibrating with certain wavelengths, which can make standing waves only at certain distances from the nucleus. Thus, the model of the atom changed from one with electrons viewed as solid particles to one that views them as vibrations on a string.

The most recently developed scientific theory was refined and expanded during the 1970s. This theory concerns the surface of Earth, and it has changed our model of what the Earth is like. At first, however, the basic idea of today's accepted theory was pure and simple conjecture. The term *conjecture* usually means an explanation or idea based on speculation, or one based on trivial grounds without any real evidence. Scientists would look at a map of Africa and South America, for example, and mull over how the two continents seem to be as pieces of a picture puzzle that had moved apart (figure 1.14). Any talk of moving continents was considered conjecture because it was not based on anything acceptable as real evidence.

Many years after the early musings about moving continents, evidence was collected from deep-sea drilling rigs that the ocean floor becomes progressively older toward the African and South American continents. This was good enough evidence to establish the "seafloor spreading hypothesis" that described the two continents moving apart.

If a hypothesis survives much experimental testing and leads, in turn, to the design of new experiments with the generation of new hypotheses that can be tested, you now have a working *theory.* A **theory** is defined as a broad, working hypothesis that is based on extensive experimental evidence. A scientific theory is a well-tested hypothesis that tells you *why* something happens. For example, the "seafloor spreading hypothesis" did survive requisite experimental testing and, together with other working hypotheses, is today found as part of the *plate tectonic theory.* The plate tectonic theory describes how the continents have moved apart, just like pieces of a picture puzzle. Is this the same idea that was once considered conjecture? Sort of, but this time it is supported by experimental evidence.

The term *scientific theory* is reserved for historic schemes of thought that have survived the test of detailed examination for long periods of time. The *atomic theory,* for example, was developed in the late 1800s and has been the subject of extensive investigation and experimentation over the last century. The atomic theory and other scientific theories form the framework of scientific thought and experimentation today. Scientific theories point to new ideas about the behavior of nature, and these ideas result in more experiments, more data to collect, and more explanations to develop. All of this may lead to a slight modification of an existing theory, a major modification, or perhaps the creation of an entirely new one. These activities continue in an ongoing attempt to satisfy the curiosity of people by understanding nature.

1.8 SCIENCE, NONSCIENCE, AND PSEUDOSCIENCE

As you can see from the discussion of the nature of science, a scientific approach to the world requires a certain way of thinking. There is an insistence on ample supporting evidence by numerous studies rather than easy acceptance of strongly stated

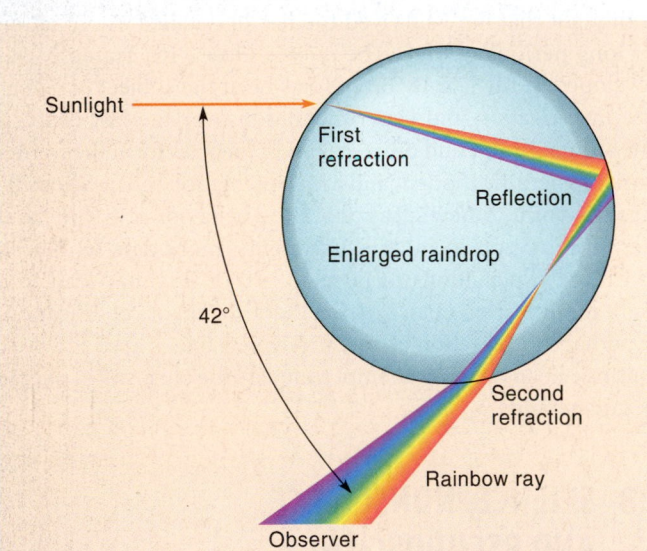
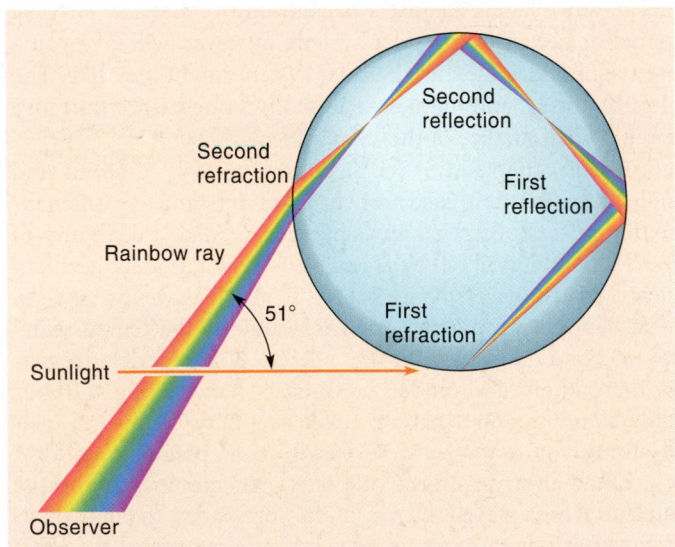

FIGURE 1.13 A model helps you visualize something that cannot be observed. You cannot observe what is making a double rainbow, for example, but models of light entering the upper and lower surface of a raindrop help you visualize what is happening. The drawings in (*B*) serve as a model that explains how a double rainbow is produced. (Also, see "The Rainbow" in chapter 7.)

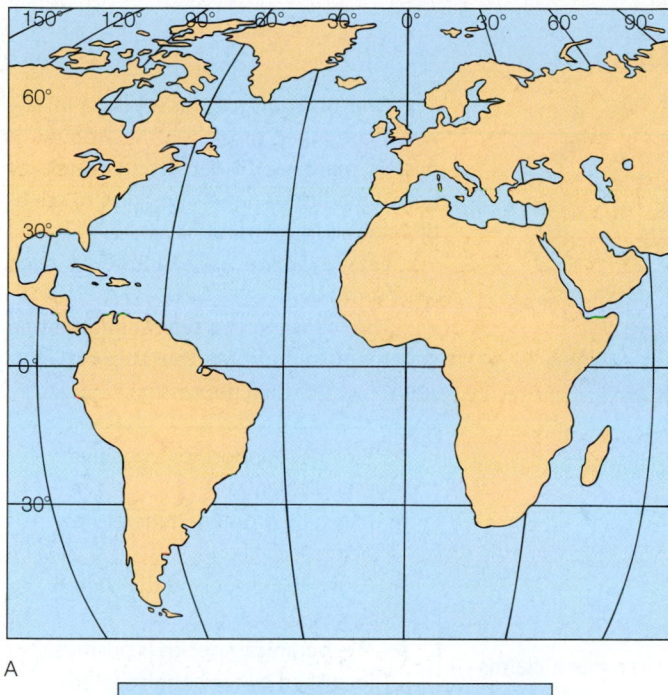

A

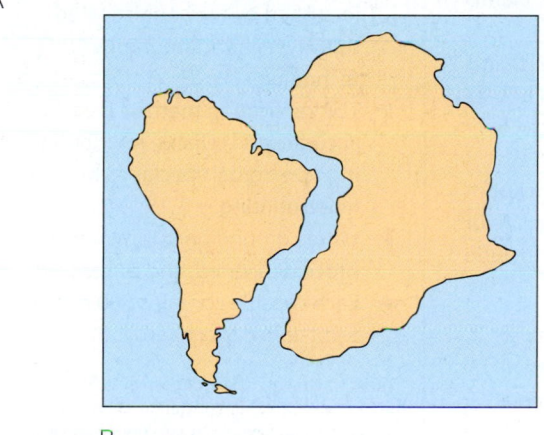

B

FIGURE 1.14 (*A*) Normal position of the continents on a world map. (*B*) A sketch of South America and Africa, suggesting that they once might have been joined together and subsequently separated by a continental drift.

opinions. Scientists must separate opinions from statements of fact. A scientist is a healthy skeptic.

Careful attention to detail is also important. Since scientists publish their findings and their colleagues examine their work, there is a strong desire to produce careful work that can be easily defended. This does not mean that scientists do not speculate and state opinions. When they do, however, they should take great care to clearly distinguish fact from opinion.

There is also a strong ethic of honesty. Scientists are not saints, but the fact that science is conducted out in the open in front of one's peers tends to reduce the incidence of dishonesty. In addition, the scientific community strongly condemns and severely penalizes those who steal the ideas of others, perform shoddy science, or falsify data. Any of these infractions could lead to the loss of one's job and reputation.

From Experimentation to Application

The scientific method has helped us to understand and control many aspects of our natural world. Some information is extremely important in understanding the structure and functioning of things in our world but at first glance appears to have little practical value. For example, understanding the life cycle of a star may be important for people who are trying to answer questions about how the universe is changing, but it seems of little practical value to the average citizen. However, as our knowledge has increased, the time between first discovery to practical application has decreased significantly.

For example, scientists known as *genetic engineers* have altered the chemical code system of small organisms (microorganisms) so that they may produce many new drugs such as antibiotics, hormones, and enzymes. The ease with which these complex chemicals are produced would not have been possible had it not been for the information gained from the basic, theoretical sciences of microbiology, molecular biology, and genetics. Our understanding of how organisms genetically control the manufacture of proteins has led to the large-scale production of enzymes. Some of these chemicals can remove stains from clothing, deodorize, clean contact lenses, remove damaged skin from burn patients, and "stone wash" denim for clothing.

Science and Nonscience

The differences between science and nonscience are often based on the assumptions and methods used to gather and organize information and, most important, the testing of these assumptions. The difference between a scientist and a nonscientist is that a scientist continually challenges and tests principles and assumptions to determine a cause-and-effect relationship, whereas a nonscientist may not feel that this is important.

Once you understand the nature of science, you will not have any trouble identifying astronomy, chemistry, physics, and biology as sciences. But what about economics, sociology, anthropology, history, philosophy, and literature? All of these fields may make use of certain central ideas that are derived in a logical way, but they are also nonscientific in some ways. Some things cannot be approached using the scientific method. Art, literature, theology, and philosophy are rarely thought of as sciences. They are concerned with beauty, human emotion, and speculative thought rather than with facts and verifiable laws. On the other hand, physics, chemistry, geology, and biology are always considered sciences.

Music is an area of study in a middle ground where scientific approaches may be used to some extent. "Good" music is certainly unrelated to science, but the study of how the human larynx generates the sound of a song is based on scientific principles. Any serious student of music will study the physics of sound and how the vocal cords vibrate to generate sound waves.

Science and Society

Herbal Medicine, Legislation, and Pseudoscience

The Dietary Supplement Health and Education Act of 1994 effectively removed herbal medicines from regulation by the U.S. Food and Drug Administration (FDA). The law, which amended the federal Food, Drug, and Cosmetic Act, defined "dietary supplements" as a separate regulatory category with very few regulations.

Following passage of the Dietary Supplement Health and Education Act,

- Manufacturers no longer needed to ask for permission from the FDA to produce and distribute products unless it is a "new product." There is no useful definition of a "new product." Since herbal medicines have been used for centuries, manufacturers could show that most of the things they would market would have been used previously somewhere and would not be new products.
- Manufacturers are responsible for determining the safety of the product. The FDA can only act if it finds a product is unsafe.
- Manufacturers determine serving size or quantity to be taken.
- Manufacturers do not need to prove that the herbal product actually works. The public is not protected from the sale of useless but harmless products.

Box Figure 1.2 This is Saint-John's-wort, an herbal medicine claimed to relieve depression.

- Manufacturers may not make claims about curing particular disease conditions but may make vague claims such as "improves mood," "boosts immune system," or "supports liver function."
- If a manufacturer makes a claim about how a product affects the structure or function of the body, it must include the following disclaimer:

These statements have not been evaluated by the Food and Drug Administration. This product is not intended to diagnose, treat, cure, or prevent any disease.

One outcome of this change in law was a continued growth in the amount of misinformation provided by marketers of herbal medicines under the guise of scientific statements—pseudoscience.

Let's examine an example of such misinformation.

The following is a typical information statement supporting the use of Saint-John's-wort to treat depression.

- Provides support for a positive mood balance.*
- May help promote general well-being.*
- Grows in Europe and the United States.
- The botanical species is positively identified by the sophisticated thin layer chromatography (TLC) technology.
- TLC verification method is as accurate and reliable for identifying true herbal species as human fingerprinting.
- Whole ground herb is minimally processed, dried, and pulverized.
- Each capsule contains 500 mg of Saint-John's-wort herb.

* These statements have not been evaluated by the Food and Drug Administration. This product is not intended to diagnose, treat, cure, or prevent any disease.

Similarly, economists use mathematical models and established economic laws to make predictions about future economic conditions. However, the regular occurrence of unpredicted economic changes indicates that economics is far from scientific, since the reliability of predictions is a central criterion of science. Anthropology and sociology are also scientific in nature in many respects, but they cannot be considered true sciences because many of the generalizations they have developed cannot be tested by repeated experimentation. They also do not show a significantly high degree of cause-and-effect, or they have poor predictive value.

Pseudoscience

Pseudoscience (*pseudo* = false) is a deceptive practice that uses the appearance or language of science to convince, confuse, or mislead people into thinking that something has scientific validity when it does not. When pseudoscientific claims are closely examined, they are not found to be supported by unbiased tests. For example, although nutrition is a respected scientific field, many individuals and organizations make claims about their nutritional products and diets that cannot be supported. Because of nutritional research, we all know that we must obtain certain nutrients such as amino acids, vitamins, and minerals from the food that we eat or we may become ill. Many scientific experiments reliably demonstrate the validity of this information. However, in most cases, it has not been proven that the nutritional supplements so vigorously promoted are as useful or desirable as advertised. Rather, selected bits of scientific information (amino acids, vitamins, and minerals are essential to good health) have been used to create the feeling that additional amounts of these nutritional supplements are necessary or

Typical Herbal Supplements and Claims

Common Name of Plant	Scientific Name of Plant	Claimed Benefit
Saint-John's-wort	*Hypericum perforatum*	Relieves mild or moderate depression
Gingko	*Ginkgo biloba*	Improves memory
Garlic	*Allium sativum*	Improves immune function; prevents atherosclerosis
Ginseng	*Panax quinquefolius* or *P. ginseng*	Elevates energy level
Green tea	*Camellia sinensis*	Prevents certain cancers, atherosclerosis, and tooth decay
Echinacea (purple coneflower)	*Echinacea pupurea*	Improves immune system
Black cohosh	*Cimicifuga racemosa* or *Actaea racemosa*	Relieves menopause symptoms or painful menstruation

How does this information exemplify pseudoscience?

1. The first two statements provide vague statements about mental health. They do not say that it is used to treat depression because that link has not been scientifically established to most scientists' satisfaction. The manufacturer relies on scientifically unverified statements in popular literature and on the Internet, and propagated by word of mouth to provide the demand for the product.

2. The two statements about thin layer chromatography are meant to suggest accuracy and purity. In fact, thin layer chromatography is a poor way to identify plants. It simply looks at the pigments present in plants. The use of this technique is unnecessary, since Saint-John's-wort is a plant that is easily identified by amateurs.

3. The statement about the herb being "minimally processed" is meant to suggest that nothing was lost from the plant during the preparation of the capsule and implies a high level of quality control and purity. Although the statement may be true, it is quite possible that the plant material collected contained dust, insects, and other contaminants. It is also highly likely that individual plants differ greatly in the amount of specific chemicals they contain.

Questions to Discuss

1. Select an herbal medicine and examine the claims made about this herb. Design a means of experimentally testing these claims. Decide with your group what would be acceptable evidence and what would not be acceptable.

2. Discuss with your group why people tend to ignore experimentally testing a claim.

3. Discuss with your group the possibility of a "placebo effect," that is, if someone strongly believes a claim, it might come true.

that they can improve your health. In reality, the average person eating a varied diet will obtain all of these nutrients in adequate amounts, and nutritional supplements are not required.

Another related example involves the labeling of products as organic or natural. Marketers imply that organic or natural products have greater nutritive value because they are organically grown (grown without pesticides or synthetic fertilizers) or because they come from nature. Although there are questions about the health effects of trace amounts of pesticides in foods, no scientific study has shown that a diet of natural or organic products has any benefit over other diets. The poisons curare, strychnine, and nicotine are all organic molecules that are produced in nature by plants that could be grown organically, but we would not want to include them in our diet.

Absurd claims that are clearly pseudoscience sometimes appear to gain public acceptance because of promotion in the media. Thus, some people continue to believe stories that psychics can really help solve puzzling crimes, that perpetual energy machines exist, or that sources of water can be found by a person with a forked stick. Such claims could be subjected to scientific testing and disposed of if they fail the test, but this process is generally ignored. In addition to experimentally testing such a claim that appears to be pseudoscience, here are some questions that you should consider when you suspect something is pseudoscience:

1. What is the background and scientific experience of the person promoting the claim?
2. How many articles have been published by the person in peer-reviewed scientific journals?
3. Has the person given invited scientific talks at universities and national professional organization meetings?

People Behind the Science

Florence Bascom (1862–1945)

Florence Bascom, a U.S. geologist, was an expert in the study of rocks and minerals and founded the geology department at Bryn Mawr College, Pennsylvania. This department was responsible for training the foremost women geologists of the early twentieth century.

Bascom's interest in geology had been sparked by a driving tour she took with her father and his friend Edward Orton, a geology professor at Ohio State University. It was an exciting time for geologists with new areas opening up all the time. Bascom was also inspired by her teachers at Wisconsin and Johns Hopkins, who were experts in the new fields of metamorphism and crystallography. Bascom's Ph.D. thesis was a study of rocks that had previously been thought to be sediments but which she proved to be metamorphosed lava flows.

While studying for her doctorate, Bascom became a popular teacher, passing on her enthusiasm and rigor to her students. She taught at the Hampton Institute for Negroes and American Indians and at Rockford College before becoming an instructor and associate professor of geology at Ohio State University from 1892 to 1895. Moving to Bryn Mawr College, where geology was considered subordinate to the other sciences, she spent two years teaching in a storeroom while building a considerable collection of fossils, rocks, and minerals. While at Bryn Mawr, she took great pride in passing on her knowledge and training to a generation of women who would become

successful. At Bryn Mawr, she rose rapidly, becoming reader (1898), associate professor (1903), professor (1906), and finally professor emerita from 1928 till her death in 1945 in Northampton, Massachusetts.

Bascom became, in 1896, the first woman to work as a geologist on the U.S. Geological Survey, spending her summers mapping formations in Pennsylvania, Maryland, and New Jersey and her winters analyzing slides. Her results were published in Geographical Society of America bulletins. In 1924, she became the first woman to be elected a fellow of the Geographical Society and went on, in 1930, to become the first woman vice president. She was associate editor of the *American Geologist* (1896–1905) and achieved a four-star place in the first edition of *American Men and Women of Science* (1906), a sign of how highly regarded she was in her field.

Bascom was the author of over forty research papers. She was an expert on the crystalline rocks of the Appalachian Piedmont, and she published her research on Piedmont geomorphology. Geologists in the Piedmont area still value her contributions, and she is still a powerful model for women seeking status in the field of geology today.

Source: Modified from the *Hutchinson Dictionary of Scientific Biography.* © RM, 2011. All rights reserved. Helicon Publishing is a division of RM.

---

4. Has the claim been researched and published by the person in a peer-reviewed scientific journal *and* have other scientists independently validated the claim?
5. Does the person have something to gain by making the claim?

Limitations of Science

By definition, science is a way of thinking and seeking information to solve problems. Therefore, scientific methods can be applied only to questions that have a factual basis. Questions concerning morals, value judgments, social issues, and attitudes cannot be answered using the scientific methods. What makes a painting great? What is the best type of music? Which wine is best? What color should I paint my car? These questions are related to values, beliefs, and tastes; therefore, scientific methods cannot be used to answer them.

Science is also limited by the ability of people to pry understanding from the natural world. People are fallible and do not always come to the right conclusions, because information is lacking or misinterpreted, but science is self-correcting. As new information is gathered, old incorrect ways of thinking must be changed or discarded. For example, at one time, people

were sure that the Sun went around Earth. They observed that the Sun rose in the east and traveled across the sky to set in the west. Since they could not feel Earth moving, it seemed perfectly logical that the sun traveled around Earth. Once they understood that Earth rotated on its axis, people began to understand that the rising and setting of the Sun could be explained in other ways. A completely new concept of the relationship between the Sun and Earth developed.

CONCEPTS APPLIED

Seekers of Pseudoscience

See what you can find out about some recent claims that might not stand up to direct scientific testing. Look into the scientific testing—or lack of testing—behind claims made in relation to cold fusion, cloning human beings, a dowser carrying a forked stick to find water, psychics hired by police departments, Bigfoot, the Bermuda Triangle, and others you might wish to investigate. One source to consider is www.randi.org/jr/archive.html.

Although this kind of study seems rather primitive to us today, this change in thinking about the Sun and Earth was a very important step in understanding the universe and how the various parts are related to one another. This background information was built upon by many generations of astronomers and space scientists, and finally led to space exploration.

People also need to understand that science cannot answer all the problems of our time. Although science is a powerful tool, there are many questions it cannot answer and many problems it cannot solve. The behavior and desires of people generate most of the problems societies face. Famine, drug abuse, and pollution are human-caused and must be resolved by humans. Science may provide some tools for social planners, politicians, and ethical thinkers, but science does not have, nor does it attempt to provide, all the answers to the problems of the human race. Science is merely one of the tools at our disposal.

SUMMARY

Science is a search for order in our surroundings. People have *concepts,* or mental images, about material *objects* and intangible *events* in their surroundings. Concepts are used for thinking and communicating. Concepts are based on *properties,* or attributes that describe a thing or event. Every property implies a *referent* that describes the property. Referents are not always explicit, and most communications require assumptions.

Measurement is a process that uses a well-defined and agreed upon *referent* to describe a *standard unit.* The unit is compared to the property being defined by an *operation* that determines the *value* of the unit by *counting.* Measurements are always reported with a *number,* or value, and a *name* for the unit.

The two major *systems* of standard units are the *English system* and the *metric system.* The English system uses standard units that were originally based on human body parts, and the metric system uses standard units based on referents found in nature. The metric system also uses a system of prefixes to express larger or smaller amounts of units. The metric standard units for length, mass, and time are the *meter, kilogram,* and *second.*

Measurement information used to describe something is called *data.* One way to extract meanings and generalizations from data is to use a *ratio,* a simplified relationship between two numbers. Density is a ratio of mass to volume, or $\rho = m/V$.

Symbols are used to represent *quantities,* or measured properties. Symbols are used in *equations,* which are shorthand statements that describe a relationship where the quantities (both number values and units) are identical on both sides of the equal sign. Equations are used to (1) *describe* a property, (2) *define* a concept, or (3) *describe* how *quantities change* together.

Quantities that can have different values at different times are called *variables.* Variables that increase or decrease together in the same ratio are said to be in *direct proportion.* If one variable increases while the other decreases in the same ratio, the variables are in *inverse* proportion. Proportionality statements are not necessarily equations. A *proportionality constant* can be used to make such a statement into an equation. Proportionality constants might have numerical value only, without units, or they might have both value and units.

Modern science began about three hundred years ago during the time of Galileo and Newton. Since that time, *scientific investigation* has been used to provide *experimental evidence* about nature. The investigations provide *accurate, specific,* and *reliable* data that are used to develop and test *explanations.* A *hypothesis* is a tentative explanation that is accepted or rejected from experimental data. An accepted hypothesis may result in a *scientific law,* an explanation concerned with important phenomena. Laws are sometimes identified with the name of a scientist and can be expressed verbally, with an equation, or with a graph.

A *model* is used to help understand something that cannot be observed directly, explaining the unknown in terms of things already understood. Physical models, mental models, and equations are all examples of models that explain how nature behaves. A *theory* is a broad, detailed explanation that guides development and interpretations of experiments in a field of study.

Science and *nonscience* can be distinguished by the kinds of laws and rules that are constructed to unify the body of knowledge. Science involves the continuous *testing* of rules and principles by the collection of new facts. If the rules are not testable, or if no rules are used, it is not science. *Pseudoscience* uses scientific appearances to mislead.

Summary of Equations

1.1
$$\text{density} = \frac{\text{mass}}{\text{volume}}$$

$$\rho = \frac{m}{V}$$

KEY TERMS

controlled experiment (p. **13**)
data (p. **7**)
density (p. **8**)
equation (p. **10**)

hypothesis (p. **13**)
measurement (p. **4**)
model (p. **15**)
properties (p. **3**)

pseudoscience (p. **18**)
quantities (p. **10**)
referent (p. **4**)
scientific law (p. **14**)

standard unit (p. **4**)
theory (p. **15**)
unit (p. **4**)
variables (p. **11**)

APPLYING THE CONCEPTS

Answers are located in appendix F.
Additional multiple choice and conversion exercise questions are on the website: **www.mhhe.com/tillery.**

1. The process of comparing a property of an object to a well-defined and agreed-upon referent is called the process of
   a. generalizing.
   b. measurement.
   c. graphing.
   d. scientific investigation.

2. The height of an average person is closest to
   a. 1.0 m.
   b. 1.5 m.
   c. 2.5 m.
   d. 3.5 m.

3. Which of the following standard units is defined in terms of an object as opposed to an event?
   a. kilogram
   b. meter
   c. second
   d. none of the above

4. One-half liter of water has a mass of
   a. 0.5 g.
   b. 5 g.
   c. 50 g.
   d. 500 g.

5. A cubic centimeter (cm^3) of water has a mass of about 1
   a. mL.
   b. kg.
   c. g.
   d. dm.

6. Measurement information that is used to describe something is called
   a. referents.
   b. properties.
   c. data.
   d. a scientific investigation.

7. The property of volume is a measure of
   a. how much matter an object contains.
   b. how much space an object occupies.
   c. the compactness of matter in a certain size.
   d. the area on the outside surface.

8. As the volume of a cube becomes larger and larger, the surface area-to-volume ratio
   a. increases.
   b. decreases.
   c. remains the same.
   d. sometimes increases and sometimes decreases.

9. If you consider a very small portion of a material that is the same throughout, the density of the small sample will be
   a. much less.
   b. slightly less.
   c. the same.
   d. greater.

10. A scientific law can be expressed as
   a. a written concept.
   b. an equation.
   c. a graph.
   d. Any of the above is correct.

QUESTIONS FOR THOUGHT

1. What is a concept?
2. What are two components of a measurement statement? What does each component tell you?
3. Other than familiarity, what are the advantages of the English system of measurement?
4. Define the metric standard units for length, mass, and time.
5. Does the density of a liquid change with the shape of a container? Explain.

6. Does a flattened pancake of clay have the same density as the same clay rolled into a ball? Explain.
7. Compare and contrast a scientific hypothesis and a scientific law.
8. What is a model? How are models used?
9. Are all theories always completely accepted or completely rejected? Explain.
10. What is pseudoscience and how can you always recognize it?

FOR FURTHER ANALYSIS

1. Select a statement that you feel might represent pseudoscience. Write an essay supporting *and* refuting your selection, noting facts that support one position or the other.
2. Evaluate the statement that science cannot solve human-produced problems such as pollution. What does it mean to say pollution is caused by humans and can only be solved by humans? Provide evidence that supports your position.
3. Make an experimental evaluation of what happens to the density of a substance at larger and larger volumes.
4. If your wage were dependent on your work-time squared, how would it affect your pay if you double your hours?

5. *Merriam-Webster's 11th Collegiate Dictionary* defines science, in part, as "knowledge or a system of knowledge covering general truths or the operation of general laws especially as obtained and tested through scientific method." How would you define science?
6. Are there any ways in which scientific methods differ from common-sense methods of reasoning?
7. The United States is the only country in the world that does not use the metric system of measurement. With this understanding, make a list of advantages and disadvantages for adopting the metric system in the United States.

INVITATION TO INQUIRY

Paper Helicopters

Construct paper helicopters and study the effects that various variables have on their flight. After considering the size you wish to test, copy the pattern shown in figure 1.15A on a sheet of notebook paper. Note that solid lines are to be cut and dashed lines are to be folded. Make three scissor cuts on the solid lines. Fold A toward you and B away from you to form the wings. Then fold C and D inward to overlap, forming the body. Finally, fold up the bottom on the dashed line and hold it together with a paper clip. Your finished product should look like the helicopter in figure 1.15B. Try a preliminary flight test by standing on a chair or stairs and dropping your helicopter.

Decide what variables you would like to study to find out how they influence the total flight time. Consider how you will hold everything else constant while changing one variable at a time. You can change the wing area by making new helicopters with more or less area in the A and B flaps. You can change the weight by adding more paper clips. Study these and other variables to find out who can design a helicopter that will remain in the air the longest. Who can design a helicopter that is most accurate in hitting a target?

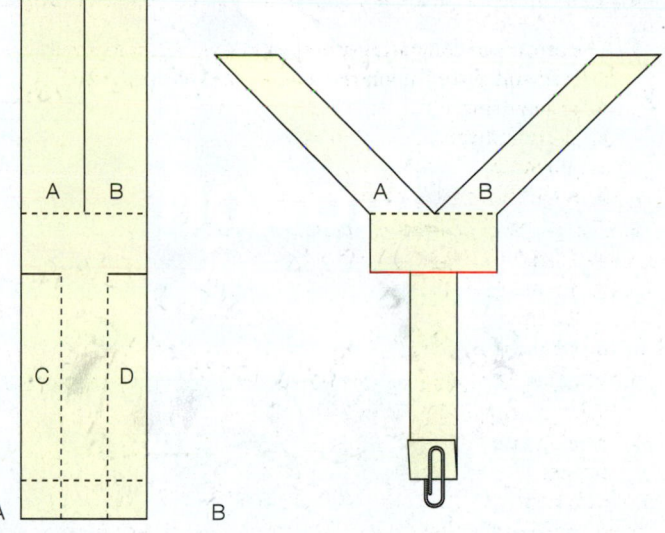

FIGURE 1.15   Pattern for a paper helicopter.

PARALLEL EXERCISES

The exercises in groups A and B cover the same concepts. Solutions to group A exercises are located in appendix G.

Group A

Note: You will need to refer to table 1.3 to complete some of the following exercises.

1. What is your height in meters? In centimeters?
2. What is the mass density of mercury if 20.0 cm³ has a mass of 272 g?
3. What is the mass of a 10.0 cm³ cube of lead?
4. What is the volume of a rock with a mass density of 3.00 g/cm³ and a mass of 600 g?
5. If you have 34.0 g of a 50.0 cm³ volume of one of the substances listed in table 1.3, which one is it?
6. What is the mass of water in a 40 L aquarium?
7. A 2.1 kg pile of aluminum cans is melted, then cooled into a solid cube. What is the volume of the cube?
8. A cubic box contains 1,000 g of water. What is the length of one side of the box in meters? Explain your reasoning.
9. A loaf of bread (volume 3,000 cm³) with a density of 0.2 g/cm³ is crushed in the bottom of a grocery bag into a volume of 1,500 cm³. What is the density of the mashed bread?
10. According to table 1.3, what volume of copper would be needed to balance a 1.00 cm³ sample of lead on a two-pan laboratory balance?

Group B

Note: You will need to refer to table 1.3 to complete some of the following exercises.

1. What is your mass in kilograms? In grams?
2. What is the mass density of iron if 5.0 cm³ has a mass of 39.5 g?
3. What is the mass of a 10.0 cm³ cube of copper?
4. If ice has a mass density of 0.92 g/cm³, what is the volume of 5,000 g of ice?
5. If you have 51.5 g of a 50.0 cm³ volume of one of the substances listed in table 1.3, which one is it?
6. What is the mass of gasoline ($\rho = 0.680$ g/cm³) in a 94.6 L gasoline tank?
7. What is the volume of a 2.00 kg pile of iron cans that are melted, then cooled into a solid cube?
8. A cubic tank holds 1,000.0 kg of water. What are the dimensions of the tank in meters? Explain your reasoning.
9. A hot dog bun (volume 240 cm³) with a density of 0.15 g/cm³ is crushed in a picnic cooler into a volume of 195 cm³. What is the new density of the bun?
10. According to table 1.3, what volume of iron would be needed to balance a 1.00 cm³ sample of lead on a two-pan laboratory balance?

2
Motion

Information about the mass of a hot air balloon and forces on the balloon will enable you to predict if it is going to move up, down, or drift across the land. This chapter is about such relationships between force, mass, and changes in motion.

CORE **CONCEPT**

A net force is required for any change in a state of motion.

OUTLINE

OVERVIEW

In chapter 1, you learned some "tools and rules" and some techniques for finding order in your surroundings. Order is often found in the form of patterns, or relationships between quantities that are expressed as equations. Equations can be used to (1) describe properties, (2) define concepts, and (3) describe how quantities change relative to one another. In all three uses, patterns are quantified, conceptualized, and used to gain a general understanding about what is happening in nature.

In the study of science, certain parts of nature are often considered and studied together for convenience. One of the more obvious groupings involves *movement*. Most objects around you appear to spend a great deal of time sitting quietly without motion. Buildings, rocks, utility poles, and trees rarely, if ever, move from one place to another. Even things that do move from time to time sit still for a great deal of time. This includes you, automobiles, and bicycles (figure 2.1). On the other hand, the Sun, the Moon, and starry heavens always seem to move, never standing still. Why do things stand still? Why do things move?

Questions about motion have captured the attention of people for thousands of years. But the ancient people answered questions about motion with stories of mysticism and spirits that lived in objects. It was during the classic Greek culture, between 600 B.C. and 300 B.C., that people began to look beyond magic and spirits. One particular Greek philosopher, Aristotle, wrote a theory about the universe that offered not only explanations about things such as motion but also offered a sense of beauty, order, and perfection. The theory seemed to fit with other ideas that people had and was held to be correct for nearly two thousand years after it was written. It was not until the work of Galileo and Newton during the 1600s that a new, correct understanding about motion was developed. The development of ideas about motion is an amazing and absorbing story. You will learn in this chapter how to describe and use some properties of motion. This will provide some basic understandings about motion and will be very helpful in understanding some important aspects of astronomy and the earth sciences, as well as the movement of living things.

2.1 DESCRIBING MOTION

Motion is one of the more common events in your surroundings. You can see motion in natural events such as clouds moving, rain and snow falling, and streams of water, all moving in a never-ending cycle. Motion can also be seen in the activities of people who walk, jog, or drive various machines from place to place. Motion is so common that you would think everyone would intuitively understand the concepts of motion, but history indicates that it was only during the past three hundred years or so that people began to understand motion correctly. Perhaps the correct concepts are subtle and contrary to common sense, requiring a search for simple, clear concepts in an otherwise complex situation. The process of finding such order in a multitude of sensory impressions by taking measurable data and then inventing a concept to describe what is happening is the activity called *science*. We will now apply this process to motion.

What is motion? Consider a ball that you notice one morning in the middle of a lawn. Later in the afternoon, you notice that the ball is at the edge of the lawn, against a fence, and you wonder if the wind or some person moved the ball. You do not know if the wind blew it at a steady rate, if many gusts of wind moved it, or even if some children kicked it all over the yard. All you know for sure is that the ball has been moved because it is in a different position after some time passed. These are

FIGURE 2.1 The motion of this snowboarder, and of other moving objects, can be described in terms of the distance covered during a certain time period.

the two important aspects of motion: (1) a change of position and (2) the passage of time.

If you did happen to see the ball rolling across the lawn in the wind, you would see more than the ball at just two locations. You would see the ball moving continuously. You could consider, however, the ball in continuous motion to be a series of individual locations with very small time intervals. Moving involves a change of position during some time period. Motion is the act or process of something changing position.

The motion of an object is usually described with respect to something else that is considered to be not moving. (Such a stationary object is said to be "at rest.") Imagine that you are traveling in an automobile with another person. You know that you

are moving across the land outside the car since your location on the highway changes from one moment to another. Observing your fellow passenger, however, reveals no change of position. You are in motion relative to the highway outside the car. You are not in motion relative to your fellow passenger. Your motion, and the motion of any other object or body, is the process of a change in position *relative* to some reference object or location. Thus, *motion* can be defined as the act or process of changing position relative to some reference during a period of time.

2.2 MEASURING MOTION

You have learned that objects can be described by measuring certain fundamental properties such as mass and length. Since motion involves (1) a change of position and (2) the passage of *time*, the motion of objects can be described by using combinations of the fundamental properties of length and time. Combinations of these measurements describe three properties of motion: *speed, velocity,* and *acceleration.*

Speed

Suppose you are in a car that is moving over a straight road. How could you describe your motion? You need at least two measurements: (1) the distance you have traveled and (2) the time that has elapsed while you covered this distance. Such a distance and time can be expressed as a ratio that describes your motion. This ratio is a property of motion called **speed,** which is a measure of how fast you are moving. Speed is defined as distance per unit of time, or

$$\text{speed} = \frac{\text{distance}}{\text{time}}$$

The units used to describe speed are usually miles/hour (mi/h), kilometers/hour (km/h), or meters/second (m/s).

Let's go back to your car that is moving over a straight highway and imagine you are driving to cover equal distances in equal periods of time (figure 2.2). If you use a stopwatch to measure the time required to cover the distance between highway mile markers (those little signs with numbers along major highways), the time intervals will all be equal. You might find, for example, that one minute elapses between each mile marker. Such a uniform straight-line motion that covers equal distances in equal periods of time is the simplest kind of motion.

If your car were moving over equal distances in equal periods of time, it would have a *constant speed.* This means that the car is neither speeding up nor slowing down. It is usually difficult to maintain a constant speed. Other cars and distractions such as interesting scenery cause you to reduce your speed. At other times you increase your speed. If you calculate your speed over an entire trip, you are considering a large distance between two places and the total time that elapsed. The increases and decreases in speed would be averaged. Therefore, most speed

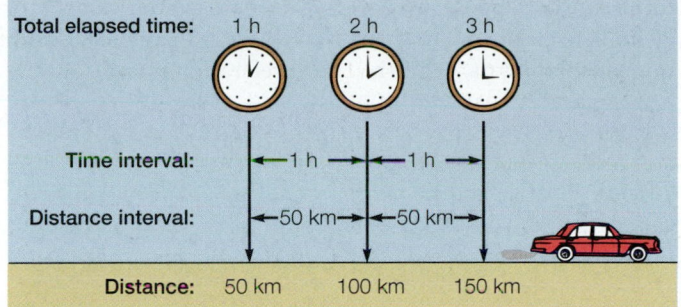

Total elapsed time:	1 h	2 h	3 h

Time interval: ←—1 h —→ ←—1 h —→

Distance interval: ←—50 km—→ ←—50 km—→

Distance:	50 km	100 km	150 km

FIGURE 2.2 If you know the value of any two of the three variables of distance, time, and speed, you can find the third. What is the average speed of this car?

calculations are for an *average speed*. The speed at any specific instant is called the *instantaneous speed*. To calculate the instantaneous speed, you would need to consider a very short time interval—one that approaches zero. An easier way would be to use the speedometer, which shows the speed at any instant.

It is easier to study the relationships between quantities if you use symbols instead of writing out the whole word. The letter v can be used to stand for speed when dealing with straight-line motion, which is the only kind of motion that will be considered in the problems in this text. A bar over the v ($\bar{v}$) is a symbol that means average (it is read "v-bar" or "v-average"). The letter d can be used to stand for distance and the letter t to stand for time. The relationship between average speed, distance, and time is therefore

$$\bar{v} = \frac{d}{t}$$

equation 2.1

This is one of the three types of equations that were discussed earlier, and in this case, the equation defines a motion property.

Constant, instantaneous, or average speeds can be measured with any distance and time units. Common units in the English system are miles/hour and feet/second. Metric units for speed are commonly kilometers/hour and meters/second. The ratio of any distance/time is usually read as distance per time, such as miles per hour.

Velocity

The word *velocity* is sometimes used interchangeably with the word *speed*, but there is a difference. **Velocity** describes the *speed and direction* of a moving object. For example, a speed might be described as 60 km/h. A velocity might be described as 60 km/h to the west. To produce a change in velocity, either the speed or the direction is changed (or both are changed). A satellite moving with a constant speed in a circular orbit around Earth does not have a constant velocity since its direction of movement is constantly changing. Velocities can be represented graphically with arrows. The lengths of the arrows are proportional to the speed, and the arrowheads indicate the direction (figure 2.3).

EXAMPLE 2.1 *(Optional)*

The driver of a car moving at 72.0 km/h drops a road map on the floor. It takes her 3.00 seconds to locate and pick up the map. How far did she travel during this time?

SOLUTION

The car has a speed of 72.0 km/h and the time factor is 3.00 s, so km/h must be converted to m/s. From inside the front cover of this book, the conversion factor is 1 km/h = 0.2778 m/s, so

$$\bar{v} = \frac{0.2778 \frac{m}{s}}{\frac{km}{h}} \times 72.0 \frac{km}{h}$$

$$= (0.2778)(72.0) \frac{m}{s} \times \frac{\cancel{h}}{\cancel{km}} \times \frac{\cancel{km}}{\cancel{h}}$$

$$= 20.0 \frac{m}{s}$$

The relationship between the three variables, $\bar{v}$, t, and d, is found in equation 2.1: $\bar{v} = d/t$

$$\bar{v} = 20.0 \frac{m}{s} \qquad \bar{v} = \frac{d}{t}$$

$$t = 3.00 \text{ s} \qquad \bar{v}t = \frac{d\cancel{t}}{\cancel{t}}$$

$$d = ? \qquad d = \bar{v}t$$

$$= \left(20.0 \frac{m}{s}\right)(3.00 \text{ s})$$

$$= (20.0)(3.00) \frac{m}{\cancel{s}} \times \frac{\cancel{s}}{1}$$

$$= \boxed{60.0 \text{ m}}$$

EXAMPLE 2.2 *(Optional)*

A bicycle has an average speed of 8.00 km/h. How far will it travel in 10.0 seconds? (Answer: 22.2 m)

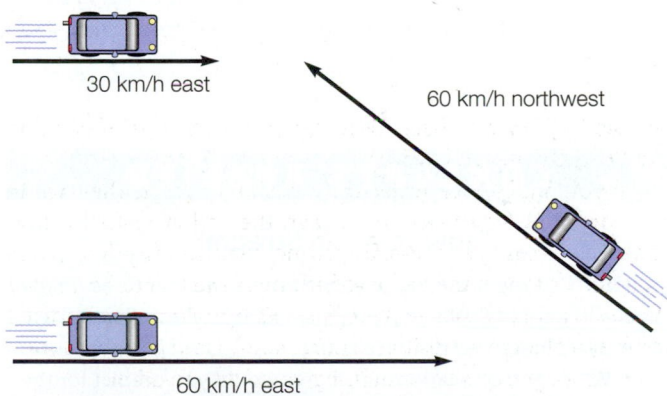

30 km/h east

60 km/h northwest

60 km/h east

FIGURE 2.3 Velocity can be presented graphically with arrows. Here are three different velocities represented by three different arrows. The length of each arrow is proportional to the speed, and the arrowhead shows the direction of travel.

Acceleration

Motion can be changed in three different ways: (1) by changing the speed, (2) by changing the direction of travel, or (3) by changing both the speed and the direction of travel. Since velocity describes both the speed and the direction of travel, any of these three changes will result in a change of velocity. You need at least one additional measurement to describe a change of motion, which is how much time elapsed while the change was taking place. The change of velocity and time can be combined to define the *rate* at which the motion was changed. This rate is called **acceleration.** Acceleration is defined as a change of velocity per unit time, or

$$\text{acceleration} = \frac{\text{change of velocity}}{\text{time elapsed}}$$

Another way of saying "change in velocity" is the final velocity minus the initial velocity, so the relationship can also be written as

$$\text{acceleration} = \frac{\text{final velocity} - \text{initial velocity}}{\text{time elapsed}}$$

Acceleration due to a change in speed only can be calculated as follows. Consider a car that is moving with a constant, straight-line velocity of 60 km/h when the driver accelerates

CONCEPTS APPLIED

Style Speeds

Observe how many different styles of walking you can identify in students walking across the campus. Identify each style with a descriptive word or phrase.

Is there any relationship between any particular style of walking and the speed of walking? You could find the speed of walking by measuring a distance, such as the distance between two trees, then measuring the time required for a student to walk the distance. Find the average speed for each identified style of walking by averaging the walking speeds of ten people.

Report any relationships you find between styles of walking and the average speed of people with each style. Include any problems you found in measuring, collecting data, and reaching conclusions.

CONCEPTS APPLIED

How Fast Is a Stream?

A stream is a moving body of water. How could you measure the speed of a stream? Would timing how long it takes a floating leaf to move a measured distance help?

What kind of relationship, if any, would you predict for the speed of a stream and a recent rainfall? Would you predict a direct relationship? Make some measurements of stream speeds and compare your findings to recent rainfall amounts.

to 80 km/h. Suppose it takes 4 s to increase the velocity of 60 km/h to 80 km/h. The change in velocity is therefore 80 km/h minus 60 km/h, or 20 km/h. The acceleration was

$$\text{acceleration} = \frac{80\,\dfrac{\text{km}}{\text{h}} - 60\,\dfrac{\text{km}}{\text{h}}}{4\text{s}}$$

$$= \frac{20\,\dfrac{\text{km}}{\text{h}}}{4\text{s}}$$

$$= 5\frac{\text{km/h}}{\text{s}} \text{ or,}$$

$$= 5 \text{ km/h/s}$$

The average acceleration of the car was 5 km/h for each ("per") second. This is another way of saying that the velocity increases an average of 5 km/h in each second. The velocity of the car was 60 km/h when the acceleration began (initial velocity). At the end of 1 s, the velocity was 65 km/h. At the end of 2 s, it was 70 km/h; at the end of 3 s, 75 km/h; and at the end of 4 s (total time elapsed), the velocity was 80 km/h (final velocity). Note how fast the velocity is changing with time. In summary,

start (initial velocity)	60 km/h
end of first second	65 km/h
end of second second	70 km/h
end of third second	75 km/h
end of fourth second (final velocity)	80 km/h

As you can see, acceleration is really a description of how fast the speed is changing; in this case, it is increasing 5 km/h each second.

Usually, you would want all the units to be the same, so you would convert km/h to m/s. A change in velocity of 5.0 km/h converts to 1.4 m/s and the acceleration would be 1.4 m/s/s. The units "m/s per s" mean what change of velocity (1.4 m/s) is occurring every second. The combination "m/s/s" is rather cumbersome, so it is typically treated mathematically to simplify the expression (to simplify a fraction, invert the divisor and multiply, or m/s × 1/s = m/s²). Remember that the expression 1.4 m/s² means the same as 1.4 m/s per s, a change of velocity in a given time period.

The relationship among the quantities involved in acceleration can be represented with the symbols a for average acceleration, v_f for final velocity, v_i for initial velocity, and t for time. The relationship is

$$a = \frac{v_f - v_i}{t} \qquad \textbf{equation 2.2}$$

There are also other changes in the motion of an object that are associated with acceleration. One of the more obvious is a change that results in a decreased velocity. Your car's brakes, for example, can slow your car or bring it to a complete stop. This is *negative acceleration,* which is sometimes called *deceleration.* Another change in the motion of an object is a change of direction. Velocity encompasses both the rate of motion as well as direction, so a change of direction is an acceleration. The satellite moving with a constant speed in a circular orbit around Earth is constantly changing its direction of movement. It is therefore constantly accelerating because of this constant change in its motion. Your automobile has three devices that could change the state of its motion. Your automobile therefore has three accelerators—the gas pedal (which can increase magnitude of velocity), the brakes (which can decrease magnitude of velocity), and the steering wheel (which can change direction of velocity). (See figure 2.4.) The important thing to remember is that acceleration results from any *change* in the motion of an object.

EXAMPLE 2.3 *(Optional)*

A bicycle moves from rest to 5 m/s in 5 s. What was the acceleration?

SOLUTION

$$v_i = 0 \text{ m/s}$$
$$v_f = 5 \text{ m/s}$$
$$t = 5 \text{ s}$$
$$a = ?$$

$$a = \frac{v_f - v_i}{t}$$

$$= \frac{5 \text{ m/s} - 0 \text{ m/s}}{5 \text{ s}}$$

$$= \frac{5}{5} \frac{\text{m/s}}{\text{s}}$$

$$= 1 \frac{\text{m}}{\text{s}} \times \frac{1}{\text{s}}$$

$$= \boxed{1 \frac{\text{m}}{\text{s}^2}}$$

CONCEPTS APPLIED

Acceleration Patterns

Suppose the radiator in your car has a leak and drops of coolant fall constantly, one every second. What pattern would the drops make on the pavement when you accelerate the car from a stoplight? What pattern would they make when you drive a constant speed? What pattern would you observe as the car comes to a stop? Use a marker to make dots on a sheet of paper that illustrate (1) acceleration, (2) constant speed, and (3) negative acceleration. Use words to describe the acceleration in each situation.

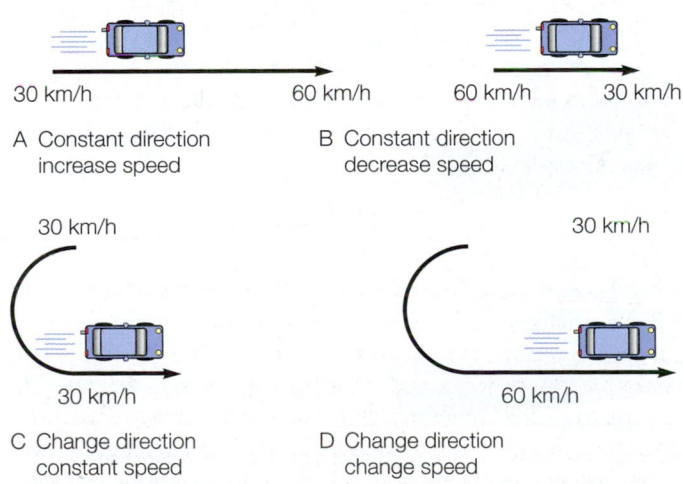

A Constant direction
 increase speed

B Constant direction
 decrease speed

C Change direction
 constant speed

D Change direction
 change speed

FIGURE 2.4   Four different ways (*A–D*) to accelerate a car.

The super-speed magnetic levitation (maglev) train is a completely new technology based on magnetically suspending a train 3 to 10 cm (about 1 to 4 in) above a monorail, then moving it along with a magnetic field that travels along the monorail guides. The maglev train does not have friction between wheels and the rails since it does not have wheels. This lack of resistance and the easily manipulated magnetic fields makes very short acceleration distances possible. For example, a German maglev train can accelerate from 0 to 300 km/h (about 185 mi/h) over a distance of just 5 km (about 3 mi). A conventional train with wheels requires about 30 km (about 19 mi) in order to reach the same speed from a standing start. The maglev is attractive for short runs because of its superior acceleration. It is also attractive for longer runs because of its high top speed—up to about 500 km/h (about 310 mi/h). Today, only an aircraft can match such a speed.

EXAMPLE 2.4   *(Optional)*

An automobile uniformly accelerates from rest at 5 m/s² for 6 s. What is the final velocity in m/s? (Answer: 30 m/s)

2.3 FORCES

The Greek philosopher Aristotle considered some of the first ideas about the causes of motion back in the fourth century B.C. However, he had it all wrong when he reportedly stated that a dropped object falls at a constant speed that is determined by its weight. He also incorrectly thought that an object moving across Earth's surface requires a continuously applied force to continue moving. These ideas were based on observing and thinking, not measurement, and no one checked to see if they were correct. It took about two thousand years before people began to correctly understand motion.

Aristotle did recognize an association between force and motion, and this much was acceptable. It is partly correct because a force is closely associated with *any* change of motion, as you will see. This section introduces the concept of a force, which will be developed more fully when the relationship between forces and motion is considered later.

A **force** is a push or a pull that is capable of changing the state of motion of an object. Consider, for example, the movement of a ship from the pushing of two tugboats (figure 2.5). Tugboats can vary the strength of the force exerted on a ship, but they can also push in different directions. What effect does direction have on two forces acting on an object? If the tugboats were side by side, pushing in the same direction, the overall force is the sum of the two forces. If they act in exactly opposite directions, one pushing on each side of the ship, the overall force is the difference between the strength of the two forces. If they have the same strength, the overall effect is to cancel each other without producing any motion. The **net force** is the sum of all the forces acting on an object. Net force means "final," after the forces are added (figure 2.6).

FIGURE 2.5 The rate of movement and the direction of movement of this ship are determined by a combination of direction and size of force from each of the tugboats. Which direction are the two tugboats pushing? What evidence would indicate that one tugboat is pushing with a greater force? If the tugboat by the numbers is pushing with a greater force and the back tugboat is keeping the back of the ship from moving, what will happen?

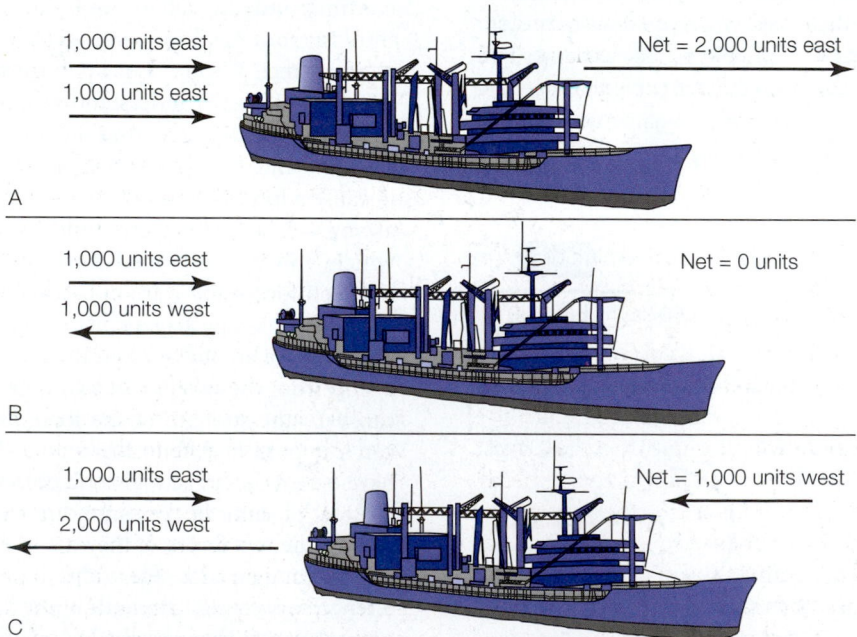

1,000 units east
1,000 units east
Net = 2,000 units east

A

1,000 units east
1,000 units west
Net = 0 units

B

1,000 units east
2,000 units west
Net = 1,000 units west

C

FIGURE 2.6 (*A*) When two parallel forces are acting on the ship in the same direction, the net force is the two forces added together. (*B*) When two forces are opposite and of equal size, the net force is zero. (*C*) When two parallel forces are not of equal size, the net force is the difference in the direction of the larger force.

When two parallel forces act in the same direction, they can be simply added. In this case, there is a net force that is equivalent to the sum of the two forces. When two parallel forces act in opposite directions, the net force is the difference between the two forces and is in the direction of the larger force. When two forces act neither in a way that is exactly together nor exactly opposite each other, the result will be like a new, different force having a new direction and strength.

Forces have a strength and direction that can be represented by force arrows. The tail of the arrow is placed on the object that feels the force, and the arrowhead points in the direction in which the force is exerted. The length of the arrow is proportional to the strength of the force. The use of force arrows helps you visualize and understand all the forces and how they contribute to the net force.

There are four **fundamental forces** that *cannot* be explained in terms of any other force. They are gravitational, electromagnetic, weak, and the strong nuclear force. Gravitational forces act between all objects in the universe—between you and Earth, between Earth and the Sun, between the planets in the solar systems, and, in fact, hold stars in large groups called galaxies. Switching scales from the very large galaxy to inside an atom, we find electromagnetic forces acting between electrically charged parts of atoms, such as electrons and protons. Electromagnetic forces are responsible for the structure of atoms, chemical change, and electricity and magnetism. Weak and strong forces act inside the nucleus of an atom, so they are not as easily observed at work as are gravitational and electromagnetic forces. The weak force is involved in certain nuclear reactions. The strong nuclear force is involved in close-range holding of the nucleus together. In general, the strong nuclear force between particles inside a nucleus is about 10^2 times stronger than the electromagnetic force and about 10^{39} times stronger than the gravitational force. The fundamental forces are responsible for everything that happens in the universe, and we will learn more about them in chapters on electricity, light, nuclear energy, chemistry, geology, and astronomy.

2.4 HORIZONTAL MOTION ON LAND

Everyday experience seems to indicate that Aristotle's idea about horizontal motion on Earth's surface is correct. After all, moving objects that are not pushed or pulled do come to rest in a short period of time. It would seem that an object keeps moving only if a force continues to push it. A moving automobile will slow and come to rest if you turn off the ignition. Likewise, a ball that you roll along the floor will slow until it comes to rest. Is the natural state of an object to be at rest, and is a force necessary to keep an object in motion? This is exactly what people thought until Galileo (figure 2.7) published his book *Two New Sciences* in 1638, which described his findings about motion.

The book had three parts that dealt with uniform motion, accelerated motion, and projectile motion. Galileo described details of simple experiments, measurements, calculations, and thought experiments as he developed definitions and concepts of motion. In one of his thought experiments, Galileo presented an argument against Aristotle's view that a force is needed to keep an object in motion. Galileo imagined an object (such as a ball) moving over a horizontal surface without the force of friction. He concluded that the object would move forever with a constant velocity as long as there was no unbalanced force acting to change the motion.

Why does a rolling ball slow to a stop? You know that a ball will roll farther across a smooth, waxed floor such as a bowling lane than it will across floor covered with carpet. The rough carpet offers more resistance to the rolling ball. The resistance of the floor friction is shown by a force arrow, F_{floor}, in figure 2.8. This force, along with the force arrow for air resistance, F_{air}, oppose the forward movement of the ball. Notice the dashed line arrow in part A of figure 2.8. There is no other force applied to the ball, so the rolling speed decreases until the ball finally comes to a complete stop. Now imagine what force you would need to exert by pushing with your hand, moving along with the ball to keep it rolling at a uniform rate. An examination of the forces in part B of figure 2.8 can help you determine the amount of force. The force

FIGURE 2.7 Galileo (left) challenged the Aristotelian view of motion and focused attention on the concepts of distance, time, velocity, and acceleration.

you apply, $F_{applied}$, must counteract the resistance forces. It opposes the forces that are slowing down the ball as illustrated by the direction of the arrows. To determine how much force you should apply, look at the arrow equation. $F_{applied}$ has the same length as the sum of the two resistance forces, but it is in the opposite direction of the resistance forces. Therefore, the overall force, F_{net}, is zero. The ball continues to roll at a uniform rate when you *balance* the force opposing its motion. It is reasonable, then, that if there were no opposing forces, you would not need to apply a force to keep it rolling. This was the kind of reasoning that Galileo did when he discredited the Aristotelian view that a force was necessary to keep an object moving. Galileo concluded that a moving object would continue moving with a constant velocity if no unbalanced forces were applied; that is, if the net force were zero.

It could be argued that the difference in Aristotle's and Galileo's views of forced motion is really a degree of analysis. After all, moving objects on Earth do come to rest unless continuously pushed or pulled. But Galileo's conclusion describes *why* they must be pushed or pulled and reveals the true nature of the motion of objects. Aristotle argued that the natural state of objects is to be at rest and attempted to explain why objects move. Galileo, on the other hand, argued that it is just as natural for an object to be moving and attempted to explain why they come to rest. The behavior of matter to persist in its state of motion is called **inertia**. Inertia is the *tendency of an object to remain in unchanging motion whether actually moving or at rest, when the net force is zero*. The development of this concept changed the way people viewed the natural state of an object and opened the way for further understandings about motion. Today, it is understood that a satellite moving through free space will continue to do so with no unbalanced forces acting on it (figure 2.9A). An unbalanced force is needed to slow the satellite (figure 2.9B), increase its speed (figure 2.9C), or change its direction of travel (figure 2.9D).

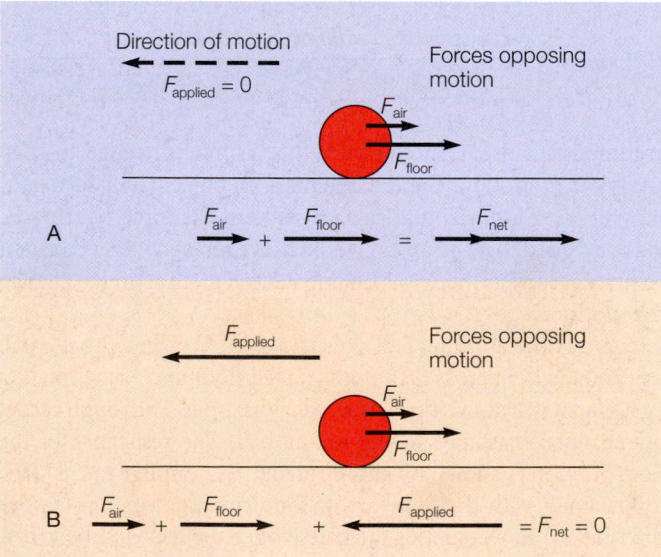

FIGURE 2.8 (*A*) This ball is rolling to your left with no forces in the direction of motion. The sum of the force of floor friction (F_{floor}) and the force of air friction (F_{air}) results in a net force opposing the motion, so the ball slows to a stop. (*B*) A force is applied to the moving ball, perhaps by a hand that moves along with the ball. The force applied ($F_{applied}$) equals the sum of the forces opposing the motion, so the ball continues to move with a constant velocity.

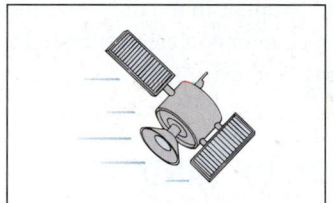

A No force—constant speed in straight line

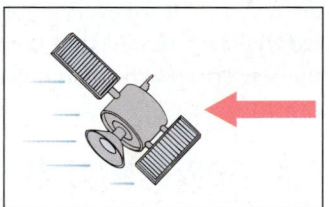

B Force applied against direction of motion

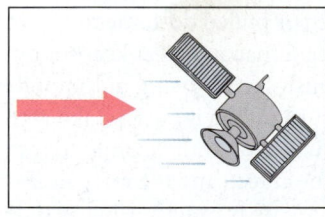

C Force applied in same direction as motion

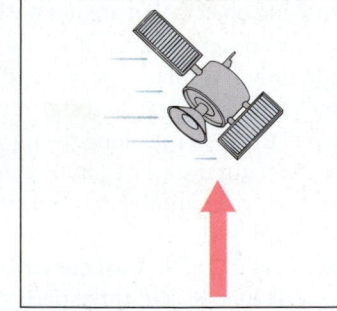

D Force applied sideways to direction of motion

FIGURE 2.9 Explain how the combination of drawings (*A–D*) illustrates inertia.

A Closer Look

A Bicycle Racer's Edge

Galileo was one of the first to recognize the role of friction in opposing motion. As shown in figure 2.8, friction with the surface and air friction combine to produce a net force that works against anything that is moving on the surface. This article is about air friction and some techniques that bike riders use to reduce that opposing force—perhaps giving them an edge in a close race.

The bike riders in box figure 2.1 are forming a single-file line, called a *paceline*, because the slip-stream reduces the air resistance for a closely trailing rider. Cyclists say that riding in the slipstream of another cyclist will save much of their energy. In fact, the cyclists will be able to move 8 km/h faster than they would expending the same energy while riding alone.

In a sense, riding in a slipstream means that you do not have to push as much air out of your way. It has been estimated that at 32 km/h a cyclist must move a little less than half a ton of air out of the way every minute.

Along with the problem of moving about a half-ton of air out of the way every

Box Figure 2.1   The object of the race is to be in the front, to finish first. If this is true, why are these racers forming single-file lines?

minute, there are two basic factors related to air resistance. These are (1) a turbulent versus a smooth flow of air and (2) the problem of frictional drag. A turbulent flow of air contributes to air resistance because it causes the air to separate slightly on the back side, which increases the pressure on the front of the moving object. This is why racing cars, airplanes, boats, and other racing vehicles are streamlined to a teardroplike shape. This shape is not as likely to have the lower-pressure-producing air turbulence behind (and resulting greater

pressure in front) because it smoothes or streamlines the air flow.

The frictional drag of air is similar to the frictional drag that occurs when you push a book across a rough tabletop. You know that smoothing the rough tabletop will reduce the frictional drag on the book. Likewise, the smoothing of a surface exposed to moving air will reduce air friction. Cyclists accomplish this "smoothing" by wearing smooth spandex clothing and by shaving hair from arm and leg surfaces that are exposed to moving air. Each hair contributes to the overall frictional drag, and removal of the arm and leg hair can thus result in seconds saved. This might provide enough of an edge to win a close race. Shaving legs and arms, together with the wearing of spandex or some other tight, smooth-fitting garments, are just a few of the things a cyclist can do to gain an edge. Perhaps you will be able to think of more ways to reduce the forces that oppose motion.

Myths, Mistakes, and Misunderstandings

Walk or Run in Rain?

Is it a mistake to run in rain if you want to stay drier? One idea is that you should run because you spend less time in the rain so you will stay drier. On the other hand, this is true only if the rain lands only on the top of your head and shoulders. If you run, you will end up running into more raindrops on the larger surface area of your face, chest, and front of your legs.

Two North Carolina researchers looked into this question with one walking and the other running over a measured distance while wearing cotton sweatsuits. They then weighed their clothing and found that the walking person's sweatsuit weighed more. This means you should run to stay drier.

2.5 FALLING OBJECTS

Did you ever wonder what happens to a falling rock during its fall? Aristotle reportedly thought that a rock falls at a uniform speed that is proportional to its weight. Thus, a heavy

rock would fall at a faster uniform speed than a lighter rock. As stated in a popular story, Galileo discredited Aristotle's conclusion by dropping a solid iron ball and a solid wooden ball simultaneously from the top of the Leaning Tower of Pisa (figure 2.10). Both balls, according to the story, hit the ground nearly at the same time. To do this, they would have to fall with the same velocity. In other words, the velocity of a falling object does not depend on its weight. Any difference in freely falling bodies is explainable by air resistance. Soon after the time of Galileo, the air pump was invented. The air pump could be used to remove the air from a glass tube. The effect of air resistance on falling objects could then be demonstrated by comparing how objects fall in the air with how they fall in an evacuated glass tube. You know that a coin falls faster than a feather when they are dropped together in the air. A feather and heavy coin will fall together in the near vacuum of an evacuated glass tube because the effect of air resistance on the feather has been removed. When objects fall toward Earth without considering air resistance, they are said to be in **free fall.** Free fall considers only gravity and neglects air resistance.

Galileo concluded that light and heavy objects fall together in free fall, but he also wanted to know the details of what was going on while they fell. He now knew that the velocity of an

Connections . . .

Sports

There are two different meanings for the term *free fall*. In physics, *free fall* means the unconstrained motion of a body in a gravitational field, without considering air resistance. Without air resistance, all objects are assumed to accelerate toward the surface at 9.8 m/s².

In the sport of skydiving, *free fall* means falling within the atmosphere without a drag-producing device such as a parachute. Air provides a resisting force that opposes the motion of a falling object, and the net force is the difference between the downward force (weight) and the upward force of air resistance. The weight of the falling object depends on the mass and acceleration from gravity, and this is the force downward. The resisting force is determined by at least two variables: (1) the area of the object exposed to the airstream and (2) the speed of the falling object. Other variables such as streamlining, air temperature, and turbulence play a role, but the greatest effect seems to be from exposed area and the increased resistance as speed increases.

A skydiver's weight is constant, so the downward force is constant. Modern skydivers typically free-fall from about 3,650 m (about 12,000 ft) above the ground until about 750 m (about 2,500 ft), where they open their parachutes. After jumping from the plane, the diver at first accelerates toward the surface, reaching speeds up to about 185–210 km/h (about 115–130 mi/h). The air resistance increases with increased speed and the net force becomes less and less. Eventually, the downward weight force will be balanced by the upward air resistance force, and the net force becomes zero. The person now falls at a constant speed, and we say the terminal velocity has been reached. It is possible to change your body position to vary your rate of fall up or down by 32 km/h (about 20 mi/h). However, by diving or "standing up" in free fall, experienced skydivers can reach speeds of up to 290 km/h (about 180 mi/h). The record free fall speed, done without any special equipment, is 517 km/h (about 321 mi/h). Once the parachute opens, a descent rate of about 16 km/h (about 10 mi/h) is typical.

FIGURE 2.10 According to a widespread story, Galileo dropped two objects with different weights from the Leaning Tower of Pisa. They reportedly hit the ground at about the same time, discrediting Aristotle's view that the speed during the fall is proportional to weight.

Falling Bodies

Galileo concluded that all objects fall together, with the same acceleration, when the upward force of air resistance is removed. It would be most difficult to remove air from the room, but it is possible to do some experiments that provide some evidence of how air influences falling objects.

1. Take a sheet of paper and your textbook and drop them side by side from the same height. Note the result.
2. Place the sheet of paper on top of the book and drop them at the same time. Do they fall together?
3. Crumple the sheet of paper into a loose ball and drop the ball and book side by side from the same height.
4. Crumple a sheet of paper into a very tight ball and again drop the ball and book side by side from the same height.

Explain any evidence you found concerning how objects fall.

object in free fall was *not* proportional to the weight of the object. He observed that the velocity of an object in free fall *increased* as the object fell and reasoned from this that the velocity of the falling object would have to be (1) somehow proportional to the *time* of fall and (2) somehow proportional to the *distance* the object fell. If the time and distance were both related to the velocity of a falling object at a given time and distance, how were they related to one another?

Galileo reasoned that a freely falling object should cover a distance *proportional to the square of the time of the fall* ($d \propto t^2$). In other words, the object should fall 4 times as far in 2 s as in 1 s ($2^2 = 4$), 9 times as far in 3 s ($3^2 = 9$), and so on. Galileo checked this calculation by rolling balls on an inclined board with a smooth groove in it. He used the inclined board to slow the motion of descent in order to measure the distance and time relationships, a necessary requirement since he lacked the accurate timing devices that exist today. He found, as predicted, that the falling balls moved through a distance proportional to the square of the time of falling. This also means that the *velocity of the falling object increased at a constant rate*. Recall that a change of velocity during some time period is called *acceleration*. In other words, a falling object *accelerates* toward the surface of Earth.

Since the velocity of a falling object increases at a constant rate, this must mean that falling objects are *uniformly accelerated* by the force of gravity. *All objects in free fall experience a constant acceleration.* During each second of fall, the object on Earth gains 9.8 m/s (32 ft/s) in velocity. This gain is the acceleration of the falling object, 9.8 m/s² (32 ft/s²).

The acceleration of objects falling toward Earth varies slightly from place to place on Earth's surface because of Earth's shape and spin. The acceleration of falling objects decreases from the poles to the equator and also varies from place to place because Earth's mass is not distributed equally. The value of 9.8 m/s² (32 ft/s²) is an approximation that is fairly close to, but not exactly, the acceleration due to gravity in any particular location. The acceleration due to gravity is important in a number of situations, so the acceleration from this force is given a special symbol, **g**.

2.6 COMPOUND MOTION

So far we have considered two types of motion: (1) the horizontal, straight-line motion of objects moving on the surface of Earth and (2) the vertical motion of dropped objects that accelerate toward the surface of Earth. A third type of motion occurs when an object is thrown, or projected, into the air. Essentially, such a projectile (rock, football, bullet, golf ball, or whatever) could be directed straight upward as a vertical projection, directed straight out as a horizontal projection, or directed at some angle between the vertical and the horizontal. Basic to understanding such compound motion is the observation that (1) gravity acts on objects *at all times,* no matter where they are, and (2) the acceleration due to gravity (*g*) is *independent of any motion* that an object may have.

Vertical Projectiles

Consider first a ball that you throw straight upward, a vertical projection. The ball has an initial velocity but then reaches a maximum height, stops for an instant, then accelerates back toward Earth. Gravity is acting on the ball throughout its climb, stop, and fall. As it is climbing, the force of gravity is continually reducing its velocity. The overall effect during the climb is deceleration, which continues to slow the ball until the instantaneous stop. The ball then accelerates back to the surface just like a ball that has been dropped. If it were not for air resistance, the ball would return with the same velocity that it had initially. The velocity arrows for a ball thrown straight up are shown in figure 2.11.

Horizontal Projectiles

Horizontal projectiles are easier to understand if you split the complete motion into vertical and horizontal parts. Consider, for example, an arrow shot horizontally from a bow. The force of gravity accelerates the arrow downward, giving it an increasing downward velocity as it moves through the air. This increasing downward velocity is shown in figure 2.12 as increasingly longer velocity arrows (v_v). There are no forces in the horizontal direction if you can ignore air resistance, so the horizontal velocity of the arrow remains the same as shown by the v_h velocity arrows. The combination of the increasing vertical (v_v) motion and the unchanging horizontal (v_h) motion causes the arrow to follow a curved path until it hits the ground.

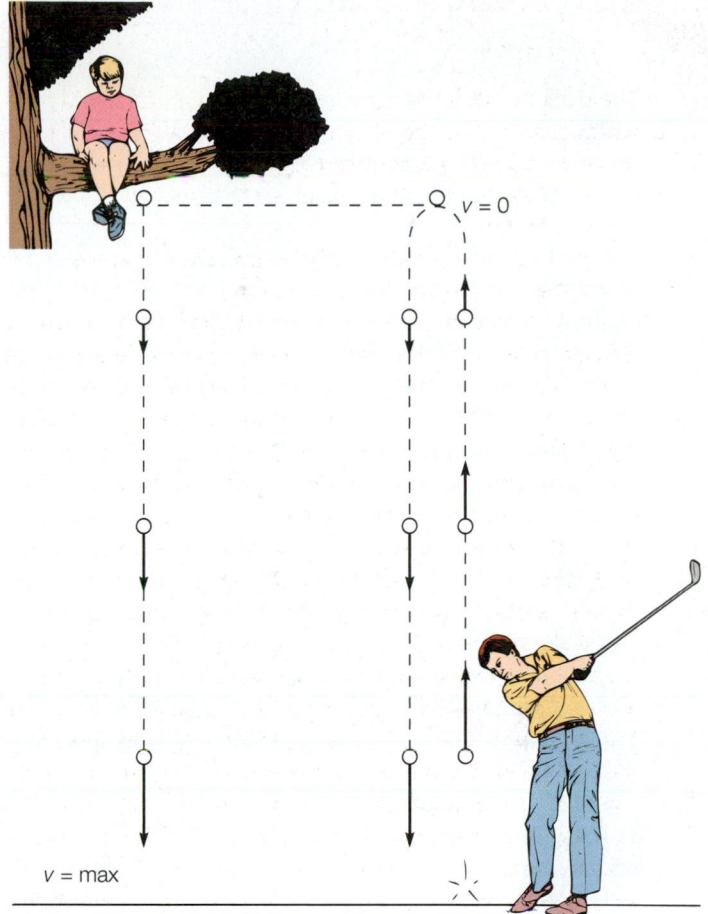

FIGURE 2.11   On its way up, a vertical projectile such as this misdirected golf ball is slowed by the force of gravity until an instantaneous stop; then it accelerates back to the surface, just as another golf ball does when dropped from the same height. The straight up- and down-moving golf ball has been moved to the side in the sketch so we can see more clearly what is happening.

An interesting prediction that can be made from the shot arrow analysis is that an arrow shot horizontally from a bow will hit the ground at the same time as a second arrow that is simply dropped from the same height (figure 2.12). Would this be true of a bullet dropped at the same time as one fired horizontally from a rifle? The answer is yes; both bullets would hit the ground at the same time. Indeed, without air resistance, all the bullets and arrows should hit the ground at the same time if dropped or shot from the same height.

Golf balls, footballs, and baseballs are usually projected upward at some angle to the horizon. The horizontal motion of these projectiles is constant as before because there are no horizontal forces involved. The vertical motion is the same as that of a ball projected directly upward. The combination of these two motions causes the projectile to follow a curved path called a *parabola,* as shown in figure 2.13. The next time you have the opportunity, observe the path of a ball that has been

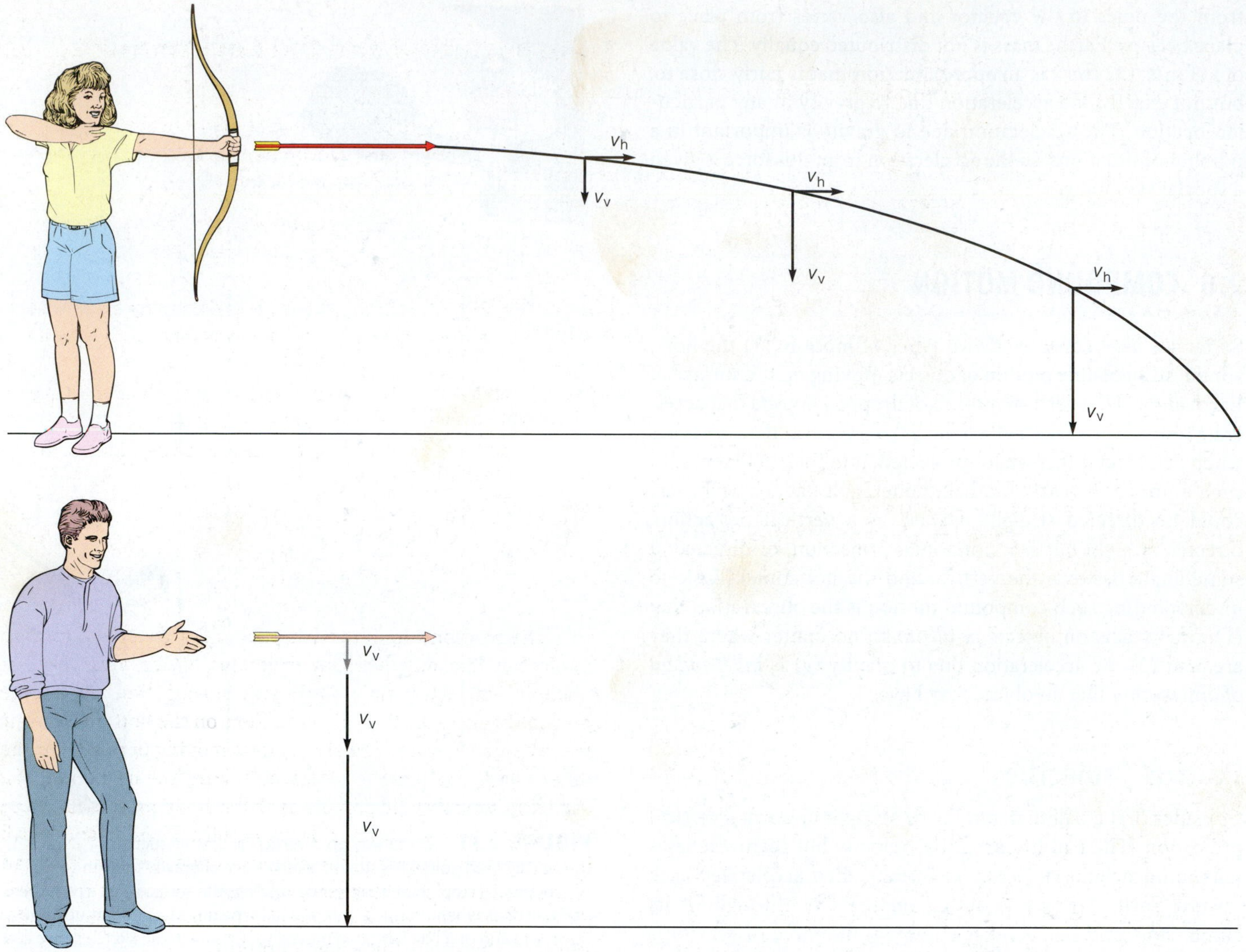

FIGURE 2.12 A horizontal projectile has the same horizontal velocity throughout the fall as it accelerates toward the surface, with the combined effect resulting in a curved path. Neglecting air resistance, an arrow shot horizontally will strike the ground at the same time as one dropped from the same height above the ground, as shown here by the increasing vertical velocity arrows.

FIGURE 2.13 A football is thrown at some angle to the horizon when it is passed downfield. Neglecting air resistance, the horizontal velocity is a constant, and the vertical velocity decreases, then increases, just as in the case of a vertical projectile. The combined motion produces a parabolic path. Contrary to statements by sportscasters about the abilities of certain professional quarterbacks, it is impossible to throw a football with a "flat trajectory" because it begins to accelerate toward the surface as soon as it leaves the quarterback's hand.

projected at some angle (figure 2.14). Note that the second half of the path is almost a reverse copy of the first half. If it were not for air resistance, the two values of the path would be exactly the same. Also note the distance that the ball travels as compared to the angle of projection. An angle of projection of 45° results in the maximum distance of travel.

2.7 THREE LAWS OF MOTION

In the previous sections, you learned how to describe motion in terms of distance, time, velocity, and acceleration. In addition, you learned about different kinds of motion, such as straight-line motion, the motion of falling objects, and the compound motion of objects projected up from the surface of Earth. You were also introduced, in general, to two concepts closely associated with motion: (1) that objects have inertia, a tendency to resist a change in motion, and (2) that forces are involved in a change of motion.

Environmental science is an interdisciplinary study of Earth's environment. The concern of this study is the overall problem of human degradation of the environment and remedies for that damage. As an example of an environmental topic of study, consider the damage that results from current human activities involving the use of transportation. Researchers estimate that overall transportation activities are responsible for about one-third of the total U.S. carbon emissions that are added to the air every day. Carbon emissions are a problem because they are directly harmful in the form of carbon monoxide. They are also indirectly harmful because of the contribution of carbon dioxide to global warming and the consequences of climate change.

Here is a list of things that people might do to reduce the amount of environmental damage from transportation:

A. Use a bike, carpool, walk, or take public transportation wherever possible.

B. Plan to combine trips to the store, mall, and work, leaving the car parked whenever possible.

C. Purchase hybrid electric or fuel cell-powered cars and vehicles whenever possible.

D. Move to a planned community that makes the use of cars less necessary and less desirable.

Questions to Discuss

Discuss with your group the following questions concerning connections between thought and feeling:

1. What are your positive or negative feelings associated with each item in the list?
2. Would you feel differently if you had a better understanding of the global problem?
3. Do your feelings mean that you have reached a conclusion?
4. What new items could be added to the list?

FIGURE 2.14 Without a doubt, this baseball player is aware of the relationship between the projection angle and the maximum distance acquired for a given projection velocity.

The relationship between forces and a change of motion is obvious in many everyday situations. When a car, bus, or plane in which you are riding starts moving, you feel a force on your back. Likewise, you feel a force on the bottoms of your feet when an elevator you are in starts moving upward. On the other hand, you seem to be forced toward the dashboard if a car stops quickly, and it feels as if the floor pulls away from your feet when an elevator drops rapidly. These examples all involve patterns between forces and motion, patterns that can be quantified, conceptualized, and used to answer questions about why things move or stand still. These patterns are the subject of Newton's three laws of motion.

Newton's First Law of Motion

Newton's first law of motion is also known as the *law of inertia* and is very similar to one of Galileo's findings about motion. Recall that Galileo used the term *inertia* to describe the tendency of an object to resist changes in motion. Newton's first law describes this tendency more directly. In modern terms (not Newton's words), the **first law of motion** is as follows:

Every object retains its state of rest or its state of uniform straight-line motion unless acted upon by an unbalanced force.

This means that an object at rest will remain at rest unless it is put into motion by an unbalanced force; that is, the net force must be greater than zero. Likewise, an object moving with uniform straight-line motion will retain that motion unless a net force causes it to speed up, slow down, or change its direction of travel. Thus, Newton's first law describes the tendency of an object to resist *any* change in its state of motion.

Think of Newton's first law of motion when you ride standing in the aisle of a bus. The bus begins to move, and you, being an independent mass, tend to remain at rest. You take a

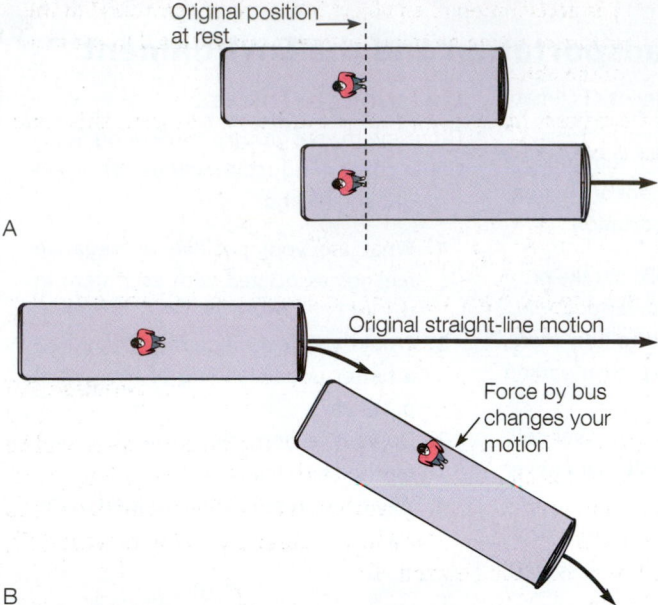

A

Original position at rest

Original straight-line motion

Force by bus changes your motion

B

FIGURE 2.15 Top view of a person standing in the aisle of a bus. (*A*) The bus is at rest and then starts to move forward. Inertia causes the person to remain in the original position, appearing to fall backward. (*B*) The bus turns to the right, but inertia causes the person to retain the original straight-line motion until forced in a new direction by the side of the bus.

few steps back as you tend to maintain your position relative to the ground outside. You reach for a seat back or some part of the bus. Once you have a hold on some part of the bus, it supplies the forces needed to give you the same motion as the bus and you no longer find it necessary to step backward. You now have the same motion as the bus, and no forces are involved, at least until the bus goes around a curve. You now feel a tendency to move to the side of the bus. The bus has changed its straight-line motion, but you, again being an independent mass, tend to move straight ahead. The side of the seat forces you into following the curved motion of the bus. The forces you feel when the bus starts moving or turning are a result of your tendency to remain at rest or follow a straight path until forces correct your motion so that it is the same as that of the bus (figure 2.15).

Newton's Second Law of Motion

Newton successfully used Galileo's ideas to describe the nature of motion. Newton's first law of motion explains that any object, once started in motion, will continue with a constant velocity in a straight line unless a force acts on the moving object. This law not only describes motion but establishes the role of a force as well. A change of motion is therefore *evidence* of the action of a net force. The association of forces and a change of motion is common in your everyday experience. You have felt forces on your back in an accelerating automobile, and you have felt other forces as the automobile turns or stops. You have also learned about gravitational forces that accelerate objects toward the surface of Earth. Unbalanced forces and acceleration are involved in any change of motion. Newton's second law of motion is a relationship between *net force, acceleration,* and *mass* that describes the cause of a change of motion.

Consider the motion of you and a bicycle you are riding. Suppose you are riding your bicycle over level ground in a straight line at 10 mi/h. Newton's first law tells you that you will continue with a constant velocity in a straight line as long as no external, unbalanced force acts on you and the bicycle. The force that you *are* exerting on the pedals seems to equal some external force that moves you and the bicycle along (more on this later). The force exerted as you move along is needed to *balance* the resisting forces of tire friction and air resistance. If these resisting forces were removed, you would not need to exert any force at all to continue moving at a constant velocity. The net force is thus the force you are applying minus the forces from tire friction and air resistance. The *net force* is therefore zero when you move at a constant speed in a straight line (figure 2.16).

If you now apply a greater force on the pedals, the *extra* force you apply is unbalanced by friction and air resistance. Hence, there will be a net force greater than zero, and you will accelerate. You will accelerate during, and *only* during, the time that the net force is greater than zero. Likewise, you will slow down if you apply a force to the brakes, another kind of resisting friction. A third way to change your velocity is to apply a force on the handlebars, changing the direction of

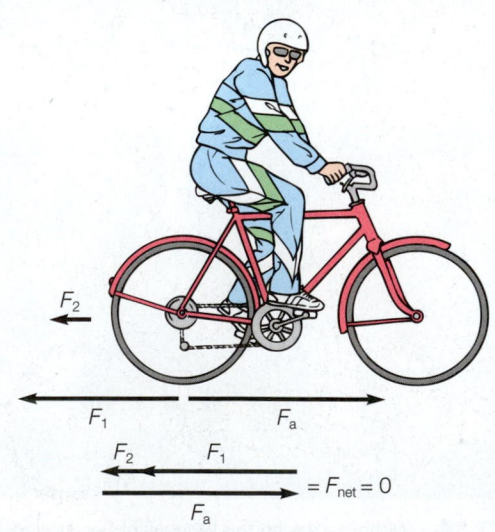

FIGURE 2.16   At a constant velocity, the force of tire friction (F_1) and the force of air resistance (F_2) have a sum that equals the force applied (F_a). The net force is therefore zero.

your velocity. Thus, *unbalanced forces* on you and your bicycle produce an *acceleration.*

Starting a bicycle from rest suggests a relationship between force and acceleration. You observe that the harder you push on the pedals, the greater your acceleration. If you double the net force, then you will also double the acceleration, reaching the same velocity in half the time. Likewise, if you triple the net force, you will increase the acceleration threefold. Recall that when quantities increase or decrease together in the same ratio, they are said to be *directly proportional.* The acceleration is therefore directly proportional to the force.

Suppose that your bicycle has two seats, and you have a friend who will ride (but not pedal) with you. Suppose also that the addition of your friend on the bicycle will double the mass of the bike and riders. If you use the same net force as before, the bicycle will undergo a much smaller acceleration. In fact, with all other factors equal, doubling the mass and applying the same extra force will produce an acceleration of only half as much (figure 2.17). An even more massive friend would reduce the acceleration even more. If you triple the mass and apply the same extra force, the acceleration will be one-third as much. Recall that when a relationship between two quantities shows that one quantity increases as another decreases, in the same ratio, the quantities are said to be *inversely proportional.*

The acceleration of an object depends on *both* the *net force applied* and the *mass* of the object. The **second law of motion** is as follows:

The acceleration of an object is directly proportional to the net force acting on it and inversely proportional to the mass of the object.

If we express force in appropriate units, we can write this statement as an equation,

$$a = \frac{F}{m}$$

By solving for *F*, we rearrange the equation into the form it is most often expressed,

$$F = ma \qquad \textbf{equation 2.3}$$

In the metric system, you can see that the units for force will be the units for mass (*m*) times acceleration (*a*). The unit for mass is kg and the unit for acceleration is m/s². The combination of these units, (kg) (m/s²) is a unit of force called the **newton** (N), in honor of Isaac Newton. So,

$$1 \text{ newton} = 1 \text{ N} = 1 \frac{\text{kg} \cdot \text{m}}{\text{s}^2}$$

Newton's second law of motion is the essential idea of his work on motion. According to this law, there is always a relationship between the acceleration, a net force, and the mass of an object. Implicit in this statement are three understandings: (1) that we are talking about the net force, meaning total external force acting on an object, (2) that the motion statement is concerned with acceleration, not velocity, and (3) that the mass does not change unless specified.

EXAMPLE 2.5   *(Optional)*

A 60 kg bicycle and rider accelerate at 0.5 m/s². How much extra force was applied?

SOLUTION

The mass (*m*) of 60 kg and the acceleration (*a*) of 0.5 m/s² are given. The problem asked for the extra force (*F*) needed to give the mass the acquired acceleration. The relationship is found in equation 2.3, $F = ma$.

$$m = 60 \text{ kg} \qquad F = ma$$
$$a = 0.5 \frac{\text{m}}{\text{s}^2} \qquad = (60 \text{ kg})\left(0.5 \frac{\text{m}}{\text{s}^2}\right)$$
$$F = ? \qquad = (60)(0.5)\text{kg} \times \frac{\text{m}}{\text{s}^2}$$
$$= 30 \frac{\text{kg} \cdot \text{m}}{\text{s}^2}$$
$$= \boxed{30 \text{ N}}$$

An extra force of 30 N beyond that required to maintain constant speed must be applied to the pedals for the bike and rider to maintain an acceleration of 0.5 m/s². (Note that the units kg·m/s² form the definition of a newton of force, so the symbol N is used.)

FIGURE 2.17   More mass results in less acceleration when the same force is applied. With the same force applied, the riders and bike with twice the mass will have half the acceleration, with all other factors constant. Note that the second rider is not pedaling.

EXAMPLE 2.6 *(Optional)*

What is the acceleration of a 20 kg cart if the net force on it is 40 N? (Answer: 2 m/s²)

Weight and Mass

What is the meaning of weight—is it the same concept as mass? Weight is a familiar concept to most people, and in everyday language, the word is often used as having the same meaning as mass. In physics, however, there is a basic difference between weight and mass, and this difference is very important in Newton's explanation of motion and the causes of motion.

Mass is defined as the property that determines how much an object resists a change in its motion. The greater the mass the greater the *inertia,* or resistance to change in motion. Consider, for example, that it is easier to push a small car into motion than to push a large truck into motion. The truck has more mass and therefore more inertia. Newton originally defined mass as the "quantity of matter" in an object, and this definition is intuitively appealing. However, Newton needed to measure inertia because of its obvious role in motion and redefined mass as a measure of inertia.

You could use Newton's second law to measure a mass by exerting a force on the mass and measuring the resulting acceleration. This is not very convenient, so masses are usually measured on a balance by comparing the force of gravity acting on a standard mass compared to the force of gravity acting on the unknown mass.

The force of gravity acting on a mass is the *weight* of an object. Weight is a force and has different units (N) from those of mass (kg). Since weight is a measure of the force of gravity acting on an object, the force can be calculated from Newton's second law of motion,

$$F = ma$$

or

downward force = (mass)(acceleration due to gravity)

or

$$\text{weight} = (\text{mass})(g)$$

$$w = mg \qquad \text{equation 2.4}$$

You learned previously that g is the symbol used to represent acceleration due to gravity. Near Earth's surface, g has an approximate value of 9.8 m/s². To understand how g is applied to an object not moving, consider a ball you are holding in your hand. By supporting the weight of the ball, you hold it stationary, so the upward force of your hand and the downward force of the ball (its weight) must add to a net force of zero. When you let go of the ball, the gravitational force is the only force acting on the ball. The ball's weight is then the net force that accelerates it at g, the acceleration due to gravity. Thus, $F_{net} = w = ma = mg$. The weight of the ball never changes in a given location, so its weight is always equal to $w = mg$, even if the ball is not accelerating.

An important thing to remember is that *pounds* and *newtons* are units of *force* (table 2.1). A *kilogram,* on the other hand, is a measure of *mass.* Thus, the English unit of 1.0 lb is comparable to the metric unit of 4.5 N (or 0.22 lb is equivalent to 1.0 N). Conversion tables sometimes show how to convert from pounds (a unit of weight) to kilograms (a unit of mass). This is possible because weight and mass are proportional in a given location on the surface of Earth. Using conversion factors from inside the front cover of this book, see if you can express your weight in pounds and newtons and your mass in kilograms.

EXAMPLE 2.7 *(Optional)*

What is the weight of a 60.0 kg person on the surface of Earth?

SOLUTION

A mass (m) of 60.0 kg is given, and the acceleration due to gravity (g) 9.8 m/s² is implied. The problem asked for the weight (w). The relationship is found in equation 2.4, $w = mg$, which is a form of $F = ma$.

$$m = 60.0 \text{ kg} \qquad w = mg$$
$$g = 9.8 \text{ m/s}^2 \qquad = (60.0 \text{ kg}) \, 9.8 \frac{\text{m}}{\text{s}^2}$$
$$w = ? \qquad = (60.0)(9.8) \text{ kg} \times \frac{\text{m}}{\text{s}^2}$$
$$= 588 \frac{\text{kg} \cdot \text{m}}{\text{s}^2}$$
$$= 588 \text{ N}$$
$$= \boxed{590 \text{ N}}$$

EXAMPLE 2.8 *(Optional)*

A 60.0 kg person weighs 100.0 N on the Moon. What is the value of g on the Moon? (Answer: 1.67 m/s²)

TABLE 2.1   Units of Mass and Weight in the Metric and English Systems of Measurement

	Mass	×	Acceleration	=	Force
Metric system	kg	×	$\dfrac{m}{s^2}$	=	$\dfrac{kg \cdot m}{s^2}$ (newton)
English system	$\left(\dfrac{lb}{ft/s^2}\right)$	×	$\dfrac{ft}{s^2}$	=	lb (pound)

Connections . . .

Weight on Different Planets*

Planet	Acceleration of Gravity	Approximate Weight(N)	Approximate Weight (lb)
Mercury	3.72 m/s²	223 N	50 lb
Venus	8.92 m/s²	535 N	120 lb
Earth	9.80 m/s²	588 N	132 lb
Mars	3.72 m/s²	223 N	50 lb
Jupiter	24.89 m/s²	1,493 N	336 lb
Saturn	10.58 m/s²	635 N	143 lb
Uranus	8.92 m/s²	535 N	120 lb
Neptune	11.67 m/s²	700 N	157 lb

*For a 60.0 kg person

Newton's Third Law of Motion

Newton's first law of motion states that an object retains its state of motion when the net force is zero. The second law states what happens when the net force is *not* zero, describing how an object with a known mass moves when a given force is applied. The two laws give one aspect of the concept of a force; that is, if you observe that an object starts moving, speeds up, slows down, or changes its direction of travel, you can conclude that an unbalanced force is acting on the object. Thus, any change in the state of motion of an object is *evidence* that an unbalanced force has been applied.

Newton's third law of motion is also concerned with forces. First, consider where a force comes from. A force is always produced by the interaction of two objects. Sometimes we do not know what is producing forces, but we do know that they always come in pairs. Anytime a force is exerted, there is always a matched and opposite force that occurs at the same time. For example, if you push on the wall, the wall pushes back with an equal and opposite force. The two forces are opposite and balanced, and you know this because $F = ma$ and neither you nor the wall accelerated. If the acceleration is zero, then you know from $F = ma$ that the net force is zero (zero equals zero). Note also that the two forces were between two different objects, you and the wall. Newton's third law always describes what happens between two different objects. To simplify the many interactions that occur on Earth, consider a satellite freely floating in space. According to Newton's second law ($F = ma$), a force must be applied to change the state of motion of the satellite. What is a possible source of such a force? Perhaps an astronaut pushes on the satellite for 1 second. The satellite would accelerate *during* the application of the force, then move away from the original position at some constant velocity. The astronaut would also move away from the original position, but in the opposite direction (figure 2.18). A *single force does not exist* by itself. There is always a matched and opposite force that occurs at the same time. Thus, the astronaut exerted a momentary force on the satellite, but the satellite evidently exerted a momentary force back on the astronaut as well, for the astronaut moved away from the original position in the opposite direction. Newton did not have astronauts and satellites to think about, but this is the kind of reasoning he did when he concluded that forces always occur in matched pairs that are equal and opposite. Thus, the **third law of motion** is as follows:

> **Whenever two objects interact, the force exerted on one object is equal in strength and opposite in direction to the force exerted on the other object.**

The third law states that forces always occur in matched pairs that act in opposite directions and on two *different* bodies. Sometimes the third law of motion is expressed as follows: "For every action there is an equal and opposite reaction," but this can be misleading. Neither force is the cause of the other. The forces are at every instant the cause of each other, and they appear and disappear at the same time. If you are going to describe the force exerted on a satellite by an astronaut, then you must realize that there is a simultaneous force exerted on the astronaut by the satellite. The forces (astronaut on satellite and satellite on astronaut) are equal in magnitude but opposite in direction.

Perhaps it would be more common to move a satellite with a small rocket. A satellite is maneuvered in space by firing a rocket in the direction opposite to the direction someone wants to move the satellite. Exhaust gases (or compressed gases) are accelerated in one direction and exert an equal but opposite force on the satellite that accelerates it in the opposite direction. This is another example of the third law.

Force of satellite on astronaut
F

Force of astronaut on satellite
F

FIGURE 2.18   Forces occur in matched pairs that are equal in strength and opposite in direction.

Consider how the pairs of forces work on Earth's surface. You walk by pushing your feet against the ground (figure 2.19). Of course you could not do this if it were not for friction. You would slide as on slippery ice without friction. But since friction does exist, you exert a backward horizontal force on the ground, and, as the third law explains, the ground exerts an equal and opposite force on you. You accelerate forward from the net force as explained by the second law. If Earth had the same mass as you, however, it would accelerate backward at the same rate that you were accelerated forward. Earth is much more massive than you, however, so any acceleration of Earth is a vanishingly small amount. The overall effect is that you are accelerated forward by the force the ground exerts on you.

Return now to the example of riding a bicycle that was discussed previously. What is the source of the *external* force that accelerates you and the bike? Pushing against the pedals is not external to you and the bike, so that force will *not* accelerate you and the bicycle forward. This force is transmitted through the bike mechanism to the rear tire, which pushes against the ground. It is the ground exerting an equal and opposite force against the system of you and the bike that accelerates you forward. You must consider the forces that act on the system of you and the bike before you can apply $F = ma$. The only forces that will affect the forward motion of the bike system are the force of the ground pushing it forward and the frictional forces that oppose the forward motion. This is another example of the third law.

2.8 MOMENTUM

Sportscasters often refer to the *momentum* of a team, and newscasters sometimes refer to an election where one of the candidates has *momentum*. Both situations describe a competition where one side is moving toward victory and it is difficult to stop. It seems appropriate to borrow this term from the physical sciences because momentum is a property of movement. It takes a longer time to stop something from moving when it has a lot of momentum. The physical science concept of momentum is closely related to Newton's laws of motion. **Momentum** (p) is defined as the product of the mass (m) of an object and its velocity (v),

$$\text{momentum} = \text{mass} \times \text{velocity}$$

or

$$p = mv \qquad \text{equation 2.5}$$

The astronaut in figure 2.20 has a mass of 60.0 kg and a velocity of 0.750 m/s as a result of the interaction with the satellite. The resulting momentum is therefore (60.0 kg) (0.750 m/s), or 45.0 kg·m/s. As you can see, the momentum would be greater if the astronaut had acquired a greater velocity or if the astronaut had a greater mass and acquired the same velocity. Momentum involves both the inertia and the velocity of a moving object.

FIGURE 2.19 The football player's foot is pushing against the ground, but it is the ground pushing against the foot that accelerates the player forward to catch a pass.

Conservation of Momentum

Notice that the momentum acquired by the satellite in figure 2.20 is *also* 45.0 kg·m/s. The astronaut gained a certain momentum in one direction, and the satellite gained the *very same momentum in the opposite direction*. Newton originally defined the second law in terms of a change of momentum being proportional to the net force acting on an object. Since the third law explains that the forces exerted on both the astronaut and satellite were equal and opposite, you would expect both objects to acquire equal momentum in the opposite direction. This result is observed any time objects in a system interact and the only forces involved are those between the interacting objects (figure 2.20). This statement leads to a

Newton's laws of motion apply to animal motion as well as that of satellites and automobiles. Consider, for example, the dilemma of a growing scallop. A scallop is the shell often seen as a logo for a certain petroleum company, a fan-shaped shell with a radiating fluted pattern (box figure 2.2). The scallop is a marine mollusk that is most unusual since it is the only clam-like mollusk that is capable of swimming. By opening and closing its shell, it is able to propel itself by forcing water from the interior of the shell in a jetlike action. The popular seafood called "scallops" is the edible muscle that the scallop uses to close its shell.

A scallop is able to swim by orienting its shell at a proper angle and maintaining a minimum acceleration to prevent sinking. For example, investigations have found that one particular species of scallop must force enough water backward to move about six body lengths per second with a 10 degree angle of attack to maintain level swimming. Such a swimming effort can be maintained for up to about 20 seconds, enabling the scallop to escape predation or some other disturbing condition.

A more massive body limits the swimming ability of the scallop, as a greater force is needed to give a greater mass the same acceleration (as you would expect from Newton's second law of motion). This prob-

lem becomes worse as the scallop grows larger and larger without developing a greater and greater jet force.

Box Figure 2.2   A scallop shell.

particular kind of relationship called a *law of conservation*. In this case, the law applies to momentum and is called the **law of conservation of momentum:**

> **The total momentum of a group of interacting objects remains the same in the absence of external forces.**

Conservation of momentum, energy, and charge are among examples of conservation laws that apply to everyday situations. These situations always illustrate two understandings: (1) each conservation law is an expression that describes a physical prin-

ciple that can be observed, and (2) each law holds regardless of the details of an interaction or how it took place. Since the conservation laws express something that always occurs, they tell us what might be expected to happen and what might be expected not to happen in a given situation. The conservation laws also allow unknown quantities to be found by analysis. The law of conservation of momentum, for example, is useful in analyzing motion in simple systems of collisions such as those of billiard balls, automobiles, or railroad cars. It is also useful in measuring action and reaction interactions, as in rocket propulsion, where the backward momentum of the exhaust gases equals the momentum given to the rocket in the opposite direction (figure 2.21). When this is done, momentum is always found to be conserved.

Impulse

Have you ever heard that you should "follow through" when hitting a ball? When you follow through, the bat is in contact with the ball for a longer period of time. The force of the hit is important, of course, but both the force and how long the force is applied determine the result. The product of the force and the time of application is called **impulse.** This quantity can be expressed as

$$\text{impulse} = Ft$$

where F is the force applied during the time of contact (t). The impulse you give the ball determines how fast the ball will move and thus how far it will travel.

Impulse is related to the change of motion of a ball of a given mass, so the change of momentum (mv) is brought about by the impulse. This can be expressed as

$$\text{Change of momentum} = (\text{applied force})(\text{time of contact})$$

$$\Delta p = Ft \qquad\qquad \textbf{equation 2.6}$$

$F = 30.0$ N
$t = 1.50$ s

$F = 30.0$ N
$t = 1.50$ s

$m = 60.0$ kg
$v = 0.750$ m/s
$p = mv$

$$= (60.0 \text{ kg})\left(0.750 \frac{m}{s}\right)$$

$$= \boxed{45.0 \frac{kg \cdot m}{s}}$$

$m = 120.0$ kg
$v = 0.375$ m/s
$p = mv$

$$= (120.0 \text{ kg})\left(0.375 \frac{m}{s}\right)$$

$$= \boxed{45.0 \frac{kg \cdot m}{s}}$$

FIGURE 2.20   Both the astronaut and the satellite receive a force of 30.0 N for 1.50 s when they push on each other. Both then have a momentum of 45.0 kg·m/s in the opposite direction. This is an example of the law of conservation of momentum.

FIGURE 2.21 According to the law of conservation of momentum, the momentum of the expelled gases in one direction equals the momentum of the rocket in the other direction in the absence of external forces.

Contact time is also important in safety. Automobile airbags, the padding in elbow and knee pads, and the plastic barrels off the highway in front of overpass supports are examples of designs intended to increase the contact time. Again, increasing the contact time reduces the force, since $\Delta p = Ft$. The impact force is reduced and so are the injuries. Think about this the next time you see a car that was crumpled and bent by a collision. The driver and passengers were probably saved from injuries that are more serious since more time was involved in stopping the car that crumpled. A car that crumples is a safer car in a collision.

where Δp is a change of momentum. You "follow through" while hitting a ball in order to increase the contact time. If the same force is used, a longer contact time will result in a greater impulse. A greater impulse means a greater change of momentum, and since the mass of the ball does not change, the overall result is a moving ball with a greater velocity. This means following through will result in more distance from hitting the ball with the same force. That's why it is important to follow through when you hit the ball.

Now consider bringing a moving object to a stop by catching it. In this case, the mass and the velocity of the object are fixed at the time you catch it, and there is nothing you can do about these quantities. The change of momentum is equal to the impulse, and the force and time of force application *can* be manipulated. For example, consider how you would catch a raw egg that is tossed to you. You would probably move your hands with the egg as you catch it, increasing the contact time. Increasing the contact time has the effect of reducing the force, since $\Delta p = Ft$. You changed the force applied by increasing the contact time, and hopefully, you reduced the force sufficiently so the egg does not break.

2.9 FORCES AND CIRCULAR MOTION

Consider a communications satellite that is moving at a uniform speed around Earth in a circular orbit. According to the first law of motion, there *must be* forces acting on the satellite, since it does *not* move off in a straight line. The second law of motion also indicates forces, since an unbalanced force is required to change the motion of an object.

Recall that acceleration is defined as a rate of change in velocity and that velocity has both strength and direction. Velocity is changed by a change in speed, direction, or both speed and direction. The satellite in a circular orbit is continuously being accelerated. This means that there is a continuously acting unbalanced force on the satellite that pulls it out of a straight-line path.

The force that pulls an object out of its straight-line path and into a circular path is called a **centripetal** (center-seeking) **force.** Perhaps you have swung a ball on the end of a string in a horizontal circle over your head. Once you have the ball moving, the only unbalanced force (other than gravity) acting on the ball is the centripetal force your hand exerts on the ball

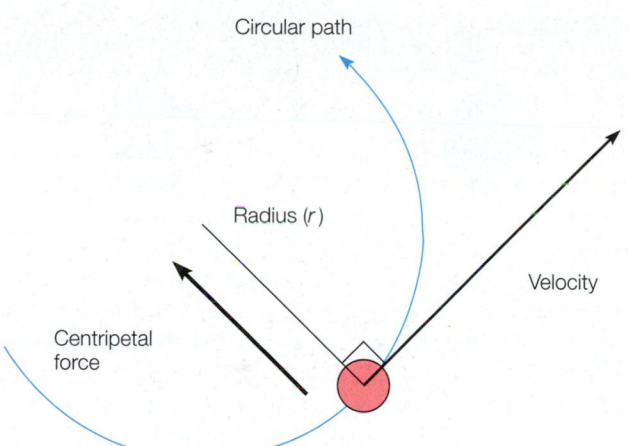

Circular path

Radius (r)

Velocity

Centripetal force

FIGURE 2.22 Centripetal force on the ball causes it to change direction continuously, or accelerate into a circular path. Without the force acting on it, the ball would continue in a straight line.

through the string. This centripetal force pulls the ball from its natural straight-line path into a circular path. There are no outward forces acting on the ball. The force that you feel on the string is a consequence of the third law; the ball exerts an equal and opposite force on your hand. If you were to release the string, the ball would move away from the circular path in a *straight line* that has a right angle to the radius at the point of release (figure 2.22). When you release the string, the centripetal force ceases, and the ball then follows its natural straight-line motion. If other forces were involved, it would follow some other path. Nonetheless, the apparent outward force has been given a name just as if it were a real force. The outward tug is called a *centrifugal force.*

The magnitude of the centripetal force required to keep an object in a circular path depends on the inertia, or mass, of the object and the acceleration of the object, just as you learned in the second law of motion. The acceleration of an object moving in a circle can be shown to be directly proportional to the square of the speed around the circle (v^2) and inversely proportional to the radius of the circle (r). (A smaller radius requires a greater acceleration.) Therefore, the acceleration of an object moving in uniform circular motion (a_c) is

$$a_c = \frac{v^2}{r}$$

equation 2.7

The magnitude of the centripetal force of an object with a constant mass (m) that is moving with a velocity (v) in a circular orbit of a radius (r) can be found by substituting equation 2.7 in $F = ma$, or

$$F = \frac{mv^2}{r}$$

equation 2.8

EXAMPLE 2.9 (Optional)

A 0.25 kg ball is attached to the end of a 0.5 m string and moved in a horizontal circle at 2.0 m/s. What net force is needed to keep the ball in its circular path?

SOLUTION

$m = 0.25$ kg
$r = 0.5$ m
$v = 2.0$ m/s
$F = ?$

$$F = \frac{mv^2}{r}$$

$$= \frac{(0.25 \text{ kg})(2.0 \text{ m/s})^2}{0.5 \text{ m}}$$

$$= \frac{(0.25 \text{ kg})(4.0 \text{ m}^2/\text{s}^2)}{0.5 \text{ m}}$$

$$= \frac{(0.25)(4.0)}{0.5} \frac{\text{kg} \cdot \text{m}^2}{\text{s}^2} \times \frac{1}{\text{m}}$$

$$= 2 \frac{\text{kg} \cdot \text{m}^2}{\text{m} \cdot \text{s}^2}$$

$$= 2 \frac{\text{kg} \cdot \text{m}}{\text{s}^2}$$

$$= \boxed{2 \text{ N}}$$

EXAMPLE 2.10 (Optional)

Suppose you make the string in example 2.9 half as long, 0.25 m. What force is now needed? (Answer: 4.0 N)

2.10 NEWTON'S LAW OF GRAVITATION

You know that if you drop an object, it always falls to the floor. You define *down* as the direction of the object's movement and *up* as the opposite direction. Objects fall because of the force of gravity, which accelerates objects on Earth at $g = 9.8$ m/s^2 (32 ft/s^2) and gives them weight, $w = mg$.

Gravity is an attractive force, a pull that exists between all objects in the universe. It is a mutual force that, just like all other forces, comes in matched pairs. Since Earth attracts you with a certain force, you must attract Earth with an exact opposite force. The magnitude of this force of mutual attraction depends on several variables. These variables were first described by Newton in *Principia*, his famous book on motion that was printed in 1687. Newton had, however, worked out his ideas much earlier, by the age of twenty-four, along with ideas about his laws of motion and the formula for centripetal acceleration. In a biography written by a friend in 1752, Newton stated that the notion of gravitation came to mind during a time of thinking that "was occasioned by the fall of an apple." He was thinking about why the Moon stays in orbit around Earth rather than moving off in a straight line as would be predicted by the first law of motion. Perhaps the same force that attracts the Moon toward Earth, he thought, attracts the apple to Earth. Newton developed a theoretical equation for gravitational force that explained not only the motion of the Moon but the motion of the whole solar system. Today, this relationship is known as the **universal law of gravitation:**

> **Every object in the universe is attracted to every other object with a force that is directly proportional to the product of their masses and inversely proportional to the square of the distances between them.**

In symbols, m_1 and m_2 can be used to represent the masses of two objects, d the distance between their centers, and G a constant of proportionality. The equation for the law of universal gravitation is therefore

$$F = G\frac{m_1 m_2}{d^2} \qquad \textbf{equation 2.9}$$

This equation gives the magnitude of the attractive force that each object exerts on the other. The two forces are oppositely directed. The constant G is a universal constant, since the law applies to all objects in the universe. It was first measured experimentally by Henry Cavendish in 1798. The accepted value today is $G = 6.67 \times 10^{-11}$ N·m^2/kg^2. Do not confuse G, the universal constant, with g, the acceleration due to gravity on the surface of Earth.

Thus, the magnitude of the force of gravitational attraction is determined by the mass of the two objects and the distance between them (figure 2.23). The law also states that *every* object is attracted to every other object. You are attracted to all the objects around you—chairs, tables, other people, and so forth. Why don't you notice the forces between you and other objects? One or both of the interacting objects must be quite massive before a noticeable force results from the interaction. That is why you do not notice the force of gravitational attraction between you and objects that are not very massive compared to Earth. The attraction between you and Earth overwhelmingly predominates, and that is all you notice.

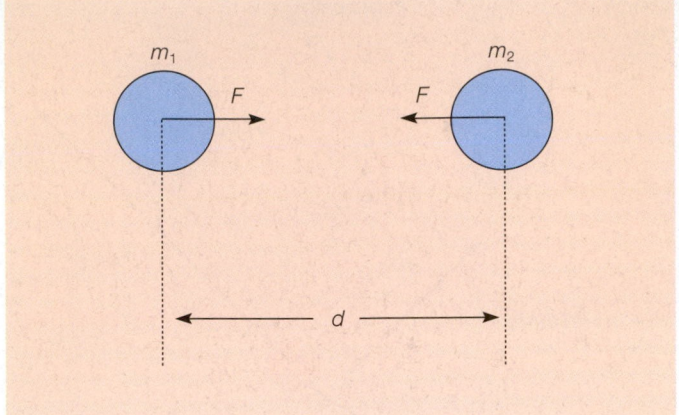

FIGURE 2.23 The variables involved in gravitational attraction. The force of attraction (F) is proportional to the product of the masses (m_1, m_2) and inversely proportional to the square of the distance (d) between the centers of the two masses.

EXAMPLE 2.11 *(Optional)*

What is the force of gravitational attraction between two 60.0 kg (132 lb) students who are standing 1.00 m apart?

SOLUTION

$G = 6.67 \times 10^{-11}$ Nm^2 / kg^2

$m_1 = 60.0$ kg

$m_2 = 60.0$ kg $\qquad F = G\dfrac{m_1 m_2}{d^2}$

$d = 1.00$ m

$F = ?$

$$= \frac{(6.67 \times 10^{-11}\,\text{Nm}^2/\text{kg}^2)(60.0\,\text{kg})(60.0\,\text{kg})}{(1.00\,\text{m})^2}$$

$$= (6.67 \times 10^{-11})(3.60 \times 10^3)\frac{\text{Nm}^2\frac{\text{kg}^2}{\text{kg}^2}}{\text{m}^2}$$

$$= 2.40 \times 10^{-7}(\text{Nm}^2)\left(\frac{1}{\text{m}^2}\right)$$

$$= 2.40 \times 10^{-7}\frac{\text{Nm}^2}{\text{m}^2}$$

$$= \boxed{2.40 \times 10^{-7}\ \text{N}}$$

(Note: A force of 2.40×10^{-7} [0.00000024] N is equivalent to a force of 5.40×10^{-8} [0.00000005] lb, a force that you would not notice. In fact, it would be difficult to measure such a small force.)

EXAMPLE 2.12 *(Optional)*

What would be the value of g if Earth were less dense, with the same mass and double the radius? (Answer: $g = 2.44$ m/s^2)

The acceleration due to gravity, g, is about 9.8 m/s^2 on Earth and is practically a constant for relatively short distances above the surface. Notice, however, that Newton's law of gravitation is an inverse square law. This means if you double the distance, the force is $1/(2)^2$ or 1/4 as great. If you triple the distance, the force is $1/(3)^2$ or 1/9 as great. In other words, the force of

People Behind the Science

Isaac Newton (1642–1727)

Isaac Newton was a British physicist who is regarded as one of the greatest scientists ever to have lived. He discovered the three laws of motion that bear his name and was the first to explain gravitation, clearly defining the nature of mass, weight, force, inertia, and acceleration. In his honor, the SI unit of force is called the newton. Newton also made fundamental discoveries in light, finding that white light is composed of a spectrum of colors and inventing the reflecting telescope.

Newton was born on January 4, 1643 (by the modern calendar). He was a premature, sickly baby born after his father's death, and his survival was not expected. When he was three, his mother remarried, and the young Newton was left in his grandmother's care. He soon began to take refuge in things mechanical, making water clocks, kites bearing fiery lanterns aloft, and a model mill powered by a mouse, as well as innumerable drawings and diagrams. When Newton was twelve, his mother withdrew him from school with the intention of making him into a farmer. Fortunately, his uncle recognized Newton's ability and managed to get him back into school to prepare for college.

Newton was admitted to Trinity College, Cambridge, and graduated in 1665, the same year that the university was closed because of the plague. Newton returned to his boyhood farm to wait out the plague, making only an occasional visit back to Cambridge. During this period, he performed his first prism experiments and thought about motion and gravitation.

Newton returned to study at Cambridge after the plague had run its course, receiving a master's degree in 1668 and becoming a professor at the age of only twenty-six. Newton remained at Cambridge almost thirty years, studying alone for the most part, though in frequent contact with other leading scientists by letter and through the Royal Society in London. These were Newton's most fertile years. He labored day and night, thinking and testing ideas with calculations.

In Cambridge, he completed what may be described as his greatest single work, *Philosophae Naturalis Principia Mathematica* (*Mathematical Principles of Natural Philosophy*). This was presented to the Royal Society in 1686, which subsequently withdrew from publishing it because of a shortage of funds. The astronomer Edmund Halley (1656–1742), a wealthy man and friend of Newton, paid for the publication of the *Principia* in 1687. In it, Newton revealed his laws of motion and the law of universal gravitation.

Newton's greatest achievement was to demonstrate that scientific principles are of universal application. In *Principia Mathematica,* he built the evidence of experiment and observation to develop a model of the universe that is still of general validity. "If I have seen further than other men," he once said, "it is because I have stood on the shoulders of giants"; and Newton was certainly able to bring together the knowledge of his forebears in a brilliant synthesis.

No knowledge can ever be total, but Newton's example brought about an explosion of investigation and discovery that has never really abated. He perhaps foresaw this when he remarked, "To myself, I seem to have been only like a boy playing on the seashore, and diverting myself in now and then finding a smoother pebble or a prettier shell than ordinary, whilst the great ocean of truth lay all undiscovered before me."

With his extraordinary insight into the workings of nature and rare tenacity in wresting its secrets and revealing them in as fundamental and concise a way as possible, Newton stands as a colossus of science. In physics, only Archimedes (287–212 B.C.) and Albert Einstein (1879–1955), who also possessed these qualities, may be compared to him.

Source: Modified from the *Hutchinson Dictionary of Scientific Biography.* © RM, 2011. All rights reserved. Helicon Publishing is a division of RM.

---

gravitational attraction and *g* decrease inversely with the square of the distance from Earth's center.

Newton was able to calculate the acceleration of the Moon toward Earth, about 0.0027 m/s². The Moon "falls" toward Earth because it is accelerated by the force of gravitational attraction. This attraction acts as a *centripetal force* that keeps the Moon from following a straight-line path as would be predicted from the first law. Thus, the acceleration of the Moon keeps it in a somewhat circular orbit around Earth. Figure 2.24 shows that the Moon would be in position A if it followed a straight-line path instead of "falling" to position B as it does. The Moon thus "falls" around Earth. Newton was able to analyze the motion of the Moon quantitatively as evidence that it is gravitational force that keeps the Moon in its orbit. The law of gravitation was extended to the Sun, other planets, and eventually the universe. The quantitative predictions of observed relationships

CONCEPTS APPLIED

Apparent Weightlessness

Use a sharp pencil to make a small hole in the bottom of a Styrofoam cup. The hole should be large enough for a thin stream of water to flow from the cup, but small enough for the flow to continue for 3 or 4 seconds. Test the water flow over a sink.

Hold a finger over the hole on the cup as you fill it with water. Stand on a ladder or outside stairwell as you hold the cup out at arm's length. Move your finger, allowing a stream of water to flow from the cup as you drop it. Observe what happens to the stream of water as the cup is falling. Explain your observations. Also predict what you would see if you were falling with the cup.

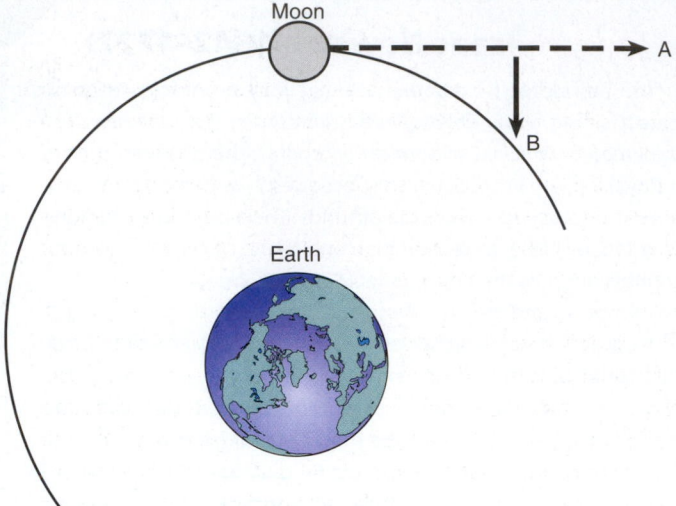

FIGURE 2.24 Gravitational attraction acts as a centripetal force that keeps the Moon from following the straight-line path shown by the dashed line to position A. It was pulled to position B by gravity (0.0027 m/s²) and thus "fell" toward Earth the distance from the dashed line to B, resulting in a somewhat circular path.

FIGURE 2.25 This photograph of Earth from space shows that it is nearly spherical.

among the planets were strong evidence that all objects obey the same law of gravitation. In addition, the law provided a means to calculate the mass of Earth, the Moon, the planets, and the Sun. Newton's law of gravitation, laws of motion, and work with mathematics formed the basis of most physics and technology for the next two centuries, as well as accurately describing the world of everyday experience.

Earth Satellites

As you can see in figure 2.25, Earth is round-shaped and nearly spherical. The curvature is obvious in photographs taken from space but not so obvious back on the surface because Earth is so large. However, you can see evidence of the curvature in places on the surface where you have unobstructed vision for long distances. For example, a tall ship appears to "sink" on the horizon as it sails away, following Earth's curvature below your line of sight. The surface of Earth curves away from your line of sight or any other line tangent to the surface, dropping at a rate of about 4.9 m for every 8 km (16 ft in 5 mi). This means that a ship 8 km away will appear to drop about 5 m below the horizon and anything less than about 5 m tall at this distance will be out of sight, below the horizon.

Recall that a falling object accelerates toward Earth's surface at g, which has an average value of 9.8 m/s². Ignoring air resistance, a falling object will have a speed of 9.8 m/s at the end of 1 second and will fall a distance of 4.9 m. If you wonder why the object did not fall 9.8 m in 1 s, recall that the object starts with an initial speed of zero and has a speed of 9.8 m/s only during the last instant. The average speed was an average of the initial and final speeds, which is 4.9 m/s. An average speed of 4.9 m/s over a time interval of 1 second will result in a distance covered of 4.9 m.

Did you know that Newton was the first to describe how to put an artificial satellite into orbit around Earth? He did not discuss rockets, however, but described in *Principia* how to put a cannonball into orbit. He described how a cannonball shot with sufficient speed straight out from a mountaintop would go into orbit around Earth. If it had less than the sufficient speed, it would fall back to Earth following the path of projectile motion, as discussed in the section on "Compound Motion." What speed does it need to go into orbit? Earth curves away from a line tangent to the surface at 4.9 m per 8 km. Any object falling from a resting position will fall a distance of 4.9 m during the first second. Thus, a cannonball shot straight out from a mountaintop with a speed of 8 km/s (nearly 18,000 mi/hr, or 5 mi/s) will fall toward the surface, dropping 4.9 m during the first second. But the surface of Earth drops, too, curving away below the falling cannonball. So the cannonball is still moving horizontally, no closer to the surface than it was a second ago. As it falls 4.9 m during the next second, the surface again curves away 4.9 m over the 8 km distance. This repeats again and again, and the cannonball stays the same distance from the surface, and we say it is now an *artificial satellite* in orbit. The satellite requires no engine or propulsion as it continues to fall toward the surface, with Earth curving away from it continuously. This assumes, of course, no loss of speed from air resistance.

Today, an artificial satellite is lofted by rocket or rockets to an altitude of more than 320 km (about 200 mi), above the air friction of the atmosphere, before being aimed horizontally. The satellite is then "injected" into orbit by giving it the correct tangential speed. This means it has attained an orbital speed of

Gravity Problems

Gravity does act on astronauts in spacecraft that are in orbit around Earth. Since gravity is acting on the astronaut and spacecraft, the term *zero gravity* is not an accurate description of what is happening. The astronaut, spacecraft, and everything in it are experiencing *apparent weightlessness* because they are continuously falling toward the surface. Everything seems to float because everything is falling together. But, strictly speaking, everything still has weight, because weight is defined as a gravitational force acting on an object ($w = mg$).

Whether weightlessness is apparent or real, however, the effects on people are the same. Long-term orbital flights have provided evidence that the human body changes from the effect of weightlessness. Bones lose calcium and other minerals, the heart shrinks to a much smaller size, and leg muscles shrink so much on prolonged flights that astronauts cannot walk when they return to the surface. These changes occur because on Earth, humans are constantly subjected to the force of gravity. The nature of the skeleton and the strength of the muscles are determined by how the body reacts to this force. Metabolic pathways and physiological processes that maintain strong bones and muscles evolved having to cope with a specific gravitational force. When we are suddenly subjected to a place where gravity is significantly different, these processes result in weakened systems. If we lived on a planet with a different gravitational force, we would have muscles and bones that were adapted to the gravity on that planet. Many kinds of organisms have been used in experiments in space to try to develop a better understanding of how their systems work without gravity.

The problems related to prolonged weightlessness must be worked out before long-term weightless flights can take place. One solution to these problems might be a large, uniformly spinning spacecraft. The astronauts would tend to move in a straight line, and the side of the turning spacecraft (now the "floor") would exert a force on them to make them go in a curved path. This force would act as an artificial gravity.

---

at least 8 km/s (5 mi/s) but less than 11 km/s (7 mi/s). At a speed less than 8 km/s, the satellite would fall back to the surface in a parabolic path. At a speed more than 11 km/s, it will move faster than the surface curves away and will escape from Earth into space. But with the correct tangential speed, and above the atmosphere and air friction, the satellite will follow a circular orbit for long periods of time without the need for any more propulsion. An orbit injection speed of more than 8 km/s (5 mi/s) would result in an elliptical rather than a circular orbit.

A satellite could be injected into orbit near the outside of the atmosphere, closer to Earth but outside the air friction that might reduce its speed. The satellite could also be injected far away from Earth, where it takes a longer time to complete one orbit. Near the outer limits of the atmosphere—that is, closer to the surface—a satellite might orbit Earth every 90 minutes or so. A satellite as far away as the Moon, on the other hand, orbits Earth in a little less than 28 days. A satellite at an altitude of 36,000 km (a little more than 22,000 mi) has a period of 1 day. In the right spot over the equator, such a satellite is called a **geosynchronous satellite,** since it turns with Earth and does not appear to move across the sky (figure 2.26). The photographs of the cloud cover you see in weather reports were taken from one or more geosynchronous weather satellites. Communications networks are also built around geosynchronous satellites. One way to locate one of these geosynchronous satellites is to note the aiming direction of backyard satellite dishes that pick up television signals.

Weightlessness

News photos sometimes show astronauts "floating" in the Space Shuttle or next to a satellite (figure 2.27). These astronauts appear to be weightless but technically are no more weightless than a skydiver in free fall or a person in a falling elevator. Recall that weight is a gravitational force, a measure of the gravitational attraction between Earth and an object (*mg*). The weight of a cup of coffee, for example, can be measured by placing the cup on a scale. The force the cup of coffee exerts against the scale

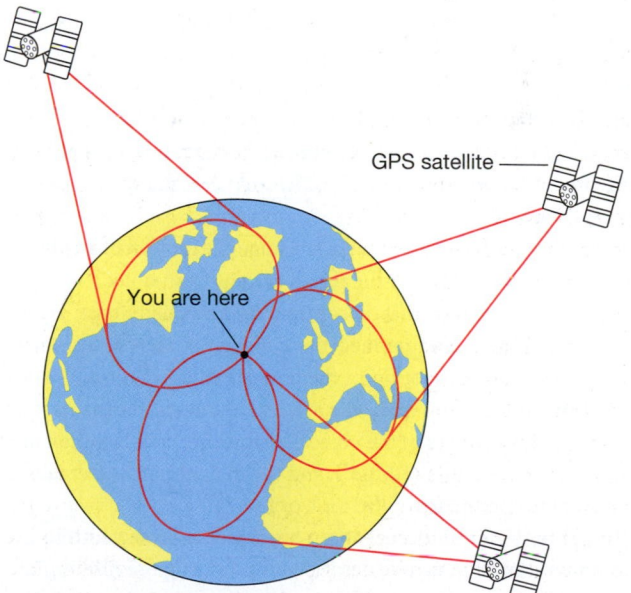

FIGURE 2.26 In the Global Positioning System (GPS), each of a fleet of orbiting satellites sends out coded radio signals that enable a receiver on the earth to determine both the exact position of the satellite in space and its exact distance from the receiver. Given this information, a computer in the receiver then calculates the circle on the Earth's surface on which the receiver must lie. Data from three satellites give three circles and the receiver must be located at the one point where all three intersect (also see page 173).

FIGURE 2.27 Astronauts in an orbiting space station may appear to be weightless. Technically, however, they are no more weightless than a skydiver in free fall or a person in a falling elevator near or on the surface of Earth.

is its weight. You also know that the scale pushes back on the cup of coffee since it is not accelerating, which means the net force is zero.

Now consider what happens if a skydiver tries to pour a cup of coffee while in free fall. Even if you ignore air resistance, you can see that the skydiver is going to have a difficult time, at best. The coffee, the cup, and the skydiver will all be falling together. Gravity is acting to pull the coffee downward, but gravity is also acting to pull the cup from under it at the same rate. The coffee, the cup, and the skydiver all fall together, and the skydiver will see the coffee appear to "float" in blobs. If the diver lets go of the cup, it too will appear to float as everything continues to fall together. However, this is only *apparent weightlessness*, since gravity is still acting on everything; the coffee, the cup, and the skydiver only *appear* to be weightless because they are all accelerating at *g*.

Astronauts in orbit are in free fall, falling toward Earth just as the skydiver, so they too are undergoing apparent weightlessness. To experience true weightlessness, the astronauts would have to travel far from Earth and its gravitational field, and far from the gravitational fields of other planets.

SUMMARY

Motion can be measured by speed, velocity, and acceleration. *Speed* is a measure of how fast something is moving. It is a ratio of the distance covered between two locations to the time that elapsed while moving between the two locations. The *average speed* considers the distance covered during some period of time, while the *instantaneous speed* is the speed at some specific instant. *Velocity* is a measure of the speed and direction of a moving object. *Acceleration* is a change of velocity per unit of time.

A *force* is a push or a pull that can change the motion of an object. The *net force* is the sum of all the forces acting on an object.

Galileo determined that a continuously applied force is not necessary for motion and defined the concept of *inertia*: an object remains in unchanging motion in the absence of a net force. Galileo also determined that falling objects accelerate toward Earth's surface independent of the weight of the object. He found the acceleration due to gravity, *g*, to be 9.8 m/s^2(32 ft/s^2), and the distance an object falls is proportional to the square of the time of free fall ($d \propto t^2$).

Compound motion occurs when an object is projected into the air. Compound motion can be described by splitting the motion into vertical and horizontal parts. The acceleration due to gravity, *g*, is a constant that is acting at all times and acts independently of any motion that an object has. The path of an object that is projected at some angle to the horizon is therefore a parabola.

Newton's *first law of motion* is concerned with the motion of an object and the lack of a net force. Also known as the *law of inertia*, the first law states that an object will retain its state of straight-line motion (or state of rest) unless a net force acts on it.

The *second law of motion* describes a relationship between net force, mass, and acceleration. A *newton* of force is the force needed to give a 1.0 kg mass an acceleration of 1.0 m/s^2.

Weight is the downward force that results from Earth's gravity acting on the mass of an object. Weight is measured in *newtons* in the metric system and *pounds* in the English system.

Newton's *third law of motion* states that forces are produced by the interaction of *two different* objects. These forces always occur in matched pairs that are equal in size and opposite in direction.

Momentum is the product of the mass of an object and its velocity. In the absence of external forces, the momentum of a group of interacting objects always remains the same. This relationship is the *law of conservation of momentum*. *Impulse* is a change of momentum equal to a force times the time of application.

An object moving in a circular path must have a force acting on it, since it does not move in a straight line. The force that pulls an object out of its straight-line path is called a *centripetal force*. The centripetal force needed to keep an object in a circular path depends on the mass of the object, its velocity, and the radius of the circle.

The *universal law of gravitation* is a relationship between the masses of two objects, the distance between the objects, and a proportionality constant. Newton was able to use this relationship to show that gravitational attraction provides the centripetal force that keeps the Moon in its orbit.

Summary of Equations

2.1
$$\text{average speed} = \frac{\text{distance}}{\text{time}}$$
$$\overline{v} = \frac{d}{t}$$

2.2
$$\text{acceleration} = \frac{\text{change of velocity}}{\text{time}}$$
$$= \frac{\text{final velocity} - \text{initial velocity}}{\text{time}}$$
$$a = \frac{v_f - v_i}{t}$$

2.3
$$\text{force} = \text{mass} \times \text{acceleration}$$
$$F = ma$$

2.4
$$\text{weight} = \text{mass} \times \text{acceleration due to gravity}$$
$$w = mg$$

2.5
$$\text{momentum} = \text{mass} \times \text{velocity}$$
$$p = mv$$

2.6
$$\text{change of momentum} = \text{force} \times \text{time}$$
$$\Delta p = Ft$$

2.7
$$\text{centripetal acceleration} = \frac{\text{velocity squared}}{\text{radius of circle}}$$
$$a_c = \frac{v^2}{r}$$

2.8
$$\text{centripetal force} = \frac{\text{mass} \times \text{velocity squared}}{\text{radius of circle}}$$
$$F = \frac{mv^2}{r}$$

2.9
$$\text{gravitational force} = \text{constant} \times \frac{\text{one mass} \times \text{another mass}}{\text{distance squared}}$$
$$F = G\frac{m_1 m_2}{d^2}$$

KEY TERMS

acceleration (p. **28**)
centripetal force (p. **44**)
first law of motion (p. **37**)
force (p. **29**)
free fall (p. **33**)
fundamental forces (p. **31**)

g (p. **35**)
geosynchronous satellite (p. **49**)
impulse (p. **43**)
inertia (p. **32**)
law of conservation of
    momentum (p. **43**)

mass (p. **40**)
momentum (p. **42**)
net force (p. **29**)
newton (p. **39**)
second law of motion (p. **39**)
speed (p. **26**)

third law of motion (p. **41**)
universal law of gravitation
    (p. **46**)
velocity (p. **27**)

APPLYING THE CONCEPTS

Answers are located in appendix F.

1. A quantity of 5 m/s^2 is a measure of
   a. metric area.
   b. acceleration.
   c. speed.
   d. velocity.

2. An automobile has how many different devices that can cause it to undergo acceleration?
   a. none
   b. one
   c. two
   d. three or more

3. Ignoring air resistance, an object falling toward the surface of Earth has a *velocity* that is
   a. constant.
   b. increasing.
   c. decreasing.
   d. acquired instantaneously but dependent on the weight of the object.

4. Ignoring air resistance, an object falling near the surface of Earth has an *acceleration* that is
   a. constant.
   b. increasing.
   c. decreasing.
   d. dependent on the weight of the object.

5. Two objects are released from the same height at the same time, and one has twice the weight of the other. Ignoring air resistance,
   a. the heavier object hits the ground first.
   b. the lighter object hits the ground first.
   c. they both hit at the same time.
   d. whichever hits first depends on the distance dropped.

6. A ball rolling across the floor slows to a stop because
   a. there is a net force acting on it.
   b. the force that started it moving wears out.
   c. the forces are balanced.
   d. the net force equals zero.

7. Considering the forces on the system of you and a bicycle as you pedal the bike at a constant velocity in a horizontal straight line,
   a. the force you are exerting on the pedal is greater than the resisting forces.
   b. all forces are in balance, with the net force equal to zero.
   c. the resisting forces of air and tire friction are less than the force you are exerting.
   d. the resisting forces are greater than the force you are exerting.

8. If you double the unbalanced force on an object of a given mass, the acceleration will be
   a. doubled.
   b. increased fourfold.
   c. increased by one-half.
   d. increased by one-fourth.

9. If you double the mass of a cart while it is undergoing a constant unbalanced force, the acceleration will be
   a. doubled.
   b. increased fourfold.
   c. half as much.
   d. one-fourth as much.

10. Doubling the distance between the center of an orbiting satellite and the center of Earth will result in what change in the gravitational attraction of Earth for the satellite?
   a. one-half as much
   b. one-fourth as much
   c. twice as much
   d. four times as much

11. If a ball swinging in a circle on a string is moved twice as fast, the force on the string will be
   a. twice as great.
   b. four times as great.
   c. one-half as much.
   d. one-fourth as much.

12. A ball is swinging in a circle on a string when the string length is doubled. At the same velocity, the force on the string will be
   a. twice as great.
   b. four times as great.
   c. one-half as much.
   d. one-fourth as much.

QUESTIONS FOR THOUGHT

1. An insect inside a bus flies from the back toward the front at 5.0 mi/h. The bus is moving in a straight line at 50.0 mi/h. What is the speed of the insect?

2. Disregarding air friction, describe all the forces acting on a bullet shot from a rifle into the air.

3. Can gravity act in a vacuum? Explain.

4. Is it possible for a small car to have the same momentum as a large truck? Explain.

5. What net force is needed to maintain the constant velocity of a car moving in a straight line? Explain.

6. How can an unbalanced force exist if every action has an equal and opposite reaction?

7. Why should you bend your knees as you hit the ground after jumping from a roof?

8. Is it possible for your weight to change as your mass remains constant? Explain.

9. What maintains the speed of Earth as it moves in its orbit around the Sun?

10. Suppose you are standing on the ice of a frozen lake and there is no friction whatsoever. How can you get off the ice? (*Hint:* Friction is necessary to crawl or walk, so that will not get you off the ice.)

11. A rocket blasts off from a platform on a space station. An identical rocket blasts off from free space. Considering everything else to be equal, will the two rockets have the same acceleration? Explain.

12. An astronaut leaves a spaceship that is moving through free space to adjust an antenna. Will the spaceship move off and leave the astronaut behind? Explain.

FOR FURTHER ANALYSIS

1. What are the significant similarities and differences between speed and velocity?

2. What are the significant similarities and differences between velocity and acceleration?

3. Compare your beliefs and your own reasoning about motion before and after learning Newton's three laws of motion.

4. Newton's law of gravitation explains that every object in the universe is attracted to every other object in the universe. Describe a conversation between yourself and another person who does not believe this law, as you persuade him or her that the law is indeed correct.

5. Why is it that your weight can change by moving from one place to another, but your mass stays the same?

6. Assess the reasoning that Newton's first law of motion tells us that centrifugal force does not exist.

INVITATION TO INQUIRY

The Domino Effect

The *domino effect* is a cumulative effect produced when one event initiates a succession of similar events. In the actual case of dominoes, a row is made by standing dominoes on their end face to face in a line. When one on the end is tipped over toward the others, it will fall into its neighbor, which falls into the next one, and so on until the whole row has fallen. How should the dominoes be spaced so the row falls with maximum speed? Should one domino strike the next one as high as possible, in the center, or as low as possible? If you accept this invitation, you will need to determine how you plan to space the dominoes as well as how you will measure the speed.

PARALLEL EXERCISES

The exercises in groups A and B cover the same concepts. Solutions to group A exercises are located in appendix G.
Note: Neglect all frictional forces in all exercises.

Group A

1. How far away was a lightning strike if thunder is heard 5.00 seconds after seeing the flash? Assume that sound traveled at 350.0 m/s during the storm.

2. What is the acceleration of a car that moves from rest to 15.0 m/s in 10.0 s?

3. What is the average speed of a truck that makes a 160-kilometer trip in 2.0 hours?

4. What force will give a 40.0 kg grocery cart an acceleration of 2.4 m/s^2?

5. An unbalanced force of 18 N will give an object an acceleration of 3 m/s^2. What force will give this very same object an acceleration of 10 m/s^2?

6. What is the weight of a 70.0 kg person?

7. What is the momentum of a 100 kg football player who is moving at 6 m/s?

8. A car weighing 13,720 N is speeding down a highway with a velocity of 91 km/h. What is the momentum of this car?

9. A 15 g bullet is fired with a velocity of 200 m/s from a 6 kg rifle. What is the recoil velocity of the rifle?

10. A net force of 5,000.0 N accelerates a car from rest to 90.0 km/h in 5.0 s. (a) What is the mass of the car? (b) What is the weight of the car?

11. How much centripetal force is needed to keep a 0.20 kg ball on a 1.50 m string moving in a circular path with a speed of 3.0 m/s?

12. On Earth, an astronaut and equipment weigh 1,960.0 N. While weightless in space, the astronaut fires a 100 N rocket backpack for 2.0 s. What is the resulting velocity of the astronaut and equipment?

Group B

1. How many meters away is a cliff if an echo is heard one-half second after the original sound? Assume that sound traveled at 343 m/s on that day.

2. What is the acceleration of a car that moves from a speed of 5.0 m/s to a speed of 15 m/s during a time of 6.0 s?

3. What is the average speed of a car that travels 400.0 kilometers in 4.5 hours?

4. What force would an asphalt road have to give a 6,000 kg truck in order to accelerate it at 2.2 m/s^2 over a level road?

5. If a space probe weighs 39,200 N on the surface of Earth, what will be the mass of the probe on the surface of Mars?

6. How much does a 60.0 kg person weigh?

7. What is the momentum of a 30.0 kg shell fired from a cannon with a velocity of 500 m/s?

8. What is the momentum of a 39.2 N bowling ball with a velocity of 7.00 m/s?

9. A 30.0 kg shell fired from a 2,000 kg cannon will have a velocity of 500 m/s. What is the resulting velocity of the cannon?

10. A net force of 3,000.0 N accelerates a car from rest to 36.0 km/h in 5.00 s. (a) What is the mass of the car? (b) What is the weight of the car?

11. What tension must a 50.0 cm length of string support in order to whirl an attached 1,000.0 g stone in a circular path at 5.00 m/s?

12. A 200.0 kg astronaut and equipment move with a velocity of 2.00 m/s toward an orbiting spacecraft. How long will the astronaut need to fire a 100.0 N rocket backpack to stop the motion relative to the spacecraft?

3

Energy

The wind can be used as a source of energy. All you need is a way to capture the energy—such as these wind turbines in California—and to live someplace where the wind blows enough to make it worthwhile.

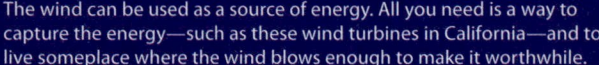

CORE **CONCEPT**

Energy is transformed through working or heating, and the total amount remains constant.

OUTLINE

Physics

▷ Heat is energy in transit (Ch. 3).
▷ Sound is a mechanical form of energy (Ch. 5).
▷ Electricity is a form of energy (Ch. 6).

Chemistry

▷ Chemical change is a form of energy (Ch. 8–10).
▷ A huge amount of energy exists in the nucleus of an atom (Ch. 11).

Earth Science

▷ Energy changes alter Earth and cause materials to be cycled (Ch. 14–18).

Life Science

▷ Living things use energy and materials in complex interactions (Ch. 19–26).

Astronomy

▷ Stars are nuclear reactors (Ch. 12).

▷ Meteorite impacts occurred (Ch. 13).

OVERVIEW

The term *energy* is closely associated with the concepts of force and motion. Naturally moving matter, such as the wind or moving water, exerts forces. You have felt these forces if you have ever tried to walk against a strong wind or stand in one place in a stream of rapidly moving water. The motion and forces of moving air and moving water are used as *energy sources* (figure 3.1). The wind is an energy source as it moves the blades of a windmill, performing useful work. Moving water is an energy source as it forces the blades of a water turbine to spin, turning an electric generator. Thus, moving matter exerts a force on objects in its path, and objects moved by the force can also be used as an energy source.

Matter does not have to be moving to supply energy; matter *contains* energy. Food supplied the energy for the muscular exertion of the humans and animals that accomplished most of the work before the twentieth century. Today in the developed world, machines do the work that was formerly accomplished by muscular exertion. Machines also use the energy contained in matter. They use gasoline, for example, as they supply the forces and motion to accomplish work.

Moving matter and matter that contains energy can be used as energy sources to perform work. The concepts of work and energy and the relationship to matter are the topics of this chapter. You will learn how energy flows in and out of your surroundings as well as a broad, conceptual view of energy.

3.1 WORK

You learned earlier that the term *force* has a special meaning in science that is different from your everyday concept of force. In everyday use, you use the term in a variety of associations such as police force, economic force, or the force of an argument. Earlier, force was discussed in a general way as a push or pull. Then a more precise scientific definition of force was developed from Newton's laws of motion—a force is a result of an interaction that is capable of changing the state of motion of an object.

The word *work* represents another one of those concepts that has a special meaning in science that is different from your everyday concept. In everyday use, work is associated with a task to be accomplished or the time spent in performing the task. You might work at understanding science, for example, or you might tell someone that science is a lot of work. You also probably associate physical work, such as lift-

ing or moving boxes, with how tired you become from the effort. The definition of mechanical work is not concerned with tasks, time, or how tired you become from doing a task. It is concerned with the application of a force to an object and the distance the object moves as a result of the force. The **work** done on the object is defined as *the product of the applied force and the parallel distance through which the force acts:*

$$\text{work} = \text{force} \times \text{distance}$$
$$W = Fd \qquad \text{equation 3.1}$$

Mechanical work is the product of a force and the distance an object moves as a result of the force. There are two important considerations to remember about this definition: (1) something *must move* whenever work is done, and (2) the movement must be in the *same direction* as the direction of the force. When you move a book to a higher shelf in a book-

FIGURE 3.1 Glen Canyon Dam on the Colorado River between Utah and Arizona is 216 m (710 ft) tall, dropping 940 m³ of water per second (33,000 ft³/s) through eight generating units.

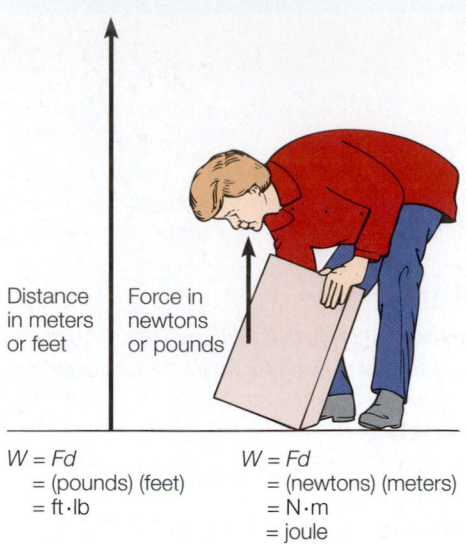

FIGURE 3.3 Work is done against gravity when lifting an object. Work is measured in joules or foot-pounds.

case, you are doing work on the book. You apply a vertically upward force equal to the weight of the book as you move it in the same direction as the direction of the applied force. The work done on the book can therefore be calculated by multiplying the weight of the book by the distance it was moved (figure 3.2).

If you simply stand there holding the book, however, you are doing no work on the book. Your arm may become tired from holding the book, since you must apply a vertically upward force equal to the weight of the book. But this force is not acting through a distance, since the book is not moving. According to equation 3.1, a distance of zero results in zero work. Only a force that results in motion in the same direction results in work.

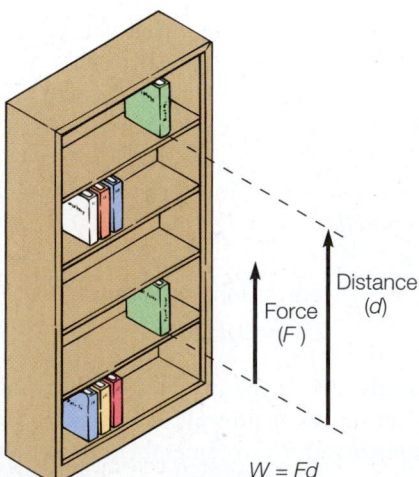

FIGURE 3.2 The force on the book moves it through a vertical distance from the second shelf to the fifth shelf, and work is done, $W = Fd$.

Units of Work

The units of work can be obtained from the definition of work, $W = Fd$. In the metric system, a force is measured in newtons (N), and distance is measured in meters (m), so the unit of work is

$$W = Fd$$
$$W = (\text{newton})(\text{meter})$$
$$W = (\text{N})(\text{m})$$

The newton-meter is therefore the unit of work. This derived unit has a name. The newton-meter is called a **joule** (J) (pronounced "jool").

$$1 \text{ joule} = 1 \text{ newton-meter}$$

The units for a newton are kg·m/s², and the unit for a meter is m. It therefore follows that the units for a joule are kg·m²/s².

In the English system, the force is measured in pounds (lb), and the distance is measured in feet (ft). The unit of work in the English system is therefore the ft·lb. The ft·lb does not have a name of its own as the N·m does (figure 3.3). (Note that although the equation is $W = Fd$, and this means = (pounds)(feet), the unit is called the ft·lb.)

EXAMPLE 3.1 *(Optional)*

How much work is needed to lift a 5.0 kg backpack to a shelf 1.0 m above the floor?

SOLUTION

The backpack has a mass (m) of 5.0 kg, and the distance (d) is 1.0 m. To lift the backpack requires a vertically upward force equal to the weight of the backpack. Weight can be calculated from $w = mg$:

—Continued on page 58

Simple Machines

Simple machines are tools that people use to help them do work. Recall that work is a force times a distance, and you can see that the simple machine helps you do work by changing a force or a distance that something is moved. The force or distance advantage you gain by using the machine is called the *mechanical advantage*. The larger the mechanical advantage, the greater the effort that you would save by using the machine.

A lever is a simple machine, and box figure 3.1 shows what is involved when a lever reduces a force needed to do work. First, note there are two forces involved. The force that you provide by using the machine is called the *effort force*. You and the machine are working against the second force, called the *resistance force*. In the illustration, a 60 N effort force is used to move a resistance force of 300 N.

There are also two distances involved in using the lever. The distance over which your effort force acts is called the *effort distance*, and the distance the resistance moves is called the *resistance distance*. You pushed down with an effort force of 60 N through an effort distance of 1 m. The 300 N rock, on the other hand, was raised a resistance distance of 0.2 m.

You did 60 N × 1 m, or 60 J, of work on the lever. The work done on the rock by the lever was 300 N × .02 m, or 60 J, of work. The work done by you on the lever is the same as the work done by the lever on the rock, so

$$\text{work input} = \text{work output}$$

Since work is force times distance, we can write this concept as

$$\text{effort force} \times \text{effort distance} =$$
$$\text{resistance force} \times \text{resistance distance}$$

Ignoring friction, the work you get out of any simple machine is the same as the work you put into it. The lever enabled you to trade force for distance, and the mechanical advantage (*MA*) can be found from a ratio of the resistance force (F_R) divided by the effort force (F_E):

$$MA = \frac{F_R}{F_E}$$

Therefore, the example lever in box figure 3.1 had a mechanical advantage of

$$MA = \frac{F_R}{F_E}$$
$$= \frac{300 \text{ N}}{60 \text{ N}}$$
$$= 5$$

You can also find the mechanical advantage by dividing the effort distance (d_E) by the resistance distance (d_R):

$$MA = \frac{d_E}{d_R}$$

For the example lever, we find

$$MA = \frac{d_E}{d_R}$$
$$= \frac{1 \text{ m}}{0.2 \text{ m}}$$
$$= 5$$

So, we can use either the forces or the distances involved in simple machines to calculate the mechanical advantage. In summary, a simple machine works for you by making it possible to apply a small force over a large distance to get a large force working over a small distance.

There are six kinds of simple machines: inclined plane, wedge, screw, lever, wheel and axle, and pulley. As you will see, the screw and wedge can be considered types of inclined planes; the wheel and axle and the pulley can be considered types of levers.

1. The *inclined plane* is a stationary ramp that is used to trade distance for force. You are using an inclined plane when you climb a stairway, drive a road that switches back and forth when going up a mountainside, or use a board to slide a heavy box up to a loading dock. Each use gives a large mechanical advantage by trading distance for force. For example, sliding a heavy box up a 10 m ramp to a 2 m high loading dock raises the box with less force through a greater distance. The mechanical advantage of this inclined plane is

$$MA = \frac{d_E}{d_R}$$
$$= \frac{10 \text{ m}}{2 \text{ m}}$$
$$= 5$$

Ignoring friction, a mechanical advantage of 5 means that a force of only 20 newtons would be needed to push a box weighing 100 newtons up the ramp.

2. The *wedge* is an inclined plane that moves. An ax is two back-to-back inclined planes that move through the wood it is used to split. Wedges are found in knives, axes, hatchets, and nails.

3. The *screw* is an inclined plane that has been wrapped around a cylinder, with the threads playing the role of the incline. A finely threaded screw has a higher mechanical advantage and

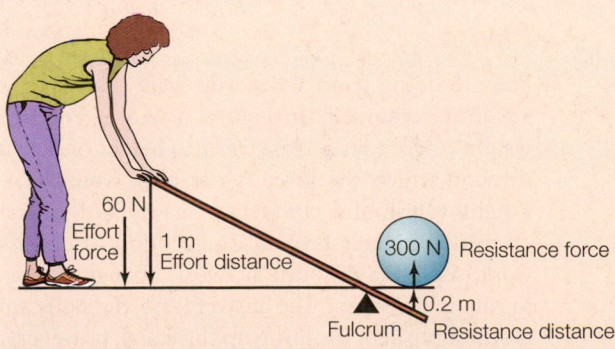

Box Figure 3.1   The lever is one of six simple machines.

60 N
Effort force
1 m
Effort distance
300 N   Resistance force
0.2 m
Fulcrum   Resistance distance

—*Continued top of next page*

Continued—

requires less force to turn, but it also requires a greater effort distance.

4. The *lever* is a bar or board that is free to pivot about a fixed point called a *fulcrum*. There are three classes of levers based on the location of the fulcrum, effort force, and resistance force (box figure 3.2). A first-class lever has the fulcrum between the effort force and the resistance force. Examples are a seesaw, pliers, scissors, crowbars, and shovels. A second-class lever has the effort resistance between the fulcrum and the effort force. Examples are nutcrackers and twist-type jar openers. A third-class lever has the effort force between the resistance force and the fulcrum. Examples are fishing rods and tweezers.

A claw hammer can be used as a first-class lever to remove nails from a board. If the hammer handle is 30 cm and the distance from the nail slot to the fulcrum is 5 cm, the mechanical advantage will be

$$MA = \frac{d_E}{d_R}$$
$$= \frac{30 \text{ cm}}{5 \text{ cm}}$$
$$= 6$$

5. A *wheel and axle* has two circles, with the smaller circle called the *axle* and the larger circle called the *wheel*. The wheel and axle can be considered to be a lever that can move in a circle. Examples are a screwdriver, door knob, steering wheel, and any application of a turning crank. The mechanical advantage is found from the radius of the wheel, where the effort is applied, to the radius of the axle, which is the distance over which the resistance moves. For example, a large screw-driver has a radius of 15 cm in the handle (the wheel) and 0.5 cm in the bit (the axle). The mechanical advantage of this screwdriver is

$$MA = \frac{d_E}{d_R}$$
$$= \frac{3 \text{ cm}}{0.5 \text{ cm}}$$
$$= 6$$

6. A *pulley* is a movable lever that rotates around a fulcrum. A single fixed pulley can only change the direction of a force. To gain a mechanical advantage, you need a fixed pulley and a movable pulley such as those found in a block and tackle. The mechanical advantage of a block and tackle can be found by comparing the length of rope or chain pulled to the distance the resistance has moved.

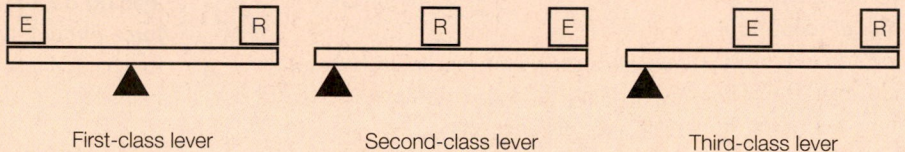

First-class lever   Second-class lever   Third-class lever

Box Figure 3.2 The three classes of levers are defined by the relative locations of the fulcrum, effort, and resistance.

—EXAMPLE 3.1 *continued*

$$m = 5.0 \text{ kg} \qquad w = mg$$
$$g = 9.8 \text{ m/s}^2$$
$$w = ? \qquad = (5.0 \text{ kg})\left(9.8\frac{\text{m}}{\text{s}^2}\right)$$
$$= (5.0)(9.8) \text{ kg} \times \frac{\text{m}}{\text{s}^2}$$
$$= 49\frac{\text{kg}\cdot\text{m}}{\text{s}^2}$$
$$= 49 \text{ N}$$

The definition of work is found in equation 3.1,

$$F = 49 \text{ N} \qquad W = Fd$$
$$d = 1.0 \text{ m} \qquad = (49 \text{ N})(1.0 \text{ m})$$
$$W = ? \qquad = (49)(1.0) \text{ N} \times \text{m}$$
$$= 49 \text{ N}\cdot\text{m}$$
$$= \boxed{49 \text{ J}}$$

EXAMPLE 3.2 *(Optional)*

How much work is required to lift a 50 lb box vertically a distance of 2 ft? (Answer: 100 ft·lb)

Power

You are doing work when you walk up a stairway, since you are lifting yourself through a distance. You are lifting your weight (force exerted) the *vertical* height of the stairs (distance through which the force is exerted). Consider a person who weighs 120 lb and climbs a stairway with a vertical distance of 10 ft. This person will do (120 lb)(10 ft) or 1,200 ft·lb of work. Will the amount of work change if the person were to run up the stairs? The answer is no, the same amount of work is accomplished. Running up the stairs, however, is more tiring than walking up the stairs. You use the same amount of energy but at a greater *rate* when running. The rate at which energy is

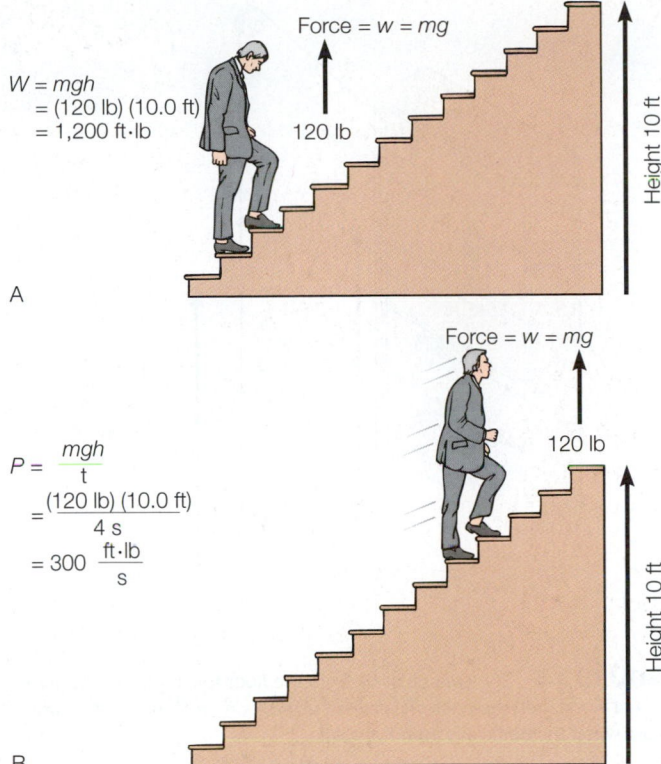

$W = mgh$
$= (120 \text{ lb}) (10.0 \text{ ft})$
$= 1,200 \text{ ft·lb}$

Force $= w = mg$

120 lb

Height 10 ft

A

$P = \dfrac{mgh}{t}$
$= \dfrac{(120 \text{ lb}) (10.0 \text{ ft})}{4 \text{ s}}$
$= 300 \dfrac{\text{ft·lb}}{\text{s}}$

Force $= w = mg$

120 lb

Height 10 ft

B

FIGURE 3.4 (A) The work accomplished in climbing a stairway is the person's weight times the vertical distance. (B) The power level is the work accomplished per unit of time.

transformed or the rate at which work is done is called **power** (figure 3.4). **Power is defined as work per unit of time,**

$$\text{power} = \frac{\text{work}}{\text{time}}$$

$$P = \frac{W}{t} \qquad \textbf{equation 3.2}$$

Considering just the work and time factors, the 120 lb person who ran up the 10 ft height of stairs in 4 seconds would have a power rating of

$$P = \frac{W}{t} = \frac{(120 \text{ lb})(10 \text{ ft})}{4 \text{ s}} = 300 \frac{\text{ft·lb}}{\text{s}}$$

If the person had a time of 3 seconds on the same stairs, the power rating would be greater, 400 ft·lb/s. This is a greater *rate* of energy use, or greater power.

When the steam engine was first invented, there was a need to describe the rate at which the engine could do work. Since people at this time were familiar with using horses to do their work, the steam engines were compared to horses. James Watt, who designed a workable steam engine, defined **horsepower** as a power rating of 550 ft·lb/s (figure 3.5A). To convert a power rating in the English units of ft·lb/s to horsepower, divide the power rating by 550 ft·lb/s/hp. For example, the 120 lb person

who had a power rating of 400 ft·lb/s had a horsepower of 400 ft·lb/s ÷ 550 ft·lb/s/hp, or 0.7 hp.

In the metric system, power is measured in joules per second. The unit J/s, however, has a name. A J/s is called a **watt** (W). The watt (figure 3.5B) is used with metric prefixes for large numbers: 1,000 W = 1 kilowatt (kW) and 1,000,000 W = 1 megawatt (MW). It takes 746 W to equal 1 horsepower. One kilowatt is equal to about 1⅓ horsepower. The electric utility company charges you for how much electrical energy you have used. Electrical energy is measured by power (kW) times the time of use (h). Thus, electrical energy is measured in kWh. We will return to kilowatts and kilowatt-hours later when we discuss electricity.

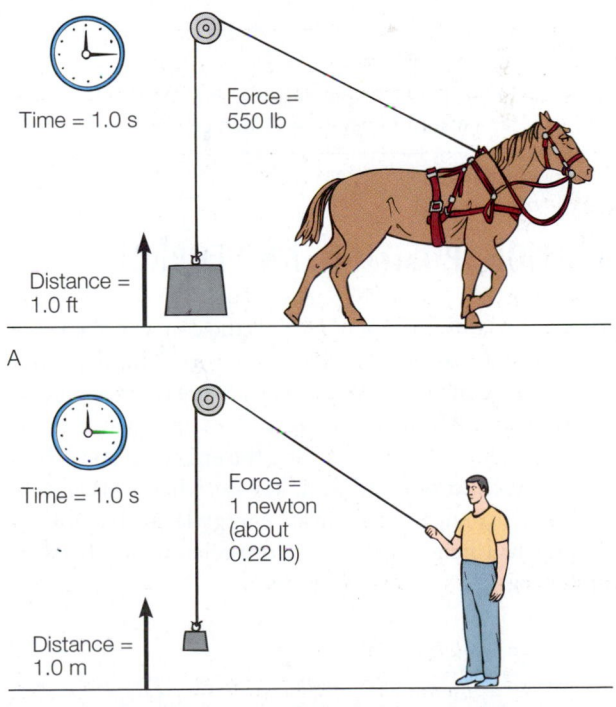

Time = 1.0 s

Force = 550 lb

Distance = 1.0 ft

A

Time = 1.0 s

Force = 1 newton (about 0.22 lb)

Distance = 1.0 m

B

FIGURE 3.5 (A) A horsepower is defined as a power rating of 550 ft·lb/s. (B) A watt is defined as a newton-meter per second, or joule per second.

EXAMPLE 3.3 (Optional)

An electric lift can raise a 500.0 kg mass a distance of 10.0 m in 5.0 s. What is the power of the lift?

SOLUTION

Power is the work per unit time ($P = W/t$), and work is force times distance ($W = Fd$). The vertical force required is the weight lifted, and $w = mg$. Therefore, the work accomplished would be $W = mgh$, and the power would be $P = mgh/t$. Note that h is for height, a vertical distance (d).

$$m = 500.0 \text{ kg}$$
$$g = 9.8 \text{ m/s}^2$$
$$h = 10.0 \text{ m}$$
$$t = 5.0 \text{ s}$$

$$P = \frac{mgh}{t}$$

$$= \frac{(500.0 \text{ kg})\left(9.8 \frac{\text{m}}{\text{s}^2}\right)(10.0 \text{ m})}{5.0 \text{ s}}$$

$$= \frac{(500.0)(9.8)(10.0)}{5.0} \frac{\text{kg} \times \frac{\text{m}}{\text{s}^2} \times \text{m}}{\text{s}}$$

$$= 9{,}800 \frac{\text{N} \cdot \text{m}}{\text{s}}$$

$$= 9{,}800 \frac{\text{J}}{\text{s}}$$

$$= 9{,}800 \text{ W}$$

$$= \boxed{9.8 \text{ kW}}$$

The power in horsepower (hp) units would be

$$9{,}800 \text{ W} \times \frac{\text{hp}}{746 \text{ W}} = 13 \text{ hp}$$

EXAMPLE 3.4 (Optional)

A 150 lb person runs up a 15 ft stairway in 10.0 s. What is the horsepower rating of the person? (Answer: 0.41 horsepower)

3.2 MOTION, POSITION, AND ENERGY

Closely related to the concept of work is the concept of **energy.** Energy can be defined as the *ability to do work.* This definition of energy seems consistent with everyday ideas about energy and physical work. After all, it takes a lot of energy to do a lot of work. In fact, one way of measuring the energy of something is to see how much work it can do. Likewise, when work is done *on* something, a change occurs in its energy level. The following examples will help clarify this close relationship between work and energy.

Potential Energy

Consider a book on the floor next to a bookcase. You can do work on the book by vertically raising it to a shelf. You can measure this work by multiplying the vertical upward force

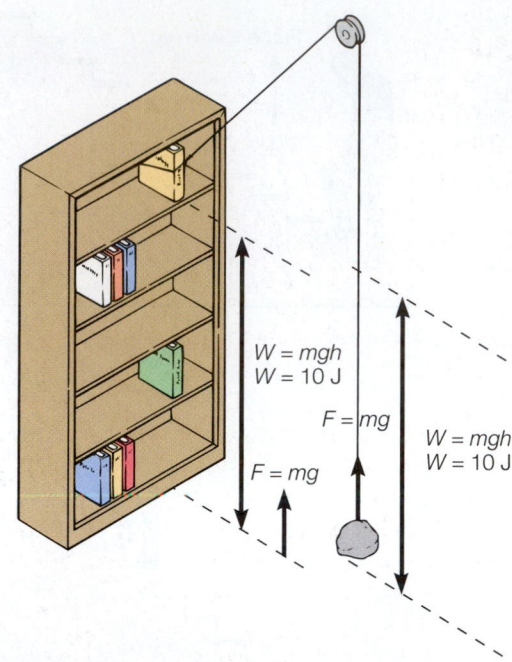

FIGURE 3.6   If moving a book from the floor to a high shelf requires 10 J of work, then the book will do 10 J of work on an object of the same mass when the book falls from the shelf.

applied times the distance that the book is moved. You might find, for example, that you did an amount of work equal to 10 J on the book.

Suppose that the book has a string attached to it, as shown in figure 3.6. The string is threaded over a frictionless pulley and attached to an object on the floor. If the book is caused to fall from the shelf, the object on the floor will be vertically lifted through some distance by the string. The falling book exerts a force on the object through the string, and the object is moved through a distance. In other words, the *book* did work on the object through the string, $W = Fd$.

The book can do more work on the object if it falls from a higher shelf, since it will move the object a greater distance. The higher the shelf, the greater the *potential* for the book to do work. The ability to do work is defined as energy. The energy that an object has because of its position is called **potential energy** (*PE*). Potential energy is defined as *energy due to position.* This type of potential energy is called *gravitational potential energy,* since it is a result of gravitational attraction. There are other types of potential energy, such as that in a compressed or stretched spring.

The gravitational potential energy of an object can be calculated from the work done *on* the object to change its position. You exert a force equal to its weight as you lift it some height above the floor, and the work you do is the product of the weight and height. Likewise, the amount of work the object *could* do because of its position is the product of its weight and height. For the metric unit of mass, weight is the product of the mass of an object times g, the acceleration due to gravity, so

gravitational potential energy = weight × height

$$PE = mgh \qquad \text{equation 3.3}$$

For English units, the pound *is* the gravitational unit of force, or weight, so equation 3.3 becomes $PE = (w)(h)$.

Under what conditions does an object have zero potential energy? Considering the book in the bookcase, you could say that the book has zero potential energy when it is flat on the floor. It can do no work when it is on the floor. But what if that floor happens to be the third floor of a building? You could, after all, drop the book out of a window. The answer is that it makes no difference. The same results would be obtained in either case since it is the *change of position* that is important

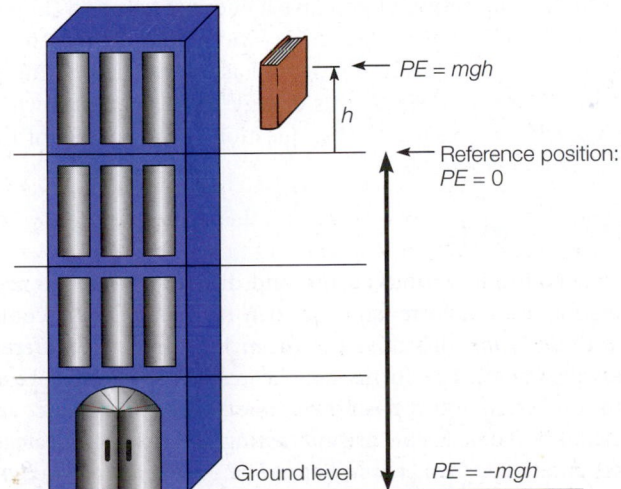

FIGURE 3.7 The zero reference level for potential energy is chosen for convenience. Here the reference position chosen is the third floor, so the book will have a negative potential energy at ground level.

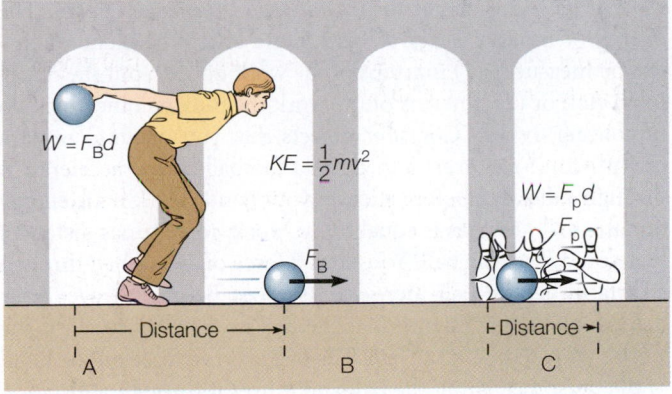

FIGURE 3.8 (A) Work is done on the bowling ball as a force (F_B) moves it through a distance. (B) This gives the ball a kinetic energy equal in amount to the work done on it. (C) The ball does work on the pins and has enough remaining energy to crash into the wall behind the pins.

in potential energy. The zero reference position for potential energy is therefore arbitrary. A zero reference point is chosen as a matter of convenience. Note that if the third floor of a building is chosen as the zero reference position, a book on ground level would have negative potential energy. This means that you would have to do work on the book to bring it back to the zero potential energy position (figure 3.7). You will learn more about negative energy levels later in the chapters on chemistry.

EXAMPLE 3.5 (*Optional*)

What is the potential energy of a 2.14 kg book that is on a bookshelf 1.0 m above the floor?

SOLUTION

Equation 3.3, $PE = mgh$, shows the relationship between potential energy (PE), weight (mg), and height (h).

$$m = 2.14 \text{ kg} \qquad PE = mgh$$
$$h = 1.0 \text{ m} \qquad\quad = (2.14 \text{ kg})(9.8 \tfrac{\text{m}}{\text{s}^2})(1.0 \text{ m})$$
$$PE = ? \qquad\qquad = (2.14)(9.8)(1.0) \frac{\text{kg} \cdot \text{m} \cdot \text{m}}{\text{s}^2}$$
$$\qquad\qquad = 21 \frac{\text{kg} \cdot \text{m}}{\text{s}^2} \times \text{m}$$
$$\qquad\qquad = 21 \text{ Nm}$$
$$\qquad\qquad = \boxed{21 \text{ J}}$$

EXAMPLE 3.6 (*Optional*)

How much work can a 5.00 kg mass do if it is 5.00 m above the ground? (Answer: 250 J)

Kinetic Energy

Moving objects have the ability to do work on other objects because of their motion. A rolling bowling ball exerts a force on the bowling pins and moves them through a distance, but the ball loses speed as a result of the interaction (figure 3.8). A moving car has the ability to exert a force on a tree and knock it down, again with a corresponding loss of speed. Objects in motion have the ability to do work, so they have energy. The energy of motion is known as **kinetic energy.** Kinetic energy can be measured (1) in terms of the work done to put the object in motion or (2) in terms of the work the moving object will do in coming to rest. Consider objects that you put into motion by throwing. You exert a force on a football as you accelerate it through a distance before it leaves your hand. The kinetic energy that the ball now has is equal to the work (force times distance) that you did on the ball. You exert a force on a baseball through a distance as the ball increases its speed before it leaves your hand. The kinetic energy that the ball now has is equal to the work that you did on the ball. The ball exerts a force on the hand of the person catching the ball and moves it through a distance. The net work done on the hand is equal to the kinetic energy that the ball had.

A baseball and a bowling ball moving with the same velocity do not have the same kinetic energy. You cannot knock down many bowling pins with a slowly rolling baseball. Obviously, the more-massive bowling ball can do much more work than a less-massive baseball with the same velocity. Is it possible for the bowling ball and the baseball to have the same kinetic energy? The answer is yes, if you can give the baseball sufficient velocity. This might require shooting the baseball from a cannon, however. Kinetic energy is proportional to the mass of a moving object, but velocity has a greater influence. Consider two balls of the same mass, but one is moving twice as fast as the other. The ball with twice the velocity will do *four* times as much work as the slower ball. A ball with three times the velocity will do *nine* times as much work as the slower ball. Kinetic energy is proportional to the square of the velocity ($2^2 = 4$; $3^2 = 9$). The kinetic energy (KE) of an object is

$$\text{kinetic energy} = \frac{1}{2}(\text{mass})(\text{velocity})^2$$

$$KE = \frac{1}{2}mv^2 \qquad \textbf{equation 3.4}$$

Kinetic energy is measured in joules, as is work ($F \times d$ or N·m) and potential energy (mgh or N·m).

FIGURE 3.9 Mechanical energy is the energy of motion, or the energy of position, of many familiar objects. This boat has energy of motion.

EXAMPLE 3.7   (OPTIONAL)

A 7.00 kg bowling ball is moving in a bowling lane with a velocity of 5.00 m/s. What is the kinetic energy of the ball?

SOLUTION

The relationship between kinetic energy (KE), mass (m), and velocity (v) is found in equation 3.4, $KE = 1/2mv^2$:

$$m = 7.00 \text{ kg} \qquad KE = \frac{1}{2}mv^2$$
$$v = 5.00 \text{ m/s}$$
$$KE = ?$$

$$= \frac{1}{2}(7.00 \text{ kg})\left(5.00\frac{m}{s}\right)^2$$

$$= \frac{1}{2}(7.00 \text{ kg})\left(25.0\frac{m^2}{s^2}\right)$$

$$= \frac{1}{2}(7.00)(25.0) \text{ kg} \times \frac{m^2}{s^2}$$

$$= \frac{1}{2} \times 175\frac{\text{kg}\cdot m^2}{s^2}$$

$$= 87.5\frac{\text{kg}\cdot m}{s^2} \times m$$

$$= 87.5 \text{ N}\cdot\text{m}$$

$$= \boxed{87.5 \text{ J}}$$

EXAMPLE 3.8   (Optional)

A 100.0 kg football player moving with a velocity of 6.0 m/s tackles a stationary quarterback. How much work was done on the quarterback? (Answer: 1,800 J)

3.3   ENERGY FLOW

The key to understanding the individual concepts of work and energy is to understand the close relationship between the two. When you do work on something, you give it energy of position (potential energy) or you give it energy of motion (kinetic energy). In turn, objects that have kinetic or potential energy can now do work on something else as the transfer of energy continues. Where does all this energy come from and where does it go? The answer to this question is the subject of this section on energy flow.

Energy Forms

Energy comes in various forms, and different terms are used to distinguish one form from another. Although energy comes in various *forms,* this does not mean that there are different *kinds* of energy. The forms are the result of the more common fundamental forces—gravitational, electromagnetic, and nuclear—and objects that are interacting. Energy can be categorized into five forms: (1) *mechanical,* (2) *chemical,* (3) *radiant,* (4) *electrical,* and (5) *nuclear.* The following is a brief discussion of each of the five forms of energy.

Mechanical energy is the form of energy of familiar objects and machines (figure 3.9). A car moving on a highway has

A

B

FIGURE 3.10 Chemical energy is a form of potential energy that is released during a chemical reaction. Both (A) wood and (B) coal have chemical energy that has been stored through the process of photosynthesis. The pile of wood might provide fuel for a small fireplace for several days. The pile of coal might provide fuel for a power plant for a hundred days.

kinetic mechanical energy. Water behind a dam has potential mechanical energy. The spinning blades of a steam turbine have kinetic mechanical energy. The form of mechanical energy is usually associated with the kinetic energy of everyday-sized objects and the potential energy that results from gravity. There are other possibilities (e.g., sound), but this description will serve the need for now.

Chemical energy is the form of energy involved in chemical reactions (figure 3.10). Chemical energy is released in the chemical reaction known as *oxidation*. The fire of burning wood is an example of rapid oxidation. A slower oxidation releases energy from food units in your body. As you will learn in the chemistry unit, chemical energy involves electromagnetic forces between the parts of atoms. Until then, consider the following comparison. Photosynthesis is carried on in green plants. The plants use the energy of sunlight to rearrange carbon dioxide and water into plant materials and oxygen. Leaving

FIGURE 3.11 This demonstration solar cell array converts radiant energy from the Sun to electrical energy, producing an average of 200,000 watts of electric power (after conversion).

out many steps and generalizing, this could be represented by the following word equation:

$$\text{energy} + \text{carbon dioxide} + \text{water} = \text{wood} + \text{oxygen}$$

The plant took energy and two substances and made two different substances. This is similar to raising a book to a higher shelf in a bookcase. That is, the new substances have more energy than the original ones did. Consider a word equation for the burning of wood:

$$\text{wood} + \text{oxygen} = \text{carbon dioxide} + \text{water} + \text{energy}$$

Notice that this equation is exactly the reverse of photosynthesis. In other words, the energy used in photosynthesis was released during oxidation. Chemical energy is a kind of potential energy that is stored and later released during a chemical reaction.

Radiant energy is energy that travels through space (figure 3.11). Most people think of light or sunlight when considering this form of energy. Visible light, however, occupies only a small part of the complete electromagnetic spectrum, as shown in figure 3.12. Radiant energy includes light and all other parts of the spectrum (see page 154). Infrared radiation is sometimes called "heat radiation" because of the association with heating when this type of radiation is absorbed. For example, you feel the interaction of infrared radiation when you hold your hand near a warm range element. However, infrared radiation is another type of radiant energy. In fact, some snakes, such as the sidewinder, have pits between their eyes and nostrils that can detect infrared radiation emitted from warm animals. Microwaves are another type of radiant energy that is used in cooking. As with other forms of energy, light, infrared, and microwaves will be considered in more detail later. For now, consider all types of radiant energy to be forms of energy that travel through space.

Electrical energy is another form of energy from electromagnetic interactions that will be considered in detail later. You

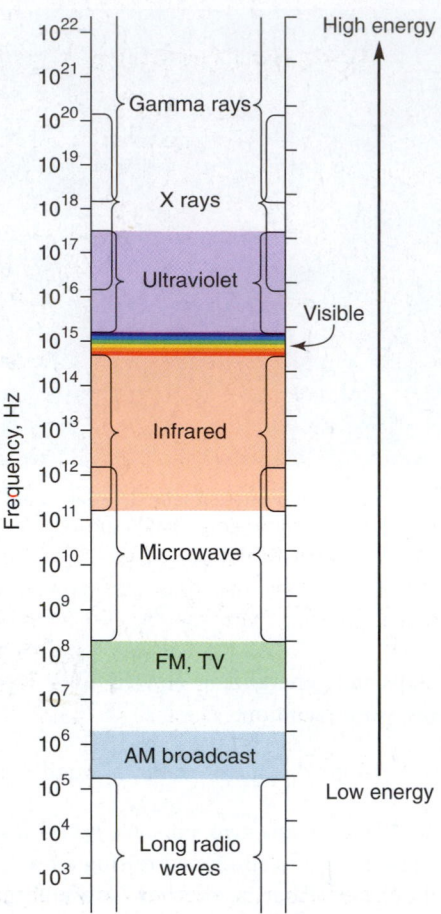

FIGURE 3.12 The electromagnetic spectrum includes many categories of radiant energy. Note that visible light occupies only a tiny part of the entire spectrum.

FIGURE 3.13 The blades of a steam turbine. In a power plant, chemical or nuclear energy is used to heat water to steam, which is directed against the turbine blades. The mechanical energy of the turbine turns an electric generator. Thus, a power plant converts chemical or nuclear energy to mechanical energy, which is then converted to electrical energy.

are familiar with electrical energy that travels through wires to your home from a power plant (figure 3.13), electrical energy that is generated by chemical cells in a flashlight, and electrical energy that can be "stored" in a car battery.

Nuclear energy is a form of energy often discussed because of its use as an energy source in power plants. Nuclear energy is another form of energy from the atom, but this time the energy involves the nucleus, the innermost part of an atom, and nuclear interactions.

Energy Conversion

Potential energy can be converted to kinetic energy and vice versa. The simple pendulum offers a good example of this conversion. A simple pendulum is an object, called a bob, suspended by a string or wire from a support. If the bob is moved to one side and then released, it will swing back and forth in an arc. At the moment that the bob reaches the top of its swing, it stops for an instant, then begins another swing. At the instant of stopping, the bob has 100 percent potential energy and no kinetic energy. As the bob starts back down through the swing, it is gaining kinetic energy and losing potential energy. At the instant the bob is at the bottom of the swing, it has 100 percent kinetic energy and no potential energy. As the bob now climbs

through the other half of the arc, it is gaining potential energy and losing kinetic energy until it again reaches an instantaneous stop at the top, and the process starts over. The kinetic energy of the bob at the bottom of the arc is equal to the potential energy it had at the top of the arc (figure 3.14). Disregarding friction, the sum of the potential energy and the kinetic energy remains constant throughout the swing.

Any *form* of energy can be converted to another form. In fact, most technological devices that you use are nothing more than *energy-form converters* (figure 3.15). A lightbulb, for example, converts electrical energy to radiant energy. A car converts chemical energy to mechanical energy. A solar cell converts radiant energy to electrical energy, and an electric motor converts electrical energy to mechanical energy. Each technological device converts some form of energy (usu-

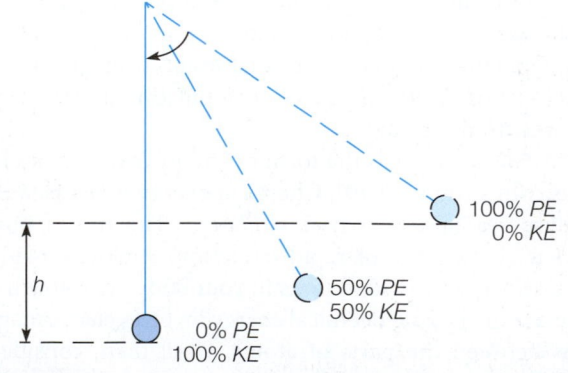

FIGURE 3.14 This pendulum bob loses potential energy (*PE*) and gains an equal amount of kinetic energy (*KE*) as it falls through a distance *h*. The process reverses as the bob moves up the other side of its swing.

Energy Converter Coaster

A roller coaster is an energy converter that swaps kinetic and potential energy back and forth. An outside energy source is used only to move the cars to the top of the first hill. The first hill is always the highest hill above the ground, and here the cars will have the most potential energy they will have for the entire ride.

When the coaster reaches the top of the first hill, it begins to move down a sloping track with increasing speed as potential energy is converted to kinetic energy. At the bottom of this hill, the track then starts up a second hill, and the cars this time convert kinetic energy back to potential energy. Ideally, all the potential energy is converted to kinetic energy as the cars move down a hill, and all the kinetic energy is converted to potential energy as the cars move up a hill. There is no perfect energy conversion, and some energy is lost to air resistance and friction. The roller coaster design allows for these losses and leaves room for a slight surplus of kinetic energy. You know this is true because the operator must apply brakes at the end of the ride.

Is the speed of the moving roller coaster at the bottom of a hill directly proportional to the height of the hill? Ignoring friction, the speed of a coaster at the bottom of a hill is proportional to the *square root of the height* of the hill. This means that to double the speed of the coaster at the bottom of the hill, you would need to build the hill more than four times higher. It would need to be more than four times higher because the coaster does not drop straight down. Doubling the height increases the theoretical speed only 40 percent.

What is the relationship between the weight of the people on a roller coaster and the speed achieved at the bottom of the hill? The answer is that the weight of the people does not matter. Ignoring air resistance and friction, a heavy roller coaster full of people and a lighter one with just a few people will both have the same speed at the bottom of the hill.

Draw a profile of a roller coaster ride to find out about potential and kinetic energy exchanges and the height of each hill. The profile should show the relative differences in height between the crest of each hill and the bottom of each upcoming dip. From such a profile, you could find where in a ride the maximum speed would occur, as well as what speed to expect. As you study a profile keep in mind that the roller coaster is designed to produce many changes of speed—accelerations—rather than high speed alone.

---

ally chemical or electrical) to another form that you desire (usually mechanical or radiant).

It is interesting to trace the *flow of energy* that takes place in your surroundings. Suppose, for example, that you are riding a bicycle. The bicycle has kinetic mechanical energy as it moves along. Where did the bicycle get this energy? It came from you, as you use the chemical energy of food units to contract your muscles and move the bicycle along. But where did your chemical energy come from? It came from your food, which consists of plants, animals who eat plants, or both plants and animals. In any case, plants are at the bottom of your food chain. Plants convert radiant energy from the Sun into chemical energy. Radiant energy comes to the plants from the Sun because of the nuclear reactions that took place in the core of the Sun. Your bicycle is therefore powered by nuclear energy that has undergone a number of form conversions!

Energy Conservation

Energy can be transferred from one object to another, and it can be converted from one form to another form. If you make a detailed accounting of all forms of energy before and after a transfer or conversion, the total energy will be *constant*. Consider your bicycle coasting along over level ground when you apply the brakes. What happened to the kinetic mechanical energy of the bicycle? It went into heating the rim and brakes of your bicycle, then eventually radiated to space as infrared radiation. All radiant energy that reaches Earth is eventually radiated back to space (figure 3.16). Thus, throughout all the form conversions and energy transfers that take place, the total sum of energy remains constant.

The total energy is constant in every situation that has been measured. This consistency leads to another one of the conservation laws of science, the **law of conservation of energy**:

Energy is never created or destroyed. Energy can be converted from one form to another but the total energy remains constant.

FIGURE 3.15   The energy forms and some conversion pathways.

Grow Your Own Fuel?

Have you heard of biodiesel? Biodiesel is a vegetable-based oil that can be used for fuel in diesel engines. It can be made from soy oils, canola oil, or even recycled deep-fryer oil from a fast-food restaurant. Biodiesel can be blended with regular diesel oil in any amount. Or it can be used in its 100 percent pure form in diesel cars, trucks, buses, or as home heating oil.

Why would we want to use vegetable oil to run diesel engines? First, it is a sustainable (or renewable) resource. It also reduces dependency on foreign oil and lessens the trade deficit. It runs smoother, produces less exhaust smoke, and reduces the health risks associated with petroleum diesel. The only negative aspect seems to occur when recycled oil from fast-food restaurants is used. People behind such a biodiesel-powered school bus complained that it smells like fried potatoes, making them hungry.

A website is maintained by some biodiesel users where you can learn how to produce your own biodiesel from algae. See www.biodieselnow.com and search for algae.

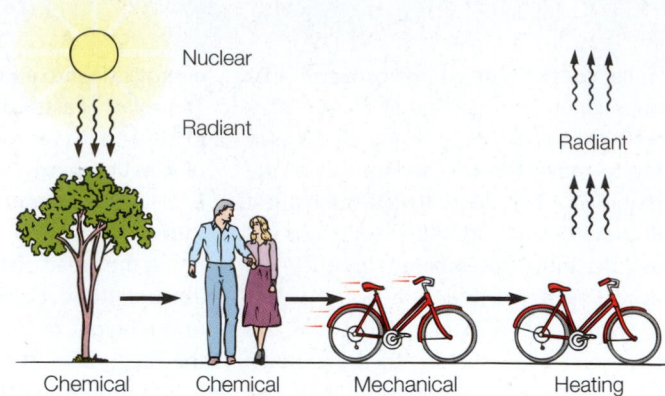

FIGURE 3.16 Energy arrives from the Sun, goes through a number of conversions, then radiates back into space. The total sum leaving eventually equals the original amount that arrived.

See also "Thermodynamics" in chapter 4. You may be wondering about the source of nuclear energy. Does a nuclear reaction create energy? Albert Einstein answered this question back in the early 1900s when he formulated his now-famous relationship between mass and energy, $E = mc^2$. This relationship will be discussed in detail in chapter 11. Basically, the relationship states that mass *is* a form of energy, and this has been experimentally verified many times.

Energy Transfer

Earlier it was stated that when you do work on something, you give it energy. The result of work could be increased kinetic mechanical energy, increased gravitational potential energy, or an increase in the temperature of an object. You could summarize this by stating that either *working* or *heating* is always involved any time energy is transformed. This is not unlike your financial situation. To increase or decrease your financial status, you need some mode of transfer, such as cash or checks, as a means of conveying assets. Just as with cash flow from one individual to another, energy flow from one object to another requires a mode of transfer. In energy matters, the mode of transfer is working or heating. Any time you see working or heating occurring, you know that an energy transfer is taking place. The next time you see heating, think about what energy form is being converted to what new energy form. (The final form is usually radiant energy.) Heating is the topic of chapter 4, where you will consider the role of heat in energy matters.

3.4 ENERGY SOURCES TODAY

Prometheus, according to ancient Greek mythology, stole fire from heaven and gave it to humankind. Fire has propelled human advancement ever since. All that was needed was something to burn—fuel for Prometheus's fire.

Any substance that burns can be used to fuel a fire, and various fuels have been used over the centuries as humans advanced. First, wood was used as a primary source for heating. Then coal fueled the Industrial Revolution. Eventually, humankind roared into the twentieth century burning petroleum. Today, petroleum is the most widely used source of energy (figure 3.17). It provides about 40 percent of the total energy used by the United States. Natural gas contributes about 23 percent of the total energy used today. The use of coal also provides about 23 percent of the total. Biomass—any material formed by photosynthesis—contributes about 3 percent of the total energy used today. Note that petroleum, coal, biomass, and natural gas are all chemical sources of energy, sources that are mostly burned for their energy. These chemical sources supply about 87 percent of the total energy consumed. About a third of this is burned for heating, and the rest is burned to drive engines or generators.

Nuclear energy and hydropower are the nonchemical sources of energy. These sources are used to generate electrical

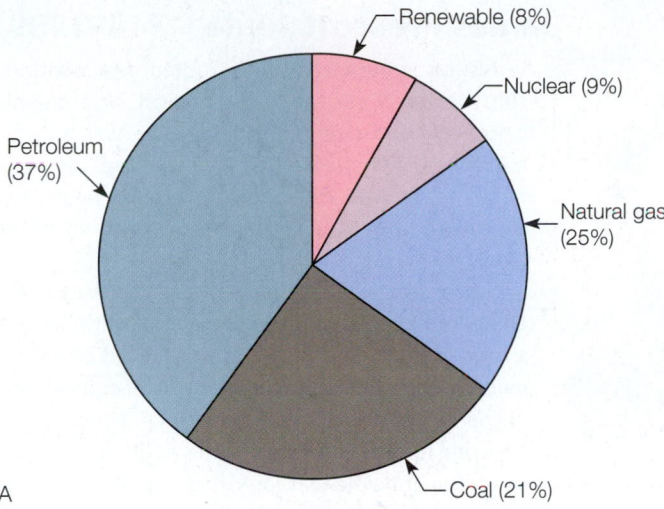

A

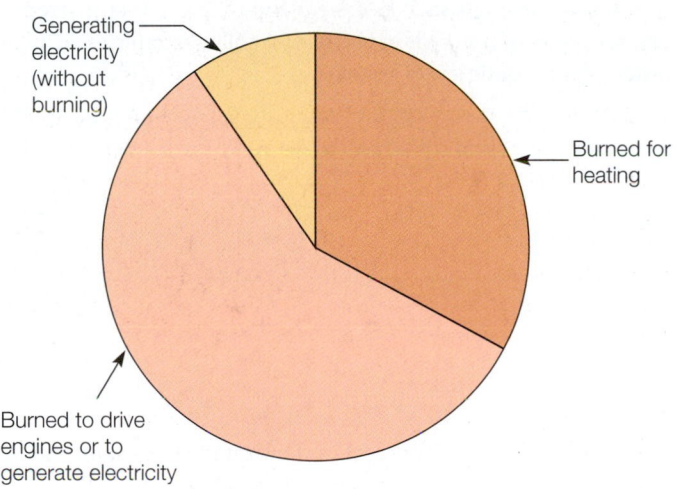

B

FIGURE 3.17 (A) Primary energy consumed in the United States, 2009,* and (B) uses of this energy.

*Source: Energy Information Administration (www.eia.doe.gov/emeu/aer/pdf/pages/sec1.pdf).

energy. The alternative sources of energy, such as solar, wind, and geothermal, provide about 1 percent of the total energy consumed in the United States today.

The energy-source mix has changed from past years, and it will change in the future. Wood supplied 90 percent of the energy until the 1850s, when the use of coal increased. Then, by 1910, coal was supplying about 75 percent of the total energy needs. Then petroleum began making increased contributions to the energy supply. Now increased economic and environmental constraints and a decreasing supply of petroleum are producing another supply shift. The present petroleum-based energy era is about to shift to a new energy era.

About 95 percent of the total energy consumed today is provided by four sources: (1) petroleum (including natural gas), (2) coal, (3) moving water, and (4) nuclear. The following is a brief introduction to these four sources.

Petroleum

The word *petroleum* is derived from the word *petra,* meaning rock, and the word *oleum,* meaning oil. Petroleum is oil that comes from oil-bearing rock. Natural gas is universally associated with petroleum and has similar origins. Both petroleum and natural gas form from organic sediments, materials that have settled out of bodies of water. Sometimes a local condition permits the accumulation of sediments that are exceptionally rich in organic material. This could occur under special conditions in a freshwater lake, or it could occur on shallow ocean basins. In either case, most of the organic material is from plankton—tiny free-floating animals and plants such as algae. It is from such accumulations of buried organic material that petroleum and natural gas are formed. The exact process by which these materials become petroleum and gas is not understood. It is believed that bacteria, pressure, appropriate temperatures, and time are all important. Natural gas is formed at higher temperatures than is petroleum. Varying temperatures over time may produce a mixture of petroleum and gas or natural gas alone.

Petroleum forms a thin film around the grains of the rock where it formed. Pressure from the overlying rock and water moves the petroleum and gas through the rock until they reach a rock type or structure that stops them. If natural gas is present, it occupies space above the accumulating petroleum. Such accumulations of petroleum and natural gas are the sources of supply for these energy sources.

Discussions about the petroleum supply and the cost of petroleum usually refer to a "barrel of oil." The *barrel* is an accounting device of 42 U.S. gallons. Such a 42-gallon barrel does not exist. When or if oil is shipped in barrels, each drum holds 55 U.S. gallons.

Petroleum is used for gasoline (about 45 percent), diesel (about 40 percent), and heating oil (about 15 percent). Petroleum is also used in making medicine, clothing fabrics, plastics, and ink.

Coal

Petroleum and natural gas formed from the remains of tiny organisms that lived millions of years ago. Coal, on the other hand, formed from an accumulation of plant materials that collected under special conditions millions of years ago. Thus, petroleum, natural gas, and coal are called **fossil fuels.** Fossil fuels contain the stored radiant energy of organisms that lived millions of years ago.

The first thing to happen in the formation of coal was for plants in swamps to die and sink. Stagnant swamp water protected the plants and plant materials from consumption by animals and decomposition by microorganisms. Over time, chemically altered plant materials collected at the bottom of pools of water in the swamp. This carbon-rich material is *peat* (not to be confused with peat moss). Peat is used as a fuel in many places in the world. The flavor of Scotch (whisky) is the result of the peat fires used to brew the liquor. Peat is still being produced naturally in swampy areas today. Under pressure and

James Prescott Joule (1818–1889)

James Joule was a British physicist who helped develop the principle of conservation of energy by experimentally measuring the mechanical equivalent of heat. In recognition of Joule's pioneering work on energy, the SI unit of energy is named the joule.

Joule was born on December 24, 1818, into a wealthy brewing family. He and his brother were educated at home between 1833 and 1837 in elementary math, natural philosophy, and chemistry, partly by the English chemist John Dalton (1766–1844).

Joule had great dexterity as an experimenter, and he could measure temperatures very precisely. At first, other scientists could not believe such accuracy and were skeptical about the theories that Joule developed to explain his results. The encouragement of Lord Kelvin from 1847 on changed these attitudes, however, and Kelvin subsequently used Joule's practical ability to great advantage. By 1850, Joule was highly regarded by other scientists and was elected a fellow of the Royal Society.

Joule realized the importance of accurate measurement very early on, and exact data became his hallmark. His most active research period was between 1837 and 1847. In a long series of experiments, he studied the relationship between electrical, mechanical, and chemical effects and heat, and in 1843, he was able to announce his determination of the amount of work required to produce a unit of heat. This is called the mechanical equivalent of heat (4.184 joules per calorie).

One great value of Joule's work was the variety and completeness of his experimental evidence. He showed that the same relationship could be examined experimentally and that the ratio of equivalence of the different forms of energy did not depend on how one form was converted into another or on the materials involved. The principle that Joule had established is that energy cannot be created or destroyed but only transformed.

Joule lives on in the use of his name to measure energy, supplanting earlier units such as the erg and calorie. It is an appropriate reflection of his great experimental ability and his tenacity in establishing a basic law of science.

Source: Modified from the *Hutchinson Dictionary of Scientific Biography*. © RM, 2011. All rights reserved. Helicon Publishing is a division of RM.

at high temperatures, peat will eventually be converted to coal. There are several stages, or *ranks,* in the formation of coal. The lowest rank is lignite (brown coal), then subbituminous, then bituminous (soft coal), and the highest rank is anthracite (hard coal).

Each rank of coal has different burning properties and a different energy content. Coal also contains impurities of clay, silt, iron oxide, and sulfur. The mineral impurities leave an ash when the coal is burned, and the sulfur produces sulfur dioxide, a pollutant.

Most of the coal mined today is burned by utilities to generate electricity (about 80 percent). The coal is ground to a face-powder consistency and blown into furnaces. This greatly increases efficiency but produces *fly ash,* ash that "flies" up the chimney. Industries and utilities are required by the federal Clean Air Act to remove sulfur dioxide and fly ash from plant emissions. About 20 percent of the cost of a new coal-fired power plant goes into air pollution control equipment. Coal is an abundant but dirty energy source.

Moving Water

Moving water has been used as a source of energy for thousands of years. It is considered a renewable energy source, inexhaustible as long as the rain falls. Today, hydroelectric plants generate about 3 percent of the United States' *total* energy consumption at about 2,400 power-generating dams across the nation. Hydropower furnished about 40 percent of the United States' electric power in 1940. Today, dams furnish 9 percent of the electric power. Energy consumption has increased, but hydropower production has not kept pace because geography limits the number of sites that can be built.

Water from a reservoir is conducted through large pipes called penstocks to a powerhouse, where it is directed against turbine blades that turn a shaft on an electric generator.

Nuclear

Nuclear power plants use nuclear energy to produce electricity. Energy is released as the nuclei of uranium and plutonium atoms split, or undergo fission. The fissioning takes place in a large steel vessel called a *reactor.* Water is pumped through the reactor to produce steam, which is used to produce electrical energy, just as in the fossil fuel power plants. The nuclear processes are described in detail in chapter 11, and the process of producing electrical energy is described in detail in chapter 6. Nuclear power plants use nuclear energy to produce electricity, but there are opponents of this process. The electric utility companies view nuclear energy as *one* energy source used to produce electricity. They state that they have no allegiance to any one energy source but are seeking to utilize the most reliable and dependable of several energy sources. Petroleum, coal, and hydropower are also presently utilized as energy sources for electric power production. The electric utility companies are concerned that petroleum and natural gas are becoming increasingly expensive, and there are questions about long-term supplies. Hydropower has limited potential for growth, and solar energy is prohibitively expensive today. Utility companies see two major energy sources that are available for growth: coal and nuclear. There are problems and advantages to each, but the utility companies feel they must use coal and nuclear power until the new technologies, such as solar power, are economically feasible.

Conserving Energy

Conservation is not a way of generating energy, but it is a way of reducing the need for additional energy consumption and it saves money for the consumer. Some conservation technologies are sophisticated, while others are quite simple. For example, if a small, inexpensive wood-burning stove were developed and used to replace open fires in the less-developed world, energy consumption in these regions could be reduced by 50 percent.

Many observers have pointed out that demanding more energy while failing to conserve is like demanding more water to fill a bathtub while leaving the drain open. To be sure, conservation and efficiency strategies by themselves will not eliminate demands for energy, but they can make the demands much easier to meet, regardless of what options are chosen to provide the primary energy. Energy efficiency improvements have significantly reduced the need for additional energy sources. Consider these facts, which are based primarily on data published by the U.S. Energy Information Administration:

- Total primary energy use per capita in the United States in 2003 was almost identical to that in 1973. Over the same thirty-year period, economic output (gross domestic product, or GDP) per capita increased 74 percent.
- National energy intensity (energy use per unit of GDP) fell 43 percent between 1973 and 2002. About 60 percent of this decline is attributable to energy efficiency improvements.
- If the United States had not dramatically reduced its energy intensity over the past thirty years, consumers and businesses would have spent at least $430 billion more on energy purchases in 2003.

Even though the United States is much more energy-efficient today than it was thirty years ago, the potential is still enormous for additional cost-effective energy savings. Some newer energy efficiency measures have barely begun to be adopted. Other efficiency measures could be developed and commercialized in coming years.

Much of the energy we consume is wasted. This statement is not meant as a reminder to simply turn off lights and lower furnace thermostats; it is a technological challenge. Our use of energy is so inefficient that most potential energy in fuel is lost as waste heat, becoming a form of environmental pollution.

The amount of energy wasted through poorly insulated windows and doors alone is about as much energy as the United States receives from the Alaskan pipeline each year. It is estimated that by using inexpensive, energy-efficient measures, the average energy bills of a single home could be reduced by 10 percent to 50 percent and could help reduce the emissions of carbon dioxide into the atmosphere.

Many conservation techniques are relatively simple and highly cost-effective. More efficient and less energy-intensive industry and domestic practices could save large amounts of energy. Improved automobile efficiency, better mass transit, and increased railroad use for passenger and freight traffic are simple and readily available means of conserving transportation energy. In response to the 1970s' oil price shocks, automobile mileage averages in the United States more than doubled, from 5.55 km/L (13 mi/gal) in 1975 to 12.3 km/L (28.8 mi/gal) in 1988. Unfortunately, the oil glut and falling fuel prices of the late 1980s discouraged further conservation. Between 1990 and 1997, the average slipped to only 11.8 km/L (27.6 mi/gal). It remains to be seen if the sharp increase of gasoline prices in the early years of the twenty-first century will translate into increased miles per gallon in new car design.

Several technologies that reduce energy consumption are now available. Highly efficient fluorescent lightbulbs (CFL) that can be used in regular incandescent fixtures give the same amount of light for 25 percent of the energy and produce less heat. Since lighting and air conditioning (which removes the heat from inefficient incandescent lighting) account for 25 percent of U.S. electricity consumption, widespread use of these lights could significantly reduce energy consumption. Low-emissive glass for windows can reduce the amount of heat entering a building while allowing light to enter. The use of this type of glass in new construction and replacement windows could have a major impact on the energy picture. Many other technologies, such as automatic light-dimming or shutoff devices, are being used in new construction.

The shift to more efficient use of energy needs encouragement. Often, poorly designed, energy-inefficient buildings and machines can be produced inexpensively. The short-term cost is low, but the long-term cost is high. The public needs to be educated to look at the long-term economic and energy costs of purchasing poorly designed buildings and appliances.

Electric utilities have recently become part of the energy conservation picture. In some states, they have been allowed to make money on conservation efforts; previously, they could

Myths, Mistakes, and Misunderstandings

Reusing or Recycling?

It is a mistake to say you are recycling plastic bags if you take them back to the store when you shop. This is reusing, not recycling. Recycling means that an item is broken down into raw materials, then made into new items. Since recycling uses energy, the reuse of materials is more environmentally desirable.

make money only by building more power plants. This encourages them to become involved in energy conservation education, because teaching their customers how to use energy more efficiently allows them to serve more people without building new power plants.

3.5 ENERGY TOMORROW

An *alternative source of energy* is one that is different from the typical sources used today. The sources used today are fossil fuels (coal, petroleum, and natural gas), nuclear, and falling water. Alternative sources could be solar, geothermal, hydrogen gas, fusion, or any other energy source that a new technology could utilize.

Solar Technologies

The term *solar energy* is used to describe a number of technologies that directly or indirectly utilize sunlight as an alternative energy source (figure 3.18). There are eight main categories of these solar technologies:

1. **Solar cells.** A solar cell is a thin crystal of silicon, gallium, or some polycrystalline compound that generates electricity when exposed to light. Also called photovoltaic devices, solar cells have no moving parts and produce electricity directly, without the need for hot fluids or intermediate conversion states. Solar cells have been used extensively in space vehicles and satellites. Here on Earth, however, use has been limited to demonstration projects, remote-site applications, and consumer specialty items such as solar-powered watches and calculators. The problem with solar cells today is that the manufacturing cost is too high (they

FIGURE 3.18 Wind is another form of solar energy. This wind turbine generates electrical energy for this sailboat, charging batteries for backup power when the wind is not blowing. In case you are wondering, the turbine cannot be used to make a wind to push the boat. In accord with Newton's laws of motion, this would not produce a net force on the boat.

are essentially handmade). Research is continuing on the development of highly efficient, affordable solar cells that could someday produce electricity for the home. See page 147 to find out how a solar cell is able to create a current.

2. **Power tower.** This is another solar technology designed to generate electricity. One type of planned power tower will have a 171 m (560 ft) tower surrounded by some 9,000 special mirrors called heliostats. The heliostats will focus sunlight on a boiler at the top of the tower where salt (a mixture of sodium nitrate and potassium nitrate) will be heated to about 566°C (about 1,050°F). This molten salt will be pumped to a steam generator, and the steam will be used to drive a generator, just like other power plants. Water could be heated directly in the power tower boiler. Molten salt is used because it can be stored in an insulated storage tank for use when the Sun is not shining, perhaps for up to 20 hours.

3. **Passive application.** In passive applications, energy flows by natural means, without mechanical devices such as motors, pumps, and so forth. A passive solar house would include considerations such as its orientation to the Sun, the size and positioning of windows, and a roof overhang that lets sunlight in during the winter but keeps it out during the summer. There are different design plans to capture, store, and distribute solar energy throughout a house, and some of these designs are described on pages 88–89.

4. **Active application.** An active solar application requires a solar collector in which sunlight heats air, water, or some liquid. The liquid or air is pumped through pipes in a house to generate electricity, or it is used directly for hot water. Solar water heating makes more economic sense today than the other solar applications.

5. **Wind energy.** The wind has been used for centuries to move ships, grind grain into flour, and pump water. The wind blows, however, because radiant energy from the Sun heats some parts of Earth's surface more than others. This differential heating results in pressure differences and the horizontal movement of air, which is called wind. Thus, wind is another form of solar energy. Wind turbines are used to generate electrical energy or mechanical energy. The biggest problem with wind energy is the inconsistency of the wind. Sometimes the wind speed is too great, and other times it is not great enough. Several methods of solving this problem are being researched. (See "Wind Power" in chapter 17 for more on wind energy.)

6. **Biomass.** Biomass is any material formed by photosynthesis, including small plants, trees, and crops, and any garbage, crop residue, or animal waste. Biomass can be burned directly as a fuel, converted into a gas fuel (methane), or converted into liquid fuels such as alcohol. The problems with using biomass include the energy expended in gathering the biomass, as well as the energy used to convert it to a gaseous or liquid fuel.

7. **Agriculture and industrial heating.** This is a technology that simply uses sunlight to dry grains, cure paint, or do anything that can be done with sunlight rather than using traditional energy sources.

8. **Ocean thermal energy conversion (OTEC).** This is an electric generating plant that would take advantage of the approximately 22°C (about 40°F) temperature difference between the surface and the depths of tropical, subtropical, and equatorial ocean waters. Basically, warm water is drawn into the system to vaporize a fluid, which expands through a turbine generator. Cold water from the depths condenses the vapor back to a liquid form, which is then cycled back to the warm-water side. The concept has been tested and found to be technically successful. The greatest interest seems to be from islands that have warm surface waters and cold depths, such as Hawaii, Puerto Rico, Guam, and the Virgin Islands.

Geothermal Energy

Geothermal energy is energy from beneath Earth's surface. The familiar geysers, hot springs, and venting steam of Yellowstone National Park are clues that this form of energy exists. There is substantially more geothermal energy than is revealed in Yellowstone, however, and geothermal resources are more widespread than once thought. Earth has a high internal temperature, and recoverable geothermal resources may underlie most states. These resources occur in four broad categories of geothermal energy: (1) dry steam, (2) hot water, (3) hot, dry rock, and (4) geopressurized resources. Together, the energy contained in these geothermal resources represents about 15,000 times more than is consumed in the United States in a given year. The only problem is getting to the geothermal energy, then using it in a way that is economically attractive.

Most geothermal energy occurs as *hot, dry rock,* which accounts for about 85 percent of the total geothermal resource. Hot, dry rock is usually in or near an area of former volcanic activity. The problem of utilizing this widespread resource is how to get the energy to the surface. Research has been conducted by drilling wells, then injecting water into one well and extracting energy from the heated water pumped from the second well. There is more interest in the less widespread but better understood geothermal systems of hot water and steam.

Geopressurized resources are trapped underground reservoirs of hot water that contain dissolved natural gas. The water temperature is higher than the boiling point, so heat could be used as a source of energy as well as the dissolved natural gas. Such geopressurized reservoirs make up about 14 percent of the total accessible geothermal energy found on Earth. They are being studied in some areas to determine whether they are large enough to be economically feasible as an energy source. More is known about recovering energy from other types of hot water and steam resources, so these seem more economically attractive.

Hot water and steam make up the smallest geothermal resource category, together comprising only about 1 percent of the total known resource. However, more is known about the utilization and recovery of these energy sources, which are estimated to contain an amount of energy equivalent to about half of the known reserves of petroleum in the United States. *Steam* is very rare, occurring in only three places in the United States. Two of these places are national parks (Lassen and Yellowstone), so this geothermal steam cannot be used as an energy source. The third place is at the Geysers, an area of fumaroles near San Francisco, California. Steam from the Geysers is used to generate a significant amount of electricity.

Hot water systems make up most of the *recoverable* geothermal resources. Heat from deep volcanic or former volcanic sources create vast, slow-moving convective patterns in groundwater. If the water circulating back near the surface is hot enough, it can be used for generating electricity, heating buildings, or many other possible applications. Worldwide, geothermal energy is used to operate pulp and paper mills, cool hotels, raise fish, heat greenhouses, dry crops, desalt water, and dozens of other things. Thousands of apartments, homes, and businesses are today heated geothermally in Oregon and Idaho in the United States, as well as in Hungary, France, Iceland, and New Zealand. Today, each British thermal unit (Btu) of heat supplied by geothermal energy does not have to be supplied by fossil fuels. Tomorrow, geothermal resources will become more and more attractive as the price and the consequences of using fossil fuels continue to increase.

Hydrogen

Hydrogen is the lightest and simplest of all the elements, occurring as a diatomic gas that can be used for energy directly in a fuel cell or burned to release heat. Hydrogen could be used to replace natural gas with a few modifications of present natural gas burners. A big plus in favor of hydrogen as a fuel is that it produces no pollutants. In addition to the heat produced, the only emission from burning hydrogen is water, as shown in the following equation:

$$\text{Hydrogen} + \text{oxygen} \rightarrow \text{water} + 68{,}300 \text{ calories}$$

The primary problem with using hydrogen as an energy source is that *it does not exist* on or under Earth's surface in any but trace amounts! Hydrogen must therefore be obtained by a chemical reaction from such compounds as water. Water is a plentiful substance on Earth, and an electric current will cause decomposition of water into hydrogen and oxygen gas. Measurement of the electric current and voltage will show that:

$$\text{Water} + 68{,}300 \text{ calories} \rightarrow \text{hydrogen} + \text{oxygen}$$

Thus, assuming 100 percent efficiency, the energy needed to obtain hydrogen gas from water is exactly equal to the energy released by hydrogen combustion. So hydrogen cannot be used to produce energy, since hydrogen gas is not available, but it can be used as a means of storing energy for later use. Indeed, hydrogen may be destined to become an effective solution to the problems of storing and transporting energy derived from solar energy sources. In addition, hydrogen might serve as the transportable source of energy, such as that needed for cars and trucks, replacing the fossil fuels. In summary, hydrogen has the potential to provide clean, alternative energy for a number of uses, including lighting, heating, cooling, and transportation.

SUMMARY

Work is defined as the product of an applied force and the distance through which the force acts. Work is measured in newton-meters, a metric unit called a *joule*. *Power* is work per unit of time. Power is measured in *watts*. One watt is 1 joule per second. Power is also measured in *horsepower*. One horsepower is 550 ft·lb/s.

Energy is defined as the ability to do work. An object that is elevated against gravity has a potential to do work. The object is said to have *potential energy*, or *energy of position*. Moving objects have the ability to do work on other objects because of their motion. The *energy of motion* is called *kinetic energy*.

Work is usually done *against inertia, fundamental forces, friction, shape,* or *combinations of these*. As a result, there is a gain of *kinetic energy, potential energy, an increased temperature,* or *any combination of these*. Energy comes in the *forms* of *mechanical, chemical, radiant, electrical,* or *nuclear*. Potential energy can be *converted* to kinetic and kinetic can be *converted* to potential. Any form of energy can be *converted* to any other form. Most technological devices are *energy-form converters* that do work for you. Energy flows into and out of the surroundings, but the amount of energy is always constant. The *law of conservation of energy* states that *energy is never created or destroyed*. Energy conversion always takes place through *heating* or *working*.

The basic energy sources today are the chemical *fossil fuels* (petroleum, natural gas, and coal), *nuclear energy,* and *hydropower*. Petroleum and *natural gas* were formed from organic material of plankton, tiny free-floating plants and animals. A barrel of petroleum is 42 U.S. gallons, but such a container does not actually exist. *Coal* formed from plants that were protected from consumption by falling into a swamp.

The decayed plant material, *peat,* was changed into the various *ranks* of coal by pressure and heating over some period of time. Coal is a dirty fuel that contains impurities and sulfur. Controlling air pollution from burning coal is costly. Moving water and nuclear energy are used for the generation of electricity. An *alternative* source of energy is one that is different from the typical sources used today. Alternative sources could be *solar, geothermal,* or *hydrogen*.

Summary of Equations

3.1
$$\text{work} = \text{force} \times \text{distance}$$
$$W = Fd$$

3.2
$$\text{power} = \frac{\text{work}}{\text{time}}$$
$$P = \frac{W}{t}$$

3.3
$$\text{potential energy} = \text{weight} \times \text{height}$$
$$PE = mgh$$

3.4
$$\text{kinetic energy} = \frac{1}{2}(\text{mass})(\text{velocity})^2$$
$$KE = \frac{1}{2}mv^2$$

KEY TERMS

energy (p. **60**)
fossil fuels (p. **67**)
geothermal energy (p. **71**)

horsepower (p. **59**)
joule (p. **56**)
kinetic energy (p. **62**)

law of conservation of
energy (p. **65**)
potential energy (p. **60**)

power (p. **59**)
watt (p. **59**)
work (p. **55**)

APPLYING THE CONCEPTS

Answers are located in appendix F.

1. The metric unit of a joule (J) is a unit of
   a. potential energy.
   b. work.
   c. kinetic energy.
   d. Any of the above is correct.

2. Power is
   a. the rate at which work is done.
   b. the rate at which energy is expended.
   c. work per unit time.
   d. Any of the above is correct.

3. Which of the following is a combination of units called a watt?
   a. N·m/s
   b. kg·m²/s²/s
   c. J/s
   d. All of the above are correct.

4. About how many watts are equivalent to 1 horsepower?
   a. 7.5   b. 75
   c. 750   d. 7500

5. A kilowatt-hour is a unit of
   a. power   b. work
   c. time   d. electrical charge

6. The potential energy of a box on a shelf, relative to the floor, is a measure of
   a. the work that was required to put the box on the shelf from the floor.
   b. the weight of the box times the distance above the floor.
   c. the energy the box has because of its position above the floor.
   d. All of the above are correct.

7. A rock on the ground is considered to have zero potential energy. In the bottom of a well, then, the rock would be considered to have
   a. zero potential energy, as before.
   b. negative potential energy.
   c. positive potential energy.
   d. zero potential energy but will require work to bring it back to ground level.

8. Which quantity has the greatest influence on the amount of kinetic energy that a large truck has while moving down the highway?
   a. mass
   b. weight
   c. velocity
   d. size

9. Most energy comes to and leaves Earth in the form of
   a. nuclear energy.
   b. chemical energy.
   c. radiant energy.
   d. kinetic energy.

10. The law of conservation of energy is basically that
    a. energy must not be used up faster than it is created, or the supply will run out.
    b. energy should be saved because it is easily destroyed.
    c. energy is never created or destroyed.
    d. you are breaking a law if you needlessly destroy energy.

11. The most widely used source of energy today is
    a. coal.
    b. petroleum.
    c. nuclear.
    d. moving water.

12. The accounting device of a "barrel of oil" is defined to hold how many U.S. gallons of petroleum?
    a. 24
    b. 42
    c. 55
    d. 100

QUESTIONS FOR THOUGHT

1. How is work related to energy?
2. What is the relationship between the work done while moving a book to a higher bookshelf and the potential energy that the book has on the higher shelf?
3. Does a person standing motionless in the aisle of a moving bus have kinetic energy? Explain.
4. A lamp bulb is rated at 100 W. Why is a time factor not included in the rating?
5. Is a kWh a unit of work, energy, power, or more than one of these? Explain.
6. If energy cannot be destroyed, why do some people worry about the energy supplies?
7. A spring clamp exerts a force on a stack of papers it is holding together. Is the spring clamp doing work on the papers? Explain.

8. Why are petroleum, natural gas, and coal called *fossil fuels*?
9. From time to time, people claim to have invented a machine that will run forever without energy input and develops more energy than it uses (perpetual motion). Why would you have reason to question such a machine?
10. Define a joule. What is the difference between a joule of work and a joule of energy?
11. Compare the energy needed to raise a mass 10 m on Earth to the energy needed to raise the same mass 10 m on the Moon. Explain the difference, if any.
12. What happens to the kinetic energy of a falling book when the book hits the floor?

FOR FURTHER ANALYSIS

1. Evaluate the requirement that something must move whenever work is done. Why is this a requirement?
2. What are the significant similarities and differences between work and power?
3. Whenever you do work on something, you give it energy. Analyze how you would know for sure that this is a true statement.
4. Simple machines are useful because they are able to trade force for distance moved. Describe a conversation between yourself and another person who believes that you do less work when you use a simple machine.
5. Use the equation for kinetic energy to prove that speed is more important than mass in bringing a speeding car to a stop.

6. Describe at least several examples of negative potential energy and how each shows a clear understanding of the concept.
7. The five forms of energy are the result of the fundamental forces—gravitational, electromagnetic, and nuclear—and objects that are interacting. Analyze which force is probably involved with each form of energy.
8. Most technological devices convert one of the five forms of energy into another. Try to think of a technological device that does not convert an energy form to another. Discuss the significance of your finding.
9. Are there any contradictions to the law of conservation of energy in any area of science?

INVITATION TO INQUIRY

New Energy Source?

Is waste paper a good energy source? There are 103 waste-to-energy plants in the United States that burn solid garbage, so waste paper would be a good source, too. The plants burn solid garbage to make steam that is used to heat buildings and generate electricity. Schools might be able to produce a pure waste paper source because waste paper accumulates near computer print stations and in offices. Collecting waste paper from such sources would yield 6,800 Btu/lb, which is about half the heat value of coal.

If you accept this invitation, start by determining how much waste paper is created per month in your school. Would this amount produce enough energy to heat buildings or generate electricity?

PARALLEL EXERCISES

The exercises in groups A and B cover the same concepts. Solutions to group A exercises are located in appendix G.
Note: Neglect all frictional forces in all exercises.

Group A

1. A force of 200 N is needed to push a table across a level classroom floor for a distance of 3 m. How much work was done on the table?

2. An 880 N box is pushed across a level floor for a distance of 5.0 m with a force of 440 N. How much work was done on the box?

3. How much work is done in raising a 10.0 kg backpack from the floor to a shelf 1.5 m above the floor?

4. If 5,000 J of work is used to raise a 102 kg crate to a shelf in a warehouse, how high was the crate raised?

5. A 60.0 kg student runs up a 5.00 m high stairway in a time of 3.92 s. (a) How many watts of power did she develop? (b) How many horsepower is this?

6. What is the kinetic energy of a 2,000 kg car moving at 72 km/h?

7. How much work is needed to stop a 1,000.0 kg car that is moving straight down the highway at 54.0 km/h?

8. A 1,000 kg car stops on top of a 51.02 m hill. (a) How much energy was used in climbing the hill? (b) How much potential energy does the car have?

9. A horizontal force of 10.0 lb is needed to push a bookcase 5 ft across the floor. (a) How much work was done on the bookcase? (b) How much did the gravitational potential energy change as a result?

10. (a) How much work is done in moving a 2.0 kg book to a shelf 2.00 m high? (b) What is the change in potential energy of the book as a result? (c) How much kinetic energy will the book have as it hits the ground as it falls?

11. A 60.0 kg jogger moving at 2.0 m/s decides to double the jogging speed. How did this change in speed change the kinetic energy?

12. A 170.0 lb student runs up a stairway to a classroom 25.0 ft above ground level in 10.0 s. (a) How much work did the student do? (b) What was the average power output in horsepower?

Group B

1. Work of 1,200 J is done while pushing a crate across a level floor for a distance of 1.5 m. What force was used to move the crate?

2. How much work is done by a hammer that exerts a 980.0 N force on a nail, driving it 1.50 cm into a board?

3. A 5.0 kg textbook is raised a distance of 30.0 cm as a student prepares to leave for school. How much work did the student do on the book?

4. An electric hoist does 196,000 J of work in raising a 250.0 kg load. How high was the load lifted?

5. What is the horsepower of a 1,500.0 kg car that can go to the top of a 360.0 m high hill in exactly one minute?

6. What is the kinetic energy of a 30.0 g bullet that is traveling at 200.0 m/s?

7. How much work will be done by a 30.0 g bullet traveling at 200 m/s?

8. A 10.0 kg box is lifted 15 m above the ground by a construction crane. (a) How much work did the crane do on the box? (b) How much potential energy does the box have relative to the ground?

9. A force of 50.0 lb is used to push a box 10.0 ft across a level floor. (a) How much work was done on the box? (b) What is the change of potential energy as a result of this move?

10. (a) How much work is done in raising a 50.0 kg crate a distance of 1. 5 m above a storeroom floor? (b) What is the change of potential energy as a result of this move? (c) How much kinetic energy will the crate have as it falls and hits the floor?

11. The driver of an 800.0 kg car decides to double the speed from 20.0 m/s to 40.0 m/s. What effect would this have on the amount of work required to stop the car, that is, on the kinetic energy of the car?

12. A 70.0 kg student runs up the stairs of a football stadium to a height of 10.0 m above the ground in 10.0 s. (a) What is the power of the student in kilowatts? (b) in horsepower?

4

Heat and Temperature

Sparks fly from a plate of steel as it is cut by an infrared laser. Today, lasers are commonly used to cut as well as weld metals, so the cutting and welding is done by light, not by a flame or electric current.

CORE **CONCEPT**

A relationship exists between heat, temperature, and the motion and position of the molecules.

OUTLINE

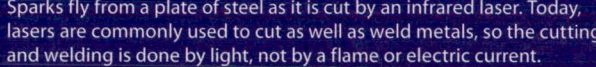

All matter is made of molecules that move and interact (p. 76).

Temperature is a measure of the average kinetic energy of molecules (p. 79).

Heat is a measure of internal energy that has been transferred or absorbed (p. 82).

The laws of thermodynamics describe a relationship between changes of internal energy, work, and heat (p. 94).

Chemistry

▷ Chemical reactions involve changes in the internal energy of molecules (Ch. 9).

Astronomy

▷ Stars form from compressional heating (Ch. 12).

Earth Science

▷ Earth's interior is hot, making possible the movement of plates on the surface (Ch. 15–16).

▷ Movements of air, air masses, and ocean currents are driven in cycles involving temperature differences (Ch. 17–18).

Life Science

▷ The process of aerobic respiration results in a release of heat (Ch. 20).

OVERVIEW

Heat has been closely associated with the comfort and support of people throughout history. You can imagine the appreciation when your earliest ancestors first discovered fire and learned to keep themselves warm and cook their food. You can also imagine the wonder and excitement about 3000 B.C., when people put certain earthlike substances on the hot, glowing coals of a fire and later found metallic copper, lead, or iron. The use of these metals for simple tools followed soon afterwards. Today, metals are used to produce complicated engines that use heat for transportation and do the work of moving soil and rock, construction, and agriculture. Devices made of heat-extracted metals are also used to control the temperature of structures, heating or cooling the air as necessary. Thus, the production and control of heat gradually built the basis of civilization today (figure 4.1).

The sources of heat are the energy forms that you learned about in chapter 3. The fossil fuels are *chemical* sources of heat. Heat is released when oxygen is combined with these fuels. Heat also results when *mechanical* energy does work against friction, such as in the brakes of a car coming to a stop. Heat also appears when *radiant* energy is absorbed. This is apparent when solar energy heats water in a solar collector or when sunlight melts snow. The transformation of *electrical* energy to heat is apparent in toasters, heaters, and ranges. *Nuclear* energy provides the heat to make steam in a nuclear power plant. Thus, all the energy forms can be converted to heat.

The relationship between energy forms and heat appears to give an order to nature, revealing patterns that you will want to understand. All that you need is some kind of explanation for the relationships—a model or theory that helps make sense of it all. This chapter is concerned with heat and temperature and their relationship to energy. It begins with a simple theory about the structure of matter and then uses the theory to explain the concepts of heat, energy, and temperature changes.

4.1 THE KINETIC MOLECULAR THEORY

The idea that substances are composed of very small particles can be traced back to certain early Greek philosophers. The earliest record of this idea was written by Democritus during the fifth century B.C. He wrote that matter was empty space filled with tremendous numbers of tiny, indivisible particles called *atoms*. This idea, however, was not acceptable to most of the ancient Greeks, because matter seemed continuous, and empty space was simply not believable. The idea of atoms was rejected by Aristotle as he formalized his belief in continuous matter composed of the earth, air, fire, and water elements.

Aristotle's belief about matter, like his beliefs about motion, predominated through the 1600s. Some people, such as Galileo and Newton, believed the ideas about matter being composed of tiny particles, or atoms, since this theory seemed to explain the behavior of matter. Widespread acceptance of the particle model did not occur, however, until strong evidence was developed through chemistry in the late 1700s and early 1800s. The experiments finally led to a collection of assumptions about the small particles of matter and the space around them. Collectively, the assumptions could be called the **kinetic molecular theory.** The following is a general description of some of these assumptions.

FIGURE 4.1 Heat and modern technology are inseparable. These glowing steel slabs, at over 1,100°C (about 2,000°F), are cut by an automatic flame torch. The slab caster converts 300 tons of molten steel into slabs in about 45 minutes. The slabs are converted to sheet steel for use in the automotive, appliance, and building industries.

Molecules

The basic assumption of the kinetic molecular theory is that all matter is made up of tiny, basic units of structure called *atoms*. Atoms are neither divided, created, nor destroyed during any type of chemical or physical change. There are similar groups of atoms that make up the pure substances known as chemical *elements*. Each element has its own kind of atom, which is different from the atoms of other elements. For example, hydrogen, oxygen, carbon, iron, and gold are chemical elements, and each has its own kind of atom.

In addition to the chemical elements, there are pure substances called *compounds* that have more complex units of structure. Pure substances, such as water, sugar, and alcohol, are composed of atoms of two or more elements that join together in definite proportions. Water, for example, has structural units that are made up of two atoms of hydrogen tightly bound to one atom of oxygen (H_2O). These units are not easily broken apart and stay together as small physical particles of which water is composed. Each is the smallest particle of water that can exist, a molecule of water. A *molecule* is generally defined as a tightly bound group of atoms in which the atoms maintain their identity. How atoms become bound together to form molecules is discussed in chapter 9.

Some elements exist as gases at ordinary temperatures, and all elements are gases at sufficiently high temperatures. At ordinary temperatures, the atoms of oxygen, nitrogen, and other gases are paired in groups of two to form *diatomic molecules*. Other gases, such as helium, exist as single, unpaired atoms at ordinary temperatures. At sufficiently high temperatures, iron, gold, and other metals vaporize to form gaseous, single,

unpaired atoms. In the kinetic molecular theory, the term *molecule* has the additional meaning of the smallest, ultimate particle of matter that can exist. Thus, the ultimate particle of a gas, whether it is made up of two or more atoms bound together or of a single atom, is conceived of as a molecule. A single atom of helium, for example, is known as a *monatomic molecule*. For now, a **molecule** is defined as the smallest particle of a compound, or a gaseous element, that can exist and still retain the characteristic properties of that substance.

Molecules Interact

Some molecules of solids and liquids interact, strongly attracting and clinging to each other. When this attractive force is between the same kind of molecules, it is called *cohesion*. It is a stronger cohesion that makes solids and liquids different from gases, and without cohesion, all matter would be in the form of gases. Sometimes one kind of molecule attracts and clings to a different kind of molecule. The attractive force between unlike molecules is called *adhesion*. Water wets your skin because the adhesion of water molecules and skin is stronger than the cohesion of water molecules. Some substances, such as glue, have a strong force of adhesion when they harden from a liquid state, and they are called adhesives.

Phases of Matter

Three phases of matter are common on Earth under conditions of ordinary temperature and pressure. These phases—or forms of existence—are solid, liquid, and gas. Each of these has a different molecular arrangement (figure 4.2). The different characteristics of each phase can be attributed to the molecular arrangements and the strength of attraction between the molecules (table 4.1).

Solids have definite shapes and volumes because they have molecules that are fixed distances apart and bound by relatively strong cohesive forces. Each molecule is a nearly fixed distance from the next, but it does vibrate and move around an equilibrium position. The masses of these molecules and the spacing between them determine the density of the solid. The hardness of a solid is the resistance of a solid to forces that tend to push its molecules further apart.

Liquids have molecules that are not confined to an equilibrium position as in a solid. The molecules of a liquid are close together and bound by cohesive forces that are not as strong as in a solid. This permits the molecules to move from place to place within the liquid. The molecular forces are strong enough to give the liquid a definite volume but not strong enough to give it a definite shape. Thus, a pint of milk is always a pint of milk (unless it is under tremendous pressure) and takes the shape of the container holding it. Because the forces between the molecules of a liquid are weaker than the forces between the molecules of a solid, a liquid cannot support the stress of a rock placed on it as a solid does. The liquid molecules *flow*, rolling over each other as the rock pushes its way between the

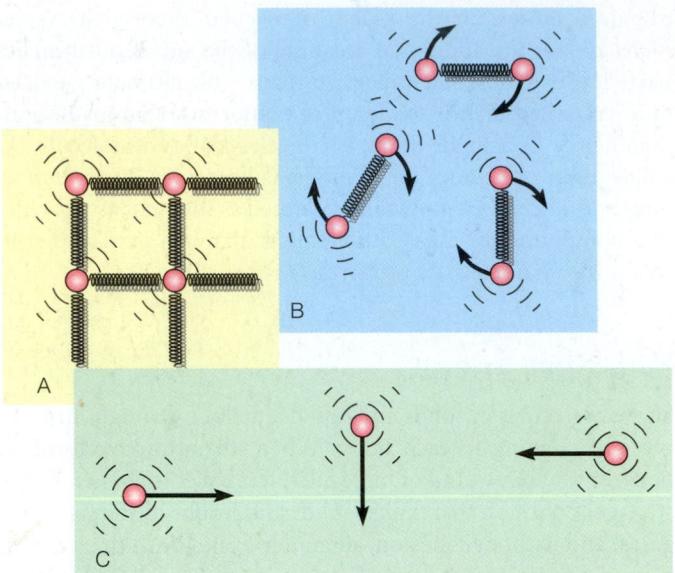

FIGURE 4.2 (*A*) In a solid, molecules vibrate around a fixed equilibrium position and are held in place by strong molecular forces. (*B*) In a liquid, molecules can rotate and roll over each other because the molecular forces are not as strong. (*C*) In a gas, molecules move rapidly in random, free paths.

TABLE 4.1   The Shape and Volume Characteristics of Solids, Liquids, and Gases Are Reflections of Their Molecular Arrangements*

	Solids	Liquids	Gases
Shape	Fixed	Variable	Variable
Volume	Fixed	Fixed	Variable

*These characteristics are what would be expected under ordinary temperature and pressure conditions on the surface of Earth.

molecules. Yet, the molecular forces are strong enough to hold the liquid together, so it keeps the same volume.

Gases are composed of molecules with weak cohesive forces acting between them. The gas molecules are relatively far apart and move freely in a constant, random motion that is changed often by collisions with other molecules. Gases therefore have neither fixed shapes nor fixed volumes.

Gases that are made up of positive ions and negative electrons are called *plasmas*. Plasmas have the same properties as gases but also conduct electricity and interact strongly with magnetic fields. Plasmas are found in fluorescent and neon lights on Earth, and in the Sun and other stars. Nuclear fusion occurs in plasmas of stars (see chapters 11–12), producing starlight as well as sunlight. Plasma physics is studied by scientists in their attempt to produce controlled nuclear fusion.

There are other distinctions between the phases of matter. The term *vapor* is sometimes used to describe a gas that is usually in the liquid phase. Water vapor, for example, is the

gaseous form of liquid water. Liquids and gases are collectively called *fluids* because of their ability to flow, a property that is lacking in most solids.

Molecules Move

Suppose you are in an evenly heated room with no air currents. If you open a bottle of ammonia, the odor of ammonia is soon noticeable everywhere in the room. According to the kinetic molecular theory, molecules of ammonia leave the bottle and bounce around among the other molecules making up the air until they are everywhere in the room, slowly becoming more evenly distributed. The ammonia molecules *diffuse*, or spread, throughout the room. The ammonia odor diffuses throughout the room faster if the air temperature is higher and slower if the air temperature is lower. This would imply a relationship between the temperature and the speed at which molecules move about.

The relationship between the temperature of a gas and the motion of molecules was formulated in 1857 by Rudolf Clausius. He showed that the temperature of a gas is proportional to the average kinetic energy of the gas molecules. This means that ammonia molecules have a greater average velocity at a higher temperature and a slower average velocity at a lower temperature. This explains why gases diffuse at a greater rate at higher temperatures. Recall, however, that kinetic energy involves the mass of the molecules as well as their velocity ($KE = 1/2\ mv^2$). It is the *average kinetic energy* that is proportional to the temperature, which involves the molecular mass as well as the molecular velocity. Whether the kinetic energy is jiggling, vibrating, rotating, or moving from place to place, the **temperature** of a substance is a *measure of the average kinetic energy of the molecules making up the substance* (figure 4.3).

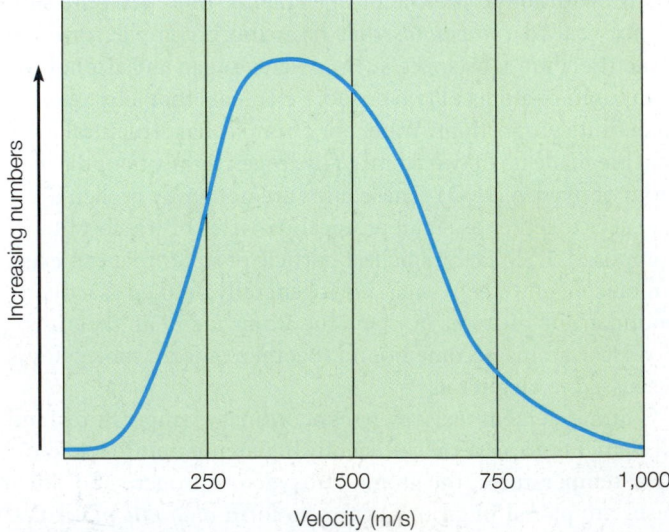

FIGURE 4.3 The number of oxygen molecules with certain velocities that you might find in a sample of air at room temperature. Notice that a few are barely moving and some have velocities over 1,000 m/s at a given time, but the *average* velocity is somewhere around 500 m/s.

4.2 TEMPERATURE

If you ask people about the temperature, they usually respond with a referent ("hotter than the summer of '89") or a number ("68°F or 20°C"). Your response, or feeling, about the referent or number depends on a number of factors, including a *relative* comparison. A temperature of 20°C (68°F), for example, might seem cold during the month of July but warm during the month of January. The 20°C temperature is compared to what is expected at the time, even though 20°C is 20°C, no matter what month it is.

When people ask about the temperature, they are really asking *how hot or how cold something is*. Without a thermometer, however, most people can do no better than *hot* or *cold*, or perhaps *warm* or *cool*, in describing a relative temperature. Even then, there are other factors that confuse people about temperature. Your body judges temperature on the basis of the net *direction* of energy flow. You sense situations in which heat is flowing into your body as *warm* and situations in which heat is flowing from your body as *cool*. Perhaps you have experienced having your hands in snow for some time, then washing your hands in cold water. The cold water feels warm. Your hands are colder than the water, energy flows into your hands, and they communicate "warm."

Thermometers

The human body is a poor sensor of temperature, so a device called a *thermometer* is used to measure the hotness or coldness of something. Most thermometers are based on the relationship between some property of matter and changes in temperature. Almost all materials expand with increasing temperatures. A strip of metal is slightly longer when hotter and slightly shorter when cooler, but the change of length is too small to be useful in a thermometer. A more useful, larger change is obtained when two metals that have different expansion rates are bonded together in a strip. The bimetallic (*bi* = two; *metallic* = metal) strip will bend toward the metal with less expansion when the strip is heated (figure 4.4). Such a bimetallic strip is formed into a coil and used in thermostats and dial thermometers (figure 4.5).

The common glass thermometer is a glass tube with a bulb containing a liquid, usually mercury or colored alcohol, that expands up the tube with increases in temperature and contracts back toward the bulb with decreases in temperature. The height of this liquid column is used with a referent scale

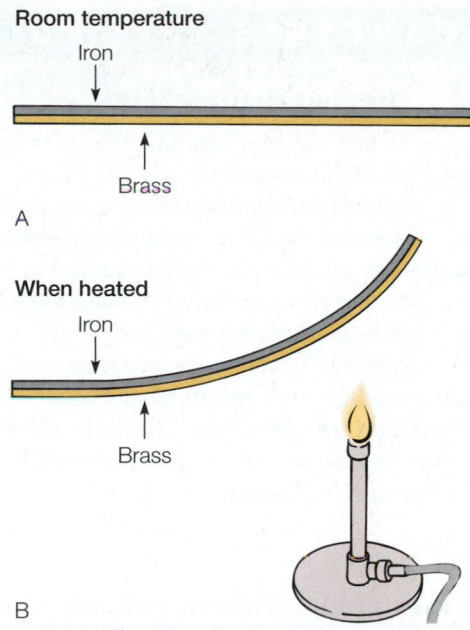

FIGURE 4.4 (*A*) A bimetallic strip is two different metals, such as iron and brass, bonded together as a single unit, shown here at room temperature. (*B*) Since one metal expands more than the other, the strip will bend when it is heated. In this example, the brass expands more than the iron, so the bimetallic strip bends away from the brass.

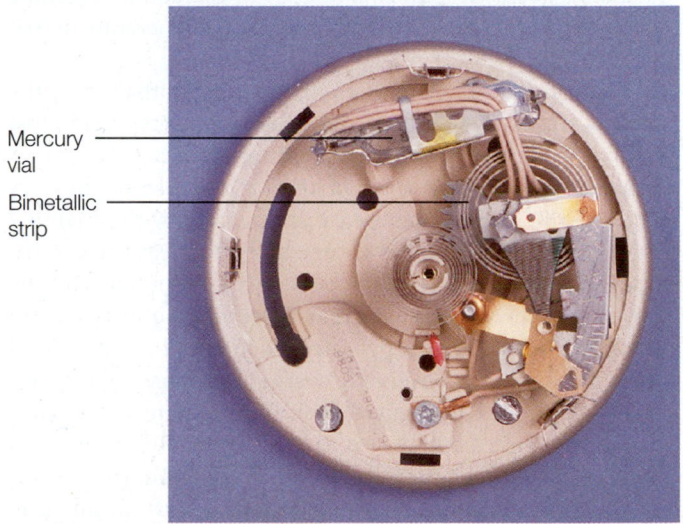

FIGURE 4.5 This thermostat has a coiled bimetallic strip that expands and contracts with changes in the room temperature. The attached vial of mercury is tilted one way or the other, and the mercury completes or breaks an electric circuit that turns the heating or cooling system on or off.

to measure temperature. Some thermometers, such as a fever thermometer, have a small constriction in the tube so the liquid cannot normally return to the bulb. Thus, the thermometer shows the highest reading, even if the temperature it measures has fluctuated up and down during the reading. The liquid must be forced back into the bulb by a small swinging motion, bulb-end down, then sharply stopping the swing with a snap of the wrist. The inertia of the mercury in the bore forces it past the constriction and into the bulb. The fever thermometer is then ready to use again.

Today, scientists have developed a different type of thermometer and a way around the problems of using a glass-mercury fever thermometer. This new approach measures the internal core temperature by quickly reading infrared radiation from the eardrum. All bodies with a temperature above absolute zero emit radiation, including your body. The intensity of the radiation is a sensitive function of body temperature, so reading the radiation emitted will tell you about the temperature of that body.

The human eardrum is close to the hypothalamus, the body's thermostat, so a temperature reading taken here must be close to the temperature of the internal core. You cannot use a mercury thermometer in the ear because of the very real danger of puncturing the eardrum, along with obtaining doubtful readings from a mercury bulb. You can use a pyroelectric material to measure the infrared radiation coming from the entrance to the ear canal, however, to quickly obtain a temperature reading. A pyroelectric material is a polarized crystal that generates an electric charge in proportion to a temperature change. The infrared fever thermometer has a short barrel that is inserted in the ear canal opening. A button opens a shutter inside the battery-powered device, admitting infrared radiation for about 300 milliseconds. Infrared radiation from the ear canal increases the temperature of a thin pyroelectric crystal, which develops an electric charge. A current spike from the pyroelectric sensor now moves through some filters and converters and into a microprocessor chip. The chip is programmed with the relationship between the body temperature and the infrared radiation emitted. Using this information, it calculates the temperature by measuring the current spike produced by the infrared radiation falling on the pyroelectric crystal. The microprocessor now sends the temperature reading to an LCD display on the outside of the device, where it can be read almost immediately.

Thermometer Scales

Several referent scales are used to define numerical values for measuring temperatures (figure 4.6). The **Fahrenheit scale** was developed by the German physicist Gabriel D. Fahrenheit in about 1715. Fahrenheit invented a mercury-in-glass thermometer with a scale based on two arbitrarily chosen reference points. The original Fahrenheit scale was based on the temperature of an ice and salt mixture for the lower reference point (0°) and the temperature of the human body as the upper reference point (about 100°). Thus, the original Fahrenheit scale was a centigrade scale with 100 divisions between the high and the low reference points. The distance between the two reference points was then divided into equal intervals called *degrees*. There were problems with identifying a "normal" human body temperature as a reference point, since body temperature naturally changes during a given day and from day to day. However, some people "normally" have a higher body temperature than others. Some may have a normal body temperature of 99.1°F, while others have a temperature of 97°F. The average for a large population is 98.6°F. The only consistent thing about the human body temperature is constant change. The standards for the Fahrenheit scale were eventually changed to something more consistent, the freezing point and the boiling point of water at normal atmospheric pressure. The original scale was retained with the new reference points, however, so the "odd" numbers of 32°F (freezing point of water) and 212°F (boiling point of water under normal pressure) came to be the reference points. There are 180 equal intervals, or degrees, between the freezing and boiling points on the Fahrenheit scale.

The **Celsius scale** was invented by Anders C. Celsius, a Swedish astronomer, in about 1735. The Celsius scale uses the freezing point and the boiling point of water at normal atmospheric pressure, but it has different arbitrarily assigned values. The Celsius scale identifies the freezing point of water as 0°C and the boiling point as 100°C. There are 100 equal intervals, or degrees, between these two reference points, so the Celsius scale is sometimes called the *centigrade* scale.

There is nothing special about either the Celsius scale or the Fahrenheit scale. Both have arbitrarily assigned numbers, and one is no more accurate than the other. The Celsius scale is more convenient because it is a decimal scale and because

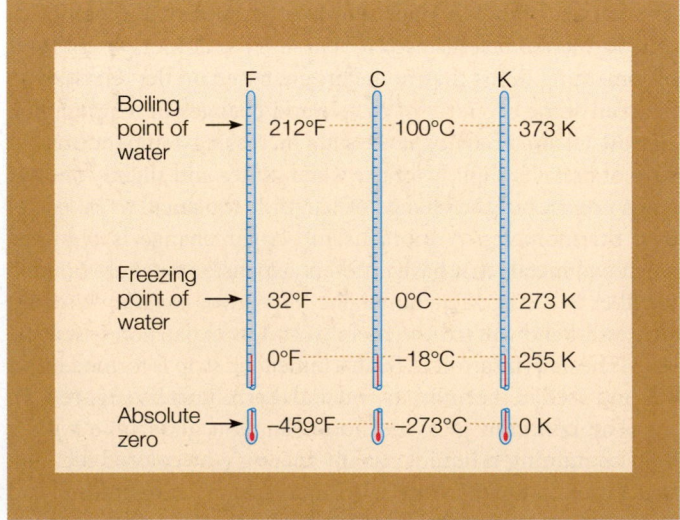

FIGURE 4.6 The Fahrenheit, Celsius, and Kelvin temperature scales.

Goose Bumps and Shivering

For an average age and minimal level of activity, many people feel comfortable when the environmental temperature is about 25°C (77°F). Comfort at this temperature probably comes from the fact that the body does not have to make an effort to conserve or get rid of heat.

Changes in the body that conserve heat occur when the temperature of the air and clothing directly next to a person becomes less than 20°C, or if the body senses rapid heat loss. First, blood vessels in the skin are constricted. This slows the flow of blood near the surface, which reduces heat loss by conduction. Constriction of skin blood vessels reduces body heat loss but may also cause the skin and limbs to become significantly cooler than the body core temperature (producing cold feet, for example).

Sudden heat loss, or a chill, often initiates another heat-saving action by the body. Skin hair is pulled upright, erected to slow heat loss to cold air moving across

the skin. Contraction of a tiny muscle attached to the base of the hair shaft makes a tiny knot, or bump, on the skin. These are sometimes called "goose bumps" or "chill bumps." Although "goose bumps" do not significantly increase insulation in humans, the equivalent response in birds and many mammals elevates hairs or feathers and greatly enhances insulation.

Further cooling after the blood vessels in the skin have been constricted results in the body taking yet another action. The body now begins to produce *more* heat, making up for heat loss through involuntary muscle contractions called "shivering." The greater the need for more body heat, the greater the activity of shivering.

If the environmental temperatures rise above about 25°C (77°F), the body triggers responses that cause it to *lose* heat. One response is to make blood vessels in the skin larger, which increases blood flow in the skin. This brings more heat from the core to be conducted through the skin,

then radiated away. It also causes some people to have a red blush from the increased blood flow in the skin. This action increases conduction through the skin, but radiation alone provides insufficient cooling at environmental temperatures above about 29°C (84°F). At about this temperature, sweating begins and perspiration pours onto the skin to provide cooling through evaporation. The warmer the environmental temperature, the greater the rate of sweating and cooling through evaporation.

The actual responses to a cool, cold, warm, or hot environment will be influenced by a person's level of activity, age, and gender, and environmental factors such as the relative humidity, air movement, and combinations of these factors. Temperature is the single most important comfort factor. However, when the temperature is high enough to require perspiration for cooling, humidity also becomes an important factor in human comfort.

---

it has a direct relationship with a third scale to be described shortly, the Kelvin scale. Both scales have arbitrarily assigned reference points and an arbitrary number line that indicates *relative* temperature changes. Zero is simply one of the points on each number line and does *not* mean that there is no temperature. Likewise, since the numbers are relative measures of temperature change, 2° is not twice as hot as a temperature of 1° and 10° is not twice as hot as a temperature of 5°. The numbers simply mean some measure of temperature *relative to the* freezing and boiling points of water under normal conditions.

You can convert from one temperature to the other by considering two differences in the scales: (1) the difference in the degree size between the freezing and boiling points on the two scales, and (2) the difference in the values of the lower reference points.

The Fahrenheit scale has 180° between the boiling and freezing points (212°F − 32°F) and the Celsius scale has 100° between the same two points. Therefore, each Celsius degree is 180/100 or 9/5 as large as a Fahrenheit degree. Each Fahrenheit degree is 100/180 or 5/9 of a Celsius degree. In addition, considering the difference in the values of the lower reference points (0°C and 32°F) gives the equations for temperature conversion.

$$T_F = \frac{9}{5}T_C + 32°$$

equation 4.1

$$T_C = \frac{5}{9}(T_F - 32°)$$

equation 4.2

EXAMPLE 4.1 *(Optional)*

The average human body temperature is 98.6°F. What is the equivalent temperature on the Celsius scale?

SOLUTION

$$T_C = \frac{5}{9}(T_F - 32°)$$

$$= \frac{5}{9}(98.6° - 32°)$$

$$= \frac{5}{9}(66.6°)$$

$$= \frac{333°}{9}$$

$$= \boxed{37°C}$$

EXAMPLE 4.2 *(Optional)*

A temperature display outside of a bank indicates 20°C (room temperature). What is the equivalent temperature on the Fahrenheit scale? (Answer: 68°F)

There is a temperature scale that does not have arbitrarily assigned reference points and zero *does* mean nothing. This is not a relative scale but an absolute temperature scale called the **absolute scale,** or **Kelvin scale.** The zero point on the absolute scale is thought to be the lowest limit of temperature. **Absolute zero** is the *lowest temperature possible,* occurring when all random motion of molecules has ceased. Absolute zero is written as 0 K. A degree symbol is not used, and the K stands for the SI standard scale unit kelvin. The absolute scale uses the same degree size as the Celsius scale and $-273°C = 0$ K. Note in figure 4.6 that 273 K is the freezing point of water and 373 K is the boiling point. You could think of the absolute scale as a Celsius scale with the zero point shifted by 273°. Thus, the relationship between the absolute and Celsius scales is

$$T_K = T_C + 273 \qquad \textbf{equation 4.3}$$

A temperature of absolute zero has never been reached, but scientists have cooled a sample of sodium to 700 nanokelvins, or 700 billionths of a kelvin above absolute zero.

EXAMPLE 4.3 *(Optional)*

A science article refers to a temperature of 300.0 K. (a) What is the equivalent Celsius temperature? (b) What is the equivalent Fahrenheit temperature?

SOLUTION

(a) The relationship between the absolute scale and Celsius scale is found in equation 4.3, $T_K = T_C + 273$. Solving this equation for Celsius yields $T_C = T_K - 273$.

$$T_C = T_K - 273$$
$$= 300.0 - 273$$
$$= \boxed{27°C}$$

(b)
$$T_F = \frac{9}{5}T_C + 32°$$
$$= \frac{9}{5}27° + 32°$$
$$= \frac{243°}{5} + 32°$$
$$= 48.6° + 32°$$
$$= \boxed{81°F}$$

4.3 HEAT

Suppose you have a bowl of hot soup or a cup of hot coffee that is too hot. What can you do to cool it? You can blow across the surface, which speeds evaporation and therefore results in cooling, but this is a slow process. If you were in a hurry, you would probably add something cooler, such as ice. Adding a cooler substance will cool the hot liquid.

You know what happens when you mix fluids or objects with a higher temperature with fluids or objects with a lower temperature. The warmer temperature object becomes cooler and the cooler temperature object becomes warmer. Eventually, both will have a temperature somewhere between the warmer and the cooler. This might suggest that something is moving between the warmer and cooler objects, changing the temperature.

The relationship that exists between energy and temperature will help explain the concept of heat, so we will consider it first. If you rub your hands together a few times, they will feel a little warmer. If you rub them together vigorously for a while, they will feel a lot warmer, maybe hot. A temperature increase takes place anytime mechanical energy causes one surface to rub against another (figure 4.7). The two surfaces could be solids, such as the two blocks, but they can also be the surface of a solid and a fluid, such as air. A solid object moving through the air encounters compression, which results in a higher temperature of the surface. A high-velocity meteor enters Earth's atmosphere and is heated so much from the compression that it begins to glow, resulting in the fireball and smoke trail of a "falling star."

To distinguish between the energy of the object and the energy of its molecules, we use the terms *external* and *internal* energy. **External energy** is the total potential and kinetic energy of an everyday-sized object. All the kinetic and potential energy considerations discussed in previous chapters were about the external energy of an object.

Internal energy is the total kinetic and potential energy of the *molecules* of an object. The kinetic energy of a molecule can be much more complicated than straight-line velocity might suggest, however, as a molecule can have many different types of

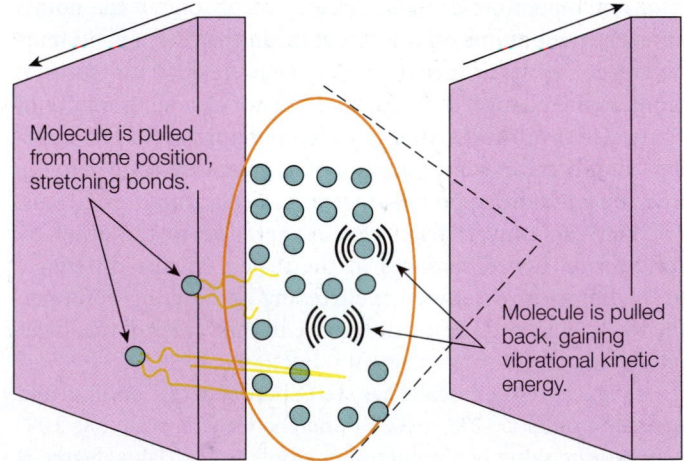

FIGURE 4.7   Here is how friction results in increased temperatures: Molecules on one moving surface will catch on another surface, stretching the molecular forces that are holding it. The molecules are pulled back to their home position with a snap, resulting in a gain of vibrational kinetic energy.

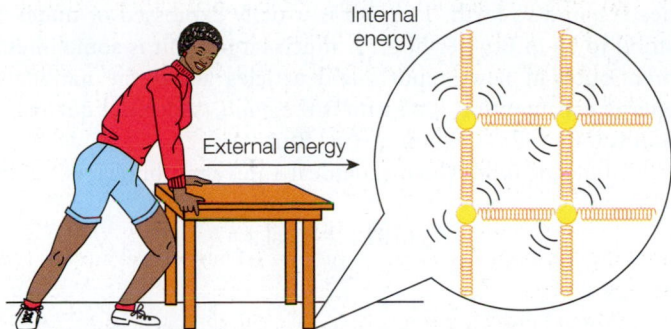

FIGURE 4.8 *External energy* is the kinetic and potential energy that you can see. *Internal energy* is the total kinetic and potential energy of molecules. When you push a table across the floor, you do work against friction. Some of the external mechanical energy goes into internal kinetic and potential energy, and the bottom surface of the legs becomes warmer.

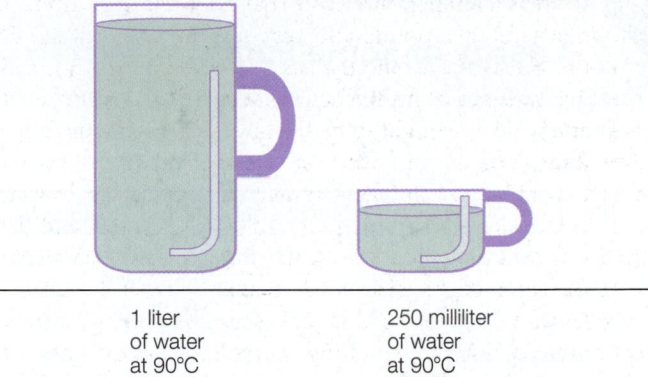

1 liter of water at 90°C

250 milliliter of water at 90°C

FIGURE 4.9 Heat and temperature are different concepts, as shown by a liter of water (1,000 mL) and a 250 mL cup of water, both at the same temperature. You know the liter of water contains more internal energy since it will require more ice cubes to cool it to, say, 25°C than will be required for the cup of water. In fact, you will have to remove 48,750 *additional* calories to cool the liter of water.

motion at the same time (pulsing, twisting, turning, etc.). Overall, internal energy is characterized by properties such as temperature, density, heat, volume, pressure of a gas, and so forth.

When you push a table across the floor, the observable *external* kinetic energy of the table is transferred to the *internal* kinetic energy of the molecules between the table legs and the floor, resulting in a temperature increase (figure 4.8). The relationship between external and internal kinetic energy explains why the heating is proportional to the amount of mechanical energy used.

Heat as Energy Transfer

Temperature is a measure of the degree of hotness or coldness of a body, a measure that is based on the average molecular kinetic energy. Heat, on the other hand, is based on the *total internal energy* of the molecules of a body. You can see one difference in heat and temperature by considering a cup of water and a pitcher of water. If both the small and the large amount of water have the same temperature, both must have the same average molecular kinetic energy. Now, suppose you wish to cool both by, say, 20°. The pitcher of water would take much longer to cool, so it must be that the large amount of water has more internal energy (figure 4.9). Heat is a measure based on the *total* internal energy of the molecules of a body, and there is more total energy in a pitcher of water than in a cup of water at the same temperature.

How can we measure heat? Since it is difficult to see molecules, internal energy is difficult to measure directly. Thus, heat is nearly always measured during the process of a body gaining or losing energy. This measurement procedure will also give us a working definition of **heat:**

> **Heat is a measure of the internal energy that has been absorbed or transferred from one body to another.**

The *process* of increasing the internal energy is called "heating," and the *process* of decreasing internal energy is called "cooling." The word *process* is italicized to emphasize that heat is energy in transit, not a material thing you can add or take away. Heat is understood to be a measure of internal energy that can be measured as energy flows in or out of an object.

There are two general ways that heating can occur. These are (1) from a temperature difference, with energy moving from the region of higher temperature, and (2) from an object gaining energy by way of an energy-form conversion.

When a *temperature difference* occurs, energy is transferred from a region of higher temperature to a region of lower temperature. Energy flows from a hot range element, for example, to a pot of cold water on a range. It is a natural process for energy to flow from a region of a higher temperature to a region of a lower temperature just as it is natural for a ball to roll downhill. The temperature of an object and the temperature of the surroundings determine if heat will be transferred to or from an object. The terms *heating* and *cooling* describe the direction of energy flow, naturally moving from a region of higher energy to one of lower energy.

Measures of Heat

Since heating is a method of energy transfer, a quantity of heat can be measured just like any quantity of energy. The metric unit for measuring work, energy, or heat is the *joule*. However, the separate historical development of the concepts of heat and the concepts of motion resulted in separate units, some based on temperature differences.

The metric unit of heat is called the **calorie** (cal). A calorie is defined as the *amount of energy (or heat) needed to increase the temperature of 1 gram of water 1 degree Celsius.* A more precise definition specifies the degree interval from 14.5°C to 15.5°C because the energy required varies slightly at different temperatures. This precise definition is not needed for a general discussion. A **kilocalorie** (kcal) is the *amount of energy (or heat) needed to increase the temperature of 1 kilogram of water 1 degree Celsius.* The measure of the energy released by the

Connections . . .

Energy of Some Foods

Note: Typical basal metabolic rate equals 1,200–2,200 Calories per day.

Food	Nutritional Calories	Kilojoules
Candy bar, various	120 to 230	502 to 963
Beer, regular	150	628
Beer, lite	100	419
Carbonated drink	150	628
Arby's roast beef	380	1,590
Whopper, w/cheese	705	2,951
DQ large cone	340	1,423
Domino's deluxe pizza	225 per slice (16")	942
Hardee's big deluxe	500	2,093
Jumbo Jack w/cheese	630	2,638
KFC breast	300	1,256
Big Mac	560	2,344
Pizza Hut supreme	140 per slice (10")	586
Taco Bell taco	160	670
Wendy's big classic	570	2,386

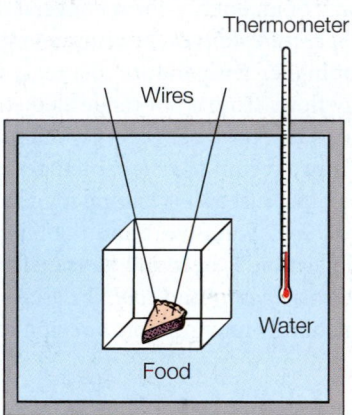

FIGURE 4.10 The Calorie value of food is determined by measuring the heat released from burning the food. If there is 10.0 kg of water and the temperature increases from 10°C to 20°C, the food contains 100 Calories (100,000 calories). The food illustrated here would release much more energy than this.

oxidation of food is the kilocalorie, but it is called the Calorie (with a capital C) by nutritionists (figure 4.10). Confusion can be avoided by making sure that the scientific calorie is never capitalized (cal) and the dieter's Calorie is always capitalized. The best solution would be to call the Calorie what it is, a kilocalorie (kcal).

The English system's measure of heating is called the **British thermal unit** (Btu). A Btu is *the amount of energy (or heat) needed to increase the temperature of 1 pound of water 1 degree Fahrenheit.* The Btu is commonly used to measure the heating or cooling rates of furnaces, air conditioners, water

heaters, and so forth. The rate is usually expressed or understood to be in Btu per hour. A much larger unit is sometimes mentioned in news reports and articles about the national energy consumption. This unit is the *quad*, which is 1 quadrillion Btu (a million billion or 10^{15} Btu).

In metric units, the mechanical equivalent of heat is

$$4.184 \text{ J} = 1 \text{ cal}$$

or

$$4,184 \text{ J} = 1 \text{ kcal}$$

The establishment of this precise proportionality means that, fundamentally, mechanical energy and heat are different forms of the same thing.

EXAMPLE 4.4   *(Optional)*

A 1,000.0 kg car is moving at 90.0 km/h (25.0 m/s). How many kilocalories are generated when the car brakes to a stop?

SOLUTION

The kinetic energy of the car is

$$KE = \frac{1}{2}mv^2$$

$$= \frac{1}{2}(1,000.0 \text{ kg})\left(25.0\frac{\text{m}}{\text{s}}\right)^2$$

$$= \frac{1}{2}(1,000.0 \text{ kg})\left(625\frac{\text{m}^2}{\text{s}^2}\right)$$

$$= (500.0)(625)\frac{\text{kg} \cdot \text{m}^2}{\text{s}^2}$$

$$= 312,500 \text{ J}$$

You can convert this to kcal by using the relationship between mechanical energy and heat:

$$(312,500 \text{ J})\frac{1 \text{ kcal}}{4,184 \text{ J}}$$

$$\frac{312,500}{4,184}\frac{\text{J} \cdot \text{kcal}}{\text{J}}$$

$$\boxed{74.7 \text{ kcal}}$$

(Note: The temperature increase from this amount of heating could be calculated from equation 4.4.)

Specific Heat

You can observe a relationship between heat and different substances by doing an experiment in "kitchen physics." Imagine that you have a large pot of liquid to boil in preparing a meal. Three variables influence how much heat you need:

1. The initial temperature of the liquid;

2. How much liquid is in the pot; and,
3. The nature of the liquid (water or soup, for example).

What this means specifically, is

1. **Temperature change.** The amount of heat needed is proportional to the temperature change. It takes more heat to raise the temperature of cool water, so this relationship could be written as $Q \propto \Delta T$.
2. **Mass.** The amount of heat needed is also proportional to the amount of the substance being heated. A larger mass requires more heat to go through the same temperature change than a smaller mass. In symbols, $Q \propto m$.
3. **Substance.** Different materials require different amounts of heat to go through the same temperature range when their masses are equal (figure 4.11). This property is called the **specific heat** of a material, which is defined as the amount of heat needed to increase the temperature of 1 gram of a substance 1 degree Celsius.

Considering all the variables involved in our kitchen physics cooking experience, we find the heat (Q) need is described by the relationship of

$$Q = mc\Delta T \qquad \text{equation 4.4}$$

where c is the symbol for specific heat.

Specific heat is responsible for the larger variation of temperature observed over land than near a large body of water. Table 4.2 gives the specific heat of soil as 0.200 cal/gC°*

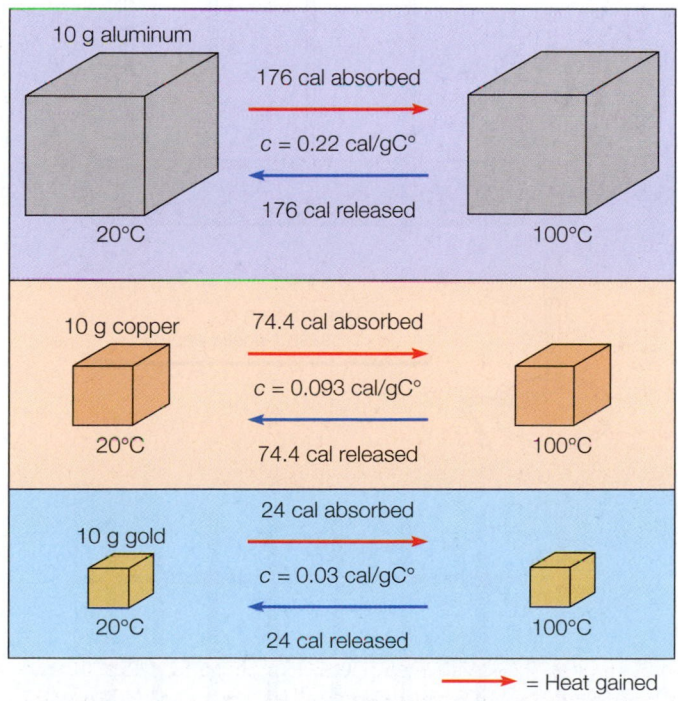

FIGURE 4.11 Of these three metals, aluminum needs the most heat per gram per degree when warmed and releases the most heat when cooled. Why are the cubes different sizes?

TABLE 4.2 **The Specific Heat of Selected Substances**

Substance	Specific Heat (cal/gC° or kcal/kgC°)
Air	0.17
Aluminum	0.22
Concrete	0.16
Copper	0.093
Glass (average)	0.160
Gold	0.03
Ice	0.500
Iron	0.11
Lead	0.0305
Mercury	0.033
Seawater	0.93
Silver	0.056
Soil (average)	0.200
Steam	0.480
Water	1.00

Note: To convert to specific heat in J/kgC°, multiply each value by 4,184. Also note that 1 cal/gC° = 1 kcal/kgC°.

CONCEPTS APPLIED

More Kitchen Physics

Consider the following information as it relates to the metals of cooking pots and pans:

1. It is easier to change the temperature of metals with low specific heats.
2. It is harder to change the temperature of metals with high specific heats.

Look at the list of metals and specific heats listed in table 4.2 and answer the following questions:

1. Considering specific heat alone, which metal could be used for making practical pots and pans that are the most energy efficient to use?
2. Again considering specific heat alone, would certain combinations of metals provide any advantages for rapid temperature changes?

and the specific heat of water as 1.00 cal/gC°. Since specific heat is defined as the amount of heat needed to increase the temperature of 1 gram of a substance 1 degree, this means 1 gram of water exposed to 1 calorie of sunlight will warm 1°C. One gram of soil exposed to 1 calorie of sunlight, on the other hand, will be warmed to 5°C since it only takes 0.2 calories to warm the soil 1°C. Thus, the temperature is more even near large bodies of water since it is harder to change the temperature of the water.

*The Celsius degree (C°) represents a temperature interval; the difference between two temperatures. 5°C + 15°C = 20C°

EXAMPLE 4.5 (Optional)

How much heat must be supplied to a 500.0 g pan to raise its temperature from 20.0°C to 100.0°C if the pan is made of (a) iron and (b) aluminum?

SOLUTION

The relationship between the heat supplied (Q), the mass (m), and the temperature change (ΔT) is found in equation 4.4. The specific heats (c) of iron and aluminum can be found in table 4.2.

(a) Iron:

$m = 500.0$ g
$c = 0.11$ cal/gC°
$T_f = 100.0°C$
$T_i = 20.0°C$
$Q = ?$

$$Q = mc\Delta T$$
$$= (500.0 \text{ g})\left(0.11\frac{\text{cal}}{\text{gC}°}\right)(80.0\,\text{C}°)$$
$$= (500.0)(0.11)(80.0)\,\text{g} \times \frac{\text{cal}}{\text{g} \cdot \text{C}°} \times \text{C}°$$
$$= 4,400\,\frac{\text{g} \cdot \text{cal} \cdot \text{C}°}{\text{g} \cdot \text{C}°}$$
$$= \boxed{4.4 \text{ kcal}}$$

(b) Aluminum:

$m = 500.0$ g
$c = 0.22$ cal/gC°
$T_f = 100.0°C$
$T_i = 20.0°C$
$Q = ?$

$$Q = mc\Delta T$$
$$= (500.0 \text{ g})\left(0.22\frac{\text{cal}}{\text{gC}°}\right)(80.0\,\text{C}°)$$
$$= (500.0)(0.22)(80.0)\,\text{g} \times \frac{\text{cal}}{\text{g} \cdot \text{C}°} \times \text{C}°$$
$$= 8,800\,\frac{\text{g} \cdot \text{cal} \cdot \text{C}°}{\text{g} \cdot \text{C}°}$$
$$= \boxed{8.8 \text{ kcal}}$$

It takes twice as much heat energy to warm the aluminum pan through the same temperature range as the iron pan. Thus, with equal rates of energy input, the iron pan will warm twice as fast as the aluminum pan. Note that using kcal rather than the cal unit results in a more compact number. Either unit is correct, but the more compact number is easier to work with, and this makes it more desirable.

EXAMPLE 4.6 (Optional)

What is the specific heat of a 2 kg metal sample if 1.2 kcal are needed to increase the temperature from 20.0°C to 40.0°C? (Answer: 0.03 kcal/kgC°)

Heat Flow

In a previous section, you learned that heating is a transfer of energy that involves (1) a temperature difference or (2) energy-form conversions. Heat transfer that takes place because of a temperature difference takes place in three different ways: by conduction, convection, or radiation.

Conduction

Anytime there is a temperature difference, there is a natural transfer of heat from the region of higher temperature to the region of lower temperature. In solids, this transfer takes place as heat is *conducted* from a warmer place to a cooler one. Recall that the molecules in a solid vibrate in a fixed equilibrium position and that molecules in a higher temperature region have more kinetic energy, on the average, than those in a lower temperature region. When a solid, such as a metal rod, is held in a flame, the molecules in the warmed end vibrate violently. Through molecular interaction, this increased energy of vibration is passed on to the adjacent, slower-moving molecules, which also begin to vibrate more violently. They, in turn, pass on more vibrational energy to the molecules next to them. The increase in activity thus moves from molecule to molecule, causing the region of increased activity to extend along the rod. This is called **conduction,** the transfer of energy from molecule to molecule (figure 4.12).

Most insulating materials are good insulators because they contain many small air spaces (figure 4.13). The small air spaces are poor conductors because the molecules of air are far apart, compared to a solid, making it more difficult to pass the increased vibrating motion from molecule to molecule. Styrofoam, glass wool, and wool cloth are good insulators because they have many small air spaces, not because of the material they are made of. The best insulator is a vacuum, since there are no molecules to pass on the vibrating motion.

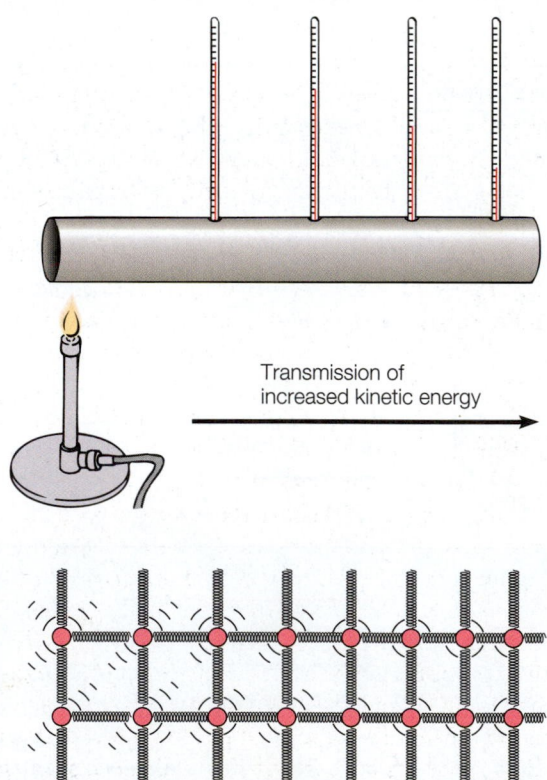

Transmission of increased kinetic energy

FIGURE 4.12 Thermometers placed in holes drilled in a metal rod will show that heat is conducted from a region of higher temperature to a region of lower temperature. The increased molecular activity is passed from molecule to molecule in the process of conduction.

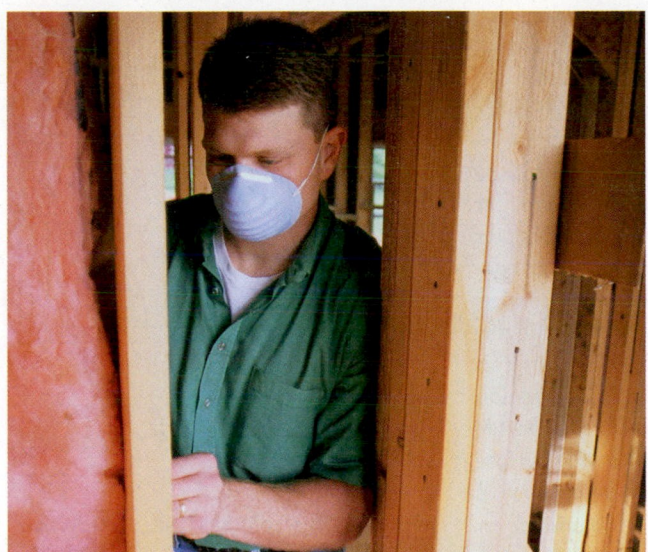

FIGURE 4.13 Fiberglass insulation is rated in terms of R-value, a ratio of the conductivity of the material to its thickness.

Require Insulation?

Can you spend too much for home insulation? Should the local government require maximum insulation in new homes to save energy? Research economic limits to insulation: when does the cost of installing insulation exceed any returns in energy saved?

Questions to Discuss

Write a script designed to inform a city planning board about the benefits, advantages, and disadvantages of requiring maximum insulation in all new homes.

Wooden and metal parts of your desk have the same temperature, but the metal parts will feel cooler if you touch them. Metal is a better conductor of heat than wood and feels cooler because it conducts heat from your finger faster. This is the same reason that a wood or tile floor feels cold to your bare feet. You use an insulating rug to slow the conduction of heat from your feet.

Convection

Convection is the transfer of heat by a large-scale displacement of groups of molecules with relatively higher kinetic energy. In conduction, increased kinetic energy is passed from molecule to molecule. In convection, molecules with higher kinetic energy are moved from one place to another place. Conduction happens primarily in solids, but convection happens only in liquids and gases, where fluid motion can carry molecules with higher kinetic energy over a distance. When molecules gain energy, they move more rapidly and push more vigorously against their surroundings. The result is an expansion as the region of heated molecules pushes outward and increases the volume. Since the same amount of matter now occupies a larger volume, the overall density has been decreased.

In fluids, expansion sets the stage for convection. Warm, less-dense fluid is pushed upward by the cooler, more-dense fluid around it. In general, cooler air is more dense; it sinks and flows downhill. Cold air, being more dense, flows out near the bottom of an open refrigerator. You can feel the cold, dense air pouring from the bottom of an open refrigerator to your toes on the floor. On the other hand, you hold your hands *over* a heater because the warm, less-dense air is pushed upward. In a room, warm air is pushed upward from a heater. The warm air spreads outward along the ceiling and is slowly displaced as newly warmed air is pushed upward to the ceiling. As the air cools it sinks over another part of the room, setting up a circulation pattern known as a *convection current* (figure 4.14). Convection

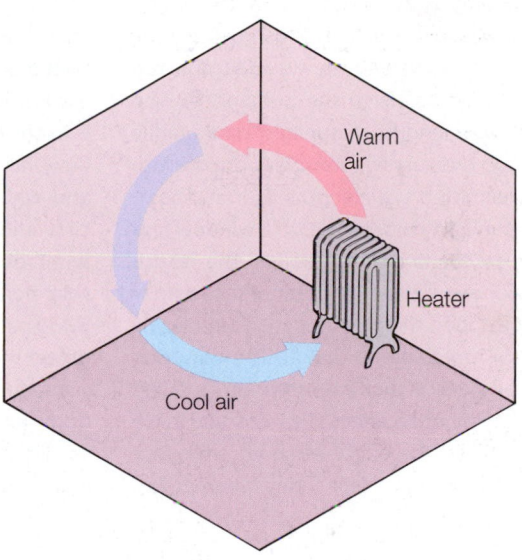

FIGURE 4.14 Convection currents move warm air throughout a room as the air over the heater becomes warmed, expands, and is moved upward by cooler air.

currents can also be observed in a large pot of liquid that is heating on a range. You can see the warmer liquid being forced upward over the warmer parts of the range element, then sink over the cooler parts. Overall, convection currents give the pot of liquid an appearance of turning over as it warms.

Radiation

The third way that heat transfer takes place because of a temperature difference is called **radiation.** Radiation involves the form of energy called *radiant energy,* energy that moves through space. As you will learn in chapter 7, radiant energy includes visible light and many other forms as well. All objects with a temperature above absolute zero give off radiant energy. The absolute temperature of the object determines the rate, intensity, and kinds of radiant energy emitted. You know that visible light is emitted if an object is heated to a certain temperature. A heating element on an electric range, for example, will glow with a reddish-orange light when at the highest setting, but it produces no visible light at lower temperatures, although

A Closer Look

Passive Solar Design

Passive solar application is an economically justifiable use of solar energy today. Passive solar design uses a structure's construction to heat a living space with solar energy. Few electric fans, motors, or other energy sources are used. The passive solar design takes advantage of free solar energy; it stores and then distributes this energy through natural conduction, convection, and radiation.

Sunlight that reaches Earth's surface is mostly absorbed. Buildings, the ground, and objects become warmer as the radiant energy is absorbed. Nearly all materials, however, reradiate the absorbed energy at longer wavelengths, wavelengths too long to be visible to the human eye. The short wavelengths of sunlight pass readily through ordinary window glass, but the longer, reemitted wavelengths cannot. Therefore, sunlight passes through a window and warms objects inside a house. The reradiated longer wavelengths cannot pass readily back through the glass but are absorbed by certain molecules in the air. The temperature of the air is thus increased. This is called the "greenhouse effect." Perhaps you have experienced the effect when you left your car windows closed on a sunny, summer day.

In general, a passive solar home makes use of the materials from which it is constructed to capture, store, and distribute solar energy to its occupants. Sunlight enters the house through large windows facing south and warms a thick layer of concrete, brick, or stone. This energy "storage mass" then releases energy during the day and, more important, during the night. This release of energy can be by direct radiation to occupants, by conduction to adjacent air, or by convection of air across the surface of the storage mass. The living space is thus heated without special plumbing or forced air circulation. As you can imagine, the key to a successful passive solar home is to consider every detail of natural energy flow, including the materials of which floors and walls are constructed, convective air circulation patterns, and the size and placement of windows. In addition, a passive solar home requires a different lifestyle and living patterns. Carpets, for example, would defeat the purpose of a storage-mass floor, since they would insulate the storage mass from sunlight. Glass is not a good insulator, so windows must have curtains or movable insulation panels to slow energy loss

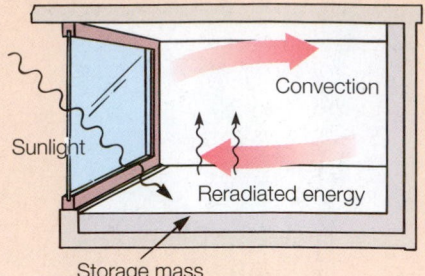

Box Figure 4.1 The direct solar gain design collects and stores solar energy in the living space.

at night. This requires the daily activity of closing curtains or moving insulation panels at night and then opening curtains and moving panels in the morning. Passive solar homes, therefore, require a high level of personal involvement by the occupants.

There are three basic categories of passive solar design: (1) direct solar gain, (2) indirect solar gain, and (3) isolated solar gain.

A *direct solar gain* home is one in which solar energy is collected in the actual living space of the home (box figure 4.1). The advantage of this design is the large, open window space with a calculated over-

---

CONCEPTS APPLIED

Candle Wax, Bubbles, and a Lightbulb

Here are three experiments you can do on heat flow:

Conduction. Use melted candle wax to stick thumbtack heads to a long metal rod. Heat one end of the rod with a flame. Record any evidence you observe that heat moves across the rod by conduction.

Convection. Choose a calm, warm day with plenty of strong sunlight. Make soap bubbles to study convection currents between a grass field and an adjacent asphalt parking lot. Find other adjacent areas where you think you might find convection currents and study them with soap bubbles, too. Record your experiments, findings, and explanations for what you observed.

Radiation. Hold your hand under an unlighted electric lightbulb and then turn on the bulb. Describe evidence that what you feel traveled to you by radiation, not conduction or convection. Describe any experiments you can think of to prove you felt radiant energy.

you feel warmth in your hand when you hold it near the element. Your hand absorbs the nonvisible radiant energy being emitted from the element. The radiant energy does work on the molecules of your hand, giving them more kinetic energy. You sense this as an increase in temperature, that is, warmth.

All objects above absolute zero emit radiant energy, but all objects also absorb radiant energy. A hot object, however, emits more radiant energy than a cold object. The hot object will emit more energy than it absorbs from the colder object, and the colder object will absorb more energy from the hot object than it emits. There is, therefore, a net energy transfer that will take place by radiation as long as there is a temperature difference between the two objects.

4.4 ENERGY, HEAT, AND MOLECULAR THEORY

The kinetic molecular theory of matter is based on evidence from different fields of science, not just one subject area. Chemists and physicists developed some convincing conclusions

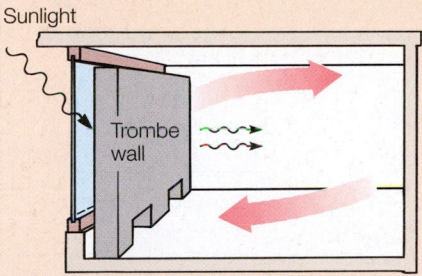

Box Figure 4.2 The indirect solar gain design uses a Trombe wall to collect, store, and distribute solar energy.

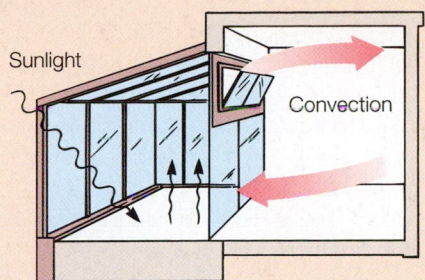

Box Figure 4.3 The isolated solar gain design uses a separate structure to collect and store solar energy.

hang that admits maximum solar energy in the winter but prevents solar gain in the summer. The disadvantage is that the occupants are living in the collection and storage components of the design and can place nothing (such as carpets and furniture) that would interfere with warming the storage mass in the floors and walls.

An *indirect solar gain* home uses a massive wall inside a window that serves as a storage mass. Such a wall, called a *Trombe wall*, is shown in box figure 4.2. The Trombe wall collects and stores solar energy, then warms the living space with radiant energy and convection currents. The disadvantage

to the indirect solar gain design is that large windows are blocked by the Trombe wall. The advantage is that the occupants are not in direct contact with the solar collection and storage area, so they can place carpets and furniture as they wish. Controls to prevent energy loss at night are still necessary with this design.

An *isolated solar gain* home uses a structure that is separated from the living space to collect and store solar energy. Examples of an isolated gain design are an attached greenhouse or sun porch (box figure 4.3). Energy flow between the attached structure and the living space can be by

conduction, convection, and radiation, which can be controlled by opening or closing off the attached structure. This design provides the best controls, since it can be completely isolated, opened to the living space as needed, or directly used as living space when the conditions are right. Additional insulation is needed for the glass at night, however, and for sunless winter days.

It has been estimated that building a passive solar home would cost about 10 percent more than building a traditional home of the same size. Considering the possible energy savings, you might believe that most homes would now have a passive solar design. They do not, however, as most new buildings require technology and large amounts of energy to maximize comfort. Yet, it would not require too much effort to consider where to place windows in relation to the directional and seasonal intensity of the Sun and where to plant trees. Perhaps in the future you will have an opportunity to consider using the environment to your benefit through the natural processes of conduction, convection, and radiation.

about the structure of matter over the past 150 years, using carefully designed experiments and mathematical calculations that explained observable facts about matter. Step by step, the detailed structure of this submicroscopic, invisible world of particles became firmly established. Today, an understanding of this particle structure is basic to physics, chemistry, biology, geology, and practically every other science subject. This understanding has also resulted in present-day technology.

Phase Change

Solids, liquids, and gases are the three common phases of matter, and each phase is characterized by different molecular arrangements. The motion of the molecules in any of the three common phases can be increased through (1) the addition of heat through a temperature difference or (2) the absorption of one of the five forms of energy, which results in heating. In either case, the temperature of the solid, liquid, or gas increases according to the specific heat of the substance, and more heating generally means higher temperatures.

More heating, however, does not always result in increased temperatures. When a solid, liquid, or gas changes from one

phase to another, the transition is called a **phase change.** A phase change always absorbs or releases energy, *a quantity of heat that is not associated with a temperature change.* Since the quantity of heat associated with a phase change is not associated with a temperature change, it is called *latent heat.* Latent heat refers to the "hidden" energy of phase changes, which is energy (heat) that goes into or comes out of *internal potential energy* (figure 4.15).

Three kinds of major phase changes can occur: (1) *solid-liquid*, (2) *liquid-gas*, and (3) *solid-gas*. In each case, the phase change can go in either direction. For example, the solid-liquid phase change occurs when a solid melts to a liquid or when a liquid freezes to a solid. Ice melting to water and water freezing to ice are common examples of this phase change and its two directions. Both occur at a temperature called the *freezing point* or the *melting point*, depending on the direction of the phase change. In either case, however, the freezing and melting points are the same temperature.

The liquid-gas phase change also occurs in two different directions. The temperature at which a liquid boils and changes to a gas (or vapor) is called the *boiling point*. The temperature at which a gas or vapor changes back to a liquid is called the

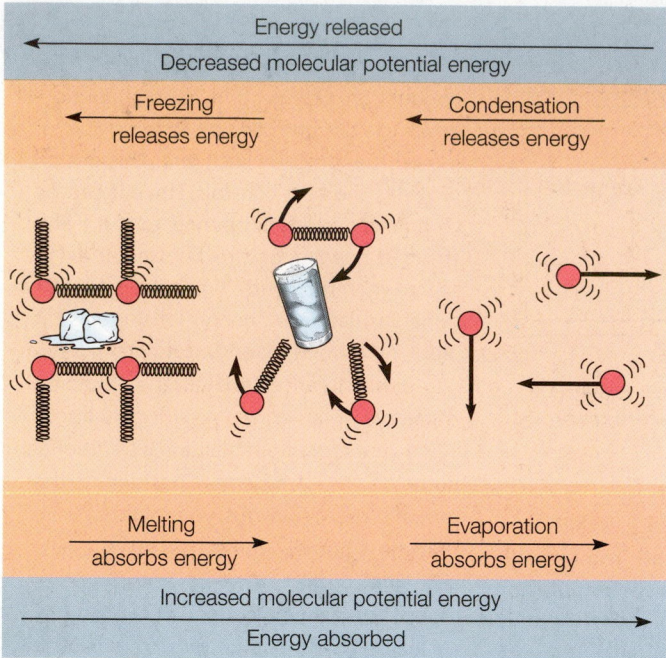

FIGURE 4.15 Each phase change absorbs or releases a quantity of latent heat, which goes into or is released from molecular potential energy.

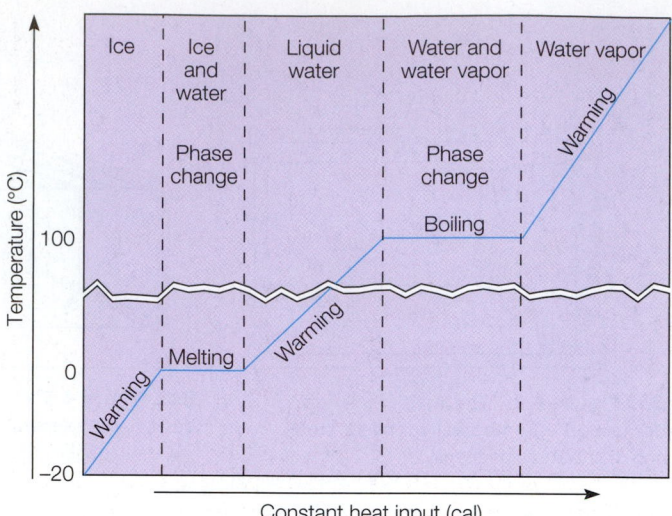

FIGURE 4.16 This graph shows three warming sequences and two phase changes with a constant input of heat. The ice warms to the melting point, then absorbs heat during the phase change as the temperature remains constant. When all the ice has melted, the now-liquid water warms to the boiling point, where the temperature again remains constant as heat is absorbed during this second phase change from liquid to gas. After all the liquid has changed to gas, continued warming increases the temperature of the water vapor.

condensation point. The boiling and condensation points are the same temperature. There are conditions other than boiling under which liquids may undergo liquid-gas phase changes, and these conditions are discussed in the next section.

You probably are not as familiar with solid-gas phase changes, but they are common. A phase change that takes a solid directly to a gas or vapor is called *sublimation.* Mothballs and dry ice (solid CO_2) are common examples of materials that undergo sublimation, but frozen water, meaning common ice, also sublimates under certain conditions. Perhaps you have noticed ice cubes in a freezer become smaller with time as a result of sublimation. The frost that forms in a freezer, on the other hand, is an example of a solid-gas phase change that takes place in the other direction. In this case, water vapor forms the frost without going through the liquid state, a solid-gas phase change that takes place in an opposite direction to sublimation.

For a specific example, consider the changes that occur when ice is subjected to a constant source of heat (figure 4.16). Starting at the left side of the graph, you can see that the temperature of the ice increases from the constant input of heat. The ice warms according to $Q = mc\Delta T$, where c is the specific heat of ice. When the temperature reaches the melting point (0°C), it stops increasing as the ice begins to melt. More and more liquid water appears as the ice melts, but the temperature *remains* at 0°C even though heat is still being added at a constant rate. It takes a certain amount of heat to melt all of the ice. Finally, when all the ice is completely melted, the temperature again increases at a constant rate between the melting and boiling points. Then, at constant temperature, the addition of heat produces another phase change, from liquid to gas. The quantity of heat involved in

this phase change is used in doing the work of breaking the molecule-to-molecule bonds in the solid, making a liquid with molecules that are now free to move about and roll over one another. Since the quantity of heat (Q) is absorbed without a temperature change, it is called the **latent heat of fusion** (L_f). The latent heat of fusion is *the heat involved in a solid-liquid phase change in melting or freezing.* For water, the latent heat of fusion is 80.0 cal/g (144.0 Btu/lb). This means that every gram of ice that melts *absorbs* 80.0 cal of heat. Every gram of water that freezes *releases* 80.0 cal. The total heat involved in a solid-liquid phase change depends on the mass of the substance involved, so

$$Q = mL_f \qquad \textbf{equation 4.5}$$

where L_f is the latent heat of fusion for the substance involved.

Refer again to figure 4.16. After the solid-liquid phase change is complete, the constant supply of heat increases the temperature of the water according to $Q = mc\Delta T$, where c is now the specific heat of liquid water. When the water reaches the boiling point, the temperature again remains constant even though heat is still being supplied at a constant rate. The quantity of heat involved in the liquid-gas phase change again goes into doing the work of overcoming the attractive molecular forces. This time the molecules escape from the liquid state to become single, independent molecules of gas. The quantity of heat (Q) absorbed or released during this phase change is called the **latent heat of vaporization** (L_v). The latent heat of vaporization is *the heat involved in a liquid-gas phase change where there is evaporation or condensation.* For water, the latent heat of vaporization is 540.0 cal/g (970.0 Btu/lb).

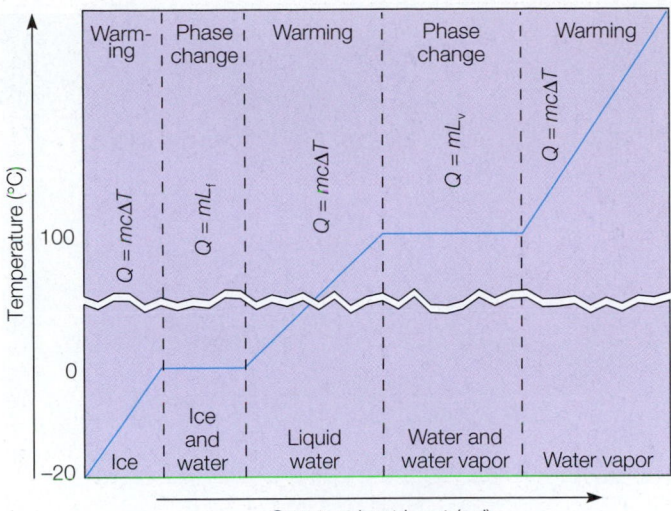

FIGURE 4.17 Compare this graph to the one in figure 4.16. This graph shows the relationships between the quantity of heat absorbed during warming and phase changes as water is warmed from ice at −20°C to water vapor at some temperature above 100°C. Note that the specific heat for ice, liquid water, and water vapor (steam) has different values.

TABLE 4.3 Some Physical Constants for Water and Heat

Specific Heat (c)

Water	$c = 1.00$ cal/gC°
Ice	$c = 0.500$ cal/gC°
Steam	$c = 0.480$ cal/gC°

Latent Heat of Fusion

L_f (water)	$L_f = 80.0$ cal/g

Latent Heat of Vaporization

L_v (water)	$L_v = 540.0$ cal/g

Mechanical Equivalent of Heat

1 kcal	4,184 J

This means that every gram of water vapor that condenses on your bathroom mirror releases 540.0 cal, which warms the bathroom. The total heating depends on how much water vapor condensed, so

$$Q = mL_v \qquad \textbf{equation 4.6}$$

where L_v is the latent heat of vaporization for the substance involved. The relationships between the quantity of heat absorbed during warming and phase changes are shown in figure 4.17. Some physical constants for water and heat are summarized in table 4.3.

EXAMPLE 4.7 *(Optional)*

How much energy does a refrigerator remove from 100.0 g of water at 20.0°C to make ice at −10.0°C?

SOLUTION

This type of problem is best solved by subdividing it into smaller steps that consider (1) the heat added or removed and the resulting temperature changes for each phase of the substance, and (2) the heat flow resulting from any phase change that occurs within the ranges of changes as identified by the problem (see figure 4.17). The heat involved in each phase change and the heat involved in the heating or cooling of each phase are identified as Q_1, Q_2, and so forth. Temperature readings are calculated with absolute values, so you ignore any positive or negative signs.

1. Water in the liquid state cools from 20.0°C to 0°C (the freezing point) according to the relationship $Q = mc\Delta T$, where c is the specific heat of water, and

$$Q_1 = mc\Delta T$$
$$= (100.0 \text{ g})\left(1.00 \frac{\text{cal}}{\text{gC°}}\right)(0°C - 20.0°C)$$
$$= (100.0)(1.00)(20.0) \text{ g} \times \frac{\text{cal}}{\text{g} \cdot \text{C°}} \times \text{C°}$$
$$= 2,000 \frac{\text{g} \cdot \text{cal} \cdot \text{C°}}{\text{g} \cdot \text{C°}}$$
$$= 2.00 \text{ kcal}$$

2. The latent heat of fusion must now be removed as water at 0°C becomes ice at 0°C through a phase change, and

$$Q_2 = mL_f$$
$$= (100.0 \text{ g})\left(80.0 \frac{\text{cal}}{\text{g}}\right)$$
$$= (100.0)(80.0) \text{ g} \times \frac{\text{cal}}{\text{g}}$$
$$= 8,000 \text{ cal}$$
$$= 8.00 \text{ kcal}$$

3. The ice is now at 0°C and is cooled to −10°C, as specified in the problem. The ice cools according to $Q = mc\Delta T$, where c is the specific heat of ice. The specific heat of ice is 0.500 cal/gC°, and

$$Q_3 = mc\Delta T$$
$$= (100.0 \text{ g})\left(0.500 \frac{\text{cal}}{\text{gC°}}\right)(10°C - 0°C)$$
$$= (100.0)(0.500)(10.0) \text{ g} \times \frac{\text{cal}}{\text{g} \cdot \text{C°}} \times \text{C°}$$
$$= 500 \text{ cal}$$
$$= 0.500 \text{ kcal}$$

4. The total energy removed is then

$$Q_T = Q_1 + Q_2 + Q_3$$
$$= (2.00 \text{ kcal}) + (8.00 \text{ kcal}) + (0.500 \text{ kcal})$$
$$= \boxed{10.50 \text{ kcal}}$$

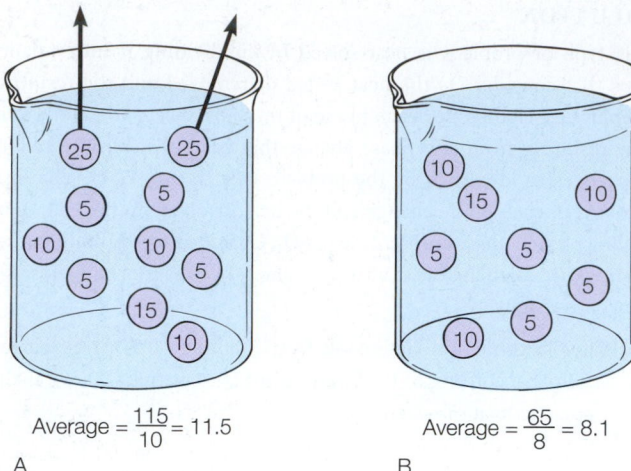

$$\text{Average} = \frac{115}{10} = 11.5$$

A

$$\text{Average} = \frac{65}{8} = 8.1$$

B

FIGURE 4.18 Temperature is associated with the average energy of the molecules of a substance. These numbered circles represent arbitrary levels of molecular kinetic energy that, in turn, represent temperature. The two molecules with the higher kinetic energy values [25 in (A)] escape, which lowers the average values from 11. 5 to 8. 1 (B). Thus, evaporation of water molecules with more kinetic energy contributes to the cooling effect of evaporation, in addition to the absorption of latent heat.

Evaporation and Condensation

Liquids do not have to be at the boiling point to change to a gas and, in fact, tend to undergo a phase change at any temperature when left in the open. The phase change occurs at any temperature but does occur more rapidly at higher temperatures. The temperature of the water is associated with the *average* kinetic energy of the water molecules. The word *average* implies that some of the molecules have a greater energy and some have less. If a molecule of water that has an exceptionally high energy is near the surface and is headed in the right direction, it may overcome the attractive forces of the other water molecules and escape the liquid to become a gas. This is the process of *evaporation.* Evaporation reduces a volume of liquid water as water molecules leave the liquid state to become water vapor in the atmosphere (figure 4.18).

Water molecules that evaporate move about in all directions, and some will return, striking the liquid surface. The same forces that they escaped from earlier capture the molecules, returning them to the liquid state. This is called the process of condensation. Condensation is the opposite of evaporation. In *evaporation,* more molecules are leaving the liquid state than are returning. In *condensation,* more molecules are returning to the liquid state than are leaving. This is a dynamic, ongoing process with molecules leaving and returning continuously. The net number leaving or returning determines if evaporation or condensation is taking place (figure 4.19).

When the condensation rate *equals* the evaporation rate, the air above the liquid is said to be *saturated.* The air immediately next to a surface may be saturated, but the condensation of water molecules is easily moved away with air movement. There is no net energy flow when the air is saturated, since the heat carried away by evaporation is returned by condensation.

FIGURE 4.19 The inside of this closed bottle is isolated from the environment, so the space above the liquid becomes saturated. While it is saturated, the evaporation rate equals the condensation rate. When the bottle is cooled, condensation exceeds evaporation and droplets of liquid form on the inside surfaces.

This is why you fan your face when you are hot. The moving air from the fanning action pushes away water molecules from the air near your skin, preventing the adjacent air from becoming saturated, thus increasing the rate of evaporation. Think about this process the next time you see someone fanning his or her face.

There are four ways to increase the rate of evaporation. (1) An increase in the temperature of the liquid will increase the average kinetic energy of the molecules and thus increase the number of high-energy molecules able to escape from the liquid state. (2) Increasing the surface area of the liquid will also increase the likelihood of molecular escape to the air. This is why you spread out wet clothing to dry or spread out a puddle you want to evaporate. (3) Removal of water vapor from near the surface of the liquid will prevent the return of the vapor molecules to the liquid state and thus increase the net rate of evaporation. This is why things dry more rapidly on a windy day. (4) *Pressure* is defined as *force per unit area,* which can be measured in lb/in^2 or N/m^2. Gases exert a pressure, which is interpreted in terms of the kinetic molecular theory.

Evaporative Coolers

Evaporative cooling is one of the earliest forms of mechanical air conditioning. An evaporative cooler (sometimes called a "swamp cooler") works by cooling outside air by evaporation, then blowing the cooler but now more humid air through a house. Usually, an evaporative cooler moves a sufficient amount of air through a house to completely change the air every two or three minutes.

An evaporative cooler is a metal or fiberglass box with louvers. Inside the box is a fan and motor, and there might be a small pump to recycle water. Behind the louvers are loose pads made of wood shavings (excelsior) or some other porous material. Water is pumped from the bottom of the cooler and trickles down through the pads, thoroughly wetting them. The fan forces air into the house and dry, warm outside air moves through the wet pads into the house in a steady flow. Windows or a door in the house must be partly open, or the air will not be able to flow through the house.

The water in the pads evaporates, robbing heat from the air moving through them. Each liter of water that evaporates could take over 540 kcal of heat from the air. The actual amount removed depends on the temperature of the water, the efficiency of the cooler, and the relative humidity of the outside air.

An electric fan cools you by evaporation, but an evaporative cooler cools the air that blows around you. Relative humidity is the main variable that determines how much an evaporative cooler will cool the air. The following data show the cooling power of a typical evaporative cooler at various relative humidities when the outside air is 38°C (about 100°F).

The advantage of an evaporative cooler is the low operating cost compared to refrigeration. The disadvantages are that it doesn't cool the air that much when the humidity is high, it adds even more humidity to the air, and outside air with its dust, pollen, and pollution is continually forced through the house. Another disadvantage is that mineral deposits left by evaporating hard water could require a frequent maintenance schedule.

At This Humidity:	Air at 38°C (100°F) Will Be Cooled to:
10%	22°C (72°F)
20%	24°C (76°F)
30%	27°C (80°F)
40%	29°C (84°F)
50%	31°C (88°F)
60%	33°C (91°F)
70%	34°C (93°F)
80%	35°C (95°F)
90%	37°C (98°F)

Atmospheric pressure is discussed in detail in chapter 17. For now, consider that the atmosphere exerts a pressure of about 10.0 N/cm² (14.7 lb/in²) at sea level. The atmospheric pressure, as well as the intermolecular forces, tend to hold water molecules in the liquid state. Thus, reducing the atmospheric pressure will reduce one of the forces holding molecules in a liquid state. Perhaps you have noticed that wet items dry more quickly at higher elevations, where the atmospheric pressure is less.

Relative Humidity

There is a relationship between evaporation-condensation and the air temperature. If the air temperature is decreased, the average kinetic energy of the molecules making up the air is decreased. Water vapor molecules condense from the air when they slow enough that molecular forces can pull them into the liquid state. Fast-moving water vapor molecules are less likely to be captured than slow-moving ones. Thus, as the air temperature increases, there is less tendency for water vapor molecules to return to the liquid state. Warm air can therefore hold more water vapor than cool air. In fact, air at 38°C (100°F) can hold five times as much water vapor as air at 10°C (50°F) (figure 4.20).

The ratio of how much water vapor *is* in the air to how much water vapor *could be* in the air at a certain temperature is called **relative humidity.** This ratio is usually expressed as a percentage, and

$$\text{relative humidity} = \frac{\text{water vapor in air}}{\text{capacity at present temperature}} \times 100\%$$

$$R.H. = \frac{\text{g/m}^3 (\text{present})}{\text{g/m}^3 (\text{max})} \times 100\% \qquad \textbf{equation 4.7}$$

Figure 4.20 shows the maximum amount of water vapor that can be in the air at various temperatures. Suppose that the air contains 10 g/m³ of water vapor at 10°C (50°F). According to figure 4.20, the maximum amount of water vapor that *can be* in the air when the air temperature is 10°C (50°F) is 10 g/m³. Therefore, the relative humidity is (10 g/m³) ÷ (10 g/m³) × 100%, or 100 percent. This air is therefore saturated. If the air held only 5 g/m³ of water vapor at 10°C, the relative humidity would be 50 percent, and 2 g/m³ of water vapor in the air at 10°C is 20 percent relative humidity.

As the air temperature increases, the capacity of the air to hold water vapor also increases. This means that if the *same amount* of water vapor is in the air during a temperature increase, the relative humidity will decrease. Thus, the relative humidity increases every night because the air temperature decreases, not because water vapor has been added to the air. The relative humidity is important because it is one of the things that controls the rate of evaporation, and the evaporation rate is one of the variables involved in how well you can cool yourself in hot weather.

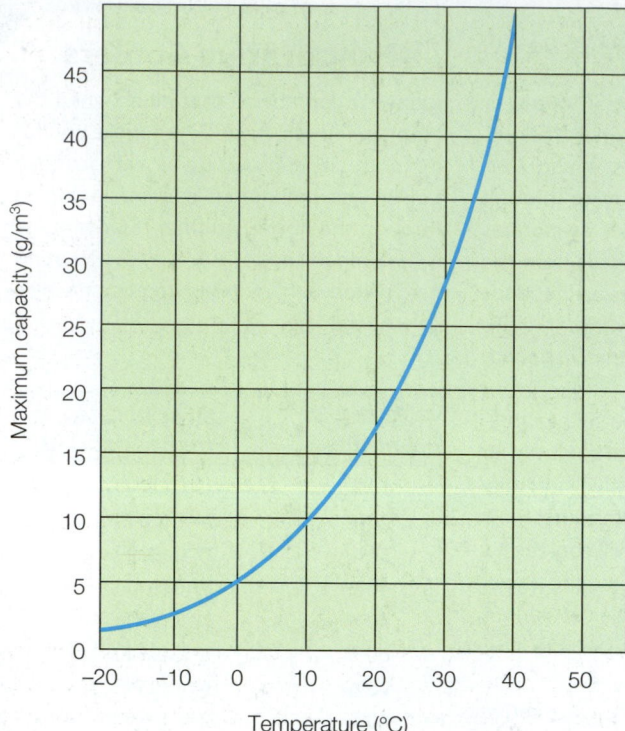

FIGURE 4.20 The curve shows the *maximum* amount of water vapor in g/m³ that can be in the air at various temperatures.

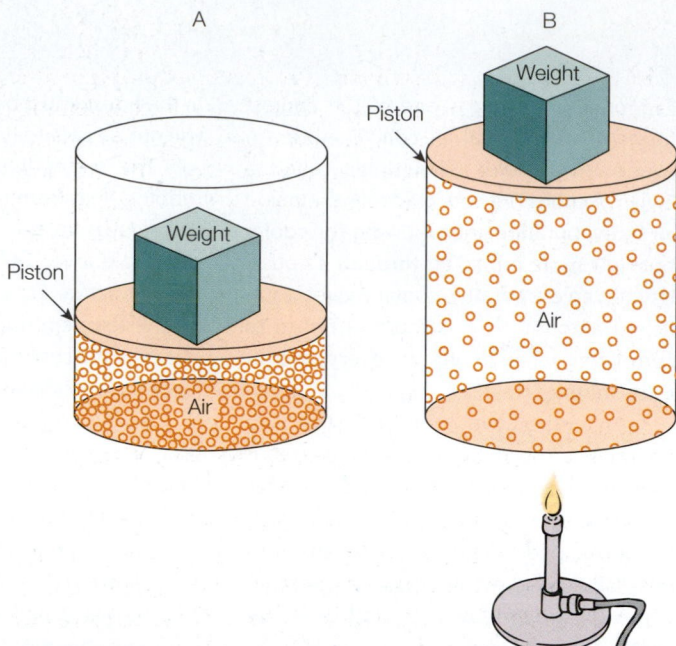

FIGURE 4.21 A very simple heat engine. The air in (*B*) has been heated, increasing the molecular motion and thus the pressure. Some of the heat is transferred to the increased gravitational potential energy of the weight as it is converted to mechanical energy.

CONCEPTS APPLIED

Why It Is Called a Pop Can!

Obtain two empty, clean pop (soft drink) cans, a container of ice water with ice cubes, and a container of boiling water. You might want to "dry run" this experiment to make sure of the procedure before actually doing it.

Place about 2 cm of water in a pop can and heat it on a stove until the water boils and you see evidence of steam coming from the opening. Using tongs, quickly invert the can half way into a container of ice water. Note how much water runs from the can as you remove it from the ice water.

Repeat this procedure, this time inverting the can half way into a container of boiling water. Note how much water runs from the can as you remove it from the boiling water.

Explain your observations in terms of the kinetic molecular theory, evaporation, and condensation. It is also important to explain any differences observed between what happened to the two pop cans.

4.5 THERMODYNAMICS

The branch of physical science called *thermodynamics* is concerned with the study of heat and its relationship to mechanical energy, including the science of heat pumps, heat engines, and the transformation of energy in all its forms. The *laws of thermodynamics* describe the relationships concerning what happens as energy is transformed to work and the reverse, also serving as useful intellectual tools in meteorology, chemistry, and biology.

Mechanical energy is easily converted to heat through friction, but a special device is needed to convert heat to mechanical energy. A *heat engine* is a device that converts heat into mechanical energy. The operation of a heat engine can be explained by the kinetic molecular theory, as shown in figure 4.21. This illustration shows a cylinder, much like a big can, with a closely fitting piston that traps a sample of air. The piston is like a slightly smaller cylinder and has a weight resting on it, supported by the trapped air. If the air in the large cylinder is now heated, the gas molecules will acquire more kinetic energy. This results in more gas molecule impacts with the enclosing surfaces, which results in an increased pressure. Increased pressure results in a net force, and the piston and weight move upward, as shown in figure 4.21B. Thus, some of the heat has now been transformed to the increased gravitational potential energy of the weight.

Thermodynamics is concerned with the *internal energy* (*U*), the total internal potential and kinetic energies of molecules making up a substance, such as the gases in the simple heat engine. The variables of temperature, gas pressure, volume, heat, and so forth characterize the total internal energy, which is called the *state* of the system. Once the system is identified, everything else is called the *surroundings*. A system can exist in a number of states since the variables that characterize a state can have any number of values and combinations of values. Any two systems that have the same values of variables that characterize internal energy are said to be in the same state.

The First Law of Thermodynamics

Any thermodynamic system has a unique set of properties that will identify the internal energy of the system. This state can be changed two ways: (1) by heat flowing into (Q_{in}) or out (Q_{out}) of the system, or (2) by the system doing work (W_{out}) or work being done on the system (W_{in}). Thus, work (W) and heat (Q) can change the internal energy of a thermodynamic system according to

$$JQ - W = U_2 - U_1 \qquad \textbf{equation 4.8}$$

where J is the mechanical equivalent of heat (J = 4.184 joule/calorie) and ($U_2 - U_1$) is the internal energy difference between two states. This equation represents the **first law of thermodynamics,** which states that the energy supplied to a thermodynamic system in the form of heat minus the work done by the system is equal to the change in internal energy. The first law of thermodynamics is an application of the *law of conservation of energy,* which applies to all energy matters. The first law of thermodynamics is concerned specifically with a thermodynamic system. As an example, consider energy conservation that is observed in the thermodynamic system of a heat engine (see figure 4.22). As the engine cycles to the original state of internal energy ($U_2 - U_1 = 0$), all the external work accomplished must be equal to all the heat absorbed in the cycle. The heat supplied to the engine from a high temperature source (Q_H) is partly converted to work (W), and the rest is rejected in the lower-temperature exhaust (Q_L). The work accomplished is therefore the difference in the heat input and the heat output ($Q_H - Q_L$), so the work accomplished represents the heat used,

$$W = J(Q_H - Q_L) \qquad \textbf{equation 4.9}$$

where J is the mechanical equivalence of heat (J = 4.184 joules/calorie). A schematic diagram of this relationship is shown in figure 4.22. You can increase the internal energy (produce heat) as long as you supply mechanical energy (or do work). The first law of thermodynamics states that the conversion of work to heat is reversible, meaning that heat can be changed to work. There are several ways of converting heat to work, for example, the use of a steam turbine or gasoline automobile engine.

The Second Law of Thermodynamics

A heat pump is the opposite of a heat engine, as shown schematically in figure 4.23. The heat pump does work (W) in compressing vapors and moving heat from a region of lower temperature (Q_L) to a region of higher temperature (Q_H). That work is required to move heat this way is in accord with

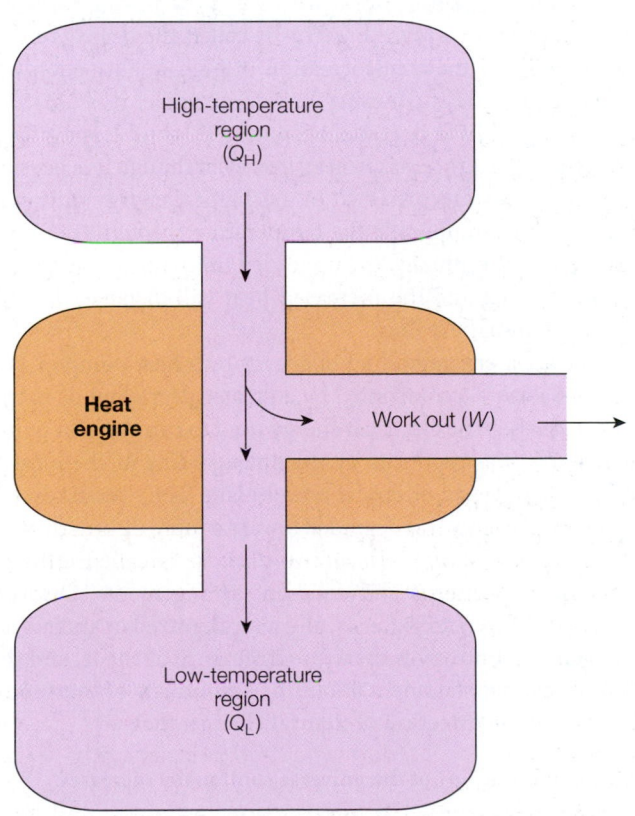

FIGURE 4.22 The heat supplied (Q_H) to a heat engine goes into the mechanical work (W), and the remainder is expelled in the exhaust (Q_L). The work accomplished is therefore the difference in the heat input and output ($Q_H - Q_L$), so the work accomplished represents the heat used, $W = J(Q_H - Q_L)$.

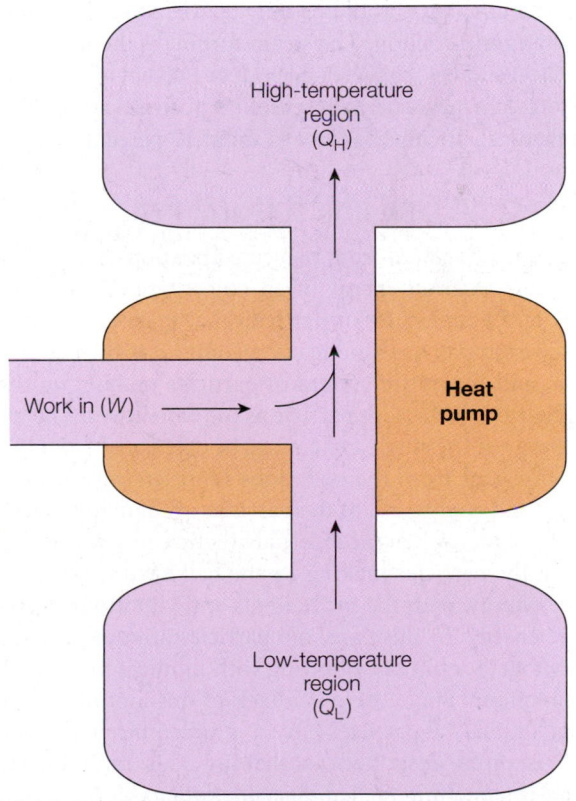

FIGURE 4.23 A heat pump uses work (W) to move heat from a low temperature region (Q_L) to a high temperature region (Q_H). The heat moved (Q_L) requires work (W), so $J(Q_H - Q_L) = W$. A heat pump can be used to chill things at the Q_L end or warm things at the Q_H end.

the observation that heat naturally flows from a region of higher temperature to a region of lower temperature. Energy is required for the opposite, moving heat from a cooler region to a warmer region. The natural direction of this process is called the **second law of thermodynamics,** which is that heat flows from objects with a higher temperature to objects with a cooler temperature. In other words, if you want heat to flow from a colder region to a warmer one, you must *cause* it to do so by using energy. And if you do, such as with the use of a heat pump, you necessarily cause changes elsewhere, particularly in the energy sources used in the generation of electricity.

Another statement of the second law is that it is impossible to convert heat completely into mechanical energy. This does not say that you cannot convert mechanical energy completely into heat, for example, in the brakes of a car when the brakes bring it to a stop. The law says that the reverse process is not possible, that you cannot convert 100 percent of a heat source into mechanical energy.

The heat used by a machine can be expressed as a measure of work, so it is possible to judge how well the machine operated by looking at the work input and the work output. The ratio of work output to work input is called mechanical *efficiency* (efficiency = work output/work input). If a machine converted all of the heat into mechanical energy, the efficiency would be 100 percent. However, the efficiency of a machine is always less than 100 percent because of the second law of thermodynamics. Some of the work put into a machine is always lost to friction. Even if it were possible to convert 100 percent of the heat to mechanical energy, the actual efficiency would be only 60 to 80 percent because of heat losses and friction. The steam turbine is the most efficient of the heat engines, converting about 40 percent of the generated heat to mechanical energy. A gasoline-powered automobile has a mechanical efficiency as low as about 17 percent.

The Second Law and Natural Processes

Energy can be viewed from two considerations of scale: (1) the observable *external energy* of an object, and (2) the *internal energy* of the molecules or particles that make up an object. A ball, for example, has kinetic energy after it is thrown through the air, and the entire system of particles making up the ball acts like a single massive particle as the ball moves. The motion and energy of the single system can be calculated from the laws of motion and from the equations representing the concepts of work and energy. All of the particles are moving together in *coherent motion* when the external kinetic energy is considered.

But the particles making up the ball have another kind of kinetic energy, with the movements and vibrations of internal kinetic energy. In this case, the particles are not moving uniformly together but are vibrating with motions in many different directions. Since there is a lack of net motion and a lack of correlation, the particles have a jumbled *incoherent motion*, which is often described as chaotic. This random, chaotic motion is sometimes called *thermal motion*.

Thus, there are two kinds of motion that the particles of an object can have: (1) a coherent motion where they move together, in step, and (2) an incoherent, chaotic motion of individual particles. These two types of motion are related to the two modes of energy transfer, working and heating. The relationship is that *work* on an object is associated with its *coherent motion,* while *heating* an object is associated with its internal *incoherent motion.*

The second law of thermodynamics implies a direction to the relationship between work (coherent motion) and heat (incoherent motion), and this direction becomes apparent as you analyze what happens to the motions during energy conversions. Some forms of energy, such as electrical and mechanical, have a greater amount of order since they involve particles moving together in a coherent motion. The term *quality of energy* is used to identify the amount of coherent motion. Energy with high order and coherence is called *high-quality energy.* Energy with less order and less coherence, on the other hand, is called *low-quality energy.* In general, high-quality energy can be easily converted to work, but low-quality energy is less able to do work.

High-quality electrical and mechanical energy can be used to do work but then become dispersed as heat through energy-form conversions and friction. The resulting heat can be converted to do more work only if there is a sufficient temperature difference. The temperature differences do not last long, however, as conduction, convection, and radiation quickly disperse the energy even more. Thus, the transformation of high-quality energy into lower-quality energy is a natural process. Energy tends to disperse, both from the conversion of an energy form to heat and from the heat flow processes of conduction, convection, and radiation. Both processes flow in one direction only and cannot be reversed. This is called the *degradation of energy,* which is the transformation of high-quality energy to lower-quality energy. In every known example, it is a natural process of energy to degrade, becoming less and less available to do work. The process is *irreversible,* even though it is possible to temporarily transform heat to mechanical energy through a heat engine or to upgrade the temperature through the use of a heat pump. Eventually, the upgraded mechanical energy will degrade to heat and the increased heat will disperse through the processes of heat flow.

The apparent upgrading of energy by a heat pump or heat engine is always accompanied by a greater degrading of energy someplace else. The electrical energy used to run the heat pump, for example, was produced by the downgrading of chemical or nuclear energy at an electrical power plant. The overall result is that the *total* energy was degraded toward a more disorderly state.

A *thermodynamic measure of disorder* is called **entropy.** Order means patterns and coherent arrangements. Disorder means dispersion, no patterns, and a randomized or spread-out arrangement. Entropy is therefore a measure of chaos, and this leads to another statement about the second law of thermodynamics and the direction of natural change, that

the total entropy of the universe continually increases.

Note the use of the words *total* and *universe* in this statement of the second law. The entropy of a system can decrease (more order); for example, when a heat pump cools and condenses the random, chaotically moving water vapor molecules into

Thermodynamics in Action

The laws of thermodynamics are concerned with changes in energy and heat. This application explores some of these relationships.

Obtain an electric blender and a thermometer. Fill the blender half way with water, then let it remain undisturbed until the temperature is constant as shown by two consecutive temperature readings.

Run the blender at the highest setting for a short time, then stop and record the temperature of the water. Repeat this procedure several times.

Explain your observations in terms of thermodynamics. See if you can think of other experiments that show relationships between changes in energy and heat.

It Makes Its Own Fuel?

Have you ever heard of a perpetual motion machine? A perpetual motion machine is a hypothetical device that would produce useful energy out of nothing. This is generally accepted as being impossible according to laws of physics. In particular, perpetual motion machines would violate either the first or the second law of thermodynamics.

A perpetual motion machine that violates the first law of thermodynamics is called a "machine of the first kind." In general, the first law says that you can never get something for nothing. This means that without energy input, there can be no change in internal energy, and without a change in internal energy, there can be no work output. Machines of the first kind typically use no fuel or make their own fuel faster than they use it. If this type of machine appears to work, look for some hidden source of energy.

A "machine of the second kind" does not attempt to make energy out of nothing. Instead, it tries to extract random molecular motion into useful work or extract useful energy from some degraded source such as outgoing radiant energy. The second law of thermodynamics says this just cannot happen any more than rocks can roll uphill on their own. This just does not happen.

The American Physical Society states, "The American Physical Society deplores attempts to mislead and defraud the public based on claims of perpetual motion machines or sources of unlimited useful free energy, unsubstantiated by experimentally tested established physical principles."

the more ordered state of liquid water. When the energy source for the production, transmission, and use of electrical energy is considered, however, the *total* entropy will be seen

as increasing. Likewise, the total entropy increases during the growth of a plant or animal. When all the food, waste products, and products of metabolism are considered, there is again an increase in *total* entropy.

Thus, the *natural process* is for a state of order to degrade into a state of disorder with a corresponding increase in entropy. This means that all the available energy of the universe is gradually diminishing, and over time, the universe should therefore approach a limit of maximum disorder called the *heat death* of the universe. The heat death of the universe is the theoretical limit of disorder, with all molecules spread far, far apart, vibrating slowly with a uniform low temperature.

The heat death of the universe seems to be a logical consequence of the second law of thermodynamics, but scientists are not certain if the second law should apply to the whole universe. What do you think? Will the universe end as matter becomes spread out with slowly vibrating molecules? As has been said, nature is full of symmetry. So why should the universe begin with a bang and end with a whisper?

EXAMPLE 4.8   *(Optional)*

A heat engine operates with 65.0 kcal of heat supplied and exhausts 40.0 kcal of heat. How much work did the engine do?

SOLUTION

Listing the known and unknown quantities:

heat input $\qquad Q_H = 65.0$ kcal

heat rejected $\qquad Q_L = 40.0$ kcal

mechanical equivalent of heat 1 kcal = 4,184 J

The relationship between these quantities is found in equation 4.9 $W = J(Q_H - Q_L)$. This equation states a relationship between the heat supplied to the engine from a high-temperature source (Q_H), which is partly converted to work (W), with the rest rejected in a lower-temperature exhaust (Q_L). The work accomplished is therefore the difference in the heat input and the heat output ($Q_H - Q_L$), so the work accomplished represents the heat used, where J is the mechanical equivalence of heat (1 kcal = 4,184 J). Therefore,

$$W = J(Q_H - Q_L)$$

$$= 4{,}184 \, \frac{J}{\text{kcal}} \, (65.0 \text{ kcal} - 40.0 \text{ kcal})$$

$$= 4{,}184 \, \frac{J}{\text{kcal}} \, (25.0 \text{ kcal})$$

$$= (4{,}184)(25.0) \, \frac{J \, \text{kcal}}{\text{kcal}}$$

$$= 104{,}600 \text{ J}$$

$$= \boxed{105 \text{ kJ}}$$

People Behind the Science

Count Rumford (Benjamin Thompson) (1753–1814)

Count Rumford was a U.S.-born physicist who first demonstrated conclusively that heat is not a fluid but a form of motion. He was born Benjamin Thompson in Woburn, Massachusetts, on March 26, 1753. At the age of nineteen, he became a schoolmaster as a result of much self-instruction and some help from local clergy.

Rumford's early work in Bavaria combined social experiments with his lifelong interests concerning heat in all its aspects. When he employed beggars from the streets to manufacture military uniforms, he faced the problem of feeding them. A study of nutrition led him to recognize the importance of water and vegetables, and Rumford decided that soups would fit his requirements. He devised many recipes and developed cheap food emphasizing the potato. Soldiers were employed in gardening to produce the vegetables. Rumford's enterprise of manufacturing military uniforms led to a study of insulation and to the conclusion that heat was lost mainly through convection. Therefore, he designed clothing to inhibit convection—sort of the first thermal clothing.

No application of heat technology was too humble for Rumford's experiments. He devised the domestic range—the "fire in a box"—and special utensils to go with it. In the interest of fuel efficiency, he devised a calorimeter to compare the heats of combustion of various fuels. Smoky fireplaces also drew his attention, and after a study of the various air movements, he produced designs incorporating all the features now considered essential in open fires and chimneys, such as the smoke shelf and damper.

His search for an alternative to alcoholic drinks led to the promotion of coffee and the design of the first percolator.

The work for which Rumford is best remembered took place in 1798. As military commander for the elector of Bavaria, he was concerned with the manufacture of cannons. These were bored from blocks of iron with drills, and it was believed that the cannons became hot because as the drills cut into the iron, heat was escaping in the form of a fluid called caloric. However, Rumford noticed that heat production increased as the drills became blunter and cut less into the metal. If a very blunt drill was used, no metal was removed, yet the heat output appeared to be limitless. Clearly, heat could not be a fluid in the metal but must be related to the work done in turning the drill. Rumford also studied the expansion of liquids of different densities and different specific heats, and showed by careful weighing that the expansion was not due to caloric taking up the extra space.

Rumford's contribution to science in demolishing the caloric theory of heat was very important, because it paved the way to the realization that heat is related to energy and work and that all forms of energy can be converted to heat. However, it took several decades to establish the understanding that caloric does not exist and there was no basis for the caloric theory of heat.

Source: Modified from the *Hutchinson Dictionary of Scientific Biography.* © RM, 2011. All rights reserved. Helicon Publishing is a division of RM.

SUMMARY

The kinetic theory of matter assumes that all matter is made up of tiny, ultimate particles of matter called *molecules.* A molecule is defined as the smallest particle of a compound or a gaseous element that can exist and still retain the characteristic properties of that substance. Molecules interact, attracting each other through a force of *cohesion.* Liquids, solids, and gases are the *phases of matter* that are explained by the molecular arrangements and forces of attraction between their molecules. A *solid* has a definite shape and volume because it has molecules that vibrate in a fixed equilibrium position with strong cohesive forces. A *liquid* has molecules that have cohesive forces strong enough to give it a definite volume but not strong enough to give it a definite shape. The molecules of a liquid can flow, rolling over each other. A *gas* is composed of molecules that are far apart, with weak cohesive forces. Gas molecules move freely in a constant, random motion.

The *temperature* of an object is related to the *average kinetic energy* of the molecules making up the object. A measure of temperature tells how hot or cold an object is on two arbitrary scales, the *Fahrenheit scale* and the *Celsius scale.* The *absolute scale,* or *Kelvin scale,* has the coldest temperature possible ($-273°C$) as zero (0 K).

The observable potential and kinetic energy of an object is the *external energy* of that object, while the potential and kinetic energy of the molecules making up the object is the *internal energy* of the object. Heat refers to the total internal energy and is a transfer of energy that takes place (1) because of a *temperature difference* between two objects or

(2) because of an *energy-form conversion*. An energy-form conversion is actually an energy conversion involving work at the molecular level, so all energy transfers involve *heating* and *working*.

A quantity of heat can be measured in *joules* (a unit of work or energy) or *calories* (a unit of heat). A *kilocalorie* is 1,000 calories, another unit of heat. A *Btu*, or *British thermal unit,* is the English system unit of heat. The *mechanical equivalent of heat* is 4,184 J = 1 kcal.

The *specific heat* of a substance is the amount of energy (or heat) needed to increase the temperature of 1 gram of a substance 1 degree Celsius. The specific heat of various substances is not the same because the molecular structure of each substance is different.

Energy transfer that takes place because of a temperature difference does so through conduction, convection, or radiation. *Conduction* is the transfer of increased kinetic energy from molecule to molecule. Substances vary in their ability to conduct heat, and those that are poor conductors are called *insulators*. Gases, such as air, are good insulators. The best insulator is a vacuum. *Convection* is the transfer of heat by the displacement of large groups of molecules with higher kinetic energy. Convection takes place in fluids, and the fluid movement that takes place because of density differences is called a *convection current*. *Radiation* is radiant energy that moves through space. All objects with an absolute temperature above zero give off radiant energy, but all objects absorb it as well. Energy is transferred from a hot object to a cold one through radiation.

The transition from one phase of matter to another is called a *phase change*. A phase change always absorbs or releases a quantity of *latent heat* not associated with a temperature change. Latent heat is energy that goes into or comes out of *internal potential energy*. The *latent heat of fusion* is absorbed or released at a solid-liquid phase change. The latent heat of fusion for water is 80.0 cal/g (144.0 Btu/lb). The *latent heat of vaporization* is absorbed or released at a liquid-gas phase change. The latent heat of vaporization for water is 540.0 cal/g (970.0 Btu/lb).

Molecules of liquids sometimes have a high enough velocity to escape the surface through the process called *evaporation*. Evaporation is a cooling process, since the escaping molecules remove the latent heat of vaporization in addition to their high molecular energy. Vapor molecules return to the liquid state through the process called *condensation*. Condensation is the opposite of evaporation and is a warming process. When the condensation rate equals the evaporation rate, the air is said to be *saturated*. The rate of evaporation can be *increased* by (1) increased temperature, (2) increased surface area, (3) removal of evaporated molecules, and (4) reduced atmospheric pressure.

Warm air can hold more water vapor than cold air, and the ratio of how much water vapor is in the air to how much could be in the air at that temperature (saturation) is called *relative humidity*.

Thermodynamics is the study of heat and its relationship to mechanical energy, and the *laws of thermodynamics* describe these relationships: The *first law of thermodynamics* is a thermodynamic statement of the law of conservation of energy. The *second law of thermodynamics* states that heat flows from objects with a higher temperature to objects with a cooler temperature. The second law implies a *degradation of energy* as *high-quality* (more ordered) energy sources undergo *degradation* to *low-quality* (less ordered) sources. *Entropy* is a thermodynamic measure of disorder, and entropy is seen as continually increasing in the universe and may result in the maximum disorder called the *heat death* of the universe.

Summary of Equations

4.1
$$T_F = \frac{9}{5}T_C + 32°$$

4.2
$$T_C = \frac{5}{9}(T_F - 32°)$$

4.3
$$T_K = T_C + 273$$

4.4 Quantity of heat = (mass)(specific heat)(temperature change)
$$Q = mc\Delta T$$

4.5 Heat absorbed or released = (mass)(latent heat off usion)
$$Q = mL_f$$

4.6 Heat absorbed or released = (mass)(latent heat of vaporization)
$$Q = mL_v$$

4.7 relative humidity = $\dfrac{\text{water vapor in air}}{\text{capacity at present temperature}} \times 100\%$

$$R.H. = \frac{g/m^3 (\text{present})}{g/m^3 (\text{max})} \times 100\%$$

4.8 (mechanical equivalence of heat) − (work) = internal energy difference between two states
$$JQ - W = U_2 - U_1$$

4.9 work = (mechanical equivalence of heat)(difference between heat input and output)
$$W = J(Q_H - Q_L)$$

KEY TERMS

absolute scale (p. **82**)
absolute zero (p. **82**)
British thermal unit (p. **84**)
calorie (p. **83**)
Celsius scale (p. **80**)
conduction (p. **86**)
convection (p. **87**)
entropy (p. **96**)

external energy (p. **82**)
Fahrenheit scale (p. **80**)
first law of thermodynamics
  (p. **95**)
heat (p. **83**)
internal energy (p. **82**)
Kelvin scale (p. **82**)
kilocalorie (p. **83**)

kinetic molecular theory (p. **76**)
latent heat of fusion (p. **90**)
latent heat of vaporization
  (p. **90**)
molecule (p. **77**)
phase change (p. **89**)
radiation (p. **87**)
relative humidity (p. **93**)

second law of thermodynamics
  (p. **96**)
specific heat (p. **85**)
temperature (p. **78**)

APPLYING THE CONCEPTS

Answers are located in appendix F.

1. The kinetic molecular theory explains the expansion of a solid material with increases of temperature as basically the result of
   a. individual molecules expanding.
   b. increased translational kinetic energy.
   c. molecules moving a little farther apart.
   d. heat taking up the spaces between molecules.

2. Using the absolute temperature scale, the freezing point of water is correctly written as
   a. 0 K.
   b. 0°K.
   c. 273 K.
   d. 273°K.

3. The metric unit of heat called a *calorie* is
   a. the specific heat of water.
   b. the energy needed to increase the temperature of 1 gram of water 1 degree Celsius.
   c. equivalent to a little over 4 joules of mechanical work.
   d. All of the above are correct.

4. Table 4.2 lists the specific heat of soil as 0.20 kcal/kgC° and the specific heat of water as 1.00 kcal/kgC°. This means that if 1 kg of soil and 1 kg of water each receive 1 kcal of energy, ideally
   a. the water will be warmer than the soil by 0.8°C.
   b. the soil will be 4°C warmer than the water.
   c. the water will be 4°C warmer than the soil.
   d. the water will warm by 1°C, and the soil will warm by 0.2°C.

5. The heat transfer that takes place by energy moving directly from molecule to molecule is called
   a. conduction.
   b. convection.
   c. radiation.
   d. None of the above is correct.

6. The heat transfer that does not require matter is
   a. conduction.
   b. convection.
   c. radiation.
   d. impossible, for matter is always required.

7. Styrofoam is a good insulating material because
   a. it is a plastic material that conducts heat poorly.
   b. it contains many tiny pockets of air.
   c. of the structure of the molecules making up the Styrofoam.
   d. it is not very dense.

8. The transfer of heat that takes place because of density difference in fluids is
   a. conduction.
   b. convection.
   c. radiation.
   d. None of the above is correct.

9. As a solid undergoes a phase change to a liquid state, it
   a. releases heat while remaining at a constant temperature.
   b. absorbs heat while remaining at a constant temperature.
   c. releases heat as the temperature decreases.
   d. absorbs heat as the temperature increases.

10. The condensation of water vapor actually
    a. warms the surroundings.
    b. cools the surroundings.
    c. sometimes warms and sometimes cools the surroundings, depending on the relative humidity at the time.
    d. neither warms nor cools the surroundings.

11. No water vapor is added to or removed from a sample of air that is cooling, so the relative humidity of this sample of air will
    a. remain the same.
    b. be lower.
    c. be higher.
    d. be higher or lower, depending on the extent of change.

12. Compared to cooler air, warm air can hold
    a. more water vapor.
    b. less water vapor.
    c. the same amount of water vapor.
    d. less water vapor, the amount depending on the humidity.

QUESTIONS FOR THOUGHT

1. What is temperature? What is heat?

2. Explain why most materials become less dense as their temperature is increased.

3. Would the tight packing of more insulation, such as glass wool, in an enclosed space increase or decrease the insulation value? Explain.

4. A true vacuum bottle has a double-walled, silvered bottle with the air removed from the space between the walls. Describe how this design keeps food hot or cold by dealing with conduction, convection, and radiation.

5. Why is cooler air found in low valleys on calm nights?

6. Why is air a good insulator?

7. A piece of metal feels cooler than a piece of wood at the same temperature. Explain why.

8. What is condensation? Explain, on a molecular level, how the condensation of water vapor on a bathroom mirror warms the bathroom.

9. Which Styrofoam cooler provides more cooling, one with 10 lb of ice at 0°C or one with 10 lb of ice water at 0°C? Explain your reasoning.

10. Explain why a glass filled with a cold beverage seems to sweat. Would you expect more sweating inside a house during the summer or during the winter? Explain.

11. Explain why a burn from 100°C steam is more severe than a burn from water at 100°C.

12. The relative humidity increases almost every evening after sunset. Explain how this is possible if no additional water vapor is added to or removed from the air.

FOR FURTHER ANALYSIS

1. Considering the criteria for determining if something is a solid, liquid, or a gas, what is table salt, which can be poured?
2. What are the significant similarities and differences between heat and temperature?
3. Gas and plasma are phases of matter, yet gas runs a car and plasma is part of your blood. Compare and contrast these terms and offer an explanation for the use of similar names.
4. Analyze the table of specific heats (table 4.2) and determine which metal would make an energy-efficient and practical pan, providing more cooking for less energy.
5. This chapter contains information about three types of passive solar home design. Develop criteria or standards of evaluation that would help someone decide which design is right for their local climate.
6. Could a heat pump move heat without the latent heat of vaporization? Explain.
7. Explore the assumptions on which the "heat death of the universe" idea is based. Propose and evaluate an alternative idea for the future of the universe.

INVITATION TO INQUIRY

Who Can Last Longest?

How can we be more energy efficient? Much of our household energy consumption goes into heating and cooling, and much is lost through walls and ceilings. This invitation is about the insulating properties of various materials and their arrangement.

The challenge of this invitation is to create an insulated container that can keep an ice cube from melting. Decide on a maximum size for the container, and then decide what materials to use. Consider how you will use the materials. For example, if you are using aluminum foil, should it be shiny side in or shiny side out? If you are using newspapers, should they be folded flat or crumpled loosely?

One ice cube should be left outside the container to use as a control. Find out how much longer your insulated ice cube will outlast the control.

PARALLEL EXERCISES

The exercises in groups A and B cover the same concepts. Solutions to group A exercises are located in appendix G.
Note: Neglect all frictional forces in all exercises.

Group A

1. The average human body temperature is 98.6°F. What is the equivalent temperature on the Celsius scale?
2. An electric current heats a 221 g copper wire from 20.0°C to 38.0°C. How much heat was generated by the current? (c_{copper} = 0.093 kcal/kgC°)
3. A bicycle and rider have a combined mass of 100.0 kg. How many calories of heat are generated in the brakes when the bicycle comes to a stop from a speed of 36.0 km/h?
4. A 15.53 kg bag of soil falls 5.50 m at a construction site. If all the energy is retained by the soil in the bag, how much will its temperature increase? (c_{soil} = 0.200 kcal/kgC°)
5. A 75.0 kg person consumes a small order of french fries (250.0 Cal) and wishes to "work off" the energy by climbing a 10.0 m stairway. How many vertical climbs are needed to use all the energy?
6. A 0.5 kg glass bowl (c_{glass} = 0.2 kcal/kgC°) and a 0.5 kg iron pan (c_{iron} = 0.11 kcal/kgC°) have a temperature of 68°F when placed in a freezer. How much heat will the freezer have to remove from each to cool them to 32°F?
7. A sample of silver at 20.0°C is warmed to 100.0°C when 896 cal is added. What is the mass of the silver? (c_{silver} = 0.056 kcal/kgC°)

Group B

1. The Fahrenheit temperature reading is 98° on a hot summer day. What is this reading on the Kelvin scale?
2. A 0.25 kg length of aluminum wire is warmed 10.0°C by an electric current. How much heat was generated by the current? ($c_{aluminum}$ = 0.22 kcal/kgC°)
3. A 1,000.0 kg car with a speed of 90.0 km/h brakes to a stop. How many calories of heat are generated by the brakes as a result?
4. A 1.0 kg metal head of a geology hammer strikes a solid rock with a velocity of 5.0 m/s. Assuming all the energy is retained by the hammer head, how much will its temperature increase? (c_{head} = 0.11 kcal/kgC°)
5. A 60.0 kg person will need to climb a 10.0 m stairway how many times to "work off" each excess Cal (kcal) consumed?
6. A 50.0 g silver spoon at 20.0°C is placed in a cup of coffee at 90.0°C. How much heat does the spoon absorb from the coffee to reach a temperature of 89.0°C?
7. If the silver spoon placed in the coffee in problem 6 causes it to cool 0.75°C, what is the mass of the coffee? (Assume c_{coffee} = 1.0 cal/gC°)

—continued

Group A

8. A 300.0 W immersion heater is used to heat 250.0 g of water from 10.0°C to 70.0°C. About how many minutes did this take?

9. A 100.0 g sample of metal is warmed 20.0°C when 60.0 cal is added. What is the specific heat of this metal?

10. How much heat is needed to change 250.0 g of water at 80.0°C to steam at 100.0°C?

11. A 100.0 g sample of water at 20.0°C is heated to steam at 125.0°C. How much heat was absorbed?

12. In an electric freezer, 400.0 g of water at 18.0°C is cooled, frozen, and the ice is chilled to −5.00°C. How much total heat was removed from the water?

Group B

8. How many minutes would be required for a 300.0 W immersion heater to heat 250.0 g of water from 20.0°C to 100.0°C?

9. A 200.0 g china serving bowl is warmed 65.0°C when it absorbs 2.6 kcal of heat from a serving of hot food. What is the specific heat of the china dish?

10. A 500.0 g pot of water at room temperature (20.0°C) is placed on a stove. How much heat is required to change this water to steam at 100.0°C?

11. Spent steam from an electric generating plant leaves the turbines at 120.0°C and is cooled to 90.0°C liquid water by water from a cooling tower in a heat exchanger. How much heat is removed by the cooling tower water for each kilogram of spent steam?

12. Lead is a soft, dense metal with a specific heat of 0.028 kcal/kgC°, a melting point of 328.0°C, and a heat of fusion of 5.5 kcal/kg. How much heat must be provided to melt a 250.0 kg sample of lead with a temperature of 20.0°C?

5

Wave Motions and Sound

Compared to the sounds you hear on a calm day in the woods, the sounds from a waterfall can carry up to a million times more energy.

CORE **CONCEPT**

Sound is transmitted as increased and decreased pressure waves that carry energy.

OUTLINE

Mechanical waves are longitudinal or transverse; sound waves are longitudinal (p. 107).

All sounds originate from vibrating matter (p. 108).

5.1 Forces and Elastic Materials
Forces and Vibrations
Describing Vibrations
5.2 Waves
Kinds of Waves
Waves in Air
Hearing Waves in Air
5.3 Describing Waves
A Closer Look: Hearing Problems
5.4 Sound Waves
Velocity of Sound in Air
Refraction and Reflection
Interference

5.5 Energy and Sound
Loudness
Resonance
5.6 Sources of Sounds
Vibrating Strings
People Behind the Science: *Johann Christian Doppler*
Sounds from Moving Sources

An external force that matches the natural frequency results in resonance (p. 116).

The Doppler effect is an apparent change of pitch brought about from motion (p. 119).

Life Science

▷ The brain perceives sounds based on signals from the ear (Ch. 24).

OVERVIEW

Sometimes you can feel the floor of a building shake for a moment when something heavy is dropped. You can also feel prolonged vibrations in the ground when a nearby train moves by. The floor of a building and the ground are solids that transmit vibrations from a disturbance. Vibrations are common in most solids because the solids are elastic, having a tendency to rebound, or snap back, after a force or an impact deforms them. Usually you cannot see the vibrations in a floor or the ground, but you sense they are there because you can feel them.

There are many examples of vibrations that you can see. You can see the rapid blur of a vibrating guitar string (figure 5.1). You can see the vibrating up-and-down movement of a bounced-upon diving board. Both the vibrating guitar string and the diving board set up a vibrating motion of air that you identify as a sound. You cannot see the vibrating motion of the air, but you sense it is there because you hear sounds.

There are many kinds of vibrations that you cannot see but can sense. Heat, as you have learned, is associated with molecular vibrations that are too rapid and too tiny for your senses to detect other than as an increase in temperature. Other invisible vibrations include electrons that vibrate, generating spreading electromagnetic radio waves or visible light. Thus, vibrations take place as an observable motion of objects but are also involved in sound, heat, electricity, and light. The vibrations involved in all these phenomena are alike in many ways and all involve energy. Therefore, many topics of science are concerned with vibrational motion. In this chapter, you will learn about the nature of vibrations and how they produce waves in general. These concepts will be applied to sound in this chapter and to electricity and electromagnetic radiation in later chapters.

5.1 FORCES AND ELASTIC MATERIALS

If you drop a rubber ball, it bounces because it is capable of recovering its shape when it hits the floor. A ball of clay, on the other hand, does not recover its shape and remains a flattened blob on the floor. An *elastic* material is one that is capable of recovering its shape after a force deforms it. A rubber ball is elastic and a ball of clay is not elastic. You know a metal spring is elastic because you can stretch it or compress it and it always recovers its shape.

A direct relationship exists between the extent of stretching or compression of a spring and the amount of force applied to it. A large force stretches a spring a lot; a small force stretches it a little. As long as the applied force does not exceed the elastic limit of the spring, it will always return to its original shape when you remove the applied force. There are three important considerations about the applied force and the response of the spring:

1. The greater the applied force, the greater the compression or stretch of the spring from its original shape.
2. The spring appears to have an *internal restoring force*, which returns it to its original shape.
3. The farther the spring is pushed or pulled, the *stronger* the restoring force that returns the spring to its original shape.

Forces and Vibrations

A **vibration** is a back-and-forth motion that repeats itself. Such a motion is not restricted to any particular direction and can be in many different directions at the same time. Almost any solid can be made to vibrate if it is elastic. To see how forces are involved in vibrations, consider the spring and mass in figure 5.2. The spring and mass are arranged so that the mass can freely move back and forth on a frictionless surface. When the mass has not been disturbed, it is at rest at an *equilibrium position* (figure 5.2A). At the equilibrium position, the spring is not compressed or stretched, so it applies no force on the mass. If, however, the mass is pulled to the right (figure 5.2B), the spring is stretched and applies a restoring force on the mass toward the left. The farther the mass is displaced, the greater the stretch of the spring and thus the greater the restoring force. The restoring force is proportional to the displacement and is in the opposite direction of the applied force.

If the mass is now released, the restoring force is the only force acting (horizontally) on the mass, so it accelerates back toward the equilibrium position. This force will continuously decrease until the moving mass arrives back at the equilibrium position, where the force is zero. The mass will have a maximum

FIGURE 5.1 Vibrations are common in many elastic materials, and you can see and hear the results of many in your surroundings. Other vibrations in your surroundings, such as those involved in heat, electricity, and light, are invisible to the senses.

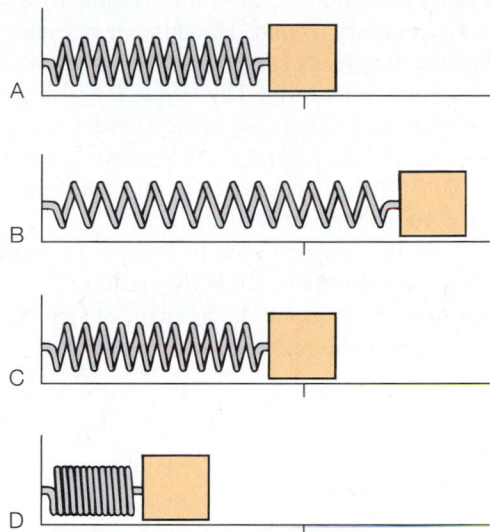

FIGURE 5.2 A mass on a frictionless surface is at rest at an equilibrium position (*A*) when undisturbed. When the spring is stretched (*B*) or compressed (*D*), then released (*C*), the mass vibrates back and forth because restoring forces pull opposite to and proportional to the displacement.

velocity when it arrives, however, so it overshoots the equilibrium position and continues moving to the left (figure 5.2C). As it moves to the left of the equilibrium position, it compresses the spring, which exerts an increasing force on the mass. The moving mass comes to a temporary halt (figure 5.2D), but now the restoring force again starts it moving back toward the equilibrium position. The whole process repeats itself again and again as the mass moves back and forth over the same path.

The periodic vibration, or oscillation, of the mass is similar to many vibrational motions found in nature called *simple harmonic motion.* Simple harmonic motion is defined as the vibratory motion that occurs when there is a restoring force opposite to and proportional to a displacement.

The vibrating mass and spring system will continue to vibrate for a while, slowly decreasing with time until the vibrations stop completely. The slowing and stopping is due to air resistance and internal friction. If these could be eliminated or compensated for with additional energy, the mass would continue to vibrate with a repeating, or *periodic,* motion.

Describing Vibrations

A vibrating mass is described by measuring several variables (figure 5.3). The extent of displacement from the equilibrium position is called the **amplitude.** A vibration that has a mass displaced a greater distance from equilibrium thus has a greater amplitude than a vibration with less displacement.

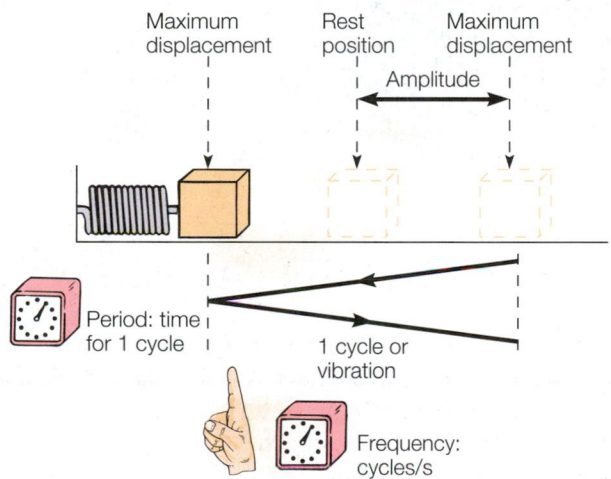

FIGURE 5.3 A vibrating mass attached to a spring is displaced from rest, or equilibrium position, and then released. The maximum displacement is called the *amplitude* of the vibration. A cycle is one complete vibration. The period is the time required for one complete cycle. The frequency is a count of how many cycles it completes in 1 s.

A complete vibration is called a **cycle.** A cycle is the movement from some point, say the far left, all the way to the far right, and back to the same point again, the far left in this example. The **period** (*T*) is simply the time required to complete one cycle. For example, suppose 0.1 s is required for an object to move through one complete cycle, to complete the back-and-forth motion from one point, then back to that point. The period of this vibration is 0.1 s.

Sometimes it is useful to know how frequently a vibration completes a cycle every second. The number of cycles per second is called the **frequency** (f). For example, a vibrating object moves through 10 cycles in 1 s. The frequency of this vibration is 10 cycles per second. Frequency is measured in a unit called a **hertz** (Hz). The unit for a hertz is 1/s because a cycle does not have dimensions. Thus, a frequency of 10 cycles per second is referred to as 10 hertz or 10 1/s.

The period and frequency are two ways of describing the time involved in a vibration. Since the period (T) is the total time involved in one cycle and the frequency (f) is the number of cycles per second, the relationship is

$$T = \frac{1}{f}$$

equation 5.1

or

$$f = \frac{1}{T}$$

equation 5.2

EXAMPLE 5.1   *(Optional)*

A vibrating system has a period of 0.5 s. What is the frequency in Hz?

SOLUTION

$T = 0.5 \text{ s}$
$f = ?$

$$f = \frac{1}{T}$$
$$= \frac{1}{0.5 \text{ s}}$$
$$= \frac{1}{0.5}\frac{1}{\text{s}}$$
$$= 2\frac{1}{\text{s}}$$
$$= \boxed{2 \text{ Hz}}$$

You can obtain a graph of a vibrating object, which makes it easier to measure the amplitude, period, and frequency. If a pen is fixed to a vibrating mass and a paper is moved beneath it at a steady rate, it will draw a curve, as shown in figure 5.4. The greater the amplitude of the vibrating mass, the greater the height of this curve. The greater the frequency, the closer together the peaks and valleys. Note the shape of this curve. This shape is characteristic of simple harmonic motion and is called a *sinusoidal*, or sine, graph. It is so named because it is the same shape as a graph of the sine function in trigonometry.

5.2 WAVES

A vibration can be a repeating, or *periodic,* type of motion that disturbs the surroundings. A *pulse* is a disturbance of a single event of short duration. Both pulses and periodic vibrations can create a physical *wave* in the surroundings. A wave is a disturbance that moves through a medium such as a solid or the air. A heavy object dropped on the floor, for example, makes a pulse that sends a mechanical wave that you feel. It might also make a sound wave in the air that you hear. In either case, the medium that transported a wave (solid floor or air) returns to its normal state after the wave has passed. The medium does not travel from place to place, the wave does. Two major considerations about a wave are that (1) a wave is a traveling disturbance and (2) a wave transports energy.

You can observe waves when you drop a rock into a still pool of water. The rock pushes the water into a circular mound as it enters the water. Since it is forcing the water through a distance, it is doing work to make the mound. The mound starts to move out in all directions, in a circle, leaving a depression behind. Water moves into the depression, and a circular wave—mound and depression—moves from the place of disturbance outward (figure 5.5). Any floating object in the path of the wave, such as a leaf, exhibits an up-and-down motion as the mound and

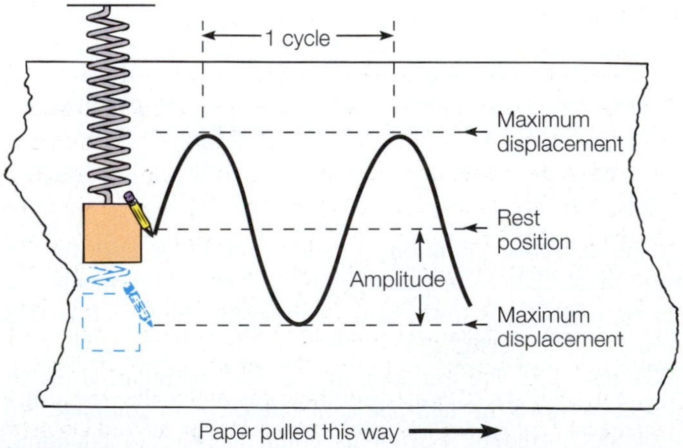

FIGURE 5.4   A graph of simple harmonic motion is described by a sinusoidal curve.

FIGURE 5.5   A water wave moves across the surface. How do you know for sure that it is energy and not water that is moving across the surface?

depression of the wave pass. But the leaf merely bobs up and down and after the wave has passed, it is much in the same place as before the wave. Thus, it was the disturbance that traveled across the water, not the water itself. If the wave reaches a leaf floating near the edge of the water, it may push the leaf up and out of the water, doing work on the leaf. Thus, the wave is a moving disturbance that transfers energy from one place to another.

Kinds of Waves

If you could see the motion of an individual water molecule near the surface as a water wave passed, you would see it trace out a circular path as it moves up and over, down and back. This circular motion is characteristic of the motion of a particle reacting to a water wave disturbance. There are other kinds of waves, and each involves particles in a characteristic motion.

A **longitudinal wave** is a disturbance that causes particles to move closer together or farther apart in the same direction that the wave is moving. If you attach one end of a coiled spring to a wall and pull it tight, you will make longitudinal waves in the spring if you grasp the spring and then move your hand back and forth parallel to the spring. Each time you move your hand toward the length of the spring, a pulse of closer-together coils will move across the spring (figure 5.6A). Each time you pull your hand back, a pulse of farther-apart coils will move across the spring. The coils move back and forth in the same direction that the wave is moving, which is the characteristic movement in reaction to a longitudinal wave.

You will make a different kind of wave in the stretched spring if you now move your hand up and down perpendicular to the length of the spring. This creates a **transverse wave.** A transverse wave is a disturbance that causes motion perpendicular to the direction that the wave is moving. Particles

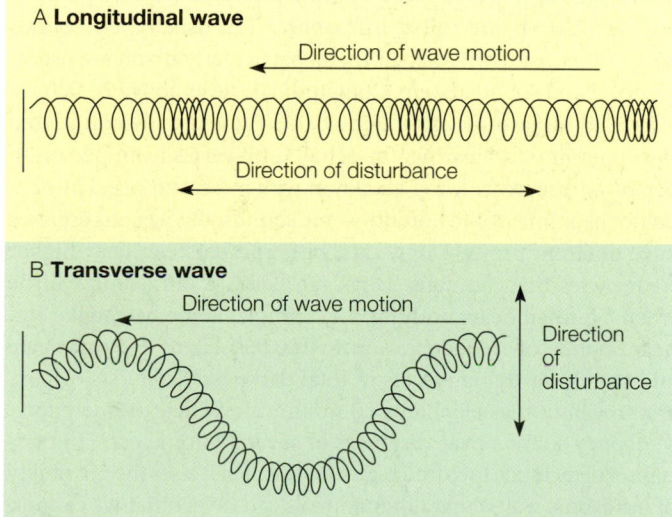

FIGURE 5.6 (A) Longitudinal waves are created in a spring when the free end is moved back and forth parallel to the spring. (B) Transverse waves are created in a spring when the free end is moved up and down.

CONCEPTS APPLIED

Making Waves

Obtain a Slinky or a long coiled spring and stretch it out on the floor. Have another person hold the opposite end stationary while you make waves move along the spring. Make longitudinal and transverse waves, observing how the disturbance moves in each case. If the spring is long enough, measure the distance, then time the movement of each type of wave. How fast were your waves?

responding to a transverse wave do not move closer together or farther apart in response to the disturbance; rather, they vibrate back and forth or up and down in a direction perpendicular to the direction of the wave motion (see figure 5.6B).

Whether you make mechanical longitudinal or transverse waves depends not only on the nature of the disturbance creating the waves but also on the nature of the medium. Mechanical transverse waves can move through a material only if there is some interaction, or attachment, between the molecules making up the medium. In a gas, for example, the molecules move about freely without attachments to one another. A pulse can cause these molecules to move closer together or farther apart, so a gas can carry a longitudinal wave. But if a gas molecule is caused to move up and then down, there is no reason for other molecules to do the same, since they are not attached. Thus, a gas will carry mechanical longitudinal waves but not mechanical transverse waves. Likewise a liquid will carry mechanical longitudinal waves but not mechanical transverse waves since the liquid molecules simply slide past one another. The surface of a liquid, however, is another story because of surface tension. A surface water wave is, in fact, a combination of longitudinal and transverse wave patterns that produce the circular motion of a disturbed particle. Solids can and do carry both mechanical longitudinal and transverse waves because of the strong attachments between the molecules.

Waves in Air

Waves that move through the air are longitudinal, so sound waves must be longitudinal waves. A familiar situation will be used to describe the nature of a longitudinal wave moving through air before considering sound specifically. The situation involves a small room with no open windows and two doors that open into the room. When you open one door into the room, the other door closes. Why does this happen? According to the kinetic molecular theory, the room contains many tiny, randomly moving gas molecules that make up the air. As you opened the door, it pushed on these gas molecules, creating a jammed-together zone of molecules immediately adjacent to the door. This jammed-together zone of air now has a greater density and pressure, which immediately spreads outward from

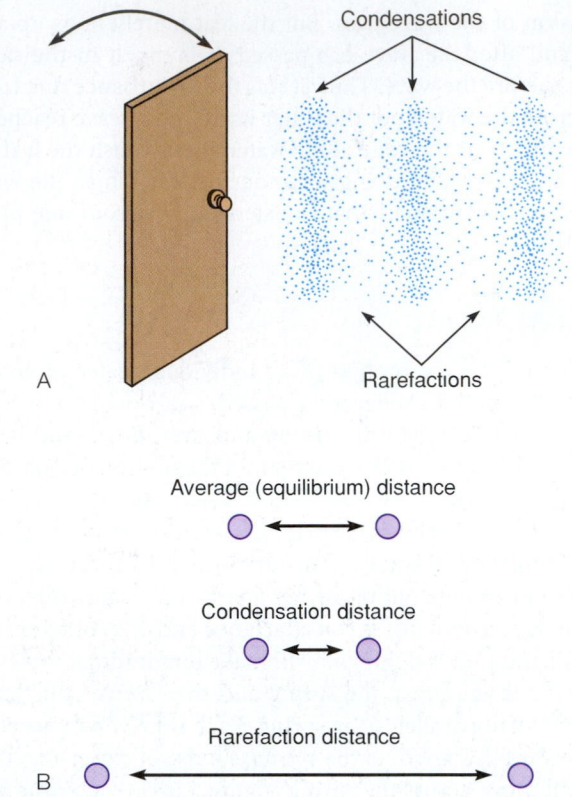

FIGURE 5.7 (A) Swinging the door inward produces pulses of increased density and pressure called *condensations*. Pulling the door outward produces pulses of decreased density and pressure called *rarefactions*. (B) In a condensation, the average distance between gas molecules is momentarily decreased as the pulse passes. In a rarefaction, the average distance is momentarily increased.

the door as a pulse. The disturbance is rapidly passed from molecule to molecule, and the pulse of compression spreads through the room. The pulse of greater density and increased pressure of air reached the door at the other side of the room, and the composite effect of the molecules impacting the door, that is, the increased pressure, caused it to close.

If the door at the other side of the room does not latch, you can probably cause it to open again by pulling on the first door quickly. By so doing, you send a pulse of thinned-out molecules of lowered density and pressure. The door you pulled quickly pushed some of the molecules out of the room. Other molecules quickly move into the region of less pressure, then back to their normal positions. The overall effect is the movement of a thinned-out pulse that travels through the room. When the pulse of slightly reduced pressure reaches the other door, molecules exerting their normal pressure on the other side of the door cause it to move. After a pulse has passed a particular place, the molecules are very soon homogeneously distributed again due to their rapid, random movement.

If you were to swing a door back and forth, it would be a vibrating object. As it vibrates back and forth, it would have a certain frequency in terms of the number of vibrations per second. As the vibrating door moves toward the room, it creates a pulse of jammed-together molecules called a **condensation** (or compression) that quickly moves throughout the room. As the vibrating door moves away from the room, a pulse of thinned-out molecules called a **rarefaction** quickly moves throughout the room. The vibrating door sends repeating pulses of condensation (increased density and pressure) and rarefaction (decreased density and pressure) through the room as it moves back and forth (figure 5.7). You know that the pulses transmit energy because they produce movement, or do work on, the other door. Individual molecules execute a harmonic motion about their equilibrium position and can do work on a movable object. Energy is thus transferred by this example of longitudinal waves.

Hearing Waves in Air

You cannot hear a vibrating door because the human ear normally hears sounds originating from vibrating objects with a frequency between 20 and 20,000 Hz. Longitudinal waves with frequencies less than 20 Hz are called **infrasonic.** You usually *feel* sounds below 20 Hz rather than hear them, particularly if you are listening to a good sound system. Longitudinal waves above 20,000 Hz are called **ultrasonic.** Although 20,000 Hz is usually considered the upper limit of hearing, the actual limit varies from person to person and becomes lower and lower with increasing age. Humans do not hear infrasonic nor ultrasonic sounds, but various animals have different limits. Dogs, cats, rats, and bats can hear higher frequencies than humans. Dogs can hear an ultrasonic whistle when a human hears nothing, for example. Some bats make and hear sounds of frequencies up to 100,000 Hz as they navigate and search for flying insects in total darkness. Scientists discovered recently that elephants communicate with extremely low-frequency sounds over distances of several kilometers. Humans cannot detect such low-frequency sounds. This raises the possibility of infrasonic waves that other animals can detect that we cannot.

A tuning fork that vibrates at 260 Hz makes longitudinal waves much like the swinging door, but these longitudinal waves are called *sound waves* because they are within the frequency

Connections . . .

The Human Ear

Box figure 5.1 shows the anatomy of the ear. The sound that arrives at the ear is first funneled by the external ear to the tympanum, also known as the eardrum. The cone-shaped nature of the external ear focuses sound on the tympanum and causes it to vibrate at the same frequency as the sound waves reaching it. Attached to the tympanum are three tiny bones known as the malleus (hammer), incus (anvil), and stapes (stirrup). The malleus is attached to the tympanum, the incus is attached to the malleus and stapes, and the stapes is attached to a small, membrane-covered opening called the oval window in a snail-shaped structure known as the cochlea. The vibration of the tympanum causes the tiny bones (malleus, incus, and stapes) to vibrate, and they, in turn, cause a corresponding vibration in the membrane of the oval window.

The cochlea of the ear is the structure that detects sound, and it consists of a snail-shaped set of fluid-filled tubes. When the oval window vibrates, the fluid in the cochlea begins to move, causing a membrane in the cochlea, called the basilar membrane, to vibrate. High-pitched, short-wavelength sounds cause the basilar membrane to vibrate at the base of the cochlea near the oval window. Low-pitched, long-wavelength sounds vibrate the basilar membrane far from the oval window. Loud sounds cause the basilar membrane to vibrate more vigorously than do faint sounds. Cells on this membrane depolarize when they are stimulated by its vibrations. Since they synapse with neurons, messages can be sent to the brain.

Because sounds of different wavelengths stimulate different portions of the cochlea, the brain is able to determine the pitch of a sound. Most sounds consist of a mixture of pitches that are heard. Louder sounds stimulate the membrane more forcefully, causing the sensory cells in the cochlea to send more nerve impulses per second. Thus, the brain is able to perceive the loudness of various sounds, as well as the pitch.

Associated with the cochlea are two fluid-filled chambers and a set of fluid-filled tubes called the semicircular canals. These structures are not involved in hearing but are involved in maintaining balance and posture. In the walls of these canals and chambers are cells similar to those found on the basilar membrane. These cells are stimulated by movements of the head and by the position of the head with respect to the force of gravity. The constantly changing position of the head results in sensory input that is important in maintaining balance.

Box Figure 5.1 A schematic sketch of the human ear.

range of human hearing. The prongs of a struck tuning fork vibrate, moving back and forth. This is more readily observed if the prongs of the fork are struck, then held against a sheet of paper or plunged into a beaker of water. In air, the vibrating prongs first move toward you, pushing the air molecules into a condensation of increased density and pressure. As the prongs then move back, a rarefaction of decreased density and pressure is produced. The alternation of increased and decreased pressure pulses moves from the vibrating tuning fork and spreads outward equally in all directions, much like the surface of a rapidly expanding balloon (figure 5.8). Your eardrum is forced in and out by the pulses it receives. It now vibrates with the same frequency as the tuning fork. The vibrations of the eardrum are transferred by three tiny bones to a fluid in a coiled chamber. Here, tiny hairs respond to the frequency and size of the disturbance, activating nerves that transmit the information to the brain. The brain interprets a frequency as a sound with a certain **pitch.** High-frequency sounds are interpreted as high-pitched musical notes, for example, and low-frequency sounds are interpreted as low-pitched musical notes. The brain then selects certain sounds from all you hear,

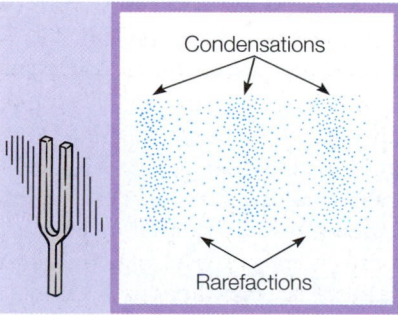

FIGURE 5.8 A vibrating tuning fork produces a series of condensations and rarefactions that move away from the tuning fork. The pulses of increased and decreased pressure reach your ear, vibrating the eardrum. The ear sends nerve signals to the brain about the vibrations, and the brain interprets the signals as sounds.

and you "tune" to certain ones, enabling you to listen to whatever sounds you want while ignoring the background noise, which is made up of all the other sounds.

5.3 DESCRIBING WAVES

A tuning fork vibrates with a certain frequency and amplitude, producing a longitudinal wave of alternating pulses of increased-pressure condensations and reduced-pressure rarefactions. A graph of the frequency and amplitude of the vibrations is shown in figure 5.9A, and a representation of the condensations and rarefactions is shown in figure 5.9B. The wave pattern can also be represented by a graph of the changing air pressure of the traveling sound wave, as shown in figure 5.9C. This graph can be used to define some interesting concepts associated with sound waves. Note the correspondence between the (1) amplitude, or displacement, of the vibrating prong, (2) the pulses of condensations and rarefactions, and (3) the changing air pressure. Note also the correspondence between the frequency of the vibrating prong and the frequency of the wave cycles.

Figure 5.10 shows the terms commonly associated with waves from a continuously vibrating source. The wave *crest* is the maximum disturbance from the undisturbed (rest) position. For a sound wave, this would represent the maximum increase of air pressure. The wave *trough* is the maximum disturbance in the opposite direction from the rest position. For a sound wave, this would represent the maximum decrease of air pressure. The *amplitude* of a wave is the displacement from rest to the crest *or* from rest to the trough. The time required for a wave to repeat itself is the *period* (*T*). To repeat itself means the time required to move through one full wave, such as from the crest of one wave to the crest of the next wave. This length in which the wave repeats itself is called the **wavelength** (the symbol is λ, which is the Greek letter lambda). Wavelength is measured in centimeters or meters just like any other length.

There is a relationship between the wavelength, period, and speed of a wave. The relationship is

$$v = \lambda f \qquad \text{equation 5.3}$$

which we will call the *wave equation*. This equation tells you that the velocity of a wave can be obtained from the product of the wavelength and the frequency. Note that it also tells you that the wavelength and frequency are inversely proportional at a given velocity.

EXAMPLE 5.2 *(Optional)*

A sound wave with a frequency of 260 Hz has a wavelength of 1.27 m. With what speed would you expect this sound wave to move?

SOLUTION

$$f = 260 \text{ Hz} \qquad v = \lambda f$$
$$\lambda = 1.27 \text{ m}$$
$$v = ? \qquad = (1.27 \text{ m}) \left(260 \frac{1}{s} \right)$$
$$= (1.27)(260) \text{ m} \times \frac{1}{s}$$
$$= \boxed{330 \frac{m}{s}}$$

EXAMPLE 5.3 *(Optional)*

In general, the human ear is most sensitive to sounds at 2,500 Hz. Assuming that sound moves at 330 m/s, what is the wavelength of sounds to which people are most sensitive? (Answer: 13 cm)

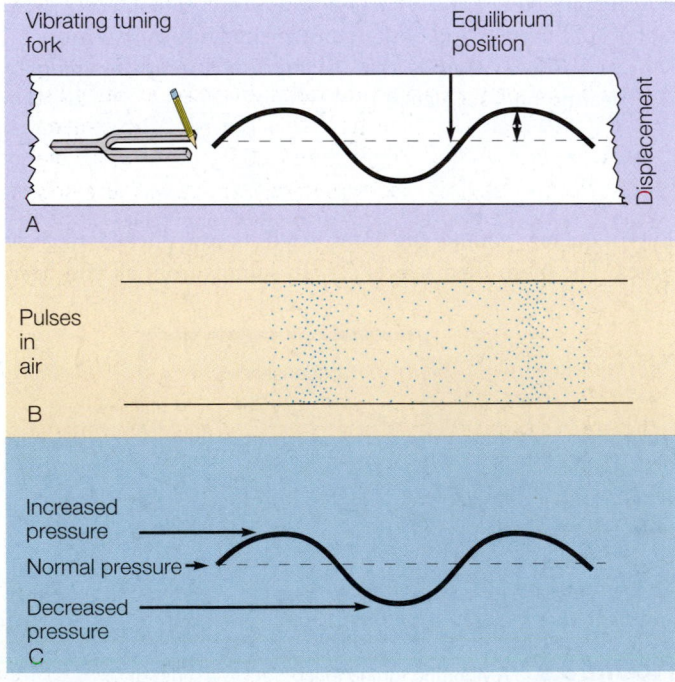

FIGURE 5.9 Compare the (*A*) back-and-forth vibrations of a tuning fork with (*B*) the resulting condensations and rarefactions that move through the air and (*C*) the resulting increases and decreases of air pressure on a surface that intercepts the condensations and rarefactions.

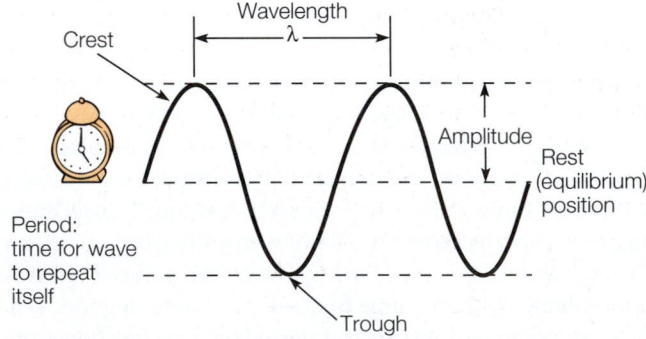

FIGURE 5.10 Here are some terms associated with periodic waves. The *wavelength* is the distance from a part of one wave to the same part in the next wave, such as from one crest to the next. The *amplitude* is the displacement from the rest position. The *period* is the time required for a wave to repeat itself, that is, the time for one complete wavelength to move past a given location.

Hearing Problems

There are three general areas of hearing problems that are related to your age and the intensity and duration of sounds falling on your ears. These are (1) middle ear infections of young children, (2) loss of ability to hear higher frequencies because of aging, and (3) ringing and other types of noise heard in the head or ear.

Middle ear infections are one of the most common illnesses of young children. The middle ear is a small chamber behind the eardrum that has three tiny bones that transfer sound vibrations from the eardrum to the inner ear (see box figure 5.1). The middle ear is connected to the throat by a small tube named the eustachian tube. This tube allows air pressure to be balanced behind the eardrum. A middle ear infection usually begins with a cold. Small children have short eustachian tubes, which become swollen from a cold, and this traps fluid in the middle ear. Fluid buildup causes pain and discomfort, as well as reduced hearing ability. This condition often clears on its own in several weeks or more. For severe and recurring cases, small drainage tubes are sometimes inserted through the eardrum to allow fluid drainage. These tubes eventually fall out, and the eardrum heals. Middle ear infections are less of a problem by the time a child reaches school age.

There is a normal loss of hearing because of aging. This loss is more pronounced for higher frequencies and is also greater in males than females. The normal loss of hearing begins in the early twenties, and then increases through the sixties and seventies. However, this process is accelerated by spending a lot of time in places with very loud sounds. Loud concerts, loud ear buds, and riding in a "boom car" are examples that will speed the loss of ability to hear higher frequencies. A "boom car" is one with loud music you can hear "booming" a half a block or more away.

Tinnitus is a sensation of sound, as a ringing or swishing that seems to originate in the ears or head and can only be heard by the person affected. There are several different causes, but the most common is nerve damage in the inner ear. Exposure to loud noises, explosions, firearms, and loud bands are common causes of tinnitus. Advancing age and certain medications can also cause tinnitus. Medications, such as aspirin, can be stopped to end the tinnitus, but there is no treatment for nerve damage to the inner ear.

5.4 SOUND WAVES

The transmission of a sound wave requires a medium, that is, a solid, liquid, or gas to carry the disturbance. Therefore, sound does not travel through the vacuum of outer space, since there is nothing to carry the vibrations from a source. The nature of the molecules making up a solid, liquid, or gas determines how well or how rapidly the substance will carry sound waves. The two variables are (1) the inertia of the molecules and (2) the strength of the interaction, if the molecules are attached to one another. Thus, hydrogen gas, with the least massive molecules with no interaction or attachments, will carry a sound wave at 1,284 m/s (4,213 ft/s) when the temperature is 0°C. More massive helium gas molecules have more inertia and carry a sound wave at only 965 m/s (3,166 ft/s) at the same temperature. A solid, however, has molecules that are strongly attached so vibrations are passed rapidly from molecule to molecule. Steel, for example, is highly elastic, and sound will move through a steel rail at 5,940 m/s (19,488 ft/s). Thus, there is a reason for the old saying, "Keep your ear to the ground," because sounds move through solids more rapidly than through a gas (table 5.1).

Velocity of Sound in Air

Most people have observed that sound takes some period of time to move through the air. If you watch a person hammering on a roof a block away, the sounds of the hammering are not in sync with what you see. Light travels so rapidly that you can consider what you see to be simultaneous with what is actually happening for all practical purposes. Sound, however travels much more slowly and the sounds arrive late in comparison to what you are seeing. This is dramatically illustrated by seeing a flash of lightning, then hearing thunder seconds later. Perhaps you know of a way to estimate the distance to a lightning flash by timing the interval between the flash and boom.

The air temperature influences how rapidly sound moves through the air. The gas molecules in warmer air have a greater kinetic energy than those of cooler air. The molecules of warmer air therefore transmit an impulse from molecule to molecule more rapidly. More precisely, the speed of a sound wave increases 0.60 m/s (2.0 ft/s) for *each* Celsius degree increase in temperature. In *dry* air at sea-level density (normal pressure) and 0°C (32°F), the velocity of sound is about 331 m/s (1,087 ft/s).

TABLE 5.1 **Speed of Sound in Various Materials**

Medium	m/s	ft/s
Carbon dioxide (0°C)	259	850
Dry air (0°C)	331	1,087
Helium (0°C)	965	3,166
Hydrogen (0°C)	1,284	4,213
Water (25°C)	1,497	4,911
Seawater (25°C)	1,530	5,023
Lead	1,960	6,430
Glass	5,100	16,732
Steel	5,940	19,488

Refraction and Reflection

When you drop a rock into a still pool of water, circular patterns of waves move out from the disturbance. These water waves are on a flat, two-dimensional surface. Sound waves, however, move in three-dimensional space like a rapidly expanding balloon. Sound waves are *spherical waves* that move outward from the source. Spherical waves of sound move as condensations and rarefactions from a continuously vibrating source at the center. If you identify the same part of each wave in the spherical waves, you have identified a *wave front*. For example, the crests of each condensation could be considered as a wave front. From one wave front to the next, therefore, identifies one complete wave or wavelength. At some distance from the source, a small part of a spherical wave front can be considered a *linear wave front* (figure 5.11).

Waves move within a homogeneous medium such as a gas or a solid at a fairly constant rate but gradually lose energy to friction. When a wave encounters a different condition (temperature, humidity, or nature of material), however, drastic changes may occur rapidly. The division between two physical conditions is called a *boundary*. Boundaries are usually encountered (1) between different materials or (2) between the same materials with different conditions. An example of a wave moving between different materials is a sound made in the next room that moves through the air to the wall and through the wall to the air in the room where you are. The boundaries are air-wall and wall-air. If you have ever been in a room with "thin walls," it is obvious that sound moved through the wall and air boundaries.

An example of sound waves moving through the same material with different conditions is found when a wave front moves through air of different temperatures. Since sound travels faster in warm air than in cold air, the wave front becomes bent. The bending of a wave front between boundaries is called *refraction*. Refraction changes the direction of travel of a wave front. Consider, for example, that on calm, clear nights, the air near Earth's surface is cooler than air farther above the surface. Air at rooftop height above the surface might be four or five degrees warmer under such ideal conditions. Sound will travel faster in the higher, warmer air than it will in the lower, cooler air close to the surface. A wave front will therefore become bent, or refracted, toward the ground on a cool night and you will be able to hear sounds from farther away than on warm nights (figure 5.12A). The opposite process occurs during the day as Earth's surface becomes warmer from sunlight (figure 5.12B). Wave fronts are refracted upward because part of the wave front travels faster in the warmer air near the surface. Thus, sound does not seem to carry as far in the summer as it does in the winter. What is actually happening is that during the summer, the wave fronts are refracted away from the ground before they travel very far.

When a wave front strikes a boundary that is parallel to the front the wave may be absorbed, transmitted, or undergo *reflection,* depending on the nature of the boundary medium, or the wave may be partly absorbed, partly transmitted, partly reflected, or any combination thereof. Some materials, such as hard, smooth surfaces, reflect sound waves more than they absorb them. Other materials, such as soft, ruffly curtains,

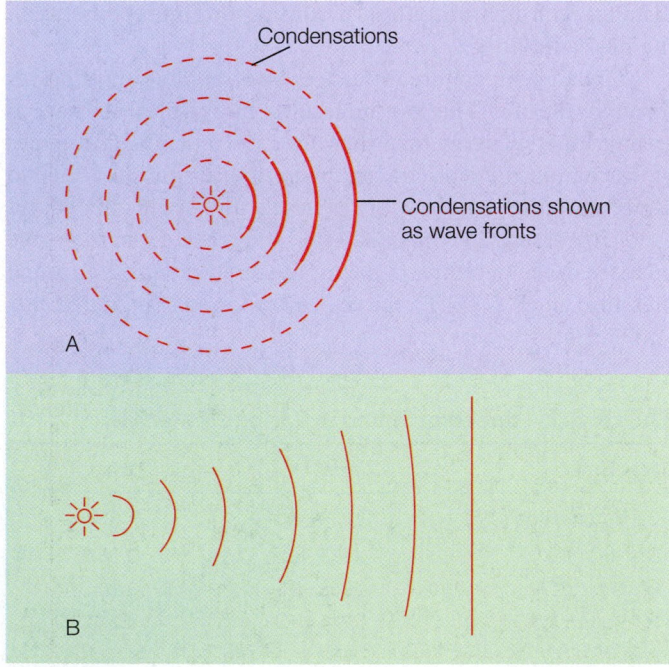

FIGURE 5.11 (*A*) Spherical waves move outward from a sounding source much as a rapidly expanding balloon. This two-dimensional sketch shows the repeating condensations as spherical wave fronts. (*B*) Some distance from the source, a spherical wave front is considered a linear, or plane, wave front.

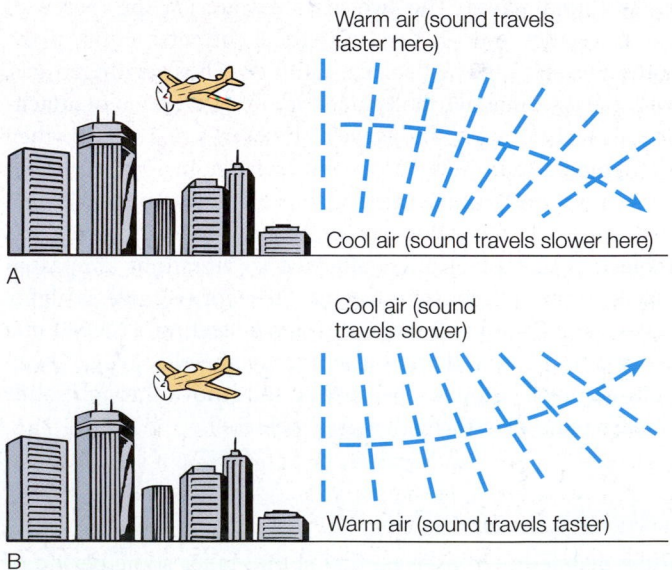

FIGURE 5.12 (*A*) Since sound travels faster in warmer air, a wave front becomes bent, or refracted, toward Earth's surface when the air is cooler near the surface. (*B*) When the air is warmer near the surface, a wave front is refracted upward, away from the surface.

FIGURE 5.13 This closed-circuit TV control room is acoustically treated by covering the walls with sound-absorbing baffles.

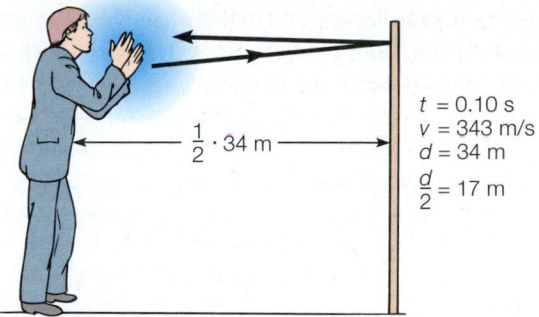

A **Echo**

$t = 0.10$ s
$v = 343$ m/s
$d = 34$ m
$\frac{d}{2} = 17$ m

$\frac{1}{2} \cdot 34$ m

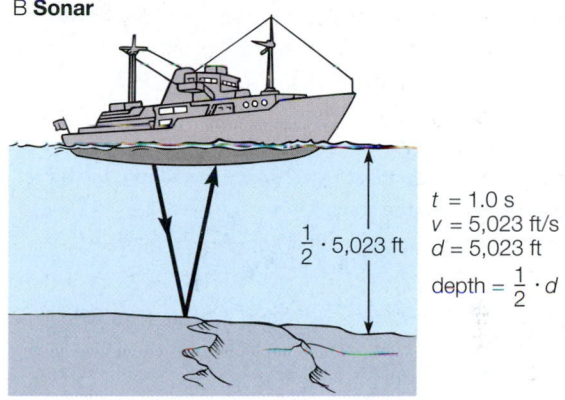

B **Sonar**

$t = 1.0$ s
$v = 5{,}023$ ft/s
$d = 5{,}023$ ft

depth $= \frac{1}{2} \cdot d$

$\frac{1}{2} \cdot 5{,}023$ ft

FIGURE 5.14 (A) At room temperature, sound travels at 343 m/s. In 0.10 s, sound would travel 34 m. Since the sound must travel to a surface and back in order for you to hear an echo, the distance to the surface is one-half the total distance. (B) Sonar measures a depth by measuring the elapsed time between an ultrasonic sound pulse and the echo. The depth is one-half the round trip.

absorb sound waves more than they reflect them. If you have ever been in a room with smooth, hard walls and with no curtains, carpets, or furniture, you know that sound waves may be reflected several times before they are finally absorbed.

Do you sing in the shower? Many people do because the tone is more pleasing than singing elsewhere. The walls of a shower are usually hard and smooth, reflecting sounds back and forth several times before they are absorbed. The continuation of many reflections causes a tone to gain in volume. Such mixing of reflected sounds with the original is called **reverberation.** Reverberation adds to the volume of a tone, and it is one of the factors that determines the acoustical qualities of a room, lecture hall, or auditorium. An open-air concert sounds flat without the reverberation of an auditorium and is usually enhanced electronically to make up for the lack of reflected sounds. Too much reverberation in a room or classroom is not good since the spoken word is not as sharp. Sound-absorbing materials are therefore used on the walls and floors where clear, distinct speech is important (figure 5.13). The carpet and drapes you see in a movie theater are not decorator items but are there to absorb sounds.

If a reflected sound arrives after 0.10 s, the human ear can distinguish the reflected sound from the original sound. A reflected sound that can be distinguished from the original is called an **echo.** Thus, a reflected sound that arrives before 0.10 s is perceived as an increase in volume and is called a reverberation, but a sound that arrives after 0.10 s is perceived as an echo.

Sound wave echoes are measured to determine the depth of water or to locate underwater objects by a *sonar* device. The word *sonar* is taken from *so*und *na*vigation *r*anging. The device generates an underwater ultrasonic sound pulse, then measures the elapsed time for the returning echo. Sound waves travel at about 1,531 m/s (5,023 ft/s) in seawater at 25°C (77°F). A 1 s lapse between the ping of the generated sound and the echo return would mean that the sound traveled 5,023 ft for the round trip. The bottom would be half this distance below the surface (figure 5.14).

EXAMPLE 5.4 (Optional)

The human ear can distinguish a reflected sound pulse from the original sound pulse if 0.10 s or more elapses between the two sounds. What is the minimum distance to a reflecting surface from which we can hear an echo (see figure 5.14A) if the speed of sound is 343 m/s?

SOLUTION

$$t = 0.10 \text{ s (minimum)} \qquad v = \frac{d}{t} \therefore d = vt$$

$$v = 343 \text{ m/s}$$

$$d = ? \qquad\qquad = \left(343 \frac{\text{m}}{\text{s}}\right)(0.10 \text{ s})$$

$$= (343)(0.10)\frac{\text{m}}{\text{s}} \times s$$

$$= 34.3 \frac{\text{m} \cdot \text{s}}{\text{s}}$$

$$= 34 \text{ m}$$

Since the sound pulse must travel from the source to the reflecting surface, then back to the source,

$$34 \text{ m} \times 1/2 = \boxed{17 \text{ m}}$$

The minimum distance to a reflecting surface from which we hear an echo when the air is at room temperature is therefore 17 m (about 56 ft).

EXAMPLE 5.5 (Optional)

An echo is heard exactly 1.00 s after a sound when the speed of sound is 1,147 ft/s. How many feet away is the reflecting surface? (Answer: 574 ft)

Interference

Waves interact with a boundary much as a particle would, reflecting or refracting because of the boundary. A moving ball, for example, will bounce from a surface at the same angle it strikes the surface, just as a wave does. A particle or a ball, however, can be in only one place at a time, but waves can be spread over a distance at the same time. You know this since many different people in different places can hear the same sound at the same time.

Another difference between waves and particles is that two or more waves can exist in the same place at the same time. When two patterns of waves meet, they pass through each other without refracting or reflecting. However, at the place where they meet, the waves interfere with each other, producing a *new* disturbance. This new disturbance has a different amplitude, which is the algebraic sum of the amplitudes of the two separate wave patterns. If the wave crests or wave troughs arrive at the same place at the same time, the two waves are said to be *in phase*. The result of two waves arriving in phase is a new disturbance with a crest and trough that has greater displacement than either of the two separate waves. This is called *constructive interference* (figure 5.15A). If the trough of one wave arrives at the same place and time as the crest of another

wave, the waves are completely *out of phase*. When two waves are completely out of phase, the crest of one wave (positive displacement) will cancel the trough of the other wave (negative displacement), and the result is zero total disturbance, or no wave. This is called *destructive interference* (figure 5.15B). If the two sets of wave patterns do not have the exact same amplitudes or wavelengths, they will be neither completely in phase nor completely out of phase. The result will be partly constructive or destructive interference, depending on the exact nature of the two wave patterns.

Suppose that two vibrating sources produce sounds that are in phase, equal in amplitude, and equal in frequency. The resulting sound will be increased in volume because of constructive interference. But suppose the two sources are slightly different in frequency, for example, 350 Hz and 352 Hz. You will hear a regularly spaced increase and decrease of sound known as **beats**. Beats occur because the two sound waves experience alternating constructive and destructive interferences (figure 5.16). The phase relationship changes because of the difference in frequency, as you can see in the illustration. These alternating constructive and destructive interference zones are moving from the source to the receiver, and the receiver hears the results as a rapidly rising and falling sound level. The beat frequency is the difference between the frequencies of the two sources. A 352 Hz source and 350 Hz source sounded together would result in a beat frequency of 2 Hz. Thus, as two frequencies are closer and closer together, fewer beats will be heard per second. You may be familiar with the phenomenon of beats if

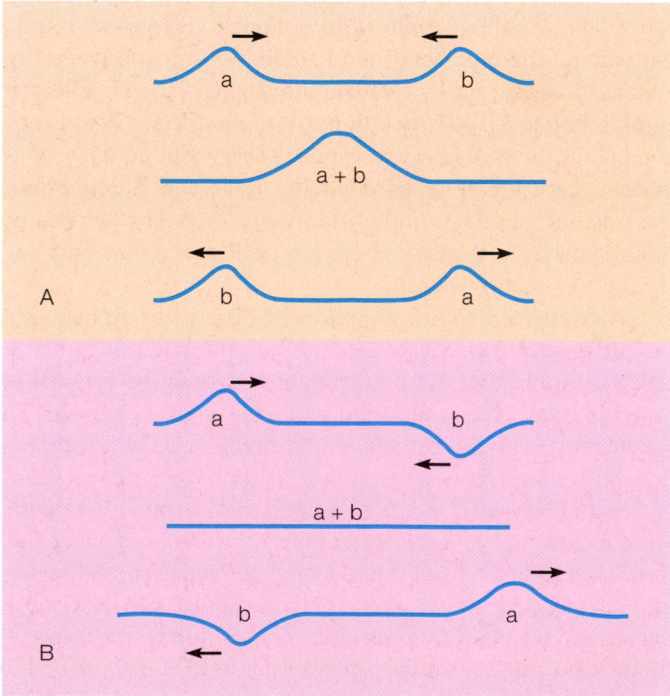

FIGURE 5.15 (A) Constructive interference occurs when two equal, in-phase waves meet. (B) Destructive interference occurs when two equal, out-of-phase waves meet. In both cases, the wave displacements are superimposed when they meet, but they then pass through one another and return to their original amplitudes.

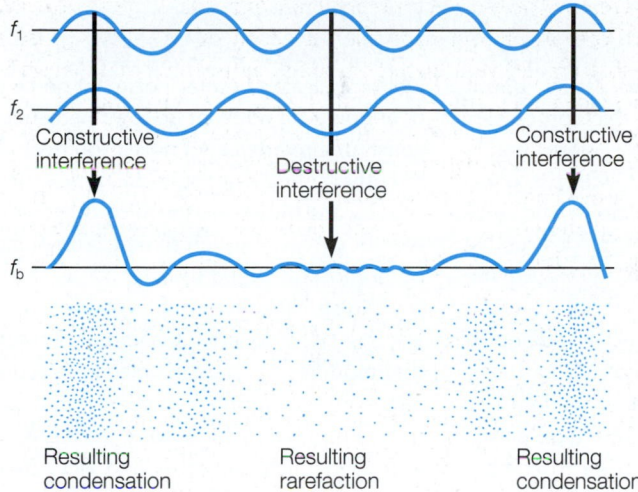

FIGURE 5.16 Two waves of equal amplitude but slightly different frequencies interfere destructively and constructively. The result is an alternation of loudness called a *beat*.

you have ever flown in an airplane with two engines. If one engine is running slightly faster than the other, you hear a slow beat. The beat frequency (f_b) is equal to the absolute difference in frequency of two interfering waves with slightly different frequencies, or

$$f_b = f_2 - f_1$$

equation 5.4

5.5 ENERGY AND SOUND

All waves involve the transportation of energy, including sound waves. The vibrating mass and spring in figure 5.2 vibrate with an amplitude that depends on how much work you did on the mass in moving it from its equilibrium position. More work on the mass results in a greater displacement and a greater amplitude of vibration. A vibrating object that is producing sound waves will produce more intense condensations and rarefactions if it has a greater amplitude. The intensity of a sound wave is a measure of the energy the sound wave is carrying (figure 5.17). **Intensity** is defined as the power (in watts) transmitted by a wave to a unit area (in square meters) that is perpendicular to the waves.

Loudness

The *loudness* of a sound is a subjective interpretation that varies from person to person. Loudness is also related to (1) the energy of a vibrating object, (2) the condition of the air the sound wave travels through, and (3) the distance between you and the vibrating source. Furthermore, doubling the amplitude of the vibrating source will quadruple the *intensity* of the resulting sound wave, but the sound will not be perceived as four times as loud. The relationship between perceived loudness and the intensity of a sound wave is not a linear relationship.

In fact, a sound that is perceived as twice as loud requires ten times the intensity, and quadrupling the loudness requires a one-hundredfold increase in intensity.

The human ear is very sensitive, capable of hearing sounds with intensities as low as 10^{-12} W/m², and is not made uncomfortable by sound until the intensity reaches about 1 W/m². The second intensity is a million million (10^{12}) times greater than the first. Within this range, the subjective interpretation of intensity seems to vary by powers of ten. This observation led to the development of the **decibel scale** to measure the intensity level. The scale is a ratio of the intensity level of a given sound to the threshold of hearing, which is defined as 10^{-12} W/m² at 1,000 Hz. In keeping with the power-of-ten subjective interpretations of intensity, a logarithmic scale is used rather than a linear scale. Originally, the scale was the logarithm of the ratio of the intensity level of a sound to the threshold of hearing. This definition set the zero point at the threshold of human hearing. The unit was named the *bel* in honor of Alexander Graham Bell. This unit was too large to be practical, so it was reduced by one-tenth and called a *decibel*. The intensity level of a sound is therefore measured in decibels (table 5.2). Compare

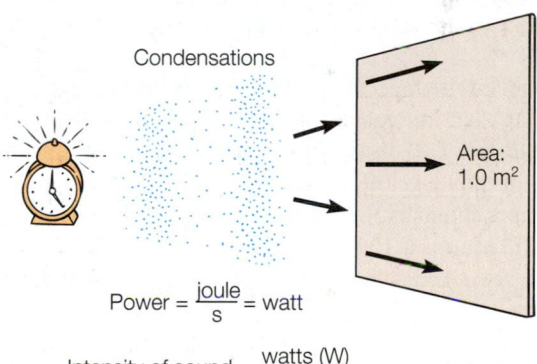

$$\text{Power} = \frac{\text{joule}}{\text{s}} = \text{watt}$$

$$\text{Intensity of sound} = \frac{\text{watts (W)}}{\text{area (m}^2)}$$

FIGURE 5.17 The intensity of a sound wave is the rate of energy transferred to an area perpendicular to the waves. Intensity is measured in watts per square meter, W/m².

TABLE 5.2 Comparison of Noise Levels in Decibels with Intensity

Example	Response	Decibels	Intensity W/m²
Least needed for hearing	Barely perceived	0	1×10^{-12}
Calm day in woods	Very, very quiet	10	1×10^{-11}
Whisper (15 ft)	Very quiet	20	1×10^{-10}
Library	Quiet	40	1×10^{-8}
Talking	Easy to hear	65	3×10^{-6}
Heavy street traffic	Conversation difficult	70	1×10^{-5}
Pneumatic drill (50 ft)	Very loud	95	3×10^{-3}
Jet plane (200 ft)	Discomfort	120	1

Connections . . .

Noise Pollution Solution?

Bells, jet planes, sirens, motorcycles, jack-hammers, and construction noises all contribute to the constant racket that has become commonplace in many areas. Such noise pollution is everywhere, and it is a rare place where you can escape the ongoing din. Earplugs help reduce the noise level, but they also block sounds that you want to hear, such as music and human voices. In addition, earplugs are more effective against high-frequency sounds than they are against lower ones such as aircraft engines or wind noise.

A better solution for noise pollution might be the relatively new "antinoise" technology that cancels sound waves before they reach your ears. A microphone detects background noise and transmits information to microprocessors about the noise waveform. The microprocessors then generate an "antinoise" signal that is 180 degrees out of phase with the background noise. When the noise and anti-noise meet, they undergo destructive interference and significantly reduce the final sound loudness.

The noise-canceling technology is available today as a consumer portable electronic device with a set of headphones, looking much like a portable tape player. The headphones can be used to combat steady, ongoing background noise such as you might hear from a spinning computer drive, a vacuum cleaner, or while inside the cabin of a jet plane. Some vendors claim their device cancels up to 40 percent of whirring air conditioner noise, 80 percent of ongoing car noise, or up to 95 percent of constant airplane cabin noise. You hear all the other sounds—people talking, warning sounds, and music—as the microprocessors are not yet fast enough to match anything but a constant sound source.

Noise-canceling microphones can also be used to limit background noise that muddles the human voice in teleconferencing or during cellular phone use from a number of high-noise places. With improved microprocessor and new digital applications, noise-canceling technology will help bring higher sound quality to voice-driven applications. It may also find extended practical use in helping to turn down the volume on our noisy world.

the decibel noise level of familiar sounds listed in table 5.2, and note that each increase of 10 on the decibel scale is matched by a *multiple* of 10 on the intensity level. For example, moving from a decibel level of 10 to a decibel level of 20 requires *ten times* more intensity. Likewise, moving from a decibel level of 20 to 40 requires a 100-fold increase in the intensity level. As you can see, the decibel scale is not a simple linear scale.

Resonance

You know that sound waves transmit energy when you hear a thunderclap rattle the windows. In fact, the sharp sounds from an explosion have been known not only to rattle but also break windows. The source of the energy is obvious when thunder-claps or explosions are involved. But sometimes energy transfer occurs through sound waves when it is not clear what is happening. A truck drives down the street, for example, and one window rattles but the others do not. A singer shatters a crystal water glass by singing a single note, but other objects remain undisturbed. A closer look at the nature of vibrating objects and the transfer of energy will explain these phenomena.

Almost any elastic object can be made to vibrate and will vibrate freely at a constant frequency after being sufficiently disturbed. Entertainers sometimes discover this fact and appear on late-night talk shows playing saws, wrenches, and other odd objects as musical instruments. All material objects have a **natural frequency** of vibration determined by the materials and shape of the objects. The natural frequencies of different wrenches enable an entertainer to use the suspended tools as if they were the bars of a glockenspiel.

If you have ever pumped a swing, you know that small forces can be applied at any frequency. If the frequency of the applied forces matches the natural frequency of the moving swing, there is a dramatic increase in amplitude. When the two frequencies match, energy is transferred very efficiently. This condition, when the frequency of an external force matches the natural frequency, is called **resonance.** The natural frequency of an object is thus referred to as the *resonant frequency*, that is, the frequency at which resonance occurs.

A silent tuning fork will resonate if a second tuning fork with the same frequency is struck and vibrates nearby (figure 5.18). You will hear the previously silent tuning fork sounding if you stop the vibrations of the struck fork by touching it. The waves of condensations and rarefactions produced by the struck tuning fork produce a regular series of impulses

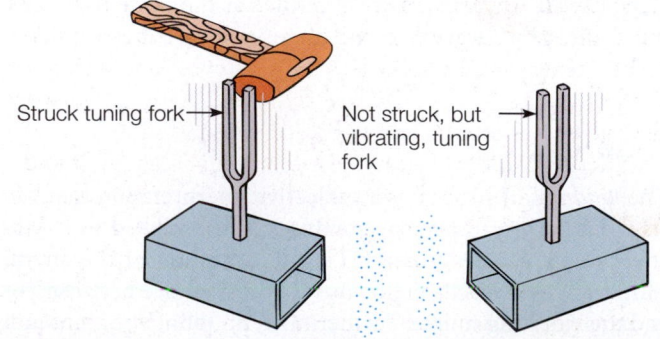

Struck tuning fork →

Not struck, but vibrating, tuning fork →

FIGURE 5.18   When the frequency of an applied force, including the force of a sound wave, matches the natural frequency of an object, energy is transferred very efficiently. The condition is called *resonance*.

Science and Society

CONCEPTS APPLIED

A Singing Glass

Did you ever hear a glass "sing" when the rim is rubbed? The trick to making the glass sing is to remove as much oil from your finger as possible. Then you lightly rub around and on the top of the glass rim at the correct speed. Without oil, your finger will imperceptibly catch on the glass as you rub the rim. With the appropriate pressure and speed, your catching finger might match the natural frequency of the glass. The resonate vibration will cause the glass to "sing" with a high-pitched note.

Laser Bug

Hold a fully inflated balloon lightly between your fingertips and talk. You will be able to feel the slight vibrations from your voice. Likewise, the sound waves from your voice will cause a nearby window to vibrate slightly. If a laser beam is bounced off the window, the reflection will be changed by the vibrations. The incoming laser beam is coherent; all the light has the same frequency and amplitude (see p. 169). The reflected beam, however, will have different frequencies and amplitudes from the window pane vibrating in and out. The changes can be detected by a receiver and converted into sound in a headphone.

You cannot see an infrared laser beam because infrared is outside the frequencies that humans can see. Any sound-sensitive target can be used by the laser bug, including a windowpane, inflated balloon, hanging picture, or the glass front of a china cabinet.

Questions to Discuss

1. Is it legal for someone to listen in on your private conversations?
2. Should the sale of technology such as the laser bug be permitted? What are the issues?

that match the natural frequency of the silent tuning fork. This illustrates that at resonance, relatively little energy is required to start vibrations.

A truck causing vibrations as it is driven past a building may cause one window to rattle while others do not. Vibrations caused by the truck have matched the natural frequency of this window but not the others. The window is undergoing resonance from the sound wave impulses that matched its natural frequency. It is also resonance that enables a singer to break a water glass. If the tone is at the resonant frequency of the glass, the resulting vibrations may be large enough to shatter it.

Resonance considerations are important in engineering. A large water pump, for example, was designed for a nuclear power plant. Vibrations from the electric motor matched the resonant frequency of the impeller blades, and they shattered after a short period of time. The blades were redesigned to have a different natural frequency when the problem was discovered. Resonance vibrations are particularly important in the design of buildings.

5.6 SOURCES OF SOUNDS

All sounds have a vibrating object as their source. The vibrations of the object send pulses or waves of condensations and rarefactions through the air. These sound waves have physical properties that can be measured, such as frequency and intensity. Subjectively, your response to frequency is to identify a certain pitch. A high-frequency sound is interpreted as a high-pitched sound, and a low-frequency sound is interpreted as a low-pitched sound. Likewise, a greater intensity is interpreted as increased loudness, but there is not a direct relationship between intensity and loudness as there is between frequency and pitch.

There are other subjective interpretations about sounds. Some sounds are bothersome and irritating to some people but go unnoticed by others. In general, sounds made by brief, irregular vibrations such as those made by a slamming door, dropped book, or sliding chair are called *noise*. Noise is characterized by sound waves with mixed frequencies and jumbled intensities (figure 5.19). On the other hand, there are sounds made by very regular, repeating vibrations such as those made by a tuning fork. A tuning fork produces a *pure tone* with a sinusoidal curved

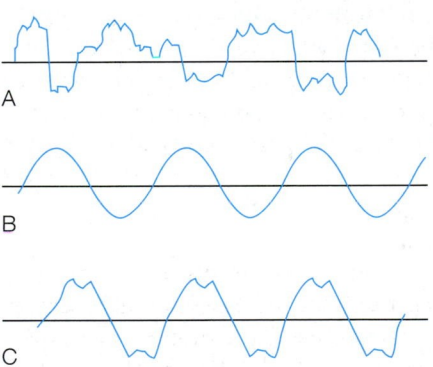

FIGURE 5.19   Different sounds that you hear include (A) noise, (B) pure tones, and (C) musical notes.

pressure variation and regular frequency. Yet a tuning fork produces a tone that most people interpret as bland. You would not call a tuning fork sound a musical note! Musical sounds from instruments have a certain frequency and loudness, as do noise and pure tones, but you can readily identify the source of the very same musical note made by two different instruments. You recognize it as a musical note, not noise and not a pure tone. You also recognize if the note was produced by a violin or a guitar. The difference is in the wave form of the sounds made by the two instruments, and the difference is called the *sound quality*. How does a musical instrument produce a sound of a characteristic quality? The answer may be found by looking at instruments that make use of vibrating strings.

Vibrating Strings

A stringed musical instrument, such as a guitar, has strings that are stretched between two fixed ends. When a string is plucked, waves of many different frequencies travel back and forth on the string, reflecting from the fixed ends. Many of these waves quickly fade away, but certain frequencies resonate, setting up patterns of waves. Before considering these resonant patterns in detail, keep in mind that (1) two or more waves can be in the same place at the same time, traveling through one another from opposite directions; (2) a confined wave will be reflected at a boundary, and the reflected wave will be inverted (a crest becomes a trough); and (3) reflected waves interfere with incoming waves of the same frequency to produce **standing waves.** Figure 5.20 is a graphic "snapshot" of what happens when reflected wave patterns meet incoming wave patterns. The incoming wave is shown as a solid line, and the reflected wave is shown as a dotted line. The result is (1) places of destructive interference, called *nodes,* which show no disturbance, and (2) loops of constructive interference, called *antinodes,* which take place where the crests and troughs of the two wave patterns produce a disturbance that rapidly alternates upward and downward. This pattern of alternating nodes and antinodes does not move along the string and is thus called a *standing wave.* Note that the standing wave for *one wavelength* will have a node at both ends and in the center, as well as two antinodes. Standing waves occur at the natural, or resonant, frequencies of the string, which are a consequence of the nature of the string, the string length, and the tension in the string. Since the standing waves are resonant vibrations, they continue as all other waves quickly fade away.

Since the two ends of the string are not free to move, the ends of the string will have nodes. The *longest* wave that can

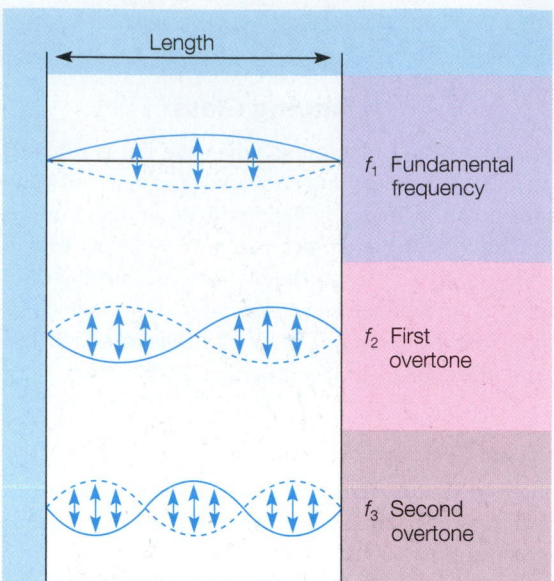

FIGURE 5.21 A stretched string of a given length has a number of possible resonant frequencies. The lowest frequency is the fundamental, f_1; the next higher frequencies, or overtones, shown are f_2 and f_3.

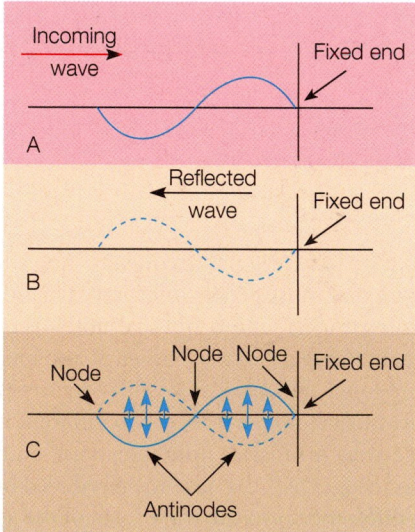

FIGURE 5.20 An incoming wave on a cord with a fixed end (*A*) meets a reflected wave (*B*) with the same amplitude and frequency, producing a standing wave (*C*). Note that a standing wave of one wavelength has three nodes and two antinodes.

make a standing wave on such a string has a wavelength (λ) that is twice the length (L) of the string. Since frequency (f) is inversely proportional to wavelength ($f = v/\lambda$ from equation 5.3), this longest wavelength has the lowest frequency possible, called the **fundamental frequency.** The fundamental frequency has one antinode, which means that the length of the string has one-half a wavelength. The fundamental frequency (f_1) determines the pitch of the *basic* musical note being sounded and is called the first harmonic. Other resonant frequencies occur at the same time, however, since other standing waves can also fit onto the string. A higher frequency of vibration (f_2) could fit two half-wavelengths between the two fixed nodes. An even higher frequency (f_3) could fit three half-wavelengths between the two fixed nodes (figure 5.21). Any whole number of halves of the wavelength will permit a standing wave to form. The frequencies ($f_2, f_3,$ etc.) of these wavelengths are called the *overtones,* or harmonics. It is the presence and strength of various overtones that give a musical note from a certain instrument its characteristic quality. The fundamental and the overtones add together to produce the characteristic *sound quality,* which is different for the same-pitched note produced by a violin and by a guitar (figure 5.22).

The vibrating string produces a waveform with overtones, so instruments that have vibrating strings are called *harmonic instruments.* Instruments that use an air column as a sound maker are also harmonic instruments. These include all the wind instruments such as the clarinet, flute, trombone, trumpet, pipe organ, and many others. The various wind instruments have different ways of making a column of air vibrate. In the flute, air vibrates as it moves over a sharp edge, while in the clarinet, saxophone, and other reed instruments, it vibrates through fluttering thin reeds. The air column in brass instru-

People Behind the Science

Johann Christian Doppler (1803–1853)

Johann Doppler was an Austrian physicist who discovered the Doppler effect, which relates the observed frequency of a wave to the relative motion of the source and the observer. The Doppler effect is readily observed in moving sound sources, producing a fall in pitch as the source passes the observer, but it is of most use in astronomy, where it is used to estimate the velocities and distances of distant bodies.

Doppler explained the effect that bears his name by pointing out that sound waves from a source moving toward an observer will reach the observer at a greater frequency than if the source is stationary, thus increasing the observed frequency and raising the pitch of the sound. Similarly, sound waves from a source moving away from the observer reach the observer more slowly, resulting in a decreased frequency and a lowering of pitch. In 1842, Doppler put forward this explanation and derived the observed frequency mathematically in Doppler's principle.

The first experimental test of Doppler's principle was made in 1845 at Utrecht in Holland. A locomotive was used to carry a group of trumpeters in an open carriage to and fro past some musicians able to sense the pitch of the notes being played. The variation of

pitch produced by the motion of the trumpeters verified Doppler's equations.

Doppler correctly suggested that his principle would apply to any wave motion and cited light as an example as well as sound. He believed that all stars emit white light and that differences in color are observed on Earth because the motion of stars affects the observed frequency of the light and hence its color. This idea was not universally true, as stars vary in their basic color. However, Armand Fizeau (1819–1896) pointed out in 1848 that shifts in the spectral lines of stars could be observed and ascribed to the Doppler effect and hence enable their motion to be determined. This idea was first applied in 1868 by William Huggins (1824–1910), who found that Sirius is moving away from the solar system by detecting a small redshift in its spectrum. With the linking of the velocity of a galaxy to its distance by Edwin Hubble (1889–1953) in 1929, it became possible to use the redshift to determine the distances of galaxies. Thus, the principle that Doppler discovered to explain an everyday and inconsequential effect in sound turned out to be of truly cosmological importance.

Source: Modified from the *Hutchinson Dictionary of Scientific Biography*. © RM, 2011. All rights reserved. Helicon Publishing is a division of RM.

---

ments, on the other hand, is vibrated by the tightly fluttering lips of the player.

The length of the air column determines the frequency, and woodwind instruments have holes in the side of a tube that are opened or closed to change the length of the air column. The resulting tone depends on the length of the air column and the resonate overtones.

Sounds from Moving Sources

When the source of a sound is stationary, equally spaced sound waves expand from a source in all directions. But if the sounding source starts moving, then successive sound waves become displaced in the direction of movement and this changes the pitch. For example, the siren of an approaching ambulance seems to change pitch when the ambulance passes you. The sound wave is "squashed" as the ambulance approaches you and you hear a higher-frequency siren than the people inside the ambulance. When the ambulance passes you, the sound waves are "stretched" and you hear a lower-frequency siren.

The overall effect of a higher pitch as a source approaches and then a lower pitch as it moves away is called the **Doppler effect.**

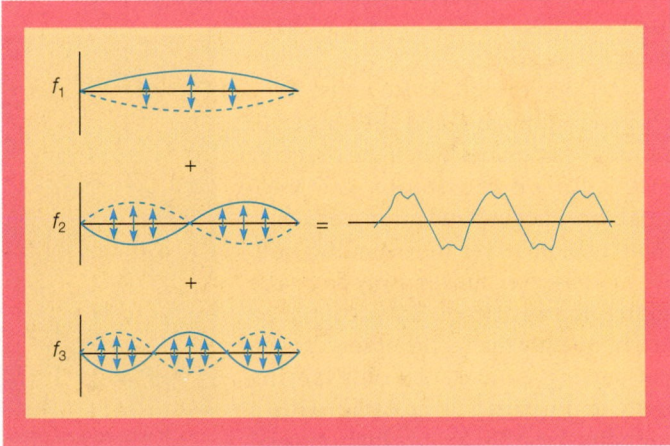

FIGURE 5.22   A combination of the fundamental and overtone frequencies produces a composite waveform with a characteristic sound quality.

Connections . . .

Doppler Radar

The Doppler effect was named after the Austrian scientist Christian Doppler, who first demonstrated the effect using sound waves back in 1842. The same principle applies to electromagnetic radiation as well as sound, but now the shifts are in frequency of the radiation. A lower frequency is observed when a source of light is moving away, and this is called a "redshift." Also, a "blueshift" toward a higher frequency occurs when a source of light is moving toward an observer. Radio waves will also experience such shifts of frequency, and weather radar that measures frequency changes as a result of motion is called *Doppler radar*.

Weather radar broadcasts short radio waves from an antenna. When directed at a storm, the waves are reflected back to the antenna by rain, snow, and hail. Reflected radar waves are electronically converted and displayed on a monitor, showing the location and intensity of precipitation. A Doppler radar also measures frequency shifts in the reflected radio waves. Waves from objects moving toward the antenna show a higher frequency, and waves from objects moving away from the antenna show a lower frequency. These shifts of frequency are measured, then displayed as the speed and direction of winds, moving raindrops, and other objects in the storm.

Weather forecasters can direct a Doppler radar machine to measure different elevations of a storm system. This shows highly accurate information that can be used to identify, for example, where and when a tornado might form, as well as the intensity of storm winds in a given area, and even provide an estimate of how much precipitation fell from the storm.

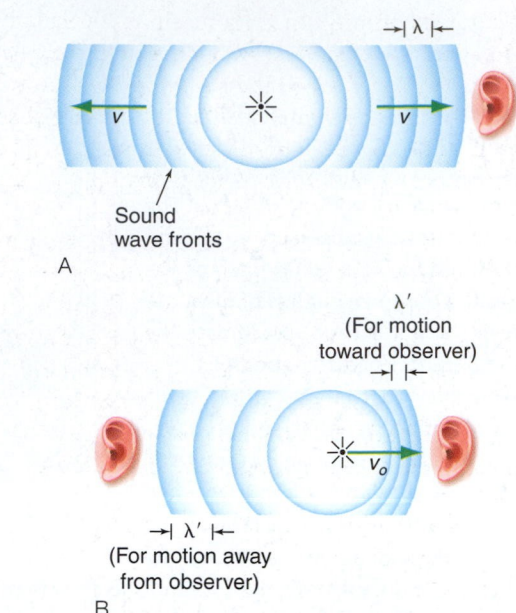

FIGURE 5.23 (*A*) Sound waves emitted by a stationary source and observed by a stationary observer. (*B*) Sound waves emitted by a source in motion toward the right. An observer on the right receives wavelengths that are shortened; an observer on the left receives wavelengths that are lengthened.

The Doppler effect is evident if you stand by a street and an approaching car sounds its horn as it drives by you. You will hear a higher-pitched horn as the car approaches, which shifts to a lower-pitched horn as the waves go by you. The driver of the car, however, will hear the continual, true pitch of the horn since the driver is moving with the source (figure 5.23).

A Doppler shift is also noted if the observer is moving and the source of sound is stationary. When the observer moves toward the source, the wave fronts are encountered more frequently than if the observer were standing still. As the observer moves away from the source, the wave fronts are encountered less frequently than if the observer were not moving. An observer on a moving train approaching a crossing with a sounding bell thus hears a high-pitched bell that shifts to a lower-pitched bell as the train whizzes by the crossing.

When an object moves through the air at the speed of sound, it keeps up

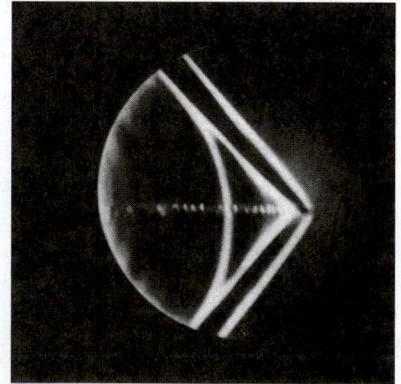

Shock waves are produced when a projectile travels faster than the speed of sound in air.

with its own sound waves. All the successive wave fronts pile up on one another, creating a large wave disturbance called a *shock wave*. The shock wave from a supersonic airplane is a cone-shaped wave of intense condensations trailing backward at an angle dependent on the speed of the aircraft. Wherever this cone of superimposed crests passes, a **sonic boom** occurs (figure 5.24). The many crests have been added together, each contributing to the pressure increase. The human ear cannot differentiate between such a pressure wave created by a supersonic aircraft and a pressure wave created by an explosion.

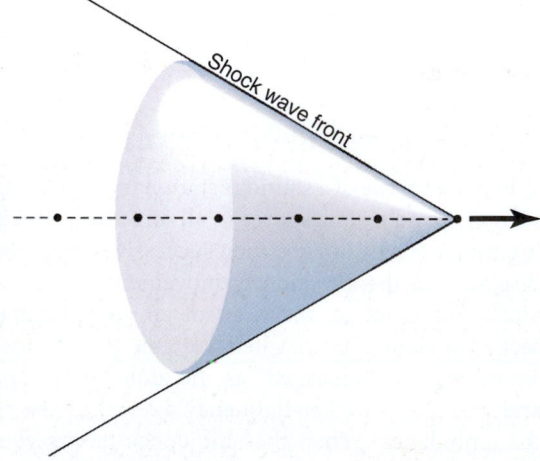

FIGURE 5.24 A sound source moves with velocity *greater* than the speed of sound in the medium. The envelope of spherical wave front forms the conical shock wave.

Does a sonic boom occur only when an airplane breaks the sound barrier? The answer is no, an airplane traveling at or faster than the speed of sound produces the shock wave continuously, and a sonic boom will be heard everywhere the plane drags its cone-shaped shock wave. Can you find evidence of shock waves associated with projections on the airplane pictured in figure 5.25?

The Austrian physicist Ernst Mach published a paper in 1877 laying out the principles of supersonics. He also came up with the idea of using a ratio of the velocity of an object to the velocity of sound. Today, this ratio is called the Mach number. A plane traveling at the speed of sound has a Mach number of 1, a plane traveling at twice the speed of sound has a Mach number of 2, and so on. Ernst Mach was also the first to describe what is happening to produce a sonic boom, and he observed the

FIGURE 5.25   A cloud sometimes forms just as a plane accelerates to break the sound barrier. Moist air is believed to form the cloud water droplets as air pressure drops behind the shock wave.

existence of a conical shock wave formed by a projectile as it approached the speed of sound.

SUMMARY

Elastic objects *vibrate,* or move back and forth, in a repeating motion when disturbed by some external force. They are able to do this because they have an *internal restoring force* that returns them to their original positions after being deformed by some external force. If the internal restoring force is opposite to and proportional to the deforming displacement, the vibration is called a *simple harmonic motion.* The extent of displacement is called the *amplitude,* and one complete back-and-forth motion is one *cycle.* The time required for one cycle is a *period.* The *frequency* is the number of cycles per second, and the unit of frequency is the *hertz.* A *graph* of the displacement as a function of time for a simple harmonic motion produces a *sinusoidal* graph.

Periodic, or repeating, vibrations or the *pulse* of a single disturbance can create *waves,* disturbances that carry energy through a medium. A wave that disturbs particles in a back-and-forth motion in the direction of the wave travel is called a *longitudinal wave.* A wave that disturbs particles in a motion perpendicular to the direction of wave travel is called a *transverse wave.* The nature of the medium and the nature of the disturbance determine the type of wave created.

Waves that move through the air are longitudinal and cause a back-and-forth motion of the molecules making up the air. A zone of molecules forced closer together produces a *condensation,* a pulse of increased density and pressure. A zone of reduced density and pressure is a *rarefaction.* A vibrating object produces condensations and rarefactions that expand outward from the source. If the frequency is between 20 Hz and 20,000 Hz, the human ear perceives the waves as *sound* of a certain *pitch.* High frequency is interpreted as high-pitched sound and low frequency as low-pitched sound.

A graph of pressure changes produced by condensations and rarefactions can be used to describe sound waves. The condensations produce *crests,* and the rarefactions produce *troughs.* The *amplitude* is the maximum change of pressure from the normal. The *wavelength* is the distance between any two successive places on a wave train,

such as the distance from one crest to the next crest. The *period* is the time required for a wave to repeat itself. The *velocity* of a wave is how quickly a wavelength passes. The *frequency* can be calculated from the *wave equation,* $v = \lambda f$.

Sound waves can move through any medium but not a vacuum. The velocity of sound in a medium depends on the molecular inertia and strength of interactions. Sound, therefore, travels most rapidly through a solid, then a liquid, then a gas. In air, sound has a greater velocity in warmer air than in cooler air because the molecules of air are moving about more rapidly, therefore transmitting a pulse more rapidly.

Sound waves are *reflected* or *refracted* from a *boundary,* which means a change in the transmitting medium. Reflected waves that are *in phase* with incoming waves undergo *constructive interference,* and waves that are *out of phase* undergo *destructive interference.* Two waves that are otherwise alike but with slightly different frequencies produce an alternating increasing and decreasing of loudness called *beats.*

The *energy* of a sound wave is called the wave *intensity,* which is measured in watts per square meter. The intensity of sound is expressed on the *decibel scale,* which relates it to changes in loudness as perceived by the human ear.

All elastic objects have *natural frequencies* of vibration that are determined by the materials they are made of and their shapes. When energy is transferred at the natural frequencies, there is a dramatic increase of amplitude called *resonance.* The natural frequencies are also called *resonant frequencies.*

Sounds are compared by pitch, loudness, and *quality.* The quality is determined by the instrument sounding the note. Each instrument has its own characteristic quality because of the resonant frequencies that it produces. The basic, or *fundamental,* frequency is the longest standing wave that it can make. The fundamental frequency determines the basic note being sounded, and other resonant frequencies, or standing waves called *overtones* or *harmonics,* combine with the fundamental to give the instrument its characteristic quality.

A moving source of sound or a moving observer experiences an apparent shift of frequency called the *Doppler effect*. If the source is moving as fast or faster than the speed of sound, the sound waves pile up into a *shock wave* called a *sonic boom*. A sonic boom sounds very much like the pressure wave from an explosion.

Summary of Equations

5.1

$$\text{period} = \frac{1}{\text{frequency}}$$

$$T = \frac{1}{f}$$

5.2

$$\text{frequency} = \frac{1}{\text{period}}$$

$$f = \frac{1}{T}$$

5.3

$$\text{velocity} = (\text{wavelength})(\text{frequency})$$

$$v = \lambda f$$

5.4

$$\text{beat frequency} = \text{one frequency} - \text{other frequency}$$

$$f_b = f_2 - f_1$$

KEY TERMS

amplitude (p. **105**)
beats (p. **114**)
condensation (p. **108**)
cycle (p. **105**)
decibel scale (p. **115**)
Doppler effect (p. **119**)
echo (p. **113**)

frequency (p. **106**)
fundamental frequency
   (p. **118**)
hertz (p. **106**)
infrasonic (p. **108**)
intensity (p. **115**)
longitudinal wave (p. **107**)

natural frequency (p. **116**)
period (p. **105**)
pitch (p. **109**)
rarefaction (p. **108**)
resonance (p. **116**)
reverberation (p. **113**)
sonic boom (p. **120**)

standing waves (p. **118**)
transverse wave (p. **107**)
ultrasonic (p. **108**)
vibration (p. **104**)
wavelength (p. **110**)

APPLYING THE CONCEPTS

Answers are located in appendix F.

1. The time required for a vibrating object to complete one full cycle is the
   a. frequency.
   b. amplitude.
   c. period.
   d. hertz.

2. The unit of cycles per second is called a
   a. hertz.
   b. lambda.
   c. wave.
   d. watt.

3. The period of a vibrating object is related to the frequency, since they are
   a. directly proportional.
   b. inversely proportional.
   c. frequently proportional.
   d. not proportional.

4. A longitudinal mechanical wave causes particles of a material to move
   a. back and forth in the same direction the wave is moving.
   b. perpendicular to the direction the wave is moving.
   c. in a circular motion in the direction the wave is moving.
   d. in a circular motion opposite the direction the wave is moving.

5. Transverse mechanical waves will move only through
   a. solids.
   b. liquids.
   c. gases.
   d. All of the above are correct.

6. Longitudinal mechanical waves will move only through
   a. solids.
   b. liquids.
   c. gases.
   d. All of the above are correct.

7. The characteristic of a wave that is responsible for what you interpret as pitch is the wave
   a. amplitude.
   b. shape.
   c. frequency.
   d. height.

8. The number of cycles that a vibrating tuning fork experiences each second is related to the resulting sound wave characteristic of
   a. frequency.
   b. amplitude.
   c. wave height.
   d. quality.

9. Sound waves travel faster in
   a. solids as compared to liquids.
   b. liquids as compared to gases.
   c. warm air as compared to cooler air.
   d. All of the above are correct.
10. Sound interference is necessary to produce the phenomenon known as
    a. resonance.
    b. decibels.
    c. beats.
    d. reverberation.
11. The efficient transfer of energy that takes place at a natural frequency is known as
    a. resonance.
    b. beats.
    c. the Doppler effect.
    d. reverberation.
12. An observer on the ground will hear a sonic boom from an airplane traveling faster than the speed of sound
    a. only when the plane breaks the sound barrier.
    b. as the plane is approaching.
    c. when the plane is directly overhead.
    d. after the plane has passed by.

QUESTIONS FOR THOUGHT

1. What is a wave?
2. Is it possible for a transverse wave to move through air? Explain.
3. A piano tuner hears three beats per second when a tuning fork and a note are sounded together and six beats per second after the string is tightened. What should the tuner do next, tighten or loosen the string? Explain.
4. Why do astronauts on the moon have to communicate by radio even when close to one another?
5. What is resonance?
6. Explain why sounds travel faster in warm air than in cool air.
7. Do all frequencies of sound travel with the same velocity? Explain your answer by using the wave equation.
8. What eventually happens to a sound wave traveling through the air?
9. What gives a musical note its characteristic quality?
10. Does a supersonic aircraft make a sonic boom only when it cracks the sound barrier? Explain.
11. What is an echo?
12. Why are fundamental frequencies and overtones also called resonant frequencies?

FOR FURTHER ANALYSIS

1. How would distant music sound if the speed of sound decreased with frequency?
2. What are the significant similarities and differences between longitudinal and transverse waves? Give examples of each.
3. Sometimes it is easier to hear someone speaking in a full room than in an empty room. Explain how this could happen.
4. Describe how you can use beats to tune a musical instrument.
5. Is sound actually destroyed in destructive interference?
6. Are vibrations the source of all sounds? Discuss if this is supported by observations or if it is an inference.
7. How can sound waves be waves of pressure changes if you can hear several people talking at the same time?
8. Why is it not a good idea for a large band to march in unison across a bridge?

INVITATION TO INQUIRY

Does a Noisy Noise Annoy You?

An old question-and-answer game played by children asks, "What annoys an oyster?"

The answer is, "A noisy noise annoys an oyster."

You could do an experiment to find out how much noise it takes to annoy an oyster, but you might have trouble maintaining live oysters, as well as measuring how annoyed they might become. So consider using different subjects, including humans. You could modify the question to, "What noise level affects how well we concentrate?"

If you choose to accept this invitation, start by determining how you are going to make the noise, how you can control different noise levels, and how you can measure the concentration level of people. A related question could be, "Does it help or hinder learning to play music while studying?"

PARALLEL EXERCISES

The exercises in groups A and B cover the same concepts. Solutions to group A exercises are located in appendix G.

Group A

1. A vibrating object produces periodic waves with a wavelength of 50 cm and a frequency of 10 Hz. How fast do these waves move away from the object?

2. The distance between the center of a condensation and the center of an adjacent rarefaction is 1.50 m. If the frequency is 112.0 Hz, what is the speed of the wave front?

3. Water waves are observed to pass under a bridge at a rate of one complete wave every 4.0 s. (a) What is the period of these waves? (b) What is the frequency?

4. A sound wave with a frequency of 260 Hz moves with a velocity of 330 m/s. What is the distance from one condensation to the next?

5. The following sound waves have what velocity?
   a. Middle C, or 256 Hz and 1.34 m λ.
   b. Note A, or 440.0 Hz and 78.0 cm λ.
   c. A siren at 750.0 Hz and λ of 45.7 cm.
   d. Note from a stereo at 2,500.0 Hz and λ of 13.7 cm.

6. You hear an echo from a cliff 4.80 s after you shout, "Hello." How many feet away is the cliff if the speed of sound is 1,100 ft/s?

7. During a thunderstorm, thunder was timed 4.63 s after lightning was seen. How many feet away was the lightning strike if the speed of sound is 1,140 ft/s?

8. If the velocity of a 440 Hz sound is 1,125 ft/s in the air and 5,020 ft/s in seawater, find the wavelength of this sound (a) in air, (b) in seawater.

Group B

1. A tuning fork vibrates 440.0 times a second, producing sound waves with a wavelength of 78.0 cm. What is the velocity of these waves?

2. The distance between the center of a condensation and the center of an adjacent rarefaction is 65.23 cm. If the frequency is 256.0 Hz, how fast are these waves moving?

3. A warning buoy is observed to rise every 5.0 s as crests of waves pass by it. (a) What is the period of these waves? (b) What is the frequency?

4. Sound from the siren of an emergency vehicle has a frequency of 750.0 Hz and moves with a velocity of 343.0 m/s. What is the distance from one condensation to the next?

5. The following sound waves have what velocity?
   a. 20.0 Hz, λ of 17.2 m
   b. 200.0 Hz, λ of 1.72 m
   c. 2,000.0 Hz, λ of 17.2 cm
   d. 20,000.0 Hz, λ of 1.72 cm

6. A ship at sea sounds a whistle blast, and an echo returns from the coastal land 10.0 s later. How many kilometers is it to the coastal land if sound travels at 337 m/s?

7. How many seconds will elapse between seeing lightning and hearing the thunder if the lightning strikes 1 mile (5,280 ft) away and sound travels at 1,151 ft/s?

8. A 600.0 Hz sound has a velocity of 1,087.0 ft/s in the air and a velocity of 4,920.0 ft/s in water. Find the wavelength of this sound (a) in the air, (b) in the water.

6

Electricity

A thunderstorm produces an interesting display of electrical discharge. Each bolt can carry over 150,000 amperes of current with a voltage of 100 million volts.

CORE CONCEPT

Electric and magnetic fields interact and can produce forces.

OUTLINE

OVERVIEW

The previous chapters have been concerned with *mechanical* concepts, explanations of the motion of objects that exert forces on one another. These concepts were used to explain straight-line motion, the motion of free fall, and the circular motion of objects on Earth as well as the circular motion of planets and satellites. The mechanical concepts were based on Newton's laws of motion and are sometimes referred to as Newtonian physics. The mechanical explanations were then extended into the submicroscopic world of matter through the kinetic molecular theory. The objects of motion were now particles, molecules that exert force on one another, and concepts associated with heat were interpreted as the motion of these particles. In a further extension of Newtonian concepts, mechanical explanations were given for concepts associated with sound, a mechanical disturbance that follows the laws of motion as it moves through the molecules of matter.

You might wonder, as did the scientists of the 1800s, if mechanical interpretations would also explain other natural phenomena such as electricity, chemical reactions, and light. A mechanical model would be very attractive since it already explained so many other facts of nature, and scientists have always looked for basic, unifying theories. Mechanical interpretations were tried, as electricity was considered as a moving fluid and light was considered as a mechanical wave moving through a material fluid. There were many unsolved puzzles with such a model, and gradually, it was recognized that electricity, light, and chemical reactions could not be explained by mechanical interpretations. Gradually, the point of view changed from a study of particles to a study of the properties of the *space* around the particles. In this chapter, you will learn about electric charge in terms of the space around particles. This model of electric charge, called the *field model*, will be used to develop concepts about electric current, the electric circuit, and electrical work and power. A relationship between electricity and the fascinating topic of magnetism is discussed next, including what magnetism is and how it is produced. The relationship is then used to explain the mechanical production of electricity (figure 6.1), how electricity is measured, and how electricity is used in everyday technological applications.

6.1 ELECTRIC CHARGE

It was a big mystery for thousands of years. No one could figure out why a rubbed piece of amber, which is fossilized tree resin, would attract small pieces of paper, thread, and hair. This unexplained attraction was called the "amber effect." Then about one hundred years ago, Joseph J. Thomson found the answer while experimenting with electric currents. From these experiments, Thomson concluded that negatively charged particles were present in all matter and in fact might be the stuff of which matter is made. The amber effect was traced to the movement of these particles, so they were called *electrons* after the Greek word for amber. The word *electricity* is also based on the Greek word for amber.

Today, we understand that the basic unit of matter is the *atom*, which is made up of electrons and other particles such as *protons* and *neutrons*. The atom is considered to have a dense center part called a *nucleus* that contains the closely situated protons and neutrons. The electrons move around the nucleus at some relatively greater distance (figure 6.2). For understanding electricity, you need only to consider the protons in the nucleus, the electrons that move around the nucleus, and the fact that electrons can be moved from an atom and caused to move to or from an object. Details on the nature of protons, neutrons, electrons, and models on how the atom is constructed will be considered in chapter 8.

Electrons and protons have a property called **electric charge.** Electrons have a *negative electric charge* and protons have a *positive electric charge*. The negative and positive description simply means that these two properties are opposite; it does not mean that one charge is better than the other. Charge is as fundamental to these subatomic particles as gravitational

FIGURE 6.1 The importance of electrical power seems obvious in a modern industrial society. What is not so obvious is the role of electricity in magnetism, light, and chemical change, and as the very basis for the structure of matter. As you will see, all matter is in fact electrical in nature.

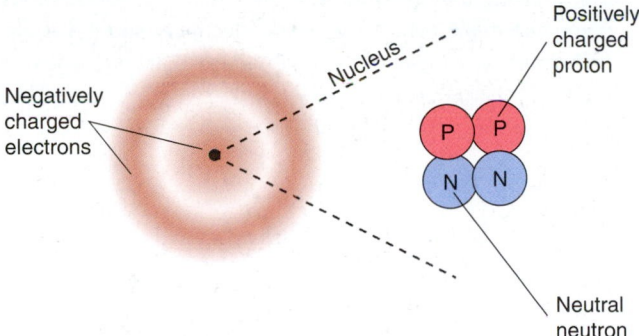

FIGURE 6.2 A very highly simplified model of an atom has most of the mass in a small, dense center called the *nucleus*. The nucleus has positively charged protons and neutral neutrons. Negatively charged electrons move around the nucleus at a much greater distance than is suggested by this simplified model. Ordinary atoms are neutral because there is balance between the number of positively charged protons and negatively charged electrons.

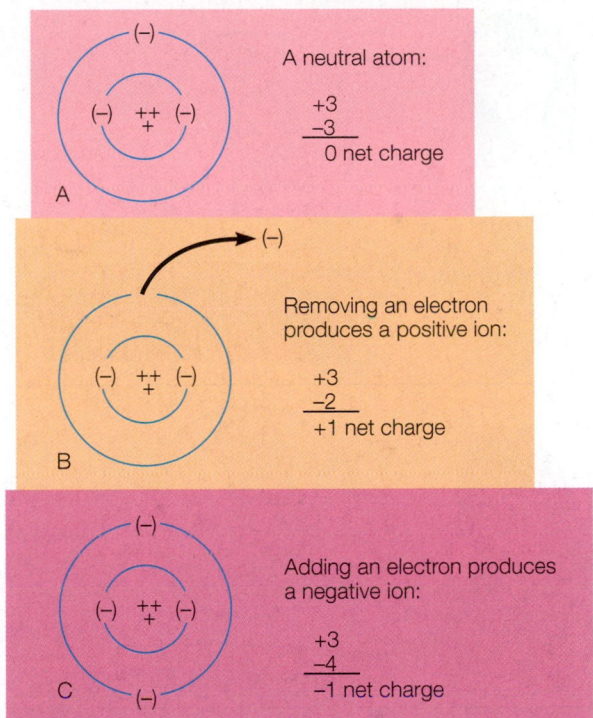

FIGURE 6.3 (A) A neutral atom has no net charge because the numbers of electrons and protons are balanced. (B) Removing an electron produces a net positive charge; the charged atom is called a positive ion. (C) The addition of an electron produces a net negative charge and a negative ion.

attraction is fundamental to masses. This means that you cannot separate gravity from a mass, and you cannot separate charge from an electron or a proton.

Electric charges interact to produce what is called the **electric force.** Like charges produce a repulsive electric force as positive repels positive and negative repels negative. Unlike charges produce an attractive electric force as positive and negative charges attract each other. You can remember how this happens with the simple rule of "*like charges repel and unlike charges attract.*"

Ordinary atoms are usually neutral because there is a balance between the number of positively charged protons and the number of negatively charged electrons. A number of different physical and chemical interactions can result in an atom gaining or losing electrons. In either case, the atom is said to be *ionized* and *ions* are produced as a result. An atom that is ionized by losing electrons results in a *positive ion* since

it has a net positive charge. An atom that is ionized by gaining electrons results in a *negative ion* since it has a net negative charge (figure 6.3).

Electrons can be moved from atom to atom to create ions. They can also be moved from one object to another by friction. Since electrons are negatively charged, an object that acquires an excess of electrons becomes a negatively charged body. The loss of electrons by another body results in a deficiency of electrons, which results in a positively charged object. Thus, *electric charges on objects result from the gain or loss of electrons.* Because the electric charge is confined to an object and is not moving, it is called an **electrostatic charge.** You probably call this charge *static electricity.* Static electricity is an accumulated electric charge at rest, that is, one that is not moving. When you comb your hair with a hard rubber comb, the comb becomes negatively charged because electrons are transferred from your hair to the comb, and your hair acquires a positive charge (figure 6.4). Both the negative charge on the comb from an excess of electrons and the positive charge on your hair from a deficiency of electrons are charges that are momentarily at rest, so they are electrostatic charges.

Once charged by friction, objects such as the rubber comb soon return to a neutral, or balanced, state by the movement of electrons. This happens more quickly on a humid day because water vapor assists with the movement of electrons to or from charged objects. Thus, static electricity is more noticeable on dry days than on humid ones.

A charged object can also exert a force of attraction on a second object that does not have a net charge. For example,

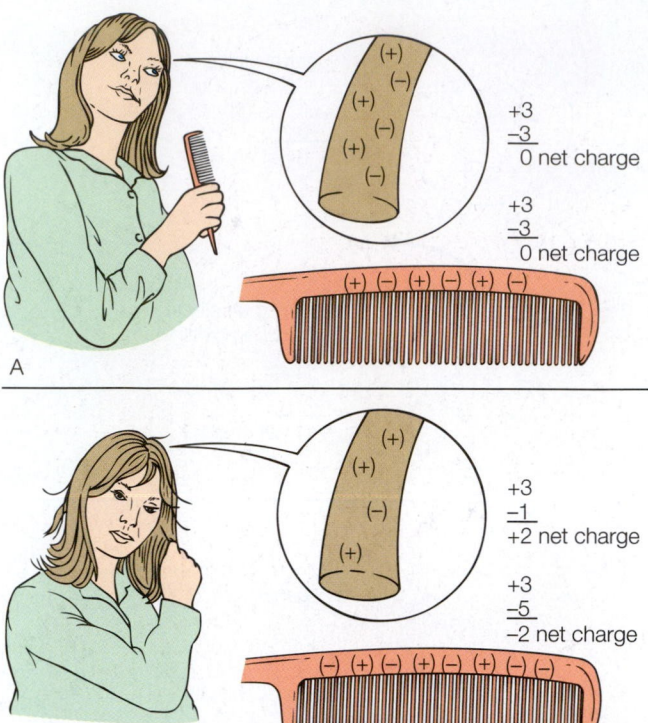

FIGURE 6.4 Arbitrary numbers of protons (+) and electrons (−) on a comb and in hair (A) before and (B) after combing. Combing transfers electrons from the hair to the comb by friction, resulting in a negative charge on the comb and a positive charge on the hair.

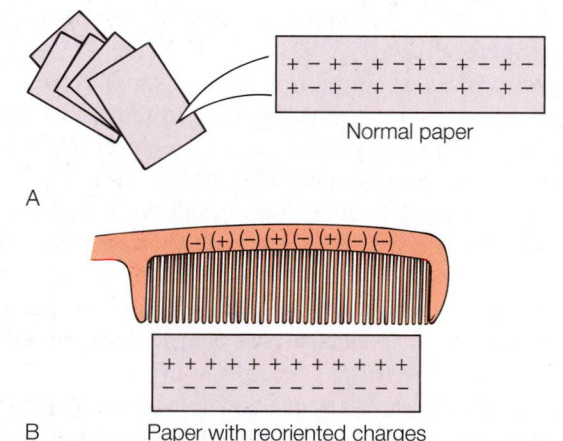

FIGURE 6.5 Charging by induction. The comb has become charged by friction, acquiring an excess of electrons. The paper (A) normally has a random distribution of (+) and (−) charges. (B) When the charged comb is held close to the paper, there is a reorientation of charges because of the repulsion of like charges. This leaves a net positive charge on the side close to the comb, and since unlike charges attract, the paper is attracted to the comb.

you can give your hard rubber comb a negative charge by combing your hair. Your charged comb will attract tiny pieces of paper that do not have a charge, pulling them to the comb. This happens because the negative charge on the comb repels electrons on the tiny bits of paper, giving the paper a positive charge on the side nearest the comb (figure 6.5). These unlike charges attract more than the like ones repel since the distance to the like ones is greater, and the paper is pulled to the comb.

Measuring Electric Charge

As you might have experienced, sometimes you receive a slight shock after walking across a carpet, and sometimes you are really zapped. You receive a greater shock when you have accumulated a greater electric charge. Since there is less electric charge at one time and more at another, it should be evident that charge occurs in different amounts, and these amounts can be measured. The magnitude of an electric charge is identified with the number of electrons that have been transferred onto or away from an object. The quantity of such a charge (q) is measured in a unit called a **coulomb** (C). The coulomb is a fundamental metric unit of measure like the meter, kilogram, and second.

Every electron has a charge of -1.60×10^{-19} C and every proton has a charge of $+1.60 \times 10^{-19}$ C. To accumulate a negative charge of 1 C, you would need to accumulate more than 6 billion billion (10^{18}) electrons.

The charge on an electron (or proton), 1.60×10^{-19} C, is the smallest common charge known (more exactly, $1.6021892 \times 10^{-19}$ C). It is the **fundamental charge** of the electron ($e^- = 1.60 \times 10^{-19}$ C) and the proton ($p^+ = 1.60 \times 10^{-19}$ C). All charged objects have multiples of this fundamental charge. An object might have a charge on the order of about 10^{-8} to 10^{-6} C.

Measuring Electric Force

Recall that two objects with like charges, (−) and (−) or (+) and (+), produce a repulsive force, and two objects with unlike charges, (−) and (+), produce an attractive force. The size of either force depends on the amount of charge of each object and on the distance between the objects. The relationship is known as **Coulomb's law**, which is

$$F = k\frac{q_1 q_2}{d^2}$$ equation 6.1

where k has the value of 9.00×10^9 newton·meters^2/coulomb^2 (9.00×10^9 N·m^2/C^2).

The force between the two charged objects is repulsive if q_1 and q_2 are the same charge and attractive if they are different (like charges repel, unlike charges attract). Whether the force is attractive or repulsive, you know that both objects feel the same force, as described by Newton's third law of motion. In addition, the strength of this force decreases if the distance between the objects increases. (A doubling of the distance reduces the force to ¼ of the original value.)

EXAMPLE 6.1   *(Optional)*

Electrons carry a negative electric charge and move about the nucleus of the atom, which carries a positive electric charge from the proton. The electron is held by the force of electrical attraction at a typical distance of 1.00×10^{-10} m. What is the force of electrical attraction between an electron and proton?

SOLUTION

$q_1 = 1.60 \times 10^{-19}$ C
$q_2 = 1.60 \times 10^{-19}$ C
$d = 1.00 \times 10^{-10}$ m
$k = 9.00 \times 10^9$ N·m^2/C^2
$F = ?$

$$F = k\frac{q_1 q_2}{d^2}$$

$$= \frac{\left(9.00 \times 10^9 \dfrac{\text{N·m}^2}{\text{C}^2}\right)(1.60 \times 10^{-19}\,\text{C})(1.60 \times 10^{-19}\,\text{C})}{(1.00 \times 10^{-10}\,\text{m})^2}$$

$$= \frac{(9.00 \times 10^9)(1.60 \times 10^{-19})(1.60 \times 10^{-19})}{1.00 \times 10^{-20}} \frac{\left(\dfrac{\text{N·m}^2}{\text{C}^2}\right)(\text{C}^2)}{\text{m}^2}$$

$$= \frac{2.30 \times 10^{-28}}{1.00 \times 10^{-20}} \frac{\text{N·m}^2}{\text{C}^2} \times \frac{\text{C}^2}{1} \times \frac{1}{\text{m}^2}$$

$$= \boxed{2.30 \times 10^{-8}\,\text{N}}$$

The electrical force of attraction between the electron and proton is 2.30×10^{-8} newton.

The model of a *field* is useful in understanding how a charge can attract or repel another charge some distance away. This model does not consider the force that one object exerts on another one through space. Instead, it considers the condition of space around a charge. The condition of space around an electric charge is changed by the presence of the charge. The charge produces a *force field* in the space around it. Since this force field is produced by an electric charge, it is called an **electric field.** Imagine a second electric charge, called a "test charge," that is far enough away from the electric charge that no forces are experienced. As you move the test charge closer and closer, it will experience an increasing force as it enters the

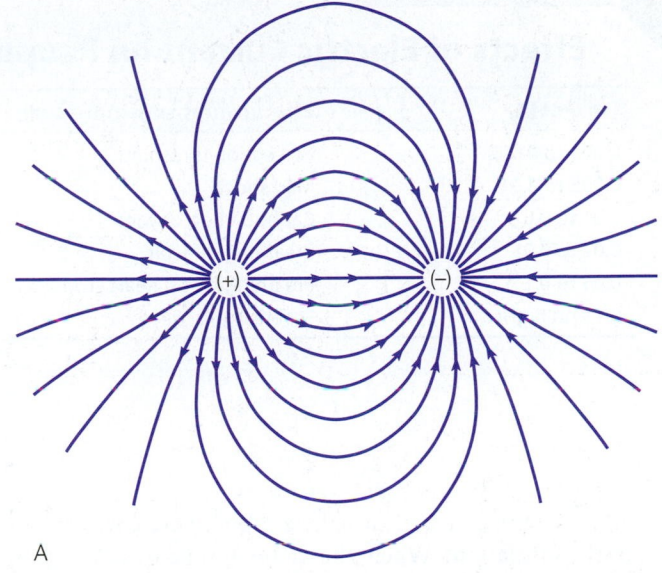

A

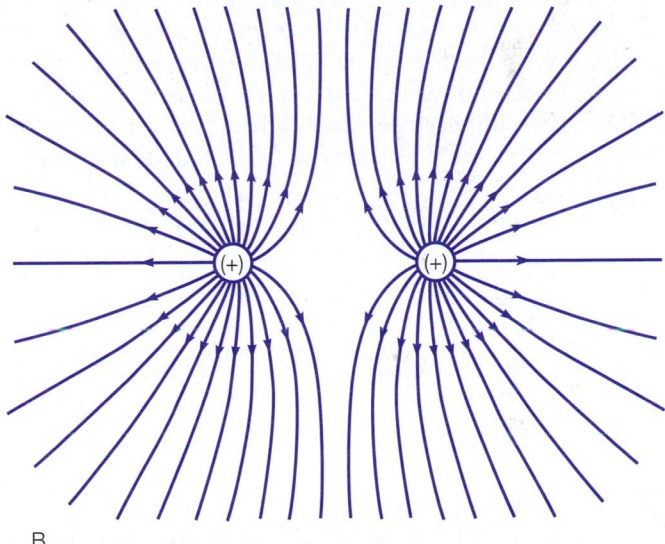

B

FIGURE 6.6   Lines of force diagrams for (*A*) a negative charge and (*B*) a positive charge when the charges have the same strength as the test charge.

electric field. The test charge can be used to identify the electric field that spreads out and around the space of an electric charge.

An electric field can be visualized by making a map of the field. The field is represented by electric field lines that indicate the strength and direction of the force the field would exert on the field of another charge. The field lines always point outward around a positively charged particle and point inward around a negatively charged particle. The spacing of the field lines shows the strength of the field. The field is stronger where the lines are closer together and weaker where they are farther apart (figure 6.6).

6.2 ELECTRIC CURRENT

Electric current means a flow of charge in the same way that "water current" means a flow of water. An electric current is flow of charge and this flow can be either negative or positive.

Connections . . .

Effects of Electric Current on People

Current (A)	Effect (varies with individual)
0.001 to 0.005	Perception threshold
0.005 to 0.01	Mild shock
0.01 to 0.02	Cannot let go of wire
0.02 to 0.05	Breathing difficult
0.05 to 0.1	Breathing stops, heart stops
0.1 and higher	Severe burns, death

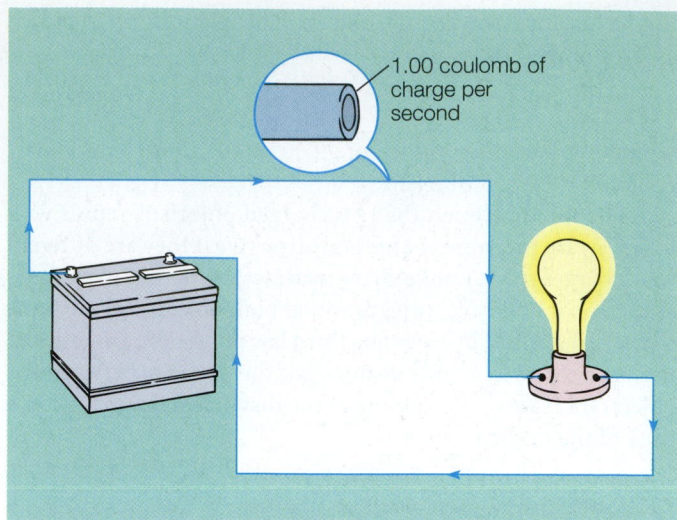

FIGURE 6.7 A simple electric circuit carrying a current of 1.00 coulomb per second through a cross section of a conductor has a current of 1.00 amp.

TABLE 6.1 **Electric Conductors and Insulators**

Conductors	Insulators
Silver	Rubber
Copper	Glass
Gold	Carbon (diamond)
Aluminum	Plastics
Carbon (graphite)	Wood
Tungsten	
Iron	
Lead	
Nichrome	

An electric wire in a car is usually a copper wire covered with a plastic insulation. When the radio is operated by using such a wire, electrons are moved from one terminal of the car battery through the radio, then back toward the battery. In a computer monitor, electrons are accelerated inside the picture tube to the back of the screen, producing an image as directed by the processor. Finally, a computer printer is plugged into a household current, which has an insulated copper wire with electrons moving back and forth. In all of these examples, there was movement of charge, also known as an electric current.

Many examples of different types of current exist, but all can be compared by the amount of electric charge that flows per second. The relationship is

$$\text{electric current} = \frac{\text{quantity of charge}}{\text{time}}$$

$$I = \frac{q}{t} \qquad \qquad \textbf{equation 6.2}$$

The unit for current is coulomb/second, which is called an **ampere** (A or **amp** for short). A current of 1 ampere in a wire is 1 coulomb of charge flowing through the wire every second. A 2 amp current is 2 coulombs of charge flowing through the wire every second, and so on (figure 6.7).

Charge can flow easily through some materials, such as metals, because they have many loosely attached electrons that can be easily moved from atom to atom. A substance that allows charges to flow easily is called a *conductor*. Materials such as plastic, wood, and rubber hold tightly to their electrons and do not readily allow for the flow of charge. A substance that does not allow charges to flow is called an *insulator* (table 6.1). Thus, metal wires are used to conduct an electric current from one place to another, and rubber, glass, and plastics are used as insulators to keep the current from going elsewhere.

There is a third class of materials, such as silicon and germanium, that sometimes conduct and sometimes insulate, depending on the conditions and how pure they are. These materials are called *semiconductors,* and their special proper-

ties make possible a number of technological devices such as computers, electrostatic copying machines, laser printers, solar cells, and so forth.

Resistance

Insulators have a property of limiting a current, and this property is called *electric resistance*. A good electric conductor has a very low electric resistance, and a good electric insulator has a very high electric resistance. The actual magnitude of electric resistance of a metal wire conductor depends on four variables (figure 6.8):

1. **Material.** Different materials have different resistances, as shown by the list of conductors in table 6.1. Silver, for example, is at the top of the list because it offers the least resistance, followed by copper, gold, then aluminum. Of the materials listed in the table, nichrome is the conductor with the greatest resistance. By definition, conductors have

Hydrogen and Fuel Cells

There is more than one way to power an electric vehicle, and a rechargeable lead-acid battery is not the best answer. A newly developing technology uses a hydrogen-powered fuel cell to power an electric vehicle. Storage batteries, dry cells, and fuel cells all operate by using a chemical reaction to produce electricity. The difference is that the fuel cell does not need charging like a storage battery or replacing like a dry cell. A fuel cell can use a fuel to run continuously, and that is why it is called a "fuel cell."

Fuel cells are able to generate electricity directly onboard an electric vehicle, so heavy batteries are not needed. A short driving range is not a problem, and time lost charging the batteries is not a problem. Instead, electricity is produced electrochemically, in a device without any moving parts. As you probably know, energy is required to separate water into its component gases of hydrogen and oxygen. Thus, as you might expect, energy is released when hydrogen and oxygen combine to form water. It is this energy that a fuel cell uses to produce an electric current.

One design of a fuel cell has two electrodes, one positive and one negative, where the reactions that produce electricity take place. A proton exchange membrane (PEM) separates the electrodes, which also use catalysts to speed chemical reactions. In general terms, hydrogen atoms enter the fuel cell at the positive terminal where a chemical reaction strips them of their electrons (box figure 6.1). The hydrogen atoms are now positive ions, which pass through the membrane and leave their electrons behind. Once through the membrane, the hydrogen ions combine with oxygen, forming water, which drains from the cell. The electrons on the original side of the membrane increase in number, building up an electrical potential difference. Once a wire connects the two sides of the membrane, the electrons are able to make a current, providing electricity.

Box Figure 6.1
A schematic of a PEM fuel cell.

As long as such a fuel cell is supplied with hydrogen and oxygen, it will generate electricity to run a car.

The silent-running, nonpolluting fuel cell with no moving parts sounds too good to be true, but the technology works. The technology has been too expensive for everyday use until recently but is now more affordable. Fuel-cell powered vehicles can operate directly on compressed hydrogen gas or liquid hydrogen, and when they do, the only emission is water vapor. Direct use of hydrogen is also very efficient, with a 50 to 60 percent efficiency compared to the typical 15 to 20 percent efficiency for automobiles that run on petroleum in an internal combustion engine. Other fuels can also be used by running them through an onboard reformer, which transforms

the fuel to hydrogen. Methanol or natural gas, for example, can be used with significantly less CO_2, CO, HC, and NO_x emissions than produced by an internal combustion engine. An added advantage to the use of methanol is that the existing petroleum fuel distribution system (tanks, pumps, etc.) can be used to distribute this liquid fuel. Liquid or compressed hydrogen, on the other hand, requires a completely new type of distribution system.

A fuel cell–powered electric vehicle gives the emission benefits of a battery-powered vehicle without the problems of constantly recharging the batteries. Before long, you may see a fuel-cell vehicle in your neighborhood. It is the car of the future, which is needed now for the environment.

less electric resistance than insulators, which have a very large electric resistance.

2. **Length.** The resistance of a conductor varies directly with the length; that is, a longer wire has more resistance and a shorter wire has less resistance. The longer the wire is, the greater the resistance.

3. **Diameter.** The resistance varies inversely with the cross-sectional area of a conductor. A thick wire has a greater

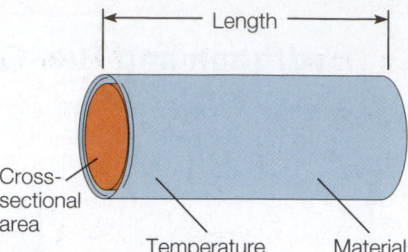

FIGURE 6.8 The four factors that influence the resistance of an electric conductor are the material the conductor is made of, the length of the conductor, the cross-sectional area of the conductor, and the temperature of the conductor.

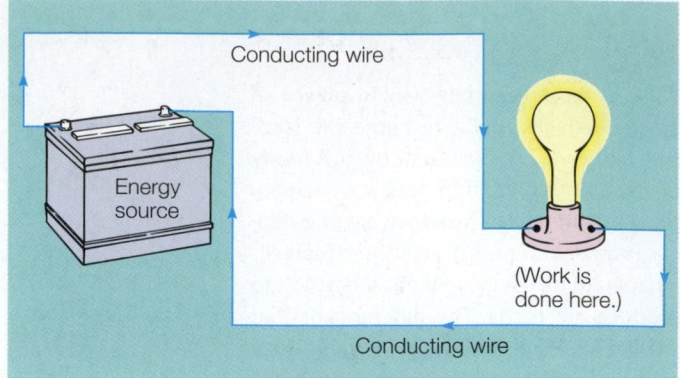

FIGURE 6.9 A simple electric circuit has an energy source (such as a generator or battery), some device (such as a lamp or motor) where work is done, and continuous pathways for the current to follow.

cross-sectional area and therefore has less resistance than a thin wire. The thinner the wire is, the greater the resistance.

4. **Temperature.** For most materials, the resistance increases with increases in temperature. This is a consequence of the increased motion of electrons and ions at higher temperatures, which increases the number of collisions. At very low temperatures (100 K or less), the resistance of some materials approaches zero, and the materials are said to be *superconductors*.

AC and DC

Another aspect of the nature of an electric current is the direction the charge is flowing. The wires in a car have currents that always move in one direction, and this is called a **direct current** (DC). Chemical batteries, fuel cells, and solar cells produce a direct current, and direct currents are utilized in electronic devices. Electric utilities and most of the electrical industry, on the other hand, use an **alternating current** (AC). An alternating current, as the name implies, moves the electrons alternately one way, then the other way. Since household electric circuits use alternating current, there is no flow of electrons from the electrical outlets through the wires. Instead an electric *field* moves back and forth through a wire at nearly the speed of light, causing electrons to jiggle back and forth. This constitutes a current that flows one way, then the other with the changing field. The current changes like this 120 times a second in a 60 hertz alternating current.

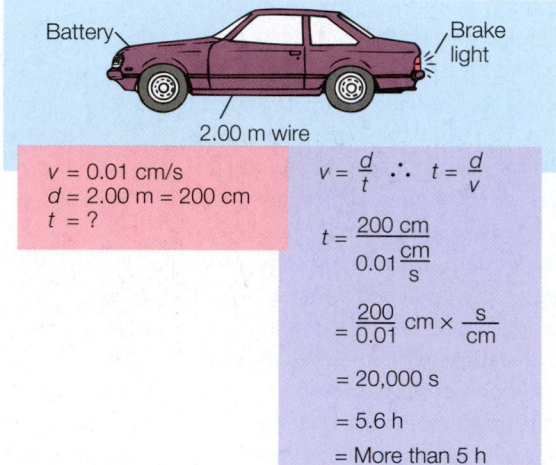

FIGURE 6.10 Electrons move very slowly in a direct current circuit. With a drift velocity of 0.01 cm/s, more than 5 hours would be required for an electron to travel 200 cm from a car battery to the brake light. It is the electric field, not the electrons, that moves at near the speed of light in an electric circuit.

6.3 THE ELECTRIC CIRCUIT

An electric current is established in a conductor when an electric field exerts a force on charges in the conductor. A car battery, for example, is able to light a bulb because the battery produces an electric field that forces electrons to move through the lightbulb filament. An **electric circuit** contains some device, such as a battery or electric generator, that acts as a source of energy as it forces charges to move out one

terminal, through the wires of the circuit, and then back in the other terminal (figure 6.9). The charges do work in another part of the circuit as they light bulbs, run motors, or provide heat. The charges flow through connecting wires to make a continuous path, and the number of coulombs of charge that leaves one terminal is the same as the number of coulombs of charge that enters the other terminal. The electric field moves through the circuit at nearly the speed of light, forcing the electrons to move along. The electrons, however, actually move through the circuit very slowly (figure 6.10).

As we learned in chapter 3, work is done when a force moves an object over a distance. In a circuit, work is done by the device that creates the electric field (battery, for example) as it exerts a force on the electrons and moves them through a distance in the circuit. Disregarding any losses due to the very

Benjamin Franklin (1706–1790)

Benjamin Franklin was the first great U.S. scientist. He made an important contribution to physics by arriving at an understanding of the nature of electric charge, introducing the terms *positive* and *negative* to describe charges. He also proved in a classic experiment that lightning is electrical in nature and went on to invent the lightning rod. In addition to being a scientist and inventor, Franklin is widely remembered as a statesman. He played a leading role in drafting the Declaration of Independence and the Constitution of the United States.

Franklin was born in Boston, Massachusetts, of British settlers on January 17, 1706. He started life with little formal instruction, and by the age of ten, he was helping his father in the tallow and soap business. Soon, apprenticed to his brother, a printer, he was launched into that trade, leaving home in 1724 to set himself up as a printer in Philadelphia.

In 1746, his business booming, Franklin turned his thoughts to electricity and spent the next seven years executing a remarkable series of experiments. Although he had little formal education, his voracious reading habits gave him the necessary background, and his practical skills, together with an analytical yet intuitive approach, enabled Franklin to put the whole topic on a very sound basis. It was said that he found electricity a curiosity and left it a science.

In 1752, Franklin carried out his famous experiments with kites. By flying a kite in a thunderstorm, he was able to produce sparks from the end of the wet string, which he held with a piece of insulating silk. The lightning rod used everywhere today owes its origin to these experiments. Furthermore, some of Franklin's last work in this area demonstrated that while most thunderclouds have negative charges, a few are positive—something confirmed in modern times.

Finally, Franklin also busied himself with such diverse topics as the first public library, bifocal lenses, population control, the rocking chair, and daylight-savings time.

Benjamin Franklin is arguably the most interesting figure in the history of science and not only because of his extraordinary range of interests, his central role in the establishment of the United States, and his amazing willingness to risk his life to perform a crucial experiment—a unique achievement in science. By conceiving of the fundamental nature of electricity, he began the process by which a most detailed understanding of the structure of matter has been achieved.

Source: Modified from the *Hutchinson Dictionary of Scientific Biography.* © RM, 2011. All rights reserved. Helicon Publishing is a division of RM.

small work done in moving electrons through a wire, the work done in some device (lamp, for example) is equal to the work done by the battery. The amount of work can be quantified by considering the work done and the size of the charge moved, and this ratio is used to define *voltage*. The voltage is defined by taking the ratio of the work done to the size of the charge that is being moved. So,

$$\text{voltage} = \frac{\text{work}}{\text{charge moved}}$$

$$V = \frac{W}{q} \qquad \textbf{equation 6.3}$$

The unit of voltage is the **volt,** which is the ratio that results when 1 joule of work is used to move 1 coulomb of charge:

$$1 \text{ volt (V)} = \frac{1 \text{ joule (J)}}{1 \text{ coulomb (C)}}$$

Thus, the voltage is the energy transfer per coulomb. The energy transfer can be measured by the work that is done to move the charge or by the work that the charge can do because of its position in the field. This is perfectly analogous to the work that must be done to give an object gravitational potential energy or to the work that the object can potentially do because of its new position. Thus, when a 12-volt battery is charging, 12 joules of work are done to transfer 1 coulomb of charge from an outside source against the electric field of the battery terminal. When the 12-volt battery is used, it does 12 joules of work for each coulomb of charge transferred from one terminal of the battery through the electrical system and back to the other terminal. Household circuits usually have a difference of potential of 120 or 240 volts. A voltage of 120 means that each coulomb of charge that moves through the circuit can do 120 joules of work in some electrical device.

The current in a circuit depends on the resistance as well as the voltage that is causing the current. If a conductor offers a small resistance, less voltage would be required to push an amp of current through the circuit. If a conductor offers more resistance, then more voltage will be required to push the same amp of current through the circuit. Resistance (R) is therefore a ratio between the voltage (V) and the resulting current (I). This ratio is

$$R = \frac{V}{I}$$

In units, this ratio is

$$1 \text{ ohm } (\Omega) = \frac{1 \text{ volt (V)}}{1 \text{ amp (A)}}$$

The ratio of volts/amps is the unit of resistance called an **ohm** (Ω) after the German physicist who discovered the relationship.

Another way to show the relationship between the voltage, current, and resistance is

$$V = IR \qquad \text{equation 6.4}$$

which is known as **Ohm's law.** This is one of three ways to show the relationship, but this way (solved for V) is convenient for easily solving the equation for other unknowns.

EXAMPLE 6.2 (*Optional*)

A lightbulb in a 120 V circuit is switched on, and a current of 0.50 A flows through the filament. What is the resistance of the bulb?

SOLUTION

The current (I) of 0.50 A is given with a potential difference (V) of 120 V. The relationship to resistance (R) is given by Ohm's law (equation 6.4)

$$
\begin{aligned}
I &= 0.50 \text{ A} & V = IR \therefore R &= \frac{V}{I} \\
V &= 120 \text{ V} \\
R &= ? & &= \frac{120 \text{ V}}{0.50 \text{ A}} \\
& & &= 240 \frac{\text{V}}{\text{A}} \\
& & &= 240 \text{ ohm} \\
& & &= \boxed{240 \ \Omega}
\end{aligned}
$$

EXAMPLE 6.3 (*Optional*)

What current would flow through an electrical device in a circuit with a potential difference of 120 V and a resistance of 30 Ω? (Answer: 4 A)

6.4 ELECTRIC POWER AND WORK

All electric circuits have three parts in common:

1. A voltage source, such as a battery or electric generator that uses some nonelectric source of energy to do work on electrons.
2. An electric device, such as a lightbulb or electric motor, where work is done by the electric field.
3. Conducting wires that maintain the current between the electric device and voltage source.

The work done by a voltage source (battery, electric generator) is equal to the work done by the electric field in an electric device (lightbulb, electric motor) plus the energy lost to resistance. Resistance is analogous to friction in a mechanical device, so low-resistance conducting wires are used to reduce this loss. Disregarding losses to resistance, electric work can therefore be measured where the voltage source does work in moving charges.

If you include a time factor with the work done in moving charges, you will be considering the *power output* of the voltage source. The power is determined by the voltage and current in the following relationship:

$$\text{power} = \text{voltage} \times \text{current}$$
$$P = VI \qquad \text{equation 6.5}$$

In units, you can see that multiplying the current ($I = \text{C/s}$) by the voltage ($V = \text{J/C}$) yields

$$\frac{\text{coulomb}}{\text{second}} \times \frac{\text{joule}}{\text{coulomb}} = \frac{\text{joule}}{\text{second}}$$

A joule/second is a unit of power called the **watt.** Therefore, electric power is measured in units of watts. Note that the relationship between power output and the current is directly proportional. Therefore, the greater the current supplied, the greater the power output.

Household electric devices are designed to operate on a particular voltage, usually 120 or 240. They therefore draw a certain current to produce the designed power. Information about these requirements is usually found somewhere on the device. A lightbulb, for example, is usually stamped with the designed power, such as 100 watts. Other electric devices may be stamped with amp and volt requirements. You can determine the power produced in these devices by using equation 6.5; that is, amps $\times$ volts = watts (figure 6.11). Another handy conversion factor to remember is that 746 watts are equivalent to 1 horsepower.

EXAMPLE 6.4 *(Optional)*

A 1,100 W hair dryer is designed to operate on 120 V. How much current does the dryer require?

SOLUTION

The power (P) produced is given in watts with a potential difference of 120 V across the dryer. The relationship between the units of amps, volts, and watts is found in equation 6.5, $P = VI$

$$P = 1,100 \text{ W} \qquad P = VI \therefore I = \frac{P}{V}$$
$$V = 120 \text{ V}$$
$$I = ?$$
$$= \frac{1,100 \frac{\text{J}}{\text{s}}}{120 \frac{\text{J}}{\text{C}}}$$
$$= \frac{1,100}{120} \frac{\text{J}}{\text{s}} \times \frac{\text{C}}{\text{J}}$$
$$= 9.2 \frac{\text{J} \cdot \text{C}}{\text{s} \cdot \text{J}}$$
$$= 9.2 \frac{\text{C}}{\text{s}}$$
$$= \boxed{9.2 \text{ A}}$$

EXAMPLE 6.5 *(Optional)*

An electric fan is designed to draw 0.5 A in a 120 V circuit. What is the power rating of the fan? (Answer: 60 W)

Connections . . .

Inside a Dry Cell

The common dry cell used in a flashlight produces electric energy from a chemical reaction between ammonium chloride and the zinc can (box figure 6.2). The reaction leaves a negative charge on the zinc and a positive charge on the carbon rod. Manganese dioxide takes care of hydrogen gas, which is a by-product of the reaction. Dry cells always produce 1.5 volts, regardless of their size. Larger voltages are produced by combinations of smaller cells (making a true "battery").

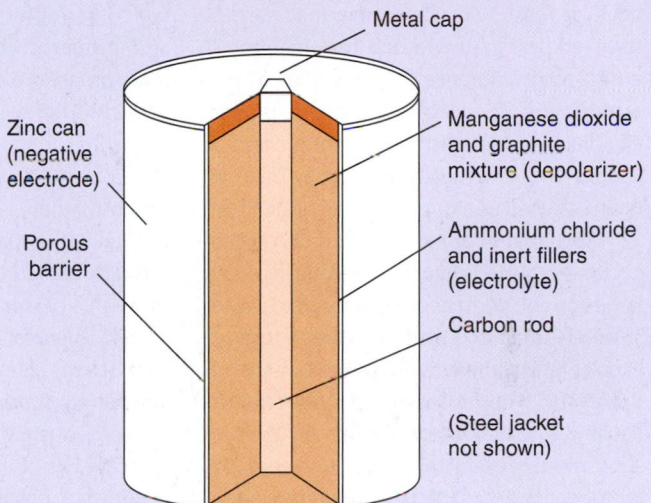

Box Figure 6.2 A schematic sketch of a dry cell.

A

B

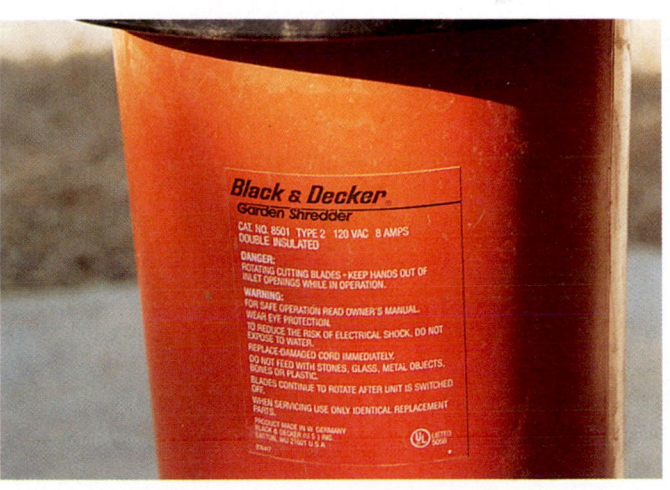

C

FIGURE 6.11 What do you suppose it would cost to run each of these appliances for one hour? (A) This lightbulb is designed to operate on a potential difference of 120 volts and will do work at the rate of 100 W. (B) The finishing sander does work at the rate of 1.6 amp × 120 volts, or 192 W. (C) The garden shredder does work at the rate of 8 amps × 120 volts, or 960 W.

Household Circuits and Safety

The example of a household circuit in box figure 6.3B shows a light and four wall outlets in a parallel circuit. The use of the term *parallel* means that a current can flow through any of the separate branches without having to first go through any of the other devices. It does not imply that the branches are necessarily lined up with each other. Lights and outlets in this circuit all have at least two wires, one that carries the electrical load and one that maintains a potential difference by serving as a system ground. The load-carrying wire is usually black (or red) and the system ground is usually white. A third wire, usually bare or green, is used as a ground for an appliance.

Too many appliances running in a circuit—or a short circuit—can result in a very large current, perhaps great enough to cause strong heating and possibly a fire. A fuse or circuit breaker prevents this by disconnecting the circuit when it reaches a preset value, usually 15 or 20 amps. A fuse contains a short piece of metal that melts when heated by too large a current, creating a gap when the current through the circuit reaches the preset rating. The gap disconnects the circuit, just as cutting the wire or using a switch. The circuit breaker has the same purpose but uses the proportional relationship between the magnitude of a current and the strength

of the magnetic field that forms around the conductor. When the current reaches a preset level, the magnetic field is strong enough to open a spring-loaded switch. The circuit breaker is reset by flipping the switch back to its original position.

Besides a fuse or circuit breaker, a modern household electric circuit has three-pronged plugs, polarized plugs, and ground fault interrupters to help protect people and property from electrical damage. A *three-pronged plug* provides a grounding wire through a (usually round) prong on the plug. The grounding wire connects the housing of an appliance directly to the ground. If there is a short circuit, the current will take the path of least resistance—through the grounding wire—rather than through a person.

A *polarized plug* has one of the two flat prongs larger than the other. An alternating current moves back and forth with a frequency of 60 Hz, and *polarized* in this case has nothing to do with positive or negative. A polarized plug in an AC circuit means that one prong always carries the load. The smaller plug is connected to the load-carrying wire and the larger one is connected to the neutral wire. An ordinary, nonpolarized plug can fit into an outlet either way, which means there is a 50–50 chance that one of the

wires will be the one that carries the load. The polarized plug always has the load-carrying wire on the same side of the circuit, so the switch can be wired in so it is always on the load-carrying side. The switch will function the same on either wire, but when it is on the ground wire, the appliance contains a load-carrying wire, just waiting for a short circuit. When the switch is on the load-carrying side, the appliance does not have this potential safety hazard.

Yet another safety device called a *ground-fault interrupter* (GFI) offers a different kind of protection. Normally, the current in the load-carrying and system ground wire is the same. If a short circuit occurs, some of the current might be diverted directly to the ground or to the appliance ground. A GFI device monitors the load-carrying and system ground wires, and if any difference is detected, it trips, opening the circuit within a fraction of a second. This is much quicker than the regular fuse or general circuit breaker can react, and the difference might be enough to prevent a fatal shock. The GFI device is usually placed in an outside or bathroom circuit, places where currents might be diverted through people with wet feet. Note the GFI can also be tripped by a line surge that might occur during an electrical thunderstorm.

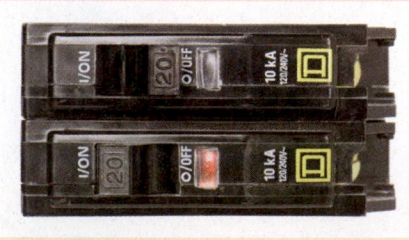

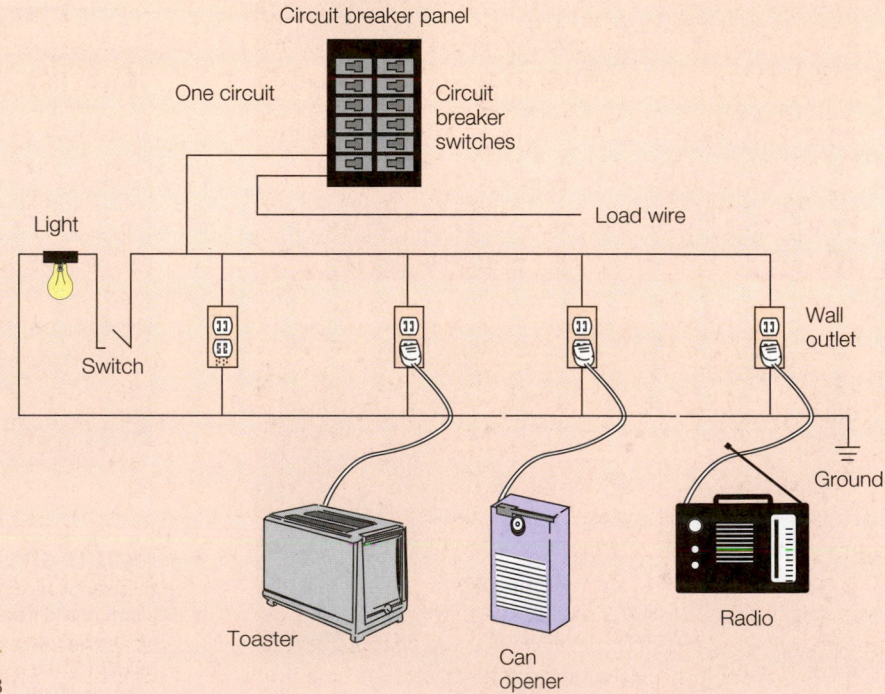

Box Figure 6.3 (*A*) One circuit breaker in this circuit breaker panel has tripped, indicating an overload or short circuit. (*B*) The tripped breaker could be circuit number 6.

The electric utility charge for the electric work done is at a rate of cents per kilowatt-hour. The rate varies from place to place across the country, depending on the cost of producing the power (typically, 5 to 15 cents per kWh). You can predict the cost of running a particular electric appliance with the following equation,

$$\text{cost} = \frac{(\text{watts})(\text{time})(\text{rate})}{1,000\ \dfrac{\text{watt}}{\text{kilowatt}}}$$

equation 6.6

If the watt power rating is not given, it can be obtained by multiplying amps times volts. Also note that since the time unit is in hours, if you want to know the cost of running an appliance for a number of minutes, the time must be converted to the decimal equivalent of an hour (x min ÷ 60 min).

EXAMPLE 6.6   *(Optional)*

What is the cost of operating a 100 W lightbulb for 1.00 h if the utility rate is $0.10 per kWh?

SOLUTION

The power rating is given as 100 W, so the volt and amp units are not needed. Therefore,

$IV = P = 100$ W
$t = 1.00$ h
rate = $0.10/kWh
cost = ?

$$\text{cost} = \frac{(\text{watts})(\text{time})(\text{rate})}{1,000\ \dfrac{\text{watts}}{\text{kilowatt}}}$$

$$= \frac{(100\ \text{W})(1.00\ \text{h})(\$0.10/\text{kWh})}{1,000\ \dfrac{\text{watts}}{\text{kilowatt}}}$$

$$= \frac{(100)(1.00)(0.10)}{1,000}\ \frac{W}{1} \times \frac{h}{1} \times \frac{\$}{kWh} \times \frac{kW}{W}$$

$$= \boxed{\$0.01}$$

The cost of operating a 100 W lightbulb at a rate of $0.10/kWh is 1¢/h.

EXAMPLE 6.7   *(Optional)*

An electric fan draws 0.5 A in a 120 V circuit. What is the cost of operating the fan if the rate is $0.10/kWh? (Answer: $0.006, which is 0.6 of a cent per hour)

As you may have noticed, the electric cord to certain appliances such as hair dryers becomes warm when the appliance is used. Heating occurs in any conductor that has any resistance, and it represents a loss of useable energy. The amount of heating depends on the resistance and the size of the current. Increasing the current or increasing any of the four resistance factors discussed earlier will cause an increase in heating. A large current in a wire of high resistance can become very hot, hot enough to melt insulation and ignite materials. This is why

all circuits have fuses or circuit breakers to "break" the circuit if the current exceeds a preset safe limit (see "A Closer Look: Household Circuits and Safety").

6.5 MAGNETISM

The ability of a certain naturally occurring rock to attract iron has been known since at least 600 B.C. This rock was called "Magnesian stone" since it was discovered near the ancient city of Magnesia in Turkey. Knowledge about the iron-attracting properties of the Magnesian stone grew slowly. About A.D. 100, the Chinese learned to magnetize a piece of iron with the stone, and sometime before A.D. 1000, they learned to use the magnetized iron or stone as a direction finder (compass). The rock that attracts iron is known today as the mineral named *magnetite*.

Magnetite is a natural magnet that strongly attracts iron and steel but also attracts cobalt and nickel. Such substances that are attracted to magnets are said to have *ferromagnetic* properties, or simply magnetic properties. Iron, cobalt, and nickel are considered to have magnetic properties, and most other common materials are considered not to have magnetic properties. Most of these nonmagnetic materials, however, are slightly attracted or slightly repelled by a strong magnet. In

Connections . . .

Magnetic Fields and Instinctive Behavior

Do animals use Earth's magnetic fields to navigate? Since animals move from place to place to meet their needs, it is useful to be able to return to a nest, water hole, den, or favorite feeding spot. This requires some sort of memory of their surroundings (a mental map) and a way of determining direction. Often it is valuable to have information about distance as well. Direction can be determined by such things as magnetic fields, identified landmarks, scent trails, or reference to the Sun or stars. If the Sun or stars are used for navigation, then some sort of time sense is also needed since these bodies move in the sky.

Instinctive behaviors are automatic, preprogrammed, and genetically determined. Such behaviors are found in a wide range of organisms from simple one-celled protozoans to complex vertebrates. These behaviors are performed correctly the first time without previous experience when the proper stimulus is given. A stimulus is some change in the internal or external environment of the organism that causes it to react. The reaction of the organism to the stimulus is called a response.

An organism can respond only to stimuli it can recognize. For example, it is difficult for us as humans to appreciate what the world seems like to a bloodhound. The bloodhound is able to identify individuals by smell, whereas we have great difficulty detecting, let alone distinguishing, many odors. Some animals, such as dogs, deer, and mice, are color-blind and are able to see only shades of gray. Others, such as honeybees, can see ultra-violet light, which is invisible to us. Some birds and other animals are able to detect the magnetic field of Earth.

There is evidence that some birds navigate by compass direction, that is, they fly as if they had a compass in their heads. They seem to be able to sense magnetic north. Their ability to sense magnetic fields has been proven at the U.S. Navy's test facility in Wisconsin. The weak magnetism radiated from this test site has changed the flight pattern of migrating birds, but it is yet to be proven that birds use the magnetism of Earth to guide their migration. Homing pigeons are famous for their ability to find their way home. They make use of a wide variety of clues, but it has been shown that one of the clues they use involves magnetism. Birds with tiny magnets glued to the sides of their heads were very poor navigators, while others with nonmagnetic objects attached to the sides of their heads did not lose their ability to navigate.

---

addition, certain rare earth elements as well as certain metal oxides exhibit strong magnetic properties.

Every magnet has two **magnetic poles,** or ends, about which the force of attraction seems to be concentrated. Iron filings or other small pieces of iron are attracted to the poles of a magnet, for example, revealing their location (figure 6.12). A magnet suspended by a string will turn, aligning itself in a north-south direction. The north-seeking pole is called the *north pole* of the magnet. The south-seeking pole is likewise named the *south pole* of the magnet. All magnets have both a north pole and a south pole, and neither pole can exist by itself. You cannot separate a north pole from a south pole. If a magnet is broken into pieces, each new piece will have its own north and south pole (figure 6.13).

You are probably familiar with the fact that two magnets exert forces on each other. For example, if you move the north pole of one magnet near the north pole of a second magnet, each will experience a repelling force. A repelling force also occurs if two south poles are moved close together. But if the north pole of one magnet is brought near the south pole of a second magnet, an attractive force occurs. The rule is *"like magnetic poles repel and unlike magnetic poles attract."*

A similar rule of "likes charges repel and unlike charges attract" was used for electrostatic charges, so you might wonder if there is some similarity between charges and poles. The answer is they are not related. A magnet has no effect on a charged glass rod and the charged glass rod has no effect on either pole of a magnet.

A magnet moved into the space near a second magnet experiences a magnetic force as it enters the **magnetic field**

FIGURE 6.12 Every magnet has ends, or poles, about which the magnetic properties seem to be concentrated. As this photo shows, more iron filings are attracted to the poles, revealing their location.

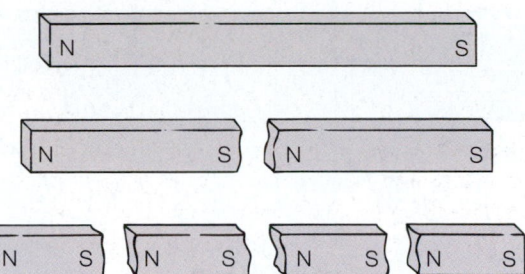

FIGURE 6.13   A bar magnet cut into halves always makes new, complete magnets with both a north and a south pole. The poles always come in pairs, and the separation of a pair into single poles, called monopoles, has never been accomplished.

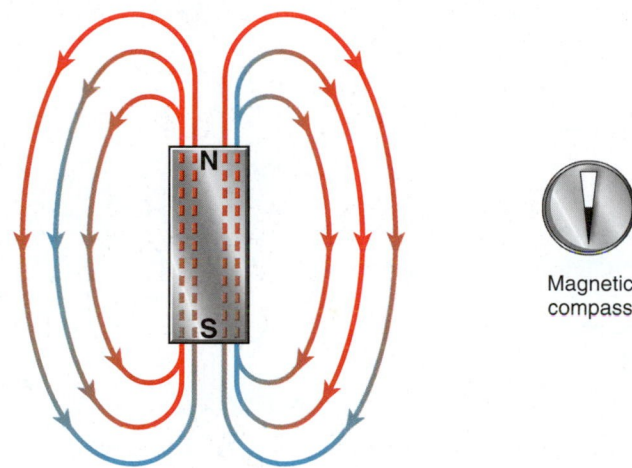

Magnetic compass

FIGURE 6.14   These lines are a map of the magnetic field around a bar magnet. The needle of a magnetic compass will follow the lines, with the north end showing the direction of the field.

The Inventor

Michael Faraday (1791–1867) is often regarded as the greatest experimental scientist of the 1800s. Among other experimental work, he invented the electric motor, electric generator, and transformer, discovered electromagnetic induction, and established the laws of electrolysis. He also coined the terms *anode*, *cathode, cation, anion, electrode,* and *electrolyte* to help explain his new inventions and laws.

Faraday was the son of a poor blacksmith and received a poor education as a child, with little knowledge of mathematics. At fourteen, he was apprenticed to a bookbinder and began to read many books on chemistry and physics. He also began attending public lectures at science societies and in general learned about science and scientific research. At twenty-one, he was appointed an assistant at the Royal Institution in London, where he liquefied gases for the first time and continued to work in electricity and magnetism. To understand electric and magnetic fields that he could not see or represent mathematically (he was not very accomplished at math), Faraday invented field lines—lines that do not exist but help us to picture what is going on.

Michael Faraday was a modest man, content to work in science without undue rewards. He declined both a knighthood and the presidency of the Royal Society. His accomplishments were recognized nonetheless, and the SI unit of capacitance, the farad, was named in his honor, as was the faraday. The faraday is the amount of electricity needed to liberate a standard amount of something through electrolysis.

of the second magnet. A magnetic field can be represented by *magnetic field lines.* By convention, magnetic field lines are drawn to indicate how the *north pole* of a tiny imaginary magnet would point when in various places in the magnetic field. Arrowheads indicate the direction that the north pole would point, thus defining the direction of the magnetic field. The strength of the magnetic field is greater where the lines are closer together and weaker where they are farther apart. Figure 6.14 shows the magnetic field lines around the familiar bar magnet.

The north end of a magnetic compass needle points north because Earth has a magnetic field. Earth's magnetic field is shaped and oriented as if there were a huge bar magnet inside Earth (figure 6.15). The geographic North Pole is the axis of Earth's rotation, and this pole is used to determine the direction of true north on maps. A magnetic compass does not point to true north because the north magnetic pole and the geographic North Pole are in two different places. The difference is called the *magnetic declination.* The map in figure 6.16 shows approximately how many degrees east or west of true north a compass needle will point in different locations. Magnetic declination must be considered when navigating with a compass.

Geographic North Pole

Magnetic north pole

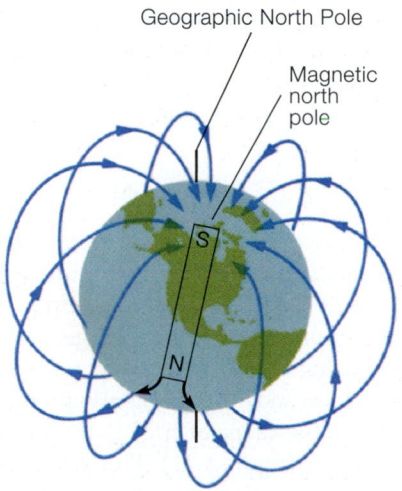

FIGURE 6.15   Earth's magnetic field. Note that the magnetic north pole and the geographic North Pole are not in the same place. Note also that the magnetic north pole acts as if the south pole of a huge bar magnet were inside Earth. You know that it must be a magnetic south pole, since the north end of a magnetic compass is attracted to it, and opposite poles attract.

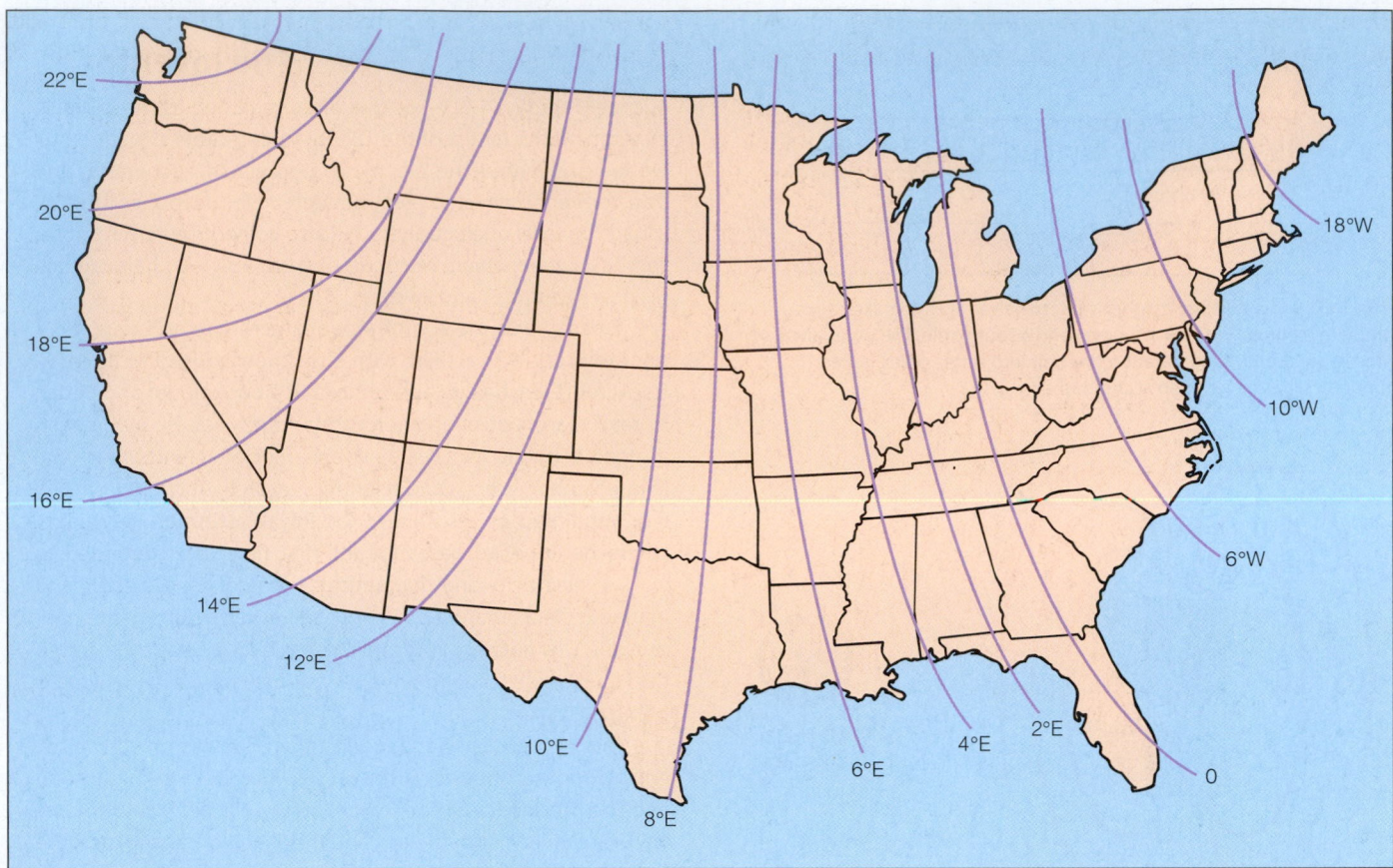

22°E

20°E

18°E

16°E

14°E

12°E

10°E

8°E

6°E

4°E

2°E

0

6°W

10°W

18°W

FIGURE 6.16   This magnetic declination map shows the approximate number of degrees east or west of the true geographic north that a magnetic compass will point in various locations.

Note in figure 6.15 that Earth's magnetic field acts as if there is a huge bar magnet inside Earth with a south pole near Earth's geographic North Pole. This is not an error. The north pole of a magnet is attracted to the south pole of a second magnet, and the north pole of a compass needle points to the north. Therefore, the bar magnet must be arranged as shown. This apparent contradiction is a result of naming the magnetic poles after their "seeking" direction.

Moving Charges and Magnetic Fields

Recall that every electric charge is surrounded by an electric field. If the charge is moving, it is surrounded by an electric field *and* a magnetic field. The magnetic field is formed in the shape of circles around the moving charge. A DC current is many charges moving through a conductor in response to an electric field, and these moving charges also produce a magnetic field. The field is in the shape of circles around the length of a current-carrying wire (figure 6.17). This relationship suggests that electricity and magnetism are two different manifestations of charges in motion. The electric field of a charge, for example, is fixed according to the fundamental charge of the particle. The magnetic field, however, changes with the velocity of the moving charge. The magnetic field does not exist at all if the charge is not moving, and the strength of the magnetic

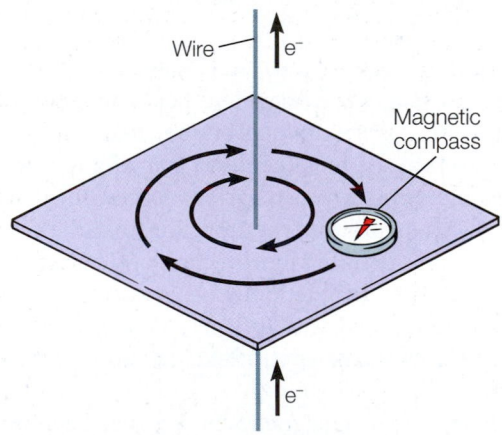

Wire   e⁻

Magnetic compass

e⁻

FIGURE 6.17   A magnetic compass shows the presence and direction of the magnetic field around a straight length of current-carrying wire.

field increases with increases in velocity. It seems clear that magnetic fields are produced by the motion of charges, or electric currents. Thus, a magnetic field is a property of the space around a moving charge.

You can see the shape of the magnetic field established by the current by running a straight wire vertically through a sheet of paper. The wire is connected to a battery, and iron filings are sprinkled on the paper. The filings will become aligned as each tiny piece of iron is moved parallel to the field. Overall, filings near the wire form a pattern of circles with the wire in the center.

A current-carrying wire that is formed into a loop has circular magnetic field lines that pass through the inside of the loop in the same direction. This has the effect of concentrating the field lines, which increases the magnetic field intensity. Since the field lines all pass through the loop in the same direction, one side of the loop will have a north pole and the other side a south pole (figure 6.18).

Many loops of wire formed into a cylindrical coil are called a **solenoid.** When a current passes through the loops of wire in a solenoid, each loop contributes field lines along the length of the cylinder (figure 6.19). The overall effect is a magnetic field around the solenoid that acts just like the magnetic field

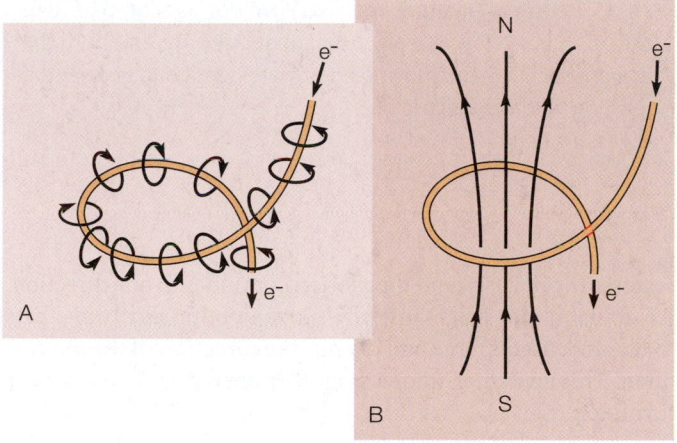

FIGURE 6.18 (A) Forming a wire into a loop causes the magnetic field to pass through the loop in the same direction. (B) This gives one side of the loop a north pole and the other side a south pole.

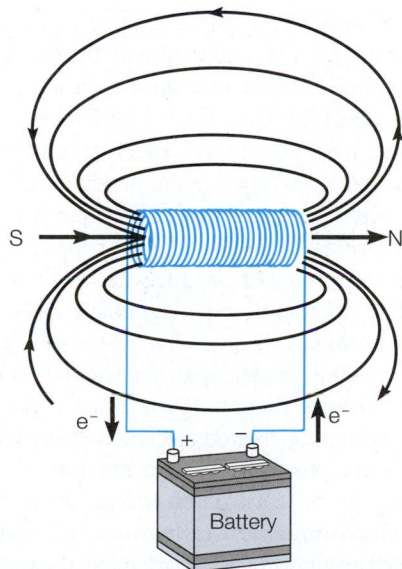

FIGURE 6.19 When a current is run through a cylindrical coil of wire, a solenoid, it produces a magnetic field like the magnetic field of a bar magnet.

of a bar magnet. This **electromagnet** can be turned on or off by turning the current on or off. In addition, the strength of the electromagnet depends on the size of the current and the number of loops. The strength of the electromagnet can also be increased by placing a piece of soft iron in the coil.

A solenoid can serve as an electrical switch or valve. It can be used as a water valve by placing a spring-loaded movable piece of iron inside the wire coil. When a current flows in such a coil, the iron is pulled into the coil by the magnetic field, turning the hot or cold water on in a washing machine or dishwasher, for example. Solenoids are also used as mechanical switches on VCRs, automobile starters, and signaling devices such as door bells and buzzers.

Electrons in atoms are moving around the nucleus, so they produce a magnetic field. Electrons also have a magnetic field associated with their spin. In most materials, these magnetic fields cancel one another and neutralize the overall magnetic effect. In other materials, such as an iron magnet, the electrons are arranged and oriented in such a way that individual magnetic fields are aligned. Each field becomes essentially a tiny magnet with a north and south pole. In an unmagnetized piece of iron, the fields are oriented in all possible directions and effectively cancel any overall magnetic effect. The net magnetism is therefore zero or near zero.

Magnetic Fields Interact

An electric charge at rest does not have a magnetic field, so it does not interact with a magnetic field from a magnet. A moving electric charge does produce a magnetic field, and this field will interact with another magnetic field, producing a force. A DC current is many charges moving through a wire, so it follows that a current-carrying wire will also interact with another magnetic field, producing a force. This force varies, depending on the size and direction of the current and the orientation of the external field. With a constant current, the force is at a maximum when the direction of the current is at a right angle to the other magnetic field.

Since you cannot measure electricity directly, it must be measured indirectly through one of the effects that it produces. Recall that the strength of the magnetic field around a current-carrying wire is proportional to the size of the current. Thus, one way to measure a current is to measure the magnetic field that it produces. A device that measures the size of a current from the size of its magnetic field is called a **galvanometer** (figure 6.20). A galvanometer has a coil of wire that can rotate on pivots in the magnetic field of a permanent magnet. When there is a current in the coil, the magnetic field produced is attracted and repelled by the field of the permanent magnet. The larger the current, the greater the force and the more the coil will rotate. The amount of movement of the coil is proportional to the current in the coil, and the pointer shows this on a scale. With certain modifications and applications, the device can be used to measure current (ammeter), voltage (voltmeter), and resistance (ohmmeter).

An electric motor is able to convert electrical energy to mechanical energy by using the force produced by interacting

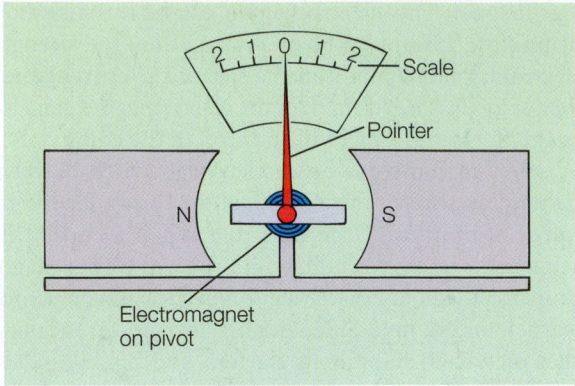

FIGURE 6.20 A galvanometer measures the direction and relative strength of an electric current from the magnetic field it produces. A coil of wire wrapped around an iron core becomes an electromagnet that rotates in the field of a permanent magnet. The rotation moves a pointer on a scale.

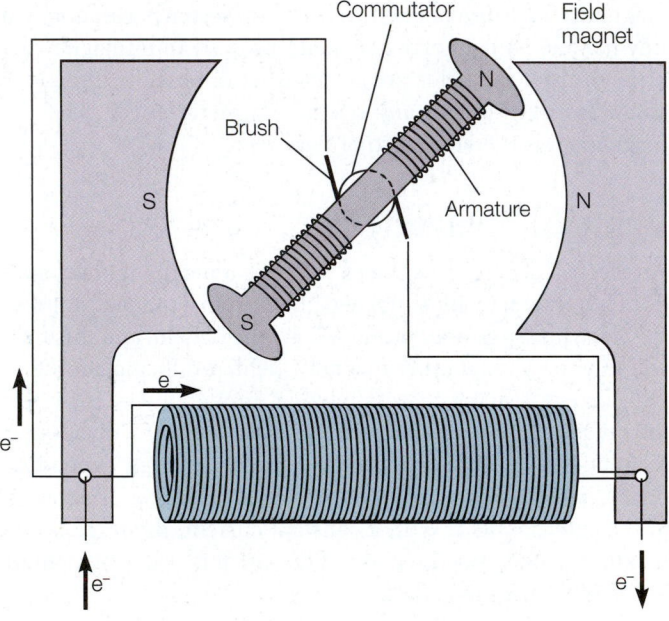

FIGURE 6.21 A schematic of a simple electric motor.

magnetic fields. Basically, a motor has two working parts, a stationary electromagnet and an electromagnet that moves. The movable electromagnet rotates in the magnetic field of the stationary one. It turns fan blades, compressors, drills, pulleys, or other devices that do mechanical work.

Different designs of electric motors are used for various applications, but the simple demonstration motor shown in figure 6.21 can be used as an example of the basic operating principle. Both the stationary and the movable electromagnets are connected to an electric current. When the current is turned on, the unlike poles of the two fields attract, rotating the movable electromagnet for a half turn. The motor has a simple device that now reverses the direction of the current. This switches the magnetic poles on the movable electromagnet, which is now repelled for another half turn. The device again

reverses the polarity, and the motion continues in one direction. An actual motor has many wire loops in both electromagnets to obtain a useful force and change the direction of the current often. This gives the motor a smoother operation with a greater turning force.

A Moving Magnet Produces an Electric Field

So far, you have learned what happens when we move a charge: (1) a moving charge and a current-carrying wire produce a magnetic field and (2) a second magnetic field exerts a force on a moving charge and exerts a force on a current-carrying wire as their magnetic fields interact. Since a moving charge produces a magnetic field, what should we expect if we move a magnet? The answer is that a *moving magnet* produces an *electric field* in the shape of circles around the path of the magnet. Here we see some similarity: A moving charge produces a magnetic field and a moving magnet produces an electric field. Thus, we observe that electricity and magnetism interact, but only when there is motion.

A moving magnet produces an electric field that is circular around the path of the magnet. If you place a coil of wire near the moving magnet, the created electric field will interact with charges in the wire, forcing them to move as a current. The process of creating an electric current with a moving magnetic field is called **electromagnetic induction.** One way to produce electromagnetic induction is to move a bar magnet into or out of a coil of wire (figure 6.22). A galvanometer shows that the induced current flows one way when the bar magnet is moved toward the coil and flows the other way when the bar magnet is

Lemon Battery

1. You can make a simple compass galvanometer that will detect a small electric current (box figure 6.4). All you need is a magnetic compass and some thin insulated wire (the thinner the better).
2. Wrap the thin insulated wire in parallel windings around the compass. Make as many parallel windings as you can, but leave enough room to see both ends of the compass needle. Leave the wire ends free for connections.
3. To use the galvanometer, first turn the compass so the needle is parallel to the wire windings. When a current passes through the coil of wire, the magnetic field produced will cause the needle to move from its north-south position, showing the presence of a current. The needle will deflect one way or the other, depending on the direction of the current.
4. Test your galvanometer with a "lemon battery." Roll a soft lemon on a table while pressing on it with the palm of your hand. Cut two slits in the lemon about 1 cm apart. Insert a 8 cm (approximate) copper wire in one slit and a same-sized length of a straightened paper clip in the other slit, making sure the metals do not touch inside the lemon. Connect the galvanometer to the two metals. Try the two metals in other fruits, vegetables, and liquids. Can you find a pattern?

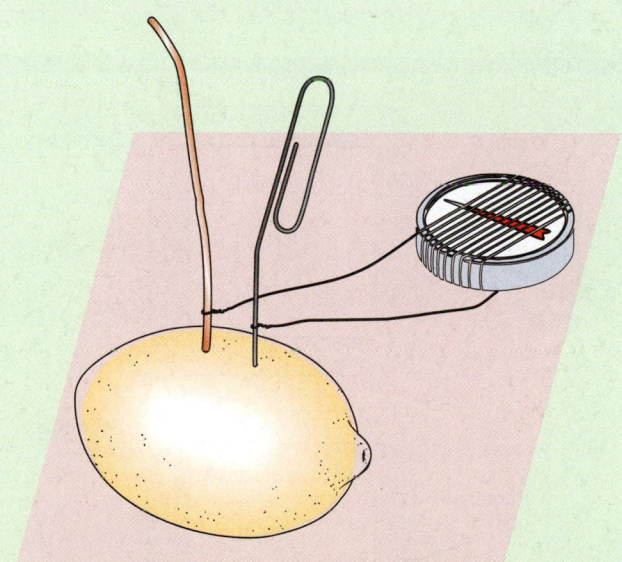

Box Figure 6.4 You can use the materials shown here to create and detect an electric current.

Swinging Coils

The interactions between moving magnets and moving charges can be easily demonstrated with two large magnets, two coils of wire, and a galvanometer.

1. Make a coil of insulated bell wire (#18 copper wire) by wraping fifty windings around a narrow jar. Tape the coil at several places so it does not come apart.
2. Connect the coil to a galvanometer (see "Concepts Applied: Lemon Battery" to make your own).
3. Move a strong bar magnet into and out of the stationary coil of wire and observe the galvanometer. Note the magnetic pole, direction of movement, and direction of current for both in and out movements.
4. Move the coil of wire back and forth over the stationary bar magnet.
5. Now make a second coil of wire from insulated bell wire and tape the coil as before.
6. Suspend the coil of wire on a ring stand or some other support on a table top. The coil should hang so it will swing with the broad circle of the coil moving back and forth. Place a large magnet on supports so it is near the center of the coil.
7. Set up an identical coil of wire, ring stand or support, and magnet on another table. Connect the two coils of wire.
8. Move one of the coils of wire and observe what happens to the second coil. The second coil should move, mirroring the movements of the first coil. (If it does not move, find some stronger magnets.)
9. Explain what happens at the first coil and the second coil.

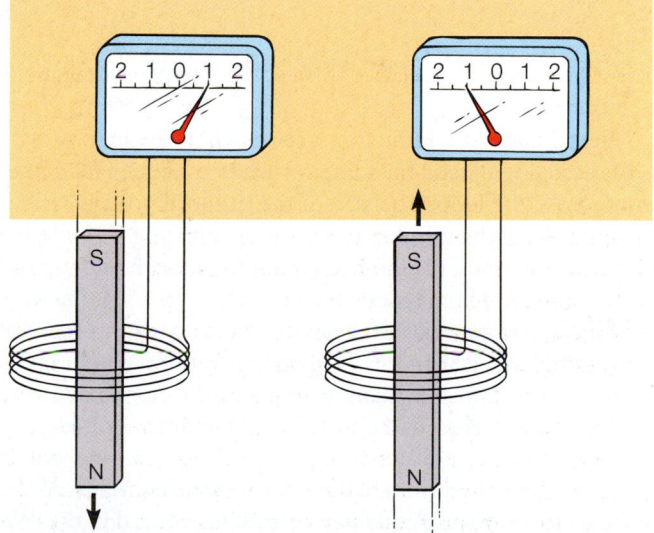

FIGURE 6.22 A current is induced in a coil of wire moved through a magnetic field. The direction of the current depends on the direction of motion.

moved away from the coil. The same effect occurs if you move the coil back and forth over a stationary magnet. Furthermore, no current is detected when the magnetic field and the coil of wire are not moving. Thus, electromagnetic induction depends on the relative motion of the magnetic field and the coil of wire. It does not matter which one moves or changes, but one

Connections . . .

Current War

Thomas Edison built the first electric generator and electrical distribution system to promote his new long-lasting lightbulbs. The DC generator and distribution system was built in lower Manhattan, New York City, and was switched on September 4, 1882. It supplied 110 volts DC to fifty-nine customers. Edison studied both AC and DC systems and chose DC because of the advantages it offered at the time. Direct current was used because batteries are DC, and batteries were used as a system backup. Also, DC worked fine with electric motors, and AC motors were not yet available.

George Westinghouse was in the business of supplying gas for gas lighting, and he could see that electric lighting would soon be replacing all the gaslights. After studying the matter, he decided that Edison's low-voltage system was not efficient enough. In 1885, he began experimenting with AC generators and transformers in Pittsburgh. In 1886, he installed a 500-volt AC system in Great Barrington, Massachusetts. The system stepped up the voltage to 3,000 volts for transmission, then back down to 110 volts, 60 Hz AC to power lightbulbs in homes and businesses.

Westinghouse's promotion of AC led to direct competition with Edison and his DC electrical systems. A "war of currents" resulted, with Edison claiming that transmission of such high voltage was dangerous. He emphasized this point by recommending the use of high-voltage AC in an electric chair as the best way to execute prisoners.

The advantages of AC were greater since you could increase the voltage, transmit for long distances at a lower cost, and then decrease the voltage to a safe level. Eventually, even Edison's own General Electric company switched to producing AC equipment. Westinghouse turned his attention to the production of large steam turbines for producing AC power and was soon setting up AC distribution systems across the nation.

"input" coil and (2) a *secondary* or "output" coil, which is close by. Both coils are often wound on a single iron core but are always fully insulated from each other. When an alternating current flows through the primary coil, a magnetic field grows around the coil to a maximum size, collapses to zero, then grows to a maximum size with an opposite polarity. This happens 120 times a second as the alternating current oscillates at 60 hertz. The growing and collapsing magnetic field moves across the wires in the secondary coil, inducing a voltage in the secondary coil.

The size of the induced voltage in the secondary coil is proportional to the number of wire loops in the two coils. Each turn or loop of the output coil has the same voltage induced in it, so more loops in an output coil means higher voltage output. If the output coil has the same number of loops as the input coil, the induced voltage in the secondary coil will be the same as the voltage in the primary coil. If the secondary coil has one-tenth as many loops as the primary coil, then the induced voltage in the secondary coil will be one-tenth the voltage in the primary coil. This is called a **step-down transformer** since the voltage was stepped down in the secondary coil. On the other hand, if the secondary coil has ten times more loops than the primary coil, then the voltage will be *increased* by a factor of 10. This is a **step-up transformer.** How much the voltage is stepped up or stepped down depends on the ratio of wire loops in the primary and secondary coils (figure 6.23).

must move or change relative to the other for electromagnetic induction to occur.

Electromagnetic induction occurs when the loop of wire moves across magnetic field lines or when magnetic field lines move across the loop. The size of the induced voltage is proportional to (1) the number of wire loops passing through the magnetic field lines, (2) the strength of the magnetic field, and (3) the rate at which magnetic field lines pass through the wire.

An **electric generator** is a device that converts mechanical energy into electrical energy. A simple generator is built much like an electric motor, with an axle with many loops in a wire coil that rotates. The coil is turned by some form of mechanical energy, such as a water turbine or a steam turbine, which uses steam generated from fossil fuels or nuclear energy. As the coil rotates in a magnetic field, a current is induced in the coil.

Current from a power plant goes to a transformer to step up the voltage. A **transformer** is a device that steps up or steps down the AC voltage. It has two basic parts: (1) a *primary* or

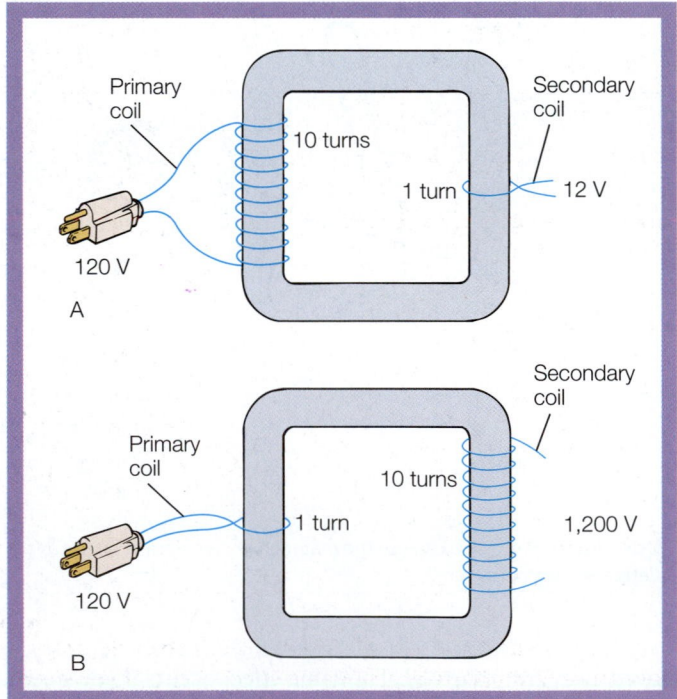

FIGURE 6.23 (A) This step-down transformer has ten turns on the primary for each turn on the secondary and reduces the voltage from 120 V to 12 V. (B) This step-up transformer increases the voltage from 120 V to 1,200 V, since there are ten turns on the secondary to each turn on the primary.

Note that the volts per wire loop are the same in each coil. The relationship is

$$\frac{\text{volts}_{\text{primary}}}{(\text{number of loops})_{\text{primary}}} = \frac{\text{volts}_{\text{secondary}}}{(\text{number of loops})_{\text{secondary}}}$$

or

$$\frac{V_p}{N_p} = \frac{V_s}{N_s} \qquad \text{equation 6.7}$$

EXAMPLE 6.8 *(Optional)*

A step-up transformer has five loops on its primary coil and twenty loops on its secondary coil. If the primary coil is supplied with an alternating current at 120 V, what is the voltage in the secondary coil?

SOLUTION

$N_p = 5$ loops

$N_s = 20$ loops

$V_p = 120$ V

$V_s = ?$

$$\frac{V_p}{N_p} = \frac{V_s}{N_s} \therefore V_s = \frac{V_p N_s}{N_p}$$

$$= \frac{(120 \text{ V})(20 \text{ loops})}{5 \text{ loops}}$$

$$= \frac{(120)(20)}{5} \frac{\text{V} \cdot \cancel{\text{loops}}}{\cancel{\text{loops}}}$$

$$= \boxed{480 \text{ V}}$$

EXAMPLE 6.9 *(Optional)*

The step-up transformer in example 6.8 is supplied with an alternating current at 120 V and a current of 10.0 A in the primary coil. What current flows in the secondary circuit?

$V_p = 120$ V

$I_p = 10.0$ A

$V_s = 480$ V

$I_s = ?$

$$V_p I_p = V_s I_s \therefore I_s = \frac{V_p I_p}{V_s}$$

$$= \frac{(120 \text{ V})(10.0 \text{ A})}{480 \text{ V}}$$

$$= \frac{(120)(10.0)}{480} \frac{\text{V} \cdot \text{A}}{\text{V}}$$

$$= \boxed{2.5 \text{ A}}$$

A step-up or step-down transformer steps up or steps down the voltage of an alternating current according to the ratio of wire loops in the primary and secondary coils. Assuming no losses in the transformer, the *power input* on the primary coil equals the *power output* on the secondary coil. Since $P = IV$, you can see that when the voltage is stepped up the current is correspondingly decreased, as

$$\text{power input} = \text{power output}$$

$$\text{watts input} = \text{watts output}$$

$$(\text{amps} \times \text{volts})_{\text{in}} = (\text{amps} \times \text{volts})_{\text{out}}$$

or

$$V_p I_p = V_s I_s \qquad \text{equation 6.8}$$

Energy losses in transmission are reduced by stepping up the voltage. Recall that electrical resistance results in an energy loss and a corresponding absolute temperature increase in the conducting wire. If the current is large, there are many collisions between the moving electrons and positive ions of the wire, resulting in a large energy loss. Each collision takes energy from the electric field, diverting it into increased kinetic energy of the positive ions and thus increased temperature of the conductor. The energy lost to resistance is therefore reduced by lowering the current, which is what a transformer does by increasing the voltage. Hence, electric power companies step up the voltage of generated power for economical transmission. A step-up transformer at a power plant, for example, might step up the voltage from 22,000 volts to 500,000 volts for transmission across the country to a city (figure 6.24A). This step up in voltage correspondingly reduces the current, lowering the resistance losses to a more acceptable 4 or 5 percent over long distances. A step-down transformer at a substation near the city reduces the voltage to several thousand volts for transmission around the city. Additional step-down transformers reduce this voltage to 120 volts for transmission to three or four houses (figure 6.24B).

A

B

FIGURE 6.24 Energy losses in transmission are reduced by increasing the voltage, so the voltage of generated power is stepped up at the power plant. (*A*) These transformers, for example, might step up the voltage from tens to hundreds of thousands of volts. After a step-down transformer reduces the voltage at a substation, still another transformer (*B*) reduces the voltage to 120 volts for transmission to three or four houses.

You may be familiar with many solid-state devices such as calculators, computers, word processors, digital watches, VCRs, digital stereos, and camcorders. All of these are called solid-state devices because they use a solid material, such as the semiconductor silicon, in an electric circuit in place of vacuum tubes. Solid-state technology developed from breakthroughs in the use of semiconductors during the 1950s, and the use of thin pieces of silicon crystal is common in many electric circuits today.

A related technology also uses thin pieces of a semiconductor such as silicon but not as a replacement for a vacuum tube. This technology is concerned with photovoltaic devices, also called *solar cells,* that generate electricity when exposed to light (box figure 6.5). A solar cell is unique in generating electricity since it produces electricity directly, without moving parts or chemical reactions, and potentially has a very long lifetime. This reading is concerned with how a solar cell generates electricity.

The conducting properties of silicon can be changed by *doping,* that is, artificially forcing atoms of other elements into the crystal lattice. Phosphorus, for example, has five electrons in its outermost shell, compared to the four in a silicon atom. When phosphorus atoms replace silicon atoms in the crystal lattice, there are extra electrons not tied up in the two electron bonds. The extra electrons move easily through the crystal lattice, carrying a charge. Since the phosphorus-doped silicon carries a negative charge, it is called an *n-type* semiconductor. The *n* means negative charge carrier.

A silicon crystal doped with boron will have atoms in the lattice with only three electrons in the outermost shell. This results in a deficiency, that is, electron "holes" that act as positive charges. A hole can move as an electron is attracted to it, but it leaves another hole elsewhere, from where it moved. Thus, a flow of electrons in one direction is equivalent to a flow of holes in the opposite direction. A hole, there-

A   B

Box Figure 6.5   Solar cells are economical in remote uses such as (*A*) navigational aids and (*B*) communications. The solar panels in both of these examples are oriented toward the south.

fore, behaves as a positive charge. Since the boron-doped silicon carries a positive charge, it is called a *p-type* semiconductor. The *p* means positive charge carrier.

The basic operating part of a silicon solar cell is typically an 8 cm wide and 3×10^{-1}mm (about one-hundredth of an inch) thick wafer cut from a silicon crystal. One side of the wafer is doped with boron to make p-silicon, and the other side is doped with phosphorus to make n-silicon. The place of contact between the two is called the p-n junction, which creates a *cell barrier.* The cell barrier forms as electrons are attracted from the n-silicon to the holes in the p-silicon. This creates a very thin zone of negatively charged p-silicon and positively charged n-silicon (box figure 6.6). Thus, an internal electric field is established at the p-n junction, and the field is the cell barrier.

The cell is thin, and light can penetrate through the p-n junction. Light impacts the p-silicon, freeing electrons. Low-energy free electrons might combine with a hole, but high-energy electrons cross the cell

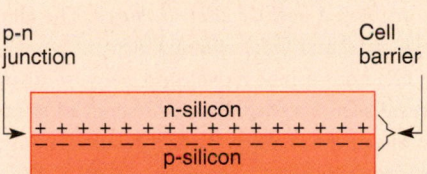

Box Figure 6.6   The cell barrier forms at the p-n junction between the n-silicon and the p-silicon. The barrier creates a "one-way" door that accumulates negative charges in the n-silicon.

barrier into the n-silicon. The electron loses some of its energy, and the barrier prevents it from returning, creating an excess negative charge in the n-silicon and a positive charge in the p-silicon. This establishes a potential that will drive a current.

Today, solar cells are essentially handmade and are economical only in remote power uses (navigational aids, communications, or irrigation pumps) and in consumer specialty items (solar-powered watches and calculators). Research continues on finding methods of producing highly efficient, highly reliable solar cells that are affordably priced.

Science and Society

Blackout Reveals Pollution

In August 2003, problems in a huge electrical grid resulted in power plants shutting down and a massive electric power blackout that affected some 50 million people. Scientists from the University of Maryland took advantage of the shutdown, measuring different levels of atmospheric air pollution while the fossil-fueled power plants in the Ohio Valley were shut down. Scooping many air samples with a small airplane 24 hours after the blackout, they found 90 percent less sulfur dioxide, 50 percent less ozone, and 70 percent fewer light-scattering particles from the air in the same area than when the power plants were running. The scientists stated that the result could come from an underestimation of emission from power plants or from unknown chemical reactions in the atmosphere.

Questions to Discuss

1. Are there atmospheric factors that might have contributed to the findings?

2. Are there factors on the ground that might have contributed to the findings?

3. Do the results mean that power plants contribute that much pollution, or what else should be considered?

4. How would you conduct a pollution-measuring experiment that would leave no room for doubt about the results?

SUMMARY

The basic unit of matter is the *atom,* which is made up of *protons, electrons,* and other particles. Protons and electrons have a property called *electric charge;* electrons have a *negative electric charge* and protons have a *positive electric charge.* The charges interact, and *like charges repel* and *unlike charges attract.*

Electrons can be moved and *electrostatic charge,* or *static electricity,* results from a surplus or deficiency of electrons.

A *quantity of charge* (q) is measured in units of *coulombs* (C), the charge equivalent to the transfer of 6.24×10^{18} charged particles such as the electron. The *fundamental charge* of an electron or proton is 1.60×10^{-19} coulomb. The *electric forces* between two charged objects can be calculated from the relationship between the quantity of charge and the distance between two charged objects. The relationship is known as *Coulomb's law.*

A flow of electric charge is called an *electric current* (I). Current (I) is measured as the *rate* of flow of charge, the quantity of charge (q) through a conductor in a period of time (t). The unit of current in coulomb/second is called an *ampere,* or *amp* for short (A).

An *electric circuit* has some device that does work in moving charges through wires to do work in another part of the circuit. The *work done* and the *size of the charge* moved defines *voltage.* A *volt* (V) is the ratio of work to charge moved, $V = W/q$. The ratio of volts/amps in a circuit is the *unit of resistance* called an *ohm.* Ohm's law is $V = IR$.

Disregarding the energy lost to resistance, the *work* done by a voltage source is equal to the work accomplished in electrical devices in a circuit. The *rate of doing work* is *power,* or *work per unit time,* $P = W/t$. *Electrical power* can be calculated from the relationship of $P = IV$, which gives the power unit of *watts.*

Magnets have two poles about which their attraction is concentrated. When free to turn, one pole moves to the north and the other to the south. The north-seeking pole is called the *north pole* and the south-seeking pole is called the *south pole. Like poles repel* one another and *unlike poles attract.*

A *current-carrying wire* has magnetic field lines of closed, *concentric circles* that are at right angles to the length of wire. The *direction* of the magnetic field depends on the direction of the current. A coil of many loops is called a *solenoid* or *electromagnet.* The electromagnet is the working part in electric meters, electromagnetic switches, and the electric motor.

When a loop of wire is moved in a magnetic field, or if a magnetic field is moved past a wire loop, a voltage is induced in the wire loop. The interaction is called *electromagnetic induction.* An electric generator is a rotating coil of wire in a magnetic field. The coil is rotated by mechanical energy, and electromagnetic induction induces a voltage, thus converting mechanical energy to electrical energy. A *transformer* steps up or steps down the voltage of an alternating current. The ratio of input and output voltage is determined by the number of loops in the primary and secondary coils. Increasing the voltage decreases the current, which makes long-distance transmission of electrical energy economically feasible.

Summary of Equations

6.1
$$\text{electric force} = (\text{constant}) \times \frac{\text{charge on one object} \times \text{charge on second object}}{\text{distance between objects squared}}$$

$$F = k\frac{q_1 q_2}{d^2}$$

where $k = 9.00 \times 10^9$ newton·meters²/coulomb²

6.2
$$\text{electrical current} = \frac{\text{quantity of charge}}{\text{time}}$$

$$I = \frac{q}{t}$$

$$6.3 \qquad \text{voltage} = \frac{\text{work}}{\text{charge moved}}$$

$$V = \frac{W}{q}$$

$$6.4 \qquad \text{voltage} = \text{current} \times \text{resistance}$$

$$V = IR$$

$$6.5 \qquad \text{power} = \text{voltage} \times \text{current}$$

$$P = VI$$

$$6.6 \qquad \text{cost} = \frac{(\text{work})(\text{time})(\text{rate})}{1{,}000 \, \dfrac{\text{watt}}{\text{kilowatt}}}$$

$$6.7 \qquad \frac{\text{volts}_{\text{primary}}}{(\text{number of loops})_{\text{primary}}} = \frac{\text{volts}_{\text{secondary}}}{(\text{number of loops})_{\text{secondary}}}$$

$$\frac{V_\text{p}}{N_\text{p}} = \frac{V_\text{s}}{N_\text{s}}$$

$$6.8 \qquad (\text{amps} \times \text{volts})_{\text{in}} = (\text{amps} \times \text{volts})_{\text{out}}$$

$$V_\text{p}I_\text{p} = V_\text{s}I_\text{s}$$

KEY TERMS

alternating current (p. **132**)	electric circuit (p. **132**)	electrostatic charge (p. **127**)	solenoid (p. **141**)
amp (p. **130**)	electric current (p. **129**)	fundamental charge (p. **128**)	step-down transformer (p. **144**)
ampere (p. **130**)	electric field (p. **129**)	galvanometer (p. **141**)	step-up transformer (p. **144**)
coulomb (p. **128**)	electric generator (p. **144**)	magnetic field (p. **138**)	transformer (p. **144**)
Coulomb's law (p. **129**)	electric force (p. **127**)	magnetic poles (p. **138**)	volt (p. **133**)
direct current (p. **132**)	electromagnet (p. **141**)	ohm (p. **134**)	watt (p. **134**)
electric charge (p. **126**)	electromagnetic induction (p. **142**)	Ohm's law (p. **134**)	

APPLYING THE CONCEPTS

Answers are located in appendix F.

1. An object that acquires an excess of electrons becomes a (an)
   a. ion.
   b. negatively charged object.
   c. positively charged object.
   d. electric conductor.

2. Which of the following is most likely to acquire an electrostatic charge?
   a. electric conductor
   b. electric nonconductor
   c. Both are equally likely.
   d. None of the above is correct.

3. A quantity of electric charge is measured in a unit called a (an)
   a. coulomb.
   b. volt.
   c. amp.
   d. watt.

4. The unit that describes the potential difference that occurs when a certain amount of work is used to move a certain quantity of charge is called the
   a. ohm.
   b. volt.
   c. amp.
   d. watt.

5. An electric current is measured in units of
   a. coulomb.
   b. volt.
   c. amp.
   d. watt.

6. In an electric current, the electrons are moving
   a. at a very slow rate.
   b. at the speed of light.
   c. faster than the speed of light.
   d. at a speed described as "Warp 8."

7. If you multiply amps $\times$ volts, the answer will be in units of
   a. resistance.
   b. work.
   c. current.
   d. power.

8. The unit of resistance is the
   a. watt.
   b. ohm.
   c. amp.
   d. volt.

9. Compared to a thick wire, a thin wire of the same length, material, and temperature has
   a. less electric resistance.
   b. more electric resistance.
   c. the same electric resistance.
   d. None of the above is correct.

10. A permanent magnet has magnetic properties because
    a. the magnetic fields of its electrons are balanced.
    b. of an accumulation of monopoles in the ends.
    c. the magnetic fields of elections are aligned.
    d. All of the above are correct.

11. A current-carrying wire has a magnetic field around it because
    a. a moving charge produces a magnetic field of its own.
    b. the current aligns the magnetic fields in the metal of the wire.
    c. the metal was magnetic before the current was established and the current enhanced the magnetic effect.
    d. None of the above is correct.

12. A step-up transformer steps up (the)
    a. power.
    b. current.
    c. voltage.
    d. All of the above are correct.

QUESTIONS FOR THOUGHT

1. Explain why a balloon that has been rubbed sticks to a wall for a while.
2. Explain what is happening when you walk across a carpet and receive a shock when you touch a metal object.
3. Why does a positively or negatively charged object have multiples of the fundamental charge?
4. Explain how you know that it is an electric field, not electrons, that moves rapidly through a circuit.
5. Is a kWh a unit of power or a unit of work? Explain.
6. What is the difference between AC and DC?
7. What is a magnetic pole? How are magnetic poles named?
8. How is an unmagnetized piece of iron different from the same piece of iron when it is magnetized?
9. Explain why the electric utility company increases the voltage of electricity for long-distance transmission.
10. Describe how an electric generator is able to generate an electric current.
11. Why does the north pole of a magnet generally point to the geographic North Pole if like poles repel?
12. Explain what causes an electron to move toward one end of a wire when the wire is moved across a magnetic field.

FOR FURTHER ANALYSIS

1. Explain how the model of electricity as electrons moving along a wire is an oversimplification that misrepresents the complex nature of an electric current.
2. What are the significant similarities and differences between AC and DC? What determines which is "better" for a particular application?
3. Transformers usually have signs warning, "Danger—High Voltage." Analyze if this is a contradiction since it is exposure to amps, not volts, that harms people.
4. Will a fuel cell be the automobile engine of the future? Identify the facts, beliefs, and theories that support or refute your answer.
5. Analyze the apparent contradiction in the statement that "solar energy is free" with the fact that solar cells are too expensive to use as an energy source.
6. What are the basic similarities and differences between an electric field and a magnetic field?

INVITATION TO INQUIRY

Earth Power?

Investigate if you can use Earth's magnetic field to induce an electric current in a conductor. Connect the ends of a 10 m wire to a galvanometer. Have a partner hold the ends of the wire on the galvanometer while you hold the end of the wire loop and swing the doubled wire like a jump rope.

If you accept this invitation, try swinging the wire in different directions. Can you figure out a way to measure how much electricity you can generate?

PARALLEL EXERCISES

The exercises in groups A and B cover the same concepts. Solutions to group A exercises are located in appendix G.

Group A

1. A rubber balloon has become negatively charged from being rubbed with a wool cloth, and the charge is measured as 1.00×10^{-14} C. According to this charge, the balloon contains an excess of how many electrons?

2. An electric current through a wire is 6.00 C every 2.00 s. What is the magnitude of this current?

3. A current of 4.00 A flows through a toaster connected to a 120.0 V circuit. What is the resistance of the toaster?

4. What is the current in a 60.0 Ω resistor when the potential difference across it is 120.0 V?

5. A lightbulb with a resistance of 10.0 Ω allows a 1.20 A current to flow when connected to a battery. (a) What is the voltage of the battery? (b) What is the power of the lightbulb?

6. A small radio operates on 3.00 V and has a resistance of 15.0 Ω. At what rate does the radio use electric energy?

7. A 1,200 W hair dryer is operated on a 120 V circuit for 15 min. If electricity costs $0.10/kWh, what was the cost of using the blow dryer?

8. An automobile starter rated at 2.00 hp draws how many amps from a 12.0 V battery?

9. An average-sized home refrigeration unit has a 1/3 hp fan motor for blowing air over the inside cooling coils, a 1/3 hp fan motor for blowing air over the outside condenser coils, and a 3.70 hp compressor motor. (a) All three motors use electric energy at what rate? (b) If electricity costs $0.10/kWh, what is the cost of running the unit per hour? (c) What is the cost for running the unit 12 hours a day for a 30-day month?

10. A 15 ohm toaster is turned on in a circuit that already has a 0.20 hp motor, three 100 W lightbulbs, and a 600 W electric iron that are on. Will this trip a 15 A circuit breaker? Explain.

11. A power plant generator produces a 1,200 V, 40 A alternating current that is fed to a step-up transformer before transmission over the high lines. The transformer has a ratio of 200 to 1 wire loops. (a) What is the voltage of the transmitted power? (b) What is the current?

12. A step-down transformer has an output of 12 V and 0.5 A when connected to a 120 V line. Assuming no losses: (a) What is the ratio of primary to secondary loops? (b) What current does the transformer draw from the line? (c) What is the power output of the transformer?

Group B

1. An inflated rubber balloon is rubbed with a wool cloth until an excess of a billion electrons is on the balloon. What is the magnitude of the charge on the balloon?

2. A wire carries a current of 2.0 A. At what rate is the charge flowing?

3. A current of 0.83 A flows through a lightbulb in a 120 V circuit. What is the resistance of this lightbulb?

4. What is the voltage across a 60.0 Ω resistor with a current of 3⅓ amp?

5. A 10.0 Ω lightbulb is connected to a 12.0 V battery. (a) What current flows through the bulb? (b) What is the power of the bulb?

6. A lightbulb designed to operate in a 120.0 V circuit has a resistance of 192 Ω. At what rate does the bulb use electric energy?

7. What is the monthly energy cost of leaving a 60 W bulb on continuously if electricity costs $0.10 per kWh?

8. An electric motor draws a current of 11.5 A in a 240 V circuit. (a) What is the power of this motor in W? (b) How many horsepower is this?

9. A swimming pool requiring a 2.0 hp motor to filter and circulate the water runs for 18 hours a day. What is the monthly electrical cost for running this pool pump if electricity costs $0.10 per kWh?

10. Is it possible for two people to simultaneously operate 1,300 W hair dryers on the same 120 V circuit without tripping a 15 A circuit breaker? Explain.

11. A step-up transformer has a primary coil with 100 loops and a secondary coil with 1,500 loops. If the primary coil is supplied with a household current of 120 V and 15 A, (a) what voltage is produced in the secondary circuit? (b) What current flows in the secondary circuit?

12. The step-down transformer in a local neighborhood reduces the voltage from a 7,200 V line to 120 V. (a) If there are 125 loops on the secondary, how many are on the primary coil? (b) What current does the transformer draw from the line if the current in the secondary is 36 A? (c) What are the power input and output?

Light

This fiber optics bundle carries pulses of light from a laser to carry much more information than could be carried by electrons moving through wires. This is part of a dramatic change under way, a change that will first find a hybrid "optoelectronics" replacing the more familiar "electronics" of electrons and wires.

CORE **CONCEPT**

Light is electromagnetic radiation—energy—that interacts with matter.

OUTLINE

Physics

▷ Light is a form of energy (Ch. 3; Ch. 7).
▷ Light energy can be converted to electricity (Ch. 6).

Chemistry

▷ Light is emitted from an atom when an electron jumps from a higher to a lower orbital (Ch. 8).
▷ Gamma radiation is electromagnetic radiation (Ch. 11).

Astronomy

▷ Light from a star will tell you about the star's temperature and motion (Ch. 12).

Life Science

▷ Sunlight provides energy flow in ecosystems (Ch. 22).

OVERVIEW

You use light and your eyes more than any other sense to learn about your surroundings. All of your other senses—touch, taste, sound, and smell—involve matter, but the most information is provided by light. Yet, light seems more mysterious than matter. You can study matter directly, measuring its dimensions, taking it apart, and putting it together to learn about it. Light, on the other hand, can only be studied indirectly in terms of how it behaves (figure 7.1). Once you understand its behavior, you know everything there is to know about light. Anything else is thinking about what the behavior means.

The behavior of light has stimulated thinking, scientific investigations, and debate for hundreds of years. The investigations and debate have occurred because light cannot be directly observed, which makes the exact nature of light very difficult to pin down. For example, you know that light moves energy from one place to another place. You can feel energy from the Sun as sunlight warms you, and you know that light has carried this energy across millions of miles of empty space. The ability of light to move energy like this could be explained (1) as energy transported by waves, just as sound waves carry energy from a source, or (2) as the kinetic energy of a stream of moving particles, which give up their energy when they strike a surface. The movement of energy from place to place could be explained equally well by a wave model of light or by a particle model of light. When two possibilities exist like this in science, experiments are designed and measurements are made to support one model and reject the other. Light, however, presents a baffling dilemma. Some experiments provide evidence that light consists of waves and not a stream of moving particles. Yet other experiments provide evidence of just the opposite, that light is a stream of particles and not a wave. Evidence for accepting a wave or particle model seems to depend on which experiments are considered.

The purpose of using a model is to make new things understandable in terms of what is already known. When these new things concern light, three models are useful in visualizing separate behaviors. Thus, the electromagnetic wave model will be used to describe how light is created at a source. Another model, a model of light as a ray, a small beam of light, will be used to discuss some common properties of light such as reflection and the refraction, or bending, of light. Finally, properties of light that provide evidence for a particle model will be discussed before ending with a discussion of the present understanding of light.

7.1 SOURCES OF LIGHT

The Sun, and other stars, lightbulbs, and burning materials all give off light. When something produces light, it is said to be **luminous.** The Sun is a luminous object that provides almost all of the *natural* light on Earth. A small amount of light does reach Earth from the stars but not really enough to see by on a moonless night. The Moon and planets shine by reflected light and do not produce their own light, so they are not luminous.

Burning has been used as a source of *artificial* light for thousands of years. A wood fire and a candle flame are luminous because of their high temperatures. When visible light is given off as a result of high temperatures, the light source is said to be **incandescent.** A flame from any burning source, an ordinary lightbulb, and the Sun are all incandescent sources because of high temperatures.

How do incandescent objects produce light? One explanation is given by the electromagnetic wave model. This model describes a relationship between electricity, magnetism, and light. The model pictures an electromagnetic wave as forming whenever an electric charge is *accelerated* by some external force. The acceleration produces a wave consisting of electrical and magnetic fields that become separated from the accelerated

FIGURE 7.1 Light, sounds, and odors can identify this pleasing environment, but light provides the most information. Sounds and odors can be identified and studied directly, but light can only be studied indirectly, that is, in terms of how it behaves. As a result, the behavior of light has stimulated thinking, scientific investigations, and debate for hundreds of years. Perhaps you have wondered about light and its behaviors. What is light?

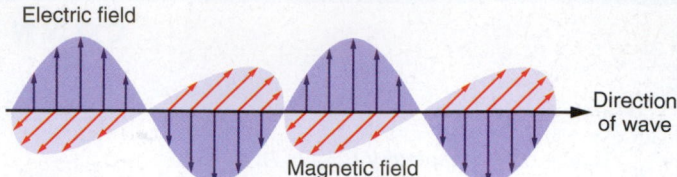

FIGURE 7.2 The electric and magnetic fields in an electromagnetic wave vary together. Here the fields are represented by arrows that indicate the strength and direction of the fields. Note the fields are perpendicular to each other and to the direction of the wave.

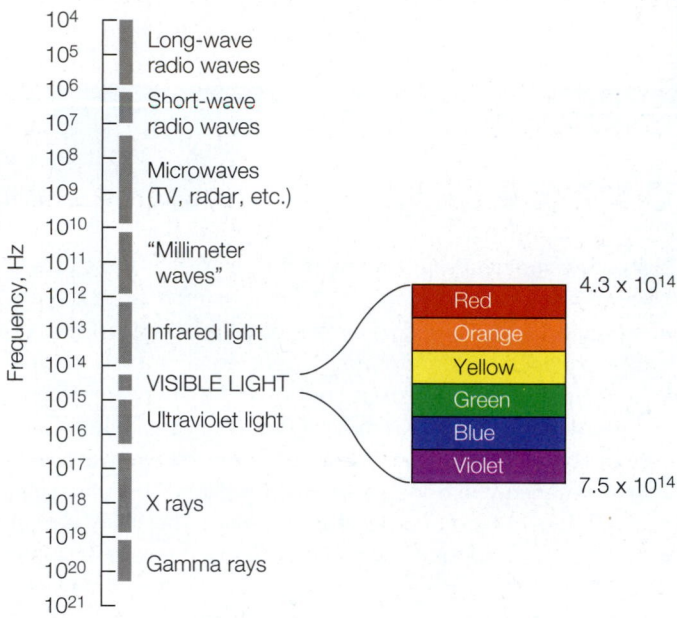

FIGURE 7.3 The electromagnetic spectrum. All electromagnetic waves have the same fundamental character and the same speed in a vacuum, but many aspects of their behavior depend on their frequency.

charge, moving off into space (figure 7.2). As the wave moves through space, the two fields exchange energy back and forth, continuing on until they are absorbed by matter and give up their energy.

The frequency of an electromagnetic wave depends on the acceleration of the charge; the greater the acceleration, the higher the frequency of the wave that is produced. The complete range of frequencies is called the *electromagnetic spectrum* (figure 7.3). The spectrum ranges from radio waves at the low frequency end of the spectrum to gamma rays at the high frequency end. Visible light occupies only a small part of the middle portion of the complete spectrum.

Visible light is emitted from incandescent sources at high temperatures, but actually electromagnetic radiation is given off from matter at *any* temperature. This radiation is called **blackbody radiation,** which refers to an idealized material (the *blackbody*) that perfectly absorbs and perfectly emits electromagnetic radiation. From the electromagnetic wave model, the radiation originates from the acceleration of charged particles near the surface of an object. The frequency of the blackbody radiation is determined by the energy available for accelerating charged particles, that is, the temperature of the object. Near

absolute zero, there is little energy available and no radiation is given off. As the temperature of an object is increased, more energy is available, and this energy is distributed over a range of values, so more than one frequency of radiation is emitted. A graph of the frequencies emitted from the range of available energy is thus somewhat bell-shaped. The steepness of the curve and the position of the peak depend on the temperature (figure 7.4). As the temperature of an object increases, there is an increase in the *amount* of radiation given off, and the peak radiation emitted progressively *shifts* toward higher and higher frequencies.

At room temperature, the radiation given off from an object is in the infrared region, invisible to the human eye. When the temperature of the object reaches about 700°C (about 1,300°F), the peak radiation is still in the infrared region, but the peak has shifted enough toward the higher frequencies that a little visible light is emitted as a dull red glow. As the temperature of the object continues to increase, the amount

Ultraviolet Light and Life

Ultraviolet (UV) light is a specific portion of the electromagnetic spectrum that has a significant impact on living things. UV light has been subdivided into different categories based on wavelength—the shorter the wavelength, the higher the energy level. Ultraviolet light causes mutations when absorbed by DNA.

▶ Ultraviolet light is important in forming vitamin D in the skin. When cholesterol in the skin is struck by ultraviolet light, some of the cholesterol is converted to vitamin D. Therefore, some exposure to ultraviolet light is good. However, since large amounts of ultraviolet light damage the skin, the normal tanning reaction seen in the skin of light-skinned people is a protective reaction to the damaging effects of ultraviolet light. The darker skin protects the deeper layers of the skin from damage.

▶ Tanning booths have mostly UV-A with a small amount of UV-B. Dosages in tanning booths are usually higher than in sunlight.

▶ Since UV-C disrupts DNA, it can be used to sterilize medical instruments, drinking water, and wastewater. A standard mercury vapor light emits much of its light in the UV-C range, but the lamp is coated to prevent light in that range from leaving the lamp. Specially manufactured lamps without the coating and ultraviolet transparent glass are used to provide UV-C for germicidal purposes.

▶ Some animals are able to see light in the ultraviolet spectrum. In particular, bees do not see the color red but are able to see ultraviolet. When flowers are photographed with filters that allow ultraviolet light to be recorded on the film, flowers often have a different appearance from what we see. The pattern of ultraviolet light assists the bees in finding the source of nectar within the flower.

Class	Wavelength (nanometers)	Characteristics	Biological Effects
UV-A	320–400	Most passes through the ozone layer; can pass through window glass	Causes skin damage and promotes skin cancer
UV-B	280–320	90% blocked by ozone layer; amount is highest at midday	Causes sunburn, skin damage, and skin cancer
UV-C	100–280	Nearly all blocked by ozone layer	Kills bacteria and viruses

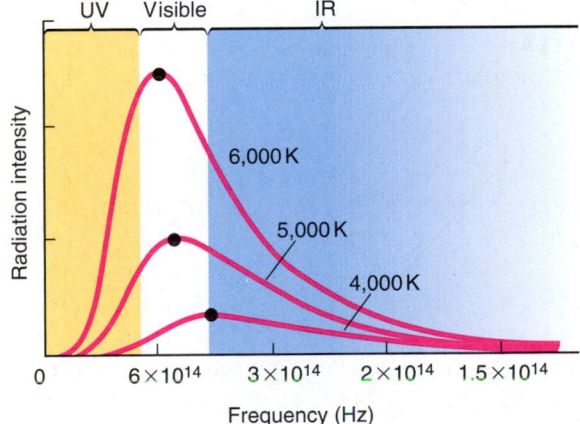

FIGURE 7.4 Blackbody spectra for several different temperatures. The frequency of the peak of the curve (shown by a dot) shifts to higher frequency at higher temperatures.

temperature is noted in the referent description of an object being "red hot," "white hot," and so forth.

The incandescent flame of a candle or fire results from the blackbody radiation of carbon particles in the flame. At a black-body temperature of 1,500°C (about 2,700°F), the carbon particles emit visible light in the red to yellow frequency range. The tungsten filament of an incandescent lightbulb is heated to about 2,200°C (about 4,000°F) by an electric current. At this temperature, the visible light emitted is in the reddish, yellow-white range.

The radiation from the Sun, or sunlight, comes from the Sun's surface, which has a temperature of about 5,700°C (about 10,000°F). As shown in figure 7.5, the Sun's radiation has a broad spectrum centered near the yellow-green frequency. Your eye is most sensitive to this frequency of sunlight. The spectrum of sunlight before it travels through Earth's atmosphere is infrared (about 51 percent), visible light (about 40 percent), and ultraviolet (about 9 percent). Sunlight originated as energy released from nuclear reactions in the Sun's core (see chapter 12). This energy requires about a million years to work its way up to the surface. At the surface, the energy from the core accelerates charged particles, which then emit light like tiny antennas. The sunlight requires about eight minutes to travel the distance from the Sun's surface to Earth.

of radiation increases, and the peak continues to shift toward shorter wavelengths. Thus, the object begins to glow brighter, and the color changes from red, to orange, to yellow, and eventually to white. The association of this color change with

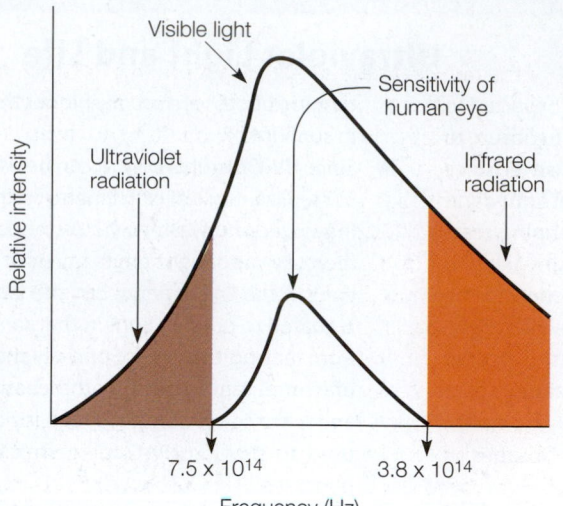

FIGURE 7.5 Sunlight is about 9 percent ultraviolet radiation, 40 percent visible light, and 51 percent infrared radiation before it travels through Earth's atmosphere.

7.2 PROPERTIES OF LIGHT

You can see luminous objects from the light they emit, and you can see nonluminous objects from the light they reflect, but you cannot see the path of the light itself. For example, you cannot see a flashlight beam unless you fill the air with chalk dust or smoke. The dust or smoke particles reflect light, revealing the path of the beam. This simple observation must be unknown to the makers of science fiction movies, since they always show visible laser beams zapping through the vacuum of space.

Some way to represent the invisible travels of light is needed in order to discuss some of its properties. Throughout history, a **light ray model** has been used to describe the travels of light. The meaning of this model has changed over time, but it has always been used to suggest that "something" travels in *straight-line paths*. The light ray is a line that is drawn to represent the straight-line travel of light. A line is drawn to represent this imaginary beam to illustrate the law of reflection (as from a mirror) and the law of refraction (as through a lens). There are limits to using a light ray for explaining some properties of light, but it works very well in explaining mirrors, prisms, and lenses.

Light Interacts with Matter

A ray of light travels in a straight line from a source until it encounters some object or particles of matter (figure 7.6). What happens next depends on several factors, including (1) the smoothness of the surface, (2) the nature of the material, and (3) the angle at which the light ray strikes the surface.

The *smoothness* of the surface of an object can range from perfectly smooth to extremely rough. If the surface is perfectly smooth, rays of light undergo *reflection,* leaving the surface parallel to each other. A mirror is a good example of a very smooth surface that reflects light this way (figure 7.7A).

If a surface is not smooth, the light rays are reflected in many random directions as *diffuse reflection* takes place (figure 7.7B). Rough and irregular surfaces and dust in the air make diffuse reflections. It is diffuse reflection that provides light in places not in direct lighting, such as under a table or a tree. Such shaded areas would be very dark without diffuse reflection of light.

Some materials allow much of the light that falls on them to move through the material without being reflected. Materials that allow transmission of light through them are called *transparent*. Glass and clear water are examples of transparent materials. Many materials do not allow transmission of any light and are called *opaque*. Opaque materials reflect light, absorb light, or some combination of partly absorbing and

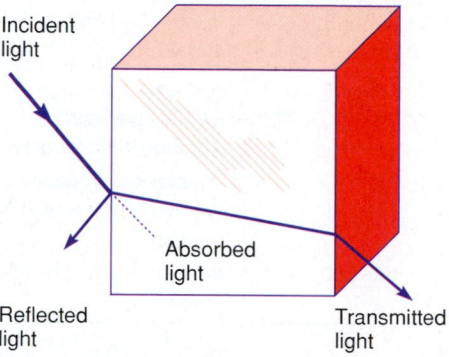

FIGURE 7.6 Light that interacts with matter can be reflected, absorbed, or transmitted through transparent materials. Any combination of these interactions can take place, but a particular substance is usually characterized by what it mostly does to light.

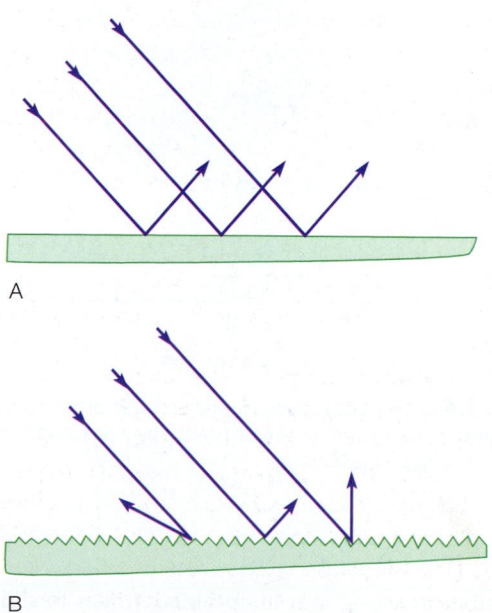

FIGURE 7.7 (A) Rays reflected from a perfectly smooth surface are parallel to each other. (B) Diffuse reflection from a rough surface causes rays to travel in many random directions.

FIGURE 7.8 Light travels in a straight line, and the color of an object depends on which wavelengths of light the object reflects. Each of these flowers absorbs the colors of white light and reflects the color that you see.

partly reflecting light (figure 7.8). The light that is reflected varies with wavelength and gives rise to the perception of color, which will be discussed shortly. Absorbed light gives up its energy to the material and may be reemitted at a different wavelength, or it may simply show up as a temperature increase.

The *angle* of the light ray to the surface and the nature of the material determine if the light is absorbed, transmitted through a transparent material, or reflected. Vertical rays of light, for example, are mostly transmitted through a transparent material with some reflection and some absorption. If the rays strike the surface at some angle, however, much more of the light is reflected, bouncing off the surface. Thus, the glare of reflected sunlight is much greater around a body of water in the late afternoon than when the Sun is directly overhead.

Light that interacts with matter is reflected, transmitted, or absorbed, and all combinations of these interactions are possible. Materials are usually characterized by which of these interactions they *mostly* do, but this does not mean that other interactions are not occurring too. For example, a window glass is usually characterized as a transmitter of light. Yet, the glass always *reflects* about 4 percent of the light that strikes it. The reflected light usually goes unnoticed during the day because of the bright light that is transmitted from the outside. When it is dark outside, you notice the reflected light as the window glass now appears to act much like a mirror. A one-way mirror is another example of both reflection and transmission occurring. A mirror is usually characterized as a reflector of light. A one-way mirror, however, has a very thin silvering that reflects most of the light but still transmits a little. In a lighted room, a one-way mirror appears to reflect light just as any other mirror does. But a person behind the mirror in a dark room can see into the lighted room by means of the transmitted light. Thus, you know that this mirror transmits as well as reflects light. One-way mirrors are used to unobtrusively watch for shoplifters in many businesses.

Reflection

Most of the objects that you see are visible from diffuse reflection. For example, consider some object such as a tree that you see during a bright day. Each *point* on the tree must reflect light in all directions, since you can see any part of the tree from any angle (figure 7.9). As a model, think of bundles of light rays entering your eye, which enable you to see the tree. This means that you can see any part of the tree from any angle because different bundles of reflected rays will enter your eye from different parts of the tree.

Light rays that are diffusely reflected move in all possible directions, but rays that are reflected from a smooth surface, such as a mirror, leave the mirror in a definite direction. Suppose you look at a tree in a mirror. There is only one place on the mirror where you look to see any one part of the tree. Light is reflecting off the mirror from all parts of the tree, but the only rays that reach your eye are the rays that are reflected at a certain angle from the place where you look. The relationship between the light rays moving from the tree and the direction in which they are reflected from the mirror to reach your eyes can be understood by drawing three lines: (1) a line representing an original ray from the tree, called the *incident ray,* (2) a line representing a reflected ray, called the *reflected ray,* and (3) a reference line that is perpendicular to the reflecting surface and is located at the point where the incident ray struck the surface. This line is called the *normal.* The angle between the incident ray and the normal is called the *angle of incidence,* θ_i, and the angle between the reflected ray and the normal is called the *angle of reflection,* θ_r (figure 7.10). The *law of reflection,* which

FIGURE 7.9 Bundles of light rays are reflected diffusely in all directions from every point on an object. Only a few light rays are shown from only one point on the tree in this illustration. The light rays that move to your eyes enable you to see the particular point from which they were reflected.

was known to the ancient Greeks, is that the *angle of incidence equals the angle of reflection,* or

$$\theta_i = \theta_r \qquad \text{equation 7.1}$$

Figure 7.11 shows how the law of reflection works when you look at a flat mirror. Light is reflected from all points on the box, and of course only the rays that reach your eyes are detected. These rays are reflected according to the law of reflection, with the angle of reflection equaling the angle of incidence. If you move your head slightly, then a different bundle of rays reaches your eyes. Of all the bundles of rays that reach your eyes, only two rays from a point are shown in the illustration. After these two rays are reflected, they continue to spread apart at the same rate that they were spreading before reflection. Your eyes and brain do not know that the rays have been reflected, and the diverging rays appear to come from behind the mirror, as the dashed lines show. The image, therefore, appears to be the same distance *behind* the mirror as the box is from the front of the mirror. Thus, a mirror image is formed where the rays of light *appear* to originate. This is called a **virtual image.** A virtual image is the result of your eyes' and brain's interpretations of light rays, not actual light rays originating from an image. Light rays that do originate from the other kind of image are called a **real image.** A real image is like the one displayed on a movie screen, with light originating from the image. A virtual image cannot be displayed on a screen, since it results from an interpretation.

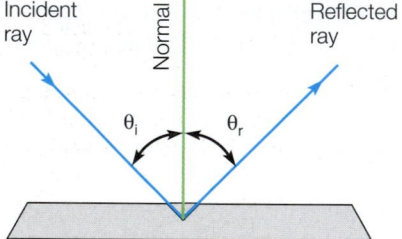

FIGURE 7.10 The law of reflection states that the angle of incidence (θ_i) is equal to the angle of reflection (θ_r). Both angles are measured from the *normal,* a reference line drawn perpendicular to the surface at the point of reflection.

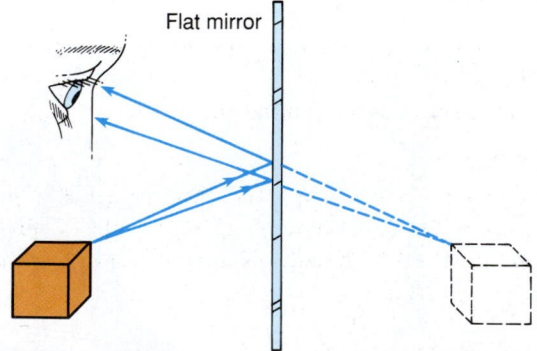

FIGURE 7.11 Light rays leaving a point on the block are reflected according to the law of reflection, and those reaching your eye are seen. After reflecting, the rays continue to spread apart at the same rate. You interpret this to be a block the same distance behind the mirror. You see a virtual image of the block, because light rays do not actually move from the image.

Curved mirrors are either *concave,* with the center part curved inward, or *convex,* with the center part bulging outward. A concave mirror can be used to form an enlarged virtual image, such as a shaving or makeup mirror, or it can be used to form a real image, as in a reflecting telescope. Convex mirrors, for example, the mirrors on the sides of trucks and vans, are often used to increase the field of vision. Convex mirrors are also placed above an aisle in a store to show a wide area.

Refraction

You may have observed that an object that is partly in the air and partly in water appears to be broken, or bent, where the air and water meet. When a light ray moves from one transparent material to another, such as from water through air, the ray undergoes a change in the direction of travel at the boundary between the two materials. This change of direction of a light ray at the boundary is called **refraction.** The amount of change can be measured as an angle from the normal, just as it was for the angle of reflection. The incoming ray is called the *incident ray,* as before, and the new direction of travel is called the *refracted ray.* The angles of both rays are measured from the normal (figure 7.12).

Refraction results from a *change in speed* when light passes from one transparent material into another. The speed of light in a vacuum is 3.00×10^8 m/s, but it is slower when moving through a transparent material. In water, for example, the speed of light is reduced to about 2.30×10^8 m/s. The speed of

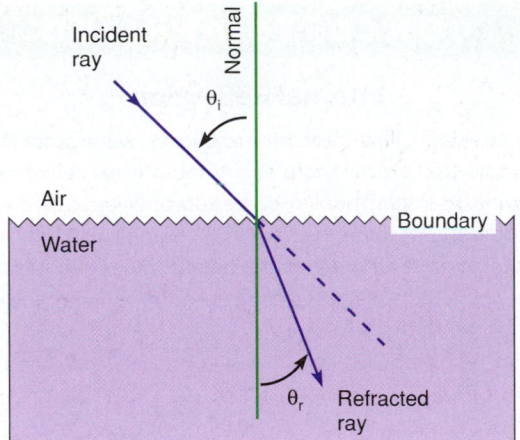

FIGURE 7.12 A ray diagram shows refraction at the boundary as a ray moves from air through water. Note that θ_i does not equal θ_r in refraction.

light has a magnitude that is specific for various transparent materials.

When light moves from one transparent material to another transparent material with a *slower* speed of light, the ray is refracted *toward* the normal (figure 7.13A). For example, light travels through air faster than through water. Light traveling from air into water is therefore refracted toward the normal as it enters the water. On the other hand, if light has a *faster* speed in the new material, it is refracted *away* from the normal. Thus, light traveling from water into the air is refracted away from the normal as it enters the air (figure 7.13B).

The magnitude of refraction depends on (1) the angle at which light strikes the surface and (2) the ratio of the speed of light in the two transparent materials. An incident ray that is perpendicular (90°) to the surface is not refracted at all. As the angle of incidence is increased, the angle of refraction is also increased. There is a limit, however, that occurs when the angle of refraction reaches 90°, or along the water surface. Figure 7.14 shows rays of light traveling from water to air at various angles. When the incident ray is about 49°, the angle of refraction that results is 90°, along the water surface. This limit to the angle of incidence that results in an angle of refraction of 90° is called the *critical angle* for a water-to-air surface (figure 7.14). At any incident angle greater than the critical angle, the light ray does not move from the water to the air but is *reflected* back from the surface as if it were a mirror. This is

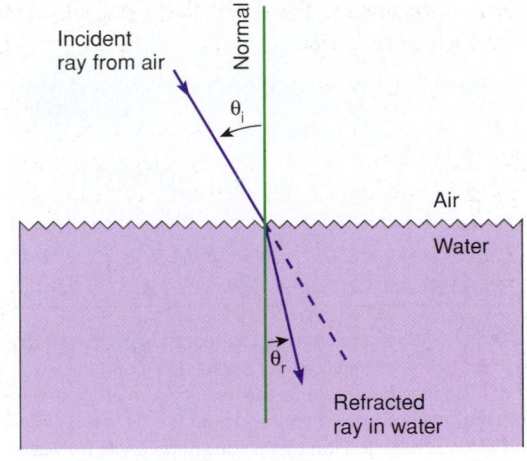

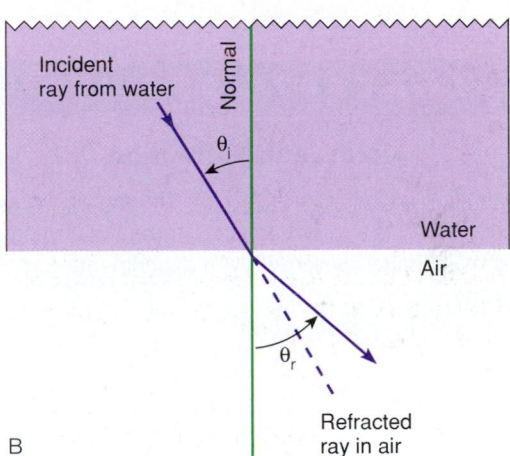

FIGURE 7.13 (A) A light ray moving to a new material with a slower speed of light is refracted toward the normal ($\theta_i > \theta_r$). (B) A light ray moving to a new material with a faster speed is refracted away from the normal ($\theta_i < \theta_r$).

called **total internal reflection** and implies that the light is trapped inside if it arrived at the critical angle or beyond. Faceted transparent gemstones such as the diamond are brilliant because they have a small critical angle and thus reflect much light internally. Total internal reflection is also important in fiber optics.

EXAMPLE 7.1 *(Optional)*

What is the speed of light in a diamond?

SOLUTION

The relationship between the speed of light in a material (v), the speed of light in a vacuum ($c = 3.00 \times 10^8$ m/s), and the index of refraction is given in equation 7.2. The index of refraction of a diamond is found in table 7.1 ($n = 2.42$).

$$n_{diamond} = 2.42 \qquad n = \frac{c}{v} \therefore v = \frac{c}{n}$$
$$c = 3.00 \times 10^8 \text{ m/s}$$
$$v = ? \qquad\qquad = \frac{3.00 \times 10^8 \text{ m/s}}{2.42}$$
$$= \boxed{1.24 \times 10^8 \text{ m/s}}$$

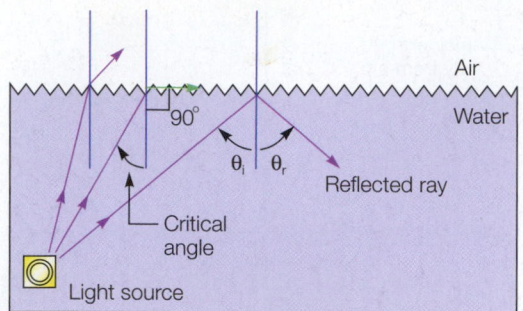

FIGURE 7.14 When the angle of incidence results in an angle of refraction of 90°, the refracted light ray is refracted along the water surface. The angle of incidence for a material that results in an angle of refraction of 90° is called the *critical angle*. When the incident ray is at this critical angle or greater, the ray is reflected internally. The critical angle for water is about 49°, and for a diamond it is about 25°.

CONCEPTS APPLIED

Colors and Refraction

A convex lens is able to magnify by forming an image with refracted light. This application is concerned with magnifying, but it is really more concerned with experimenting to find an explanation.

Here are three pairs of words:

SCIENCE BOOK
RAW HIDE
CARBON DIOXIDE

Hold a cylindrical solid glass rod over the three pairs of words, using it as a magnifying glass. A clear, solid, and transparent plastic rod or handle could also be used as a magnifying glass.

Notice that some words appear inverted but others do not. Does this occur because red letters are refracted differently than blue letters?

Make some words with red and blue letters to test your explanation. What is your explanation for what you observed?

As was stated earlier, refraction results from a change in speed when light passes from one transparent material into another. The ratio of the speeds of light in the two materials determines the magnitude of refraction at any given angle of incidence. The greatest speed of light possible, according to current theory, occurs when light is moving through a vacuum. The speed of light in a vacuum is accurately known to nine decimals but is usually rounded to 3.00×10^8 m/s for general discussion. The speed of light in a vacuum is a very important constant in physical science, so it is given a symbol of its own, c. The ratio of c to the speed of light in some transparent material, v, is called the **index of refraction,** n, of that material or

$$n = \frac{c}{v}$$

equation 7.2

CONCEPTS APPLIED

Internal Reflection

Seal a flashlight in a clear plastic bag to waterproof it, then investigate the critical angle and total internal reflection in a swimming pool, play pool, or large tub of water. In a darkened room or at night, shine the flashlight straight up from beneath the water, then at different angles until it shines almost horizontally beneath the surface. Report your observation of the critical angle for the water used.

The indexes of refraction for some substances are listed in table 7.1. The values listed are constant physical properties and can be used to identify a specific substance. Note that a larger value means a greater refraction at a given angle. Of the materials listed, diamond refracts light the most and air the least. The index for air is nearly 1, which means that light is slowed only slightly in air.

Note that table 7.1 shows that colder air at 0°C (32°F) has a higher index of refraction than warmer air at 30°C (86°F), which means that light travels faster in warmer air. This difference explains the "wet" highway that you sometimes see at a distance in the summer. The air near the road is hotter on a clear, calm day. Light rays traveling toward you in this hotter air are refracted upward as they enter the cooler air. Your brain interprets this refracted light as *reflected* light, but no reflection is taking place. Light traveling downward from other cars is also refracted upward toward you, and you think you are seeing cars "reflected" from the wet highway (figure 7.15).

TABLE 7.1   Index of Refraction

Substance	$n = c/v$
Glass	1.50
Diamond	2.42
Ice	1.31
Water	1.33
Benzene	1.50
Carbon tetrachloride	1.46
Ethyl alcohol	1.36
Air (0°C)	1.00029
Air (30°C)	1.00026

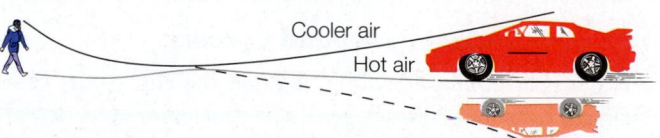

FIGURE 7.15 Mirages are caused by hot air near the ground refracting, or bending, light rays upward into the eyes of a distant observer. The observer believes he is seeing an upside-down image reflected from water on the highway.

Historians tell us there are many early stories and legends about the development of ancient optical devices. The first glass vessels were made about 1500 B.C., so it is possible that samples of clear, transparent glass were available soon after. One legend claimed that the ancient Chinese invented eyeglasses as early as 500 B.C. A burning glass (lens) was mentioned in a Greek play written about 424 B.C. Several writers described how Archimedes saved his hometown of Syracuse with a burning glass in about 214 B.C. Syracuse was besieged by Roman ships when Archimedes supposedly used the burning glass to focus sunlight on the ships, setting them on fire. It is not known if this story is true or not, but it is known that the Romans indeed did have burning glasses. Glass spheres, which were probably used to start fires, have been found in Roman ruins, including a convex lens recovered from the ashes of Pompeii.

Today, lenses are no longer needed to start fires, but they are common in cameras, scanners, optical microscopes, eyeglasses, lasers, binoculars, and many other optical devices. Lenses are no longer just made from glass, and today many are made from a transparent, hard plastic that is shaped into a lens.

The primary function of a lens is to form an image of a real object by refracting incoming parallel light rays. Lenses have two basic shapes, with the center of a surface either bulging in or bulging out. The outward bulging shape is thicker at the center than around the outside edge and is called a *convex lens* (box figure 7.1A). The other basic lens shape is just the opposite—thicker around the outside edge than at the center—and is called a *concave lens* (box figure 7.1B).

Convex lenses are used to form images in magnifiers, cameras, eyeglasses, projectors, telescopes, and microscopes (box figure 7.2). Concave lenses are used in some eyeglasses and in combinations with the convex lens to correct for defects. The convex lens is the most commonly used lens shape.

Your eyes are optical devices with convex lenses. Box figure 7.3 shows the basic

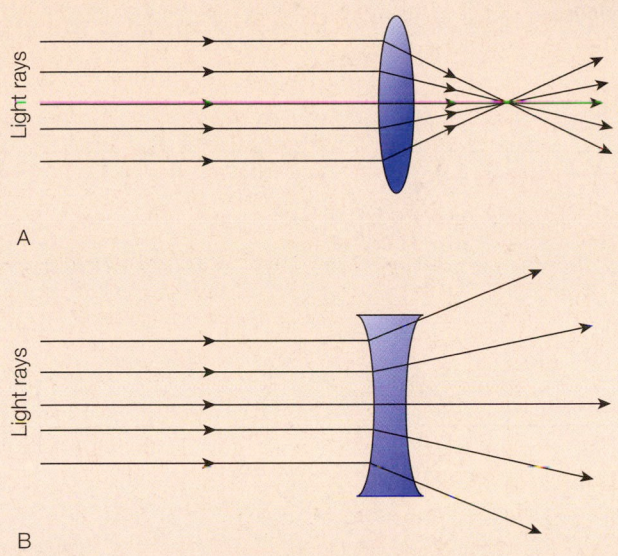

Box Figure 7.1 (*A*) Convex lenses are called converging lenses since they bring together, or converge, parallel rays of light. (*B*) Concave lenses are called diverging lenses since they spread apart, or diverge, parallel rays of light.

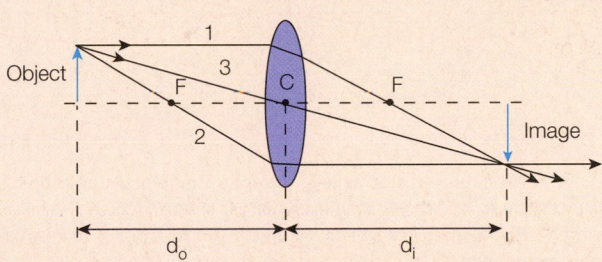

Box Figure 7.2 A convex lens forms an inverted image from refracted light rays of an object outside the focal point. Convex lenses are mostly used to form images in cameras, film or overhead projectors, magnifying glasses, and eyeglasses.

structure. First, a transparent hole called the *pupil* allows light to enter the eye. The size of the pupil is controlled by the *iris*, the colored part that is a muscular diaphragm. The *lens* focuses a sharp image on the back surface of the eye, the *retina*. The retina is made up of millions of light-sensitive structures, and nerves carry signals (impulses) from the retina through the optic nerve to the brain.

The lens is a convex, pliable material held in place and changed in shape by the attached *ciliary muscle*. When the eye is focused on a distant object, the ciliary muscle is completely relaxed. Looking at a closer object requires the contraction of the ciliary muscles to change the curvature of the lens. This adjustment of focus by action of the

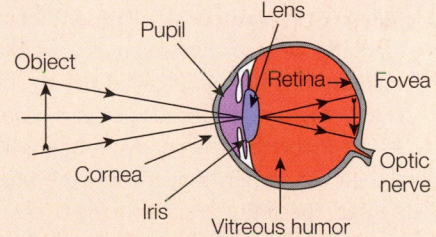

Box Figure 7.3 Light rays from a distant object are focused by the lens onto the retina, a small area on the back of the eye.

ciliary muscle is called *accommodation*. The closest distance an object can be seen without a blurred image is called the *near point*, and this is the limit to accommodation.

The near point moves outward with age as the lens becomes less pliable. By

—Continued top of next page

Continued—

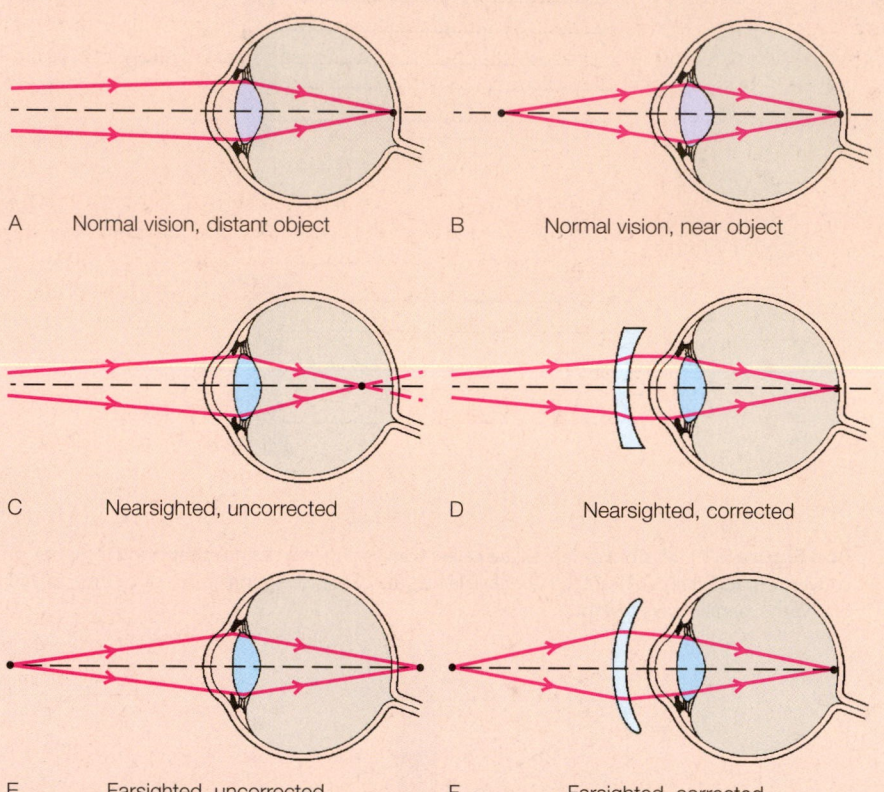

A   Normal vision, distant object

B   Normal vision, near object

C   Nearsighted, uncorrected

D   Nearsighted, corrected

E   Farsighted, uncorrected

F   Farsighted, corrected

Box Figure 7.4 (*A*) The relaxed, normal eye forms an image of distant objects on the retina. (*B*) For close objects, the lens of the normal eye changes shape to focus the image on the retina. (*C*) In a nearsighted eye, the image of a distant object forms in front of the retina. (*D*) A diverging lens corrects for nearsightedness. (*E*) In a farsighted eye, the image of a nearby object forms beyond the retina. (*F*) A converging lens corrects for farsightedness.

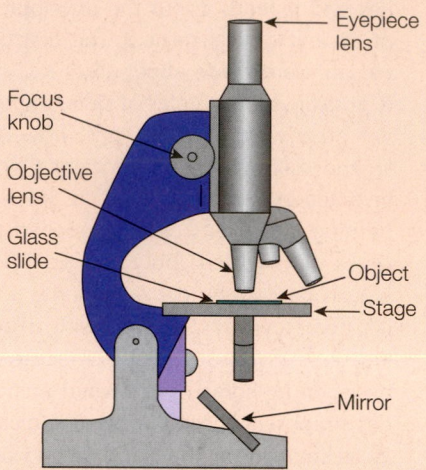

Box Figure 7.5 A simple microscope uses a system of two lenses, which are an objective lens that makes an enlarged image of the specimen and an eyepiece lens that makes an enlarged image of that image.

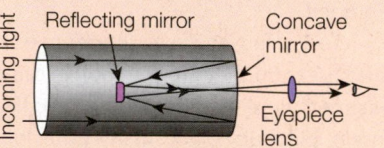

Box Figure 7.6 This illustrates how the path of light moves through a simple reflecting astronomical telescope. Several different designs and placement of mirrors are possible.

middle age, the near point may be twice this distance or greater, creating the condition known as farsightedness. The condition of farsightedness, or *hyperopia,* is a problem associated with aging (called presbyopia). Hyperopia can be caused at an early age by an eye that is too short or by problems with the cornea or lens that focuses the image behind the retina. Farsightedness can be corrected with a convex lens as shown in box figure 7.4A.

Nearsightedness, or *myopia,* is a problem caused by an eye that is too long or problems with the cornea or lens that focus the image in front of the retina. Nearsightedness can be corrected with a concave lens, as shown in box figure 7.4B.

The microscope is an optical device used to make things look larger. It is essentially a system of two lenses, one to produce an image of the object being studied and the other to act as a magnifying glass and enlarge that image. The power of the microscope is basically determined by the *objective lens,* which is placed close to the specimen on the stage of the microscope. Light is projected up through the specimen, and the objective lens makes an enlarged image of the specimen inside the tube between the two lenses. The *eyepiece lens* is positioned so that it makes a sharp enlarged image of the image produced by the objective lens (box figure 7.5).

Telescopes are optical instruments used to provide enlarged images of near and distant objects. There are two major types of telescopes, *refracting* telescopes that use two lenses and *reflecting* telescopes that use combinations of mirrors or a mirror and a lens. The refracting telescope has two lenses, with the objective lens forming a reduced image, which is viewed with an eyepiece lens to enlarge that image. In reflecting telescopes, mirrors are used instead of lenses to collect the light (box figure 7.6).

Finally, the *digital camera* is one of the more recently developed light-gathering and photograph-taking optical instruments. This camera has a group of small photocells, with perhaps thousands lined up on the focal plane behind a converging lens. An image falls on the array, and each photocell stores a charge that is proportional to the amount of light falling on the cell. A microprocessor measures the amount of charge registered by each photocell and considers it as a pixel, a small bit of the overall image. A shade of gray or a color is assigned to each pixel, and the image is ready to be enhanced, transmitted to a screen, printed, or magnetically stored for later use.

When you reach the place where the "water" seemed to be, it disappears, only to appear again farther down the road (also see refraction of sound on page 112).

Sometimes convection currents produce a mixing of warmer air near the road with the cooler air just above. This mixing refracts light one way, then the other, as the warmer and cooler air mix. This produces a shimmering or quivering that some people call "seeing heat." They are actually seeing changing refraction, which is a *result* of heating and convection. In addition to causing distant objects to quiver, the same effect causes the point source of light from stars to appear to twinkle. The light from closer planets does not twinkle because the many light rays from the disklike sources are not refracted together as easily as the fewer rays from the point sources of stars. The light from planets will appear to quiver, however, if the atmospheric turbulence is great.

Dispersion and Color

Electromagnetic waves travel with the speed of light with a whole spectrum of waves of various frequencies and wavelengths. The speed of electromagnetic waves (c) is related to the wavelength (λ) and the frequency (f) by a form of the wave equation, or

$$c = \lambda f \qquad \text{equation 7.3}$$

Visible light is the part of the electromagnetic spectrum that your eyes can detect, a narrow range of wavelength from about 7.90×10^{-7} m to 3.90×10^{-7} m. In general, this range of visible light can be subdivided into ranges of wavelengths that you perceive as colors (figure 7.16). These are the colors of the rainbow, and there are six distinct colors that blend one into another. These colors are *red, orange, yellow, green, blue,* and *violet.* The corresponding ranges of wavelengths and frequencies of these colors are given in table 7.2.

In general, light is interpreted to be white if it has the same mixture of colors as the solar spectrum. That sunlight is made up of component colors was first investigated in detail by Isaac Newton. While a college student, Newton became interested in grinding lenses, light, and color. At the age of twenty-three, Newton visited a local fair and bought several triangular glass prisms and proceeded to conduct a series of experiments with a beam of sunlight in his room. In 1672, he reported the results of his experiments with prisms and color, concluding that white light is a mixture of all the independent colors. Newton found that a beam of sunlight falling on a glass prism in a darkened room produced a band of colors he called a *spectrum.* Further, he found that a second glass prism would not subdivide each separate color but would combine all the colors back into white sunlight. Newton concluded that sunlight consists of a mixture of the six colors.

EXAMPLE 7.2   *(Optional)*

The colors of the spectrum can be measured in units of wavelength, frequency, or energy, which are alternative ways of describing colors of light waves. The human eye is most sensitive to light with a wavelength of 5.60×10^{-7} m, which is a yellow-green color. What is the frequency of this wavelength?

SOLUTION

The relationship between the wavelength (λ), frequency (f), and speed of light in a vacuum (c) is found in equation 7.3, $c = \lambda f$.

$$\lambda = 5.60 \times 10^{-7} \text{ m} \qquad c = \lambda f \therefore f = \frac{c}{\lambda}$$

$$c = 3.00 \times 10^8 \text{ m/s}$$

$$f = ?$$

$$= \frac{3.00 \times 10^8 \, \frac{\text{m}}{\text{s}}}{5.60 \times 10^{-7} \text{ m}}$$

$$= \frac{3.00 \times 10^8}{5.60 \times 10^{-7}} \frac{\text{m}}{\text{s}} \times \frac{1}{\text{m}}$$

$$= 5.40 \times 10^{14} \frac{1}{\text{s}}$$

$$= \boxed{5.40 \times 10^{14} \text{ Hz}}$$

FIGURE 7.16   The flowers appear to be red because they reflect light in the 7.9×10^{-7} m to 6.2×10^{-7} m range of wavelengths.

TABLE 7.2   Range of Wavelengths and Frequencies of the Colors of Visible Light

Color	Wavelength (in meters)	Frequency (in hertz)
Red	7.9×10^{-7} to 6.2×10^{-7}	3.8×10^{14} to 4.8×10^{14}
Orange	6.2×10^{-7} to 6.0×10^{-7}	4.8×10^{14} to 5.0×10^{14}
Yellow	6.0×10^{-7} to 5.8×10^{-7}	5.0×10^{14} to 5.2×10^{14}
Green	5.8×10^{-7} to 4.9×10^{-7}	5.2×10^{14} to 6.1×10^{14}
Blue	4.9×10^{-7} to 4.6×10^{-7}	6.1×10^{14} to 6.6×10^{14}
Violet	4.6×10^{-7} to 3.9×10^{-7}	6.6×10^{14} to 7.7×10^{14}

Connections . . .

Diamond Quality

The beauty of a diamond depends largely on optical properties, including degree of refraction, but how well the cutter shapes the diamond determines the brilliance and reflection of light from a given stone. The "finish" describes the precision of facet placement on a cut stone, but overall quality involves much more. Four factors are generally used to determine the quality of a cut diamond. These are sometimes called the four C's of carat, color, clarity, and cut.

Carat is a measure of the weight of a diamond. When measuring diamonds (as opposed to gold), 1 metric carat is equivalent to 0.2 g, so a 5 carat diamond would have a mass of 1 g. Diamonds are weighed to the nearest hundredth of a carat, and 100 points equal 1 carat. Thus, a 3 point diamond is 3/100 of a carat, with a mass of about 0.006 g.

Color is a measure of how much (or how little) color a diamond has. The color grade is based on the color of the body of a diamond, not the color of light dispersion from the stone. The color grades range from colorless, near colorless, faint yellow, very light yellow, and down to light yellow. It is measured by comparing a diamond to a standard of colors, but the color of a diamond weighing one-half carat or less is often difficult to determine.

Clarity is a measure of inclusions (grains or crystals), cleavages, or other blemishes. Such inclusions and flaws may affect transparency, brilliance, and even the durability of a diamond. The clarity grade ranges from flawless down to industrial-grade stones, which may appear milky from too many tiny inclusions.

Cut involves the relationships between the sizes of a stone's major physical features and their various angles. They determine, above all else, the limit to which a diamond will accomplish its optical potential. There are various types of faceted cuts, including the round brilliant cut, with a flat top, a flat bottom, and about fifty-eight facets. Facet shapes of square, triangular, diamond-shaped, and trapezoidal cuts might be used.

The details of a given diamond's four C's determine how expensive the diamond is going to be, and a top grade in all factors would be very expensive. No two stones are alike, and it is possible to find a brilliant diamond with a few lower grades in some of the factors that is a good value.

A glass prism separates sunlight into a spectrum of colors because the index of refraction is different for different wavelengths of light. The same processes that slow the speed of light in a transparent substance have a greater effect on short wavelengths than they do on longer wavelengths. As a result, violet light is refracted most, red light is refracted least, and the other colors are refracted between these extremes. This results in a beam of white light being separated, or dispersed, into a spectrum when it is refracted. Any transparent material in which the index of refraction varies with wavelength has the property of *dispersion*. The dispersion of light by ice crystals sometimes produces a colored halo around the Sun and the Moon.

7.3 EVIDENCE FOR WAVES

The nature of light became a topic of debate toward the end of the 1600s as Isaac Newton published his *particle theory* of light. He believed that the straight-line travel of light could be better explained as small particles of matter that traveled at great speed from a source of light. Particles, reasoned Newton, should follow a straight line according to the laws of motion. Waves, on the other hand, should bend as they move, much as water waves on a pond bend into circular shapes as they move away from a disturbance. About the same time that Newton developed his particle theory of light, Christian Huygens (pronounced *hi-ganz*) was concluding that light is not a stream of particles but rather a longitudinal wave.

Both theories had advocates during the 1700s, but the majority favored Newton's particle theory. By the beginning of the 1800s, new evidence was found that favored the wave theory, evidence that could not be explained in terms of anything but waves.

Interference

In 1801, Thomas Young published evidence of a behavior of light that could only be explained in terms of a wave model of light. Young's experiment is illustrated in figure 7.17A. Light from a single source is used to produce two beams of light that are in phase, that is, having their crests and troughs together as they move away from the source. This light falls on a card with two slits, each less than a millimeter in width. The light moves out from each slit as an expanding arc. Beyond the card, the light from one slit crosses over the light from the other slit to produce a series of bright lines on a screen. Young had produced a phenomenon of light called **interference,** and interference can only be explained by waves!

Young found all of the experimental data such as this in full agreement with predictions from a wave theory of light. About fifteen years later, A. J. Fresnel (pronounced *fray-nel*) demonstrated mathematically that behaviors of light could be fully explained with the wave theory. In 1821, Fresnel determined that the wavelength of red light was about 8×10^{-7} m and of violet light about 4×10^{-7} m, with other colors in between these two extremes. The work of Young and Fresnel seemed to resolve the issue of considering light to be a stream of particles or a wave, and it was generally agreed that light must be waves.

Why Is the Sky Blue?

Sunlight entering our atmosphere is scattered, or redirected by interactions with air molecules. Sunlight appears to be white to the human eye but is actually a mixture of all the colors of the rainbow. The blue and violet part of the spectrum has shorter wavelengths than the red and orange part. Shorter wavelength light—blue and violet light—is scattered more strongly than red and orange light. When you look at the sky, you see the light that was redirected by the atmosphere into your line of sight. Since blue and violet light is scattered more efficiently than red and orange light, the sky appears blue. When viewing a sunrise or sunset, you see only light that has not been scattered in other directions. The red and orange part of sunlight travels through a maximum length of the atmosphere, and the blue and violet has been scattered away, so a sunrise or sunset appears to be more orange and reddish.

Polarization

Huygens' wave theory and Newton's particle theory could explain some behaviors of light satisfactorily, but there were some behaviors that neither (original) theory could explain. Both theories failed to explain some behaviors of light, such as light moving through certain transparent crystals. For example, a slice of the mineral tourmaline transmits what appears to be a low-intensity greenish light. But if a second slice of tourmaline is placed on the first and rotated, the transmitted light passing through both slices begins to dim. The transmitted light is practically zero when the second slice is rotated 90°. Newton suggested that this behavior had something to do with "sides" or "poles" and introduced the concept of what is now called the *polarization* of light.

The waves of Huygens' wave theory were longitudinal, moving like sound waves, with wave fronts moving in the direction of travel. A longitudinal wave could not explain the polarization behavior of light. In 1817, Young modified Huygens' theory by describing the waves as *transverse*, vibrating at right angles to the direction of travel. This modification helped explain the polarization behavior of light transmitted through the two crystals and provided firm evidence that light is a transverse wave. As shown in figure 7.18A, **unpolarized light** is assumed to consist of transverse waves vibrating in all conceivable random directions. Polarizing materials, such as the tourmaline crystal, transmit light that is vibrating in one direction only, such as the vertical direction in figure 7.18B. Such a wave is said to be **polarized,** or *plane-polarized*, since it vibrates only in one plane. The single crystal polarized light by transmitting only waves that vibrate parallel to a certain direction while selectively absorbing waves that vibrate in all other directions. Your eyes cannot tell the difference between unpolarized and polarized light, so the light transmitted through a single crystal looks just like any other light. When a second crystal is placed on the first, the amount of light transmitted depends on the alignment of the two crystals (figure 7.19). When the two crystals are *aligned*, the polarized light from the first crystal passes through the second with little absorption. When the crystals are *crossed* at 90°, the light transmitted by the first is vibrating in a plane that is absorbed

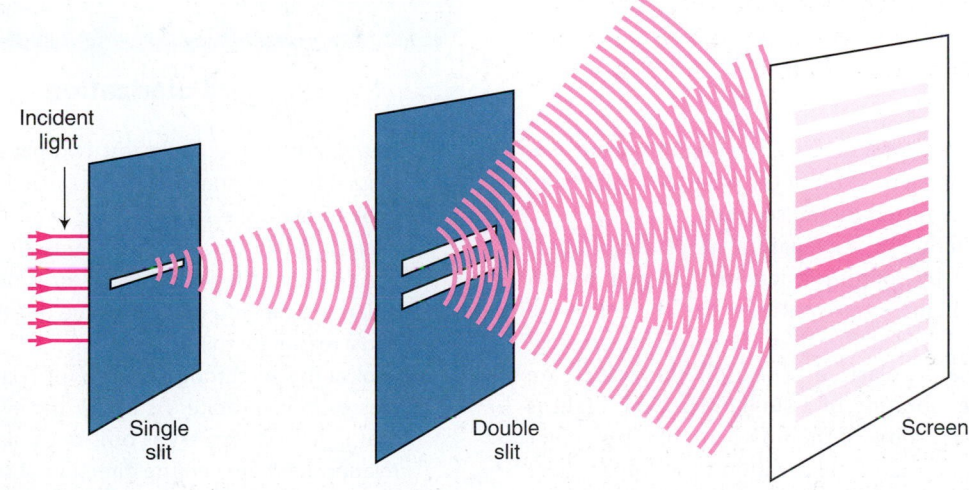

FIGURE 7.17   (*A*) The arrangement for Young's double-slit experiment. Sunlight passing through the first slit is coherent and falls on two slits close to each other. Light passing beyond the two slits produces an interference pattern on a screen.

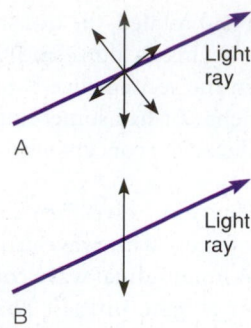

FIGURE 7.18 (*A*) Unpolarized light has transverse waves vibrating in all possible directions perpendicular to the direction of travel. (*B*) Polarized light vibrates only in one plane. In this illustration, the wave is vibrating in a vertical direction only.

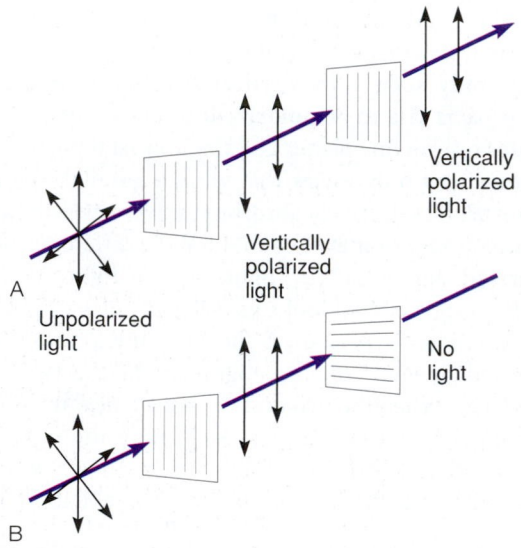

FIGURE 7.19 (*A*) Two crystals that are aligned both transmit vertically polarized light that looks like any other light. (*B*) When the crystals are crossed, no light is transmitted.

by the second crystal, and practically all the light is absorbed. At some other angle, only a fraction of the polarized light from the first crystal is transmitted by the second.

You can verify whether or not a pair of sunglasses is made of polarizing material by rotating a lens of one pair over a lens of a second pair. Light is transmitted when the lenses are aligned but mostly absorbed at 90° when the lenses are crossed.

Light is completely polarized when all the waves are removed except those vibrating in a single direction. Light is partially polarized when some of the waves are in a particular orientation, and any amount of polarization is possible. There are several means of producing partially or completely polarized light, including (1) selective absorption, (2) reflection, and (3) scattering.

Selective absorption is the process that takes place in certain crystals, such as tourmaline, where light in one plane is transmitted and all the other planes are absorbed. A method of manufacturing a polarizing film was developed in the 1930s

by Edwin H. Land. The film is called *Polaroid*. Today, Polaroid is made of long chains of hydrocarbon molecules that are aligned in a film. The long-chain molecules ideally absorb all light waves that are parallel to their lengths and transmit light that is perpendicular to their lengths. The direction that is *perpendicular* to the oriented molecular chains is thus called the polarization direction or the *transmission axis*.

Reflected light with an angle of incidence between 1° and 89° is partially polarized as the waves parallel to the reflecting surface are reflected more than other waves. Complete polarization, with all waves parallel to the surface, occurs at a particular angle of incidence. This angle depends on a number of variables, including the nature of the reflecting material. Figure 7.20 illustrates polarization by reflection. Polarizing sunglasses reduce the glare of reflected light because they have vertically oriented transmission axes. This absorbs the horizontally oriented reflected light. If you turn your head from side to side so as to rotate your sunglasses while looking at a reflected glare, you will see the intensity of the reflected light change. This means that the reflected light is partially polarized.

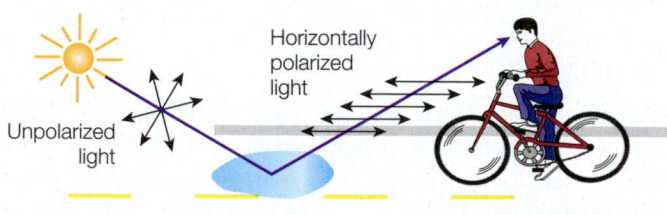

FIGURE 7.20 Light that is reflected becomes partially or fully polarized in a horizontal direction, depending on the incident angle and other variables.

CONCEPTS APPLIED

Polarization

1. Describe how you could test two pairs of polarizing sunglasses to make sure they are polarized.
2. Look through the glass of a car windshield while rotating a lens of a pair of polarizing sunglasses. What evidence do you find that the windshield is or is not polarized? If it is polarized, can you determine the direction of polarization and a reason for this?
3. Look at the sky through one lens of a pair of polarizing sunglasses as you rotate the lens. What evidence do you find that light from the sky is or is not polarized? Is there any relationship between the direction of polarization and the position of the Sun?
4. Position yourself with a wet street or puddle of water between you and the Sun when it is low in the sky. Rotate the lens of a pair of polarizing sunglasses as you look at the glare reflected off the water. Explain how these sunglasses are able to eliminate reflected glare.

A rainbow is a spectacular, natural display of color that is supposed to have a pot of gold under one end. Understanding the why and how of a rainbow requires information about water droplets and knowledge of how light is reflected and refracted. This information will also explain why the rainbow seems to move when you move—making it impossible to reach the end to obtain that mythical pot of gold.

First, note the pattern of conditions that occurs when you see a rainbow. It usually appears when the Sun is shining low in one part of the sky and rain is falling in the opposite part. With your back to the Sun, you are looking at a zone of raindrops that are all showing red light, another zone that are all showing violet light, with zones of the other colors between (ROYGBV). For a rainbow to form like this requires a surface that refracts and reflects the sunlight, a condition met by spherical raindrops.

Water molecules are put together in such a way that they have a positive side and a negative side, and this results in strong molecular attractions. It is the strong attraction of water molecules for one another that results in the phenomenon of surface tension. Surface tension is the name given to the surface of water acting as if it is covered by an ultrathin elastic membrane that is contracting. It is surface tension that pulls raindrops into a spherical shape as they fall through the air.

Box figure 7.7 shows one thing that can happen when a ray of sunlight strikes a single spherical raindrop near the top of the drop. At this point, some of the sunlight is reflected, and some is refracted into the raindrop. The refraction disperses the light into its spectrum colors, with the violet

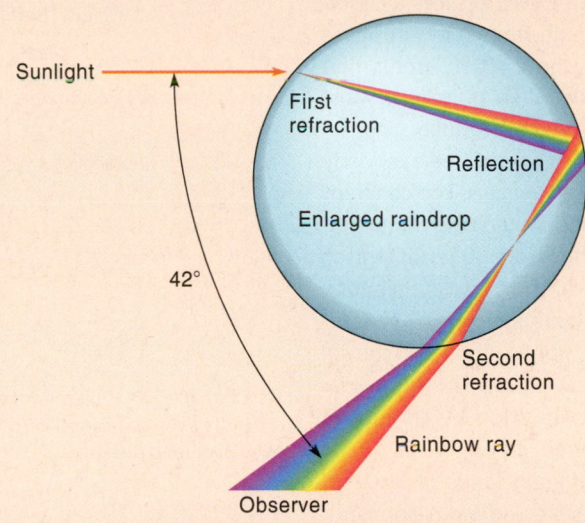

Box Figure 7.7 Light is refracted when it enters a raindrop and when it leaves. The part that leaves the front surface of the raindrop is the source of the light in thousands upon thousands of raindrops from which you see zones of color—a rainbow.

light being refracted most and red the least. The refracted light travels through the drop to the opposite side, where some of it might be reflected back into the drop. The reflected part travels back through the drop again, leaving the front surface of the raindrop. As it leaves, the light is refracted for a second time. The combined refraction, reflection, and second refraction is the source of the zones of colors you see in a rainbow. This also explains why you see a rainbow in the part of the sky opposite from the Sun.

The light from any one raindrop is one color, and that color comes from all drops on the arc of a circle that is a certain angle between the incoming sunlight and the refracted light. Thus, the raindrops in the red region refract red light toward your

eyes at an angle of 42°, and all other colors are refracted over your head by these drops. Raindrops in the violet region refract violet light toward your eyes at an angle of 40° and the red and other colors toward your feet. Thus, the light from any one drop is seen as one color, and all drops showing this color are on the arc of a circle. An arc is formed because the angle between the sunlight and the refracted light of a color is the same for each of the spherical drops.

Sometimes a fainter secondary rainbow, with colors reversed, forms from sunlight entering the bottom of the drop, reflecting twice, and then refracting out the top. The double reflection reverses the colors, and the angles are 50° for the red and 54° for the violet (see figure 1.13 on page 16).

The phenomenon called *scattering* occurs when light is absorbed and reradiated by particles about the size of gas molecules that make up the air. Sunlight is initially unpolarized. When it strikes a molecule, electrons are accelerated and vibrate horizontally and vertically. The vibrating charges reradiate polarized light. Thus, if you look at the blue sky with a pair of polarizing sunglasses and rotate them, you will observe that light from the sky is polarized. Bees are believed to be able to detect polarized skylight and use it to orient the direction of

their flights. Violet and blue light have the shortest wavelengths of visible light, and red and orange light have the longest. The violet and blue rays of sunlight are scattered the most. At sunset, the path of sunlight through the atmosphere is much longer than when the Sun is more directly overhead. Much of the blue and violet have been scattered away as a result of the longer path through the atmosphere at sunset. The remaining light that comes through is mostly red and orange, so these are the colors you see at sunset.

Connections . . .

Lasers

A laser is a device that produces a coherent beam of single-frequency, in-phase light. The beam comes from atoms that have been stimulated by electricity. Most ordinary light sources produce incoherent light; light that is emitted randomly and at different frequencies. The coherent light from a laser has the same frequency, phase, and direction, so it does not tend to spread out and can be very intense (box figure 7.8). This has made possible a number of specialized applications, and the list of uses continues to grow. The word *laser* is from *l*ight *a*mplification by *s*timulated *e*mission of *r*adiation.

There are different kinds of lasers in use and new ones are under development. One common type of laser is a gas-filled tube with mirrors at both ends. The mirror at one end is only partly silvered, which allows light to escape as the laser beam. The distance between the mirrors matches the resonate frequency of the light produced, so the trapped light will set up an optical standing wave. An electric discharge produces fast electrons that raise the energy level of the electrons of the specific gas atoms in the tube. The electrons of the energized gas atoms emit a particular

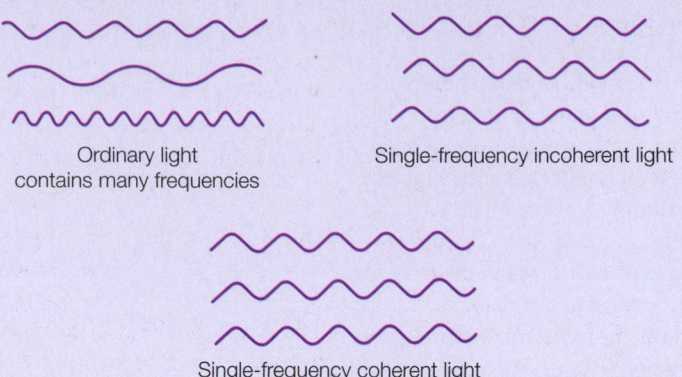

Ordinary light contains many frequencies

Single-frequency incoherent light

Single-frequency coherent light

Box Figure 7.8 A laser produces a beam of light whose waves all have the same frequency and are in step with one another (coherent). The beam is also very narrow and spreads out very little even over long distances.

frequency of light as they drop back to the original level, and this emitted light sets up the standing wave. The standing wave stimulates other atoms of the gas, resulting in the emission of more light at the same frequency and phase.

Lasers are everywhere today and have connections with a wide variety of technologies. At the supermarket, a laser and detector unit reads the bar code on each grocery item. The laser sends the pattern to a computer, which sends a price to the register while also tracking the store inventory. A low-powered laser and detector also reads your CD music disk or MP3 disk and can be used to make a three-dimensional image. Most laser printers use a laser, and a laser makes the operational part of a fiber optics communication system. Stronger lasers are used for cutting, drilling, and welding. Lasers are used extensively in many different medical procedures, from welding a detached retina to bloodless surgery.

7.4 EVIDENCE FOR PARTICLES

In 1850, J. L. Foucault was able to prove that light travels much slower in transparent materials than it does in air. This was in complete agreement with the wave theory and completely opposed to the particle theory. By the end of the 1800s, James Maxwell's theoretical concept of electric and magnetic fields changed the concept of light from mechanical waves to waves of changing electric and magnetic fields. Further evidence removed the necessity for ether, the material supposedly needed for waves to move through. Light was now seen as electromagnetic waves that could move through empty space. By this time, it was possible to explain all behaviors of light moving through empty space or through matter with a wave theory. Yet, there were nagging problems that the wave theory could not explain. In general, these problems concerned light that is absorbed by or emitted from matter.

Photoelectric Effect

Light is a form of energy, and it gives its energy to matter when it is absorbed. Usually, the energy of absorbed

light results in a temperature increase, such as the warmth you feel from absorbed sunlight. Sometimes, however, the energy from absorbed light results in other effects. In some materials, the energy is acquired by electrons, and some of the electrons acquire sufficient energy to jump out of the material. The movement of electrons as a result of energy acquired from light is known as the **photoelectric effect.** The photoelectric effect is put to a practical use in a solar cell, which transforms the energy of light into an electric current (figure 7.21).

The energy of light can be measured with great accuracy. The kinetic energy of electrons after they absorb light can also be measured with great accuracy. When these measurements were made of the light and electrons involved in the photoelectric effect, some unexpected results were observed. Monochromatic light, that is, light of a single, fixed frequency, was used to produce the photoelectric effect. First, a low-intensity, or dim, light was used, and the numbers and energy of the ejected electrons were measured. Then a high-intensity light was used, and the numbers and energy of the ejected electrons were again measured. Measurement showed that (1) low-intensity light caused fewer electrons to be ejected, and high-intensity light caused many to be ejected, and (2) all electrons ejected from low- or high-intensity light ideally had the *same* kinetic energy. Surprisingly, the kinetic energy of the ejected electrons was found to be *independent* of the light intensity. This was contrary to what the wave theory of light would predict, since a stronger light should mean that waves with more energy have more energy to give to the electrons. Here is a behavior involving light that the wave theory could not explain.

Quantization of Energy

In addition to the problem of the photoelectric effect, there were problems with blackbody radiation, light emitted from hot objects. The experimental measurements of light emitted through blackbody radiation did not match predictions made from theory. In 1900, Max Planck (pronounced *plonk*), a German physicist, found that he could fit the experimental measurements and theory together by assuming that the vibrating molecules that emitted the light could only have a *discrete amount* of energy. Instead of energy existing through a continuous range of amounts, Planck found that the vibrating molecules could only have energy in multiples of energy in certain amounts, or **quanta** (meaning "discrete amounts"; *quantum* is singular and *quanta*, plural).

Planck's discovery of quantized energy states was a radical, revolutionary development, and most scientists, including Planck, did not believe it at the time. Planck, in fact, spent considerable time and effort trying to disprove his own discovery. It was, however, the beginning of the quantum theory, which was eventually to revolutionize physics.

Five years later, in 1905, Albert Einstein applied Planck's quantum concept to the problem of the photoelectric effect. Einstein described the energy in a light wave as quanta of energy called **photons.** Each photon has an energy E that is related to the frequency f of the light through Planck's constant h, or

$$E = hf \qquad \text{equation 7.4}$$

The value of Planck's constant is 6.63×10^{-34} J·s. This relationship says that higher-frequency light (e.g., blue light at 6.50×10^{14} Hz) has more energy than lower-frequency light (e.g., red light at 4.00×10^{14} Hz). The energy of such high- and low-frequency light can be verified by experiment.

The photon theory also explained the photoelectric effect. According to this theory, light is a stream of moving photons. It is the number of photons in this stream that determines if the light is dim or intense. A high-intensity light has many, many photons, and a low-intensity light has only a few photons. At any particular fixed frequency, all the photons would have the same energy, the product of the frequency and Planck's constant (hf). When a photon interacts with matter, it is absorbed and gives up all of its energy. In the photoelectric effect, this interaction takes place between photons and electrons. When an intense light is used, there are more photons to interact with the electrons, so more electrons are ejected. The energy given up by each photon is a function of the frequency of the light, so at a fixed frequency, the energy of each photon, hf, is the same, and the acquired kinetic energy of each ejected electron is the same. Thus, the photon theory explains the measured experimental results of the photoelectric effect.

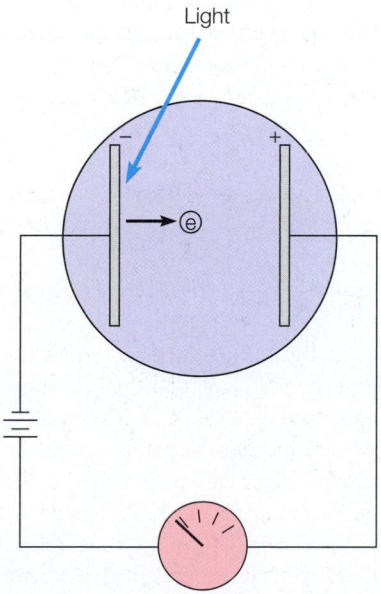

FIGURE 7.21   A setup for observing the photoelectric effect. Light strikes the negatively charged plate, and electrons are ejected. The ejected electrons move to the positively charged plate and can be measured as a current in the circuit.

EXAMPLE 7.3   *(Optional)*

What is the energy of a photon of red light with a frequency of 4.00×10^{14} Hz?

SOLUTION

The relationship between the energy of a photon (E) and its frequency (f) is found in equation 7.4. Planck's constant is given as 6.63×10^{-34} J·s.

$$f = 4.00 \times 10^{14} \text{ Hz} \qquad E = hf$$
$$h = 6.63 \times 10^{-34} \text{ J·s}$$
$$E = ? \qquad = (6.63 \times 10^{-34} \text{ J·s})\left(4.00 \times 10^{14}\,\frac{1}{\text{s}}\right)$$

$$= (6.63 \times 10^{-34})(4.00 \times 10^{14})\,\text{J·s} \times \frac{1}{\text{s}}$$

$$= 2.65 \times 10^{-19}\,\frac{\text{J·s}}{\text{s}}$$

$$= \boxed{2.65 \times 10^{-19} \text{ J}}$$

EXAMPLE 7.4   *(Optional)*

What is the energy of a photon of violet light with a frequency of 7.00×10^{14} Hz? (Answer: 4.64×10^{-19} J)

The photoelectric effect is explained by considering light to be photons with quanta of energy, not a wave of continuous energy. This is not the only evidence about the quantum nature of light, and more will be presented in chapter 8. But, as you can see, there is a dilemma. The electromagnetic wave theory and the photon theory seem incompatible. Some experiments cannot be explained by the wave theory and seem to support the photon theory. Other experiments are contradictions, providing seemingly equal evidence to reject the photon theory in support of the wave theory.

7.5 THE PRESENT THEORY

Today, light is considered to have a dual nature, sometimes acting like a wave and sometimes acting like a particle. A wave model is useful in explaining how light travels through space and how it exhibits such behaviors as refraction, interference, and diffraction. A particle model is useful in explaining how light is emitted from and absorbed by matter, exhibiting such behaviors as blackbody radiation and the photoelectric effect. Together, both of these models are part of a single theory of light, a theory that pictures light as having both particle and wave properties. Some properties are more useful when explaining some observed behaviors, and other properties are more useful when explaining other behaviors.

Frequency is a property of a wave, and the energy of a photon is a property of a particle. Both frequency and the energy of a photon are related in equation 7.4, $E = hf$. It is thus possible to describe light in terms of a frequency (or wavelength) or in terms of a quantity of energy. Any part of the electromagnetic spectrum can thus be described by units of frequency, wavelength, or energy, which are alternative means of describing light. The radio radiation parts of the spectrum are low-frequency, low-energy, and long-wavelength radiations. Radio radiations have more wave properties and practically no particle properties, since the energy levels are low. Gamma radiation, on the other hand, is high-frequency, high-energy, and short wavelength radiation. Gamma radiation has more particle properties, since the extremely short wavelengths have very high energy levels. The more familiar part of the spectrum, visible light, is between these two extremes and exhibits both wave and particle properties, but it never exhibits both properties at the same time in the same experiment.

Part of the problem in forming a concept or mental image of the exact nature of light is understanding this nature in terms of what is already known. The things you already know about are observed to be particles, or objects, or they are observed to be waves. You can see objects that move through the air, such as baseballs or footballs, and you can see waves on water or in a field of grass. There is nothing that acts like a moving object in some situations but acts like a wave in other situations. Objects are objects, and waves are waves, but objects do not become waves, and waves do not become objects. If this dual nature did exist, it would seem very strange. Imagine, for example, holding a

book at a certain height above a lake (figure 7.22). You can make measurements and calculate the kinetic energy the book will have when dropped into the lake. When it hits the water, the book disappears, and water waves move away from the point of impact in a circular pattern that moves across the water. When the waves reach another person across the lake, a book identical to the one you dropped pops up out of the water as the waves disappear. As it leaves the water across the lake, the book has the same kinetic energy that your book had when it hit the water in front of you. You and the other person could measure things about either book, and you could measure things about the waves, but you could not measure both at the same time. You might say that this behavior is not only strange but impossible. Yet, it is an analogy to the observed behavior of light.

As stated, light has a dual nature, sometimes exhibiting the properties of a wave and sometimes exhibiting the properties of moving particles but never exhibiting both properties at the same time. Both the wave and the particle nature are accepted as being part of one model today, with the understanding that the exact nature of light is not describable in terms of anything that is known to exist in the everyday-sized world. Light is an extremely small-scale phenomenon that must be different, without a sharp distinction between a particle and a wave. Evidence about this strange nature of an extremely small-scale phenomenon will be considered again in chapter 8 as a basis for introducing the quantum theory of matter.

7.6 RELATIVITY

The electromagnetic wave model brought together and explained electric and magnetic phenomena, and explained that light can be thought of as an electromagnetic wave (see figure 7.2). There remained questions, however, that would not be answered until Albert Einstein developed a revolutionary new theory. Even at the age of seventeen, Einstein

FIGURE 7.22   It would seem very strange if a book fell into and jumped out of water with the same kinetic energy. Yet this appears to be the nature of light.

The Compact Disc (CD)

A compact disc (CD) is a laser-read (also called *optically read*) data storage device. There are a number of different formats in use today, including music CDs, DVD movies, Blu-Ray DVD, and CDs for storing computer data. All of these utilize the general working principles described below, but some have different refinements. Some, for example, can fit much more data on a disc by utilizing smaller recording tracks.

The CD disc rotates between 200 and 500 revolutions per minute, but the drive changes speed to move the head at a constant linear velocity over the recording track, faster near the inner hub and slower near the outer edge of the disc. Furthermore, the drive reads from the inside out, so the disc will slow as it is played.

The CD disc is a 12 cm diameter, 1.3 mm thick sandwich of a hard plastic core, a mirrorlike layer of metallic aluminum, and a tough, clear plastic overcoating that protects the thin layer of aluminum. The CD records digitized data: music, video, or computer data that have been converted into a string of binary numbers. First, a master disc is made. The binary numbers are translated into a series of pulses that are fed to a laser. The laser is focused onto a photosensitive material on a spinning master disc. Whenever there is a pulse in the signal, the laser burns a small oval pit into the surface, making a pattern of pits and bumps on the track of the master disc. The laser beam is incredibly small, making marks about a micron or so in diameter. A micron is one-millionth of a meter, so you can fit a tremendous number of data tracks onto the disc, which has each track spaced 1.6 microns apart. Next, commercial CD discs are made by using the master disc as a mold. Soft plastic is pressed against the master disc in a vacuum-forming machine so the small physical marks—the pits and bumps made by the laser—are pressed into the plastic. This makes a record of the strings of binary numbers that were etched into the master disc by the strong but tiny laser beam. During playback, a low-powered laser beam is reflected off the track to read the binary marks on it. The optical sensor head contains a tiny diode laser, a lens, mirrors, and tracking devices that can move the head in three directions. The head moves side to side to keep the head over a single track (within 1.6 microns), it moves up and down to keep the laser beam in focus, and it moves forward and backward as a fine adjustment to maintain a constant linear velocity.

The disadvantage of the commercial CD is the lack of ability to do writing or rewriting. Writing and rewritable optical media are available, and these are called CD-R and CD-RW.

A CD-R records data to a disc by using a laser to burn spots into an organic dye. Such a "burned" spot reflects less light than an area that was not heated by the laser. This is designed to mimic the way light reflects from pits and bumps of a commercial CD, except this time the string of binary numbers are burned (nonreflective) and not burned areas (reflective). Since this is similar to how data on a commercial CD is represented, a CD-R disc can generally be used in a CD player as if it were a commercial CD. The dyes in a CD-R disc are photosensitive organic compounds that are similar to those used in making photographs. The color of a CD-R disc is a result of the kind of dye that was used in the recording layer combined with the type of reflective coating used. Some of these dye and reflective coating combinations appear green, some appear blue, and others appear to be gold. Once a CD-R disc is burned, it cannot be rewritten or changed.

The CD-RW is designed to have the ability to do writing or rewriting. It uses a different technology but again mimics the way light reflects from the pits and bumps of a pressed commercial CD. Instead of a dye-based recording layer, the CD-RW uses a compound made from silver, indium, antimony, and tellurium. This layer has a property that permits rewriting the information on a disc. The nature of this property is that when it is heated to a certain temperature and cooled, it becomes crystalline. However, when it is heated to a higher temperature and cooled, it becomes noncrystalline. A crystalline surface reflects a laser beam while a noncrystalline surface absorbs the laser beam. The CD-RW is again designed to mimic the way light reflects from the pits and bumps of a commercial CD, except this time the string of binary numbers are noncrystalline (nonreflective) and crystalline areas (reflective). In order to write, erase, and read, the CD-RW recorder must have three different laser powers. It must have (1) a high power to heat spots to about 600°C, which cool rapidly and make noncrystalline spots that are less reflective. It must have (2) a medium power to erase data by heating the media to about 200°C, which allows the media to crystallize and have a uniform reflectivity. Finally, it must have (3) a low setting that is used for finding and reading nonreflective and the more reflective areas of a disc. The writing and rewriting of a CD-RW can be repeated hundreds of times.

---

was already thinking about ideas that would eventually lead to his new theory. For example, he wondered about chasing a beam of light if you were also moving at the speed of light. Would you see the light as an oscillating electric and magnetic field at rest? He realized there was no such thing, either on the basis of experience or according to Maxwell's theory of electromagnetic waves.

In 1905, at the age of twenty-six, Albert Einstein published an analysis of how space and time are affected by motion between an observer and what is being measured. This analysis is called the *special theory of relativity*. Eleven years later, Einstein published an interpretation of gravity as distortion of the structure of space and time. This analysis is called the *general theory of relativity*. A number of remarkable predictions have been made based on this theory, and all have been verified by many experiments.

Special Relativity

The **special theory of relativity** is concerned with events as observed from different points of view, or different "reference frames." Here is an example: You are on a bus traveling straight down a highway at a constant 100 km/h. An insect is observed to fly from the back of the bus at 5 km/h. With respect to the bus, the insect is flying at 5 km/h. To someone observing from the ground, however, the speed of the insect is 100 km/h plus 5 km/h, or 105 km/h. If the insect is flying toward the back of the bus, its speed is 100 km/h minus 5 km/h, or 95 km/h with respect to Earth. Generally, the reference frame is understood to be Earth, but this is not always stated. Nonetheless, we must specify a reference frame whenever a speed or velocity is measured.

Einstein's special theory is based on two principles. The first concerns frames of reference and the fact that all motion is relative to a chosen frame of reference. This principle could be called the **consistent law principle:**

> **The laws of physics are the same in all reference frames that are moving at a constant velocity with respect to each other.**

Ignoring vibrations, if you are in a windowless bus, you will not be able to tell if the bus is moving uniformly or if it is not moving at all. If you were to drop something—say, your keys—in a moving bus, they would fall straight down, just as they would in a stationary bus. The keys fall straight down with respect to the bus in either case. To an observer outside the bus, in a different frame of reference, the keys would appear to take a curved path because they have an initial velocity. Moving objects follow the same laws in a uniformly moving bus or any other uniformly moving frame of reference (figure 7.23).

The second principle concerns the speed of light and could be called the **constancy of speed principle:**

> **The speed of light in empty space has the same value for all observers regardless of their velocity.**

The speed of light in empty space is 3.00×10^8 m/s (186,000 mi/s). An observer traveling toward a source would measure the speed of light in empty space as 3.00×10^8 m/s. An observer not moving with respect to the source would measure this very same speed. This is not like the insect moving in a bus—you do not add or subtract the velocity of the source from the velocity of light. The velocity is always 3.00×10^8 m/s for all observers, regardless of the velocity of the observers and regardless of the velocity of the source of light. Light behaves differently than anything in our everyday experience.

The special theory of relativity is based solely on the consistent law principle and the constancy of speed principle. Together, these principles result in some very interesting results if you compare measurements from the ground of the length, time, and mass of a very fast airplane with measurements made by someone moving with the airplane. You, on the ground, would find that

FIGURE 7.23   All motion is relative to a chosen frame of reference. Here the photographer has turned the camera to keep pace with one of the cyclists. Relative to him, both the road and the other cyclists are moving. There is no fixed frame of reference in nature and, therefore, no such thing as "absolute motion"; all motion is relative.

- The length of an object is shorter when it is moving.
- Moving clocks run more slowly.
- Moving objects have increased mass.

The special theory of relativity shows that measurements of length, time, and mass are different in different moving reference frames. Einstein developed equations that describe each of the changes described above. These changes have been verified countless times with elementary particle experiments, and the data always fit Einstein's equations with predicted results.

General Relativity

Einstein's **general theory of relativity** could also be called Einstein's geometric theory of gravity. According to Einstein, a gravitational interaction does not come from some mysterious force called gravity. Instead, the interaction is between a mass and the geometry of space and time where the mass is located. Space and time can be combined into a fourth-dimensional "spacetime" structure. A mass is understood to interact with the spacetime, telling it how to curve. Spacetime also interacts with a mass, telling it how to move. A gravitational interaction is considered to be a local event of movement along a geodesic (shortest distance between two points on a curved surface) in curved spacetime (figure 7.24). This different viewpoint has led to much more accurate measurements and has been tested by many events in astronomy (see page 299 for one example).

People Behind the Science

James Clerk Maxwell (1831–1879)

James Maxwell was a British physicist who discovered that light consists of electromagnetic waves and established the kinetic theory of gases. He also proved the nature of Saturn's rings and demonstrated the principles governing color vision.

Maxwell's development of the electromagnetic theory of light took many years. It began with the paper *On Faraday's Lines of Force*, in which Maxwell built on the views of Michael Faraday (1791–1867) that electric and magnetic effects result from fields of lines of force that surround conductors and magnets. Maxwell drew an analogy between the behavior of the lines of force and the flow of an incompressible liquid, thereby deriving equations that represented known electric and magnetic effects. The next step toward the electromagnetic theory took place with the publication of the paper *On Physical Lines of Force* (1861–1862). In it, Maxwell developed a model for the medium in which electric and magnetic effects could occur. In *A Dynamical Theory of the Electromagnetic Field* (1864), Maxwell developed the fundamental equations that describe the electromagnetic field. These showed that light is propagated in two waves, one magnetic and the other electric, which vibrate perpendicular to each other and to the direction of propagation. This was confirmed in Maxwell's *Note on the Electromagnetic Theory of Light* (1868), which used an electrical derivation of the theory instead of the dynamical formulation, and

Maxwell's whole work on the subject was summed up in *Treatise on Electricity and Magnetism* in 1873.

The treatise also established that light has a radiation pressure and suggested that a whole family of electromagnetic radiations must exist, of which light was only one. This was confirmed in 1888 with the sensational discovery of radio waves by Heinrich Hertz (1857–1894). Sadly, Maxwell did not live long enough to see this triumphant vindication of his work.

Maxwell is generally considered to be the greatest theoretical physicist of the 1800s, as his forebear Michael Faraday was the greatest experimental physicist. His rigorous mathematical ability was combined with great insight to enable him to achieve brilliant syntheses of knowledge in the two most important areas of physics at that time. In building on Faraday's work to discover the electromagnetic nature of light, Maxwell not only explained electromagnetism but also paved the way for the discovery and application of the whole spectrum of electromagnetic radiation that has characterized modern physics.

Source: Modified from the *Hutchinson Dictionary of Scientific Biography*. © RM, 2011. All rights reserved. Helicon Publishing is a division of RM.

Relativity Theory Applied

You use the Global Positioning System (GPS) when you use the locator in your cell phone or trace your route on a car navigation system. The GPS consists of a worldwide network of 24 satellites, each with an atomic clock that keeps accurate time to within three nanoseconds (0.000000003 of a second). The satellites broadcast a radio signal with the position and time of transmission. A receiver on the surface of Earth, for example your navigation system, uses signals from four satellites to determine location, speed, and time. A computer chip in the receiver uses such data from the satellites to calculate latitude and longitude, which can be used in a mapping application.

GPS satellites move with high velocity at a high altitude above the surface. This results in a combination of errors from the satellite velocity (special relativity error) and from the high location in Earth's gravitational field (general relativity error). Clocks moving with high velocity run slower compared to an identical clock on Earth's surface. For a satellite moving at 14,000 km/h this amounts to a slowing of 7,200 nanoseconds/day.

Clocks located at a higher altitude run faster than an identical clock on Earth's surface. For a satellite at 26,000 km this amounts to running fast by 45,900 nanoseconds/day. Combining special and general relativity errors results in a GPS clock running fast by 38,700 nanoseconds/day. This would result in a position error of more than 10 km/day. GPS satellite clocks correct relativity errors by adjusting the rate so the fast moving, high altitude clocks tick at the same rate as an identical clock on Earth's surface.

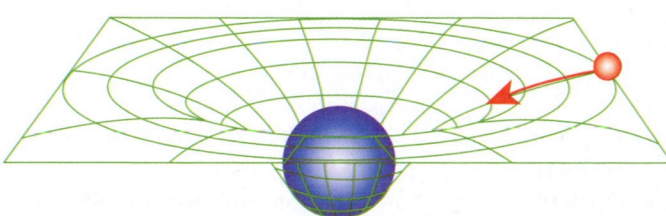

FIGURE 7.24 General relativity pictures gravity as a warping of the structure of space and time due to the presence of a body of matter. An object nearby experiences an attractive force as a result of this distortion in spacetime, much as a marble rolls toward the bottom of a saucer-shaped hole in the ground.

SUMMARY

Electromagnetic radiation is emitted from all matter with a temperature above absolute zero, and as the temperature increases, more radiation and shorter wavelengths are emitted. Visible light is emitted from matter hotter than about 700°C, and this matter is said to be *incandescent*. The Sun, a fire, and the ordinary lightbulb are incandescent sources of light.

The behavior of light is shown by a light ray model that uses straight lines to show the straight-line path of light. Light that interacts with matter is *reflected* with parallel rays, moves in random directions by *diffuse reflection* from points, or is *absorbed*, resulting in a temperature increase. Matter is *opaque*, reflecting light, or *transparent*, transmitting light.

In reflection, the incoming light, or *incident ray*, has the same angle as the *reflected ray* when measured from a perpendicular from the point of reflection, called the *normal*. That the two angles are equal is called the *law of reflection*. The law of reflection explains how a flat mirror forms a *virtual image*, one from which light rays do not originate. Light rays do originate from the other kind of image, a *real image*.

Light rays are bent, or *refracted*, at the boundary when passing from one transparent media to another. The amount of refraction depends on the *incident angle* and the *index of refraction*, a ratio of the speed of light in a vacuum to the speed of light in the media. When the refracted angle is 90°, *total internal reflection* takes place. This limit to the angle of incidence is called the *critical angle*, and all light rays with an incident angle at or beyond this angle are reflected internally.

Each color of light has a range of wavelengths that forms the *spectrum* from red to violet. A glass prism has the property of *dispersion*, separating a beam of white light into a spectrum. Dispersion occurs because the index of refraction is different for each range of colors, with short wavelengths refracted more than larger ones.

A wave model of light can be used to explain diffraction, interference, and polarization, all of which provide strong evidence for the wavelike nature of light. *Interference* occurs when light passes through two small slits or holes and produces an *interference pattern* of bright lines and dark zones. *Polarized light* vibrates in one direction only, in a plane. Light can be polarized by certain materials, by reflection, or by scattering. Polarization can only be explained by a transverse wave model.

A wave model fails to explain observations of light behaviors in the *photoelectric effect* and *blackbody radiation*. Max Planck found that he could modify the wave theory to explain blackbody radiation by assuming that vibrating molecules could only have discrete amounts, or *quanta*, of energy and found that the quantized energy is related to the frequency and a constant known today as *Planck's constant*. Albert Einstein applied Planck's quantum concept to the photoelectric effect and described a light wave in terms of quanta of energy called *photons*. Each photon has an energy that is related to the frequency and Planck's constant.

Today, the properties of light are explained by a model that incorporates both the wave and the particle nature of light. Light is considered to have both wave and particle properties and is not describable in terms of anything known in the everyday-sized world.

The *special theory of relativity* is an analysis of how space and time are affected by motion between an observer and what is being measured. The *general theory of relativity* relates gravity to the structure of space and time.

Summary of Equations

7.1   angle of incidence = angle of reflection
$$\theta_i = \theta_r$$

7.2   index of refraction $= \dfrac{\text{speed of light in vacuum}}{\text{speed of light in material}}$
$$n = \frac{c}{v}$$

7.3   speed of light in vacuum = (wavelength)(frequency)
$$c = \lambda f$$

7.4 $\begin{array}{l}\text{energy of} \\ \text{photon}\end{array} = \left(\begin{array}{l}\text{Planck's} \\ \text{constant}\end{array}\right)(\text{frequency})$
$$E = hf$$

KEY TERMS

blackbody radiation (p. **154**)
consistent law principle (p. **172**)
constancy of speed principle (p. **172**)
general theory of relativity (p. **173**)
incandescent (p. **153**)
index of refraction (p. **160**)
interference (p. **164**)
light ray model (p. **156**)
luminous (p. **153**)
photoelectric effect (p. **168**)
photons (p. **169**)
polarized (p. **165**)
quanta (p. **169**)
real image (p. **158**)
refraction (p. **158**)
special theory of relativity (p. **172**)
total internal reflection (p. **159**)
unpolarized light (p. **165**)
virtual image (p. **158**)

APPLYING THE CONCEPTS

Answers are located in appendix F.

1. An object is hot enough to emit a dull red glow. When this object is heated even more, it will emit
   a. shorter-wavelength, higher-frequency radiation.
   b. longer-wavelength, lower-frequency radiation.
   c. the same wavelengths as before but with more energy.
   d. more of the same wavelengths with more energy.

2. The difference in the light emitted from a candle, an incandescent lightbulb, and the Sun is basically from differences in
   a. energy sources.
   b. materials.
   c. temperatures.
   d. phases of matter.

3. Before it travels through Earth's atmosphere, sunlight is mostly
   a. infrared radiation.
   b. visible light.
   c. ultraviolet radiation.
   d. blue light.

4. You are able to see in shaded areas, such as under a tree, because light has undergone
   a. refraction.
   b. incident bending.
   c. a change in speed.
   d. diffuse reflection.

5. The ratio of the speed of light in a vacuum to the speed of light in some transparent materials is called
   a. the critical angle.
   b. total internal reflection.
   c. the law of reflection.
   d. the index of refraction.

6. Any part of the electromagnetic spectrum, including the colors of visible light, can be measured in units of
   a. wavelength.
   b. frequency.
   c. energy.
   d. Any of the above is correct.

7. A prism separates the colors of sunlight into a spectrum because
   a. each wavelength of light has its own index of refraction.
   b. longer wavelengths are refracted more than shorter wavelengths.
   c. red light is refracted the most, violet the least.
   d. All of the above are correct.

8. Which of the following can only be explained by a wave model of light?
   a. reflection
   b. refraction
   c. interference
   d. photoelectric effect

9. The polarization behavior of light is best explained by considering light to be
   a. longitudinal waves.
   b. transverse waves.
   c. particles.
   d. particles with ends, or poles.

10. Max Planck made the revolutionary discovery that the energy of vibrating molecules involved in blackbody radiation existed only in
   a. multiples of certain fixed amounts.
   b. amounts that smoothly graded one into the next.
   c. the same, constant amount of energy in all situations.
   d. amounts that were never consistent from one experiment to the next.

11. Einstein applied Planck's quantum discovery to light and found
   a. a direct relationship between the energy and frequency of light.
   b. that the energy of a photon divided by the frequency of the photon always equaled a constant known as Planck's constant.
   c. that the energy of a photon divided by Planck's constant always equaled the frequency.
   d. All of the above are correct.

12. Today, light is considered to be
   a. tiny particles of matter that move through space, having no wave properties.
   b. electromagnetic waves only, with no properties of particles.
   c. a small-scale phenomenon without a sharp distinction between particle and wave properties.
   d. something that is completely unknown.

QUESTIONS FOR THOUGHT

1. What determines if an electromagnetic wave emitted from an object is a visible light wave or a wave of infrared radiation?
2. What model of light does the polarization of light support? Explain.
3. Which carries more energy, red light or blue light? Should this mean anything about the preferred color of warning and stop lights? Explain.
4. What model of light is supported by the photoelectric effect? Explain.
5. What happens to light that is absorbed by matter?
6. One star is reddish, and another is bluish. Do you know anything about the relative temperatures of the two stars? Explain.

7. When does total internal reflection occur? Why does this occur in the diamond more than other gemstones?
8. Why does a highway sometimes appear wet on a hot summer day when it is not wet?
9. How can you tell if a pair of sunglasses is polarizing or not?
10. Explain why the intensity of reflected light appears to change if you tilt your head from side to side while wearing polarizing sunglasses.
11. What was so unusual about Planck's findings about blackbody radiation? Why was this considered revolutionary?
12. Why are both the photon model and the electromagnetic wave model accepted today as a single theory? Why was this so difficult for people to accept at first?

FOR FURTHER ANALYSIS

1. Clarify the distinction between light reflection and light refraction by describing clear, obvious examples of each.
2. Describe how you would use questions alone to help someone understand that the shimmering they see above a hot pavement is not heat.
3. Use a dialogue as you "think aloud" in considering the evidence that visible light is a wave, a particle, or both.

4. Compare and contrast the path of light through a convex and a concave lens. Give several uses for each lens, and describe how the shape of the lens results in that particular use.
5. Analyze how the equation $E = hf$ could mean that visible light is a particle and a wave at the same time.
6. How are visible light and a radio wave different? How are they the same?

INVITATION TO INQUIRY

Best Sunglasses?

Obtain several different types of sunglasses. Design experiments to determine which combination of features will be found in the best pair of sunglasses. First, design an experiment to determine which type reduces reflected glare the most. Find out how sunglasses are able to block ultraviolet radiation. According to your experiments and research, describe the "best" pair of sunglasses.

PARALLEL EXERCISES

The exercises in groups A and B cover the same concepts. Solutions to group A exercises are located in appendix G.

Group A

1. What is the speed of light while traveling through (a) water and (b) ice?

2. How many minutes are required for sunlight to reach Earth if the Sun is 1.50×10^8 km from Earth?

3. How many hours are required before a radio signal from a space probe near the dwarf planet Pluto reaches Earth, 6.00×10^9 km away?

4. A light ray is reflected from a mirror with an angle 10° to the normal. What was the angle of incidence?

5. Light travels through a transparent substance at 2.20×10^8 m/s. What is the substance?

6. The wavelength of a monochromatic light source is measured to be 6.00×10^{-7} m in an experiment. (a) What is the frequency? (b) What is the energy of a photon of this light?

7. At a particular location and time, sunlight is measured on a 1 m² solar collector with a power of 1,000.0 W. If the peak intensity of this sunlight has a wavelength of 5.60×10^{-7} m, how many photons are arriving each second?

8. A light wave has a frequency of 4.90×10^{14} cycles per second. (a) What is the wavelength? (b) What color would you observe (see table 7.2)?

9. What is the energy of a gamma photon of frequency 5.00×10^{20} Hz?

10. What is the energy of a microwave photon of wavelength 1.00 mm?

Group B

1. (a) What is the speed of light while traveling through a vacuum? (b) While traveling through air at 30°C? (c) While traveling through air at 0°C?

2. How much time is required for reflected sunlight to travel from the Moon to Earth if the distance between Earth and the Moon is 3.85×10^5 km?

3. How many minutes are required for a radio signal to travel from Earth to a space station on Mars if Mars is 7.83×10^7 km from Earth?

4. An incident light ray strikes a mirror with an angle of 30° to the surface of the mirror. What is the angle of the reflected ray?

5. The speed of light through a transparent substance is 2.00×10^8 m/s. What is the substance?

6. A monochromatic light source used in an experiment has a wavelength of 4.60×10^{-7} m. What is the energy of a photon of this light?

7. In black-and-white photography, a photon energy of about 4.00×10^{-19} J is needed to bring about the changes in the silver compounds used in the film. Explain why a red light used in a darkroom does not affect the film during developing.

8. The wavelength of light from a monochromatic source is measured to be 6.80×10^{-7} m. (a) What is the frequency of this light? (b) What color would you observe?

9. How much greater is the energy of a photon of ultraviolet radiation ($\lambda = 3.00 \times 10^{-7}$ m) than the energy of an average photon of sunlight ($\lambda = 5.60 \times 10^{-7}$ m)?

10. At what rate must electrons in a wire vibrate to emit microwaves with a wavelength of 1.00 mm?

8

Atoms and Periodic Properties

This is a picture of pure zinc, one of the eighty-nine naturally occurring elements found on Earth.

CORE **CONCEPT**

Different fields of study contributed to the model of the atom.

OUTLINE

The electron was discovered from experiments with electricity (p. 180).

8.1 Atomic Structure Discovered
  o Discovery of the Electron
  o The Nucleus
8.2 The Bohr Model
    The Quantum Concept
    Atomic Spectra
  o Bohr's Theory

The nucleus and proton were discovered from experiments with radioactivity (p. 181).

Experiments with light and line spectra and application of the quantum concept led to the Bohr model of the atom (pp. 184–86).

8.3 Quantum Mechanics o
8.4 The Periodic Table
A Closer Look: The Rare Earths
8.5 Metals, Nonmetals, and Semiconductors
People Behind the Science: Dmitri Ivanovich Mendeleyev

Application of the wave properties of electrons led to the quantum mechanics model of the atom (pp. 186–88).

Chemistry

▷ Electron configuration can be used to explain how atoms join together (Ch. 9).

▷ Water and its properties can be explained by considering the electron structure of hydrogen and oxygen atoms (Ch. 10).

▷ The nature of the atomic nucleus will explain radioactivity and nuclear energy (Ch. 11).

Astronomy

▷ The energy from stars originates in nuclear reactions in their core (Ch. 12).

▷ Stars are nuclear reactors (Ch. 12).

Earth Science

▷ Materials that make up Earth can be understood by considering their atomic structures (Ch. 15–16).

▷ Earth cycles materials through change (Ch. 15–18).

Life Science

▷ Living things use energy and materials in complex interactions (Ch. 19–26).

OVERVIEW

The development of the modern atomic model illustrates how modern scientific understanding comes from many different fields of study. For example, you will learn how studies of electricity led to the discovery that atoms have subatomic parts called *electrons*. The discovery of radioactivity led to the discovery of more parts, a central nucleus that contains protons and neutrons. Information from the absorption and emission of light was used to construct a model of how these parts are put together, a model resembling a miniature solar system with electrons circling the nucleus. The solar system model had initial, but limited, success and was inconsistent with other understandings about matter and energy. Modifications of this model were attempted, but none solved the problems. Then the discovery of wave properties of matter led to an entirely new model of the atom (figure 8.1).

The atomic model will be put to use in later chapters to explain the countless varieties of matter and the changes that matter undergoes. In addition, you will learn how these changes can be manipulated to make new materials, from drugs to ceramics. In short, you will learn how understanding the atom and all the changes it undergoes not only touches your life directly but shapes and affects all parts of civilization.

8.1 ATOMIC STRUCTURE DISCOVERED

Did you ever wonder how scientists could know about something so tiny that you cannot see it, even with the most powerful optical microscope? The atom is a tiny unit of matter, so small that 1 gram of hydrogen contains about 600,000,000,000,000,000,000,000 (six-hundred-thousand-billion billion or 6×10^{23}) atoms. Even more unbelievable is that atoms are not individual units but are made up of even smaller particles. How is it possible that scientists are able to tell you about the parts of something so small that it cannot be seen? The answer is that these things cannot be observed directly, but their existence can be inferred from experimental evidence. The following story describes the evidence and how scientists learned about the parts—electrons, the nucleus, protons, neutrons, and others— and how they are all arranged in the atom.

The atomic concept is very old, dating back to ancient Greek philosophers some 2,500 years ago. The ancient Greeks also reasoned about the way that pure substances are put together. A glass of water, for example, appears to be the same throughout. Is it the same? Two plausible, but conflicting, ideas were possible as an intellectual exercise. The water could have a *continuous* structure; that is, it could be completely homogeneous throughout. The other idea was that the water only appears to be continuous but is actually *discontinuous*. This means that if you continue to divide the water into smaller and smaller volumes, you would eventually reach a limit to this dividing, a particle that could not be further subdivided. The Greek philosopher Democritus (460–362 B.C.) developed this model in the fourth century B.C., and he called the indivisible particle an *atom*, from a Greek word meaning "uncuttable." However, neither Plato nor Aristotle accepted the atomic theory of matter, and it was not until about two thousand years later that the atomic concept of matter was reintroduced. In the early 1800s, the English chemist John Dalton brought back the ancient Greek idea of hard, indivisible atoms to explain

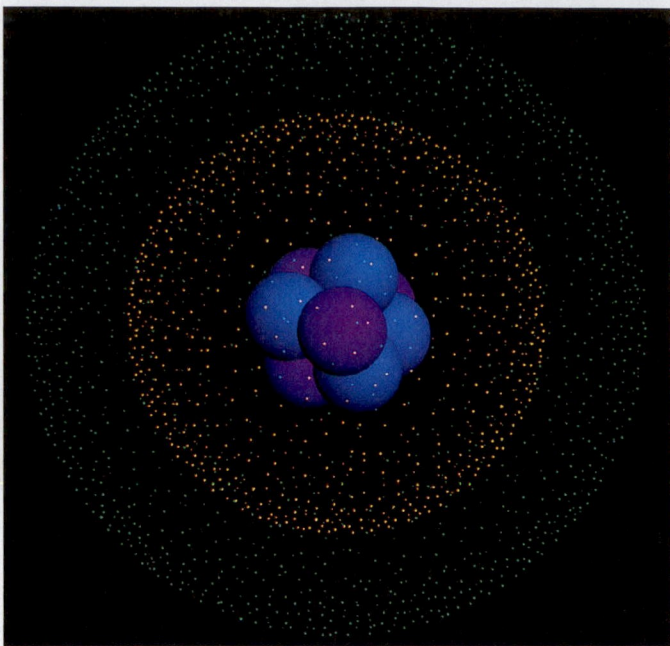

FIGURE 8.1 This is a computer-generated model of a beryllium atom, showing the nucleus and electron spacing relationships. This configuration can also be predicted from information on a periodic table. (Not to scale.)

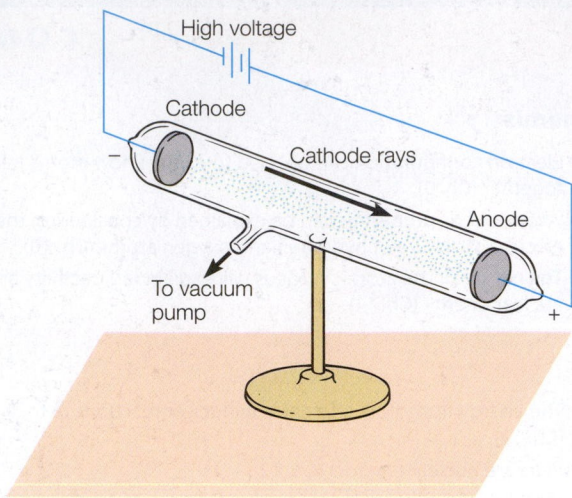

FIGURE 8.2 A vacuum tube with metal plates attached to a high-voltage source produces a greenish beam called *cathode rays*. These rays move from the cathode (negative charge) to the anode (positive charge).

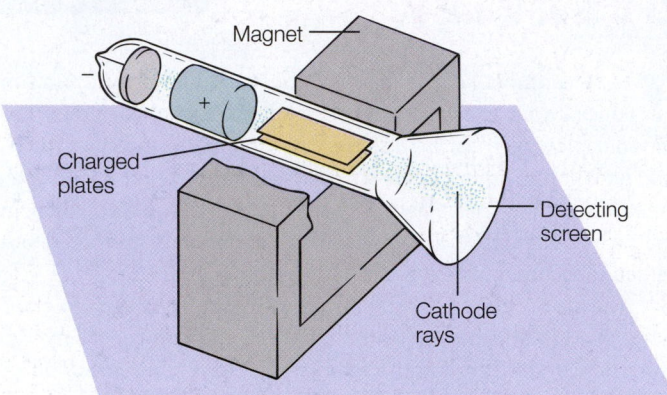

FIGURE 8.3 A cathode ray passed between two charged plates is deflected toward the positively charged plate. The ray is also deflected by a magnetic field. By measuring the deflection by both, J. J. Thomson was able to calculate the ratio of charge to mass. He was able to measure the deflection because the detecting screen was coated with zinc sulfide, a substance that produces a visible light when struck by a charged particle.

chemical reactions. Five statements will summarize his theory. As you will soon see, today we know that statement 2 is not strictly correct:

1. Indivisible minute particles called atoms make up all matter.
2. All the atoms of an element are exactly alike in shape and mass.
3. The atoms of different elements differ from one another in their masses.
4. Atoms chemically combine in definite whole-number ratios to form chemical compounds.
5. Atoms are neither created nor destroyed in chemical reactions.

During the 1800s, Dalton's concept of hard, indivisible atoms was familiar to most scientists. Yet, the existence of atoms was not generally accepted by all scientists. There was skepticism about something that could not be observed directly. Strangely, full acceptance of the atom came in the early 1900s with the discovery that the atom was not indivisible after all. The atom has parts that give it an internal structure. The first part to be discovered was the *electron,* a part that was discovered through studies of electricity.

Discovery of the Electron

Scientists of the late 1800s were interested in understanding the nature of the recently discovered electric current. To observe a current directly, they tried to produce one by itself, away from wires, by removing the air from a tube and then running electricity through the vacuum. When metal plates inside a tube

were connected to the negative and positive terminals of a high-voltage source (figure 8.2), a greenish beam was observed that seemed to move from the cathode (negative terminal) through the empty tube and collect at the anode (positive terminal). Since this mysterious beam seemed to come out of the cathode, it was said to be a *cathode ray*.

The English physicist J. J. Thomson figured out what the cathode ray was in 1897. He placed charged metal plates on each side of the beam (figure 8.3) and found that the beam was deflected away from the negative plate. Since it was known that like charges repel, this meant that the beam was composed of negatively charged particles.

The cathode ray was also deflected when caused to pass between the poles of a magnet. By balancing the deflections made by the magnet with the deflections made by the electric field, Thomson could determine the ratio of the charge to mass for an individual particle. Today, the charge-to-mass ratio is considered to be 1.7584×10^{11} coulomb/kilogram. A significant

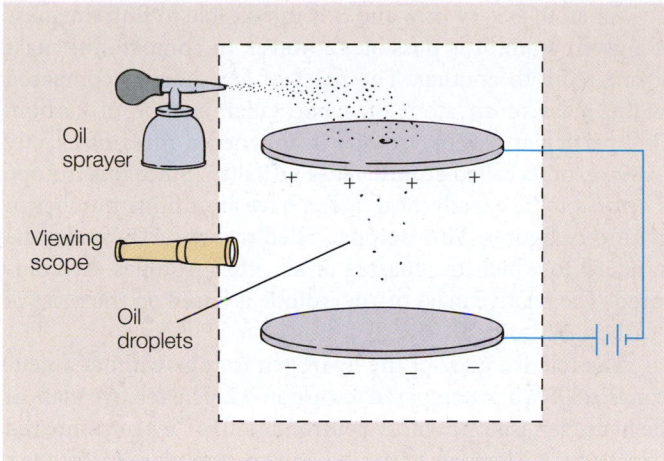

FIGURE 8.4 Millikan measured the charge of an electron by balancing the pull of gravity on oil droplets with an upward electrical force. Knowing the charge-to-mass ratio that Thomson had calculated, Millikan was able to calculate the charge on each droplet. He found that all the droplets had a charge of 1.60×10^{-19} coulomb or multiples of that charge. The conclusion was that this had to be the charge of an electron.

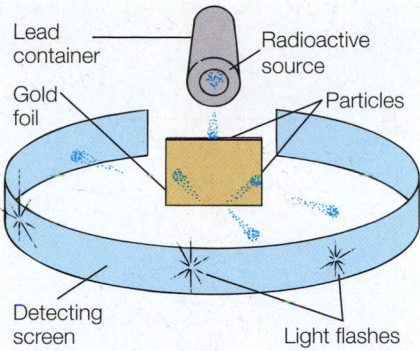

FIGURE 8.5 Rutherford and his co-workers studied particle scattering from a thin metal foil. The particles struck the detecting screen, producing a flash of visible light. Measurements of the angles between the flashes, the metal foil, and the source of the particles showed that they were scattered in all directions, including straight back toward the source.

part of Thomson's experiments was that he found the charge-to-mass ratio was the same no matter what gas was in the tube and of what materials the electrodes were made. Thomson had discovered the **electron,** a fundamental particle of matter.

A method for measuring the charge and mass of the electron was worked out by an American physicist, Robert A. Millikan, around 1906. Millikan used an apparatus like the one illustrated in figure 8.4 to measure the charge on tiny droplets of oil. Millikan found that none of the droplets had a charge less than one particular value (1.60×10^{-19} coulomb) and that larger charges on various droplets were always multiples of this unit of charge. Since all of the droplets carried the single unit of charge or multiples of the single unit, the unit of charge was understood to be the charge of a single electron.

Knowing the charge of a single electron and knowing the charge-to-mass ratio that Thomson had measured now made it possible to calculate the mass of a single electron. The mass of an electron was thus determined to be about 9.11×10^{-31} kg, or about 1/1,840 of the mass of the lightest atom, hydrogen.

Thomson had discovered the negatively charged electron, and Millikan had measured the charge and mass of the electron. But atoms themselves are electrically neutral. If an electron is part of an atom, there must be something else that is positively charged, canceling the negative charge of the electron. The next step in the sequence of understanding atomic structure would be to find what is neutralizing the negative charge and to figure out how all the parts are put together.

Thomson had proposed a model for what was known about the atom at the time. He suggested that an atom could be a blob of massless, positively charged stuff in which electrons were stuck like "raisins in plum pudding." If the mass of a hydrogen atom is due to the electrons embedded in a massless, positively charged matrix, then 1,840 electrons would be needed together with sufficient positive matter to make the atom electrically neutral.

The Nucleus

The nature of radioactivity and matter were the research interests of a British physicist, Ernest Rutherford (see page 242). In 1907, Rutherford was studying the scattering of radioactive particles directed toward a thin sheet of metal. As shown in figure 8.5, the particles from a radioactive source were allowed to move through a small opening in a lead container, so only a narrow beam of the massive, fast-moving particles would penetrate a very thin sheet of gold. The particles were detected by plates, which produced a small flash of light when struck by the particles.

Rutherford found that most of the particles went straight through the foil. However, he was astounded to find that some were deflected at very large angles and some were even reflected backward. He could account for this only by assuming that the massive, positively charged particles were repelled by a massive positive charge concentrated in a small region of the atoms in the gold foil (figure 8.6). He concluded that an atom must have a tiny, massive, and positively charged **nucleus** surrounded by electrons.

From measurements of the scattering, Rutherford estimated electrons must be moving around the nucleus at a distance 100,000 times the radius of the nucleus. This means the volume of an atom is mostly empty space. A few years later, Rutherford was able to identify the discrete unit of positive charge, which we now call a **proton.** Rutherford also speculated about the existence of a neutral particle in the nucleus, a neutron. The **neutron** was eventually identified in 1932 by James Chadwick.

Today, the number of protons in the nucleus of an atom is called the **atomic number.** All the atoms of a particular element have the same number of protons in their nucleus, so all atoms of an element have the same atomic number. Hydrogen has an atomic number of 1, so any atom that has one proton in its nucleus is an atom of the element hydrogen. Today, scientists

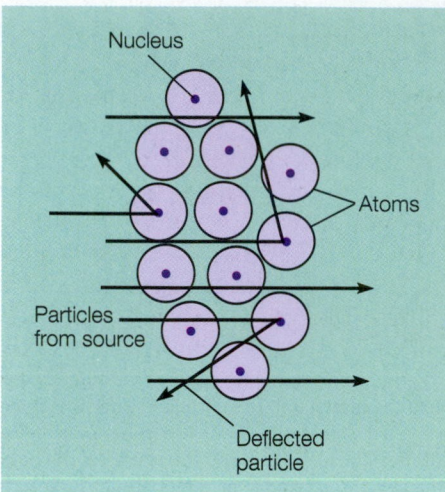

FIGURE 8.6 Rutherford's nuclear model of the atom explained the particle-scattering results as positive particles experiencing a repulsive force from the positive nucleus. Measurements of the percent of particles passing straight through and of the various angles of scattering of those coming close to the nuclei gave Rutherford a means of estimating the size of the nucleus.

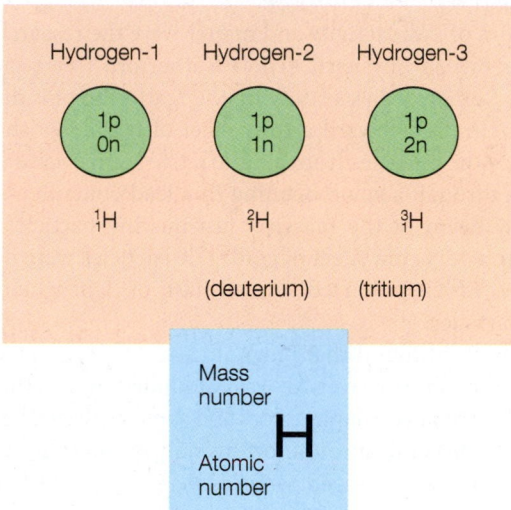

FIGURE 8.7 The three isotopes of hydrogen have the same number of protons but different numbers of neutrons. Hydrogen-1 is the most common isotope. Hydrogen-2, with an additional neutron, is named *deuterium*, and hydrogen-3 is called *tritium*.

have identified 117 different kinds of elements, each with a different number of protons.

The neutrons of the nucleus, along with the protons, contribute to the mass of an atom. Although all the atoms of an element must have the same number of protons in their nuclei, the number of neutrons may vary. Atoms of an element that have different numbers of neutrons are called **isotopes.** There are three isotopes of hydrogen illustrated in figure 8.7. All three isotopes have the same number of protons and electrons, but one isotope has no neutrons, one isotope has 1 neutron (deuterium), and one isotope has 2 neutrons (tritium).

An atom is very tiny, and it is impossible to find the mass of a given atom. It is possible, however, to compare the mass of one atom to another. The mass of any atom is compared to the mass of an atom of a particular isotope of carbon. This particular carbon isotope is assigned a mass of exactly 12.00 . . . units called **atomic mass units** (u). Since this isotope is *defined* to be exactly 12 u, it can have an infinite number of significant figures. This isotope, called *carbon-12,* provides the standard to which the masses of all other isotopes are compared. The relative mass of any isotope is based on the mass of a carbon-12 isotope.

The relative mass of the hydrogen isotope without a neutron is 1.007 when compared to carbon-12. The relative mass of the hydrogen isotope with 1 neutron is 2.0141 when compared to carbon-12. Elements occur in nature as a mixture of isotopes, and the contribution of each is calculated in the atomic weight. **Atomic weight** is a weighted average of the isotopes based on their mass compared to carbon-12 and their relative abundance found on Earth. Of all the hydrogen isotopes, for example, 99.985 percent occurs as the isotope without a neutron and 0.015 percent is the isotope with 1 neutron (the other isotope is not considered because it is radioactive). The fractional part of occurrence is multiplied by the relative atomic mass for each isotope, and the results are summed to obtain the atomic weight. Table 8.1 gives the atomic weight of hydrogen as 1.0079 as a result of this calculation.

The sum of the number of protons and neutrons in a nucleus of an atom is called the **mass number** of that atom. Mass numbers are used to identify isotopes. A hydrogen atom with 1 proton and 1 neutron has a mass number of $1 + 1$, or 2, and is referred to as hydrogen-2. A hydrogen atom with 1 proton and 2 neutrons has a mass number of $1 + 2$, or 3, and is referred to as hydrogen-3. Using symbols, hydrogen-3 is written as

$$^{3}_{1}H$$

where H is the chemical symbol for hydrogen, the subscript to the bottom left is the atomic number, and the superscript to the top left is the mass number.

How are the electrons moving around the nucleus? It might occur to you, as it did to Rutherford and others, that an atom might be similar to a miniature solar system. In this analogy, the

TABLE 8.1 Selected Atomic Weights Calculated from Mass and Abundance of Isotopes

Stable Isotopes	Mass of Isotope Compared to C-12		Abundance		Atomic Weight
$^{1}_{1}H$	1.007	×	99.985%		
$^{2}_{1}H$	2.0141	×	0.015%	=	1.0079
$^{9}_{4}Be$	9.01218	×	100%	=	9.01218
$^{14}_{7}N$	14.00307	×	99.63%		
$^{15}_{7}N$	15.00011	×	0.37%	=	14.0067

nucleus is in the role of the Sun, electrons in the role of moving planets in their orbits, and electrical attractions between the nucleus and electrons in the role of gravitational attraction. There are, however, big problems with this idea. If electrons were moving in circular orbits, they would continually change their direction of travel and would therefore be accelerating (see p. 154). According to the Maxwell model of electromagnetic radiation, an accelerating electric charge emits electromagnetic radiation such as light. If an electron gave off light, it would lose energy. The energy loss would mean that the electron could not maintain its orbit, and it would be pulled into the oppositely charged nucleus. The atom would collapse as electrons spiraled into the nucleus. Since atoms do not collapse like this, there is a significant problem with the solar system model of the atom.

8.2 THE BOHR MODEL

Niels Bohr was a young Danish physicist who visited Rutherford's laboratory in 1912 and became very interested in questions about the solar system model of the atom. He wondered what determined the size of the electron orbits and the energies of the electrons. He wanted to know why orbiting electrons did not give off electromagnetic radiation. Seeking answers to questions such as these led Bohr to incorporate the *quantum concept* of Planck and Einstein with Rutherford's model to describe the electrons in the outer part of the atom. This quantum concept will be briefly reviewed before proceeding with the development of Bohr's model of the hydrogen atom.

The Quantum Concept

In 1900, Max Planck introduced the idea that matter emits and absorbs energy in discrete units that he called **quanta.** Planck had been trying to match data from spectroscopy experiments with data that could be predicted from the theory of electromagnetic radiation. In order to match the experimental findings with the theory, he had to assume that specific, discrete amounts of energy were associated with different frequencies of radiation. In 1905, Albert Einstein extended the quantum concept to light, stating that light consists of discrete units of energy that are now called **photons.** The energy of a photon is directly proportional to the frequency of vibration, and the higher the frequency of light, the greater the energy of the individual photons. In addition, the interaction of a photon with matter is an "all-or-none" affair; that is, matter absorbs an entire photon or none of it. The relationship between frequency (f) and energy (E) is

$$E = hf \qquad \text{equation 8.1}$$

where h is the proportionality constant known as *Planck's constant* (6.63×10^{-34} J·s). This relationship means that higher-frequency light, such as ultraviolet, has more energy than lower-frequency light, such as red light.

EXAMPLE 8.1 *(Optional)*

What is the energy of a photon of red light with a frequency of 4.60×10^{14} Hz?

SOLUTION

$f = 4.60 \times 10^{14}$ Hz $E = hf$

$h = 6.63 \times 10^{-34}$ J·s

$E = ?$

$\qquad = (6.63 \times 10^{-34} \text{ J·s})\left(4.60 \times 10^{14} \dfrac{1}{\text{s}}\right)$

$\qquad = (6.63 \times 10^{-34})(4.60 \times 10^{14}) \text{J·s} \times \dfrac{1}{\text{s}}$

$\qquad = \boxed{3.05 \times 10^{-19} \text{ J}}$

EXAMPLE 8.2 *(Optional)*

What is the energy of a photon of violet light with a frequency of 7.30×10^{14} Hz? (Answer: 4.84×10^{-19} J)

Atomic Spectra

Planck was concerned with hot solids that emit electromagnetic radiation. The nature of this radiation, called *blackbody radiation,* depends on the temperature of the source. When this light is passed through a prism, it is dispersed into a *continuous spectrum,* with one color gradually blending into the next as in a rainbow. Today, it is understood that a continuous spectrum comes from solids, liquids, and dense gases because the atoms interact, and all frequencies within a temperature-determined range are emitted. Light from an incandescent gas, on the other hand, is dispersed into a **line spectrum,**

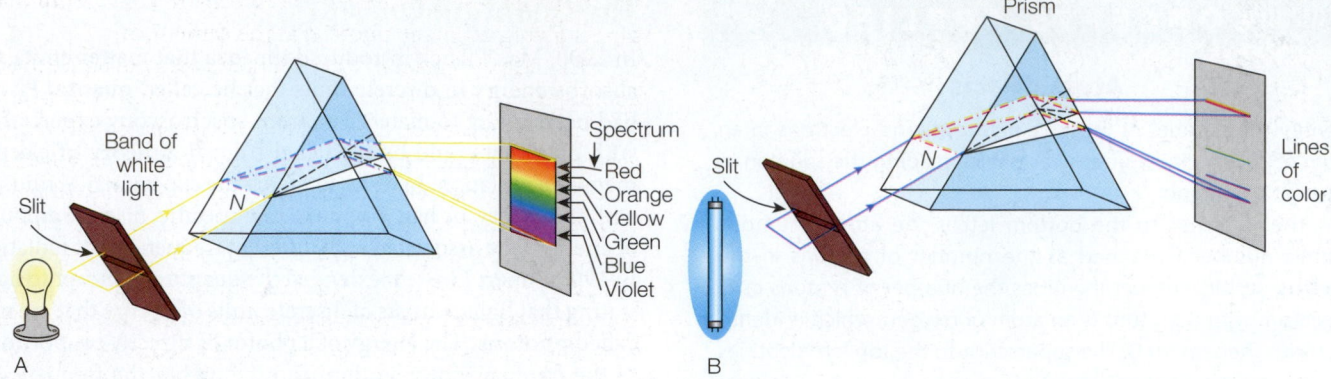

FIGURE 8.8 (A) Light from incandescent solids, liquids, or dense gases produces a continuous spectrum as atoms interact to emit all frequencies of visible light. (B) Light from an incandescent gas produces a line spectrum as atoms emit certain frequencies that are characteristic of each element.

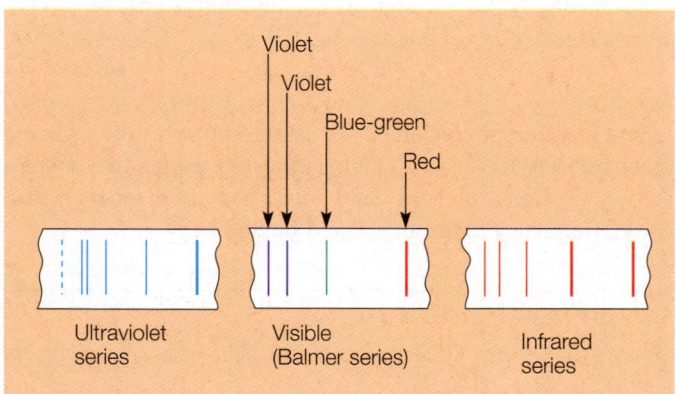

FIGURE 8.9 Atomic hydrogen produces a series of characteristic line spectra in the ultraviolet, visible, and infrared parts of the total spectrum. The visible light spectra always consist of two violet lines, a blue-green line, and a bright red line.

narrow lines of colors with no light between the lines (figure 8.8). The atoms in the incandescent gas are able to emit certain characteristic frequencies, and each frequency is a line of color that represents a definite value of energy. The line spectra are specific for a substance, and increased or decreased temperature changes only the intensity of the lines of colors. Thus, hydrogen always produces the same colors of lines in the same position. Helium has its own specific set of lines, as do other substances. Line spectra are a kind of fingerprint that can be used to identify a gas. A line spectrum might also extend beyond visible light into ultraviolet, infrared, and other electromagnetic regions.

In 1885, a Swiss mathematics teacher named J. J. Balmer was studying the regularity of spacing of the hydrogen line spectra. Balmer was able to develop an equation that fit all the visible lines. These four lines became known as the *Balmer series*. Other series were found later, outside the visible part of the spectrum (figure 8.9).

Such regularity of observable spectral lines must reflect some unseen regularity in the atom. At this time it was known that hydrogen had only one electron. How could one electron produce series of spectral lines with such regularity?

Bohr's Theory

An acceptable model of the hydrogen atom would have to explain the characteristic line spectra and their regularity as described by Balmer. In fact, a successful model should be able to predict the occurrence of each color line as well as account for its origin. By 1913, Bohr was able to do this by applying the quantum concept to a solar system model of the atom. He began by considering the single hydrogen electron to be a single "planet" revolving in a circular orbit around the nucleus. Three sets of rules described this electron:

1. **Allowed orbits.** An electron can revolve around an atom only in specific allowed orbits. Bohr considered the electron to be a particle with a known mass in motion around the nucleus and used Newtonian mechanics to calculate the distances of the allowed orbits. According to the Bohr model, electrons can exist only in one of these allowed orbits and nowhere else.

2. **Radiationless orbits.** An electron in an allowed orbit does not emit radiant energy as long as it remains in the orbit. According to Maxwell's theory of electromagnetic radiation, an accelerating electron should emit an electromagnetic wave, such as light, which would move off into space from the electron. Bohr recognized that electrons moving in a circular orbit are accelerating, since they are changing direction continuously. Yet, hydrogen atoms did not emit light in their normal state. Bohr decided that the situation must be different for orbiting electrons and that electrons could stay in their allowed orbits and not give off light. He postulated this rule as a way to make his theory consistent with other scientific theories.

3. **Quantum leaps.** An electron gains or loses energy only by moving from one allowed orbit to another (figure 8.10). In the Bohr model, the energy an electron has depends on which allowable orbit it occupies. The only way that an electron can change its energy is to jump from one allowed orbit to another in quantum "leaps." An electron must acquire energy to jump from a lower orbit to a higher one. Likewise, an electron gives up energy when jumping from a higher orbit to a lower one. Such jumps must be all at once, not part way and not gradual. An electron

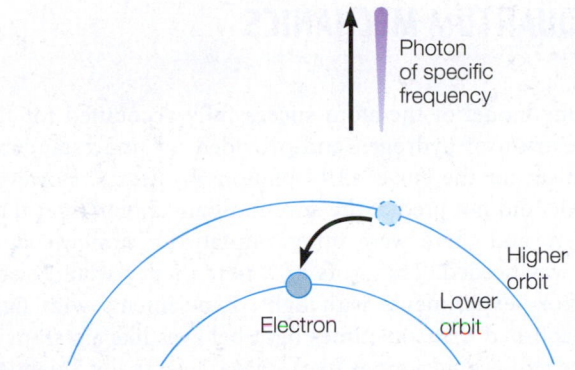

Photon
of specific
frequency

Higher
orbit

Lower
orbit

Electron

FIGURE 8.10 Each time an electron makes a "quantum leap," moving from a higher-energy orbit to a lower-energy orbit, it emits a photon of a specific frequency and energy value.

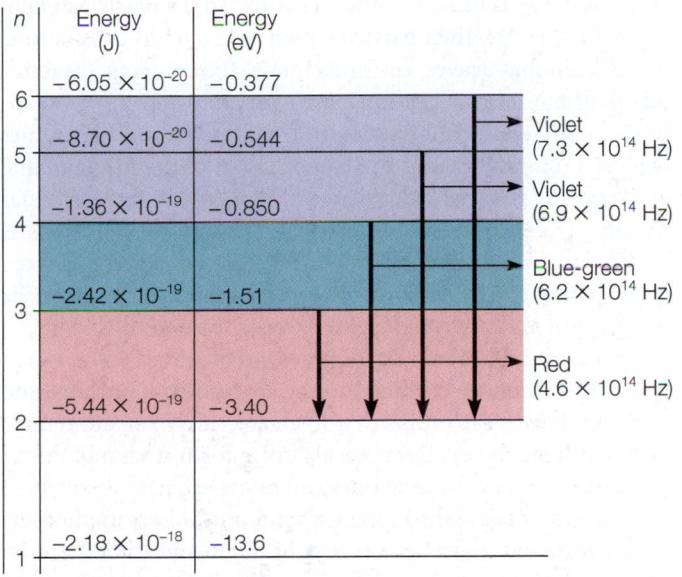

n	Energy (J)	Energy (eV)	
6	-6.05×10^{-20}	-0.377	
5	-8.70×10^{-20}	-0.544	Violet (7.3×10^{14} Hz)
4	-1.36×10^{-19}	-0.850	Violet (6.9×10^{14} Hz)
3	-2.42×10^{-19}	-1.51	Blue-green (6.2×10^{14} Hz)
2	-5.44×10^{-19}	-3.40	Red (4.6×10^{14} Hz)
1	-2.18×10^{-18}	-13.6	

FIGURE 8.11 An energy level diagram for a hydrogen atom, not drawn to scale. The energy levels (n) are listed on the left side, followed by the energies of each level in J and eV. The color and frequency of the visible light photons emitted are listed on the right side, with the arrow showing the orbit moved from and to.

acquires energy from high temperatures or from electrical discharges to jump to a higher orbit. An electron jumping from a higher to a lower orbit gives up energy in the form of light. A single photon is emitted when a downward jump occurs, and the energy of the photon is *exactly* equal to the difference in the energy level of the two orbits.

The energy level diagram in figure 8.11 shows the energy states for the orbits of a hydrogen atom. The lowest energy state is the **ground state** (or normal state). The higher states are the **excited states.** The electron in a hydrogen atom would normally occupy the ground state, but high temperatures or electric discharge can give the electron sufficient energy to jump to one of the excited states. Once in an excited state, the electron immediately jumps back to a lower state, as shown by the arrows in the figure. The length of the arrow represents the frequency of the photon that the electron emits in the process. A hydrogen atom can give off only one photon at a time, and

the many lines of a hydrogen line spectrum come from many atoms giving off many photons at the same time.

The reference level for the potential energy of an electron is considered to be zero when the electron is *removed* from an atom. The electron, therefore, has a lower and lower potential energy at closer and closer distances to the nucleus and has a negative value when it is in some allowed orbit. By way of analogy, you could consider ground level as a reference level where the potential energy of some object equals zero. But suppose there are two basement levels below the ground. An object on either basement level would have a gravitational potential energy less than zero, and work would have to be done on each object to bring it back to the zero level. Thus, each object would have a negative potential energy. The object on the lowest level would have the largest negative value of energy, because more work would have to be done on it to bring it back to the zero level. Therefore, the object on the lowest level would have the *least* potential energy, and this would be expressed as the *largest negative value.*

Just as the objects on different basement levels have negative potential energy, the electron has a definite negative potential energy in each of the allowed orbits. Bohr calculated the energy of an electron in the orbit closest to the nucleus to be -2.18 ×10^{-18} J, which is called the energy of the lowest state. The energy of electrons can be expressed in units of the electron volt (eV). An electron volt is defined as the energy of an electron moving through a potential of 1 volt. Because this energy is charge times voltage (from equation 6.3, $V = W/q$), 1.00 eV is equivalent to 1.60×10^{-19} J. Therefore, the energy of the electron in the innermost orbit is its energy in joules divided by 1.60×10^{-19} J/eV, or -13.6 eV.

Bohr found that the energy of each of the allowed orbits could be found from the simple relationship of

$$E_n = \frac{E_1}{n^2}$$

equation 8.2

where E_1 is the energy of the innermost orbit (-13.6 eV) and n is the quantum number for an orbit, or 1, 2, 3, and so on. Thus, the energy for the second orbit ($n = 2$) is $E_2 = -13.6$ eV/4 = -3.40 eV. The energy for the third orbit out ($n = 3$) is $E_3 = -13.6$ eV/9 = -1.51 eV, and so forth (figure 8.11). Thus, the energy of each orbit is quantized, occurring only as a definite value.

In the Bohr model, the energy of the electron is determined by which allowable orbit it occupies. The only way that an electron can change its energy is to jump from one allowed orbit to another in quantum "jumps." An electron must acquire energy to jump from a lower orbit to a higher one. Likewise, an electron gives up energy when jumping from a higher orbit to a lower one. Such jumps must be all at once, not partway and not gradual. By way of analogy, this is very much like the gravitational potential energy that you have on the steps of a staircase. You have the lowest potential on the bottom step and the greatest amount on the top step. Your potential energy is quantized because you can increase or decrease it by going up or down a number of steps, but you cannnot stop between the steps.

An electron acquires energy from high temperatures or from electrical discharges to jump to a higher orbit. An electron jumping from a higher to a lower orbit gives up energy in the form of light. A single photon is emitted when a downward jump occurs, and the energy of the photon is exactly equal to the difference in the energy level of the two orbits. If E_L represents the lower-energy level (closest to the nucleus) and E_H represents a higher-energy level (farthest from the nucleus), the energy of the emitted photon is

$$hf = E_H - E_L \qquad \text{equation 8.3}$$

where h is Planck's constant and f is the frequency of the emitted light.

As you can see, the energy level diagram in figure 8.11 shows how the change of known energy levels from known orbits results in the exact frequencies of the color lines in the Balmer series. Bohr's theory did explain the lines in the hydrogen spectrum with a remarkable degree of accuracy. However, the model did not have much success with larger atoms. The Bohr model could not explain the spectra line of larger atoms with its single quantum number. A German physicist, A. Sommerfeld, tried to modify Bohr's model by adding elliptical orbits in addition to Bohr's circular orbits. It soon became apparent that the "patched up" model, too, was not adequate. Bohr had made the rule that there were radiationless orbits without an explanation, and he did not have an explanation for the quantized orbits. There was something fundamentally incomplete about the model.

EXAMPLE 8.3   (Optional)

An electron in a hydrogen atom jumps from the excited energy level $n = 4$ to $n = 2$. What is the frequency of the emitted photon?

SOLUTION

The frequency of an emitted photon can be calculated from equation 8.3, $hf = E_H - E_L$. The values for the two energy levels can be obtained from figure 8.11. (Note: E_H and E_L must be in joules. If the values are in electron volts, they can be converted to joules by multiplying by the ratio of joules per electron volt, or (eV) $(1.60 \times 10^{-19}$ J/eV) = joules.)

$$E_H = -1.36 \times 10^{-19} \text{ J}$$
$$E_L = -5.44 \times 10^{-19} \text{ J}$$
$$h = 6.63 \times 10^{-34} \text{ J} \cdot \text{s}$$
$$f = ?$$

$$hf = E_H - E_L \therefore f = \frac{E_H - E_L}{h}$$

$$= \frac{(-1.36 \times 10^{-19} \text{J}) - (-5.44 \times 10^{-19} \text{J})}{6.63 \times 10^{-34} \text{J} \cdot \text{s}}$$

$$= \frac{4.08 \times 10^{-19}}{6.63 \times 10^{-34}} \frac{\text{J}}{\text{J} \cdot \text{s}}$$

$$= 6.15 \times 10^{14} \frac{1}{\text{s}}$$

$$= 6.15 \times 10^{14} \text{Hz}$$

This is approximately the blue-green line in the hydrogen line spectrum.

8.3   QUANTUM MECHANICS

The Bohr model of the atom successfully accounted for the line spectrum of hydrogen and provided an understandable mechanism for the emission of photons by atoms. However, the model did not predict the spectra of any atom larger than hydrogen, and there were other limitations. A new, better theory was needed. The roots of a new theory would again come from experiments with light. Experiments with light had established that sometimes light behaves like a stream of particles and at other times like a wave. Eventually, scientists began to accept that light has both wave properties and particle properties, which is now referred to as the *wave-particle duality of light*. In 1923, Louis de Broglie, a French physicist, reasoned that symmetry is usually found in nature, so if a particle of light has a dual nature, then particles such as electrons should too.

Recall that waves confined on a fixed string establish resonant modes of vibration called *standing waves* (see chapter 5). Only certain fundamental frequencies and harmonics can exist on a string, and the combination of the fundamental and overtones gives the stringed instrument its particular quality. The same result of resonant modes of vibrations is observed in *any* situation where waves are confined to a fixed space. Characteristic standing wave patterns depend on the wavelength and wave velocity for waves formed on strings, in enclosed columns of air, or for any kind of wave in a confined space. Electrons are confined to the space near a nucleus, and electrons have wave properties, so an electron in an atom must be a confined wave. Does an electron form a characteristic wave pattern? This was the question being asked in about 1925 when Heisenberg, Schrödinger, Dirac, and others applied the wave nature of the electron to develop a new model of the atom based on the mechanics of electron waves. The new theory is now called *wave mechanics,* or quantum mechanics.

Erwin Schrödinger, an Austrian physicist, treated the atom as a three-dimensional system of waves to derive the *Schrödinger equation*. Instead of the simple circular planetary orbits of the Bohr model, solving the Schrödinger equation results in a description of three-dimensional shapes of the patterns that develop when electron waves are confined by a nucleus. Schrödinger first considered the hydrogen atom, calculating the states of vibration that would be possible for an electron wave confined by a nucleus. He found that the frequency of these vibrations, when multiplied by Planck's constant, matched exactly, to the last decimal point, the observed energies of the quantum states of the hydrogen atom. The conclusion is that the wave nature of the electron is the important property to consider for a successful model of the atom.

The quantum mechanics theory of the atom proved to be very successful; it confirmed all the known experimental facts and predicted new discoveries.

Quantum mechanics does share at least one idea with the Bohr model. This is that an electron emits a photon when jumping from a higher state to a lower one. The Bohr model, however, considered the *particle* nature of an electron moving in a circular orbit with a definitely assigned position at a given time. Quantum mechanics considers the *wave* nature, with the electron as a confined wave with well-defined shapes and frequencies. A wave is not localized like a particle and is spread out in space. The quantum mechanics model is, therefore, a series of orbitlike smears, or fuzzy statistical representations, of where the electron might be found.

The quantum mechanical theory is not an extension or refinement of the Bohr model. The Bohr model considered electrons as particles in circular orbits that could be only certain distances from the nucleus. The quantum mechanical model, on the other hand, considers the electron as a wave and considers the energy of its harmonics, or modes, of standing waves. In the Bohr model, the location of an electron was certain—in an orbit. In the quantum mechanical model, the electron is a spread-out wave.

Quantum mechanics describes the energy state of an electron wave with four *quantum numbers:*

1. **Distance from the nucleus.** The *principal quantum number* describes the *main energy level* of an electron in terms of its most probable distance from the nucleus. The lowest energy state possible is closest to the nucleus and is assigned the principal quantum number of 1 ($n = 1$). Higher states are assigned progressively higher positive whole numbers of $n = 2$, $n = 3$, $n = 4$, and so on. Electrons with higher principal quantum numbers have higher energies and are located farther from the nucleus.

2. **Energy sublevel.** The *angular momentum quantum number* defines energy sublevels within the main energy levels. Each sublevel is identified with a letter. The first four of these letters, in order of increasing energy, are s, p, d, and f. The letter s represents the lowest sublevel, and the letter f represents the highest sublevel. A principal quantum number and a letter indicating the angular momentum quantum number are combined to identify the main energy state and energy sublevel of an electron. For an electron in the lowest main energy level, $n = 1$, and in the lowest sublevel, s, the number and letter are 1s (read as "one-s"). Thus, 1s indicates an electron that is as close to the nucleus as possible in the lowest energy sublevel possible.

There are limits to how many sublevels can occupy each of the main energy levels. Basically, the lowest main energy level can have only the lowest sublevel, and another sublevel is added as you move up through the main energy levels. Thus, the lowest main energy level, $n = 1$, can have only the s sublevel. The $n = 2$ can have s and p sublevels. The $n = 3$ main energy level can have the s, p, and the d sublevels. Finally, the $n = 4$ main energy level can have all four sublevels, with s, p, d, and f. Therefore, the number of possible sublevels is the same as the principal quantum number.

The Bohr model considered the location of an electron as certain, like a tiny shrunken marble in an orbit. The quantum mechanics model considers the electron as a wave, and knowledge of its location is very uncertain. The **Heisenberg uncertainty principle** states that you cannot measure the exact position of a wave because a wave is spread out. One cannot specify the position and the momentum of a spread-out electron. The location of the electron can be described only in terms of *probabilities*

of where it might be at a given instant. The probability of location is described by a fuzzy region of space called an **orbital.** An orbital defines the space where an electron is likely to be found. Orbitals have characteristic three-dimensional shapes and sizes and are identified with electrons of characteristic energy levels (figure 8.12). An orbital shape represents where an electron could probably be located at any particular instant. This "probability cloud" could likewise have any particular orientation in space, and the direction of this orientation is uncertain.

3. **Orientation in space.** An external magnetic field applied to an atom produces different energy levels that are related to the orientation of the orbital to the magnetic field. The orientation of an orbital in space is described by the *magnetic quantum number.* This number is related to the energies of orbitals as they are oriented in space relative to an external magnetic field, a kind of energy sub-sublevel. In general, the lowest-energy sublevel (s) has only one orbital orientation (figure 8.12A). The next higher-energy sublevel (p) can have three orbital orientations (figure 8.12B). The d sublevel can have five orbital orientations, and the highest sublevel, f, can have a total of seven different orientations (figure 8.12C) (also see table 8.2).

4. **Direction of spin.** Detailed studies have shown that an electron spinning one way, say clockwise, in an external magnetic field would have a different energy than one spinning the other way, say counterclockwise. The *spin quantum number* describes these two spin orientations (figure 8.13).

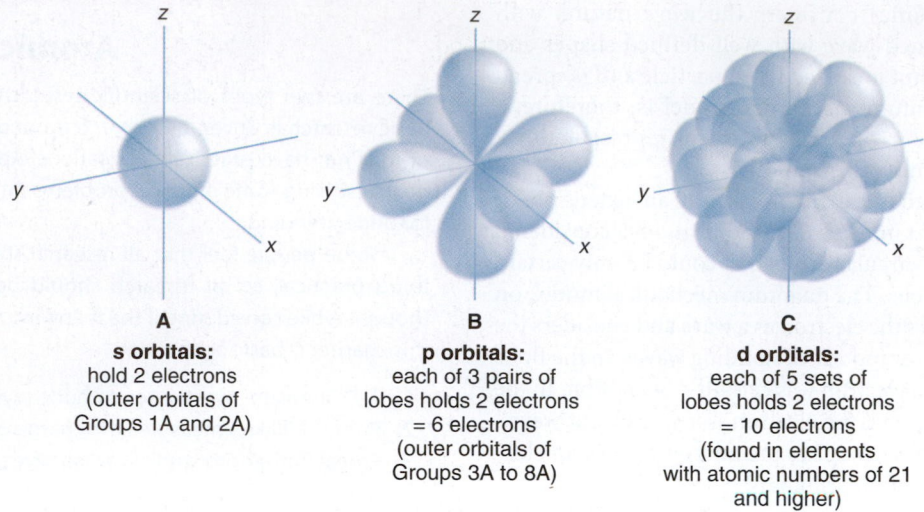

A
s orbitals:
hold 2 electrons
(outer orbitals of
Groups 1A and 2A)

B
p orbitals:
each of 3 pairs of
lobes holds 2 electrons
= 6 electrons
(outer orbitals of
Groups 3A to 8A)

C
d orbitals:
each of 5 sets of
lobes holds 2 electrons
= 10 electrons
(found in elements
with atomic numbers of 21
and higher)

FIGURE 8.12 The general shapes of s, p, and d orbitals, the regions of space around the nuclei of atoms in which electrons are likely to be found. (The f orbital is too difficult to depict.)

TABLE 8.2 **Quantum Numbers and Electron Distribution to $n = 4$**

Main Energy Level	Energy Sublevels	Maximum Number of Electrons	Maximum Number of Electrons per Main Energy Level
$n = 1$	s	2	2
$n = 2$	s	2	
	p	6	8
$n = 3$	s	2	
	p	6	
	d	10	18
$n = 4$	s	2	
	p	6	
	d	10	
	f	14	32

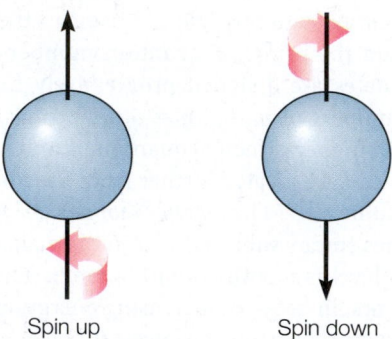

Spin up   Spin down

FIGURE 8.13 Experimental evidence supports the concept that electrons can be considered to spin one way or the other as they move about an orbital under an external magnetic field.

Electron spin is an important property that helps determine the electronic structure of an atom. As it turns out, two electrons spinning in opposite directions produce unlike magnetic fields that are attractive, balancing some of the normal repulsion from two like charges. Two electrons of opposite spin, called an electron pair, can thus occupy the same orbital. Wolfgang Pauli, a German physicist, summarized this idea in 1924. His summary, now known as the **Pauli exclusion principle,** states that *no two electrons in an atom can have the same four quantum numbers.* This principle provides the key for understanding the electron structure of atoms. (For a detailed discussion of electron configuration, see the chapter 8 resources on www.mhhe.com/tillery.)

8.4 THE PERIODIC TABLE

The periodic table is made up of rows and columns of cells, with each element having its own cell in a specific location.

The cells are not arranged symmetrically. The arrangement has a meaning, both about atomic structure and about chemical behaviors. It will facilitate your understanding of the code if you refer frequently to a periodic table during the following discussion (figure 8.14).

An element is identified in each cell with its chemical symbol. The number above the symbol is the atomic number of the element, and the number below the symbol is the rounded atomic weight of the element.* Horizontal rows of elements run from left to right with increasing atomic numbers. Each row is called a *period.* The periods are numbered from 1 to 7 on the left side. A vertical column of elements is called a *family* (or group) of elements. Elements in families have similar properties, but this is more true of some families than others. The table

---

Isotopes are identified using a chemical symbol, atomic number, and mass number as shown in figure 8.7. These units are also used in writing nuclear equations. *Elements* are compared in the periodic table using a chemical symbol, atomic number, and atomic weight as shown in figure 8.14. The table is used to find atomic weights in chemical calculations.

Compounds of the ra
identified when they
uncommon minerals i
elements are very reac
chemical properties, s
ognized as elements
later. Thus, they were
earths, that is, non-me
in fact they are met
were also considered
that time, they were kn
uncommon minerals.
elements are known t
in the Earth than gol
tungsten. The rarest o
lium, is twice as abund
earth elements are ne
and they are importa
electronic, and metall

You can identify
two lowermost rows
These rows contain tw

with an **electron do**
symbol with dots aro
electrons. Electron do
elements in figure 8.1
the noble gases are in
eight outer electrons
have one dot, all the
pattern will explain
third group of in-bet

There are sever
properties. One exa
to the physical prop
conductivity, malleal

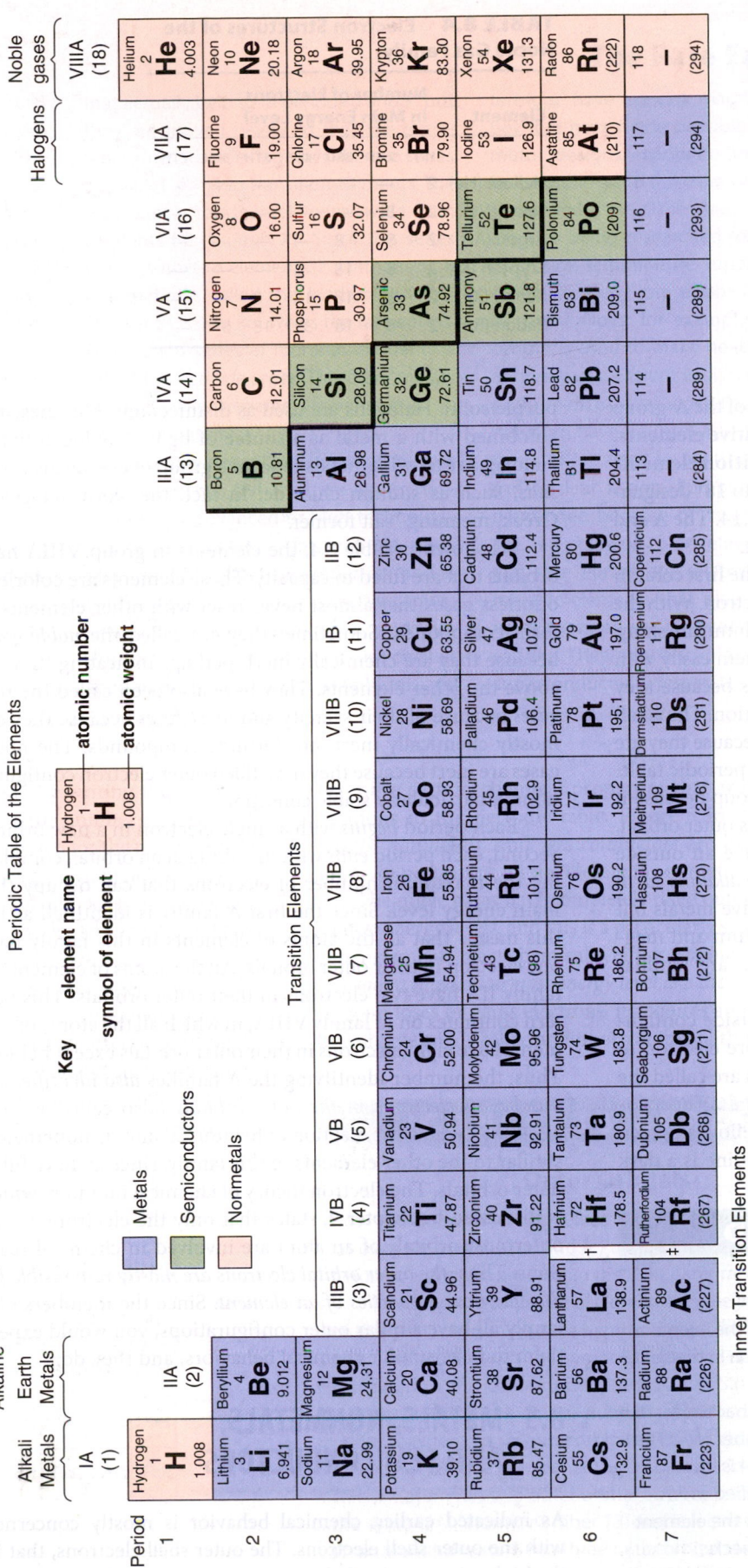

FIGURE 8.15 Elect

Periodic Table of the Elements

Key

element name → Hydrogen
atomic number → 1
symbol of element → **H**
atomic weight → 1.008

Metals
Semiconductors
Nonmetals

Values in parentheses are the mass numbers of the most stable or best-known isotopes.

Names and symbols for elements 113–118 are under review.

Transition Elements

Inner Transition Elements

†Lanthanides 6

‡Actinides 7

TABLE 8.3 Electr
Alkal

Element	Numb in Mai
	1ST 2
Lithium (Li)	2
Sodium (Na)	2
Potassium (K)	2
Rubidium (Rb)	2
Cesium (Cs)	2
Francium (Fr)	2

is subdivided into A a
families are called th
The members of the
(or metals). Some sci
tion for the A and B g
B designations will b

As shown in tabl
have an outside electr
exception of hydroger
low-density metals th
a knife. These metals
react violently with w
metals do not occur i
so reactive. Hydroger
It is not an alkali me
seems to fit there bec

The elements in
configuration of two
metals. The alkaline
not as reactive or sof
nesium, in the form
examples of this grou

The elements in
ration of seven elect
completely fill the ou
halogens. The haloge
gens fluorine and chl
gases. Bromine is a r

CONC

Identify the periodic
silicon. Write your ans
paragraph.

According to the
this text, silicon has th
square with the symb
the third period (third

Now, can you ide
iron? Compare your a

Connections . . .

Nutrient Elements

All molecules required to support living things are called nutrients. Some nutrients are inorganic molecules such as calcium, iron, or potassium, and others are organic molecules such as carbohydrates, proteins, fats, and vitamins. All minerals are elements and cannot be synthesized by the body. Because they are elements, they cannot be broken down or destroyed by metabolism or cooking. They commonly occur in many foods and in water. Minerals retain their characteristics whether they are in foods or in the body, and each plays a different role in metabolism.

Minerals* can function as regulators, activators, transmitters, and controllers of various enzymatic reactions. For example, sodium ions (Na^+) and potassium ions (K^+) are important in the transmission of nerve impulses, while magnesium ions (Mg^{2+}) facilitate energy release during reactions involving ATP. Without iron, not enough hemoglobin would be formed to transport oxygen, a condition called anemia, and a lack of calcium may result in osteoporosis. Osteoporosis is a condition that results from calcium loss leading to painful, weakened bones. There are many minerals that are important in your diet. In addition to those just mentioned, you need chlorine, cobalt, copper, iodine, phosphorus, potassium, sulfur, and zinc to remain healthy. With few exceptions, adequate amounts of minerals are obtained in a normal diet. Calcium and iron supplements may be necessary, particularly in women.

Osteoporosis is a nutritional deficiency disease that results in a change in the density of the bones as a result of the loss of bone mass. Bones that have undergone this change look lacy or like Swiss cheese, with larger than normal holes. A few risk factors found to be associated with this disease are being female and fair skinned; having a sedentary lifestyle; using alcohol, caffeine, and tobacco; being anorexic; and having reached menopause.

*Note that the term *mineral* has a completely different meaning in geology. See page 330 in chapter 15.

has three protons (plus charges) and three electrons (negative charges). If it loses the outermost electron, it now has an outer filled orbital structure like helium, a noble gas. It is also now an ion, since it has three protons (3+) and two electrons (2−), for a net charge of 1+. A lithium ion thus has a 1+ charge.

Elements with one, two, or three outer electrons tend to lose electrons to form positive ions. The metals lose electrons like this, and the *metals are elements that lose electrons to form positive ions* (figure 8.17). Nonmetals, on the other hand, are elements with five to seven outer electrons that tend to acquire electrons to fill their outer orbitals. *Nonmetals are elements that gain electrons to form negative ions.* In general, elements located in the left two-thirds or so of the periodic table are metals. The nonmetals are on the right side of the table.

The dividing line between the metals and nonmetals is a steplike line from the left top of group IIIA down to the bottom left of group VIIA. This is not a line of sharp separation between the metals and nonmetals, and elements *along* this line sometimes act like metals, sometimes like nonmetals, and sometimes like both. These hard-to-classify elements are called **semiconductors** (or *metalloids*). Silicon, germanium, and arsenic have physical properties of nonmetals; for example, they are brittle materials that cannot be hammered into a new shape. Yet these elements conduct electric currents under certain conditions. The ability to conduct an electric current is a property of a metal, and nonmalleability is a property of nonmetals, so as you can see, these semiconductors have the properties of both metals and nonmetals.

Semiconductors play a major role in our lives, and they now make computers, cell phones, CDs, and recent innovations in television possible. Tomorrow we will find semiconductors playing an even larger role with increased uses in solar cells and light emitting diodes (LEDs). Solar cell use will increase as we turn more and more to this clean way of producing electricity. LED use will increase as LEDs make possible more innovations,

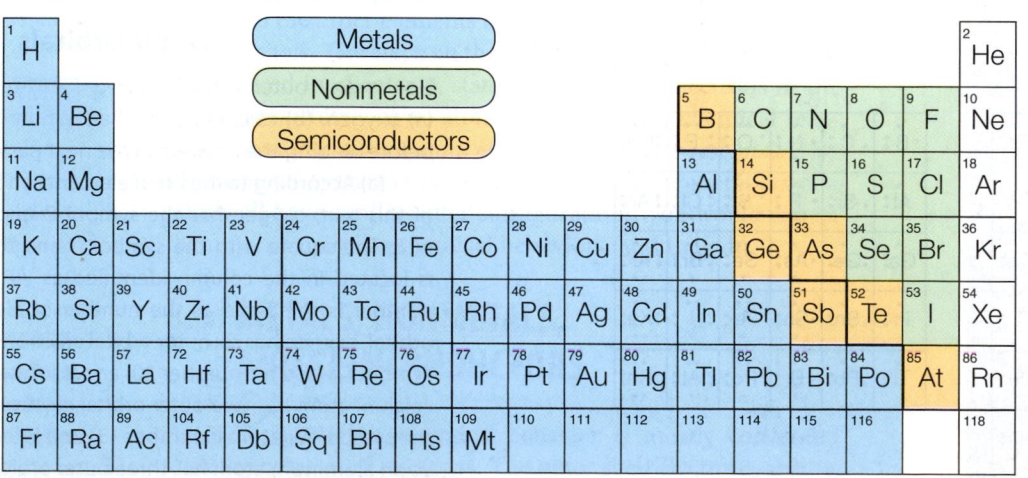

FIGURE 8.16 The location of metals, nonmetals, and semiconductors in the periodic table.

People Behind the Science

Dmitri Ivanovich Mendeleyev (1834–1907)

Dmitri Mendeleyev was a Russian chemist whose name will always be linked with his outstanding achievement, the development of the periodic table. He was the first chemist to understand that all elements are related members of a single, ordered system. He converted what had been a highly fragmented and speculative branch of chemistry into a true, logical science. The spelling of his name has been a source of confusion for students and frustration for editors for more than a century, and the forms Mendeléeff, Mendeléev, and even Mendelejeff can all be found in print.

Before Mendeleyev produced his periodic law, understanding of the chemical elements had long been an elusive and frustrating task. According to Mendeleyev, the properties of the elements are periodic functions of their atomic weights. In 1869, he stated that "the elements arranged according to the magnitude of atomic weights show a periodic change of properties." Other chemists, notably Lothar Meyer in Germany, had meanwhile come to similar conclusions, with Meyer publishing his findings independently.

Mendeleyev compiled the first true periodic table, listing all the sixty-three elements then known. Not all elements would "fit"

properly using the atomic weights of the time, so he altered indium from 76 to 114 (modern value 114.8) and beryllium from 13.8 to 9.2 (modern value 9.013). In 1871, he produced a revisionary paper showing the correct repositioning of seventeen elements.

To make the table work, Mendeleyev also had to leave gaps, and he predicted that further elements would eventually be discovered to fill them. These predictions provided the strongest endorsement of the periodic law. Three were discovered in Mendeleyev's lifetime: gallium (1871), scandium (1879), and germanium (1886), all with properties that tallied closely with those he had assigned to them.

Farsighted though Mendeleyev was, he had no notion that the periodic recurrences of similar properties in the list of elements reflected anything in the structures of their atoms. It was not until the 1920s that it was realized that the key parameter in the periodic system is not the atomic weight but the atomic number of the elements—a measure of the number of protons in the atom. Since then, great progress has been made in explaining the periodic law in terms of the electronic structures of atoms and molecules.

Source: Modified from the *Hutchinson Dictionary of Scientific Biography.* © RM, 2011. All rights reserved. Helicon publishing is a division of RM.

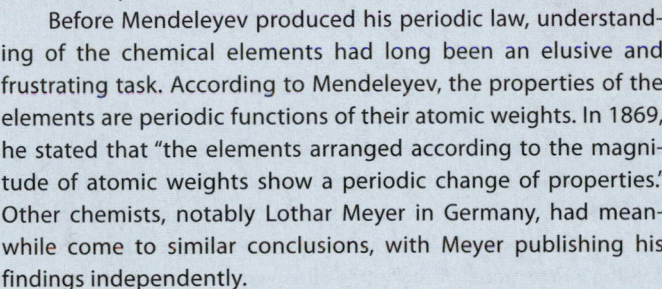

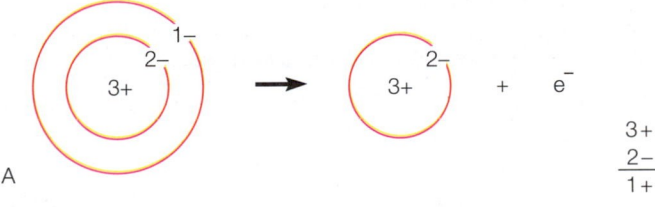

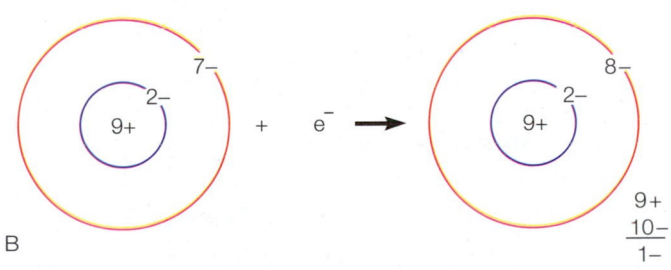

FIGURE 8.17 (*A*) Metals lose their outer electrons to acquire a noble gas structure and become positive ions. Lithium becomes a 1+ ion as it loses its one outer electron. (*B*) Nonmetals gain electrons to acquire an outer noble gas structure and become negative ions. Fluorine gains a single electron to become a 1− ion.

CONCEPTS APPLIED

Metals and Charge

Is strontium a metal, nonmetal, or semiconductor? What is the charge on a strontium ion?

The list of elements inside the back cover identifies the symbol for strontium as Sr (atomic number 38). In the periodic table, Sr is located in family IIA, which means that an atom of strontium has two electrons in its outer shell. For several reasons, you know that strontium is a metal: (1) An atom of strontium has two electrons in its outer shell, and atoms with one, two, or three outer electrons are identified as metals; (2) strontium is located in the IIA family, the alkaline earth metals; and (3) strontium is located on the left side of the periodic table, and in general, elements located in the left two-thirds of the table are metals.

Elements with one, two, or three outer electrons tend to lose electrons to form positive ions. Since strontium has an atomic number of 38, you know that it has thirty-eight protons (38+) and thirty-eight electrons (38−). When it loses its two outer shell electrons, it has 38+ and 36− for a charge of 2+ .

Chapter 8 *Atoms and Periodic Properties*   **193**

but also as they replace more conventional sources of light. The huge TV screens found at concerts and football stadiums are examples of innovated uses of LEDs. LEDs are replacing conventional incandescent bulbs because they last thousands of hours compared to the hundreds of hours of incandescent bulbs. They are also more efficient because electricity is not wasted as heat as it is in incandescent bulbs.

The transition elements, which are all metals, are located in the B-group families. Unlike the representative elements, which form vertical families of similar properties, the transition elements tend to form horizontal groups of elements with similar properties. Iron (Fe), cobalt (Co), and nickel (Ni) in group VIIIB, for example, are three horizontally arranged metallic elements that show magnetic properties.

A family of representative elements all form ions with the same charge. Alkali metals, for example, all lose an electron to form a 1+ ion. The transition elements have *variable charges*. Some transition elements, for example, lose their one outer electron to form 1+ ions (copper, silver). Copper, because of its special configuration, can also lose an additional electron to form a 2+ ion. Thus, copper can form either a 1+ ion or a 2+ ion. Most transition elements have two outer s orbital electrons and lose them both to form 2+ ions (iron, cobalt, nickel), but some of these elements also have special configurations that permit them to lose more of their electrons. Thus, iron and cobalt, for example, can form either a 2+ ion or a 3+ ion. Much more can be interpreted from the periodic table, and more generalizations will be made as the table is used in the following chapters.

SUMMARY

Attempts at understanding matter date back to ancient Greek philosophers, who viewed matter as being composed of *elements,* or simpler substances. Two models were developed that considered matter to be (1) *continuous,* or infinitely divisible, or (2) *discontinuous,* made up of particles called *atoms.*

In the early 1800s, Dalton published an *atomic theory,* reasoning that matter was composed of hard, indivisible atoms that were joined together or dissociated during chemical change.

When a good air pump to provide a vacuum was invented in 1885, *cathode rays* were observed to move from the negative terminal in an evacuated glass tube. The nature of cathode rays was a mystery. The mystery was solved in 1887 when Thomson discovered they were negatively charged particles now known as *electrons.* Thomson had discovered the first elementary particle of which atoms are made and measured their charge-to-mass ratio.

Rutherford developed a solar system model based on experiments with alpha particles scattered from a thin sheet of metal. This model had a small, massive, and positively charged *nucleus* surrounded by moving electrons. These electrons were calculated to be at a distance from the nucleus of 100,000 times the radius of the nucleus, so the volume of an atom is mostly empty space. Later, Rutherford proposed that the nucleus contained two elementary particles: *protons* with a positive charge and *neutrons* with no charge. The *atomic number* is the number of protons in an atom. Atoms of elements with different numbers of neutrons are called *isotopes.* The mass of each isotope is compared to the mass of carbon-12, which is assigned a mass of exactly 12.00 *atomic mass units.* The mass contribution of the isotopes of an element according to their abundance is called the *atomic weight* of an element. Isotopes are identified by their *mass number,* which is the sum of the number of protons and neutrons in the nucleus. Isotopes are identified by their chemical symbol, with the atomic number as a subscript and the mass number as a superscript.

Bohr developed a model of the hydrogen atom to explain the characteristic *line spectra* emitted by hydrogen. His model specified that (1) electrons can move only in allowed orbits, (2) electrons do not emit radiant energy when they remain in an orbit, and (3) electrons move from one allowed orbit to another when they gain or lose energy. When an electron jumps from a higher orbit to a lower one, it gives up energy in the form of a single photon. The energy of the photon corresponds to the difference in energy between the two levels. The Bohr model worked well for hydrogen but not for other atoms.

Schrödinger and others used the wave nature of the electron to develop a new model of the atom called *wave mechanics,* or *quantum mechanics.* This model was found to confirm exactly all the experimental data as well as predict new data. The quantum mechanical model describes the energy state of the electron in terms of quantum numbers based on the wave nature of the electron. The quantum numbers defined the *probability* of the location of an electron in terms of fuzzy regions of space called *orbitals.*

The *periodic table* has horizontal rows of elements called *periods* and vertical columns of elements called *families.* Members of a given family have the same outer orbital electron configurations, and it is the electron configuration that is mostly responsible for the chemical properties of an element.

Summary of Equations

8.1 $\quad$ energy = (Planck's constant)(frequency)

$$E = hf$$

where $h = 6.63 \times 10^{-34} \text{ J} \cdot \text{s}$

8.2 $\quad \dfrac{\text{energy state of orbit number}}{} = \dfrac{\text{energy state of innermost orbit}}{\text{number squared}}$

$$E_n = \frac{E_1}{n^2}$$

where $E_1 = -13.6 \text{ eV}$ and $n = 1, 2, 3 \ldots$

8.3 $\quad \dfrac{\text{energy of photon}}{} = \left(\begin{array}{c}\text{energy state of higher orbit}\end{array}\right) - \left(\begin{array}{c}\text{energy state of lower orbit}\end{array}\right)$

$$hf = E_H - E_L$$

where $h = 6.63 \times 10^{-34} \text{ J} \cdot \text{s}$ and E_H and E_L must be in joules

KEY TERMS

atomic mass units (p. **182**)
atomic number (p. **181**)
atomic weight (p. **182**)
electron (p. **181**)
electron dot notation (p. **191**)
excited states (p. **185**)

ground state (p. **185**)
Heisenberg uncertainty principle
   (p. **187**)
ion (p. **191**)
isotope (p. **182**)
line spectrum (p. **183**)

mass number (p. **182**)
neutron (p. **181**)
nucleus (p. **181**)
orbital (p. **187**)
Pauli exclusion principle (p. **188**)
photons (p. **183**)

proton (p. **181**)
quanta (p. **183**)
representative elements (p. **190**)
semiconductors (p. **192**)
transition elements (p. **190**)

APPLYING THE CONCEPTS

Answers are located in appendix F.

1. The electron was discovered through experiments with
   a. radioactivity.
   b. light.
   c. matter waves.
   d. electricity.

2. Thomson was convinced that he had discovered a subatomic particle, the electron, from the evidence that
   a. the charge-to-mass ratio was the same for all materials.
   b. cathode rays could move through a vacuum.
   c. electrons were attracted toward a negatively charged plate.
   d. the charge was always 1.60×10^{-19} coulomb.

3. The existence of a tiny, massive, and positively charged nucleus was deduced from the observation that
   a. fast, massive, and positively charged alpha particles all move straight through metal foil.
   b. alpha particles were deflected by a magnetic field.
   c. some alpha particles were deflected by metal foil.
   d. None of the above is correct.

4. According to Rutherford's calculations, the volume of an atom is mostly
   a. occupied by protons and neutrons.
   b. filled with electrons.
   c. occupied by tightly bound protons, electrons, and neutrons.
   d. empty space.

5. Hydrogen, with its one electron, produces a line spectrum in the visible light range with
   a. one color line.
   b. two color lines.
   c. three color lines.
   d. four color lines.

6. According to the Bohr model, an electron gains or loses energy only by
   a. moving faster or slower in an allowed orbit.
   b. jumping from one allowed orbit to another.
   c. being completely removed from an atom.
   d. jumping from one atom to another atom.

7. When an electron in a hydrogen atom jumps from an orbit farther from the nucleus to an orbit closer to the nucleus, it
   a. emits a single photon with an energy equal to the energy difference of the two orbits.
   b. emits four photons, one for each of the color lines observed in the line spectrum of hydrogen.
   c. emits a number of photons dependent on the number of orbit levels jumped over.
   d. None of the above is correct.

8. The quantum mechanics model of the atom is based on
   a. the quanta, or measured amounts of energy of a moving particle.
   b. the energy of a standing electron wave that can fit into an orbit.
   c. calculations of the energy of the three-dimensional shape of a circular orbit of an electron particle.
   d. Newton's laws of motion but scaled down to the size of electron particles.

9. The Bohr model of the atom described the energy state of electrons with one quantum number. The quantum mechanics model uses how many quantum numbers to describe the energy state of an electron?
   a. one
   b. two
   c. four
   d. ten

10. The space in which it is probable that an electron will be found is described by a(an)
    a. circular orbit.
    b. elliptical orbit.
    c. orbital.
    d. geocentric orbit.

11. Two different isotopes of the same element have
    a. the same number of protons, neutrons, and electrons.
    b. the same number of protons and neutrons but different numbers of electrons.
    c. the same number of protons and electrons but different numbers of neutrons.
    d. the same number of neutrons and electrons but different numbers of protons.

12. If you want to know the number of neutrons in an atom of a given element, you
    a. round the atomic weight to the nearest whole number.
    b. add the mass number and the atomic number.
    c. subtract the atomic number from the mass number.
    d. add the mass number and the atomic number, then divide by 2.

QUESTIONS FOR THOUGHT

1. What was the experimental evidence that Thomson had discovered the existence of a subatomic particle when working with cathode rays?
2. Describe the experimental evidence that led Rutherford to the concept of a nucleus in an atom.
3. What is the main problem with a solar system model of the atom?
4. Compare the size of an atom to the size of its nucleus.
5. What does *atomic number* mean? How does the atomic number identify the atoms of a particular element? How is the atomic number related to the number of electrons in an atom?
6. An atom has 11 protons in the nucleus. What is the atomic number? What is the name of this element? What is the electron configuration of this atom?
7. How is the atomic weight of an element determined?
8. Describe the three main points in the Bohr model of the atom.
9. Why do the energies of electrons in an atom have negative values? (*Hint:* It is *not* because of the charge of the electron.)
10. Which has the lowest energy, an electron in the first energy level ($n = 1$) or an electron in the third energy level ($n = 3$)? Explain.
11. What is similar about the Bohr model of the atom and the quantum mechanical model? What are the fundamental differences?
12. What is the difference between a hydrogen atom in the ground state and one in the excited state?

FOR FURTHER ANALYSIS

1. Evaluate Millikan's method for finding the charge of an electron. Are there any doubts about the results of using this technique?
2. What are the significant similarities and differences between the isotopes of a particular element?
3. Thomson's experiments led to the discovery of the electron. Analyze how you know for sure that he discovered the electron.
4. Describe a conversation between yourself and another person as you correct his or her belief that atomic weight has something to do with gravity.
5. Analyze the significance of the observation that matter only emits and absorbs energy in discrete units.
6. Describe at least several basic differences between the Bohr and quantum mechanics models of the atom.

INVITATION TO INQUIRY

Too Small to See?

As Rutherford knew when he conducted his famous experiment with radioactive particles and gold foil, the structure of an atom is too small to see, so it must be inferred from other observations. To illustrate this process, pour 50 mL of 95 percent ethyl alcohol (or some other almost pure alcohol) into a graduated cylinder. In a second graduated cylinder, measure 50 mL of water. Mix the alcohol and water together thoroughly and record the volume of the combined liquids. Assuming no evaporation took place, is the result contrary to your expectation? What question could be asked about the result?

Answers to questions about things that you cannot see often require a model that can be observed. For example, a model might represent water and alcohol molecules by using beans for alcohol molecules and sand for water molecules. Does mixing 50 mL of beans and 50 mL of sand result in 100 mL of mixed sand and beans? Explain the result of mixing alcohol and water based on your observation of mixing beans and sand.

To continue this inquiry, fill a water glass to the brim with water. Add a small amount of salt to the water. You know that two materials cannot take up the same space at the same time, so what happened to the salt? How can you test your ideas about things that are too tiny to see?

PARALLEL EXERCISES

The exercises in groups A and B cover the same concepts. Solutions to group A exercises are located in appendix G.

Group A

1. How much energy is needed to move an electron in a hydrogen atom from $n = 2$ to $n = 6$? (See figure 8.11 for needed values.)
2. What frequency of light is emitted when an electron in a hydrogen atom jumps from $n = 6$ to $n = 2$? What color would you see?
3. How much energy is needed to completely remove the electron from a hydrogen atom in the ground state?
4. Thomson determined the charge-to-mass ratio of the electron to be -1.76×10^{11} coulomb/kilogram. Millikan determined the charge on the electron to be -1.60×10^{-19} coulomb. According to these findings, what is the mass of an electron?

Group B

1. How much energy is needed to move an electron in a hydrogen atom from the ground state ($n = 1$) to $n = 3$?
2. What frequency of light is emitted when an electron in a hydrogen atom jumps from $n = 2$ to the ground state ($n = 1$)?
3. How much energy is needed to completely remove an electron from $n = 2$ in a hydrogen atom?
4. If the charge-to-mass ratio of a proton is 9.58×10^7 coulomb/kilogram and the charge is 1.60×10^{-19} coulomb, what is the mass of the proton?

9

Chemical Reactions

A chemical change occurs when iron rusts, and rust is a different substance with physical and chemical properties different from iron. This rusted old car makes a colorful display in an abandoned small town.

CORE **CONCEPT**

Electron structure will explain how and why atoms join together in certain numbers.

OUTLINE

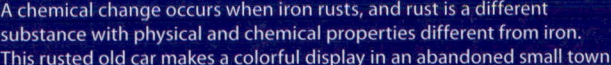

Chemical reactions are changes in matter in which different substances are created by forming or breaking chemical bonds (p. 200).

A chemical bond is an attractive force that holds atoms together in a compound (p. 203).

An ionic bond is a chemical bond of electrostatic attraction between ions (p. 204).

A covalent bond is a chemical bond formed by the sharing of electrons (p. 205).

Chemistry

▷ The electron structure of water will explain its properties and the properties of solutions (Ch. 10).

Life Science

▷ Organic chemistry and biochemistry consider chemical changes in the important organic compounds of life (Ch. 19).

Earth Science

▷ Chemical change will explain one way solid rocks on Earth's surface are weathered (Ch. 16).

OVERVIEW

In chapter 8, you learned how the modern atomic theory is used to describe the structures of atoms of different elements. The electron structures of different atoms successfully account for the position of elements in the periodic table as well as for groups of elements with similar properties. On a large scale, all metals were found to have a similarity in electron structure, as were nonmetals. On a smaller scale, chemical families such as the alkali metals were found to have the same outer electron configurations. Thus, the modern atomic theory accounts for observed similarities between elements in terms of atomic structure.

So far, only individual, isolated atoms have been discussed; we have not considered how atoms of elements join together to produce compounds. There is a relationship between the electron structure of atoms and the reactions they undergo to produce specific compounds. Understanding this relationship will explain the changes that matter itself undergoes. For example, hydrogen is a highly flammable, gaseous element that burns with an explosive reaction. Oxygen, on the other hand, is a gaseous element that supports burning. As you know, hydrogen and oxygen combine to form water. Water is a liquid that neither burns nor supports burning. Here are some questions to consider about this reaction:

1. What happens when atoms of elements such as hydrogen and oxygen join to form molecules such as water?
2. Why do atoms join and why do they stay together?
3. Why does water have properties different from the elements that combine to produce it?
4. Why is water H_2O and not H_3O or H_4O?

Answers to questions about why and how atoms join together in certain numbers are provided by considering the electronic structures of the atoms. Chemical substances are formed from the interactions of electrons as their structures merge, forming new patterns that result in molecules with new properties. It is the new electron pattern of the water molecule that gives water properties that are different from the oxygen or hydrogen from which it formed (figure 9.1). Understanding how electron structures of atoms merge to form new patterns is understanding the changes that matter itself undergoes, the topic of this chapter.

9.1 COMPOUNDS

Matter that you see around you seems to come in a wide variety of sizes, shapes, forms, and kinds. Are there any patterns in the apparent randomness that will help us comprehend matter? Yes, there are many patterns, and one of the more obvious is that all matter occurs as either a mixture or as a pure substance (figure 9.2). A *mixture* has unlike parts and a composition that varies from sample to sample. For example, sand from a beach is a variable mixture of things such as bits of rocks, minerals, and sea shells.

There are two distinct ways in which mixtures can have unlike parts with a variable composition. A *heterogeneous mixture* has physically distinct parts with different properties.

Beach sand, for example, is usually a heterogeneous mixture of tiny pieces of rocks and tiny pieces of shells that you can see. It is said to be heterogeneous because any two given samples will have a different composition with different kinds of particles. A solution of salt dissolved in water also meets the definition of a mixture since it has unlike parts and can have a variable composition. A solution, however, is different from a sand mixture since it is a *homogeneous mixture,* meaning it is the same throughout a given sample. A homogeneous mixture, or solution, is the same throughout. The key to understanding that a solution is a mixture is found in its variable composition; that is, a given solution might be homogeneous, but one solution can vary from the next. Thus, you can have a salt solution with a 1 percent concentration, another with a 7 percent concentra-

FIGURE 9.1 Water is the most abundant liquid on Earth and is necessary for all life. Because of water's great dissolving properties, any sample is a solution containing solids, other liquids, and gases from the environment. This stream also carries suspended, ground-up rocks, called *rock flour*, from a nearby glacier.

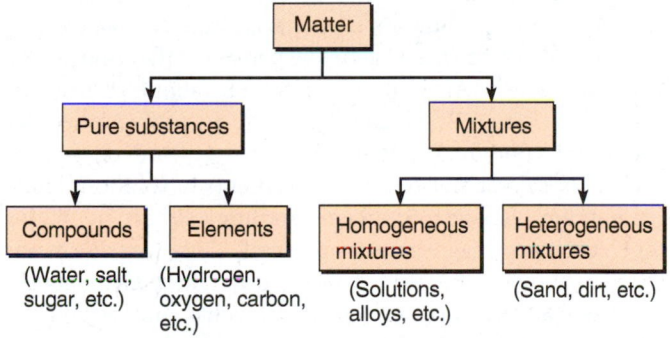

FIGURE 9.2 A classification scheme for matter.

 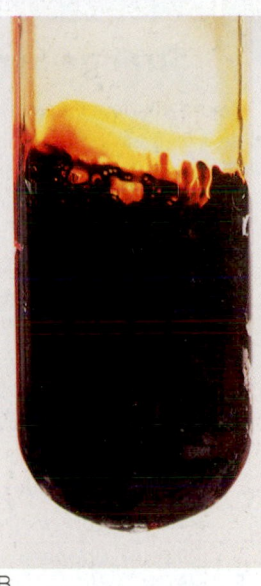

A   B

FIGURE 9.3 Sugar (*A*) is a compound that can be easily decomposed to simpler substances by heating. (*B*) One of the simpler substances is the black element carbon, which cannot be further decomposed by chemical or physical means.

tion, yet another with a 10 percent concentration, and so on. Solutions, mixed gases, and metal alloys are all homogeneous mixtures since they are made of unlike parts and do not have a fixed, definite composition.

Mixtures can be separated into their component parts by physical means. For example, you can physically separate the parts making up a sand mixture by using a magnifying glass and tweezers to move and isolate each part. A solution of salt in water is a mixture since the amount of salt dissolved in water can have a variable composition. But how do you separate the parts of a solution? One way is to evaporate the water, leaving the salt behind. There are many methods for separating mixtures, but all involve a **physical change.** A physical change does not alter the *identity* of matter. When water is evaporated from the salt solution, for example, it changes to a different state of matter (water vapor) but is still recognized as water. Physical changes involve physical properties only; no new substances

are formed. Examples of physical changes include evaporation, condensation, melting, freezing, and dissolving, as well as reshaping processes such as crushing or bending.

Mixtures can be physically separated into *pure substances,* materials that are the same throughout and have a fixed, definite composition. If you closely examine a sample of table salt, you will see that it is made up of hundreds of tiny cubes. Any one of these cubes will have the same properties as any other cube, including a salty taste. Sugar, like table salt, has all of its parts alike. Unlike salt, sugar grains have no special shape or form, but each grain has the same sweet taste and other properties as any other sugar grain.

If you heat salt and sugar in separate containers, you will find very different results. Salt, like some other pure substances, undergoes a physical change and melts, changing back to a solid upon cooling with the same properties that it had originally. Sugar, however, changes to a black material upon heating, while it gives off water vapor. The black material does not change back to sugar upon cooling. The sugar has *decomposed* to a new substance, while the salt did not. The new substance is carbon, and it has properties completely different from sugar. The sugar has gone through a **chemical change.** A chemical change alters the identity of matter, producing new substances with different properties. In this case, the chemical change was one of decomposition. Heat produced a chemical change by decomposing the sugar into carbon and water vapor.

The decomposition of sugar always produces the same mass ratio of carbon to water, so sugar has a fixed, definite composition. A **compound** is a pure substance that can be decomposed by a chemical change into simpler substances with a fixed mass ratio. This means that sugar is a compound (figure 9.3).

A pure substance that cannot be broken down into anything simpler by chemical or physical means is an **element.**

Connections . . .

Strange Symbols

Some elements have symbols that are derived from the earlier use of their Latin names. For example, the symbol Au is used for gold because the metal was earlier known by its Latin name of *aurum*, meaning "shining dawn." There are ten elements with symbols derived from Latin names and one with a symbol from a German name. These eleven elements are listed below, together with the sources of their names and their symbols.

Atomic Number	Name	Source of Symbol	Symbol
11	Sodium	Latin: *Natrium*	Na
19	Potassium	Latin: *Kalium*	K
26	Iron	Latin: *Ferrum*	Fe
29	Copper	Latin: *Cuprum*	Cu
47	Silver	Latin: *Argentum*	Ag
50	Tin	Latin: *Stannum*	Sn
51	Antimony	Latin: *Stibium*	Sb
74	Tungsten	German: *Wolfram*	W
79	Gold	Latin: *Aurum*	Au
80	Mercury	Latin: *Hydrargyrum*	Hg
82	Lead	Latin: *Plumbum*	Pb

TABLE 9.1 Elements Making up 99 Percent of Earth's Crust

Element (Symbol)	Percent by Weight
Oxygen (O)	46.6
Silicon (Si)	27.7
Aluminum (Al)	8.1
Iron (Fe)	5.0
Calcium (Ca)	3.6
Sodium (Na)	2.8
Potassium (K)	2.6
Magnesium (Mg)	2.1

9.2 ELEMENTS

Elements are not equally abundant and only a few are common. In table 9.1, for example, you can see that only eight elements make up about 99 percent of the solid surface of Earth. Oxygen is most abundant, making up about 50 percent of the weight of Earth's crust. Silicon makes up more than 25 percent, so these two nonmetals alone make up about 75 percent of Earth's solid surface. Almost all the rest is made up of just six metals, as shown in the table.

The number of common elements is limited elsewhere, too. Only two elements make up about 99 percent of the atmospheric air around Earth. Air is mostly nitrogen (about 78 percent) and oxygen (about 21 percent), with traces of five other elements and compounds. Water is hydrogen and oxygen, of course, but seawater also contains elements in solution. These elements are chlorine (55 percent), sodium (31 percent), sulfur (8 percent), and magnesium (4 percent). Only three elements make up about 97 percent of your body. These elements are hydrogen (60 percent), oxygen (26 percent), and carbon (11 percent). Generally, all of this means that the elements are not equally distributed or equally abundant in nature (figure 9.4).

9.3 CHEMICAL CHANGE

The air you breathe, the liquids you drink, and all the things around you are elements, compounds, or mixtures. Most are compounds, however, and very few are pure elements. Water, sugar, gasoline, and chalk are examples of compounds. Each can be broken down into the elements that make it up. Recall that elements are basic substances that cannot be broken down into simpler substances. Examples of elements are hydrogen, carbon, and calcium. Why and how these elements join together in different ways to form different compounds is the subject of this chapter.

You have already learned that elements are made up of atoms that can be described by the modern atomic theory. You can also consider an **atom** to be *the smallest unit of an element that can exist alone or in combination with other elements.* Compounds are formed when atoms are held together by an

Sugar is decomposed by heating into carbon and water vapor. Carbon cannot be broken down further, so carbon is an element. It has been known since about 1800 that water is a compound that can be broken down by electrolysis into hydrogen and oxygen, two gases that cannot be broken down to anything simpler. So sugar is a compound made from the elements of carbon, hydrogen, and oxygen.

But what about the table salt? Is table salt a compound? Table salt is a stable compound that is not decomposed by heating. It melts at a temperature of about 800°C and then returns to the solid form with the same salty properties upon cooling. Electrolysis—the splitting of a compound by means of electricity—can be used to decompose table salt into the elements sodium and chlorine, positively proving that it is a compound.

Pure substances are either compounds or elements. Decomposition through heating and decomposition through electrolysis are two means of distinguishing between compounds and elements. If a substance can be decomposed into something simpler, you know for sure that it is a compound. If the substance cannot be decomposed, it might be an element, or it might be a stable compound that resists decomposition. More testing would be necessary before you can be confident that you have identified an element. Most pure substances are compounds. There are millions of different compounds but only 117 known elements at the present time. These elements are the fundamental materials of which all matter is made.

FIGURE 9.4 The elements of aluminum, iron, oxygen, and silicon make up about 88 percent of Earth's solid surface. Water on the surface and in the air as clouds and fog is made up of hydrogen and oxygen. The air is 99 percent nitrogen and oxygen. Hydrogen, oxygen, and carbon make up 97 percent of a person. Thus, almost everything you see in this picture is made up of just six elements.

FIGURE 9.5 Magnesium is an alkaline earth metal that burns brightly in air, releasing heat and light. As chemical energy is released, a new chemical substance is formed. The new chemical material is magnesium oxide, a soft powdery material that forms an alkaline solution in water (called *milk of magnesia*).

attractive force called a *chemical bond*. The chemical bond binds individual atoms together in a compound. A molecule is generally thought of as a tightly bound group of atoms that maintains its identity. More specifically, a **molecule** is defined as *the smallest particle of a compound, or a gaseous element, that can exist and still retain the characteristic chemical properties of a substance.* Compounds with one type of chemical bond, as you will see, have molecules that are electrically neutral groups of atoms held together strongly enough to be considered independent units. For example, water is a compound. The smallest unit of water that can exist alone is an electrically neutral unit made up of two hydrogen atoms and one oxygen atom held together by chemical bonds. The concept of a molecule will be expanded as chemical bonds are discussed.

Compounds occur naturally as gases, liquids, and solids. Many common gases occur naturally as molecules made up of two or more atoms. For example, at ordinary temperatures, hydrogen gas occurs as molecules of two hydrogen atoms bound together. Oxygen gas also usually occurs as molecules of two oxygen atoms bound together. Both hydrogen and oxygen occur naturally as *diatomic molecules* (*di-* means "two"). Oxygen sometimes occurs as molecules of three oxygen atoms bound together. These *triatomic molecules* (*tri-* means "three") are called *ozone.*

When molecules of any size are formed or broken down into simpler substances, new materials with new properties are produced. This kind of a change in matter is called a chemical change, and the process is called a chemical reaction. A **chemical reaction** is defined as

a change in matter in which different chemical substances are created by forming or breaking chemical bonds.

In general, chemical bonds are formed when atoms of elements are bound together to form compounds. Chemical bonds are broken when a compound is decomposed into simpler substances. Chemical bonds are electrical in nature, formed by electrical attractions, as discussed in chapter 6.

Chemical reactions happen all the time, all around you. A growing plant, burning fuels, and your body's utilization of food all involve chemical reactions. These reactions produce different chemical substances with greater or smaller amounts of internal potential energy (see chapter 4 for a discussion of internal potential energy). Energy is *absorbed* to produce new chemical substances with more internal potential energy. Energy is *released* when new chemical substances are produced with less internal potential energy (figure 9.5). In general, changes in internal potential energy are called **chemical energy.** For example, new chemical substances are produced in green plants through the process called *photosynthesis.* A green plant uses radiant energy (sunlight), carbon dioxide, and water to produce new chemical materials and oxygen. These new chemical materials, the stuff that leaves, roots, and wood are made of, contain more chemical energy than the carbon dioxide and water they were made from.

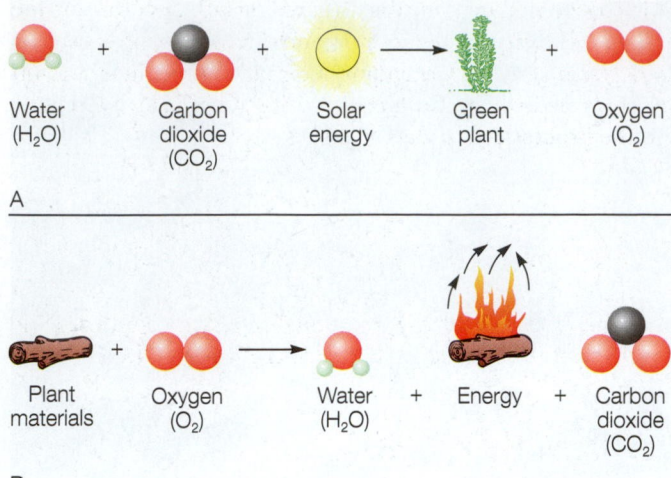

Water
(H_2O)
+
Carbon
dioxide
(CO_2)
+
Solar
energy
→
Green
plant
+
Oxygen
(O_2)

A

Plant
materials
+
Oxygen
(O_2)
→
Water
(H_2O)
+
Energy
+
Carbon
dioxide
(CO_2)

B

FIGURE 9.6 (*A*) New chemical bonds are formed as a green plant makes new materials and stores solar energy through the photosynthesis process. (*B*) The chemical bonds are later broken, and the same amount of energy and the same original materials are released. The same energy and the same materials are released rapidly when the plant materials burn, and they are released slowly when the plant decomposes.

A **chemical equation** is a way of describing what happens in a chemical reaction. The chemical reaction of photosynthesis can be described by using words in an equation:

$$\text{energy (sunlight)} + \text{carbon dioxide molecules} + \text{water molecules} \rightarrow \text{plant material molecules} + \text{oxygen molecules}$$

The substances that are changed are on the left side of the word equation and are called *reactants*. The reactants are carbon dioxide molecules and water molecules. The equation also indicates that energy is absorbed, since the term *energy* appears on the left side. The arrow means *yields*. The new chemical substances are on the right side of the word equation and are called *products*. Reading the photosynthesis reaction as a sentence, you would say, "Carbon dioxide and water absorb energy to react, yielding plant materials and oxygen."

The plant materials produced by the reaction have more internal potential energy, also known as *chemical energy*, than the reactants. You know this from the equation because the term *energy* appears on the left side but not the right. This means that the energy on the left went into internal potential energy on the right. You also know this because the reaction can be reversed to release the stored energy (figure 9.6). When plant materials (such as wood) are burned, the materials react with oxygen, and chemical energy is released in the form of radiant energy (light) and high kinetic energy of the newly formed gases and vapors. In words,

$$\text{plant material molecules} + \text{oxygen molecules} \rightarrow \text{carbon dioxide molecules} + \text{water molecules} + \text{energy}$$

If you compare the two equations, you will see that burning is the opposite of the process of photosynthesis! The energy released in burning is exactly the same amount of solar energy that was stored as internal potential energy by the plant. Such chemical changes, in which chemical energy is stored in one reaction and released by another reaction, are the result of the making, then the breaking, of chemical bonds. Chemical bonds were formed by utilizing energy to produce new chemical substances. Energy was released when these bonds were broken then re-formed to produce the original substances. In this example, chemical reactions and energy flow can be explained by the making and breaking of chemical bonds. Chemical bonds can be explained in terms of changes in the electron structures of atoms. Thus, the place to start in seeking understanding about chemical reactions is the electron structure of the atoms themselves.

9.4 VALENCE ELECTRONS AND IONS

As discussed in chapter 8, it is the number of electrons in the outermost orbital that usually determines the chemical properties of an atom. These outer electrons are called **valence electrons,** and it is the valence electrons that participate in chemical bonding. The inner electrons are in stable, fully occupied orbitals and do not participate in chemical bonds. The representative elements (the A-group families) have valence electrons in the outermost orbitals, which contain from one to eight valence electrons. Recall that you can easily find the number of valence electrons by referring to a periodic table. The number at the top of each representative family is the same as the number of outer orbital electrons (with the exception of helium).

The noble gases have filled outer orbitals and do not normally form compounds. Apparently, half-filled and filled orbitals are particularly stable arrangements. Atoms have a tendency to seek such a stable, filled outer orbital arrangement such as the one found in the noble gases. For the representative elements, this tendency is called the **octet rule.** The octet rule states that *atoms attempt to acquire an outer orbital with eight electrons* through chemical reactions. This rule is a generalization, and a few elements do not meet the requirement of eight electrons but do seek the same general trend of stability. There are a few other exceptions, and the octet rule should be considered a generalization that helps keep track of the valence electrons in most representative elements.

The family number of the representative element in the periodic table tells you the number of valence electrons and what the atom must do to reach the stability suggested by the octet rule. For example, consider sodium (Na). Sodium is in family IA, so it has one valence electron. If the sodium atom can get rid of this outer valence electron through a chemical reaction, it will have the same outer electron configuration as an atom of the noble gas neon (Ne) (compare figure 9.7B and 9.7C).

When a sodium atom (Na) loses an electron to form a sodium ion (Na^+), it has the same, stable outer electron configuration as a neon atom (Ne). The sodium ion (Na^+) is still

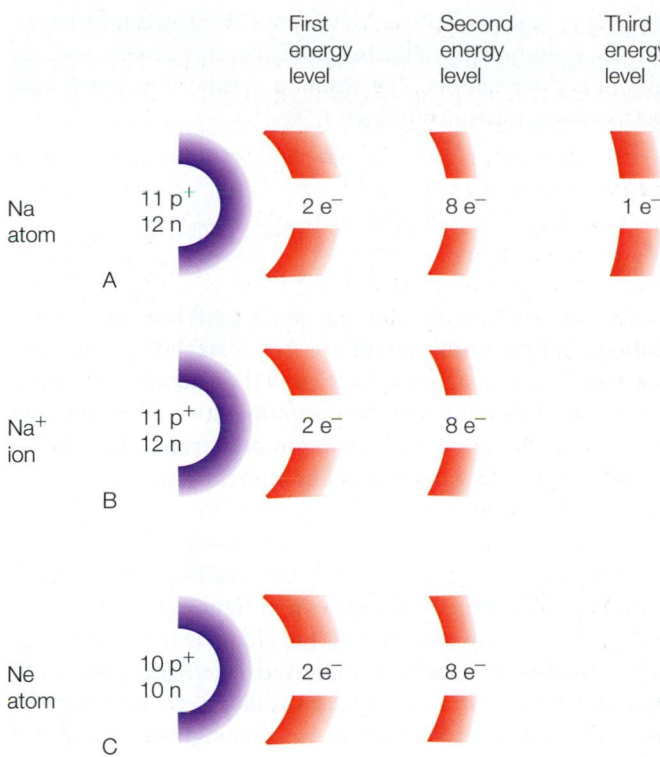

First energy level	Second energy level	Third energy level

Na atom
11 p$^+$
12 n
2 e$^-$ 8 e$^-$ 1 e$^-$

A

Na$^+$ ion
11 p$^+$
12 n
2 e$^-$ 8 e$^-$

B

Ne atom
10 p$^+$
10 n
2 e$^-$ 8 e$^-$

C

FIGURE 9.7　(*A*) A sodium atom has two electrons in the first energy level, eight in the second energy level, and one in the third level. (*B*) When it loses its one outer, or valence, electron, it becomes a positively charged sodium ion with the same electron structure as an atom of neon (*C*).

a form of sodium since it still has eleven protons. But it is now a sodium *ion*, not a sodium *atom,* since it has eleven protons (eleven positive charges) and now has ten electrons (ten negative charges) for a total of

$$11 + \text{(protons)}$$
$$\underline{10 - \text{(electrons)}}$$
$$1 + \text{(net charge on sodium ion)}$$

CONCEPTS APPLIED

Calcium

What is the symbol and charge for a calcium ion?

　From the list of elements on the inside back cover, the symbol for calcium is Ca, and the atomic number is 20. The periodic table tells you that Ca is in family IIA, which means that calcium has two valence electrons. According to the octet rule, the calcium ion must lose two electrons to acquire the stable outer arrangement of the noble gases. Since the atomic number is 20, a calcium atom has twenty protons (20 +) and twenty electrons (20 −). When it is ionized, the calcium ion will lose two electrons for a total charge of (20 +) + (18 −), or 2 +. The calcium ion is represented by the chemical symbol for calcium and the charge shown as a superscript: Ca^{2+}. What is the symbol and charge for an aluminum ion? (Answer: Al^{3+})

This charge is shown on the chemical symbol of Na$^+$ for the *sodium ion.* Note that the sodium nucleus and the inner orbitals do not change when the sodium atom is ionized. The sodium ion is formed when a sodium atom loses its valence electron, and the process can be described by

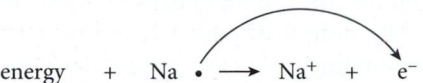

energy　+　Na ·　⟶　Na$^+$　+　e$^-$

where Na is the electron dot symbol for sodium, and the e$^-$ is the electron that has been pulled off the sodium atom.

9.5 CHEMICAL BONDS

Atoms gain or lose electrons through a chemical reaction to achieve a state of lower energy, the stable electron arrangement of the noble gas atoms. Such a reaction results in a **chemical bond,** an *attractive force that holds atoms together in a compound.* There are three general classes of chemical bonds: (1) ionic bonds, (2) covalent bonds, and (3) metallic bonds.

Ionic bonds are formed when atoms *transfer* electrons to achieve the noble gas electron arrangement. Electrons are given up or acquired in the transfer, forming positive and negative ions. The electrostatic attraction between oppositely charged ions forms ionic bonds, and ionic compounds are the result. In general, ionic compounds are formed when a metal from the left side of the periodic table reacts with a nonmetal from the right side.

Covalent bonds result when atoms achieve the noble gas electron structure by *sharing* electrons. Covalent bonds are generally formed between the nonmetallic elements on the right side of the periodic table.

Metallic bonds are formed in solid metals such as iron, copper, and the other metallic elements that make up about 80 percent of all the elements. The atoms of metals are closely packed and share many electrons in a "sea" that is free to move throughout the metal, from one metal atom to the next. Metallic bonding accounts for metallic properties such as high electrical conductivity.

　Ionic, covalent, and metallic bonds are attractive forces that hold atoms or ions together in molecules and crystals. There are two ways to describe what happens to the electrons when one of these bonds is formed: by considering (1) the new patterns formed when atomic orbitals overlap to form a combined orbital called a *molecular orbital* or (2) the atoms in a molecule as *isolated* atoms with changes in their outer shell arrangements. The molecular orbital description considers that the electrons belong to the whole molecule and form a molecular orbital with its own shape, orientation, and energy levels. The isolated atom description considers the electron energy levels as if the atoms in the molecule were alone, isolated from the molecule. The isolated atom description is less accurate than the molecular orbital description, but it is less complex and more easily understood. Thus, the following details about chemical bonding will mostly consider individual atoms and ions in compounds.

Ionic Bonds

An **ionic bond** is defined as the *chemical bond of electrostatic attraction* between negative and positive ions. Ionic bonding occurs when an atom of a metal reacts with an atom of a nonmetal. The reaction results in a transfer of one or more valence electrons from the metal atom to the valence shell of the non-metal atom. The atom that loses electrons becomes a positive ion, and the atom that gains electrons becomes a negative ion. Oppositely charged ions attract one another, and when pulled together, they form an ionic solid with the ions arranged in an orderly geometric structure (figure 9.8). This results in a crystalline solid that is typical of salts such as sodium chloride (figure 9.9).

As an example of ionic bonding, consider the reaction of sodium (a soft reactive metal) with chlorine (a pale yellow-green gas). When an atom of sodium and an atom of chlorine collide, they react violently as the valence electron is transferred from the sodium to the chlorine atom. This produces a sodium ion and a chlorine ion. The reaction can be illustrated with electron dot symbols as follows:

$$Na\bullet \ + \ \bullet\ddot{\underset{..}{Cl}}: \ \longrightarrow \ Na^+ \ (:\ddot{\underset{..}{Cl}}:)^-$$

As you can see, the sodium ion transferred its valence electron, and the resulting ion now has a stable electron configuration. The chlorine atom accepted the electron in its outer orbital to acquire a stable electron configuration. Thus, a stable positive ion and a stable negative ion are formed. Because of opposite electrical charges, the ions attract each other to produce an ionic bond. When many ions are involved, each Na^+ ion is surrounded by six Cl^- ions, and each Cl^- ion is surrounded by six Na^+ ions. This gives the resulting solid NaCl its crystalline cubic structure, as shown in figure 9.9. In the solid state, all the sodium ions and all the chlorine ions are bound together in one giant unit. Thus, the term *molecule* is not really appropriate for ionic solids such as sodium chloride. But the term is sometimes used anyway, since any given sample will have the same number of Na^+ ions as Cl^- ions.

The sodium-chlorine reaction can be represented with electron dot notation as occurring in three steps:

1. $\quad$ energy $\ + \ Na\bullet \ \longrightarrow \ Na^+ \ + \ e^-$

2. $\quad \bullet\ddot{\underset{..}{Cl}}: \ + \ e^- \ \longrightarrow \ (:\ddot{\underset{..}{Cl}}:)^- \ +$ energy

3. $\quad Na^+ \ + \ (:\ddot{\underset{..}{Cl}}:)^- \ \longrightarrow \ Na^+ \ (:\ddot{\underset{..}{Cl}}:)^- \ +$ energy

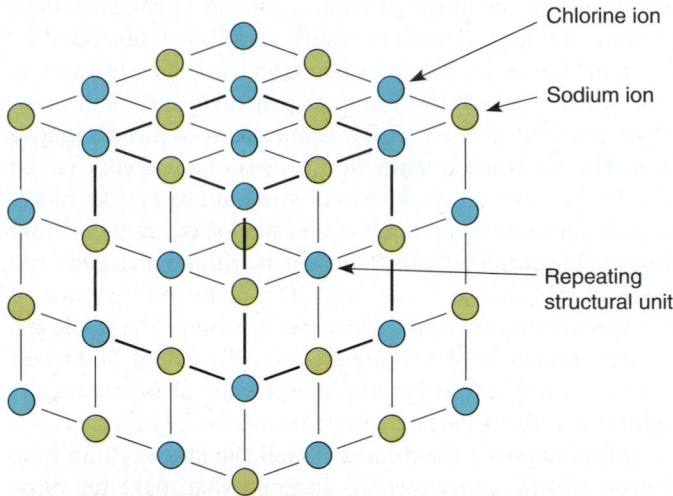

FIGURE 9.8 Sodium chloride crystals are composed of sodium and chlorine ions held together by electrostatic attraction. A crystal builds up, giving the sodium chloride crystal a cubic structure.

Chlorine ion

Sodium ion

Repeating structural unit

The energy released in steps 2 and 3 is greater than the energy absorbed in step 1, and an ionic bond is formed. The energy released is called the **heat of formation.** It is also the amount of energy required to decompose the compound (sodium chloride) into its elements. The reaction does not take place in steps as described, however, but occurs all at once. Note again, as in the photosynthesis-burning reactions described earlier, that the total amount of chemical energy is conserved. The energy released by the formation of the sodium chloride compound is the *same* amount of energy needed to decompose the compound.

Ionic bonds are formed by electron transfer, and electrons are conserved in the process. This means that electrons are not created or destroyed in a chemical reaction. The same total number of electrons exists after a reaction that existed before the reaction. There are two rules you can use for keeping track of electrons in ionic bonding reactions:

1. Ions are formed as atoms gain or lose valence electrons to achieve the stable noble gas structure.

FIGURE 9.9 You can clearly see the cubic structure of these ordinary table salt crystals because they have been magnified about ten times.

2. There must be a balance between the number of electrons lost and the number of electrons gained by atoms in the reaction.

The sodium-chlorine reaction follows these two rules. The loss of one valence electron from a sodium atom formed a stable sodium ion. The gain of one valence electron by the chlorine atom formed a stable chlorine ion. Thus, both ions have noble gas configurations (rule 1), and one electron was lost and one was gained, so there is a balance in the number of electrons lost and the number gained (rule 2).

The **formula** of a compound *describes what elements are in the compound and in what proportions.* Sodium chloride contains one positive sodium ion for each negative chlorine ion. The formula of the compound sodium chloride is NaCl. If there are no subscripts at the lower right part of each symbol, it is understood that the symbol has a number "1." Thus, NaCl indicates a compound made up of the elements sodium and chlorine, and there is one sodium atom for each chlorine atom.

Calcium (Ca) is an alkaline metal in family IIA, and fluorine (F) is a halogen in family VIIA. Since calcium is a metal and fluorine is a nonmetal, you would expect calcium and fluorine atoms to react, forming a compound with ionic bonds. Calcium must lose two valence electrons to acquire a noble gas configuration. Fluorine needs one valence electron to acquire a noble gas configuration. So calcium needs to lose two electrons and fluorine needs to gain one electron to achieve a stable configuration (rule 1). Two fluorine atoms, each acquiring one electron, are needed to balance the number of electrons lost and the number of electrons gained. The compound formed from the reaction, calcium fluoride, will therefore have a calcium ion with a charge of plus two for every two fluorine ions with a charge of minus one. Recalling that electron dot symbols show only the outer valence electrons, you can see that the reaction is

which shows that a calcium atom transfers two electrons, one each to two fluorine atoms. Now showing the results of the reaction, a calcium ion is formed from the loss of two electrons (charge 2+) and two fluorine ions are formed by gaining one electron each (charge 1−):

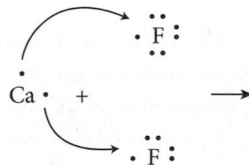

The formula of the compound is therefore CaF_2, with the subscript 2 for fluorine and the understood subscript 1 for calcium. This means that there are two fluorine atoms for each calcium atom in the compound.

Sodium chloride (NaCl) and calcium fluoride (CaF_2) are examples of compounds held together by ionic bonds. Such compounds are called **ionic compounds.** Ionic compounds of the representative elements are generally white, crystalline solids that form colorless solutions. Sodium chloride, the most common example, is common table salt. Many of the transition elements form colored compounds that make colored solutions. Ionic compounds dissolve in water, producing a solution of ions that can conduct an electric current.

In general, the elements in families IA and IIA of the periodic table tend to form positive ions by losing electrons. The ion charge for these elements equals the family number of these elements. The elements in families VIA and VIIA tend to form negative ions by gaining electrons. The ion charge for these elements equals their family number minus 8. The elements in families IIIA and VA have less of a tendency to form ionic compounds, except for those in higher periods. Common ions of some representative elements are given in table 9.2. The transition elements form positive ions of several different charges. Common ions of some transition elements are listed in table 9.3.

Covalent Bonds

Most substances do not have the properties of ionic compounds since they are not composed of ions. Most substances are molecular, composed of electrically neutral groups of atoms that are tightly bound together. As noted earlier, many gases are diatomic, occurring naturally as two atoms bound together as an electrically neutral molecule. Hydrogen, for example, occurs as molecules of H_2 and no ions are involved. The hydrogen atoms are held together by a covalent bond. A **covalent bond** is a *chemical bond formed by the sharing of at least a pair of electrons.* In the diatomic hydrogen molecule, each hydrogen atom contributes a single electron to the shared pair. Both hydrogen atoms count the shared pair of electrons in achieving

TABLE 9.2 **Common Ions of Some Representative Elements**

Element	Symbol	Ion
Lithium	Li	1+
Sodium	Na	1+
Potassium	K	1+
Magnesium	Mg	2+
Calcium	Ca	2+
Barium	Ba	2+
Aluminum	Al	3+
Oxygen	O	2−
Sulfur	S	2−
Hydrogen	H	1+, 1−
Fluorine	F	1−
Chlorine	Cl	1−
Bromine	Br	1−
Iodine	I	1−

TABLE 9.3 **Common Ions of Some Transition Elements**

Single-Charge Ions		
Element	**Symbol**	**Charge**
Zinc	Zn	2+
Tungsten	W	6+
Silver	Ag	1+
Cadmium	Cd	2+

Variable-Charge Ions		
Element	**Symbol**	**Charge**
Chromium	Cr	2+, 3+, 6+
Manganese	Mn	2+, 4+, 7+
Iron	Fe	2+, 3+
Cobalt	Co	2+, 3+
Nickel	Ni	2+, 3+
Copper	Cu	1+, 2+
Tin	Sn	2+, 4+
Gold	Au	1+, 3+
Mercury	Hg	1+, 2+
Lead	Pb	2+, 4+

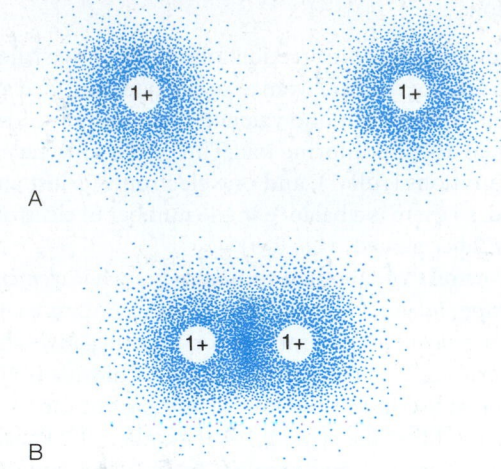

A

B

FIGURE 9.10 (*A*) Two hydrogen atoms, each with its own probability distribution of electrons about the nucleus. (*B*) When the hydrogen atoms bond, a new electron distribution pattern forms around the entire molecule, and both atoms share the two electrons.

their noble gas configuration. Hydrogen atoms both share one pair of electrons, but other elements might share more than one pair to achieve a noble gas structure.

Consider how the covalent bond forms between two hydrogen atoms by imagining two hydrogen atoms moving toward one another. Each atom has a single electron. As the atoms move closer and closer together, their orbitals begin to overlap. Each electron is attracted to the oppositely charged nucleus of the other atom and the overlap tightens. Then the repulsive forces from the like-charged nuclei will halt the merger. A state of stability is reached between the two nuclei and two electrons, and an H_2 molecule has been formed. The two electrons are now shared by both atoms, and the attraction of one nucleus for the other electron and vice versa holds the atoms together (figure 9.10).

Electron dot notation can be used to represent the formation of covalent bonds. For example, the joining of two hydrogen atoms to form an H_2 molecule can be represented as

$$H\cdot + H\cdot \longrightarrow H:H$$

Since an electron pair is *shared* in a covalent bond, the two electrons move throughout the entire molecular orbital. Since each hydrogen atom now has both electrons on an equal basis, each can be considered now to have the noble gas configuration of helium. A dashed circle around each symbol shows that both atoms have two electrons:

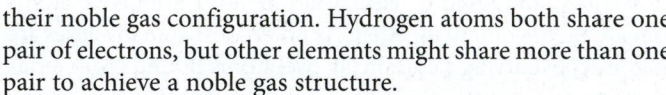

Hydrogen and fluorine react to form a covalent molecule (how this is known will be discussed shortly), and this bond can be represented with electron dots. Fluorine is in the VIIA family, so you know an atom of fluorine has seven valence electrons in the outermost energy level. The reaction is

$$H\cdot + \cdot\ddot{\underset{..}{F}}: \longrightarrow \left(H\underset{..}{\overset{..}{:}}F:\right)$$

Each atom shares a pair of electrons to achieve a noble gas configuration. Hydrogen achieves the helium configuration, and fluorine achieves the neon configuration. All the halogens have seven valence electrons, and all need to gain one electron (ionic bond) or share an electron pair (covalent bond) to achieve a noble gas configuration. This also explains why the halogen gases occur as diatomic molecules. Two fluorine atoms can achieve a noble gas configuration by sharing a pair of electrons:

$$\cdot\ddot{\underset{..}{F}}: + \cdot\ddot{\underset{..}{F}}: \longrightarrow \left(:\ddot{\underset{..}{F}}\overset{..}{:}\ddot{\underset{..}{F}}:\right)$$

The diatomic hydrogen (H_2) and fluorine (F_2), hydrogen fluoride (HF), and water (H_2O) are examples of compounds held together by covalent bonds. A compound held together by covalent bonds is called a **covalent compound.** In general, covalent compounds form from nonmetallic elements on the right side of the periodic table. For elements in families IVA through VIIA, the number of unpaired electrons (and thus the number of covalent bonds formed) is eight minus the family number. You can get a lot of information from the periodic table from generalizations like this one. For another generalization, compare table 9.4 with the periodic table. The table gives the structures of nonmetals combined with hydrogen and the resulting compounds.

Two dots can represent a lone pair of valence electrons, or they can represent a bonding pair, a single pair of electrons being shared by two atoms. Bonding pairs of electrons are often represented by a simple line between two atoms. For example,

$$H : H \quad \text{is shown as} \quad H - H$$

and

$$:\overset{..}{\underset{H}{O}}: \quad \text{is shown as} \quad \overset{O}{\diagup \diagdown}_{H \quad H}$$

Note that the line between the two hydrogen atoms represents an electron pair, so each hydrogen atom has two electrons in the outer shell, as does helium. In the water molecule, each hydrogen atom has two electrons as before. The oxygen atom has two lone pairs (a total of four electrons) and two bonding pairs (a total of four electrons) for a total of eight electrons. Thus, oxygen has acquired a stable octet of electrons.

A covalent bond in which a single pair of electrons is shared by two atoms is called a *single covalent bond,* or simply a **single bond.** Some atoms have two unpaired electrons and can share more than one electron pair. A **double bond** is a covalent bond formed when *two pairs* of electrons are shared by two atoms. This happens mostly in compounds involving atoms of the elements C, N, O, and S. Ethylene, for example, is a gas given off from ripening fruit. The electron dot formula for ethylene is

$$H \quad \quad H \\ :C :: C: \quad \text{or} \quad \overset{H}{\diagdown} C = C \overset{H}{\diagup} \\ H \quad \quad H \quad\quad\quad\quad H \diagup \quad \diagdown H$$

The ethylene molecule has a double bond between two carbon atoms. Since each line represents two electrons, you can simply count the lines around each symbol to see if the octet rule has been satisfied. Each H has one line, so each H atom is sharing two electrons. Each C has four lines so each C atom has eight electrons, satisfying the octet rule.

A **triple bond** is a covalent bond formed when *three pairs* of electrons are shared by two atoms. Triple bonds occur mostly in compounds with atoms of the elements C and N. Acetylene, for example, is a gas often used in welding torches (figure 9.11). The electron dot formula for acetylene is

$$H : C ::: C : H \quad \text{or} \quad H - C \equiv C - H$$

TABLE 9.4 **Structures and Compounds of Nonmetallic Elements Combined with Hydrogen**

Nonmetallic Element	Elements (E represents any element of family)	Compound
Family IVA: C, Si, Ge	$\cdot \overset{.}{\underset{.}{E}} \cdot$	H:E:H (with H above and below)
Family VA: N, P, As, Sb	$\cdot \overset{.}{\underset{.}{E}} \cdot$	H:E:H (with H below)
Family VIA: O, S, Se, Te	$\cdot \overset{..}{E} \cdot$	H:E:H
Family VIIA: F, Cl, Br, I	$\cdot \overset{..}{\underset{..}{E}} :$	H:E:

A Closer Look

Name That Compound

Ionic Compound Names

Ionic compounds formed by metal ions are named by stating the name of the metal (positive ion), then the name of the nonmetal (negative ion). Ionic compounds formed by variable-charge ions have an additional rule to identify which variable-charge ion is involved. There was an old way of identifying the charge on the ion by adding either -ic or -ous to the name of the metal. The suffix -ic meant the higher of two possible charges, and the suffix -ous meant the lower of two possible charges. For example, iron has two possible charges, 2+ or 3+. The old system used the Latin name for the root. The Latin name for iron is *ferrum,* so a higher charged iron ion (3+) was named a ferric ion. The lower charged iron ion (2+) was called a ferrous ion. You still hear the old names sometimes, but chemists now have a better way to identify the variable-charge ion. The newer system uses the English name of the metal with Roman numerals in parentheses to indicate the charge number. Thus, an iron ion with a charge of 2+ is called an iron(II) ion and an iron ion with a charge of 3+ is an iron(III) ion. Box table 9.1 gives some of the modern names for variable-charge ions. These names are used with the name of a nonmetal ending in -ide, just like the single-charge ions in ionic compounds made up of two different elements.

Some ionic compounds contain three or more elements, and so are more complex than a combination of a metal ion and a nonmetal ion. This is possible because they have **polyatomic ions,** groups of two or more atoms that are bound together tightly and behave very much like a single monatomic ion (box table 9.2). For example, the OH^- ion is an oxygen atom bound to a hydrogen atom with a net charge of $1-$. This polyatomic ion is called a *hydroxide ion.* The hydroxide compounds make up one of the main groups of ionic compounds, the *metal hydroxides.* A metal hydroxide is an ionic compound consisting of a metal with the hydroxide ion. Another main group consists of the salts with polyatomic ions.

Box Table 9.1

Modern Names of Some Variable-Charge Ions

Ion	Name of Ion
Fe^{2+}	Iron(II) ion
Fe^{3+}	Iron(III) ion
Cu^+	Copper(I) ion
Cu^{2+}	Copper(II) ion
Pb^{2+}	Lead(II) ion
Pb^{4+}	Lead(IV) ion
Sn^{2+}	Tin(II) ion
Sn^{4+}	Tin(IV) ion
Cr^{2+}	Chromium(II) ion
Cr^{3+}	Chromium(III) ion
Cr^{6+}	Chromium(VI) ion

The metal hydroxides are named by identifying the metal first and the term *hydroxide* second. Thus, NaOH is named sodium hydroxide and KOH is potassium hydroxide. The salts are similarly named, with the metal (or ammonium ion) identified first, then the name of the polyatomic ion. Thus, $NaNO_3$ is named sodium nitrate and $NaNO_2$ is sodium nitrite. Note that the suffix -ate means the polyatomic ion with one more oxygen atom than the -ite ion. For example, the chlor*ate* ion is $(ClO_3)^-$ and the chlor*ite* ion is $(ClO_2)^-$. Sometimes more than two possibilities exist, and more oxygen atoms are identified with the prefix *per-* and less with the prefix *hypo-*. Thus, the *per*chlor*ate* ion is $(ClO_4)^-$ and the *hypo*chlor*ite* ion is $(ClO)^-$.

Covalent Compound Names

Covalent compounds are molecular, and the molecules are composed of two *nonmetals,* as opposed to the metal and nonmetal elements that make up ionic compounds. The combinations of nonmetals do not present simple names as the ionic compounds did, so a different set of rules for naming and formula writing is needed.

Ionic compounds were named by stating the name of the positive metal ion, then the name of the negative nonmetal ion with an -ide ending. This system is not adequate for naming the covalent com-

Box Table 9.2

Some Common Polyatomic Ions

Ion Name	Formula
Acetate	$(C_2H_3O_2)^-$
Ammonium	$(NH_4)^+$
Borate	$(BO_3)^{3-}$
Carbonate	$(CO_3)^{2-}$
Chlorate	$(ClO_3)^-$
Chromate	$(CrO_4)^{2-}$
Cyanide	$(CN)^-$
Dichromate	$(Cr_2O_7)^{2-}$
Hydrogen carbonate (or bicarbonate)	$(HCO_3)^-$
Hydrogen sulfate (or bisulfate)	$(HSO_4)^-$
Hydroxide	$(OH)^-$
Hypochlorite	$(ClO)^-$
Nitrate	$(NO_3)^-$
Nitrite	$(NO_2)^-$
Perchlorate	$(ClO_4)^-$
Permanganate	$(MnO_4)^-$
Phosphate	$(PO_4)^{3-}$
Phosphite	$(PO_3)^{3-}$
Sulfate	$(SO_4)^{2-}$
Sulfite	$(SO_3)^{2-}$

pounds. To begin, covalent compounds are composed of two or more nonmetal atoms that form a molecule. It is possible for some atoms to form single, double, or even triple bonds with other atoms, including atoms of the same element. The net result is that the same two elements can form more than one kind of covalent compound. Carbon and oxygen, for example, can combine to form the gas released from burning and respiration, carbon dioxide (CO_2). Under certain conditions, the very same elements combine to produce a different gas, the poisonous carbon monoxide (CO). Similarly, sulfur and oxygen can combine differently to produce two different covalent compounds. A successful system for naming covalent compounds must therefore provide a means of identifying different compounds made of the same elements. This is accomplished by using a system of Greek prefixes (see box table 9.3). The rules are as follows:

1. The first element in the formula is named first with a prefix indicating the number of atoms if the number is greater than one.
2. The stem name of the second element in the formula is next. A prefix is used with the stem if two elements form more than one compound. The suffix -ide is again used to indicate a compound of only two elements. For example, CO is carbon monoxide and CO_2 is carbon dioxide. The compound BF_3 is boron trifluoride and N_2O_4 is dinitrogen tetroxide. Knowing the for-

mula and the prefix and stem information in box table 9.3, you can write the name of any covalent compound made up of two elements by ending it with -ide. Conversely, the name will tell you the formula. However, there are a few polyatomic ions with -ide endings that are compounds made up of more than just two elements (hydroxide and cyanide). Compounds formed with the ammonium ion will also have an -ide ending, and these are also made up of more than two elements.

Box Table 9.3

Prefixes and Element Stem Names

Prefixes		Stem Names	
PREFIX	MEANING	ELEMENT	STEM
Mono-	1	Hydrogen	Hydr-
Di-	2	Carbon	Carb-
Tri-	3	Nitrogen	Nitr-
Tetra-	4	Oxygen	Ox-
Penta-	5	Fluorine	Fluor-
Hexa-	6	Phosphorus	Phosph-
Hepta-	7	Sulfur	Sulf-
Octa-	8	Chlorine	Chlor-
Nona-	9	Bromine	Brom-
Deca-	10	Iodine	Iod-

Note: The a or o ending on the prefix is often dropped if the stem name begins with a vowel, for example, "tetroxide," not "tetraoxide."

FIGURE 9.11 Acetylene is a hydrocarbon consisting of two carbon atoms and two hydrogen atoms held together by a triple covalent bond between the two carbon atoms. When mixed with oxygen gas (the tank to the right), the resulting flame is hot enough to cut through most metals.

The acetylene molecule has a triple bond between two carbon atoms. Again, note that each line represents two electrons. Each C atom has four lines, so the octet rule is satisfied.

9.6 COMPOSITION OF COMPOUNDS

As you can imagine, there are literally millions of different chemical compounds from all the possible combinations of over ninety natural elements held together by ionic or covalent bonds. Each of these compounds has its own name, so there are millions of names and formulas for all the compounds. In the early days, compounds were given *common names* according to how they were used, where they came from, or some other means of identifying them. Thus, sodium carbonate was called soda, and closely associated compounds were called baking soda (sodium bicarbonate), washing soda (sodium carbonate), and caustic soda (sodium hydroxide), and the bubbly drink made by reacting soda

Myths, Mistakes, and Misunderstandings

Ban DHMO?

"Dihydrogen monoxide (DHMO) is colorless, odorless, tasteless, and kills uncounted thousands of people every year. Most of these deaths are caused by accidental inhalation of DHMO, but the dangers do not end there. Prolonged exposure to its solid form causes severe tissue damage. Symptoms of DHMO ingestion can include excessive sweating and urination, and possibly a bloated feeling, nausea, vomiting, and body electrolyte imbalance. For those who have become dependent, DHMO withdrawal means certain death."

The preceding statement is part of a hoax that was recently circulated on the Internet. The truth is that dihydrogen monoxide is the chemical name of H_2O—water.

CONCEPTS APPLIED

Household Chemicals

Pick a household product that has a list of ingredients with names of covalent compounds or of ions you have met in this chapter. Write the brand name of the product and the type of product (example: Sani-Flush; toilet-bowl cleaner), then list the ingredients as given on the label, writing them one under the other (column 1). Beside each name put the formula, if you can figure out what it should be (column 2). Also, in a third column, put whatever you know or can guess about the function of that substance in the product. (Example: This is an acid; helps dissolve mineral deposits.)

FIGURE 9.12 These substances are made up of sodium and some form of a carbonate ion. All have common names with the term *soda* for this reason. Soda water (or "soda pop") was first made by reacting soda (sodium carbonate) with an acid, so it was called "soda water."

with acid was called soda water, later called soda pop (figure 9.12). Potassium carbonate was extracted from charcoal by soaking in water and came to be called potash. Such common names are colorful, and some are descriptive, but it was impossible to keep up with the names as the number of known compounds grew. So a systematic set of rules was developed to determine the name and formula of each compound. Once you know the rules, you can write the formula when you hear the name. Conversely, seeing the formula will tell you the systematic name of the compound. This can be an interesting intellectual activity and can also be important when reading the list of ingredients to understand the composition of a product.

A different set of systematic rules is used with ionic compounds and covalent compounds, but there are a few rules in common. For example, a compound made of only two different elements always ends with the suffix *-ide*. So when you hear the name of a compound ending with *-ide*, you automatically know that the compound is made up of only two elements. Sodium chlor*ide* is an ionic compound made of sodium and chlorine ions. Carbon diox*ide* is a covalent compound with carbon and oxygen atoms. Thus, the systematic name tells you what elements are present in a compound with an *-ide* ending.

9.7 CHEMICAL EQUATIONS

Chemical reactions occur when bonds between the outermost parts of atoms are formed or broken. Bonds are formed, for example, when a green plant uses sunlight—a form of energy—to create molecules of sugar, starch, and plant fibers. Bonds are broken and energy is released when you digest the sugars and starches or when plant fibers are burned. Chemical reactions thus involve changes in matter, the creation of new materials with new properties, and energy exchanges. So far, you have considered chemical symbols as a concise way to represent elements and formulas as a concise way to describe what a compound is made of. There is also a concise way to describe a chemical reaction, the *chemical equation*.

Word equations are useful in identifying what has happened before and after a chemical reaction. The substances that existed before a reaction are called *reactants*, and the substances that exist after the reaction are called the *products*. The equation has a general form of

$$\text{reactants} \rightarrow \text{products}$$

where the arrow signifies a separation in time; that is, it identifies what existed before the reaction and what exists after the reaction. For example, the charcoal used in a barbecue grill is carbon (figure 9.13). The carbon reacts with oxygen while burning, and the reaction (1) releases energy and (2) forms carbon dioxide. The reactants and products for this reaction can be described as

FIGURE 9.13 The charcoal used in a grill is basically carbon. The carbon reacts with oxygen to yield carbon dioxide. The chemical equation for this reaction, $C + O_2 \rightarrow CO_2$, contains the same information as the English sentence but has quantitative meaning as well.

How to Write a Chemical Formula

Ionic Compound Formulas

The formulas for ionic compounds are easy to write. There are two rules:

1. **The symbols.** Write the symbol for the positive element first, followed by the symbol for the negative element (same order as in the name).
2. **The subscripts.** Add subscripts to indicate the numbers of ions needed to produce an electrically neutral compound.

As an example, let us write the formula for the compound calcium chloride. The name tells you that this compound consists of positive calcium ions and negative chlorine ions. The suffix -ide tells you there are only two elements present. Following rule 1, the symbols would be CaCl.

For rule 2, note the calcium ion is Ca^{2+} and the chlorine ion is Cl^-. You know the calcium is plus two and chlorine is negative one by applying the atomic theory, knowing their positions in the periodic table, or by using a table of ions and their charges. To be electrically neutral, the compound must have an equal number of pluses and minuses. Thus, you will need two negative chlorine ions for every calcium ion with its 2+ charge. Therefore, the formula is $CaCl_2$. The total charge of two chlorines is thus 2−, which balances the 2+ charge on the calcium ion.

One easy way to write a formula showing that a compound is electrically neutral is to cross over the absolute charge numbers (without plus or minus signs) and use them as subscripts. For example, the symbols for the calcium ion and the chlorine ion are

$$Ca^{2+}Cl^{1-}$$

Crossing the absolute numbers as subscripts, as follows

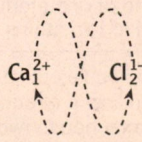

and then dropping the charge numbers gives

$$Ca_1\,Cl_2$$

No subscript is written for 1; it is understood. The formula for calcium chloride is thus

$$CaCl_2$$

The crossover technique works because ionic bonding results from a transfer of electrons, and the net charge is conserved. A calcium ion has a 2+ charge because the atom lost two electrons and two chlorine atoms gain one electron each, for a total of two electrons gained. Two electrons lost equals two electrons gained, and the net charge on calcium chloride is zero, as it has to be.

When using the crossover technique, it is sometimes necessary to reduce the ratio to the lowest common multiple. Thus, Mg_2O_2 means an equal ratio of magnesium and oxygen ions, so the correct formula is MgO.

The formulas for variable-charge ions are easy to write, since the Roman numeral tells you the charge number. The formula for tin(II) fluoride is written by crossing over the charge numbers (Sn^{2+}, F^{1-}), and the formula is SnF_2.

The formulas for ionic compounds with polyatomic ions are written from combinations of positive metal ions or the ammonium ion with the polyatomic ions, as listed in box table 9.2. Since the polyatomic ion is a group of atoms that has a charge and stays together in a unit, it is sometimes necessary to indicate this with parentheses. For example, magnesium hydroxide is composed of Mg^{2+} ions and $(OH)^{1-}$ ions. Using the crossover technique to write the formula, you get

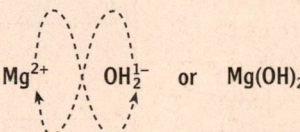

The parentheses are used and the subscript is written *outside* the parenthesis to show that the entire hydroxide unit is taken twice. The formula $Mg(OH)_2$ means

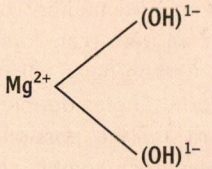

which shows that the pluses equal the minuses. Parentheses are not used, however, when only one polyatomic ion is present. Sodium hydroxide is NaOH, not $Na(OH)_1$.

Covalent Compound Formulas

The systematic name tells you the formula for a covalent compound. The gas that dentists use as an anesthetic, for example, is dinitrogen monoxide. This tells you there are two nitrogen atoms and one oxygen atom in the molecule, so the formula is N_2O. A different molecule composed of the very same elements is nitrogen dioxide. Nitrogen dioxide is the pollutant responsible for the brownish haze of smog. The formula for nitrogen dioxide is NO_2. Other examples of formulas from systematic names are carbon dioxide (CO_2) and carbon tetrachloride (CCl_4).

Formulas of covalent compounds indicate a pattern of how many atoms of one element combine with atoms of another. Carbon, for example, combines with no more than two oxygen atoms to form carbon dioxide. Carbon combines with no more than four chlorine atoms to form carbon tetrachloride. Electron dot formulas show these two molecules as

$$\ddot{O}::C::\ddot{O} \qquad :\!\ddot{\underset{..}{Cl}}\!:\!\overset{\displaystyle :\ddot{Cl}:}{\underset{\displaystyle :\ddot{Cl}:}{C}}\!:\!\ddot{\underset{..}{Cl}}\!:$$

Using a dash to represent bonding pairs, we have

$$O=C=O \qquad \overset{\displaystyle Cl}{\underset{\displaystyle Cl}{Cl-\!C\!-Cl}}$$

—Continued top of next page

In both of these compounds, the carbon atom forms four covalent bonds with another atom. The number of covalent bonds that an atom can form is called its *valence*. Carbon has a valence of four and can form single, double, or triple bonds. Here are the possibilities for a single carbon atom (combining elements not shown):

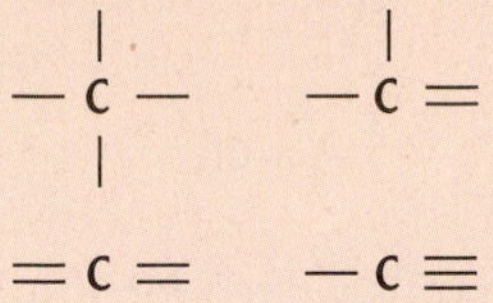

$$-\overset{|}{\underset{|}{C}}- \qquad -\overset{|}{C}=$$

$$=C= \qquad -C\equiv$$

Hydrogen has only one unshared electron, so the hydrogen atom has a valence of

one. Oxygen has a valence of two, and nitrogen has a valence of three. Here are the possibilities for hydrogen, oxygen, and nitrogen:

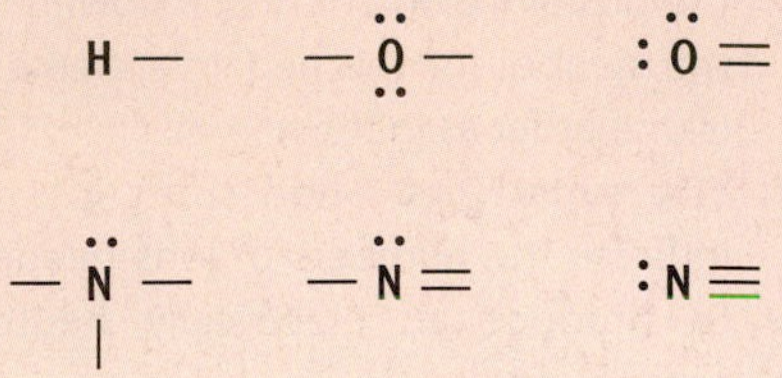

$$H- \qquad -\overset{..}{\underset{..}{O}}- \qquad :\overset{..}{O}=$$

$$-\overset{..}{\underset{|}{N}}- \qquad -\overset{..}{N}= \qquad :N\equiv$$

---

carbon + oxygen → carbon dioxide

The arrow means *yields,* and the word equation is read as, "Carbon reacts with oxygen to yield carbon dioxide." This word equation describes what happens in the reaction but says nothing about the quantities of reactants or products.

Chemical symbols and formulas can be used in the place of words in an equation and the equation will have a whole new meaning. For example, the equation describing carbon reacting with oxygen to yield carbon dioxide becomes

$$C + O_2 \rightarrow CO_2 \qquad \textbf{(balanced)}$$

The new, added meaning is that one atom of carbon (C) reacts with one molecule of oxygen (O_2) to yield one molecule of carbon dioxide (CO_2). Note that the equation also shows one atom of carbon and two atoms of oxygen (recall that oxygen occurs as a diatomic molecule) as reactants on the left side and one atom of carbon and two atoms of oxygen as products on the right side. Since the same number of each kind of atom appears on both sides of the equation, the equation is said to be *balanced.*

You would not want to use a charcoal grill in a closed room because there might not be enough oxygen. An insufficient supply of oxygen produces a completely different product, the poisonous gas carbon monoxide (CO). An equation for this reaction is

$$C + O_2 \rightarrow CO \qquad \textbf{(not balanced)}$$

As it stands, this equation describes a reaction that violates the **law of conservation of mass,** that matter is neither created nor destroyed in a chemical reaction. From the point of view of an equation, this law states that

mass of reactants = mass of products

Mass of reactants here means all that you start with, including some that might not react. Thus, elements are neither created nor destroyed, and this means the elements present and their mass. In any chemical reaction, the kind and mass of the reactive elements are identical to the kind and mass of the product elements.

From the point of view of atoms, the law of conservation of mass means that *atoms are neither created nor destroyed in the chemical reaction.* A chemical reaction is the making or breaking of chemical bonds between atoms or groups of atoms.

Atoms are not lost or destroyed in the process, nor are they changed to a different kind. The equation for the formation of carbon monoxide has two oxygen atoms in the reactants (O_2) but only one in the product (CO). An atom of oxygen has disappeared somewhere, and that violates the law of conservation of mass. You cannot fix the equation by changing the CO to a CO_2, because this would change the identity of the compounds. Carbon monoxide is a poisonous gas that is different from carbon dioxide, a relatively harmless product of burning and respiration. *You cannot change the subscript in a formula* because that would change the formula. A different formula means a different composition and thus a different compound.

You cannot change the subscripts of a formula, but you can place a number called a *coefficient* in *front* of the formula. Changing a coefficient changes the *amount* of a substance, not the identity. Thus 2 CO means two molecules of carbon monoxide and 3 CO means three molecules of carbon monoxide. If there is no coefficient, 1 is understood, as with subscripts. The meaning of coefficients and subscripts is illustrated in figure 9.14.

C	means	One atom of carbon
O	means	One atom of oxygen
O_2	means	 One molecule of oxygen consisting of two atoms of oxygen
CO	means	 One molecule of carbon monoxide consisting of one atom of carbon attached to one atom of oxygen
CO_2	means	 One molecule of carbon dioxide consisting of one atom of carbon attached to two atoms of oxygen
3 CO_2	means	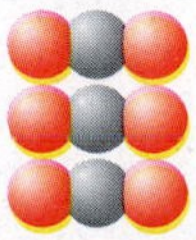 Three molecules of carbon dioxide, each consisting of one atom of carbon attached to two atoms of oxygen

FIGURE 9.14 The meaning of subscripts and coefficients used with a chemical formula. The subscripts tell you how many atoms of a particular element are in a compound. The coefficient tells you about the quantity, or number, of molecules of the compound.

On Balancing Equations

Natural gas is mostly methane, CH_4, which burns by reacting with oxygen (O_2) to produce carbon dioxide (CO_2) and water vapor (H_2O). Write a balanced chemical equation for this reaction by following a procedure of four steps.

Step 1. Write the correct formulas for the reactants and products in an unbalanced equation. For the burning of methane, the unbalanced, but otherwise correct, formula equation would be

$$CH_4 + O_2 \rightarrow CO_2 + H_2O \quad \textbf{(not balanced)}$$

Step 2. Inventory the number of each kind of atom on both sides of the unbalanced equation. In the example, there are

Reactants	Products
1 C	1 C
4 H	2 H
2 O	3 O

This step shows that the H and O are unbalanced.

Step 3. Determine where to place coefficients in front of formulas to balance the equation. It is often best to focus on the simplest thing you can do with whole number ratios. The H and the O are unbalanced, for example, and there are 4 H atoms on the left and 2 H atoms on the right. Placing a coefficient 2 in front of H_2O will balance the H atoms:

$$CH_4 + O_2 \rightarrow CO_2 + 2\,H_2O \quad \textbf{(not balanced)}$$

Now take a second inventory:

Reactants	Products
1 C	1 C
4 H	4 H
2 O	4 O ($O_2 + 2\,O$)

This shows the O atoms are still unbalanced with 2 on the left and 4 on the right. Placing a coefficient of 2 in front of O_2 will balance the O atoms.

$$CH_4 + 2\,O_2 \rightarrow CO_2 + 2\,H_2O \quad \textbf{(balanced)}$$

Step 4. Take another inventory to determine if the numbers of atoms on both sides are now equal. If they are, determine if the coefficients are in the lowest possible whole-number ratio. The inventory is now

Reactants	Products
1 C	1 C
4 H	4 H
4 O	4 O

The number of each kind of atom on each side of the equation is the same, and the ratio of $1:2 \rightarrow 1:2$ is the lowest possible whole-number ratio. The equation is balanced, which is illustrated with sketches of molecules in box figure 9.1.

Balancing chemical equations is mostly a trial-and-error procedure. But with practice, you will find there are a few generalized "role models" that can be useful in balancing equations for many simple reactions. The key to success at balancing equations is to think it out step-by-step while remembering the following:

1. Atoms are neither lost nor gained nor do they change their identity in a chemical reaction. The same kind and number of atoms in the reactants must appear in the products, meaning atoms are conserved.

2. A correct formula of a compound cannot be changed by altering the number or placement of subscripts. Changing subscripts changes the identity of a compound and the meaning of the entire equation.

3. A coefficient in front of a formula multiplies everything in the formula by that number.

Here are a few generalizations that can be helpful in balancing equations:

1. Look first to formulas of compounds with the most atoms and try to balance the atoms or compounds from which they were formed or to which they decomposed.

2. You should treat polyatomic ions that appear on both sides of the equation as independent units with a charge. That is, consider the polyatomic ion as a unit while taking an inventory rather than the individual atoms making up the polyatomic ion. This will save time and simplify the procedure.

3. Both the "crossover technique" and the use of "fractional coefficients" can be useful in finding the least common multiple to balance an equation.

4. The physical state of reactants and products in a reaction is often identified by the symbols (*g*) for gas, (*l*) for liquid, (*s*) for solid, and (*aq*) for an aqueous solution, which means water. If a gas escapes, this is identified with an arrow pointing up (↑). A solid formed from a solution is identified with an arrow pointing down (↓). The Greek symbol delta (Δ) is often used under or over the yield sign to indicate a change of temperature or other physical values.

Reaction:
    Methane reacts with oxygen to yield carbon dioxide and water

Balanced equation:
$$CH_4 + 2\,O_2 \longrightarrow CO_2 + 2\,H_2O$$

Sketches representing molecules:

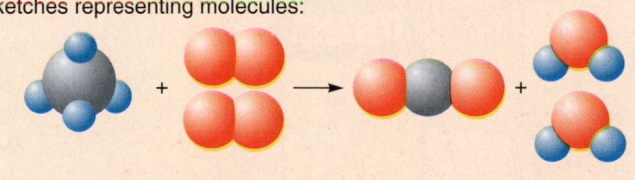

Meaning:

1 molecule of methane	+	2 molecules of oxygen	→	1 molecule of carbon dioxide	+	2 molecules of water

Box Figure 9.1 Compare the numbers of each kind of atom in the balanced equation with the numbers of each kind of atom in the sketched representation. Both the equation and the sketch have the same number of atoms in the reactants and in the products.

Placing a coefficient of 2 in front of the C and a coefficient of 2 in front of the CO in the equation will result in the same numbers of each kind of atom on both sides:

$$2\,C + O_2 \rightarrow 2\,CO$$

Reactants: 2 C	Products: 2 C
2 O	2 O

The equation is now balanced.

Generalizing from groups of chemical reactions also makes it possible to predict what will happen in similar reactions. For example, the combustion of methane (CH_4), propane (C_3H_8), and octane (C_8H_{18}) all involve a *hydrocarbon,* a compound of the elements hydrogen and carbon. Each hydrocarbon reacts with O_2, yielding CO_2 and releasing the energy of combustion. Generalizing from these reactions, you could predict that the combustion of *any* hydrocarbon would involve the combination of atoms of the hydrocarbon molecule with O_2 to produce CO_2 and H_2O with the release of energy. Such reactions could be analyzed by chemical experiments, and the products could be identified by their physical and chemical properties. You would find your predictions based on similar reactions would be correct, thus justifying predictions from such generalizations. Butane, for example, is a hydrocarbon with the formula C_4H_{10} The balanced equation for the combustion of butane is

$$2\,C_4H_{10}(g) + 13\,O_2(g) \rightarrow 8\,CO_2(g) + 10\,H_2O(g)$$

You could extend the generalization further, noting that the combustion of compounds containing oxygen as well as carbon and hydrogen also produces CO_2 and H_2O (figure 9.15). These compounds are *carbohydrates,* composed of carbon and water. Glucose, for example, is a compound with the formula $C_6H_{12}O_6$.

FIGURE 9.15 *Hydrocarbons* are composed of the elements hydrogen and carbon. Propane (C_3H_8) and gasoline, which contains a mixture of molecules such as octane (C_8H_{18}), are examples of hydrocarbons. *Carbohydrates* are composed of the elements of hydrogen, carbon, and oxygen. Table sugar, for example, is the carbohydrate $C_{12}H_{22}O_{11}$. Generalizing, all hydrocarbons and carbohydrates react completely with oxygen to yield CO_2 and H_2O.

Glucose combines with oxygen to produce CO_2 and H_2O, and the balanced equation is

$$C_6H_{12}O_6(s) + 6\,O_2(g) \rightarrow 6\,CO_2(g) + 6\,H_2O(g)$$

Note that three molecules of oxygen were not needed from the O_2 reactant since the other reactant, glucose, contains six oxygen atoms per molecule. An inventory of atoms will show that the equation is thus balanced.

Combustion is a rapid reaction with O_2 that releases energy, usually with a flame. A very similar, although much slower, reaction takes place in plant and animal respiration. In respiration, carbohydrates combine with O_2 and release energy used for biological activities. This reaction is slow compared to combustion and requires enzymes to proceed at body temperature. Nonetheless, CO_2 and H_2O are the products.

9.8 TYPES OF CHEMICAL REACTIONS

The reactions involving hydrocarbons and carbohydrates with oxygen are examples of an important group of chemical reactions called *oxidation-reduction* reactions. When the term *oxidation* was first used, it specifically meant reactions involving the combination of oxygen with other atoms. But fluorine, chlorine, and other nonmetals were soon understood to have similar reactions as oxygen, so the definition was changed to one concerning the shifts of electrons in the reaction.

An **oxidation-reduction reaction** (or **redox reaction**) is broadly defined as a reaction in which electrons are transferred from one atom to another. As is implied by the name, such a reaction has two parts and each part tells you what happens to the electrons. *Oxidation* is the part of a redox reaction in which there is a loss of electrons by an atom. *Reduction* is the part of a redox reaction in which there is a gain of electrons by an atom. The name also implies that in any reaction in which oxidation

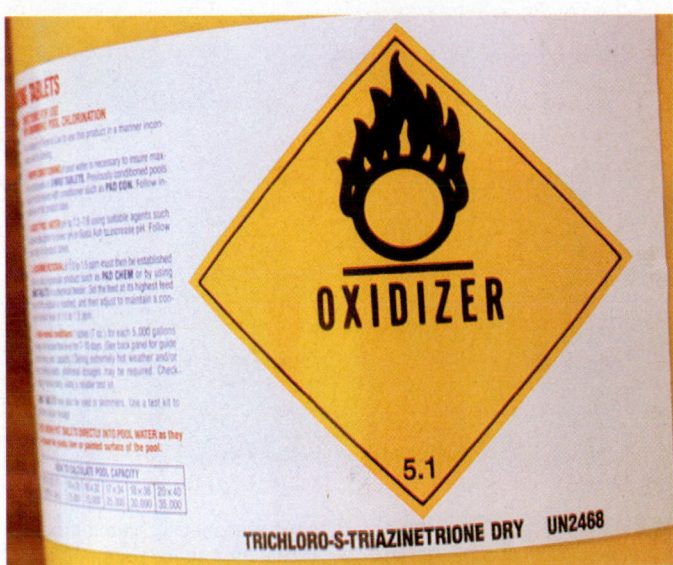

FIGURE 9.16 Oxidizing agents take electrons from other substances that are being oxidized. Oxygen and chlorine are commonly used strong oxidizing agents.

occurs, reduction must take place, too. One cannot take place without the other.

Substances that take electrons from other substances are called **oxidizing agents.** Oxidizing agents take electrons from the substances being oxidized. Oxygen is the most common oxidizing agent, and several examples have already been given about how it oxidizes foods and fuels. Chlorine is another commonly used oxidizing agent, often for the purposes of bleaching or killing bacteria (figure 9.16).

A **reducing agent** supplies electrons to the substance being reduced. Hydrogen and carbon are commonly used reducing agents. Carbon is commonly used as a reducing agent to extract metals from their ores. For example, carbon (from coke, which is coal that has been baked) reduces Fe_2O_3, an iron ore, in the reaction

$$2\,Fe_2O_3(s) + 3\,C(s) \rightarrow 4\,Fe(s) + 3\,CO_2\uparrow$$

The Fe in the ore gained electrons from the carbon, the reducing agent in this reaction.

Many chemical reactions can be classified as redox or nonredox reactions. Another way to classify chemical reactions is to consider what is happening to the reactants and products. This type of classification scheme leads to four basic categories of chemical reactions, which are (1) *combination,* (2) *decomposition,* (3) *replacement,* and (4) *ion exchange* reactions. The first three categories are subclasses of redox reactions. It is in the ion exchange reactions that you will find the first example of a reaction that is not a redox reaction.

Combination Reactions

A **combination reaction** is a synthesis reaction in which two or more substances combine to form a single compound. The combining substances can be (1) elements, (2) compounds, or (3) combinations of elements and compounds. In generalized form, a combination reaction is

$$X + Y \rightarrow XY$$

Many redox reactions are combination reactions. For example, metals are oxidized when they burn in air, forming a metal oxide. Consider magnesium, which gives off a bright white light as it burns:

$$2\,Mg(s) + O_2(g) \rightarrow 2\,MgO(s)$$

The rusting of metals is oxidation that takes place at a slower pace than burning, but metals are nonetheless oxidized in the process. Again noting the generalized form of a combination reaction, consider the rusting of iron:

$$4\,Fe(s) + 3\,O_2(g) \rightarrow 2\,Fe_2O_3(s)$$

Nonmetals are also oxidized by burning in air, for example, when carbon burns with a sufficient supply of O_2:

$$C(s) + O_2(g) \rightarrow CO_2(g)$$

Note that all the combination reactions follow the generalized form of $X + Y \rightarrow XY$.

Decomposition Reactions

A **decomposition reaction,** as the term implies, is the opposite of a combination reaction. In decomposition reactions, a compound is broken down (1) into the elements that make up the compound, (2) into simpler compounds, or (3) into elements and simpler compounds. Decomposition reactions have a generalized form of

$$XY \rightarrow X + Y$$

Decomposition reactions generally require some sort of energy, which is usually supplied in the form of heat or electrical energy. An electric current, for example, decomposes water into hydrogen and oxygen:

$$2\,H_2O(l) \xrightarrow{\text{electricity}} 2\,H_2(g) + O_2(g)$$

Mercury(II) oxide is decomposed by heat, an observation that led to the discovery of oxygen:

$$2\,HgO(s) \xrightarrow{\Delta} 2\,Hg(s) + O_2\uparrow$$

Note that all the decomposition reactions follow the generalized form of $XY \rightarrow X + Y$.

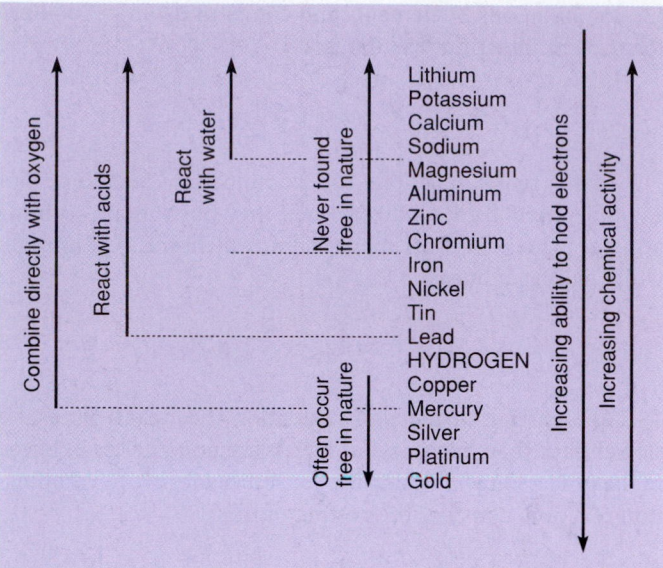

FIGURE 9.17 The activity series for common metals, together with some generalizations about the chemical activities of the metals. The series is used to predict which replacement reactions will take place and which reactions will not occur. (Note that hydrogen is not a metal and is placed in the series for reference to acid reactions.)

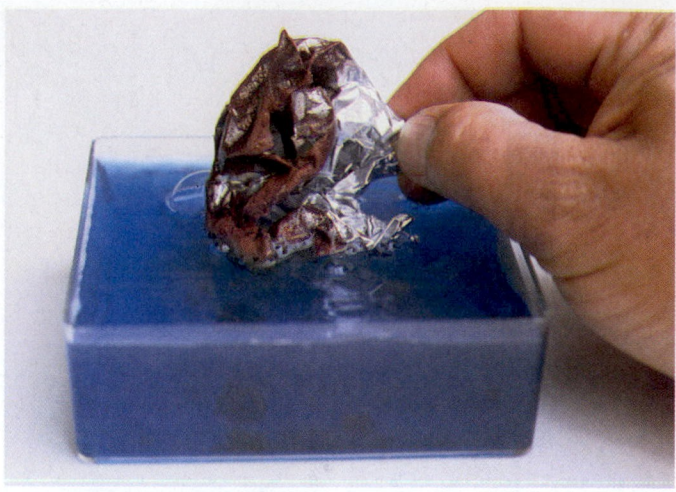

FIGURE 9.18 This shows a reaction between metallic aluminum and the blue solution of copper(II) chloride. Aluminum is above copper in the activity series, and aluminum replaces the copper ions from the solution as copper is deposited as a metal. The aluminum loses electrons to the copper and forms aluminum ions in solution.

Replacement Reactions

In a **replacement reaction,** an atom or polyatomic ion is replaced in a compound by a different atom or polyatomic ion. The replaced part can be either the negative or positive part of the compound. In generalized form, a replacement reaction is

$$XY + Z \rightarrow XZ + Y$$

(negative part replaced)

or

$$XY + A \rightarrow AY + X$$

(positive part replaced)

Replacement reactions occur because some elements have a stronger electron-holding ability than other elements. Elements that have the least ability to hold on to their electrons are the most chemically active. Figure 9.17 shows a list of chemical activity of some metals, with the most chemically active at the top. Hydrogen is included because of its role in acids. Take a few minutes to look over the generalizations listed in figure 9.17. The generalizations apply to combination, decomposition, and replacement reactions.

Replacement reactions take place as more active metals give up electrons to elements lower on the list with a greater electron-holding ability. For example, aluminum is higher on the activity series than copper. When aluminum foil is placed in a solution of copper(II) chloride, aluminum is oxidized, losing electrons to the copper. The loss of electrons from metallic aluminum forms aluminum ions in solution, and the copper comes out of solution as a solid metal (figure 9.18).

$$2 \, \text{Al}(s) + 3 \, \text{CuCl}_2(aq) \rightarrow 2 \, \text{AlCl}_3(aq) + 3 \, \text{Cu}(s)$$

A metal will replace any metal ion in solution that it is above in the activity series. If the metal is listed below the metal ion in solution, no reaction occurs. For example, $\text{Ag}(s) + \text{CuCl}_2(aq) \rightarrow$ no reaction.

The very active metals (lithium, potassium, calcium, and sodium) react with water to yield metal hydroxides and hydrogen. For example,

$$2 \, \text{Na}(s) + 2 \, \text{H}_2\text{O}(l) \rightarrow 2 \, \text{NaOH}(aq) + \text{H}_2 \uparrow$$

Acids yield hydrogen ions in solution, and metals above hydrogen in the activity series will replace hydrogen to form a metal salt. For example,

$$\text{Zn}(s) + \text{H}_2\text{SO}_4(aq) \rightarrow \text{ZnSO}_4(aq) + \text{H}_2 \uparrow$$

In general, the energy involved in replacement reactions is less than the energy involved in combination or decomposition reactions.

Ion Exchange Reactions

An **ion exchange reaction** is a reaction that takes place when the ions of one compound interact with the ions of another compound, forming (1) a solid that comes out of solution (a precipitate), (2) a gas, or (3) water.

A water solution of dissolved ionic compounds is a solution of ions. For example, solid sodium chloride dissolves in water to become ions in solution,

$$\text{NaCl}(s) \rightarrow \text{Na}^+(aq) + \text{Cl}^-(aq)$$

If a second ionic compound is dissolved with a solution of another, a mixture of ions results. The formation of a precipitate, a gas, or water, however, removes ions from the solution,

Science and Society

The Catalytic Converter

The modern automobile produces two troublesome products in the form of (1) nitrogen monoxide and (2) hydrocarbons from the incomplete combustion of gasoline. These products from the exhaust enter the air to react in sunlight, eventually producing an irritating haze known as photochemical smog. To reduce photochemical smog, modern automobiles are fitted with a catalytic converter as part of their exhaust system (box figure 9.2).

Molecules require a certain amount of energy to change chemical bonds. This certain amount of energy is called the *activation energy*, and it represents an energy barrier that must be overcome before a chemical reaction can take place. This explains why chemical reactions proceed at a faster rate at higher temperatures. At higher temperatures, molecules have greater average kinetic energies; thus, they already have part of the minimum energy needed for a reaction to take place.

The rate at which a chemical reaction proceeds is affected by a *catalyst*, a material that speeds up a chemical reaction without being permanently changed by the reaction. A catalyst appears to speed a chemical reaction by lowering the activation energy. Molecules become temporarily attached to the surface of the catalyst, which weakens the chemical bonds holding the molecule together. The weakened molecule is easier to break apart and the activation energy is lowered. Some catalysts do this better with some specific compounds than others, and extensive chemical research programs are devoted to finding new and more effective catalysts.

Automobile catalytic converters use metals such as platinum and transition metal oxides such as copper(II) oxide and chromium(III) oxide. Catalytic reactions that occur in the converter can reduce or oxidize about 90 percent of the hydrocarbons, 85 percent of the carbon monoxide, and 40 percent of the nitrogen monoxide from exhaust gases. Other controls, such as exhaust gas recirculation, are used to reduce further nitrogen monoxide formation.

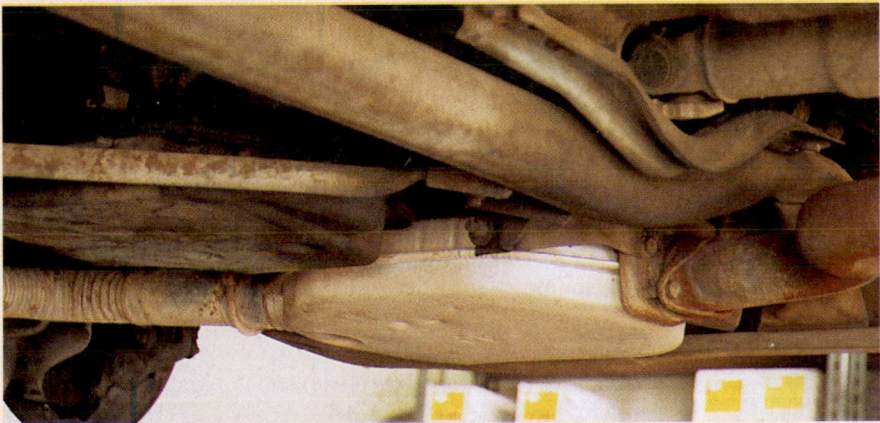

Box Figure 9.2 This silver-colored canister is the catalytic converter. The catalytic converter is located between the engine and the muffler, which is farther back toward the rear of the car.

---

and this must occur before you can say that an ionic exchange reaction has taken place. For example, water being treated for domestic use sometimes carries suspended matter that is removed by adding aluminum sulfate and calcium hydroxide to the water. The reaction is

$$3\,Ca(OH)_2(aq) + Al_2(SO_4)_3(aq) \rightarrow 3\,CaSO_4(aq) + 2\,Al(OH)_3 \downarrow$$

The aluminum hydroxide is a jellylike solid that traps the suspended matter for sand filtration. The formation of the insoluble aluminum hydroxide removed the aluminum and hydroxide ions from the solution, so an ion exchange reaction took place.

In general, an ion exchange reaction has the form

$$AX + BY \rightarrow AY + BX$$

where one of the products removes ions from the solution. The calcium hydroxide and aluminum sulfate reaction took place as the aluminum and calcium ions traded places. A solubility table such as the one in appendix B will tell you if an ionic exchange reaction has taken place. Aluminum hydroxide is

insoluble, according to the table, so the reaction did take place. No ionic exchange reaction occurred if the new products are both soluble.

Another way for an ion exchange reaction to occur is if a gas or water molecule forms to remove ions from the solution. When an acid reacts with a base (an alkaline compound), a salt and water are formed

$$HCl(aq) + NaOH(aq) \rightarrow NaCl(aq) + H_2O(l)$$

The reactions of acids and bases are discussed in chapter 10.

People Behind the Science

Linus Carl Pauling (1901–1994)

Linus Pauling was a U.S. theoretical chemist and biologist whose achievements ranked among the most important of any in twentieth-century science. His main contribution was to the understanding of molecular structure and chemical bonding. He was one of the very few people to have been awarded two Nobel prizes: he received the 1954 Nobel Prize for chemistry (for his work on intermolecular forces) and the 1962 Peace Prize. Throughout his career, his work was noted for the application of intuition and inspiration, assisted by his phenomenal memory; he often carried over principles from one field of science to another.

In 1931, Pauling published a classic paper, "The Nature of the Chemical Bond," in which he used quantum mechanics to explain that an electron-pair bond is formed by the interaction of two unpaired electrons, one from each of two atoms, and that once paired, these electrons cannot take part in the formation of other bonds. It was followed by the book *Introduction to Quantum Mechanics* (1935), of which he was coauthor. He was a pioneer in the application of quantum mechanical principles to the structures of molecules.

It was Pauling who introduced the concept of hybrid orbitals in molecules to explain the symmetry exhibited by carbon atoms in most of its compounds. Pauling also investigated electronegativity of atoms and polarization in chemical bonds. He assigned electronegativities on a scale up to 4.0. A pair of electrons in a bond is pulled preferentially toward an atom with a higher electronegativity. In hydrogen chloride (HCl), for example, hydrogen has an electronegativity of 2.1 and chlorine of 3.5. The bonding electrons are pulled toward the chlorine atom, giving it a small excess negative charge (and leaving the hydrogen atom with a small excess positive charge), polarizing the hydrogen-chlorine bond.

Pauling's ideas on chemical bonding are fundamental to modern theories of molecular structure. Much of this work was consolidated in his book *The Nature of the Chemical Bond, The Structure of Molecules and Crystals* (1939). In the 1940s, Pauling turned his attention to the chemistry of living tissues and systems. He applied his knowledge of molecular structure to the complexity of life, principally to proteins in blood. With Robert Corey, he worked on the structures of amino acids and polypeptides. They proposed that many proteins have structures held together with hydrogen bonds, giving them helical shapes. This concept assisted Francis Crick and James Watson in their search for the structure of DNA, which they eventually resolved as a double helix.

In his researches on blood, Pauling investigated immunology and sickle-cell anemia. Later work confirmed his hunch that the disease is genetic and that normal hemoglobin and the hemoglobin in abnormal "sickle" cells differ in electrical charge. Throughout the 1940s, he studied living materials; he also carried out research on anesthesia. At the end of this period, he published two textbooks, *General Chemistry* (1948) and *College Chemistry* (1950), which became best-sellers.

Source: Modified from the *Hutchinson Dictionary of Scientific Biography.* © RM, 2011. All rights reserved. Helicon Publishing is a division of RM.

SUMMARY

Mixtures are made up of *unlike parts* with a *variable composition*. *Pure substances* are the *same throughout* and have a *definite composition*. Mixtures can be separated into their components by *physical changes*, changes that do not alter the identity of matter. Some pure substances can be broken down into simpler substances by a *chemical change*, a change that *alters the identity of matter as it produces new substances with different properties*. A pure substance that can be decomposed by chemical change into simpler substances with a definite composition is a *compound*. A pure substance that cannot be broken down into anything simpler is an *element*.

A *chemical change* produces new substances by making or breaking chemical bonds. The process of chemical change is called a *chemical reaction*. During a chemical reaction, different chemical substances with greater or lesser amounts of internal potential energy are produced. *Chemical energy* is the change of internal potential energy during a chemical reaction. A *chemical equation* is a shorthand way of describing a chemical reaction. An equation shows the substances that are changed, the *reactants*, on the left side and the new substances produced, the *products*, on the right side.

Chemical reactions involve *valence electrons*, the electrons in the outermost energy level of an atom. Atoms tend to lose or acquire electrons to achieve the configuration of the noble gases with stable, filled outer orbitals. This tendency is generalized as the *octet rule*, that atoms lose or gain electrons to acquire the noble gas structure of eight electrons in the outer orbital. Atoms form negative or positive *ions* in the process.

A chemical bond is an attractive force that holds atoms together in a compound. Chemical bonds formed when atoms transfer electrons to become ions are *ionic bonds*. An ionic bond is an electrostatic attraction between oppositely charged ions. Chemical bonds formed when ions share electrons are *covalent bonds*.

Ionic bonds result in *ionic compounds* with a crystalline structure. The energy released when an ionic compound is formed is called the *heat of formation*. It is the same amount of energy that is required to decompose the compound into its elements.

A *formula* of a compound uses symbols to tell what elements are in a compound and in what proportions. Ions of representative elements have a single, fixed charge, but many transition elements have variable charges. Electrons are conserved when ionic compounds are formed, and the ionic compound is electrically neutral. A formula shows the overall balance of charges.

Covalent compounds are molecular, composed of electrically neutral groups of atoms bound together by *covalent bonds*. The sharing of a pair of electrons, with each atom contributing a single electron to the shared pair, forms a *single covalent bond*. Covalent bonds formed when two pairs of electrons are shared are called *double bonds*, and a *triple bond* is the sharing of three pairs of electrons.

Compounds are named with different rules for ionic and covalent compounds. Both ionic and covalent compounds that are made up of only two different elements always end with an *-ide* suffix, but there are a few exceptions.

The rule for naming variable-charge ions states the English name and gives the charge with Roman numerals in parentheses. Ionic compounds are electrically neutral, and formulas must show a balance of charge. The *crossover technique* is an easy way to write formulas that show a balance of charge.

Covalent compounds are molecules of two or more nonmetal atoms held together by a covalent bond. The system for naming covalent compounds uses Greek prefixes to identify the numbers of atoms, since more than one compound can form from the same two elements (CO and CO_2, for example).

A concise way to describe a chemical reaction is to use formulas in a *chemical equation*. A chemical equation with the same number of each kind of atom on both sides is called a *balanced equation*. A balanced equation is in accord with the *law of conservation of mass*, which states that atoms are neither created nor destroyed in a chemical reaction. To balance a chemical equation, *coefficients* are placed in front of chemical formulas. Subscripts of formulas may not be changed since this would change the formula, meaning a different compound.

One important group of chemical reactions is called *oxidation reduction reactions*, or *redox* reactions for short. Redox reactions are reactions where shifts of electrons occur. The process of losing electrons is called *oxidation*, and the substance doing the losing is said to be *oxidized*. The process of gaining electrons is called *reduction*, and the substance doing the gaining is said to be *reduced*. Substances that take electrons from other substances are called *oxidizing agents*. Substances that supply electrons are called *reducing agents*.

Chemical reactions can also be classified as (1) *combination*, (2) *decomposition*, (3) *replacement*, or (4) *ion exchange*. The first three of these are redox reactions, but ion exchange is not.

KEY TERMS

atom (p. **200**)	covalent compound (p. **206**)	ionic compounds (p. **205**)	polyatomic ion (p. **208**)
chemical bond (p. **203**)	decomposition reaction	law of conservation of mass	redox reaction (p. **214**)
chemical change (p. **199**)	(p. **215**)	(p. **212**)	reducing agent (p. **215**)
chemical energy (p. **201**)	double bond (p. **207**)	molecule (p. **201**)	replacement reaction (p. **216**)
chemical equation (p. **202**)	element (p. **199**)	octet rule (p. **202**)	single bond (p. **207**)
chemical reaction (p. **201**)	formula (p. **205**)	oxidation-reduction reaction	triple bond (p. **207**)
combination reaction (p. **215**)	heat of formation (p. **204**)	(p. **214**)	valence electrons (p. **202**)
compound (p. **199**)	ion exchange reaction (p. **216**)	oxidizing agents (p. **215**)	
covalent bond (p. **205**)	ionic bond (p. **204**)	physical change (p. **199**)	

APPLYING THE CONCEPTS

Answers are located in appendix F.

1. Which of the following represents a chemical change?
   a. heating a sample of ice until it melts
   b. tearing a sheet of paper into tiny pieces
   c. burning a sheet of paper
   d. All of the above are correct.

2. A pure substance that cannot be decomposed into anything simpler by chemical or physical means is called a (an)
   a. element.
   b. compound.
   c. mixture.
   d. isotope.

3. The electrons that participate in chemical bonding are (the)
   a. valence electrons.
   b. electrons in fully occupied orbitals.
   c. stable inner electrons.
   d. All of the above are correct.

4. Which type of chemical bond is formed between two atoms by the sharing of electrons?
   a. ionic
   b. covalent
   c. metallic
   d. None of the above is correct.

5. Which combination of elements forms crystalline solids that will dissolve in water, producing a solution of ions that conduct an electric current?
   a. metal and metal
   b. metal and nonmetal
   c. nonmetal and nonmetal
   d. All of the above are correct.

6. An inorganic compound made of only two different elements has a name that always ends with the suffix
   a. -ite.
   c. -ide.
   b. -ate.
   d. -ous.

7. Dihydrogen monoxide is a compound with the common name of
   a. laughing gas.
   c. smog.
   b. water.
   d. rocket fuel.

8. A chemical equation is balanced by changing (the)
   a. subscripts.
   b. superscripts.
   c. coefficients.
   d. any of the above as necessary to achieve a balance.

9. Since wood is composed of carbohydrates, you should expect what gases to exhaust from a fireplace when complete combustion takes place?
   a. carbon dioxide, carbon monoxide, and pollutants
   b. carbon dioxide and water vapor
   c. carbon monoxide and smoke
   d. It depends on the type of wood being burned.

10. When carbon burns with an insufficient supply of oxygen, carbon monoxide is formed according to the following equation: $2C + O_2 \rightarrow 2CO$. What category of chemical reaction is this?
    a. combination
    b. ion exchange
    c. replacement
    d. None of the above is correct.

11. Of the elements listed below, the one with the greatest chemical activity is
    a. aluminum.
    b. zinc.
    c. iron.
    d. mercury.

12. A balanced chemical equation has
    a. the same number of molecules on both sides of the equation.
    b. the same kinds of molecules on both sides of the equation.
    c. the same number of each kind of atom on both sides of the equation.
    d. All of the above are correct.

QUESTIONS FOR THOUGHT

1. What is the difference between a chemical change and a physical change? Give three examples of each.

2. Describe how the following are alike and how they are different: (a) a sodium atom and a sodium ion, and (b) a sodium ion and a neon atom.

3. What is the difference between an ionic and covalent bond? What do atoms forming the two bond types have in common?

4. What is the octet rule and why is it important?

5. What is a polyatomic ion? Give the names and formulas for several common polyatomic ions.

6. Write the formula for magnesium hydroxide. Explain what the parentheses mean.

7. What is the basic difference between a single bond and a double bond?

8. What is the law of conservation of mass? How do you know if a chemical equation is in accord with this law?

9. Describe in your own words how a chemical equation is balanced.

10. How is the activity series for metals used to predict if a replacement reaction will occur or not?

11. What must occur in order for an ion exchange reaction to take place? What is the result if this does not happen?

12. Predict the products for the following reactions: (a) the combustion of ethyl alcohol (C_2H_5OH), and (b) the rusting of aluminum (Al).

FOR FURTHER ANALYSIS

1. What are the significant similarities and differences between a physical change and a chemical change?

2. Analyze how you would know for sure that a pure substance you have is a compound and not an element.

3. Analyze how you would know for sure that a pure substance you have is an element and not a compound.

4. Make up an explanation for why ionic compounds are formed when a metal from the left side of the periodic table reacts with a nonmetal from the right side, but covalent bonds are formed between nonmetallic elements on the right side of the table.

5. What are the advantages and disadvantages to writing a chemical equation with chemical symbols and formulas rather than just words?

6. Provide several examples of each of the four basic categories of chemical reactions and describe how each illustrates a clear representation of the category.

7. Summarize for another person the steps needed for successfully writing a balanced chemical equation.

INVITATION TO INQUIRY

Rate of Chemical Reactions

Temperature is one of the more important factors that influence the rate of a chemical reaction. You can use a "light stick" or "light tube" to study how temperature can influence a chemical reaction. Light sticks and tubes are devices that glow in the dark and are very popular on July 4 and at other times when people might be outside after sunset. They work from a chemical reaction that is similar to the chemical reaction that produces light in a firefly. Design an experiment that uses light sticks to find out the effect of temperature on the brightness of light and how long the device will continue providing light. Perhaps you will be able to show by experimental evidence that use at a particular temperature produces the most light for the longest period of time.

PARALLEL EXERCISES

The exercises in groups A and B cover the same concepts. Solutions to group A exercises are located in appendix G.

Group A

1. How many outer orbital electrons are found in an atom of:
   a. Li
   b. N
   c. F
   d. Cl
   e. Ra
   f. Be

2. Write electron dot notations for the following elements:
   a. Boron
   b. Bromine
   c. Calcium
   d. Potassium
   e. Oxygen
   f. Sulfur

3. Identify the charge on the following ions:
   a. Boron
   b. Bromine
   c. Calcium
   d. Potassium
   e. Oxygen
   f. Nitrogen

4. Name the following polyatomic ions:
   a. $(OH)^-$
   b. $(SO_3)^{2-}$
   c. $(ClO)^-$
   d. $(NO_3)^-$
   e. $(CO_3)^{2-}$
   f. $(ClO_4)^-$

5. Use the crossover technique to write formulas for the following compounds:
   a. Iron(III) hydroxide
   b. Lead(II) phosphate
   c. Zinc carbonate
   d. Ammonium nitrate
   e. Potassium hydrogen carbonate
   f. Potassium sulfite

6. Write formulas for the following covalent compounds:
   a. Carbon tetrachloride
   b. Dihydrogen monoxide
   c. Manganese dioxide
   d. Sulfur trioxide
   e. Dinitrogen pentoxide
   f. Diarsenic pentasulfide

Group B

1. How many outer orbital electrons are found in an atom of:
   a. Na
   b. P
   c. Br
   d. I
   e. Te
   f. Sr

2. Write electron dot notations for the following elements:
   a. Aluminum
   b. Fluorine
   c. Magnesium
   d. Sodium
   e. Carbon
   f. Chlorine

3. Identify the charge on the following ions:
   a. Aluminum
   b. Chlorine
   c. Magnesium
   d. Sodium
   e. Sulfur
   f. Hydrogen

4. Name the following polyatomic ions:
   a. $(C_2H_3O_2)^-$
   b. $(HCO_3)^-$
   c. $(SO_4)^{2-}$
   d. $(NO_2)^-$
   e. $(MnO_4)^-$
   f. $(CO_3)^{2-}$

5. Use the crossover technique to write formulas for the following compounds:
   a. Aluminum hydroxide
   b. Sodium phosphate
   c. Copper(II) chloride
   d. Ammonium sulfate
   e. Sodium hydrogen carbonate
   f. Cobalt(II) chloride

6. Write formulas for the following covalent compounds:
   a. Silicon dioxide
   b. Dihydrogen sulfide
   c. Boron trifluoride
   d. Dihydrogen dioxide
   e. Carbon tetrafluoride
   f. Nitrogen trihydride

—continued

PARALLEL EXERCISES—*Continued*

Group A

7. Name the following covalent compounds:
   a. CO
   b. CO_2
   c. CS_2
   d. N_2O
   e. P_4S_3
   f. N_2O_3

8. Write balanced chemical equations for each of the following unbalanced reactions:
   a. $SO_2 + O_2 \rightarrow SO_3$
   b. $P + O_2 \rightarrow P_2O_5$
   c. $Al + HCl \rightarrow AlCl_3 + H_2$
   d. $NaOH + H_2SO_4 \rightarrow Na_2SO_4 + H_2O$
   e. $Fe_2O_3 + CO \rightarrow Fe + CO_2$
   f. $Mg(OH)_2 + H_3PO_4 \rightarrow Mg_3(PO_4)_2 + H_2O$

9. Identify the following as combination, decomposition, replacement, or ion exchange reactions:
   a. $NaCl(aq) + AgNO_3(aq) \rightarrow NaNO_3(aq) + AgCl\downarrow$
   b. $H_2O(l) + CO_2(g) \rightarrow H_2CO_3(l)$
   c. $2\ NaHCO_3(s) \rightarrow Na_2CO_3(s) + H_2O(g) + CO_2(g)$
   d. $2\ Na(s) + Cl_2(g) \rightarrow 2\ NaCl(s)$
   e. $Cu(s) + 2\ AgNO_3(aq) \rightarrow Cu(NO_3)_2(aq) + 2\ Ag(s)$
   f. $CaO(s) + H_2O(l) \rightarrow Ca(OH)_2(aq)$

10. Write complete, balanced equations for each of the following reactions:
    a. $C_5H_{12}(g) + O_2(g) \rightarrow$
    b. $HCl(aq) + NaOH(aq) \rightarrow$
    c. $Al(s) + Fe_2O_3(s) \rightarrow$
    d. $Fe(s) + CuSO_4(aq) \rightarrow$
    e. $MgCl_2(aq) + Fe(NO_3)_2(aq) \rightarrow$
    f. $C_6H_{10}O_5(s) + O_2(g) \rightarrow$

Group B

7. Name the following covalent compounds:
   a. N_2O
   b. SO_2
   c. SiC
   d. PF_5
   e. $SeCl_6$
   f. N_2O_4

8. Write balanced chemical equations for each of the following unbalanced reactions:
   a. $NO + O_2 \rightarrow NO_2$
   b. $KClO_3 \rightarrow KCl + O_2$
   c. $NH_4Cl + Ca(OH)_2 \rightarrow CaCl_2 + NH_3 + H_2O$
   d. $NaNO_3 + H_2SO_4 \rightarrow Na_2SO_4 + HNO_3$
   e. $PbS + H_2O_2 \rightarrow PbSO_4 + H_2O$
   f. $Al_2(SO_4)_3 + BaCl_2 \rightarrow AlCl_3 + BaSO_4$

9. Identify the following as combination, decomposition, replacement, or ion exchange reactions:
   a. $ZnCO_3(s) \rightarrow ZnO(s) + CO_2\uparrow$
   b. $2\ NaBr(aq) + Cl_2(g) \rightarrow 2\ NaCl(aq) + Br_2(g)$
   c. $2\ Al(s) + 3\ Cl_2(g) \rightarrow 2\ AlCl_3(s)$
   d. $Ca(OH)_2(aq) + H_2SO_4(aq) \rightarrow CaSO_4(aq) + 2\ H_2O(l)$
   e. $Pb(NO_3)_2(aq) + H_2S(g) \rightarrow 2\ HNO_3(aq) + PbS\downarrow$
   f. $C(s) + ZnO(s) \rightarrow Zn(s) + CO\uparrow$

10. Write complete, balanced equations for each of the following reactions:
    a. $C_3H_6(g) + O_2(g) \rightarrow$
    b. $H_2SO_4(aq) + KOH(aq) \rightarrow$
    c. $C_6H_{12}O_6(s) + O_2(g) \rightarrow$
    d. $Na_3PO_4(aq) + AgNO_3(aq) \rightarrow$
    e. $NaOH(aq) + Al(NO_3)_3(aq) \rightarrow$
    f. $Mg(OH)_2(aq) + H_3PO_4(aq) \rightarrow$

10

Water and Solutions

Water is often referred to as the *universal solvent* because it makes so many different kinds of solutions. Eventually, moving water can dissolve solid rock, carrying it away in solution.

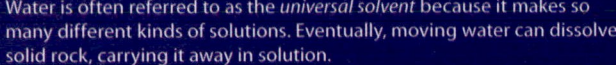

CORE **CONCEPT**

Water and solutions of water have unique properties.

OUTLINE

Earth Science

▷ Water cycles in and out of the atmosphere (Ch. 17).
▷ Less than 1 percent of all water on Earth is fit for human consumption or agriculture (Ch. 18).

Life Science

▷ Metabolic processes require water for all aspects (Ch. 20).
▷ Water and temperature ranges determine biomes (Ch. 23).

OVERVIEW

What do you think about when you see a stream (figure 10.1)? Do you wonder about the water quality and what might be dissolved in the water? Do you wonder where the stream comes from and if it will ever run out of water?

Many people can look at a stream, but they might think about different things. A farmer might think about how the water could be diverted and used for his crops. A city planner might wonder if the water is safe for domestic use, and if not, what it would cost to treat the water. Others might wonder if the stream has large fish they could catch. Many large streams can provide water for crops, domestic use, and recreation, and still meet the requirements for a number of other uses.

It is the specific properties of water that make it important for agriculture, domestic use, and recreation. Living things evolved in a watery environment, so water and its properties are essential to life on Earth. Some properties of water, such as the ability to dissolve almost anything, also make water very easy to pollute. This chapter is concerned with some of the unique properties of water, water solutions, and household use of water.

10.1 HOUSEHOLD WATER

Water is an essential resource, not only because it is required for life processes but also because of its role in a modern society. (See chapter 18.) Water is used in the home for drinking and cooking (2 percent), washing dishes (6 percent), laundry (11 percent), bathing (23 percent), toilets (29 percent), and maintaining lawns and gardens (29 percent).

The water supply is obtained from streams, lakes, and reservoirs on the surface or from groundwater pumped from below the surface. Surface water contains more sediments, bacteria, and possible pollutants than water from a well because it is exposed to the atmosphere and water runs off the land into streams and rivers. Surface water requires filtering to remove suspended particles, treatment to kill bacteria, and sometimes processing to remove pollution. Well water is generally cleaner but still might require treatment to kill bacteria and remove pollution that has seeped through the ground from waste dumps, agricultural activities, or industrial sites.

Most pollutants are usually too dilute to be considered a significant health hazard, but there are exceptions. There are five types of contamination found in U.S. drinking water that are responsible for the most widespread danger, and these are listed in table 10.1. In spite of these general concerns and other occasional local problems, the U.S. water supply is considered to be among the cleanest in the world.

Demand for domestic water sometimes exceeds the immediate supply in some growing metropolitan areas. This is most common during the summer, when water demand is high

FIGURE 10.1   A freshwater stream has many potential uses.

Who Has the Right?

As the population grows and new industries develop, more and more demands are placed on the water supply. This raises some issues about how water should be divided among agriculture, industries, and city domestic use. Agricultural interests claim they should have the water because they produce the food and fibers that people must have. Industrial interests claim they should have the water because they create the jobs and the products that people must have. Cities, on the other hand, claim that domestic consumption is the most important because people cannot survive without water. Yet, others claim that no group has a right to use water when it is needed to maintain habitats.

Questions to Discuss:

1. Who should have the first priority for water use?
2. Who should have the last priority for water use?
3. What determined your answers to questions 1 and 2?

TABLE 10.1 Possible Pollution Problems in the U.S. Water Supply

Pollutant	Source	Risk
Lead	Lead pipes in older homes; solder in copper pipes, brass fixtures	Nerve damage, miscarriage, birth defects, high blood pressure, hearing problems
Chlorinated solvents	Industrial pollution	Cancer
Trihalomethanes	Chlorine disinfectant reacting with other pollutants	Liver damage, kidney damage, possibly cancer
PCBs	Industrial waste, older transformers	Liver damage, possibly cancer
Bacteria and viruses	Septic tanks, outhouses, overflowing sewer lines	Gastrointestinal problems, serious diseases

and rainfall is often low. Communities in these areas often have public education campaigns designed to help reduce the demand for water. For example, did you know that taking a tub bath can use up to 135 liters (about 36 gal) of water compared to only 95 liters (about 25 gal) for a regular shower? Even more water is saved by a shower that does not run continuously—wetting down, soaping up, and rinsing off uses only 15 liters (about 4 gal) of water. You can also save about 35 liters (about 9 gal) of water by not letting the water run continuously while brushing your teeth.

It is often difficult to convince people to conserve water when it is viewed as an inexpensive, limitless supply. However, efforts to conserve water increase dramatically as the cost to the household consumer increases.

The issues involved in maintaining a safe water supply are better understood by considering some of the properties of water and water solutions. These are the topics of the following sections.

10.2 PROPERTIES OF WATER

Water is essential for life since living organisms are made up of cells filled with water and a great variety of dissolved substances. Foods are mostly water, with fruits and vegetables containing up to 95 percent water and meat consisting of about 50 percent water. Your body is over 70 percent water by weight. Since water is such a large component of living things, understanding the properties of water is important to understanding life. One important property is water's unusual ability to act as a solvent. Water is called a "universal solvent" because of its ability to dissolve most molecules. In living things, these dissolved molecules can be transported from one place to another by diffusion or by some kind of a circulatory system.

The usefulness of water does not end with its unique abilities as a solvent and transporter; it has many more properties that are useful, although unusual. For example, unlike other liquids, water in its liquid phase has a greater density than solid water (ice). This important property enables solid ice to float on the surface of liquid water, insulating the water below and permitting fish and other water organisms to survive the winter. If ice were denser than water, it would sink, freezing all lakes and rivers from the bottom up. Fish and most organisms that live in water would not be able to survive in a lake or river of solid ice.

As described in chapter 4, water is also unusual because it has a high specific heat. The same amount of sunlight falling on equal masses of soil and water will warm the soil 5°C for each 1°C increase in water temperature. Thus, it will take five times more sunlight to increase the temperature of the water to as much as the soil. This enables large bodies of water to moderate the temperature, making it more even.

A high latent heat of vaporization is yet another unusual property of water. This particular property enables people to dissipate large amounts of heat by evaporating a small amount of water. Since people carry this evaporative cooling system with them, they can survive some very warm desert temperatures, for example.

Finally, other properties of water are not crucial for life but are interesting nonetheless. For example, why do all snowflakes have six sides? Is it true that no two snowflakes are alike? The unique structure of the water molecule will explain water's unique solvent abilities, why solid water is less dense than liquid water, its high specific heat, its high latent heat of vaporization, and perhaps why no two snowflakes seem to be alike.

Structure of the Water Molecule

In chapter 9, you learned that atoms combine two ways. Atoms from opposite sides of the periodic table form ionic bonds after transferring one or more electrons. Atoms from the right side of the periodic table form covalent bonds by sharing one or more pairs of electrons. This distinction is clear-cut in many compounds but not in water. The way atoms share electrons in a water molecule is not exactly covalent, but it is not ionic either.

In a molecule of water, an oxygen atom shares a pair of electrons with two hydrogen atoms. Oxygen has six outer electrons and needs two more to satisfy the octet rule, achieving the noble gas structure of eight. Each hydrogen atom needs one more electron to fill its outer orbital with two. Therefore, one oxygen atom bonds with two hydrogen atoms, forming H_2O. Both oxygen and hydrogen are more stable with the outer orbital configuration of the noble gases (neon and helium in this case).

Electrons are shared in a water molecule but not equally. Oxygen, with its eight positive protons, has a greater attraction for the shared electrons than do either of the hydrogens with a single proton. Therefore, the shared electrons spend more time around the oxygen part of the molecule than they do around the hydrogen part. This results in the oxygen end of the molecule being more negative than the hydrogen end. When electrons in a covalent bond are not equally shared, the molecule is said to be polar. A **polar molecule** has a *dipole* (*di* = two; *pole* = side or end), meaning it has a positive end and a negative end.

A water molecule has a negative center at the oxygen end and a positive center at the hydrogen end. The positive charges on the hydrogen end are separated, giving the molecule a bent arrangement rather than a straight line. Figure 10.2A shows a model of a water molecule showing its polar nature.

It is the polar structure of the water molecule that is responsible for many of the unique properties of water. Polar molecules of any substances have attractions between the positive end of a molecule and the negative end of another molecule. When the polar molecule has hydrogen at one end and fluorine, oxygen, or nitrogen on the other part, the attractions are strong enough to make a type of bonding called **hydrogen bonding.** Hydrogen bonding is a bond that occurs between the hydrogen end of a molecule and the fluorine, oxygen, or nitrogen end of other similar molecules. A better name for this would be a hydrogen-fluorine bond, a hydrogen-oxygen bond, or a hydrogen-nitrogen bond. However, for brevity, the second part of the bond is not named and all the hydrogen-something bonds are simply known of as "hydrogen" bonds. The dotted line between the hydrogen and oxygen molecules in figure 10.2B represents a hydrogen bond. A dotted line is used to represent a bond that is not as strong as the bond represented by the solid line of a covalent compound.

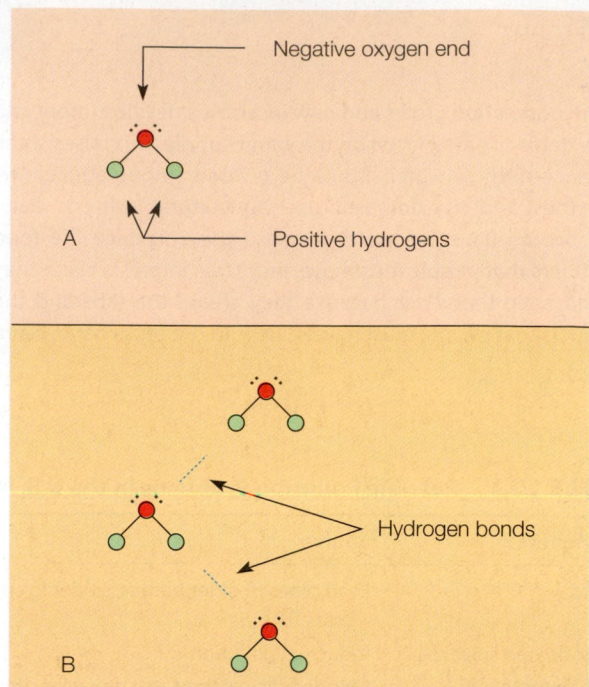

FIGURE 10.2 (*A*) The water molecule is polar, with centers of positive and negative charges. (*B*) Attractions between these positive and negative centers establish hydrogen bonds between adjacent molecules.

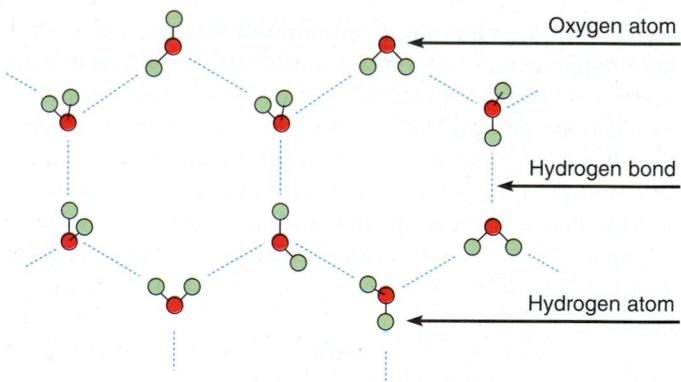

FIGURE 10.3 The hexagonal structure of ice. Hydrogen bonding between the oxygen atom and two hydrogen atoms of other water molecules results in an arrangement, which forms the open, hexagonal structure of ice. Note that the angles of the water molecules do not change but have different orientations.

Hydrogen bonding accounts for the physical properties of water, including its unusual density changes with changes in temperature. Figure 10.3 shows the hydrogen-bonded structure of ice. Water molecules form a six-sided hexagonal structure that extends out for billions of molecules. The large channels, or holes, in the structure result in ice being less dense than water. The shape of the hexagonal arrangement also suggests why snowflakes always have six sides. Why does it seem like no two snowflakes are alike? Perhaps the answer can be found in the almost infinite variety of shapes that can be built from billions and billions of tiny hexagons of ice crystals.

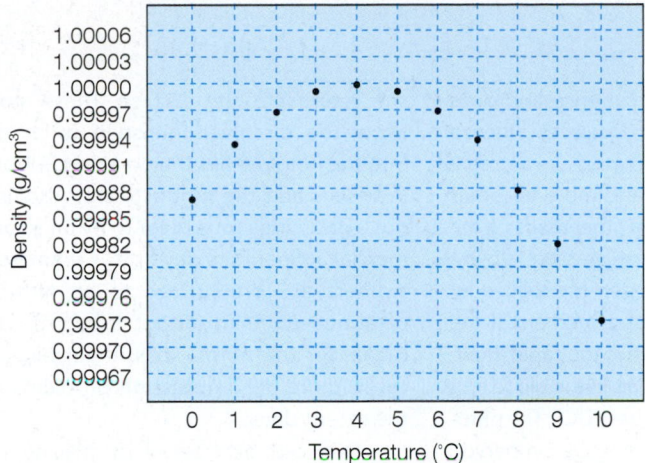

FIGURE 10.4 The density of water from 0°C to 10°C. The density of water is at a maximum at 4°C, becoming less dense as it is cooled or warmed from this temperature. Hydrogen bonding explains this unusual behavior.

When ice is warmed, the increased vibrations of the molecules begin to expand and stretch the hydrogen bond structure. When ice melts, about 15 percent of the hydrogen bonds break and the open structure collapses into the more compact arrangement of liquid water. As the liquid water is warmed from 0°C, still more hydrogen bonds break down, and the density of the water steadily increases. At 4°C, the expansion of water from the increased molecular vibrations begins to predominate, and the density decreases steadily with further warming (figure 10.4). Thus, water has its greatest density at a temperature of 4°C.

The heat of fusion, specific heat, and heat of vaporization of water are unusually high when compared to other chemically similar substances. These high values are accounted for by the additional energy needed to break hydrogen bonds.

The Dissolving Process

A **solution** is a homogeneous mixture of ions or molecules of two or more substances. *Dissolving* is the process of making a solution. During dissolving, the different components that make up the solution become mixed. For example, when sugar dissolves in water, molecules of sugar become uniformly dispersed throughout the molecules of water. The uniform taste of sweetness of any part of the sugar solution is a result of this uniform mixing.

The general terms of *solvent* and *solute* identify the components of a solution. The solvent is the component present in the larger amount. The solute is the component that dissolves in the solvent. Atmospheric air, for example, is about 78 percent nitrogen, so nitrogen is considered the solvent. Oxygen (about 21 percent), argon (about 0.9 percent), and other gases make up the solutes. If one of the components of a solution is a liquid, it is usually identified as the solvent. An *aqueous solution* is a solution of a solid, a liquid, or a gas in water.

A solution is formed when the molecules or ions of two or more substances become homogeneously mixed. But the process of dissolving must be more complicated than the simple mixing together of particles because (1) solutions become saturated, meaning there is a limit on solubility, and (2) some substances are *insoluble*, not dissolving at all or at least not noticeably. In general, the forces of attraction between molecules or ions of the solvent and solute determine if something will dissolve and if there are limits on the solubility. These forces of attraction and their role in the dissolving process will be considered in the following examples.

First, consider the dissolving process in gaseous and liquid solutions. In a gas, the intermolecular forces are small, so gases can mix in any proportion. Fluids that can mix in any proportion like this are called **miscible fluids.** Fluids that do not mix are called *immiscible fluids.* Air is a mixture of gases, so gases (including vapors) are miscible.

Liquid solutions can dissolve a gas, another liquid, or a solid. Gases are miscible in liquids, and a carbonated beverage (your favorite cola) is the common example, consisting of carbon dioxide dissolved in water. Whether or not two given liquids form solutions depends on some similarities in their molecular structures. The water molecule, for example, is a polar molecule with a negative end and a positive end. On the other hand, carbon tetrachloride (CCl_4) is a molecule with polar bonds that are symmetrically arranged. Because of the symmetry, CCl_4 has no negative or positive ends, so it is nonpolar. Thus, some liquids have polar molecules, and some have nonpolar molecules. The general rule for forming solutions is *like dissolves like.* A nonpolar compound, such as carbon tetrachloride, will dissolve oils and greases because they are nonpolar compounds. Water, a polar compound, will not dissolve the nonpolar oils and greases. Carbon tetrachloride was at one time used as a cleaning solvent because of its oil and grease dissolving abilities. Its use is no longer recommended because it causes liver damage.

Some molecules, such as soap, have a part of the molecule that is polar and a part that is nonpolar. Washing with water alone will not dissolve oils because water and oil are immiscible. When soap is added to the water, however, the polar end of the soap molecule is attracted to the polar water molecules, and the nonpolar end is absorbed into the oil. A particle (larger than a molecule) is formed, and the oil is washed away with the water.

The "like dissolves like" rule applies to solids and liquid solvents as well as liquids and liquid solvents. Polar solids, such as salt, will readily dissolve in water, which has polar molecules,

A Closer Look

Decompression Sickness

Decompression sickness (DCS) is a condition caused by the formation of nitrogen bubbles in the blood and tissues of a scuba diver who surfaces too quickly, causing a rapid drop in pressure. This condition is usually marked by joint pain but can include chest pain, skin irritation, and muscle cramps. This may be barely noticed in mild cases, but severe cases can be fatal. The joint pain is often called *bends*.

DCS can occur in any situation where a person is subjected to a higher air pressure for some time and then experiences a rapid decompression. Here is what happens: Air is about 79 percent nitrogen and 21 percent oxygen. Nitrogen is inert to the human body, and what we inhale is exhaled, but some becomes dissolved in the blood. Dissolved nitrogen is a normal occurrence and does not present a problem. However, when a person is breathing higher-pressure air, more nitrogen is breathed in and more becomes dissolved in the blood and other tissues. If the person returns to normal pressure slowly, the extra dissolved nitrogen is expelled by the lungs, and this is not a problem. If the return to normal pressure is too rapid, however, nitrogen bubbles form in the blood and other tissues, and this causes DCS. The nitrogen bubbles can cause pressure on nerves, block circulation, and cause joint pain. These symptoms usually appear when the diver returns to the surface or within eight hours of returning to normal air pressure.

One way to prevent DCS is for a diver to make "decompression stops" as he or she returns to the surface. These stops allow the dissolved nitrogen to diffuse into the lungs rather than making bubbles in tissues and the bloodstream. Also, it is not a good idea for a diver to fly on a commercial airliner for a day or so. Commercial airliners are pressurized to an altitude of about 2,500 m (about 8,200 ft). This is safe for the normal passenger but can cause more decompression problems for the recent diver.

DCS can be treated by placing the affected person in a special sealed chamber. Pressure is slowly increased in the chamber to cause nitrogen bubbles to go back into solution, and then slowly decreased to allow the dissolved nitrogen to be expelled by the lungs.

---

but do not dissolve readily in oil, grease, or other nonpolar solvents. Polar water readily dissolves salt because the charged polar water molecules are able to exert an attraction on the ions, pulling them away from the crystal structure. Thus, ionic compounds dissolve in water.

Ionic compounds vary in their solubilities in water. This difference is explained by the existence of two different forces involved in an ongoing "tug of war." One force is the attraction between an ion on the surface of the crystal and a water molecule, an *ion-polar molecule force*. When solid sodium chloride and water are mixed together, the negative ends of the water molecules (the oxygen ends) become oriented toward the positive sodium ions on the crystal. Likewise, the positive ends of water molecules (the hydrogen ends) become oriented toward the negative chlorine ions. The attraction of water molecules for ions is called **hydration.** If the force of hydration is greater than the attraction between the ions in the solid, they are pulled away from the solid, and dissolving occurs (figure 10.5). Considering sodium chloride only, the equation is

$$Na^+Cl^-(s) \rightarrow Na^+(aq) + Cl^-(aq)$$

which shows that the ions were separated from the solid to become a solution of ions. In other compounds, the attraction between the ions in the solid might be greater than the energy of hydration. In this case, the ions of the solid would win the "tug of war," and the ionic solid is insoluble.

The saturation of soluble compounds is explained in terms of hydration eventually occupying a large number of the polar water molecules. Fewer available water molecules means less attraction on the ionic solid, with more solute ions being pulled back to the surface of the solid. The tug of war continues back and forth as an equilibrium condition is established. (For a detailed discussion of different measures of solution concentration, see the chapter 10 Resources on www.mhhe.com/tillery.)

Solubility

Gases and liquids appear to be soluble in all proportions, but there is a limit to how much solid can be dissolved in a liquid. You may have noticed that a cup of hot tea will dissolve several

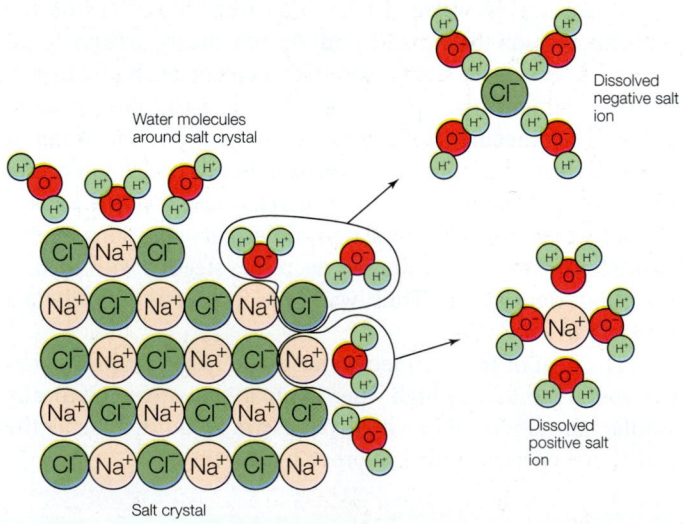

FIGURE 10.5 An ionic solid dissolves in water because the number of water molecules around the surface is greater than the number of other ions of the solid. The attraction between polar water molecules and a charged ion enables the water molecules to pull ions away from the crystal so the salt crystals dissolve in the water.

teaspoons of sugar, but the limit of solubility is reached quickly in a glass of iced tea. The limit of how much sugar will dissolve seems to depend on the temperature of the tea. More sugar added to the cold tea after the limit is reached will not dissolve, and solid sugar granules begin to accumulate at the bottom of the glass. At this limit, the sugar and tea solution is said to be *saturated.* Dissolving does not actually stop when a solution becomes saturated and undissolved sugar continues to enter the solution. However, dissolved sugar is now returning to the undissolved state at the same rate as it is dissolving. The overall equilibrium condition of sugar dissolving as sugar is coming out of solution is called a **saturated solution.** A saturated solution is a *state of equilibrium that exists between dissolving solute and solute coming out of solution.* You actually cannot see the dissolving and coming out of solution that occur in a saturated solution because the exchanges are taking place with particles the size of molecules or ions.

Not all compounds dissolve as sugar does, and more or less of a given compound may be required to produce a saturated solution at a particular temperature. In general, the difficulty of dissolving a given compound is referred to as *solubility.* More specifically, the **solubility** of a solute is defined as the *concentration that is reached in a saturated solution at a particular temperature.* Solubility varies with the temperature, as the sodium and potassium salt examples show in figure 10.6. These solubility curves describe the amount of solute required to reach the saturation equilibrium at a particular temperature. In general, the solubilities of most ionic solids increase with

temperature, but there are exceptions. In addition, some salts release heat when dissolved in water, and other salts absorb heat when dissolved. The "instant cold pack" used for first aid is a bag of water containing a second bag of ammonium nitrate (NH_4NO_3). When the bag of ammonium nitrate is broken, the compound dissolves and absorbs heat.

You can usually dissolve more of a solid, such as salt or sugar, as the temperature of the water is increased. Contrary to what you might expect, gases usually become *less* soluble in water as the temperature increases. As a glass of water warms, small bubbles collect on the sides of the glass as dissolved air comes out of solution. The first bubbles that appear when warming a pot of water to boiling are also bubbles of dissolved air coming out of solution. This is why water that has been boiled usually tastes "flat." The dissolved air has been removed by the heating. The "normal" taste of water can be restored by pouring the boiled water back and forth between two glasses. The water dissolves more air during this process, restoring the usual taste.

Changes in pressure have no effect on the solubility of solids in liquids but greatly affect the solubility of gases. The release of bubbles (fizzing) when a bottle or can of soda is opened occurs because pressure is reduced on the beverage and dissolved carbon dioxide comes out of solution. In general, *gas solubility decreases with temperature and increases with pressure.* As usual, there are exceptions to this generalization.

10.3 PROPERTIES OF WATER SOLUTIONS

Pure solvents have characteristic physical and chemical properties that are changed by the addition of a solute. Following are some of the more interesting changes.

Electrolytes

Water solutions of ionic substances will conduct an electric current, so they are called **electrolytes.** Ions must be present and free to move in a solution to carry the charge, so electrolytes are solutions containing ions. Pure water will not conduct an electric current as it is a covalent compound, which ionizes only very slightly. Water solutions of sugar, alcohol, and most other covalent compounds are nonconductors, so they are called *nonelectrolytes.* Nonelectrolytes are covalent compounds that form molecular solutions, so they cannot conduct an electric current.

Some covalent compounds are nonelectrolytes as pure liquids but become electrolytes when dissolved in water. Pure hydrogen chloride (HCl), for example, does not conduct an electric current, so you can assume that it is a molecular substance. When dissolved in water, hydrogen chloride does conduct a current, so it must now contain ions. Evidently, the hydrogen chloride has become *ionized* by the water. The process of forming ions from molecules is called *ionization.* Hydrogen chloride, just as water, has polar molecules. The positive hydrogen atom on the HCl molecule is attracted to the negative

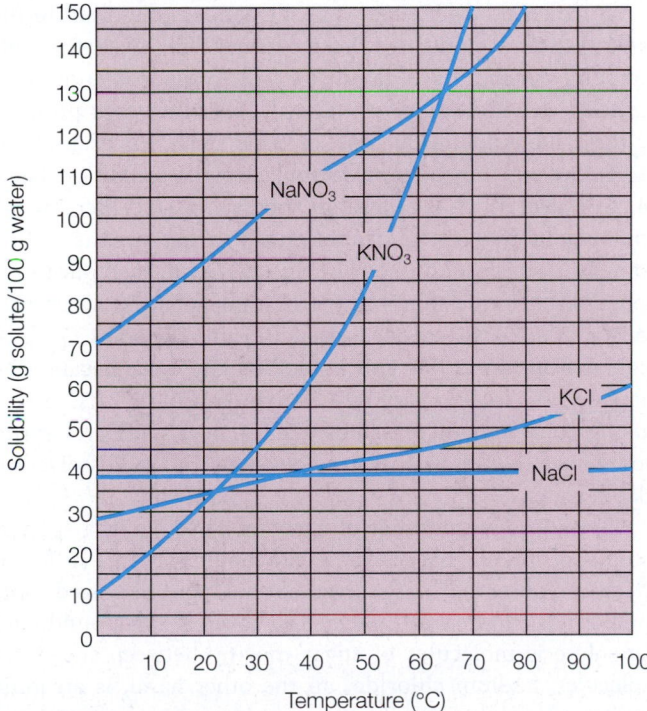

FIGURE 10.6 Approximate solubility curves for sodium nitrate, potassium nitrate, potassium chloride, and sodium chloride.

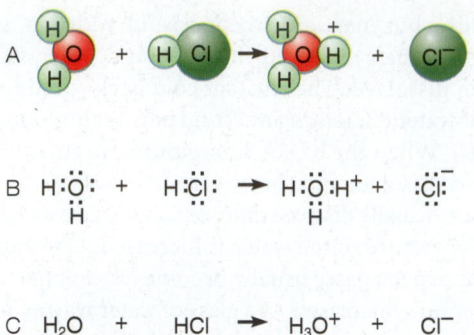

FIGURE 10.7 Three representations of water and hydrogen chloride in an ionizing reaction. (*A*) Sketches of molecules involved in the reaction. (*B*) Electron dot equation of the reaction. (*C*) The chemical equation for the reaction. Each of these representations shows the hydrogen being pulled away from the chlorine atom to form H_3O^+, the hydronium ion.

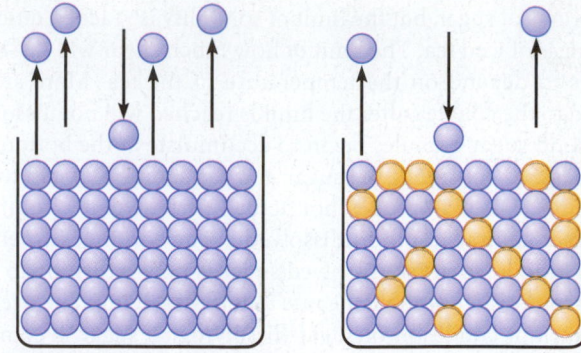

FIGURE 10.8 The rate of evaporation, and thus the vapor pressure, is less for a solution than for a solvent in the pure state. The greater the solute concentration, the less the vapor pressure.

oxygen end of a water molecule, and the force of attraction is strong enough to break the hydrogen-chlorine bond, forming charged particles (figure 10.7). The reaction is

$$H_2O(l) + HCl(l) \rightarrow H_3O^+(aq) + Cl^-(aq)$$

The H_3O^+ ion is called a **hydronium ion.** A hydronium ion is basically a molecule of water with an attached hydrogen ion. The presence of the hydronium ion gives the solution new chemical properties; the solution is no longer hydrogen chloride but is *hydrochloric acid.* Hydrochloric acid, and other acids, will be discussed shortly.

Boiling Point

Boiling occurs when the pressure of the vapor escaping from a liquid (vapor pressure) is equal to the atmospheric pressure on the liquid. The *normal* boiling point is defined as the temperature at which the vapor pressure is equal to the average atmospheric pressure at sea level. For pure water, this temperature is 100°C (212°F). It is important to remember that boiling is a purely physical process. No bonds within water molecules are broken during boiling.

The vapor pressure over a solution is *less* than the vapor pressure over the pure solvent at the same temperature. Molecules of a liquid can escape into the air only at the surface of the liquid, and the presence of molecules of a solute means that fewer solvent molecules can be at the surface to escape. Thus, the vapor pressure over a solution is less than the vapor pressure over a pure solvent (figure 10.8).

Because the vapor pressure over a solution is less than that over the pure solvent, the solution boils at a higher temperature. A higher temperature is required to increase the vapor pressure to that of the atmospheric pressure. Some cooks have been observed to add a "pinch" of salt to a pot of water before boiling. Is this to increase the boiling point and therefore cook the food more quickly? How much does a pinch of salt increase the boiling temperature? The answers are found in the relationship

between the concentration of a solute and the boiling point of the solution.

It is the number of solute particles (ions or molecules) at the surface of a solution that increases the boiling point. A mole is a measure that can be defined as a number of particles called Avogadro's number. Since the number of particles at the surface is proportional to the ratio of particles in the solution, the concentration of the solute will directly influence the increase in the boiling point. In other words, the boiling point of any dilute solution is increased proportional to the concentration of the solute. For water, the boiling point is increased 0.521°C for every mole of solute dissolved in 1,000 g of water. Thus, any water solution will boil at a higher temperature than pure water.

It makes no difference what substance is dissolved in the water; 1 mole of solute in 1,000 g of water will elevate the boiling point by 0.521°C. A mole contains Avogadro's number of particles, so a mole of any solute will lower the vapor pressure by the same amount. Sucrose, or table sugar, for example, is $C_{12}H_{22}O_{11}$ and has a gram-formula weight of 342 g. Thus, 342 g of sugar in 1,000 g of water (about a liter) will increase the boiling point by 0.521°C. Therefore, if you measure the boiling point of a sugar solution, you can determine the concentration of sugar in the solution. For example, pancake syrup that boils at 100.261°C (sea-level pressure) must contain 171 g of sugar dissolved in 1,000 g of water. You know this because the increase of 0.261°C over 100°C is one-half of 0.521°C. If the boiling point were increased by 0.521°C over 100°C, the syrup would have the full gram-formula weight (342 g) dissolved in a kg of water.

Since it is the number of particles of solute in a specific sample of water that elevates the boiling point, different effects are observed in dissolved covalent and dissolved ionic compounds (figure 10.9). Sugar is a covalent compound, and the solute is molecules of sugar moving between the water molecules. Sodium chloride, on the other hand, is an ionic compound and dissolves by the separation of ions, or

$$Na^+Cl^-(s) \rightarrow Na^+(aq) + Cl^-(aq)$$

This equation tells you that 1 mole of NaCl separates into 1 mole of sodium ions and 1 mole of chlorine ions for a total of 2 moles of solute. The boiling point elevation of a solution made from 1 mole of NaCl (58.5 g) is therefore multiplied by two, or $2 \times 0.521°C = 1.04°C$. The boiling point of a solution made by adding 58.5 g of NaCl to 1,000 g of water is therefore 101.04°C at normal sea-level pressure.

Now back to the question of how much a pinch of salt increases the boiling point of a pot of water. Assuming the pot contains about a liter of water (about a quart) and assuming that a "pinch" of salt has a mass of about 0.2 gram, the boiling point will be increased by 0.0037°C. Thus, there must be some reason other than increasing the boiling point that a cook adds a pinch of salt to a pot of boiling water. Perhaps the salt is for seasoning?

Freezing Point

Freezing occurs when the kinetic energy of molecules has been reduced sufficiently so the molecules can come together, forming the crystal structure of the solid. Reduced kinetic energy of the molecules, that is, reduced temperature, results in a specific freezing point for each pure liquid. The *normal* freezing point for pure water, for example, is 0°C (32°F) under normal pressure. The presence of solute particles in a solution interferes with the water molecules as they attempt to form the six-sided hexagonal structure. The water molecules cannot get by the solute particles until the kinetic energy of the solute particles is reduced, that is, until the temperature is below the normal freezing point. Thus, the presence of solute particles lowers the freezing point, and solutions freeze at a lower temperature than the pure solvent.

The freezing-point depression of a solution has a number of interesting implications for solutions such as seawater. When seawater freezes, the water molecules must work their way around the salt particles as was described earlier. Thus, the solute particles are *not* normally included in the hexagonal structure of ice. Ice formed in seawater is practically pure water. Since the solute was *excluded* when the ice formed, the freezing of seawater increases the salinity. Increased salinity means increased concentration, so the freezing point of seawater is further depressed and more ice forms only at a lower temperature. When this additional ice forms, more pure water is removed and the process goes on. Thus, seawater does not have a fixed freezing point but has a lower and lower freezing point as more and more ice freezes.

The depression of the freezing point by a solute has a number of interesting applications in colder climates. Salt, for example, is spread on icy roads to lower the freezing point (and thus the melting point) of the ice. Calcium chloride, $CaCl_2$, is a salt that is often used for this purpose. Water in a car radiator would also freeze in colder climates if a solute, called antifreeze, were not added to the radiator water. Methyl alcohol has been used as an antifreeze because it is soluble in water and does not damage the cooling system. Methyl alcohol, however, has a low boiling point and tends to boil away. Ethylene glycol has a higher boiling point, so it is called a "permanent" antifreeze.

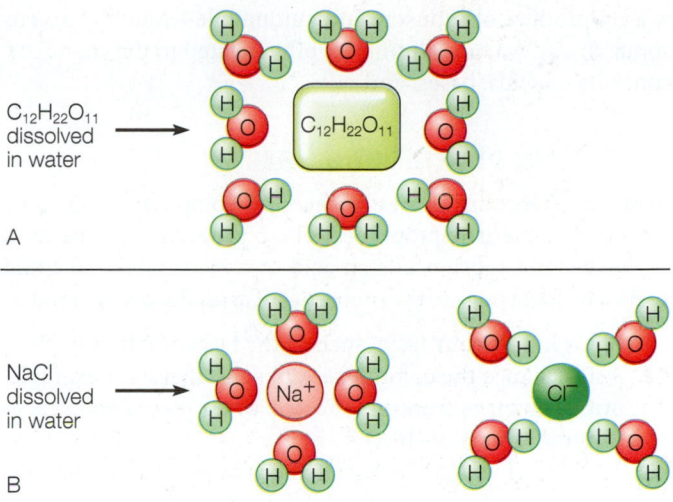

A

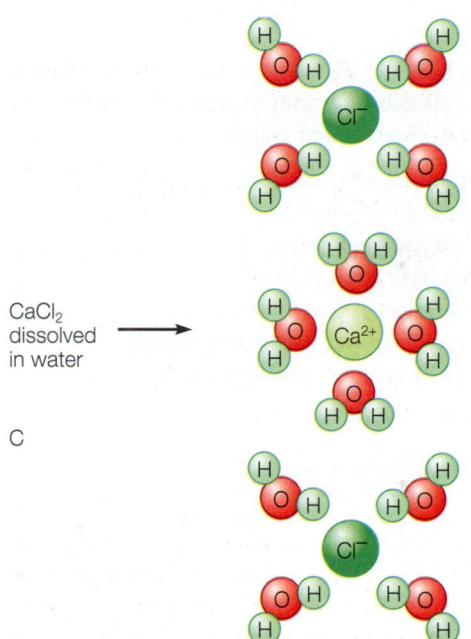

C

FIGURE 10.9 Since ionic compounds dissolve by the separation of ions, they provide more particles in solution than molecular compounds. (*A*) A mole of sugar provides Avogadro's number of particles. (*B*) A mole of NaCl provides two times Avogadro's number of particles. (*C*) A mole of $CaCl_2$ provides three times Avogadro's number of particles.

Like other solutes, ethylene glycol also raises the boiling point, which is an added benefit for summer driving.

10.4   ACIDS, BASES, AND SALTS

The electrolytes known as *acids, bases,* and *salts* are evident in environmental quality, foods, and everyday living. Environmental quality includes the hardness of water, which is determined by the presence of certain salts, the acidity of soils, which determines how well plants grow, and acid rain, which is a by-product of industry and automobiles. Many concerns about air and water pollution are often related to the chemistry concepts of acids, bases, and salts.

Properties of Acids and Bases

Acids and bases are classes of chemical compounds that have certain characteristic properties. These properties can be used to identify if a substance is an acid or a base (tables 10.2 and 10.3). The following are the properties of *acids* dissolved in water:

1. Acids have a sour taste, such as the taste of citrus fruits.
2. Acids change the color of certain substances; for example, litmus changes from blue to red when placed in an acid solution (figure 10.10A).
3. Acids react with active metals, such as magnesium or zinc, releasing hydrogen gas.
4. Acids *neutralize* bases, forming water and salts from the reaction.

Likewise, *bases* have their own characteristic properties. Bases are also called alkaline substances, and the following are the properties of bases dissolved in water:

1. Bases have a bitter taste, for example, the taste of caffeine.
2. Bases reverse the color changes that were caused by acids. Red litmus is changed back to blue when placed in a solution containing a base (figure 10.10B).
3. Basic solutions feel slippery on the skin. They have a *caustic* action on plant and animal tissue, converting tissue into soluble materials. A strong base, for example, reacts with

fat to make soap and glycerine. This accounts for the slippery feeling on the skin.
4. Bases *neutralize* acids, forming water and salts from the reaction.

Tasting an acid or base to see if it is sour or bitter can be hazardous, since some are highly corrosive or caustic. Many organic acids are not as corrosive and occur naturally in foods. Citrus fruit, for example, contains citric acid, vinegar is a solution of acetic acid, and sour milk contains lactic acid. The stings or bites of some insects (bees, wasps, and ants) and some plants (stinging nettles) are painful because an organic acid, formic acid, is injected by the insect or plant. Your stomach contains a solution of hydrochloric acid. In terms of relative strength, the hydrochloric acid in your stomach is about ten times stronger than the carbonic acid (H_2CO_3) of carbonated beverages.

Examples of bases include solutions of sodium hydroxide (NaOH), which has a common name of lye or caustic soda, and potassium hydroxide (KOH), which has a common name of caustic potash. These two bases are used in products known as drain cleaners. They open plugged drains because of their caustic action, turning grease, hair, and other organic "plugs" into soap and other soluble substances that are washed away. A weaker base is a solution of ammonia (NH_3), which is often used as a household cleaner. A solution of magnesium hydroxide, $Mg(OH)_2$, has a common name of milk of magnesia and is sold as an antacid and laxative.

Many natural substances change color when mixed with acids or bases. You may have noticed that tea changes color slightly, becoming lighter, when lemon juice (which contains citric acid) is added. Some plants have flowers of one color when grown in acidic soil and flowers of another color when grown in basic soil. A vegetable dye that changes color in the presence of acids or bases can be used as an *acid-base indicator*. An indicator is simply a vegetable dye that is used to distinguish between acid and base solutions by a color change. Litmus, for example, is an acid-base indicator made from a dye extracted from certain species of lichens. The dye is applied to paper strips, which turn red in acidic solutions and blue in basic solutions.

TABLE 10.2   Some Common Acids

Name	Formula	Comment
Acetic acid	CH_3COOH	A weak acid found in vinegar
Boric acid	H_3BO_3	A weak acid used in eyedrops
Carbonic acid	H_2CO_3	The weak acid of carbonated beverages
Formic acid	HCOOH	Makes the sting of insects and certain plants
Hydrochloric acid	HCl	Also called muriatic acid; used in swimming pools, soil acidifiers, and stain removers
Lactic acid	$CH_3CHOHCOOH$	Found in sour milk, sauerkraut, and pickles; gives tart taste to yogurt
Nitric acid	HNO_3	A strong acid
Phosphoric acid	H_3PO_4	Used in cleaning solutions; added to carbonated beverages for tartness
Sulfuric acid	H_2SO_4	Also called oil of vitriol; used as battery acid and in swimming pools

TABLE 10.3   Some Common Bases

Name	Formula	Comment
Sodium hydroxide	NaOH	Also called lye or caustic soda; a strong base used in oven cleaners and drain cleaners
Potassium hydroxide	KOH	Also called caustic potash; a strong base used in drain cleaners
Ammonia	NH_3	A weak base used in household cleaning solutions
Calcium hydroxide	$Ca(OH)_2$	Also called slaked lime; used to make brick mortar
Magnesium hydroxide	$Mg(OH)_2$	Solution is called milk of magnesia; used as antacid and laxative

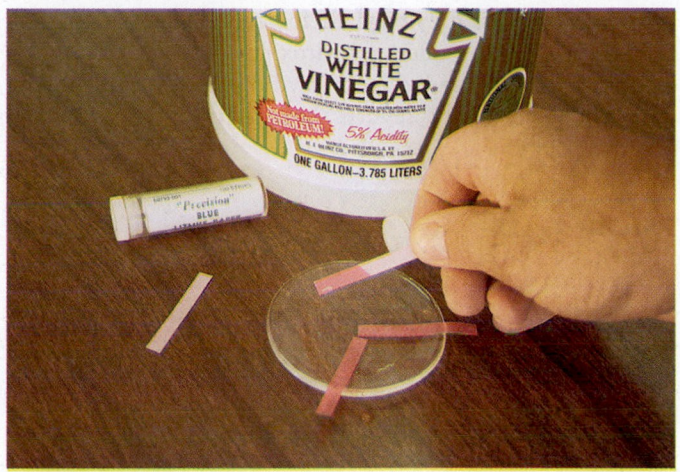

A

B

FIGURE 10.10 (A) Acid solutions will change the color of blue litmus to red. (B) Solutions of bases will change the color of red litmus to blue.

Explaining Acid-Base Properties

Comparing the lists in tables 10.2 and 10.3, you can see that acids and bases appear to be chemical opposites. Notice in table 10.2 that the acids all have an H, or hydrogen atom, in their formulas. In table 10.3, most of the bases have a hydroxide ion, OH^-, in their formulas. Could this be the key to acid-base properties?

The modern concept of an acid considers the properties of acids in terms of the hydronium ion, H_3O^+. As was mentioned earlier, the hydronium ion is a water molecule to which an H^+ ion is attached. Since a hydrogen ion is a hydrogen atom without its single electron, it could be considered as an ion consisting of a single proton. Thus, the H^+ ion can be called a *proton*. An **acid** is defined as any substance that is a *proton donor* when dissolved in water, increasing the hydronium ion concentration. For example, hydrogen chloride dissolved in water has the following reaction:

$$\text{(H)}Cl(aq) + H_2O(l) \longrightarrow H_3O^+(aq) + Cl^-(aq)$$

Cabbage Indicator

To see how acids and bases change the color of certain vegetable dyes, consider the dye that gives red cabbage its color. Shred several leaves of red cabbage and boil them in a pan of water to extract the dye. After you have a purple solution, squeeze the juice from the cabbage into the pan and allow the solution to cool. Add vinegar in small amounts as you stir the solution, continuing until the color changes. Add ammonia in small amounts, stirring until the color changes again. Reverse the color change again by adding vinegar in small amounts. Will this purple cabbage acid-base indicator tell you if other substances are acids or bases?

The dotted circle and arrow were added to show that the hydrogen chloride donated a proton to a water molecule. The resulting solution contains H_3O^+ ions and has acid properties, so the solution is called hydrochloric acid. It is the H_3O^+ ion that is responsible for the properties of an acid.

The bases listed in table 10.3 all appear to have a hydroxide ion, OH^-. Water solutions of these bases do contain OH^- ions, but the definition of a base is much broader. A **base** is defined as any substance that is a *proton acceptor* when dissolved in water, increasing the hydroxide ion concentration. For example, ammonia dissolved in water has the following reaction:

$$NH_3(g) + \text{(H}_2\text{)}O(l) \longrightarrow (NH_4)^+ + OH^-$$

The dotted circle and arrow show that the ammonia molecule accepted a proton from a water molecule, providing a hydroxide ion. The resulting solution contains OH^- ions and has basic properties, so a solution of ammonium hydroxide is a base.

Carbonates, such as sodium carbonate (Na_2CO_3), form basic solutions because the carbonate ion reacts with water to produce hydroxide ions.

$$(CO_3)^{2-}(aq) + H_2O(l) \rightarrow (HCO_3)^-(aq) + OH^-(aq)$$

Thus, sodium carbonate produces a basic solution.

Acids could be thought of as simply solutions of hydronium ions in water, and bases could be considered solutions of hydroxide ions in water. The proton donor and proton acceptor definition is much broader, and it does include the definition of acids and bases as hydronium and hydroxide compounds. The broader, more general definition covers a wider variety of reactions and is therefore more useful.

The modern concept of acids and bases explains why the properties of acids and bases are **neutralized,** or lost, when acids and bases are mixed together. For example, consider the

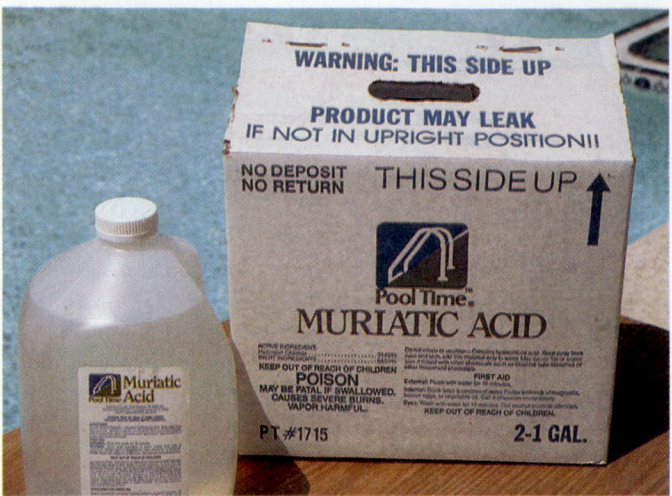

FIGURE 10.11 Hydrochloric acid (HCl) has the common name of *muriatic* acid. Hydrochloric acid is a strong acid used in swimming pools, soil acidifiers, and stain removers.

hydronium ion produced in the hydrochloric acid solution and the hydroxide ion produced in the ammonia solution. When these solutions are mixed together, the hydronium ion reacts with the hydroxide ion, and

$$H_3O^+(aq) + OH^+(aq) \rightarrow H_2O(l) + H_2O(l)$$

Thus, a proton is transferred from the hydronium ion (an acid), and the proton is accepted by the hydroxide ion (a base). Water is produced, and both the acid and base properties disappear or are neutralized.

Strong and Weak Acids and Bases

Acids and bases are classified according to their degree of ionization when placed in water. *Strong acids* ionize completely in water, with all molecules dissociating into ions. Nitric acid, for example, reacts completely in the following equation:

$$HNO_3(aq) + H_2O(l) \rightarrow H_3O^+(aq) + (NO_3)^-(aq)$$

Nitric acid, hydrochloric acid (figure 10.11), and sulfuric acid are common strong acids.

Acids that ionize only partially produce fewer hydronium ions, so they are *weak acids*. Vinegar, for example, contains acetic acid that reacts with water in the following reaction:

$$HC_2H_3O_2 + H_2O \rightarrow H_3O^+ + (C_2H_3O_2)^-$$

Only about 1 percent or less of the acetic acid molecules ionize, depending on the concentration.

Bases are also classified as strong or weak. A *strong base* is completely ionized in solution. Sodium hydroxide, or lye, is the most common example of a strong base. It dissolves in water to form a solution of sodium and hydroxide ions:

$$Na^+OH^-(s) \rightarrow Na^+(aq) + OH^-(aq)$$

A *weak base* is only partially ionized. Ammonia, magnesium hydroxide, and calcium hydroxide are examples of weak bases. Magnesium and calcium hydroxide are only slightly soluble in water, and this reduces the *concentration* of hydroxide ions in a solution.

The pH Scale

The strength of an acid or a base is usually expressed in terms of a range of values called a **pH scale.** The pH scale is based on the concentration of the hydronium ion (in moles/L) in an acidic or a basic solution. To understand how the scale is able to express both acid and base strength in terms of the hydronium ion, first note that pure water is very slightly ionized in the reaction:

$$H_2O(l) + H_2O(l) \rightarrow H_3O^+(aq)\ OH^-(aq)$$

The amount of self-ionization by water has been determined through measurements. In pure water at 25°C or any neutral water solution at that temperature, the H_3O^+ concentration is 1×10^{-7} moles/L, and the OH^- concentration is also 1×10^{-7} moles/L. Since both ions are produced in equal numbers, then the H_3O^+ concentration equals the OH^- concentration, and pure water is neutral, neither acidic nor basic.

In general, adding an acid substance to pure water increases the H_3O^+ concentration. Adding a base substance to pure water increases the OH^- concentration. Adding a base also *reduces* the H_3O^+ concentration as the additional OH^- ions are able to combine with more of the hydronium ions to produce unionized water. Thus, at a given temperature, an increase in OH^- concentration is matched by a *decrease* in H_3O^+ concentration. The concentration of the hydronium ion can be used as a measure of acidic, neutral, and basic solutions. In general, (1) acidic solutions have H_3O^+ concentrations above 1×10^{-7} moles/L, (2) neutral solutions have H_3O^+ concentrations equal to 1×10^{-7} moles/L, and (3) basic solutions have H_3O^+ concentrations less than 1×10^{-7} moles/L. These three statements lead directly to the pH scale, which is named from the French *pouvoir hydrogene*, meaning "hydrogen power." Power refers to the exponent of the hydronium ion concentration, and the pH is a *power of ten notation that expresses the H_3O^+ concentration.*

A neutral solution has a pH of 7.0. Acidic solutions have pH values below 7, and smaller numbers mean greater acidic properties. Increasing the OH^- concentration decreases the

CONCEPTS APPLIED

Acid or Base?

Pick some household product that probably has an acid or base character (Example: pH increaser for aquariums). On a separate paper, write the listed ingredients and identify any you believe would be distinctly acidic or basic in a water solution. Tell whether you expect the product to be an acid or a base. Describe your findings of a litmus paper test.

H_3O^+ Concentration (moles/liters)	pH	Meaning
1×10^{-0} (=1)	0	
1×10^{-1}	1	
1×10^{-2}	2	Increasing acidity
1×10^{-3}	3	
1×10^{-4}	4	
1×10^{-5}	5	
1×10^{-6}	6	
1×10^{-7}	7	Neutral
1×10^{-8}	8	
1×10^{-9}	9	
1×10^{-10}	10	Increasing basicity
1×10^{-11}	11	
1×10^{-12}	12	
1×10^{-13}	13	
1×10^{-14}	14	

FIGURE 10.12 The pH scale.

TABLE 10.4 The Approximate pH of Some Common Substances

Substance	pH (or pH Range)
Hydrochloric acid (4%)	0
Gastric (stomach) solution	1.6–1.8
Lemon juice	2.2–2.4
Vinegar	2.4–3.4
Carbonated soft drinks	2.0–4.0
Grapefruit	3.0–3.2
Oranges	3.2–3.6
Acid rain	4.0–5.5
Tomatoes	4.2–4.4
Potatoes	5.7–5.8
Natural rainwater	5.6–6.2
Milk	6.3–6.7
Pure water	7.0
Seawater	7.0–8.3
Blood	7.4
Sodium bicarbonate solution	8.4
Milk of magnesia	10.5
Ammonia cleaning solution	11.9
Sodium hydroxide solution	13.0

H_3O^+ concentration, so the strength of a base is indicated on the same scale with values greater than 7. Note that the pH scale is logarithmic, so a pH of 2 is ten times as acidic as a pH of 3. Likewise, a pH of 10 is one hundred times as basic as a pH of 8. Figure 10.12 is a diagram of the pH scale, and table 10.4 compares the pH of some common substances (figure 10.13).

Properties of Salts

Salt is produced by a neutralization reaction between an acid and a base. A **salt** is defined as any ionic compound except those with hydroxide or oxide ions. Table salt, NaCl, is but one example of this large group of ionic compounds. As an example of a salt produced by a neutralization reaction, consider the reaction of HCl (an acid in solution) with $Ca(OH)_2$ (a base in solution). The reaction is

$$2\,HCl(aq) + Ca(OH)_2(aq) \rightarrow CaCl_2(aq) + 2\,H_2O(l)$$

This is an ionic exchange reaction that forms molecular water, leaving Ca^{2+} and Cl^- in solution. As the water is evaporated, these ions begin forming ionic crystal structures as the solution concentration increases. When the water is all evaporated, the white crystalline salt of $CaCl_2$ remains.

If sodium hydroxide had been used as the base instead of calcium hydroxide, a different salt would have been produced:

$$HCl(aq) + NaOH(aq) \rightarrow NaCl(aq) + H_2O(l)$$

Salts are also produced when elements combine directly, when an acid reacts with a metal, and by other reactions.

Salts are essential in the diet both as electrolytes and as a source of certain elements, usually called *minerals* in this context. Plants must have certain elements that are derived from water-soluble salts. Potassium, nitrates, and phosphate salts are often used to supply the needed elements. There is no

FIGURE 10.13 The pH increases as the acidic strength of these substances decreases from left to right. Did you know that lemon juice is more acidic than vinegar? That a soft drink is more acidic than orange juice or grapefruit juice?

scientific evidence that plants prefer to obtain these elements from natural sources, as compost, or from chemical fertilizers. After all, a nitrate ion is a nitrate ion, no matter what its source. Table 10.5 lists some common salts and their uses.

Hard and Soft Water

Salts vary in their solubility in water, and a solubility chart appears in appendix B. Table 10.6 lists some generalizations

TABLE 10.5 Some Common Salts and Their Uses

Common Name	Formula	Use
Alum	$KAl(SO_4)_2$	Medicine, canning, baking powder
Baking soda	$NaHCO_3$	Fire extinguisher, antacid, deodorizer, baking powder
Bleaching powder (chlorine tablets)	$CaOCl_2$	Bleaching, deodorizer, disinfectant in swimming pools
Borax	$Na_2B_4O_7$	Water softener
Chalk	$CaCO_3$	Antacid tablets, scouring powder
Chile saltpeter	$NaNO_3$	Fertilizer
Cobalt chloride	$CoCl_2$	Hygrometer (pink in damp weather, blue in dry weather)
Epsom salt	$MgSO_4 \cdot 7\,H_2O$	Laxative
Fluorspar	CaF_2	Metallurgy flux
Gypsum	$CaSO_4 \cdot 2\,H_2O$	Plaster of Paris, soil conditioner
Lunar caustic	$AgNO_3$	Germicide and cauterizing agent
Niter (or saltpeter)	KNO_3	Meat preservative, makes black gunpowder (75 parts KNO_3, 15 of carbon, 10 of sulfur)
Potash	K_2CO_3	Makes soap, glass
Rochelle salt	$KNaC_4H_4O_6$	Baking powder ingredient
TSP	Na_3PO_4	Water softener, fertilizer

TABLE 10.6 Generalizations About Salt Solubilities

Salts	Solubility	Exceptions
Sodium Potassium Ammonium	Soluble	None
Nitrate Acetate Chlorate	Soluble	None
Chlorides	Soluble	Ag and Hg(I) are insoluble
Sulfates	Soluble	Ba, Sr, and Pb are insoluble
Carbonates Phosphates Silicates	Insoluble	Na, K, and NH_4 are soluble
Sulfides	Insoluble	Na, K, and NH_4 are soluble: Mg, Ca, Sr, and Ba decompose

concerning the various common salts. Some of the salts are dissolved by water that will eventually be used for domestic supply. When the salts are soluble calcium or magnesium compounds, the water will contain calcium or magnesium ions in solution. A solution of Ca^{2+} or Mg^{2+} ions is said to be *hard water* because it is hard to make soap lather in the water. "Soft" water, on the other hand, makes a soap lather easily. The difficulty occurs because soap is a sodium or potassium compound that is soluble in water. The calcium or magnesium ions, when present, replace the sodium or potassium ions in the soap compound, forming an insoluble compound. It is this insoluble compound that forms a "bathtub ring" and also collects on clothes being washed, preventing cleansing.

The key to "softening" hard water is to remove the troublesome calcium and magnesium ions. If the hardness is caused by magnesium or calcium *bicarbonates*, the removal is accomplished by simply heating the water. Upon heating, they decompose, forming an insoluble compound that effectively removes the ions from solution. The decomposition reaction for calcium bicarbonate is

$$Ca^{2+}(HCO_3)_2^-(aq) \rightarrow CaCO_3(s) + H_2O(l) + CO_2\uparrow$$

The reaction is the same for magnesium bicarbonate. As the solubility chart in appendix B shows, magnesium and calcium

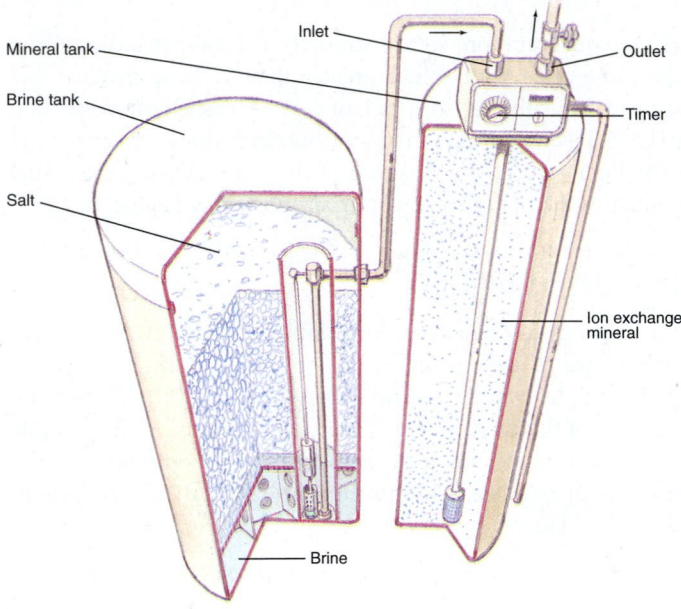

FIGURE 10.14 A water softener exchanges sodium ions for the calcium and magnesium ions of hard water. Thus, the water is now soft, but it contains the same number of ions as before.

carbonates are insoluble, so the ions are removed from solution in the solid that is formed. Perhaps you have noticed such a white compound forming around faucets if you live where bicarbonates are a problem. Commercial products to remove such deposits usually contain an acid, which reacts with the carbonate to make a new, soluble salt that can be washed away.

Water hardness is also caused by magnesium or calcium *sulfate*, which requires a different removal method. Certain chemicals such as sodium carbonate (washing soda), trisodium phosphate (TSP), and borax will react with the troublesome ions,

Acid rain is a general term used to describe any acidic substances, wet or dry, that fall from the atmosphere. Wet acidic deposition could be in the form of rain, but snow, sleet, and fog could also be involved. Dry acidic deposition could include gases, dust, or any solid particles that settle out of the atmosphere to produce an acid condition.

Pure, unpolluted rain is naturally acidic. Carbon dioxide in the atmosphere is absorbed by rainfall, forming carbonic acid (H_2CO_3). Carbonic acid lowers the pH of pure rainfall to a range of 5.6 to 6.2. Decaying vegetation in local areas can provide more CO_2, making the pH even lower. A pH range of 4.5 to 5.0, for example, has been measured in remote areas of the Amazon jungle. Human-produced exhaust emissions of sulfur and nitrogen oxides can lower the pH of rainfall even more, to a 4.0 to 5.5 range. This is the pH range of acid rain.

The sulfur and nitrogen oxides that produce acid rain come from exhaust emissions of industries and electric utilities that burn coal and from the exhaust of cars, trucks, and buses (box figure 10.1). The emissions are sometimes called "SO_x" and "NO_x" which is read "socks" and "knox." The x subscript implies the variable presence of any or all of the oxides, for example, nitrogen monoxide (NO), nitrogen dioxide (NO_2), and dinitrogen tetroxide (N_2O_4) for NO_x.

SO_x and NO_x are the raw materials of acid rain and are not themselves acidic. They react with other atmospheric chemicals to form sulfates and nitrates, which combine with water vapor to form sulfuric acid (H_2SO_4) and nitric acid (HNO_3). These are the chemicals of concern in acid rain.

Many variables influence how much and how far SO_x and NO_x are carried in the atmosphere and if they are converted to acid rain or simply return to the surface as a dry gas or particles. During the 1960s and 1970s, concerns about local levels of pollution led to the replacement of short smokestacks of about 60 m (about 200 ft) with taller smokestacks of about 200 m (about 650 ft). This reduced the local levels of pollution by dumping the exhaust higher in the atmosphere where winds could carry it away. It also set the stage for longer-range transport of SO_x and NO_x and their eventual conversion into acids.

There are two main reaction pathways by which SO_x and NO_x are converted to acids: (1) reactions in the gas phase and (2) reactions in the liquid phase, such as in water droplets in clouds and fog. In the gas phase, SO_x and NO_x are oxidized to acids, mainly by hydroxyl ions and ozone, and the acid is absorbed by cloud droplets and precipitated as rain or snow. Most of the nitric acid in acid rain and about one-fourth of the sulfuric acid is formed in gas-phase reactions. Most of the liquid-phase reactions that produce sulfuric acid involve the absorbed SO_x and hydrogen peroxide (H_2O_2), ozone, oxygen, and particles of carbon, iron oxide, and manganese oxide particles. These particles also come from the exhaust of fossil fuel combustion.

Acid rain falls on land, bodies of water, forests, crops, buildings, and people. Concerns about acid rain center on its environmental impact on lakes, forests, crops, materials, and human health. Lakes in different parts of the world, for example, have been increasing in acidity over the past fifty years. Lakes in northern New England, the Adirondacks, and parts of Canada now have a pH of less than 5.0, and correlations have been established between lake acidity and decreased fish populations. Trees, mostly conifers, are dying at unusually rapid rates in the northeastern United States. Red spruce in Vermont's Green Mountains and the mountains of New York and New Hampshire have been affected by acid rain, as have pines in New Jersey's Pine Barrens. It is believed

Box Figure 10.1 Natural rainwater has a pH of 5.6 to 6.2. Exhaust emissions of sulfur and nitrogen oxides can lower the pH of rainfall to a range of 4.0 to 5.5. The exhaust emissions come from industries, electric utilities, and automobiles. Not all emissions are as visible as those pictured in this illustration.

that acid rain leaches essential nutrients, such as calcium, from the soil and mobilizes aluminum ions. The aluminum ions disrupt the water equilibrium of fine root hairs, and when the root hairs die, so do the trees.

Human-produced emissions of sulfur and nitrogen oxides from burning fossil fuels are the cause of acid rain. The heavily industrialized northeastern part of the United States, from the Midwest through New England, releases sulfur and nitrogen emissions that result in a precipitation pH of 4.0 to 4.5. This region is the geographic center of the nation's acid rain problem. The solution to the problem is found in (1) using fuels other than fossil fuels and (2) reducing the thousands of tons of SO_x and NO_x that are dumped into the atmosphere per day when fossil fuels are used.

---

forming an insoluble solid that removes them from solution. For example, washing soda and calcium sulfate react as follows:

$$Na_2CO_3(aq) + CaSO_4(aq) \rightarrow Na_2SO_4(aq) + CaCO_3\downarrow$$

Calcium carbonate is insoluble; thus, the calcium ions are removed from solution before they can react with the soap.

Many laundry detergents have Na_2CO_3, TSP, or borax ($Na_2B_4O_7$) added to soften the water. TSP causes other problems, however, as the additional phosphates in the waste water can act as a fertilizer, stimulating the growth of algae to such an extent that other organisms in the water die.

A water softener unit is an ion exchanger (figure 10.14). The unit contains a mineral that exchanges sodium ions for calcium and

People Behind the Science

Johannes Nicolaus Brönsted (1879–1947)

Johannes Brönsted was a Danish physical chemist whose work in solution chemistry, particularly electrolytes, resulted in a new theory of acids and bases.

Brönsted was born on February 22, 1879, in Varde, Jutland, the son of a civil engineer. He was educated at local schools before going to study chemical engineering at the Technical Institute of the University of Copenhagen in 1897. He graduated two years later and then turned to chemistry, in which he qualified in 1902. After a short time in industry, he was appointed an assistant in the university's chemical laboratory in 1905, becoming professor of physical and inorganic chemistry in 1908.

Brönsted's early work was wide-ranging, particularly in the fields of electrochemistry, the measurement of hydrogen ion concentrations, amphoteric electrolytes, and the behavior of indicators. He discovered a method of eliminating potentials in the measurement of hydrogen ion concentrations and devised a simple equation that connects the activity and osmotic coefficients of an electrolyte, as well as another that relates activity coefficients to reaction velocities. From the absorption spectra of chromium(III) salts, he concluded that strong electrolytes are completely dissociated and that the changes of molecular conductivity and freezing point that accompany changes in concentration are caused by the electrical forces between ions in solution.

In 1887, Svante Arrhenius had proposed a theory of acidity that explained its nature on an atomic level. He defined an acid as a compound that could generate hydrogen ions in aqueous solution and an alkali as a compound that could generate hydroxyl ions. A strong acid is completely ionized (dissociated) and produces many hydrogen ions, whereas a weak acid is only partly dissociated and produces few hydrogen ions. Conductivity measurements confirm the theory, as long as the solutions are not too concentrated.

In 1923, Brönsted published (simultaneously with Thomas Lowry in Britain) a new theory of acidity, which has certain important advantages over that of Arrhenius. Brönsted defined an acid as a proton donor and a base as a proton acceptor. The definition applies to all solvents, not just water. It also explains the different behavior of pure acids in solution. Pure liquid sulfuric acid or acetic acid does not change the color of indicators nor does it react with carbonates or metals. But as soon as water is added, all of these reactions occur.

Source: Modified from the *Hutchinson Dictionary of Scientific Biography.* © RM, 2011. All rights reserved. Helicon Publishing is a division of RM.

magnesium ions as water is run through it. The softener is regenerated periodically by flushing with a concentrated sodium chloride solution (brine). The sodium ions replace the calcium and magnesium ions, which are carried away in the rinse water. The softener is then ready for use again. The frequency of renewal cycles depends on the water hardness, and each cycle can consume from 4 to 20 pounds of sodium chloride per renewal cycle. In general, water with less than 75 ppm calcium and magnesium ions is called soft water; with greater concentrations, it is called hard water. The greater the concentration above 75 ppm, the harder the water.

SUMMARY

A water molecule consists of two hydrogen atoms and an oxygen atom with bonding and electron pairs in a tetrahedral arrangement. Electrons spend more time around the oxygen, producing a *polar molecule,* with centers of negative and positive charge. Polar water molecules interact. The force of attraction is called a *hydrogen bond.* The hydrogen bond accounts for the decreased density of ice, the high heat of fusion, and the high heat of vaporization of water. The hydrogen bond is also involved in the *dissolving* process.

A *solution* is a homogeneous mixture of ions or molecules of two or more substances. The substance present in the large amount is the *solvent,* and the *solute* is dissolved in the solvent. If one of the components is a liquid, however, it is called the solvent.

Fluids that mix in any proportion are called *miscible fluids,* and *immiscible fluids* do not mix. Polar substances dissolve in polar solvents but not nonpolar solvents, and the general rule is *like dissolves like.* Thus, oil, a nonpolar substance, is immiscible in water, a polar substance.

A limit to dissolving solids in a liquid occurs when the solution is *saturated.* A *saturated solution* is one with equilibrium between solute dissolving and solute coming out of solution. The *solubility* of a solid is the concentration of a saturated solution at a particular temperature.

Water solutions that carry an electric current are called *electrolytes,* and nonconductors are called *nonelectrolytes.* In general, ionic substances make electrolyte solutions, and molecular substances make nonelectrolyte solutions. Polar molecular substances may be *ionized* by polar water molecules, however, making an electrolyte from a molecular solution.

The *boiling point of a solution* is greater than the boiling point of the pure solvent, and the increase depends only on the concentration of the solute (at a constant pressure). For water, the boiling point is increased 0.521°C for each mole of solute in each kg of water. The *freezing point of a solution* is lower than the freezing point of the pure solvent, and the depression also depends on the concentration of the solute.

Acids, bases, and salts are chemicals that form ionic solutions in water, and each can be identified by simple properties. These properties are accounted for by the modern concepts of each. *Acids* are *proton donors* that form *hydronium ions* (H_3O^+) in water solutions. *Bases* are *proton acceptors* that form *hydroxide ions* (OH^-) in water solutions. The strength of an acid or base is measured on the *pH scale*, a power of ten notation of the hydronium ion concentration. On the scale, numbers from 0 up to 7 are acids, 7 is neutral, and numbers above 7 and up to 14 are bases. Each unit represents a tenfold increase or decrease in acid or base properties.

A *salt* is any ionic compound except those with hydroxide or oxide ions. Salts provide plants and animals with essential elements. The solubility of salts varies with the ions that make up the compound. Solutions of magnesium or calcium produce *hard water,* water in which it is hard to make soap lather. Hard water is softened by removing the magnesium and calcium ions.

KEY TERMS

acid (p. **233**)
base (p. **233**)
electrolytes (p. **229**)
hydration (p. **228**)

hydrogen bonding (p. **226**)
hydronium ion (p. **230**)
miscible fluids (p. **227**)
neutralized (p. **233**)

pH scale (p. **234**)
polar molecule (p. **226**)
salt (p. **235**)
saturated solution (p. **229**)

solubility (p. **229**)
solution (p. **227**)

APPLYING THE CONCEPTS

Answers are located in appendix F.

1. Which of the following is *not* a solution?
   a. seawater
   b. carbonated water
   c. sand
   d. brass

2. Atmospheric air is a homogeneous mixture of gases that is mostly nitrogen gas. The nitrogen is therefore (the)
   a. solvent.
   b. solution.
   c. solute.
   d. None of the above is correct.

3. Water has the greatest density at what temperature?
   a. 100°C
   b. 20°C
   c. 4°C
   d. 0°C

4. A solid salt is insoluble in water, so the strongest force must be the
   a. ion-water molecule force.
   b. ion-ion force.
   c. force of hydration.
   d. polar molecule force.

5. The ice that forms in freezing seawater is
   a. pure water.
   b. the same salinity as liquid seawater.
   c. more salty than liquid seawater.
   d. more dense than liquid seawater.

6. Which of the following would have a pH of *less* than 7?
   a. a solution of ammonia
   b. a solution of sodium chloride
   c. pure water
   d. carbonic acid

7. Which of the following would have a pH of *more* than 7?
   a. a solution of ammonia
   b. a solution of sodium chloride
   c. pure water
   d. carbonic acid

8. Substance A has a pH of 2 and substance B has a pH of 3. This means that
   a. substance A has more basic properties than substance B.
   b. substance B has more acidic properties than substance A.
   c. substance A is ten times more acidic than substance B.
   d. substance B is ten times more acidic than substance A.

9. As the temperature of water *decreases,* the solubility of carbon dioxide gas in the water
   a. increases.
   b. decreases.
   c. remains the same.
   d. increases or decreases, depending on the specific temperature.

10. Hard water is a solution of
    a. Na^+ and Cl^- ions.
    b. Na^+ and K^+ ions.
    c. Ca^{2+} and Mg^{2+} ions.
    d. Ba^{2+} and Cl^- ions.

11. The solubility of most ionic salts in water
    a. increases with temperature increase.
    b. decreases with temperature increase.
    c. depends on the amount of salt.
    d. increases with stirring.

12. The heat of fusion, specific heat, and heat of vaporization of water are high compared to similar substances such as hydrogen sulfide, H_2S, because
    a. ionic bonds form in water molecules.
    b. hydrogen bonds form between water molecules.
    c. covalent bonds form between water molecules.
    d. covalent bonds form in water molecules.

QUESTIONS FOR THOUGHT

1. How is a solution different from other mixtures?
2. Explain why some ionic compounds are soluble while others are insoluble in water.
3. Explain why adding salt to water increases the boiling point.
4. A deep lake in Minnesota is covered with ice. What is the water temperature at the bottom of the lake? Explain your reasoning.
5. Explain why water has a greater density at 4°C than at 0°C.
6. What is hard water? How is it softened?
7. According to the definition of an acid and the definition of a base, would the pH increase, decrease, or remain the same when NaCl is added to pure water? Explain.
8. What is a hydrogen bond? Explain how a hydrogen bond forms.
9. What feature of a soap molecule gives it cleaning ability?
10. What ion is responsible for (a) acidic properties? (b) for basic properties?

FOR FURTHER ANALYSIS

1. Analyze the basic reason that water is a universal solvent, becomes less dense when it freezes, has a high heat of fusion, has a high specific heat, and has a high heat of vaporization.
2. What is the same and what is different between a salt that will dissolve in water and one that is insoluble?
3. There are at least three ways to change the boiling point of water, so describe how you know for sure that 100°C (212°F) is the boiling point.
4. What are the significant similarities and differences between an acid, a base, and a salt?
5. Describe how you would teach someone why the pH of an acid is a low number (less than 7), while the pH of a base is a larger number (greater than 7).
6. Describe at least four different examples of how you could make hard water soft.

INVITATION TO INQUIRY

Which Freezes Faster?

Is it true that hot water that has boiled freezes faster than fresh cold water from the tap? Investigate freezing of hot, boiled water and fresh cold water. Make sure you control all the variables in your experimental design. Compare your findings to those of classmates.

11

Nuclear Reactions

With the top half of the steel vessel and control rods removed, fuel rod bundles can be replaced in the water-flooded nuclear reactor.

CORE **CONCEPT**

Nuclear reactions involve changes in the nucleus of the atom.

OUTLINE

Natural radioactivity is the spontaneous emission of particles or energy from a disintegrating nucleus (p. 242).

Nuclear instability results from an imbalance between the attractive nuclear force and the repulsive electromagnetic force (pp. 244–46).

An unstable nucleus becomes more stable by emitting alpha, beta, or gamma radiation from the nucleus (p. 246–49).

The relationship between energy and mass changes is $E = mc^2$ (p. 251–52).

OVERVIEW

The ancient alchemist dreamed of changing one element into another, such as lead into gold. The alchemist was never successful, however, because such changes were attempted with chemical reactions. Chemical reactions are reactions that involve only the electrons of atoms. Electrons are shared or transferred in chemical reactions, and the internal nucleus of the atom is unchanged. Elements thus retain their identity during the sharing or transferring of electrons. This chapter is concerned with a different kind of reaction, one that involves the *nucleus* of the atom. In nuclear reactions, the nucleus of the atom is often altered, changing the identity of the elements involved. The ancient alchemist's dream of changing one element into another was actually a dream of achieving a nuclear change, that is, a nuclear reaction.

Understanding nuclear reactions is important because although fossil fuels are the major source of energy today, there are growing concerns about (1) air pollution from fossil fuel combustion, (2) increasing levels of CO_2 from fossil fuel combustion, which contributes to the warming of Earth (the greenhouse effect), and (3) the dwindling fossil fuel supply, which cannot last forever. Energy experts see nuclear energy as a means of meeting rising energy demands in an environmentally acceptable way. However, the topic of nuclear energy is controversial, and discussions of it often result in strong emotional responses. Decisions about the use of nuclear energy require some understandings about nuclear reactions and some facts about radioactivity and radioactive materials (figure 11.1). These understandings and facts are the topics of this chapter.

11.1 NATURAL RADIOACTIVITY

Natural **radioactivity** is the spontaneous emission of particles or energy from an atomic nucleus as it disintegrates. It was discovered in 1896 by Henri Becquerel, a French scientist who was very interested in the recent discovery of X rays. Becquerel was experimenting with fluorescent minerals, minerals that give off visible light after being exposed to sunlight. He wondered if fluorescent minerals emitted X rays in addition to visible light. From previous work with X rays, Becquerel knew that they would penetrate a wrapped, light-tight photographic plate, exposing it as visible light exposes an unprotected plate. Thus, Becquerel decided to place a fluorescent uranium mineral on a protected photographic plate while the mineral was exposed to sunlight. Sure enough, he found a silhouette of the mineral on the plate when it was developed. Believing the uranium mineral emitted X rays, he continued his studies until the weather turned cloudy. Storing a wrapped, protected photographic plate and the uranium mineral together during the cloudy weather, Becquerel returned to the materials later and developed the photographic plate to again find an image of the mineral (figure 11.2). He concluded that the mineral was emitting an "invisible radiation" that was not induced by sunlight. The emission of invisible radiation was later named

radioactivity. Materials that have the property of radioactivity are called *radioactive* materials.

Becquerel's discovery led to the beginnings of the modern atomic theory and to the discovery of new elements. Ernest Rutherford studied the nature of radioactivity and found that there are three kinds, which are today known by the first three letters of the Greek alphabet—alpha (α), beta (β), and gamma (γ). These Greek letters were used at first before the nature of the radiation was known. Today, an **alpha particle** (sometimes called an alpha ray) is known to be the nucleus of a helium atom, that is, two protons and two neutrons. A **beta particle** (or beta ray) is a high-energy electron. A **gamma ray** is electromagnetic radiation, as is light, but of very short wavelength (figure 11.3).

It was Rutherford's work with alpha particles that resulted in the discovery of the nucleus and the proton (see chapter 8). At Becquerel's suggestion, Marie Curie searched for other radioactive materials and, in the process, discovered two new elements, polonium and radium. More radioactive elements have been discovered since that time, and, in fact, all the isotopes of all the elements with an atomic number greater than 83 (bismuth) are radioactive. As a result of radioactive disintegration, the nucleus of an atom often undergoes a change of identity, becoming a simpler nucleus. The spontaneous

FIGURE 11.1 Decisions about nuclear energy require some understanding of nuclear reactions and the nature of radioactivity. This is one of the three units of the Palo Verde Nuclear Generating Station in Arizona. With all three units running, enough power is generated to meet the electrical needs of nearly 4 million people.

disintegration of a given nucleus is a purely natural process and cannot be controlled or influenced. The natural spontaneous disintegration or decomposition of a nucleus is also called **radioactive decay**. Although it is impossible to know *when* a given nucleus will undergo radioactive decay, as you will see later, it is possible to deal with the *rate* of decay for a given radioactive material with precision.

Nuclear Equations

There are two main subatomic particles in the nucleus, the proton and the neutron. The proton and neutron are called

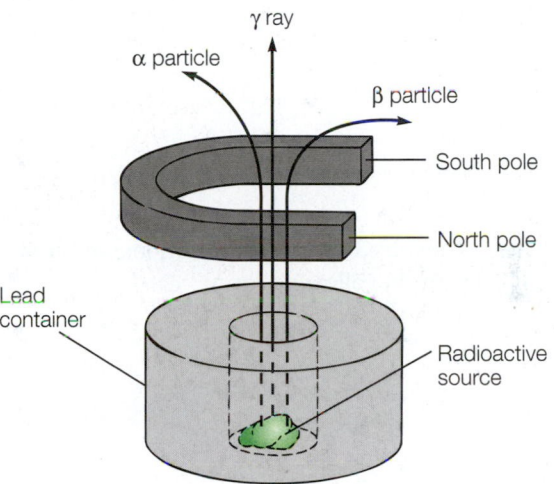

FIGURE 11.3 Radiation passing through a magnetic field shows that massive, positively charged alpha particles are deflected one way and less massive beta particles with their negative charge are greatly deflected in the opposite direction. Gamma rays, like light, are not deflected.

A

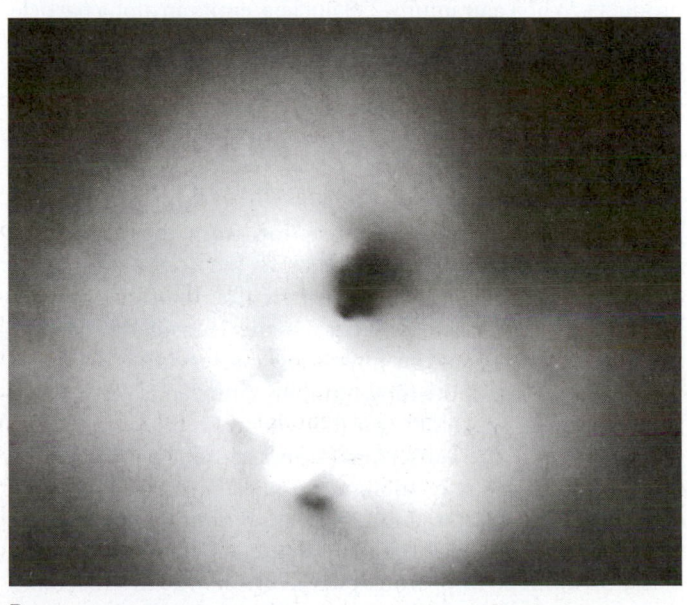

B

FIGURE 11.2 Radioactivity was discovered by Henri Becquerel when he exposed a light-tight photographic plate to a radioactive mineral, then developed the plate. (*A*) A photographic film is exposed to a uraninite ore sample. (*B*) The film, developed normally after a four-day exposure to uraninite. Becquerel found an image like this one and deduced that the mineral gave off invisible radiation.

nucleons. Recall that the number of protons, the *atomic number,* determines what element an atom is and that all atoms of a given element have the same number of protons. The number of neutrons varies in *isotopes,* which are atoms with the same atomic number but different numbers of neutrons. The number of protons and neutrons together determines the *mass number,* so different isotopes of the same element are identified with their mass numbers. Thus, the two most common, naturally occurring isotopes of uranium are referred to as uranium-238 and uranium-235, and the 238 and 235 are the mass numbers of these isotopes. Isotopes are also represented by the following symbol:

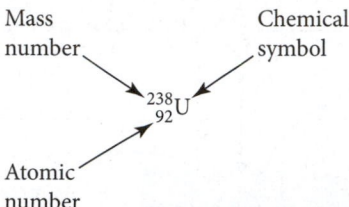

Subatomic particles involved in nuclear reactions are represented by symbols with the following form:

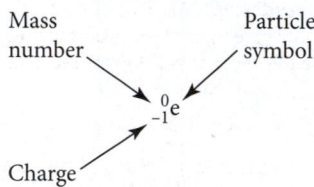

Symbols are used in an equation for a nuclear reaction that is written much like a chemical reaction with reactants and products. When a uranium-238 nucleus emits an alpha particle (^4_2He), for example, it loses two protons and two neutrons. The nuclear reaction is written in equation form as

$$^{238}_{92}\text{U} \rightarrow \, ^{234}_{90}\text{Th} + \, ^4_2\text{He}$$

The *products* of this nuclear reaction from the decay of a uranium-238 nucleus are (1) the alpha particle (^4_2He) given off and (2) the nucleus, which remains after the alpha particle leaves the original nucleus. What remains is easily determined since all nuclear equations must show conservation of charge and conservation of the total number of nucleons. Therefore, (1) the number of protons (positive charge) remains the same, and the sum of the subscripts (atomic number, or numbers of protons) in the reactants must equal the sum of the subscripts in the products; and (2) the total number of nucleons remains the same, and the sum of the superscripts (atomic mass, or number of protons plus neutrons) in the reactants must equal the sum of the superscripts in the products. The new nucleus remaining after the emission of an alpha particle, therefore, has an atomic number of 90 ($92 - 2 = 90$). According to the table of atomic numbers on the inside back

TABLE 11.1 Names, Symbols, and Properties of Subatomic Particles in Nuclear Equations

Name	Symbol	Mass Number	Charge
Proton	^1_1H (or ^1_1p)	1	1+
Electron	$^{\,0}_{-1}\text{e}$ (or $^{\,0}_{-1}\beta$)	0	1−
Neutron	^1_0n	1	0
Gamma photon	$^0_0\gamma$	0	0

cover of this text, this new nucleus is thorium (Th). The mass of the thorium isotope is 238 minus 4, or 234. The emission of an alpha particle thus decreases the number of protons by 2 and the mass number by 4. From the subscripts, you can see that the total charge is conserved ($92 = 90 + 2$). From the superscripts, you can see that the total number of nucleons is also conserved ($238 = 234 + 4$). The mass numbers (superscripts) and the atomic numbers (subscripts) are *balanced* in a correctly written nuclear equation. Such nuclear equations are considered to be independent of any chemical form or chemical reaction. Nuclear reactions are independent and separate from chemical reactions, whether or not the atom is in the pure element or in a compound. Each particle that is involved in nuclear reactions has its own symbol with a superscript indicating mass number and a subscript indicating the charge. These symbols, names, and numbers are given in table 11.1.

The Nature of the Nucleus

The modern atomic theory does not picture the nucleus as a group of stationary protons and neutrons clumped together by some "nuclear glue." The protons and neutrons are understood to be held together by a *nuclear force,* a strong fundamental force of attraction that is functional only at very short distances, on the order of 10^{-15} m or less. At distances greater than about 10^{-15} m, the nuclear force is negligible, and the weaker *electromagnetic force,* the force of repulsion between like charges, is the operational force. Thus, like-charged protons experience a repulsive force when they are farther apart than about 10^{-15} m. When closer together than 10^{-15} m, the short-range, stronger nuclear force predominates, and the protons experience a strong attractive force. This explains why the like-charged protons of the nucleus are not repelled by their like electric charges.

Observations of radioactive decay reactants and products and experiments with nuclear stability have led to a *shell model of the nucleus.* This model considers the protons and neutrons moving in energy levels in the nucleus analogous to the orbital structure of electrons in the outermost part of the atom. As in the electron orbitals, there are certain configurations of nuclear shells that have a greater stability than others. Considering electrons, filled and half-filled orbitals are more stable than other arrangements, and maximum stability occurs with the noble gases and their 2, 10, 18, 36, 54, and 86 electrons. Considering the nucleus, atoms with 2, 8, 20, 28, 50, 82, or 126 protons or

Plutonium Decay Equation

A plutonium-242 nucleus undergoes radioactive decay, emitting an alpha particle. Write the nuclear equation for this nuclear reaction.

Step 1: The table of atomic weights on the inside back cover gives the atomic number of plutonium as 94. Plutonium-242 therefore has a symbol of $^{242}_{94}Pu$. The symbol for an alpha particle is (^4_2He), so the nuclear equation so far is

$$^{242}_{94}Pu \rightarrow {}^4_2He + ?$$

Step 2: From the subscripts, you can see that $94 = 2 + 92$, so the new nucleus has an atomic number of 92. The table of atomic weights identifies element 92 as uranium with a symbol of U.

Step 3: From the superscripts, you can see that the mass number of the uranium isotope formed is $242 - 4 = 238$, so the product nucleus is $^{238}_{92}U$ and the complete nuclear equation is

$$^{242}_{94}Pu \rightarrow {}^4_2He + {}^{238}_{92}U$$

Step 4: Checking the subscripts ($94 = 2 + 92$) and the superscripts ($242 = 4 + 238$), you can see that the nuclear equation is balanced.

What is the product nucleus formed when radium emits an alpha particle? (Answer: Radon-222, a chemically inert, radioactive gas.)

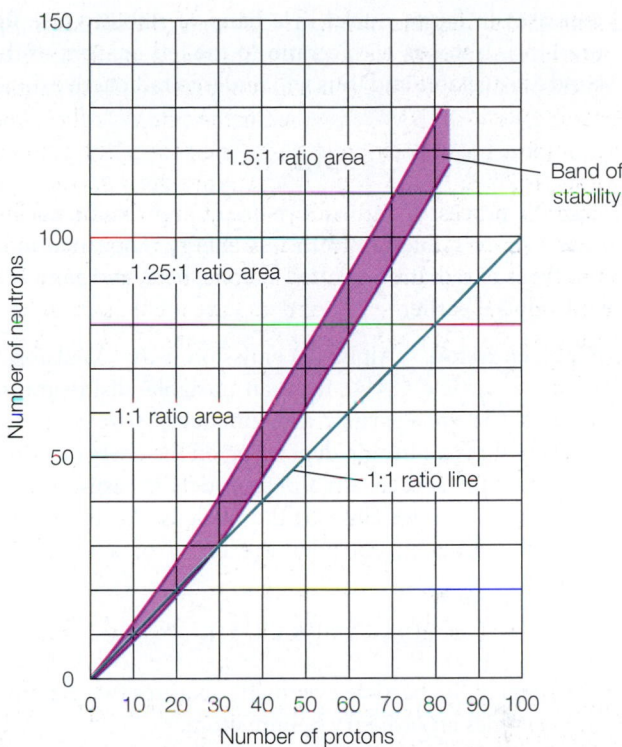

FIGURE 11.4 The shaded area indicates stable nuclei, which group in a band of stability according to their neutron-to-proton ratio. As the size of nuclei increases, so does the neutron-to-proton ratio that represents stability. Nuclei outside this band of stability are radioactive.

neutrons have a maximum nuclear stability. The stable numbers are not the same for electrons and nucleons because of differences in nuclear and electromagnetic forces.

Isotopes of uranium, radium, and plutonium, as well as other isotopes, emit an alpha particle during radioactive decay to a simpler nucleus. The alpha particle is a helium nucleus, (^4_2He). The alpha particle contains two protons as well as two neutrons, which is one of the nucleon numbers of stability, so you would expect the helium nucleus (or alpha particle) to have a stable nucleus, and it does. *Stable* means it does not undergo radioactive decay. Pairs of protons and pairs of neutrons have increased stability, just as pairs of electrons in a molecule do. As a result, nuclei with an *even number* of both protons and neutrons are, in general, more stable than nuclei with odd numbers of protons and neutrons. There are a little more than 150 stable isotopes with an even number of protons and an even number of neutrons, but there are only 4 stable isotopes with odd numbers of each. Just as in the case of electrons, other factors come into play as the nucleus becomes larger and larger with increased numbers of nucleons.

The results of some of these factors are shown in figure 11.4, which is a graph of the number of neutrons versus the number of protons in nuclei. As the number of protons increases, the neutron-to-proton ratio of the *stable nuclei* also

increases in a **band of stability.** Within the band, the neutron-to-proton ratio increases from about 1:1 at the bottom left to about 1½:1 at the top right. The increased ratio of neutrons is needed to produce a stable nucleus as the number of protons increases. Neutrons provide additional attractive *nuclear* (not electrical) forces, which counter the increased electrical repulsion from a larger number of positively charged protons. Thus, more neutrons are required in larger nuclei to produce a stable nucleus. However, there is a limit to the additional attractive forces that can be provided by more and more neutrons, and all isotopes of all elements with more than 83 protons are unstable and thus undergo radioactive decay.

The generalizations about nuclear stability provide a means of predicting if a particular nucleus is radioactive. The generalizations are as follows:

1. All isotopes with an atomic number greater than 83 have an unstable nucleus.
2. Isotopes that contain 2, 8, 20, 28, 50, 82, or 126 protons or neutrons in their nucleus occur in more stable isotopes than those with other numbers of protons or neutrons.
3. Pairs of protons and pairs of neutrons have increased stability, so isotopes that have nuclei with even numbers of both protons and neutrons are generally more stable than nuclei with odd numbers of both protons and neutrons.
4. Isotopes with an atomic number less than 83 are stable when the ratio of neutrons to protons in the nucleus is about 1:1 in isotopes with up to 20 protons, but the ratio

increases in larger nuclei in a band of stability (see figure 11.4). Isotopes with a ratio to the left or right of this band are unstable and thus will undergo radioactive decay.

Types of Radioactive Decay

Through the process of radioactive decay, an unstable nucleus becomes a more stable one with less energy. The three more familiar types of radiation emitted—alpha, beta, and gamma—were introduced earlier.

1. **Alpha emission.** Alpha (α) emission is the expulsion of an alpha particle (^4_2He) from an unstable, disintegrating nucleus. The alpha particle, a helium nucleus, travels from 2 to 12 cm through the air, depending on the energy of emission from the source. An alpha particle is easily stopped by a sheet of paper close to the nucleus. As an example of alpha emission, consider the decay of a radon-222 nucleus,

$$^{222}_{86}Rn \rightarrow\ ^{218}_{84}Po +\ ^4_2He$$

The spent alpha particle eventually acquires two electrons and becomes an ordinary helium atom.

2. **Beta emission.** Beta (β^-) emission is the expulsion of a different particle, a beta particle, from an unstable disintegrating nucleus. A beta particle is simply an electron ($^0_{-1}e$) ejected from the nucleus at a high speed. The emission of a beta particle *increases the number of protons* in a nucleus. It is as if a neutron changed to a proton by emitting an electron, or

$$^1_0n \rightarrow\ ^1_1p +\ ^0_{-1}e$$

Carbon-14 is a carbon isotope that decays by beta emission:

$$^{14}_6C \rightarrow\ ^{14}_7N +\ ^0_{-1}e$$

Note that the number of protons increased from six to seven, but the mass number remained the same. The mass number is unchanged because the mass of the expelled electron (beta particle) is negligible.

Beta particles are more penetrating than alpha particles and may travel several hundred centimeters through the air. They can be stopped by a thin layer of metal close to the emitting nucleus, such as a 1 cm thick piece of aluminum. A spent beta particle may eventually join an ion to become part of an atom, or it may remain a free electron.

3. **Gamma emission.** Gamma (γ) emission is a high-energy burst of electromagnetic radiation from an excited nucleus. It is a burst of light (photon) of a wavelength much too short to be detected by the eye. Other types of radioactive decay, such as alpha or beta emission, sometimes leave the nucleus with an excess of energy, a condition called an *excited state*. As in the case of excited electrons, the nucleus returns to a lower energy state by emitting electromagnetic radiation. From a nucleus, this radiation is in the high-energy portion of the electromagnetic spectrum. Gamma is the most penetrating of the three common types of nuclear radiation. Like X rays, gamma rays can pass completely through a person, but all gamma radiation can be stopped by a 5 cm thick piece of lead close to the source. As with other types of electromagnetic radiation, gamma radiation is absorbed by and gives its energy to materials. Since the product nucleus changed from an excited state to a lower energy state, there is no change in the number of nucleons. For example, radon-222 is an isotope that emits gamma radiation:

$$^{222}_{86}Rn^* \rightarrow\ ^{222}_{86}Rn +\ ^0_0\gamma$$

(* denotes excited state)

Radioactive decay by alpha, beta, and gamma emission is summarized in table 11.2, which also lists the unstable nuclear conditions that lead to the particular type of emission. Just as electrons seek a state of greater stability, a nucleus undergoes radioactive decay to achieve a balance between nuclear attractions, electromagnetic repulsions, and a low quantum of nuclear shell energy. The key to understanding the types of reactions that occur is found in the band of stable nuclei illustrated in figure 11.4. The isotopes within this band have achieved the state of stability, and other isotopes above, below, or beyond the band are unstable and thus radioactive.

Nuclei that have a neutron-to-proton ratio beyond the upper right part of the band are unstable because of an imbalance between the proton-proton electromagnetic repulsions and all the combined proton and neutron nuclear attractions. Recall that the neutron-to-proton ratio increases from about 1:1 to about 1½:1 in the larger nuclei. The additional neutron provided additional nuclear attractions to hold the nucleus together, but atomic number 83 appears to be the upper limit to this additional stabilizing contribution. Thus, all nuclei with an atomic number greater than 83 are outside the upper right limit of the band of stability. Emission of an alpha particle reduces the number of protons by two and the number of neutrons by two, moving the nucleus more toward the band of stability. Thus, you can expect a nucleus that lies beyond the upper right part of the band of stability to be an alpha emitter (figure 11.5).

A nucleus with a neutron-to-proton ratio that is too large will be on the left side of the band of stability. Emission of a

TABLE 11.2 Radioactive Decay

Unstable Condition	Type of Decay	Emitted	Product Nucleus
More than 83 protons	Alpha emission	^4_2He	Lost 2 protons and 2 neutrons
Neutron-to-proton ratio too large	Beta emission	$^0_{-1}e$	Gained 1 proton, no mass change
Excited nucleus	Gamma emission	$^0_0\gamma$	No change
Neutron-to-proton ratio too small	Other emission	^0_1e	Lost 1 proton, no mass change

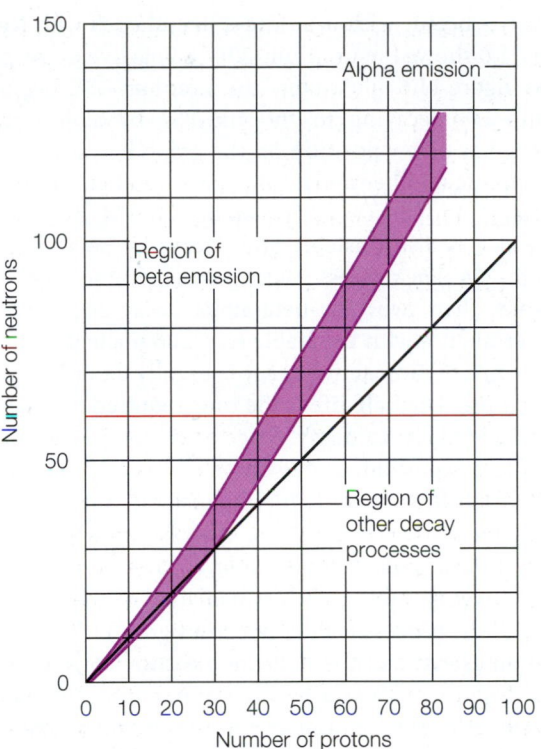

FIGURE 11.5 Unstable nuclei undergo different types of radioactive decay to obtain a more stable nucleus. The type of decay depends, in general, on the neutron-to-proton ratio, as shown.

beta particle decreases the number of neutrons and increases the number of protons, so a beta emission will lower the neutron-to-proton ratio. Thus, you can expect a nucleus with a large neutron-to-proton ratio, that is, one to the left of the band of stability, to be a beta emitter.

A nucleus that has a neutron-to-proton ratio that is too small will be on the right side of the band of stability. These nuclei can increase the number of neutrons and reduce the number of protons in the nucleus by other types of radioactive decay. As usual when dealing with broad generalizations and trends, there are exceptions to the summarized relationships between neutron-to-proton ratios and radioactive decay.

Radioactive Decay Series

A radioactive decay reaction produces a simpler and eventually more stable nucleus than the reactant nucleus. As discussed in the previous section, large nuclei with an atomic number greater than 83 decay by alpha emission, giving up two protons and two neutrons with each alpha particle. A nucleus with an atomic number greater than 86, however, will emit an alpha particle and *still* have an atomic number greater than 83, which means the product nucleus will also be radioactive. This nucleus will also undergo radioactive decay, and the process will continue through a series of decay reactions until a stable nucleus is achieved. Such a series of decay reactions that (1) begins with one radioactive nucleus, which (2) decays to a second nucleus, which (3) then decays to a third nucleus, and so on until

(4) a stable nucleus is reached is called a *radioactive decay series.* There are three naturally occurring radioactive decay series. One begins with thorium-232 and ends with lead-208, another begins with uranium-235 and ends with lead-207, and the

third series begins with uranium-238 and ends with lead-206. Figure 11.6 shows the uranium-238 radioactive decay series.

As figure 11.6 illustrates, the uranium-238 begins with uranium-238 decaying to thorium-234 by alpha emission. Thorium has a new position on the graph because it now has a new atomic number and a new mass number. Thorium-234 is unstable and decays to protactinium-234 by beta emission, which is also unstable and decays by beta emission to uranium-234. The process continues with five sequential alpha emissions, then two beta-beta-alpha decay steps before the series terminates with the stable lead-206 nucleus.

The rate of radioactive decay is usually described in terms of its *half-life*. The **half-life** is the time required for one-half of the unstable nuclei to decay. Since each isotope has a characteristic decay constant, each isotope has its own characteristic half-life. Half-lives of some highly unstable isotopes are measured in fractions of seconds, and other isotopes have half-lives measured in seconds, minutes, hours, days, months, years, or billions of years. Table 11.3 lists half-lives of some of the isotopes, and the process is illustrated in figure 11.7.

As an example of the half-life measure, consider a hypothetical isotope that has a half-life of one day. The half-life is independent of the amount of the isotope being considered, but suppose you start with a 1.0 kg sample of this element with

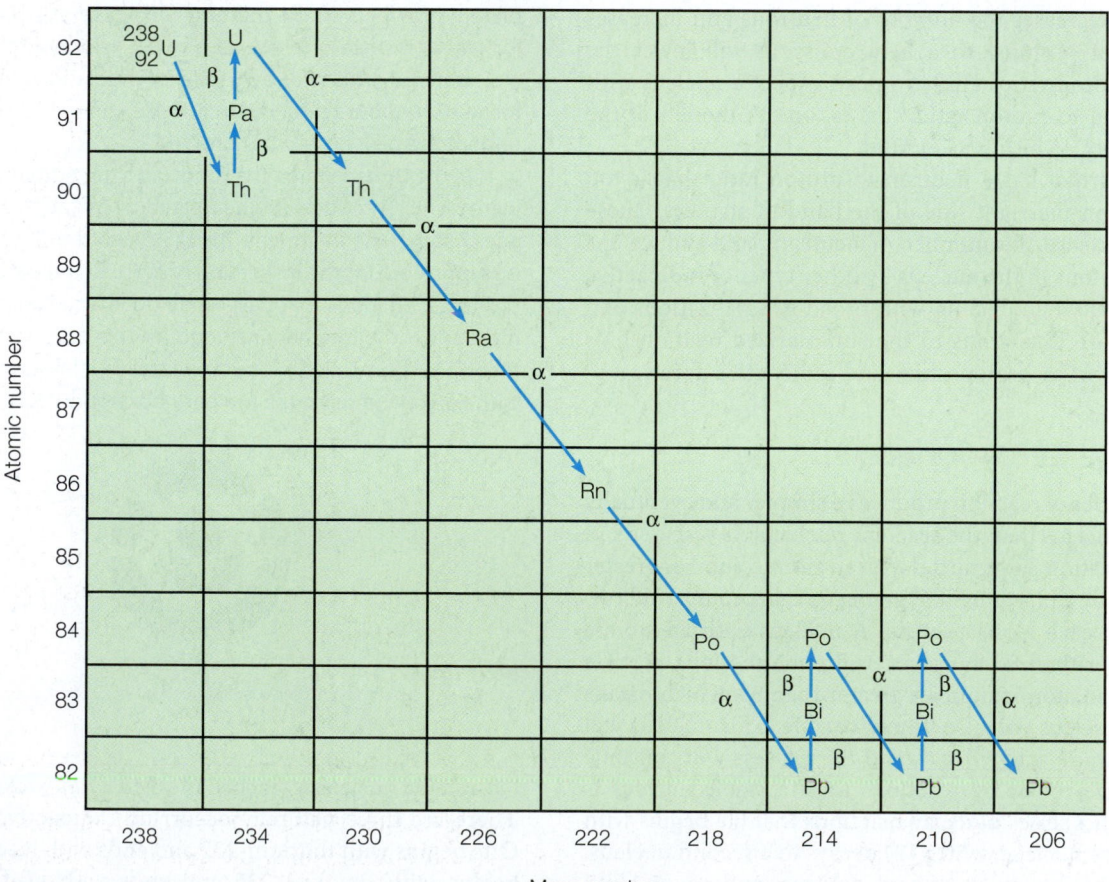

FIGURE 11.6 The radioactive decay series for uranium-238. This is one of three naturally occurring series.

TABLE 11.3 Half-Lives of Some Radioactive Isotopes

Isotope	Half-Life	Mode of Decay
$^{3}_{1}H$ (tritium)	12.26 years	Beta
$^{14}_{6}C$	5,730 years	Beta
$^{90}_{38}Sr$	28 years	Beta
$^{131}_{53}I$	8 days	Beta
$^{133}_{54}Xe$	5.27 days	Beta
$^{238}_{92}U$	4.51×10^{9} years	Alpha
$^{242}_{94}Pu$	3.79×10^{5} years	Alpha
$^{240}_{94}Pu$	6,760 years	Alpha
$^{239}_{94}Pu$	24,360 years	Alpha

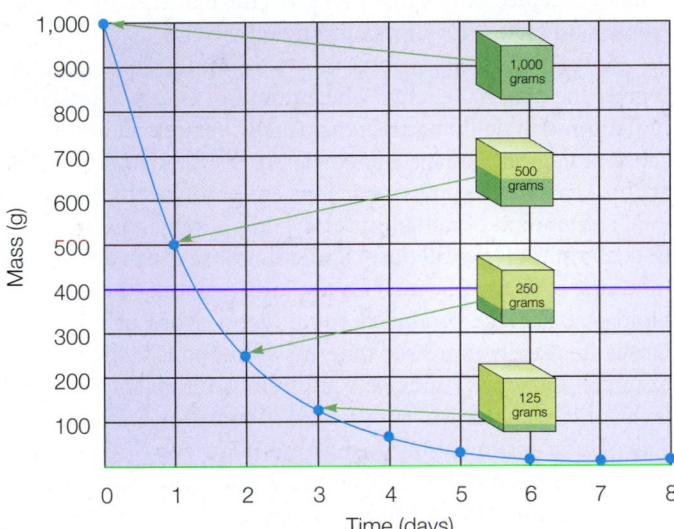

FIGURE 11.7 Radioactive decay of a hypothetical isotope with a half-life of one day. The sample decays each day by one-half to some other element. Actual half-lives may be in seconds, minutes, or any time unit up to billions of years.

a half-life of one day. One day later, you will have half of the original sample, or 500 g. The other half did not disappear, but it is now the decay product, that is, some new element. During the next day, half of the remaining nuclei will disintegrate, and only 250 g of the initial sample is still the original element. One-half of the remaining sample will disintegrate each day until the original sample no longer exists.

11.2 MEASUREMENT OF RADIATION

The measurement of radiation is important in determining the half-life of radioactive isotopes. Radiation measurement is

How Is Half-Life Determined?

It is not possible to predict when a radioactive nucleus will decay because it is a random process. It is possible, however, to deal with nuclear disintegration statistically since the rate of decay is not changed by any external conditions of temperature, pressure, or any chemical state. When dealing with a large number of nuclei, the ratio of the rate of nuclear disintegration per unit of time to the total number of radioactive nuclei is a constant, or

$$\text{radioactive decay constant} = \frac{\text{decay rate}}{\text{number of nuclei}}$$

The radioactive decay constant is a specific constant for a particular isotope, and each isotope has its own decay constant that can be measured. For example, a 238 g sample of uranium-238 (1 mole) that has 2.93×10^{6} disintegrations per second would have a decay constant of

$$\text{radioactive decay constant} = \frac{\text{decay rate}}{\text{number of nuclei}}$$
$$= \frac{2.93 \times 10^{6} \text{ nuclei/s}}{6.01 \times 10^{23} \text{ nuclei}}$$
$$= 4.87 \times 10^{-18} \text{ 1/s}$$

The half-life of a radioactive nucleus is related to its radioactive decay constant by

$$\text{half-life} = \frac{\text{a mathematical constant}}{\text{decay constant}}$$

The half-life of uranium-238 is therefore

$$\text{half-life} = \frac{\text{a mathematical constant}}{\text{decay constant}}$$
$$= \frac{0.693}{4.87 \times 10^{-18} \text{ 1/s}}$$
$$= 1.42 \times 10^{17} \text{ s}$$

This is the half-life of uranium-238 in seconds. There are $60 \times 60 \times 24 \times 365$, or 3.15×10^{7} s in a year, so

$$\frac{1.42 \times 10^{17} \text{ s}}{3.15 \times 10^{7} \text{ s/yr}} = 4.5 \times 10^{9} \text{ yr}$$

The half-life of uranium-238 is thus 4.5 billion years.

also important in considering biological effects, which will be discussed in the next section. As is the case with electricity, it is not possible to make direct measurements on things as small as electrons and other parts of atoms. Indirect measurement methods are possible, however, by considering the effects of the radiation.

Measurement Methods

As Becquerel discovered, radiation affects photographic film, exposing it as visible light does. Since the amount of film exposure is proportional to the amount of radiation, photographic film can be used as an indirect measure of radiation. Today, people who work around radioactive materials or X rays carry light-tight film badges. The film is replaced periodically and developed. The optical density of the developed film provides a record of the worker's exposure to radiation since the darkness of the developed film is proportional to the exposure.

There are also devices that indirectly measure radiation by measuring an effect of the radiation. An *ionization counter* is one type of device that measures ions produced by radiation. A second type of device is called a *scintillation counter*. *Scintillate* is a word meaning "sparks or flashes," and a scintillation counter measures the flashes of light produced when radiation strikes a phosphor.

The most common example of an ionization counter is known as a *Geiger counter*. The working components of a Geiger counter are illustrated in figure 11.8. Radiation is received in a metal tube filled with an inert gas, such as argon. An insulated wire inside the tube is connected to the positive terminal of a direct current source. The metal cylinder around the insulated wire is connected to the negative terminal. There is not a current between the center wire and the metal cylinder because the gas acts as an insulator. When radiation passes through the window, however, it ionizes some of the gas atoms, releasing free electrons. These electrons are accelerated by the field between the wire and cylinder, and the accelerated electrons ionize more gas molecules, which results in an *avalanche* of free electrons. The avalanche creates a pulse of current that is amplified and then measured. More radiation means more avalanches, so the pulses are an indirect means of measuring radiation. When connected to a speaker or earphone, each avalanche produces a "pop" or "click."

Some materials are *phosphors,* substances that emit a flash of light when excited by radiation. Zinc sulfide, for example, gives off a tiny flash of light when struck by radiation from a disintegrating radium nucleus. A scintillation counter measures the flashes of light through the photoelectric effect, producing free electrons that are accelerated to produce a pulse of current. Again, the pulses of current are used as an indirect means to measure radiation.

Radiation Units

You have learned that *radioactivity* is a property of isotopes with unstable, disintegrating nuclei and *radiation* is emitted particles (alpha or beta) or energy traveling in the form of photons (gamma). Radiation can be measured (1) at the source of radioactivity or (2) at a place of reception, where the radiation is absorbed.

The *activity* of a radioactive source is a measure of the number of nuclear disintegrations per unit of time. The unit of activity at the source is called a **curie** (Ci), which is defined as 3.70×10^{10} nuclear disintegrations per second. Activities are usually expressed in terms of fractions of curies, for example, a *picocurie* (pCi), which is a millionth of a millionth of a curie. Activities are sometimes expressed in terms of so many picocuries per liter (pCi/L).

The International System of Units (SI) unit for radioactivity is the *Becquerel* (Bq), which is defined as one nuclear disintegration per second. The unit for reporting radiation in the United States is the curie, but the Becquerel is the internationally accepted unit. Table 11.4 gives the names, symbols, and conversion factors for units of radioactivity.

As radiation from a source moves out and strikes a material, it gives the material energy. The amount of energy released by radiation striking living tissue is usually very small, but it can cause biological damage nonetheless because chemical bonds are broken and free polyatomic ions are produced by radiation.

The amount of radiation received by a human is expressed in terms of radiological dose. Radiation dose is usually written in units of a **rem,** which takes into account the possible biological damage produced by different types of radiation. Doses are usually expressed in terms of fractions of the rem, for example, a *millirem* (mrem). A millirem is 1/1,000 of a rem and is the unit of choice when low levels of radiation are discussed. The SI unit for radiation dose is the *millisievert* (mSv). Both the millirem and the millisievert relate ionizing radiation and biological effect to humans. The natural radiation that people receive from nature in one day is less than 1 millirem (0.01 millisievert). A single dose of 100,000 to 200,000 millirem (1,000 to 2,000 millisievert) can cause radiation sickness in humans (table 11.5). A single dose of 500,000 millirem (5,000 millisievert) results in death about 50 percent of the time.

Another measure of radiation received by a material is the **rad.** The term *rad* is from radiation *a*bsorbed *d*ose. The SI unit

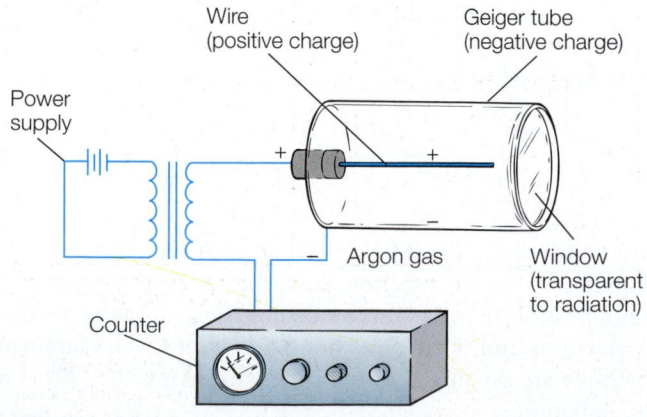

FIGURE 11.8   The working parts of a Geiger counter.

TABLE 11.4   Names, Symbols, and Conversion Factors for Radioactivity

Name	Symbol	To Obtain	Multiply By
becquerel	Bq	Ci	2.7×10^{11}
gray	Gy	rad	100
sievert	Sv	rem	100
curie	Ci	Bq	3.7×10^{10}
rem	rem	Sv	0.01
millirem	millirem	rem	0.001
rem	rem	millirem	1,000

TABLE 11.5	Approximate Single-Dose, Whole-Body Effects of Radiation Exposure
Level	Comment
0.240 rem	Average annual exposure to natural background radiation
0.500 rem	Upper limit of annual exposure to general public
25.0 rem	Threshold for observable effects such as blood count changes
100.0 rem	Fatigue and other symptoms of radiation sickness
200.0 rem	Definite radiation sickness, bone marrow damage, possibility of developing leukemia
500.0 rem	Lethal dose for 50 percent of individuals
1,000.0 rem	Lethal dose for all

for radiation received by a material is the *gray* (Gy). One gray is equivalent to an exposure of 100 rad.

Radiation Exposure

Natural radioactivity is a part of your environment, and you receive between 100 and 500 millirems each year from natural sources. This radiation from natural sources is called **background radiation.** Background radiation comes from outer space in the form of cosmic rays and from unstable isotopes in the ground, building materials, and foods. Many activities and situations will increase your yearly exposure to radiation. For example, the atmosphere absorbs some of the cosmic rays from space, so the less atmosphere above you, the more radiation you will receive. You are exposed to one additional millirem per year for each 100 feet you live above sea level. You receive approximately 0.3 millirem for each hour spent on a jet flight. Airline crews receive an additional 300 to 400 millirems per year because they spend so much time high in the atmosphere. Additional radiation exposure comes from medical X rays and television sets. In general, the worldwide average background radiation exposure for the average person is about 240 millirems per year.

What are the consequences of radiation exposure? Radiation can be a hazard to living organisms because it produces ionization along its path of travel. This ionization can (1) disrupt chemical bonds in essential macromolecules such as DNA and (2) produce molecular fragments, which are free polyatomic ions that can interfere with enzyme action and other essential cell functions. Tissues with highly active cells are more vulnerable to radiation damage than others, such as blood-forming tissue. Thus, one of the symptoms of an excessive radiation exposure is an altered blood count. Table 11.5 compares the estimated results of various levels of acute radiation exposure.

Radiation is not a mysterious, unique health hazard. It is a hazard that should be understood and neither ignored nor exaggerated. Excessive radiation exposure should be avoided, just as you avoid excessive exposure to other hazards such as certain chemicals, electricity, or even sunlight. Everyone agrees that *excessive* radiation exposure should be avoided, but there is some controversy about long-term, low-level exposure and its possible role in cancer. Some claim that tolerable low-level

exposure does not exist because that is not possible. Others point to many studies comparing high and low background radioactivity with cancer mortality data. For example, no cancer mortality differences could be found between people receiving 500 or more millirems a year and those receiving fewer than 100 millirems a year. The controversy continues, however, because of lack of knowledge about long-term exposure. Two models of long-term, low-level radiation exposure have been proposed: (1) a linear model and (2) a threshold model. The *linear model* proposes that any radiation exposure above zero is damaging and can produce cancer and genetic damage. The *threshold model* proposes that the human body can repair damage and get rid of damaging free polyatomic ions up to a certain exposure level called the threshold (figure 11.9). The controversy over long-term, low-level radiation exposure will probably continue until there is clear evidence about which model is correct. Whichever is correct will not lessen the need for rational risks versus cost-benefit analyses of all energy alternatives.

11.3 NUCLEAR ENERGY

Some nuclei are unstable because they are too large or because they have an unstable neutron-to-proton ratio. These unstable nuclei undergo radioactive decay, eventually forming products

Myths, Mistakes, and Misunderstandings

Antiradiation Pill?

It is a myth that an antiradiation pill exists that will protect you from ionizing radiation. There is a pill, an iodine supplement, that is meant to saturate your thyroid with a nonradioactive isotope of iodine. Once saturated, your thyroid will not absorb radioactive isotopes for storage in the gland, which could be dangerous.

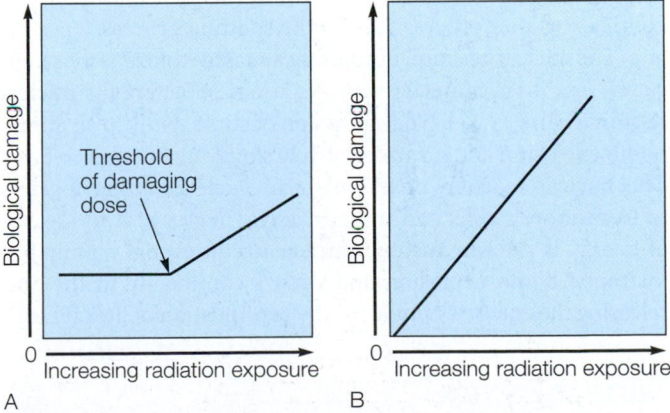

FIGURE 11.9 Graphic representation of the (*A*) threshold model and (*B*) linear model of low-level radiation exposure. The threshold model proposes that the human body can repair damage up to a threshold. The linear model proposes that any radiation exposure is damaging.

of greater stability. An example of this radioactive decay is the alpha emission reaction of uranium-238 to thorium-234,

$$^{238}_{92}U \rightarrow \, ^{234}_{90}Th + \, ^{4}_{2}He$$

$$238.0003 \text{ u} \rightarrow 233.9942 \text{ u} + 4.00150 \text{ u}$$

The numbers below the nuclear equation are the *nuclear* masses (u) of the reactant and products. As you can see, there seems to be a loss of mass in the reaction,

$$233.9942 + 4.00150 - 238.0003 = -0.0046 \text{ u}$$

This change in mass is related to the energy change according to the relationship that was formulated by Albert Einstein in 1905. The relationship is

$$E = mc^2 \qquad \textbf{equation 11.1}$$

where E is a quantity of energy, m is a quantity of mass, and c is a constant equal to the speed of light in a vacuum, 3.00×10^8 m/s. According to this relationship, matter and energy are the same thing, and energy can be changed to matter and vice versa.

The relationship between mass and energy explains why the mass of a nucleus is always *less* than the sum of the masses of the individual particles of which it is made.

The difference between the mass of the individual nucleons making up a nucleus and the actual mass of the nucleus is called the **mass defect** of the nucleus. The explanation for the mass defect is again found in $E = mc^2$. When nucleons join to make a nucleus, energy is released as the more stable nucleus is formed.

The energy equivalent released when a nucleus is formed is the same as the **binding energy,** the energy required to break the nucleus into individual protons and neutrons. The binding energy of the nucleus of any isotope can be calculated from the mass defect of the nucleus.

The ratio of binding energy to nucleon number is a reflection of the stability of a nucleus (figure 11.10). The greatest binding energy per nucleon occurs near mass number 56, then decreases for both more massive and less massive nuclei. This means that more massive nuclei can gain stability by splitting into smaller nuclei with the release of energy. It also means that less massive nuclei can gain stability by joining together with the release of energy. The slope also shows that more energy is released in the coming-together process than in the splitting process.

The nuclear reaction of splitting a massive nucleus into more stable, less massive nuclei with the release of energy is **nuclear fission** (figure 11.11). Nuclear fission occurs rapidly in an atomic bomb explosion and occurs relatively slowly in a nuclear reactor. The nuclear reaction of less massive nuclei, coming together to form more stable, and more massive, nuclei with the release of energy is **nuclear fusion.** Nuclear fusion occurs rapidly in a hydrogen bomb explosion and occurs continually in the Sun, releasing the energy essential for the continuation of life on Earth.

Nuclear Fission

Nuclear fission was first accomplished in the late 1930s when researchers were attempting to produce isotopes by bombard-

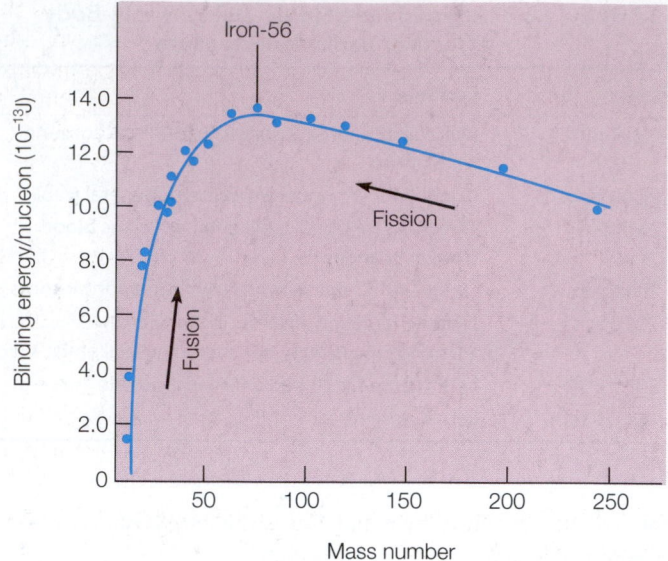

FIGURE 11.10   The maximum binding energy per nucleon occurs around mass number 56, then decreases in both directions. As a result, fission of massive nuclei and fusion of less massive nuclei both release energy.

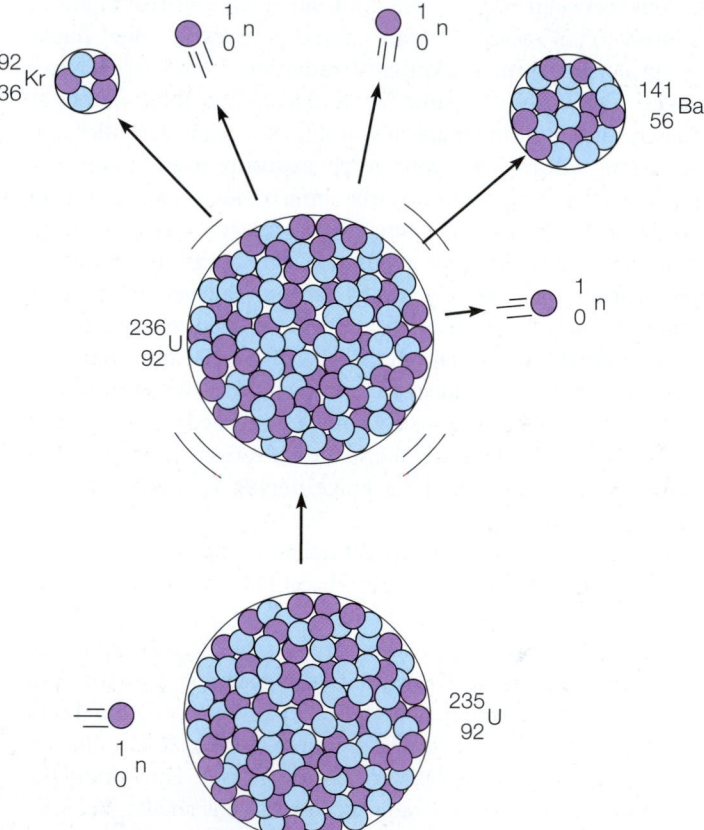

FIGURE 11.11   The fission reaction occurring when a neutron is absorbed by a uranium-235 nucleus. The deformed nucleus splits any number of ways into lighter nuclei, releasing neutrons in the process.

ing massive nuclei with neutrons. In 1938, two German scientists, Otto Hahn and Fritz Strassman, identified the element barium in a uranium sample that had been bombarded with neutrons. Where the barium came from was a puzzle at the time, but soon afterward, Lise Meitner, an associate who had moved to Sweden, deduced that the uranium nuclei had split, producing barium. The reaction might have been

$$_0^1\text{n} + {}_{92}^{235}\text{U} \rightarrow {}_{56}^{141}\text{Ba} + {}_{36}^{92}\text{Kr} + 3\,_0^1\text{n}$$

The phrase "might have been" is used because a massive nucleus can split in many different ways, producing different products. About thirty-five different, less massive elements have been identified among the fission products of uranium-235. Some of these products are fission fragments, and some are produced by unstable fragments that undergo radioactive decay. Selected fission fragments are listed in table 11.6 below, together with their major modes of radioactive decay and half-lives. Some of the isotopes are the focus of concern about nuclear wastes, the topic of the "Science and Society" reading at the end of this chapter.

The fission of a uranium-235 nucleus produces two or three neutrons along with other products. These neutrons can each collide with other uranium-235 nuclei, where they are absorbed, causing fission with the release of more neutrons, which collide with other uranium-235 nuclei to continue the process. A reaction where the products are able to produce more reactions in a self-sustaining series is called a **chain reaction.** A chain reaction is self-sustaining until all the

uranium-235 nuclei have fissioned or until the neutrons fail to strike a uranium-235 nucleus (figure 11.12).

You might wonder why all the uranium in the universe does not fission in a chain reaction. Natural uranium is mostly uranium-238, an isotope that does not fission easily. Only about 0.7 percent of natural uranium is the highly fissionable uranium-235. This low ratio of readily fissionable uranium-235 nuclei makes it unlikely that a stray neutron would be able to achieve a chain reaction.

To achieve a chain reaction, there must be (1) a sufficient mass with (2) a sufficient concentration of fissionable nuclei. When the mass and concentration are sufficient to sustain a chain reaction, the amount is called a **critical mass.** Likewise, a mass too small to sustain a chain reaction is called a *subcritical mass.* A mass of sufficiently pure uranium-235 (or plutonium-239) that is large enough to produce a rapidly accelerating chain reaction is called a *supercritical mass.* An atomic bomb is simply a device that uses a small, conventional explosive to push subcritical masses of fissionable material into a supercritical mass. Fission occurs almost instantaneously in the supercritical mass, and tremendous energy is released in a violent explosion.

Nuclear Power Plants

The nuclear part of a nuclear power plant is the *nuclear reactor,* a steel vessel in which a controlled chain reaction of fissionable material releases energy (figure 11.13). In the most

TABLE 11.6 Fragments and Products from Nuclear Reactors Using Fission of Uranium-235

Isotope	Major Mode of Decay	Half-Life	Isotope	Major Mode of Decay	Half-Life
Tritium	Beta	12.26 years	Cerium-144	Beta, gamma	285 days
Carbon-14	Beta	5,730 years	Promethium-147	Beta	2.6 years
Argon-41	Beta, gamma	1.83 hours	Samarium-151	Beta	90 years
Iron-55	Electron capture	2.7 years	Europium-154	Beta, gamma	16 years
Cobalt-58	Beta, gamma	71 days	Lead-210	Beta	22 years
Cobalt-60	Beta, gamma	5.26 years	Radon-222	Alpha	3.8 days
Nickel-63	Beta	92 years	Radium-226	Alpha, gamma	1,620 years
Krypton-85	Beta, gamma	10.76 years	Thorium-229	Alpha	7,300 years
Strontium-89	Beta	5.4 days	Thorium-230	Alpha	26,000 years
Strontium-90	Beta	28 years	Uranium-234	Alpha	2.48×10^5 years
Yttrium-91	Beta	59 days	Uranium-235	Alpha, gamma	7.13×10^8 years
Zirconium-93	Beta	9.5×10^5 years	Uranium-238	Alpha	4.51×10^9 years
Zirconium-95	Beta, gamma	65 days	Neptunium-237	Alpha	2.14×10^6 years
Niobium-95	Beta, gamma	35 days	Plutonium-238	Alpha	89 years
Technetium-99	Beta	2.1×10^5 years	Plutonium-239	Alpha	24,360 years
Ruthenium-106	Beta	1 year	Plutonium-240	Alpha	6,760 years
Iodine-129	Beta	1.6×10^7 years	Plutonium-241	Beta	13 years
Iodine-131	Beta, gamma	8 days	Plutonium-242	Alpha	3.79×10^5 years
Xenon-133	Beta, gamma	5.27 days	Americium-241	Alpha	458 years
Cesium-134	Beta, gamma	2.1 years	Americium-243	Alpha	7,650 years
Cesium-135	Beta	2×10^6 years	Curium-242	Alpha	163 days
Cesium-137	Beta	30 years	Curium-244	Alpha	18 years
Cerium-141	Beta	32.5 days			

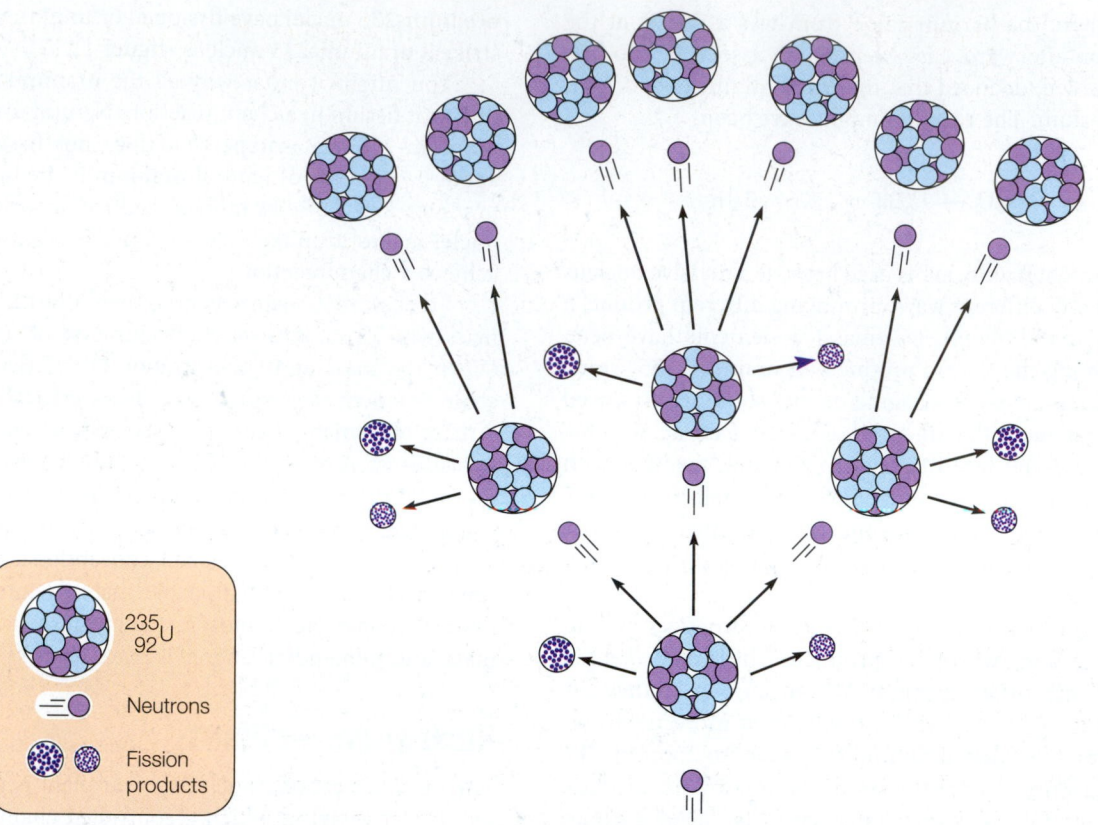

FIGURE 11.12 A schematic representation of a chain reaction. Each fissioned nucleus releases neutrons, which move out to fission other nuclei. The number of neutrons can increase quickly with each series.

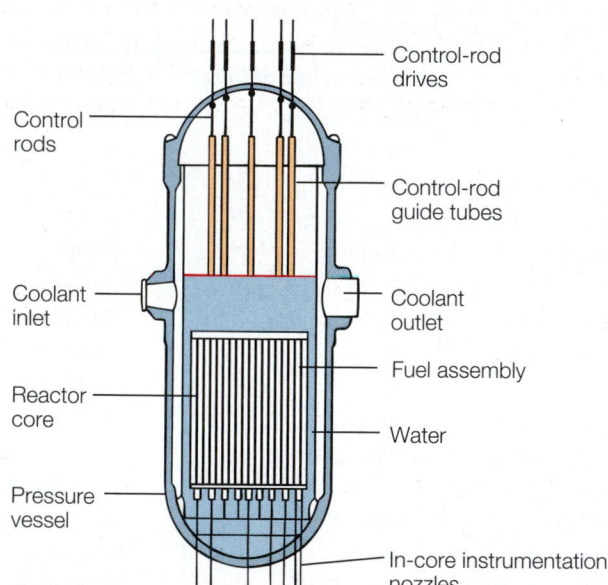

FIGURE 11.13 A schematic representation of the basic parts of a nuclear reactor. The largest commercial nuclear power plant reactors are 23 to 28 cm (9 to 11 in) thick steel vessels with a stainless steel liner, standing about 12 m (40 ft) high with a diameter of about 5 m (16 ft). Such a reactor has four pumps, which move 1,665,581 liters (440,000 gallons) of water per minute through the primary loop.

popular design, called a pressurized light-water reactor, the fissionable material is enriched 3 percent uranium-235 and 97 percent uranium-238 that has been fabricated in the form of

small ceramic pellets (figure 11.14A). The pellets are encased in a long zirconium alloy tube called a *fuel rod*. The fuel rods are locked into a *fuel rod assembly* by locking collars, arranged to permit pressurized water to flow around each fuel rod (figure 11.14B) and to allow the insertion of *control rods* between the fuel rods. *Control rods* are constructed of materials, such as cadmium, that absorb neutrons. The lowering or raising of control rods within the fuel rod assemblies slows or increases the chain reaction by varying the amount of neutrons absorbed. When they are lowered completely into the assembly, enough neutrons are absorbed to stop the chain reaction.

It is physically impossible for the low-concentration fuel pellets to form a supercritical mass. A nuclear reactor in a power plant can only release energy at a comparatively slow rate, and it is impossible for a nuclear power plant to produce a nuclear explosion. In a pressurized water reactor, the energy released is carried away from the reactor by pressurized water in a closed pipe called the *primary loop* (figure 11.15). The water is pressurized at about 150 atmospheres (about 2,200 lb/in^2) to keep it from boiling, since its temperature may be 350°C (about 660°F).

In the pressurized light-water (ordinary water) reactor, the circulating pressurized water acts as a coolant, carrying heat away from the reactor. The water also acts as a *moderator*, a substance that slows neutrons so they are more readily absorbed by uranium-235 nuclei. Other reactor designs use heavy water (deuterium dioxide) or graphite as a moderator.

Water from the closed primary loop is circulated through a heat exchanger called a *steam generator* (figure 11.15). The

A

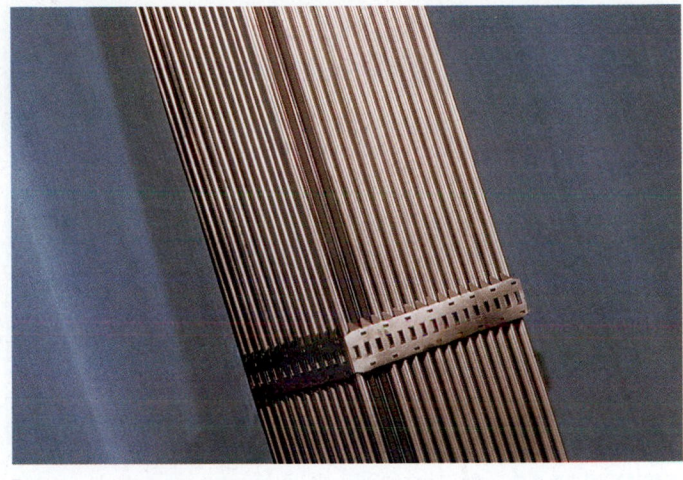

B

FIGURE 11.14 (*A*) These are uranium oxide fuel pellets that are stacked inside fuel rods, which are then locked together in a fuel rod assembly. (*B*) A fuel rod assembly. See also figure 11.17, which shows a fuel rod assembly being loaded into a reactor.

pressurized high-temperature water from the reactor moves through hundreds of small tubes inside the generator as *feedwater* from the *secondary loop* flows over the tubes. The water in the primary loop heats feedwater in the steam generator and then returns to the nuclear reactor to become heated again. The feedwater is heated to steam at about 235°C (455°F) with a pressure of about 68 atmospheres (1,000 lb/in²). This steam is piped to the turbines, which turn an electric generator (figure 11.16).

After leaving the turbines, the spent steam is condensed back to liquid water in a second heat exchanger receiving water from the cooling towers. Again, the cooling water does not mix with the closed secondary loop water. The cooling-tower water enters the condensing heat exchanger at about 32°C (90°F) and

leaves at about 50°C (about 120°F) before returning to a cooling tower, where it is cooled by evaporation. The feedwater is preheated, then recirculated to the steam generator to start the cycle over again. The steam is condensed back to liquid water because of the difficulty of pumping and reheating steam.

After a period of time, the production of fission products in the fuel rods begins to interfere with effective neutron transmission, so the reactor is shut down annually for refueling. During refueling, about one-third of the fuel that had the longest exposure in the reactor is removed as "spent" fuel. New fuel rod assemblies are inserted to make up for the part removed (figure 11.17). However, only about 4 percent of the "spent" fuel is unusable waste, about 94 percent is uranium-238,

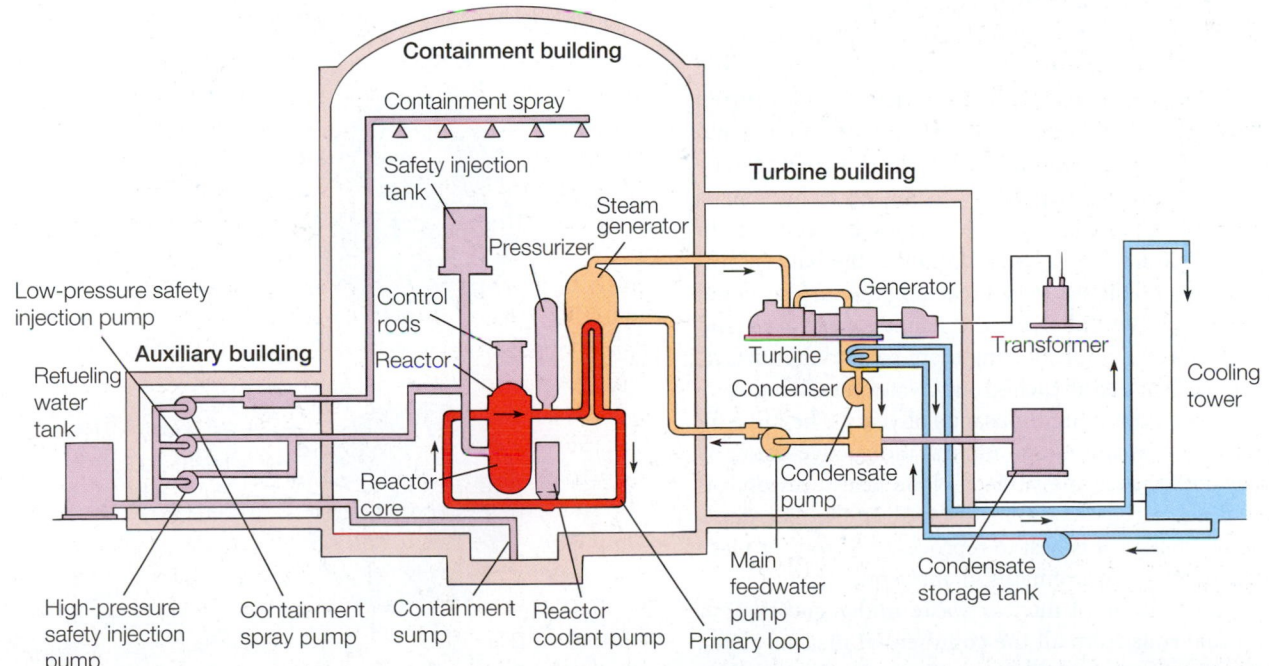

FIGURE 11.15 A schematic general system diagram of a pressurized water nuclear power plant, not to scale. The containment building is designed to withstand an internal temperature of 149°C (300°F) at a pressure of 414 kilopascals (60 lbs/in²) and still maintain its leak-tight integrity.

FIGURE 11.16 The turbine deck of a nuclear generating station. One large generator is in line with four steam turbines in this non-nuclear part of the plant. The large silver tanks are separators that remove water from the steam after it has left the high-pressure turbine and before it is recycled back into the low-pressure turbines.

0.8 percent is uranium-235, and about 0.9 percent is plutonium (figure 11.18). Thus, "spent" fuel rods contain an appreciable amount of usable uranium and plutonium. For now, spent reactor fuel rods are mostly stored in cooling pools at the nuclear plant sites. In the future, a decision will be made either to reprocess the spent fuel, recovering the uranium and plutonium through chemical reprocessing, or put the fuel in terminal storage. Concerns about reprocessing are based on the fact that plutonium-239 and uranium-235 are fissionable and could possibly be used by terrorist groups to construct nuclear explosive devices. Six other countries do have reprocessing plants, however, and the spent fuel rods represent a significant energy source. Some energy experts say that it would be inappropriate to dispose of such an energy source.

The technology to dispose of fuel rods exists if the decision is made to do so. The longer half-life waste products are mostly alpha emitters. These metals could be converted to oxides, mixed with powdered glass (or a ceramic), melted, and then poured into stainless steel containers. The solidified canisters would then be buried in a stable geologic depository. The glass technology is used in France for disposal of high-level wastes. Buried at 610 to 914 m (2,000–3,000 ft) depths in solid granite, the only significant means of the radioactive wastes reaching the surface would be through groundwater dissolving the stainless steel, glass, and waste products and then transporting them back to the surface. Many experts believe that if such groundwater dissolving were to take place, it would require thousands of years. The radioactive isotopes would thus undergo natural radioactive decay by the time they could reach the surface. Nonetheless, research is continuing on nuclear waste and its disposal. In the meantime, the question of whether it is best to reprocess fuel rods or place them in permanent storage remains unanswered.

What is the volume of nuclear waste under question? If all the spent fuel rods from all the commercial nuclear plants accumulated since coming online were reprocessed, then mixed with glass, the total amount of glassified waste would make a pile on one football field an estimated 4 m (about 13 ft) high.

Nuclear Fusion

As the graph of nuclear binding energy versus mass numbers shows (see figure 11.10), nuclear energy is released when (1) massive nuclei such as uranium-235 undergo fission and (2) when less massive nuclei come together to form more massive nuclei through nuclear fusion. Nuclear fusion is responsible for the energy released by the Sun and other stars. At the present halfway point in the Sun's life—with about 5 billion years to go—the core is now 35 percent hydrogen and 65 percent helium. Through fusion, the Sun converts about 650 million tons of hydrogen to 645 million tons of helium every second. The other roughly 5 million tons of matter are converted into energy.

FIGURE 11.17 Spent fuel rod assemblies are removed and new ones are added to a reactor head during refueling. This shows an initial fuel load to a reactor, which has the upper part removed and set aside for the loading.

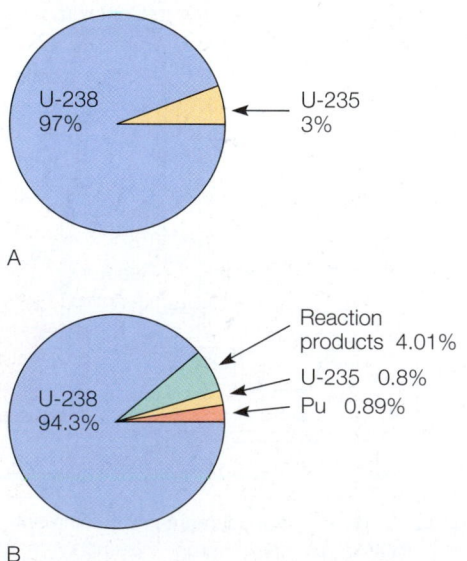

FIGURE 11.18 The composition of the nuclear fuel in a fuel rod (A) before and (B) after use over a three-year period in a nuclear reactor.

Three Mile Island, Chernobyl, and Fukushima I

Three Mile Island (downriver from Harrisburg, Pennsylvania), Chernobyl (former USSR, now Ukraine), and Fukushima I (north of Tokyo) are three nuclear power plants that became famous because of accidents. Here is a brief accounting of what happened.

Three Mile Island

It was March 28, 1979, and the 880-megawatt Three Mile Island Nuclear Plant, operated by Metropolitan Edison Company, was going full blast. At 4:00 A.M. that morning the main feedwater pump that pumps water to the steam generator failed for some unexplained reason (follow this description in figure 11.15). Backup feedwater pumps kicked in, but the valves that should have been open were closed for maintenance, blocking the backup source of water for the steam generator. All of a sudden there was not a source of water for the steam generator that removes heat from the primary loop. Events began to happen quickly at that point.

The computer sensors registered that the steam generator was not receiving water, and the computer began to follow a shutdown procedure. First, the turbine was shut down as steam was vented from the steam line out through the turbine building, sounding much like a large jet plane. Within six seconds, the reactor was shut down (scrammed) with the control rods dropping down between the fuel rods in the reactor vessel. Fissioning began to slow, but the reactor was still hot.

Between three and six seconds, a significant event occurred when a pressure relief valve opened on the primary loop, relieving the excess pressure that was generated because feedwater was not entering the steam generator to remove heat from the primary loop. The valve should have closed when the excess pressure was released, but it did not. It was stuck in an open position. Pressurized water and steam were pouring from the primary loop into the containment building. As water was lost from the primary loop, temperatures inside the reactor began to climb. The loss of pressure resulted in high-pressure water flashing into steam. If an operator had

pressed the right button in the control room, it would have closed the open valve, but this did not happen until thirty-two minutes later.

At this point, the reactor could have recovered from the events. Two minutes after the initial shutdown, the computer sensors noted the loss of pressure from the open valve and kicked in the emergency core cooling system, which pumps more water into the reactor vessel. However, for some unknown reason, the control room operators shut down one pump four and a half minutes after the initial event and the second pump six minutes later.

Water continued to move through the open pressure relief valve into the containment building. At seven and a half minutes after the start of the event, the radioactive water on the floor was 2 feet deep and the sump pumps started pumping water into tanks in the auxiliary building. This water would become the source of the radioactivity that did escape the plant. It escaped because the pump seals leaked and spilled radioactive water. The filters had been removed from the auxiliary building air vents, allowing radioactive gases to escape.

Eleven minutes after the start of the event, an operator restarted the emergency core cooling system that had been turned off. With the cooling water flowing again, the pressure in the reactor stopped falling. The fuel rods, some thirty-six thousand in this reactor, had not yet suffered any appreciable damage. This would be taken care of by the next incredible event.

What happened next was that operators began turning off the emergency cooling pumps, perhaps because they were vibrating too much. In any case, with the pumps off, the water level in the reactor fell again, this time uncovering the fuel rods. Within minutes, the temperature was high enough to rupture fuel rods, dropping radioactive oxides into the bottom of the reactor vessel. The operators now had a general emergency.

Eleven hours later, the operators decided to start the main reactor coolant pump. This pump had shut down at the beginning of the series of events. Water again covered

the fuel rods, and the pressure and temperature stabilized.

The consequences of this series of events were as follows:

1. Local residences did receive some radiation from the release of gases. They received 10 millirem (0.1 millisievert) in a low-exposure area and up to 25 millirem (0.25 millisievert) in a high-exposure area.
2. Cleaning up the damaged reactor vessel and core required more than ten years to cut it up, pack it into canisters, and ship everything to the Federal Nuclear Reservation at Idaho Falls.
3. The cost of the cleanup was more than $1 billion.
4. Changes were implemented at other nuclear power plants as a consequence: Pressure relief valves have been removed, operators can no longer turn off the emergency cooling system, and operators must now spend about one-fourth of their time in training.

Chernobyl

The Soviet-designed Chernobyl reactor was a pressurized water reactor with individual fuel channels, which is very different from the pressurized water reactors used in the United States. The Chernobyl reactor was constructed with each fuel assembly in an individual pressure tube with circulating pressurized water. This heated water was circulated to a steam separator, and the steam was directed to turbines, which turned a generator to produce electricity. Graphite blocks surrounded the pressure tubes, serving as moderators to slow the neutrons involved in the chain reaction. The graphite was cooled by a mixture of helium and nitrogen. The reactor core was located in a concrete bunker that acted as a radiation shield. The top part was a steel cap and shield that supported the fuel assemblies. There were no containment buildings around the Soviet reactors as there are in the United States.

—Continued top of next page

Continued

The Chernobyl accident was the result of a combination of a poorly engineered reactor design, poorly trained reactor operators, and serious mistakes made by the operators on the day of the accident.

The reactor design was flawed because at low power, steam tended to form pockets in the water-filled fuel channels, creating a condition of instability. Instability occurred because (1) steam is not as efficient at cooling as is liquid water, and (2) liquid water acts as a moderator and neutron absorber while steam does not. Excess steam therefore leads to overheating and increased power generation. Increased power can lead to increased steam generation, which leads to further increases in power. This coupled response is very difficult to control because it feeds itself.

On April 25, 1986, the operators of Chernobyl unit 4 undertook a test to find out how long the turbines would spin and supply power following the loss of electrical power. The operators disabled the automatic shutdown mechanisms and then started the test early on April 26. The plan was to stabilize the reactor at 1,000 MW, but an error was made and the power fell to about 30 MW and pockets of steam became a problem. They tried to increase the power by removing all the control rods. At 1:00 A.M., they were able to stabilize the reactor at 200 MW. Then instability returned and the operators were making continuous adjustments to maintain a constant power. They reduced the feedwater to maintain steam pressure, and this created even more steam voids in the fuel channels. Power surged to a very high level and fuel elements ruptured. A steam explosion moved the reactor cap, exposing individual fuel channels and releasing fission products to the environment. A second explosion knocked a hole in the roof, exposed more of the reactor core, and the reactor graphite burst into flames. The graphite burned for nine days, releasing about 324 million Ci (12×10^{18} Bq) into the environment.

The fiery release of radioactivity was finally stopped by using a helicopter to drop sand, boron, lead, and other materials onto the burning graphite reactor. After the fire was out, the remains of the reactor were covered with a large concrete shelter.

In addition to destroying the reactor, the accident killed 30 people, 28 of whom died from radiation exposure. Another 134 people were treated for acute radiation poisoning and all recovered from the immediate effects.

Cleanup crews over the next year received about 10 rem (100 millisievert) to 25 rem (250 millisievert) and some received as much as 50 rem (500 millisievert). In addition to this direct exposure, large expanses of Belarus, Ukraine, and Russia were contaminated by radioactive fallout from the reactor fire. Hundreds of thousands of people have been resettled into less contaminated areas. The World Health Organization and other international agencies have studied the data to understand the impact of radiation-related disease. These studies do confirm a rising incidence of thyroid cancer but no increases in leukemia so far.

Fukushima I

It was 2:46 PM local time time on March, 11, 2011, when the largest earthquake to ever strike Japan hit with a magnitude of 9.0. The Fukushima I nuclear power plant is a 4.7-gigawatt plant located on the coast, about 220 km (140 mi) north of Tokyo. The plant has six boiling water reactors that are owned and operated by the Tokyo Electric Power Company (TEPCO).

When the earthquake struck, reactors 1, 2, and 3 were running and reactors 4, 5, and 6 were shut down for maintenance. Reactors 1, 2, and 3 were immediately shut down (scrammed) with control rods in place between the fuel rods in the reactor vessel. The reactor still had decay heat from the short-lived decay products. Several days of cooling are needed for a cold shutdown, and this cooling normally takes place by circulating water. However, a source of electricity is needed to run the cooling pumps, and the source of electricity was lost when the reactors shut down. Backup diesel generators are used to generate the power needed to run the cooling pumps. The diesel generators started, just as they were supposed to, and they ran for about an hour before the tsunami knocked them out. Batteries are used to back up the diesel generators, but the batteries provide power for only eight hours.

With all backup sources of power out of commission, or exhausted, there was no way to provide cooling. The reactor normally runs at 285°C (545°F), and the used fuel rods are cooled to about 60°C (140°F). Without cooling, both the reactor rods and the stored rods begin to undergo an increase in temperature. At a temperature of 1,200°C (2,192°F), the zirconium cladding of the fuel rods (see figure 11.14) reacts with water, producing zirconium oxide and hydrogen gas. The hydrogen gas is highly explosive, and a number of explosions occurred before the reactor building could be vented. The presence of cesium and iodine isotopes in the vented gas suggest that the metal cladding had been damaged, allowing isotope products to escape. Water in the reactor and in the spent fuel rod storage pool was boiling away, so the operators tried to add water by dropping it from helicopters and by using fire hoses. Finally, they were able to pump seawater to cover the fuel rods. Seawater is highly corrosive, but it did decrease the temperature to a nondamaging level. However, it also means that the reactors will be damaged by corrosion. The uranium fuel is a ceramic and has a melting temperature of 2800°C (5072°F). This is a very high temperature, but there is evidence that fuel rod melting nonetheless occurred in all three reactors that were running.

It might take years to cool, stabilize, and contain the reactor fuel rods and spent fuel rods. Contaminated water will need to be isolated, treated, and stored. It might take another eight to ten years to defuel and clean up the reactors. What is left of the reactors will need to be isolated and stored for a very long time.

A Closer Look

Nuclear Waste

There are two general categories of nuclear wastes: (1) low-level wastes and (2) high-level wastes. *Low-level wastes* are produced by the normal operation of a nuclear reactor. Radioactive isotopes sometimes escape from fuel rods in the reactor and in the spent fuel storage pools. These isotopes are removed from the water by ion-exchange resins and from the air by filters. The used resins and filters will contain the radioactive isotopes and will become low-level wastes. In addition, any contaminated protective clothing, tools, and discarded equipment also become low-level wastes.

Low-level liquid wastes are evaporated, mixed with cement, then poured into 55 gallon steel drums. Solid wastes are compressed and placed in similar drums. The drums are currently disposed of by burial in government-licensed facilities. In general, low-level waste has an activity of less than 1.0 curie per cubic foot. Contact with the low-level waste could expose a person to up to 20 millirems per hour of contact.

High-level wastes from nuclear power plants are spent nuclear fuel rods. At the present time, most of the commercial nuclear power plants have these rods in temporary storage at the plant site. These rods are "hot" in the radioactive sense, producing about 100,000 curies per cubic foot. They are also hot in the thermal sense, continuing to generate heat for months after removal from the reactor. The rods are cooled by heat exchangers connected to storage pools; they could otherwise achieve an internal temperature as high as 800°C for several decades. In the future, these spent fuel rods will be reprocessed or disposed of through terminal storage.

Agencies of the U.S. government have also accumulated millions of gallons of high-level wastes from the manufacture of nuclear weapons and nuclear research programs. These liquid wastes are stored in million-gallon stainless steel containers that are surrounded by concrete. The future of this large amount of high-level wastes may be evaporation to a solid form or mixture with a glass or ceramic matrix, which is melted and poured into stainless steel containers. These containers would be buried in solid granite rock in a stable geologic depository. Such high-level wastes must be contained for thousands of years as they undergo natural radioactive decay (box figure 11.2). Burial at a depth of 610 to 914 m (2,000–3,000 ft) in solid granite

Box Figure 11.2 This is a standard warning sign for a possible radioactive hazard. Such warning signs would have to be maintained around a nuclear waste depository for thousands of years.

would provide protection from exposure by explosives, meteorite impact, or erosion. One major concern about this plan is that a hundred generations later, people might lose track of what is buried in the nuclear garbage dump.

Even at this rate, the Sun has enough hydrogen to continue the process for an estimated 5 billion years. Several fusion reactions take place between hydrogen and helium isotopes, including the following:

$$\frac{1}{1}H + \frac{1}{1}H \rightarrow \frac{2}{1}H + \frac{0}{1}e$$

$$\frac{2}{1}H + \frac{2}{1}H \rightarrow \frac{3}{2}He + \frac{1}{0}n$$

$$\frac{3}{2}He + \frac{3}{2}He \rightarrow \frac{4}{2}He + 2\frac{1}{1}H$$

The fusion process would seem to be a desirable energy source on Earth because (1) two isotopes of hydrogen, deuterium ($\frac{2}{1}H$) and tritium ($\frac{3}{1}H$), undergo fusion at a relatively low temperature; (2) the supply of deuterium is practically unlimited, with each gallon of seawater containing about a teaspoonful of deuterium dioxide; and (3) enormous amounts of energy are released with no radioactive by-products.

The oceans contain enough deuterium to generate electricity for the entire world for millions of years, and tritium can be constantly produced by a fusion device. Researchers know what needs to be done to tap this tremendous energy source. The problem is *how* to do it in an economical, continuous energy-

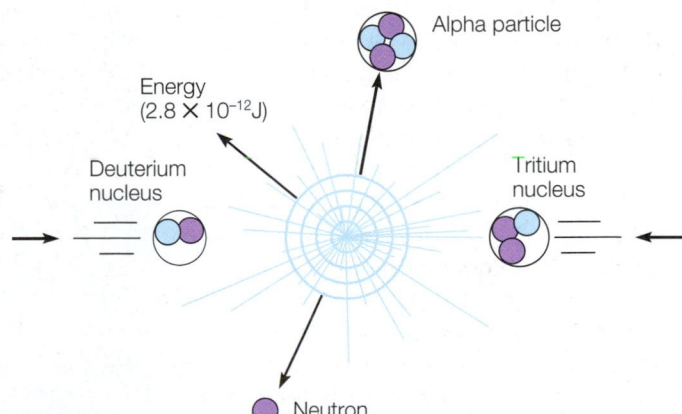

FIGURE 11.19 A fusion reaction between a tritium nucleus and a deuterium nucleus requires a certain temperature, density, and time of containment to take place.

producing fusion reactor. The problem, one of the most difficult engineering tasks ever attempted, is meeting three basic fusion reaction requirements of (1) temperature, (2) density, and (3) time (figure 11.19):

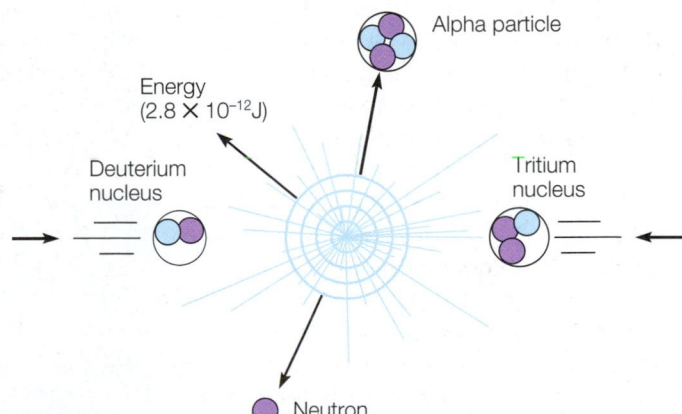

Science and Society

High-Level Nuclear Waste

In 1982, the U.S. Congress established a national policy to solve the problem of nuclear waste disposal. This policy is a federal law called the Nuclear Waste Policy Act. Congress based this policy on what most scientists worldwide agree is the best way to dispose of nuclear waste.

The Nuclear Waste Policy Act made the U.S. Department of Energy (DOE) responsible for finding a site, building, and operating an underground disposal facility called a geologic repository.

In 1983, the DOE selected nine locations in six states for consideration as potential repository sites. This selection was based on data collected for nearly ten years. The nine sites were studied and results of these preliminary studies were reported in 1985. Based on these reports, President Ronald Reagan approved three sites for intensive scientific study called site characterization. The three sites were Hanford, Washington; Deaf Smith County, Texas; and Yucca Mountain, Nevada.

In 1987, Congress amended the Nuclear Waste Policy Act and directed the Department of Energy to study only Yucca Mountain. On July 9, 2002, the U.S. Senate cast the final legislative vote approving the development of a repository at Yucca Mountain.

The current administration has determined that Yucca Mountain is not a workable option for a long-term repository for spent nuclear fuel and high-level radioactive waste. Accordingly, on February 1, 2010, DOE filed a motion with the Nuclear Regulatory Commission to stay the pending proceedings on DOE's Yucca Mountain repository license application filed June 2008 until DOE's planned motion to withdraw is resolved. The motion to stay the proceedings was granted on February 16, 2010. On March 3, 2010, DOE filed a motion to withdraw the license application with prejudice.*

Press Release: http://energy.gov/sites/prod/files/edg/media/DOE_Motion_to_Withdraw.pdf

People Behind the Science

Marie Curie (1867–1934)

Marie Curie was a Polish-born French scientist who, with her husband, Pierre Curie (1859–1906), was an early investigator of radioactivity. From 1896, the Curies worked together, building on the results of Henri Becquerel, who had discovered radioactivity from uranium salts. Marie Curie discovered that thorium also emits radiation and found that the mineral pitchblende was even more radioactive than could be accounted for by any uranium and thorium content. The Curies then carried out an exhaustive search and in July 1898 announced the discovery of polonium, followed in December of that year with the discovery of radium. They shared the 1903 Nobel Prize for physics with Becquerel for the discovery of radioactivity. The Curies did not participate in Becquerel's discovery but investigated radioactivity and gave the phenomenon its name. Marie Curie went on to study the chemistry and medical applications of radium, and was awarded the 1911 Nobel Prize for chemistry in recognition of her work in isolating the pure metal.

At the outbreak of World War I in 1914, Marie Curie helped to equip ambulances with X-ray equipment and drove the ambulances to the front lines. The International Red Cross made her head of its Radiological Service. She taught medical orderlies and doctors how to use the new technique. By the late 1920s, her health began to deteriorate: continued exposure to high-energy radiation had given her leukemia. She entered a sanatorium and died on July 4, 1934.

Throughout much of her life, Marie Curie was poor, and the painstaking radium extractions were carried out in primitive conditions. The Curies refused to patent any of their discoveries, wanting them to benefit everyone freely. They used the Nobel Prize money and other financial rewards to finance further research. One of the outstanding applications of their work has been the use of radiation to treat cancer, one form of which cost Marie Curie her life.

Source: Modified from the *Hutchinson Dictionary of Scientific Biography.* © RM, 2011. All rights reserved. Helicon Publishing is a division of RM.

1. **Temperature.** Nuclei contain protons and are positively charged, so they experience the electromagnetic repulsion of like charges. This force of repulsion can be overcome, moving the nuclei close enough to fuse together, by giving the nuclei sufficient kinetic energy. The fusion reaction of deuterium and tritium, which has the lowest temperature requirements of any fusion reaction known at the present time, requires temperatures on the order of 100 million degrees Celsius.
2. **Density.** There must be a sufficiently dense concentration of heavy hydrogen nuclei, on the order of $10^{14}/cm^3$, so many reactions occur in a short time.
3. **Time.** The nuclei must be confined at the appropriate density up to a second or longer at pressures of at least 10 atmospheres to permit a sufficient number of reactions to take place.

The temperature, density, and time requirements of a fusion reaction are interrelated. A short time of confinement, for example, requires an increased density, and a longer confinement time requires less density. The primary problems of fusion research are the high-temperature requirements and confinement. No material in the world can stand up to a temperature of 100 million degrees Celsius, and any material container would be instantly vaporized. Thus, research has centered on meeting the fusion reaction requirements without a material container. Two approaches are being tested, *magnetic confinement* and *inertial confinement.*

Magnetic confinement utilizes a very hot *plasma,* a gas consisting of atoms that have been stripped of their electrons because of the high kinetic energies. The resulting positively and negatively charged particles respond to electrical and magnetic forces, enabling researchers to develop a "magnetic bottle," that is, magnetic fields that confine the plasma and avoid the problems of material containers that would vaporize. A magnetically confined plasma is very unstable, however, and researchers have compared the problem to trying to carry a block of Jell-O on a pair of rubber bands. Different magnetic field geometries and magnetic "mirrors" are the topics of research in attempts to stabilize the hot, wobbly plasma. Electric currents, injection of fast ions, and radio frequency (microwave) heating methods are also being studied.

Inertial confinement is an attempt to heat and compress small frozen pellets of deuterium and tritium with energetic laser beams or particle beams, producing fusion. The focus of this research is new and powerful lasers, light ion and heavy ion beams. If successful, magnetic or inertial confinement will provide a long-term solution for future energy requirements.

SUMMARY

Radioactivity is the spontaneous emission of particles or energy from an unstable atomic nucleus. The modern atomic theory pictures the nucleus as protons and neutrons held together by a short-range *nuclear force* that has moving *nucleons* (protons and neutrons) in *energy shells* analogous to the shell structure of electrons. A graph of the number of neutrons to the number of protons in a nucleus reveals that stable nuclei have a certain neutron-to-proton ratio in a *band of stability.* Nuclei that are above or below the band of stability and nuclei that are beyond atomic number 83 are radioactive and undergo *radioactive decay.*

Three common examples of radioactive decay involve the emission of an *alpha particle,* a *beta particle,* and a *gamma ray.* An alpha particle is a helium nucleus, consisting of two protons and two neutrons. A beta particle is a high-speed electron that is ejected from the nucleus. A gamma ray is a short-wavelength electromagnetic radiation from an excited nucleus. In general, nuclei with an atomic number of 83 or larger become more stable by alpha emission. Nuclei with a neutron-to-proton ratio that is too large become more stable by beta emission. Gamma ray emission occurs from a nucleus that was left in a high-energy state by the emission of an alpha or beta particle.

Each radioactive isotope has its own specific *radioactive decay rate.* This rate is usually described in terms of *half-life,* the time required for one-half the unstable nuclei to decay.

Radiation is measured by (1) its effects on photographic film, (2) the number of ions it produces, or (3) the flashes of light produced on a phosphor. It is measured at a source in units of a *curie,* defined as 3.70×10^{10} nuclear disintegrations per second. It is measured where received in units of a *rad.* A *rem* is a measure of radiation that takes into account the biological effectiveness of different types of radiation damage. In general, the natural environment exposes everyone to 100 to 500 millirems per year, an exposure called *background radiation.* Lifestyle and location influence the background radiation received, but the worldwide average is 240 millirems per year.

Energy and mass are related by Einstein's famous equation of $E = mc^2$, which means that *matter can be converted to energy and energy to matter.* The mass of a nucleus is always less than the sum of the masses of the individual particles of which it is made. This *mass defect* of a nucleus is equivalent to the energy released when the nucleus was formed according to $E = mc^2$. It is also the *binding energy,* the energy required to break the nucleus apart into nucleons.

When the binding energy is plotted against the mass number, the greatest binding energy per nucleon is seen to occur for an atomic number near that of iron. More massive nuclei therefore release energy by fission, or splitting to more stable nuclei. Less massive nuclei release energy by fusion, the joining of less massive nuclei to produce a more stable, more massive nucleus. Nuclear fission provides the energy for atomic explosions and nuclear power plants. Nuclear fusion is the energy source of the Sun and other stars and also holds promise as a future energy source for humans.

Summary of Equations

11.1   energy = mass × the speed of light squared

$$E = mc^2$$

KEY TERMS

alpha particle (p. 242)
background radiation (p. 251)
band of stability (p. 245)
beta particle (p. 242)
binding energy (p. 252)

chain reaction (p. 253)
critical mass (p. 253)
curie (p. 250)
gamma ray (p. 242)
half-life (p. 248)

mass defect (p. 252)
nuclear fission (p. 252)
nuclear fusion (p. 252)
nucleons (p. 244)
rad (p. 250)

radioactive decay (p. 243)
radioactivity (p. 242)
rem (p. 250)

APPLYING THE CONCEPTS

Answers are located in appendix F.

1. A high-speed electron ejected from a nucleus during radioactive decay is called a (an)
   a. alpha particle.
   b. beta particle.
   c. gamma ray.
   d. None of the above is correct.

2. The ejection of an alpha particle from a nucleus results in
   a. an increase in the atomic number by one.
   b. an increase in the atomic mass by four.
   c. a decrease in the atomic number by two.
   d. None of the above is correct.

3. An atom of radon-222 loses an alpha particle to become a more stable atom of
   a. radium.
   b. bismuth.
   c. polonium.
   d. radon.

4. A sheet of paper will stop a (an)
   a. alpha particle.
   b. beta particle.
   c. gamma ray.
   d. None of the above is correct.

5. The most penetrating of the three common types of nuclear radiation is the
   a. alpha particle.
   b. beta particle.
   c. gamma ray.
   d. All have equal penetrating ability.

6. An atom of an isotope with an atomic number greater than 83 will probably emit a (an)
   a. alpha particle.
   b. beta particle.
   c. gamma ray.
   d. None of the above is correct.

7. An atom of an isotope with a large neutron-to-proton ratio will probably emit a (an)
   a. alpha particle.
   b. beta particle.
   c. gamma ray.
   d. None of the above is correct.

8. The rate of radioactive decay can be increased by increasing the
   a. temperature.
   b. pressure.
   c. size of the sample.
   d. None of the above is correct.

9. Isotope A has a half-life of seconds, and isotope B has a half-life of millions of years. Which isotope is more radioactive?
   a. It depends on the sample size.
   b. isotope A
   c. isotope B
   d. Unknown from the information given.

10. A measure of radioactivity at the *source* is (the)
    a. curie.
    b. rad.
    c. rem.
    d. Any of the above is correct.

11. A measure of radiation received that considers the biological effect resulting from the radiation is (the)
    a. curie.
    b. rad.
    c. rem.
    d. Any of the above is correct.

12. Used fuel rods from a nuclear reactor contain about
    a. 96% usable uranium and plutonium.
    b. 33% usable uranium and plutonium.
    c. 4% usable uranium and plutonium.
    d. 0% usable uranium and plutonium.

QUESTIONS FOR THOUGHT

1. How is a radioactive material different from a material that is not radioactive?

2. What is radioactive decay? Describe how the radioactive decay rate can be changed if this is possible.

3. Describe three kinds of radiation emitted by radioactive materials. Describe what eventually happens to each kind of radiation after it is emitted.

4. How are positively charged protons able to stay together in a nucleus since like charges repel?

5. What is half-life? Give an example of the half-life of an isotope, describing the amount remaining and the time elapsed after five half-life periods.

6. Would you expect an isotope with a long half-life to be more, the same, or less radioactive than an isotope with a short half-life? Explain.

7. What is meant by background radiation? What is the normal radiation dose for the average person from background radiation?

8. Why is there controversy about the effects of long-term, low levels of radiation exposure?

9. What is a mass defect? How is it related to the binding energy of a nucleus? How can both be calculated?

10. Compare and contrast nuclear fission and nuclear fusion.

FOR FURTHER ANALYSIS

1. What are the significant differences between a radioactive isotope and an isotope that is not radioactive?

2. Analyze the different types of radioactive decay to explain how each is a hazard to living organisms.

3. Make up a feasible explanation for why some isotopes have half-lives of seconds, yet other kinds of isotopes have half-lives in the billions of years.

4. Suppose you believe the threshold model of radiation exposure is correct. Describe a conversation between yourself and another person who feels strongly that the linear model of radiation exposure is correct.

5. Explain how the fission of heavy elements and the fusion of light elements both release energy.

6. Write a letter to your congressional representative describing why used nuclear fuel rods should be reprocessed rather than buried as nuclear waste.

7. What are the similarities and differences between a nuclear fission power plant and a nuclear fusion power plant?

INVITATION TO INQUIRY

How Much Radiation?

Ionizing radiation is understood to be potentially harmful if certain doses are exceeded. How much ionizing radiation do you acquire from the surroundings where you live, from your lifestyle, and from medical procedures? Investigate radiation from cosmic sources, the Sun, television sets, time spent in jet airplanes, and dental or other X-ray machines. What are other sources of ionizing radiation in your community? How difficult is it to find relevant information and make recommendations? Does any agency monitor the amount of radiation that people receive? What are the problems and issues with such monitoring?

PARALLEL EXERCISES

The exercises in groups A and B cover the same concepts. Solutions to group A exercises are located in appendix G.

Group A

Note: You will need the table of atomic weights inside the back cover of this text.

1. Give the number of protons and the number of neutrons in the nucleus of each of the following isotopes:
   a. cobalt-60
   b. potassium-40
   c. neon-24
   d. lead-208

2. Write the nuclear symbols for each of the nuclei in exercise 1.

3. Predict if the nuclei in exercise 1 are radioactive or stable, giving your reasoning behind each prediction.

4. Write a nuclear equation for the decay of the following nuclei as they give off a beta particle:
   a. $^{56}_{26}Fe$
   b. $^{24}_{11}Na$
   c. $^{64}_{29}Cu$
   d. $^{24}_{11}Na$
   e. $^{214}_{82}Pb$
   f. $^{32}_{15}P$

Group B

Note: You will need the table of atomic weights inside the back cover of this text.

1. Give the number of protons and the number of neutrons in the nucleus of each of the following isotopes:
   a. aluminum-25
   b. technetium-95
   c. tin-120
   d. mercury-200

2. Write the nuclear symbols for each of the nuclei in exercise 1.

3. Predict if the nuclei in exercise 1 are radioactive or stable, giving your reasoning behind each prediction.

4. Write a nuclear equation for the beta emission decay of each of the following:
   a. $^{14}_{6}C$
   b. $^{60}_{27}Co$
   c. $^{24}_{11}Na$
   d. $^{241}_{94}Pu$
   e. $^{131}_{53}I$
   f. $^{210}_{82}Pb$

PARALLEL EXERCISES—*Continued*

5. Write a nuclear equation for the decay of the following nuclei as they undergo alpha emission:

   a. $^{235}_{92}U$

   b. $^{226}_{88}Ra$

   c. $^{239}_{94}Pu$

   d. $^{214}_{83}Bi$

   e. $^{230}_{90}Th$

   f. $^{210}_{84}Po$

6. The half-life of iodine-131 is 8 days. How much of a 1.0 oz sample of iodine-131 will remain after 32 days?

7. How much energy must be supplied to break a single iron-56 nucleus into separate protons and neutrons? (The mass of an iron-56 nucleus is 55. 9206 u, one proton is 1.00728 u, and one neutron is 1.00867 u.)

5. Write a nuclear equation for each of the following alpha emission decay reactions:

   a. $^{241}_{95}Am$

   b. $^{232}_{90}Th$

   c. $^{223}_{88}Ra$

   d. $^{234}_{92}U$

   e. $^{242}_{96}Cm$

   f. $^{237}_{93}Np$

6. If the half-life of cesium-137 is 30 years, how much time will be required to reduce a 1.0 kg sample to 1.0 g?

7. How much energy is needed to separate the nucleons in a single lithium-7 nucleus? (The mass of a lithium-7 nucleus is 7.01435 u, one proton is 1.00728 u, and one neutron is 1.00867 u.)

12

The Universe

The large Whirlpool Galaxy (left) is known for its sharply defined spiral arms. Their prominence could be the result of the Whirlpool's gravitational tug-of-war with its smaller companion galaxy (right).

CORE **CONCEPT**

The night sky is filled with billions of stars, and the Sun is an ordinary star with an average brightness.

OUTLINE

Stars are very large accumulations of gases that release energy from nuclear fusion reactions (p. 267).

Stars vary in brightness and temperature (p. 268).

The Hertzsprung-Russell diagram is a plot of temperature versus brightness of stars (p. 269).

Stars are organized in galaxies, the basic unit of the universe (p. 273).

OVERVIEW

Astronomy is an exciting field of science that has fascinated people since the beginnings of recorded history. Ancient civilizations searched the heavens in wonder, some recording on clay tablets what they observed. Many religious and philosophical beliefs were originally based on interpretations of these ancient observations. Today, we are still awed by space, but now we are fascinated with ideas of space travel, black holes, and the search for extraterrestrial life. Throughout history, people have speculated about the universe and their place in it and watched the sky and wondered (figure 12.1). What is out there and what does it all mean? Are there other people on other planets, looking at the star in their sky that is our Sun, wondering if we exist?

Until about thirty years ago, progress in astronomy was limited to what could be observed and photographed. Developments in technology then began to provide the details of what is happening in the larger expanses of space away from Earth. This included new data made available from the development of infrared, radio, and X-ray telescopes and the Hubble Space Telescope. New data and the discovery of pulsars, neutron stars, and black holes began to fit together like the pieces of a puzzle. Theoretical models emerged about how stars, galaxies, and the universe have evolved. This chapter is concerned with these topics and how the stars are arranged in space. The chapter concludes with theoretical models of how the universe began and what may happen to it in the future.

12.1 THE NIGHT SKY

Early civilizations had a much better view of the night sky before city lights, dust, and pollution obscured much of it. Today, you must travel far from cities, perhaps to a remote mountaintop, to see a clear night sky as early people observed it. Back then, people could clearly see the motion of the moon and stars night after night, observing recurring cycles of motion. These cycles became important as people associated them with certain events. Thus, watching the Sun, Moon, and star movements became a way to identify when to plant crops, when to harvest, and when it was time to plan for other events. Observing the sky was an important activity, and many early civilizations built observatories with sighting devices to track and record astronomical events. Stonehenge, for example, was an ancient observatory built in England around 2600 B.C. by Neolithic people (figure 12.2).

Light from the stars and planets must pass through Earth's atmosphere to reach you, and this affects the light. Stars appear as point *sources* of light, and each star generates its own light. The stars seem to twinkle because density differences in the atmosphere refract the point of starlight one way, then the other, as the air moves. The result is the slight dancing about and change in intensity called twinkling. The points of starlight are much steadier when viewed on a calm night or from high in the mountains, where there is less atmosphere for the starlight to pass through. Astronauts outside the atmosphere see no twinkling, and the stars appear as steady point sources of light.

Back at ground level, within the atmosphere, the *reflected* light from a planet does not seem to twinkle. A planet appears as a disk of light rather than a point source, so refraction from moving air of different densities does not affect the image as much. Sufficient air movement can cause planets to appear to shimmer, however, just as a road appears to shimmer on a hot summer day.

How far away is a star? When you look at the sky, it appears that all the stars are at the same distance. It seems impossible to know anything about the actual distance to any given star. Standard referent units of length such as kilometers or miles

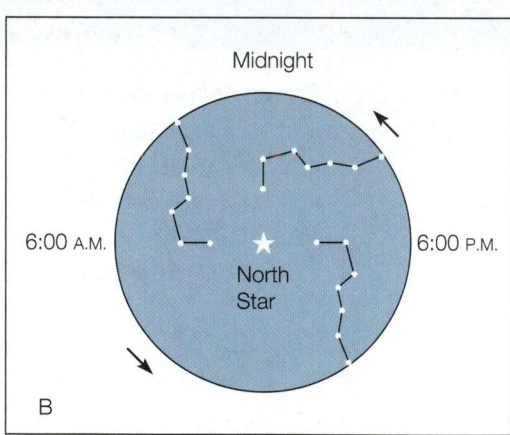

FIGURE 12.1 Ancient civilizations used celestial cycles of motion as clocks and calendars. (*A*) This photograph shows the path of stars around the North Star. (*B*) A "snapshot" of the position of the Big Dipper over a period of twenty-four hours as it turns around the North Star. This shows how the Big Dipper can be used to help you keep track of time.

have little meaning in astronomy since there are no referent points of comparison. Distance without a referent point can be measured in terms of angles or time. The unit of astronomical distance that uses time is the **light-year** (ly). A light-year is

FIGURE 12.2 The stone pillars of Stonehenge were positioned so that they could be used to follow the movement of the Sun and Moon with the seasons of the year.

the distance that light travels in one year, about 9.5×10^{12} km (about 6×10^{12} mi).

If you could travel by spaceship a few hundred light-years from Earth, you would observe the Sun shrink to a bright point of light among the billions and billions of other stars. The Sun is just an ordinary star with an average brightness. Like the other stars, the Sun is a massive, dense ball of gases with a surface heated to incandescence by energy released from fusion reactions deep within. Since the Sun is an average star, it can be used as a reference for understanding all the other stars.

12.2 ORIGIN OF STARS

Theoretically, stars are born from swirling clouds of hydrogen gas in the deep space between other stars. Such interstellar (between stars) clouds are called **nebulae.** These clouds consist of random, swirling atoms of gases that have little gravitational attraction for one another because they have little mass. Complex motion of stars, however, can produce a shock wave that causes particles to move closer together, making local compressions. Their mutual gravitational attraction then begins to pull them together into a cluster. The cluster grows as more atoms are pulled in, which increases the mass and thus the gravitational attraction, and still more atoms are pulled in from farther away. Theoretical calculations indicate that on the order of 1×10^{57} atoms are necessary, all within a distance of 3 trillion km (about 1.9 trillion mi). When these conditions occur, the cloud of gas atoms begins to condense by gravitational attraction to a **protostar,** an accumulation of gases that will become a star.

Gravitational attraction pulls the average protostar from a cloud with a diameter of trillions of kilometers (trillions of miles) down to a dense sphere with a diameter of 2.5 million km (1.6 million mi) or so. As gravitational attraction accelerates the atoms toward the center, they gain kinetic energy, and the interior temperature increases. Over a period of some 10 million years of contracting and heating,

the temperature and density conditions at the center of the protostar are sufficient to start nuclear fusion reactions. Pressure from hot gases and energy from increasing fusion reactions begin to balance the gravitational attraction over the next 17 million years, and the newborn, average star begins its stable life, which will continue for the next 10 billion years.

The interior of an average star, such as the Sun, is modeled after the theoretical pressure, temperature, and density conditions that would be necessary to produce the observed energy and light from the surface. This model describes the interior as a set of three shells: (1) the core, (2) a radiation zone, and (3) the convection zone (figure 12.3).

Our model describes the **core** as a dense, very hot region where nuclear fusion reactions release gamma and X-ray radiation. The density of the core is about twelve times that of solid lead. Because of the exceedingly hot conditions, however, the core remains in a plasma state even at this density.

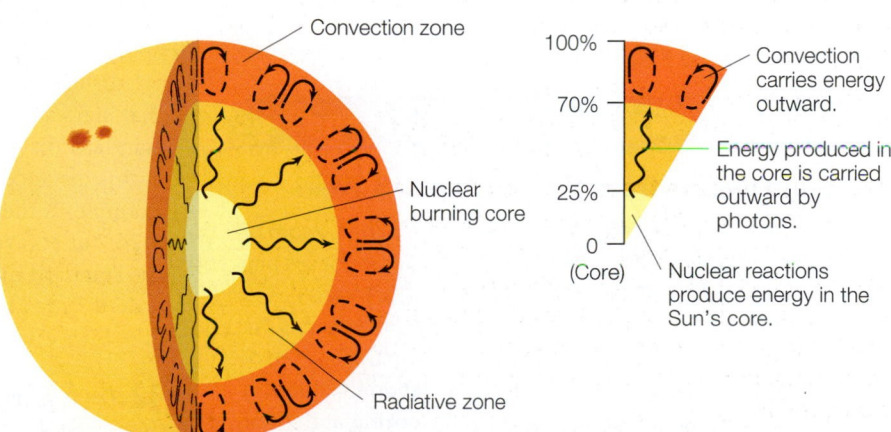

FIGURE 12.3 Energy-producing nuclear reactions occur only within the inner 25 percent of the Sun's radius. The energy produced by these reactions is carried outward by photons to 70 percent of the Sun's radius. From that distance outward, convection carries most of the Sun's energy.

The model describes the **radiation zone** as less dense than the core, having a density about the same as the density of water. Energy in the form of gamma and X rays from the core is absorbed and reemitted by collisions with atoms in this zone. The radiation slowly diffuses outward because of the countless collisions over a distance comparable to the distance between Earth and the Moon. It could take millions of years before this radiation finally escapes the radiation zone.

In the model, the **convection zone** begins about seven-tenths of the way to the surface, where the density of the plasma is about 1 percent the density of water. Plasma at the bottom of this zone is heated by radiation from the radiation zone below, expands from the heating, and rises to the surface by convection. At the surface, the plasma emits energy in the form of visible light, ultraviolet radiation, and infrared radiation, which moves out into space. As it loses energy, the plasma contracts in volume and sinks back to the radiation zone to become heated again, continuously carrying energy from the radiation zone to the surface in convection cells. The surface is continuously heated by the convection cells as it gives off energy to space, maintaining a temperature of about 5,800 K (about 5,500°C).

As an average star, the Sun converts about 1.4×10^{17} kg of matter to energy every year as hydrogen nuclei are fused to produce helium. The Sun was born about 5 billion years ago and has sufficient hydrogen in the core to continue shining for another 4 or 5 billion years. Other stars, however, have masses that are much greater or much less than the mass of the Sun so they have different life spans. More massive stars generate higher temperatures in the core because they have a greater gravitational contraction from their greater masses. Higher temperatures mean increased kinetic energy, which results in increased numbers of collisions between hydrogen nuclei with the end result an increased number of fusion reactions. Thus, a more massive star uses up its hydrogen more rapidly than a less massive star. On the other hand, stars that are less massive than the Sun use their hydrogen at a slower rate so they have longer life spans. The life spans of the stars range from a few million years for large, massive stars, to 10 billion years for average stars such as the Sun, to trillions of years for small, less massive stars.

12.3 BRIGHTNESS OF STARS

Stars generate their own light, but some stars appear brighter than others in the night sky. As you can imagine, this difference in brightness could be related to (1) the amount of light produced by the stars, (2) the size of each star, or (3) the distance to a particular star. A combination of these factors is responsible for the brightness of a star as it appears to you in the night sky. A classification scheme for different levels of brightness that you see is called the **apparent magnitude** scale (table 12.1).

The apparent magnitude scale is based on a system established by the Greek astronomer Hipparchus over two thousand years ago. Hipparchus made a catalog of stars he could see and assigned a numerical value to each to identify its relative brightness. The brightness values ranged from 1 to 6, with the number 1 assigned to the brightest star and the number 6 assigned to the faintest star that could be seen. Later, some stars were found to be brighter than the apparent magnitude of +1, which extended the scale into negative numbers. The brightest star in the night sky is Sirius, for example, with an apparent magnitude of −1.42 (see table 12.1).

The apparent magnitude of a star depends on how far away stars are in addition to differences in the stars themselves. Stars at a farther distance will appear fainter, and those closer will appear brighter, just as any other source of light. To compensate for distance differences, astronomers calculate the brightness that stars would appear to have if they were all at a defined, standard distance (32.6 light-years). The brightness of a star at this distance is called the **absolute magnitude.** The Sun, for example, is the closest star and has an apparent magnitude of −26.7 at an average distance from Earth. When viewed from the standard distance, the Sun would have an absolute magnitude of +4.8, which is about the brightness of a faint star.

12.4 STAR TEMPERATURE

If you observe the stars on a clear night, you will notice that some are brighter than others, and you will also notice some color differences. Some stars have a reddish color, some have

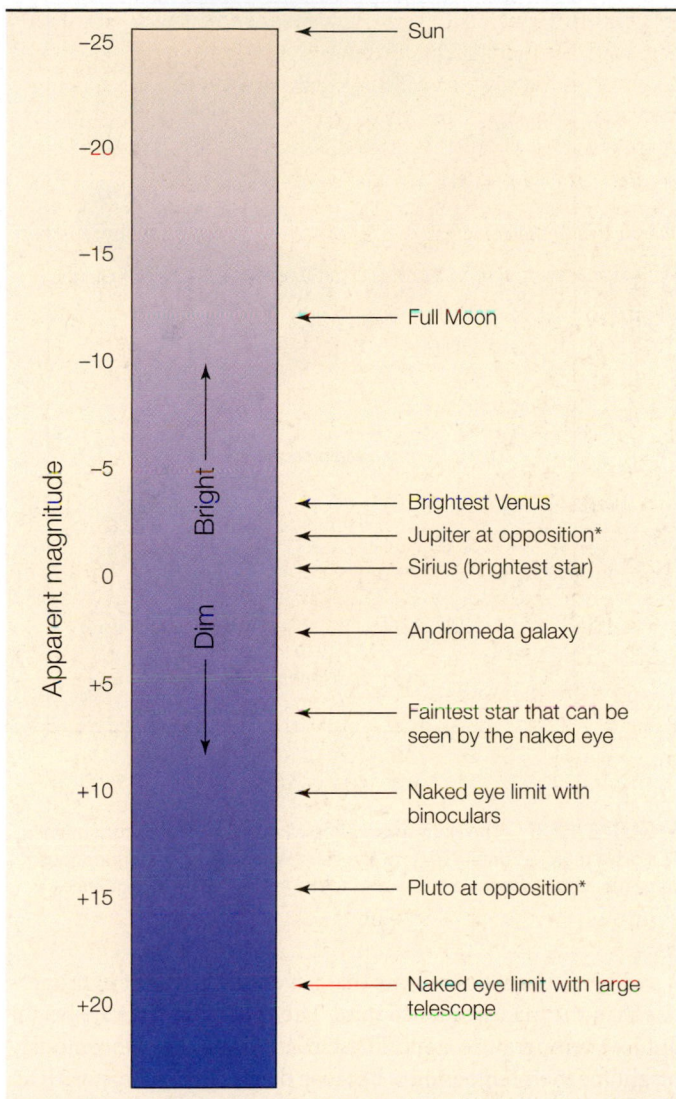

*When the planet is opposite the Sun in the sky.

the group with the strongest hydrogen line spectrum, B for slightly weaker lines, and on to the last group with the faintest lines. Later, astronomers realized that the star temperature was the important variable, so they rearranged the categories according to decreasing temperatures. The original letter categories were retained, however, resulting in classes of stars with the hottest temperature first and the coolest last with the sequence O B A F G K M. Table 12.2 compares the color, temperature ranges, and other features of the stellar spectra classification scheme.

12.5 STAR TYPES

In 1910, Henry Russell in the United States and Ejnar Hertzsprung in Denmark independently developed a scheme to classify stars with a temperature-luminosity graph. The graph is called the **Hertzsprung-Russell diagram,** or the *H-R diagram* for short. The diagram is a plot with temperature indicated by spectral types, and the true brightness indicated by absolute magnitude. The diagram, as shown in figure 12.5, plots temperature by spectral types sequenced O through M types, so the temperature decreases from left to right. The hottest, brightest stars are thus located at the top left of the diagram, and the coolest, faintest stars are located at the bottom right.

Each dot is a data point representing the surface temperature and brightness of a particular star. The Sun, for example, is a type G star with an absolute magnitude of about +5, which places the data point for the Sun almost in the center of the diagram. This means that the Sun is an ordinary, average star with respect to both surface temperature and true brightness.

Most of the stars plotted on an H-R diagram fall in or close to a narrow band that runs from the top left to the lower right. This band is made up of **main sequence stars.** Stars along the main sequence band are normal, mature stars that are using

a bluish white color, and others have a yellowish color. This color difference is understood to be a result of the relationship that exists between the color and the temperature of a glowing object. The colors of the various stars are a result of the temperatures of the stars (figure 12.4). You see a cooler star as reddish in color and comparatively hotter stars as bluish white. Stars with in-between temperatures, such as the Sun, appear to have a yellowish color.

Astronomers use information about the star temperature and spectra as the basis for a star classification scheme. Originally, the classification scheme was based on sixteen categories according to the strength of the hydrogen line spectra (see line spectrum in chapter 8). The groups were identified alphabetically with A for

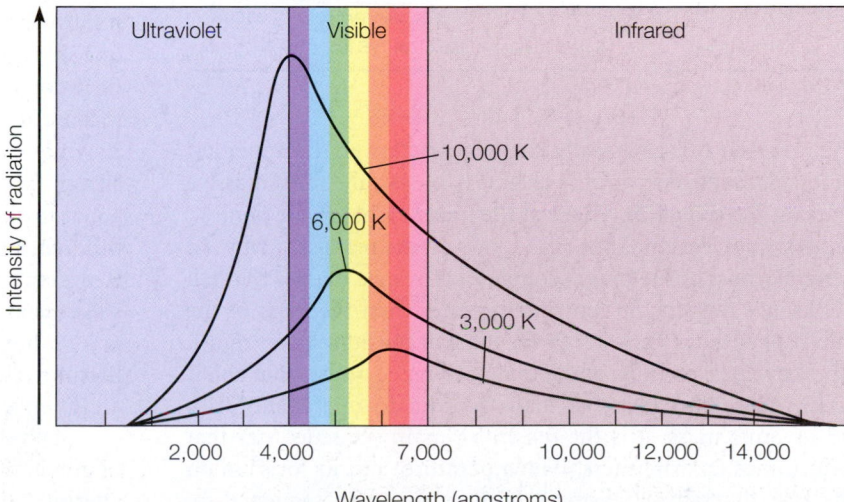

FIGURE 12.4 The distribution of radiant energy emitted is different for stars with different surface temperatures. Note that the peak radiation of a cooler star is more toward the red part of the spectrum and the peak radiation of a hotter star is more toward the blue part of the spectrum.

Seeing Spectra

Inexpensive diffraction grating on plastic film is available from many scientific materials supply houses. If available, view the light from gas discharge tubes to see bright line spectra. Use the grating to examine light from incandescent lightbulbs of different wattages, fluorescent lights, lighted "neon" signs of different colors, and street lights. Describe the type of spectrum each produces. If it has lines, see if you can identify the elements present.

TABLE 12.2 Major Stellar Spectral Types and Temperatures

Type	Color	Temperature (K)	Comment
O	Bluish	30,000–80,000	Spectrum with ionized helium and hydrogen but little else; short-lived and rare stars
B	Bluish	10,000–30,000	Spectrum with neutral helium, none ionized
A	Bluish	7,500–10,000	Spectrum with no helium, strongest hydrogen, some magnesium and calcium
F	White	6,000–7,500	Spectrum with ionized calcium, magnesium, neutral atoms of iron
G	Yellow	5,000–6,000	The spectral type of the Sun. Spectrum shows sixty-seven elements in the Sun
K	Orange-red	3,500–5,000	Spectrum packed with lines from neutral metals
M	Reddish	2,000–3,500	Band spectrum of molecules, e.g., titanium oxide; other related spectral types (R, N, and S) are based on other molecules present in each spectral type

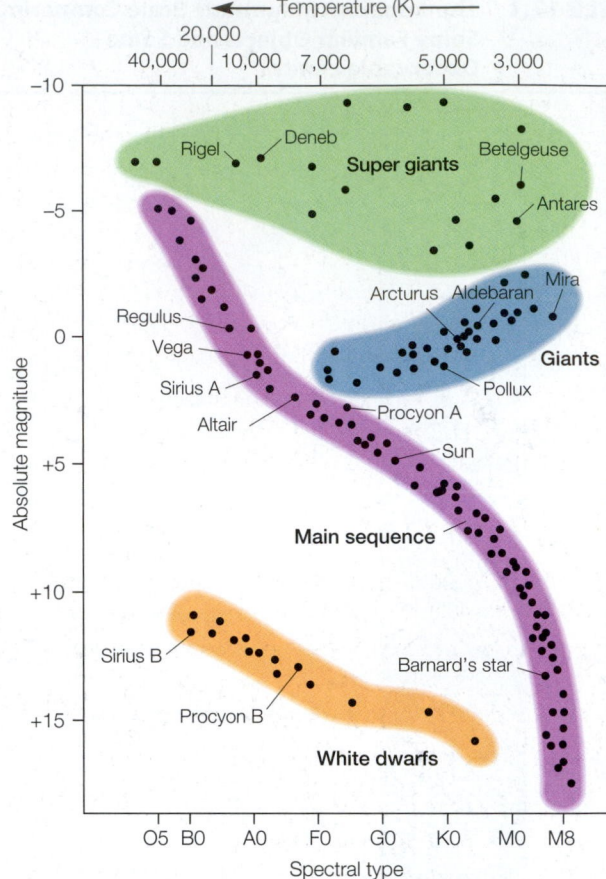

FIGURE 12.5 The Hertzsprung-Russell diagram. The main sequence and giant regions contain most of the stars, whereas hot underluminous stars, the white dwarfs, lie below and to the left of the main sequence.

their nuclear fuel at a steady rate. Those stars on the upper left of the main sequence are the brightest, bluest, and most massive stars on the sequence. Those at the lower right are the faintest, reddest, and least massive of the stars on the main sequence. In general, most of the main sequence stars have masses that fall between a range from ten times greater than the mass of the Sun (upper left) to one-tenth the mass of the Sun (lower right). The extremes, or ends, of the main sequence range from about sixty times more massive than the Sun to one-twenty-fifth of the Sun's mass. It is the *mass* of a main sequence star that determines its brightness, its temperature, and its location on the H-R diagram. High-mass stars on the main sequence are brighter, hotter, and have shorter lives than low-mass stars. These relationships do not apply to the other types of stars in the H-R diagram.

There are groups of stars that have a different set of properties than the main sequence stars. The **red giant stars** are bright but low-temperature stars. These reddish stars are enormously bright for their temperature because they are very large, with an enormous surface area giving off light. A red giant might be one hundred times larger but have the same mass as the Sun. These low-density red giants are located in the upper right part of the H-R diagram. The **white dwarf stars,** on the other hand, are located at the lower left because they are faint, white-hot stars. A white dwarf is faint because it is small, perhaps twice the size of Earth. It is also very dense, with a mass approximately equal to the Sun's. During its lifetime, a star will be found in different places on the H-R diagram as it undergoes changes. Red giants and white dwarfs are believed to be evolutionary stages that aging stars pass through, and the path a star takes across the diagram is called an evolutionary track. During the lifetime of the Sun, it will be a main sequence star, a red giant, and then a white dwarf.

Stars such as the Sun emit a steady light because the force of gravitational contraction is balanced by the outward flow of energy. *Variable stars,* on the other hand, are stars that change in brightness over a period of time. A **Cepheid variable** is a bright variable star that is used to measure distances. There is a general relationship between the period and the brightness: the longer the time needed for one pulse, the greater the apparent brightness of that star. The period-brightness relationship to distance was calibrated by comparing the apparent brightness with the absolute magnitude (true brightness) of a Cepheid at a known distance with a known period. Using the period to predict how bright the star would appear at various distances allowed astronomers to calculate the distance to a Cepheid given its apparent brightness.

Edwin Hubble used the Cepheid period-brightness relationship to find the distances to other galaxies and discovered yet another relationship, that the greater the distance to a galaxy, the greater a shift in spectral lines toward the red end of the spectrum (redshift). This relationship is called Hubble's law, and it forms the foundation for understanding our expanding universe. Measuring redshift provides another means of establishing distances to other, far-out galaxies.

12.6 THE LIFE OF A STAR

A star is born in a gigantic cloud of gas and dust in interstellar space, then spends billions of years calmly shining while it fuses hydrogen nuclei in the core. How long a star shines and what happens to it when it uses up the hydrogen in the core depends on the mass of the star. Of course, no one has observed a life cycle of over billions of years. The life cycle of a star is a theoretical outcome based on what is known about nuclear reactions. The predicted outcomes seem to agree with observations of stars today, with different groups of stars that can be plotted on the H-R diagram. Thus, the groups of stars on the diagram—main sequence, red giants, and white dwarfs, for example—are understood to be stars in various stages of their lives.

Protostar Stage. The first stage in the theoretical model of the life cycle of a star is the formation of the protostar. As gravity pulls the gas of a protostar together, the density, pressure, and temperature increase from the surface down to the center. Eventually, the conditions are right for nuclear fusion reactions to begin in the core, which requires a temperature of 10 million kelvins. The initial fusion reaction essentially combines four hydrogen nuclei to form a helium nucleus with the release of much energy. This energy heats the core beyond the temperature reached by gravitational contraction, eventually to 16 million kelvins. Since the star is plasma, the increased temperature expands the volume of the star. The outward pressure of expansion balances the inward pressure from gravitational collapse, and the star settles down to a balanced condition of calmly converting hydrogen to helium in the core, radiating the energy released into space (figure 12.6). The theoretical time elapsed from the initial formation and collapse of the protostar to the main sequence is about 50 million years for a star of a solar mass.

Main Sequence Stage. Where the star is located on the main sequence and what happens to it next depend only on how massive it is. The more massive stars have higher core temperatures and use up their hydrogen more rapidly as they shine at higher surface temperatures (O type stars). Less massive stars shine at lower surface temperatures (M type stars) as they use their fuel at a slower rate. The overall life span on the main sequence ranges from millions of years for O type stars to trillions of years for M type stars. An average one-solar-mass star will last about 10 billion years.

Red Giant Stage. The next stage in the theoretical life of a star begins when much of the hydrogen in the core has been fused into helium. With fewer hydrogen fusion reactions, less energy is released and less outward balancing pressure is produced, so the star begins to collapse. The collapse heats the core, which now is composed primarily of helium, and the surrounding shell where hydrogen still exists. The increased temperature causes the hydrogen in the shell to undergo fusion, and the increased release of energy causes the outer layers of the star to expand. With an increased surface area, the amount of radiation emitted per unit area is less, and the star acquires the properties of a brilliant red giant. Its position on the H-R diagram changes since it now has different brightness and temperature properties. (The star has not physically *moved*. The changing properties move its temperature brightness data point, not the star, to a new position.)

Back Toward Main Sequence. After about 500 million years as a red giant, the star now has a surface temperature of about 4,000 kelvins compared to its main sequence surface temperature of 6,000 kelvins. The radius of the red giant is now a thousand times greater, a distance that will engulf Earth when the Sun reaches this stage, assuming Earth is in the same position as it is today. Even though the surface temperature has decreased from the expansion, the helium core is continually heating and eventually reaches a temperature of 100 million kelvins, the critical temperature necessary for the helium nuclei to undergo fusion to produce carbon. The red giant now has helium fusion reactions in the core and hydrogen fusion reactions in a shell around the core. This changes the radius, the surface temperature, and the brightness with the overall result depending on the composition of the star. In general,

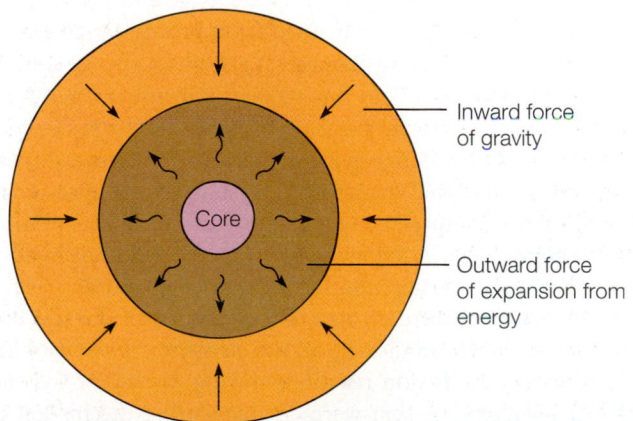

Inward force of gravity

Core

Outward force of expansion from energy

FIGURE 12.6 A star becomes stable when the outward forces of expansion from the energy released in nuclear fusion reactions balance the inward forces of gravity.

the radius and brightness decrease when this stage is reached, moving the star back toward the main sequence (figure 12.7).

Beginning of the End for Less Massive Stars. After millions of years of helium fusion reactions, the core is gradually converted to a carbon core and helium fusion begins in the shell surrounding the core. The core reactions decrease as the star now has a helium-fusing shell surrounded by a second, hydrogen-fusing shell. This releases additional energy, and the star again expands to a red giant for the second time. A star the size of the Sun or less massive may cool enough at this point that nuclei at the surface become neutral atoms rather than a plasma. As neutral atoms, they can absorb radiant energy coming from within the star, heating the outer layers. Changes in temperature produce changes in pressure, which change the balance between the temperature, pressure, and the internal energy generation rate. The star begins to expand outward from heating. The expanded gases are cooled by the expansion process, however, and are pulled back to the star by gravity, only to be heated and expand outward again. In other words, the outer layers of the star begin to pulsate in and out. Finally, a violent expansion blows off the outer layers of the star, leaving the hot core. Such blown-off outer layers of a star form circular nebulae called *planetary nebulae* (figure 12.8). The nebulae continue moving away from the core, eventually adding to the dust and gases between the stars. The remaining carbon core and helium-fusing shell begin gravitationally to contract to a small, dense *white dwarf* star. A star with the original mass of the Sun or less slowly cools from white to red, then to a black lump of carbon in space (figure 12.9).

Beginning of the End for Massive Stars. A more massive star will have a different theoretical ending than the slow cooling of a white dwarf. A massive star will contract, just as the less massive stars, after blowing off its outer shells. In a more massive star, however, heat from the contraction may reach the critical temperature of 600 million kelvins to begin carbon fusion reactions. Thus, a more massive star may go through a carbon fusing stage and other fusion reaction stages that will continue to produce new elements until the element iron is reached. (See binding energy and figure 11.10 on page 252.) After iron, energy is no longer released by the fusion process, and the star has used up all of its energy sources. Lacking an energy source, the star is no longer able to maintain its internal temperature. The star loses the outward pressure of expansion from the high temperature, which had previously balanced the inward pressure from gravitational attraction. The star thus collapses, then rebounds like a compressed spring into a catastrophic explosion called a **supernova.** A supernova produces a brilliant light in the sky that may last for months before it begins to dim as the new elements that were created during the life of the star diffuse into space. These include all the elements up to iron that were produced by fusion reactions during the life of the star and heavier elements that were created during the instant of the explosion. All the elements heavier than iron were created as some less massive nuclei disintegrated in the explosion, joining with each other and with lighter nuclei to produce the

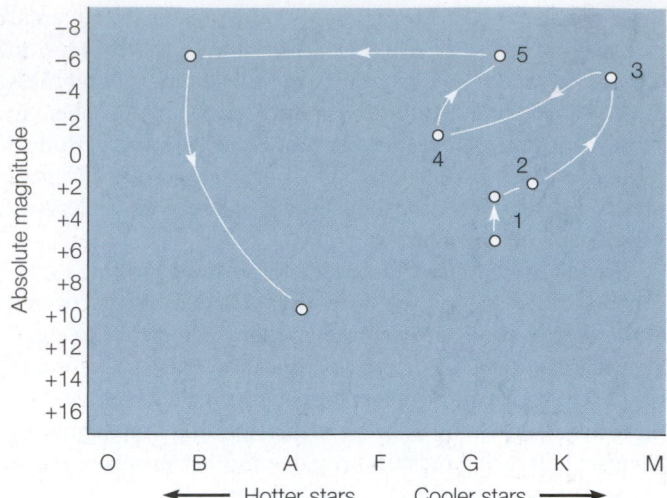

FIGURE 12.7 The evolution of a star of solar mass as it depletes hydrogen in the core (1), fuses hydrogen in the shell to become a red giant (2 to 3), becomes hot enough to produce helium fusion in the core (3 to 4), then expands to a red giant again as helium and hydrogen fusion reactions move out into the shells (4 to 5). It eventually becomes unstable and blows off the outer shells to become a white dwarf star.

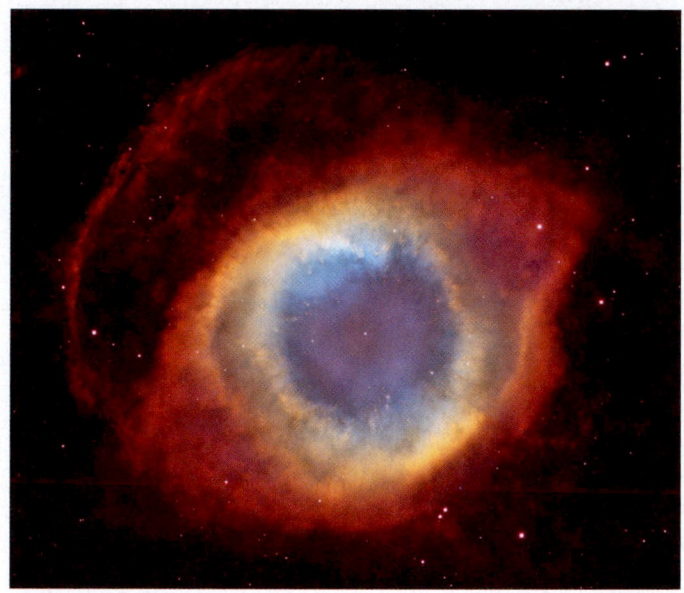

FIGURE 12.8 The blown-off outer layers of stars form ringlike structures called *planetary nebulae.*

nuclei of the elements from iron to uranium. As you will see in chapter 13, these newly produced, scattered elements will later become the building blocks for new stars and planets such as the Sun and Earth.

The remains of the compressed core after the supernova have yet another fate if the core has a remaining mass greater than 1.4 solar masses. The gravitational forces on the remaining matter, together with the compressional forces of the supernova explosion, are great enough to collapse nuclei, forcing protons and electrons together into neutrons, forming the core of a **neutron star.** A neutron star is the very small (10 to 20 km

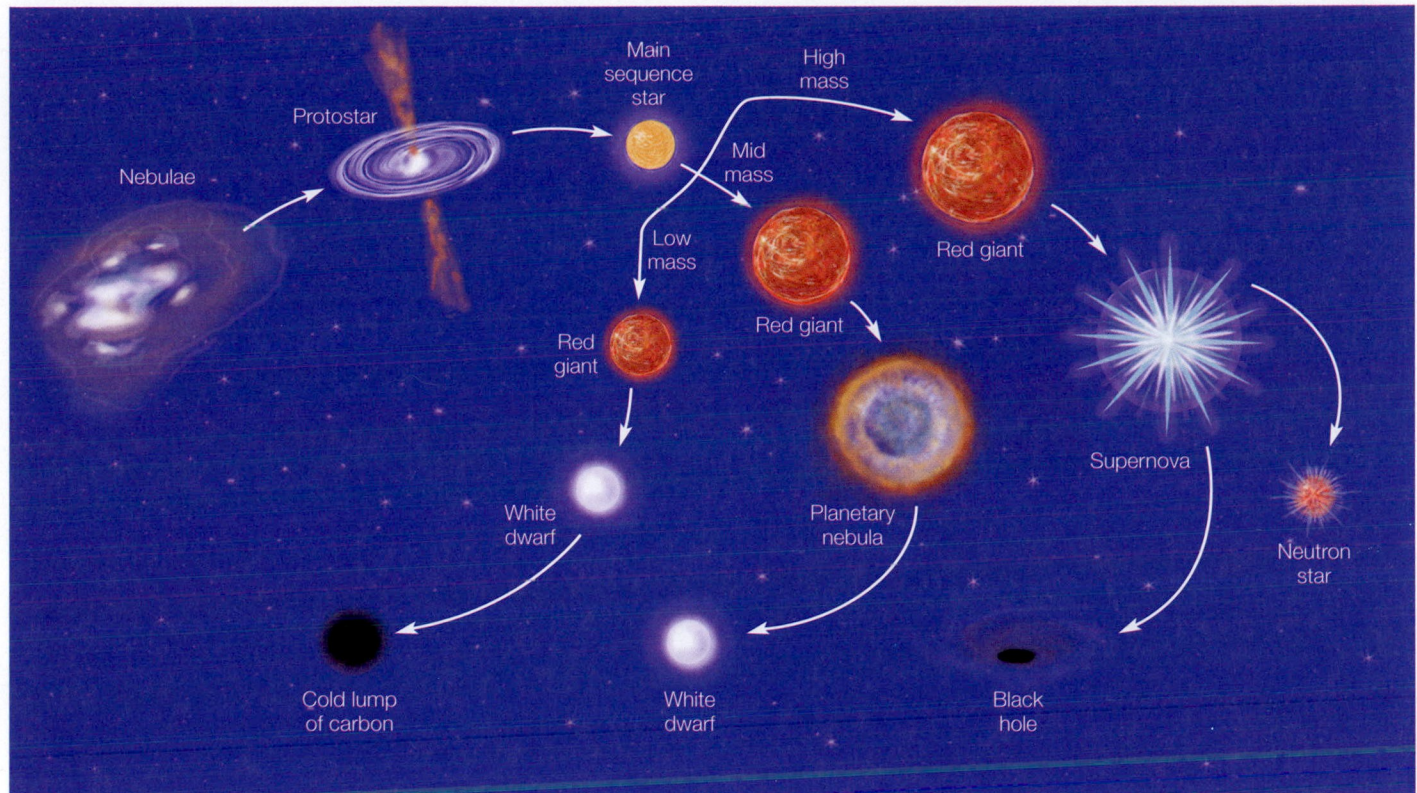

FIGURE 12.9 This shows some of the possible stages in the birth and aging of a star. The differences are determined by the mass of the star.

diameter), superdense (10^{11} kg/cm^3 or greater) remains of a supernova with a center core of pure neutrons.

Because it is a superdense form of matter, the neutron star also has an extremely powerful magnetic field, capable of becoming a pulsar. A **pulsar** is a very strongly magnetized neutron star that emits a uniform series of equally spaced electromagnetic pulses. Evidently, the magnetic field of a rotating neutron star makes it a powerful electric generator, capable of accelerating charged particles to very high energies. These accelerated charges are responsible for emitting a beam of electromagnetic radiation, which sweeps through space with amazing regularity (figure 12.10). The pulsating radio signals from a pulsar were a big mystery when first discovered. For a time, extraterrestrial life was considered as the source of the signals, so they were jokingly identified as LGM (for "little green men"). Over three hundred pulsars have been identified, and most emit radiation in the form of radio waves. Two, however, emit visible light, two emit beams of gamma radiation, and one emits X-ray pulses.

Another theoretical limit occurs if the remaining core has a mass of about 3 solar masses or more. At this limit, the force of gravity overwhelms *all* nucleon forces, including the repulsive forces between like charged particles. If this theoretical limit is reached, nothing can stop the collapse, and the collapsed star will become so dense that even light cannot escape. The star is now a **black hole** in space. Since nothing can stop the collapsing star, theoretically a black hole would continue to collapse to a pinpoint and then to a zero radius called a *singularity*. This event seems contrary to anything that can be directly observed in the physical universe, but it does agree with the general theory of relativity and concepts about the curvature of space produced by such massively dense objects. Black holes are theoretical and none has been seen, of course, because a black hole theoretically pulls in radiation of all wavelengths and emits nothing. Evidence for the existence of a black hole is sought by studying X rays that would be given off by matter as it is accelerated into a black hole.

Another form of evidence for the existence of a black hole has now been provided by the Hubble Space Telescope. Hubble pictured a disk of gas only about 60 light-years out from the center of a galaxy (M87), moving at more than 1.6 million km/h (about 1 million mi/h). The only known possible explanation for such a massive disk of gas moving with this velocity at the distance observed would require the presence of a 1–2 billion solar-mass black hole. This gas disk could only be resolved by the Hubble Space Telescope, so this telescope has provided the first observational evidence of a black hole.

12.7 GALAXIES

Stars are associated with other stars on many different levels, from double stars that orbit a common center of mass, to groups of tens or hundreds of stars that have gravitational links and a common origin, to the billions and billions of stars that form the basic unit of the universe, a **galaxy.** The Sun is but one of an estimated 400 billion stars that are held together by gravitational attraction in the Milky Way galaxy (figure 12.11).

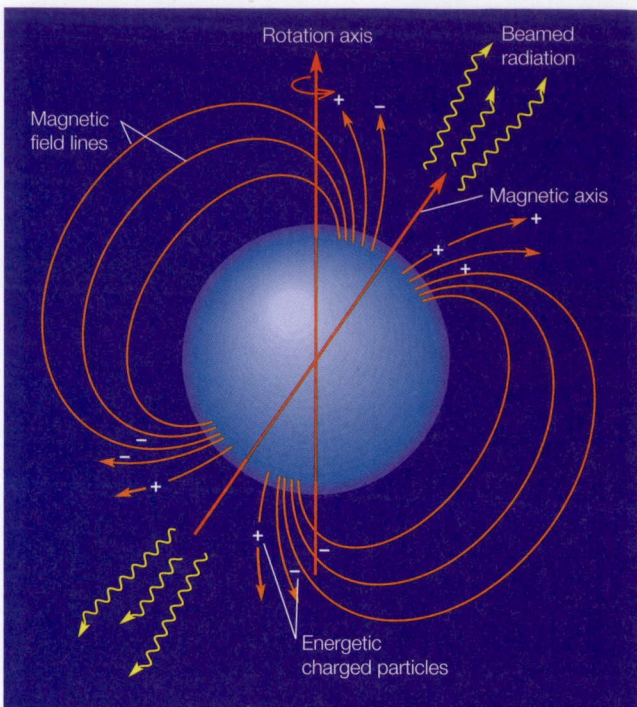

FIGURE 12.10 The magnetic axis of the pulsar is inclined with respect to the rotation axis. Rapidly moving electrons in the regions near the magnetic poles emit radiation in a beam pointed outward. When the beam sweeps past Earth, a pulse is detected.

The numbers of stars and vastness of the Milky Way galaxy alone seem almost beyond comprehension, but there is more to come. The Milky Way is but one of *billions* of galaxies that are associated with other galaxies in clusters, and these clusters are associated with one another in superclusters. Through a large telescope, you can see more galaxies than individual stars in any direction, each galaxy with its own structure of billions of stars. Yet, there are similarities that point to a common origin. Some of the similarities and associations of stars will be introduced in this section along with the Milky Way galaxy, the vast, flat, spiraling arms of stars, gas, and dust where the Sun is located.

The Milky Way Galaxy

Away from city lights, you can clearly see the faint, luminous band of the Milky Way galaxy on a moonless night. Through a telescope or a good pair of binoculars, you can see that the luminous band is made up of countless numbers of stars. You may also be able to see the faint glow of nebulae, concentrations of gas and dust. There are dark regions in the Milky Way that also give an impression of something blocking starlight, such as dust. You can also see small groups of stars called **galactic clusters.** Galactic clusters are gravitationally bound subgroups of as many as one thousand stars that move together within the Milky Way. Other clusters are more symmetrical and tightly packed, containing as many as a million stars, and are known as **globular clusters.**

Viewed from a distance in space, the Milky Way would appear to be a huge, flattened cloud of spiral arms radiating out from the center. There are three distinct parts: (1) the spherical concentration of stars at the center of the disk called the *galactic nucleus;* (2) the rotating *galactic disk,* which contains most of the bright, blue stars along with much dust and gas; and (3) a spherical *galactic halo,* which contains some 150 globular clusters located outside the galactic disk (figure 12.12). The Sun is located in one of the arms of the galactic disk, some 25,000 to 30,000 light-years from the center. The galactic disk rotates, and the Sun completes one full rotation every 200 million years.

The diameter of the *galactic disk* is about 100,000 light-years. Yet, in spite of the recent estimate of 400 billion stars in the Milky Way, it is mostly full of emptiness. By way of analogy, imagine reducing the size of the Milky Way disk until stars like the Sun were reduced to the size of tennis balls. The distance between two of these tennis-ball-sized stars would now compare to the distance across the state of Texas. The space between the stars is not actually empty since it contains a thin concentration of gas, dust, and molecules of chemical compounds. The gas particles outnumber the dust particles about 10^{12} to 1. The gas is mostly hydrogen, and the dust is mostly solid iron, carbon, and silicon compounds. Over forty different chemical molecules have been discovered in the space between the stars, including many organic molecules. Some nebulae consist of clouds of molecules with a maximum density of about 10^6 molecules/cm^3. The gas, dust, and chemical compounds make up part of the mass of the galactic disk, and the stars make up the remainder. The gas plays an important role in the formation of new stars, and the dust and chemical compounds play an important role in the formation of planets.

Other Galaxies

Outside the Milky Way is a vast expanse of emptiness, lacking even the few molecules of gas and dust spread thinly through the galactic nucleus. There is only the light from faraway galaxies

![Wide-angle view of the Milky Way galaxy]

FIGURE 12.11   A wide-angle view toward the center of the Milky Way galaxy. Parts of the white, milky band are obscured from sight by gas and dust clouds in the galaxy. The white streak is a meteorite.

and the time that it takes for this light to travel across the vast vacuum of intergalactic space. How far away is the nearest galaxy? Recall that the Milky Way is so large that it takes light 100,000 years to travel the length of its diameter.

The nearest galactic neighbor to the Milky Way is a dwarf spherical galaxy only 80,000 light-years from the solar system. The nearby galaxy is called a dwarf because it has a diameter of only about 1,000 light-years. It is apparently in the process of being pulled apart by the gravitational pull of the Milky Way, which now is known to have eleven satellite galaxies.

The nearest galactic neighbor similar to the Milky Way is Andromeda, about 2 million light-years away. Andromeda is similar to the Milky Way in size and shape, with about 400 billion

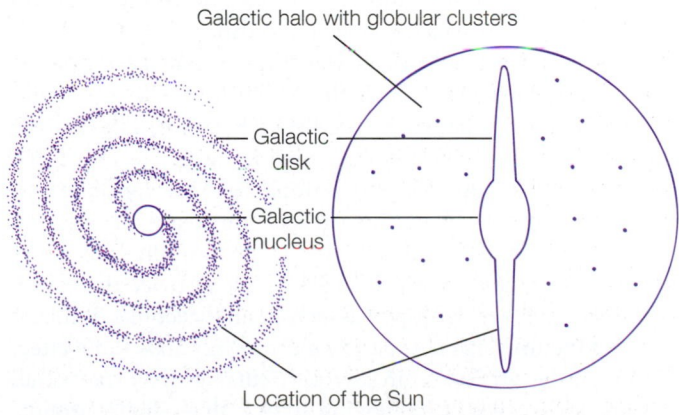

FIGURE 12.12   The structure of the Milky Way galaxy.

stars, gas, and dust turning in a giant spiral pinwheel (figure 12.13). Other galaxies have other shapes and other characteristics. The American astronomer Edwin Hubble developed a classification scheme for the structure of galaxies based on his 1926 study of some six hundred different galaxies. The basic galactic structures were identified as elliptical, spiral, barred, and irregular.

The Life of a Galaxy

Hubble's classification of galaxies into distinctly different categories of shape was an exciting accomplishment because it suggested that some relationship or hidden underlying order might exist in the shapes. Finding underlying order is important because it leads to the discovery of the physical laws that govern the universe. Soon after Hubble published his classification results in 1926, two models of galactic evolution were proposed. One model, which was suggested by Hubble, had extremely slowly spinning spherical galaxies forming first, which gradually flattened out as their rate of spin increased while they condensed. This is a model of spherical galaxies flattening out to increasingly elliptical shapes, eventually spinning off spirals until they finally broke up into irregular shapes over a long period of time.

Among many uses, the Hubble Space Telescope is used to study young galaxies and galaxies on collision courses. Based on these studies, astronomers today recognize that the different shapes of galaxies do not represent an evolutionary sequence. Their shapes are understood to be a result of the various conditions under which the galaxies were formed.

The current model of how galaxies form is based on the **big bang theory** of the creation of the universe. The big bang theory considers the universe to have had an explosive beginning. According to this theory, all matter in the universe was located together in an arbitrarily dense state from which it began to expand, an expansion that continues today. Evidence that supports the big bang theory comes from Albert Einstein's theory of general relativity as well as three areas of physical observations:

1. **Expansion of the universe.** The initial evidence for the big bang theory came from Edwin Hubble and his earlier work with galaxies. Hubble had determined the distances to some of the galaxies that had redshifted spectra. From this expansive redshift, it was known that these galaxies were moving away from the Milky Way. Hubble found a relationship between the distance to a galaxy and the velocity with which it was moving away. He found the velocity to be directly proportional to the distance; that is, the greater the distance to a galaxy, the greater the velocity. This means that a galaxy twice as far from the Milky Way is moving away from the Milky Way at twice the speed. Since this relationship was seen in all directions, it meant that the universe is expanding uniformly. The same effect would be viewed from any particular galaxy; that is, all the other galaxies are moving away with a velocity propor-

FIGURE 12.13 The Andromeda galaxy, which is believed to be similar in size, shape, and structure to the Milky Way galaxy.

tional to the distance to the other galaxies. This points to a common beginning, a time when all matter in the universe was together.

2. **Background radiation.** The big bang occurred as some unstable form of energy expanded and cooled, eventually creating matter and space. The initial temperature was 10 billion kelvins or so, and began to cool as the universe expanded. The afterglow of the big bang is called cosmic background radiation. A measurement of cosmic background radiation today agrees with the radiation that should be present according to an expanding model of the universe.

3. **Abundance of elements.** The proportion of helium in the universe should be about 24 percent, based on an expanding model of the universe. A measurement of the abundance of helium verifies the proportion as predicted by the big bang theory.

Cosmic Background Explorer (COBE) spacecraft studied diffuse cosmic background radiation to help answer such ques-

A Closer Look

Extraterrestrials?

Extraterrestrial is a descriptive term, meaning a thing or event outside Earth or its atmosphere. The term is also used to describe a being, a life-form that originated away from Earth. This reading is concerned with the search for extraterrestrials, intelligent life that might exist beyond Earth and outside the solar system.

Why do people believe that extraterrestrials might exist? The affirmative answer comes from a mixture of current theories about the origin and development of stars, statistical odds, and faith. Considering the statistical odds, note that our Sun is one of the some 400 billion stars that make up our Milky Way galaxy. The Milky Way galaxy is one of some 10 billion galaxies in the observable universe. Assuming an average of about 400 billion stars per galaxy, this means there are some 400 billion times 10 billion stars, or 4×10^{21} stars in the observable universe. There is nothing special or unusual about our Sun, and astronomers believe it to be quite ordinary among all the other stars (all 4,000,000,000,000,000,000,000 or so).

So the Sun is an ordinary star, but what about our planet? Not too long ago, most people, including astronomers, thought our solar system with its life-supporting planet (Earth) to be unique. Evidence collected over the past decade or so, however, has strongly suggested that this is not so. Planets are now believed to be formed as a natural part of the star-forming process. Evidence of planets around other stars has also been detected by astronomers. One of the stars with planets is "only" 53 light-years from Earth.

Even with a very low probability of planetary systems forming with the development of stars, a population of 4×10^{21} stars means there are plenty of planetary

Box Figure 12.1   A radio telescope.

systems in existence, some with the conditions necessary to support life. (Note: If 1 percent have planetary systems, this means 4×10^{19} stars have planets.) Thus, it is a statistical observation that suitable planets for life are very likely to exist. In addition, radio astronomers have found that many organic molecules exist, even in the space between the stars. Based on statistics alone, there should be life on other planets, life that may have achieved intelligence and developed into a technological civilization.

If extraterrestrial life exists, why have we not detected them or why have they not contacted us? The answer to this question is found in the unbelievable distances involved in interstellar space. For example, a logical way to search for extraterrestrial intelligence is to send radio signals to space and analyze those coming from space. Modern radio telescopes can send powerful radio beams, and present-day computers and data processing techniques can now search through incoming radio

signals for the patterns of artificially generated radio signals (box figure 12.1). Radio signals, however, travel through space at the speed of light. The diameter of our Milky Way galaxy is about 100,000 light-years, which means 100,000 years would be required for a radio transmission to travel across the galaxy. For this discussion, assume Earth is located on one edge of the Milky Way. If we were to transmit a super, super strong radio beam from Earth, it would travel at the speed of light and cross the distance of our galaxy in 100,000 years. If some extraterrestrials on the other side of the Milky Way galaxy did detect the message and send a reply, it could not arrive at Earth until 200,000 years after the message was sent. Now consider the fact that of all the 10 billion other galaxies in the observable universe, our nearest galactic neighbor similar to the Milky Way is Andromeda. Andromeda is 2 million light-years from the Milky Way galaxy.

In addition to problems with distance and time, there are questions about in which part of the sky you should send and look for radio messages, questions about which radio frequency to use, and problems with the power of present-day radio transmitters and detectors. Realistically, the hope for any exchange of radio-transmitted messages would be restricted to within several hundred light-years of Earth.

Considering all the limitations, what sort of signals should we expect to receive from extraterrestrials? Probably a series of pulses that somehow indicate counting, such as 1, 2, 3, 4, and so on, repeated at regular intervals. This is the most abstract, while at the same time the simplest, concept that an intelligent being anywhere would have. It could provide the foundation for communications between the stars.

---

tions as how matter is distributed in the universe, whether the universe is uniformly expanding, and how and when galaxies first formed.

The 2003 results from NASA's orbiting *Wilkinson Microwave Anisotropy Probe (WMAP)* produced a precision map of the remaining cosmic microwave background from the big

bang. *WMAP* surveyed the entire sky for a whole year with a resolution some 40 times greater than *COBE*. Analysis of *WMAP* data revealed that the universe is 13.7 billion years old, with a 1 percent margin of error. The *WMAP* data found strong support for the big bang and expanding universe theories. It also revealed that the content of the universe includes 4 percent

A Closer Look

Redshift and Hubble's Law

As described in chapter 5, the Doppler effect tells us that the frequency of a wave depends on the relative motion of the source and observer. When the source and observer are moving toward each other, the frequency appears to be higher. If the source and observer are moving apart, the frequency appears to be lower.

Light from a star or galaxy is changed by the Doppler effect, and the frequency of the observed spectral lines depends on the relative motion. The Doppler effect changes the frequency from what it would be if the star or galaxy were motionless relative to the observer. If the star or galaxy is moving toward the observer, a shift occurs in the spectral lines toward a higher frequency (blueshift). If the star or galaxy is moving away from the observer, a shift occurs in the spectral lines toward a lower frequency (redshift). Thus, a redshift or blueshift in the spectral lines will tell you if a star or galaxy is moving toward or away from you.

One of the first measurements of the distance to other galaxies was made by Edwin Hubble at Mount Wilson Observatory in California. When Hubble compared the distance figures with the observed redshifts, he found that the recession speeds were proportional to the distance. Farther-away galaxies were moving away from the Milky Way, but galaxies that are more distant are moving away faster than closer galaxies. This proportional relationship between galactic speed and distances was discovered in 1929 by Hubble and today is known as *Hubble's law*. The conclusion was that all the galaxies are moving away from one another and an observer on any given galaxy would have the impression that all galaxies were moving away in all directions. In other words, the universe is expanding with component galaxies moving farther and farther apart.

---

ordinary matter, 23 percent of an unknown type of dark matter, and 73 percent of a mysterious dark energy.

How old is the universe? As mentioned earlier, astronomical and physical "clocks" indicate that the universe was created in a "big bang" some 13.7 billion years ago, expanding as an intense and brilliant fireball with a temperature of some 10 billion kelvins. This estimate of the age is based on precise measurements of the rate at which galaxies are moving apart, an expansion that started with the big bang. Astronomers used data on the expansion to back-calculate the age, much like running a movie backward, to arrive at the age estimate. At first, this technique found that the universe was about 18 to 20 billion years old. Then astronomers found that the universe is not expanding at a constant rate. Instead, the separation of galaxies is actually accelerating, pushed by a mysterious force known as "**dark energy.**" By adding in calculations for this poorly understood force, the estimate of 13 to 14 billion years was developed.

Other age-estimating techniques agree with this ballpark age. Studies of white dwarfs, for example, established their rate of cooling. By looking at the very faintest and oldest white dwarfs with the Hubble Space Telescope, astronomers were able to use the cooling rate to estimate the age of the universe. The result found the dimmest of the white dwarfs to be about 13 billion years old, plus or minus about half a billion years. More recently, data from the orbiting *WMAP* spacecraft has produced a precision map of the cosmic microwave background that shows the universe to be 13.7 billion years old. An age of 13.7 billion years may not be the final answer for the age of the universe, but when three independent measurement techniques agree closely, the answer becomes more believable.

An often-used analogy for the movement of galaxies after the big bang is a loaf of rising raisin bread. Consider galaxies as raisins in a loaf of raisin bread dough as it rises (figure 12.14).

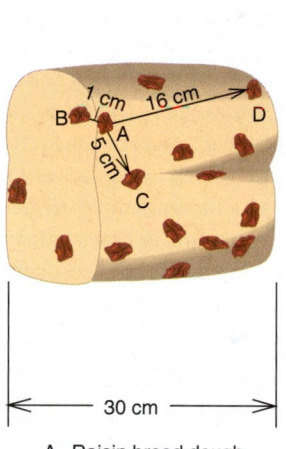

A Raisin bread dough before rising

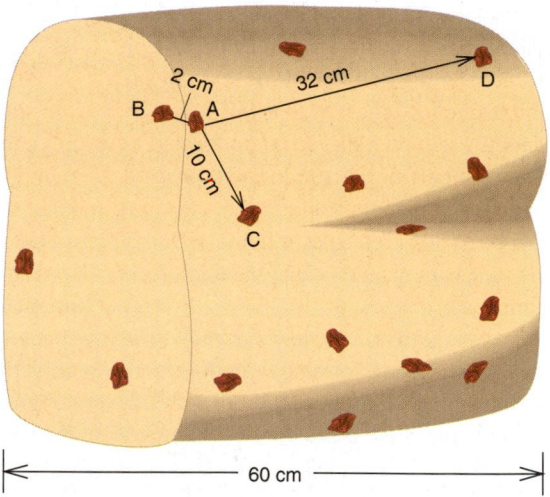

B Raisin bread dough after rising

FIGURE 12.14 As the universe (here represented by a loaf of raisin bread) expands, the expansion carries galaxies (represented by the raisins) away from each other at speeds that are proportional to their distances from each other. It doesn't matter within which galaxy an astronomer resides; the other galaxies all appear to be moving away.

Dark Matter

Will the universe continue to expand forever, or will it be pulled back together into a really big crunch? Whether the universe will continue expanding or gravity will pull it back together again depends on the *critical density* of the universe. If there is enough matter, the actual density of the universe will be above the critical density, and gravity will stop the current expansion and pull everything back together again. On the other hand, if the actual density of the universe is less than or equal to the critical density, the universe will be able to escape its own gravity and continue expanding. Is the actual density small enough for the universe to escape itself?

All the detailed calculations of astrophysicists point to an actual density that is *less* than the critical density, meaning the universe will continue to expand forever. However, these calculations also show that there is more matter in the universe than can be accounted for in the stars and galaxies. This means there must be matter in the universe that is not visible, or not shining at least, so we cannot see it. Three examples follow that show why astrophysicists believe more matter exists than we can see.

1. There is a relationship between the light emitted from a star and the mass of that star. Thus, you can indirectly measure the amount of matter in a galaxy by measuring the light output of that galaxy. Clusters of galaxies have been observed moving, and the motion of galaxies within a cluster does not agree with matter information from the light. The motion suggests that galaxies are attracted by a gravitational force from about ten times more matter than can be accounted for by the light coming from the galaxies.

2. There are mysterious variations in the movement of stars within an individual galaxy. The rate of rotation of stars about the center of rotation of a galaxy is related to the distribution of matter in that galaxy. The outer stars are observed to rotate too fast for the amount of matter that can be seen in the galaxy. Again, measurements indicate there is about ten times more matter in each galaxy than can be accounted for by the light coming from the galaxies.

3. Finally, there are estimates of the total matter in the universe that can be made by measuring ratios such as deuterium to ordinary hydrogen, lithium-7 to helium-3, and helium to hydrogen. In general, theoretical calculations all seem to account for only about 10 percent of all the matter in the universe.

Calculations from a variety of sources all seem to agree that 90 percent or more of the matter that makes up the universe is missing. The missing matter and the mysterious variations in the orbits of galaxies and stars can be accounted for by the presence of **dark matter,** which is invisible and unseen. What is the nature of this dark matter? You could speculate that dark matter is simply normal matter that has been overlooked, such as dark galaxies, brown dwarfs, or planetary material such as rock, dust, and the like. You could also speculate that dark matter is dark because it is in the form of subatomic particles, too small to be seen. As you can see, scientists have reasons to believe that dark matter exists, but they can only speculate about its nature.

The nature of dark matter represents one of the major unsolved problems in astronomy today. There are really two dark matter problems: (1) the nature of the dark matter and (2) how much dark matter contributes to the actual density of the universe. Not everyone believes that dark matter is simply normal matter that has been overlooked. There are at least two schools of thought on the nature of dark matter and what should be the focus of research. One school of thought focuses on *particle physics* and contends that dark matter consists primarily of exotic undiscovered particles or known particles such as the neutrino. Neutrinos are electrically neutral, stable subatomic particles. This school of thought is concerned with forms of dark matter called WIMPs (after *w*eakly *i*nteracting *m*assive *p*articles). In spite of the name, particles under study are not always massive. It is entirely possible that some low-mass species of neutrino—or some as-of-yet unidentified WIMP—will be experimentally discovered and found to have the mass needed to account for the dark matter and thus close the universe.

The other school of thought about dark matter focuses on *astrophysics* and contends that dark matter is ordinary matter in the form of brown dwarfs, unseen planets outside the solar system, and galactic halos. This school of thought is concerned with forms of dark matter called MACHOs (after *m*assive *a*strophysical *c*ompact *h*alo *o*bjects). Dark matter considerations in this line of reasoning consider massive objects, but ordinary matter (protons and neutrons) is also considered as the material making up galactic halos. Protons and neutrons belong to a group of subatomic particles called the *baryons*, so they are sometimes referred to as *baryonic dark matter*. Some astronomers feel that baryonic dark matter is the most likely candidate for halo dark matter.

In general, astronomers can calculate the probable cosmological abundance of WIMPs, but there is no proof of their existence. By contrast, MACHOs dark matter candidates are known to exist, but there is no way to calculate their abundance. MACHOs astronomers assert that halo dark matter and probably all dark matter may be baryonic. Do you believe WIMPs or MACHOs will provide an answer about the future of the universe? What is the nature of dark matter, how much dark matter exists, and what is the fate of the universe? Answers to these and more questions await further research.

Dark Energy

The first evidence of dark energy came from astronomers trying to measure how fast the expanding universe is slowing. At the time, the generally accepted model of the universe was that it was still expanding but had been slowing from gravitational attraction since the big bang. The astronomers intended to measure the slowing rate to determine the average density of matter in the universe, providing a clue about the extent of dark matter. Their idea was to compare light traveling from a supernova in the distant universe with light traveling from a much closer supernova. Light from the distant universe is just now reaching us since it was emitted when the universe was very young. The brightness of the older and younger supernova would provide information about distance, and expansive redshift data would provide information about their speed of expansion. By comparing these two light sources, they would be able to calculate the expansion rate for now and in the distant past.

To their surprise, the astronomers did not find a rate of slowing expansion. Rather, they found that the expansion is speeding up. There was no explanation for this speeding up other than to assume that some unknown antigravity force was at work. This unknown force must be pushing the galaxies farther apart at the same time that gravity is trying to pull them together. The unknown repulsive force became known as "dark energy."

The idea of an unknown repulsive force is not new. It was a new idea when Einstein added a "cosmological constant" to the equations in the theory of general relativity. This constant represents a force that opposes gravity, and the force grows as a function of space. This means there was not much—shall we say, dark energy—when the universe was smaller, and gravity slowed the expansion. As the distance between galaxies increased, there was an increase of dark energy and the expansion accelerated. Einstein removed this constant when Edwin Hubble reported evidence that the universe is expanding. Removing the constant may have been a mistake. One of the problems in understanding what was happening to the early universe was a lack of information about the distant past as is represented by the distant universe of the present. Today, this is changing as more new technology is helping astronomers study the distant universe. Will they find the meaning of dark energy? What is dark energy? The answer is that no one knows. Stay tuned . . . this story is to be continued.

Worth the Cost?

People are fascinated with ideas of space travel as well as new discoveries about the universe and how it formed. Unmanned space probes, new kinds of instruments, and new kinds of telescopes have resulted in new information. We are learning what is happening in outer space away from Earth, as well as about the existence of other planets.

Few would deny that the space program has provided valuable information. However, some people wonder if it is worth the cost. They point to many problems here on Earth, such as growing energy and water needs, pollution problems, and ongoing health problems. They say that the money spent on exploring space could be better used for helping resolve problems on Earth.

In addition to new information and understanding coming out of the space program, supporters of space exploration point to new technology that helps people living on Earth. Satellites now provide valuable information for agriculture, including land use and weather monitoring. Untold numbers of lives have been saved thanks to storm warnings provided by weather satellites. There are many other spin-offs from the space program, including improvements in communication systems.

Questions to Discuss

What do you think: have the gain of knowledge and the spin-offs merited the expense? Or would the funds be better spent on solving other problems?

As the dough expands in size, it carries along the raisins, which are moved farther and farther apart by the expansion. If you were on a raisin, you would see all the other raisins moving away from you, and the speed of this movement would be directly proportional to their distance away. It would not matter which raisin you were on since all the raisins would appear to be moving away from all the other raisins. In other words, the galaxies are not expanding into space that is already there. It is space itself that is expanding, and the galaxies move with the expanding space (figure 12.15).

People Behind the Science

Jocelyn (Susan) Bell Burnell (1943–)

Jocelyn Bell is a British astronomer who discovered pulsating radio stars—pulsars—an important astronomical discovery of the 1960s.

Bell spent two years in Cambridge building a radio telescope that was specially designed to track quasars—her Ph.D. research topic. The telescope that she and her team built had the ability to record rapid variations in signals. It was also nearly 2 hectares (about 5 acres) in area, equivalent to a dish of 150 m (about 500 ft) in diameter, making it an extremely sensitive instrument.

The sky survey began when the telescope was finally completed in 1967, and Bell was given the task of analyzing the signals received. One day, while scanning the charts of recorded signals, she noticed a rather unusual radio source that had occurred during the night and had been picked up in a part of the sky that was opposite in direction to the Sun. This was curious because strong variations in the signals from quasars are caused by solar wind and are usually weak during the night. At first, she thought that the signal might be due to a local interference. After a month of further observations, it became clear that the position of the peculiar signals remained fixed with respect to the stars, indicating that it was neither terrestrial nor solar in origin. A more detailed examination of the signal showed that it was in fact composed of a rapid set of pulses that occurred precisely every 1.337 seconds. The pulsed signal was as regular as the most regular clock on Earth.

One attempted explanation of this curious phenomenon was that it represented an interstellar beacon sent out by extraterrestrial life on another star, so initially it was nicknamed LGM, for "little green men." Within a few months of noticing this signal, however, Bell located three other similar sources. They too pulsed at an extremely regular rate, but their periods varied over a few fractions of a second, and they all originated from widely spaced locations in our galaxy. Thus, it seemed that a more likely explanation of the signals was that they were being emitted by a special kind of star—a pulsar.

Since the astonishing discovery was announced, other observatories have searched the heavens for new pulsars. Some three hundred are now known to exist, their periods ranging from hundredths of a second to four seconds. It is thought that neutron stars are responsible for the signal. These are tiny stars, only about 7 km (about 4.3 mi) in diameter, but they are incredibly massive. The whole star and its associated magnetic field are spinning at a rapid rate, and the rotation produces the pulsed signal.

Source: Modified from the *Hutchinson Dictionary of Scientific Biography.* © RM, 2011. All rights reserved. Helicon Publishing is a division of RM.

The ultimate fate of the universe will strongly depend on the mass of the universe and if the expansion is slowing or continuing. All the evidence tells us it began by expanding with a big bang. Will it continue to expand, becoming increasingly cold and diffuse? Will it slow to a halt, then start to contract to a fiery finale of a big crunch (figure 12.16)? Researchers continue searching for answers with experimental data and theoretical models, looking for clues in matches between the data and models.

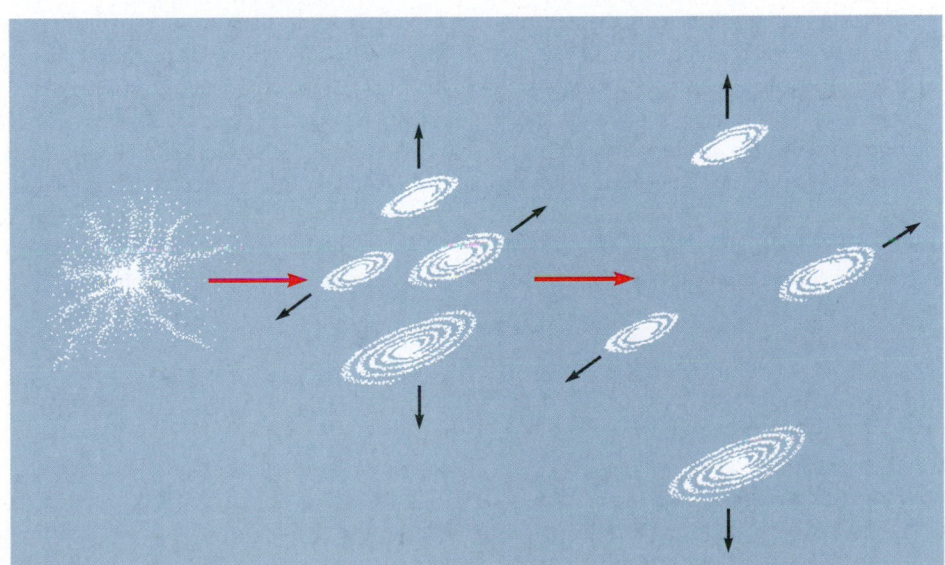

FIGURE 12.15 Will the universe continue expanding as the dust and gas in galaxies become locked up in white dwarf stars, neutron stars, and black holes?

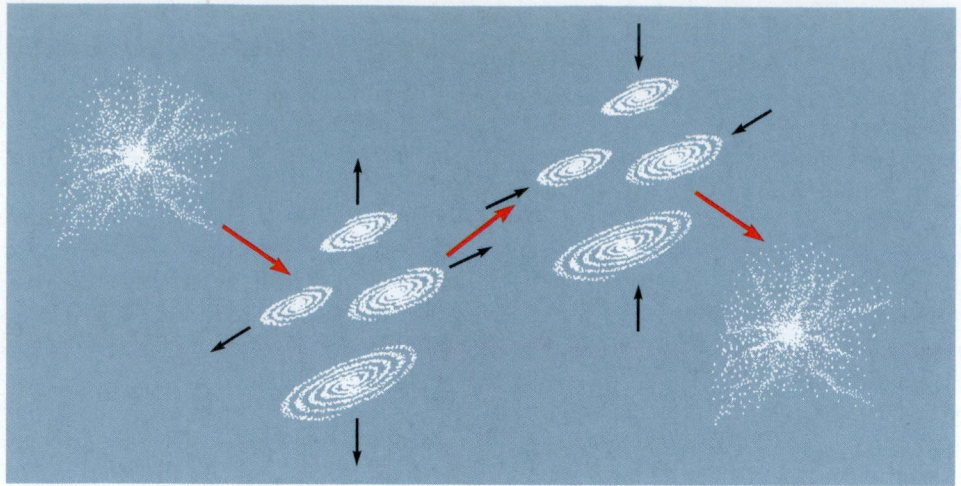

FIGURE 12.16 One theory of the universe assumes that the space between the galaxies is expanding, as does the big bang theory, but the galaxies gradually come back together to begin all over in another big bang.

SUMMARY

Stars are theoretically born in clouds of hydrogen gas and dust in the space between other stars. Gravity pulls huge masses of hydrogen gas together into a *protostar,* a mass of gases that will become a star. The protostar contracts, becoming increasingly hotter at the center, eventually reaching a temperature high enough to start *nuclear fusion* reactions between hydrogen atoms. Pressure from hot gases balances the gravitational contraction, and the average newborn star will shine quietly for billions of years. The average star has a dense, hot *core* where nuclear fusion releases radiation, a less dense *radiation zone* where radiation moves outward, and a thin *convection zone* that is heated by the radiation at the bottom, then moves to the surface to emit light to space.

The brightness of a star is related to the amount of energy and light it is producing, the size of the star, and the distance to the star. The *apparent magnitude* is the brightness of a star as it appears to you. To compensate for differences in brightness due to distance, astronomers calculate the brightness that stars would have at a standard distance. This standard-distance brightness is called the *absolute magnitude.*

Stars appear to have different colors because they have different surface temperatures. A graph of temperature by spectral types and brightness by absolute magnitude is called the *Hertzsprung-Russell diagram,* or H-R diagram for short. Such a graph shows that normal, mature stars fall on a narrow band called the *main sequence* of stars. Where a star falls on the main sequence is determined by its brightness and temperature, which in turn are determined by the mass of the star. Other groups of stars on the H-R diagram have different sets of properties that are determined by where they are in their evolution.

The life of a star consists of several stages, the longest of which is the *main sequence* stage after a relatively short time as a *protostar.* After using up the hydrogen in the core, a star with an average mass expands to a *red giant,* then blows off the outer shell to become a *white dwarf star,* which slowly cools to a black lump of carbon. The blown-off outer shell forms a *planetary nebula,* which disperses over time to become the gas and dust of interstellar space. More massive stars collapse into *neutron stars* or *black holes* after a violent *supernova* explosion.

Galaxies are the basic units of the universe. The Milky Way galaxy has three distinct parts: (1) the *galactic nucleus,* (2) a rotating *galactic disk,* and (3) a *galactic halo.* The galactic disk contains subgroups of stars that move together as *galactic clusters.* The halo contains symmetrical and tightly packed clusters of millions of stars called *globular clusters.*

Evidence from astronomical and physical "clocks" indicates that the galaxies formed some 13.7 billion years ago, expanding ever since from a common origin in a *big bang.* The *big bang theory* describes how the universe began by expanding.

KEY TERMS

absolute magnitude (p. **268**)
apparent magnitude (p. **268**)
big bang theory (p. **276**)
black hole (p. **273**)
Cepheid variable (p. **270**)
convection zone (p. **268**)

core (p. **267**)
dark energy (p. **278**)
dark matter (p. **279**)
galactic clusters (p. **274**)
galaxy (p. **273**)
globular clusters (p. **274**)

Hertzsprung-Russell diagram (p. **269**)
light-year (p. **266**)
main sequence stars (p. **269**)
nebulae (p. **267**)
neutron star (p. **272**)

protostar (p. **267**)
pulsar (p. **273**)
radiation zone (p. **268**)
red giant stars (p. **270**)
supernova (p. **272**)
white dwarf stars (p. **270**)

APPLYING THE CONCEPTS

Answers are located in appendix F.

1. Stars twinkle and planets do not twinkle because
   a. planets shine by reflected light, and stars produce their own light.
   b. all stars are pulsing light sources.
   c. stars appear as point sources of light, and planets are disk sources.
   d. All of the above are correct.

2. Which of the following stars would have the longer life spans?
   a. the less massive
   b. between the more massive and the less massive
   c. the more massive
   d. All have the same life span.

3. A bright blue star on the main sequence is probably
   a. very massive.
   b. less massive.
   c. between the more massive and the less massive.
   d. None of the above is correct.

4. The basic property of a main sequence star that determines most of its other properties, including its location on the H-R diagram, is
   a. brightness.
   b. color.
   c. temperature.
   d. mass.

5. All the elements that are more massive than the element iron were formed in a
   a. nova.
   b. white dwarf.
   c. supernova.
   d. black hole.

6. If the core remaining after a supernova has a mass between 1.5 and 3 solar masses, it collapses to form a
   a. white dwarf.
   b. neutron star.
   c. red giant.
   d. black hole.

7. The basic unit of the universe is a
   a. star.
   b. solar system.
   c. galactic cluster.
   d. galaxy.

8. The greater the distance to a galaxy, the greater a redshift in its spectral lines. This is known as
   a. Doppler's law.
   b. Cepheid's law.
   c. Hubble's law.
   d. Kepler's law.

9. Dark energy calculations and the age of cooling white dwarfs indicate that the universe is about how old?
   a. 6,000 years
   b. 4.5 billion years
   c. 13.7 billion years
   d. 100,000 billion years

10. Whether the universe will continue to expand or will collapse back into another big bang seems to depend on what property of the universe?
    a. the density of matter in the universe
    b. the age of galaxies compared to the age of their stars
    c. the availability of gases and dust between the galaxies
    d. the number of black holes

11. Stars that are faint, very dense, white hot, and close to the end of their lifetime are
    a. red giant stars.
    b. novas.
    c. white dwarf stars.
    d. Cepheid variable stars.

12. Which of the following elements form only in a supernova explosion of a dying star?
    a. hydrogen
    b. carbon
    c. nitrogen
    d. nickel

QUESTIONS FOR THOUGHT

1. What is a light-year, and how is it defined?
2. Why are astronomical distances not measured with standard referent units of distance such as kilometers or miles?
3. Explain why a protostar heats up internally as it gravitationally contracts.
4. What is the Hertzsprung-Russell diagram? What is the significance of the diagram?
5. Describe in general the structure and interior density, pressure, and temperature conditions of an average star such as the Sun.
6. Describe, in general, the life history of a star with an average mass like the Sun.
7. What is a nova? What is a supernova?
8. Describe the theoretical physical circumstances that lead to the creation of (a) a white dwarf star, (b) a red giant, (c) a neutron star, (d) a black hole, and (e) a supernova.
9. What is meant by the main sequence of the H-R diagram? What one thing determines where a star is plotted on the main sequence?
10. Which size of star has the longest life span, a star 60 times more massive than the Sun, one just as massive as the Sun, or a star that has a mass of one-twenty-fifth that of the Sun? Explain.
11. What is the difference between apparent magnitude and absolute magnitude?
12. What does the color of a star indicate about the surface temperature of the star? What is the relationship between the temperature of a star and the spectrum of the star? Describe in general the spectral classification scheme based on temperature and stellar spectra.
13. Describe the two forces that keep a star in a balanced, stable condition while it is on the main sequence. Explain how these forces are able to stay balanced for a period of billions of years or longer.
14. What is the source of all the elements in the universe that are more massive than helium but less massive than iron? What is the source of all the elements in the universe that are more massive than iron?
15. Why must the internal temperature of a star be hotter for helium fusion reactions than for hydrogen fusion reactions?
16. When does a protostar become a star? Explain.

17. What is a red giant star? Explain the conditions that lead to the formation of a red giant. How can a red giant become brighter than it was as a main sequence star if it now has a lower surface temperature?

18. Why is an average star like the Sun unable to have carbon fusion reactions in its core?

19. If the universe is expanding, are the galaxies becoming larger? Explain.

20. What is the evidence that supports a big bang theory of the universe?

FOR FURTHER ANALYSIS

1. A star is 520 light-years from Earth. During what event in history did the light now arriving at Earth leave the star?

2. What are the significant differences between the life and eventual fate of a massive star and an average-sized star such as the Sun?

3. Analyze when apparent magnitude is a better scale of star brightness and when absolute magnitude is a better scale of star brightness.

4. What is the significance of the Hertzsprung-Russell diagram?

5. The Milky Way galaxy is a huge, flattened cloud of spiral arms radiating out from the center. Describe several ideas that explain why it has this shape. Identify which idea you favor and explain why.

INVITATION TO INQUIRY

It Keeps Going, and Going, and . . .

Pioneer 10 was the first space probe to visit an outer planet of our solar system. It was launched March 2, 1972, and successfully visited Jupiter on June 13, 1983. After transmitting information and relatively close-up pictures of Jupiter, *Pioneer 10* continued on its trajectory, eventually becoming the first space probe to leave the solar system. It continued to move silently into deep space and sent the last signal on January 22, 2003, when it was 12.2 billion km (7.6 billion mi) from Earth. It will now continue to drift for the next 2 million years toward the star Aldebaran in the constellation Taurus.

As the first human-made object out of the solar system, *Pioneer 10* carries a gold-plated plaque with the image shown in box figure 12.2. Perhaps intelligent life will find the plaque and decipher the image to learn about us. What information is in the image? Try to do your own deciphering to reveal the information. When you have exhausted your efforts, see grin.hq.nasa.gov/ABSTRACTS/GPN-2000-001623.html.

For more on the *Pioneer 10* mission, see nssdc.gsfc.nasa.gov/nmc/mastercatalog.do?sc=1972-012A.

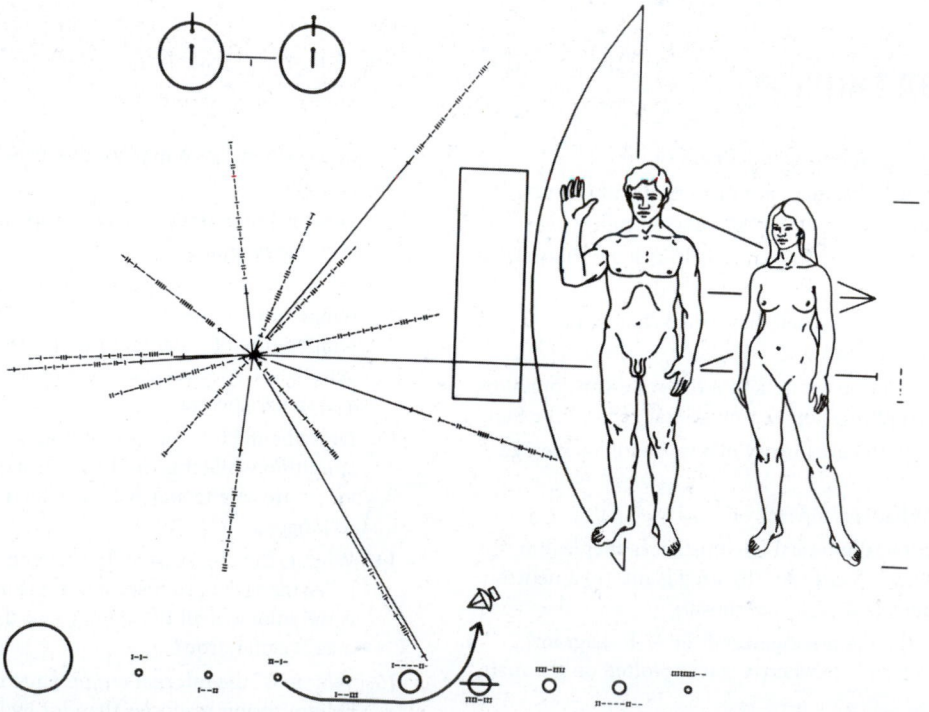

Box Figure 12.2 *Pioneer 10* plaque symbology.

13

The Solar System

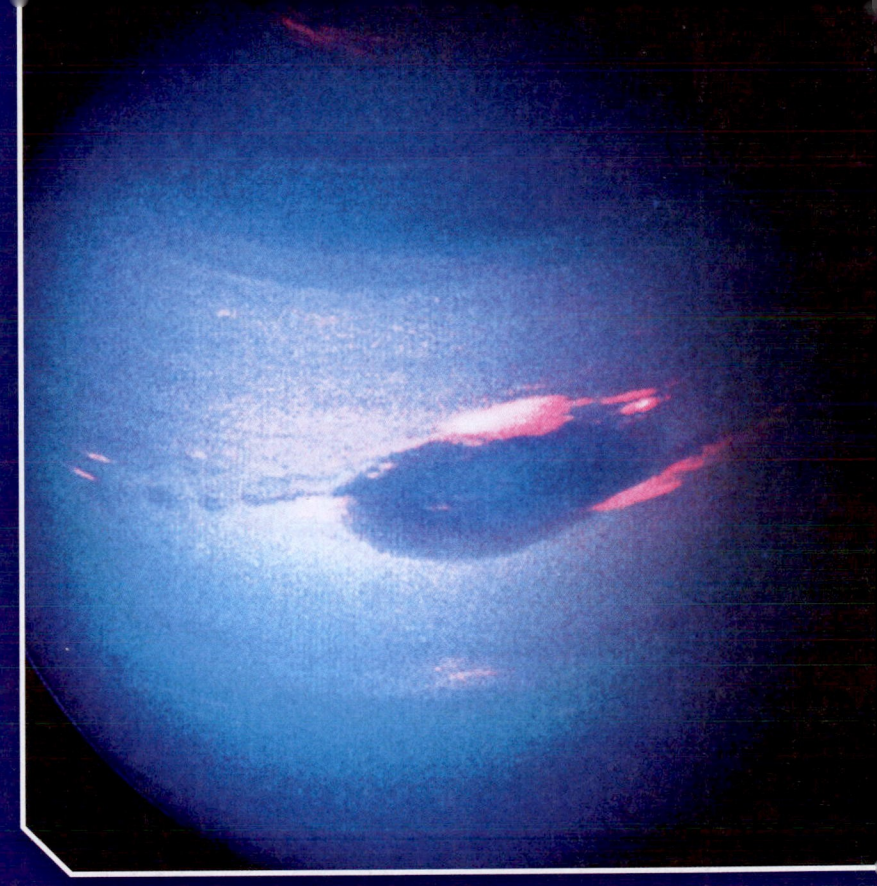

Neptune, the most distant and smallest of the gas giant planets, is a cold and interesting place. It has a Great Dark Spot, as you can see in this photograph made by *Voyager*. This spot is about the size of Earth and is similar to the Great Red Spot on Jupiter. Neptune has the strongest winds of any planet of the solar system—up to 2,000 km/h (1,200 mi/h). Clouds were observed by *Voyager* to be "scooting" around Neptune every sixteen hours or so. *Voyager* scientists called these fast-moving clouds "scooters."

CORE **CONCEPT**

The solar system is composed of the Sun, a system of related planets, moons, comets, and asteroids.

OUTLINE

The interior planets of Mercury, Venus, Earth, and Mars are composed of rocky materials with a metallic nickel and iron core (pp. 287–94).

The outer planets of Jupiter, Saturn, and Uranus are giant planets mostly composed of hydrogen, helium, and methane (pp. 294–97).

The Sun, planets, moons, asteroids, and comets are believed to have formed from gas, dust, and chemical elements of previously existing stars (p. 303).

Physics

▷ Newton's laws of motion describe the relationships between forces, mass, and motion (Ch. 2).

▷ Newton's universal law of gravitation describes how every object in the universe is attracted to every other object in the universe (Ch. 2).

Astronomy

▷ A massive star ages to become a supernova, which spreads elements and gas in space (Ch. 12).

OVERVIEW

For generations, people have observed the sky in awe, wondering about the bright planets moving across the background of stars, but they could do no more than wonder. You are among the first generations on Earth to see close-up photographs of the planets, comets, and asteroids and to see Earth as it appears from space. Spacecrafts have now made thousands of photographs of the other planets and their moons, measured properties of the planets, and in some cases, studied their surfaces with landers. Astronauts have left Earth and visited the Moon, bringing back rock samples, data, and photographs of Earth as seen from the Moon (figure 13.1). All of these photographs and findings have given a new perspective of Earth, the planets, and the moons, comets, and asteroids that make up the solar system.

Viewed from the Moon, Earth is a spectacular blue globe with expanses of land and water covered by huge changing patterns of white clouds. Viewed from a spacecraft, other planets present a very different picture, each unique in its own way. Mercury has a surface that is covered with craters, looking very much like the surface of Earth's Moon. Venus is covered with clouds of sulfuric acid over an atmosphere of mostly carbon dioxide, which is under great pressures with surface temperatures hot enough to melt lead. The surface of Mars has great systems of canyons, inactive volcanoes, dry riverbeds and tributaries, and ice beneath the surface. The giant planets Jupiter and Saturn have orange, red, and white bands of organic and sulfur compounds and storms with gigantic lightning discharges much larger than anything ever seen on Earth. One moon of Jupiter has active volcanoes spewing out liquid sulfur and gaseous sulfur dioxide. The outer giant planets Uranus and Neptune have moons and particles in rings that appear to be covered with powdery, black carbon.

These and many more findings, some fascinating surprises and some expected, have stimulated the imagination as well as added to our comprehension of the frontiers of space. The new information about the Sun's impressive system of planets, moons, comets, and asteroids has also added to speculations and theories about the planets and how they evolved over time in space. This information, along with the theories and speculations, will be presented in this chapter to give you a picture of the solar system.

13.1 PLANETS, MOONS, AND OTHER BODIES

The International Astronomical Union (IAU) is the governing authority over names of celestial bodies. At the August 24, 2006, meeting of the IAU, the definitions of planets, dwarf planets, and small solar system bodies were clarified and approved. To be a classical **planet,** an object must be orbiting the Sun, must be nearly spherical, and must be large enough to clear all matter from its orbital zone. A **dwarf planet** is defined as an object that is orbiting the Sun, is nearly spherical but has *not* cleared matter from its orbital zone, and is *not* a satellite. All other objects orbiting the Sun are referred to collectively as **small solar system bodies.**

Some astronomers had dismissed Pluto as a true planet for years because it has properties that do not fit with those of the other planets. The old definition of a planet was anything spherical that orbits the Sun, which resulted in nine planets. But in 2003, a new astronomical body was discovered. This body was named "Eris" after the Greek goddess of discord. Eris is larger than Pluto, round, and circles the Sun. Is it a planet? If so, the asteroid Ceres would also be a planet as would the fifty or so large, icy bodies believed to be orbiting the Sun far beyond Pluto. The idea of fifty or sixty planets in the solar system spurred astronomers to clarify the definition of a planet. However, Pluto does not clear its orbital zone, so it was downgraded to a dwarf planet. Today, there are eight planets, five dwarf planets (Pluto, Eris, Ceres, Makemake, and Haumea), and many, many small solar system bodies. These definitions may change again in the future as more is learned about our solar system.

FIGURE 13.1 This view of rising Earth was seen by the *Apollo 11* astronauts after they entered the orbit around the Moon. Earth is just above the lunar horizon in this photograph.

In this chapter, we will visit each of the planets, Earth's Moon, and other bodies of the solar system. (figure 13.2).

The Sun has seven hundred times the mass of all the planets, moons, and minor members of the solar system together. It is the force of gravitational attraction between the comparatively

massive Sun and the rest of our solar system that holds it all together. The distance from Earth to the Sun is known as one **astronomical unit** (AU). One AU is about 1.5×10^8 km (about 9.3×10^7 mi). The astronomical unit is used to describe distances in the solar system, for example, Earth is 1 AU from the Sun.

Table 13.1 compares the basic properties of the eight planets. From this table, you can see that the planets can be classified into two major groups based on size, density, and nature of the atmosphere. The interior planets of Mercury, Venus, and Mars have densities and compositions similar to those of Earth, so these planets, along with Earth, are known as the **terrestrial planets.**

Outside the orbit of Mars are four **giant planets,** which are similar in density and chemical composition. The terrestrial planets are mostly composed of rocky materials and metallic nickel and iron. The giant planets of Jupiter, Saturn, Uranus, and Neptune, on the other hand, are massive giants mostly composed of hydrogen, helium, and methane. The density of the giant planets suggests the presence of rocky materials and iron as a core surrounded by a deep layer of compressed gases beneath a deep atmosphere of vapors and gases. Note that the terrestrial planets are separated from the giant planets by the asteroid belt.

We will start with the planet closest to the Sun and work our way outward, moving farther and farther from the Sun as we learn about our solar system.

Mercury

Mercury is the innermost planet, moving rapidly in a highly elliptical orbit that averages about 0.4 astronomical unit, or about 0.4 of the average distance of Earth from the Sun. Mercury is the smallest planet and is slightly larger than Earth's Moon. Mercury is very bright because it is so close to the Sun, but it is difficult to observe because it only appears briefly for a few hours

FIGURE 13.2 The order of the planets out from the Sun. The planets are (1) Mercury, (2) Venus, (3) Earth, (4) Mars, (5) Jupiter, (6) Saturn, (7) Uranus, and (8) Neptune. The orbits and the planet sizes are not drawn to scale, and not all rings or moons are shown. Also, the planets are not in a line as shown.

TABLE 13.1 Properties of the Planets

	Mercury	Venus	Earth	Mars	Jupiter	Saturn	Uranus	Neptune
Average Distance from the Sun:								
in 10^6 km	58	108	150	228	778	1,400	3,000	4,497
in AU	0.38	0.72	1.0	1.5	5.2	9.5	19.2	30.1
Inclination to Ecliptic	7°	3.4°	0°	1.9°	1.3°	2.5°	0.8°	1.8°
Revolution Period								
(Earth years)	0.24	0.62	1.00	1.88	11.86	29.46	84.01	164.8
Rotation Period*	59 days	−243 days	23 h	24 h	9 h	10 h	−17 h	16 h
(Earth days,			56 min	37 min	50 min	39 min	14 min	6.7 min
min, and s)			4 s	23 s	30 s			
Mass (Earth = 1)	0.05	0.82	1.00	0.11	317.9	95.2	14.6	17.2
Equatorial Dimensions:								
diameter in km	4,880	12,104	12,756	6,787	142,984	120,536	57,118	49,528
in Earth radius = 1	0.38	0.95	1.00	0.53	11	9	4	4
Density (g/cm³)	5.43	5.25	5.52	3.95	1.33	0.69	1.29	1.64
Atmosphere								
(major compounds)	None	CO_2	N_2, O_2	CO_2	H_2, He	H_2, He	H_2, He, CH_4	H_2, He, CH_4
Solar Energy Received								
(cal/cm²/s)	13.4	3.8	2.0	0.86	0.08	0.02	0.006	0.002

*A negative value means spin is opposite to motion in orbit.

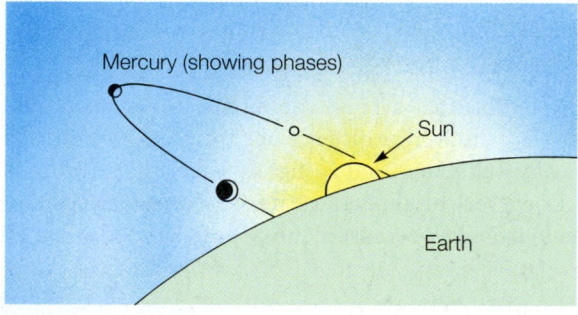

FIGURE 13.3 Mercury is close to the Sun and is visible from Earth only briefly before or after sunrise or sunset, showing phases. Mercury actually appears much smaller in an orbit that is not tilted as much as shown in this figure.

immediately after sunset or before sunrise. This appearance, low on the horizon, means that Mercury must be viewed through more of Earth's atmosphere, making the study of such a small object difficult at best (figure 13.3).

Mercury moves around the Sun in about three Earth months, giving Mercury the shortest "year" of all the planets. With the highest orbital velocity of all the planets, Mercury was appropriately named after the mythical Roman messenger of speed. Oddly, however, this speedy planet has a rather long day in spite of its very short year. With respect to the stars, Mercury rotates once every fifty-nine days. This means that Mercury rotates on its axis three times every two orbits.

The long Mercury day with a nearby large, hot Sun means high temperatures on the surface facing the Sun. High tem-

peratures mean higher gas kinetic energies, and with a low gravity, gases easily escape from Mercury so it has only trace gases for an atmosphere. The lack of an atmosphere to even the heat gains from the long days and heat losses from the long nights results in some very large temperature differences. The temperature of the surface of Mercury ranges from above the melting point of lead on the sunny side to below the temperature of liquid oxygen on the dark side.

Mercury has been visited by *Mariner 10,* which flew by three times in 1973 and 1974. The photographs transmitted by *Mariner 10* revealed that the surface of Mercury is covered with craters and very much resembles the surface of Earth's Moon. There are large craters, small craters, superimposed craters, and craters with lighter colored rays coming from them just like the craters on the Moon. Also as on Earth's Moon, there are hills and smooth areas with light and dark colors that were covered by lava in the past, some time after most of the impact craters were formed (figure 13.4).

Spacecraft MESSENGER was launched August 3, 2004, on a looping 7.9 billion km (about 4.9 billion mi) trip through the solar system. It traveled for six and a half years before becoming the first spacecraft to orbit Mercury. MESSENGER used gravity assists from Earth, Venus, and Mercury to achieve the correct orientation and speed for orbit insertion, which was achieved March 17, 2011. Mercury has a high surface temperature, and the spacecraft instruments are protected from high radiant energy from Mercury and the Sun by a sun-shade of heat-resistant ceramic fabric. (There is a NASA mission page at www.nasa.gov/mission_pages/messenger/main/index.html.)

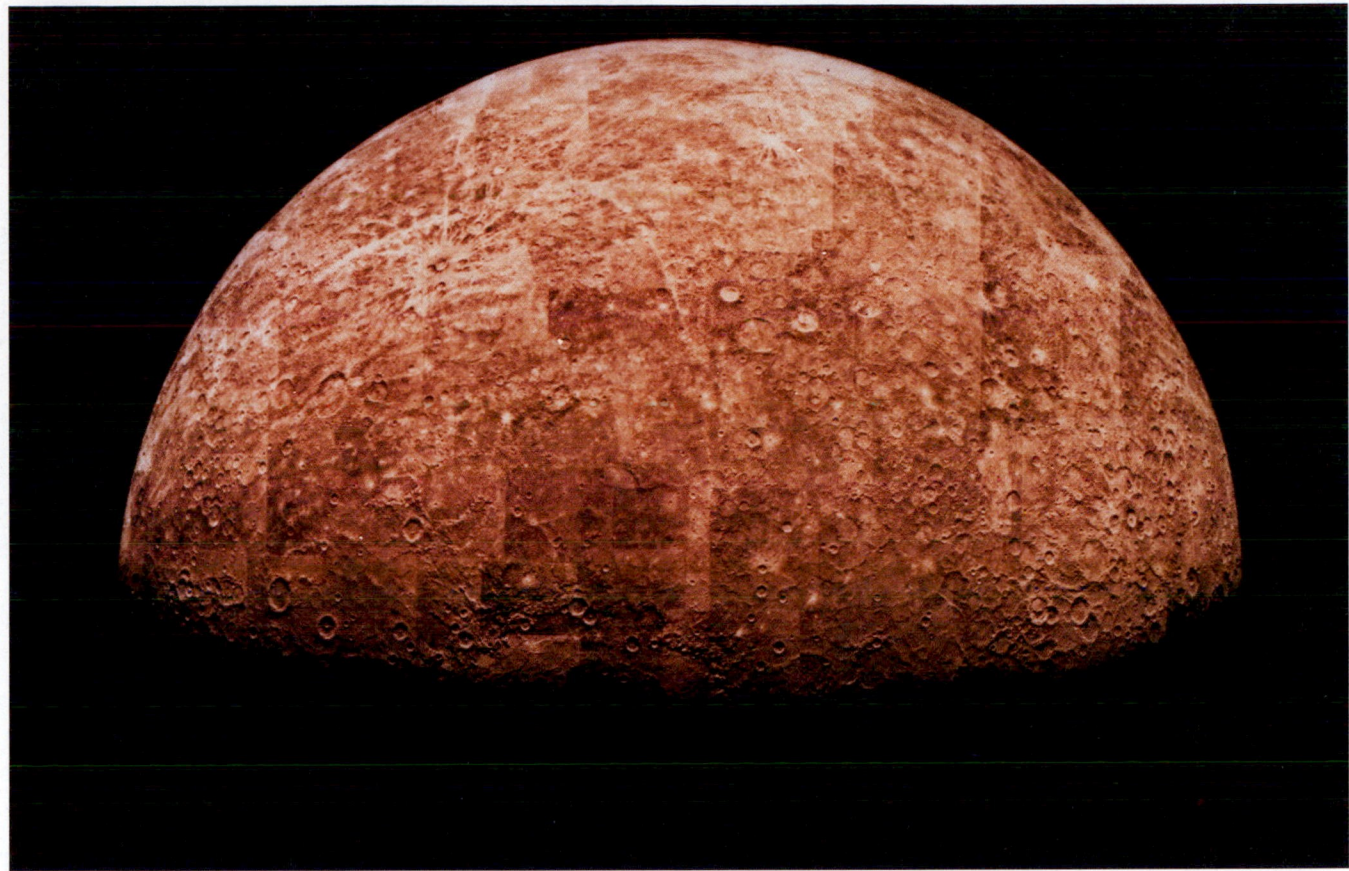

FIGURE 13.4 A photomosaic of Mercury made from pictures taken by the *Mariner 10* spacecraft. The surface of Mercury is heavily cratered, looking much like the surface of Earth's Moon. All the interior planets and the Moon were bombarded early in the life of the solar system.

Mercury has no natural satellites or moons, it has a weak magnetic field, and it has an average density more similar to that of Venus or Earth than to the Moon's. The presence of the magnetic field and the relatively high density for such a small body probably means that Mercury has a relatively large core of iron with at least part of the core molten. Because of its high density, Mercury is believed to have lost much of its less dense, outer layer of rock materials sometime during its formation.

Venus

Venus is the brilliant evening and morning "star" that appears near sunrise or sunset, sometimes shining so brightly that you can see it while it is still daylight. Venus orbits the Sun at an average distance of about 0.7 AU. Venus is sometimes to the left of the Sun, appearing as the evening star, and sometimes to the right of the Sun, appearing as the morning star. Venus also has phases just as the Moon does. When Venus is in the full phase, it is small and farthest away from Earth. A crescent Venus appears much larger and thus the brightest because it is closest to Earth when it is in its crescent phase. You can see the phases of Venus with a good pair of binoculars.

Venus shines brightly because it is covered with clouds that reflect about 80 percent of the sunlight, making it the brightest object in the sky after the Sun and the Moon. These same clouds prevented any observations of the surface of Venus until the early 1960s, when radar astronomers were able to penetrate the clouds and measure the planet's spin rate. Venus was found to spin slowly, so slowly that each day on Venus is longer than a Venus year! Also a surprise, Venus was found to spin in the *opposite* direction to its direction of movement in its orbit. On Venus, you would observe the Sun to rise in the west and set in the east, if you could see it, that is, through all the clouds.

In addition to early studies by radio astronomers, Venus has been the target of many American and Soviet spacecraft probes (table 13.2).

Venus has long been called Earth's sister planet since its mass, size, and density are very similar. That is where the similarities with Earth end, however, as Venus has been found to have a very hostile environment. Spacecraft probes found a hot, dry surface under tremendous atmospheric pressure (figure 13.5). The atmosphere consists mostly of carbon dioxide, a few percent of nitrogen, and traces of water vapor and other gases. The atmospheric pressure at the surface of Venus is almost one hundred times the pressure at the surface of Earth, a pressure many times beyond what a human could tolerate. The average surface temperature is comparable to the surface temperature on Mercury, which is hot enough to melt lead. The hot temperature on Venus, which is nearly twice the distance from the Sun as Mercury, is a result of the greenhouse effect.

TABLE 13.2   Spacecraft Missions to Venus

Date	Name	Owner	Remark
Feb 12, 1961	Venera 1	USSR	Flyby
Aug 27, 1962	Mariner 2	U.S.	Flyby
Apr 2, 1964	Zond 1	USSR	Flyby
Nov 12, 1965	Venera 2	USSR	Flyby
Nov 16, 1965	Venera 3	USSR	Crashed on Venus
June 12, 1967	Venera 4	USSR	Impacted Venus
June 14, 1967	Mariner 5	U.S.	Flyby
Jan 5, 1969	Venera 5	USSR	Impacted Venus
Jan 10, 1969	Venera 6	USSR	Impacted Venus
Aug 17, 1970	Venera 7	USSR	Venus landing
Mar 27, 1972	Venera 8	USSR	Venus landing
Nov 3, 1973	Mariner 10	U.S.	Venus, Mercury flyby photos
June 8, 1975	Venera 9	USSR	Lander/orbiter
June 14, 1975	Venera 10	USSR	Lander/orbiter
May 20, 1978	Pioneer 12 (also called Pioneer Venus 1 or Pioneer Venus)	U.S.	Orbital studies of Venus
Aug 8, 1978	Pioneer 13 (also called Pioneer Venus 2 or Pioneer Venus)	U.S.	Orbital studies of Venus
Sept 9, 1978	Venera 11	USSR	Lander; sent photos
Sept 14, 1978	Venera 12	USSR	Lander; sent photos
Oct 30, 1981	Venera 13	USSR	Lander; sent photos
Nov 4, 1981	Venera 14	USSR	Lander; sent photos
June 2, 1983	Venera 15	USSR	Radar mapper
June 7, 1983	Venera 16	USSR	Radar mapper
Dec 15, 1984	Vega 1	USSR	Venus/Comet Halley probe
Dec 21, 1984	Vega 2	USSR	Venus/Comet Halley probe
May 4, 1989	Magellan	U.S.	Orbital radar mapper
Oct 18, 1989	Galileo	U.S.	Flyby measurements and photos
Apr 11, 2006	Venus Express	ESA	Orbital spacecraft
June 5, 2007	Messenger	U.S.	Flyby photos

FIGURE 13.5   This is an image of an 8 km (5 mile) high volcano on the surface of Venus. The image was created by a computer using *Magellan* radar data, simulating a viewpoint elevation of 1.7 km (1 mi) above the surface. The lava flows extend for hundreds of kilometers across the fractured plains shown in the foreground. The simulated colors are based on color images recorded by the Soviet *Venera 13* and *14* spacecraft.

The surface of Venus is mostly a flat, rolling plain. There are several raised areas, or "continents," on about 5 percent of the surface. There is also a mountain larger than Mount Everest, a great valley deeper and wider than the Grand Canyon, and many large, old impact craters. In general, the surface of Venus appears to have evolved much as did the surface of Earth but without the erosion caused by ice, rain, and running water.

Venus, like Mercury, has no satellites. Venus also does not have a magnetic field, as might be expected. Two conditions seem to be necessary in order for a planet to generate a magnetic field: a molten center and a relatively rapid rate of rotation. Since Venus takes 243 days to complete one rotation, the slowest of all the planets, it does not have a magnetic field even if some of the interior of Venus is still liquid as in Earth.

The European Space Agency's (ESA) orbiting *Venus Express* is conducting an in-depth observation of the structure and chemistry of the atmosphere of Venus. One of the early findings was hot, extensive 21 km (13 mi) deep clouds of sulfuric acid moved by 334 km/h (220 mph) winds. The mission also plans to study the Venus greenhouse effect and volcanic activity of the past and present, if any.

Earth's Moon

Next to the Sun, the Moon is the largest, brightest object in the sky. The Moon is Earth's nearest neighbor and surface features can be observed with the naked eye. With the aid of a telescope or a good pair of binoculars, you can see light-colored mountainous regions called the *lunar highlands,* smooth dark areas called *maria,* and many sizes of craters, some with bright streaks extending from them (figure 13.6). The smooth dark areas are called maria after a Latin word meaning "sea." They

Sunlight filters through the atmosphere of Venus, warming the surface. The surface reemits the energy in the form of infrared radiation, which is absorbed by the almost pure carbon dioxide atmosphere. Carbon dioxide molecules absorb the infrared radiation, increasing their kinetic energy and the temperature.

FIGURE 13.6 You can easily see the light-colored lunar highlands, smooth and dark maria, and many craters on the surface of Earth's nearest neighbor in space.

acquired this name from early observers who thought the dark areas were oceans and the light areas were continents. Today, the maria are understood to have formed from ancient floods of molten lava that poured across the surface and solidified to form the "seas" of today. There is no water or atmosphere on the Moon.

Many facts known about the Moon were established during the *Apollo* missions, the first human exploration of a place away from Earth. A total of twelve *Apollo* astronauts walked on the Moon, taking thousands of photographs, conducting hundreds of experiments, and returning to Earth with over 380 kg (about 840 lb) of moon rocks. In addition, instruments were left on the Moon that continued to radio data back to Earth after the *Apollo* program ended in 1972. As a result of the *Apollo* missions, many questions were answered about the Moon, but unanswered questions still remain.

The *Apollo* astronauts found that the surface of the Moon is covered by a 3 m (about 10 ft) layer of fine gray dust that contains microscopic glass beads. The dust and beads were formed from millions of years of continuous bombardment of micrometeorites. These very small meteorites generally burn up in Earth's atmosphere. The Moon does not have an atmosphere, so meteorites have continually fragmented and pulverized the surface in a slow, steady rain. The glass beads are believed to have formed when larger meteorite impacts melted part of the surface, which was immediately forced into a fine spray that cooled rapidly while above the surface.

The rocks on the surface of the Moon were found to be mostly *basalts,* a type of rock formed on Earth from the cooling and solidification of molten lava. The dark-colored rocks from the maria are similar to Earth's basalts (see chapter 15) but contain greater amounts of titanium and iron oxides. The light-colored rocks from the highlands are mostly *breccias,* a kind of rock made up of rock fragments that have been compacted together. On the Moon, the compacting was done by meteorite impacts. The rocks from the highlands contain more aluminum and less iron than the maria basalts and thus have a lower density than the darker rocks.

All the moon rocks contained a substantial amount of radioactive elements, which made it possible to precisely measure their age. The light-colored rocks from the highlands were formed about 4 billion years ago. The dark-colored rocks from the maria were much younger, with ages ranging from 3.1 to 3.8 billion years. This indicates a period of repeated volcanic eruptions and lava flooding over a 700-million-year period that ended about 3 billion years ago.

The moon rocks brought back to Earth, the results of the lunar seismographs, and all the other data gathered through the *Apollo* missions have increased our knowledge about the Moon, leading to new understandings of how it formed. This model pictures the present Moon developing through four distinct stages.

The *origin stage* describes how the moon originally formed. The Moon is believed to have originated from the impact of Earth with a very large object, perhaps as large as Mars or larger. The Moon formed from ejected material produced by this collision. The collision is believed to have vaporized the colliding body as well as part of Earth. Some of the debris condensed away from Earth to form the Moon.

The collision that resulted in the Moon took place after Earth's iron core formed, so there is not much iron in moon rocks. The difference in moon and earth rocks can be accounted for by the presence of materials from the impacting body.

The *molten surface stage* occurred during the first 200 million years after the collision. Heating from a number of sources could have been involved in the melting of the entire lunar surface 100 km (about 60 mi) or so deep. The heating required to melt the surface is believed to have resulted from the impacts of rock fragments, which were leftover debris from the formation of the solar system that intensely bombarded the Moon. After a time, there were fewer fragments left to bombard the Moon, and the molten outer layer cooled and solidified to solid rock. The craters we see on the Moon today are the result of meteorites hitting the Moon between 3.9 and 4.2 billion years ago after the crust formed.

The *molten interior stage* involved the melting of the interior of the Moon. Radioactive decay had been slowly heating the interior, and 3.8 billion years ago, or about a billion years after the Moon formed, sufficient heat accumulated to melt the interior. The light and heavier rock materials separated during this period, perhaps producing a small iron core. Molten lava flowed into basins on the surface during this period, forming the smooth, darker maria seen today. The lava flooding continued for about 700 million years, ending about 3.1 billion years ago.

The *cold and quiet stage* began 3.1 billion years ago as the last lava flow cooled and solidified. Since that time, the surface of the Moon has been continually bombarded by micrometeorites and a few larger meteorites. With the exception of a few new craters, the surface of the Moon has changed little in the last 3 billion years.

Mars

Mars has always attracted attention because of its unique, bright reddish color. The properties and surface characteristics have also attracted attention, particularly since Mars seems to have similarities to Earth. It orbits the Sun at an average distance of about 1.5 AU. It makes a complete orbit every 687 days, about twice the time that Earth takes. Mars rotates on its axis in twenty-four hours, thirty-seven minutes, so the length of a day on Mars is about the same as the length of a day on Earth. The observations that Mars has an atmosphere, light and dark regions that appear to be greenish and change colors with the seasons, and white polar caps that grow and shrink with the seasons led to early speculations (and many fantasies) about the possibilities of life on Mars. These speculations increased dramatically in 1877 when Schiaparelli, an Italian astronomer, reported seeing "channels" on the Martian surface. Other astronomers began interpreting the dark greenish regions as vegetation and the white polar caps as ice caps as Earth has. In the early part of the twentieth century, the American astronomer who founded the Lowell Observatory in Arizona, Percival Lowell, published a series of popular books showing a network of hundreds of canals on Mars. Lowell and other respectable astronomers interpreted what they believed to be canals as evidence of intelligent life on Mars. Other astronomers, however, interpreted the greenish colors and the canals to be illusions, imagined features of astronomers working with the limited telescopes of that time. Since canals never appeared in photographs, said the skeptics, the canals were the result of the human tendency to see patterns in random markings where no patterns actually exist.

This speculation ended in the late 1960s and early 1970s with extensive studies and probes by spacecraft (table 13.3). Limited photographs by *Mariner* flybys in 1965 and 1969 had provided some evidence that the surface of Mars was much like the Moon, with no canals, vegetation, or much of anything else. Then in 1971, *Mariner 9* became the first spacecraft to orbit Mars, photographing the entire surface as well as making extensive measurements of the Martian atmosphere, temperature ranges, and chemistry. For about a year, *Mariner 9* sent a flood of new and surprising information about Mars back to Earth.

Mariner 9 found the surface of Mars not to be a crater-pitted surface as is found on the Moon. Mars has had a geologically active past and has four provinces, or regions, of related surface features. There are (1) volcanic regions with inactive volcanoes, one larger than any found on Earth, (2) regions with systems of canyons, some larger than any found on Earth, (3) regions of terraced plateaus near the poles, and (4) flat regions pitted with impact craters. Surprisingly, dry channels suggesting former water erosion were discovered near the cratered regions. These are sinuous, dry riverbed features with dry tributaries. Liquid water may have been present on Mars in the past, but none is to be found today.

The atmosphere of Mars is very thin, exerting an average pressure at the surface that is only 0.6 percent of the average atmospheric pressure on Earth's surface. Moreover, this thin Martian atmosphere is about 95 percent carbon dioxide, and 20 percent of this freezes as dry ice at the Martian South Pole every winter.

Does life exist on Mars? Two *Viking* spacecraft were sent to Mars in 1975 to search for signs of life. The two *Viking* spacecraft were identical, each consisting of an orbiter and a lander. After eleven months of travel time, *Viking 1* entered an orbit around Mars in June 1976 and spent a month sending high-resolution images of the surface back to Earth. From these images, a land-ing site was selected for the *Viking 1* lander. Using retrorockets, parachutes, and descent rockets, the *Viking 1* lander arrived on a dusty, rocky slope in the southern hemisphere on July 20, 1976. The *Viking 2* lander arrived forty-five days later but farther to the north. The *Viking* landers contained a mechanical soil-retrieving arm and a miniature computerized lab to analyze the soil for evidence of metabolism, respiration, and other life processes. Neither lander detected any evidence of life processes or any organic compounds that would indicate life now or in the past.

The *Viking* spacecraft continued sending images and weather data back to Earth until 1982. During their six-year life, the orbiters sent about fifty-two thousand images and mapped about 97 percent of the Martian surface. The landers sent an additional forty-five hundred images, recorded a major "Marsquake," and recorded data about regular dust storms that occur on Mars with seasonal changes.

The *Mars Exploration Rovers*, named *Spirit* and *Opportunity*, landed on Mars on January 4 and January 25, 2004,

TABLE 13.3 **Completed Spacecraft Missions to Mars**

Date	Name	Owner	Remark
Nov 5, 1964	*Mariner 3*	U.S.	Flyby
Nov 28, 1964	*Mariner 4*	U.S.	First photos
Feb 24, 1969	*Mariner 6*	U.S.	Flyby
Mar 27, 1969	*Mariner 7*	U.S.	Flyby
May 19, 1971	*Mars 2*	USSR	Lander
May 28, 1971	*Mars 3*	USSR	Orbiter/lander
May 30, 1971	*Mariner 9*	U.S.	Orbiter
Jul 21, 1973	*Mars 4*	USSR	Probe
Jul 25, 1973	*Mars 5*	USSR	Orbiter
Aug 5, 1973	*Mars 6*	USSR	Lander
Aug 9, 1973	*Mars 7*	USSR	Flyby/lander
Aug 20, 1975	*Viking 1*	U.S.	Lander/orbiter
Sept 9, 1975	*Viking 2*	U.S.	Lander/orbiter
July 7, 1988	*Phobos 1*	USSR	Orbiter/Phobos lander
July 12, 1988	*Phobos 2*	USSR	Orbiter/Phobos lander
Nov 7, 1996	*Global Surveyor*	U.S.	Orbiter
Sept 11, 1997	*Mars Global Surveyor*	U.S.	Orbiter
Dec 4, 1997	*Pathfinder*	U.S.	Lander/surface rover
Oct 23, 2001	*2001 Mars Odyssey*	U.S.	Orbiter
Dec 25, 2003	*Mars Express*	ESA	Orbiter/lander
Jan 4, 2004	*Spirit*	U.S.	Lander/surface rover
Jan 25, 2004	*Opportunity*	U.S.	Lander/surface rover
Aug 26, 2006	*Mars Reconnaissance Orbiter*	U.S.	Orbiter
Mar 10, 2006	*Mars Reconnaissance Orbiter*	U.S.	Orbiter
May 25, 2008	*Phoenix Mars Lander*	U.S.	Lander

respectively, to answer questions about the history of water on Mars. The spacecraft were sent to sites on opposite sides of Mars that appear to have been affected by liquid water in the past. After parachute and airbag landings, the 185 kg (408 lb) rovers charged their solar-powered batteries. They then began driving to different locations to perform on-site scientific investigations over the course of their mission (figure 13.7).

What did the rovers find? They found that Mars is made of basalt rock (see chapter 15) and groundwater that is dilute sulfuric acid. The acid interacts with the rock, dissolving things out of it, and then evaporates and leaves sulfur-rich salts. The *Spirit* and *Opportunity* rover results confirmed that sufficient amounts of water to alter the rocks have been present in the past (figure 13.8). The results also confirm the premission interpretations of remote-sensing data. This provides evidence that other present and future remote-sensing data is accurate.

The presence of past or present life on Mars remains an open question. Scientists already knew there was liquid water in the past, and water and life go together. Beyond that, the *Rover* mission has really not changed the prospect of finding evidence of past or present life, so the search goes on.

Jupiter

Jupiter is the largest of all the planets, with a mass equivalent to some 318 Earths and, in fact, is more than twice as massive as all the other planets combined. This massive planet is located an average 5 AU from the Sun in an orbit that takes about twelve Earth years for one complete trip around the Sun. The internal heating from gravitational contraction was tremendous when this giant formed, and today it still radiates twice the energy that it receives from the Sun. The source of this heat is the slow

FIGURE 13.7   Researchers used the rover *Spirit*'s rock abrasion tool to help them study a rock dubbed "Uchben" in the "Colombia Hills" of Mars. The tool ground into the rock, creating a shallow hole 4.5 cm (1.8 in) in diameter in the central upper portion of this image. It also used wire bristles to brush a portion of the surface below and to the right of the hole. *Spirit* used its panoramic camera during the rover's 293rd martian day (October 29, 2004) to take the frames combined into this approximately true-color image.
Source: NASA/JPL/Cornell.

FIGURE 13.8   A rock dubbed "Palenque" in the "Colombia Hills" of Mars has contrasting textures in upper and lower portions. This view of the rock combines two frames taken by the panoramic camera on NASA's Mars Exploration Rover *Spirit* during the rover's 278th martian day (October 14, 2004). The layers meet each other at an angular unconformity that may mark a change in environmental conditions between the formation of the two portions of the rock. Scientists would have liked the rover to take a closer look, but Palenque is not on a north-tilted slope, which is the type of terrain needed to keep the rover's solar panels tilted toward the winter sun. The exposed portion of the rock is about 100 cm (39 in) long.
Source: NASA/JPL/Cornell.

Planets and Astrology

Do you read the astrology forecasts in daily newspapers or on Internet sites? Below is a brief background of astrology as developed by the Babylonians, followed by some questions intended for class or small group discussions.

As early as 2000 B.C., the Babylonians began keeping track of time by dividing the year into 12 months, with 7 days to a week and 360 days to a year. They maintained observatories, noting for example that the Sun, the Moon, and five planets known at the time (Mercury, Venus, Mars, Jupiter, and Saturn) moved across the sky only along a certain path, which is today called the ecliptic. This movement was independent of the stars, which followed the motions of the seven celestial bodies but kept the same position relative to each other. The Sun appeared to move completely around the ecliptic each year.

To keep track of the Sun and the time of year, the Babylonians imagined the arrangements of certain stars to be the shapes of gods, objects, or animals. These patterns, today called constellations, were used to identify twelve equal divisions of the path the Sun followed for a year. The twelve constellations are called the zodiac, so there are twelve signs of the zodiac in a year. By 540 B.C., the Babylonians had fully developed the art of studying the zodiac, the Sun, and planets as a guide to human affairs, and this activity today is known as the pseudoscience of astrology.

First, consider astrology forecasts as they are made today. A daily horoscope might include a forecast for those with a birthday on this day and forecasts for people born during each of the twelve signs of the zodiac. In the horoscope, you can find such predictions as "your computer could crash today," "focus on your ability to tear down before you rebuild," and "take a chance on romance with a Virgo." Discuss such forecasts with your group. Consider, for example, the population of a nation and how many people are forecasted to have their computer crash. How many computers would crash without the forecast?

Next, discuss why your passage through the birth canal (your birthday) is so important. Perhaps the time when your embryo formed might be more important if you were going to be "marked" by the planets for certain things to happen to you.

Finally, discuss the topic of how Earth's axis has a slow wobble, called precession, which causes it to swing in a slow circle like the wobble of a spinning top. The axis takes about 26,000 years to complete one turn, or wobble. The moving pole changes over time in which particular signs of the zodiac appear, for example, with the spring equinox. Because of precession, the occurrence of the spring equinox has been moving backward through the zodiac constellations at about 1 degree every 72 years. So, 3,000 years ago, the Sun entered the constellation of Virgo in August, which is still the basis for horoscopes today. However, the Sun is now in the Leo constellation in August.

---

gravitational compression of the planet, not from nuclear reactions as in the Sun. Jupiter would have to be about eighty times as massive to create the internal temperatures needed to start nuclear fusion reactions, or in other words, to become a star itself. Nonetheless, the giant Jupiter and its system of satellites seem almost like a smaller version of a planetary system within the solar system.

Jupiter has an average density that is about a quarter of the density of Earth. This low density indicates that Jupiter is mostly made of light elements, such as hydrogen and helium, but does contain a percentage of heavier rocky substances. The model of Jupiter's interior (figure 13.9) is derived from this and other information from spectral studies, studies of spin rates, and measurements of heat flow. The model indicates a solid, rocky core that is more than twice the size of Earth. Surrounding this core is a thick layer of liquid hydrogen, compressed so tightly by millions of atmospheres of pressure that it is able to conduct electric currents. Liquid hydrogen with this property is called *metallic hydrogen* because it has the conductive ability of metals. Above the layer of metallic hydrogen is a thick layer of ordinary liquid hydrogen, which is under less pressure. The outer layer, or atmosphere, of Jupiter is a 500 km or so (about 300 mi) zone with hydrogen, helium, ammonia gas, crystalline compounds, and a mixture of ice and water. It is the uppermost ammonia clouds, perhaps mixed with sulphur and

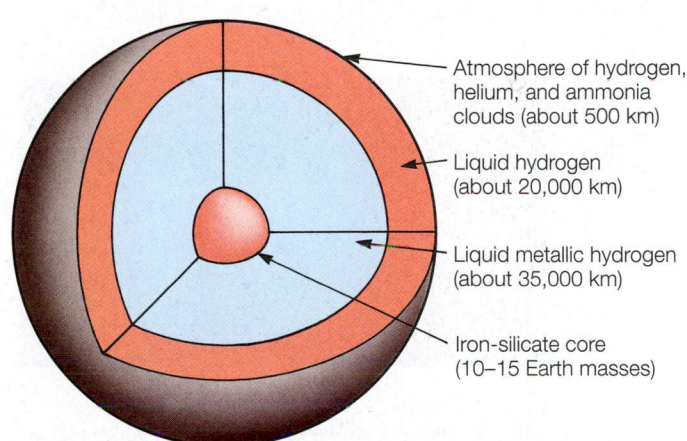

FIGURE 13.9 The interior structure of Jupiter.

organic compounds, that form the bright orange, white, and yellow bands around the planet. The banding is believed to be produced by atmospheric convection, in which bright, hot gases are forced to the top where they cool, darken in color, and sink back to the surface.

Jupiter's famous Great Red Spot is located near the equator. This permanent, deep, red oval feature was first observed by

A

B

FIGURE 13.10 Photos of Jupiter taken by *Voyager 1*. (*A*) From a distance of about 36 million km (about 22 million mi). (*B*) A closer view, from the Great Red Spot to the South Pole, showing organized cloud patterns. In general, dark features are warmer and light features are colder. The Great Red Spot soars about 25 km (about 15 mi) above the surrounding clouds and is the coldest place on the planet.

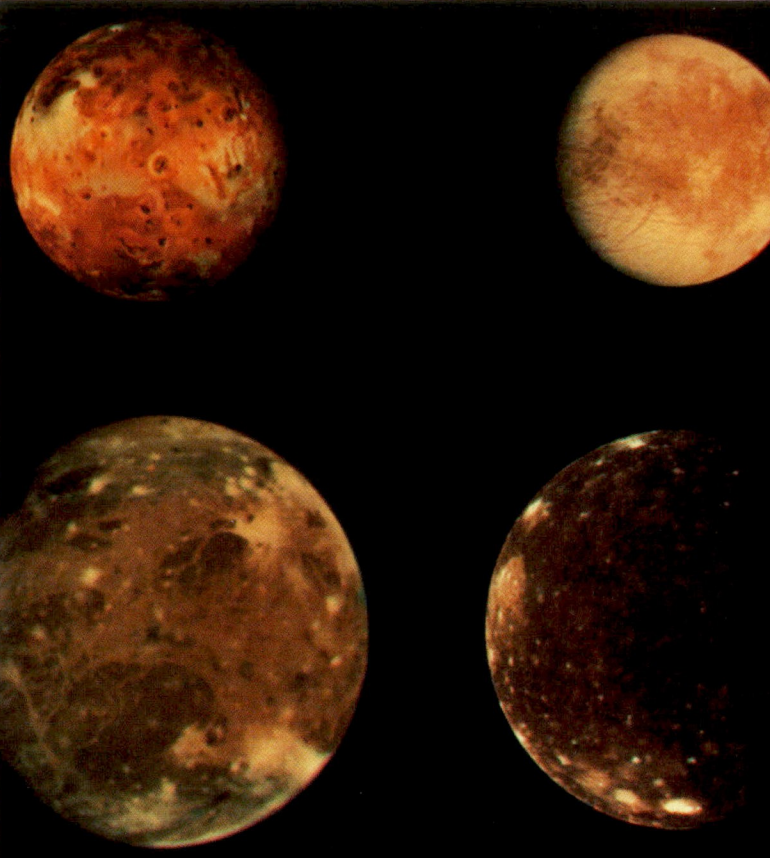

FIGURE 13.11 The four Galilean moons pictured by *Voyager 1*. Clockwise from upper left, Io, Europa, Ganymede, and Callisto. Io and Europa are about the size of Earth's Moon; Ganymede and Callisto are larger than Mercury.

Robert Hooke in the 1600s and has generated much speculation over the years. The red oval, some 40,000 km (about 25,000 mi) long, has been identified by infrared observations to be a high-pressure region, with higher and colder clouds, that has lasted for at least three hundred years. The energy source for such a huge, long-lasting feature is unknown (figure 13.10).

Jupiter has sixty-four satellites, and the four brightest and largest can be seen from Earth with a good pair of binoculars. These four are called the *Galilean moons* because they were discovered by Galileo in 1610. The Galilean moons are named Io, Europa, Ganymede, and Callisto (figure 13.11).

Observations by the *Pioneer* and *Voyager* spacecrafts revealed some fascinating and intriguing information about the moons of Jupiter. Io, for example, was discovered to have active volcanoes that eject enormous plumes of molten sulfur and sulfur dioxide gas. Europa is covered with a 19.3 km (about 12 mi) thick layer of smooth water ice, which has a network of long, straight, dark cracks. Ganymede has valleys, ridges, folded mountains, and other evidence of an active geologic history. Callisto, the most distant of the Galilean moons, was found to be the most heavily cratered object in the solar system.

These impact events are still occurring. In 1994, the comet Shoemaker-Levy 9 broke apart into a "string of pearls," (figure 13.12A) and then produced a once-in-a-lifetime spectacle as it proceeded to leave its imprint on Jupiter as well as people

UFOs and You

Unidentified flying objects (UFOs) are observed around the world. UFOs are generally sighted near small towns and out in the country, often near a military installation. Statistically, most sightings occur during the month of July at about 9:00 P.M., then at 3:00 A.M. UFOs can be grouped into three categories:

1. **Natural phenomena.** Most sightings of UFOs can be explained as natural phenomena. For example, a bright light that flashes red and green on the horizon could be a star viewed through atmospheric refraction. The vast majority of all UFO sightings are natural phenomena such as atmospheric refraction, ball lightning, swamp gas, or something simple such as a drifting weather balloon or military flares on parachutes, high in the atmosphere and drifting across the sky. This category of UFOs should also include exaggeration and fraud—such as balloons released with burning candles as a student prank.
2. **Aliens.** The idea that UFOs are alien spacecraft from other planets is very popular. However, there is no authentic, unambiguous evidence that would prove the existence of aliens. In fact, most unidentified flying objects are not objects at all. They are lights that can eventually be identified.
3. **Psychological factors.** This group of sightings includes misperceptions of natural phenomena resulting from unusual conditions that influence a person's perception and psychological health. This includes people who claim to receive information from aliens by "channeling" messages.

Questions to Discuss

Divide your group into three subgroups, with each subgroup selecting one of the three categories above. After preparing for a few minutes, have each group present reasons why we should understand UFOs to be natural phenomena, aliens from other planets, or psychological misperceptions. Then have the entire group discuss the three categories and try to come to a consensus.

Jupiter in Ultraviolet

Hubble Space Telescope Wide Field Planetary Camera 2

A

B

FIGURE 13.12 (*A*) This image, made by the Hubble Space Telescope, clearly shows the large impact site made by fragments of former comet Shoemaker-Levy 9 when it collided with Jupiter. (*B*) This is a picture of comet Shoemaker-Levy 9 after it broke into twenty-two pieces, lined up in this row, then proceeded to plummet into Jupiter during July 1994. The picture was made by the Hubble Space Telescope.

of Earth watching from the sidelines. The string of twenty-two comet fragments fell onto Jupiter during July 1994, creating a show eagerly photographed by viewers using telescopes around the world (figure 13.12B). The fragments impacted the upper atmosphere of Jupiter, producing visible, energetic fireballs. The aftereffects of these fireballs were visible for about a year. There are chains of craters on two of the Galilean moons that may have been formed by similar events.

Saturn

Saturn is slightly smaller and substantially less massive than Jupiter, and has similar features to Jupiter (see figure 13.9), but it is readily identified by its unique, beautiful system of rings. Saturn's rings consist of thousands of narrow bands of particles. Some rings are composed of particles large enough to be measured in meters and some have particles that are dust-sized (figure 13.13). Saturn is about 9.5 AU from the Sun, but its system of rings is easily spotted with a good pair of binoculars. Saturn also has the lowest average density of any of the planets, about 0.7 the density of water.

FIGURE 13.13 A part of Saturn's system of rings, pictured by *Voyager 2* from a distance of about 3 million km (about 2 million mi). More than sixty bright and dark ringlets are seen here; different colors indicate different surface compositions.

The surface of Saturn, like Jupiter's surface, has bright and dark bands that circle the planet parallel to the equator. Saturn also has a smaller version of Jupiter's Great Red Spot, but in general, the bands and spot are not as highly contrasted or brightly colored as they are on Jupiter. Saturn has sixty-two satellites, including Titan, the only moon in the solar system with a substantial atmosphere.

Titan is covered with clouds and impossible to observe. It is larger than the planet Mercury and is covered with a deep layer of reddish clouds. Titan's atmosphere is mostly nitrogen, with some hydrocarbons. This could be similar to Earth's atmosphere before life began adding oxygen to the atmosphere. The pressure on the surface is 1.5 atmospheres, but with a surface temperature of about −180°C (about −290°F), it is doubtful that life has developed on Titan.

The international *Cassini-Huygens* mission entered orbit around Saturn on July 1, 2004, after a 3.5 billion km (2.2 billion mi), seven-year voyage from Earth. Establishing orbit was the first step of a four-year study of Saturn and its rings and moons. So far, the mission has found that Saturn's largest moon, Titan, has a surface shaped by rock fracturing, winds, and erosion. Among the new discoveries is a 1,500 km (930 mi) long river of liquid methane. Titan's atmosphere is rich in organic (meaning carbon-containing) molecules. Most of the clouds on Titan occur over the south pole, and scientists believe this is where a cycle of methane rain, runoff with channel carving, and liquid methane evaporation is most active.

On January 14, 2005, *Cassini* delivered a detachable probe called *Huygens* to the moon Titan. The probe dropped by parachute, sampling the chemical composition of the atmosphere for the 2.5 hour descent and for 90 minutes after landing. It landed with more of a "splat" than a thud or splash, indicating that even though the surface temperature was −180°C, the surface was more like a thick organic stew than a solid or liquid.

Cassini has also discovered that Saturn's icy moon Enceladus has a significant atmosphere. Before this discovery, Titan was the only moon in the solar system known to have an atmosphere. Enceladus is too small to hold on to an atmosphere, so it must have a continuous source that may be volcanic, geysers, or gases escaping from the interior.

Cassini also provided a test of Einstein's general theory of relativity (p. 172). According to this theory, a massive object such as the Sun should cause spacetime to curve. This curve means that it should take longer for anything to travel by the Sun since a curved path is a greater distance. The general theory of relativity was tested by analyzing radio waves that traveled by the Sun on their way from *Cassini* to Earth. The researchers found that the radio waves traveling by the Sun were indeed delayed, and the amount of delay agreed precisely with predictions made from Einstein's theory.

The *Cassini-Huygens* mission is a cooperative project of NASA, the European Space Agency, and the Italian Space Agency.

Uranus and Neptune

Uranus and Neptune are two more giant planets that are far, far away from Earth. Uranus revolves around the Sun at an average distance of over 19 AU, taking about 84 years to circle the Sun once. Neptune is an average 30 AU from the Sun and takes about 165 years for one complete orbit. Uranus is about twice as far away from the Sun as Saturn, and Neptune is three times as far away. To give you an idea of these tremendous distances, consider that the time required for radio signals to travel from Uranus to Earth is more than 2.5 hours! It would be most difficult to carry on a conversation by radio with someone such a distance away. Even farther away, a radio signal from a

FIGURE 13.14 This is a photo image of Neptune taken by *Voyager 2*. Neptune has a turbulent atmosphere over a very cold surface of frozen hydrogen and helium.

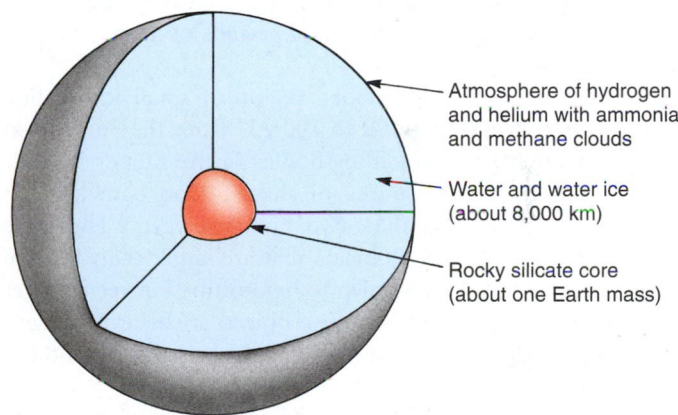

Atmosphere of hydrogen and helium with ammonia and methane clouds

Water and water ice (about 8,000 km)

Rocky silicate core (about one Earth mass)

FIGURE 13.15 The interior structure of Uranus and Neptune.

transmitter near Neptune would require over 4 hours to reach Earth, which means 8 hours would be required for two people just to say "Hello" to each other!

Uranus and Neptune are more similar to each other than Saturn is to Jupiter (figure 13.14). Both are the smallest of the giant planets, with a diameter of about 50,000 km (about 30,000 mi) and about a third the mass of Jupiter. Both planets are thought to have similar interior structures (figure 13.15), which consist of water and water ice surrounding a rocky core with an atmosphere of hydrogen and helium. Because of their great distances from the Sun, both have very low average surface temperatures.

13.2 SMALL BODIES OF THE SOLAR SYSTEM

Comets, asteroids, and meteorites are the leftovers from the formation of the Sun and planets. Presently, the total mass of all these leftovers in and around the solar system may account for a significant fraction of the mass of the solar system, perhaps as much as two-thirds of the total mass. It must have been much greater in the past, however, as evidenced by the intense bombardment that took place on the Moon and other planets up to some 4 billion years ago.

Comets

A **comet** is known to be a relatively small, solid body of frozen water, carbon dioxide, ammonia, and methane, along with dusty and rocky bits of materials mixed in. Until the 1950s, most astronomers believed that comet bodies were mixtures of sand and gravel. Fred Whipple proposed what became known as the *dirty-snowball cometary model,* which was recently verified when spacecraft probes observed Halley's comet in 1986 (table 13.4).

Based on calculation of their observed paths, comets are estimated to originate some 30 AU to a light-year or more from the Sun. Here, according to other calculations and estimates, is a region of space containing billions and billions of objects. There is a spherical "cloud" of the objects beyond the orbit of Pluto from about 30,000 AU out to a light-year or more from the Sun, called the **Oort cloud** (figure 13.16). The icy aggregates of the Oort cloud are understood to be the source of long-period comets, with orbital periods of more than two hundred years.

There is also a disk-shaped region of small icy bodies, which ranges from about 30 to 100 AU from the Sun, called the **Kuiper Belt.** The small icy bodies in the Kuiper Belt are understood to be the source of short-period comets, with orbital periods of less than two hundred years. There are thousands of Kuiper Belt objects that are larger than 100 km in diameter, and six are known to be orbiting between Jupiter and Neptune. Called *centaurs,* these objects are believed to have escaped the Kuiper Belt. Centaurs might be small, icy bodies similar to Pluto.

The current theory of the origin of comets was developed by the Dutch astronomer Jan Oort in 1950. According to the theory, the huge cloud and belt of icy, dusty aggregates are leftovers from the formation of the solar system and have been slowly circling the solar system ever since it formed. Something, perhaps a gravitational nudge from a passing star, moves one of the icy bodies enough that it is pulled toward the Sun in what will become an extremely elongated elliptical orbit. The icy, dusty body forms the only substantial part of a comet, and the body is called the comet *nucleus.*

Observations by the *Vega* and *Giotto* spacecrafts found the nucleus of Halley's comet to be an elongated mass of about 8 by 11 km (about 5 by 7 mi) with an overall density less than one-fourth that of solid water ice. As the comet nucleus moves toward the Sun, it warms from the increasing intense solar radiation. Somewhere between Jupiter and Mars, the ice and

TABLE 13.4 Completed Spacecraft Missions to Study Comets or Asteroids

Date	Name	Owner	Remark
Sep 11, 1985	*ISSE 3* (or *ICE*)	U.S.	Studies of electric and magnetic fields around Giacobini-Zinner comet from 7,860 km (4,880 mi)
Mar 6, 1986	*Vega 1*	USSR	Photos and studies of nucleus of Halley's comet from 8,892 km (5,525 mi)
Mar 8, 1986	*Suisei*	Japan	Studied hydrogen halo of Halley's comet from 151,000 km (93,800 mi)
Mar 9, 1986	*Vega 2*	USSR	Photos and studies of nucleus of Halley's comet from 8,034 km (4,992 mi)
Mar 11, 1986	*Sakigake*	Japan	Studied solar wind in front of Halley's comet from 7.1 million km (4.4 million mi)
Mar 28, 1986	*Giotto*	ESA	Photos and studies of Halley's comet from 541 km (336 mi)
Mar 28, 1986	*ISSE 3* (or *ICE*)	U.S.	Studies of electric and magnetic fields around Halley's comet from 32 million km (20 million mi)
Feb 17, 1996	*NEAR Shoe-maker*	U.S.	Studied the asteroid Eros
Oct 24, 1998	*Deep Space 1*	U.S.	Flyby of asteroid Braille and comet Borrelly
Feb 7, 1999	*Stardust*	U.S.	Studied comet Wild 2 and returned a sample of cosmic dust to Earth
Mar 2, 2004	*Rosetta*	ESA	Two probes will be launched into comet 67PChruyumov-Gerasimenko on Nov 2014
Jan 12, 2005	*Deep Impact*	U.S.	Studied comet Tempel 1 by sending impact probe on July 4, 2005
Jul 4, 2005	*Deep Impact*	U.S.	Projectile shot into comet Tempel 1
Sept 27, 2007	*Dawn*	U.S.	Study the two most massive asteroids Vesta (2011) and Ceres (2015)
Nov 4, 2010	*EPOXI*	U.S.	Used Deep Impact spacecraft to extend study of comet Hartley 2.
Jun 13, 2010	*Asteroid sample return*	JAXA*	Spacecraft *Hayabusa* returned 1,500 dust grains from asteroid Itokawa
Feb 14, 2011	*Stardust-NExT*	U.S.	Study result of projectile shot into comet Tempel 1 after one orbit of Sun

*Japan Aerospace Exploration Agency

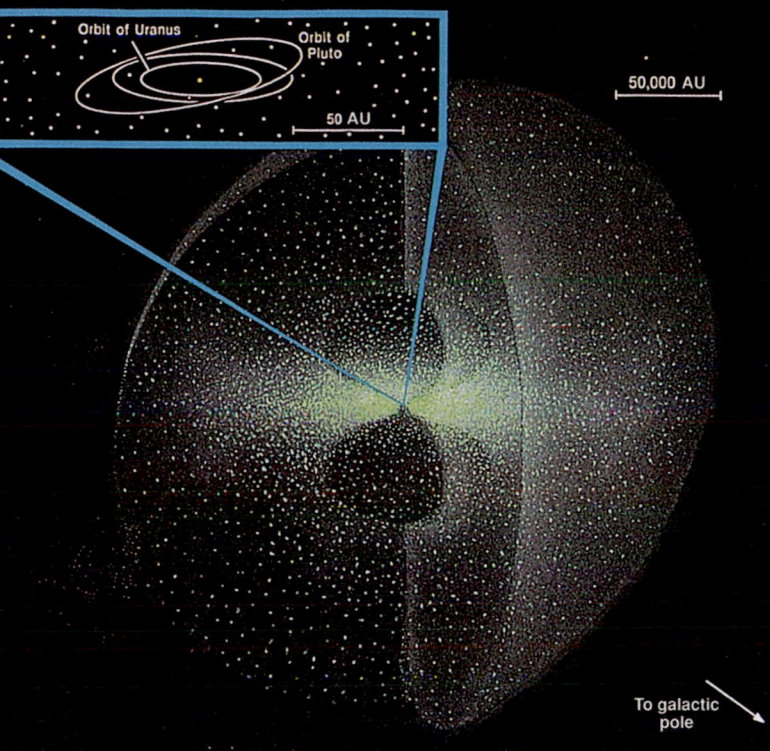

FIGURE 13.16 Although there may be as many as 1 trillion comets in the Oort cloud, the volume of space that the Oort cloud occupies is so immense that the comets are separated from one another by distances that are typically about 10 AU.

frozen gases begin to vaporize, releasing both grains of dust and evaporated ices. These materials form a large hazy head around the comet called a *coma*. The coma grows larger with increased vaporization, perhaps several hundred or thousands of kilometers across. The coma reflects sunlight as well as producing its own light, making it visible from Earth. The coma generally appears when a comet is within about 3 AU of the Sun. It reaches its maximum diameter about 1.5 AU from the Sun. The nucleus and coma together are called the *head* of the comet. In addition, a large cloud of invisible hydrogen gas surrounds the head, and this hydrogen *halo* may be hundreds of thousands of kilometers across.

As the comet nears the Sun, the solar wind and solar radiation ionize gases and push particles from the coma, pushing both into the familiar visible *tail* of the comet. Comets may have two types of tails: (1) ionized gases and (2) dust. The dust is pushed from the coma by the pressure from sunlight. It is visible because of reflected sunlight. The ionized gases are pushed into the tail by magnetic fields carried by the solar wind. The ionized gases of the tail are fluorescent, emitting visible light because they are excited by ultraviolet radiation from the Sun. The tail generally points away from the Sun, so it follows the comet as it approaches the Sun but leads the comet as it moves away from the Sun (figure 13.17).

Comets are not very massive or solid, and the porous, snow-like mass has a composition more similar to the giant planets than to the terrestrial planets in comparison. Each time a comet passes near the Sun, it loses some of its mass through evaporation of gases and loss of dust to the solar wind. After passing the Sun, the surface forms a thin, fragile crust covered with carbon and other dust particles. Each pass by the Sun means a loss of matter, and the coma and tail are dimmer with each succeeding pass. About 20 percent of the approximately six hundred comets that are known have orbits that return them to the Sun within a two-hundred-year period, some of which return as often as every five or ten years. The other 80 percent have long elliptical orbits that return them at intervals exceeding two hundred years. The famous Halley's comet has a smaller elliptical orbit and returns about every seventy-six years. Halley's comet, like all other comets, may eventually break up into a trail of gas and dust particles that orbit the Sun.

Asteroids

Between the orbits of Mars and Jupiter is a belt, or circular region of thousands of small rocky bodies called **asteroids** (figure 13.18). This belt contains thousands of *asteroids* that

FIGURE 13.17 As a comet nears the Sun it grows brighter, with the tail always pointing away from the Sun.

range in size from 1 km or less up to the largest asteroid, named Ceres, which has a diameter of about 1,000 km (over 600 mi). The asteroids are thinly distributed in the belt, 1 million km or so apart (about 600,000 mi), but there is evidence of collisions occurring in the past. Most asteroids larger than 50 km (about 30 mi) have been studied by analyzing the sunlight reflected from their surfaces. These spectra provide information about the composition of the asteroids. Asteroids on the inside of the belt are made of stony materials, and those on the outside of the belt are dark with carbon minerals. Still other asteroids are metallic, containing iron and nickel. These spectral composition studies, analyses of the orbits of asteroids, and studies of meteorites that have fallen to Earth all indicate that the asteroids are not the remains of a planet or planets that were broken up. The asteroids are now believed to have formed some 4.6 billion years ago from the original solar nebula. During their formation, or shortly thereafter, their interiors were partly melted, perhaps from the heat of short-lived radioactive decay reactions. Their location close to Jupiter, with its gigantic gravitational field, prevented the slow gravitational clumping-together process that would have formed a planet.

Jupiter's gigantic gravitational field also captured some of the asteroids, pulling them into its orbit. Today, there are two groups of asteroids, called the *Trojan asteroids,* which lead and follow Jupiter in its orbit. They lead and follow at a distance where the gravitational forces of Jupiter and the Sun balance to keep them in the orbit. A third group of asteroids, called the *Apollo asteroids,* has orbits that cross the orbit of Earth. It is possible that one of the Apollo asteroids could collide with Earth. One theory about what happened to the dinosaurs is based on evidence that such a collision indeed did occur some 65 million years ago. The chemical and physical properties of the two satellites of Mars, Phobos and Deimos, are more similar to the asteroids than to Mars. It is probable that the Martian satellites are captured asteroids.

Meteors and Meteorites

Comets leave trails of dust and rock particles after encountering the heat of the Sun, and collisions between asteroids in the past have ejected fragments of rock particles into space. In space, the remnants of comets and asteroids are called **meteoroids.** When a meteoroid encounters Earth moving through space, it accelerates toward the surface with a speed that depends on its direction of travel and the relative direction that Earth is moving. It soon begins to heat from air compression, melting into a visible trail of light and smoke. The streak of light and smoke in the sky is called a **meteor.** The "falling star" or "shooting star" is a meteor. Most meteors burn up or evaporate completely within seconds after reaching an altitude of about 100 km (about 60 mi) because they are nothing more than a speck of dust. A **meteor shower** occurs when Earth passes through a stream of particles left by a comet in its orbit. Earth might meet the stream of particles concentrated in such an orbit on a regular basis as it travels around the Sun, resulting in predictable meteor showers (table 13.5). In the third

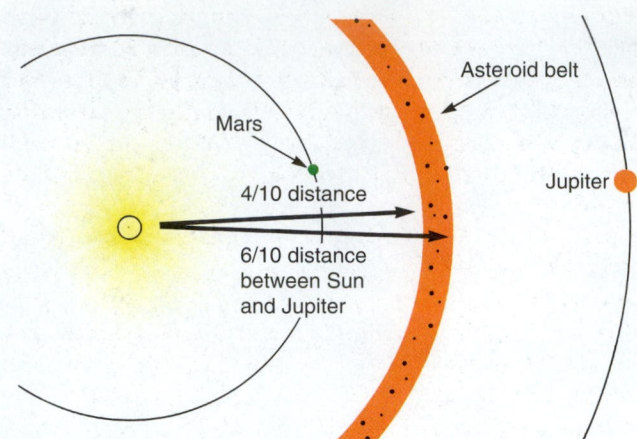

FIGURE 13.18 Most of the asteroids in the asteroid belt are about halfway between the Sun and Jupiter.

TABLE 13.5 **Some Annual Meteor Showers**

Name	Date of Maximum	Hour Rate
Quadrantid	January 3	30
Aquarid	May 4	5
Perseid	August 12	40
Orionid	October 22	15
Taurids	November 1, 16	5
Leonid	November 17	5
Geminid	December 12	55

week of October, for example, Earth crosses the orbital path of Halley's comet, resulting in a shower of some ten to fifteen meteors per hour. Meteor showers are named for the constellation in which they appear to originate. The October meteor shower resulting from an encounter with the orbit of Halley's comet, for example, is called the Orionid shower because it appears to come from the constellation Orion.

Did you know that atom-bomb-sized meteoroid explosions often occur high in Earth's atmosphere? Most smaller meteors melt into the familiar trail of light and smoke. Larger meteors may fragment upon entering the atmosphere, and the smaller fragments will melt into multiple light trails. Still larger meteors may actually explode at altitudes of about 32 km (about 20 mi) or so. Military satellites that watch Earth for signs of rockets blasting off or nuclear explosions record an average of eight meteor explosions a year. These are big explosions, with an energy equivalent estimated to be similar to a small nuclear bomb. Actual explosions, however, may be ten times larger than the estimation. Based on statistical data, scientists have estimated that every 10 million years, Earth should be hit by a very, very large meteor. The catastrophic explosion and aftermath would devastate life over much of the planet, much like the theoretical dinosaur-killing impact of 65 million years ago.

If a meteoroid survives its fiery trip through the atmosphere to strike the surface of Earth, it is called a **meteorite.** Most meteors are from fragments of comets, but most meteorites generally come from particles that resulted from collisions between asteroids that occurred long ago. Meteorites are classified into three basic groups according to their composition: (1) *iron meteorites,* (2) *stony meteorites,* and (3) *stony-iron meteorites* (figure 13.19). The most common meteorites are stony, composed of the same minerals that make up rocks on Earth. The stony meteorites are further subdivided into two groups according to their structure, the *chondrites* and the *achondrites.* Chondrites have a structure of small spherical lumps of silicate minerals or glass, called *chondrules,* held together by a fine-grained cement. The achondrites do not have the chondrules, as their name implies, but have a homogeneous texture more like volcanic rocks such as basalt that cooled from molten rock.

The iron meteorites are about half as abundant as the stony meteorites. They consist of variable amounts of iron and nickel, with traces of other elements. In general, there is proportionally much more nickel than is found in the rocks of Earth. When cut, polished, and etched, beautiful crystal patterns are observed on the surface of the iron meteorite. The patterns mean that the iron was originally molten, then cooled very slowly over millions of years as the crystal patterns formed.

A meteorite is not, as is commonly believed, a ball of fire that burns up the landscape where it lands. The iron or rock has been in the deep freeze of space for some time, and it travels rapidly through Earth's atmosphere. The outer layers become hot enough to melt, but there is insufficient time for this heat to be conducted to the inside. Thus, a newly fallen iron meteorite will be hot since metals are good heat conductors, but it will not be hot enough to start a fire. A stone meteorite is a poor conductor of heat so it will be merely warm.

13.3 ORIGIN OF THE SOLAR SYSTEM

Any model of how the solar system originated presents a problem in testing or verification. This problem is that the solar system originated a long time ago, some 5 billion years ago, according to a number of different independent sources of evidence. Also, there are no other planetary systems that can be directly observed for comparison either in existence or in the process of being formed. At the distance they occur, even the Hubble Space Telescope would not be able to directly observe planets around their suns. Astronomers have identified hundreds of extrasolar planets. They identify the presence of planets by measuring the very slight wobble of a central star and then use the magnitude of this motion to determine the presence of orbiting planets, the size and shape of their orbits, and their mass. The technique works only for larger planets and cannot detect those much smaller than about half the mass of Saturn. The technique does not provide a

A

B

FIGURE 13.19 (*A*) A stony meteorite. The smooth, black surface was melted by friction with the atmosphere. (*B*) An iron meteorite that has been cut, polished, and etched with acid. The pattern indicates that the original material cooled from a molten material over millions of years.

visual image of the planets but only measures the gravitational effect of the planets on the star.

The basic idea behind how the solar system formed was put forth by both Immanuel Kant and Pierre Laplace back in the eighteenth century. Today this idea is called the **protoplanet nebular model.** A *protoplanet* is the earliest stage in the formation of a planet. The model can be considered in stages, which are not really a part of the model but are simply a convenient way to organize the total picture (figure 13.20).

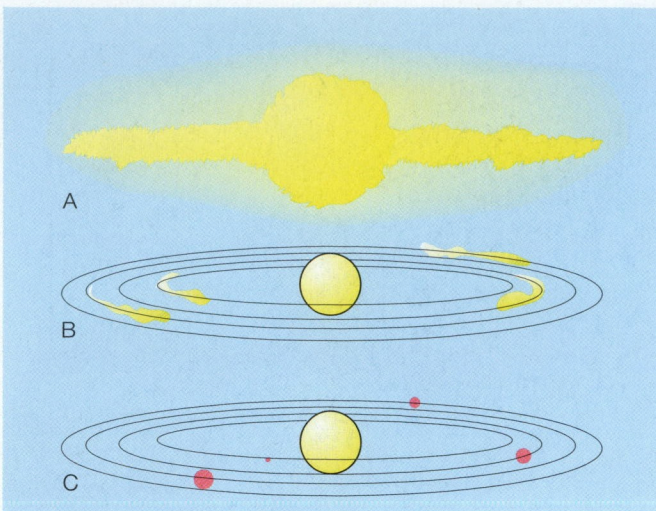

FIGURE 13.20 Formation of the solar system according to the protoplanet nebular model, not drawn to scale. (*A*) The process starts with a nebula of gas, dust, and chemical elements from previously existing stars. (*B*) The nebula is pulled together by gravity, collapsing into the protosun and protoplanets. (*C*) As the planets form, they revolve around the Sun in newly formed orbits.

Stage A

The first important event in the formation of our solar system involves stars that disappeared billions of years ago, long before the Sun was born. Earth, the other planets, and all the members of the solar system are composed of elements that were manufactured by these former stars. In a sequence of nuclear reactions, hydrogen fusion in the core of large stars results in the formation of the elements up to iron. Elements heavier than iron are formed in rare supernova explosions of dying massive stars. Thus, *stage A* of the formation of the solar system consisted of the formation of elements heavier than hydrogen in many, many previously existing stars, including the supernovas of more massive stars. Many stars had to live out their life cycles to provide the raw materials of the solar system. The death of each star, including supernovas, added newly formed elements to the accumulating gases and dust in interstellar space. Over a long period of time, these elements began to concentrate in one region of space as dust, gases, and chemical compounds, but hydrogen was still the most abundant element in the nebula that was destined to become the solar system.

Stage B

During *stage B,* the hydrogen gas, dust, elements, and chemical compounds from former stars began to form a large, slowly rotating nebula that was much, much larger than the present solar system. Under the influence of gravity, the large but diffuse, slowly rotating nebula began to contract, increasing its rate of spin. The largest mass pulled together in the center, contracting to the protostar, which eventually would become the Sun. The remaining gases, elements, and dusts formed an enormous, fat, bulging disk called an *accretion disk,* which would eventually form the planets and smaller bodies. The fragments

of dust and other solid matter in the disk began to stick together in larger and larger accumulations from numerous collisions over the first million years or so. All of the present-day elements of the planets must have been present in the nebula along with the most abundant elements of hydrogen and helium. The elements and the familiar chemical compounds accumulated into basketball-sized or larger chunks of matter.

Did the planets have an icy slush beginning? Over a period of time, perhaps 100 million years or so, huge accumulations of frozen water, frozen ammonia, and frozen crystals of methane began to accumulate, together with silicon, aluminum, and iron oxide plus other metals in the form of rock and mineral grains. Such a slushy mixture would no doubt have been surrounded by an atmosphere of hydrogen, helium, and other vapors thinly interspersed with smaller rocky grains of dust. Local concentrations of certain minerals might have occurred throughout the whole accretion disk, with a greater concentration of iron, for example, in the disk where the protoplanet Mars was forming compared to where the protoplanet Earth was forming. Evidence for this part of the model is found in Mars today, with its greater abundance of iron. The abundant iron oxides are what gives the planet Mars its red color.

All of the protoplanets might have started out somewhat similarly as huge accumulations of a slushy mixture with an atmosphere of hydrogen and helium gases. Gravitational attraction must have compressed the protoplanets as well as the protosun. During this period of contraction and heating, gravitational adjustments continue, and about a fifth of the disk nearest to the protosun must have been pulled into the central body of the protosun, leaving a larger accumulation of matter in the outer part of the accretion disk.

Stage C

During *stage C,* the warming protosun became established as a star, perhaps undergoing an initial flare-up that has been observed today in other newly forming stars. Such a flare-up might have been of such a magnitude that it blasted away the hydrogen and helium atmospheres of the interior planets (Mercury, Venus, Earth, and Mars) out past Mars, but it did not reach far enough out to disturb the hydrogen and helium atmospheres of the outer planets. The innermost of the outer planets, Jupiter and Saturn, might have acquired some of the matter blasted away from the inner planets, becoming the giants of the solar system by comparison. This is just speculation, however, and the two giants may have simply formed from greater concentrations of matter in that part of the accretion disk.

The evidence, such as separation of heavy and light mineral matter, shows that the protoplanets underwent heating early in their formation. Much of the heating may have been provided by gravitational contraction, the same process that gave the protosun sufficient heat to begin its internal nuclear fusion reactions. Heat was also provided from radioactive decay processes inside the protoplanets, and the initial greater heating from the Sun may have played a role in the protoplanet heating process. Larger bodies were able to retain this heat better than smaller ones, which radiated it to space more readily. Thus, the

People Behind the Science

Percival Lowell (1855–1916)

Percival Lowell was an American astronomer, mathematician, and the founder of an important observatory in the United States whose main field of research was the planets of the solar system. Responsible for the popularization in his time of the theory of intelligent life on Mars, he also predicted the existence of a planet beyond Neptune that was later discovered and named Pluto. Today, Pluto is recognized as a dwarf planet.

Lowell was born in Boston, Massachusetts, on March 13, 1855. His interest in astronomy began to develop during his early school years. In 1876, he graduated from Harvard University, where he had concentrated on mathematics, and then traveled for a year before entering his father's cotton business. Six years later, Lowell left the business and went to Japan. He spent most of the next ten years traveling around the Far East, partly for pleasure, partly to serve business interests, but also holding a number of minor diplomatic posts.

Lowell returned to the United States in 1893 and soon afterward decided to concentrate on astronomy. He set up an observatory at Flagstaff, Arizona, at an altitude more than 2,000 m (6,564 ft) above sea level, on a site chosen for the clarity of its air and its favorable atmospheric conditions. He first used borrowed telescopes of 12- and 18-inch (30 and 45 cm) diameters to study Mars, which at that time was in a particularly suitable position. In 1896, he acquired a larger telescope and studied Mars by night and Mercury and Venus during the day. Overwork led to deterioration in Lowell's health, and from 1897 to 1901, he could do little research, although he was able to participate in an expedition to Tripoli in 1900 to study a solar eclipse.

He was made nonresident professor of astronomy at the Massachusetts Institute of Technology in 1902 and gave several lecture series in that capacity. He led an expedition to the Chilean Andes in 1907 that produced the first high-quality photographs of Mars. The author of many books and the holder of several honorary degrees, Lowell died in Flagstaff on November 12, 1916.

The planet Mars was a source of fascination for Lowell. Influenced strongly by the work of Giovanni Schiaparelli (1835–1910)—and possibly misled by the current English translation of "canals" for the Italian *canali* ("channels")—Lowell set up his observatory at Flagstaff originally with the sole intention of confirming the presence of advanced life forms on the planet. Thirteen years later, the expedition to South America was devoted to the study and photography of Mars. Lowell "observed" a complex and regular network of canals and believed that he detected regular seasonal variations that strongly indicated agricultural activity. He found darker waves that seemed to flow from the poles to the equator and suggested that the polar caps were made of frozen water. (The waves were later attributed to dust storms, and the polar caps are now known to consist not of ice but mainly of frozen carbon dioxide. Lowell's canal system also seems to have arisen mostly out of wishful thinking; part of the system does indeed exist, but it is not artificial and is apparent only because of the chance apposition of dark patches on the Martian surface.)

Lowell is remembered as a scientist of great patience and originality. He contributed to the advancement of astronomy through his observations and his establishment of a fine research center, and he did much to bring the excitement of the subject to the general public.

Source: Modified from the *Hutchinson Dictionary of Scientific Biography.* © RM, 2011. All rights reserved. Helicon Publishing is a division of RM.

---

larger bodies underwent a more thorough heating and melting, perhaps becoming completely molten early in their history. In the larger bodies, the heavier elements, such as iron, were pulled to the center of the now molten mass, leaving the lighter elements near the surface. The overall heating and cooling process took millions of years as the planets and smaller bodies were formed. Gases from the hot interiors formed secondary atmospheres of water vapor, carbon dioxide, and nitrogen on the larger interior planets.

Interestingly, the asteroid belt was discovered from a prediction made by the German astronomer Bode at the end of the eighteenth century. Bode had noticed a pattern of regularity in the spacing of the planets that were known at the time. He found that by expressing the distances of the planets from the Sun in astronomical units, these distances could be approximated by the relationship $(n + 4)/10$, where n is a number in the sequence 0, 3, 6, 12, and so on where each number (except the first) is doubled in succession. When these calculations were done, the distances turned out to be very close to the distances of all the planets known at that time, but the numbers also predicted a planet between Mars and Jupiter where there

Myths, Mistakes, and Misunderstandings

Solar Mistakes

The following are common misunderstandings about the Sun:

- The Sun is not a star; it is a unique object of our solar system.
- The Sun is solid.
- The Sun does not rotate.
- Sunlight is from burning gas on the Sun.
- The Sun will last forever.

was none. Later, a belt of asteroids was found where the Bode numbers predicted there should be a planet. This suggested to some people that a planet had existed between Mars and Jupiter in the past and somehow this planet was broken into pieces, perhaps by a collision with another large body (table 13.6).

Such patterns of apparent regularity in the spacing of planetary orbits were of great interest because, if a true pattern existed, it could hold meaning about the mechanism that determined the location of planets at various distances from the Sun. Many attempts have been made to explain the mechanism of planetary spacing and why a belt of asteroids exists where the Bode numbers predict there should be a planet. The most successful explanations concern Jupiter and the influence of its gigantic gravitational field on the formation of clumps of matter at certain distances from the Sun. In other words, a planet does not exist today between Mars and Jupiter because there never was a planet there. The gravitational influence of Jupiter prevented the clumps of matter from joining together to form a planet, and a belt of asteroids formed instead. There is no theoretical evidence that explains the Bode numbers, and they are believed to be simply a mathematical coincidence.

TABLE 13.6 Distances from the Sun to Planets Known in the 1790s

Planet	n	Distance Predicted by $(n + 4)/10$ (AU)	Actual Distance (AU)
Mercury	0	0.4	0.39
Venus	3	0.7	0.72
Earth	6	1.0	1.0
Mars	12	1.6	1.5
(Asteroid belt)	24	2.8	—
Jupiter	48	5.2	5.2
Saturn	96	10.0	9.5
Uranus	192	19.6	19.2

SUMMARY

The planets can be classified into two major groups: (1) the *terrestrial planets* of Mercury, Venus, Mars, and Earth and (2) the *giant planets* of Jupiter, Saturn, Uranus, and Neptune.

Comets are porous aggregates of water ice, frozen methane, frozen ammonia, dry ice, and dust. The solar system is surrounded by the *Kuiper Belt* and the *Oort cloud* of these objects. Something nudges one of the icy bodies and it falls into a long elliptical orbit around the Sun. As it approaches the Sun, increased radiation evaporates ices and pushes ions and dust into a long visible tail. *Asteroids* are rocky or metallic bodies that are mostly located in a belt between Mars and Jupiter. The remnants of comets, fragments of asteroids, and dust are called *meteoroids*. A meteoroid that falls through Earth's atmosphere and melts to a visible trail of light and smoke is called a *meteor*. A meteoroid that survives the trip through the atmosphere to strike the surface of Earth is called a *meteorite*. Most meteors are fragments and pieces of dust from comets. Most meteorites are fragments that resulted from collisions between asteroids.

The *protoplanet nebular model* is the most widely accepted theory of the origin of the solar system, and this theory can be considered as a series of events, or stages. *Stage A* is the creation of all the elements heavier than hydrogen in previously existing stars. *Stage B* is the formation of a nebula from the raw materials created in stage A. The nebula contracts from gravitational attraction, forming the *protosun* in the center with a fat, bulging *accretion disk* around it. The Sun will form from the protosun, and the planets will form in the accretion disk. *Stage C* begins as the protosun becomes established as a star. The icy remains of the original nebula are the birthplace of *comets*. *Asteroids* are other remains that did undergo some melting.

KEY TERMS

asteroids (p. **301**)
astronomical unit (p. **287**)
comet (p. **300**)
dwarf planet (p. **286**)

giant planets (p. **287**)
Kuiper Belt (p. **300**)
meteor (p. **302**)
meteorite (p. **303**)

meteoroids (p. **302**)
meteor shower (p. **302**)
Oort cloud (p. **300**)
planet (p. **286**)

protoplanet nebular model (p. **303**)
small solar system bodies
  (p. **286**)
terrestrial planets (p. **287**)

APPLYING THE CONCEPTS

Answers are located in appendix F.

1. Earth, other planets, and all the members of the solar system
   a. have always existed.
   b. formed thousands of years ago from elements that have always existed.
   c. formed millions of years ago, when the elements and each body were created at the same time.
   d. formed billions of years ago from elements that were created in many previously existing stars.

2. The belt of asteroids between Mars and Jupiter is probably
   a. the remains of a planet that exploded.
   b. clumps of matter that condensed from the accretion disk but never got together as a planet.
   c. the remains of two planets that collided.
   d. the remains of a planet that collided with an asteroid or comet.

3. Which of the following planets would be mostly composed of hydrogen, helium, and methane and have a density of less than 2 g/cm^3?
   a. Uranus
   b. Mercury
   c. Mars
   d. Venus

4. Which of the following planets probably still has its original atmosphere?
   a. Mercury
   b. Venus
   c. Mars
   d. Jupiter

5. Venus appears the brightest when it is in the
   a. full phase.
   b. half phase.
   c. quarter phase.
   d. crescent phase.

6. The largest planet is
   a. Saturn.
   b. Jupiter.
   c. Uranus.
   d. Neptune.

7. The small body with a composition and structure closest to the materials that condensed from the accretion disk is a (an)
   a. asteroid.   b. meteorite.
   c. comet.   d. None of the above is correct.
8. A small body from space that falls on the surface of Earth is a
   a. meteoroid.   b. meteor.
   c. meteor shower.   d. meteorite.
9. A day on which planet is longer than a year on that planet?
   a. Mercury   c. Neptune
   b. Venus   d. Jupiter
10. The day on which planet is about the same time period as a day on Earth?
   a. Mercury   c. Mars
   b. Venus   d. Jupiter

11. Which of the following planets would be mostly composed of hydrogen, helium, and methane and have a density of less than 2 g/cm³?
   a. Uranus   c. Mars
   b. Mercury   d. Venus
12. Which of the following planets probably still has its original atmosphere?
   a. Mercury   c. Mars
   b. Venus   d. Jupiter

QUESTIONS FOR THOUGHT

1. Describe the protoplanet nebular model of the origin of the solar system. Which part or parts of this model seem least credible to you? Explain. What information could you look for today that would cause you to accept or modify this least credible part of the model?
2. What are the basic differences between the terrestrial planets and the giant planets? Describe how the protoplanet nebular model accounts for these differences.
3. Describe the surface and atmospheric conditions on Mars.
4. What evidence exists that Mars at one time had abundant liquid water? If Mars did have liquid water at one time, what happened to it and why?
5. Describe the internal structure of Jupiter and Saturn.
6. What are the rings of Saturn?
7. Describe some of the unusual features found on the moons of Jupiter.
8. What are the similarities and the differences between the Sun and Jupiter?
9. Give one idea about why the Great Red Spot exists on Jupiter. Does the existence of a similar spot on Saturn support or not support this idea? Explain.
10. What is so unusual about the motions and orbits of Venus and Uranus?
11. What evidence exists today that the number of rocks and rock particles floating around in the solar system was much greater in the past soon after the planets formed?
12. Using the properties of the planets other than Earth, discuss the possibilities of life on each of the other planets.
13. What are "shooting stars"? Where do they come from? Where do they go?
14. What is an asteroid? What evidence indicates that asteroids are parts of a broken-up planet? What evidence indicates that asteroids are not parts of a broken-up planet?
15. Where do comets come from? Why are astronomers so interested in studying the physical and chemical structure of a comet?
16. What is a meteor? What is the most likely source of meteors?
17. What is a meteorite? What is the most likely source of meteorites?
18. Technically speaking, what is wrong with calling a rock that strikes the surface of the Moon a meteorite? Again speaking technically, what should you call a rock that strikes the surface of the Moon (or any planet other than Earth)?
19. If a comet is an icy, dusty body, explain why it appears bright in the night sky.

FOR FURTHER ANALYSIS

1. What are the significant similarities and differences between the terrestrial and giant planets? Speculate why these similarities and differences exist.
2. Draw a sketch showing the positions of the Earth, Sun, and Venus when it appears as the morning star. Draw a second sketch showing the positions when Venus appears as the evening star.
3. Evaluate the statement that Venus is Earth's sister planet.
4. Describe the possibility and probability of life on each of the other planets.
5. Provide arguments that Pluto should be considered a planet. Counter this with arguments that it should not be classified as a planet.
6. Describe and analyze why it would be important to study the nucleus of a comet.

INVITATION TO INQUIRY

What's Your Sign?

Form a team to investigate horoscope forecasts in a newspaper or on an Internet site. Each team member should select one birthday and track what is forecast to happen and what actually happens each day for a week. Analyze the way the forecasts are written that may make them "come true." Compare the prediction, the actual results, and the analysis for each team member.

14

Earth in Space

This shows part of Earth as seen from space, with the Salton Sea in the right center of the photo. Could you tell someone where on Earth the Salton Sea is located? One topic of this chapter is identifying places on Earth, which should help you describe the location of any part of Earth's surface.

CORE **CONCEPT**

The way Earth moves in space is used to define time and describe location, and causes other recurrent phenomena.

OUTLINE

Earth's axis is inclined 23.5° and keeps the same orientation all year (p. 312).

Earth moves around the Sun in a yearly revolution (p. 312).

Earth spins around its axis in a daily rotation (p. 313).

Earth's axis serves as a reference point for direction and location on the entire surface (p. 315).

OVERVIEW

Earth is a common object in the solar system, one of eight planets that goes around the Sun once a year in an almost circular orbit. Earth is the third planet out from the Sun, it is fifth in mass and diameter, and it has the greatest density of all the planets (figure 14.1). Earth is unique because of its combination of an abundant supply of liquid water, a strong magnetic field, and a particular atmospheric composition. In addition to these physical properties, Earth has a unique set of natural motions that humans have used for thousands of years as a frame of reference to mark time and to identify the events of their lives. These references to Earth's motions are called the day, the month, and the year.

Eventually, about three hundred years ago, people began to understand that their references for time came from an Earth that spins like a top as it circles the Sun. It was still difficult, however, for them to understand Earth's place in the universe. The problem was not unlike that of a person trying to comprehend the motion of a distant object while riding a moving merry-go-round being pulled by a cart. Actually, the combined motions of Earth are much more complex than a simple moving merry-go-round being pulled by a cart. Imagine trying to comprehend the motion of a distant object while undergoing a combination of Earth's more conspicuous motions, which are as follows:

1. A daily rotation of 1,670 km/h (about 1,040 mi/h) at the equator and less at higher latitudes.
2. A monthly revolution of Earth around the Earth-Moon center of gravity at about 50 km/h (about 30 mi/h).
3. A yearly revolution around the Sun at about an average 106,000 km/h (about 66,000 mi/h).
4. A motion of the solar system around the core of the Milky Way at about 370,000 km/h (about 230,000 mi/h).
5. A motion of the local star group that contains the Sun as compared to other star clusters of about 1,000,000 km/h (about 700,000 mi/h).
6. Movement of the entire Milky Way galaxy relative to other, remote galaxies at about 580,000 km/h (about 360,000 mi/h).
7. Minor motions such as cycles of change in the size and shape of Earth's orbit and the tilt of Earth's axis. In addition to these slow changes, there is a gradual slowing of the rate of Earth's daily rotation.

Basically, Earth is moving through space at fantastic speeds, following the Sun in a spiral path of a giant helix as it spins like a top (figure 14.2). This ceaseless and complex motion in space is relative to various frames of reference, however, and the limited perspective from Earth's surface can result in some very different ideas about Earth and its motions. This chapter is about the more basic, or fundamental, motions of Earth and its Moon. In addition to conceptual understandings and evidences for the motions, some practical human uses of the motions will be discussed.

14.1 SHAPE AND SIZE OF EARTH

The most widely accepted theory about how the solar system formed pictures the planets forming in a disk-shaped nebula with a turning, swirling motion. The planets formed from separate accumulations of materials within this disk-shaped, turning nebula, so the orbit of each planet was established along with its rate of rotation as it formed. Thus, all the planets move around the Sun in the same direction in elliptical orbits that are nearly circular. The flatness of the solar system results in the observable planets moving in, or near, the plane of Earth's orbit, which is called the **plane of the ecliptic.**

When viewed from Earth, the planets appear to move only within a narrow band across the sky as they move in the plane of the ecliptic. The Sun also appears to move in the center of this band, which is called the *ecliptic.* As viewed from Earth, the Sun appears to move across the background of stars, completely around the ecliptic each year.

Today, almost everyone has seen pictures of Earth from space, and it is difficult to deny that it has a rounded shape. During the fifth and sixth centuries B.C., the ancient Greeks decided that Earth must be round because (1) philosophically, they considered the sphere to be the perfect shape and they considered Earth to be perfect, so therefore Earth must be a sphere, (2) Earth was observed to cast a circular shadow on the Moon during a lunar eclipse, and (3) ships were observed to slowly disappear below the horizon as they sailed off into the distance. More abstract evidence of a round Earth was found in the observation that the altitude of the North Star above the horizon appeared to increase as a person traveled northward. This established that Earth's surface was curved, at least, which seemed to fit with other evidence.

The shape and size of Earth have been precisely measured by artificial satellites circling Earth. These measurements have found that Earth is not a perfectly round sphere as believed by the ancient Greeks. It is flattened at the poles and has an

FIGURE 14.1   Illustration of eight planets in the solar system in order.

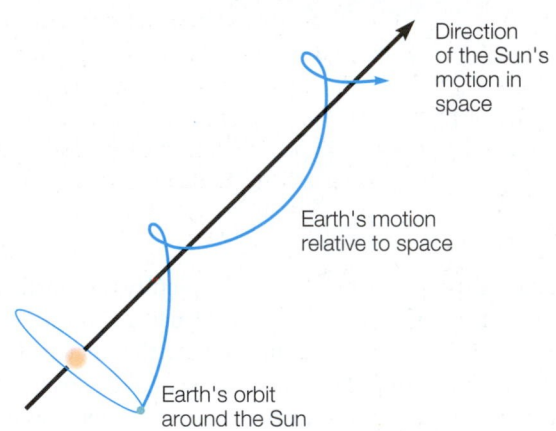

FIGURE 14.2   Earth undergoes many different motions as it moves through space. There are seven more conspicuous motions, three of which are more obvious on the surface. Earth follows the path of a gigantic helix, moving at fantastic speeds as it follows the Sun and the galaxy through space.

Direction of the Sun's motion in space

Earth's motion relative to space

Earth's orbit around the Sun

FIGURE 14.3   Earth as seen from space.

equatorial bulge, as do many other planets. In fact, you can observe through a telescope that both Jupiter and Saturn are considerably flattened at the poles. A shape that is flattened at the poles has a greater distance through the equator than through the poles, which is described as an *oblate* shape. Earth, like a water-filled, round balloon resting on a table, has an oblate shape. It is not perfectly symmetrically oblate, however, since the North Pole is slightly higher and the South Pole is slightly lower than the average surface. In addition, it is not perfectly circular around the equator, with a lump in the Pacific and a depression in the Indian Ocean. The shape of the earth is a slightly pear-shaped, slightly lopsided *oblate spheroid*. All the elevations and depressions are less than 85 m (about 280 ft), however, which is practi-

To locate the ecliptic, planets, or anything else in the sky, you need something to refer to, a referent system. A referent system is easily established by first imagining the sky to be a celestial sphere just as the ancient Greeks did. A coordinate system of lines can be visualized on this sphere. Imagine that you could inflate Earth until its surface touched the celestial sphere. If you now transfer lines to the celestial sphere, you will have a system of sky coordinates. From the surface of Earth, you can see that the *celestial equator* is a line on the celestial sphere directly above Earth's equator and the *north celestial pole* is a point directly above the North Pole of Earth. Likewise, the *south celestial pole* is a point directly above the South Pole of Earth.

You can only see half of the overall celestial sphere from any one place on the surface of Earth. Imagine a point on the celestial sphere directly above where you are located. An imaginary line that passes through this point, then passes north through the north celestial pole, continuing all the way around through the south celestial pole and back to the point directly above you makes a big circle called the *celestial meridian* (box figure 14.1). Note that the celestial meridian location is determined by where you are on Earth.

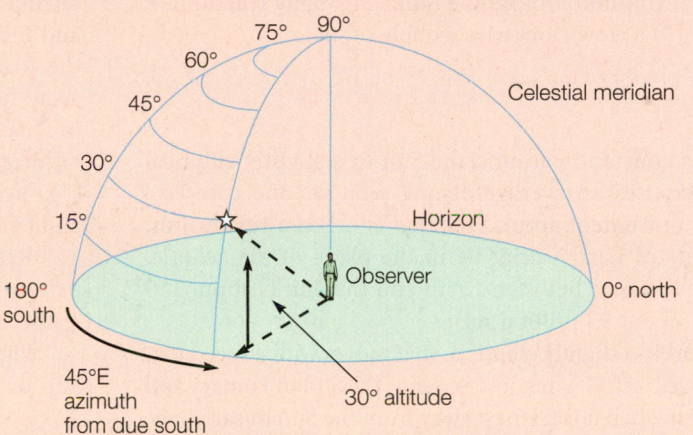

Box Figure 14.1 Once you have established the celestial equator, the celestial poles, and the celestial meridian, you can use a two-coordinate horizon system to locate positions in the sky. One popular method of using this system identifies the altitude angle (in degrees) from the horizon up to an object on the celestial sphere and the azimuth angle (again in degrees) the object on the celestial sphere is east or west of due south, where the celestial meridian meets the horizon. The illustration shows an altitude of 30° and an azimuth of 45° east of due south.

The celestial equator and the celestial poles, on the other hand, are always in the same place no matter where you are.

Overall, the celestial sphere appears to spin, turning on an axis through the celestial poles. A photograph made by pointing a camera at the north celestial pole and leaving the shutter open for several hours will show the apparent motion of the celestial sphere with star trails (see figure 12.1A).

The moderately bright star near the center is the North Star, Polaris. Polaris is almost, but not exactly, at the north celestial pole. If you observe the celestial sphere night after night, you will see that the stars maintain their positions relative to one another as they turn counterclockwise around Polaris. Those near Polaris pivot around it and are called "circumpolar." Those farther out rise in the east, move in an arc, then set in the west.

cally negligible compared to the size of Earth (figure 14.3). Thus, Earth is very close to, but not exactly, an oblate spheroid. The significance of this shape will become apparent when Earth's motions are discussed next (figure 14.4).

14.2 MOTIONS OF EARTH

Ancient civilizations had a fairly accurate understanding of the size and shape of Earth but had difficulty accepting the idea that Earth moves for at least two reasons: (1) they could not sense any motion of Earth, and (2) they had ideas about being at the center of a universe that was created for them. It was not until the 1700s that the concept of an Earth in motion became generally accepted. Today, Earth is understood to move

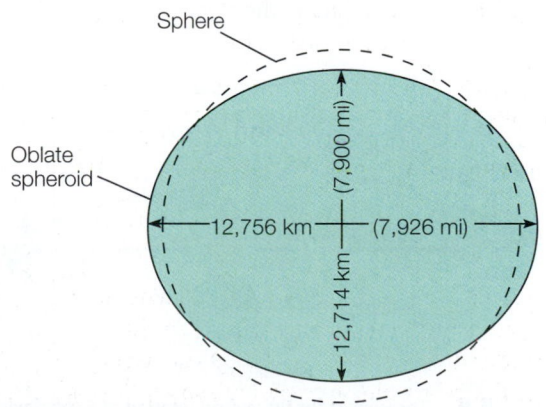

FIGURE 14.4 Earth has an irregular, slightly lopsided, slightly pear-shaped form. In general, it is considered to have the shape of an oblate spheroid, departing from a perfect sphere as shown here.

a number of different ways, seven of which were identified in the introduction to this chapter. Three of these motions are independent of motions of the Sun and the galaxy. These are (1) a yearly revolution around the Sun, (2) a daily rotation on its axis, and (3) a slow clockwise wobble of its axis.

Revolution

Earth moves constantly around the Sun in a slightly elliptical orbit that requires an average of one year for one complete circuit. The movement around the Sun is called a **revolution,** and all points of Earth's orbit lie in the plane of the ecliptic. The average distance between Earth and the Sun is about 150 million km (about 93 million mi).

Earth's orbit is slightly elliptical, so it moves with a speed that varies. It moves fastest when it is closer to the Sun in January and moves slowest when it is farthest away from the Sun in early July. Earth is about 2.5 million km (about 1.5 million mi) closer to the Sun in January and about the same distance farther away in July than it would be if the orbit were a circle. This total difference of about 5 million km (about 3 million mi) results in a January Sun with an apparent diameter that is 3 percent larger than the July Sun, and Earth as a whole receives about 6 percent more solar energy in January. The effect of being closer to the Sun is much less than the effect of some other relationships, and winter occurs in the Northern Hemisphere when Earth is closest to the Sun. Likewise, summer occurs in the Northern Hemisphere when the Sun is at its greatest distance from Earth (figure 14.5).

The important directional relationships that override the effect of Earth's distance from the Sun involve the daily **rotation,** or spinning, of Earth around an imaginary line through the geographic poles called Earth's *axis.* The important directional relationships are a constant inclination of Earth's axis to the plane of the ecliptic and a constant orientation of the axis to the stars. The *inclination of Earth's axis* to the plane of the ecliptic is about 66.5° (or 23.5° from a line perpendicular to the plane). This relationship between the plane of Earth's orbit and the tilt of its axis is considered to be the same day after day throughout the year, even though small changes do occur in the inclination over time. Likewise, the *orientation of Earth's axis* to the stars is considered to be the same throughout the year as

Earth moves through its orbit. Again, small changes do occur in the orientation over time. Thus, in general, the axis points in the same direction, remaining essentially parallel to its position during any day of the year. The essentially constant orientation and inclination of the axis result in the axis pointing toward the Sun as Earth moves in one part of its orbit, then pointing away from the Sun six months later. The generally constant inclination and orientation of the axis, together with Earth's rotation and revolution, combine to produce three related effects: (1) days and nights that vary in length, (2) changing seasons, and (3) climates that vary with latitude.

Read about Milankovitch cycles on page 408 to learn how patterns of changes in the orientation of Earth's tilt, orbital shape, and axis wobble affect Earth's climate.

Figure 14.5 shows how the North Pole points toward the Sun on June 21 or 22, then away from the Sun on December 22 or 23 as it maintains its orientation to the stars. When the North Pole is pointed toward the Sun, it receives sunlight for a full twenty-four hours, and the South Pole is in Earth's shadow for a full twenty-four hours. This is summer in the Northern Hemisphere, with the longest daylight periods and the Sun at its maximum noon height in the sky. Six months later, on December 22 or 23, the orientation is reversed with winter in the Northern Hemisphere, the shortest daylight periods, and the Sun at its lowest noon height in the sky.

The beginning of a season can be recognized from any one of the three related observations: (1) the length of the daylight period, (2) the altitude of the Sun in the sky at noon, or (3) the length of a shadow from a vertical stick at noon. All of these observations vary with changes in the direction of Earth's axis of rotation relative to the Sun. On about June 22 and December 22, the Sun reaches its highest and lowest noon altitudes as Earth moves to point the North Pole directly toward the Sun (June 21 or 22) and directly away from the Sun (December 22 or 23). Thus, the Sun appears to stop increasing or decreasing its altitude in the sky, stop, then reverse its movement twice a year. These times are known as **solstices** after the Latin meaning "Sun stand still." The Northern Hemisphere's **summer solstice** occurs on about June 22 and identifies the beginning of the summer season. At the summer solstice, the Sun at noon has the highest altitude, and the shadow from a vertical stick is shorter than on any other day of the year. The Northern Hemisphere's **winter solstice** occurs on about December 22 and identifies the beginning of the winter season. At the winter solstice, the Sun at noon has the lowest altitude, and the shadow from a vertical stick is longer than on any other day of the year (figure 14.6).

As Earth moves in its orbit between pointing its North Pole toward the Sun on about June 22 and pointing it away on about December 22, there are two times when it is halfway between. At these times, Earth's axis is perpendicular to a line between the center of the Sun and Earth, and daylight and night are of equal length. These are called the **equinoxes** after the Latin meaning "equal nights." The **spring equinox** (also called the **vernal equinox**) occurs on about March 21 and identifies the beginning of the spring season. The **autumnal equinox** occurs on about September 23 and identifies the beginning of the fall season.

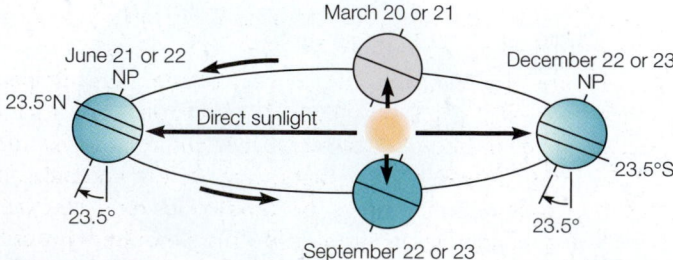

FIGURE 14.5 The consistent tilt and orientation of Earth's axis as it moves around its orbit is the cause of the seasons. The North Pole is pointing toward the Sun during the summer solstice and away from the Sun during the winter solstice.

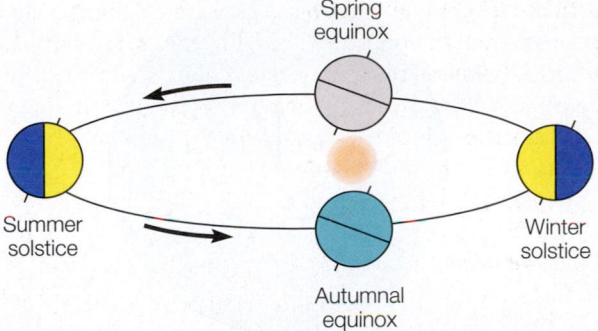

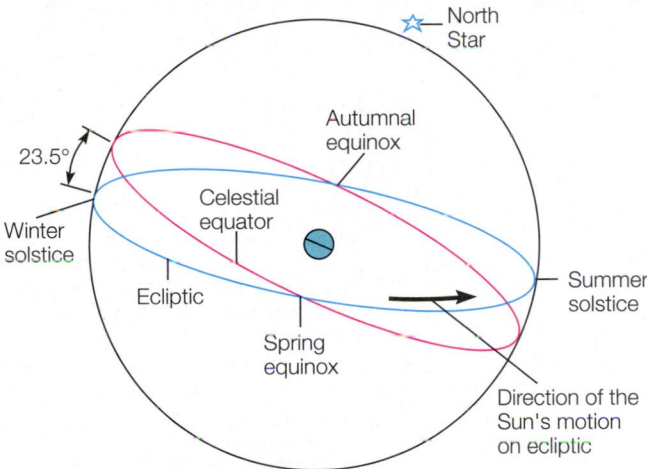

FIGURE 14.6 The length of daylight during each season is determined by the relationship of Earth's shadow to the tilt of the axis. At the equinoxes, the shadow is perpendicular to the latitudes, and day and night are of equal length everywhere. At the summer solstice, the North Pole points toward the Sun and is completely out of the shadow for a twenty-four-hour day. At the winter solstice, the North Pole is in the shadow for a twenty-four-hour night. The situation is reversed for the South Pole.

FIGURE 14.7 The position of the Sun on the celestial sphere at the solstices and the equinoxes.

The relationship between the apparent path of the Sun on the celestial sphere and the seasons is shown in figure 14.7. The celestial equator is a line on the celestial sphere directly above Earth's equator. The equinoxes are the points on the celestial sphere where the ecliptic, the path of the Sun, crosses the celestial equator. Note also that the summer solstice occurs when the ecliptic is 23.5° north of the celestial equator, and the winter solstice occurs when it is 23.5° south of the celestial equator.

Rotation

Observing the apparent turning of the celestial sphere once a day and seeing the east-to-west movement of the Sun, Moon, and stars, it certainly seems as if it is the heavenly bodies and not Earth doing the moving. You cannot sense any movement,

and there is little apparent evidence that Earth indeed moves. Evidence of a moving Earth comes from at least three different observations: (1) the observation that the other planets and the Sun rotate, (2) the observation of the changing plane of a long, heavy pendulum at different latitudes on Earth, and (3) the observation of the direction of travel of something moving across, but above, Earth's surface, such as a rocket.

Other planets, such as Jupiter, and the Sun can be observed to rotate by keeping track of features on the surface such as the Great Red Spot on Jupiter and sunspots on the Sun. While such observations are not direct evidence that Earth also rotates, they do show that other members of the solar system spin on their axes. As described earlier, Jupiter is also observed to be oblate, flattened at its poles with an equatorial bulge. Since Earth is also oblate, this is again indirect evidence that it rotates, too.

The most easily obtained and convincing evidence about Earth's rotation comes from a *Foucault pendulum*, a heavy mass swinging from a long wire. This pendulum is named after the French physicist Jean Foucault, who first used a long pendulum in 1851 to prove that Earth rotates. Foucault started a long, heavy pendulum moving just above the floor, marking the plane of its back-and-forth movement. Over some period of time, the pendulum appeared to slowly change its position, smoothly shifting its plane of rotation. Science museums often show this shifting plane of movement by setting up small objects for the pendulum to knock down. Foucault demonstrated that the pendulum actually maintains its plane of movement in space (inertia) while Earth rotates eastward (counterclockwise) under the pendulum. It is Earth that turns under the pendulum, causing the pendulum to appear to change its plane of rotation. It is difficult to imagine the pendulum continuing to move in a fixed direction in space while Earth, and everyone on it, turns under the swinging pendulum.

Figure 14.8 illustrates the concept of the Foucault pendulum. A pendulum is attached to a support on a stool that is free to rotate. If the stool is slowly turned while the pendulum is swinging, you will observe that the pendulum maintains its plane of rotation while the stool turns under it. If you were much smaller and looking from below the pendulum, it would appear to turn as you rotate with the turning stool. This is what happens on Earth. Such a pendulum at the North Pole would make a complete turn in about twenty-four hours. Moving south from the North Pole, the change decreases with latitude

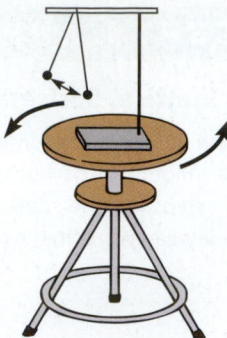

FIGURE 14.8 The Foucault pendulum swings back and forth in the same plane while a stool is turned beneath it. Likewise, a Foucault pendulum on Earth's surface swings back and forth in the same plane while Earth turns beneath it. The amount of turning observed depends on the latitude of the pendulum.

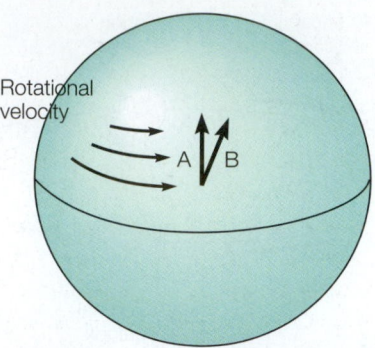

FIGURE 14.9 Earth has a greater rotational velocity at the equator and less toward the poles. As an object moves north or south (A), it passes over land with a different rotational velocity, which produces a deviation to the right in the Northern Hemisphere (B) and to the left in the Southern Hemisphere.

until, at the equator, the pendulum would not appear to turn at all. At higher latitudes, the plane of the pendulum appears to move clockwise in the Northern Hemisphere and counterclockwise in the Southern Hemisphere.

More evidence that Earth rotates is provided by objects that move above and across Earth's surface. As shown in figure 14.9, Earth has a greater rotational velocity at the equator than at the poles. As an object leaves the surface and moves north or south, the surface has a different rotational velocity, so it rotates beneath the object as it proceeds in a straight line. This gives the moving object an apparent deflection to the right of the direction of movement in the Northern Hemisphere and to the left in the Southern Hemisphere. The apparent deflection caused by Earth's rotation is called the **Coriolis effect.** The Coriolis effect will explain Earth's prevailing wind systems as well as the characteristic direction of wind in areas of high pressure and areas of low pressure (see chapter 17).

Precession

If Earth were a perfect spherically shaped ball, its axis would always point to the same reference point among the stars. The reaction of Earth to the gravitational pull of the Moon and the Sun on its equatorial bulge, however, results in a slow wobbling of Earth as it turns on its axis. This slow wobble of Earth's axis, called **precession,** causes it to swing in a slow circle like the wobble of a spinning top (figure 14.10). It takes Earth's axis about twenty-six thousand years to complete one turn, or wobble. Today, the axis points very close to the North Star, Polaris, but is slowly moving away to point to another star. In about twelve thousand years, the star Vega will appear to be in the position above the North Pole, and Vega will be the new North Star. The moving pole also causes changes over time in which particular signs of the zodiac appear with the spring equinox. Because of precession, the occurrence of the spring equinox has been moving backward (westward) through the zodiac constellations at about 1 degree every seventy-two years. Thus, after about twenty-six thousand years, the spring

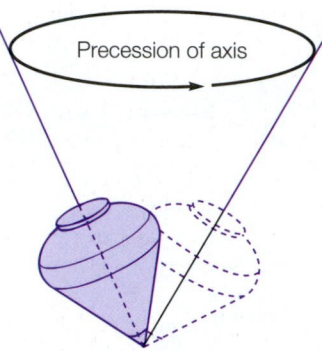

FIGURE 14.10 A spinning top wobbles as it spins, and the axis of the top traces out a small circle. The wobbling of the axis is called *precession.*

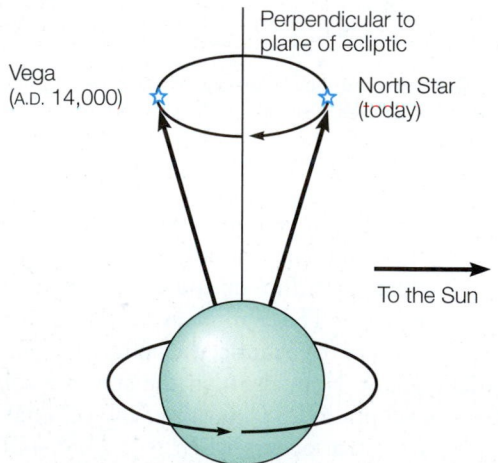

FIGURE 14.11 The slow, continuous precession of Earth's axis results in the North Pole pointing around a small circle over a period of about twenty-six thousand years.

equinox will have moved through all the constellations and will again approach the constellation of Aquarius for the next "age of Aquarius" (figure 14.11).

14.3 PLACE AND TIME

The continuous rotation and revolution of Earth establishes an objective way to determine direction, location, and time on Earth. If Earth were an unmoving sphere, there would be no side, end, or point to provide a referent for direction and location. Earth's rotation, however, defines an axis of rotation, which serves as a reference point for determination of direction and location on the entire surface. Earth's rotation and revolution together define cycles, which define standards of time. The following describes how Earth's movements are used to identify both place and time.

Identifying Place

A system of two straight lines can be used to identify a point, or position, on a flat, two-dimensional surface. The position of the letter *X* on this page, for example, can be identified by making a line a certain number of measurement units from the top of the page and a second line a certain number of measurement units from the left side of the page. Where the two lines intersect will identify the position of the letter *X*, which can be recorded or communicated to another person (figure 14.12).

A system of two straight lines can also be used to identify a point, or position, on a sphere, except this time the lines are circles. The reference point for a sphere is not as simple as in the flat, two-dimensional case, however, since a sphere does not have a top or side edge. Earth's axis provides the north-south reference point. The equator is a big circle around Earth that is exactly halfway between the two ends, or poles, of the rotational axis. An infinite number of circles are imagined to run around Earth parallel to the equator as shown in figure 14.13. The east- and west-running parallel circles are called **parallels.** Each parallel is the same distance between the equator and one of the poles all the way around Earth. The distance from the equator to a point on a parallel is called the **latitude** of that point. Latitude tells you how far north or south a point is from the equator by telling you the parallel the point is located on. The distance is measured northward from the equator (which is 0°) to the North Pole (90° north) or southward from the equator (0°) to the South Pole (90° south) (figure 14.14). If you are somewhere at a latitude of 35° north, you are somewhere on Earth on the 35° latitude line north of the equator.

Since a parallel is a circle, a location of 40°N latitude could be anyplace on that circle around Earth. To identify a location, you need another line, this time one that runs pole to pole and perpendicular to the parallels. These north-south running arcs that intersect at both poles are called **meridians** (figure 14.15). There is no naturally occurring, identifiable meridian that can be used as a point of reference such as the equator serves for parallels, so one is identified as the referent by international agreement. The referent meridian is the one that passes through the Greenwich Observatory near London, England, and this meridian is called the **prime meridian.** The distance from the prime meridian east or west is called the **longitude.** The degrees of longitude of a point on a parallel are measured to the east or to the west from

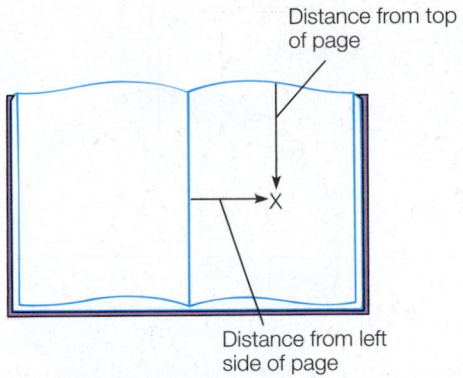

FIGURE 14.12 Any location on a flat, two-dimensional surface is easily identified with two references from two edges. This technique does not work on a motionless sphere because there are no reference points.

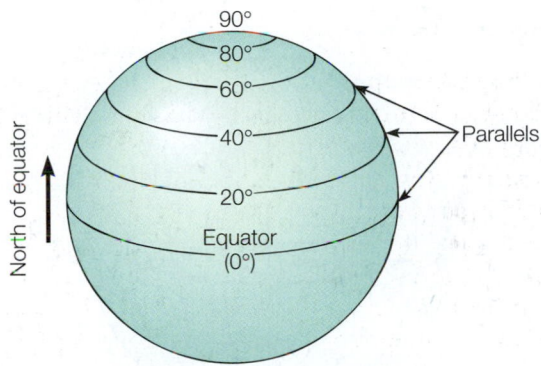

FIGURE 14.13 A circle that is parallel to the equator is used to specify a position north or south of the equator. A few of the possibilities are illustrated here.

the prime meridian up to 180° (figure 14.16). New Orleans, Louisiana, for example, has a latitude of about 30°N of the equator and a longitude of about 90°W of the prime meridian. The location of New Orleans is therefore described as 30°N, 90°W.

Locations identified with degrees of latitude north or south of the equator and degrees of longitude east or west of the prime meridian are more precisely identified by dividing each degree of latitude into subdivisions of 60 minutes (60') per degree, and each minute into 60 seconds (60"). On the other hand, latitudes near the equator are sometimes referred to in general as the *low latitudes,* and those near the poles are sometimes called the *high latitudes.*

In addition to the equator (0°) and the poles (90°), the parallels of 23.5°N and 23.5°S from the equator are important references for climatic consideration. The parallel of 23.5°N is called the **tropic of Cancer,** and 23.5°S is called the **tropic of Capricorn.** These two parallels identify the limits toward the poles within which the Sun appears directly overhead during the course of a year. The parallel of 66.5°N is called the **Arctic Circle,** and the parallel of 66.5°S is called the **Antarctic Circle.** These two parallels identify the limits toward the equator within which the Sun appears above the horizon all day during the summer (figure 14.17). This starts with six months of

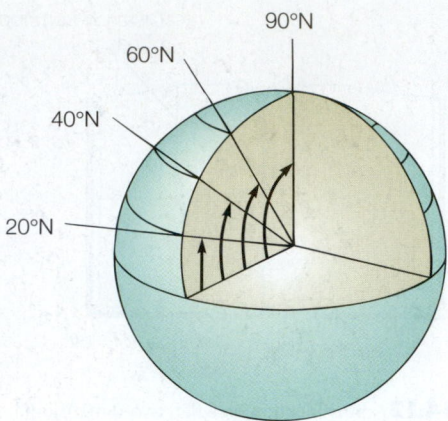

FIGURE 14.14 If you could see to Earth's center, you would see that latitudes run from 0° at the equator north to 90° at the North Pole (or to 90° south at the South Pole).

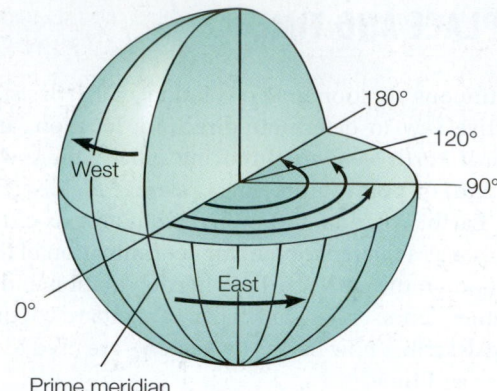

FIGURE 14.16 If you could see inside Earth, you would see 360° around the equator and 180° of longitude east and west of the prime meridian.

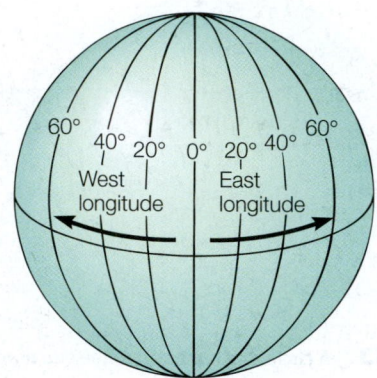

FIGURE 14.15 Meridians run pole to pole perpendicular to the parallels and provide a reference for specifying east and west directions.

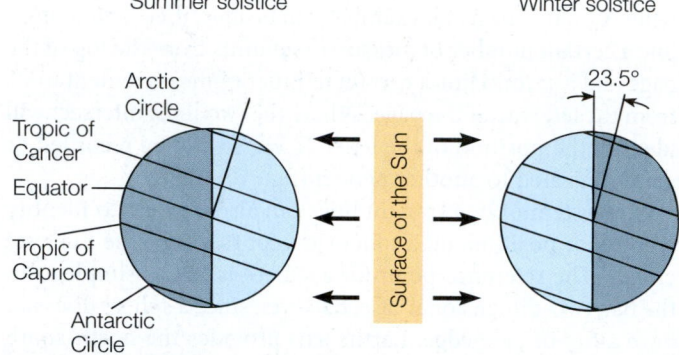

FIGURE 14.17 At the summer solstice, the noon Sun appears directly overhead at the tropic of Cancer (23.5°N) and twenty-four hours of daylight occurs north of the Arctic Circle (66.5°N). At the winter solstice, the noon Sun appears overhead at the tropic of Capricorn (23.5°S) and twenty-four hours of daylight occurs south of the Antarctic Circle (66.5°S).

daylight every day at the pole, then decreases as you get fewer days of full light until reaching the limit of one day of twenty-four-hour daylight at the 66.5° limit.

Measuring Time

Standards of time are determined by intervals between two successive events that repeat themselves in a regular way. Since ancient civilizations, many of the repeating events used to mark time have been recurring cycles associated with the rotation of Earth on its axis and its revolution around the Sun. Thus, the day, month, season, and year are all measures of time based on recurring natural motions of Earth. All other measures of time are based on other events or definitions of events. There are, however, several different ways to describe the day, month, and year, and each depends on a different set of events. These events are described in the following section.

Daily Time

The technique of using astronomical motions for keeping time originated some four thousand years ago with the Babylonian culture. The Babylonians marked the yearly journey of the Sun

against the background of the stars, which was divided into twelve periods, or months, after the signs of the zodiac. Based on this system, the Babylonian year was divided into twelve months with a total of 360 days. In addition, the Babylonians invented the week and divided the day into hours, minutes, and seconds. The week was identified as a group of seven days, each based on one of the seven heavenly bodies that were known at the time. The hours, minutes, and seconds of a day were determined from the movement of the shadow around a straight, vertical rod.

As seen from a place in space above the North Pole, Earth rotates counterclockwise turning toward the east. On Earth, this motion causes the Sun to appear to rise in the east, travel across the sky, and set in the west. The changing angle between the tilt of Earth's axis and the Sun produces an apparent shift of the Sun's path across the sky, northward in the summer season and southward in the winter season. The apparent movement of the Sun across the sky was the basis for the ancient as well as the modern standard of time known as the day.

Today, everyone knows that Earth turns as it moves around the Sun, but it is often convenient to regard space and astronomical motions as the ancient Greeks did, as a celestial sphere that turns around a motionless Earth. Recall that the celestial meridian is a great circle on the celestial sphere that passes directly overhead where you are and continues around Earth through both celestial poles. The movement of the Sun across the celestial meridian identifies an event of time called **noon**. As the Sun appears to travel west, it crosses meridians that are farther and farther west, so the instant identified as noon moves west with the Sun. The instant of noon at any particular longitude is called the **apparent local noon** for that longitude because it identifies noon from the apparent position of the Sun in the sky. The morning hours before the Sun crosses the meridian are identified as *ante meridiem* (A.M.) hours, which is Latin for "before meridian." Afternoon hours are identified as *post meridiem* (P.M.) hours, which is Latin for "after the meridian."

There are several ways to measure the movement of the Sun across the sky. The ancient Babylonians, for example, used a vertical rod called a *gnomon* to make and measure a shadow that moved as a result of the apparent changes of the Sun's position. The gnomon eventually evolved into a *sundial*, a vertical or slanted gnomon with divisions of time marked on a horizontal plate beneath the gnomon. The shadow from the gnomon indicates the **apparent local solar time** at a given place and a given instant from the apparent position of the Sun in the sky. If you have ever read the time from a sundial, you know that it usually does not show the same time as a clock or a watch (figure 14.18). In addition, sundial time is nonuniform, fluctuating throughout the course of a year, sometimes running ahead of clock time and sometimes running behind clock time.

A sundial shows the apparent local solar time, but clocks are set to measure a uniform standard time based on **mean solar time.** Mean solar time is a uniform time averaged from the apparent solar time. The apparent solar time is nonuniform, fluctuating because (1) Earth moves sometimes faster and sometimes slower in its elliptical orbit around the Sun and (2) the equator of Earth is inclined to the ecliptic. The combined consequence of these two effects is a variable, nonuniform sundial time as compared to the uniform mean solar time, otherwise known as clock time.

A day is defined as the length of time required for Earth to rotate once on its axis. There are different ways to measure this rotation, however, which result in different definitions of the day. A **sidereal day** is the interval between two consecutive crossings of the celestial meridian by a particular star (*sidereal* means "star"). This interval of time depends only on the time Earth takes to rotate 360° on its axis. One sidereal day is practically the same length as any other sidereal day because Earth's rate of rotation is constant for all practical purposes.

An **apparent solar day** is the interval between two consecutive crossings of the celestial meridian by the Sun, for example, from one local solar noon to the next solar noon. Since Earth is moving in orbit around the Sun, it must turn a little bit farther to compensate for its orbital movement, bringing the Sun back to local solar noon (figure 14.19). As a consequence, the apparent solar day is about four minutes longer than the sidereal day. This additional time accounts for the observation that the stars and constellations of the zodiac rise about four minutes earlier every night, appearing higher in the sky at the

FIGURE 14.18 A sundial indicates the apparent local solar time at a given instant in a given location. The time read from a sundial, which is usually different from the time read from a clock, is based on an average solar time.

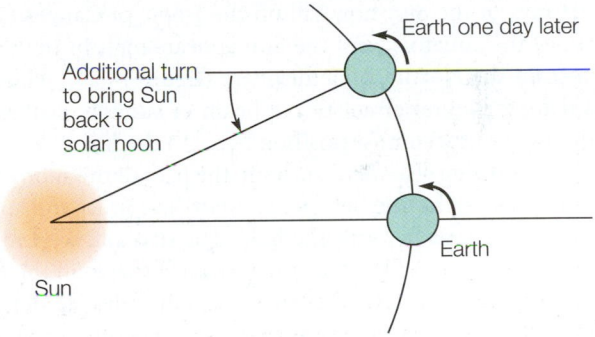

FIGURE 14.19 Because Earth is moving in orbit around the Sun, it must rotate an additional distance each day, requiring about four minutes to bring the Sun back across the celestial meridian (local solar noon). This explains why the stars and constellations rise about four minutes earlier every night.

same clock time until they complete a yearly cycle. A sidereal day is twenty-three hours, fifty-six minutes, and four seconds long. A **mean solar day** is twenty-four hours long, averaged from the mean solar time to keep clocks in closer step with the Sun than would be possible using the variable apparent solar day. Just how out of synchronization the apparent solar day can become with a clock can be illustrated with another ancient way of keeping track of the Sun's motions in the sky, the "hole in the wall" sun calendar and clock.

Variations of the "hole in the wall" sun calendar were used all over the world by many different ancient civilizations, including the early Native Americans of the American Southwest. More than one ancient Native American ruin has small holes in the western wall aligned in such a way as to permit sunlight to enter a chamber only on the longest and shortest days of the year. This established a basis for identifying the turning points in the yearly cycle of seasons.

A hole in the roof can be used as a sun clock, but it will require a whole year to establish the meaning of a beam of sunlight shining on the floor. Imagine a beam of sunlight passing through a small hole to make a small spot of light on the floor. For a year, you mark the position of the spot of light on the floor *each day* when your clock tells you the *mean solar time is noon*. You trace out an elongated, lopsided figure eight with the small end pointing south and the larger end pointing north (figure 14.20A). Note by following the monthly markings shown in figure 14.20B that the figure-eight shape is actually traced out by the spot of sunlight making two **S** shapes as the Sun changes its apparent position in the sky. Together, the two **S** shapes make the shape of the figure eight.

Why did the sunbeam trace out a figure eight over a year? The two extreme north-south positions of the figure are easy to understand because by December, Earth is in its orbit with the North Pole tilted away from the Sun. At this time, the direct rays of the Sun fall on the tropic of Capricorn (23.5° south of the equator), and the Sun appears low in the sky as seen from the Northern Hemisphere. Thus, on this date, the winter solstice, a beam of sunlight strikes the floor at its northernmost position beneath the hole. By June, Earth has moved halfway around its orbit, and the North Pole is now tilted toward the Sun. The direct rays of the Sun now fall on the tropic of Cancer (23.5° north of the equator), and the Sun appears high in the sky as seen from the Northern Hemisphere (figure 14.21). Thus, on this date, the summer solstice, a beam of sunlight strikes the floor at its southernmost position beneath the hole.

If everything else were constant, the path of the spot would trace out a straight line between the northernmost and southernmost positions beneath the hole. The east and west movements of the point of light as it makes an **S** shape on the floor must mean, however, that the Sun crosses the celestial meridian (noon) earlier one part of the year and later the other part. This

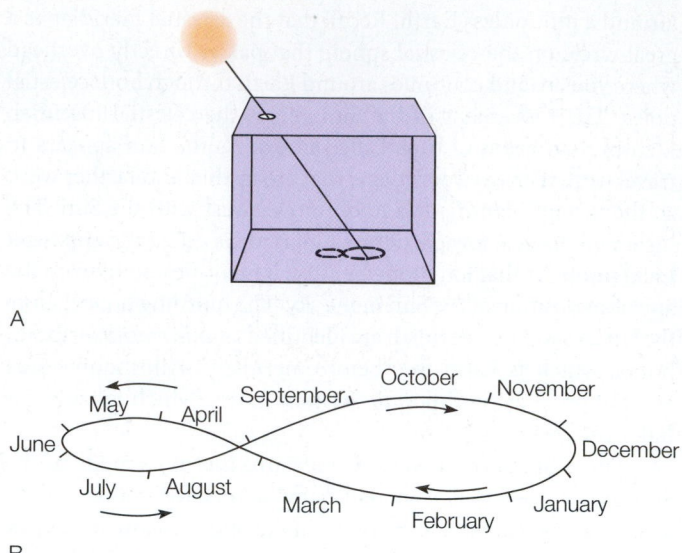

A

B

FIGURE 14.20 (*A*) During a year, a beam of sunlight traces out a lopsided figure eight on the floor if the position of the light is marked at noon every day. (*B*) The location of the point of light on the figure eight during each month.

early and late arrival is explained in part by Earth moving at different speeds in its orbit.

If changes in orbital speed were the only reason that the Sun does not cross the sky at the same rate during the year, the spot of sunlight on the floor would trace out an oval rather than a figure eight. The plane of the ecliptic, however, does not coincide with the plane of Earth's equator, so the Sun appears at different angles in the sky, and this makes it appear to change

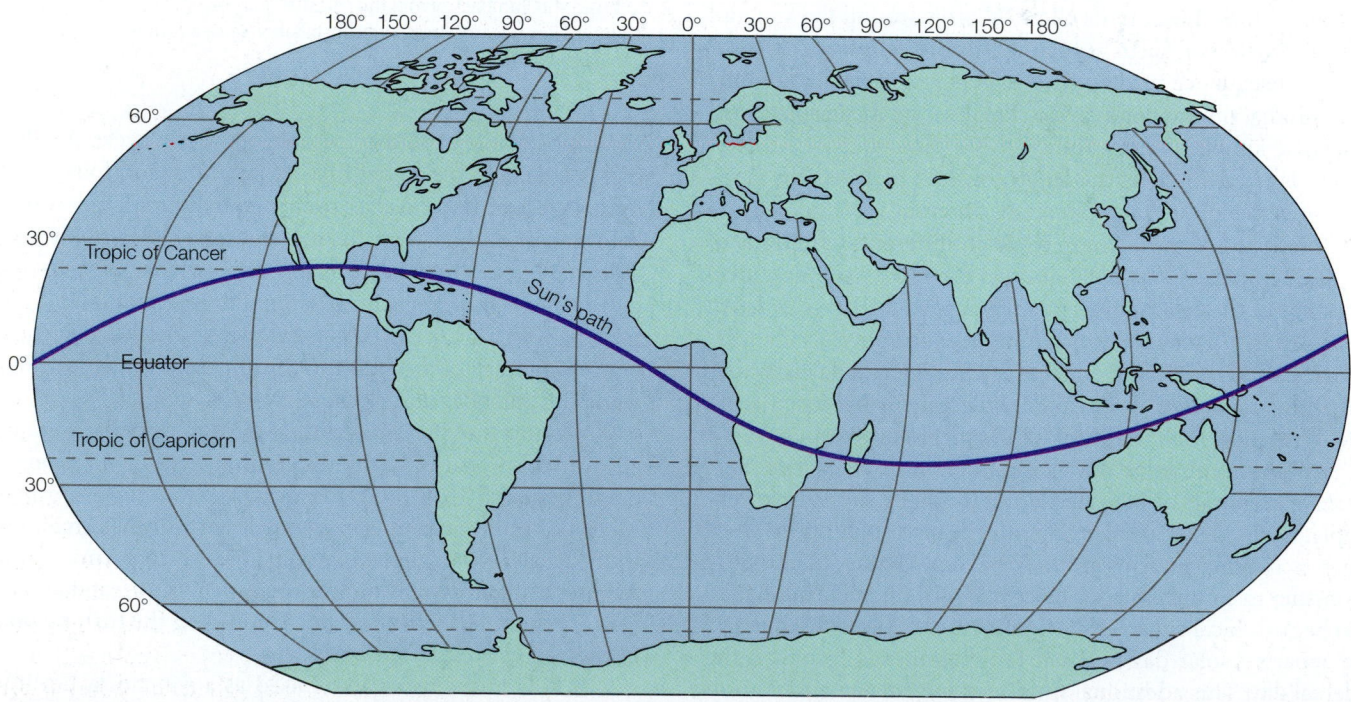

FIGURE 14.21 The path of the Sun's direct rays during a year. The Sun is directly over the tropic of Cancer at the summer solstice and high in the Northern Hemisphere sky. At the winter solstice, the Sun is directly over the tropic of Capricorn and low in the Northern Hemisphere sky.

The purpose of daylight saving time is to make better use of daylight during the summer by moving an hour of daylight from the morning to the evening. In the United States, daylight saving time is observed from the second Sunday in March to the first Sunday in November. Clocks are changed on these Sundays according to the saying, "Spring ahead, fall back." Arizona and Hawaii choose not to participate and stay on standard time all year.

Americans who say they like daylight saving time say they like it because it gives them more light in the evenings and it saves energy. Some people do not like daylight saving time because it requires them to reset all their clocks and adjust their sleep schedule twice a year. They also complain that the act of changing the clock is not saving daylight at all but it is sending them to bed an hour earlier. Farmers also complain that plants and animals are regulated by the Sun, not the clock, so they have to plan all their nonfarm interactions on a different schedule.

Questions to Discuss:

Divide your group into two subgroups, one representing those who like daylight saving time and the other representing those who do not. After a few minutes of preparation, have a short debate about the advantages and disadvantages of daylight saving time.

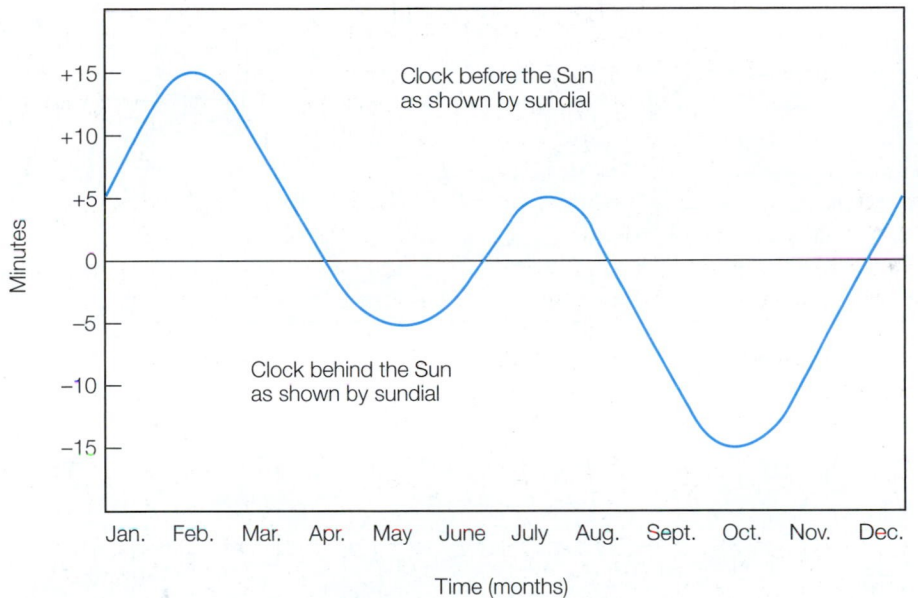

FIGURE 14.22 The equation of time, which shows how many minutes sundial time is faster or slower than clock time during different months of the year.

its speed during different times of the year. This effect changes the length of the apparent solar day by making the Sun up to ten minutes later or earlier than the mean solar time four times a year between the solstices and equinoxes.

The two effects add up to a cumulative variation between the apparent local solar time (sundial time) and the mean solar time (clock time) (figure 14.22). This cumulative variation is known as the **equation of time,** which shows how many minutes sundial time is faster or slower than clock time during different days of the year. The equation of time is often shown on globes in the figure-eight shape called an *analemma,* which also can be used to determine the latitude of direct solar radiation for any day of the year.

Since the local mean time varies with longitude, every place on an east-west line around Earth could possibly have clocks that were a few minutes ahead of those to the west and a few minutes behind those to the east. To avoid the confusion that would result from many clocks set to local mean solar time, Earth's surface is arbitrarily divided into one-hour **standard time zones** (figure 14.23). Since there are 360° around Earth and 24 hours in a day, this means that each time zone is 360° divided by 24, or 15° wide. These 15° zones are adjusted so that whole states are in the same time zone or for other political reasons. The time for each zone is defined as the mean solar time at the middle of each zone. When you cross a boundary between two zones, the clock is set ahead one hour if you are traveling east and back one hour if you are traveling west. Most states adopt **daylight saving time** during the summer, setting clocks ahead one hour in the spring and back one hour in the fall ("spring ahead and fall back"). Daylight saving time results in an extra hour of daylight during summer evenings.

The 180° meridian is arbitrarily called the **international date line,** an imaginary line established to compensate for cumulative time zone changes (figure 14.24). A traveler crossing

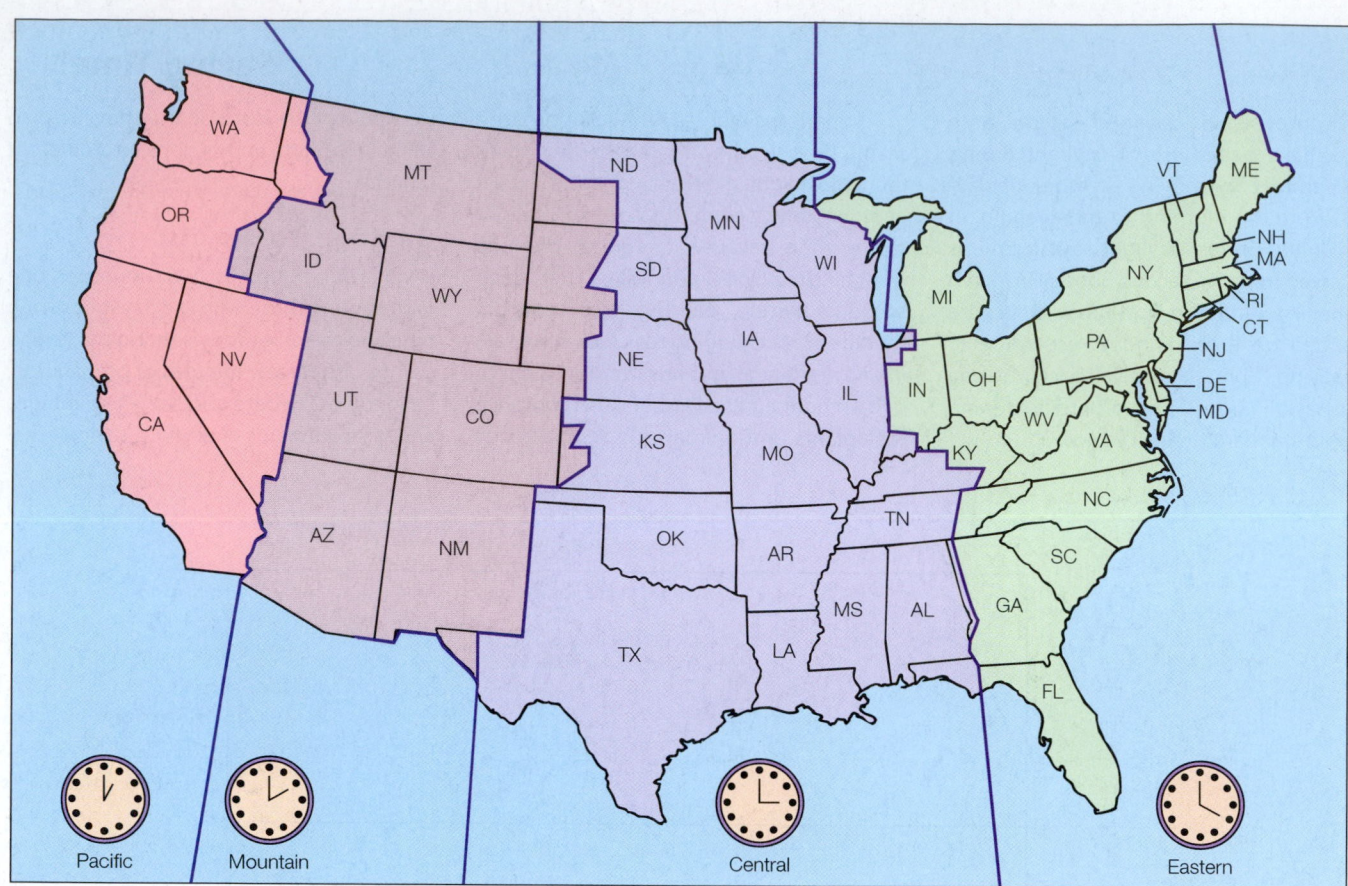

FIGURE 14.23 The standard time zones.

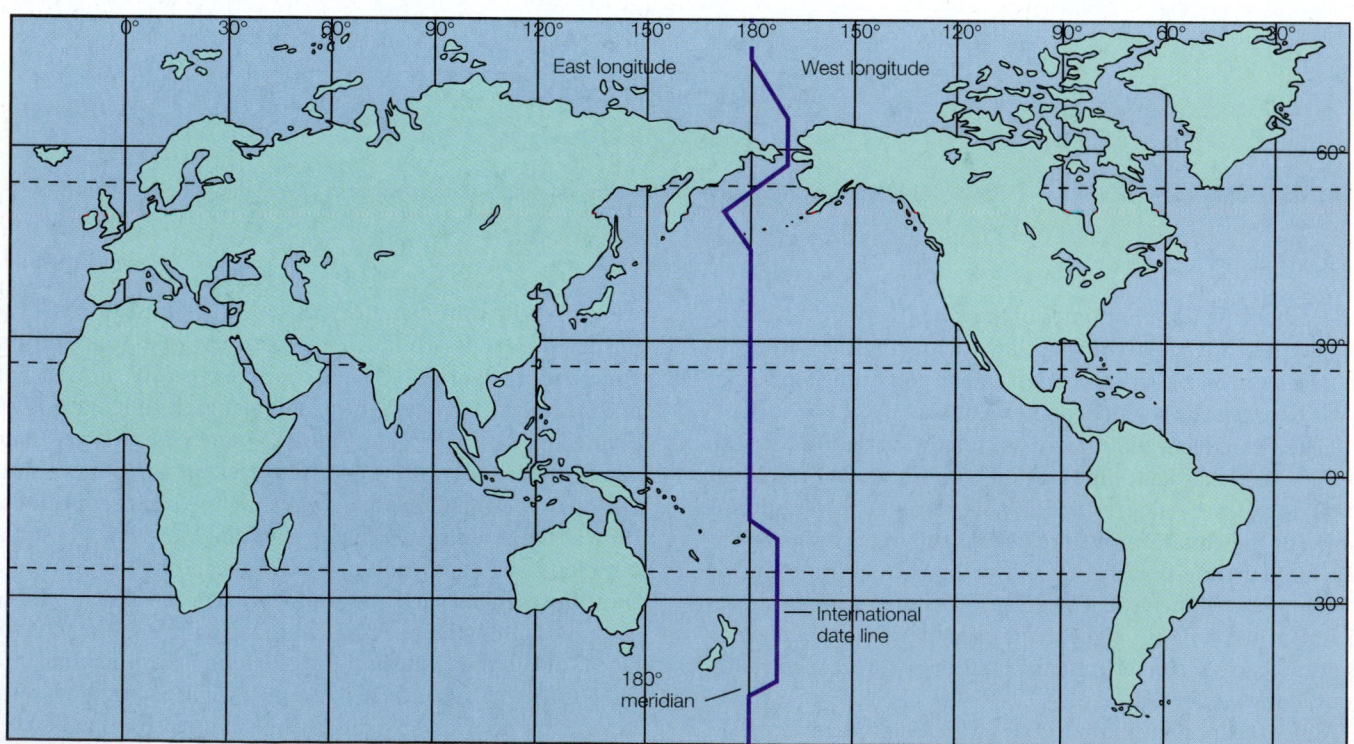

FIGURE 14.24 The international date line follows the 180° meridian but is arranged in a way that land areas and island chains have the same date.

the date line gains or loses a day just as crossing a time zone boundary results in the gain or loss of an hour. A person moving across the line while traveling westward gains a day; for example, the day after June 2 would be June 4. A person crossing the line while traveling eastward repeats a day; for example, the day after June 6 would be June 6. Note that the date line is curved around land masses to avoid local confusion.

Yearly Time

A *year* is generally defined as the interval of time required for Earth to make one complete revolution in its orbit. As was the case for definitions of a day, there are different definitions of what is meant by a year. The most common definition of a year is the interval between two consecutive spring equinoxes, which is known as the **tropical year** (*trope* is Greek for "turning"). The tropical year is 365 days, 5 hours, 48 minutes, and 46 seconds, or 365.24220 mean solar days.

A **sidereal year** is defined as the interval of time required for Earth to move around its orbit so the Sun is again in the same position relative to the stars. The sidereal year is slightly longer than the tropical year because Earth rotates more than 365.25 times during one revolution. Thus, the sidereal year is 365.25636 mean solar days, which is about 20 minutes longer than the tropical year.

The tropical and sidereal years would be the same interval of time if Earth's axis pointed in a consistent direction. The precession of the axis, however, results in the axis pointing in a slightly different direction with time. This shift of direction over the course of a year moves the position of the spring equinox westward, and the equinox is observed twenty minutes before the orbit has been completely circled. The position of the spring equinox against the background of the stars thus moves westward by some fifty seconds of arc per year.

It is the *tropical year* that is used as a standard time interval to determine the calendar year. Earth does not complete an exact number of turns on its axis while completing one trip around the Sun, so it becomes necessary to periodically adjust the calendar so it stays in step with the seasons. The calendar system that was first designed to stay in step with the seasons was devised by the ancient Romans. Julius Caesar reformed the calendar, beginning in 46 B.C., to have a 365-day year with a 366-day year (leap year) every fourth year. Since the tropical year of 365.24220 mean solar days is very close to 365¼ days, the system, called a *Julian calendar*, accounted for the ¼ day by adding a full day to the calendar every fourth year. The Julian calendar was very similar to the one now used, except the year began in March, the month of the spring equinox. The month of July was named in honor of Julius Caesar, and the following month was later named after his successor, Augustus.

There was a slight problem with the Julian calendar because it was longer than the tropical year by 365.25 minus 365.24220, or 0.0078 day per year. This small interval (which is 11 minutes, 14 seconds) does not seem significant when compared to the time in a whole year. But over the years, the error of minutes and seconds grew to an error of days. By 1582, when Pope Gregory XIII revised the calendar, the error had grown to 13 days but was corrected for 10 days of error. This revision resulted in the *Gregorian calendar,* which is the system used today. Since the accumulated error of 0.0078 day per year is almost 0.75 day per century, it follows that four centuries will have 0.75 times 4, or 3 days of error. The Gregorian system corrects for the accumulated error by dropping the additional leap year day three centuries out of every four. Thus, the century year of 2000 was a leap year with 366 days, but the century years of 2100, 2200, and 2300 will not be leap years. You will note that this approximation still leaves an error of 0.0003 day per century, so another calendar revision will be necessary in a few thousand years to keep the calendar in step with the seasons.

Monthly Time

In ancient times, people often used the Moon to measure time intervals that were longer than a day but shorter than a year. The word *month,* in fact, has its origins in the word *moon* and its period of revolution. The Moon revolves around Earth in an orbit that is inclined to the plane of Earth's orbit, the plane of the ecliptic, by about 5°. The Moon is thus never more than about ten apparent diameters from the ecliptic. It revolves in this orbit in about 27⅓ days as measured by two consecutive crossings of any star. This period is called a **sidereal month.** The Moon rotates in the same period as the time of revolution, so the sidereal month is also the time required for one rotation. Because the rotation and revolution rates are the same, you always see the same side of the Moon from Earth.

The ancient concept of a month was based on the **synodic month,** the interval of time from new moon to new moon (or any two consecutive identical phases). The synodic month is longer than a sidereal month at a little more than 29½ days. The Moon's phases (see "Phases of the Moon" in this chapter) are determined by the relative positions of Earth, Moon, and Sun. As shown in figure 14.25, the Moon moves with Earth in its orbit around the Sun. During one sidereal month, the Moon has to revolve a greater distance before the same phase is observed on Earth, and this greater distance requires 2.2 days. This makes the synodic month about 29½ days long, only a little less than 1/12 of a year, or the period of time the present calendar identifies as a "month."

14.4 THE EARTH-MOON SYSTEM

Earth and its Moon are unique in the solar system because of the size of the Moon. It is not the largest satellite, but the ratio of its mass to Earth's mass is greater than the mass ratio of any other moon to its planet. The Moon has a diameter of 3,476 km (about 2,159 mi), which is about one-fourth the diameter of Earth, and a mass of about 1/81 of Earth's mass. This is a small fraction of Earth's mass, but it is enough to affect Earth's motion as it revolves around the Sun.

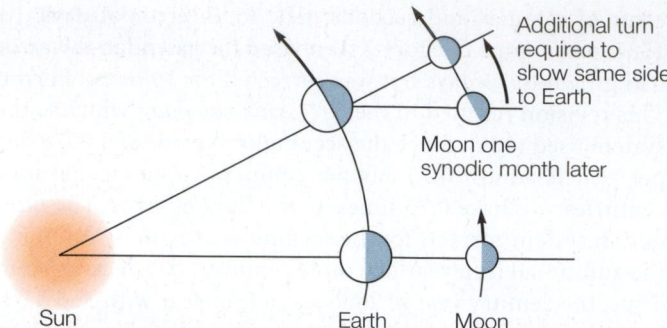

FIGURE 14.25 As the Moon moves in its orbit around Earth, it must revolve a greater distance to bring the same part to face Earth. The additional turning requires about 2.2 days, making the synodic month longer than the sidereal month.

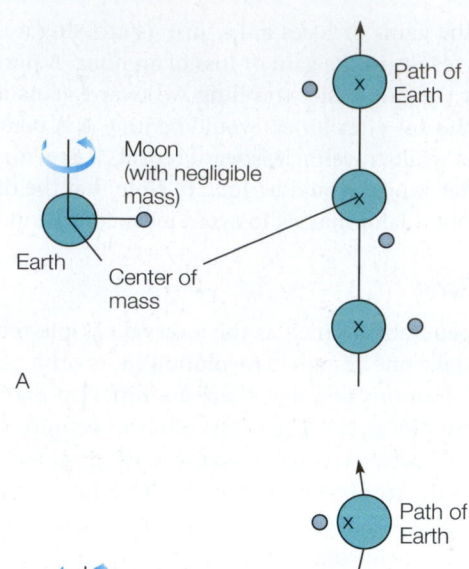

A

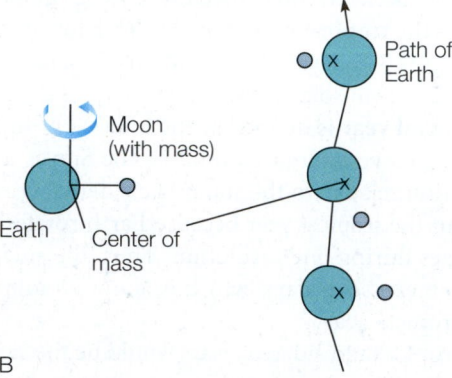

B

If the Moon had a negligible mass, it would circle Earth with a center of rotation (center of mass) located at the center of Earth. In this situation, the center of Earth would follow a smooth path around the Sun (figure 14.26A). The mass of the Moon, however, is great enough to move the center of rotation away from Earth's center toward the Moon. As a result, both bodies act as a system, moving around a center of mass. The center of mass between Earth and the Moon follows a smooth orbit around the Sun. Earth follows a slightly wavy path around the Sun as it slowly revolves around the common center of mass (figure 14.26B).

FIGURE 14.26 (A) If the Moon had a negligible mass, the center of gravity between the Moon and Earth would be Earth's center, and Earth would follow a smooth orbit around the Sun. (B) The actual location of the center of mass between Earth and Moon results in a slightly in and out, or wavy, path around the Sun.

Phases of the Moon

The phases of the Moon are a result of the changing relative positions of Earth, the Moon, and the Sun as the Earth-Moon system moves around the Sun. Sunlight always illuminates half of the Moon, and half is always in shadow. As the Moon's path takes it between Earth and the Sun, then to the dark side of Earth, you see different parts of the illuminated half called *phases* (figure 14.27). When the Moon is on the dark side of Earth, you see the entire illuminated half of the Moon called the **full moon** (or the full phase). Halfway around the orbit, the lighted side of the Moon now faces away from Earth, and the unlighted side now faces Earth. This dark appearance is called the **new moon** (or the new phase). In the new phase, the Moon is not *directly* between Earth and the Sun, so it does not produce an eclipse (see "Eclipses of the Sun and Moon" in this chapter).

As the Moon moves from the new phase in its orbit around Earth, you will eventually see half the lighted surface, which is known as the **first quarter.** Often the unlighted part of the Moon shines with a dim light of reflected sunlight from Earth called *earthshine.* Note that the division between the lighted and unlighted part of the Moon's surface is curved in an arc. A straight line connecting the ends of the arc is perpendicular to the direction of the Sun (figure 14.28). After the first quarter, the Moon moves to its full phase, then to the **last quarter** (see figure 14.27). The period of time between two consecutive phases, such as new moon to new moon, is the synodic month, or about 29.5 days.

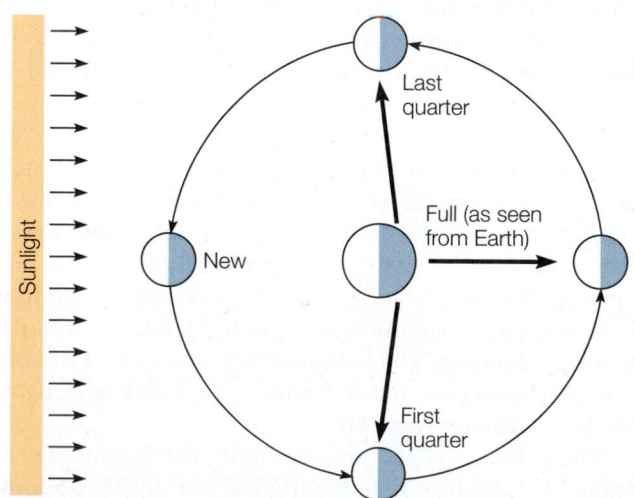

FIGURE 14.27 Half of the Moon is always lighted by the Sun, and half is always in the shadow. The Moon phases result from the view of the lighted and dark parts as the Moon revolves around Earth.

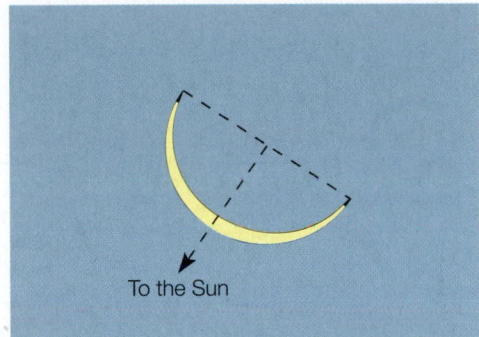

FIGURE 14.28 The cusps, or horns, of the Moon always point away from the Sun. A line drawn from the tip of one cusp to the other is perpendicular to a straight line between the Moon and the Sun.

Eclipses of the Sun and Moon

Sunlight is not visible in the emptiness of space because there is nothing to reflect the light, so the long conical shadow behind each spherical body is not visible, either. One side of Earth and one side of the Moon are always visible because they reflect sunlight. The shadow from Earth or from the Moon becomes noticeable only when it falls on the illuminated surface of the other body. This event of Earth's or the Moon's shadow falling on the other body is called an **eclipse** (table 14.1). Most of the time, eclipses do not occur because the plane of the Moon's orbit is inclined to Earth's orbit about 5° (figure 14.29). As a result, the shadow from the Moon or the shadow from Earth usually falls above or below the other body, too high or too low to produce an eclipse. An eclipse occurs only when the Sun, Moon, and Earth are in a line with each other.

The shadow from Earth and the shadow from the Moon are long cones that point away from the Sun. Both cones have two parts, an inner cone of a complete shadow called the **umbra** and

TABLE 14.1 Total Eclipses in U.S.—2008–2025*

Total Solar Eclipses	
Aug 21, 2017	Path: Oregon to South Carolina
Apr 8, 2024	Path: Texas to Maine
Total Lunar Eclipses	
Feb 21, 2008	All visible eastern; moonrise western
Dec 21, 2010	All visible
Dec 10, 2011	Visible at moonset
Apr 15, 2014	Visible at moonset
Oct 8, 2014	All visible western; at moonset eastern
Apr 4, 2015	Visible at moonset
Sept 28, 2015	Western visible at moonrise; all eastern
Jan 31, 2018	Visible at moonset
Jan 21, 2019	All visible
May 26, 2021	Visible at moonset
May 16, 2022	Visible moonrise western; all eastern
Mar 14, 2025	All visible

*Eclipse predictions by Fred Espenak, NASA/GSFC

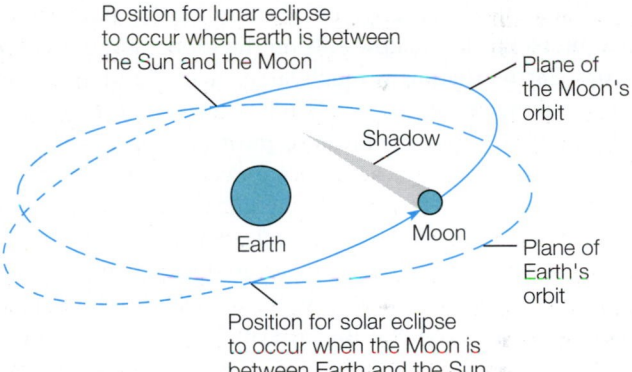

FIGURE 14.29 The plane of the Moon's orbit is inclined to the plane of Earth's orbit by about 5°. An eclipse occurs only where the two planes intersect and Earth, the Moon, and the Sun are in a line.

an outer cone of partial shadow called the **penumbra.** When and where the umbra of the Moon's shadow falls on Earth, people see a **total solar eclipse.** During a total solar eclipse, the Moon completely covers the disk of the Sun. The total solar eclipse is preceded and followed by a partial eclipse, which is seen when the observer is in the penumbra. If the observer is in a location where only the penumbra passes, then only a partial eclipse will be observed (figure 14.30). More people see partial than full solar eclipses because the penumbra covers a larger area. The occurrence of a total solar eclipse is a rare event in a given location, occurring once every several hundred years and then lasting for less than seven minutes.

The Moon's cone-shaped shadow averages a length of 375,000 km (about 233,000 mi), which is less than the average distance between Earth and the Moon. The Moon's elliptical orbit brings it sometimes closer to and sometimes farther from Earth. A total solar eclipse occurs only when the Moon is close

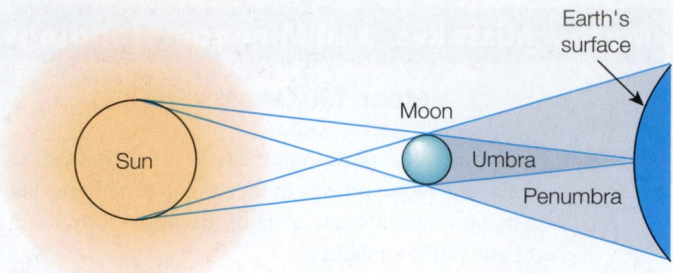

FIGURE 14.30 People in a location where the tip of the umbra falls on the surface of Earth see a total solar eclipse. People in locations where the penumbra falls on Earth's surface see a partial solar eclipse.

FIGURE 14.31 Gravitational attraction pulls on Earth's waters on the side of Earth facing the Moon, producing a tidal bulge. A second tidal bulge on the side of Earth opposite the Moon is produced when Earth, which is closer to the Moon, is pulled away from the waters.

enough so at least the tip of its umbra reaches the surface of Earth. If the Moon's umbra fails to reach Earth, an **annular eclipse** occurs. Annular means "ring-shaped," and during this eclipse, the edge of the Sun is seen to form a bright ring around the Moon. As before, people located in the area where the penumbra falls will see a partial eclipse. The annular eclipse occurs more frequently than the total solar eclipse.

When the Moon is full and the Sun, Moon, and Earth are lined up so Earth's shadow falls on the Moon, a **lunar eclipse** occurs. Earth's shadow is much larger than the Moon's diameter, so a lunar eclipse is visible to everyone on the night side of Earth. This larger shadow also means a longer eclipse that may last for hours. As the umbra moves over the Moon, the darkened part takes on a reddish, somewhat copper-colored glow from light refracted and scattered into the umbra by Earth's atmosphere. This light passes through the thickness of Earth's atmosphere on its way to the eclipsed Moon, and it acquires the reddish color for the same reason that a sunset is red: much of the blue light has been removed by scattering in Earth's atmosphere.

Tides

If you live near or have ever visited a coastal area of the ocean, you are familiar with the periodic rise and fall of sea level known as **tides.** The relationship between the motions of the Moon and the magnitude and timing of tides has been known and studied since very early times. These relationships are that (1) the greatest range of tides occurs at full and new moon phases, (2) the least range of tides occurs at quarter moon phases, and (3) in most oceans, the time between two high tides or between two low tides is an average of twelve hours and twenty-five minutes. The period of twelve hours and twenty-five minutes is half the average time interval between consecutive passages of the Moon across the celestial meridian. A location on the surface of Earth is directly under the Moon when it crosses the meridian, and directly opposite it on the far side of Earth an average twelve hours and twenty-five minutes later. There are two *tidal bulges* that follow the Moon as it moves around Earth, one on the side facing the Moon and one on the opposite side. In general, tides are a result of these bulges moving westward around Earth.

A simplified explanation of the two tidal bulges involves two basic factors: the gravitational attraction of the Moon and the motion of the Earth-Moon system (figure 14.31). Water on Earth's surface is free to move, and the Moon's gravitational attraction pulls the water to the tidal bulge on the side of Earth facing the Moon. This tide-raising force directed toward the Moon bulges the water in mid-ocean some .75 m (about 2.5 ft), but it also bulges the land, producing a land tide. Since Earth is much more rigid than water, the land tide is much smaller at about 12 cm (about 4.5 in). Since all parts of the land bulge together, this movement is not evident without measurement by sensitive instruments.

The tidal bulge on the side of Earth opposite the Moon occurs as Earth is pulled away from the ocean by the Earth-Moon gravitational interaction. Between the tidal bulge facing the Moon and the tidal bulge on the opposite side, sea level is depressed across the broad surface. The depression is called a *tidal trough,* even though it does not actually have the shape of a trough. The two tidal bulges, with the trough between, move slowly eastward following the Moon. Earth turns more rapidly on its axis, however, which forces the tidal bulge to stay in front of the Moon moving through its orbit. Thus, the tidal axis is not aligned with the Earth-Moon gravitational axis.

The tides do not actually appear as alternating bulges that move around Earth. There are a number of factors that influence the making and moving of the bulges in complex interactions that determine the timing and size of a tide at a given time in a given location. Some of these factors include (1) the relative positions of Earth, Moon, and Sun, (2) the elliptical orbit of the Moon, which sometimes brings it closer to Earth, and (3) the size, shape, and depth of the basin holding the water.

The relative positions of Earth, Moon, and Sun determine the size of a given tide because the Sun, as well as the Moon, produces a tide-raising force. The Sun is much more massive than the Moon, but it is so far away that its tide-raising force is about half that of the closer Moon. Thus, the Sun basically modifies lunar tides rather than producing distinct tides of its own. For example, Earth, Moon, and Sun are nearly in line during the full and new moon phases. At these times, the lunar and

People Behind the Science

Carl Edward Sagan (1934–1996)

Carl Edward Sagan was an American astronomer and popularizer of astronomy whose main research was on planetary atmospheres, including that of the primordial Earth. His most remarkable achievement was to provide valuable insights into the origin of life on our planet.

In the early 1960s, Sagan's first major research was into the planetary surface and atmosphere of Venus. At the time, although intense emission of radiation had shown that the dark-side temperature of Venus was nearly 600K, it was thought that the surface itself remained relatively cool—leaving open the possibility that there was some form of life on the planet. Various hypotheses were put forward to account for the strong emission actually observed: perhaps it was due to interactions between charged particles in Venus's dense upper atmosphere; perhaps it was glow discharge between positive and negative charges in the atmosphere; or perhaps emission was due to a particular radiation from charged particles trapped in the Venusian equivalent of a Van Allen Belt. Sagan showed that each of these hypotheses was incompatible with other observed characteristics or with implications of these characteristics. The positive part of Sagan's proposal was to show that all the observed characteristics were compatible with the straightforward hypothesis that the surface of Venus was very hot. On the basis of radar and optical observations, the distance between surface and clouds was calculated to be between 44 km (27 mi) and 65 km (40 mi); given the cloud-top temperature and Sagan's expectation of a "greenhouse effect" in the atmosphere, surface temperature on Venus was computed to be between 500K (227°C/440°F) and 800K (527°C/980°F)—the range that would also be expected on the basis of emission rate.

Sagan then turned his attention to the early planetary atmosphere of Earth, with regard to the origins of life. One way of understanding how life began is to try to form the compounds essential to life in conditions analogous to those of the primeval atmosphere. Before Sagan, Stanley Miller and Harold Urey had used a mixture of methane, ammonia, water vapor, and hydrogen, sparked by a corona discharge that simulated the effect of lightning, to produce amino and hydroxy acids of the sort found in life-forms. Later experiments used ultraviolet light or heat as sources of energy, and even these had less energy than would have been available in Earth's primordial state. Sagan followed a similar method and, by irradiating a mixture of methane, ammonia, water, and hydrogen sulfide, was able to produce amino acids—and, in addition, glucose, fructose, and nucleic acids. Sugars can be made from formaldehyde under alkaline conditions and in the presence of inorganic catalysts. These sugars include five-carbon sugars, which are essential to the formation of nucleic acids, glucose, and fructose—all common metabolites found as constituents of present-day life-forms. Sagan's simulated primordial atmosphere not only showed the presence of those metabolites, it also contained traces of adenosine triphosphate (ATP)—the foremost agent used by living cells to store energy.

Source: Modified from the *Hutchinson Dictionary of Scientific Biography.* © RM, 2011. All rights reserved. Helicon Publishing is a division of RM.

---

solar tide-producing forces act together, producing tides that are unusually high and corresponding low tides that are unusually low. The unusually high and low tides are called **spring tides** (figure 14.32A). Spring tides occur every two weeks and have nothing to do with the spring season. When the Moon is in its quarter phases, the Sun and Moon are at right angles to one another and the solar tides occur between the lunar tides, causing unusually less pronounced high and low tides called **neap tides** (figure 14.32B). Neap tides also occur every two weeks.

The size of the lunar-produced tidal bulge varies as the Moon's distance from Earth changes. The Moon's elliptical orbit brings it closest to Earth at a point called **perigee** and farthest from Earth at a point called **apogee.** At perigee, the Moon is about 44,800 km (about 28,000 mi) closer to Earth than at apogee, so its gravitational attraction is much greater. When perigee coincides with a new or full moon, especially high spring tides result.

The open basins of oceans, gulfs, and bays are all connected but have different shapes and sizes and have bordering landmasses in all possible orientations to the westward-moving tidal bulges. Water in each basin responds differently to the tidal forces, responding as periodic resonant oscillations that move back and forth much like the water in a bowl shifts when carried. Thus, coastal regions on open seas may experience tides that range between about 1 and 3 m (about 3 to 10 ft), but mostly enclosed basins such as the Gulf of Mexico have tides less than about 1/3 m (about 1 ft). The Gulf of Mexico, because of its size, depth, and limited connections with the open ocean, responds only to the stronger tidal attractions and has only one high and one low tide per day. Even lakes and ponds respond to tidal attractions, but the result is too small to be noticed. Other basins, such as the Bay of Fundy in Nova Scotia, are funnel-shaped and undergo an unusually high tidal range. The Bay

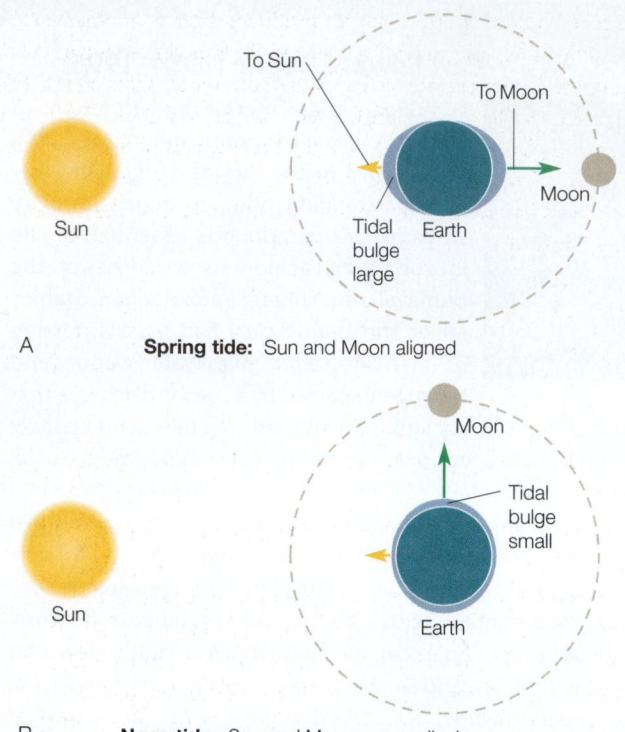

A   **Spring tide:** Sun and Moon aligned

B   **Neap tide:** Sun and Moon perpendicular

FIGURE 14.32   (A) When the Sun and Moon lie in the same or in opposite directions as seen from Earth, the tidal bulges they cause coincide, producing a large *spring* tide. (B) When the directions to the Sun and Moon are perpendicular, the tides of the Sun and Moon cancel to some extent, producing a weak *neap* tide.

of Fundy has experienced as much as a 15 m (about 50 ft) tidal range.

As the tidal bulges are pulled against a rotating Earth, friction between the moving water and the ocean basin tends to slow Earth's rotation over time. This is a very small slowing effect of 1.4 ms/day/century. Evidence for this slowing comes from a number of sources, including records of ancient solar eclipses. The solar eclipses of two thousand years ago occurred three hours earlier than would be expected by using today's time but were on the mark if a lengthening day is considered. Fossils of a certain species of coral still living today provide further evidence of a lengthening day. This particular coral adds daily growth rings, and five-hundred-million-year-old fossils show that the day was about twenty-one hours long at that time. Finally, the Moon is moving away from Earth at a rate of about 4 cm (about 1.5 in) per year. This movement out to a larger orbit is a necessary condition to conserve angular momentum as Earth slows. As the Moon moves away from Earth, the length of the month increases. Sometime in the distant future, both the day and the month will be equal, about fifty of the present days long.

SUMMARY

Earth is an *oblate spheroid* that undergoes three basic motions: (1) a yearly *revolution* around the Sun, (2) a daily *rotation* on its axis, and (3) a slow wobble of its axis called *precession*.

As Earth makes its yearly *revolution* around the Sun, it maintains a generally *constant inclination of its axis* to the *plane of the ecliptic* of 66.5°, or 23.5° from a line perpendicular to the plane. In addition, Earth maintains a generally constant *orientation of its axis* to the stars, which always points in the same direction. The constant inclination and orientation of the axis, together with Earth's rotation and revolution, produce three effects: (1) days and nights that vary in length, (2) seasons that change during the course of a year, and (3) climates that vary with latitude. When Earth is at a place in its orbit so the axis points toward the Sun, the Northern Hemisphere experiences the longest days and the summer season. This begins on June 21 or 22, which is called the *summer solstice*. Six months later, the axis points away from the Sun and the Northern Hemisphere experiences the shortest days and the winter season. This begins on December 22 or 23 and is called the *winter solstice*. On March 20 or 21, Earth is halfway between the solstices and has days and nights of equal length, which is called the *spring* (or *vernal*) *equinox*. On September 22 or 23, the *autumnal equinox*, another period of equal nights and days, identifies the beginning of the fall season.

Precession is a slow wobbling of the axis as Earth spins. Precession is produced by the gravitational tugs of the Sun and Moon on Earth's equatorial bulge.

Lines around Earth that are parallel to the equator are circles called *parallels*. The distance from the equator to a point on a parallel is called the *latitude* of that point. North and south arcs that intersect at the poles are called *meridians*. The meridian that runs through the Greenwich Observatory is a reference line called the *prime meridian*. The distance of a point east or west of the prime meridian is called the *longitude* of that point.

The event of time called *noon* is the instant the Sun appears to move across the celestial meridian. The instant of noon at a particular location is called the *apparent local noon*. The time at a given place that is determined by a sundial is called the *apparent local solar time*. It is the basis for an averaged, uniform standard time called the *mean solar time*. Mean solar time is the time used to set clocks.

A *sidereal day* is the interval between two consecutive crossings of the celestial meridian by a star. An *apparent solar day* is the interval between two consecutive crossings of the celestial meridian by the Sun, from one apparent solar noon to the next. A *mean solar day* is twenty-four hours as determined from mean solar time. The *equation of time* shows how the local solar time is faster or slower than the clock time during different days of the year.

Earth's surface is divided into one-hour *standard time zones* that are about 15° of meridian wide. The *international date line* is the 180° meridian; you gain a day if you cross this line while traveling westward and repeat a day if you are traveling eastward.

A *tropical year* is the interval between two consecutive spring equinoxes. A *sidereal year* is the interval of time between two consecutive crossings of a star by the Sun. It is the tropical year that is used as a standard time interval for the calendar year. A *sidereal month* is the interval of time between two consecutive crossings of a star by the Moon. The *synodic month* is the interval of time from a new moon to the next new moon. The synodic month is about 29½ days long, which is about ¹/₁₂ of a year.

Earth and the Moon act as a system, with both bodies revolving around a common center of mass located under Earth's surface. This combined motion around the Sun produces three phenomena: (1) as the Earth-Moon system revolves around the Sun, different parts of the illuminated lunar surface, called *phases*, are visible from Earth; (2) a *solar eclipse* is observed where the Moon's shadow falls on Earth, and a *lunar eclipse* is observed where Earth's shadow falls on the Moon; and (3) the *tides*, a periodic rising and falling of sea level, are produced by gravitational attractions of the Moon and Sun and by the movement of the Earth-Moon system.

KEY TERMS

annular eclipse (p. **324**)	first quarter (p. **322**)	parallels (p. **315**)	spring tides (p. **325**)
Antarctic Circle (p. **315**)	full moon (p. **322**)	penumbra (p. **323**)	standard time zones (p. **319**)
apogee (p. **325**)	international date line (p. **319**)	perigee (p. **325**)	summer solstice (p. **312**)
apparent local noon (p. **317**)	last quarter (p. **322**)	plane of the ecliptic (p. **309**)	synodic month (p. **321**)
apparent local solar time (p. **317**)	latitude (p. **315**)	precession (p. **314**)	tides (p. **324**)
apparent solar day (p. **317**)	longitude (p. **315**)	prime meridian (p. **315**)	total solar eclipse (p. **323**)
Arctic Circle (p. **315**)	lunar eclipse (p. **324**)	revolution (p. **312**)	tropic of Cancer (p. **315**)
autumnal equinox (p. **312**)	mean solar day (p. **317**)	rotation (p. **312**)	tropic of Capricorn (p. **315**)
Coriolis effect (p. **314**)	mean solar time (p. **317**)	sidereal day (p. **317**)	tropical year (p. **321**)
daylight saving time (p. **319**)	meridians (p. **315**)	sidereal month (p. **321**)	umbra (p. **323**)
eclipse (p. **323**)	neap tides (p. **325**)	sidereal year (p. **321**)	vernal equinox (p. **312**)
equation of time (p. **319**)	new moon (p. **322**)	solstices (p. **312**)	winter solstice (p. **312**)
equinoxes (p. **312**)	noon (p. **317**)	spring equinox (p. **312**)	

APPLYING THE CONCEPTS

Answers are located in appendix F.

1. If you are located at 20°N latitude, when will the Sun appear directly overhead?
   a. never
   b. once a year
   c. twice a year
   d. four times a year

2. If you are located on the equator (0° latitude), when will the Sun appear directly overhead?
   a. never
   b. once a year
   c. twice a year
   d. four times a year

3. If you are located at 40°N latitude, when will the Sun appear directly overhead?
   a. never
   b. once a year
   c. twice a year
   d. four times a year

4. In about twelve thousand years, the star Vega will be the North Star, not Polaris, because of Earth's
   a. uneven equinox.
   b. tilted axis.
   c. precession.
   d. recession.

5. The time as read from a sundial is the same as the time read from a clock
   a. all the time.
   b. only once a year.
   c. twice a year.
   d. four times a year.

6. You are traveling west by jet and cross three time zone boundaries. If your watch reads 3:00 P.M. when you arrive, you should reset it to
   a. 12:00 noon.
   b. 6:00 P.M.
   c. 12:00 midnight.
   d. 6:00 A.M.

7. If it is Sunday when you cross the international date line while traveling westward, the next day is
   a. Wednesday.
   b. Sunday.
   c. Tuesday.
   d. Saturday.

8. If you see a full moon, an astronaut on the Moon looking back at Earth at the same time would see a
   a. full Earth.
   b. new Earth.
   c. first quarter Earth.
   d. last quarter Earth.

9. A lunar eclipse can occur only during the moon phase of
   a. full moon.
   b. new moon.
   c. first quarter.
   d. last quarter.

10. A total solar eclipse can occur only during the moon phase of
    a. full moon.
    b. new moon.
    c. first quarter.
    d. last quarter.

11. A lunar eclipse does not occur every month because
    a. the plane of the Moon's orbit is inclined to the ecliptic.
    b. of precession.
    c. Earth moves faster in its orbit when closest to the Sun.
    d. Earth's axis is tilted with respect to the Sun.

12. The smallest range between high and low tides occurs during
    a. full moon.
    b. new moon.
    c. quarter moon phases.
    d. an eclipse.

QUESTIONS FOR THOUGHT

1. Use sketches with brief explanations to describe how the constant inclination and constant orientation of Earth's axis produce (a) a variation in the number of daylight hours and (b) a variation in seasons throughout a year.

2. Where on Earth are you if you observe the following at the instant of apparent local noon on September 23? (a) The shadow from a vertical stick points northward. (b) There is no shadow on a clear day. (c) The shadow from a vertical stick points southward.

3. What is the meaning of the word *solstice*? What causes solstices? On about what dates do solstices occur?

4. What is the meaning of *equinox*? What causes equinoxes? On about what dates do equinoxes occur?

5. Briefly describe how Earth's axis is used as a reference for a system that identifies locations on Earth's surface.

6. Use a map or a globe to identify the latitude and longitude of your present location.

7. The tropic of Cancer, tropic of Capricorn, Arctic Circle, and Antarctic Circle are parallels that are identified with specific names. What parallels do the names represent? What is the significance of each?

8. What is the meaning of (a) noon, (b) A.M. and (c) P.M.?

9. Explain why standard time zones were established. In terms of longitude, how wide is a standard time zone? Why was this width chosen?

10. When it is 12 noon in Texas, what time is it (a) in Jacksonville, Florida; (b) in Bakersfield, California; (c) at the North Pole?

11. Explain why a lunar eclipse is not observed once a month.

12. Using sketches, briefly describe the positions of Earth, Moon, and Sun during each of the major moon phases.

13. If you were on the Moon as people on Earth observed a full moon, in what phase would you observe Earth?

14. What made all the craters that can be observed on the Moon? When did this happen?

15. What phase is the Moon in if it rises at sunset? Explain your reasoning.

16. Why doesn't an eclipse of the Sun occur at each new moon when the Moon is between Earth and the Sun?

17. Does an eclipse of the Sun occur during any particular moon phase? Explain.

18. Identify the moon phases that occur with (a) a spring tide and (b) a neap tide.

19. Explain why there are two tidal bulges on opposite sides of Earth.

FOR FURTHER ANALYSIS

1. What are the significant similarities and differences between a solstice and an equinox?

2. On what date is Earth the closest to the Sun? What season is occurring in the Northern Hemisphere at this time? Explain this apparent contradiction.

3. Explain why an eclipse of the Sun does not occur at each new moon phase when the Moon is between Earth and the Sun.

4. Explain why sundial time is often different than the time as shown by a clock.

5. Analyze why the time between two consecutive tides is twelve hours and twenty-five minutes rather than twelve hours. Explore many different explanations that you imagine, then select the best and explain how it could be tested.

INVITATION TO INQUIRY

Hello, Moon!

Observe where the Moon is located relative to the landscape where you live each day or night that it is visible. Make a sketch of the outline of the landscape. For each date, draw an accurate location of where the Moon is located and note the time. Also, sketch the moon phase for each date.

Continue your observations until you have enough information for analysis. Analyze the data for trends that would enable you to make predictions. For each day, draw some circles to represent the relative positions of Earth, the Moon, and the Sun.

15

Earth

These rose-red rhodochrosite crystals are a naturally occurring form of manganese carbonate. Rhodochrosite is but one of about twenty-five hundred minerals that are known to exist, making up the solid materials of Earth's crust.

CORE **CONCEPT**

Earth is a dynamic body that cycles materials on its surface and between its surface and interior.

OUTLINE

Life Science

▷ Changes in Earth's surface create barriers that lead to the development of new species (Ch. 21).

Earth Science

▷ Changes in magnetic fields are used to determine the sequence of past geologic events (Ch. 15).

▷ Organisms trapped in sediments form fossils in sedimentary rock (Ch. 22).

OVERVIEW

The separation of the earth sciences into independent branches—such as geology, oceanography, and meteorology—was traditionally done for convenience. This made it easier to study a large and complex Earth. In the past, scientists in each branch studied their field without considering Earth as an interacting whole. Today, most earth scientists consider changes in Earth as taking place in an overall dynamic system. The parts of Earth's interior, the rocks on the surface, the oceans, the atmosphere, and the environmental conditions are today understood to be parts of a complex, interacting system with a cyclic movement of materials from one part to another.

How can materials cycle through changes from the interior to the surface and to the atmosphere and back? The answer to this question is found in the unique combination of fluids of Earth. No other known planet has Earth's combination of (1) an atmosphere consisting mostly of nitrogen and oxygen, (2) a surface that is mostly covered with liquid water, and (3) an interior that is partly fluid, partly semi-fluid, and partly solid. Earth's atmosphere is unique in terms of both its composition and interactions with the liquid water surface (figure 15.1). These interactions have cycled materials, such as carbon dioxide, from the atmosphere to the land and oceans of Earth. The internal flow of rock materials, on the other hand, produces the large-scale motion of Earth's continents and the associated phenomena of earthquakes and volcanoes. Volcanoes cycle carbon dioxide back into the atmosphere and the movement of land cycles rocks from Earth's interior to the surface and back to the interior again. Altogether, Earth's atmosphere, liquid water, and motion of its landmasses make up a dynamic cycling system that is found only on the planet Earth.

Earth also seems to be unique because there is life on Earth but apparently not on the other planets. The cycling of atmospheric gases and vapors, waters of the surface, and flowing interior rock materials sustain a wide diversity of life on Earth. There are millions of different species of plants, animals, and other kinds of organisms. Yet, there is no evidence of even one species of life existing outside Earth. The existence of life on Earth must be related to Earth's unique, dynamic system of interacting fluids. This chapter is concerned with earth materials, the internal structure of Earth, the movement of landmasses across the surface of Earth, and how earth materials are recycled.

15.1 EARTH MATERIALS

Earth, like all other solid matter in the universe, is made up of chemical elements. The different elements are not distributed equally throughout the mass of Earth, nor are they equally abundant. Chemical analysis of thousands of rocks from Earth's surface found that only eight elements make up about 98.6 percent of the crust. All the other elements make up the remaining 1.4 percent of the crust. Oxygen is the most abundant element, making up about 50 percent of the weight of the crust. Silicon makes up over 25 percent, so oxygen and silicon alone make up about 75 percent of Earth's solid surface. Figure 15.2 shows the eight most abundant elements that occur as elements or combine to form the chemical compounds of Earth's crust. Earth was apparently molten during an early stage in its development, as evidenced by distribution of the elements. During the molten stage, the heavier abundant elements, such as iron and nickel, apparently sank to the deep interior of Earth, leaving a thin layer of lighter elements on the surface. This relatively thin layer

is called the *crust*. The rocks and rock materials that you see on the surface and the materials sampled in the deepest mines and well holes are all materials of Earth's crust. The bulk of Earth's mass lies below the crust and has not been directly sampled.

Minerals

In everyday usage, the word *mineral* can have several different meanings. It can mean something your body should have (vitamins and minerals), something a fertilizer furnishes for a plant (nitrogen, potassium, and phosphorus), or sand, rock, and coal taken from Earth for human use (mineral resources). In the earth sciences, a **mineral** is defined as a naturally occurring, inorganic solid element or compound with a constant chemical composition in a crystalline structure (figure 15.3). This definition means that the element or compound cannot be synthetic (must be naturally occurring), cannot be made of organic molecules (see chapter 19), and must have atoms arranged in a regular, repeating pattern (a crystal structure). Note that the

FIGURE 15.1 No other planet in the solar system has the unique combination of fluids of Earth. Earth has a surface that is mostly covered with liquid water, water vapor in the atmosphere, and both frozen and liquid water on the land.

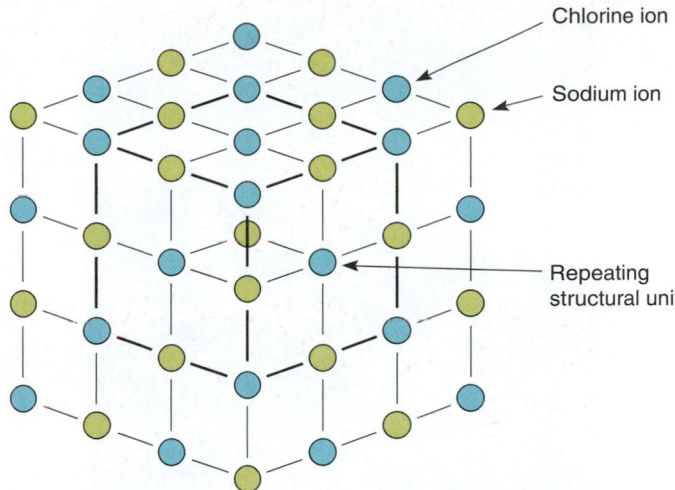

Chlorine ion

Sodium ion

Repeating structural unit

FIGURE 15.3 A crystal is composed of a structural unit that is repeated in three dimensions. This is the basic structural unit of a crystal of sodium chloride, the mineral halite.

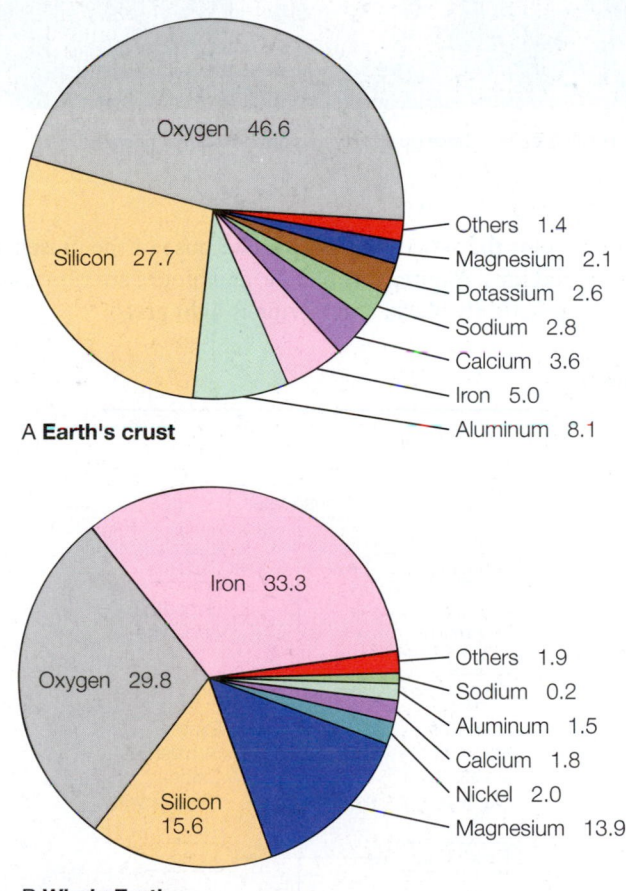

Oxygen 46.6

Silicon 27.7

Others 1.4
Magnesium 2.1
Potassium 2.6
Sodium 2.8
Calcium 3.6
Iron 5.0
Aluminum 8.1

A Earth's crust

Iron 33.3

Oxygen 29.8

Silicon 15.6

Others 1.9
Sodium 0.2
Aluminum 1.5
Calcium 1.8
Nickel 2.0
Magnesium 13.9

B Whole Earth

FIGURE 15.2 (A) The percentage by weight of the elements that make up Earth's crust. (B) The percentage by weight of the elements that make up the whole Earth.

crystal structure of a mineral can be present on the microscopic scale and is not necessarily obvious to the unaided eye. Even crystals that could be observed with the unaided eye are sometimes not noticed (figure 15.4).

The crystal structure of a mineral can be made up of atoms of one or more kinds of elements. Diamond, for example, is a mineral with only carbon atoms in a strong crystal structure. Quartz, on the other hand, is a mineral with atoms of silicon and oxygen in a different crystal structure (figure 15.5). No matter how many kinds of atoms are present, each mineral has its own defined chemical composition or range of chemical compositions. A range of chemical composition is possible because the composition of some minerals can vary with the substitution of chemically similar elements. For example, some atoms of magnesium might be substituted for some chemically similar atoms of calcium. Such substitutions might slightly alter some properties but not enough to make a different mineral.

Silicon and oxygen are the most abundant elements in Earth's crust and, as you would expect, the most common minerals contain these two elements. All minerals are classified on the basis of whether the mineral structure contains these two elements or not. The two main groups are thus called the *silicates* and the *nonsilicates* (table 15.1). Note, however, that the silicates can contain some other elements in addition to silicon and oxygen. The silicate minerals are by far the most abundant, making up about 92 percent of Earth's crust. When an ion of silicon (Si^{-4}) combines with four oxygen ions (O^{-2}), a tetrahedral structure of $(SiO_4)^{-4}$ forms (see figure 15.6). All **silicates** have a basic silicon-oxygen tetrahedral unit either isolated or joined together in the crystal structure. The structure has a total of four unattached electrons on the oxygen atoms that can combine with metallic ions such as iron or magnesium. They can also combine with the silicon atoms of *other* tetrahedral units. Some silicate minerals are thus made up of single tetrahedral units combined with metallic ions. Other silicate minerals are combinations of tetrahedral units combined in single chains, double chains, or sheets (figure 15.7).

The silicate minerals can be conveniently subdivided into two groups based on the presence of iron and magnesium. The basic tetrahedral structure joins with ions of iron, magnesium,

FIGURE 15.4 The structural unit for a crystal of table salt, sodium chloride, is cubic, as you can see in the individual grains.

FIGURE 15.5 These quartz crystals are hexagonal prisms.

calcium, and other elements in the *ferromagnesian silicates.* Examples of ferromagnesian silicates are *olivine, augite, hornblende,* and *biotite* (figure 15.8). They have a greater density and a darker color than the other silicates because of the presence of the metal ions. Augite, hornblende, and biotite are very dark in color, practically black, and olivine is light green.

TABLE 15.1   Classification Scheme of Some Common Minerals

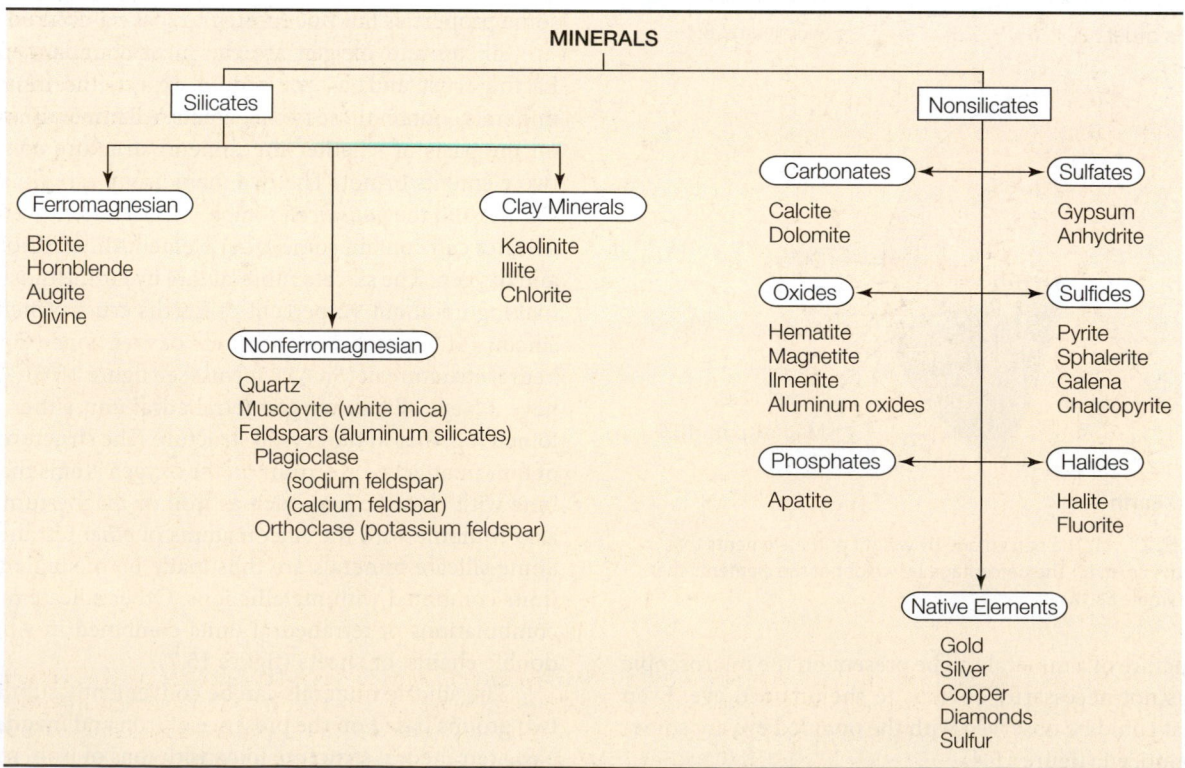

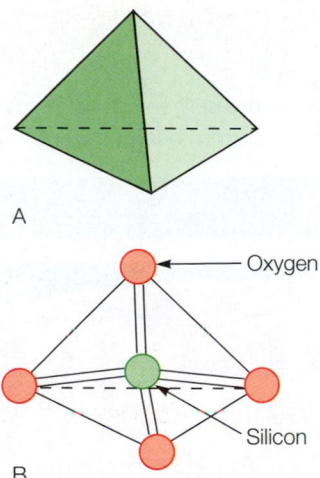

FIGURE 15.6 (A) The geometric shape of a tetrahedron with four equal sides. (B) A silicon and four oxygen atoms are arranged in the shape of a tetrahedron with the silicon in the center. This is the basic building block of all silicate minerals.

CONCEPTS APPLIED

Grow Your Own

Experiment with growing crystals from solutions of alum, copper sulfate, salt, or potassium permanganate. Write a procedure that will tell others what the important variables are for growing large, well-formed crystals.

The *nonferromagnesian silicates* have a light color and a low density compared to the ferromagnesians. This group includes the minerals *muscovite (white mica)*, the *feldspars*, and *quartz* (figure 15.9).

The remaining 8 percent of minerals making up Earth's crust that do not have silicon-oxygen tetrahedrons in their crystal structure are called *nonsilicates*. There are eight subgroups of nonsilicates: (1) carbonates, (2) sulfates, (3) oxides, (4) sulfides, (5) halides, (6) phosphates, (7) hydroxides, and (8) native elements. Some of these are identified in table 15.1. The carbonates are the most abundant of the nonsilicates, but others are important as fertilizers, sources of metals, and sources of industrial chemicals.

Rocks

Elements are *chemically* combined to make minerals. Minerals are *physically* combined to make rocks. A **rock** is defined as an aggregation of one or more minerals and perhaps other materials that have been brought together into a cohesive solid. These materials include volcanic glass, a silicate that is not considered a mineral because it lacks a crystalline structure. Thus, a rock can consist of one or more kinds of minerals that are somewhat

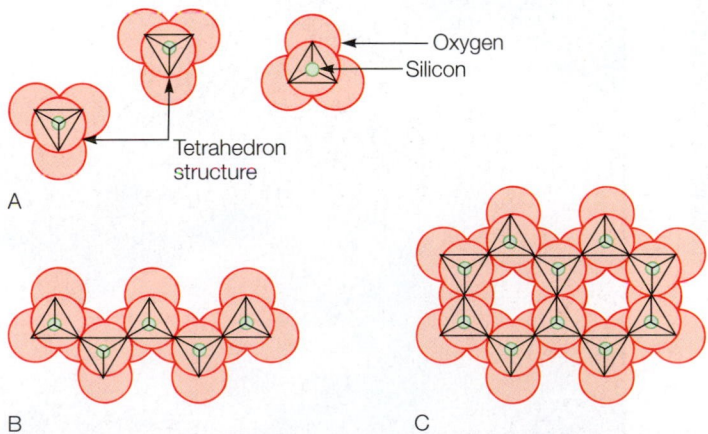

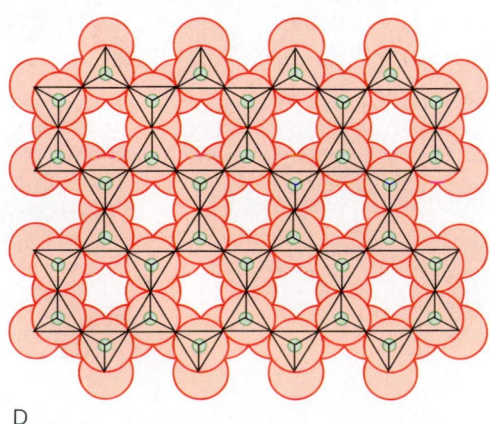

FIGURE 15.7 (A) Isolated silicon-oxygen tetrahedra do not share oxygens. This structure occurs in the mineral olivine. (B) Single chains of tetrahedra are formed by each silicon ion having two oxygens all to itself and sharing two with other silicons at the same time. This structure occurs in augite. (C) Double chains of tetrahedra are formed by silicon ions sharing either two or three oxygens. This structure occurs in hornblende. (D) The sheet structure in which each silicon shares three oxygens occurs in the micas, resulting in layers that pull off easily because of cleavage between the sheets.

A Closer Look

Asbestos

Asbestos is a common name for any of several minerals that can be separated into fireproof fibers, fibers that will not melt or ignite. The fibers can be woven into a fireproof cloth, used directly as fireproof insulation material, or mixed with plaster or other materials. People now have a fear of all asbestos because it is presumed to be a health hazard (box figure 15.1). However, there are about six commercial varieties of asbestos. Five of these varieties are made from an amphibole mineral and are commercially called "brown" and "blue" asbestos. The other variety is made from *chrysotile*, a *serpentine* family of minerals, and is commercially called "white" asbestos. White asbestos is the asbestos mined and most commonly used in North America. It is only the amphibole asbestos (brown and blue asbestos) that has been linked to cancer, even for a short exposure time. There is, however, no evidence that exposure to white asbestos results in an increased health hazard. It makes sense to ban the use of and remove all the existing amphibole asbestos from public buildings. It does not make sense to ban or remove the serpentine asbestos since it is not a proven health hazard.

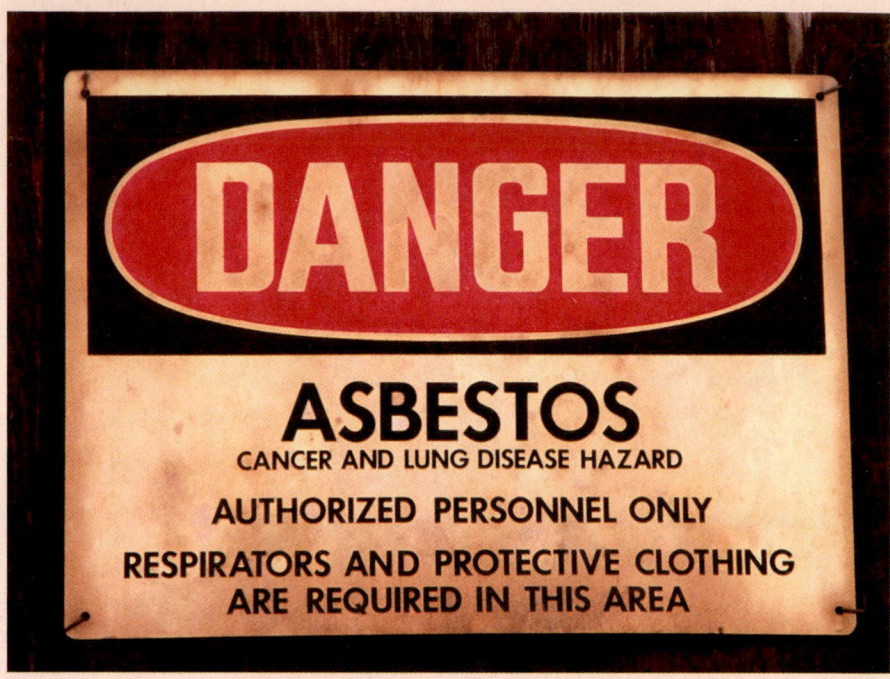

Box Figure 15.1 Did you know there are different kinds of asbestos? Are all kinds of asbestos a health hazard?

FIGURE 15.8 Compare the dark colors of the ferromagnesian silicates augite (right), hornblende (left), and biotite to the light-colored nonferromagnesian silicates in figure 15.9.

FIGURE 15.9 Compare the light colors of the nonferromagnesian silicates mica (front center), white and pink orthoclase (top and center), and quartz (left center) to the dark-colored ferromagnesian silicates in figure 15.8.

Most people understand the fact that our mineral resources are limited, and when we use them, they are gone. Of course, some mineral resources can be recycled, reducing the need to mine increasingly more minerals. For example, aluminum can be recycled and used over and over again repeatedly. Glass, copper, iron, and other metals can similarly be recycled repeatedly. Other critical resources, however, cannot be recycled and, unfortunately, cannot be replaced. Crude oil, for example, is a dwindling resource that will eventually become depleted. Oil is not recyclable once it is burned, and no new supplies are being created, at least not at a rate that would make them available in the immediate future. Even if Earth were a hollow vessel completely filled with oil, it would eventually become depleted, perhaps sooner than you might think.

There is also another of our mineral resources that is critically needed for our survival but will probably soon be depleted. That resource is phosphorus derived from phosphate rock. Phosphorus is an essential nutrient required for plant growth, and if its concentration in soils is too low, plants grow poorly, if at all. Most agricultural soils are artificially fertilized with phosphate. Without this amendment, their productivity would decline and, in some cases, cease altogether.

Phosphate occurs naturally as the mineral apatite. Deposits of apatite were formed where ocean currents carried cold, deep water rich in dissolved phosphate ions to the upper continental slope and outer continental shelf. Here, phosphate ions replaced the carbonate ions in limestone, forming the mineral apatite. Apatite also occurs as a minor accessory mineral in most igneous, sedimentary, and metamorphic rock types. Some igneous rocks serve as a source of phosphate fertilizer, but most phosphate is mined from formerly submerged coastal areas of limestone, such as those found in Florida.

Trends in phosphate production and use over the past forty years suggest that the proved world reserves of phosphate rock will be exhausted within a few decades. New sources might be discovered, but sometime soon, phosphate rock will be mined out and no longer available. When this happens, the food supply will have to be grown on lands that are naturally endowed with an adequate phosphate supply. Estimates are that this worldwide existing land area with adequate phosphate supply will supply food for only two billion people on the entire Earth. Phosphate is an essential element for all life on Earth, and no other element can function in its place.

Questions to Discuss:

Discuss with your group the following questions concerning the use of mineral resources:

1. Should the mining industry be permitted to exhaust an important mineral resource? Provide reasons with your answer.
2. What are the advantages and disadvantages of a controlled mining industry?
3. If phosphate mineral supplies become exhausted, who should be responsible for developing new supplies or substitutes: the mining industry or governments?

"glued" together by other materials such as glass. Most rocks are composed of silicate minerals, as you might expect since most minerals are silicates.

There is a classification scheme that is based on the way the rocks were formed. There are three main groups: (1) *igneous rocks* formed as a hot, molten mass of rock materials cooled and solidified; (2) *sedimentary rocks* formed from particles or dissolved materials from previously existing rocks; and (3) *metamorphic rocks* formed from rocks that were subjected to high temperatures and pressures that deformed or recrystallized the rock without complete melting.

Igneous Rocks

The word *igneous* comes from the Latin *ignis,* which means "fire." This is an appropriate name for **igneous rocks** that are defined as rocks that formed from a hot, molten mass of melted rock materials. The first step in forming igneous rocks is the creation of a very high temperature that is hot enough to melt rocks. A mass of melted rock materials is called *magma.* Magma may cool and crystallize to solid igneous rock either below or on the surface of Earth. Earth has a history of beginning as a molten material, all rocks of Earth were at one time igneous rocks. Today, about two-thirds of the outer layer, or crust, is made up of igneous rocks. This is not apparent in many loca-tions because the surface is covered by other kinds of rocks and rock materials (sand, soil, etc.). When they are found, igneous rocks are usually granite or basalt. *Granite* is a light-colored igneous rock that is primarily made up of three silicate minerals: quartz, mica, and feldspar. You can see the grains of these three minerals in a freshly broken surface of most samples of granite (figure 15.10). *Basalt* is a dark, more dense igneous rock with mineral grains too tiny to see. Basalt is the most common volcanic rock found on the surface of Earth.

Sedimentary Rocks

Sedimentary rocks are rocks that formed from particles or dissolved materials from previously existing rocks. Chemical reactions with air and water tend to break down and dissolve the less chemically stable parts of existing rocks, freeing more stable particles and grains in the process. The remaining particles are usually transported by moving water and deposited as sediments. *Sediments* are accumulations of silt, clay, sand, or other weathered materials. Weathered rock fragments and dissolved rock materials both contribute to sediment deposits (table 15.2).

Most sediments are deposited as many separate particles that accumulate as loose materials. Such accumulations of rock fragments, chemical deposits, or animal shells must become consolidated into a solid, coherent mass to become sedimentary rock.

FIGURE 15.10 Detailed view of granite that clearly shows intergrown mineral crystals of potassium feldspar, quartz, biotite, and plagioclase.

TABLE 15.2 Classification Scheme for Sedimentary Rocks

Sediment Type	Particle or Composition	Rock
Fragment	Larger than sand	Conglomerate or breccia
Fragment	Sand	Sandstone
Fragment	Silt and clay	Siltstone, claystone, or shale
Chemical	Calcite	Limestone
Chemical	Dolomite	Dolomite
Chemical	Gypsum	Gypsum
Chemical	Halite (sodium chloride)	Salt

There are two main parts to this *lithification,* or rock-forming process: (1) compaction and (2) cementation (figure 15.11).

The weight of an increasing depth of overlying sediments causes an increasing pressure on the sediments below. This pressure squeezes the deeper sediments together, gradually reducing the pore space between the individual grains. This *compaction* of the grains reduces the thickness of a sediment deposit, squeezing out water as the grains are packed more tightly together. Compaction alone is usually not enough to make loose sediment into solid rock. Cementation is needed to hold the compacted grains together.

In *cementation,* the spaces between the sediment particles are filled with a chemical deposit. The chemical deposit binds the particles together into the rigid, cohesive mass of a sedimentary rock. Compaction and cementation may occur at the same time, but the cementing agent must have been introduced before compaction restricts the movement of the fluid through the open spaces. Many soluble minerals can serve as cementing agents, and calcite (calcium carbonate) and silica (silicon dioxide) are common.

Sediments accumulate from rocks that are in various stages of being broken down, so there is a wide range of sizes of particles in sediments. The size and shape of the particles are used as criteria to name the sedimentary rocks (table 15.3).

Metamorphic Rocks

The third group of rocks is called metamorphic. **Metamorphic rocks** are previously existing rocks that have been changed by heat, pressure, or hot solutions into a distinctly different rock. The heat, pressure, or hot solutions that produced the changes are associated with geologic events of (1) movement of the crust, which will be discussed in chapter 16, and with (2) heating and hot solutions from magma. Pressures from movement of the crust can change the rock texture by flattening, deforming, or realigning mineral grains. Temperatures from magma must be just right to

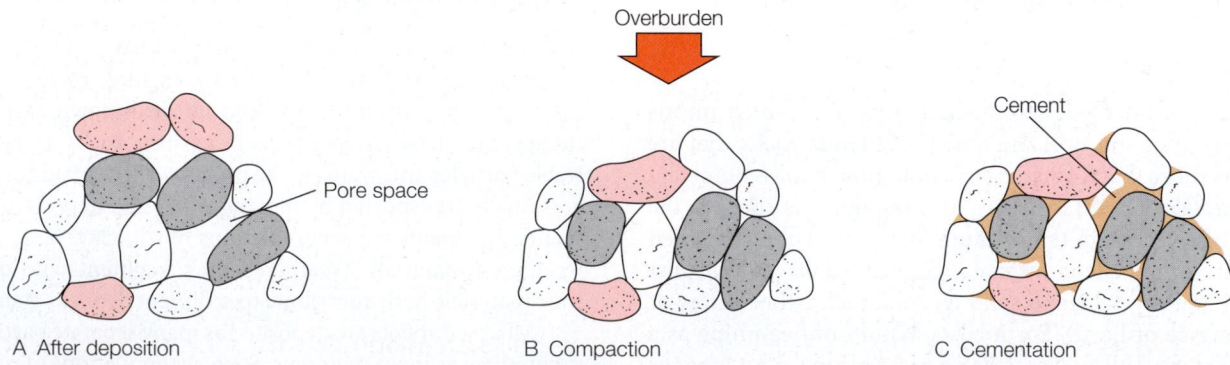

Overburden

Pore space

A After deposition

B Compaction

Cement

C Cementation

FIGURE 15.11 Cementation of sand grains to become sandstone. (A) Loose sand grains are deposited with open pore space between the grains. (B) The weight of overburden compacts the sand into a tighter arrangement, reducing pore space. (C) Precipitation of cement in the pores by groundwater binds the sand into the rock sandstone, which has a clastic texture.

TABLE 15.3 Simplified Classification Scheme for Sediment Fragments and Rocks

Sediment Name	Size Range	Rock
Boulder	Over 256 mm (10 in)	
Gravel	2 to 256 mm (0.08—10 in)	Conglomerate or breccia*
Sand	1/16 to 2 mm (0.025—0.08 in)	Sandstone
Silt (or dust)	1/256 to 1/16 mm (0.00015—0.025 in)	Siltstone**
Clay (or dust)	Less than 1/256 mm (less than 0.00015 in)	Claystone**

*Conglomerate has a rounded fragment; breccia has an angular fragment.
**Both are also known as mudstone; called shale if it splits along parallel planes.

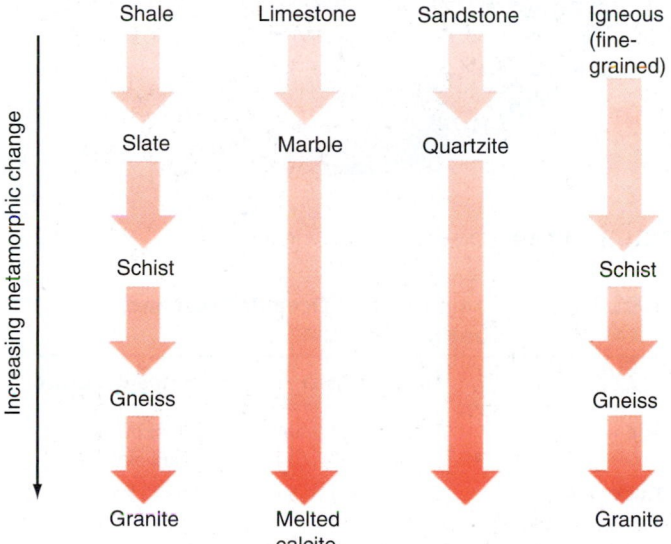

FIGURE 15.12 Increasing metamorphic change occurs with increasing temperatures and pressures. If the melting point is reached, the change is no longer metamorphic, and igneous rocks are formed.

CONCEPTS APPLIED

Sand Sort

Collect dry sand from several different locations. Use a magnifying glass to determine the minerals found in each sample.

---

produce a metamorphic rock. They must be high enough to disrupt the crystal structures to cause them to recrystallize but not high enough to melt the rocks and form igneous rocks (figure 15.12).

The Rock Cycle

Earth is a dynamic planet with a constantly changing surface and interior. As you will see in chapters 16, 17, and 18, internal changes alter Earth's surface by moving the continents and, for example, building mountains that are eventually worn away by weathering and erosion. Seas advance and retreat over the continents as materials are cycled from the atmosphere to the

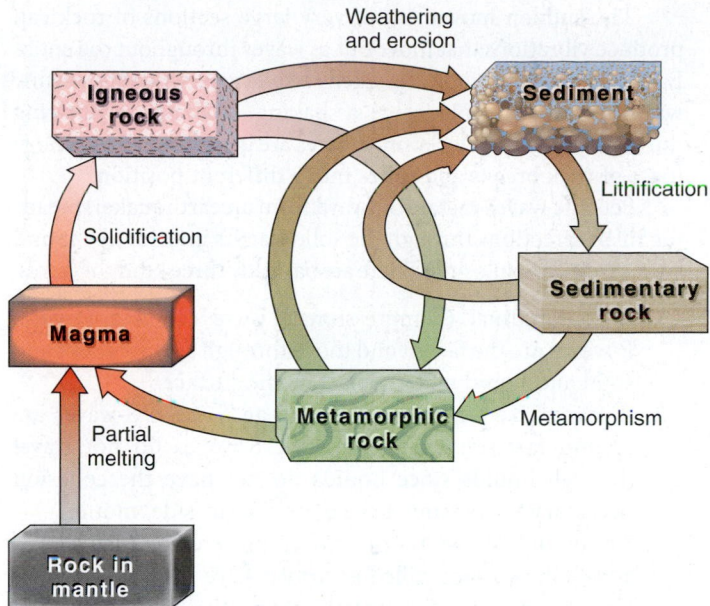

FIGURE 15.13 The rock cycle.

land and from the surface to the interior of Earth and then back again. Rocks are transformed from one type to another through this continual change. There is not a single rock on Earth's surface today that has remained unchanged through Earth's long history. The concept of continually changing rocks through time is called the *rock cycle* (figure 15.13). The rock cycle concept views an igneous, a sedimentary, or a metamorphic rock as the present but temporary stage in the ongoing transformation of rocks to new types. Any particular rock sample today has gone through countless transformations in the 4.6-billion-year history of Earth and will continue to do so in the future.

Chapter 22, The History of Life on Earth, is about how rocks are studied to find information about the history of Earth and the organisms on it. This is one of the main reasons that geologists are interested in rocks. For example, sedimentary rocks can be studied to learn about how the environment and the kinds of organisms found on Earth have changed over time.

15.2 EARTH'S INTERIOR

Earth is not a solid, stable sphere that is "solid as a rock." All rocks and rock materials can be made to flow, behaving as warm wax or putty that can be molded. Pressure and temperature just a few kilometers below the surface can be high enough so that rock flows very slowly. Rock flowage takes place as hot, buoyant material deep within Earth moves slowly upward toward the cooler surface. Elsewhere, cold, denser material moves downward.

Earth's internal heat and rock movement are related to some things that happen on the surface, such as moving continents, earthquakes, and volcanoes. Before considering how this happens, we will look at Earth's interior and some internal structures. Information about Earth's interior is deduced mostly from a study of earthquake waves as they move through different parts of Earth. Earthquake waves furnish a sort of "X-ray" machine to view Earth's interior.

The sudden movement of very large sections of rock can produce vibrations that move out as waves throughout the entire Earth. These vibrations are called *seismic waves*. Strong seismic waves are what people feel as a shaking, quaking, or vibrating during an earthquake. Seismic waves are generated when a huge mass of rock breaks and slides into a different position.

Seismic waves radiate outward from an earthquake, spreading in all directions through the solid Earth's interior like sound waves from an explosion. There are basically three kinds of waves:

1. A longitudinal (compressional) wave called a *P-wave*. P-waves are the fastest and move through surface rocks and solid and liquid materials below the surface.
2. A transverse (shear) wave called an *S-wave*. S-waves are second fastest after the P-wave. S-waves do not travel through liquids since liquids do not have the cohesion necessary to transmit a shear, or side-to-side, motion.
3. An up-and-down (crest and trough) wave that travels across the surface called a *surface wave* that is much like a wave on water that moves across the solid surface of Earth. Surface waves are the slowest and occur where S- or P-waves reach the surface. There are two important types of surface waves: *Love waves* and *Rayleigh waves*. Love waves are horizontal S-waves that move side to side. This motion knocks buildings off their foundations and can also destroy bridges and overpasses. Rayleigh waves are more like rolling water waves. Rayleigh waves are more destructive because they produce more up, down, and sideways ground movement for a longer time.

Using data from seismic waves, scientists were able to determine that the interior of Earth can be broken down into three parts (figure 15.14). The *crust* is the outer layer of rock that forms a thin shell around Earth. Below the crust is the *mantle*, a much thicker shell than the crust. The mantle separates the crust from the center part, which is called the *core*. The following section starts on Earth's surface, at the crust, and then digs deeper and deeper into Earth's interior.

The Crust

Earth's **crust** is a thin layer of rock that covers the entire Earth, existing below the oceans as well as making up the continents. According to seismic waves, there are differences in the crust making up the continents and the crust beneath the oceans (table 15.4). These differences are (1) the oceanic crust is much thinner than the continental crust and (2) seismic waves move through the oceanic crust faster than they do through the continental crust. The two types of crust vary because they are made up of different kinds of rock.

The boundary between the crust and the mantle is marked by a sharp increase in the velocity of seismic waves as they pass from the crust to the mantle. Today, this boundary is called the *Mohorovicic discontinuity*, or the "Moho" for short. The boundary is a zone where seismic P-waves increase in velocity because of changes in the composition of the materials. The increase occurs because the composition on both sides of the boundary is different. The mantle is richer in ferromagnesian minerals and poorer in silicon than the crust.

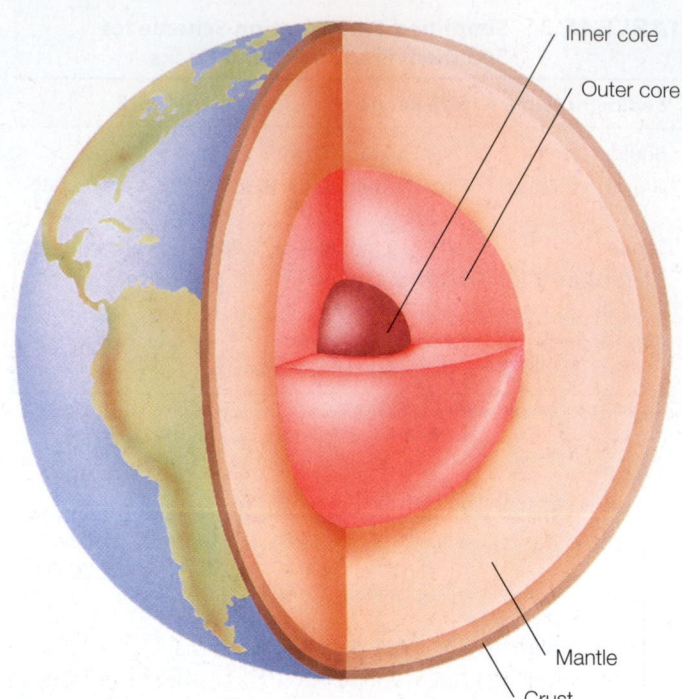

FIGURE 15.14   The structure of Earth's interior.

TABLE 15.4   **Comparison of Oceanic Crust and Continental Crust**

	Oceanic Crust	**Continental Crust**
Age	Less than 200 million years	Up to 3.8 billion years old
Thickness	5 to 8 km (3 to 5 mi)	10 to 70 km (6 to 45 mi)
Density	3.0 g/cm³	2.7 g/cm³
Composition	Basalt	Granite, schist, gneiss

Studies of the Moho show that the crust varies in thickness around Earth's surface. It is thicker under the continents and much thinner under the oceans.

The age of rock samples from Earth's continents has been compared with the age samples of rocks taken from the seafloor by oceanographic ships. This sampling has found the continental crust to be much older, with parts up to 3.8 billion years old. By comparison, the oldest oceanic crust is less than 200 million years old.

Comparative sampling also found that continental crust is a less dense, granite-type rock with a density of about 2.7 g/cm³. Oceanic crust, on the other hand, is made up of basaltic rock with a density of about 3.0 g/cm³. The less dense crust behaves as if it were floating on the mantle, much as less dense ice floats on water. There are exceptions, but in general, the thicker, less dense continental crust "floats" in the mantle above sea level and the thin, dense oceanic crust "floats" in the mantle far below sea level (figure 15.15).

The Mantle

The middle part of Earth's interior is called the **mantle.** The mantle is a thick shell between the core and the crust. This

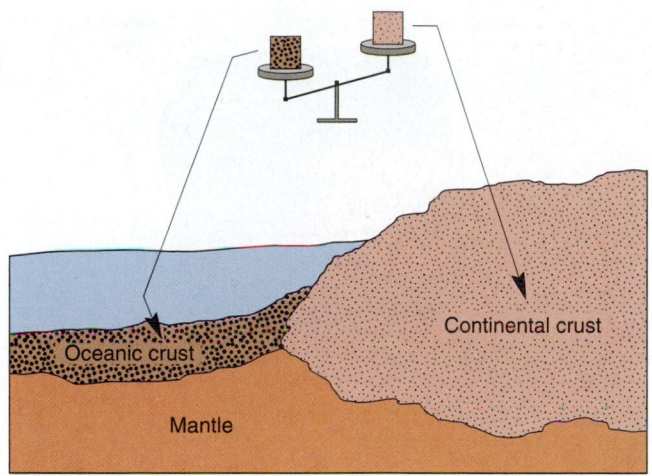

FIGURE 15.15 Continental crust is less dense, granite-type rock, while the oceanic crust is more dense, basaltic rock. Both types of crust behave as if they were floating on the mantle, which is more dense than either type of crust.

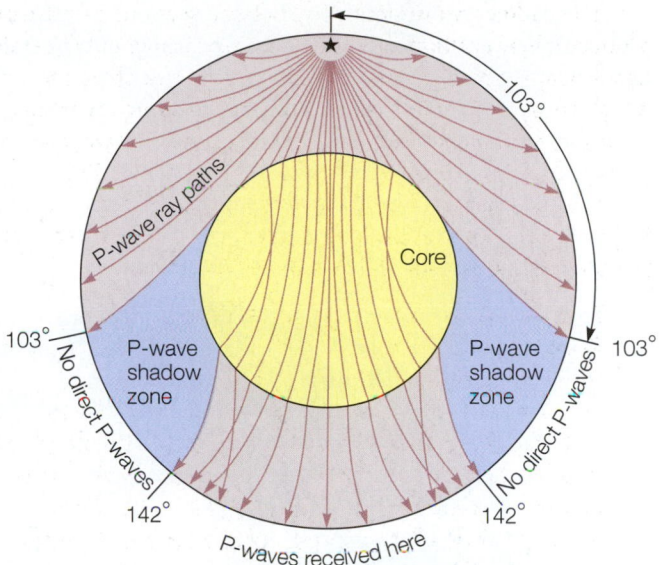

FIGURE 15.16 The P-wave shadow zone, caused by refraction of P-waves within Earth's core.

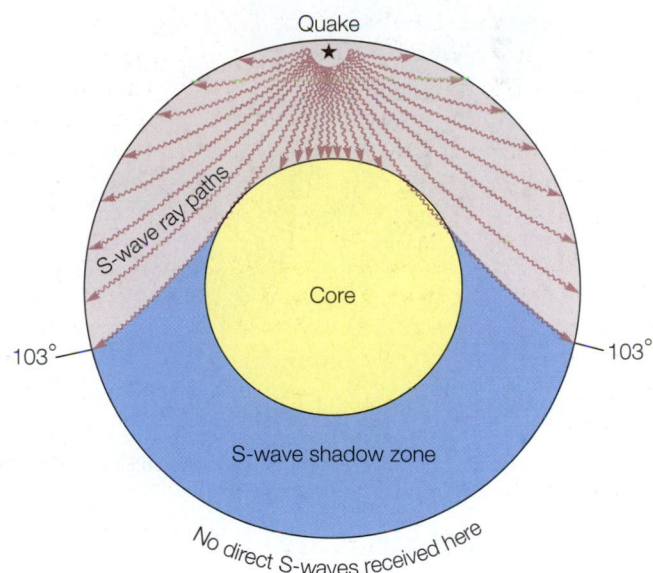

FIGURE 15.17 The S-wave shadow zone. Since S-waves cannot pass through a liquid, at least part of the core is either a liquid or has some of the same physical properties as a liquid.

shell takes up about 80 percent of the total volume of Earth and accounts for about two-thirds of Earth's total mass. Information about the composition and nature of the mantle comes from (1) studies of seismological data, (2) studies of the nature of meteorites, and (3) studies of materials from the mantle that have been ejected to Earth's surface by volcanoes. The evidence from these separate sources all indicates that the mantle is composed of silicates, predominantly the ferromagnesian silicate *olivine.* Meteorites, as mentioned in chapter 13, are basically either iron meteorites or stony meteorites. Most of the stony meteorites are silicates with a composition that would produce the chemical composition of olivine if they were melted and the heavier elements were separated by gravity. This chemical composition also agrees closely with the composition of basalt, the most common volcanic rock found on the surface of Earth.

The Core

Information about the nature of the **core,** the center part of Earth, comes from several sources of information. Seismological data provide the primary evidence for the structure of the core of Earth. Seismic P-waves spread through Earth from a large earthquake. Figure 15.16 shows how the P-waves spread out, soon arriving at seismic measuring stations all around the world. However, there are places between 103° and 142° of arc from the earthquake that do not receive P-waves. This region is called the *P-wave shadow zone.* The P-wave shadow zone is explained by P-waves being refracted by the core, leaving a shadow. The paths of P-waves can be accurately calculated, so the size and shape of Earth's core can also be accurately calculated.

Seismic S-waves leave a different pattern at seismic receiving stations around Earth. S- (sideways or transverse) waves can travel only through solid materials. An *S-wave shadow zone* also exists and is larger than the P-wave shadow zone (figure 15.17). S-waves are not recorded in the entire region more than 103° away from the epicenter. The S-wave shadow zone seems to indicate that S-waves do not travel through the core at all. If this is true, it implies that the core of Earth is a liquid, or at least acts like a liquid.

Analysis of P-wave data suggests that the core has two parts: a *liquid outer core* and a *solid inner core.* Both the P-wave and S-wave data support this conclusion. Overall, the core makes up about 15 percent of Earth's total volume and about one-third of its mass.

Evidence from the nature of meteorites indicates that Earth's core is mostly iron. Earth has a strong magnetic field that has its sources in the turbulent flow of the liquid part of its core. To produce such a field, the material of the core would have to be an electrical conductor, that is, a metal such as iron. There are two general kinds of meteorites that fall to Earth: (1) stony meteorites that are made of silicate minerals and (2) iron meteorites that are made of iron or of a nickel-iron alloy. Since Earth has a silicate-rich crust and mantle, by analogy, Earth's core must consist of iron or a nickel and iron alloy.

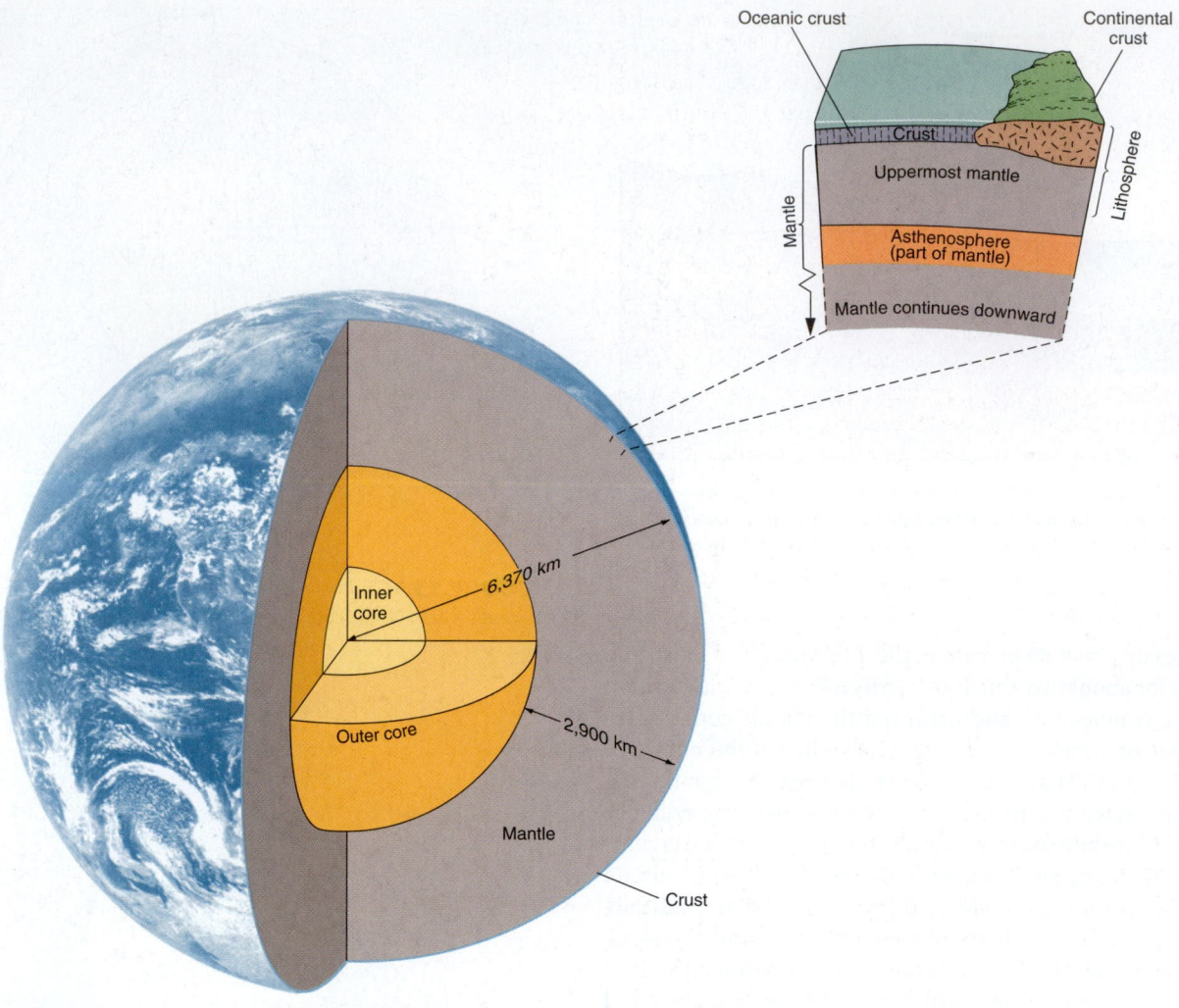

FIGURE 15.18 Earth's interior, showing the weak, plastic layer called the *asthenosphere*. The rigid, solid layer above the asthenosphere is called the *lithosphere*. The lithosphere is broken into plates that move on the upper mantle like giant ice sheets floating on water. This arrangement is the foundation for plate tectonics, which explains many changes that occur on Earth's surface such as earthquakes, volcanoes, and mountain building.

A More Detailed Structure

There is strong evidence that Earth has a layered structure with a core, mantle, and crust. This description of the structure is important for historical reasons and for understanding how Earth evolved over time. There is another, more detailed structure that can be described. This structure is far more important in understanding the history and present appearance of Earth's surface, including the phenomena of earthquakes and volcanoes.

The important part of this different detailed description of Earth's interior was first identified from seismic data. There is a thin zone in the mantle where seismic waves undergo a sharp decrease in velocity (figure 15.18). This low-velocity zone is evidently a hot, elastic semiliquid layer that extends around the entire Earth. It is called the **asthenosphere,** after the Greek word for "weak shell." The asthenosphere is weak because it is plastic, mobile, and yields to stresses. In some regions, the asthenosphere contains pockets of magma.

The rocks above and below the asthenosphere are rigid, solid, and brittle. The solid layer above the asthenosphere is called the **lithosphere,** after the Greek word for "stone shell."

The lithosphere is also known as the "strong layer" in contrast to the "weak layer" of the asthenosphere. The lithosphere includes the entire crust, the Moho, and the upper part of the mantle. As you will see in the section on plate tectonics, the asthenosphere is one important source of magma that reaches Earth's surface. It is also a necessary part of the mechanism involved in the movement of the crust. The lithosphere is made up of comparatively rigid plates that are moving, floating in the upper mantle like giant ice sheets floating in the ocean.

15.3 PLATE TECTONICS

If you observe the shape of the continents on a world map or a globe, you will notice that some of the shapes look as if they would fit together like the pieces of a puzzle. The most obvious is the eastern edge of North and South America, which seems to fit the western edge of Europe and Africa in a slight S-shaped curve. Such patterns between continental shapes seem to suggest that the continents were at one time together, breaking apart and moving to their present positions some time in the past (figure 15.19).

In the early 1900s, a German scientist named Alfred Wegener became enamored with the idea that the continents had shifted positions and published papers on the subject for nearly two decades. Wegener supposed that at one time there was a single large landmass that he called "Pangaea," which is from the Greek word meaning "all lands." He pointed out that similar fossils found in landmasses on both sides of the Atlantic Ocean today must be from animals and plants that lived in Pangaea, which later broke up and split into smaller continents. Wegener's concept came to be known as *continental drift,* the idea that individual continents could shift positions on Earth's surface. Some people found the idea of continental drift plausible, but most had difficulty imagining how a huge and massive continent could "drift" around on a solid Earth. Since Wegener had provided no good explanation of why or how continents might do this, most scientists found the concept unacceptable. The concept of continental drift was dismissed as an interesting but odd idea. Then new evidence indicated that the continents had indeed moved. The first of this evidence came from the bottom of the ocean and led to a new, broader theory about movement of Earth's crust.

Evidence from Earth's Magnetic Field

Earth's magnetic field is probably created by electric currents within the slowly circulating liquid part of the iron core. However, there is nothing static about Earth's magnetic poles. Geophysical studies have found that the magnetic poles are moving slowly around the geographic poles. Studies have also found that Earth's magnetic field occasionally undergoes magnetic reversal. Magnetic reversal is the flipping of polarity of Earth's magnetic field. During a magnetic reversal, the north magnetic pole and the south magnetic pole exchange positions. The present magnetic field orientation has persisted for the past 700,000 years and, according to the evidence, is now preparing for another reversal. The evidence, such as the magnetized iron particles found in certain Roman ceramic artifacts, shows that the magnetic field was 40 percent stronger two thousand years ago than it is today. If the present rate of weakening were to continue, Earth's magnetic field would be near zero by the end of the next two thousand years—if it weakens that far before reversing orientation and then increases to its usual value.

Many igneous rocks contain a record of the strength and direction of Earth's magnetic field at the time the rocks formed. Iron minerals, such as magnetite (Fe_3O_4), crystallize in a cooling magma and become magnetized and oriented to Earth's magnetic field, like tiny compass needles, at the time they were formed. Such rocks thus provide evidence of the direction and distance to Earth's ancient magnetic poles. The study of ancient magnetism, called *paleomagnetics,* provides the information that Earth's magnetic field has undergone twenty-two magnetic reversals during the past 4.5 million years (figure 15.20).

The record shows the time between pole flips is not consistent, sometimes reversing in as little as ten thousand years and sometimes taking as long as twenty-five million years. Once a reversal starts, however, it takes about five thousand years to complete the process.

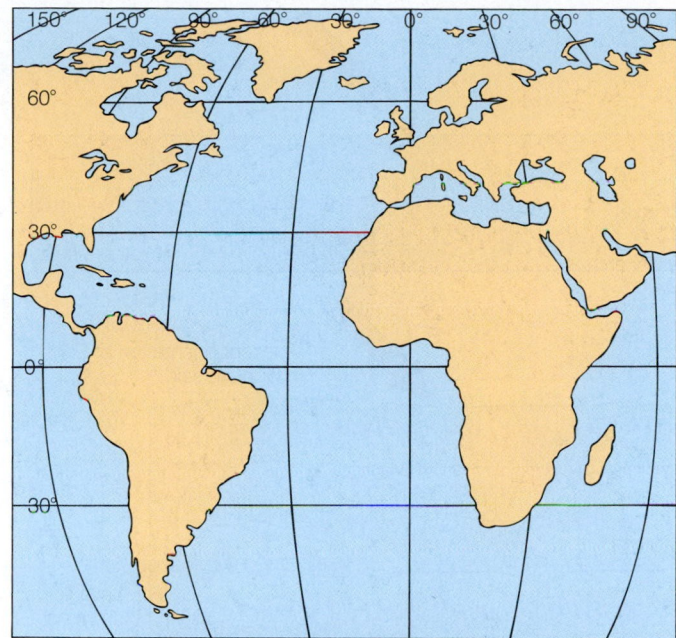

A

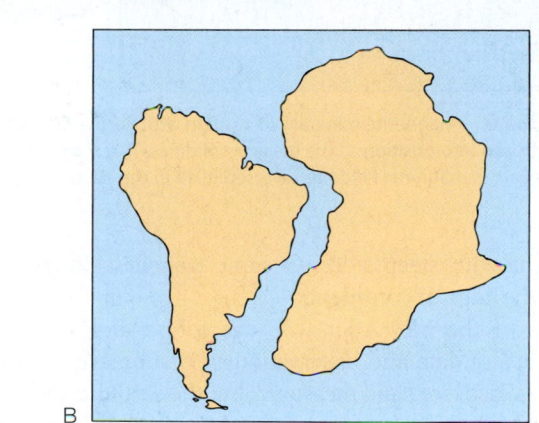

B

FIGURE 15.19 (A) Normal position of the continents on a world map. (B) A sketch of South America and Africa, showing how they once might have been joined together and subsequently separated by a continental drift.

Evidence from the Ocean

The first important studies concerning the movement of continents came from studies of the ocean basin, the bottom of the ocean floor. The basins are covered by 4 to 6 km (about 3 to 4 mi) of water and were not easily observed during Wegener's time. It was not until the development and refinement of sonar and other new technologies that scientists began to learn about the nature of the ocean basin. They found that it was not the flat, featureless plain that many had imagined. There are valleys, hills, mountains, and mountain ranges. Long, high, and continuous chains of mountains that seem to run clear around Earth were discovered, and these chains are called **oceanic ridges.** The *Mid-Atlantic Ridge* is one such oceanic ridge, which is located in the center of the Atlantic Ocean basin. The Mid-Atlantic Ridge divides the Atlantic Ocean into two nearly equal parts. Where it is high enough to reach sea level, it makes oceanic islands such as Iceland (figure 15.21). The basins also contain **oceanic trenches.** These trenches are long, narrow, and

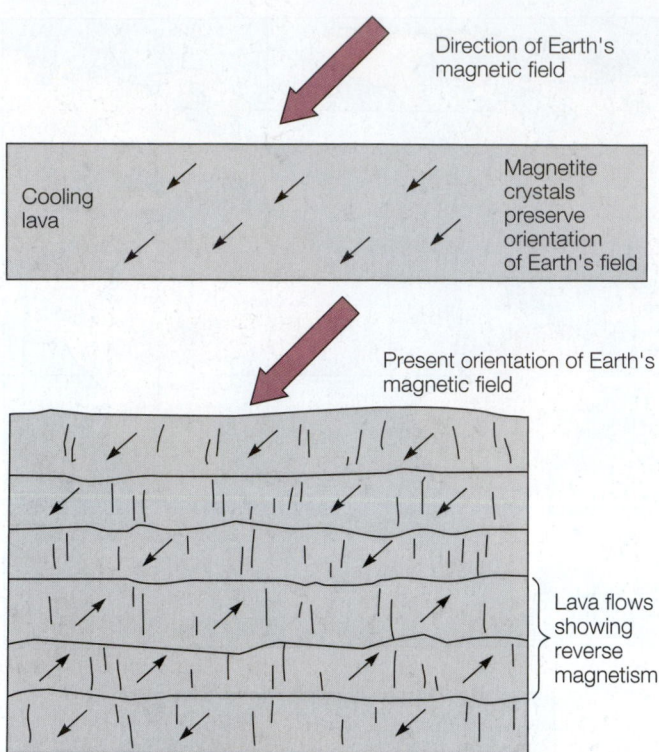

Direction of Earth's magnetic field

Cooling lava

Magnetite crystals preserve orientation of Earth's field

Present orientation of Earth's magnetic field

Lava flows showing reverse magnetism

FIGURE 15.20 Magnetite mineral grains align with Earth's magnetic field and are frozen into position as the magma solidifies. This magnetic record shows Earth's magnetic field has reversed itself in the past.

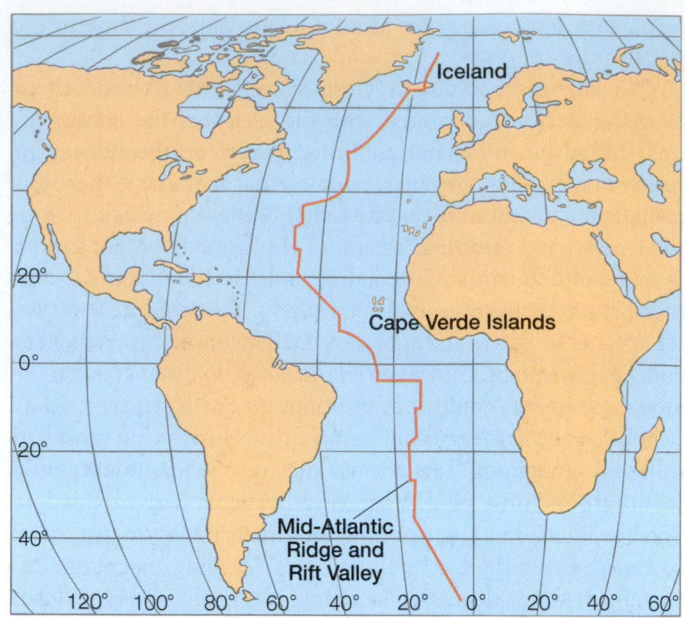

Iceland

Cape Verde Islands

Mid-Atlantic Ridge and Rift Valley

120° 100° 80° 60° 40° 20° 0° 20° 40° 60°

FIGURE 15.21 The Mid-Atlantic Ridge divides the Atlantic Ocean into two nearly equal parts. Where the ridge reaches above sea level, it makes oceanic islands, such as Iceland.

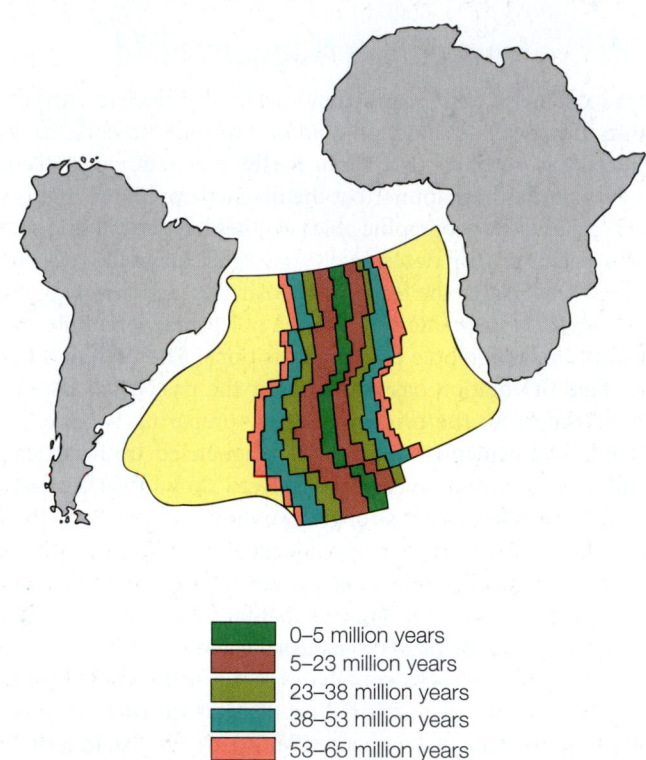

- 0–5 million years
- 5–23 million years
- 23–38 million years
- 38–53 million years
- 53–65 million years
- 65–135 million years

FIGURE 15.22 The pattern of seafloor ages on both sides of the Mid-Atlantic Ridge reflects seafloor spreading activity. Younger rocks are found closer to the ridge.

deep troughs with steep sides. Oceanic trenches always run parallel to the edge of continents.

Studies of the Mid-Atlantic Ridge found at least three related groups of data and observations: (1) submarine earthquakes were discovered and measured, but the earthquakes were all observed to occur mostly in a narrow band under the crest of the Mid-Atlantic Ridge; (2) a long, continuous **rift** was observed to run along the crest of the Mid-Atlantic Ridge for its length; and (3) a large amount of heat was found to be escaping from the rift. One explanation of the related groups of findings is that the rift might be a crack in Earth's crust, a fracture through which lava flowed to build up the ridge. The evidence of excessive heat flow, earthquakes along the crest of the ridge, and the very presence of the ridge all led to a **seafloor spreading** hypothesis. This hypothesis explained that hot, molten rock moved up from the interior of Earth to emerge along the rift, flowing out in both directions to create new rocks along the ridge. The creation of new rock like this would tend to spread the seafloor in both directions, thus the name. The test of this hypothesis would come from further studies on the ages and magnetic properties of the seafloor along the ridge (figure 15.22).

Evidence of the age of sections of the seafloor was obtained by drilling into the ocean floor from a research ship. From these drillings, scientists were able to obtain samples of fossils and sediments at progressive distances outward from the Mid-Atlantic Ridge. They found thin layers of sediments near the ridge that became progressively thicker toward the continents. This is a pattern you would expect if the seafloor were spreading, because older layers would have more time to accumulate greater depths of sediments. The fossils and sediments in the bottom of

the layer were also progressively older at increasing distances from the ridge. The oldest, which were about 150 million years old, were near the continents. This would seem to indicate that the Atlantic Ocean did not exist until 150 million years ago. At that time, a fissure formed between Africa and South America, and new materials have been continuously flowing, adding new lithosphere to the edges of the fissure.

More convincing evidence for the support of seafloor spreading came from the paleomagnetic discovery of patterns of magnetic strips in the rocks of the ocean floor. Earth's magnetic field has been reversed many times in the last 150 million years. The periods of time between each reversal were not equal, ranging from thousands to millions of years. Since iron minerals in molten basalt formed, became magnetized, then frozen in the orientation they had when the rock cooled, they made a record of reversals in Earth's ancient magnetic field (figure 15.23). Analysis of the magnetic pattern in the rocks along the Mid-Atlantic Ridge found identical patterns of magnetic bands on both sides of the ridge. This is just what you would expect if molten rock flowed out of the rift, cooled to solid basalt, then moved away from the rift on both sides. The pattern of magnetic bands also matched patterns of reversals measured elsewhere, providing a means of determining the age of the basalt. This showed that the oceanic crust is like a giant conveyor belt that is moving away from the Mid-Atlantic Ridge in both directions. It is moving at an average 5 cm (about 2 in) a year, which is about how fast your fingernails grow. This means that in fifty years, the seafloor will have moved 5 cm/yr × 50 yr, or 2.5 m (about 8 ft). This slow rate is why most people do not recognize that the seafloor—and the continents—move.

Lithosphere Plates and Boundaries

The strong evidence for seafloor spreading soon led to the development of a new theory called **plate tectonics.** According to plate tectonics, the lithosphere is broken into a number of fairly rigid plates that move on the asthenosphere. Some plates, as shown in figure 15.24, contain continents and part of an ocean basin, while other plates contain only ocean basins. The plates move, and the movement is helping to explain why mountains form where they do, the occurrence of earthquakes and volcanoes, and in general, the entire changing surface of Earth.

Earthquakes, volcanoes, and most rapid changes in Earth's crust occur at the edge of a plate, which is called a *plate boundary.* Three general kinds of plate boundaries describe how one plate moves relative to another: divergent, convergent, and transform boundaries.

Divergent boundaries occur between two plates moving away from each other. Magma forms as the plates separate, decreasing pressure on the mantle below. This molten material from the asthenosphere rises, cools, and adds new crust to the edges of the separating plates. The new crust tends to move horizontally from both sides of the divergent boundary, usually known as an oceanic ridge. A divergent boundary is thus a **new crust zone.** Most new crust zones are presently on the seafloor, producing seafloor spreading (figure 15.25).

The Mid-Atlantic Ridge is a divergent boundary between the South American and African Plates, extending north between the North American and Eurasian Plates (see figure 15.24). This ridge is one segment of the global mid-ocean ridge system that encircles Earth. The results of divergent plate movement can be seen in Iceland, where the Mid-Atlantic Ridge runs as it separates the North American and Eurasian Plates. In the northeastern part of Iceland, ground cracks are widening, often accompanied by volcanic activity. The movement was

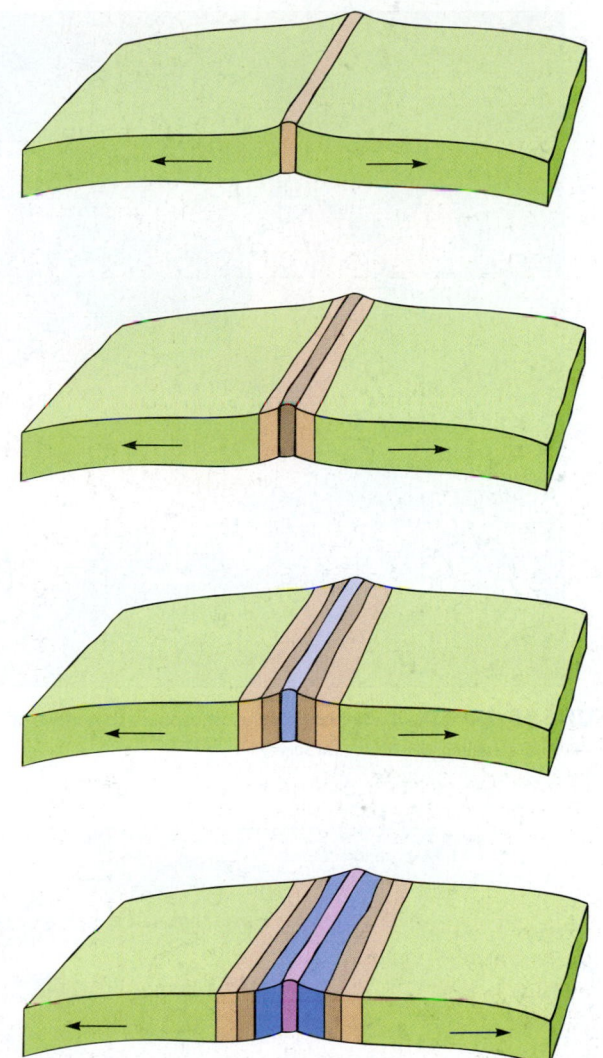

FIGURE 15.23 Formation of magnetic strips on the seafloor. As each new section of seafloor forms at the ridge, iron minerals become magnetized in a direction that depends on the orientation of Earth's field at that time. This makes a permanent record of reversals of Earth's magnetic field.

measured extensively between 1975 and 1984, when displacements caused a total separation of about 7 m (about 23 ft).

The measured rate of spreading along the Mid-Atlantic Ridge ranges from 1 to 6 centimeters per year. This may seem slow, but the process has been going on for millions of years and has caused a tiny inlet of water between the continents of Europe, Africa, and the Americas to grow into the vast Atlantic Ocean that exists today.

Another major ocean may be in the making in East Africa, where a divergent boundary has already moved Saudi Arabia away from the African continent, forming the Red Sea. If this spreading between the African Plate and the Arabian Plate continues, the Indian Ocean will flood the area and the easternmost corner of Africa will become a large island.

Convergent boundaries occur between two plates moving toward each other. The creation of new crust at a divergent boundary means that old crust must be destroyed somewhere else at the same rate, or else Earth would have a continuously expanding diameter. Old crust is destroyed by returning to the

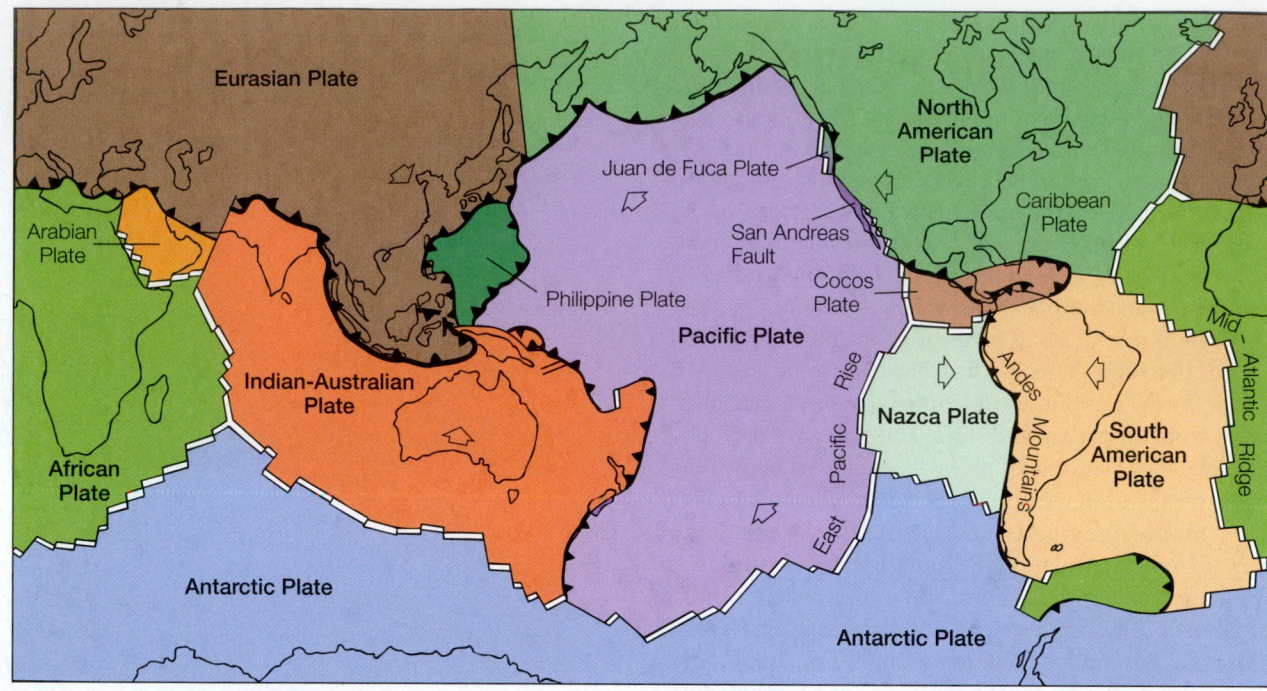

FIGURE 15.24 The major plates of the lithosphere that move on the asthenosphere.

Source: After W. Hamilton, U.S. Geological Survey.

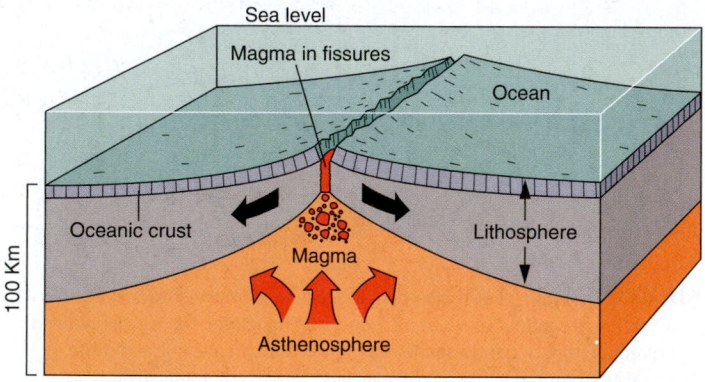

FIGURE 15.25 A diverging boundary at a mid-oceanic ridge. Hot asthenosphere wells upward beneath the ridge crest. Magma forms and squirts into fissures. Solid material that does not melt remains as mantle in the lower part of the lithosphere. As the lithosphere moves away from the spreading axis, it cools, becomes denser, and sinks to a lower level.

asthenosphere at convergent boundaries. The collision produces an elongated belt of down-bending called a **subduction zone.** The lithosphere of one plate, which contains the crust, is subducted beneath the second plate and partially melts, then becoming part of the mantle. The more dense components of this may become igneous materials that remain in the mantle. Some of it may eventually migrate to a spreading ridge to make new crust again. The less dense components may return to the surface as a silicon, potassium, and sodium-rich lava, forming volcanoes on the upper plate, or it may cool below the surface to form a body of granite. Thus, the oceanic lithosphere is being recycled through this process, which explains why ancient seafloor rocks do not exist. Convergent boundaries produce related characteristic geologic features depending on the nature of the materials in the plates, and there are three

general possibilities: (1) converging continental and oceanic plates, (2) converging oceanic plates, and (3) converging continental plates.

As an example of *ocean-continent plate convergence,* consider the plate containing the South American continent (the South American Plate) and its convergent boundary with an oceanic plate (the Nazca Plate) along its western edge. Continent-oceanic plate convergence produces a characteristic set of geologic features as the oceanic plate of denser basaltic material is subducted beneath the less dense granite-type continental plate (figure 15.26). The subduction zone is marked by an oceanic trench (the Peru-Chile Trench), deep-seated earthquakes, and volcanic mountains on the continent (the Andes Mountains). The trench is formed from the down-bending associated with subduction and the volcanic mountains from subducted and melted crust that rise up through the overlying plate to the surface. The earthquakes are associated with the movement of the subducted crust under the overlying crust.

Ocean-ocean plate convergence produces another set of characteristics and related geologic features (figure 15.27). The northern boundary of the oceanic Pacific Plate, for example, converges with the oceanic part of the North American Plate near the Bering Sea. The Pacific Plate is subducted, forming the Aleutian oceanic trench with a zone of earthquakes that are shallow near the trench and progressively more deep-seated toward the continent. The deeper earthquakes are associated with the movement of more deeply subducted crust into the mantle. The Aleutian Islands are typical **island arcs,** curving chains of volcanic islands that occur over the belt of deep-seated earthquakes. These islands form where the melted subducted material rises up through the overriding plate above sea level. The Japanese, Marianas, and Indonesians are similar groups of arc islands associated with converging oceanic-oceanic plate boundaries.

During *continent-continent plate convergence*, subduction does not occur as the less dense, granite-type materials tend to resist subduction (figure 15.28). Instead, the colliding plates pile up into a deformed and thicker crust of the lighter material. Such a collision produced the thick, elevated crust known as the Tibetan Plateau and the Himalayan Mountains.

Transform boundaries occur between two plates sliding by each other. Crust is neither created nor destroyed at transform boundaries as one plate slides horizontally past another along a long, vertical fault. The movement is neither smooth nor equal along the length of the fault, however, as short segments move independently with sudden jerks that are separated by periods without motion. The Pacific Plate, for example, is moving slowly to the northwest, sliding past the North American Plate. The San Andreas fault is one boundary along the California coastline. Vibrations from plate movements along this boundary are the famous California earthquakes.

Present-Day Understandings

The theory of plate tectonics, developed during the late 1960s and early 1970s, is new compared to most major scientific

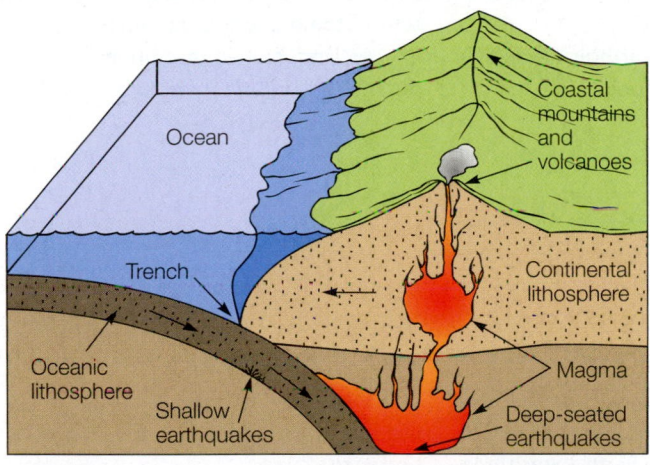

FIGURE 15.26 Ocean-continent plate convergence. This type of plate boundary accounts for shallow and deep-seated earthquakes, an oceanic trench, volcanic activity, and mountains along the coast.

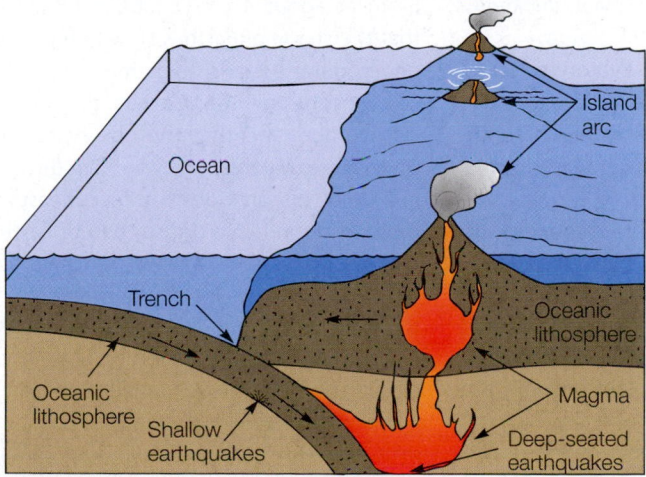

FIGURE 15.27 Ocean-ocean plate convergence. This type of plate convergence accounts for shallow and deep-focused earthquakes, an oceanic trench, and a volcanic arc above the subducted plate.

theories. Measurements are still being made, evidence is being gathered and evaluated, and the exact number of plates and their boundaries are yet to be determined with certainty. The major question that remains to be answered is what drives the plates, moving them apart, together, and past each other? One explanation is that slowly turning *convective cells* in the plastic asthenosphere drive the plates (figure 15.29). According to this hypothesis, hot fluid materials rise at the diverging boundaries. Some of the material escapes to form new crust, but most of it spreads out beneath the lithosphere. As it moves beneath the lithosphere, it drags the overlying plate with it. Eventually, it cools and sinks back inward under a subduction zone.

There is uncertainty about the existence of convective cells in the asthenosphere and their possible role because of a lack of clear evidence. Seismic data is not refined enough to show convective cell movement beneath the lithosphere. In addition, deep-seated earthquakes occur to depths of about 700 km

CONCEPTS APPLIED

New Earthquakes and Volcanoes

Locate and label the major plates of the lithosphere on an outline map of the world according to the most recent findings in plate tectonics. Show all types of boundaries and associated areas of volcanoes and earthquakes.

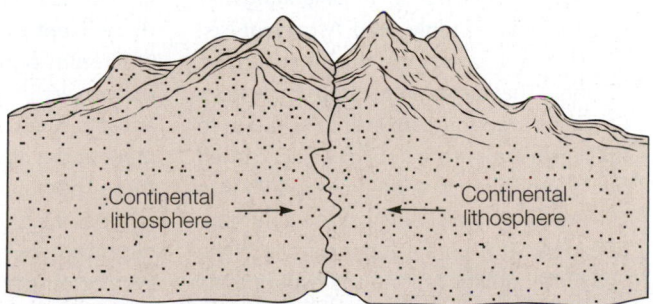

FIGURE 15.28 Continent-continent plate convergence. Rocks are deformed, and some lithosphere thickening occurs, but neither plate is subducted to any great extent.

A Closer Look

Seismic Tomography

The CAT scan is a common diagnostic imaging procedure that combines the use of X rays with computer technology. CAT stands for "computed axial tomography" and the word *tomography* means "drawing slices." The medical CAT scan shows organs of interest by making X-ray images from many different angles as the source of the X rays moves around the patient. CAT scan images are assembled by a computer into a three-dimensional picture that can show organs, bones, and tissues in great detail.

The CAT scan is applied to Earth's interior in a *seismic tomography* procedure. It works somewhat like the medical CAT scan, but seismic tomography uses seismic waves instead of X rays. The velocities of S- and P-seismic waves vary with depth, changing with density, temperature, pressure, and the composition of Earth's interior. Interior differences in temperature, pressure, and composition are what cause the motion of tectonic plates on Earth's surface. Thus,

a picture of Earth's interior can be made by mapping variations in seismic wave speeds.

Suppose an earthquake occurs and a number of seismic stations record when the S- and P- waves arrive. From these records, you could compare the seismic velocity data and identify if there were low seismic velocities between the source and some of the receivers. The late arrival of seismic waves could mean some difference in the structure of Earth between the source and the observing station. When an earthquake occurs in a new location, more data will be collected at different observing stations, which provides additional information about the shape and structure of whatever is slowing the seismic waves. Now, imagine repeating these measurements for many new earthquakes until you have enough data to paint a picture of what is beneath the surface.

Huge amounts of earthquake data—perhaps 10 million data points in 5 million

groups—are needed to construct a picture of Earth's interior. A really fast computer may take days to process this much data and construct cross sections through some interesting place, such as a subduction zone where an oceanic plate dives into the mantle.

Seismic tomography has also identified massive plumes of molten rock rising toward the surface from deep within Earth. These *superplumes* originate from the base of the mantle, rising to the lithosphere. The hot material was observed to spread out horizontally under the lithosphere toward midocean ridges. This may contribute to tectonic plate movement. Regions above the superplumes tend to bulge upward, and other indications of superplumes, such as variations in gravity, have been measured. Scientists continue to work with new seismic tomography data with more precise images and higher resolutions to better describe the interior structure of Earth.

A Closer Look

Measuring Plate Movement

According to the theory of plate tectonics, Earth's outer shell is made up of moving plates. The plates making up the continents are about 100 km (62 mi) thick and are gradually drifting at a rate of about 0.5 to 10 cm (0.2 to 4 in) per year. This reading is about one way that scientists know that Earth's plates are moving and how this movement is measured.

The very first human lunar landing mission took place in July 1969. Astronaut Neil Armstrong stepped onto the lunar surface, stating, "That's one small step for a man, one giant leap for mankind." In addition to fulfilling a dream, the *Apollo* project carried out a program of scientific experiments. The *Apollo 11* astronauts placed a number of experiments on the lunar surface in the Sea of Tranquility. Among the experiments was the first Laser Ranging Retro-Reflector

Experiment, which was designed to reflect pulses of laser light from Earth. Later, three more reflectors were placed on the Moon, including two by other *Apollo* astronauts and one by an unmanned Soviet *Lunakhod 2* lander.

The McDonald Observatory in Texas, the Lure Observatory on the island of Maui, Hawaii, and a third observatory in southern France have regularly sent laser beams through optical telescopes to the reflectors. The return signals, which are too weak to be seen with the unaided eye, are detected and measured by sensitive detection equipment at the observatories. The accuracy of these measurements, according to NASA reports, is equivalent to determining the distance between a point on the east and a point on the west coasts of the United States to an accuracy of 0.5 mm, about

the size of the period at the end of this sentence.

Reflected laser light experiments have found that the Moon is pulling away from Earth at about 4 cm/yr (about 1.6 in/yr), that the shape of Earth is slowly changing, undergoing isostatic adjustment from the compression by the glaciers during the last ice age, and that the observatory in Hawaii is slowly moving away from the one in Texas. This provides a direct measurement of the relative drift of two of Earth's tectonic plates. Thus, one way that changes on the surface of Earth are measured is through lunar ranging experiments. Results from lunar ranging, together with laser ranging to the artificial satellites in Earth's orbit, have revealed the small but constant drift rate of the plates making up Earth's dynamic surface.

(435 mi), which means that descending materials—parts of a subducted plate—must extend to that depth. This could mean that a convective cell might operate all the way down to the core-mantle boundary some 2,900 km (about 1,800 mi) below the surface. This presents another kind of problem because little is known about the lower mantle and how it interacts with the upper mantle. Theorizing without information is called speculation, and that is the best that can be done with existing data. The full answer may include the role of heat and the role of gravity.

Heat and gravity are important in a proposed mechanism of plate motion called "*ridge-push*" (figure 15.30). This idea has a plate cooling and thickening as it moves away from a divergent boundary. As it subsides, it cools asthenospheric mantle to lithospheric mantle, forming a sloping boundary between the lithosphere and the asthenosphere. The plate slides down this boundary.

Another proposed mechanism of plate motion is called "*slab-pull*" (figure 15.31). In this mechanism, the subducting plate is colder and therefore denser than the surrounding hot mantle, so it pulls the surface part of the plate downward. Density of the slab may also be increased by loss of water and reforming of minerals into more dense forms.

What is generally accepted about the plate tectonic theory is the understanding that the solid materials of Earth are engaged in a continual cycle of change. Oceanic crust is subducted, melted, then partly returned to the crust as volcanic

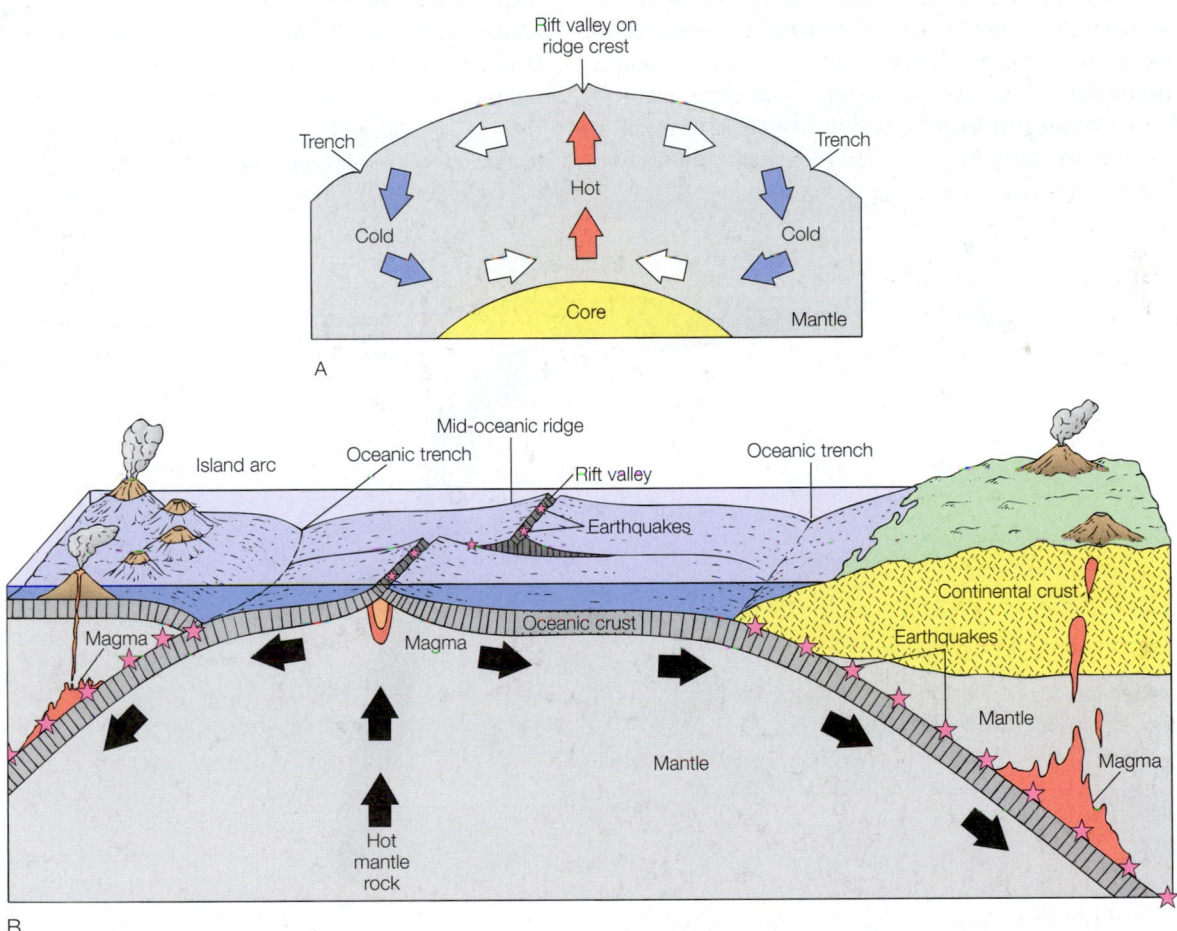

FIGURE 15.29   Not to scale. One idea about convection in the mantle has a convection cell circulating from the core to the lithosphere, dragging the overlying lithosphere laterally away from the oceanic ridge.

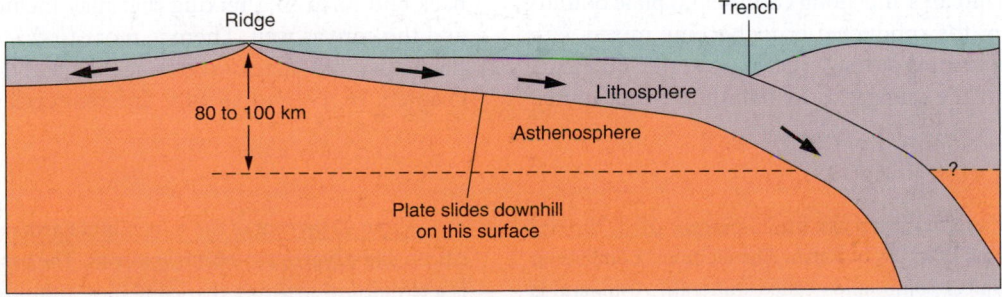

FIGURE 15.30   Ridge-push. A plate may slide downhill on the sloping boundary between the lithosphere and the asthenosphere at the base of the plate.

People Behind the Science

Frederick John Vine (1939–1988)

Frederick Vine was an English geophysicist whose work was an important contribution to the development of the theory of plate tectonics.

Vine was born in Brentford, Essex, England, and was educated at Latymer Upper School, London, and at St. John's College, Cambridge. From 1967 to 1970, he worked as assistant professor in the Department of Geological and Geophysical Sciences at Princeton University, before returning to England to become reader (1970) and then professor (1974) in the School of Environmental Sciences at the University of East Anglia.

In Cambridge in 1963, Vine collaborated with his supervisor, Drummond Hoyle Matthews (1931–), and wrote a paper, "Magnetic Anomalies over Ocean Ridges," that provided additional evidence for Harry H. Hess's (1962) seafloor spreading hypothesis. Alfred Wegener's original 1912 theory of continental drift (Wegener's hypothesis) had been met with hostility at the time because he could not explain why the continents had drifted apart, but Hess had continued his work developing the theory of seafloor spreading to explain the fact that as the oceans grew wider, the continents drifted apart.

Following the work of Brunhes and Matonori Matuyama in the 1920s on magnetic reversals, Vine and Matthews predicted that new rock emerging from the oceanic ridges would intermittently produce material of opposing magnetic polarity to the old rock. They applied paleomagnetic studies to the ocean ridges in the North Atlantic and were able to argue that parallel belts of different magnetic polarities existed on either side of the ridge crests. This evidence was vital proof of Hess's hypothesis. Studies on ridges in other oceans also showed the existence of these magnetic anomalies. Vine and Matthews's hypothesis was widely accepted in 1966 and confirmed Dietz and Hess's earlier work. Their work was crucial to the development of the theory of plate tectonics, and it revolutionized the earth sciences.

Source: Modified from the *Hutchinson Dictionary of Scientific Biography.* © RM, 2011. All rights reserved. Helicon Publishing is a division of RM.

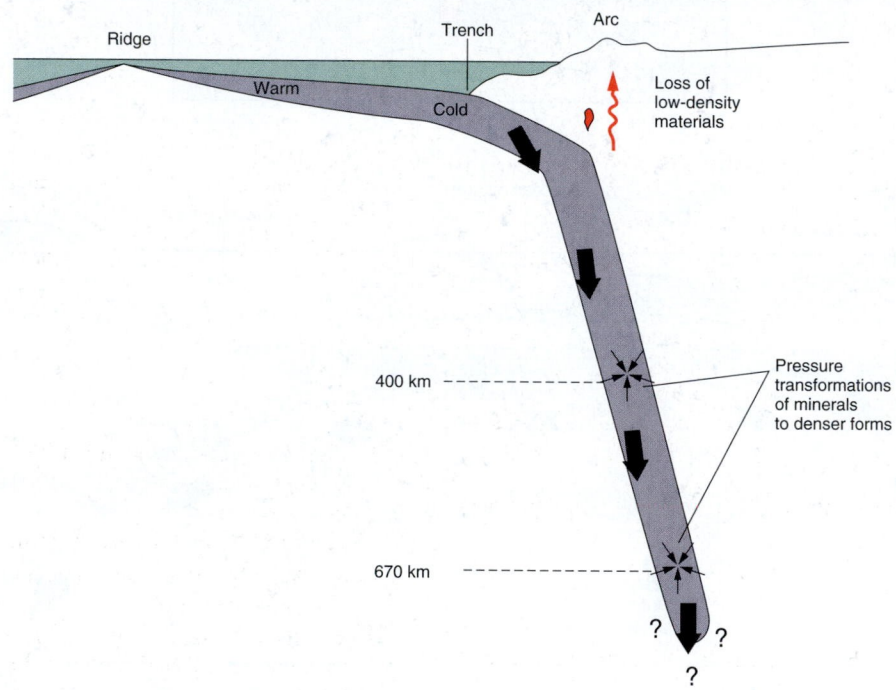

FIGURE 15.31 Slab-pull. The dense, leading edge of a subducting plate pulls the rest of the plate along. Plate density increases due to cooling, loss of low-density material, and pressure transformation of minerals to denser forms.

igneous rocks in island arcs and along continental plate boundaries. Other parts of the subducted crust become mixed with the upper mantle, returning as new crust at diverging boundaries. The materials of the crust and the mantle are thus cycled back and forth in a mixing that may include the deep mantle and the core as well. There is more to this story of a dynamic Earth that undergoes a constant change. The story continues in chapters 16, 17, and 18 with different cycles to consider.

SUMMARY

The elements silicon and oxygen make up 75 percent of all the elements in the outer layer, or *crust,* of Earth. The elements combine to make crystalline chemical compounds called minerals. A *mineral* is defined as a naturally occurring, inorganic solid element or compound with a crystalline structure.

About 92 percent of the minerals of Earth's crust are composed of silicon and oxygen, the *silicate minerals.* The basic unit of the silicates is a *tetrahedral structure* that combines with positive metallic ions or with other tetrahedral units to form chains, sheets, or an interlocking framework. The *ferromagnesian silicates* are tetrahedral structures combined with ions of iron, magnesium, calcium, and other elements. The ferromagnesian silicates are darker in color and more dense than other silicates. The *nonferromagnesian silicates* do not have irons or magnesium ions, and they are lighter in color and less dense than the ferromagnesians. The *nonsilicate minerals* do not contain silicon and are carbonates, sulfates, oxides, halides, sulfides, and native elements.

A *rock* is defined as an aggregation of one or more minerals that have been brought together into a cohesive solid. *Igneous rocks* formed as hot, molten *magma* cooled and crystallized to firm, hard rocks. *Sedimentary rocks* are formed from *sediments*, accumulations of weathered rock materials that settle out of the atmosphere or out of water. Sediments become sedimentary rocks through a rock-forming process that involves both the *compaction* and *cementation* of the sediments. *Metamorphic rocks* are previously existing rocks that have been changed by heat, pressure, or hot solution into a different kind of rock without melting. The *rock cycle* is a concept that an igneous, a sedimentary, or a metamorphic rock is a temporary stage in the ongoing transformation of rocks to new types.

Earth has a layered interior that formed as Earth's materials underwent *differentiation,* the separation of materials while in the molten state. The center part, or *core,* is predominantly iron with a solid inner part and a liquid outer part. The core makes up about 15 percent of Earth's total volume and about a third of its total mass. The

mantle is the middle part of Earth's interior that accounts for about two-thirds of Earth's total mass and about 80 percent of its total volume. The *Mohorovicic discontinuity* separates the outer layer, or *crust,* of Earth from the mantle. The crust of the *continents* is composed mostly of less dense granite-type rock. The crust of the *ocean basins* is composed mostly of the more dense basaltic rocks.

Another way to consider Earth's interior structure is to consider the weak layer in the upper mantle, the *asthenosphere* that extends around the entire Earth. The rigid, solid, and brittle layer above the asthenosphere is called the *lithosphere.* The lithosphere includes the entire crust, the Moho, and the upper part of the mantle.

Evidence from the ocean floor revived interest in the idea that continents could move. The evidence for *seafloor spreading* came from related observations concerning oceanic ridge systems, sediment and fossil dating of materials outward from the ridge, and magnetic patterns of seafloor rocks. Confirmation of seafloor spreading led to the *plate tectonic theory.* According to plate tectonics, new basaltic crust is added at *diverging boundaries* of plates, and old crust is *subducted* at *converging boundaries.* Mountain building, volcanoes, and earthquakes are seen as *related geologic features* that are caused by plate movements. The force behind the movement of plates is uncertain, but it may involve *convection* in the deep mantle.

KEY TERMS

asthenosphere (p. **340**)	island arcs (p. **344**)	oceanic ridges (p. **341**)	sedimentary rocks (p. **335**)
convergent boundaries (p. **343**)	lithosphere (p. **340**)	oceanic trenches (p. **341**)	silicates (p. **331**)
core (p. **339**)	mantle (p. **338**)	plate tectonics (p. **343**)	subduction zone (p. **344**)
crust (p. **338**)	metamorphic rocks (p. **336**)	rift (p. **342**)	transform boundaries (p. **345**)
divergent boundaries (p. **343**)	mineral (p. **330**)	rock (p. **333**)	
igneous rocks (p. **335**)	new crust zone (p. **343**)	seafloor spreading (p. **342**)	

APPLYING THE CONCEPTS

Answers are located in appendix F.

1. Sedimentary rocks are formed by the processes of compaction and
   a. pressurization.
   b. melting.
   c. cementation.
   d. heating but not melting.

2. Which type of rock probably existed first, starting the rock cycle?
   a. metamorphic
   b. igneous
   c. sedimentary
   d. All of the above are correct.

3. From seismological data, Earth's shadow zone indicates that part of Earth's interior must be
   a. liquid.
   b. solid throughout.
   c. plastic.
   d. hollow.

4. The Mohorovicic discontinuity is a change in seismic wave velocity that is believed to take place because of
   a. structural changes in minerals of the same composition.
   b. changes in the composition on both sides of the boundary.
   c. a shift in the density of minerals of the same composition.
   d. changes in the temperature with depth.

5. The oldest rocks are found in (the)
   a. continental crust.
   b. oceanic crust.
   c. neither, since both are the same age.

6. The least dense rocks are found in (the)
   a. continental crust.
   b. oceanic crust.
   c. neither, since both are the same density.

7. The idea of seafloor spreading along the Mid-Atlantic Ridge was supported by evidence from
   a. changes in magnetic patterns and ages of rocks moving away from the ridge.
   b. faulting and volcanoes on the continents.
   c. the observation that there was no relationship between one continent and another.
   d. All of the above are correct.

8. According to the plate tectonics theory, seafloor spreading takes place at a
   a. convergent boundary.
   b. subduction zone.
   c. divergent boundary.
   d. transform boundary.

9. The presence of an oceanic trench, a chain of volcanic mountains along the continental edge, and deep-seated earthquakes is characteristic of a (an)
   a. ocean-ocean plate convergence.
   b. ocean-continent plate convergence.
   c. continent-continent plate convergence.
   d. None of the above is correct.

10. The presence of an oceanic trench with shallow earthquakes and island arcs with deep-seated earthquakes is characteristic of a (an)
    a. ocean-ocean plate convergence.
    b. ocean-continent plate convergence.
    c. continent-continent plate convergence.
    d. None of the above is correct.

11. The ongoing occurrence of earthquakes without seafloor spreading, oceanic trenches, or volcanoes is most characteristic of a
    a. convergent boundary between plates.
    b. subduction zone.
    c. divergent boundary between plates.
    d. transform boundary between plates.

12. The evidence that Earth's core is part liquid or acts like a liquid comes from (the)
    a. P-wave shadow zone.
    b. S-wave shadow zone.
    c. meteorites.
    d. All of the above are correct.

QUESTIONS FOR THOUGHT

1. What is a rock?
2. Describe the concept of the rock cycle.
3. Briefly explain the basic differences among the three major kinds of rocks based on the way they were formed.
4. Which major kind of rock, based on the way it is formed, would you expect to find most of in Earth's crust? Explain.
5. What is the difference between magma and lava?
6. What is meant by the "texture" of an igneous rock? What does the texture of igneous rock tell you about its cooling history?
7. What are the basic differences between basalt and granite, the two most common igneous rocks of Earth's crust? In what part of Earth's crust are basalt and granite most common? Explain.
8. Is the igneous rock basalt *always* fine-grained? Explain.
9. Briefly describe the rock-forming process that changes sediments into solid rock.
10. What are metamorphic rocks? What limits the maximum temperatures possible in metamorphism? Explain.
11. What evidence provides information about the nature of Earth's core?

12. What is the asthenosphere? Why is it important in modern understandings of Earth?
13. Describe the origin of the magnetic strip patterns found in the rocks along an oceanic ridge.
14. Explain why very old rocks are not found on the seafloor.
15. Describe the three major types of plate boundaries and what happens at each.
16. Briefly describe the theory of plate tectonics and how it accounts for the existence of certain geologic features.
17. What is an oceanic trench? What is its relationship to major plate boundaries? Explain this relationship.
18. Describe the probable source of all the earthquakes that occur in southern California.
19. The northwestern coast of the United States has a string of volcanoes running along it. According to plate tectonics, what does this mean about this part of the North American Plate? What geologic feature would you expect to find on the seafloor off the northwestern coast? Explain.
20. Explain how the crust of Earth is involved in a dynamic, ongoing recycling process.

FOR FURTHER ANALYSIS

1. Is ice a mineral? Describe reasons to argue that ice is a mineral. Describe reasons to argue that ice is not a mineral.
2. What are the significant similarities and differences between igneous, sedimentary, and metamorphic rocks?
3. Why are there no active volcanoes in the eastern United States or Canada? Explain why you would or would not expect volcanoes there in the future.
4. Describe cycles that occur on Earth's surface and cycles that occur between the surface and the interior. Explain why these cycles do not exist on other planets of the solar system.

5. The rock cycle describes how igneous, metamorphic, and sedimentary rocks are changed into each other. If this is true, analyze why most of the rocks on Earth's surface are sedimentary.
6. If ice is a mineral, is a glacier a rock? Describe reasons to support or oppose calling a glacier a rock according to the definition of a rock.
7. Discuss evidence that would explain why plate tectonics occurs on Earth but not on other planets.

INVITATION TO INQUIRY

Measuring Plate Motion

Tectonic plate motion can be measured with several relatively new technologies, including satellite laser ranging and use of the Global Positioning System. Start your inquiry by visiting the Tectonic Plate Motion website at cddisa.gsfc.nasa.gov/926/slrtecto.html. Study the regional plate motion of the North American Plate, for example. Note the scale of 50 mm/yr, then measure to find the rate of movement at the different stations shown. Is this recent data on plate motion consistent with other available information on plate motion? How can you account for the different rates of motion at adjacent stations?

16

Earth's Surface

Folding, faulting, and lava flows, such as the one you see here, tend to build up, or elevate, Earth's surface.

CORE **CONCEPT**

The surface of the Earth is involved in an ongoing cycle of destruction and renewal.

OUTLINE

Rocks are subjected to forces associated with plate tectonics and other forces (p. 353).

Mountain ranges are features of folding and faulting on a very large scale (p. 360).

Weathering and erosion wear down the surface of Earth (p. 363).

Gravity, streams, glaciers, and wind are agents of erosion (p. 366–72).

Earth Science

▷ Earth's surface is made up of a number of rigid plates that are moving (Ch. 15).

Life Science

▷ Erosion of rock releases phosphorus for use by plants (Ch. 23).

▷ Sedimentation traps organisms and forms fossils (Ch. 22).

▷ Layers of sedimentary rock reveal the history of life on Earth (Ch. 22).

OVERVIEW

The central idea of plate tectonics, which was discussed in chapter 15, is that Earth's surface is made up of rigid plates that are slowly moving. Since the plates and the continents riding on them are in constant motion, any given map of the world is only a snapshot that shows the relative positions of the continents at a given time. The continents occupied different positions in the distant past. They will occupy different positions in the distant future. The surface of Earth, which seems so solid and stationary, is in fact mobile.

Plate tectonics has changed the accepted way of thinking about the solid, stationary nature of Earth's surface and ideas about the permanence of the surface as well. The surface of Earth is no longer viewed as having a permanent nature but is understood to be involved in an ongoing cycle of destruction and renewal. Old crust is destroyed as it is plowed back into the mantle through subduction, becoming mixed with the mantle. New crust is created as molten materials move from the mantle through seafloor spreading and volcanoes. Over time, much of the crust must cycle into and out of the mantle.

The movement of plates, the crust-mantle cycle, and the rock cycle all combine to produce a constantly changing surface. There are basically two types of surface changes: (1) changes that originate within Earth, resulting in a building up of the surface (figure 16.1) and (2) changes that occur when rocks are exposed to the atmosphere and water, resulting in a sculpturing and tearing down of the surface. This chapter is about changes in the land. The concepts of this chapter will provide you with something far more interesting about Earth's surface than the scenic aspect. The existence of different features (such as mountains, folded hills, islands) and the occurrence of certain events (such as earthquakes, volcanoes, faulting) are all related. The related features and events also have a story to tell about Earth's past, a story about the here and now, and yet another story about the future.

16.1 INTERPRETING EARTH'S SURFACE

Because many geologic changes take place slowly, it is difficult for a person to see significant change occur to mountains, canyons, and shorelines in the brief span of a lifetime. Given a mental framework based on a lack of appreciation of change over geologic time, how do you suppose people interpreted the existence of features such as mountains and canyons? Some believed, as they had observed in their lifetimes, that the mountains and canyons had "always" been there. Statements such as "unchanging as the hills" or "old as the hills" illustrate this lack of appreciation of change over geologic time. Others did not believe the features had always been there but believed they were formed by a sudden, single catastrophic event (figure 16.2). A catastrophe created a feature of Earth's surface all at once, with little or no change occurring since that time. The Grand Canyon, for example, was not interpreted as the result of incomprehensibly slow river erosion but as the result of a giant crack or rip that appeared in the surface. The canyon that you see today was interpreted as forming when Earth split open

and the Colorado River fell into the split. This interpretation was used to explain the formation of major geologic features based on the little change observed during a person's lifetime.

About two hundred years ago, the idea of unchanging, catastrophically formed landscapes was challenged by James Hutton, a Scottish physician. Hutton, who is known today as the founder of modern geology, traveled widely throughout the British Isles. Hutton was a keen observer of rocks, rock structures, and other features of the landscape. He noted that sandstone, for example, was made up of rock fragments that appeared to be (1) similar to the sand being carried by rivers and (2) similar to the sand making up the beaches next to the sea. He also noted fossil shells of sea animals in sandstone on the land, while the living relatives of these animals were found in the shallow waters of the sea. This and other evidence led Hutton to realize that rocks were being ground into fragments, then carried by rivers to the sea. He surmised that these particles would be reformed into rocks later, then lifted and shaped into the hills and mountains of the land. He saw all this as quiet, orderly change that required only *time* and the

FIGURE 16.1 An aerial view from the south of the eruption of Mount St. Helens volcano on May 18, 1980.

FIGURE 16.2 Would you believe that this rock island has "always" existed where it is? Would you believe it was formed by a sudden, single event? What evidence would it take to convince you that the rock island formed ever so slowly, starting as a part of southern California and moving very slowly, at a rate of centimeters per year, to its present location near the coast of Alaska?

ongoing work of the water and some forces to make the sediments back into rocks. With Hutton's logical conclusion came the understanding that Earth's history could be interpreted by tracing it backward, from the present to the past. This tracing required a frame of reference of slow, uniform change, not the catastrophic frame of reference of previous thinkers. The frame of reference of uniform changes is today called the **principle of uniformity** (also called *uniformitarianism*). The principle of uniformity is often represented by a statement that "the present is the key to the past." This statement means that the geologic processes you see changing rocks today are the very same processes that changed them in the ancient past, although not necessarily at the same rate. The principle of uniformity does not *exclude* the happening of sudden or catastrophic events on the surface of Earth. A violent volcanic explosion, for example, is a catastrophic event that most certainly modifies the surface of Earth. What the principle of uniformity does state is that the physical and chemical laws we observe today operated exactly the same in the past. The rates of operation may or may not have been the same in the past, but the events you see occurring today are the same events that occurred in the past. Given enough time, you can explain the formation of the structures of Earth's surface with known events and concepts.

The principle of uniformity has been used by geologists since the time of Hutton. The concept of how the constant changes occur has evolved with the development of plate tectonics, but the basic frame of reference is the same. You will see how the principle of uniformity is applied by first considering what can happen to rocks and rock layers that are deeply buried.

16.2 PROCESSES THAT BUILD UP THE SURFACE

All the possible movements of Earth's plates, including drift toward or away from other plates, and any process that deforms Earth's surface are included in the term *diastrophism*. Diastrophism is the process of deformation that changes Earth's surface. It produces many of the basic structures you see on Earth's surface, such as plateaus, mountains, and folds in the crust. The movement of magma is called *vulcanism* or *volcanism*. Diastrophism, volcanism, and earthquakes are closely related, and their occurrence can usually be explained by events involving plate tectonics. The results of diastrophism are discussed in the section on stress and strain, which is followed by a discussion of earthquakes, volcanoes, and mountain chains.

Stress and Strain

Any solid material responds to a force in a way that depends on the extent of coverage (force per unit area, or pressure), the nature of the material, and other variables such as the temperature. Consider, for example, what happens if you place the point of a ballpoint pen on the side of an aluminum pop (soda) can and apply an increasing pressure. With increasing pressure, you can observe at least four different and separate responses:

1. At first, the metal successfully resists a slight pressure and *nothing happens*.
2. At a somewhat greater pressure, you will be able to deform, or bend, the metal into a concave surface. The metal will return to its original shape, however, when the pressure is removed. This is called an *elastic deformation* since the metal was able to spring back into its original shape.
3. At a still greater pressure, the metal is deformed to a concave surface, but this time, the metal does not return to its original shape. This means the *elastic limit* of the metal has been exceeded, and it has now undergone a *plastic deformation*. Plastic deformation permanently alters the shape of a material.
4. Finally, at some great pressure, the metal will rupture, resulting in a *break* in the material.

Many materials, including rocks, respond to increasing pressures in this way, showing (1) no change, (2) an elastic change with recovery, (3) a plastic change with no recovery, and (4) finally breaking from the pressure.

A **stress** is a force that tends to compress, pull apart, or deform a rock. Rocks in Earth's solid outer crust are subjected to forces as Earth's plates move into, away from, or alongside each other. However, not all stresses are generated directly by plate interaction. Three types of forces from plate interaction that cause rock stress are:

1. *Compressive stress* is caused by two plates moving together or by one plate pushing against another plate that is not moving.
2. *Tensional stress* is the opposite of compressional stress. It occurs when one part of a plate moves away, for example, and another part does not move.
3. *Shear stress* is produced when two plates slide past one another or by one plate sliding past another plate that is not moving.

Just like the metal in the soda can, a rock is able to withstand stress up to a limit. Then it might undergo elastic deformation, plastic deformation, or breaking with progressively greater pressures. The adjustment to stress is called **strain**. A rock unit might respond to stress by changes in volume, by changes in shape, or by breaking. Thus, there are three types of strain: elastic, plastic, and fracture.

1. In *elastic strain,* rock units recover their original shape after the stress is released.
2. In *plastic strain,* rock units are molded or bent under stress and do not return to their original shape after the stress is released.
3. In *fracture strain,* rock units crack or break, as the name suggests.

The relationship between stress and strain, that is, exactly how the rock responds, depends on at least four variables. They are (1) the nature of the rock, (2) the temperature of the rock, (3) how slowly or quickly the stress is applied over time, and (4) the confining pressure on the rock. The temperature and con-

fining pressure are generally a function of how deeply the rock is buried. In general, rocks are better able to withstand compressional rather than pulling-apart stresses. Cold rocks are more likely to break than warm rocks, which tend to undergo plastic deformation. In addition, a stress that is applied quickly tends to break the rock, where stress applied more slowly over time, perhaps thousands of years, tends to result in plastic strain.

In general, rocks at great depths are under great pressure at higher temperatures. These rocks tend to undergo plastic deformation, then plastic flow, so rocks at great depths are bent and deformed extensively. Rocks closer to the surface can also bend, but they have a lower elastic limit and break more readily (figure 16.3). Rock deformation often results in recognizable surface features called folds and faults, the topics of the next sections.

Folding

Sediments that form most sedimentary rocks are deposited in nearly flat, horizontal layers at the bottom of a body of water. Conditions on the land change over time, and different mixtures of sediments are deposited in distinct layers of varying thickness. Thus, most sedimentary rocks occur naturally as structures of horizontal layers, or beds (figure 16.4).

A sedimentary rock layer that is not horizontal may have been subjected to some kind of compressive stress. The source of such a stress could be from colliding plates, from the intrusion of magma from below, or from a plate moving over a rising superplume of magma. Seismic tomography has identified massive plumes of molten rock rising toward the surface from deep within Earth. These superplumes originate from the base of the mantle, rising to the lithosphere. The hot material was observed to spread out horizontally, causing bulges as it exerts forces under the lithosphere.

Stress on buried layers of horizontal rocks can result in plastic strain, resulting in a wrinkling of the layers into *folds.* **Folds** are bends in layered bedrock (figure 16.5). They are

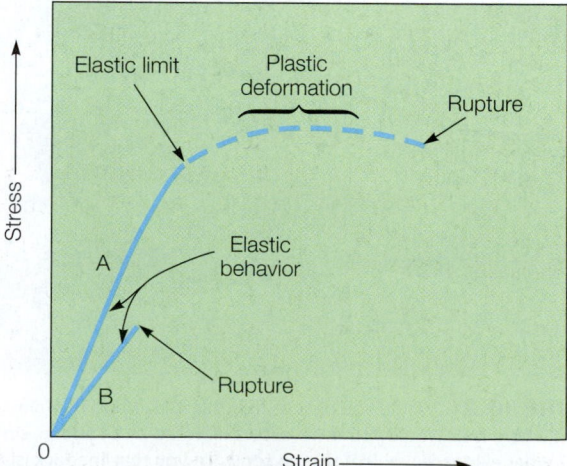

FIGURE 16.3 Stress and deformation relationships for deeply buried, warm rocks under high pressure (A) and for cooler rocks near the surface (B). Breaking occurs when stress exceeds rupture strength.

A

B

FIGURE 16.4   (*A*) Rock bedding on a grand scale in the Grand Canyon. (*B*) A closer example of rock bedding can be seen in this roadcut.

FIGURE 16.5   These folded rock layers are near Dorset, United Kingdom. Can you figure out what might have happened here to fold flat rock layers like this?

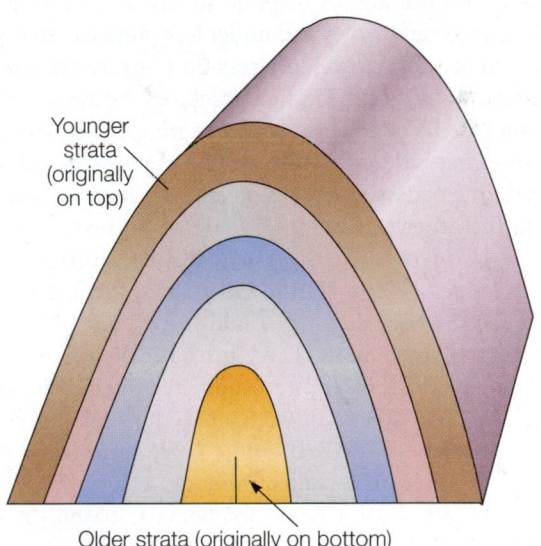

Younger strata (originally on top)

Older strata (originally on bottom)

FIGURE 16.6   An anticline, or arching fold, in layered sediments. Note that the oldest strata are at the center.

analogous to layers of rugs or blankets that were stacked horizontally, then pushed into a series of arches and troughs. Folds in layered bedrock of all shapes and sizes can occur from plastic strain, depending generally on the regional or local nature of the stress and other factors. Of course, when the folding occurred, the rock layers were in a plastic condition and were probably under considerable confining pressure from deep burial. However, you see the results of the folding when the rocks are under very different conditions at the surface.

The most common regional structures from deep plastic deformation are arch-shaped and trough-shaped folds. In general, an arch-shaped fold is called an **anticline** (figure 16.6). The corresponding trough-shaped fold is called a **syncline** (figure 16.7). Anticlines and synclines sometimes alternate

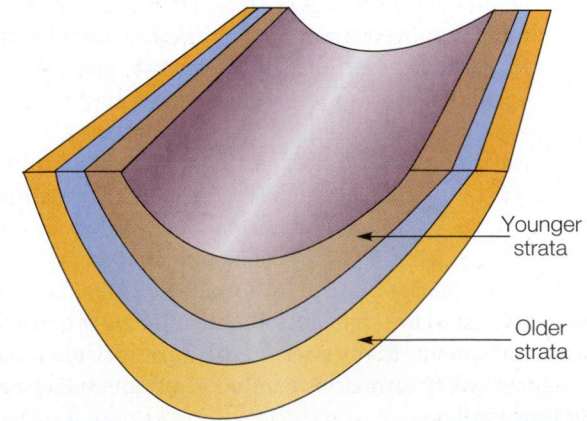

Younger strata

Older strata

FIGURE 16.7   A syncline, showing the reverse age pattern.

across the land like waves on water. You can imagine that a great compressional stress must have been involved over a wide region to wrinkle the land like this.

Anticlines, synclines, and other types of folds are not always visible as such on Earth's surface. The ridges of anticlines are constantly being weathered into sediments. The sediments, in turn, tend to collect in the troughs of synclines, filling them in. The Appalachian Mountains have ridges of rocks that are more resistant to weathering, forming hills and mountains. The San Joaquin Valley, on the other hand, is a very large syncline in California.

Note that any kind of rock can be folded. Sedimentary rocks are usually the best example of folding, however, since the fold structures of rock layers are easy to see and describe. Folding is much harder to see in igneous or metamorphic rocks that are blends of minerals without a layered structure.

Faulting

Rock layers do not always respond to stress by folding. Rocks near the surface are cooler and under less pressure, so they tend to be more brittle. A sudden stress on these rocks may reach the rupture point, resulting in a cracking and breaking of the rock structure. When there is relative movement between the rocks on either side of a fracture, the crack is called a **fault.** When faulting occurs, the rocks on one side move relative to the rocks on the other side along the surface of the fault, which is called the *fault plane.* Faults are generally described in terms of (1) the steepness of the fault plane, that is, the angle between the plane and imaginary horizontal plane, and (2) the direction of relative movement. There are basically three ways that rocks on one side of a fault can move relative to the rocks on the other side: (1) up and down (called "dip"), (2) horizontally, or sideways (called "strike"), and (3) with elements of both directions of movement (called "oblique").

One classification scheme for faults is based on an orientation referent borrowed from mining (many ore veins are associated with fault planes). Imagine a mine with a fault plane running across a horizontal shaft. Unless the plane is perfectly vertical, a miner would stand on the mass of rock below the fault plane and look up at the mass of rock above. Therefore, the mass of rock below is called the *footwall* and the mass of rock above is called the *hanging wall* (figure 16.8). How the footwall and hanging wall have moved relative to one another describes three basic classes of faults: (1) normal, (2) reverse, and (3) thrust faults. A **normal fault** is one in which the hanging wall has moved downward relative to the footwall. This seems "normal" in the sense that you would expect an upper block to slide *down* a lower block along a slope (figure 16.9A). Sometimes a huge block of rock bounded by normal faults will drop down, creating a *graben* (figure 16.9B). The opposite of a graben is a *horst,* which is a block bounded by normal faults that is uplifted (figure 16.9C). A very large block lifted sufficiently becomes a faultblock mountain. Many parts of the western United States are characterized by numerous faultblock mountains separated by adjoining valleys.

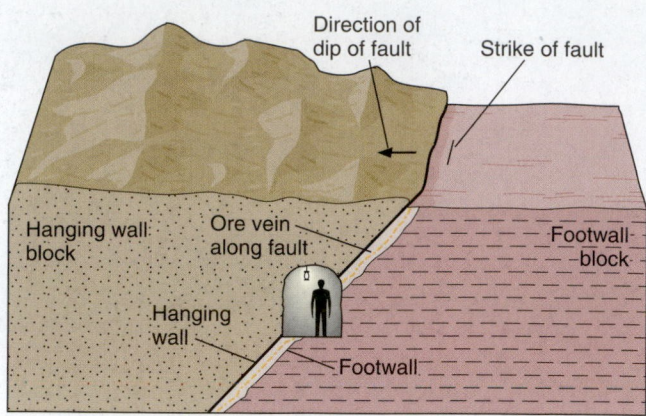

FIGURE 16.8 (A) Relationship between the hanging wall block and footwall block of a fault. (B) A photo of a fault near Kingman, Arizona, showing how the hanging wall has moved relative to the footwall.

In a **reverse fault,** the hanging wall block has moved upward relative to the footwall block. As illustrated in figure 16.10A, a reverse fault is probably the result of horizontal compressive stress.

A reverse fault with a low-angle fault plane is also called a **thrust fault** (figure 16.10B). In some thrust faults, the hanging wall block has completely overridden the lower footwall for 10 to 20 km (6 to 12 mi). This is sometimes referred to as an "over-thrust."

As shown in figures 16.9 and 16.10, the relative movement of blocks of rocks along a fault plane provides information about the stresses that produced the movement. Reverse and thrust faulting result from compressional stress in the direction of the movement. Normal faulting, on the other hand, results from a pulling-apart stress that might be associated with diverging plates. It might also be associated with the stretching and bulging up of the crust over a hot spot.

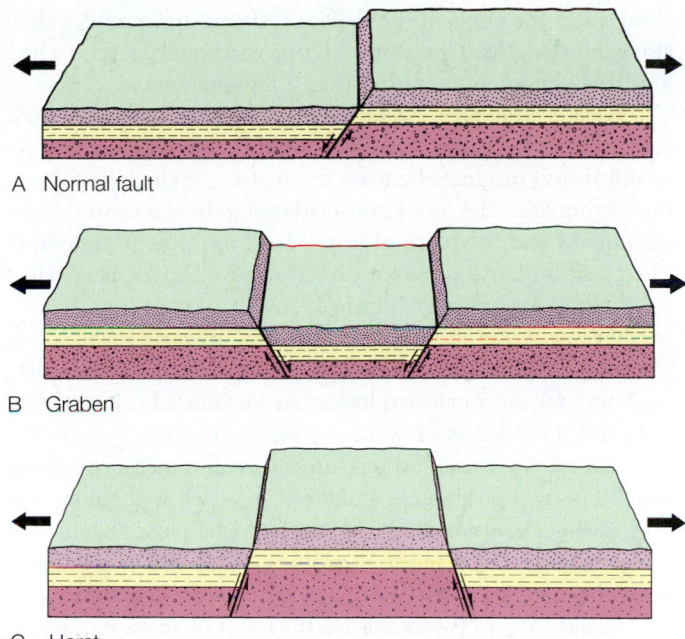

A   Normal fault

B   Graben

C   Horst

FIGURE 16.9 How tensional stress could produce (*A*) a normal fault, (*B*) a graben, and (*C*) a horst.

16.3 EARTHQUAKES

What is an earthquake? An **earthquake** is a quaking, shaking, vibrating, or upheaval of the ground. Earthquakes are the result of the sudden release of energy that comes from *stress* on rock beneath Earth's surface. In the section on folding, you learned that rock units can bend and become deformed in response to stress, but there are limits as to how much stress rock can take before it fractures. When it does fracture, the sudden movement of blocks of rock produces vibrations that move out as waves throughout Earth. These vibrations are called **seismic waves.** It is strong seismic waves that people feel as a shaking, quaking, or vibrating during an earthquake.

Seismic waves are generated when a huge mass of rock breaks and slides into a different position. As you learned in the section on folding, the plane between two rock masses that have moved into new relative positions is called a *fault.* Major earthquakes occur along existing fault planes or when a new fault is formed by the fracturing of rock. In either case, most earthquakes occur along a fault plane when there is displacement of one side relative to the other.

Most earthquakes occur along a fault plane and near Earth's surface. You might expect this to happen since the rocks near the surface are brittle and those deeper are more plastic from increased temperature and pressure. Shallow earthquakes are typical of those that occur at the boundary of the North American Plate, which is moving against the Pacific Plate. In California, the boundary between these two plates is known as the *San Andreas fault* (figure 16.11). The San Andreas fault runs north-south through California, with the Pacific Plate moving on one side and the North American Plate moving on the other.

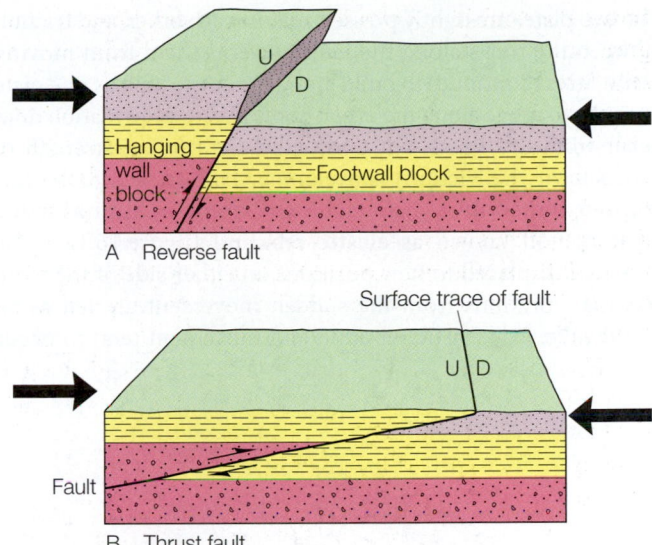

A   Reverse fault

B   Thrust fault

FIGURE 16.10 How compressive stress could produce (*A*) a reverse fault and (*B*) a thrust fault.

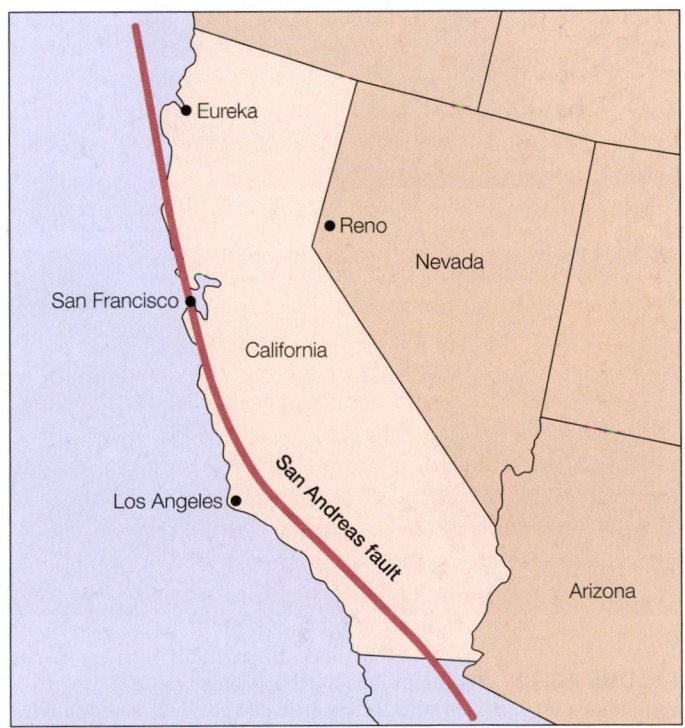

FIGURE 16.11 The San Andreas fault runs north-south for about 1,300 km (800 mi) through California.

The two plates are tightly pressed against each other, and friction between the rocks along the fault prevents them from moving easily. Stress continues to build along the entire fault as one plate attempts to move along the other. Some elastic deformation does occur from the stress, but eventually, the rupture strength of the rock (or the friction) is overcome. The stressed rock, now released of the strain, snaps suddenly into new positions in the phenomenon known as **elastic rebound** (figure 16.12). The rocks are displaced to new positions on either side of the fault, and the vibrations from the sudden movement are felt as an earthquake. The elastic rebound and movement tend to occur along short segments of the fault at different times rather than along long lengths. Thus, the resulting earthquake tends to be a localized phenomenon rather than a regional one.

Most earthquakes occur near plate boundaries. They do happen elsewhere, but those are rare. The actual place where seismic waves originate beneath the surface is called the *focus* of the earthquake. The focus is considered to be the center of the earthquake and the place of initial rock movement on a fault. The point on Earth's surface directly above the focus is called the earthquake *epicenter* (figure 16.13).

Seismic waves radiate outward from an earthquake focus, spreading in all directions through the solid Earth's interior like the sound waves from an explosion. As introduced in chapter 15, there are three kinds of waves: *P-waves*, *S-waves*, and *surface waves*. A *seismometer* is an instrument used to measure seismic waves. Information about S- and P-waves as well as about the size of an earthquake can be read from seismometer recordings. S- and P-waves provide information about the location and magnitude of an earthquake as well as information about Earth's interior.

Seismic S- and P-waves leave the focus of an earthquake at essentially the same time. As they travel away from the focus, they gradually separate because the P-waves travel faster than the S-waves. To locate an epicenter, at least three recording stations measure the time lag between the arrival of the first P-waves and the first slower S-waves. The difference in the speed between the two waves is a constant. Therefore, the farther they travel, the greater the time lag between the arrival of the faster P-waves and the slower S-waves (figure 16.14A). By measuring the time lag and knowing the speed of the two waves, it is possible to calculate the distance to their source. However, the calculated distance provides no information about the direction or location of the source of the waves. The location is found by first using the calculated distance as the radius of a circle drawn on a map. The place where the circles from the three recording stations intersect is the location of the source of the waves (figure 16.14B).

Earthquakes range from the many that are barely detectable to the few that cause widespread damage. The *intensity* of an earthquake is a measure of the effect on people and buildings. Destruction is caused by the seismic waves, which cause

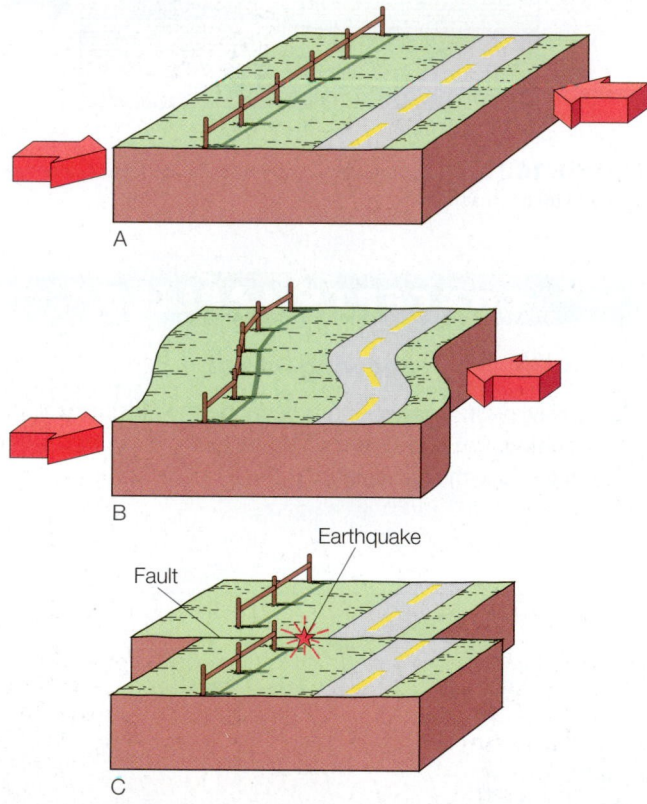

FIGURE 16.12 The elastic rebound theory of the cause of earthquakes. (*A*) Rock with stress acting on it. (*B*) Stress has caused strain in the rock. Strain builds up over a long period of time. (*C*) Rock breaks suddenly, releasing energy, with rock movement along a fault. Horizontal motion is shown; rocks can also move vertically. (*D*) Horizontal offset of rows in a lettuce field, 1979, El Centro, California.

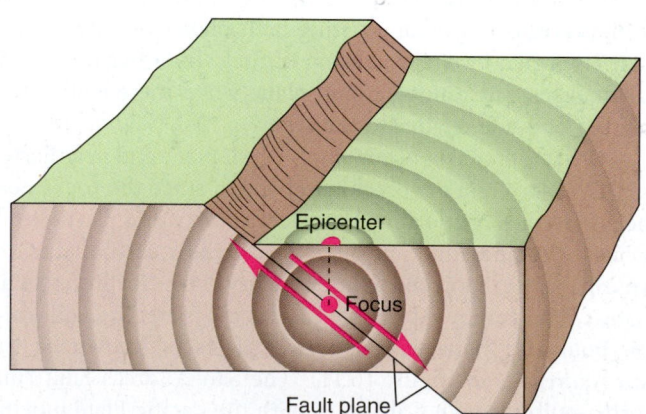

FIGURE 16.13 Simplified diagram of a fault, illustrating component parts and associated earthquake terminology.

the land and buildings to vibrate or shake. Vibrations during small quakes can crack windows and walls, but vibrations in strong quakes can topple bridges and buildings. Injuries and death are usually the result of falling debris, crumbling buildings, or falling bridges. Fire from broken gas pipes was a problem in the 1906 and 1989 earthquakes in San Francisco and the 1994 earthquake in Los Angeles. Broken water mains made it difficult to fight the 1906 San Francisco fires, but in 1989, fireboats and fire hoses using water pumped from the bay were able to extinguish the fires.

Other effects of earthquakes include landslides, displacement of the land surface, and tsunamis. Vertical, horizontal, or both vertical and horizontal displacement of the land can occur during a quake. People sometimes confuse cause and effect when they see a land displacement saying things like, "Look what the earthquake did!" The fact is that the movement of the land probably produced the seismic waves (the earthquake). The seismic waves did not produce the land displacement. Displacements from a single earthquake can be up to 10 to 15 m (about 30 to 50 ft), but such displacements rarely happen.

The size of an earthquake (if it was a "big one" or a "little one") can be measured in terms of vibrations, in terms of displacement, or in terms of the amount of energy released at the site of the earthquake. The larger the quake, the larger the waves recorded on a seismometer. From these recorded waves, scientists assign a number called the *magnitude*. Magnitude is a measure of the energy released during an earthquake. Earthquake magnitude is often reported using the *Richter scale* (table 16.1). The Richter scale was developed by Charles Richter, a seismologist at the California Institute of Technology in the early 1930s. The scale was based on the widest swing in the back-and-forth line traces of seismograph recording. The higher the magnitude of an earthquake, as measured by the Richter scale, the greater the (1) severity of the ground-shaking vibrations and (2) energy released by the earthquake. An increase of one on the Richter scale means that the amount of movement of the ground increased by a factor of ten, and the amount of energy released increased by a factor of thirty. An earthquake measuring below 3 on the scale is usually not felt by people near the epicenter. The largest earthquake measured so far had a magnitude over 9, but there is actually no upper limit to the scale. Today, professional seismologists rate the size of earthquakes in different ways, depending on what they are comparing and why, but each way results in logarithmic scales similar to the Richter scale.

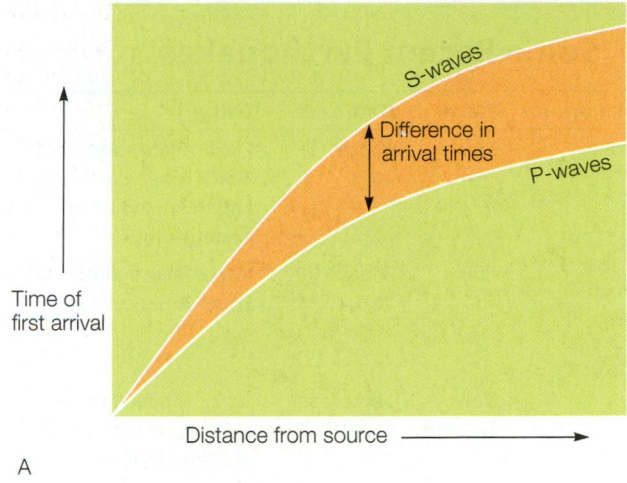

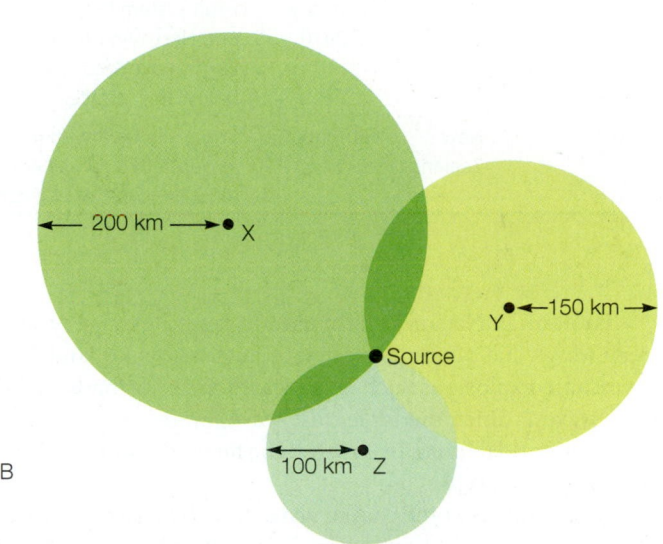

FIGURE 16.14 Use of seismic waves in locating the source of a disturbance. (*A*) Difference in times of first arrivals of P-waves and S-waves is a function of the distance from the source. (*B*) Triangulation using data from several seismograph stations allows location of the source.

TABLE 16.1 Effects of Earthquakes of Various Magnitudes

Richter Magnitudes	Description
0–2	Smallest detectable earthquake
2–3	Detected and measured but not generally felt
3–4	Felt as small earthquake but no damage occurs
4–5	Minor earthquake with local damage
5–6	Moderate earthquake with structural damage
6–7	Strong earthquake with destruction
7–8	Major earthquake with extensive damage and destruction
8–9	Great earthquake with total destruction

A Closer Look

Some Recent Earthquakes

Magnitude	Place	Date	Cause	Damage	Note
9.0	Japan	March 11, 2011	Pacific Plate subducted beneath North American Plate 6 mi. (9.7 km) deep and 80 miles (129 km) off eastern shore.	14,600 buildings destroyed; 118,000 damaged; more than 18,000 fatalities.	Movement of ocean floor created 23 ft (7 m) tsunami. More than 500 aftershocks during the first two weeks, 30 with a magnitude of 6 or greater.
8.8	Chile	February 27, 2010	Nazca Plate subducted beneath the South American Plate.	370,000 homes damaged; 486 fatalities.	Movement of ocean floor created tsunami up to 7.7 ft (2.3 m). More than 730 aftershocks during the first month, with 13 having a magnitude greater than 6.
7.1	New Zealand	September 4, 2010	Strike-slip fault within Pacific Plate at depth of 6 mi. (9.7 km) and west of Christchurch.	Widespread damage; no fatalities	No tsunami since epicenter was on land.
7.0	Haiti	February 12, 2010	Fault system between Caribbean and North American Plates slipped. Earthquake focus was 6.2 mi (10 km) below the surface.	97,300 houses destroyed; 222,600 fatalities.	Local tsunami caused four fatalities.
6.3	New Zealand	February 22, 2011	Strike-slip fault within Pacific Plate at depth of 3 mi (4.8 km) and 25 mi (40.2 km) west of Christchurch.	Widespread damage; 182 fatalities.	No tsunami since epicenter was on land. This is a different fault than the earthquake that occurred 6 months earlier.

Tsunami is a Japanese term used to describe the very large ocean waves that can be generated by an earthquake, landslide, or volcanic explosion. Such large waves were formerly called "tidal waves." Since the large, fast waves were not associated with tides or tidal forces in any way, the term *tsunami* or *seismic sea wave* is preferred.

A tsunami, like other ocean waves, is produced by a disturbance. Most common ocean waves are produced by winds, travel at speeds of 90 km/h (55 mi/h), and produce a wave height of 0.6 to 3 m (2 to 10 ft) when they break on the shore. A tsunami, on the other hand, is produced by some strong disturbance in the seafloor, travels at speeds of 725 km/h (450 mi/h), and can produce a wave height of over 8 m (about 25 ft) when it breaks on the shore. A tsunami may have a very long wavelength of up to 200 km (about 120 mi) compared to the wavelength of ordinary wind-generated waves of 400 m (1,300 ft). Because of its great wavelength, a tsunami does not just break on the shore, then withdraw. Depending on the seafloor topography, the water from a tsunami may continue to rise for five to ten minutes, flooding the coastal region before the wave withdraws. A gently sloping seafloor and a funnel-shaped bay can force tsunamis to great heights as they break on the shore. The size of a particular tsunami depends on how the seafloor was disturbed. Generally, an earthquake that causes the seafloor suddenly to rise or fall favors the generation of a large tsunami.

On December 26, 2004, an earthquake with a magnitude of 9.0 occurred west of the Indonesian island of Sumatra. This was the largest quake worldwide in four decades. The focus was about 10 km (about 6 mi) beneath the ocean floor at the interface of the India and Burma Plates and was caused by the release of stresses that develop as the India Plate subducts beneath the overriding Burma Plate. A 100 km (62 mi) wide rupture occurred in the

ocean floor, about 1,200 km (746 mi) long and parallel to the Sunda trench. The ocean floor was suddenly uplifted over 2 m (about 7 ft), with a movement about 10 m (about 33 ft) to the west-southwest. This displacement acted like a huge paddle at the bottom of the ocean, vertically displacing billions of tons of water and triggering a tsunami. The tsunami created a path of destruction across the 4,500 km (about 2,800 mi) wide Indian Ocean over the next seven hours. A series of very large waves struck the coast of the Indian Ocean, resulting in an estimated death toll of about 295,000 and over a million left homeless.

The world's largest recorded earthquakes have all occurred where one tectonic plate subducts beneath another. These include the magnitude-9.5 1960 Chile earthquake, the magnitude-9.2 1964 Prince William Sound, Alaska, earthquake, the magnitude-9.1 1957 Andreanof Islands, Alaska, earthquake, the magnitude-9.0 1952 Kamchatka earthquake, and the magnitude-9.0 2011 Japan earthquake. Such earthquakes often create large tsunamis that result in death and destruction over a wide area.

16.4 ORIGIN OF MOUNTAINS

Folding and faulting have created most of the interesting features of Earth's surface, and the most prominent of these features are **mountains.** Mountains are elevated parts of Earth's crust that rise abruptly above the surrounding surface. Most mountains do not occur in isolation but rather in chains or belts. These long, thin belts are generally found along the edges of continents rather than in the continental interior. There are a number of complex processes involved in the origin of mountains and mountain chains, and no two mountains are exactly alike. For convenience, however, mountains can be classified

FIGURE 16.15 The folded structure of the Appalachian Mountains, revealed by weathering and erosion, is obvious in this *Skylab* photograph of the Virginia-Tennessee-Kentucky boundary area. The clouds are over the Blue Ridge Mountains.

Volcanoes Change the World

A volcanic eruption changes the local landscape; that much is obvious. What is not so obvious are the worldwide changes that can happen just because of the eruption of a single volcano. Perhaps the most discussed change brought about by a volcano occurred back in 1815–16 after the eruption of Tambora in Indonesia. The Tambora eruption was massive, blasting huge amounts of volcanic dust, ash, and gas high into the atmosphere. Most of the ash and dust fell back to Earth around the volcano, but some dust particles and sulfur dioxide gas were pushed high into the stratosphere.

It is known today that the sulfur dioxide from explosive volcanic eruptions reacts with water vapor in the stratosphere, forming tiny droplets of diluted sulfuric acid. In the stratosphere, there is no convection, so the droplets of acid and dust from the volcano eventually formed a layer around the entire globe. This formed a haze that remains in the stratosphere for years, reflecting and scattering sunlight.

What were the effects of haze from Tambora? There were fantastic, brightly colored sunsets from the added haze in the stratosphere. On the other hand, it was also cooler than usual, presumably because of the reflected sunlight that did not reach Earth's surface. It snowed in New England in June 1816, and the cold continued into July. Crops failed, and 1816 became known as the "year without summer."

More information is available about the worldwide effects of present-day volcanic eruptions because there are now instruments to make more observations in more places. However, it is still necessary to do a great deal of estimating because of the relative inaccessibility of the worldwide stratosphere. It was estimated, for example, that the 1982 eruption of El Chichon in Mexico created enough haze in the stratosphere to reflect 5 percent of the solar radiation away from Earth. Researchers also estimated that the effects of the El Chichon eruption cooled the global temperatures by a few tenths of a degree for two or three years. The cooling did take place, but the actual El Chichon contribution to the cooling is not clear because of other interactions. Earth may have been undergoing global warming from the greenhouse effect, for example, so the El Chichon cooling effect could have actually been much greater. Other complicating factors such as the effects of El Niño make changes difficult to predict.

In June 1991, the Philippine volcano Mount Pinatubo erupted, blasting twice as much gas and dust into the stratosphere as El Chichon had about a decade earlier. The haze from such eruptions has the potential to cool the climate about 0.5°C (1°F). The overall result, however, will always depend on a possible greenhouse effect, a possible El Niño effect, and other complications.

according to three basic origins: (1) folding, (2) faulting, and (3) volcanic activity.

The major mountain ranges of Earth—the Appalachian, Rocky, and Himalayan Mountains, for example—have a great height that involves complex folding on a very large scale. The crust was thickened in these places as compressional forces produced tight, almost vertical folds. Thus, folding is a major feature of the major mountain ranges, but faulting and igneous intrusions are invariably also present. Differential weathering of different rock types produced the parallel features of the Appalachian Mountains that are so prominent in satellite photographs (figure 16.15). The folded sedimentary rocks of the Rockies are evident in the almost upright beds along the flanks of the front range.

A broad arching fold, which is called a dome, produced the Black Hills of South Dakota. The sedimentary rocks from the top of the dome have been weathered away, leaving a somewhat circular area of more resistant granite hills surrounded by upward-tilting sedimentary beds (figure 16.16). The Adirondack Mountains of New York are another example of this type of mountain formed from folding, called domed mountains.

Compression and relaxation of compressional forces on a regional scale can produce large-scale faults, shifting large crustal blocks up or down relative to one another. Huge blocks of rocks can be thrust to mountainous heights, creating a series of faultblock mountains. Faultblock mountains rise sharply from the surrounding land along the steeply inclined fault plane. The mountains are not in the shape of blocks, however, as weathering has carved them into their familiar mountainlike shapes (figure 16.17). The Teton Mountains of Wyoming and the Sierra Nevadas of California are classic examples of

faultblock mountains that rise abruptly from the surrounding land. The various mountain ranges of Nevada, Arizona, Utah, and southeastern California have large numbers of faultblock mountains that generally trend north and south.

Lava and other materials from volcanic vents can pile up to mountainous heights on the surface. These accumulations

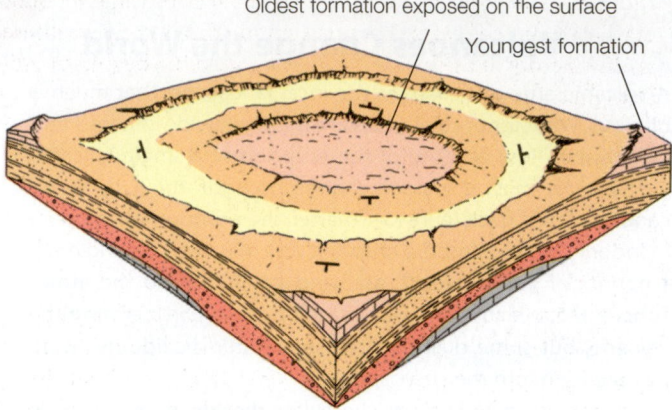

Oldest formation exposed on the surface

Youngest formation

FIGURE 16.16 A sketch of an eroded structural dome where all the rock layers dip away from the center.

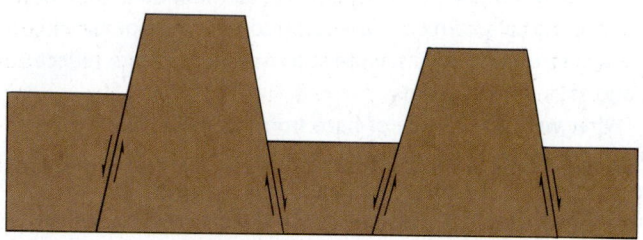

Original sharply bounded fault blocks softened by erosion and sedimentation

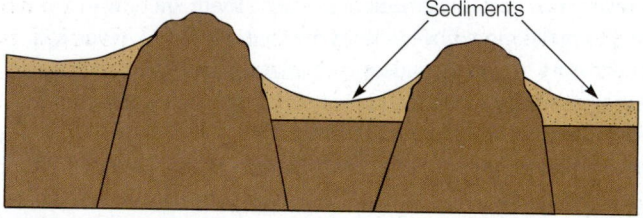

Sediments

FIGURE 16.17 Faultblock mountains are weathered and eroded as they are elevated, resulting in a rounded shape and sedimentation rather than sharply edged fault blocks.

can form local volcano-formed mountains near mountains produced by folding or faulting. Such mixed-origin mountains are common in northern Arizona, New Mexico, and western Texas. The Cascade Mountains of Washington and Oregon are a series of towering volcanic peaks, most of which are not active today. As a source of mountains, volcanic activity has an overall limited impact on the continents. The major mountains built by volcanic activity are the mid-oceanic ridges formed at diverging plate boundaries.

Deep within Earth, previously solid rock melts at high temperatures to become *magma,* a pocket of molten rock. Magma is not just melted rock alone, however, as the melt contains variable mixtures of minerals (resulting in different types of lava flows). It also includes gases such as water vapor, sulfur dioxide, hydrogen sulfide, carbon dioxide, and hydrochloric acid. You can often smell some of these gases around volcanic vents and hot springs. Hydrogen sulfide smells like rotten eggs or sewer gas. The sulfur smells like a wooden match that has just been struck.

FIGURE 16.18 This is the top of Mount St. Helens several years after the 1980 explosive eruption.

The gases dissolved in magma play a major role in forcing magma out of the ground. As magma nears the surface, it comes under less pressure, and this releases some of the dissolved gases from the magma. The gases help push the magma out of the ground. This process is similar to releasing the pressure on a can of warm soda, which releases dissolved carbon dioxide.

Magma works its way upward from its source below to Earth's surface, here to erupt into a lava flow or a volcano. A **volcano** is a hill or mountain formed by the extrusion of lava or rock fragments from magma below. Some lavas have a lower viscosity than others, are more fluid, and flow out over the land rather than forming a volcano. Such *lava flows* can accumulate into a plateau of basalt, the rock that the lava formed as it cooled and solidified. The Columbia Plateau of the states of Washington, Idaho, and Oregon is made up of layer after layer of basalt that accumulated from lava flows. Individual flows of lava formed basalt layers up to 100 m (about 330 ft) thick, covering an area of hundreds of square kilometers. In places, the Columbia Plateau is up to 3 km (about 2 mi) thick from the accumulation of many individual lava flows.

The hill or mountain of a volcano is formed by ejected material that is deposited in a conical shape. The materials are deposited around a central *vent,* an opening through which an eruption takes place. The *crater* of a volcano is a basinlike depression over a vent at the summit of the cone. Figure 16.18 is an aerial view of Mount St. Helens, looking down into the crater at a volcanic dome that formed as magma periodically welled upward into the floor of the crater. This photo was taken several years after Mount St. Helens erupted in May 1980. The volcano was quiet between that time and late 2004, when thousands of small earthquakes preceded the renewed growth of the lava dome inside the crater. According to the U.S. Geological Survey, the new dome grew to more than 152 m (500 ft) above the old dome before the volcano entered a relatively quiet state. Go to http://vulcan.wr.usgs.gov/News?framework.html for a fact sheet on the latest activity of Mount St. Helens and links to current information.

A volcano forms when magma breaks out at Earth's surface. Only a small fraction of all the magma generated actually reaches Earth's surface. Most of it remains below the ground, cooling and solidifying to form igneous rocks that were described in chapter 15. A large amount of magma that has crystallized below the surface is known as a *batholith*. A small protrusion from a batholith is called a *stock*. By definition, a stock has less than 100 km^2 (40 mi^2) of exposed surface area, and a batholith is larger. Both batholiths and stocks become exposed at the surface through erosion of the overlying rocks and rock materials, but not much is known about their shape below. The sides seem to angle away with depth, suggesting that they become larger with depth. The intrusion of a batholith sometimes tilts rock layers upward, forming a *hogback,* a ridge with equal slopes on its sides (figure 16.19). Other forms of intruded rock were formed as moving magma took the paths of least resistance, flowing into joints, faults, and planes between sedimentary bodies of rock. An intrusion that has flowed into a joint or fault that cuts across rock bodies is called a *dike.* A dike is usually tabular in shape, sometimes appearing as a great wall when exposed at the surface. One dike can occur by itself, but frequently dikes occur in great numbers, sometimes radiating out from a batholith like spokes around a wheel. If the intrusion flowed into the plane of contact between sedimentary rock layers, it is called a *sill.* A *laccolith* is similar to a sill but has an arched top where the intrusion has raised the overlying rock into a blisterlike uplift (figure 16.19).

Overall, the origin of mountain systems and belts of mountains such as the Cascades involves a complex mixture of volcanic activity as well as folding and faulting. An individual mountain, such as Mount St. Helens, can be identified as having a volcanic origin. The overall picture is best seen, however, from generalizations about how the mountains have grown along the edge of plates that are converging. Such converging boundaries are the places of folding, faulting, and associated earthquakes. They are also the places of volcanic activities, events that build and thicken Earth's crust. Thus, plate tectonics explains that mountains are built as the crust thickens at a convergent boundary between two plates. These mountains are slowly weathered and worn down as the next belt of mountains begins to build at the new continental edge.

16.5 PROCESSES THAT TEAR DOWN THE SURFACE

Sculpturing of Earth's surface takes place through agents and processes acting so gradually that humans are usually not aware that it is happening. Sure, some events such as a landslide or the movement of a big part of a beach by a storm are noticed. But the continual, slow, downhill drift of all the soil on a slope or the constant shift of grains of sand along a beach are outside the awareness of most people. People do notice the muddy water moving rapidly downstream in the swollen river after a storm, but few are conscious of the slow, steady dissolution of limestone by acid rain percolating through it. Yet, it is the processes of slow-moving, shifting grains and bits of rocks, and slow dissolving that will wear down the mountains, removing all the features of the landscape that you can see.

A mountain of solid granite on the surface of Earth might appear to be a very solid, substantial structure, but it is always undergoing slow and gradual changes. Granite on Earth's surface is exposed to and constantly altered by air, water, and other agents of change. It is altered both in appearance and in composition, slowly crumbling, and then dissolving in water. Smaller rocks and rock fragments are moved downhill by gravity or streams, exposing more granite that was previously deeply buried. The process continues, and ultimately—over much time—a mountain of solid granite is reduced to a mass of loose rock fragments and dissolved materials. The photograph in figure 16.20 is a snapshot of a mountain-sized rock mass in a stage somewhere between its formation and its eventual destruction to rock fragments. Can you imagine the length of time that such a process requires?

Weathering

The slow changes that result in the breaking up, crumbling, and destruction of any kind of solid rock are called **weathering.** The term implies changes in rocks from the action of the weather, but it actually includes chemical, physical, and biological processes. These weathering processes are important and necessary in (1) the rock cycle, (2) the formation of soils, and (3) the movement of rock materials over Earth's surface. Weathering is important in the rock cycle because it produces sediment, the raw materials for new rocks. It is important in the formation of soils because soil is an accumulation of rock fragments and organic matter. Because weathering reduces the sizes of rock particles, it is important in preparing the particles to be transported by wind or moving water.

Weathering breaks down rocks physically and chemically, and this breaking down can occur while the rocks are stationary or while they are moving. The process of physically removing

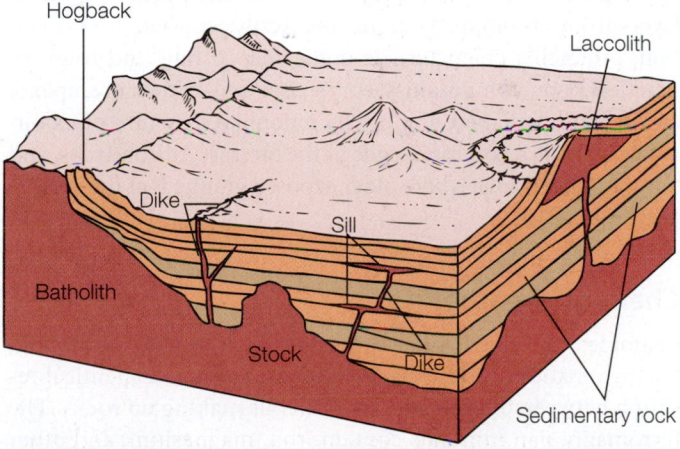

FIGURE 16.19 Here are the basic intrusive igneous bodies that form from volcanic activity.

FIGURE 16.20 The piles of rocks and rock fragments around a mass of solid rock is evidence that the solid rock is slowly crumbling away. This solid rock that is crumbling to rock fragments is in the Grand Canyon, Arizona.

weathered materials is called **erosion.** *Weathering* prepares the way for erosion by breaking solid rock into fragments. The fragments are then *eroded,* physically picked up by an agent such as a stream or a glacier. After they are eroded, the materials are then removed by *transportation.* Transportation is the movement of eroded materials by agents such as rivers, glaciers, wind, or waves. The weathering process continues during transportation. A rock being tumbled downstream, for example, is physically worn down as it bounces from rock to rock. It may be chemically altered as well as it is bounced along by the moving water. Overall, the combined action of weathering and erosion wears away and lowers the elevated parts of Earth and sculpts their surfaces.

There are two basic kinds of weathering that act to break down rocks: *mechanical weathering* and *chemical weathering.* *Mechanical weathering* is the physical breaking up of rocks without any changes in their chemical composition. Mechanical weathering results in the breaking up of rocks into smaller and

smaller pieces, so it is also called *disintegration.* If you smash a sample of granite into smaller and smaller pieces, you are mechanically weathering the granite. *Chemical weathering* is the alteration of minerals by chemical reactions with water, gases of the atmosphere, or solutions. Chemical weathering results in the dissolving or breaking down of the minerals in rocks, so it is also called *decomposition.* If you dissolve a sample of limestone in a container of acid, you are chemically weathering the limestone.

Mechanical Weathering

Examples of mechanical weathering in nature include the disintegration of rocks caused by (1) *wedging effects* and (2) the *effects of reduced pressure.* Wedging effects are often caused by the repeated freezing and thawing of water in the pores and small cracks of otherwise solid rock. If you have ever seen what happens when water in a container freezes, you know that freezing water expands and exerts a pressure on the sides of its container. As water in a pore or a crack of a rock freezes, it also expands, exerting a pressure on the walls of the pore or crack, making it slightly larger. The ice melts and the enlarged pore or crack again becomes filled with water for another cycle of freezing and thawing. As the process is repeated many times, small pores and cracks become larger and larger, eventually forcing pieces of rock to break off. This process is called **frost wedging** (figure 16.21A). It is an important cause of mechanical weathering in mountains and other locations where repeated cycles of water freezing and thawing occur. The roots of trees and shrubs can also mechanically wedge rocks apart as they grow into cracks. You may have noticed the work of roots when trees or shrubs have grown next to a sidewalk for some period of time (figure 16.22).

The other example of mechanical weathering is believed to be caused by the reduction of pressure on rocks. As more and more weathered materials are removed from the surface, the downward pressure from the weight of the material on the rock below becomes less and less. The rock below begins to expand upward, fracturing into concentric sheets from the effect of reduced pressure. These curved, sheetlike plates fall away later in the mechanical weathering process called *exfoliation* (figure 16.21B). *Exfoliation* is the term given to the process of spalling off of layers of rock, somewhat analogous to peeling layers from an onion. Granite commonly weathers by exfoliation, producing characteristic dome-shaped hills and rounded boulders. Stone Mountain, Georgia, is a well-known example of an exfoliation-shaped dome. The onionlike structure of exfoliated granite is a common sight in the Sierras, Adirondacks, and any mountain range where older exposed granite is at the surface (figure 16.23).

Chemical Weathering

Examples of chemical weathering include (1) oxidation, (2) carbonation, and (3) hydration. *Oxidation* is a chemical reaction between oxygen and the minerals making up rocks. The ferromagnesian minerals contain iron, magnesium, and other metal ions in a silicate structure. Iron can react with oxygen to produce several different iron oxides, each with its own char-

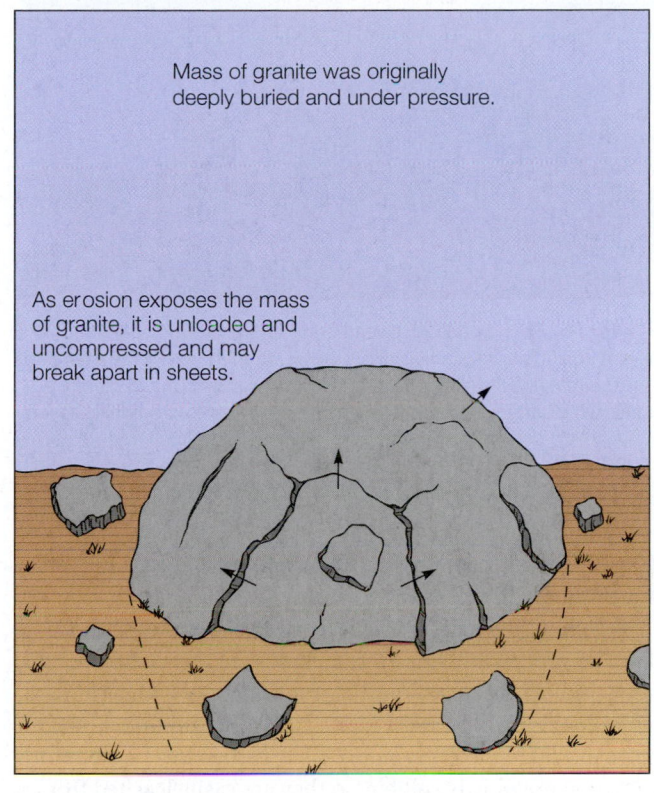

A

B

FIGURE 16.21 (*A*) Frost wedging and (*B*) exfoliation are two examples of mechanical weathering, or disintegration, of solid rock.

acteristic color. The most common iron oxide (hematite) has a deep red color. Other oxides of iron are brownish to yellow-brownish. It is the presence of such iron oxides that colors many sedimentary rocks and soils. The red soils of Oklahoma, Georgia, and many other places are colored by the presence of iron oxides produced by chemical weathering.

Carbonation is a chemical reaction between carbonic acid and the minerals making up rocks. Rainwater is naturally somewhat acidic because it dissolves carbon dioxide from the air. This forms a weak acid known as carbonic acid (H_2CO_3), the same acid that is found in carbonated soda. Carbonic acid rain falls on the land, seeping into cracks and crevices where it reacts with minerals. Limestone, for example, is easily weathered to a soluble form by carbonic acid. The limestone caves of Missouri, Kentucky, New Mexico, and elsewhere were produced by the chemical weathering of limestone by carbonation (figure 16.24). Minerals containing calcium, magnesium, sodium, potassium, and

A

B

FIGURE 16.22 Growing trees can break, separate, and move solid rock. (*A*) Note how this tree has raised the sidewalk. (*B*) This tree is surviving by growing roots into tiny joints and cracks, which become larger as the tree grows.

FIGURE 16.23 Spheroidal weathering of granite. The edges and corners of an angular rock are attacked by weathering from more than one side and retreat faster than flat rock faces. The result is rounded granite boulders, which often shed partially weathered minerals in onionlike layers.

A

iron are chemically weathered by carbonation to produce salts that are soluble in water.

Hydration is a reaction between water and the minerals of rocks. The process of hydration includes (1) the dissolving of a mineral and (2) the combining of water directly with a mineral. Some minerals, for example, halite (which is sodium chloride), dissolve in water to form a solution. The carbonates formed from carbonation are mostly soluble, so they are easily leached from a rock by dissolving. Water also combines directly with some minerals to form new, different minerals. The feldspars, for example, undergo hydration and carbonation to produce (1) water-soluble potassium carbonate, (2) a chemical product that combines with water to produce a clay mineral, and (3) silica. The silica, which is silicon dioxide (SiO_2), may appear as a suspension of finely divided particles or in solution.

Mechanical and chemical weathering are interrelated, working together in breaking up and decomposing solid rocks of Earth's surface. In general, mechanical weathering results in cracks in solid rocks and broken-off coarse fragments. Chemical weathering results in finely pulverized materials and ions in solution, the ultimate decomposition of a solid rock. Consider, for example, a mountain of solid granite, the most common rock found on continents. In general, granite is made up of 65 percent feldspars, 25 percent quartz, and about 10 percent ferromagnesian minerals. Mechanical weathering begins the destruction process as exfoliation and frost wedging create cracks in the solid mass of granite. Rainwater, with dissolved oxygen and carbon dioxide, flows and seeps into the cracks and reacts with ferromagnesian minerals to form soluble carbonates and metal oxides. Feldspars undergo carbonation and hydration, forming clay minerals and soluble salts, which are washed away. Quartz is less susceptible to chemical weathering and remains mostly unchanged to form sand grains. The end products of the complete weathering of granite are quartz sand, clay minerals, metal oxides, and soluble salts.

B

FIGURE 16.24 Limestone caves develop when slightly acidic groundwater dissolves limestone along joints and bedding planes, carrying away rock components in solution. (*A*) Joints and bedding planes in a limestone bluff. (*B*) This stream has carried away less-resistant rock components, forming a cave under the ledge.

Erosion

Weathering has prepared the way for erosion and for some agent of transportation to move or carry away the fragments,

FIGURE 16.25 The slow creep of soil is evidenced by the strange growth pattern of these trees.

FIGURE 16.26 Moving streams of water carry away dissolved materials and sediments as they slowly erode the land.

clays, and solutions that have been produced from solid rock. The weathered materials can be moved to a lower elevation by the direct result of gravity acting alone. They can also be moved to a lower elevation by gravity acting through some intermediate agent, such as running water, wind, or glaciers. The erosion of weathered materials as a result of gravity alone will be considered first.

Mass Movement

Gravity constantly acts on every mass of materials on the surface of Earth, pulling parts of elevated regions toward lower levels. Rocks in the elevated regions are able to temporarily resist this constant pull through their cohesiveness with a main rock mass or by the friction between the rock and the surface of the slope. Whenever anything happens to reduce the cohesiveness or to reduce the friction, gravity pulls the freed material to a lower elevation. Thus, gravity acts directly on individual rock fragments and on large amounts of surface materials as a mass, pulling all to a lower elevation. Erosion caused by gravity acting directly is called **mass movement** (also called mass wasting). Mass movement can be so slow that it is practically imperceptible. *Creep,* the slow downhill movement of soil down a steep slope, for example, is detectable only from the peculiar curved growth patterns of trees growing in the slowly moving soil (figure 16.25). At the other extreme, mass movement can be as sudden and swift as a rock bounding and clattering down a slope below a cliff. A *landslide* is a generic term used to describe any slow to rapid movement of any type or mass of materials, from the short slump of a hillside to the slide of a whole mountainside. Either slow or sudden, mass movement is a small victory for gravity in the ongoing process of leveling the landmass of Earth.

Running Water

Running water is the most important of all the erosional agents of gravity that remove rock materials to lower levels. Erosion by running water begins with rainfall. Each raindrop impacting the soil moves small rock fragments about but also begins to dissolve some of the soluble products of weathering. If the rainfall is heavy enough, a shallow layer, or sheet, of water forms on the surface, transporting small fragments and dissolved materials across the surface. This *sheet erosion* picks up fragments and dissolved material, then transports them to small streams at lower levels (figure 16.26). The small streams move to larger channels, and the running water transports materials three different ways: (1) as dissolved rock materials carried in solution, (2) as clay minerals and small grains carried in suspension, and (3) as sand and larger rock fragments that are rolled, bounced, and slid along the bottom of the streambed. Just how much material is eroded and transported by the stream depends on the volume of water, its velocity, and the load that it is already carrying.

Streams and major rivers are at work, for the most part, twenty-four hours a day every day of the year moving rock fragments and dissolved materials from elevated landmasses to the oceans. Any time you see mud, clay, and sand being transported by a river, you know that the river is at work moving mountains, bit by bit, to the ocean. It has been estimated that rivers remove enough dissolved materials and sediments to lower the whole surface of the United States flat in a little over 20 million years, a very short time compared to the 4.6-billion-year age of Earth.

FIGURE 16.27 A river usually stays in its channel, but during a flood, it spills over and onto the adjacent flat land called the *floodplain*.

In addition to transporting materials that were weathered and eroded by other agents of erosion, streams do their own erosive work. Streams can dissolve soluble materials directly from rocks and sediments. They also quarry and pluck fragments and pieces of rocks from beds of solid rock by hydraulic action. Most of the erosion accomplished directly by streams, however, is done by the more massive fragments that are rolled, bounced, and slid along the streambed and against each other. This results in a grinding action on the fragments and a wearing away of the streambed.

As a stream cuts downward into its bed, other agents of erosion such as mass movement begin to widen the channel as materials on the stream bank tumble into the moving water. The load that the stream carries is increased by this activity, which slows the stream. As the stream slows, it begins to develop bends, or *meanders* along the channel. Meanders have a dramatic effect on stream erosion because the water moves faster around an outside bank than it does around the inside bank downstream. This difference in stream velocity means that the stream has a greater erosion ability on the outside, downstream side and less on the sheltered area inside of curves. The stream begins to widen the floor of the valley through which it runs by eroding on the outside of the meander, then depositing the eroded material on the inside of another bend downstream. The stream thus begins to erode laterally, slowly working its way across the land. Sometimes two bends in the stream meet, forming a cutoff meander called an *oxbow lake*.

A stream, along with mass movement, develops a valley on a widening floodplain. A **floodplain** is the wide, level floor of a valley built by a stream (figure 16.27). It is called a floodplain because this is where the stream floods when it spills out of its channel. The development of a stream channel into a widening floodplain seems to follow a general, idealized aging pattern (figure 16.28). When a stream is on a recently uplifted landmass, it has a steep gradient, a vigorous, energetic ability to erode the land, and characteristic features known as the stage of youth. *Youth* is characterized by a steep gradient, a V-shaped valley without a floodplain, and the presence of features that inter-

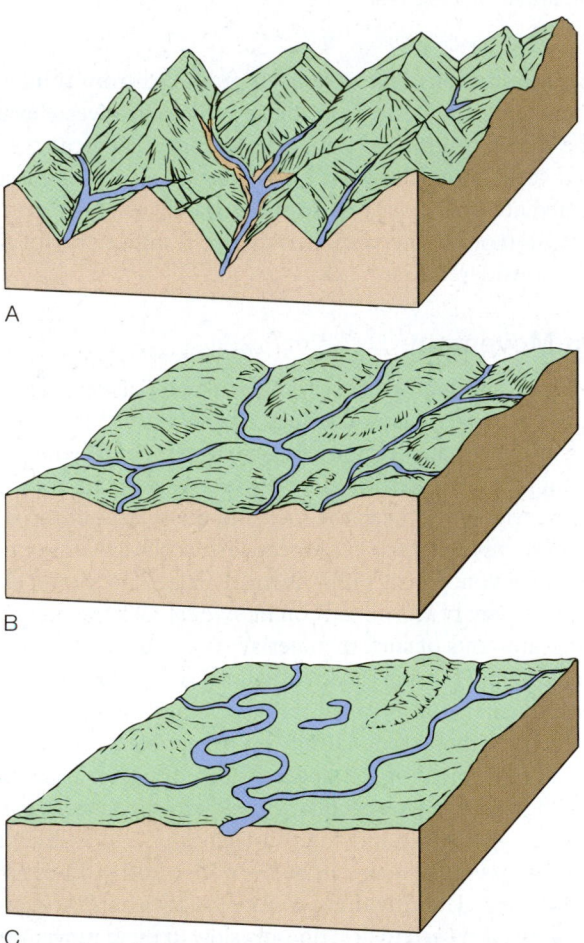

FIGURE 16.28 Three stages in the aging and development of a stream valley: (*A*) youth, (*B*) maturity, and (*C*) old age.

rupt its smooth flow such as boulders in the streambed, rapids, and waterfalls (figure 16.29). Stream erosion during youth is predominantly downward. The stream eventually erodes its way into *maturity* by eroding away the boulders, rapids, and waterfalls, and in general smoothing and lowering the stream gradient. During maturity, meanders form over a wide floodplain that now occupies the valley floor. The higher elevations are now more sloping hills at the edge of the wide floodplain rather than steep-sided walls close to the river channel. *Old age* is marked by a very low gradient in extremely broad, gently sloping valleys. The stream now flows slowly in broad meanders over the wide floodplain. Floods are more common in old age since the stream is carrying a full load of sediments and flows sluggishly.

Many assumptions are made in any generalized scheme of the erosional aging of a stream. Streams and rivers are dynamic systems that respond to local conditions, so it is possible to find an "old age feature" such as meanders in an otherwise youthful valley. This is not unlike finding a gray hair on an eighteen-year-old youth, and in this case, the presence of the gray hair does not mean old age. In general, old age characteristics are observed near the *mouth* of a stream, where it flows into an ocean, lake, or another stream. Youthful characteristics are observed at the *source*, where the water collects to first form the stream channel. As the stream slowly lowers the land, the old age characteristics will move slowly but surely upstream from the mouth toward the source.

When the stream flows into the ocean or a lake, it loses all of its sediment-carrying ability. It drops the sediments, forming a deposit at the mouth called a **delta** (figure 16.30). Large

FIGURE 16.29 The waterfall and rapids on the Yellowstone River in Wyoming indicate that the river is actively downcutting. Note the V-shaped cross-profile and lack of floodplain, characteristics of a young stream valley.

A

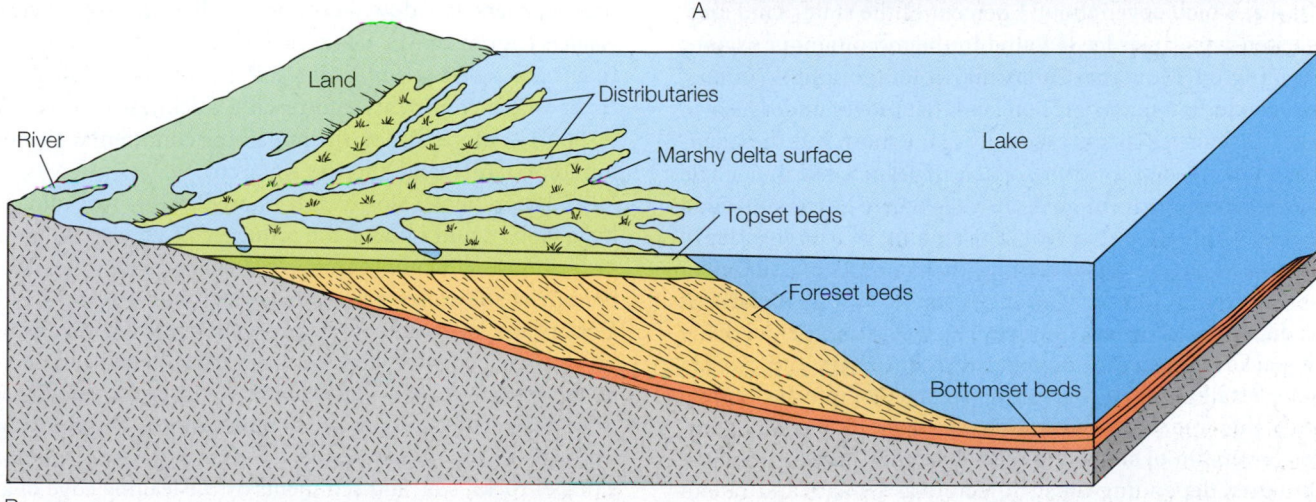

B

FIGURE 16.30 (*A*) Satellite image of the Mississippi River Delta. Note the sediment-laden water and how the land is being built outward by river sedimentation. (*B*) Cross section showing how a small delta might form. Large deltas are more complicated than this.

FIGURE 16.31 Valley glacier located in Antarctica.

rivers such as the Mississippi River have large and extensive deltas that actually extend the landmass more and more over time. In a way, you could think of the Mississippi River delta as being formed from pieces and parts of the Rocky Mountains, the Ozark Mountains, and other elevated landmasses that the Mississippi has carried there over time.

Glaciers

Glaciers presently cover only about 10 percent of Earth's continental land area, and since much of this is at higher latitudes, it might seem that glaciers would not have much of an overall effect in eroding the land. However, ice has sculptured much of the present landscape, and features attributed to glacial episodes are found over about three-quarters of the continental surface. Only a few tens of thousands of years ago, sheets of ice covered major portions of North America, Europe, and Asia. Today, the most extensive glaciers in the United States are those of Alaska, which cover about 3 percent of the state's land area. Less extensive glacier ice is found in the mountainous regions of Washington, Montana, California, Colorado, and Wyoming.

A **glacier** is a mass of ice on land that moves under its own weight. Glacier ice forms gradually from snow, but the quantity of snow needed to form a glacier does not fall in a single winter. Glaciers form in cold climates where some snow and ice persist throughout the year. The amount of winter snowfall must exceed the summer melting to accumulate a sufficient mass of snow to form a glacier. As the snow accumulates, it is gradually transformed into ice. The weight of the overlying snow packs it down, driving out much of the air, and causing it to recrystallize into a coarser, denser mass of interlocking ice crystals that appears to have a blue to deep blue color. Complete conversion of snow into glacial ice may take from 500 to 3,500 years, depending on such factors as climate and rate of snow accumulation at the top of the pile. Eventually, the mass of ice will become large enough that it begins to flow, spreading out from the accumulated mass. Glaciers that form at high

elevations in mountainous regions are called *alpine glaciers*. If these glaciers flow down into a valley they are also called *valley glaciers* (figure 16.31). Glaciers that cover a large area of a continent are called *continental glaciers*. Continental glaciers can cover whole continents and reach a thickness of 1 km (about 3,280 ft) or more. Today, the remaining continental glaciers are found on Greenland and the Antarctic.

Glaciers move slowly and unpredictably, spreading like a huge blob of putty under the influence of gravity. As an alpine glacier moves downhill through a V-shaped valley, the sides and bottom of the valley are eroded wider and deeper. When the glacier later melts, the V-shaped valley has been transformed into a U-shaped valley that has been straightened and deepened by the glacial erosion. The glacier does its erosional work using three different techniques: (1) by bulldozing, (2) by plucking, and (3) by abrasion. *Bulldozing,* as the term implies, is the pushing along of rocks, soil, and sediments by the leading edge of an advancing glacier. Deposits of bulldozed rocks and other materials that remain after the ice melts are called *moraines. Plucking* occurs as water seeps into cracked rocks and freezes, becoming a part of

People Behind the Science

James Hutton (1726–1797)

James Hutton was a Scottish natural philosopher who pioneered uniformitarian geology. The son of an Edinburgh merchant, Hutton studied at Edinburgh University, Paris, and Leiden, training first for the law but taking his doctorate in medicine in 1749 (though he never practiced). He spent the next two decades traveling and farming in the southeast of Scotland. During this time, he cultivated a love of science and philosophy, developing a special taste for geology. About 1768, he returned to his native Edinburgh. A friend of Joseph Black, William Cullen, and James Watt, Hutton shone as a leading member of the scientific and literary establishment, playing a large role in the early history of the Royal Society of Edinburgh and in the Scottish Enlightenment.

Hutton wrote widely on many areas of natural science, including chemistry (where he opposed Lavoisier), but he is best known for his geology, set out in his *Theory of the Earth*, of which a short version appeared in 1788, followed by the definitive statement in 1795. In that work, Hutton attempted (on the basis of both theoretical considerations and personal fieldwork) to demonstrate that Earth formed a steady-state system in which terrestrial causes had always been of the same kind as at present, acting with comparable intensity (the principle later known as uniformitarianism). In Earth's economy, in the imperceptible creation and devastation of landforms, there was no vestige of a beginning nor prospect of an end. Continents were continually being gradually eroded by rivers and weather. Denuded debris accumulated on the seabed, to be consolidated into strata and subsequently thrust upward to form new continents, thanks to the action of Earth's central heat. Nonstratified rocks such as granite were of igneous origin. All Earth's processes were exceptionally leisurely, and hence, Earth must be incalculably old.

Though supported by the experimental findings of Sir James Hall, Hutton's theory was vehemently attacked in its day, partly because it appeared to point to an eternal Earth and hence to atheism. It found more favor when popularized by Hutton's friend, John Playfair, and later by Charles Lyell. The notion of uniformitarianism still forms the groundwork for much geological reasoning.

Source: Modified from the *Hutchinson Dictionary of Scientific Biography.* © RM, 2011. All rights reserved. Helicon Publishing is a division of RM.

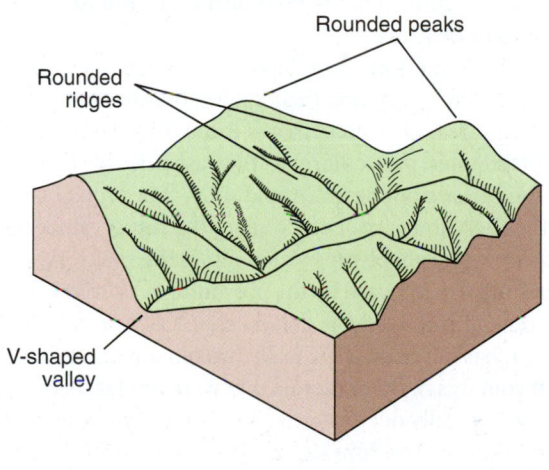

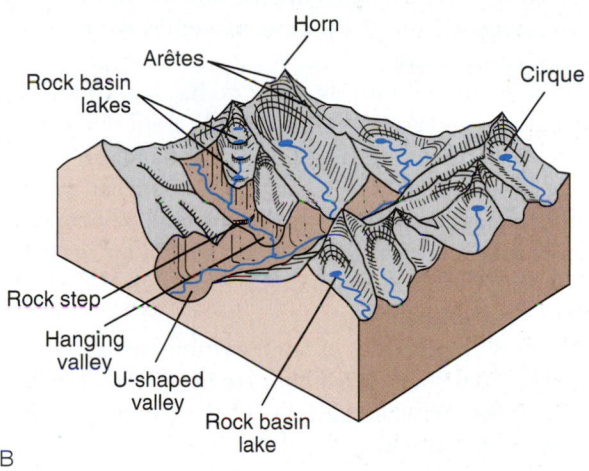

FIGURE 16.32 (*A*) A stream-carved mountainside before glaciation. (*B*) The same area after glaciation, with some of the main features of mountain glaciation labeled.

the solid glacial mass. As the glacier moves on, it pulls the fractured rock apart and plucks away chunks of it. The process is accelerated by the frost-wedging action of the freezing water. Plucking at the uppermost level of an alpine glacier, combined with weathering of the surrounding rocks, produces a rounded or bowl-like depression known as a *cirque* (figure 16.32). *Abrasion* occurs as the rock fragments frozen into the moving glacial ice scratch, polish, and grind against surrounding rocks at the base and along the valley walls. The result of this abrasion is the pulverizing of rock into ever finer fragments, eventually producing a powdery, silt-sized sediment called *rock flour*. Suspended rock flour in meltwater from a glacier gives the water a distinctive gray to blue-gray color.

Glaciation is continuously at work eroding the landscape in Alaska and many mountainous regions today. The glaciation that formed the landscape features in the Rockies, the Sierras, and across the northeastern United States took place thousands of years ago.

Wind

Like running water and moving glaciers, wind also acts as an agent shaping the surface of the land. It can erode, transport, and deposit materials. However, wind is considerably less efficient than ice or water in modifying the surface. Air is much less dense and does not have the eroding or carrying power of water or ice. In addition, a stream generally flows most of the time, but the wind blows only occasionally in most locations. Thus, on a worldwide average, winds move only a few percent as much material as do streams. Wind also lacks the ability to attack rocks chemically as water does through carbonation and

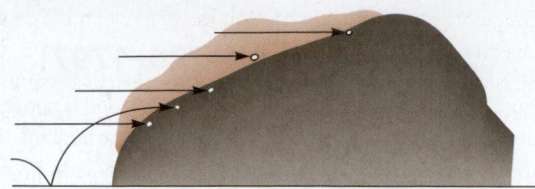

If wind is predominantly from one direction, rocks will be planed off or flattened on the upwind side.

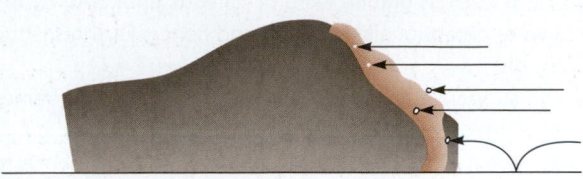

With a persistent shift in wind direction, additional facets are cut in the rock.

FIGURE 16.33 Ventifact formation by abrasion from one or several directions.

other processes, and the wind cannot carry dissolved sediments in solution. Even in many deserts, more sediment is moved during the brief periods of intense surface runoff following the occasional rainstorms than is moved by wind during the prolonged dry periods.

Flowing air and moving water do have much in common as agents of erosion since both are fluids. Both can move larger particles by rolling them along the surface and can move finer particles by carrying them in suspension. Both can move larger and more massive particles with increased velocities. Water is denser and more viscous than air, so it is more efficient at transporting quantities of material than is the wind, but the processes are quite similar.

Two major processes of wind erosion are called (1) abrasion and (2) deflation. *Wind abrasion* is a natural sandblasting process that occurs when the particles carried along by the wind break off small particles and polish what they strike. Generally, the harder mineral grains such as quartz sand accomplish this best near the ground where the wind is bouncing them along. Wind abrasion can strip paint from a car exposed to the moving particles of a dust storm, eroding the paint along with rocks on the surface. Rocks and boulders exposed to repeated wind

storms where the wind blows consistently from one or a few directions may be planed off from the repeated action of this natural sandblasting. Rocks sculptured by wind abrasion are called *ventifacts,* after the Latin meaning "wind-made" (figure 16.33).

Deflation, after the Latin meaning "to blow away," is the widespread picking up of loose materials from the surface. Deflation is naturally most active where winds are unobstructed and the materials are exposed and not protected by vegetation. These conditions are often found on deserts, beaches, and unplanted farmland between crops. During the 1930s, several years of drought killed the native vegetation in the Plains states during a period of increased farming activity. Unusually strong winds eroded the unprotected surface, removing and transporting hundreds of millions of tons of soil. This period of prolonged drought, dust storms, and general economic disaster for farmers in the area is known as the Dust Bowl episode.

The most common wind-blown deposits are (1) dunes and (2) loess. A **dune** is a low mound or ridge of sand or other sediments. Dunes form when sediment-bearing wind encounters an obstacle that reduces the wind velocity. With a slower velocity, the wind cannot carry as large a load, so sediments are deposited on the surface. This creates a windbreak, which results in a growing obstacle, a dune. Once formed, a dune tends to migrate, particularly if the winds blow predominantly from one direction. Dunes are commonly found in semiarid areas or near beaches.

Another common wind deposit is called *loess* (pronounced "luss"). Loess is a very fine dust, or silt, that has been deposited over a large area. One such area is located in the central part of the United States, particularly to the east sides of the major rivers of the Mississippi basin. Apparently, this deposit originated from the rock flour produced during the last great ice age. The rock flour was probably deposited along the major river valleys and later moved eastward by the prevailing westerly winds. Since rock flour is produced by the mechanical grinding action of glaciers, it has not been chemically broken down. Thus, the loess deposit contains many minerals that were not leached out of the deposit as typically occurs with chemical weathering. It also has an open, porous structure since it does not have as much of the chemically produced clay minerals. The good moisture-holding capacity from this open structure, together with the presence of minerals that serve as plant nutrients, make farming soils formed from loess deposits particularly productive.

SUMMARY

The *principle of uniformity* is the frame of reference that the same geologic processes you see changing rocks today are the same processes that changed them in the past.

Diastrophism is the process of deformation that changes Earth's surface, and the movement of magma is called *vulcanism*. Diastrophism, *vulcanism,* and *earthquakes* are closely related, and their occurrence can be explained most of the time by events involving *plate tectonics.*

Stress is a force that tends to compress, pull apart, or deform a rock, and the adjustment to stress is called *strain.* Rocks respond to

stress by (1) withstanding the stress without change, (2) undergoing *elastic strain,* (3) undergoing *plastic strain,* or (4) by *breaking in fracture strain.* Exactly how a particular rock responds to stress depends on (1) the nature of the rock, (2) the temperature, and (3) how quickly the stress is applied over time.

Deeply buried rocks are at a higher temperature and tend to undergo *plastic deformation,* resulting in a wrinkling of the layers into *folds.* The most common are an arch-shaped fold called an *anticline* and a trough-shaped fold called a *syncline.* Anticlines and synclines are most easily observed in sedimentary rocks because they have bedding planes, or layers.

Rocks near the surface tend to break from a sudden stress. A break with movement on one side of the break relative to the other side is called a *fault*. The vibrations that move out as waves from the movement of rocks are called an *earthquake*. The actual place where an earthquake originates is called its *focus*. The place on the surface directly above a focus is called an *epicenter*. There are three kinds of waves that travel from the focus: *S-, P-,* and *surface waves*. The magnitude of earthquake waves is measured on the *Richter scale*.

Folding and faulting produce prominent features on the surface called mountains. *Mountains* can be classified as having an origin of *folding, faulting,* or *volcanic*. In general, mountains that occur in long narrow belts called *ranges* have an origin that can be explained by *plate tectonics*.

Weathering is the breaking up, crumbling, and destruction of any kind of solid rock. The process of physically picking up weathered rock materials is called *erosion*. After the eroded materials are picked up, they are removed by *transportation* agents. The combined action of weathering, erosion, and transportation wears away and lowers the surface of Earth.

The physical breaking up of rocks is called *mechanical weathering*. Mechanical weathering occurs by *wedging effects* and the *effects of reduced pressure*. *Frost wedging* is a wedging effect that occurs from repeated cycles of water freezing and thawing. The process of spalling off of curved layers of rock from reduced pressure is called *exfoliation*.

The breakdown of minerals by chemical reactions is called *chemical weathering*. Examples include *oxidation,* a reaction between oxygen and the minerals making up rocks; *carbonation,* a reaction between carbonic acid (carbon dioxide dissolved in water) and minerals making up rocks; and *hydration,* the dissolving or combining of a mineral with water. When the end products of complete weathering of rocks are removed directly by gravity, the erosion is called *mass movement*. Erosion and transportation also occur through the agents of *running water, glaciers,* or *wind.* Each creates their own characteristic features of erosion and deposition.

KEY TERMS

anticline (p. **355**)
delta (p. **369**)
dune (p. **372**)
earthquake (p. **357**)
elastic rebound (p. **358**)
erosion (p. **364**)

fault (p. **356**)
floodplain (p. **368**)
folds (p. **354**)
frost wedging (p. **364**)
glacier (p. **370**)
mass movement (p. **367**)

mountains (p. **360**)
normal fault (p. **356**)
principle of uniformity (p. **353**)
reverse fault (p. **356**)
seismic waves (p. **357**)
strain (p. **354**)

stress (p. **354**)
syncline (p. **355**)
thrust fault (p. **356**)
tsunami (p. **360**)
volcano (p. **362**)
weathering (p. **363**)

APPLYING THE CONCEPTS

Answers are located in appendix F.

1. The basic difference in the frame of reference called the principle of uniformity and the catastrophic frame of reference used by previous thinkers is
   a. the energy for catastrophic changes is much less.
   b. the principle of uniformity requires more time.
   c. catastrophic changes have a greater probability of occurring.
   d. None of the above is correct.

2. The difference between elastic deformation and plastic deformation of rocks is that plastic deformation
   a. permanently alters the shape of a rock layer.
   b. always occurs just before a rock layer breaks.
   c. returns to its original shape after the pressure is removed.
   d. All of the above are correct.

3. Whether a rock layer subjected to stress undergoes elastic deformation, plastic deformation, or rupture depends on
   a. the temperature of the rock.
   b. the confining pressure on the rock.
   c. how quickly or slowly the stress is applied over time.
   d. All of the above are correct.

4. When subjected to stress, rocks buried at great depths are under great pressure at high temperatures, so they tend to undergo
   a. no change because of the pressure.
   b. elastic deformation because of the high temperature.
   c. plastic deformation.
   d. breaking or rupture.

5. Earthquakes that occur at the boundary between two tectonic plates moving against each other occur along
   a. the entire length of the boundary at once.
   b. short segments of the boundary at different times.
   c. the entire length of the boundary at different times.
   d. None of the above is correct.

6. Each higher number of the Richter scale
   a. increases with the magnitude of an earthquake.
   b. means ten times more ground movement.
   c. indicates about thirty times more energy released.
   d. means all of the above.

7. Other than igneous activity, all mountain ranges have an origin resulting from
   a. folding.
   b. faulting.
   c. stresses.
   d. sedimentation.

8. The preferred name for the very large ocean waves that are generated by an earthquake, landslide, or volcanic explosion is
   a. tidal wave.
   b. tsunami.
   c. tidal bore.
   d. Richter wave.

9. Freezing water exerts pressure on the wall of a crack in a rock mass, making the crack larger. This is an example of
   a. mechanical weathering.
   b. chemical weathering.
   c. exfoliation.
   d. hydration.

10. Which of the following would have the greatest overall effect in lowering the elevation of a continent such as North America?
   a. continental glaciers
   b. alpine glaciers
   c. wind
   d. running water

11. Broad meanders on a very wide, gently sloping floodplain with oxbow lakes are characteristics you would expect to find in a river valley during what stage?
   a. newborn
   b. youth
   c. maturity
   d. old age

12. A glacier forms when
   a. the temperature does not rise above freezing.
   b. snow accumulates to form ice, which begins to flow.
   c. a summer climate does not occur.
   d. a solid mass of snow moves downhill under the influence of gravity.

QUESTIONS FOR THOUGHT

1. What is the principle of uniformity? What are the underlying assumptions of this principle?

2. Describe the responses of rock layers to increasing compressional stress when it (a) increases slowly on deeply buried, warm layers; (b) it increases slowly on cold rock layers; and (c) it is applied quickly to rock layers of any temperature.

3. Describe the difference between a syncline and an anticline, using sketches as necessary.

4. What does the presence of folded sedimentary rock layers mean about the geologic history of an area?

5. Describe the conditions that would lead to faulting as opposed to folding of rock layers.

6. How would plate tectonics explain the ocurrence of normal faulting? Reverse faulting?

7. What is an earthquake? What produces an earthquake?

8. Where would the theory of plate tectonics predict that earthquakes would occur?

9. Describe how the location of an earthquake is identified by a seismic recording station.

10. Granite is the most common rock found on continents. What are the end products after granite has been completely weathered? What happens to these weathering products?

11. Describe three ways in which a river erodes its channel.

12. What is a floodplain?

13. Describe the characteristic features associated with stream erosion as the stream valley passes through the stages of youth, maturity, and old age.

14. What is rock flour and how is it produced?

15. Could a glacier erode the land lower than sea level? Explain.

16. Explain why glacial erosion produces a U-shaped valley, but stream erosion produces a V-shaped valley.

17. Compare the features caused by stream erosion, wind erosion, and glacial erosion.

18. Compare the materials deposited by streams, wind, and glaciers.

19. Why would mechanical weathering speed up chemical weathering in a humid climate but not in a dry climate?

20. Discuss all the reasons you can in favor of and in opposition to clearing away and burning tropical rainforests for agricultural purposes.

FOR FURTHER ANALYSIS

1. Evaluate the statement "the present is the key to the past" as it represents the principle of uniformity. What evidence supports this principle?

2. Does the theory of plate tectonics support or not support the principle of uniformity? Provide evidence to support your answer.

3. What are the significant similarities and differences between elastic deformation and plastic deformation?

4. Explain the combination of variables that results in solid rock layers folding rather than faulting.

5. Analyze why you would expect most earthquakes to occur as localized, shallow occurrences near a plate boundary.

6. What are the significant similarities and differences between weathering and erosion?

7. Speculate if the continents will ever be weathered and eroded flat at sea level. Provide evidence to support your speculation.

8. Is it possible for any agent of erosion to erode the land to below sea level? Provide evidence or some observation to support your answer.

INVITATION TO INQUIRY

Building Rocks

Survey the use of rocks used in building construction in your community. Compare the type of rocks that are used for building interiors and those that are used for building exteriors. Where were the rocks quar-ried? Are any trends apparent for buildings constructed in the past and those built more recently? If so, are there reasons (cost, shipping, other limitations) underlying a trend, or is it simply a matter of style?

17

Earth's Weather

This cloud forms a thin covering over the mountaintop. Likewise, Earth's atmosphere forms a thin shell around Earth, with 99 percent of the mass within 32 km (about 20 mi) of the surface

CORE **CONCEPT**

Solar radiation drives cycles in Earth's atmosphere, and some of these cycles determine weather and climate.

OUTLINE

Water cycles into Earth's atmosphere as water vapor and out as condensation and precipitation (p. 385–91).

A uniform body of air is called an air mass (p. 392).

Movement of air masses brings about rapid changes in the weather (p. 392–93).

Climate is the general pattern of weather (p. 401).

OVERVIEW

Earth's atmosphere has a unique composition because of the cyclic flow of materials. Some of these cycles involve the movement of materials in and out of Earth's atmosphere. Carbon dioxide, for example, is a very minor part of Earth's atmosphere. It has been maintained as a minor component in a mostly balanced state for about the past 570 million years, cycling into and out of the atmosphere.

Water is also involved in a global cyclic flow between the atmosphere and the surface. Water on the surface is mostly in the ocean, with lesser amounts in lakes, streams, and underground. Not much water is found in the atmosphere at any one time on a worldwide basis, but billions of tons are constantly evaporating into the atmosphere each year and returning as precipitation in an ongoing cycle.

The cycling of carbon dioxide and water to and from the atmosphere takes place in a dynamic system that is energized by the Sun. Radiant energy from the Sun heats some parts of Earth more than others. Winds redistribute this energy with temperature changes, rain, snow, and other changes that are generally referred to as the *weather*.

Understanding and predicting the weather is the subject of *meteorology*. Meteorology is the science of the atmosphere and weather phenomena, from understanding everyday rain and snow to predicting not-so-common storms and tornadoes (figure 17.1). Understanding weather phenomena depends on a knowledge of the atmosphere and the role of radiant energy on a spinning Earth that is revolving around the Sun. This chapter is concerned with understanding the atmosphere of Earth, its cycles, and the influence of radiant energy on the atmosphere.

17.1 THE ATMOSPHERE

The atmosphere is a relatively thin shell of gases that surrounds the solid Earth. If you could see the molecules making up the atmosphere, you would see countless numbers of rapidly moving particles, all undergoing a terrific jostling from the billions of collisions occurring every second. Since this jostling mass of tiny particles is pulled toward Earth by gravity, more are found near the surface than higher up. Thus, the atmosphere thins rapidly with increasing distance above the surface, gradually merging with the very diffuse medium of outer space.

To understand how rapidly the atmosphere thins with altitude, imagine a very tall stack of open boxes. At any given instant, each consecutively higher box would contain fewer of the jostling molecules than the box below it. Molecules in the lowest box on the surface, at sea level, might be able to move a distance of only 1×10^{-8} m (about 3×10^{-6} in) before

colliding with another molecule. A box moved to an altitude of 80 km (about 50 mi) above sea level would have molecules that could move perhaps 10^{-2} m (about 1/2 in) before colliding with another molecule. At 160 km (about 100 mi), the distance traveled would be about 2 m (about 7 ft). As you can see, the distance between molecules increases rapidly with increasing altitude. Since air density is defined by the number of molecules in a unit volume, the density of the atmosphere decreases rapidly with increasing altitude (figure 17.2).

It is often difficult to imagine a distance above the surface because there is nothing visible in the atmosphere for comparison. Recall our stack of boxes from the previous example. Imagine that this stack of boxes is so tall that it reaches from the surface past the top of the atmosphere. Now imagine that this tremendously tall stack of boxes is tipped over and carefully laid out horizontally on the surface of Earth. How far would you have to move along these boxes to reach the box that was in

FIGURE 17.1 The probability of a storm can be predicted, but nothing can be done to stop or slow a storm. Understanding the atmosphere may help in predicting weather changes, but it is doubtful that weather will ever be controlled on a large scale.

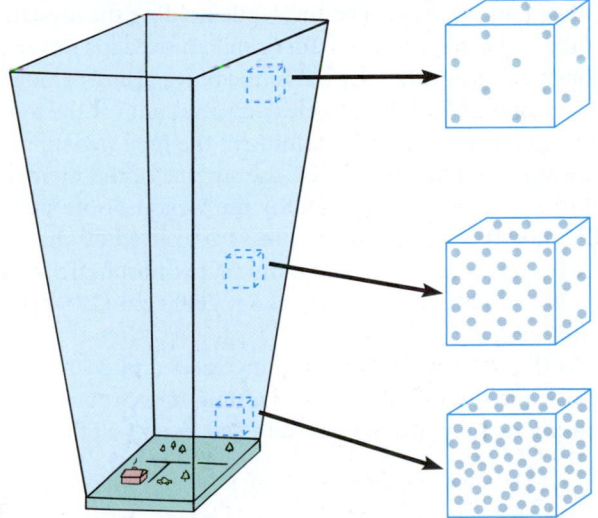

FIGURE 17.2 At greater altitudes, the same volume contains fewer molecules of the gases that make up the air. This means that the density of air decreases with increasing altitude.

outer space, outside of the atmosphere? From the bottom box, you would cover a distance of only 5.6 km (about 3.5 mi) to reach the box that was above 50 percent of the mass of Earth's atmosphere. At 12 km (about 7 mi), you would reach the box that was above 75 percent of Earth's atmosphere. At 16 km (about 10 mi), you would reach the box that was above about 90 percent of the atmosphere. And, after only 32 km (about 20 mi), you would reach the box that was above 99 percent of Earth's atmosphere. The significance of these distances might be better appreciated if you can imagine the distances to some familiar locations; for example, from your campus to a store 16 km (about 10 mi) away would place you above 90 percent of the atmosphere if you were to travel this same distance straight up.

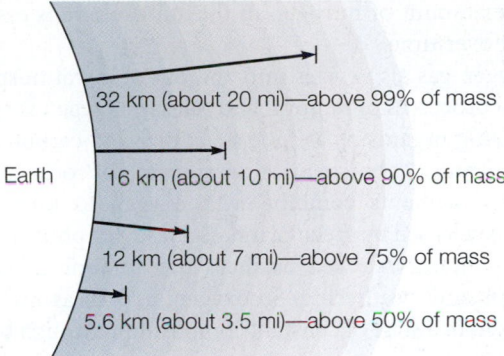

32 km (about 20 mi)—above 99% of mass

Earth   16 km (about 10 mi)—above 90% of mass

12 km (about 7 mi)—above 75% of mass

5.6 km (about 3.5 mi)—above 50% of mass

FIGURE 17.3 Earth's atmosphere thins rapidly with increasing altitude and is much closer to Earth's surface than most people realize.

Since the average radius of the solid Earth is about 6,373 km (3,960 mi), you can see that the atmosphere is a very thin shell with 99 percent of the mass within 32 km (about 20 mi) by comparison. The outer edge of the atmosphere is much closer to Earth than most people realize (figure 17.3).

Composition of the Atmosphere

A sample of pure, dry air is colorless, odorless, and composed mostly of the molecules of just three gases, nitrogen (N_2), oxygen (O_2), and argon (Ar). Nitrogen is the most abundant (about 78 percent of the total volume), followed by oxygen (about 21 percent), then argon (about 1 percent). The molecules of these three gases are well mixed, and this composition is nearly constant everywhere near Earth's surface (figure 17.4).

Nitrogen does not readily enter into chemical reactions with rocks, so it has accumulated in the atmosphere. Some nitrogen is removed from the atmosphere by certain bacteria in the soil and by lightning. The compounds formed are absorbed by plants and are consequently utilized by the food chain. Eventually, the nitrogen is returned to the atmosphere through the decay of plant and animal matter. Overall, these processes of nitrogen removal and release must be in balance

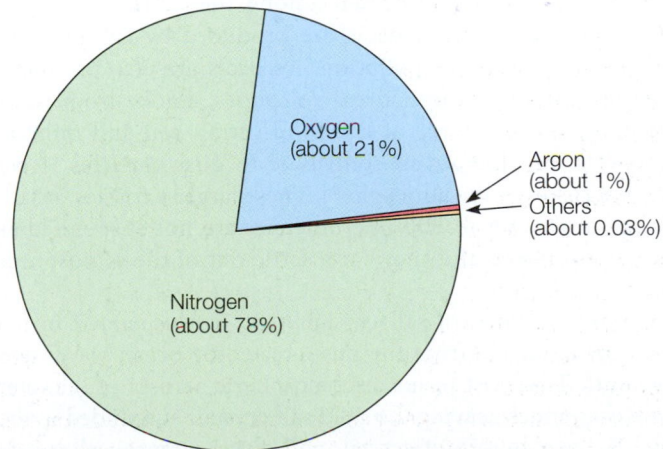

Oxygen (about 21%)

Argon (about 1%)

Others (about 0.03%)

Nitrogen (about 78%)

FIGURE 17.4 Earth's atmosphere has a unique composition of gases when compared to that of the other planets in the solar system.

since the amount of nitrogen in the atmosphere is essentially constant over time.

Oxygen gas also cycles into and out of the atmosphere in balanced processes of removal and release. Oxygen is removed (1) by living organisms as food is oxidized to carbon dioxide and water and (2) by chemical weathering of rocks as metals and other elements combine with oxygen to form oxides. Oxygen is released by green plants as a result of photosynthesis, and the amount released balances the amount removed by organisms and weathering. So oxygen, as well as nitrogen, is maintained in a state of constant composition through balanced chemical reactions.

The third major component of the atmosphere, argon, is inert and does not enter into any chemical reactions or cycles. It is produced as a product of radioactive decay and, once released, remains in the atmosphere as an inactive filler.

In addition to the relatively fixed amounts of nitrogen, oxygen, and argon, the atmosphere contains variable amounts of water vapor. Water vapor is the invisible, molecular form of water in the gaseous state, which should not be confused with fog or clouds. Fog and clouds are tiny droplets of liquid water, not water in the single molecular form of water vapor. The amount of water vapor in the atmosphere can vary from a small fraction of a percent composition by volume in cold, dry air to about 4 percent in warm, humid air. This small, variable percentage of water vapor is essential in maintaining life on Earth. It enters the atmosphere by evaporation, mostly from the ocean, and leaves the atmosphere as rain or snow.

Apart from the variable amounts of water vapor, the relatively fixed amounts of nitrogen, oxygen, and argon make up about 99.97 percent of the volume of a sample of dry air. The remaining gases are mostly carbon dioxide (CO_2) and traces of inert gases. Carbon dioxide makes up only 0.03 percent of the atmosphere, but it is important to life on Earth and it is constantly cycled through the atmosphere by the process of photosynthesis that removes CO_2, and respiration that releases CO_2.

In addition to gases and water vapor, the atmosphere contains particles of dust, smoke, salt crystals, and tiny solid or liquid particles called *aerosols*. These particles become suspended and are dispersed among the molecules of the atmospheric gases. Aerosols are produced by combustion, often resulting in air pollution. Aerosols are also produced by volcanoes and forest fires. Volcanoes, smoke from combustion, and the force of the wind lifting soil and mineral particles into the air all contribute to dust particles larger than aerosols in the atmosphere. These larger particles, which range in size up to 500 micrometers, are not suspended as the aerosols are, and they soon settle out of the atmosphere as dust and soot.

Tiny particles of salt crystals that are suspended in the atmosphere come from the mist created by ocean waves and the surf. This mist forms an atmospheric aerosol of seawater that evaporates, leaving the solid salt crystals suspended in the air. The aerosol of salt crystals and dust becomes well mixed in the lower atmosphere around the globe, playing a large and important role in the formation of clouds.

Atmospheric Pressure

At Earth's surface (sea level), the atmosphere exerts a force of about 10.0 newtons on each square centimeter (14.7 lb/in²). As you go to higher altitudes above sea level the pressure rapidly decreases with increasing altitude. At an altitude of about 5.6 km (about 3.5 mi), the air pressure is about half of what it is at sea level, about 5.0 newtons/cm² (7.4 lb/in²). At 12 km (about 7 mi), the air pressure is about 2.5 newtons/cm² (3.7 lb/in²). Compare this decreasing air pressure at greater elevations to figure 17.3. Again, you can see that most of the atmosphere is very close to Earth, and it thins rapidly with increasing altitude. Even a short elevator ride takes you high enough that the atmospheric pressure on your eardrum is reduced. You equalize the pressure by opening your mouth, allowing the air under greater pressure inside the eardrum to move through the eustachian tube. This makes a "pop" sound that most people associate with changes in air pressure.

Atmospheric pressure is measured by an instrument called a **barometer.** The mercury barometer was invented in 1643 by an Italian named Torricelli. He closed one end of a glass tube and then filled it with mercury. The tube was then placed, open end down, in a bowl of mercury while holding the mercury in the tube with a finger. When Torricelli removed his finger with the open end below the surface in the bowl, a small amount of mercury moved into the bowl leaving a vacuum at the top end of the tube. The mercury remaining in the tube was supported by the atmospheric pressure on the surface of the mercury in the bowl. The pressure exerted by the weight of the mercury in the tube thus balanced the pressure exerted by the atmosphere. At sea level, Torricelli found that atmospheric pressure balanced a column of mercury about 76.00 cm (29.92 in) tall (figure 17.5).

As the atmospheric pressure increases and decreases, the height of the supported mercury column moves up and down. Atmospheric pressure can be expressed in terms of the height

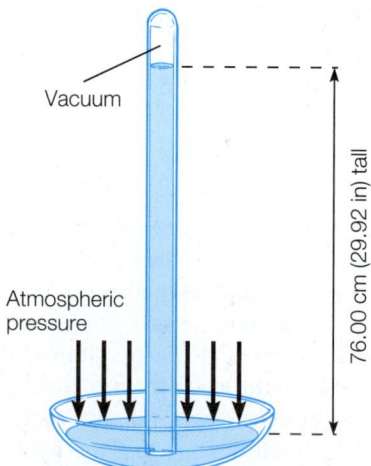

FIGURE 17.5 The mercury barometer measures the atmospheric pressure from the balance between the pressure exerted by the weight of the mercury in a tube and the pressure exerted by the atmosphere. As the atmospheric pressure increases and decreases, the mercury rises and falls. This sketch shows the average height of the column at sea level.

of such a column of mercury. Public weather reports give the pressure by referring to such a mercury column; for example, "The pressure is 30 inches (about 76 cm) and rising." If the atmospheric pressure at sea level is measured many times over long periods of time, an average value of 76.00 cm (29.92 in) of mercury is obtained. This average measurement is called the **standard atmospheric pressure** and is sometimes referred to as the *normal pressure*. It is also called *one atmosphere of pressure*.

Warming the Atmosphere

Radiation from the Sun must pass through the atmosphere before reaching Earth's surface. The atmosphere filters, absorbs, and reflects incoming solar radiation, as shown in figure 17.6. On the average, Earth as a whole reflects about 30 percent of the total radiation back into space, with two-thirds of the reflection occurring from clouds. The amount reflected at any one time depends on the extent of cloud cover, the amount of dust in the atmosphere, and the extent of snow and vegetation on the surface. Substantial changes in any of these influencing variables could increase or decrease the reflectivity, leading to increased heating or cooling of the atmosphere.

As figure 17.6 shows, only about one-half of the incoming solar radiation reaches Earth's surface. The reflection and selective filtering by the atmosphere allow a global average of about 240 watts per square meter to reach the surface. Wide variations from the average occur with latitude as well as with the season.

The incoming solar radiation that does reach Earth's surface is absorbed. Rocks, soil, water, and the ground become warmer as a result. These materials emit the absorbed solar energy as infrared radiation, wavelengths longer than the visible part of the electromagnetic spectrum. This longer-wavelength infrared radiation has a frequency that matches some of the natural frequencies of vibration of carbon dioxide and water molecules. This match means that carbon dioxide and water molecules readily absorb infrared radiation that is emitted from the surface of Earth. The absorbed infrared energy shows

up as an increased kinetic energy of the molecules, which is indicated by an increase in temperature. Carbon dioxide and water vapor molecules in the atmosphere now emit infrared radiation of their own, this time in all directions. Some of this reemitted radiation is again absorbed by other molecules in the atmosphere, some is emitted to space, and significantly, some is absorbed by the surface to start the process all over again. The net result is that less of the energy from the Sun escapes immediately to space after being absorbed and emitted as infrared. It is retained through the process of being redirected to the surface, increasing the surface temperature more than it would have otherwise been. The more carbon dioxide that is present in the atmosphere, the more energy that will be bounced around and redirected back toward the surface, increasing the temperature near the surface. The process of heating the atmosphere in the lower parts by the absorption of solar radiation and reemission of infrared radiation is called the **greenhouse effect.** It is called the greenhouse effect because greenhouse glass allows the short wavelengths of solar radiation to pass into the greenhouse but does not allow all of the longer infrared radiation to leave. This analogy is misleading, however, because carbon dioxide and water vapor molecules do not "trap" infrared radiation, but they are involved in a dynamic absorption and downward reemission process that increases the surface temperature. The more carbon dioxide molecules that are involved in this dynamic process, the more infrared radiation will be directed back to Earth and the more the temperature will increase. More layers of glass on a greenhouse will not increase the temperature significantly. The significant heating factor in a real greenhouse is the blockage of convection by the glass, a process that does not occur from the presence of carbon dioxide and water vapor in the atmosphere.

Structure of the Atmosphere

Convection currents and the repeating absorption and reemission processes of the greenhouse effect tend to heat the atmosphere from the ground up. In addition, the higher-altitude gases lose radiation to space more readily than the lower-altitude gases. Thus, the lowest part of the atmosphere is warmer, and the temperature decreases with increasing altitude. On the average, the temperature decreases about 6.5°C for each kilometer of altitude (3.5°F/1,000 ft). This change of temperature with altitude is called the *observed lapse rate*. The observed lapse rate applies only to air that is not rising or sinking, and the actual change with altitude can be very different from this average value. For example, a stagnant mass of very cold air may settle over an area, producing colder temperatures near the surface than in the air layers above. Such a layer where the temperature increases with height is called an **inversion** (figure 17.7). Inversions often occur on calm winter days after the arrival of a cold front. They also occur on calm, clear, and cool nights ("C" nights), when the surface rapidly loses radiant energy to space. In either case, the situation results in a "cap" of cooler, more dense air overlying the warmer air beneath. This often leads to an increase of air pollution because the inversion prevents dispersion of the pollutants.

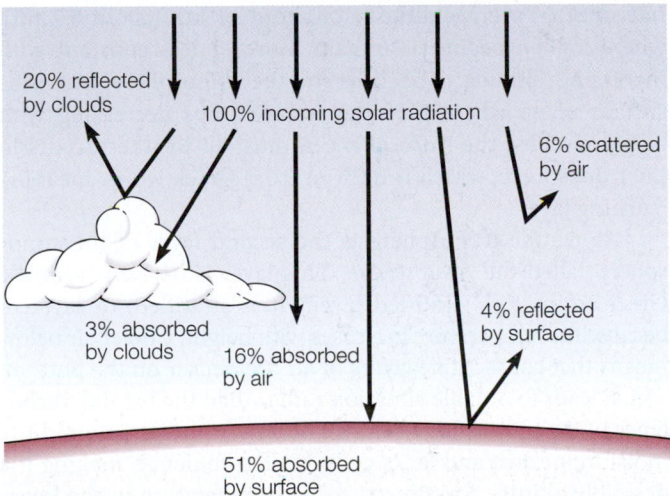

FIGURE 17.6 On the average, Earth's surface absorbs only 51 percent of the incoming solar radiation after it is filtered, absorbed, and reflected.

A Closer Look

Hole in the Ozone Layer?

Ozone is triatomic oxygen (O_3) that is concentrated mainly in the upper portions of the stratosphere. Diatomic molecules of oxygen (O_2) are concentrated in the troposphere, and monatomic molecules of oxygen (O) are found in the outer edges of the atmosphere. Although the amount of ozone present in the stratosphere is not great, its presence is vital to life on Earth's surface. Ultraviolet (UV) radiation causes mutations to the DNA of organisms. Without the stratospheric ozone, much more ultraviolet radiation would reach the surface of Earth, causing mutations in all kinds of organisms. In humans, ultraviolet radiation is known to cause skin cancer and damage the cornea of eyes. It is believed that the incidence of skin cancer would rise dramatically without the protection offered by the ozone.

Here is how stratospheric ozone shields Earth from ultraviolet radiation. The ozone concentration is not static because there is an ongoing process of ozone formation and destruction. For ozone to form, diatomic oxygen (O_2) must first be broken down into the monatomic form (O). Shortwave ultraviolet radiation is absorbed by diatomic oxygen, breaking it down into the single-atom form. This reaction is significant because of (1) the high-energy ultraviolet radiation that is removed from the sunlight and (2) the monatomic oxygen that is formed, which will combine with diatomic oxygen to make triatomic oxygen (ozone) that will absorb even more ultraviolet radiation. This initial reaction is

$$O_2 + UV \rightarrow O + O$$

When the O molecule collides with an O_2 molecule and any third, neutral molecule (NM), the following reaction takes place:

$$O_2 + O + NM \rightarrow O_3 + NM$$

When O_3 is exposed to ultraviolet radiation, the ozone absorbs the UV radiation and breaks down to two forms of oxygen in the following reaction:

$$O_3 + UV \rightarrow O_2 + O$$

The monatomic molecule that is produced combines with an ozone molecule to produce two diatomic molecules,

$$O + O_3 \rightarrow 2\,O_2$$

and the process starts all over again.

Much concern exists about Freon (CF_2Cl_2) and other similar chemicals that make their way to the stratosphere. These chemicals are broken down by UV radiation, releasing chlorine (Cl), which reacts with ozone. The reaction might be

$$CF_2Cl_2 + UV \rightarrow CF_2Cl{-}^* + Cl{-}^*$$
$$Cl{-}^* + O_3 \rightarrow {}^*{-}ClO + O_2$$
$${}^*{-}ClO + O \rightarrow O_2 + Cl{-}^*$$

$-*$ is an unattached bond

Note that ozone is decomposed in the second step, so it is removed from the UV radiation–absorbing process. Furthermore, the second and third steps repeat, so the destruction of one molecule of Freon can result in the decomposition of many molecules of ozone.

A regional zone of decreased ozone availability in the stratosphere is referred to as a "hole in the ozone layer," since much high-energy UV radiation can now reach the surface below. Concerns about the impact chlorine-containing chemicals such as Freon have on the ozone layer led to an international agreement known as the Montreal Protocol. The agreement has resulted in a phasing out of the use of Freon and similar compounds in most of the world. Subsequently, it appears that the sizes of the "holes in the ozone layer" are getting smaller.

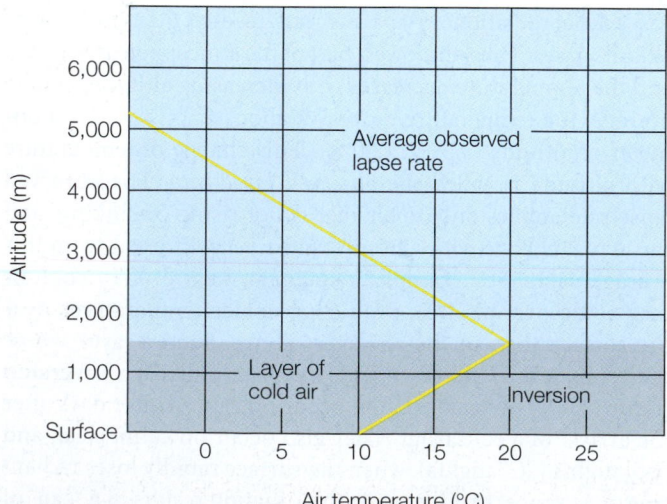

FIGURE 17.7 On the average, the temperature decreases about 6.5°C/km, which is known as the *observed lapse rate*. An inversion is a layer of air in which the temperature increases with height.

Temperature decreases with height at the observed lapse rate until an average altitude of about 11 km (about 6.7 mi), where it then begins to remain more or less constant with increasing altitude. The layer of the atmosphere from the surface up to where the temperature stops decreasing with height is called the *troposphere*. Almost all weather occurs in the troposphere, which is derived from Greek words meaning "turning layer."

Above the troposphere is the second layer of the atmosphere called the *stratosphere*. This layer is derived from the Greek terms for "stratified layer." It is stratified, or layered, because the temperature increases with height. Cooler air below means that consecutive layers of air are denser on the bottom, which leads to a stable situation rather than the turning turbulence of the troposphere below. The stratosphere contains little moisture or dust and lacks convective turbulence, making it a desirable altitude for aircraft to fly. Temperature in the lower stratosphere increases gradually with increasing altitude to a height of about 48 km (about 30 mi) (figure 17.8).

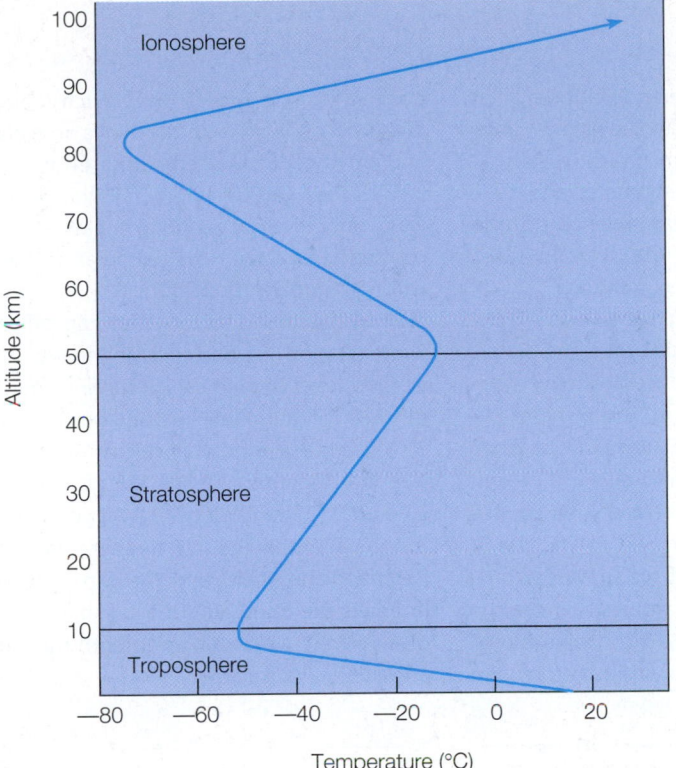

FIGURE 17.8 The structure of the atmosphere based on temperature differences.

The outermost layer is where the molecules merge with the diffuse vacuum of space. Molecules of this layer that have sufficient kinetic energy are able to escape and move off into space. This outer layer is sometimes called the *ionosphere* because of the free electrons and ions at this altitude. The electrons and ions here are responsible for reflecting radio waves around Earth and for the northern lights.

17.2 THE WINDS

The troposphere is heated from the bottom up as the surface of Earth absorbs sunlight. Uneven heating of Earth's surface sets the stage for *convection*. As a local region of air becomes heated, the air expands, reducing its density. This less dense air is pushed upward by nearby cooler, more dense air. This results in three general motions of air: (1) the upward movement of air over a region of greater heating, (2) the sinking of air over a cooler region, and (3) a horizontal air movement between the cooler and warmer regions. In general, a horizontal movement of air is called **wind,** and the direction of a wind is defined as the direction from which it blows.

Air in the troposphere rises, moves as the wind, and sinks. All three of these movements are related, and all occur at the same time over regions of the landscape. During a day with gentle breezes on the surface, the individual, fluffy clouds you see are forming over areas where the air is moving upward. The clear air between the clouds is over areas where the air

is moving downward. On a smaller scale, air can be observed moving from a field of cool grass toward an adjacent asphalt parking lot on a calm, sunlit day. Soap bubbles or smoke will often reveal the gentle air movement of this localized convection.

Local Wind Patterns

Considering average conditions, there are two factors that are important for a generalized model to help you understand local wind patterns. These factors are (1) the relationship between air temperature and air density, and (2) the relationship between air pressure and the movement of air.

The upward and downward movement of air leads to the second part of the generalized model, that (1) the upward movement produces a "lifting" effect on the surface that results in an area of lower atmospheric pressure and (2) the downward movement produces a "piling up" effect on the surface that results in an area of higher atmospheric pressure. On the surface, air is seen to move from the "piled up" area of higher pressure horizontally to the "lifted" area of lower pressure (figure 17.9). In other words, air generally moves from an area of higher pressure to an area of lower pressure. The movement of air and the pressure differences occur together, and neither is the cause of the other. This is an important relationship in a working model of air movement that can be observed and measured on a very small scale, such as between an asphalt parking lot and a grass field. It can also be observed and measured for local, regional wind patterns and for worldwide wind systems.

Adjacent areas of the surface can have different temperatures because of different heating or cooling rates. The difference is very pronounced between adjacent areas of land and water. Under identical conditions of incoming solar radiation, the temperature changes experienced by the water will be much less than the changes experienced by the adjacent land. There are three principal reasons for this difference: (1) The specific

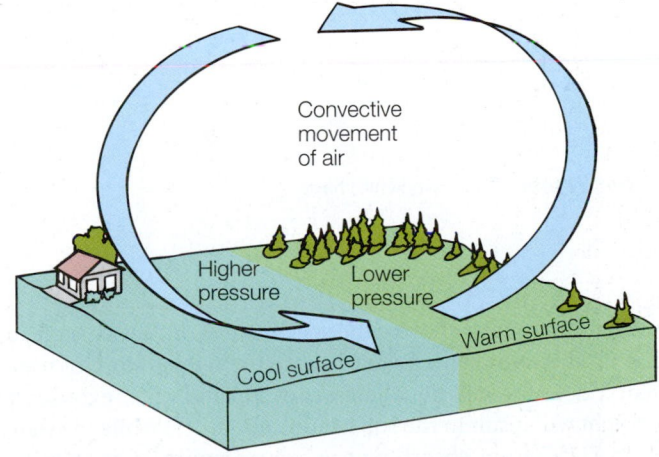

FIGURE 17.9 A model of the relationships between differential heating, the movement of air, and pressure difference in a convective cell. Cool air pushes the less-dense, warm air upward, reducing the surface pressure. As the uplifted air cools and becomes more dense, it sinks, increasing the surface pressure.

A Closer Look

The Wind Chill Factor

The term *wind chill* is attributed to the Antarctic explorer Paul A. Siple. During the 1940s, Siple and Charles F. Passel conducted experiments on how long it took a can of water to freeze at various temperatures and wind speeds at a height of 10 m (33 ft) above the ground (typical height of an anemometer). They found that the time depended on the air temperature and the wind speed. From these data, an equation was developed to calculate the **wind chill factor** for humans.

In 2001, the National Weather Service changed to a new method of computing wind chill temperatures. The new wind chill formula was developed by a year-long cooperative effort between the U.S. and Canadian governments and university scientists. The new standard is based on wind speeds at an average height of 1.5 m (5 ft) above the ground, which is closer to the height of a human face rather than the height of an anemometer and takes advantage of advances in science, technology, and computers. The new chart also highlights the danger of frostbite.

Here is the reason the wind chill factor is an important consideration. The human body constantly produces heat to maintain a core temperature, and some of this heat is radiated to the surroundings. When the wind is not blowing (and you are not moving), your body heat is also able to warm some of the air next to your body. This warm blanket of air provides some insulation, protecting your skin from the colder air farther away. If the wind blows, however, it moves this air away from your body and you feel cooler. How much cooler depends on how fast the air is moving and on the outside temperature—which is what the wind chill factor tells you. Thus, wind chill is an attempt to measure the combined effect of low temperature and wind on humans (box figure 17.1). It is just one of the many factors that can affect winter comfort. Others include the type of clothes, level of physical exertion, amount of sunshine, humidity, age, and body type.

There is a wind chill calculator at the National Weather Service website, www.nws.noaa.gov/om/windchill/. All you need to do is enter the air temperature (in degrees Fahrenheit) and the wind speed (in miles per hour), and the calculator will give you the wind chill using both the old and the new formula.

Wind (mph)	Temperature (°F)																	
	40	35	30	25	20	15	10	5	0	−5	−10	−15	−20	−25	−30	−35	−40	−45
5	36	31	25	19	13	7	1	−5	−11	−16	−22	−28	−34	−40	−46	−52	−57	−63
10	34	27	21	15	9	3	−4	−10	−16	−22	−28	−35	−41	−47	−53	−59	−66	−72
15	32	25	19	13	6	0	−7	−13	−19	−26	−32	−39	−45	−51	−58	−64	−71	−77
20	30	24	17	11	4	−2	−9	−15	−22	−29	−35	−42	−48	−55	−61	−68	−74	−81
25	29	23	16	9	3	−4	−11	−17	−24	−31	−37	−44	−51	−58	−64	−71	−78	−84
30	28	22	15	8	1	−5	−12	−19	−26	−33	−39	−46	−53	−60	−67	−73	−80	−87
35	28	21	14	7	0	−7	−14	−21	−27	−34	−41	−48	−55	−62	−69	−76	−82	−89
40	27	20	13	6	−1	−8	−15	−22	−29	−36	−43	−50	−57	−64	−71	−78	−84	−91
45	26	19	12	5	−2	−9	−16	−23	−30	−37	−44	−51	−58	−65	−72	−79	−86	−93
50	26	19	12	4	−3	−10	−17	−24	−31	−38	−45	−52	−60	−67	−74	−81	−88	−95
55	25	18	11	4	−3	−11	−18	−25	−32	−39	−46	−54	−61	−68	−75	−82	−89	−97
60	25	17	10	3	−4	−11	−19	−26	−33	−40	−48	−55	−62	−69	−76	−84	−91	−98

Frostbite times: ■ 30 minutes ■ 10 minutes ■ 5 minutes

Box Figure 17.1 Wind chill chart.

heat of water is about twice the specific heat of soil. This means that it takes more energy to increase the temperature of water than it does for soil. Equal masses of soil and water exposed to sunlight will result in the soil heating about 1°C while the water heats 1/2°C from absorbing the same amount of solar radiation. (2) Water is a transparent fluid that is easily mixed, so the incoming solar radiation warms a body of water throughout, spreading out the heating effect. Incoming solar radiation on land, on the other hand, warms a relatively thin layer on the top, concentrating the heating effect. (3) The water is cooled by evaporation, which helps keep a body of water at a lower temperature than an adjacent landmass under identical conditions of incoming solar radiation.

A local wind pattern may result from the resulting temperature differences between a body of water and adjacent landmasses. If you have ever spent some time along a coast, you may have observed that a cool, refreshing gentle breeze blows from the water toward the land during the summer. During

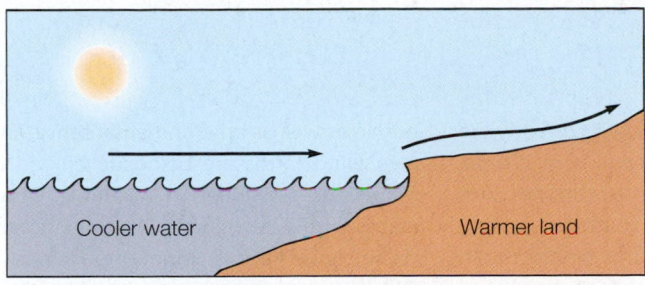

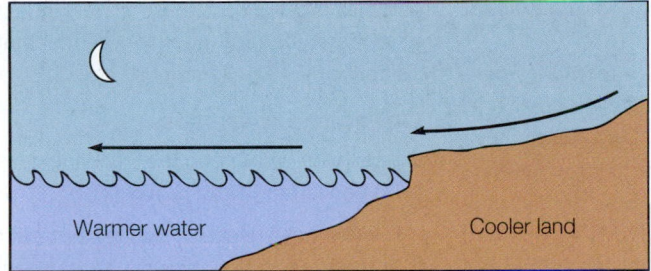

FIGURE 17.10 The land warms and cools more rapidly than an adjacent large body of water. During the day, the land is warmer, and air over the land expands and is buoyed up by cooler, more dense air from over the water. During the night, the land cools more rapidly than the water, and the direction of the breeze is reversed.

the day, the temperature of the land increases more rapidly than the water temperature. The air over the land is therefore heated more, expands, and becomes less dense. Cool, dense air from over the water moves inland under the air over the land, buoying it up. The air moving from the sea to the land is called a *sea breeze*. The sea breeze along a coast may extend inland several miles during the hottest part of the day in the summer. The same pattern is sometimes observed around the Great Lakes during the summer, but this breeze usually does not reach more than several city blocks inland. During the night, the land surface cools more rapidly than the water, and a breeze blows from the land to the sea (figure 17.10).

Another pattern of local winds develops in mountainous regions. If you have ever visited a mountain in the summer, you may have noticed that there is usually a breeze or wind blowing up the mountain slope during the afternoon. This wind pattern develops because the air over the mountain slope is heated more than the air in a valley. As shown in figure 17.11, the air over the slope becomes warmer because it receives more direct sunlight than the valley floor. Sometimes this air movement is so gentle that it would be unknown except for the evidence of clouds that form over the peaks during the day and evaporate at night. During the night, the air on the slope cools as the land loses radiant energy to space. As the air cools, it becomes denser and flows downslope, forming a reverse wind pattern to the one observed during the day.

During cooler seasons, cold, dense air may collect in valleys or over plateaus, forming a layer or "puddle" of cold air. Such an accumulation of cold air often results in some very cold night-time temperatures for cities located in valleys, temperatures that are much colder than anywhere in the surrounding region. Some weather disturbance, such as an approaching front, can

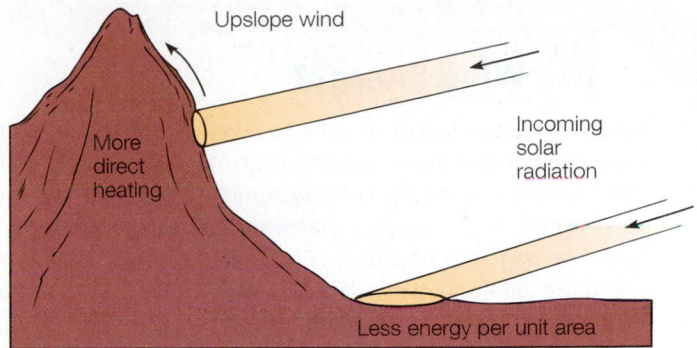

FIGURE 17.11 Incoming solar radiation falls more directly on the side of a mountain, which results in differential heating. The same amount of sunlight falls on the areas shown in this illustration, with the valley floor receiving a more spread-out distribution of energy per unit area. The overall result is an upslope mountain breeze during the day. During the night, dense cool air flows downslope for a reverse wind pattern.

disturb such an accumulation of cold air and cause it to pour out of its resting place and through canyons or lower valleys. Air moving from a higher altitude like this becomes compressed as it moves to lower elevations under increasing atmospheric pressure. Compression of air increases the temperature by increasing the kinetic energy of the molecules. This creates a wind called a *Chinook,* which is common to mountainous and adjacent regions. A Chinook is a wind of compressed air with sharp temperature increases that can melt away any existing snow cover in a single day. The *Santa Ana* is a well-known compressional wind that occurs in southern California.

Global Wind Patterns

Local wind patterns tend to mask the existence of the overall global wind pattern that is also present. The global wind pattern is not apparent if the winds are observed and measured for a particular day, week, or month. It does become apparent when the records for a long period of time are analyzed. These records show that Earth has a large-scale pattern of atmospheric circulation that varies with latitude. There are belts in which the winds average an overall circulation in one direction, belts of higher atmospheric pressure averages, and belts of lower atmospheric pressure averages. This has led to a generalized pattern of atmospheric circulation and a global atmospheric model. This model, as you will see, today provides the basis for the daily weather forecast for local and regional areas.

As with local wind patterns, it is temperature imbalances that drive the global circulation of the atmosphere. Earth receives more direct solar radiation in the equatorial region than it does at higher latitudes (figure 17.12). As a result, the temperatures of the lower troposphere are generally higher in the equatorial region, decreasing with latitude toward both poles. The lower troposphere from 10°N to 10°S of the equator is heated, expands, and becomes less dense. Hot air rises in this belt around the equator, known as the *intertropical convergence zone.* The rising air cools because it expands as it rises, resulting in heavy average precipitation. The tropical rainforests of Earth occur in this zone of high temperatures and heavy

Science and Society

Use Wind Energy?

Millions of windmills were installed in rural areas of the United States between the late 1800s and the late 1940s. These windmills used wind energy to pump water, grind grain, or generate electricity. Some are still in use today, but most were no longer needed after inexpensive electric power became generally available in those areas.

In the 1970s, wind energy began making a comeback as a clean, renewable energy alternative to fossil fuels. The windmills of the past were replaced by wind turbines of today. A wind turbine with blades that are rotated by the wind is usually mounted on a tower. The blades' rotary motion drives a generator that produces electricity. A location should have average yearly wind speeds of at least 19 km/h (12 mi/h) to provide enough wind energy for a turbine, and a greater yearly average means more energy is available. Farms, homes, and businesses in these locations can use smaller turbines, which are generally 50 kilowatts or less. Large turbines of 500 kilowatts or more are used in "wind farms," which are large clusters of interconnected wind turbines connected to a utility power grid.

The Thanet Wind Farm is currently the world's largest offshore wind farm. It has a hundred 380 ft (116 m) tall turbines situated 7 mi (11 km) off the southeast coast of England. At full capacity, this wind farm can generate 300 megawatts of wind-powered electricity, which is enough to supply about 90,000 households.

Many areas of the United States have a high potential for wind-power use. North Dakota, South Dakota, and Texas have enough wind resources to provide electricity for the entire nation. Today, only California has extensively developed wind farms, with more than thirteen thousand wind turbines on three wind farms in the Altamont Pass region (east of San Francisco), Tehachapi (southeast of Bakersfield), and San Gorgonio (near Palm Springs). With a total of 2,361 megawatts of installed capacity in California, wind energy generates enough electricity to more than meet the needs of a city the size of San Francisco. The wind farms in California have a rated capacity that is comparable to two large coal-fired power plants but without the pollution and limits of this nonrenewable energy source. Wind energy makes eco-

nomic as well as environmental sense, and new wind farms are being developed in Minnesota, Oregon, and Wyoming. Other states with a strong wind power potential include Kansas, Montana, Nebraska, Oklahoma, Iowa, Colorado, Michigan, and New York. All of these states, in fact, have a greater wind-energy potential than California.

Questions to Discuss

Discuss with your group the following questions concerning wind power:

1. Why have electric utilities not used much wind power as an energy source?

2. Should governments provide a tax break to encourage people to use wind power? Why or why not?

3. What are the advantages and disadvantages of using wind power in the place of fossil fuels?

4. What are the advantages and disadvantages of the government building huge wind farms in North Dakota, South Dakota, and Texas to supply electricity for the entire nation?

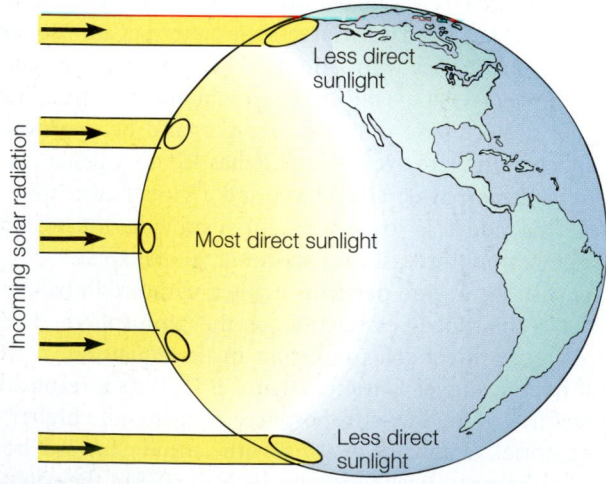

FIGURE 17.12 On a global, yearly basis, the equatorial region of Earth receives more direct incoming solar radiation than the higher latitudes. As a result, average temperatures are higher in the equatorial region and decrease with latitude toward both poles. This sets the stage for worldwide patterns of prevailing winds, high and low areas of atmospheric pressure, and climatic patterns.

rainfall. As the now dry, rising air reaches the upper parts of the troposphere, it begins to spread toward the north and the south, sinking back toward Earth's surface (figure 17.13). The descending air reaches the surface to form a high-pressure belt that is centered about 30°N and 30°S of the equator. Air moving on the surface away from this high-pressure belt produces the prevailing northeast trade winds and the prevailing westerly winds of the Northern Hemisphere. The great deserts of Earth are also located in this high-pressure belt of descending dry air.

Poleward of the belt of high pressure, the atmospheric circulation is controlled by a powerful belt of wind near the top of the troposphere called a **jet stream.** Jet streams are sinuous, meandering loops of winds that tend to extend all the way around Earth, moving generally from the west in both hemispheres at speeds of 160 km/h (about 100 mi/h) or more. A jet stream may occur as a single belt, or loop, of wind, but sometimes it divides into two or more parts. The jet stream develops north and south loops of waves much like the waves you might make on a very long rope. These waves vary in size, sometimes beginning as a small ripple but then growing slowly as the wave moves eastward. Waves that form on the jet stream

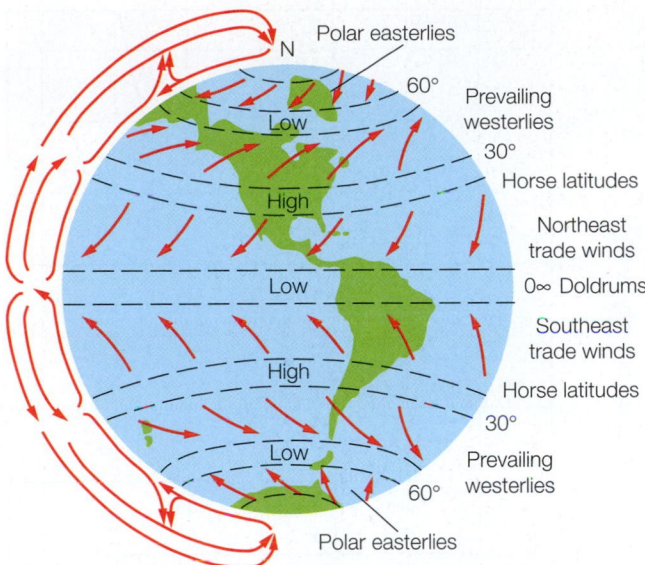

FIGURE 17.13 Simplified pattern of horizontal and vertical circulation in the actual atmosphere. Regions of high and low pressure are indicated.

bulge toward the poles (called a crest) or toward the equator (called a trough). Warm air masses move toward the poles ahead of a trough, and cool air masses move toward the equator behind a trough as it moves eastward. The development of a wave in the jet stream is understood to be one of the factors that influences the movement of warm and cool air masses, a movement that results in weather changes on the surface.

The intertropical convergence zone, the 30° belt of high pressure, and the northward and southward migration of a meandering jet stream all shift toward or away from the equator during the different seasons of the year. The troughs of the jet stream influence the movement of alternating cool and warm air masses over the belt of the prevailing westerlies, resulting in frequent shifts of fair weather to stormy weather, then back again. The average shift during the year is about 6° of latitude, which is sufficient to control the overall climate in some locations. The influence of this shift of the global circulation of Earth's atmosphere will be considered as a climatic influence after considering the roles of water and air masses in frequent weather changes.

17.3 WATER AND THE ATMOSPHERE

Water exists on Earth in all three states: (1) as a liquid when the temperature is generally above the freezing point of 0°C (32°F), (2) as a solid in the form of ice, snow, or hail when the temperature is generally below the freezing point, and (3) as the invisible, molecular form of water in the gaseous state, which is called *water vapor.*

More than 98 percent of all the water on Earth exists in the liquid state, mostly in the ocean, and only a small, variable amount of water vapor is in the atmosphere at any given time. Since so much water seems to fall as rain or snow at times, it may be a surprise that the overall atmosphere really does not contain very much water vapor. If the average amount of water vapor in Earth's atmosphere were condensed to liquid form, the vapor *and* all the droplets present in clouds would form a uniform layer around Earth only 3 cm (about 1 in) thick. Nonetheless, it is this small amount of water vapor that is eventually responsible for (1) contributing to the greenhouse effect, which helps make Earth a warmer planet, (2) serving as one of the principal agents in the weathering and erosion of the land, which creates soils and sculptures the landscape, and (3) maintaining life, for all organisms (bacteria, protozoa, algae, fungi, plants, and animals) cannot survive without water. It is the ongoing cycling of water vapor into and out of the atmosphere that makes all this possible. Understanding this cycling process and the energy exchanges involved is also closely related to understanding Earth's weather patterns.

Evaporation and Condensation

Water tends to undergo a liquid-to-gas or a gas-to-liquid phase change at any temperature. The phase change can occur in either direction at any temperature.

In *evaporation,* more molecules are leaving the liquid state than are returning. In *condensation,* more molecules are returning to the liquid state than are leaving. This is a dynamic, ongoing process with molecules leaving and returning continuously (figure 17.14). If the air were perfectly dry and still, more molecules would leave (evaporate) the liquid state than would return (condense). Eventually, however, an equilibrium would be reached with as many molecules returning to the liquid state per unit of time as are leaving. An equilibrium condition between evaporation and condensation occurs in **saturated air.** Saturated air occurs when the processes of evaporation and condensation are in balance.

Air will remain saturated as long as (1) the temperature remains constant and (2) the processes of evaporation and condensation remain balanced. Temperature influences the

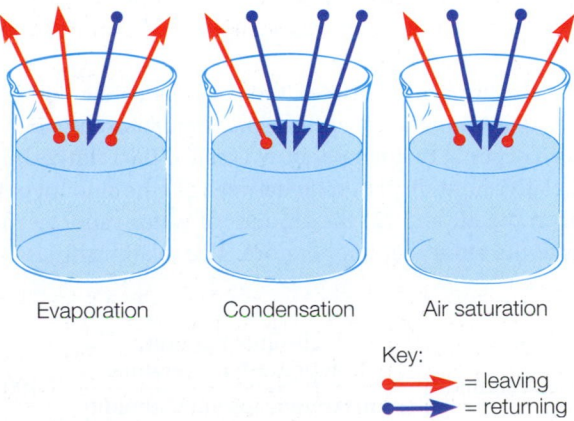

Evaporation   Condensation   Air saturation

Key:
→ = leaving
→ = returning

FIGURE 17.14 Evaporation and condensation are occurring all the time. If the number of molecules leaving the liquid state exceeds the number returning, the water is evaporating. If the number of molecules returning to the liquid state exceeds the number leaving, the water vapor is condensing. If both rates are equal, the air is saturated; that is, the relative humidity is 100 percent.

equilibrium condition of saturated air because increases or decreases in the temperature mean increases or decreases in the kinetic energy of water vapor molecules. Water vapor molecules usually undergo condensation when attractive forces between the molecules can pull them together into the liquid state. Lower temperature means lower kinetic energies, and slow-moving water vapor molecules spend more time close to one another and close to the surface of liquid water. Spending more time close together means an increased likelihood of attractive forces pulling the molecules together. On the other hand, higher temperature means higher kinetic energies, and molecules with higher kinetic energy are less likely to be pulled together. As the temperature increases, there is therefore less tendency for water molecules to return to the liquid state. If the temperature is increased in an equilibrium condition, more water vapor must be added to the air to maintain the saturated condition. Warm air can therefore hold more water vapor than cooler air. In fact, warm air on a typical summer day can hold five times as much water vapor as cold air on a cold winter day.

Humidity

The amount of water vapor in the air is referred to generally as **humidity.** Damp, moist air is more likely to have condensation than evaporation, and this air is said to have a *high humidity.* Dry air is more likely to have evaporation than condensation, on the other hand, and this air is said to have a *low humidity.* A measurement of the amount of water vapor in the air at a particular time is called the **absolute humidity** (figure 17.15). At room temperature, for example, humid air might contain 15 grams of water vapor in each cubic meter of air. At the same temperature, air of low humidity might have an absolute humidity of only 2 grams per cubic meter. The absolute humidity can range from near zero up to a maximum that is determined by the temperature at the time. Since the temperature of the water vapor present in the air is the same as the temperature of the air, the maximum absolute humidity is usually said to be determined by the air temperature. What this really means is that the maximum absolute humidity is determined by the temperature of the water vapor, that is, the average kinetic energy of the water vapor.

The relationship between the *actual* absolute humidity at a particular temperature and the *maximum* absolute humidity that can occur at that temperature is called the **relative humidity.** Relative humidity is a ratio between (1) the amount of water vapor in the air and (2) the amount of water vapor needed to saturate the air at that temperature. The relationship is

$$\text{relative humidity} = \frac{\begin{array}{c}\text{absolute humidity}\\\text{at present temperature}\end{array}}{\begin{array}{c}\text{maximum absolute humidity}\\\text{at present temperature}\end{array}} \times 100\%$$

For example, suppose a measurement of the water vapor in the air at 10°C (50°F) finds an absolute humidity of 5.0 g/m³. According to figure 17.15, the maximum amount of water vapor

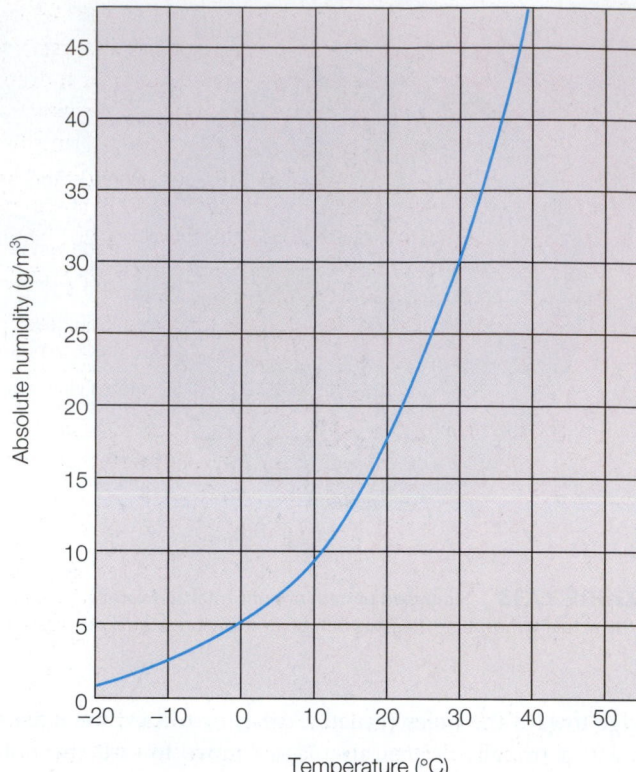

FIGURE 17.15 The maximum amount of water vapor that can be in the air at different temperatures. The amount of water vapor in the air at a particular temperature is called the *absolute humidity*.

that can be in the air when the temperature is 10°C is about 10 g/m³. The relative humidity is then

$$\frac{5.0 \text{ g/m}^3}{10 \text{ g/m}^3} \times 100\% = 50\%$$

If the absolute humidity had been 10 g/m³, then the air would have all the water vapor it could hold, and the relative humidity would be 100 percent. A humidity of 100 percent means that the air is saturated at the present temperature.

The important thing to understand about relative humidity is that the capacity of air to hold water vapor changes with the temperature. Cold air cannot hold as much water vapor, and warming the air will increase its capacity. With the capacity *increased,* the relative humidity *decreases* because you can

CONCEPTS APPLIED

Humidity Factors

Compare the relative humidity in a classroom, over a grass lawn, over a paved parking lot, and other places you might be during a particular day. What do the findings mean?

now add more water vapor to the air than you could before. Just warming the air, for example, can reduce the relative humidity from 50 percent to 3 percent. Lower relative humidity results because warming the air increases the capacity of air to hold water vapor. This explains the need to humidify a home in the winter. Evaporation occurs very rapidly when the humidity is low. Evaporation is a cooling process since the molecules with higher kinetic energy are the ones to escape, lowering the average kinetic energy as they evaporate. Dry air will therefore cause you to feel cool even though the air temperature is fairly high. Adding moisture to the air will enable you to feel warmer at lower air temperatures and thus lower your fuel bill.

The relationship between the capacity of air to hold water vapor and temperature also explains why the relative humidity increases in the evening after the Sun goes down. A cooler temperature means less capacity of air to hold water vapor. With the same amount of vapor in the air, a reduced capacity means a higher relative humidity.

The Condensation Process

Condensation depends on two factors: (1) the relative humidity and (2) the temperature of the air. During condensation, molecules of water vapor join together to produce liquid water on the surface as dew or in the air as the droplets of water making up fog or clouds. Water molecules may also join together to produce solid water in the form of frost or snow. Before condensation can occur, however, the air must be saturated, which means that the relative humidity must be 100 percent with the air containing all the vapor it can hold at the present temperature. A parcel of air can become saturated as a result of (1) water vapor being added to the air from evaporation, (2) cooling, which reduces the capacity of the air to hold water vapor and therefore increases the relative humidity, or (3) a combination of additional water vapor with cooling.

The process of condensation of water vapor explains a number of common observations. You are able to "see your breath" on a cold day, for example, because the high moisture content of your exhaled breath is condensed into tiny water droplets by cold air. The small fog of water droplets evaporates as it spreads into the surrounding air with a lower moisture content. The white trail behind a high-flying jet aircraft is also a result of condensation of water vapor. Water is one of the products of combustion, and the white trail is condensed water vapor, a trail of tiny droplets of water in the cold upper atmosphere. The trail of water droplets is called a *contrail* after "condensation trail." Back on the surface, a cold glass of beverage seems to "sweat" as water vapor molecules near the outside of the glass are cooled, moving more slowly. Slowly moving water vapor molecules spend more time closer together, and the molecular forces between the molecules pull them together, forming a thin layer of liquid water on the outside of the cold glass. This same condensation process sometimes results in a small stream of water from the cold air conditioning coils of an automobile or home mechanical air conditioner.

As air is cooled, its capacity to hold water vapor is reduced to lower and lower levels. Even without water vapor being added to the air, a temperature will eventually be reached at which saturation, 100 percent humidity, occurs. Further cooling below this temperature will result in condensation. The temperature at which condensation begins is called the **dew point temperature.** If the dew point is above 0°C (32°F), the water vapor will condense on surfaces as a liquid called **dew.** If the temperature is at or below 0°C, the vapor will condense on surfaces as a solid called **frost.** Note that dew and frost form on the tops, sides, and bottoms of objects. Dew and frost condense directly on objects and do not "fall out" of the air. Note also that the temperature that determines if dew or frost forms is the temperature of the object where they condense. This temperature near the open surface can be very different from the reported air temperature, which is measured more at eye level in a sheltered instrument enclosure.

Observations of where and when dew and frost form can lead to some interesting things to think about. Dew and frost, for example, seem to form on "C" nights, nights that can be described by the three "C" words of *clear, calm,* and *cool.* Dew and frost also seem to form more (1) in open areas rather than under trees or other shelters, (2) on objects such as grass rather than on the flat, bare ground, and (3) in low-lying areas before they form on slopes or the sides of hills. What is the meaning of these observations?

Dew and frost are related to clear nights and open areas because these are the conditions best suited for the loss of infrared radiation. Air near the surface becomes cooler as infrared radiation is emitted from the grass, buildings, streets, and everything else that absorbed the shorter-wavelength radiation of incoming solar radiation during the day. Clouds serve as a blanket, keeping the radiation from escaping to space so readily. So a clear night is more conducive to the loss of infrared radiation and therefore to cooling. On a smaller scale, a tree serves the same purpose, holding in radiation and therefore retarding the cooling effect. Thus, an open area on a clear, calm night would have cooler air near the surface than would be the case on a cloudy night or under the shelter of a tree.

The observation that dew and frost form on objects such as grass before forming on flat, bare ground is also related to loss of infrared radiation. Grass has a greater exposed surface area than the flat, bare ground. A greater surface area means a greater area from which infrared radiation can escape, so grass blades cool more rapidly than the flat ground. Other variables, such as specific heat, may be involved, but overall, frost and dew are more likely to form on grass and low-lying shrubs before they form on the flat, bare ground.

Dew and frost form in low-lying areas before forming on slopes and the sides of hills because of the density differences of cool and warm air. Cool air is more dense than warm air and is moved downhill by gravity, pooling in low-lying areas. You may have noticed the different temperatures of low-lying areas if you have ever driven across hills and valleys on a clear, calm, and cool evening. Citrus and other orchards are often located on slopes of hills rather than on valley floors because of the gravity drainage of cold air.

FIGURE 17.16 Fans like this one are used to mix the warmer, upper layers of air with the cooling air in the orchard on nights when frost is likely to form.

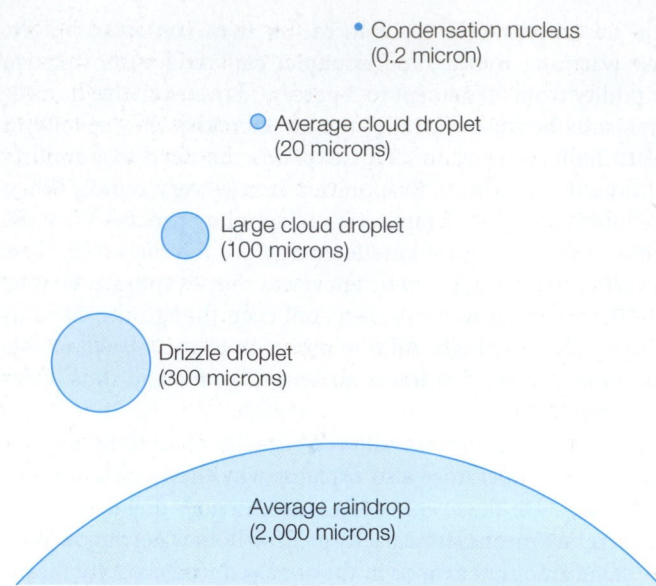

• Condensation nucleus
(0.2 micron)

◉ Average cloud droplet
(20 microns)

Large cloud droplet
(100 microns)

Drizzle droplet
(300 microns)

Average raindrop
(2,000 microns)

FIGURE 17.17 This figure compares the size of the condensation nuclei to the size of typical condensation droplets. Note that 1 micron is 1/1,000 mm.

It is air near the surface that is cooled first by the loss of radiation from the surface. Calm nights favor dew or frost formation because the wind mixes the air near the surface that is being cooled with warmer air above the surface. If you have ever driven near a citrus orchard, you may have noticed the huge, airplanelike fans situated throughout the orchard on poles. These fans are used on "C" nights when frost is likely to form to mix the warmer, upper layers of air with the cooling air in the orchard (figure 17.16).

Condensation occurs on the surface as frost or dew when the dew point is reached. When does condensation occur in the air? Water vapor molecules in the air are constantly colliding and banging into each other, but they do not just join together to form water droplets, even if the air is saturated. The water molecules need something to condense upon. Condensation of water vapor into fog or cloud droplets takes place on tiny particles present in the air. The particles are called **condensation nuclei.** There are hundreds of tiny dust, smoke, soot, and salt crystals suspended in each cubic centimeter of the air that serve as condensation nuclei. Tiny salt crystals, however, are particularly effective condensation nuclei because salt crystals attract water molecules. You may have noticed that salt in a salt shaker becomes moist on a humid day because of the way it attracts water molecules. Tiny salt crystals suspended in the air act the same way, serving as nuclei that attract water vapor into tiny droplets of liquid water.

After water vapor molecules begin to condense on a condensation nucleus, other water molecules will join the liquid water already formed, and the tiny droplet begins to increase in volume. The water droplets that make up a cloud are about fifteen hundred times larger than a condensation nuclei, and these droplets can condense out of the air in a matter of minutes. As the volume increases, however, the process slows, and

hours and days are required to form the even larger droplets and drops. For comparison to the sizes shown in figure 17.17, consider that the average human hair is about 100 microns in diameter. This is about the same diameter as the large cloud droplet of water. Large raindrops have been observed falling from clouds that formed only a few hours previously, so it must be some process or processes other than the direct condensation of raindrops that form precipitation. These processes are discussed in the section on precipitation.

Fog and Clouds

Fog and clouds are both accumulations of tiny droplets of water that have been condensed from the air. These water droplets are very small, and a very slight upward movement of the air will keep them from falling. If they do fall, they usually evaporate. Fog is sometimes described as a cloud that forms at or near the surface. A fog, as a cloud, forms because air containing water vapor and condensation nuclei has been cooled to the dew point. Some types of fog form under the same "C" night conditions favorable for dew or frost to form, that is, on clear, cool, and calm nights when the relative humidity is high. Sometimes this type of fog forms only in valleys and low-lying areas where cool air accumulates (figure 17.18). This type of fog is typical of inland fogs, those that form away from bodies of water. Other types of fog may form somewhere else, such as in the humid air over an ocean, and then move inland. Many fogs that occur along coastal regions were formed over the ocean and then carried inland by breezes. A third type of fog looks much like a steamy mist rising from melting snow on a street, over a body of water into cold air, or a steamy mist over streets after a summer rain shower. These are examples of a temporary fog that forms as a lot of water vapor is added to cool air. This is a cool fog, like other fogs, and is not hot as the steamlike appearance may lead you to believe.

Sometimes a news report states something about the Sun "burning off" a fog. A fog does not burn, of course, because it is made up of droplets of water. What the reporter really means is that the Sun's radiation will increase the temperature, which increases the air capacity to hold water vapor. With an increased capacity to hold water, the relative humidity drops, and the fog simply evaporates back to the state of invisible water vapor molecules.

Clouds, like fogs, are made up of tiny droplets of water that have been condensed from the air. Luke Howard, an English weather observer, made one of the first cloud classification schemes in 1803. He used the Latin terms *cirrus* (curly), *cumulus* (piled up), and *stratus* (spread out) to identify the basic shapes of clouds (figure 17.19). The clouds usually do not occur just in these basic cloud shapes but in combinations of the different shapes. Later, Howard's system was modified by expanding the different shapes of clouds into ten classes by using the basic cloud shapes and altitude as criteria. Clouds give practical hints about the approaching weather. The relationship between the different cloud shapes and atmospheric conditions and what clouds can mean about the coming weather are discussed in the section on precipitation.

Clouds form when a mass of air above the surface is cooled to its dew point temperature. In general, the mass of air is cooled because something has given it an upward push, moving it to higher levels in the atmosphere. There are three major causes of upward air movement: (1) *convection* resulting from differential heating, (2) mountain ranges that serve as *barriers* to moving air masses, and (3) the meeting of *moving air masses* with different densities, for example, a cold, dense mass of air meeting a warm, less dense mass of air.

The three major causes of uplifted air sometimes result in clouds, but just as often they do not. Whether clouds form or not depends on the condition of the atmosphere at the time. As a parcel of warm air is moved upward, it tends to stay together, mixing very little with the surrounding air. As it is forced upward, it becomes cooler because it is expanding. Similarly, the temperature of a gas increases when it is compressed. So rising air is cooled and descending air is warmed.

What happens to a parcel of air that is pushed upward depends on the difference in density between the parcel and the surrounding air. Air temperature will tell you about air density since the density of air is determined by its temperature. Instruments attached to a weather balloon can measure the change of temperature with altitude. By comparing this change with the rate of cooling by expansion, the state of atmospheric stability can be determined. There are many different states of atmospheric stability, and the following is a simplified description of just a few of the possible states, first considering dry air only.

The atmosphere is in a state of *stability* when a lifted parcel of air is cooler than the surrounding air. Being cooler, the parcel of air will be more dense than the surroundings. If it is moved up to a higher level and released in a stable atmosphere, it will move back to its former level. A lifted parcel of air always returns to its original level when the atmosphere is stable. Any clouds that do develop in a stable atmosphere are usually arranged in the horizontal layers of stratus-type clouds.

A

B

FIGURE 17.18 (A) An early morning aerial view of fog that developed close to the ground in cool, moist air on a clear, calm night. (B) Air moves from a warm current, then over a cool current, forming fog. The fog often moves inland at night.

The atmosphere is in a state of *instability* when a lifted parcel of air is warmer than the surrounding air. Being warmer, the parcel of air will be less dense than the surroundings. If it is moved up to a higher level, it will continue moving after the uplifting force is removed. Cumulus clouds usually develop in an unstable atmosphere, and the rising parcels of air, called thermals, can result in a very bumpy airplane ride.

So far, only dry air has been considered. As air moves upward and cools from expansion, sooner or later the dew point is reached and the air becomes saturated. As some of the water vapor in the rising parcel condenses to droplets, the latent heat of vaporization is released. The rising parcel of air

FIGURE 17.19 (*A*) Cumulus clouds. (*B*) Stratus and stratocumulus. Note the small stratocumulus clouds forming from increased convection over each of the three small islands. (*C*) An aerial view between the patchy cumulus clouds below and the cirrus and cirrostratus above (the patches on the ground are clear-cut forests). (*D*) Altocumulus. (*E*) A rain shower at the base of a cumulonimbus. (*F*) Stratocumulus.

now cools at a slower rate because of the release of this latent heat of vaporization. The release of latent heat warms the air in the parcel and decreases the density even more, accelerating the ascent. This leads to further condensation and the formation of towering cumulus clouds, often leading to rain.

Precipitation

Water that returns to the surface of Earth, in either the liquid or solid form, is called **precipitation** (figure 17.20). Note that dew and frost are not classified as precipitation because they form directly on the surface and do not fall through the air. Precipitation seems to form in clouds by one of two processes: (1) the *coalescence* of cloud droplets or (2) the *growth of ice crystals*. It would appear difficult for cloud droplets to merge, or coalesce, with one another since any air movement would seem to move them all at the same time, not bring them together. Condensation nuclei come in different sizes, however, and cloud droplets of many different sizes form on these different-sized nuclei. Larger cloud droplets are slowed less by air friction as they drift downward, and they collide and merge with smaller droplets as they fall. They may merge, or coalesce, with a million other droplets before they fall from the cloud as raindrops. This *coalescence process* of forming precipitation is thought to take place in warm cumulus clouds that form near the ocean in the tropics. These clouds contain giant salt condensation nuclei and have been observed to produce rain within about twenty minutes after forming.

Clouds at middle latitudes, away from the ocean, also produce precipitation, so there must be a second way that precipitation forms. The *ice-crystal process* of forming precipitation is

important in clouds that extend high enough in the atmosphere to be above the freezing point of water. Water molecules are more strongly bonded to each other in an ice crystal than in liquid water. Thus, an ice crystal can capture water molecules and grow to a larger size while neighboring water droplets are evaporating. As they grow larger and begin to drift toward the surface, they may coalesce with other ice crystals or droplets of water, soon falling from the cloud. During the summer, they fall through warmer air below and reach the ground as raindrops. During the winter, they fall through cooler air below and reach the ground as snow.

Tiny water droplets do not freeze as readily as a larger mass of liquid water, and many droplets do not freeze until the temperature is below about −40°C (−40°F). Water that is still in the liquid state when the temperature is below the freezing temperature is said to be *supercooled*. Supercooled clouds of water droplets are common between the temperatures of −40°C and 0°C (−40°F and 32°F), a range of temperatures that is often found in the upper atmosphere. The liquid droplets at these temperatures need solid particles called **ice-forming nuclei** to freeze upon. Generally, dust from the ground serves as ice-forming nuclei that start the ice-crystal process of forming precipitation. Artificial rainmaking has been successful by (1) dropping crushed dry ice, which is cooler than −40°C, on top of a supercooled cloud and (2) introducing "seeds" of ice-forming nuclei in supercooled clouds. Tiny crystals from the burning of silver iodide are effective ice-forming nuclei, producing ice crystals at temperatures as high as −4.0°C (about 25°F). Attempts at ground-based cloud seeding with silver iodide in the mountains of the western United States have suggested up to 15 percent more snowfall, but it is difficult to know how much snowfall would have resulted without the seeding.

In general, the basic form of a cloud has meaning about the general type of precipitation that can occur as well as the coming weather. Cumulus clouds usually produce showers or thunderstorms that last only brief periods of time. Longer periods of drizzle, rain, or snow usually occur from stratus clouds. Cirrus clouds do not produce precipitation of any kind, but they may have meaning about the coming weather, which is discussed in the section on weather producers.

17.4 WEATHER PRODUCERS

The idealized model of the general atmospheric circulation starts with the poleward movement of warm air from the tropics. The region between 10°N and 10°S of the equator receives more direct radiation, on the average, than other regions of Earth's surface. The air over this region is heated more, expands, and becomes less dense as a consequence of the heating. This less dense air is buoyed up by convection to heights up to 20 km (about 12 mi). As it rises, it is cooled by radiation to less than −73°C (about −110°F). This accumulating mass of cool, dry air spreads north and south toward both poles (see figure 17.13), then sinks back toward the surface at about 30°N and 30°S. The descending air is warmed by compression and is warm and dry by the time it reaches the surface. Part of the sinking air then

FIGURE 17.20 Precipitation is water in liquid or solid form that returns to the surface of Earth. The precipitation you see here is liquid, and each raindrop is made from billions of the tiny droplets that make up the clouds. The tiny droplets of clouds become precipitation by merging to form larger droplets or by the growth of ice crystals that melt while falling.

moves back toward the equator across the surface, completing a large convective cell. This giant cell has a low-pressure belt over the equator and high-pressure belts over the subtropics near latitudes of 30°N and 30°S. The other part of the sinking air moves poleward across the surface, producing belts of westerly winds in both hemispheres to latitudes of about 60°.

The overall pattern of pressure belts and belts of prevailing winds is seen to shift north and south with the seasons, resulting in a seasonal shift in the types of weather experienced at a location. This shift of weather is related to three related weather producers: (1) the movement of large bodies of air, called *air masses,* that have acquired the temperature and moisture conditions where they have been located, (2) the leading *fronts* of air masses when they move, and (3) the local *high- and low-pressure* patterns that are associated with air masses and fronts. These are the features shown on almost all daily weather maps, and they are the topics of this section.

Air Masses

An **air mass** is defined as a large, more or less uniform body of air with nearly the same temperature and moisture conditions. An air mass forms when a large body of air, perhaps covering millions of square kilometers, remains over a large area of land or water for an extended period of time. While it is stationary, it acquires the temperature and moisture characteristics of the land or water through the heat transfer processes of conduction, convection, and radiation, and through the moisture transfer processes of evaporation and condensation. For example, a

large body of air that remains over the cold, dry, snow-covered surface of Siberia for some time will become cold and dry. A large body of air that remains over a warm tropical ocean, on the other hand, will become warm and moist. Knowledge about the condition of air masses is important because they tend to retain the acquired temperature and moisture characteristics when they finally break away, sometimes moving long distances. An air mass that formed over Siberia can bring cold, dry air to your location, while an air mass that formed over a tropical ocean will bring warm, moist air.

Air masses are classified according to the temperature and moisture conditions where they originate. There are two temperature extreme possibilities, a *polar air mass* from a cold region and a *tropical air mass* from a warm region. There are also two moisture extreme possibilities, a moist *maritime air mass* from over the ocean and a generally dry *continental air mass* from over the land. Thus, there are four main types of air masses that can influence the weather where you live: (1) continental polar, (2) maritime polar, (3) continental tropical, and (4) maritime tropical. Figure 17.21 shows the general direction in which these air masses usually move over the mainland United States.

Once an air mass leaves its source region, it can move at speeds of up to 800 km (about 500 mi) per day while mostly retaining the temperature and moisture characteristics of the source region (figure 17.22). If it slows and stagnates over a new location, however, the air may again begin to acquire a new temperature and moisture equilibrium with the surface. When a location is under the influence of an air mass, the location is having a period of *air mass weather.* This means that the weather

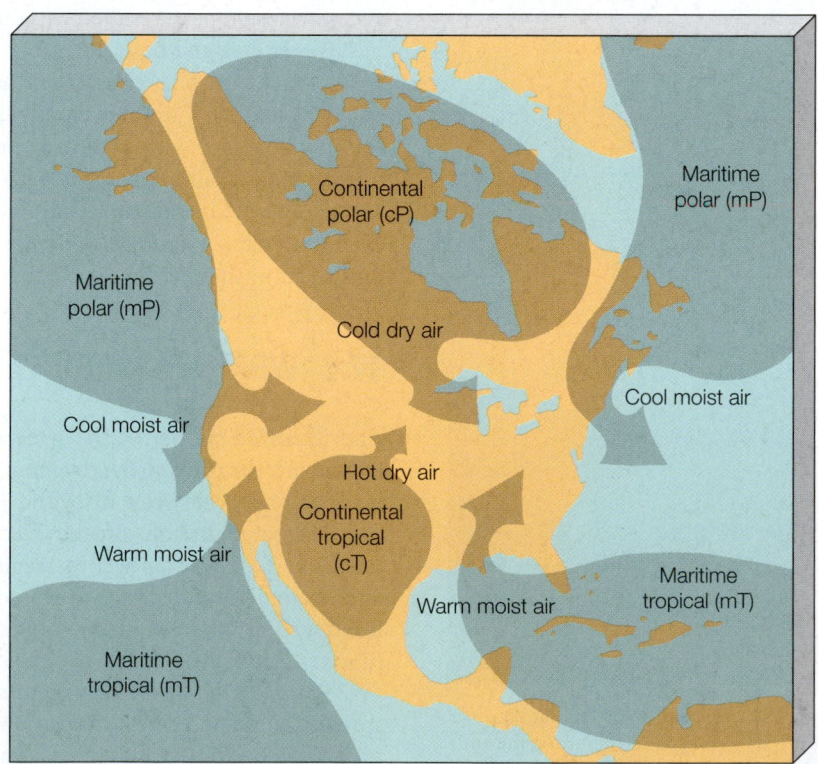

FIGURE 17.21 The air masses that affect weather in North America. The importance of the various air masses depends on the season. In winter, for instance, the continental tropical air mass disappears and the continental polar air mass exerts its greatest influence.

FIGURE 17.22 This satellite photograph shows the result of a polar air mass moving southeast over the southern United States. Clouds form over the warmer waters of the Gulf of Mexico and the Atlantic Ocean, showing the state of atmospheric instability from the temperature differences.

conditions will generally remain the same from day to day with slow, gradual changes. Air mass weather will remain the same until a new air mass moves in or until the air mass acquires the conditions of the new location. This process may take days or several weeks, and the weather conditions during this time depend on the conditions of the air mass and conditions at the new location. For example, a polar continental air mass arriving over a cool, dry land area may produce a temperature inversion with the air colder near the surface than higher up. When the temperature increases with height, the air is stable and cloudless, and cold weather continues with slow, gradual warming. The temperature inversion may also result in hazy periods of air pollution in some locations. A polar continental air mass arriving

over a generally warmer land area, on the other hand, results in a condition of instability. In this situation, each day will start clear and cold, but differential heating during the day develops cumulus clouds in the unstable air. After sunset, the clouds evaporate, and a clear night results because the thermals during the day carried away the dust and air pollution. Thus, a dry, cold air mass can bring different weather conditions, each depending on the properties of the air mass and the land it moves over.

Weather Fronts

The boundary between air masses of different temperatures is called a **front.** A front is actually a thin transition zone

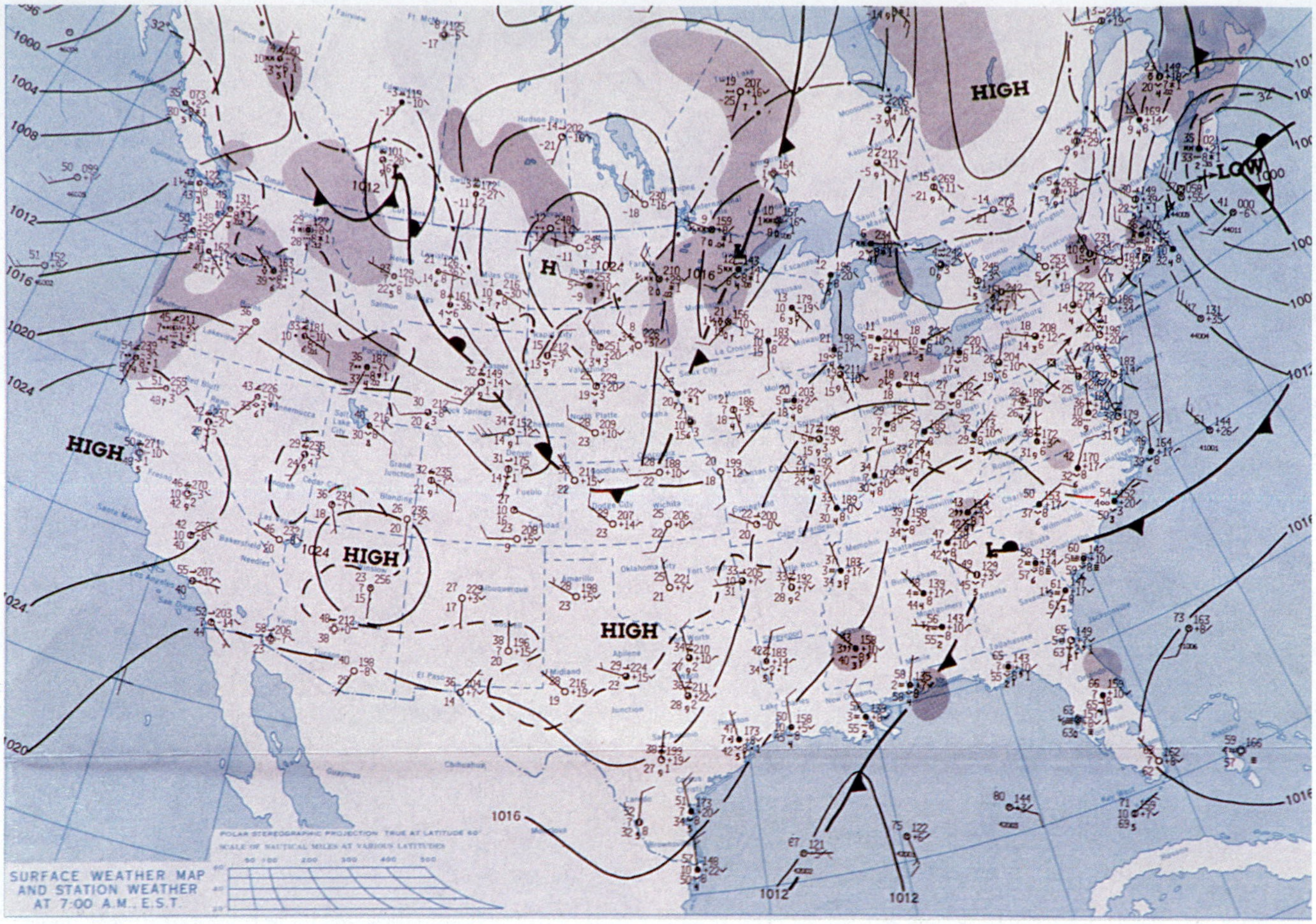

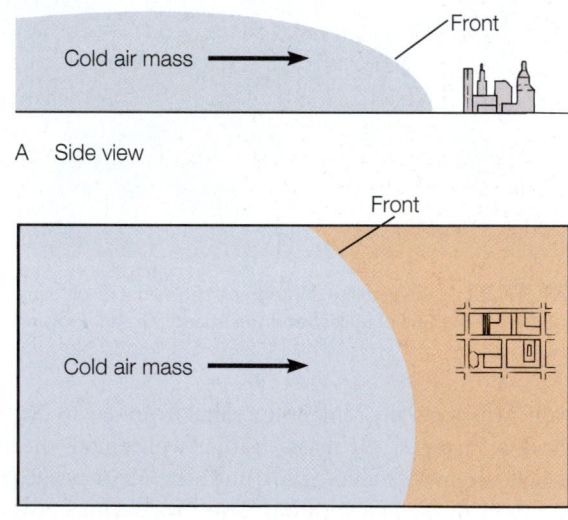

FIGURE 17.23 This weather map of the United States shows two fronts with associated low-pressure areas and five areas with high pressure.

between two air masses that ranges from about 5 to 30 km (about 3 to 20 mi) wide, and the air masses do not mix other than in this narrow zone. The density differences between the two air masses prevent any general mixing since the warm, less dense air mass is forced upward by the cooler, more dense air moving under it. You may have noticed on a daily weather map that fronts are usually represented with a line bulging outward in the direction of cold air mass movement (figure 17.23). A cold air mass is much like a huge, flattened bubble of air that moves across the land (figure 17.24). The line on a weather map represents the place where the leading edge of this huge, flattened bubble of air touches the surface of Earth.

A **cold front** is formed when a cold air mass moves into warmer air, displacing it in the process. A cold front is generally steep, and when it runs into the warmer air, it forces it to rise quickly. If the warm air is moist, it is quickly cooled to the dew point temperature, resulting in large, towering cumulus clouds and thunderclouds along the front (figure 17.25). You may have observed that thunderstorms created by an advancing cold front often form in a line along the front. These thunderstorms can be intense but are usually over quickly, soon followed by a rapid drop in temperature from the cold air mass moving past your location. The passage of the cold front is also marked by a rapid shift in the wind direction and a rapid increase in the barometric pressure. Before the cold front arrives, the wind is generally moving toward the front as warm, less dense

FIGURE 17.24 (A) A cold air mass is similar to a huge, flattened bubble of cold air that moves across the land. The front is the boundary between two air masses, a narrow transition zone of mixing. (B) A front is represented by a line on a weather map, which shows the location of the front at ground level.

air is forced upward by the cold, more dense air. The lowest barometric pressure reading is associated with the lifting of the warm air at the front. After the front passes your location,

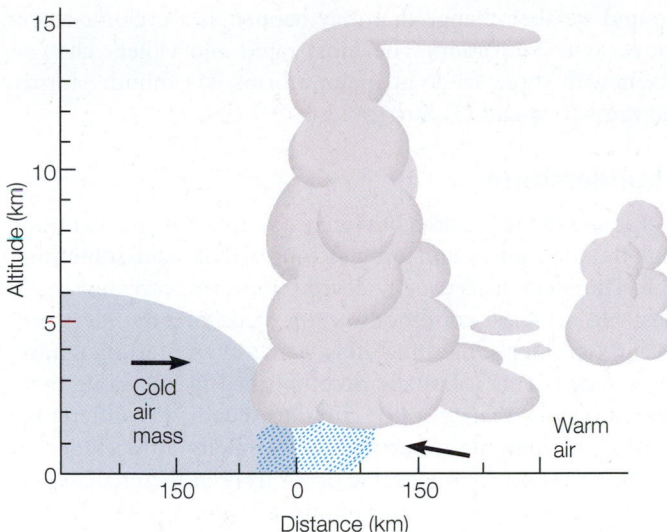

FIGURE 17.25 An idealized cold front, showing the types of clouds that might occur when an unstable cold air mass moves through unstable warm air. Stable air would result in more stratus clouds rather than cumulus clouds.

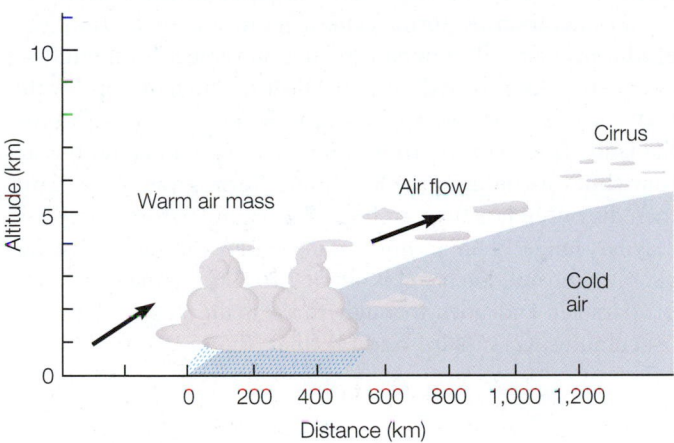

FIGURE 17.26 An idealized warm front, showing a warm air mass overriding and pushing cold air in front of it. Notice that the overriding warm air produces a predictable sequence of clouds far in advance of the moving front.

you are in the cooler, more dense air that is settling outward, so the barometric pressure increases and the wind shifts with the movement of the cold air mass.

A **warm front** forms when a warm air mass advances over a mass of cooler air. Since the advancing warm air is less dense than the cooler air it is displacing, it generally overrides the cooler air, forming a long, gently sloping front. Because of this, the overriding warm air may form clouds far in advance of the ground-level base of the front (figure 17.26). This may produce high cirrus clouds a day or more in advance of the front, which are followed by thicker and lower stratus clouds as the front advances. Usually these clouds result in a broad band of drizzle, fog, and the continuous light rain associated with stratus clouds. This light rain (and snow in the winter) may last for days as the warm front passes.

Sometimes the forces influencing the movement of a cold or warm air mass lessen or become balanced, and the front stops

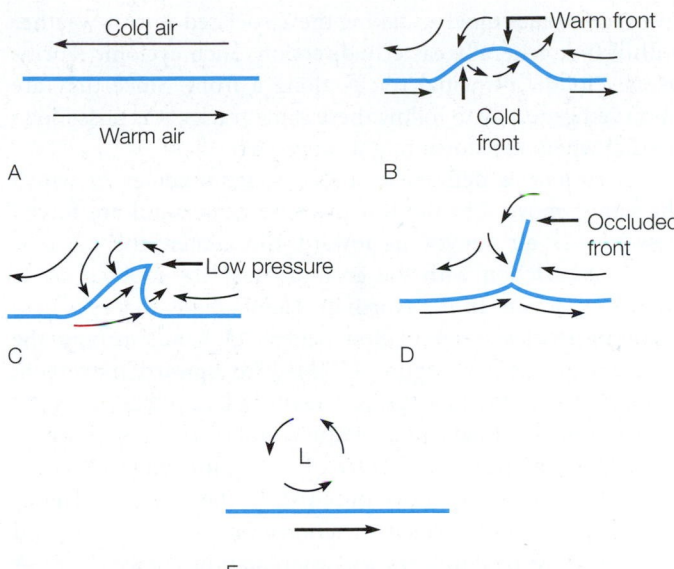

FIGURE 17.27 The development of a low-pressure center, or cyclonic storm, along a stationary front as seen from above. (*A*) A stationary front with cold air on the north side and warm air on the south side. (*B*) A wave develops, producing a warm front moving northward on the rightside and a cold front moving southward on the left side. (*C*) The cold front lifts the warm front off the surface at the apex, forming a low-pressure center. (*D*) When the warm front is completely lifted off the surface, an occluded front is formed. (*E*) The cyclonic storm is now a fully developed low-pressure center.

advancing. When this happens, a stream of cold air moves along the north side of the front, and a stream of warm air moves along the south side in an opposite direction. This is called a **stationary front** because the edge of the front is not advancing. A stationary front may sound as if it is a mild frontal weather maker because it is not moving. Actually, a stationary front represents an unstable situation that can result in a major atmospheric storm. This type of storm is discussed in the section on waves and cyclones.

Waves and Cyclones

A slowly advancing cold front and a stationary front often develop a bulge, or *wave,* in the boundary between cool and warm air moving in opposite directions (figure 17.27B). The wave grows as the moving air is deflected, forming a warm front moving northward on the right side and a cold front moving southward on the left side. Cold air is more dense than warm air, and the cold air moves faster than the slowly moving warm front. As the faster moving cold air catches up with the slower moving warm air, the cold air underrides the warm air, lifting it upward. This lifting action produces a low-pressure area at the point where the two fronts come together (figure 17.27C). The lifted air expands, cools, and reaches the dew point. Clouds form and precipitation begins from the lifting and cooling action. Within days after the wave first appears, the cold front completely overtakes the warm front, forming an occlusion (figure 17.27D). An **occluded front** is one that has been lifted completely off the ground into the atmosphere. The disturbance is now a *cyclonic storm* with a fully developed low-pressure center. After forming, the low-pressure cyclonic storm continues moving, taking the associated stormy weather with it in a generally easterly direction. Such cyclonic storms usually follow principal tracks along a front. Since they are observed generally to follow these same tracks, it is possible to predict where the storm might move next.

A *cyclone* is defined as a low-pressure center in which the winds move into the low-pressure center and are forced upward. As air moves in toward the center, the Coriolis effect and friction with the ground cause the moving air to change direction. In the Northern Hemisphere, this produces a counterclockwise circulation pattern of winds around the low-pressure center (figure 17.28). The upward movement associated with the low-pressure center of a cyclone cools the air, resulting in clouds, precipitation, and stormy conditions.

Air is sinking in the center of a region of high pressure, producing winds that move outward. In the Northern Hemisphere, the Coriolis effect and frictional forces deflect this wind to the right, producing a clockwise circulation (figure 17.28). A high-pressure center is called an *anticyclone,* or simply a *high.* Since air in a high-pressure zone sinks, it is warmed, and the relative humidity is lowered. Thus, clear, fair weather is usually associated with a high. By observing the barometric pressure, you can watch for decreasing pressure, which can mean the coming of a cyclone and its associated stormy weather. You can also watch for increasing pressure, which means a high and its associated fair weather are coming. Consulting a daily weather map makes such projections a much easier job, however.

Major Storms

A wide range of weather changes can take place as a front passes, because there is a wide range of possible temperature, moisture, stability, and other conditions between the new air mass and the air mass that it is displacing. The changes that accompany some fronts may be so mild that they go unnoticed. Others are noticed only as a day with breezes or gusty winds. Still other fronts are accompanied by a rapid and violent weather change called a **storm.** A snowstorm, for example, is

a rapid weather change that may happen as a cyclonic storm moves over a location. The most rapid and violent changes occur with three kinds of major storms: (1) thunderstorms, (2) tornadoes, and (3) hurricanes.

Thunderstorms

A **thunderstorm** is a brief but intense storm with rain, lightning and thunder, gusty and often strong winds, and sometimes hail. Thunderstorms usually develop in warm, very moist, and unstable air. These conditions set the stage for a thunderstorm to develop when something lifts a parcel of air, starting it moving upward. This is usually accomplished by the same three general causes that produce cumulus clouds: (1) differential heating, (2) mountain barriers, or (3) along an occluded or cold front. Thunderstorms that occur from differential heating usually occur during warm, humid afternoons after the Sun has had time to establish convective thermals. In the Northern Hemisphere, most of these convective thunderstorms occur during the month of July. Frontal thunderstorms, on the other hand, can occur any month and any time of the day or night that a front moves through warm, moist, and unstable air.

Frontal thunderstorms generally move with the front that produced them. Thunderstorms that developed in mountains or over flat lands from differential heating can move miles after they form, sometimes appearing to wander aimlessly across the land. These storms are not just one big rain cloud but are sometimes made up of cells that are born, grow to maturity, then die out in less than an hour. The thunderstorm, however, may last longer than an hour because new cells are formed as old ones die out. Each cell is about 2 to 8 km (about 1 to 5 mi) in diameter and goes through three main stages in its life: (1) cumulus, (2) mature, and (3) final (figure 17.29).

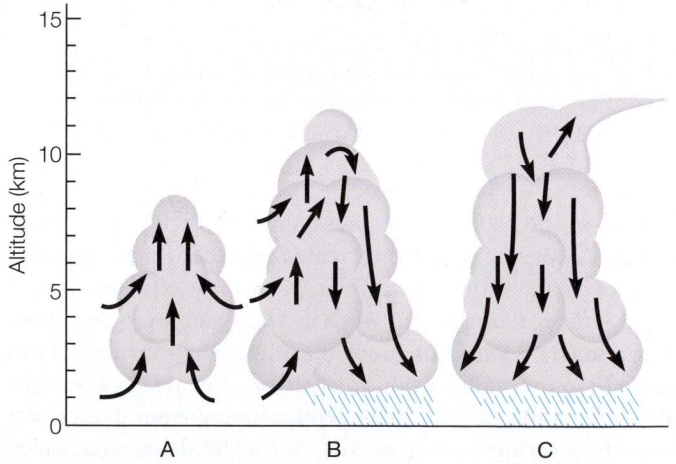

FIGURE 17.29 Three stages in the life of a thunderstorm cell. (*A*) The cumulus stage begins as warm, moist air is lifted in an unstable atmosphere. All the air movement is upward in this stage. (*B*) The mature stage begins when precipitation reaches the ground. This stage has updrafts and downdrafts side by side, which create violent turbulence. (*C*) The final stage begins when all the updrafts have been cut off and only downdrafts exist. This cuts off the supply of moisture, and the rain decreases as the thunderstorm dissipates. The anvil-shaped top is a characteristic sign of this stage.

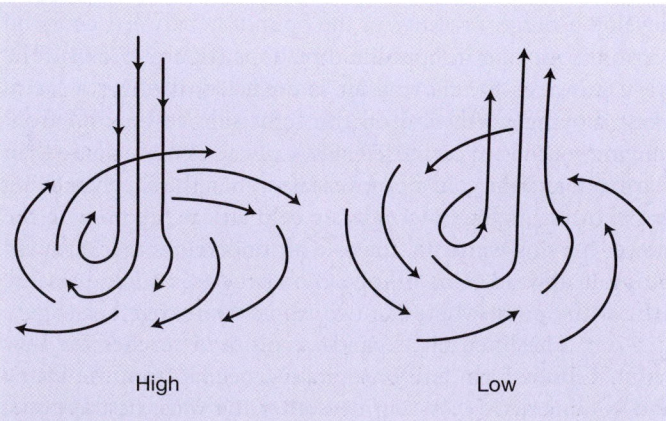

FIGURE 17.28 Air sinks over a high-pressure center and spreads over the surface in a clockwise pattern in the Northern Hemisphere. Air over the surface moves toward a low-pressure center in a counterclockwise pattern in the Northern Hemisphere.

Damage from a thunderstorm is usually caused by the associated lightning, strong winds, or hail. As illustrated in figure 17.29, the first stage of a thunderstorm begins as convection, mountains, or a dense air mass slightly lifts a mass of warm, moist air in an unstable atmosphere. The lifted air mass expands and cools to the dew point temperature, and a cumulus cloud forms. The latent heat of vaporization released by the condensation process accelerates the upward air motion, called an *updraft,* and the cumulus cloud continues to grow to towering heights. Soon the upward-moving, saturated air reaches the freezing level and ice crystals and snowflakes begin to form. When they become too large to be supported by the updraft, they begin to fall toward the surface, melting into raindrops in the warmer air they fall through. When they reach the surface, this marks the beginning of the mature stage. As the raindrops fall through the air, friction between the falling drops and the cool air produces a downdraft in the region of the precipitation. The cool air accelerates toward the surface at speeds up to 90 km/h (about 55 mi/h), spreading out on the ground when it reaches the surface. In regions where dust is raised by the winds, this spreading mass of cold air from the thunderstorm has the appearance of a small cold front with a steep, bulging leading edge. This miniature cold front may play a role in lifting other masses of warm, moist air in front of the thunderstorm, leading to the development of new cells. This stage in the life of a thunderstorm has the most intense rainfall, winds, and possibly hail. As the downdraft spreads throughout the cloud, the supply of new moisture from the updrafts is cut off and the thunderstorm enters the final, dissipating stage. The entire life cycle, from cumulus cloud to the final stage, lasts for about an hour as the thunderstorm moves across the surface. During the mature stage of powerful updrafts, the top of the thunderstorm may reach all the way to the top of the troposphere, forming a cirrus cloud that is spread into an anvil shape by the strong winds at this high altitude.

The updrafts, downdrafts, and falling precipitation separate tremendous amounts of electric charges that accumulate in different parts of the thundercloud. Large drops of water tend to carry negative charges, and cloud droplets tend to lose them. The upper part of the thunderstorm develops an accumulation of positive charges as cloud droplets are uplifted, and the middle portion develops an accumulation of negative charges from larger drops that fall. The voltage of these charge centers builds to the point that the electrical insulating ability of the air between them is overcome and a giant electrical discharge called *lightning* occurs (figure 17.30). Lightning discharges occur from the cloud to the ground, from the ground to a cloud, from one part of the cloud to another part, or between two different clouds. The discharge takes place in a fraction of a second and may actually consist of a number of strokes rather than one big discharge. The discharge produces an extremely high temperature around the channel, which may be only 6 cm (about 2 in) or so wide. The air it travels through is heated quickly, expanding into a sudden pressure wave that you hear as *thunder.* A nearby lightning strike produces a single, loud crack. Farther away strikes sound more like a rumbling boom as the sound from the separate strokes become separated over distance. Echoing of the thunder produced at farther distances also adds

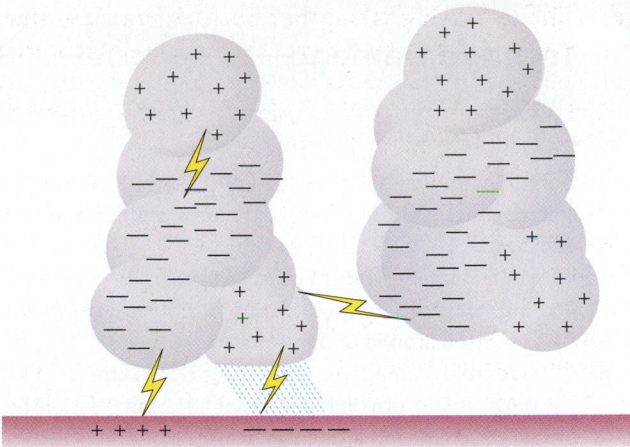

FIGURE 17.30 Different parts of a thunderstorm cloud develop centers of electric charge. Lightning is a giant electric spark that discharges the accumulated charges.

FIGURE 17.31 These hailstones fell from a thunderstorm in Iowa, damaging automobiles, structures, and crops.

to the rumbling sounds. Lightning can present a risk for people in the open, near bodies of water, or under a single, isolated tree during a thunderstorm. The safest place to be during a thunderstorm is inside a car or a building with a metal frame.

Updrafts are also responsible for **hail,** a frozen form of precipitation that can be very destructive to crops, automobiles, and other property. Hailstones can be irregular, somewhat spherical, or flattened forms of ice that range from the size of a BB to the size of a softball (figure 17.31). Most hailstones, however, are less than 2 cm (about 1 in) in diameter. The larger hailstones have alternating layers of clear and opaque, cloudy ice. These layers are believed to form as the hailstone goes through cycles of falling then being returned to the upper parts of the thundercloud by updrafts. The clear layers are believed to form as the hailstone moves through heavy layers of supercooled water droplets, which accumulate quickly on the

Connections . . .

FIGURE 17.32 A tornado might be small, but it is the most violent storm that occurs on Earth. This tornado, moving across an open road, eventually struck Dallas, Texas.

hailstone but freeze slowly because of the release of the latent heat of fusion. The cloudy layers are believed to form as the hailstone accumulates snow crystals or moves through a part of the cloud with less supercooled water droplets. In either case, rapid freezing traps air bubbles, which result in the opaque, cloudy layer. Thunderstorms with hail are most common during the month of May in Colorado, Kansas, and Nebraska.

Tornadoes

A **tornado** is the smallest, most violent weather disturbance that occurs on Earth (figure 17.32). Tornadoes occur with intense thunderstorms and resemble a long, narrow funnel or ropelike structure that drops down from a thundercloud and may or may not touch the ground. This ropelike structure is a rapidly whirling column of air, usually 100 to 400 m (about 330 to 1,300 ft) in diameter. An average tornado will travel 6 to 8 km (about 4 to 5 mi) on the ground, sometimes skipping into the air, then back down again. The bottom of the column moves across the ground at speeds that average about 50 km/h (about 30 mi/h). The speed of the whirling air in the column has been estimated to be up to about 480 km/h (about 300 mi/h), but most tornadoes have winds of less than 180 km/h (112 mi/h). The destruction is produced by the powerful winds, the sudden drop in atmospheric pressure that occurs at the center of the funnel, and the debris that is flung through the air like projectiles. A passing tornado sounds like very loud, continuous rumbling thunder with cracking and hissing noises that are punctuated by the crashing of debris projectiles.

On the average, several hundred tornadoes are reported in the United States every year. These occur mostly during spring and early summer afternoons over the Great Plains states. Texas, Oklahoma, Kansas, and Iowa have such a high occurrence of tornadoes that the region is called "tornado alley." During the spring and early summer, this region has maritime tropical air from the Gulf of Mexico at the surface. Above this warm, moist layer is a layer of dry, unstable air that has just crossed the Rocky Mountains, moved along rapidly by the jet stream. The stage is now set for some event, such as a cold air mass moving in from the north, to shove the warm, moist air upward, and the result will be violent thunderstorms with tornadoes.

Hurricanes

What is the difference between a tropical depression, tropical storm, and a hurricane? In general, they are all storms with strong upward atmospheric motion and a cyclonic surface wind circulation (figure 17.33). They are born over tropical or subtropical waters and are not associated with a weather front.

FIGURE 17.33 This is a satellite photo of hurricane John, showing the eye and counterclockwise motion.

Hurricane Damage

Hurricanes are classified according to category and damage to be expected. Here is the classification scheme:

Category	Damage	Winds	
1	minimal	120–153 km/h	(75–95 mi/h)
2	moderate	154–177 km/h	(96–110 mi/h)
3	extensive	178–210 km/h	(111–130 mi/h)
4	extreme	211–250 km/h	(131–155 mi/h)
5	catastrophic	>250 km/h	(>155 mi/h)

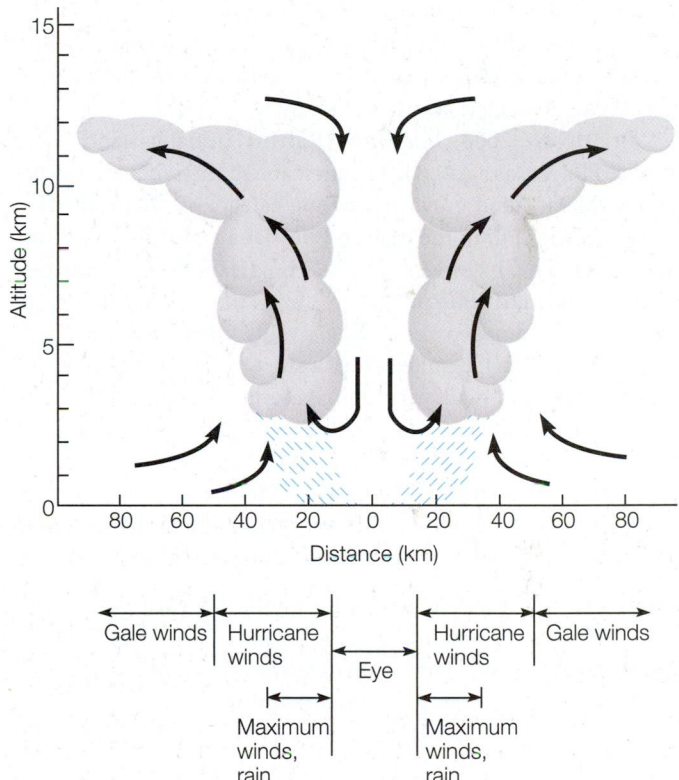

FIGURE 17.34 Cross section of a hurricane.

The varieties of storm intensities are classified according to the *speed* of the maximum sustained surface winds.

A *tropical depression* is an area of low pressure around which the winds are generally moving 55 km/h (about 35 mi/h) or less. The tropical depression might dissolve into nothing, or it might develop into a more intense disturbance. A *tropical storm* is a more intense low pressure area with winds between 56 and 120 km/h (about 35 to 75 mi/h). A **hurricane** is a very intense low-pressure area with winds greater than 120 km/h (about 75 mi/h). A strong storm of this type is called a hurricane if it occurs over the Atlantic Ocean or the Pacific Ocean east of the international date line. It is called a **typhoon** if it occurs over the North Pacific Ocean west of the international date line.

A tropical cyclone is similar to the wave cyclone of the mid-latitudes because both have low-pressure centers with a counterclockwise circulation in the Northern Hemisphere. They are different because a wave cyclone is usually about 2,500 km (about 1,500 mi) wide, has moderate winds, and receives its energy from the temperature differences between two air masses. A tropical cyclone, on the other hand, is often less than 200 km (about 125

mi) wide, has very strong winds, and receives its energy from the latent heat of vaporization released during condensation.

A fully developed hurricane has heavy bands of clouds, showers, and thunderstorms that rapidly rotate around a relatively clear, calm eye (figure 17.34). As a hurricane approaches a location, the air seems unusually calm as a few clouds appear, then thicken as the wind begins to gust. Over the next six hours or so, the overall wind speed increases as strong gusts and intense rain showers occur. Thunderstorms, perhaps with tornadoes, and the strongest winds occur just before the winds suddenly die down and the sky clears with the arrival of the eye of the hurricane. The eye is an average of 10 to 15 km (about 6 to 9 mi) across, and it takes about an hour or so to cross a location. When the eye passes, the intense rain showers, thunderstorms, and hurricane-speed winds begin again, this time blowing from the opposite direction. The whole sequence of events may be over in a day or two, but hurricanes are unpredictable and sometimes stall in one location for days. In general, they move at a rate of 15 to 50 km/h (about 10 to 30 mi/h).

Most of the damage from hurricanes results from strong winds, flooding, and the occasional tornado. Flooding occurs from the intense, heavy rainfall but also from the increased sea level that results from the strong, constant winds blowing seawater toward the shore. The sea level can be raised some 5 m (about 16 ft) above normal, with storm waves up to 15 m (about 50 ft) high on top of this elevated sea level. Overall, large inland areas can be flooded with extensive property damage. A single hurricane moving into a populated coastal region has caused billions of dollars of damage and the loss of hundreds of lives in the past. Today, the National Weather Service tracks hurricanes by weather satellites. Warnings of hurricanes, tornadoes, and severe thunderstorms are broadcast locally over special weather alert stations located across the country.

In August 2005, hurricane Katrina initially struck near Miami, Florida, as a category 1 hurricane. It then moved into the Gulf of Mexico and grew to a strong category 3 hurricane that moved up the Gulf to the eastern Louisiana and western Mississippi coast. This massive storm had hurricane-force winds that extended outward 190 km (120 mi) from the center, resulting in severe storm damage over a wide area as it struck the coast on August 29. Damage resulted from a storm surge that exceeded 7 m (25 ft), heavy rainfall, wind damage, and the failure of the levee system in New Orleans. Overall, this resulted in an estimated $81 billion in damages and more than 1,836 fatalities.

Katrina had sustained winds of 200 km/h (125 mi/h) as it struck the shore, but smaller hurricanes have had stronger sustained winds when they struck the shore. These include

- August 17, 1969: Hurricane Camille hit Mississippi with 306 km/h (190 mi/h) sustained winds.
- August 24, 1992: Hurricane Andrew hit south Florida with 266 km/h (165 mi/h) sustained winds.
- August 13, 2004: Hurricane Charley hit Punta Gorda, Florida, with 240 km/h (150 mi/h) sustained winds.*

17.5 WEATHER FORECASTING

Today, weather predictions are based on information about the characteristics, location, and rate of movement of air masses and associated fronts and pressure systems. This information is summarized as average values, then fed into a computer model of the atmosphere. The model is a scaled-down replica of the real atmosphere, and changes in one part of the model result in changes in another part of the model just as they do in the real atmosphere. Underlying the computer model are the basic scientific laws concerning solar radiation, heat, motion, and the gas laws. All these laws are written as a series of mathematical equations, which are applied to thousands of data points in a three-dimensional grid that represents the atmosphere. The computer is given instructions about the starting conditions

*Hurricane data from National Climatic Data Center

at each data point, that is, the average values of temperature, atmospheric pressure, humidity, wind speed, and so forth. The computer is then instructed to calculate the changes that will take place at each data point, according to the scientific laws, within a very short period of time. This requires billions of mathematical calculations when the program is run on a worldwide basis. The new calculated values are then used to start the process all over again, and it is repeated some 150 times to obtain a one-day forecast (figure 17.35).

A problem with the computer model of the atmosphere is that small-scale events are inadequately treated, and this introduces errors that grow when predictions are attempted for farther and farther into the future. Small eddies of air, for example, or gusts of wind in a region have an impact on larger-scale atmospheric motions such as those larger than a cumulus cloud. But all of the small eddies and gusts cannot be observed without filling the atmosphere with measuring instruments. This lack of ability to observe small events that can change the large-scale events introduces uncertainties in the data, which, over time, will increasingly affect the validity of a forecast.

To find information about the accuracy of a forecast, the computer model can be run several different times, with each run having slightly different initial conditions. If the results of all the runs are close to each other, the forecasters can feel confident that the atmosphere is in a predictable condition, and this means the forecast is probably accurate. In addition, multiple computer runs can provide forecasts in the form of probabilities. For example, if eight out of ten forecasts indicate rain, the "averaged" forecast might call for an 80 percent chance of rain.

The use of new computer technology has improved the accuracy of next-day forecasts tremendously, and the forecasts up to three days are fairly accurate, too. For forecasts of more than five days, however, the number of calculations and the effect of uncertainties increase greatly. It has been estimated that the reductions of observational errors could increase the range of accurate forecasting up to two weeks. The ultimate

FIGURE 17.35 Computers make routine weather forecasts possible by solving mathematical equations that describe changes in a mathematical model of the atmosphere.

range of accurate forecasting will require a better understanding—and thus, an improved model—of patterns of changes that occur in the ocean as well as in the atmosphere. All of this increased understanding and reduction of errors leads to an estimated ultimate future forecast of three weeks, beyond which any pinpoint forecast would be only slightly better than a wild guess. In the meantime, regional and local daily weather forecasts are fairly accurate, and computer models of the atmosphere now provide the basis for extending the forecasts for up to about a week.

17.6 CLIMATE

Changes in the atmospheric condition over a brief period of time, such as a day or a week, are referred to as changes in the *weather*. These changes are part of a composite, larger pattern called **climate.** Climate is the general pattern of the weather that occurs for a region over a number of years. Among other things, the climate determines what types of vegetation grow in a particular region, resulting in characteristic groups of plants associated with the region (figure 17.36). For example, orange, grapefruit, and palm trees grow in a region that has a climate with warm monthly temperatures throughout the year. On the other hand, blueberries, aspen, and birch trees grow in a region that has cool temperature patterns throughout the year. Climate determines what types of plants and animals live in a location, the types of houses that people build, and the lifestyles of people. Climate also influences the processes that shape the landscape, the type of soils that form, the suitability of the region for different types of agriculture, and how productive the agriculture will be in a region. This section is about climate, what determines the climate of a region, and how climate patterns are classified.

Major Climate Groups

Earth's atmosphere is heated directly by incoming solar radiation and by absorption of infrared radiation from the surface. The amount of heating at any particular latitude on the surface depends primarily on two factors: (1) the *intensity* of the incoming solar radiation, which is determined by the angle at which the radiation strikes the surface, and (2) the amount of *time* that the radiation is received at the surface, that is, the number of daylight hours compared to the number of hours of night.

Earth is so far from the Sun that all rays of incoming solar radiation reaching Earth are essentially parallel. Earth, however, has a mostly constant orientation of its axis with respect to the stars as it moves around the Sun in its orbit. Since the inclined axis points toward the Sun part of the year and away from the Sun the other part, radiation reaches different latitudes at different angles during different parts of the year. The orientation of Earth's axis to the Sun during different parts of the year also results in days and nights of nearly equal length in the equatorial region but increasing differences at increasing latitudes to the poles. During the polar winter months, the night is twenty-four hours long, which means no solar radiation is received at all. The equatorial region receives more solar radiation during

Cities Change the Weather

Using satellite measurements, NASA researchers have confirmed that the presence of a city does influence the weather. First, because of all the concrete, asphalt, and buildings, cities are 0.6°C to 5.6°C (about 1°F to 10°F) warmer than the surrounding country. The added heat of this "urban heat island" tends to make the air less stable. Less stable air brings about the second influence of more rain falling downwind of a city. Evidently, the less stable air and rougher surfaces cause air moving toward a city to rise, and the turbulence mixes in city pollutants that add nuclei for water vapor to condense upon. The result is an average 28 percent increase in the amount of rainfall 30 to 60 km (about 19 to 37 mi) downwind of cities. The NASA research team satellite measurements were verified by data from a large array of ground-based thermometers and rain gauges. The verification of these satellite-based findings are important for urban planning, water resource management, and decisions about where to farm the land. It may also show that local surface environments are more important in computer weather forecast models than had previously been believed.

Myths, Mistakes, and Misunderstandings

Rain Forecast Mistakes

It is a mistake to believe that a 20 percent chance of rain means it will rain over 20 percent of the forecast area. The correct meaning is that it will rain on 20 percent of the days when they have the same atmospheric conditions.

FIGURE 17.36 The climate determines what types of plants and animals live in a location, the types of houses that people build, and the lifestyles of people. This orange tree, for example, requires a climate that is relatively frost-free, yet it requires some cool winter nights to produce a sweet fruit.

a year, and the amount received decreases toward the poles as a result of (1) yearly changes in intensity and (2) yearly changes in the number of daylight hours.

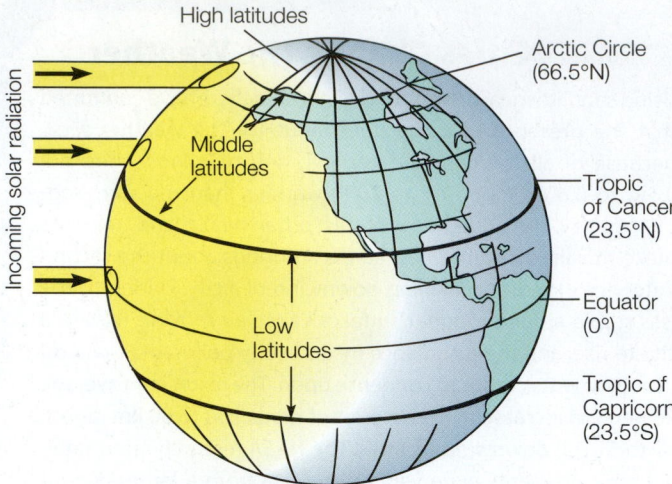

High latitudes

Arctic Circle
(66.5°N)

Middle
latitudes

Incoming solar radiation

Tropic
of Cancer
(23.5°N)

Equator
(0°)

Low
latitudes

Tropic of
Capricorn
(23.5°S)

FIGURE 17.37 Latitude groups based on incoming solar radiation. The low latitudes receive vertical solar radiation at noon some time of the year, the high latitudes receive no solar radiation at noon during some time of the year, and the middle latitudes are in between.

To generalize about the amount of radiation received at different latitudes, some means of organizing, or grouping, the latitudes is needed (figure 17.37). For this purpose, the latitudes are organized into three groups:

1. The *low latitudes,* those that some time of the year receive *vertical* solar radiation at noon.
2. The *high latitudes,* those that some time of the year receive *no* solar radiation at noon.
3. The *middle latitudes,* which are between the low and high latitudes.

This definition of low, middle, and high latitudes means that the low latitudes are between the tropics of Cancer and Capricorn (between 23.5°N and 23.5°S latitudes) and that the high latitudes are above the Arctic and Antarctic Circles (above 66.5°N and above 66.5°S latitudes).

In general:

1. The low latitudes receive a high amount of incoming solar radiation that varies little during a year. Temperatures are high throughout the year, varying little from month to month.
2. The middle latitudes receive a higher amount of incoming radiation during one part of the year and a lower amount during the other part. Overall temperatures are cooler than in the low latitudes and have a wide seasonal variation.
3. The high latitudes receive a maximum amount of radiation during one part of the year and none during another part. Overall temperatures are low, with the highest range of annual temperatures.

The low, middle, and high latitudes provide a basic framework for describing Earth's climates. These climates are associated with the low, middle, and high latitudes illustrated in figure 17.37, but they are defined in terms of yearly temperature averages. Temperature and moisture are the two most important climate factors, and temperature will be considered first.

The principal climate zones are defined in terms of yearly temperature averages, which occur in broad regions (figure 17.38). They are

1. The *tropical climate zone* of the low latitudes (figure 17.39).
2. The *polar climate zone* of the high latitudes (figure 17.40).
3. The *temperate climate zone* of the middle latitudes (figure 17.41).

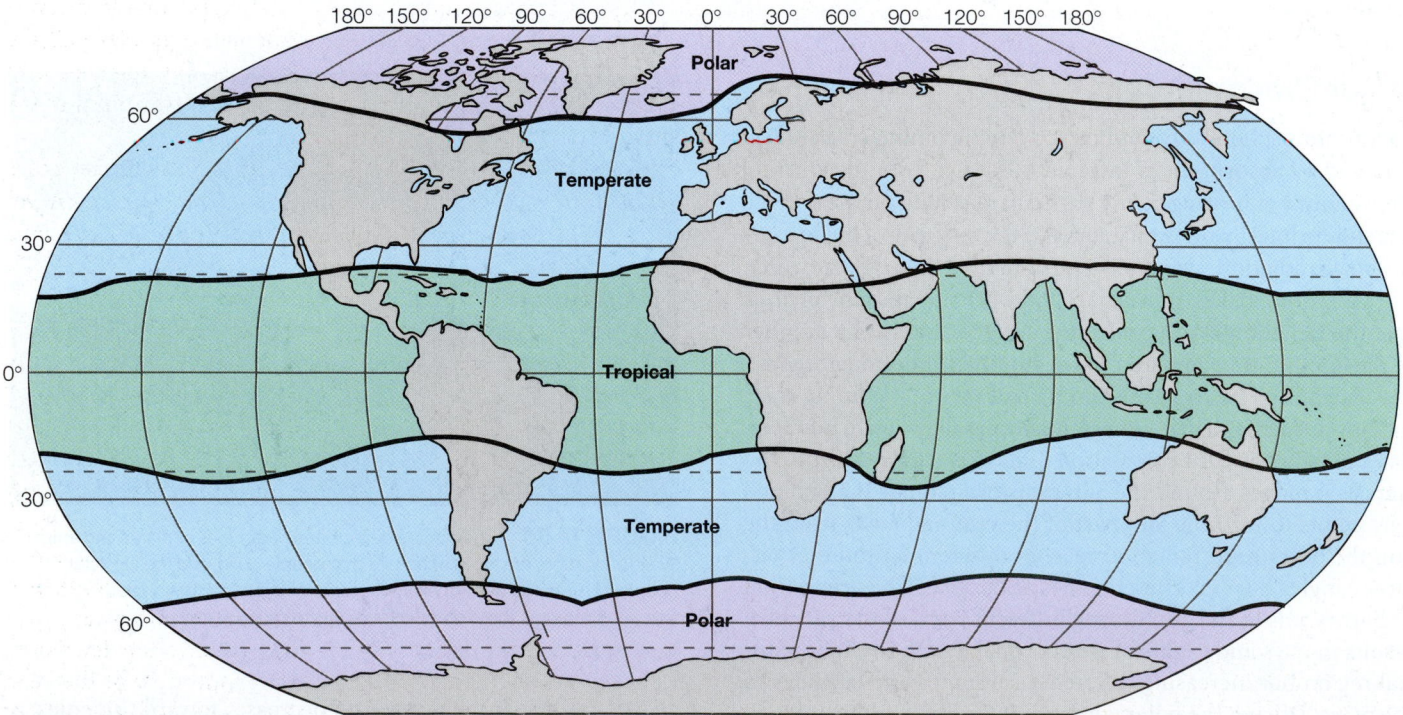

FIGURE 17.38 The principal climate zones are defined in terms of yearly temperature averages, which are determined by the amount of solar radiation received at the different latitude groups.

FIGURE 17.39   A wide variety of plant life can grow in a tropical climate, as you can see here.

FIGURE 17.40   Polar climates occur at high elevations as well as high latitudes. This mountain location has a highland polar climate and tundra vegetation but little else.

FIGURE 17.41   This temperate-climate deciduous forest responds to seasonable changes in autumn with a show of color.

The tropical climate zone is near the equator and receives the greatest amount of sunlight throughout the year. Overall, the tropical climate zone is hot. Average monthly temperatures stay above 18°C (64°F), even during the coldest month of the year.

The other extreme is found in the polar climate zone, where the sun never sets during some summer days and never rises during some winter days. Overall, the polar climate zone is cold. Average monthly temperatures stay below 10°C (50°F), even during the warmest month of the year.

The temperate climate zone is between the polar and tropical zones, with average temperatures that are neither very cold nor very hot. Average monthly temperatures stay between 10°C and 18°C (50°F and 64°F) throughout the year.

General patterns of precipitation and winds are also associated with the low, middle, and high latitudes. An idealized model of the global atmospheric circulation and pressure patterns was described in the section on weather forecasting. Recall that this model described a huge convective movement of air in the low latitudes, with air being forced upward over the equatorial region. This air expands, cools to the dew point, and produces abundant rainfall throughout most of the year. On the other hand, air is slowly sinking over 30°N and 30°S of the equator, becoming warm and dry as it is compressed. Most of the great deserts of the world are near 30°N or 30°S latitude

El Niño and La Niña

The term *El Niño* was originally used to describe an occurrence of warm, above-normal ocean temperatures off the South American coast. Fishermen along this coast learned long ago to expect associated changes in fishing patterns about every three to seven years, which usually lasted for about eighteen months. They called this event El Niño, which is Spanish for "the boy child" or "Christ child," because it typically began near Christmas. The El Niño event occurs when the trade winds along the equatorial Pacific become reduced or calm, allowing sea surface temperatures to increase much above normal. The warm water drives the fish to deeper waters or farther away from usual fishing locations.

Today, El Niño is understood to be much more involved than just a warm ocean in the Pacific. It is more than a local event, and the bigger picture is sometimes called the "El Niño–Southern Oscillation," or ENSO. In addition to the warmer tropical Pacific water of El Niño, "the boy," the term *La Niña,* "the girl," has been used to refer to the times when the water of the tropical Pacific is *colder* than normal. The "Southern Oscillation" part of the name comes from observations that atmospheric pressure around Australia seems to be inversely linked to the atmospheric pressure in Tahiti. They seem to be linked because when the pressure is low in Australia, it is high in Tahiti. Conversely, when the atmospheric pressure is high in Australia, it is low in Tahiti. The strength of this Southern Oscillation is measured by the Southern Oscillation Index (SOI), which is defined as the pressure at Darwin, Australia, subtracted from that at Tahiti. Negative values of SOI are usually associated with El Niño events, so the Southern Oscillation and the El Niño are obviously linked. How ENSO can impact the weather in other parts of the world has only recently become better understood.

The atmosphere is a system that responds to incoming solar radiation, the spinning Earth, and other factors, such as the amount of water vapor present. The ocean and atmospheric systems undergo changes by interacting with each other,

most visibly in the tropical cyclone. The ocean system supplies water vapor, latent heat, and condensation nuclei, which are the essential elements of a tropical cyclone as well as everyday weather changes and climate. The atmosphere, on the other hand, drives the ocean with prevailing winds, moving warm or cool water to locations where they affect the climate on the land. There is a complex, interdependent relationship between the ocean and the atmosphere, and it is probable that even small changes in one system can lead to bigger changes in the other.

Normally, during non–El Niño times, the Pacific Ocean along the equator has established systems of prevailing wind belts, pressure systems, and ocean currents. In July, these systems push the surface seawater offshore from South America, westward along the equator, and toward Indonesia. During El Niño times, the trade winds weaken and the warm water moves back eastward, across the Pacific to South America, where it then spreads north and south along the coast. Why the trade winds weaken and become calm is unknown and is the subject of ongoing research.

Warmer waters along the coast of South America bring warmer, more humid air and the increased possibility of thunderstorms. Thus, the possibility of towering thunderstorms, tropical storms, or hurricanes increases along the Pacific coast of South America as the warmer waters move north and south. This creates the possibility of weather changes not only along the western coast but elsewhere, too. The towering thunderstorms reach high into the atmosphere, adding tropical moisture and creating changes in prevailing wind belts. These wind belts carry or steer weather systems across the middle latitudes of North America, so typical storm paths are shifted. This shifting can result in

▶ Increased precipitation in California during the fall to spring season.
▶ A wet winter with more and stronger storms along regions of the southern United States and Mexico.

▶ A warmer and drier-than-average winter across the northern regions of Canada and the United States.
▶ A variable effect on central regions of the United States, ranging from reduced snowfall to no effect at all.
▶ Other changes in the worldwide global complex of ocean and weather events, such as droughts in normally wet climates and heavy rainfall in normally dry climates.

One major problem in these predictions is a lack of understanding of what causes many of the links and a lack of consistency in the links themselves. For example, southern California did not always have an unusually wet season every time an El Niño occurred and in fact experienced a drought during one event.

Scientists have continued to study the El Niño since the mid-1980s, searching for patterns that will reveal consistent cause-and-effect links. Part of the problem may be that other factors, such as a volcanic eruption, may influence part of the linkage but not another part. Another part of the problem may be the influence of unknown factors, such as the circulation of water deep beneath the ocean surface, the track taken by tropical cyclones, or the energy released by tropical cyclones in one year compared to the next.

The results so far have indicated that atmosphere-ocean interactions are much more complex than early theoretical models had predicted. Sometimes a new model will predict some weather changes that occur with El Niño, but no model is yet consistently correct in predicting the conditions that lead to the event and the weather patterns that result. All this may someday lead to a better understanding of how the ocean and the atmosphere interact on this dynamic planet.

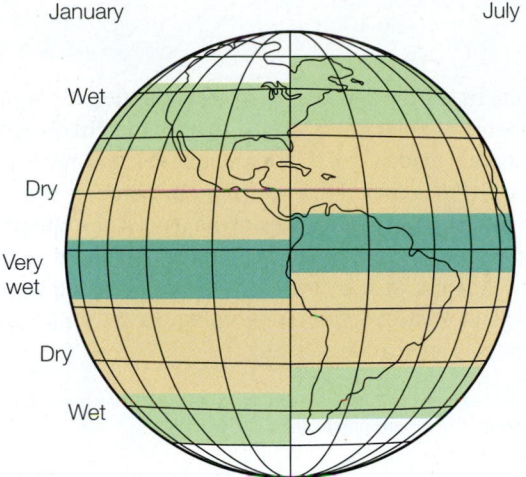

January   July

Wet

Dry

Very
wet

Dry

Wet

FIGURE 17.42 The idealized general rainfall patterns over Earth change with seasonal shifts in the wind and pressure areas of the planet's general atmospheric circulation patterns.

for this reason. There is another wet zone near 60° latitudes and another dry zone near the poles. These wet and dry zones are shifted north and south during the year with the changing seasons. This results in different precipitation patterns in each season. Figure 17.42 shows where the wet and dry zones are in winter and in summer seasons.

Regional Climatic Influence

Latitude determines the basic tropical, temperate, and polar climatic zones, and the wet and dry zones move back and forth over the latitudes with the seasons. If these were the only factors influencing the climate, you would expect to find the same climatic conditions at all locations with the same latitude. This is not what is found, however, because there are four major factors that affect a regional climate. These are (1) altitude, (2) mountains, (3) large bodies of water, and (4) ocean currents. The following describes how these four factors modify the climate of a region.

The first of the four regional climate factors is *altitude.* The atmosphere is warmed mostly by the greenhouse effect from the surface upward, and air at higher altitudes increasingly radiates more and more of its energy to space. Average air temperatures therefore decrease with altitude, and locations with higher altitudes will have lower average temperatures. This is why the tops of mountains are often covered with snow when none is found at lower elevations. St. Louis, Missouri, and Denver, Colorado, are located almost at the same latitude (within 1° of 39°N), so you might expect the two cities to have about the same average temperature. Denver, however, has an altitude of 1,609 m (5,280 ft), and the altitude of St. Louis is 141 m (465 ft). The yearly average temperature for Denver is about 10°C (about 50°F) and for St. Louis it is about 14°C (about 57°F). In general, higher altitude means lower average temperature.

The second of the regional climate factors is the influence of *mountains.* In addition to the temperature change caused

by the altitude of the mountain, mountains also affect the conditions of a passing air mass. The western United States has mountainous regions along the coast. When a moist air mass from the Pacific meets these mountains, it is forced upward and cools. Water vapor in the moist air mass condenses, clouds form, and the air mass loses much of its moisture as precipitation falls on the western side of the mountains. Air moving down the eastern slope is compressed and becomes warm and dry. As a result, the western slopes of these mountains are moist and have forests of spruce, redwood, and fir trees. The eastern slopes are dry and have grassland or desert vegetation.

The third of the regional climate factors is the presence of a large body of *water.* Water, as discussed previously, has a higher specific heat than land material, is transparent, and loses energy through evaporation. All of these affect the temperature of a landmass located near a large body of water, making the temperatures more even from day to night and from summer to winter. San Diego, California, and Dallas, Texas, for example, are at about the same latitude (both almost 33°N), but San Diego is at a seacoast and Dallas is inland. Because of its nearness to water, San Diego has an average summer temperature about 7°C (about 13°F) cooler and an average winter temperature about 5°C (about 9°F) warmer than the average temperatures in Dallas. Nearness to a large body of water keeps the temperature more even at San Diego.

The fourth of the regional climate factors is *ocean currents.* In addition to the evenness brought about by being near the ocean, currents in the ocean can bring water that has a temperature different from the land. For example, currents can move warm water northward or they can move cool water southward (figure 17.43). This can influence the temperatures of air masses that move from the water to the land and, thus, the temperatures of the land. For example, the North Pacific current brings warm waters to the western coast of North America, which results in warmer temperatures for cities near the coast.

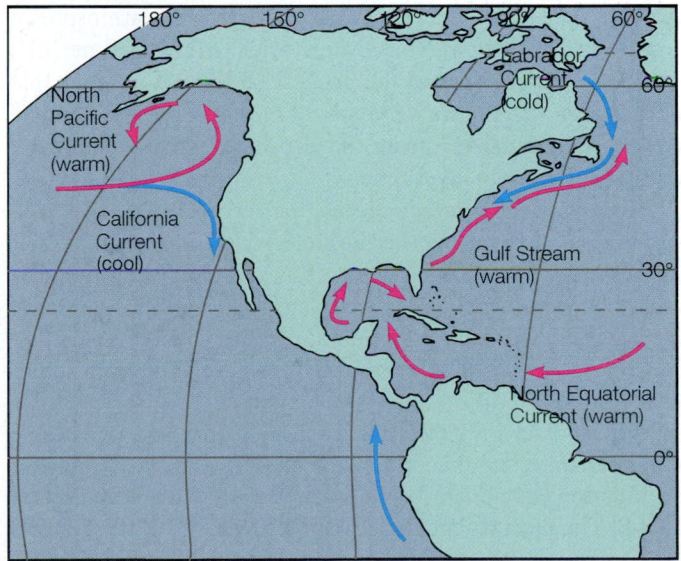

FIGURE 17.43 Ocean currents can move large quantities of warm or cool water to influence the air temperatures of nearby landmasses.

Vilhelm Firman Koren Bjerknes (1862–1951)

Vilhelm Bjerknes was the Norwegian scientist who created modern meteorology.

Bjerknes came from a talented family. His father was professor of mathematics at the Christiania (now Oslo) University and a highly influential geophysicist who clearly shaped his son's studies. Bjerknes held chairs at Stockholm and Leipzig before founding the Bergen Geophysical Institute in 1917. Bjerknes made momentous contributions that transformed meteorology into an accepted science. Not least, he showed how weather prediction could be put on a statistical basis, dependent on the use of mathematical models.

During World War I, Bjerknes instituted a network of weather stations throughout Norway; coordination of the findings from such stations led him and his coworkers to develop the highly influential theory of polar fronts, on the basis of the discovery that the atmosphere is made up of discrete air masses displaying dissimilar features. Bjerknes coined the word *front* to delineate the boundaries between such air masses. One of the many contributions of the "Bergen frontal theory" was to explain the generation of cyclones over the Atlantic, at the junction of warm and cold air wedges. Bjerknes's work gave modern meteorology its theoretical tools and methods of investigation.

Source: Modified from the *Hutchinson Dictionary of Scientific Biography*. © RM, 2011. All rights reserved. Helicon Publishing is a division of RM.

Describing Earth's climates presents a problem because there are no sharp boundaries that exist naturally between two adjacent regions with different climates. Even if two adjacent climates are very different, one still blends gradually into the other. For example, if you are driving from one climate zone to another, you might drive for miles before becoming aware that the vegetation is now different than it was an hour ago. Since the vegetation is very different from what it was before, you know that you have driven from one regional climate zone to another. Chapter 23 discusses the typical plants and animals associated with the various climate types.

Actually, no two places on Earth have exactly the same climate. Some plants will grow on the north or south side of a building, for example, but not on the other side. The two sides of the building could be considered as small, local climate zones within a larger, major climate zone.

Climate Change

As stated earlier, *weather* describes changes in the atmospheric condition (rain, wind, etc.) over a brief period of time (day, week). *Climate* describes the general pattern of weather that occurs over a region for a number of months or years. **Climate change** is a departure from the expected average pattern of climate for a region over time.

Scientists measured climate patterns during the past 120 years or so by using thermometers, rain gauges, and other instruments. Climate patterns before 120 years ago are inferred by analyzing evidence found in tree rings, lake-bottom sediments, ice cores drilled from a glacier, and other sources. A natural source used to infer temperature change, rainfall, or some other past climate condition is called **proxy data.**

Proxy data indicates that Earth's climate has undergone major changes in the past, with cold *ice ages* and glaciers dominating the climate for the past several million years (figure 17.44). The most recent ice age covered almost a third of Earth's land surface with up to 4 km (2.5 mi) ice sheets during its maximum extent. About half of the states in the United States were covered by ice, some completely and others partially. A large amount of water was locked up in glaciers, and this caused the sea level to drop some 90 m (about 300 ft), exposing a land bridge between Siberia and Alaska, among other things. The ice sheets of the most recent ice age advanced and retreated at least four times, with the sea level fluctuating with each advance and retreat.

About every 100,000 years Earth enters an *interglacial warming period* before returning to another ice age. We are currently in such a warming period, and the current period began about 18,000 years ago. About 15,000 years ago Earth's climate had warmed enough to stop the advance of glaciers, and since that time the glaciers have been retreating. By about 4000 B.C. the average temperatures had warmed to a few degrees warmer than those of today. The proxy record indicates that such interglacial periods last from 15,000 to 20,000 years before beginning a new glacial period. If the cycle continues as it has in

FIGURE 17.44 Much of Earth's surface is covered with ice sheets and glaciers during an ice age. Both are formed from accumulations of compacted snow.

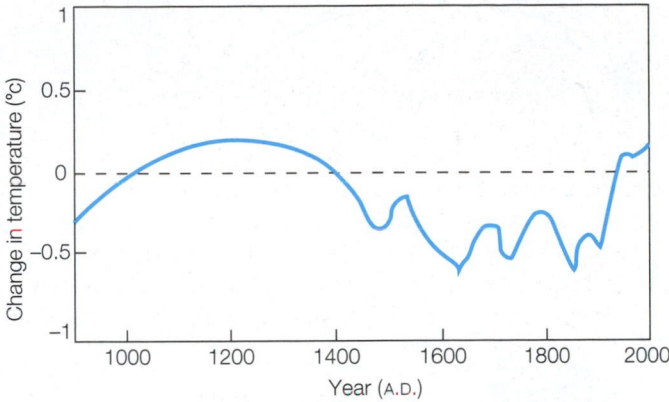

FIGURE 17.45 According to proxy data, Earth has experienced a period of warming and a period of cooling over the past 1,000 years.
Source: www.geocraft.com/WVFossils/ice_ages.html

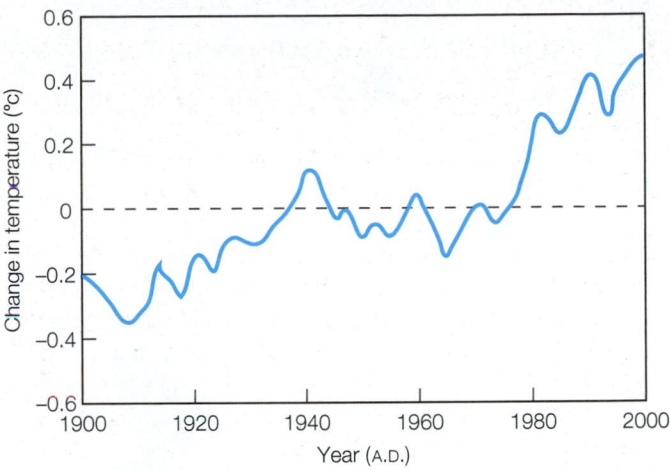

FIGURE 17.46 According to land-based weather stations, Earth has experienced a period of warming, a period of cooling, and most recently, a period of warming over the past 100 years.
Source: www.ncdc.noaa.gov

the past, we are near the end of the current interglacial period, nearing the beginning of the next ice age.

Data about ice ages and interglacial warming show that periods of warming and cooling occur in large cycles. During the past 1,000 years Earth has also undergone a number of smaller warming and cooling cycles (figure 17.45). For example it was warm—about like today—from 1000 up to 1400. Then from 1400 to 1860 there was a period of cooling called the "little ice age." The little ice age is why the Vikings left Greenland after successfully farming there from the tenth century until the thirteenth.

During the past 100 years, Earth has had two warming cycles separated by one period of cooling. The temperature increased from the early 1800s—before the Industrial Revolution—because we were coming out of the little ice age. Then the temperature decreased from about 1940 to the late 1970s (figure 17.46). And then a slight warming cycle began in the 1980s, continuing through today.

Recording stations began reading small but steady increases in temperatures near the Earth's surface, averaging about 0.25°C (0.4°F) in the past twenty-five years. Sea level has increased 1 to 2 mm per year for the past one hundred years, and this increase is due to thermal expansion of seawater in addition to melting ice.

The recent temperature increase is not uniform around the globe since the southeastern United States has cooled over the past 100 years. Artic sea ice has decreased since the early 1970s, but sea ice in the Antarctic has *increased* during the same period. Nonetheless, some people became concerned that the temperature increase was caused by the release of greenhouse gases—mostly carbon dioxide—that would result in a runaway greenhouse effect (see page 379 for information on the greenhouse effect).

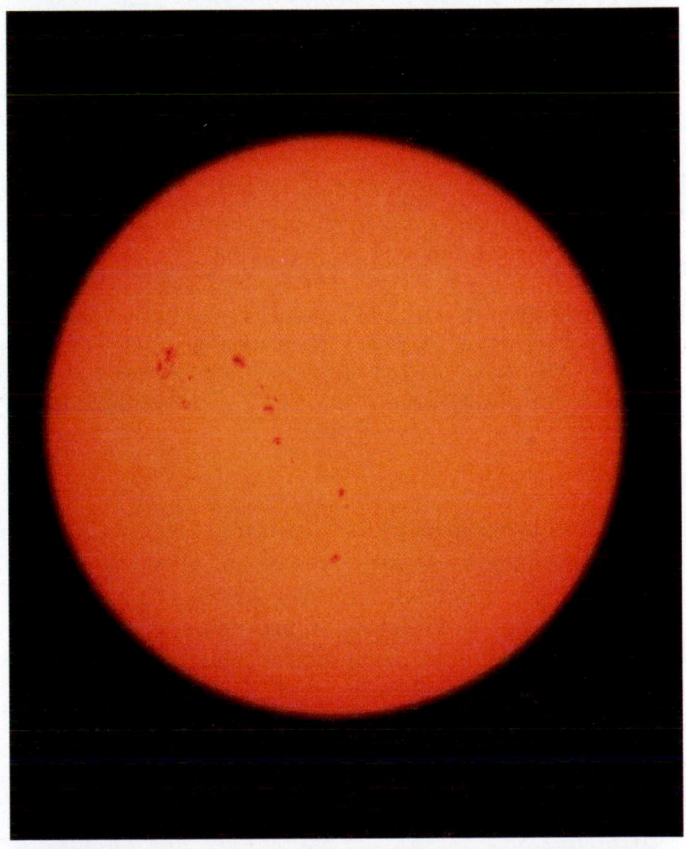

FIGURE 17.47 Sunspots appear on the Sun's surface as dark spots because they have a much cooler temperature than their surroundings. Sunspots appear in cycles that vary in number, averaging 11.1 years.

Causes of Global Climate Change

Climate change is brought about by a complex interaction of a number of factors. Some of these factors could be astronomical, and some could be factors occurring in the atmosphere. Solar energy is fundamentally responsible for weather and climate, and changes in the Sun's energy output can change the climate. The Sun's output of energy changes with *sunspots,* dark spots that appear to move around the surface of the Sun (figure 17.47).

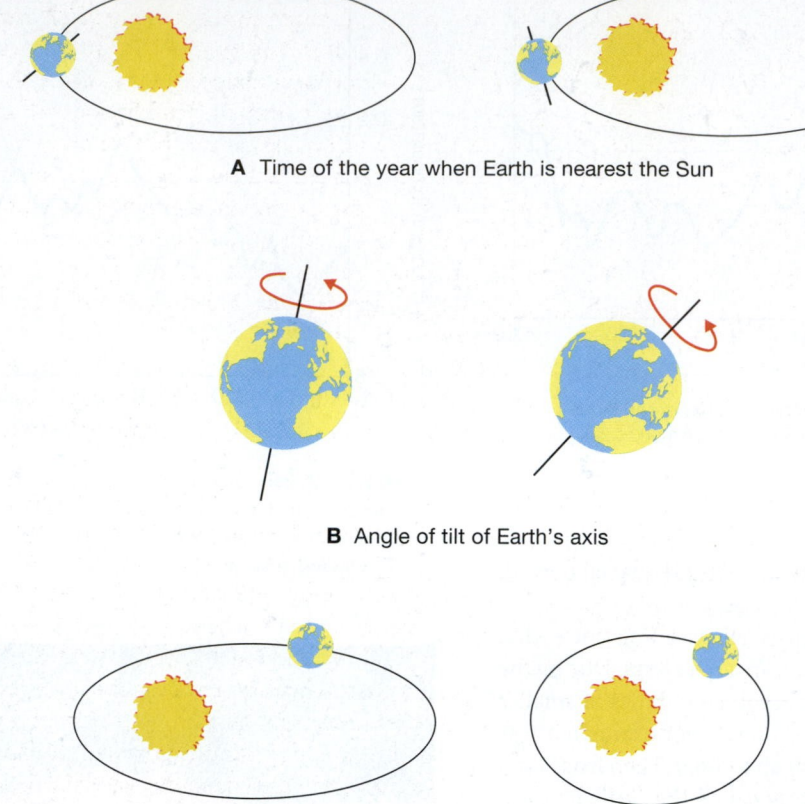

A Time of the year when Earth is nearest the Sun

B Angle of tilt of Earth's axis

C Shape of Earth's orbit

FIGURE 17.48 Three variations in Earth's motion that may be responsible for causing ice ages. (A) The time of year when Earth is nearest the Sun varies with a period of about 23,000 years. (B) The angle of tilt of Earth's axis of rotation varies with a period of about 41,000 years. (C) The shape of Earth's elliptical orbit varies with a period of about 100,000 years. These variations have relatively little effect on the total sunlight reaching Earth but a considerable effect on the sunlight reaching the polar regions in summer. The ellipticity of Earth's orbit is vastly exaggerated here; the orbit is actually less than 2 percent away from a perfect circle.

The number of sunspots varies from year to year, and there are years when sunspots are rare or absent, and years when a peak is reached. The maximum number seems to occur in a cycle that averages 11.1 years. Evidently, the amount of solar energy increases as the number of sunspots increases. More sunspots deliver more energy to Earth's surface, which increases the temperature. Estimates are that the Sun's energy output varies by up to 0.2 percent with each sunspot cycle. If this increase continues into the mid-twenty-first century, Earth's surface temperature will increase by about 0.5°C (about 1°F). However, a relationship between the sunspot cycle *length* and temperatures also comes into play. Higher than normal temperatures tend to occur with shorter cycles, and lower temperatures occur with longer cycles.

In addition to changes in the Sun, changes in the orientation of Earth's tilt, orbital shape, and axis wobble change our orientation to the Sun in predictable cycles. These are called Milankovitch cycles after Milutin Milankovitch (1879–1958), a Yugoslav physicist who calculated how the cycles would affect the climate by altering the amount of solar energy received by Earth (figure

17.48). As it works out, the amount of energy received by high latitudes varies up to 20 percent. A shorter summer allows ice to accumulate, making an ice age. The timing of the ice age and cycles of warmer or cooler average temperatures fits with periods of Earth's orbital variations.

There are also atmospheric factors that can cause climatic changes. The greenhouse gases, for example, can reradiate heat in the atmosphere, producing a warmer climate. The most abundant greenhouse gas is water vapor, which is also the dominate gas in terms of increasing the temperature. Water vapor is followed by carbon dioxide, methane, and then some trace gases (table 17.1). The natural greenhouse effect is a good thing, for without it the average Earth surface temperature would be −18°C (0.4°F). Thanks to the greenhouse effect, the global average is 14°C (57°F).

Carbon dioxide has been increasing in the atmosphere since the late 1800s (figure 17.49) and some believe the burning of fossil fuels is responsible. Precise measurements of atmosphere carbon dioxide concentration have been made since 1958. This found that the year-to-year concentration

TABLE 17.1 Man-made Contribution to the "Greenhouse Effect," Expressed as % of Total

Based on concentrations (ppb) adjusted for heat retention characteristics	% of All Greenhouse Gases	% Natural	%Man-made
Water vapor	95.000%	94.999%	0.001%
Carbon Dioxide (CO$_2$)	3.618%	3.502%	0.117%
Methane (CH$_4$)	0.360%	0.294%	0.066%
Nitrous Oxide (N$_2$O)	0.950%	0.903%	0.047%
Misc. gases (CFC's, etc.)	0.072%	0.025%	0.047%
Total	100.00%	99.72	0.28%

*From Internet article *"Global Warming: A Closer Look at the Numbers"* by Monte Hieb. Revised 01/03.

varied but had an average increase of 1.5 parts per million by volume. Carbon dioxide is naturally released and absorbed by plants and animals as well as Earth's oceans. The overall concentration varies because green plants convert carbon dioxide to plant materials. Carbon dioxide is also absorbed by ocean waters, used by marine organisms to make shells, and is converted to mineral deposits such as limestone. Also, note that carbon dioxide can go into solution, and warm ocean water dissolves less carbon dioxide than cool ocean water. Just like a glass of soda, warmer water will release carbon dioxide. Tropical deforestation is reducing one means of removing carbon dioxide from the atmosphere, and this adds up to an estimated 40 percent as much carbon dioxide as the burning of fossil fuels. How much land, the ocean, and plant and animals remove and add carbon dioxide to the atmosphere is still highly uncertain.

example, a 2002 report predicts increased global temperatures during the next 100 years somewhere between 1.4°C (2.5°F) to 5.8°C (10.4°F) as a result of increased carbon dioxide in the atmosphere. Not all the experts agree with this prediction because it is based on an estimated carbon dioxide concentration that grows exponentially, increasing 1 percent per year. These experts run their own climate model based on the actual increases in carbon dioxide concentration for the past thirty years—which did not grow exponentially—and project additional warming of only 0.5°C (0.9°F) in the next fifty years.

It may be a surprise to find such disagreements between experts. Some argue that changes of the climate are natural but not well understood. Earth has had a history of warming periods, and this was long before activities of people emitted carbon dioxide. Furthermore, not all the "evidence" points to an increase in global temperatures. A retreating glacier, for example, can be a result of less snow in the winter, or the result of a longer summer, not necessarily warmer temperatures.

Do we have increased temperatures and retreating glaciers because of increased carbon dioxide, or are these natural variations in climate cycles that will continue to change? The answer will not come from computer simulation of the global climate because the computer model will always be simpler than the climate itself. Scientists can only weigh the evidence and make a professional judgment, and this can result in disagreement. Today there is no consensus about the cause of the supposed slight warming observed during the past forty years. At least the underlying science of climate change should be understood. Then a critical analysis of climate change should be completed before any future policies about human-made carbon dioxide are implemented.

Global Warming

How do you predict the climate of the future? Scientists use mathematical models to calculate the evolving state of the atmosphere in response to changes in factors. The model is run on a large computer that uses current climatic data from sunlight, land and atmosphere interactions, and interactions with the ocean. A change in some factor is then introduced to calculate a hypothetical future. For example, the changing factor could be an increased concentration of carbon dioxide in the atmosphere. Climate models are imperfect because they do not include all the important factors; for example, they leave out the cycling of carbon dioxide into and out of the atmosphere.

The United Nations established the Intergovernmental Panel on Climate Change (IPCC) in 1988 to address climate change. The IPCC releases reports on the state of the global climate and makes projections about climate change based on changes in climate models. For

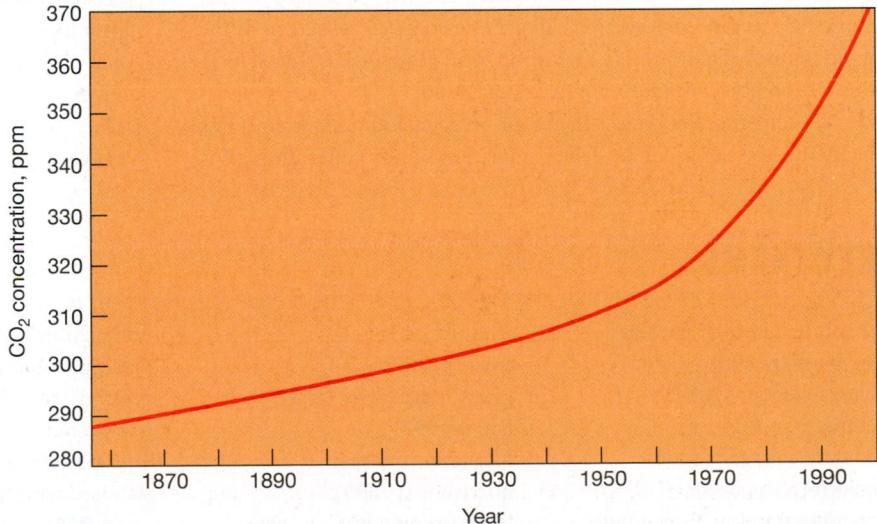

FIGURE 17.49 Carbon dioxide concentration in the atmosphere since 1860, in parts per million (ppm).

SUMMARY

Earth's *atmosphere* thins rapidly with increasing altitude. Pure, dry air is mostly *nitrogen, oxygen,* and *argon,* with traces of *carbon dioxide* and other gases. Atmospheric air also contains a variable amount of *water vapor.* Water vapor cycles into and out of the atmosphere through evaporation and precipitation.

Atmospheric pressure is measured with a *mercury barometer.* At sea level, the atmospheric pressure will support a column of mercury about 76.00 cm (about 29.92 in) tall. This is the average pressure at sea level, and it is called the *standard atmospheric pressure, normal pressure,* or *one atmosphere of pressure.*

Materials on Earth's surface absorb sunlight, emitting more and more *infrared radiation* as they are warmed. *Carbon dioxide* and *molecules* of water vapor in the atmosphere absorb infrared radiation, which then reemit the energy many times before it reaches outer space again. The overall effect warms the lower atmosphere from the bottom up in a process called the *greenhouse effect.*

The layer of the atmosphere from the surface up to where the temperature stops decreasing with height is called the *troposphere.* The *stratosphere* is the layer above the troposphere. Temperatures in the stratosphere increase because of the interaction between ozone (O_3) and ultraviolet radiation from the Sun.

The surface of Earth is not heated uniformly by sunlight. This results in a *differential heating,* which sets the stage for *convection.* The horizontal movement of air on the surface from convection is called *wind.* A generalized model for understanding why the wind blows involves (1) the relationship between *air temperature and air density* and (2) the relationship between *air pressure and the movement of air.* This model explains local wind patterns and wind patterns observed on a global scale.

The amount of water vapor in the air at a particular time is called the *absolute humidity.* The *relative humidity* is a ratio between the amount of water vapor that is in the air and the amount needed to saturate the air at the present temperature.

When the air is saturated, condensation can take place. The temperature at which this occurs is called the *dew point temperature.* If the dew point temperature is above freezing, *dew* will form. If the temperature is below freezing, *frost* will form. Both dew and frost form directly on objects and do not fall from the air.

Water vapor condenses in the air on *condensation nuclei.* If this happens near the ground, the accumulation of tiny water droplets is called a *fog. Clouds* are accumulations of tiny water droplets in the air above the ground. In general, there are three basic shapes of clouds: *cirrus, cumulus,* and *stratus.* These basic cloud shapes have meaning about the atmospheric conditions and about the coming weather conditions.

Water that returns to Earth in liquid or solid form falls from the clouds as *precipitation.* Precipitation forms in clouds through two processes: (1) the *coalescence* of cloud droplets or (2) the *growth of ice crystals* at the expense of water droplets.

Weather changes are associated with the movement of large bodies of air called *air masses,* the leading *fronts* of air masses when they move, and local *high-* and *low-pressure* patterns that accompany air masses or fronts. Examples of air masses include (1) *continental polar,* (2) *maritime polar,* (3) *continental tropical,* and (4) *maritime tropical.*

When a location is under the influence of an air mass, the location is having *air mass weather* with slow, gradual changes. More rapid changes take place when the *front,* a thin transition zone between two air masses, passes a location.

A *stationary front* often develops a bulge, or *wave,* that forms into a moving cold front and a moving warm front. The faster-moving cold front overtakes the warm front, lifting it into the air to form an *occluded front.* The lifting process forms a low-pressure center called a *cyclone.* Cyclones are associated with heavy clouds, precipitation, and stormy conditions because of the lifting action.

A *thunderstorm* is a brief, intense storm with rain, lightning and thunder, gusty and strong winds, and sometimes hail. A *tornado* is the smallest, most violent weather disturbance that occurs on Earth. A *hurricane* is a *tropical cyclone,* a large, violent circular storm that is born over warm tropical waters near the equator.

The general pattern of the weather that occurs for a region over a number of years is called *climate.* The three principal climate zones are (1) the *tropical climate zone,* (2) the *polar climate zone,* and (3) the *temperate climate zone.* The climate in these zones is influenced by four factors that determine the local climate: (1) *altitude,* (2) *mountains,* (3) *large bodies of water,* and (4) *ocean currents.* The climate for a given location is described by first considering the principal climate zone, then looking at subdivisions within each that result from local influences. Earth's climate has undergone major changes in the past, caused by changes in the Sun, changes in the Earth's orbit, and/or changes in the atmosphere.

KEY TERMS

absolute humidity (p. **386**)
air mass (p. **392**)
barometer (p. **378**)
climate (p. **401**)
climate change (p. **406**)
cold front (p. **394**)
condensation nuclei (p. **388**)
dew (p. **387**)
dew point temperature (p. **387**)

front (p. **393**)
frost (p. **387**)
greenhouse effect (p. **379**)
hail (p. **397**)
humidity (p. **386**)
hurricane (p. **399**)
ice-forming nuclei (p. **391**)
inversion (p. **379**)
jet stream (p. **384**)

occluded front (p. **395**)
precipitation (p. **391**)
proxy data (p. **406**)
relative humidity (p. **386**)
saturated air (p. **385**)
standard atmospheric pressure (p. **379**)
stationary front (p. **395**)
storm (p. **396**)

thunderstorm (p. **396**)
tornado (p. **398**)
typhoon (p. **399**)
warm front (p. **395**)
wind (p. **381**)
wind chill factor (p. **382**)

APPLYING THE CONCEPTS

Answers are located in appendix F.

1. Without adding or removing any water vapor, a sample of air experiencing an increase in temperature will have
   a. a higher relative humidity.
   b. a lower relative humidity.
   c. the same relative humidity.
   d. a changed absolute humidity.

2. Cooling a sample of air results in a (an)
   a. increased capacity to hold water vapor.
   b. decreased capacity to hold water vapor.
   c. unchanged capacity to hold water vapor.

3. On a clear, calm, and cool night, dew or frost is most likely to form
   a. under trees or other shelters.
   b. on bare ground on the side of a hill.
   c. under a tree on the side of a hill.
   d. on grass in an open, low-lying area.

4. Longer periods of drizzle, rain, or snow usually occur from which basic form of a cloud?
   a. stratus
   b. cumulus
   c. cirrus
   d. None of the above is correct.

5. Brief periods of showers are usually associated with which type of cloud?
   a. stratus
   b. cumulus
   c. cirrus
   d. None of the above is correct.

6. The type of air mass weather that results from the arrival of polar continental air is
   a. frequent snowstorms with rapid changes.
   b. clear and cold with gradual changes.
   c. unpredictable but with frequent and rapid changes.
   d. much the same from day to day, the conditions depending on the air mass and the local conditions.

7. The appearance of high cirrus clouds followed by thicker, lower stratus clouds and then continuous light rain over several days probably means which of the following air masses has moved to your area?
   a. continental polar
   b. maritime tropical
   c. continental tropical
   d. maritime polar

8. A fully developed cyclonic storm is most likely to form
   a. on a stationary front.
   b. in a high-pressure center.
   c. from differential heating.
   d. over a cool ocean.

9. The basic difference between a tropical storm and a hurricane is
   a. size.
   b. location.
   c. wind speed.
   d. amount of precipitation.

10. Most of the great deserts of the world are located
    a. near the equator.
    b. 30° north or south latitude.
    c. 60° north or south latitude.
    d. anywhere, as there is no pattern to their location.

11. The average temperature of a location is made more even by the influence of
    a. a large body of water.
    b. elevation.
    c. nearby mountains.
    d. dry air.

12. The climate of a specific location is determined by
    a. its latitude.
    b. how much sunlight it receives.
    c. its altitude and nearby mountains and bodies of water.
    d. All of the above are correct.

QUESTIONS FOR THOUGHT

1. Explain the greenhouse effect. Is a greenhouse a good analogy for Earth's atmosphere? Explain.

2. Describe how the ozone layer protects living things on Earth's surface. Why is there some concern about this ozone layer?

3. What is wind? What is the energy source for wind?

4. Explain the relationship between air temperature and air density.

5. Why does heated air rise?

6. Explain the meaning of the following expression: "It's not the heat, it's the humidity."

7. Explain why frost is more likely to form on a clear, calm, and cool night than on nights with other conditions.

8. What is a cloud? Describe how a cloud forms.

9. What is atmospheric stability? What does this have to do with the type of clouds that may form on a given day?

5. Describe two ways that precipitation may form from the water droplets of a cloud.

6. What is an air mass?

7. What kinds of clouds and weather changes are usually associated with the passing of (a) a warm front and (b) a cold front?

8. Describe the wind direction, pressure, and weather conditions that are usually associated with (a) low-pressure centers and (b) high-pressure centers.

9. In which of the four basic types of air masses would you expect to find afternoon thunderstorms? Explain.

10. Describe the three main stages in the life of a thunderstorm cell, identifying the events that mark the beginning and end of each stage.

11. What is a tornado? When and where do tornadoes usually form?

12. What is a hurricane? Describe how the weather conditions change as a hurricane approaches, passes directly over, then moves away from a location.

13. How is climate different from the weather?

14. Identify the four major factors that influence the climate of a region, and explain how each does its influencing.

15. Heated air rises, so why is snow found on top of a mountain and not at lower elevations?

FOR FURTHER ANALYSIS

1. Describe how you could use a garden hose and a bucket of water to make a barometer. How high a column of water would standard atmospheric pressure balance in a water barometer?

2. If heated air rises, why is there snow on top of a mountain and not at the bottom?

3. According to the U.S. National Oceanic and Atmospheric Administration, the atmospheric concentration of CO_2 has been increasing, global surface temperatures have increased about 0.2°C (0.4°F) over the past twenty-five years, and sea level has been rising 1 to 2 mm/yr since the late nineteenth century. Describe what evidence you would look for to confirm these increases are due to human activity rather than changes in the Sun's output or energy, or changes in Earth's orbit.

4. Evaluate the requirement that differential heating must take place before wind will blow. Do any winds exist without differential heating?

5. Given the current air temperature and relative humidity, explain how you could use the graph in figure 17.15 to find the dew point temperature.

6. Explain why dew is not considered to be a form of precipitation.

7. What are the significant similarities and differences between air mass weather and frontal weather?

8. Analyze and compare the potential damage caused by a hurricane to the potential damage caused by a tornado.

9. Describe several examples of regional climate factors completely overriding the expected weather in a given principal climatic zone. Explain how this happens.

INVITATION TO INQUIRY

Microclimate Experiments

The local climate of a small site or habitat is called a *microclimate*. Certain plants will grow within one microclimate but not another. For example, a north-facing slope of a hill is cooler and loses snow later than a south-facing slope. A different microclimate exists on each side of the hill, and some plants grow better on one side than the other.

Investigate the temperature differences in the microclimate on the north side of a building and on the south side of the same building. Determine the heights, distances, and time of day that you will take temperature readings. Remember to shade the thermometer from direct sunlight for each reading. In your report, discuss the variables that may be influencing the temperature of a microclimate. Design experiments to test your hypotheses about each variable.

18

Earth's Waters

Earth's vast oceans cover more than 70 percent of the surface of Earth. Freshwater is generally abundant on the land because the supply is replenished from ocean waters through the hydrologic cycle.

CORE **CONCEPT**

The hydrologic cycle is water evaporating from the ocean, transport by moving air masses, precipitation on the land, and the movement of water back to the ocean.

OUTLINE

Most *water on Earth is stored in Earth's oceans* (p. 414).

Water fit for human consumption is replenished by the hydrologic cycle (p. 415).

Precipitation either evaporates, runs across the surface, or soaks in to become groundwater (p. 415).

Earth Science

▷ Changes in atmospheric conditions bring precipitation (Ch. 17).

OVERVIEW

Throughout history, humans have diverted rivers and reshaped the land to ensure a supply of freshwater. There is evidence, for example, that ancient civilizations along the Nile River diverted water for storage and irrigation some five thousand years ago. The ancient Greeks and Romans built systems of aqueducts to divert streams to their cities some two thousand years ago. Some of these aqueducts are still standing today. More recent water diversion activities were responsible for the name of Phoenix, Arizona. Phoenix was named after a mythical bird that arose from its ashes after being consumed by fire. The city was given this name because it is built on a system of canals that were first designed and constructed by ancient Native Americans, then abandoned hundreds of years before settlers reconstructed the ancient canal system (figure 18.1). Water is and always has been an essential resource. Where water is in short supply, humans have historically turned to extensive diversion and supply projects to meet their needs.

Precipitation is the basic source of the water supply found today in streams, lakes, and beneath Earth's surface. Much of the precipitation that falls on the land, however, evaporates back into the atmosphere before it has a chance to become a part of this supply. The water that does not evaporate mostly moves directly to rivers and streams, flowing back to the ocean, but some soaks into the land. The evaporation of water, condensation of water vapor, and the precipitation-making processes were introduced in chapter 17 as important weather elements. They are also part of the generalized *hydrologic cycle* of evaporation from the ocean, transport through the atmosphere by moving air masses, precipitation on the land, and movement of water back to the ocean. Only part of this cycle was considered previously, however, and this was the part from evaporation through precipitation. This chapter is concerned with the other parts of the hydrologic cycle, that is, what happens to the water that falls on the land and makes it back to the ocean. It begins with a discussion of how water is distributed on Earth and a more detailed look at the hydrologic cycle. Then the travels of water across and into the land will be considered as streams, wells, springs, and other sources of usable water are discussed as limited resources. The tracing of the hydrologic cycle will be completed as the water finally makes it back to the ocean. This last part of the cycle will consider the nature of the ocean floor, the properties of seawater, and how waves and currents are generated. The water is now ready to evaporate, starting another one of Earth's never-ending cycles.

FIGURE 18.1 This is one of the water canals of the present-day system in Phoenix, Arizona. These canals were reconstructed from a system that was built by American Indians, then abandoned. Phoenix is named after a mythical bird that was consumed by fire and then arose from its ashes.

18.1 WATER ON EARTH

Some water is tied up in chemical bonds deep in Earth's interior, but free water is the most abundant chemical compound near the surface. Water is five or six times more abundant than the most abundant mineral in the outer 6 km (about 4 mi) of Earth, so it should be no surprise that water covers about 70 percent of the surface. On average, about 98 percent of this water exists in the liquid state in depressions on the surface and in sediments. Of the remainder, about 2 percent exists in the solid state as snow and ice on the surface in colder locations. Only a fraction of a percent exists as a variable amount of water vapor in the atmosphere at a given time. Water is continually moving back and forth between these "reservoirs," but the percentage found in each is assumed to be essentially constant.

As shown in figure 18.2, over 97 percent of Earth's water is stored in Earth's oceans. This water contains a relatively high level of dissolved salts, which will be discussed in the section on the nature of seawater. These dissolved salts make ocean water unfit for human consumption and for most agricultural purposes. All other water, which is fit for human consumption and agriculture, is called **freshwater.** About two-thirds of Earth's freshwater supply is locked up in the ice caps of Greenland and the Antarctic and in glaciers. This leaves less than 1 percent of all the water found on Earth as available freshwater. There is a generally abundant supply, however, because the freshwater supply is continually replenished.

Evaporation of water from the ocean is an important part of the replenishing process because (1) water vapor leaves the dissolved salts behind, forming precipitation that is freshwater, and (2) the gaseous water vapor is easily transported in the atmosphere from one part of Earth to another. Over a year, this natural desalination process produces and transports enough freshwater to cover the entire Earth with a layer about 85 cm (about 33 in.) deep. Precipitation is not evenly distributed like this, of course, and some places receive much more, while other places receive almost none. Considering global averages, more water is evaporated from the ocean than returns directly to it by precipitation. On the other hand, more water is precipitated over the land than evaporates from the land surface back to the atmosphere. The net amount evaporated and precipitated over the land and over the ocean is balanced by the return of water to the ocean by streams and rivers. This cycle of evaporation, precipitation, and return of water to the oceans is known as the *hydrologic cycle* (figure 18.3).

Freshwater

The basic source of freshwater is precipitation, but not all precipitation ends up as part of the freshwater supply. Liquid water is always evaporating, even as it falls. In arid climates, rain sometimes evaporates completely before reaching the surface, even from a fully developed thunderstorm. Evaporation continues from the water that does reach the surface. Puddles and standing water on the hard surface of city parking lots and streets, for example, gradually evaporate back to the atmosphere after a rain and the surface is soon dry. Many factors determine how much of a particular rainfall evaporates, but in general, more than two-thirds of the rain eventually returns to the atmosphere. The remaining amount either (1) flows downhill across the surface of the land toward a lower place or (2) soaks into the ground. Water moving across the surface is called **runoff.** Runoff begins as rain accumulates in thin sheets of water that move across the surface of the land. These sheets collect into a small body of running water called a *stream.* A stream is defined as any body of water that is moving across the land, from one so small that you could step across it to the widest river. Water that soaks into the ground moves downward to a saturated zone and is now called *groundwater.* Groundwater moves through sediments and rocks beneath the surface, slowly moving downhill. Streams carry the runoff of a recent rainfall or melting snow, but otherwise most of the flow comes from groundwater that seeps into the stream channel. This explains how a permanent stream is able to continue flowing when it is not being fed by runoff or melting snow (figure 18.4). Where or when the source of groundwater is in low supply, a stream may flow only part of the time, and it is designated as an *intermittent stream.*

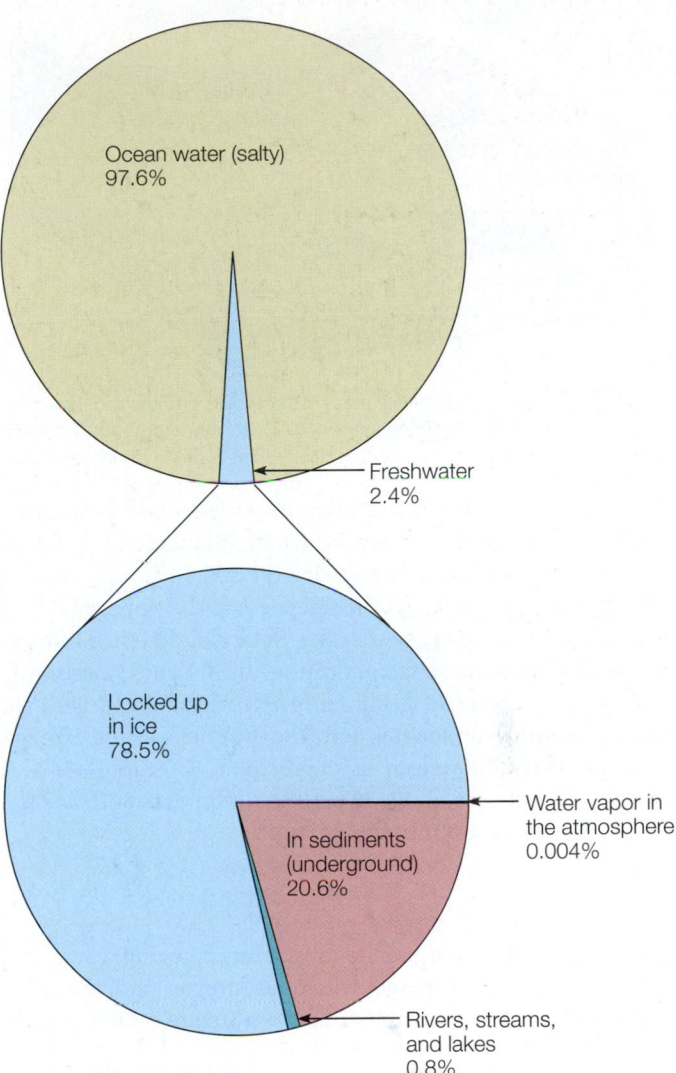

FIGURE 18.2   Estimates of the distribution of all the water found on Earth's surface.

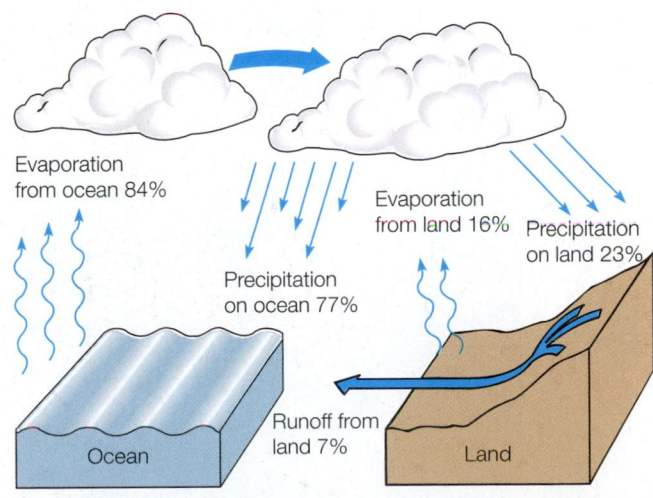

100% is based on a global average of 85 cm/yr precipitation.

FIGURE 18.3   On the average, more water is evaporated from the ocean than is returned by precipitation. More water is precipitated over the land than evaporates. The difference is returned to the ocean by rivers and streams.

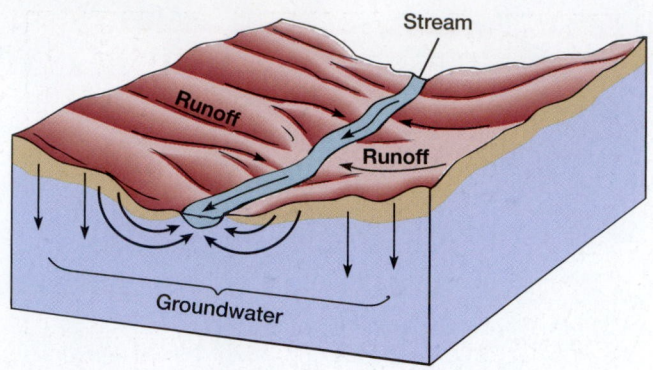

FIGURE 18.4 Some of the precipitation soaks into the ground to become groundwater. Groundwater slowly moves underground, and some of it emerges in streambeds, keeping the streams running during dry spells.

The amount of a rainfall that becomes runoff or groundwater depends on a number of factors, including (1) the type of soil on the surface, (2) how dry the soil is, (3) the amount and type of vegetation, (4) the steepness of the slope, and (5) if the rainfall is a long, gentle one or a cloudburst. Different combinations of these factors can result in from 5 percent to almost 100 percent of a rainfall event running off, with the rest evaporating or soaking into the ground. On the average, however, about 70 percent of all precipitation evaporates back into the atmosphere, about 30 percent becomes runoff, and less than 1 percent soaks into the ground.

Surface Water

If you could follow the smallest of streams downhill, you would find that it eventually merges with other streams until they form a major river. The land area drained by a stream is known as the stream's drainage basin, or **watershed.** Each stream has its own watershed, but the watershed of a large river includes all the watersheds of the smaller streams that feed into the larger river. Figure 18.5 shows the watersheds of the Columbia River, the Colorado River, and the Mississippi River. Note that the water from the Columbia River and the Colorado River watersheds empties into the Pacific Ocean. The Mississippi River watershed drains into the Gulf of Mexico, which is part of the Atlantic Ocean.

Two adjacent watersheds are separated by a line called a *divide.* Rain that falls on one side of a divide flows into one watershed, and rain that falls on the other side flows into the other watershed. A *continental divide* separates river systems that drain into opposite sides of a continent. The North American continental divide trends northwestward through the Rocky Mountains. Imagine standing over this line with a glass of water in each hand, then pouring the water to the ground. The water from one glass will eventually end up in the Atlantic Ocean, and the water from the other glass will end up in the Pacific Ocean. Sometimes the Appalachian Mountains are considered to be an eastern continental divide, but water from both sides of this divide ends up on the same side of the continent, in the Atlantic Ocean.

Water moving downhill is sometimes stopped by a depression in a watershed, a depression where water temporarily collects as a standing body of freshwater. A smaller body of standing water is usually called a *pond,* and one of much larger size is called a *lake.* A pond or lake can occur naturally in a depression, or it can be created by building a dam on a stream. A natural pond, a natural lake, or a pond or lake created by building a dam is called a *reservoir* if it is used for (1) water storage, (2) flood control, or (3) generating electricity. A reservoir can be used for one or two of these purposes but not generally for all three. A reservoir built for water storage, for example, is kept

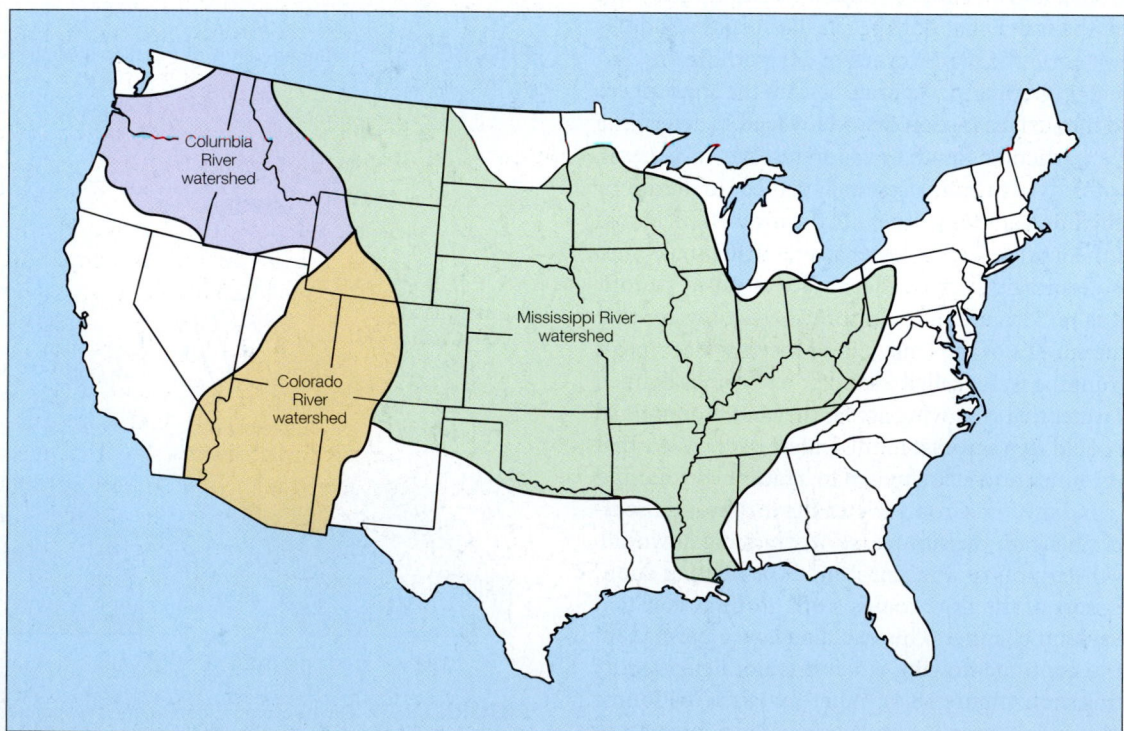

FIGURE 18.5 The approximate watersheds of the Columbia, Colorado, and Mississippi Rivers.

as full as possible to store water. This use is incompatible with use for flood control, which would require a low water level in the reservoir in order to catch runoff, preventing waters from flooding the land. In addition, extensive use of reservoir water to generate electricity requires the release of water, which could be incompatible with water storage. The water of streams, ponds, lakes, and reservoirs is collectively called *surface water,* and all serve as sources of freshwater. The management of surface water, as you can see, can present some complicated problems.

Groundwater

Precipitation soaks into the ground, or *percolates* slowly downward until it reaches an area, or zone, where the open spaces between rock and soil particles are completely filled with water. Water from such a saturated zone is called **groundwater.** There is a tremendous amount of water stored as groundwater, which makes up a supply about twenty-five times larger than all the surface water on Earth. Groundwater is an important source of freshwater for human consumption and for agriculture. Groundwater is often found within 100 m (about 330 ft) of the surface, even in arid regions where little surface water is found. Groundwater is the source of water for wells in addition to being the source that keeps streams flowing during dry periods.

Water is able to percolate down to a zone of saturation because sediments contain open spaces between the particles called *pore spaces.* The more pore space a sediment has, the more water it will hold. The total amount of pore spaces in a given sample of sediment is a measure of its *porosity.* Sand and gravel sediments, for example, have grains that have large pore spaces between them, so these sediments have a high porosity. In order for water to move through a sediment, however, the pore spaces must be connected. The ability of a given sample of sediment to transmit water is a measure of its *permeability.* Sand and gravel have a high permeability because the grains do not fit tightly together, allowing water to move from one pore space to the next. Sand and gravel sediments thus have a high porosity as well as a high permeability. Clay sediments, on the other hand, have small, flattened particles that fit tightly together. Clay thus has a low permeability, and when saturated or compressed, clay becomes *impermeable,* meaning water cannot move through it at all (figure 18.6).

The amount of groundwater available in a given location depends on a number of factors, such as the present and past climate, the slope of the land, and the porosity and permeability of the sediments beneath the surface. Generally, sand and gravel sediments, along with solid sandstone, have the best porosity and permeability for transmitting groundwater. Other solid rocks, such as granite, can also transmit groundwater if they are sufficiently fractured by joints and cracks. In any case, groundwater will percolate downward until it reaches an area where pressure and other conditions have eliminated all pores, cracks, and joints. Above this impermeable layer, it collects in all available spaces to form a *zone of saturation.* Water from the zone of saturation is considered to be groundwater. Water from the zone above is not considered to be groundwater. The surface of the boundary between the zone of saturation and

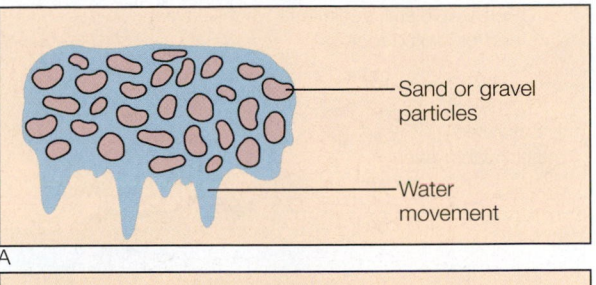

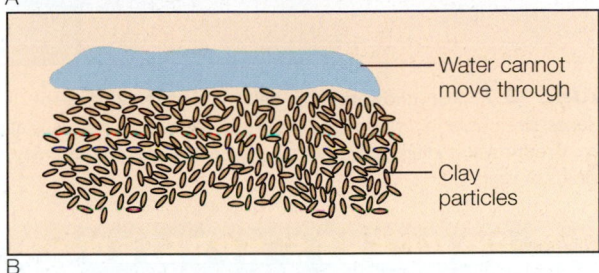

FIGURE 18.6 (*A*) Sand and gravel have large, irregular particles with large pore spaces, so they have a high porosity. Water can move from one pore space to the next, so they also have a high permeability. (*B*) Clay has small, flat particles, so it has a low porosity and is practically impermeable because water cannot move from one pore to the next.

the zone above is called the **water table.** The surface of a water table is not necessarily horizontal, but it tends to follow the topography of the surface in a humid climate. A hole that is dug or drilled through the surface to the water table is called a well. The part of the well that is below the water table will fill with groundwater, and the surface of the water in the well is generally at the same level as the water table.

Precipitation falls on the land and percolates down to the zone of saturation, then begins to move laterally, or sideways, to lower and lower elevations until it finds its way back to the surface. This surface outflowing could take place at a stream, pond, lake, swamp, or spring (figure 18.7). Groundwater flows gradually and very slowly through the tiny pore spaces, moving at a rate that ranges from kilometers (miles) per day to meters (feet) per year. Surface streams, on the other hand, move much faster, at rates up to about 30 km per hour (about 20 mi/h).

An **aquifer** is a layer of sand, gravel, sandstone, or other highly permeable material beneath the surface that is capable of producing water in usable quantities. In some places, an aquifer carries water from a higher elevation, resulting in a pressure on water trapped by impermeable layers at lower elevations. Groundwater that is under such a confining pressure is in an *artesian* aquifer. *Artesian* refers to the pressure, and groundwater from an artesian well rises above the top of the aquifer but not necessarily to the surface. Some artesian wells are under sufficient pressure to produce a fountainlike flow or spring (figure 18.8). Some people call groundwater from any deep well "artesian water," which is technically incorrect.

Freshwater as a Resource

Water is an essential resource, not only because it is required for life processes but also because of its role in a modern industrialized society (see chapter 10). Water is used in the home for drinking,

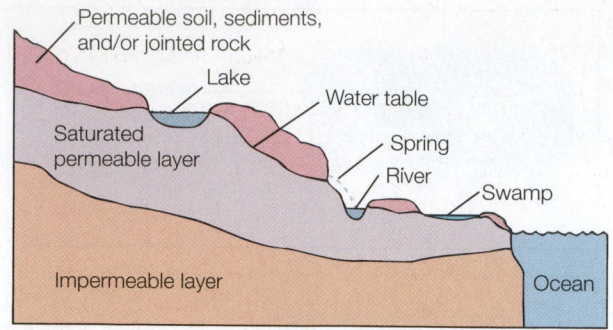

FIGURE 18.7 Groundwater from below the water table seeps into lakes, streams, and swamps and returns to the surface naturally at a spring. Groundwater eventually returns to the ocean, but the trip may take hundreds of years.

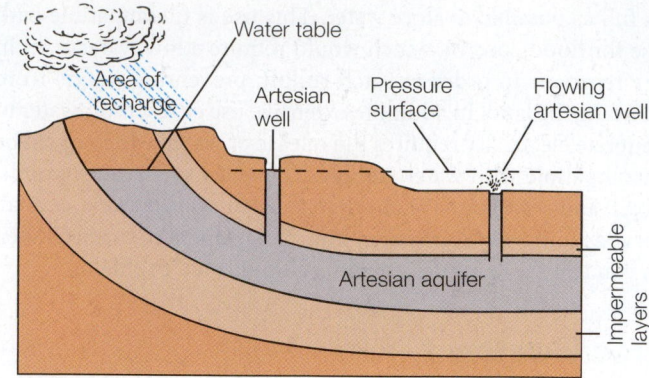

FIGURE 18.8 An artesian aquifer has groundwater that is under pressure because the groundwater is confined between two impermeable layers and has a recharge area at a higher elevation. The pressure will cause the water to rise in a well drilled into the aquifer, becoming a flowing well if the pressure is sufficiently high.

cooking, and cleaning, as a carrier to remove wastes, and for maintaining lawns and gardens. These domestic uses lead to an equivalent consumption of about 570 liters per person each day (about 150 gal/person/day), but this is only about 10 percent of the total consumed. Average daily use of water in the United States amounts to some 5,700 liters per person each day (about 1,500 gal/person/day), or about enough water to fill a small swimming pool once a week. The bulk of the water is used by agriculture (about 40 percent), for the production of electricity (about 40 percent), and for industrial purposes (about 10 percent). These overall percentages of use vary from one region of the country to another.

Most of the water supply is obtained from the surface water resources of streams, lakes, and reservoirs, and 37 percent of the municipal water supply comes from groundwater. If you then add farms, villages, and many suburban areas, the percentage of groundwater used by humans is well above 40 percent. Surface water contains more sediments, bacteria, and possible pollutants than groundwater. This means that surface water requires filtering to remove suspended particles, treatment to kill bacteria, and sometimes processing to remove dissolved chemicals. Groundwater is naturally filtered as it moves through the pore spaces of an aquifer, so it is usually relatively free of suspended particles and bacteria. Thus, the processing or treatment of groundwater is usually not necessary (figure 18.9). But groundwater, on the other hand, will cost more to use as a resource because it must be pumped to the surface. The energy required for this pumping can be very expensive. In addition, groundwater generally contains more dissolved minerals (hard water), which may require additional processing or chemical treatment to remove the troublesome minerals.

The use of surface water as a source of freshwater means that the supply depends on precipitation. When a drought occurs, low river and lake levels may require curtailing water

consumption. In some parts of the western United States, such as the Colorado River watershed, *all* of the surface water is already being used, with certain percentages allotted for domestic, industrial, and irrigation uses. Groundwater is also used in this watershed, and in some locations, it is being pumped from the ground faster than it is being replenished by precipitation (figure 18.10). As the population grows and new industries develop, more and more demands are placed on the surface water supply, which has already been committed to other uses, and on the diminishing supply of groundwater. This raises some very controversial issues about how freshwater should be divided among agriculture, industries, and city domestic use. Agricultural interests claim they should have the water because they produce the food and fibers that people must have. Industrial interests claim they should have the water because they create the jobs and the products that people must have. Cities, on the other hand, claim that domestic consumption is the most important because people cannot survive without water. Yet, others claim that no group has a right to use water when it is needed to maintain habitats. Who should have the first priority for water use in such cases?

Some have suggested that people should not try to live and grow food in areas that have a short water supply, that plenty of freshwater is available elsewhere. Others have suggested that humans have historically moved rivers and reshaped the land to obtain water, so perhaps one answer to the problem is to find new sources of freshwater. Possible sources include the recycling of wastewater and turning to the largest supply of water in the world, the ocean. About 90 percent of the water used by industries is presently dumped as a waste product. In some areas, treated city waste water is already being recycled for use in power plants and for watering parks. A practically limitless supply of freshwater could be available by desalting ocean water, something which occurs naturally in the hydrologic cycle. The treatment of seawater to obtain a new supply of freshwater is presently too expensive because of the cost of energy to accomplish the task. New technologies, perhaps ones that use solar energy, may make this more practical in the future. In the meantime, the best sources of extending the

What do you think of when you see a stream? Do you wonder how deep it is? Do you think of something to do with the water such as swimming or fishing? As you might imagine, not all people look at a stream and think about the same thing. A city engineer, for example, might wonder if the stream has enough water to serve as a source to supplement the city water supply. A rancher or farmer might wonder how the stream could be easily diverted to serve as a source of water for irrigation. An electric utility planner, on the other hand, might wonder if the stream could serve as a source of power.

Water in a stream is a resource that can be used many different ways, but using it requires knowing about the water quality as well as quantity. We need to know if the quality of the water is good enough for the intended use—and different uses have different requirements. Water fit for use in an electric power plant, for example, might not be suitable for use as a city water supply.

Indeed, water fit for use in a power plant might not be suitable for irrigation. Water quality is determined by the kinds and amounts of substances dissolved and suspended in the water and the consequences for users. Whether a source of water can be used for drinking water or not, for example, is regulated by stringent rules and guidelines about what cannot be in the water. These rules are designed to protect human health but do not call for pure water.

The water of even the healthiest stream is not absolutely pure. All water contains many naturally occurring substances such as ions of bicarbonates, calcium, and magnesium. A pollutant is not naturally occurring; it is usually a waste material that contaminates air, soil, or water. There are basically two types of water pollutants: degradable and persistent. Examples of degradable pollutants include sewage, fertilizers, and some industrial wastes. As the term implies, degradable pollutants can be broken down

into simple, nonpolluting substances such as carbon dioxide and nitrogen. Examples of persistent pollutants include some pesticides, petroleum and petroleum products, plastic materials, leached chemicals from landfill sites, oil-based paints, heavy metals and metal compounds such as lead, mercury, and cadmium, and certain radioactive materials. The damage they cause is either irreversible or reparable only over long periods of time.

Questions to Discuss

Discuss with your group the following questions concerning water quality:

1. Which water use requires the purest water? Which requires the least pure?
2. Create a hierarchy of possible water uses; for example, can water used in a power plant later be used for agriculture? For domestic use?
3. What can an individual do to improve water quality?

FIGURE 18.9   The filtering beds of a city water treatment facility. Surface water contains more sediments, bacteria, and other suspended materials because it is on the surface and exposed to the atmosphere. This means that surface water must be filtered and treated when used as a domestic resource. Such processing is not required when groundwater is used as the resource.

FIGURE 18.10   This is groundwater pumped from the ground for irrigation. In some areas, groundwater is being removed from the ground faster than it is being replaced by precipitation, resulting in a water table that is falling. It is thus possible that the groundwater resource will soon become depleted in some areas.

supply of freshwater appear to be control of pollution, recycling of wastewater, and conservation of the existing supply.

18.2 SEAWATER

More than 70 percent of the surface of Earth is covered by seawater, with an average depth of 3,800 m (about 12,500 ft). The land areas cover 30 percent, less than a third of the surface, with

A Closer Look

Wastewater Treatment

One of the most common forms of pollution control in the United States is wastewater treatment. The United States has a vast system of sewer pipes, pumping stations, and treatment plants, and about 74 percent of all Americans are served by such wastewater systems. Sewer pipelines collect the wastewater from homes, businesses, and many industries and deliver it to treatment plants. Most of these plants were designed to make wastewater fit for discharge into streams or other receiving waters.

The basic function of a waste treatment plant is to speed up the natural process of purifying the water. There are two basic stages in the treatment, called the *primary stage* and the *secondary stage*. The primary stage physically removes solids from the wastewater. The secondary stage uses biological processes to further purify wastewater. Sometimes, these stages are combined into one operation.

As raw sewage enters a treatment plant, it first flows through a screen to remove large floating objects such as rags and sticks that might cause clogs. After this initial screening, it passes into a grit chamber where cinders, sand, and small stones settle to the bottom (box figure 18.1). A grit chamber is particularly important in communities with combined sewer systems where sand or gravel may wash into the system along with rain, mud, and other stuff, all with the storm water.

After screening and grit removal, the sewage is basically a mixture of organic and inorganic matter along with other suspended solids. The solids are minute particles that can be removed in a sedimentation tank. The speed of the flow through the larger sedimentation tank is slower, and suspended solids gradually sink to the bottom of the tank. They form a mass of solids called *raw primary sludge,* which is usually removed from the tank by pumping. The sludge may be further treated for use as a fertilizer or disposed of through incineration, if necessary.

Once the effluent has passed through the primary treatment process, which is primarily physical filtering and settling, it enters a secondary stage that involves bio-

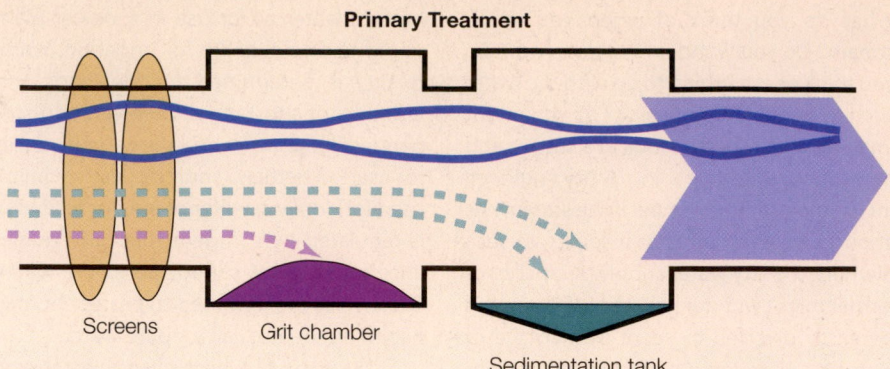

Primary Treatment

Screens

Grit chamber

Sedimentation tank

Box Figure 18.1

logical activities of microorganisms. The secondary stage of treatment removes about 85 percent of the organic matter in sewage by making use of the bacteria that are naturally a part of the sewage. There are two principal techniques used to provide secondary treatment: (1) trickling filters or (2) activated sludge. A trickling filter is simply a bed of stones from 1 to 2 m (3 to 6 ft) deep through which the effluent from the sedimentation tank flows. Interlocking pieces of corrugated plastic or other synthetic media have also been used in trickling beds, but the important part is that it provides a place for bacteria to live and grow. Bacteria grow on the stones or synthetic media and consume most of the organic matter flowing by in the effluent. The now cleaner water trickles out through pipes to another sedimentation tank to remove excess bacteria. Disinfection of the effluent with chlorine is generally used to complete this secondary stage of basic treatment.

The trend today is toward the use of an activated sludge process instead of trickling filters. The activated sludge process speeds up the work of the bacteria by bringing air and sludge heavily laden with bacteria into close contact with the effluent (box figure 18.2). After the effluent leaves the sedimentation tank in the primary stage, it is pumped into an aeration tank, where it is mixed with air and sludge loaded with bacteria and allowed to remain for several hours. During this time, the bacteria break down the organic matter into harmless by-products.

The sludge, now activated with additional millions of bacteria, can be used again by returning it to the aeration tank for mixing with new effluent and ample amounts of air. As with trickling, the final step is generally the addition of chlorine to the effluent, which kills more than 99 percent of the harmful bacteria. Some municipalities are now manufacturing chlorine solution on site to avoid the necessity of transporting and storing large amounts of chlorine, sometimes in a gaseous form. Alternatives to chlorine disinfection, such as ultraviolet light or ozone, are also being used in situations where chlorine in sewage effluents can be harmful to fish and other aquatic life.

New pollution problems have placed additional burdens on wastewater treatment systems. Today's pollutants may be more difficult to remove from water. Increased demands on the water supply only aggravate the problem. These challenges are being met through better and more complete methods of removing pollutants at treatment plants or through prevention of pollution at the source. Pretreatment of industrial waste, for example, removes many troublesome pollutants at the beginning, rather than at the end, of the pipeline.

The increasing need to reuse water calls for better and better wastewater treatment. Every use of water—whether at home, in the factory, or on a farm—results in some change in its quality. New methods for removing pollutants are being developed to return water of more usable quality to receiving lakes and streams. Advanced waste treatment techniques in use or under development range from biological treatment capable of removing nitrogen and phosphorus to physical-chemical separation techniques such as filtration, carbon adsorption, distillation, and reverse osmosis. These activities typically follow secondary treatment and are known as tertiary treatment.

These wastewater treatment processes, alone or in combination, can achieve almost any degree of pollution control desired. As waste effluents are purified to higher degrees by such treatment, the effluent water can be used for industrial, agricultural, or recreational purposes, or even drinking water supplies.

Drawings and some text from *How Wastewater Treatment Works . . . The Basics*, U.S. Environmental Protection Agency, Office of Water, http://www.epa.gov/ebtpages/watewastewatertreatment.html.

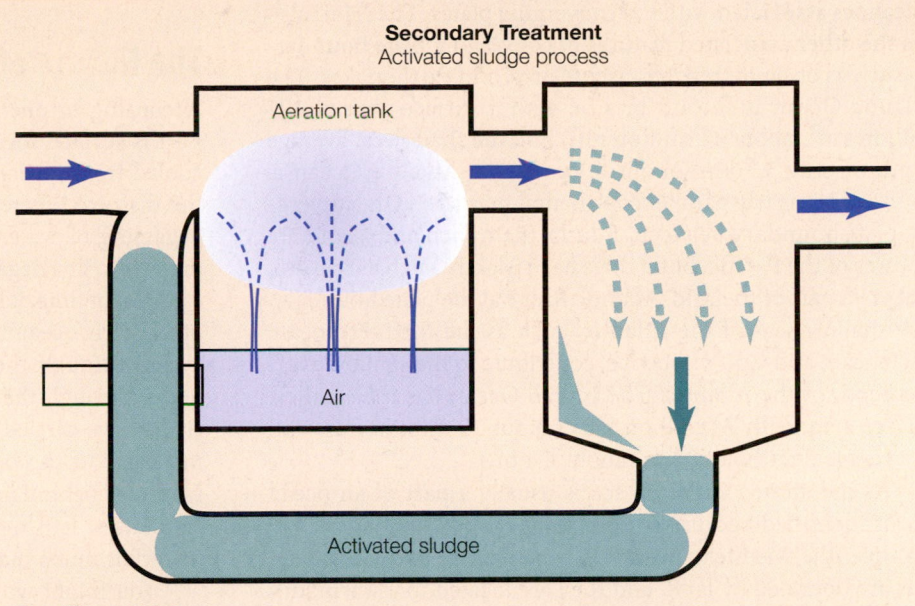

Secondary Treatment
Activated sludge process

Aeration tank

Air

Activated sludge

Box Figure 18.2

---

an average elevation of only about 830 m (about 2,700 ft). With this comparison, you can see that humans live on and fulfill most of their needs by drawing from a small part of the total Earth. As populations continue to grow and as resources of the land continue to diminish, the ocean will be looked at more as a resource rather than a convenient place for dumping wastes. The ocean already provides some food and is a source of some minerals, but it can possibly provide freshwater, new sources of food, new sources of important minerals, and new energy sources in the future. There are vast deposits of phosphorus and manganese minerals on the ocean bottom, for example, that can provide valuable resources. Phosphorus is an element that can be used to manufacture an important fertilizer needed in agriculture, and the land supplies are becoming depleted. Manganese nodules, which occur in great abundance on the ocean bottom, can be a source of manganese, iron, copper, cobalt, and nickel. Seawater contains enough deuterium to make it a feasible source of energy. One gallon of seawater contains about a spoonful of deuterium, with the energy equivalent of 300 gallons of gasoline. It has been estimated there is sufficient deuterium in the oceans to supply power at one hundred times the present consumption for the next ten billion years. The development of controlled nuclear fusion is needed, however, to utilize this potential energy source. The sea may provide new sources of food through *aquaculture,* the farming of the sea the way that the land is presently farmed. Some aquaculture projects have already started with the farming of oysters, clams, lobsters, shrimp, and certain fishes, but these projects have barely begun to utilize the full resources that are possible.

Part of the problem of utilizing the ocean is that the ocean has remained mostly unexplored and a mystery until recent times. Only now are scientists beginning to understand the complex patterns of the circulation of ocean waters, the nature of the chemical processes at work in the ocean, and the interactions of the ocean and the atmosphere, and to chart the topography of the ocean floor.

Oceans and Seas

The vast body of salt water that covers more than 70 percent of Earth's surface is usually called the *ocean* or the *sea*. In general, the **ocean** is a single, continuous body of salt water on the surface of Earth. Although there is really only one big ocean on Earth, specific regions have been given names for convenience in describing locations. For this purpose, three principal regions are recognized: the (1) Atlantic Ocean, (2) Indian Ocean, and (3) Pacific Ocean, as shown in figure 18.11. Specific regions (Atlantic, Indian, and Pacific) are often subdivided further into North Atlantic Ocean, South Atlantic Ocean, and so on.

A *sea* is usually a smaller part of the ocean, a region with characteristics that distinguish it from the larger ocean of which it is a part. Often the term *sea* is used in the name of certain inland bodies of salty water.

The Pacific Ocean is the largest of the three principal ocean regions. It has the largest surface area, covering 180 million km^2 (about 70 million mi^2), and has the greatest average depth of 3.9 km (about 2.4 mi). The Pacific is circled by active converging plate boundaries, so it is sometimes described as being circled by a "rim of fire." It is called this because of the volcanoes associated with the converging plates. The "rim" also has the other associated features of converging plate boundaries such as oceanic trenches, island arcs, and earthquakes. The Atlantic Ocean is second in size, with a surface area of 107 million km² (about 41 million mi²) and the shallowest average depth of only 3.3 km (about 2.1 mi). The Atlantic Ocean is bounded by nearly parallel continental margins with a diverging plate boundary between. It lacks the trench and island arc features of the Pacific, but it does have islands, such as Iceland, that are a part of the Mid-Atlantic Ridge at the plate boundary. The shallow seas of the Atlantic, such as the Mediterranean, Caribbean, and Gulf of Mexico, contribute to the shallow average depth of the Atlantic. The Indian Ocean has the smallest surface area, with 74 million km² (about 29 million mi²) and an average depth of 3.8 km (about 2.4 mi).

As mentioned earlier, a sea is usually a part of an ocean that is identified because some characteristic sets it apart. For example, the Mediterranean, Gulf of Mexico, and Caribbean seas are bounded by land, and they are located in a warm, dry climate. Evaporation of seawater is greater than usual at these locations, which results in the seawater being saltier. Being bounded by land and having saltier seawater characterizes these locations as being different from the rest of the Atlantic. The Sargasso Sea, on the other hand, is a part of the Atlantic that is not bounded by land and has a normal concentration of sea salts. This sea is characterized by having an abundance of floating brown seaweeds that accumulate in this region because of the global wind and ocean current patterns. The Arctic Sea, which is also sometimes called the Arctic Ocean, is a part of the North Atlantic Ocean that is less salty. Thus, the terms *ocean* and *sea* are really arbitrary terms that are used to describe different parts of Earth's one continuous ocean.

The Nature of Seawater

According to one theory, the ocean is an ancient feature of Earth's surface, forming at least three billion years ago as Earth cooled from its early molten state. The seawater and much of the dissolved materials are believed to have formed from the degassing of water vapor and other gases from molten rock materials. The degassed water vapor soon condensed, and over a period of time, it began collecting as a liquid in the depression of the early ocean basin. Ever since, seawater has continuously cycled through the hydrologic cycle, returning water to the ocean through the world's rivers. For millions of years, these rivers have carried large amounts of suspended and dissolved materials to the ocean. These dissolved materials, including salts, stay behind in the seawater as the water again evaporates, condenses, falls on the land, and then brings more dissolved materials much like a continuous conveyor belt.

You might wonder why the ocean basin has not become filled in by the continuous supply of sediments and dissolved materials that would accumulate over millions of years. The basin has not filled in because (1) accumulated sediments have been recycled to Earth's interior through plate tectonics and (2) dissolved materials are removed by natural processes just as fast as they are supplied by the rivers. Some of the dissolved materials, such as calcium and silicon, are removed by organisms to make solid shells, bones, and other hard parts. Other dissolved materials, such as iron, magnesium, and phosphorus, form solid deposits directly and also make sediments that settle to the ocean floor. Hard parts of organisms and solid deposits

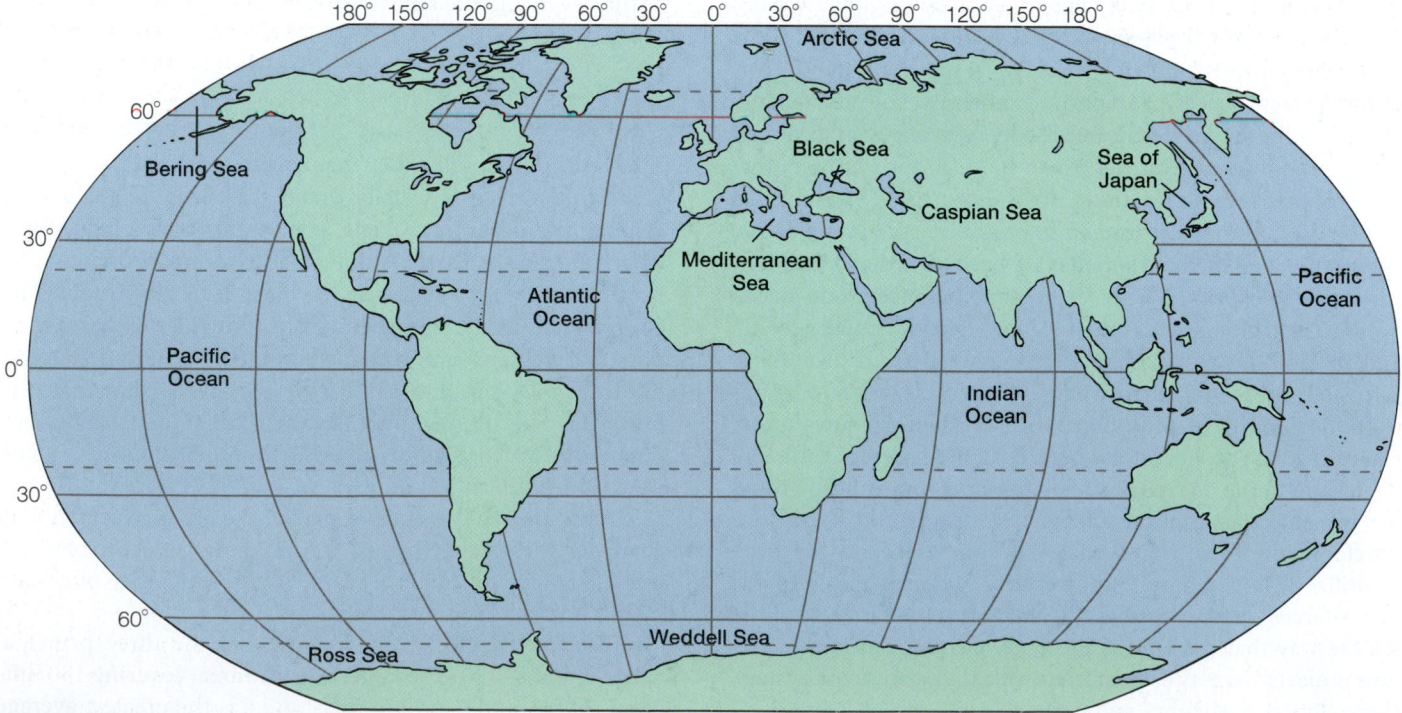

FIGURE 18.11 Distribution of the oceans and major seas on Earth's surface. There is really only one ocean; for example, where is the boundary between the Pacific, Atlantic, and Indian oceans in the Southern Hemisphere?

are cycled to Earth's interior along with suspended sediments that have settled out of the seawater. Studies of fossils and rocks indicate that the composition of seawater has changed little over the past six hundred million years.

The dissolved materials of seawater are present in the form of ions because of the strong dissolving ability of water molecules. Almost all of the chemical elements are present, but only six ions make up more than 99 percent of any given sample of seawater. As shown in table 18.1, chlorine and sodium are the most abundant ions. These are the elements of sodium chloride, or common table salt. As a sample of seawater evaporates, the positive metal ions join with the different negative ions to form a complex mixture of ionic compounds known as *sea salt*. Sea salt is mostly sodium chloride, but it also contains salts of the four metal ions (sodium, magnesium, calcium, and potassium) combined with the different negative ions of chlorine, sulfate, bicarbonate, and so on.

The amount of dissolved salts in seawater is measured as **salinity.** Salinity is defined as the mass of salts dissolved in 1.0 kg, or 1,000 g, of seawater. Since the salt content is reported in parts per thousand, the symbol ‰ is used (% means parts per hundred). Thus, 35 ‰ means that 1,000 g of seawater contains 35 g of dissolved salts (and 965 g of water). This is the same concentration as a 3.5 percent salt solution (figure 18.12). Oceanographers use the salinity measure because the mass of a sample of seawater does not change with changes in the water temperature. Other measures of concentration are based on the volume of a sample, and the volume of a liquid does vary as it expands and contracts with changes in the temperature. Thus, by using the salinity measure, any corrections due to temperature differences are eliminated.

The average salinity of seawater is about 35 ‰, but the concentration varies from a low of about 32 ‰ in some locations up to a high of about 36 ‰ in other locations. The salinity of seawater in a given location is affected by factors that tend to increase or decrease the concentration. The concentration is increased by two factors, evaporation and the formation of sea ice. Evaporation increases the concentration because it is water vapor only that evaporates, leaving the dissolved salts behind in a greater concentration. Ice that forms from freezing seawater increases the concentration because when ice forms, the salts are excluded from the crystal structure. Thus, sea ice is freshwater, and the removal of this water leaves the dissolved

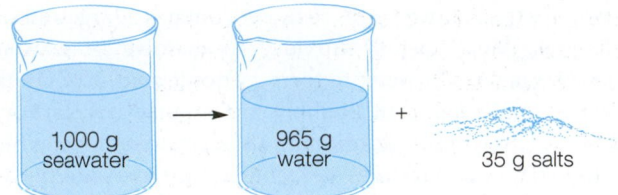

FIGURE 18.12 Salinity is defined as the mass of salts dissolved in 1.0 kg of seawater. Thus, if a sample of seawater has a salinity of 35 ‰, a 1,000 g sample would evaporate 965 g of water and leave 35 g of sea salts behind.

salts behind in a greater concentration. The salinity of seawater is decreased by three factors: heavy precipitation, the melting of ice, and the addition of freshwater by a large river. All three of these factors tend to dilute seawater with freshwater, which lowers the concentration of salts.

Note that increases or decreases in the salinity of seawater are brought about by the addition or removal of freshwater. This changes only the amount of water present in the solution. The *kinds* or *proportions* of the ions present (table 18.1) in seawater do not change with increased or decreased amounts of freshwater. The same proportion, meaning the same chemical composition, is found in seawater of any salinity of any sample taken from any location anywhere in the world, from any depth of the ocean, or taken any time of the year. Seawater has a remarkably uniform composition that varies only in concentration. This means that the ocean is well mixed and thoroughly stirred around the entire Earth. How seawater becomes so well mixed and stirred on a worldwide basis is discussed in the section on movement of seawater.

If you have ever allowed a glass of tap water to stand for a period of time, you may have noticed tiny bubbles collecting as the water warms. These bubbles are atmospheric gases, such as nitrogen and oxygen, that were dissolved in the water (figure 18.13). Seawater contains dissolved gases in addition to the dissolved salts. Near the surface, seawater contains mostly nitrogen and oxygen in similar proportions to the mixture that is found in the atmosphere. There is more carbon dioxide than

TABLE 18.1 Major Dissolved Materials in Seawater

Ion	Percent (by weight)
Chloride (Cl^-)	55.05
Sodium (Na^+)	30.61
Sulfate (SO_4^{-2})	7.68
Magnesium (Mg^{-2})	3.69
Calcium (Ca^{+2})	1.16
Potassium (K^+)	1.10
Bicarbonate (HCO_3^-)	0.41
Bromine (Br^-)	0.19
Total	99.89

FIGURE 18.13 Air will dissolve in water, and cooler water will dissolve more air than warmer water. The bubbles you see here are bubbles of carbon dioxide that came out of solution as the soda became warmer.

you would expect, however, as seawater contains a large amount of this gas. More carbon dioxide can dissolve in seawater because it reacts with water to form carbonic acid, H_2CO_3, the same acid that is found in a bubbly cola. In seawater, carbonic acid breaks down into bicarbonate and carbonate ions, which tend to remain in solution. Water temperature and salinity have an influence on how much gas can be dissolved in seawater, and increasing either or both will reduce the amount of gases that can be dissolved. Cold, lower-salinity seawater in colder regions will dissolve more gases than the warm, higher-salinity seawater in tropical locations. Abundant seaweeds (large algae) and phytoplankton (microscopic algae) in the upper, sunlit water tends to reduce the concentration of carbon dioxide and increase the concentration of dissolved oxygen through the process of photosynthesis. With increasing depth, less light penetrates the water, and below about 80 m (about 260 ft), there is insufficient light for photosynthesis. Thus, more algae and phytoplankton and more dissolved oxygen are found above this depth. Below this depth, there is more dissolved carbon dioxide and less dissolved oxygen. The oxygen-poor, deep ocean water does eventually circulate back to the surface, but the complete process may take several thousand years.

Movement of Seawater

Consider the enormity of Earth's ocean, which has a surface area of some 361 million km^2 (about 139 million mi^2) and a volume of 1,370 million km^3 (about 328 million mi^3) of seawater. There must be a terrific amount of stirring in such an enormous amount of seawater to produce the well-mixed, uniform chemical composition that is found in seawater throughout the world. The amount of mixing required is more easily imagined if you consider the long history of the ocean, the very long period of time over which the mixing has occurred. Based on investigations of the movement of seawater, it has been estimated that there is a complete mixing of all Earth's seawater about every 2,000 years or so. With an assumed age of 3 billion years, this means that Earth's seawater has been mixed $3,000,000,000 \div 2,000$, or 1.5 million times. With this much mixing, you would be surprised if seawater were *not* identical all around Earth.

How does seawater move to accomplish such a complete mixing? Seawater is in a constant state of motion, both on the surface and far below the surface. The surface has two types of motion: (1) *waves,* which have been produced by some disturbance, such as the wind, and (2) *currents,* which move water from one place to another. Waves travel across the surface as a series of wrinkles. Waves crash on the shore as booming breakers and make the surf. This produces local currents as water moves along the shore and back out to sea. There are also permanent, worldwide surface currents that move 10,000 times more water across the ocean than all the water moving in all the large rivers on the land. Beneath the surface, there are currents that move water up in some places and move it down in other places. Finally, there are enormous deep ocean currents that move tremendous volumes of seawater. The overall movement of many of the currents on the surface and their relationship to

the deep ocean currents are not yet fully mapped or understood. The surface waves are better understood. The general trend and cause of permanent, worldwide currents in the ocean can also be explained. The following is a brief description and explanation of waves, currents, and the deep ocean movements of seawater.

Waves

Any slight disturbance will create ripples that move across a water surface. For example, if you gently blow on the surface of water in a glass, you will see a regular succession of small ripples moving across the surface. These ripples, which look like small, moving wrinkles, are produced by the friction of the air moving across the water surface. The surface of the ocean is much larger, but a gentle wind produces patches of ripples in a similar way. These patches appear, then disappear as the wind begins to blow over calm water. If the wind continues to blow, larger and longer-lasting ripples are made, and the moving air can now push directly on the side of the ripples. A ripple may eventually grow into an **ocean wave,** a moving disturbance that travels across the surface of the ocean. In its simplest form, each wave has a ridge, or mound, of water called a *crest,* which is followed by a depression called a *trough.* Ocean waves are basically repeating series of these crests and troughs that move across the surface like wrinkles (figure 18.14).

The simplest form of an ocean wave can be described by measurements of three distinct characteristics: (1) the *wave height,* which is the vertical distance between the top of a crest and the bottom of the next trough, (2) the *wavelength,* which is the horizontal distance between two successive crests (or other successive parts of the wave), and (3) the *wave period,* which is the time required for two successive crests (or other successive parts) of the wave to pass a given point (figure 18.15).

FIGURE 18.14 The surface of the ocean is rarely, if ever, still. Any disturbance can produce a wave, but most waves on the open ocean are formed by a local wind.

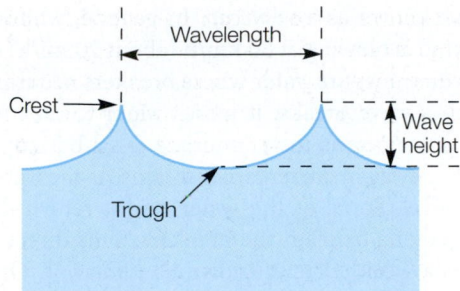

FIGURE 18.15 The simplest form of ocean waves, showing some basic characteristics. Most waves do not look like this representation because most are complicated mixtures of superimposed waves with a wide range of sizes and speeds.

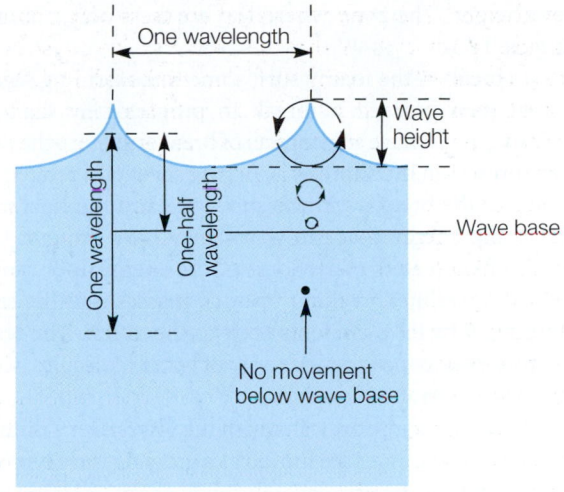

FIGURE 18.16 Water particles are moved in a circular motion by a wave passing in the open ocean. On the surface, a water particle traces out a circle with a diameter that is equal to the wave height. The diameters of the circles traced out by water particles decrease with depth to a depth that is equal to one-half the wavelength of the ocean wave. Bottom sediment cannot be moved by waves when the sediment is below the wave base.

The characteristics of an ocean wave formed by the wind depend on three factors: (1) the wind speed, (2) the length of time that the wind blows, and (3) the *fetch,* which is the distance the wind blows across the open ocean. As you can imagine, larger waves are produced by strong winds that blow for a longer time over a long fetch. In general, longer-blowing, stronger winds produce waves with greater wave heights, longer wavelengths, and longer periods, but a given wind produces waves with a wide range of sizes and speeds. In addition, the wind does not blow in just one direction, and shifting winds produce a chaotic pattern of waves of many different heights and wavelengths. Thus, the surface of the ocean in the area of a storm or strong wind has a complicated mixture of many sizes and speeds of superimposed waves. The smaller waves soon die out from friction within the water, and the larger ones grow as the wind pushes against their crests. Ocean waves range in height from a few centimeters up to more than 30 m (about 100 ft), but giant waves more than 15 m (about 50 ft) are extremely rare.

The larger waves of the chaotic, superimposed mixture of waves in a storm area last longer than the winds that formed them, and they may travel for hundreds or thousands of kilometers from their place of origin. The longer-wavelength waves travel faster and last longer than the shorter-wavelength waves, so the longer-wavelength waves tend to outrun the shorter-wavelength waves as they die out from energy losses to water friction. Thus, the irregular, superimposed waves created in the area of a storm become transformed as they travel away from the area. They become regular groups of long-wavelength waves with low wave height that are called **swell.** The regular waves of swell that you might observe near a shore may have been produced by a storm that occurred days before thousands of kilometers across the ocean.

The regular crests and troughs of swell carry energy across the ocean, but they do not transport water across the open ocean. If you have ever been in a boat that is floating in swell, you know that you move in a regular pattern of up and forward on each crest, then backward and down on the following trough. The boat does not move along with the waves unless it is moved along by a wind or by some current. Likewise, a particle of water on the surface moves upward and forward with each wave crest, then backward and down on the following trough,

tracing out a nearly circular path through this motion. The particle returns to its initial position, without any forward movement while tracing out the small circle. Note that the diameter of the circular path is equal to the wave height (figure 18.16). Water particles farther below the surface also trace out circular paths as a wave passes. The diameters of these circular paths below the surface are progressively smaller with increasing depth. Below a depth equal to about half the wavelength (the wave base), there is no circular movement of the particles. Thus, you can tell how deeply the passage of a wave disturbs the water below if you measure the wavelength.

As swell moves from the deep ocean to the shore, the waves pass over shallower and shallower water depths. When a depth is reached that is equal to about half the wavelength, the circular motion of the water particles begins to reach the ocean bottom. The water particles now move across the ocean bottom, and the friction between the two results in the waves moving slower as the wave height increases. These important modifications result in a change in the direction of travel and in an increasingly unstable situation as the wave height increases.

Most waves move toward the shore at some angle. As the wave crest nearest the shore starts to slow, the part still over deep water continues on at the same velocity. The slowing at the shoreward side *refracts,* or bends, the wave so it is more parallel to the shore. Thus, waves always appear to approach the shore head-on, arriving at the same time on all parts of the shore.

After the waves reach water that is less than one-half the wavelength, friction between the bottom and the circular motion of the water particles progressively slow the bottom part of the wave. The wave front becomes steeper and steeper as the top overruns the bottom part of the wave. When the wave front becomes too steep, the top part breaks forward and the wave is now called a *breaker* (figure 18.17). In general, this occurs where the water depth is about one and one-third times the wave height. The zone where the breakers occur is called **surf** (figure 18.18).

Waves break in the foamy surf, sometimes forming smaller waves that then proceed to break in progressively shallower water. The surf may have several sets of breakers before the water is finally thrown on the shore as a surging sheet of seawater. The turbulence of the breakers in the surf zone and the final surge expend all the energy that the waves may have brought from thousands of kilometers away. Some of the energy does work in eroding the shoreline, breaking up rock masses into the sands that are carried by local currents back to the ocean. The rest of the energy goes into the kinetic energy of water molecules, which appears as a temperature increase.

Swell does not transport water with the waves over a distance, but small volumes of water are moved as a growing wave is pushed to greater heights by the wind over the open ocean. A strong wind can topple such a wave on the open ocean, producing a foam-topped wave known as a *whitecap*. In general, whitecaps form when the wind is blowing at 30 km/h (about 20 mi/h) or more.

Waves do transport water where breakers occur in the surf zone. When a wave breaks, it tosses water toward the shore, where the water begins to accumulate. This buildup of water tends to move away in currents, or streams, as the water returns to a lower level. Some of the water might return directly to the sea by moving beneath the breakers. This direct return of water forms a weak current known as *undertow*. Other parts of the accumulated water might be pushed along by the waves, producing a *longshore current* that moves parallel to the shore in the surf zone. This current moves parallel to the shore until it finds a lower place or a channel that is deeper than the adjacent bottom. Where the current finds such a channel, it produces a *rip current*, a strong stream of water that bursts out against the waves and returns water through the surf to the sea (figure 18.19). The rip current usually extends beyond

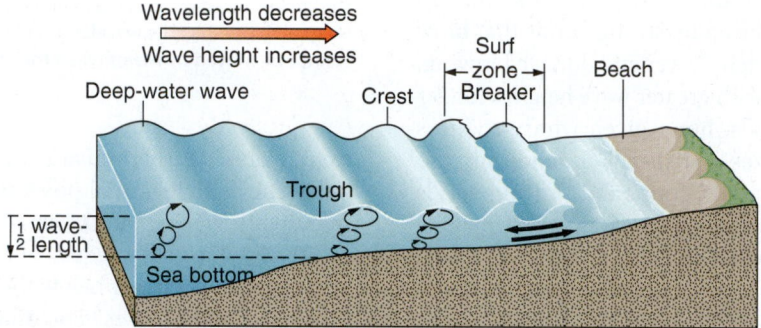

FIGURE 18.17 As a pattern of swell approaches a gently sloping beach, friction between the circular motion of the water particles and the bottom slows the wave and the wave front becomes steeper and steeper. When the depth is about one and one-third times the wave height, the wave breaks forward, moving water toward the beach.

FIGURE 18.18 The white foam is in the surf zone, which is where the waves grow taller and taller, then break forward into a froth of turbulence. Do you see any evidence of rip currents in this picture (see figure 18.19)?

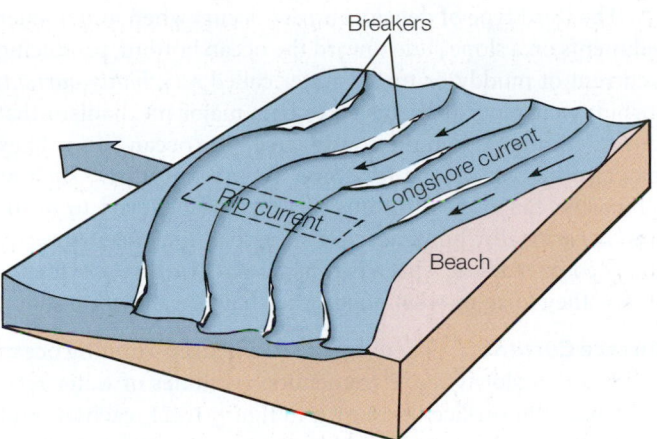

FIGURE 18.19 Breakers result in a buildup of water along the beach that moves as a long-shore current. Where it finds a shore bottom that allows it to return to the sea, it surges out in a strong flow called a *rip current*.

the surf zone and then diminishes. A rip current, or where rip currents are occurring, can usually be located by looking for the combination of (1) a lack of surf, (2) darker-looking water, which means a deeper channel, and (3) a turbid, or muddy, streak of water that extends seaward from the channel indicated by the darker water that lacks surf. See chapter 16 for information on how earthquakes can also produce waves.

Ocean Currents

Waves generated by the winds, earthquakes, and tidal forces keep the surface of the ocean in a state of constant motion. Local, temporary currents associated with this motion, such as rip currents or tidal currents, move seawater over a short distance. Seawater also moves in continuous **ocean currents,** streams of water that stay in about the same path as they move through other seawater over large distances. Ocean currents can be difficult to observe directly since they are surrounded by water that looks just like the water in the current. Wind is likewise difficult to observe directly since the moving air looks just like the rest of the atmosphere. Unlike the wind, an ocean current moves *continuously* in about the same path, often carrying water with different chemical and physical properties than the water it is moving through. Thus, an ocean current can be identified and tracked by measuring the physical and chemical characteristics of the current and the surrounding water. This shows where the current is coming from and where in the world it is going. In general, ocean currents are produced by (1) density differences in seawater and (2) winds that blow persistently in the same direction.

Density Currents. The density of seawater is influenced by three factors: (1) the water temperature, (2) salinity, and (3) suspended sediments. Cold water is generally more dense than warm water, thus sinking and displacing warmer water. Seawater of a high salinity has a higher relative density than less salty water, so it sinks and displaces water of less salinity. Likewise, seawater with a larger amount of suspended sediments has a higher relative density than clear water, so it sinks and

Rogue Waves

A rogue wave is an unusually large wave that appears with smaller waves. The rogue wave has also been called a "freak" wave. Whatever the name, it is generally one wave or a group of two or three waves that are more than twice the size of the normal surrounding waves. The rogue wave is one or several very large "walls of water" and has unpredictable behavior, not following the wind direction, for example. Large rogue waves have been reported from 21 to 35 m (69 to 114 ft) tall, and have been observed to almost capsize large ships.

It is believed that a rogue wave is an extreme storm wave. It probably forms during a storm from constructive interference (see page 114) between smaller waves when the crest and troughs happen to match, coming together to form a mountainous wave that lasts several minutes before subsiding. Other processes, such as wave focusing by the shape of the coast or movement by currents, may play a part in forming rogue waves. The source of rogue waves continues to be a mystery and an active topic of research. This much is known—rogue waves do exist.

displaces clear water. The following describes how these three ways of changing the density of seawater result in the ocean current known as a *density current,* which is an ocean current that flows because of density differences.

Earth receives more incoming solar radiation in the tropics than it does at the poles, which establishes a temperature difference between the tropical and polar oceans. The surface water in the polar ocean is often at or below the freezing point of freshwater, while the surface water in the tropical ocean averages about 26°C (about 79°F). Seawater freezes at a temperature below that of freshwater because the salt content lowers the freezing point. Seawater does not have a set freezing point, however, because as it freezes, the salinity is increased as salt is excluded from the ice structure. Increased salinity lowers the freezing point more, so the more ice that freezes from seawater, the lower the freezing point for the remaining seawater. Cold seawater near the poles is therefore the densest, sinking and creeping slowly as a current across the ocean floor toward the equator. Where and how such a cold, dense bottom current moves is influenced by the shape of the ocean floor, the rotation of Earth, and other factors. The size and the distance that cold bottom currents move can be a surprise. Cold, dense water from the Arctic, for example, moves in a 200 m (about 660 ft) diameter current on the ocean bottom between Greenland and Iceland. This current carries an estimated 5 million m^3/s (about 177 million ft^3/s) of seawater to the 3.5 km (about 2.1 mi) deep water of the North Atlantic Ocean. This is a flow rate about 250 times larger than that of the Mississippi River. At about 30°N, the cold Arctic waters meet even denser water that has moved in currents all the way from the Antarctic to the deepest part of the North Atlantic Basin (figure 18.20).

A second type of density current results because of differences in salinity. The water in the Mediterranean, for example,

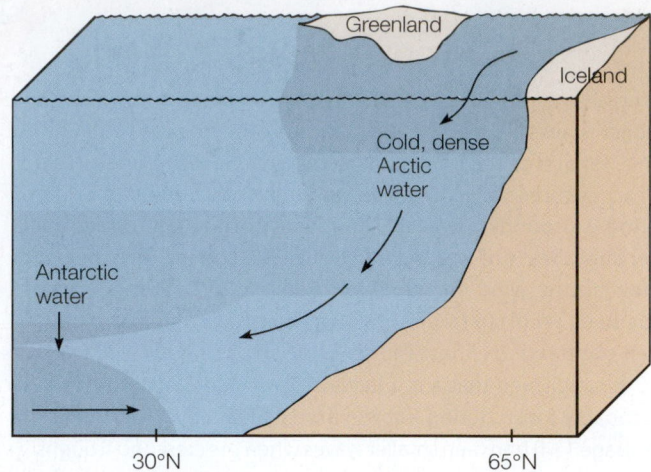

FIGURE 18.20 A cold density current carries about 250 times more water than the Mississippi River from the Arctic and between Greenland and Iceland to the deep Atlantic Ocean. At about 30°N latitude, it meets water that has moved by cold density currents all the way from the Antarctic.

has a high salinity because it is mostly surrounded by land in a warm, dry climate. The Mediterranean seawater, with its higher salinity, is more dense than the seawater in the open Atlantic Ocean. This density difference results in two separate currents that flow in opposite directions between the Mediterranean and the Atlantic. The greater-density seawater flows from the bottom of the Mediterranean into the Atlantic, while the less dense Atlantic water flows into the Mediterranean near the surface. The dense Mediterranean seawater sinks to a depth of about 1,000 m (about 3,300 ft) in the Atlantic, where it spreads over a large part of the North Atlantic Ocean. This increases the salinity of this part of the ocean, making it one of the more saline areas in the world.

The third type of density current occurs when underwater sediments on a slope slide toward the ocean bottom, producing a current of muddy or turbid water called a *turbidity current*. Turbidity currents are believed to be a major mechanism that moves sediments from the continents to the ocean basin. They may also be responsible for some undersea features, such as submarine canyons. Turbidity currents are believed to occur only occasionally, however, and none has ever been directly observed or studied. There is thus no data or direct evidence of how they form or what effects they have on the ocean floor.

Surface Currents. There are broad and deep-running ocean currents that slowly move tremendous volumes of water relatively near the surface. As shown in figure 18.21, each current is actually part of a worldwide system, or circuit, of currents. This system of ocean currents is very similar to the worldwide system of prevailing winds. This similarity exists because it is the friction of the prevailing winds on the seawater surface that drives the ocean currents. The currents are modified by other factors, such as the rotation of Earth and the shape of the ocean basins, but they are basically maintained by the wind systems.

Each ocean has a great system of moving water called a **gyre** that is centered in the mid-latitudes. The gyres rotate to the right in the Northern Hemisphere and to the left in the Southern Hemisphere. The movement of water around these systems, or gyres, plus some smaller systems, forms the surface circulation system of the world ocean. Each part of the system has a separate name, usually based on its direction of flow. All are called "currents" except one that is called a "stream" (the Gulf Stream) and those that are called "drifts." Both the Gulf Stream and the drifts are currents that are part of the connected system.

The major surface currents are like giant rivers of seawater that move through the ocean near the surface. You know that all the currents are connected, for a giant river of water cannot

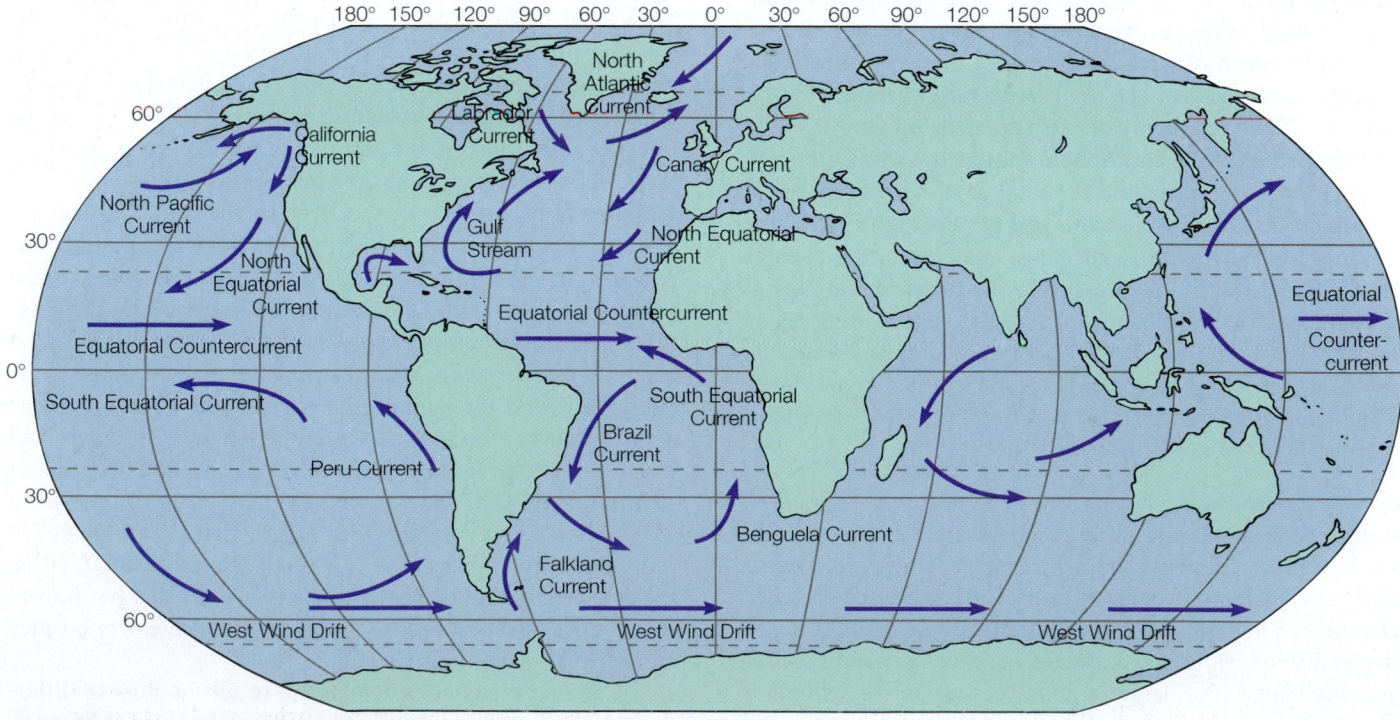

FIGURE 18.21 Earth's system of ocean currents.

Thermohaline Circulation

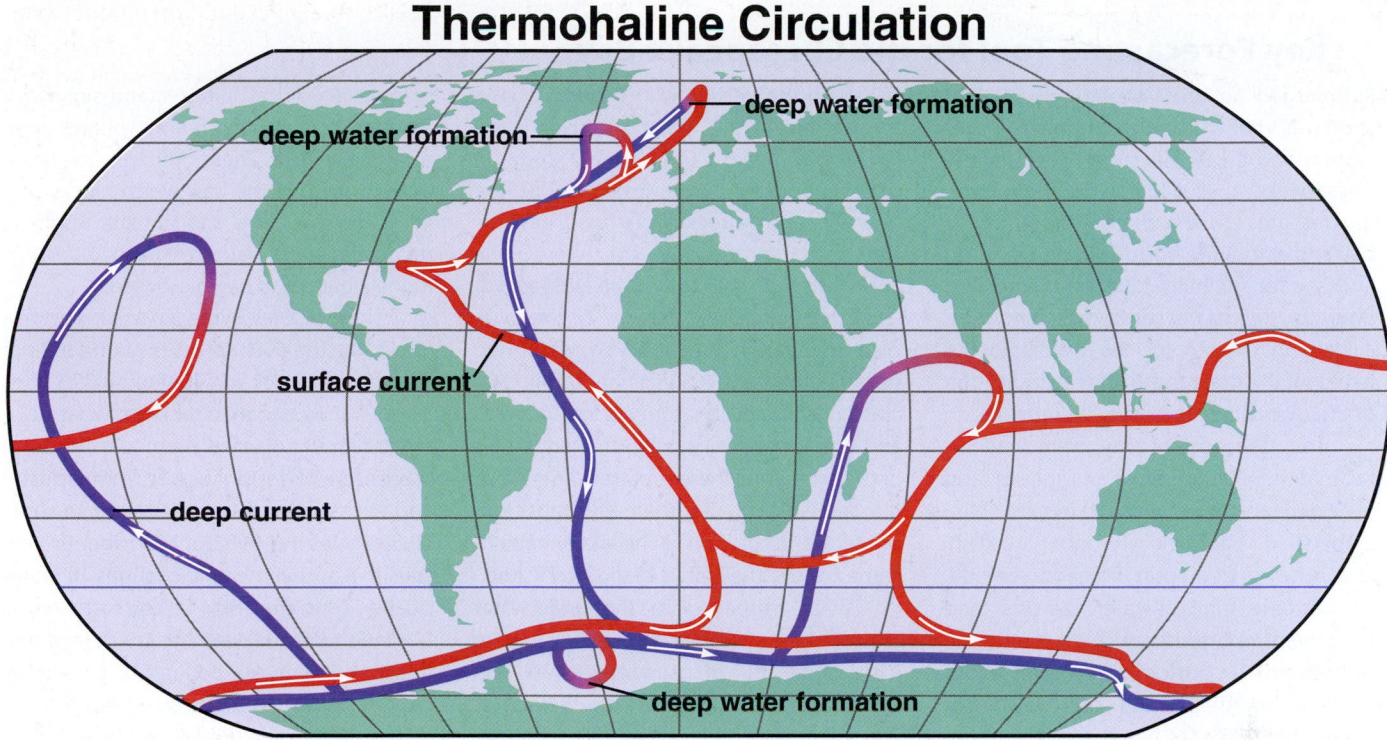

FIGURE 18.22 A summary of the path of the thermohaline circulation. Blue paths represent deep-water currents, while red paths represent surface currents..

just start moving in one place, then stop in another. The Gulf Stream, for example, is a current about 100 km (about 60 mi) wide that may extend to a depth of 1 km (about 0.6 mi) below the surface, moving more than 75 million m³ of water per second (about 2.6 billion ft³/s). The Gulf Stream carries more than 370 times more water than the Mississippi River. The California Current is weaker and broader, carrying cool water southward at a relatively slow rate. The flow rate of all the currents must be equal, however, since all the ocean basins are connected and the sea level is changing very little, if at all, over long periods of time.

The Great Ocean Conveyor Belt. Density currents and surface currents combine to create **thermohaline circulation,** which occurs deep within the ocean as well as on the surface (figure 18.22). This circulation acts like a conveyor belt as ocean water absorbs and redistributes solar energy, or heat, around Earth, so it is also called "the great ocean conveyor belt." The transport of heat by this conveyor belt influences the climate around the Earth. Ocean water moves slowly on this worldwide transportation system, eventually taking over 2,000 years to complete one trip around Earth.

18.3 THE OCEAN FLOOR

Some of the features of the ocean floor were discussed in chapter 15 because they were important in developing the theory of plate tectonics. Many features of the present ocean basins were created from the movement of large crustal plates, according to plate tectonics theory, and in fact, some ocean basins are thought to

have originated with the movement of these plates. There is also evidence that some features of the ocean floor were modified during the ice ages of the past. During an ice age, much water becomes locked up in glacial ice, which lowers the sea level. The sea level dropped as much as 140 m (about 460 ft) during the most recent major ice age, exposing the margins of the continents to erosion. Today, these continental margins are flooded with seawater, forming a zone of relatively shallow water called the **continental shelf** (figure 18.23). The continental shelf is considered to be a part of the continent and not the ocean, even though it is covered with an average depth of about 130 m (about 425 ft) of seawater. The shelf slopes gently away from the shore for an average of 75 km (about 47 mi), but it is much wider on the edge of some parts of continents than other parts.

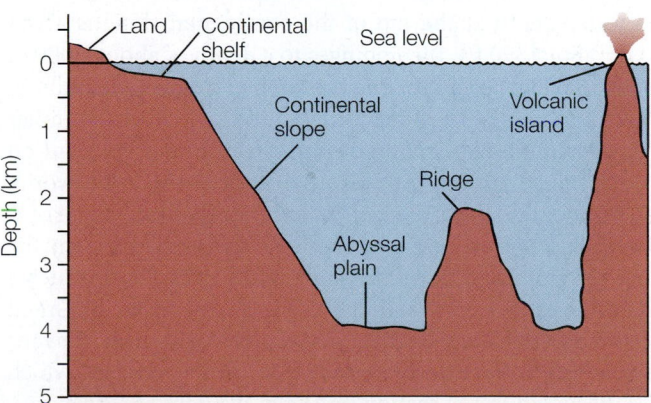

FIGURE 18.23 Some features of the ocean floor. Note that the inclination of the features is not as steep as this vertical exaggeration would suggest.

A Closer Look

Key Forecasting Tool for the Chesapeake Bay

Submerged aquatic vegetation—SAV, for short—is vital to the Chesapeake Bay ecosystem and a key measuring stick for the Chesapeake's overall health. Eel and wigeon grasses grow in the shallows around the bay and across the bottom, providing food and shelter for baby blue crabs and fish, filtering pollutants, and providing life-sustaining oxygen for the water. So the health of the grasses provides an indication of the overall health of the Chesapeake.

The Chesapeake Bay once had an estimated 600,000 acres of grasses and provided an abundance of oysters, blue crab, shad, haddock, sturgeon, rockfish, and other prized sport fish and seafood. At that time, the bay water was clear, and watermen reported they could clearly see grasses on the bottom some 6 m (about 20 ft) below their boats. Then the water clouded, the grasses began to die, and the ecosystem of the bay began to decline. The low point was reached in 1984, when less than 40,000 acres of grasses could be found in the now murky waters of the Chesapeake.

With the decline of the grasses came the decline of the aquatic species living in the bay. Shad, haddock, and sturgeon once supported large fisheries but are now scarce. The rockfish (striped bass) as well as the prized blue crab were once abundant but seem to decline and recover over the years. The decline of the blue crab population has shocked watermen because the crab is amazingly fertile. The crabs reach sexual maturity in about a year and one female bears millions of eggs. Part of the problem is the loss of the underwater grasses, which shelter the baby crabs from predators. In addition to the loss of habitat, some believe the crab is being overfished.

What happened to the underwater grasses of the Chesapeake? Scientists believe it is a combination of natural erosion and pollutants that wash from farms in the watershed. The pollutants include nutrient-rich fertilizers, chemical residues, and overflow from sewage treatment plants. There are some 6,000 chicken houses around the Chesapeake, raising more than 600 million chickens a year and producing 750,000 tons of manure. The

manure is used as fertilizer, and significant amounts of nitrogen and phosphates wash from these fields. These nutrients accelerate algae growth in the bay, which blocks sunlight. The grass dies and the loss results in muddying of the water, making it impossible for new grasses to begin growing.

There is some evidence to support this idea since the 600,000 acres of underwater grasses died back to a low of 40,000 acres in 1984, then began to rebound when sewage treatment plants were modernized and there were fewer pollutants from industry and farms. The grasses recovered to some 63,500 acres by 1999 but are evidently very sensitive to even slight changes in water quality. Thus, they may be expected to die back with unusual weather conditions that might bring more pollutants or muddy conditions but continue to rebound when the conditions are right. The abundance of blue crabs and other aquatic species can be expected to fluctuate with the health of the grasses. The trends of underwater grass growth do indeed provide a key measure for the health of the Chesapeake Bay.

The continental shelf is a part of the continent that happens to be flooded by seawater at the present time. It still retains some of the general features of the adjacent land that is above water, such as hills, valleys, and mountains, but these features were smoothed off by the eroding action of waves when the sea level was lower. Today, a thin layer of sediments from the adjacent land covers these smoothed-off features.

Beyond the gently sloping continental shelf is a steeper feature called the **continental slope.** The continental slope is the transition between the continent and the deep ocean basin. The water depth at the top of the continental slope is about 120 m (about 390 ft), then plunges to a depth of about 3,000 m (about 10,000 ft) or more. The continental slope is generally 20 to 40 km (about 12 to 25 mi) wide, so the inclination is similar to that encountered driving down a steep mountain road on an interstate highway. At various places around the world, the continental slopes are cut by long, deep, and steep-sided *submarine canyons.* Some of these canyons extend from the top of the slope and down the slope to the ocean basin. Such a submarine canyon can be similar in size and depth to the Grand Canyon on the Colorado River of Arizona. Submarine canyons are believed to have been eroded by turbidity currents, which were discussed in the section on the movement of seawater.

Beyond the continental slope is the bottom of the ocean floor, the **ocean basin.** Ocean basins are the deepest part of the

ocean, covered by about 4 to 6 km (about 2 to 4 mi) of seawater. The basin is mostly a practically level plain called the **abyssal plain** and long, rugged mountain chains called *ridges* that rise thousands of meters above the abyssal plain. The Atlantic Ocean and Indian Ocean basins have ridges that trend north and south near the center of the basin. The Pacific Ocean basin has its ridge running north and south near the eastern edge. The Pacific Ocean basin also has more *trenches* than the Atlantic Ocean or the Indian Ocean basins. A trench is a long, relatively narrow, steep-sided trough that occurs along the edges of the ocean basins. Trenches range in depth from about 8 to 11 km (about 5 to 7 mi) deep below sea level.

The ocean basin and ridges of the ocean cover more than half of Earth's surface, accounting for more of the total surface than all the land of the continents. The plain of the ocean basin alone, in fact, covers an area about equal to the area of the land. Scattered over the basin are more than ten thousand steep volcanic peaks called *seamounts*. By definition, seamounts rise more than 1 km (about 0.6 mi) above the ocean floor, sometimes higher than the sea-level surface of the ocean. A seamount that sticks above the water level makes an island. The Hawaiian Islands are examples of such giant volcanoes that have formed islands. Most seamount-formed islands are in the Pacific Ocean. Most islands in the Atlantic, on the other hand, are the tops of volcanoes of the Mid-Atlantic Ridge.

Rachel Louise Carson (1907–1964)

Rachel Louise Carson was a U.S. biologist, conservationist, and campaigner. Her writings on conservation and the dangers and hazards that many modern practices imposed on the environment inspired the creation of the modern environmental movement.

Carson was born in Springdale, Pennsylvania, on May 27, 1907, and educated at the Pennsylvania College for Women, studying English to achieve her ambition for a literary career. A stimulating biology teacher diverted her toward the study of science, and she went to Johns Hopkins University, graduating in zoology in 1929. She received her master's degree in zoology in 1932 and was then appointed to the department of zoology at the University of Maryland, spending her summers teaching and researching at the Woods Hole Marine Biological Laboratory in Massachusetts. Family commitments to her widowed mother and orphaned nieces forced her to abandon her academic career, and she worked for the U.S. Bureau of Fisheries, writing in her spare time articles on marine life and fish, and producing her first book on the sea just before the Japanese attack on Pearl Harbor. During World War II, she wrote fisheries information bulletins for the U.S. government and reorganized the publications department of what became known after the war as the U.S. Fish and Wildlife Service. In 1949, she was appointed chief biologist and editor of the service. She also became occupied with fieldwork and regularly wrote freelance articles on the natural world. During this period, she was also working on *The Sea Around Us*. Upon its publication in 1951, this book became an immediate best-seller, was translated into several languages, and won several literary awards. Given a measure of financial independence by this success, Carson resigned from her job in 1952 to become a professional writer. Her second book, *The Edge of the Sea* (1955), an ecological exploration of the seashore, further established her reputation as a writer on biological subjects. Her most famous book, *The Silent Spring* (1962), was a powerful indictment of the effects of the chemical poisons, especially DDT, with which humans were destroying the Earth, sea, and sky. Despite denunciations from the influential agrochemical lobby, one immediate effect of Carson's book was the appointment of a presidential advisory committee on the use of pesticides. By this time, Carson was seriously incapacitated by ill health, and she died in Silver Spring, Maryland, on April 14, 1964. On a larger canvas, *The Silent Spring* alerted and inspired a new worldwide movement of environmental concern. While writing about broad scientific issues of pollution and ecological exploitation, Carson also raised important issues about the reckless squandering of natural resources by an industrial world.

Source: Modified from the *Hutchinson Dictionary of Scientific Biography.* © RM, 2011. All rights reserved. Helicon Publishing is a division of RM.

SUMMARY

Precipitation that falls on the land either evaporates, flows across the surface, or soaks into the ground. Water moving across the surface is called *runoff.* Water that moves across the land as a small body of running water is called a *stream.* A stream drains an area of land known as the stream drainage basin or *watershed.* The watershed of one stream is separated from the watershed of another by a line called a *divide.* Water that collects as a small body of standing water is called a *pond,* and a larger body is called a *lake.* A *reservoir* is a natural pond, a natural lake, or a lake or pond created by building a dam for water management or control. The water of streams, ponds, lakes, and reservoirs is collectively called *surface water.*

Precipitation that soaks into the ground *percolates* downward until it reaches a *zone of saturation.* Water from the saturated zone is called *groundwater.* The amount of water that a material will hold depends on its *porosity,* and how well the water can move through the material depends on its *permeability.* The surface of the zone of saturation is called the *water table.*

The *ocean* is the single, continuous body of salt water on the surface of Earth. A *sea* is a smaller part of the ocean with different characteristics. The dissolved materials in seawater are mostly the ions of six substances, but sodium ions and chlorine ions are the most abundant. *Salinity* is a measure of the mass of salts dissolved in 1,000 g of seawater.

An *ocean wave* is a moving disturbance that travels across the surface of the ocean. In its simplest form, a wave has a ridge called a *crest* and a depression called a *trough.* Waves have a characteristic *wave height, wavelength,* and *wave period.* The characteristics of waves made by the wind depend on the wind *speed,* the *time* the wind blows, and the *fetch.* Regular groups of low-profile, long-wavelength waves are called *swell.* When swell approaches a shore, the wave slows and increases in wave height. This slowing *refracts,* or bends, the waves so they approach the shore head-on. When the wave height becomes too steep, the top part breaks forward, forming *breakers* in the *surf zone.* Water accumulates at the shore from the breakers and returns to the sea as *undertow,* as *longshore currents,* or in *rip currents.*

Ocean currents are streams of water that move through other seawater over large distances. Some ocean currents are *density currents,* which are caused by differences in *water temperature, salinity,* or *suspended sediments.* Each ocean has a great system of moving water called a *gyre* that is centered in mid-latitudes. Different parts of a gyre are given different names such as the *Gulf Stream* or the *California Current.*

The ocean floor is made up of the *continental shelf,* the *continental slope,* and the *ocean basin.* The ocean basin has two main parts, the *abyssal plain* and mountain chains called *ridges.*

KEY TERMS

abyssal plain (p. **430**)

aquifer (p. **417**)

continental shelf (p. **429**)

continental slope (p. **430**)

freshwater (p. **415**)

groundwater (p. **417**)

gyre (p. **428**)

ocean (p. **421**)

ocean basin (p. **430**)

ocean currents (p. **427**)

ocean wave (p. **424**)

runoff (p. **415**)

salinity (p. **423**)

surf (p. **426**)

swell (p. **425**)

thermohaline circulation (p. **429**)

watershed (p. **416**)

water table (p. **417**)

APPLYING THE CONCEPTS

Answers are located in appendix F.

1. The most abundant chemical compound at the surface of Earth is
   a. silicon dioxide.
   b. nitrogen gas.
   c. water.
   d. minerals of iron, magnesium, and silicon.

2. Of the total supply, the amount of water that is available for human consumption and agriculture is
   a. 97 percent.
   b. about two-thirds.
   c. about 3 percent.
   d. less than 1 percent.

3. In a region of abundant rainfall, a layer of extensively cracked but otherwise solid granite could serve as a limited source of groundwater because it has
   a. limited permeability and no porosity.
   b. average porosity and average permeability.
   c. no permeability and no porosity.
   d. limited porosity and no permeability.

4. How many different oceans are actually on Earth's surface?
   a. 14
   b. 7
   c. 3
   d. 1

5. The largest of the three principal ocean regions of Earth is the
   a. Atlantic Ocean.
   b. Pacific Ocean.
   c. Indian Ocean.
   d. South American Ocean.

6. The Gulf of Mexico is a shallow sea of the
   a. Atlantic Ocean.
   b. Pacific Ocean.
   c. Indian Ocean.
   d. South American Ocean.

7. Measurement of the salts dissolved in seawater taken from various locations throughout the world show that seawater has a
   a. uniform chemical composition and a variable concentration.
   b. variable chemical composition and a variable concentration.
   c. uniform chemical composition and a uniform concentration.
   d. variable chemical composition and a uniform concentration.

8. The salinity of seawater is *increased* locally by
   a. the addition of water from a large river.
   b. heavy precipitation.
   c. the formation of sea ice.
   d. None of the above is correct.

9. Considering only the available light and the dissolving ability of gases in seawater, more abundant life should be found in a
   a. cool, relatively shallow ocean.
   b. warm, very deep ocean.
   c. warm, relatively shallow ocean.
   d. cool, very deep ocean.

10. The regular, low-profile waves called swell are produced from
    a. constant, prevailing winds.
    b. small, irregular waves becoming superimposed.
    c. longer wavelengths outrunning and outlasting shorter wavelengths.
    d. all wavelengths becoming transformed by gravity as they travel any great distance.

11. If the wavelength of swell is 10.0 m, then you know that the fish below the surface feel the waves to a depth of
    a. 5.0 m.
    b. 10.0 m.
    c. 20.0 m.
    d. however deep it is to the bottom.

QUESTIONS FOR THOUGHT

1. Describe in general all the things that happen to the water that falls on the land.
2. Explain how a stream can continue to flow even during a dry spell.
3. What is the water table? What is the relationship between the depth to the water table and the depth that a well must be drilled? Explain.
4. Compare the advantages and disadvantages of using (a) surface water and (b) groundwater as a source of freshwater.
5. Prepare arguments for (a) agriculture, (b) industries, and (c) cities each having first priority in the use of a limited water supply. Identify one of these arguments as being the "best case" for first priority, then justify your choice.
6. Discuss some possible ways of extending the supply of freshwater.
7. The world's rivers and streams carry millions of tons of dissolved materials to the ocean each year. Explain why this does not increase the salinity of the ocean.
8. What is swell and how does it form?
9. Why do waves always seem to approach the shore head-on?
10. What factors determine the size of an ocean wave made by the wind?
11. Describe how a breaker forms from swell. What is surf?
12. Describe what you would look for to avoid where rip current occurs at a beach.
13. How are the waters of Earth distributed as a solid, a liquid, and a gas at a given time? How much of the water is salt water and how much is freshwater?
14. What are the requirements for an artesian well?
15. Describe the required wavelength and wave height that are needed before sediments will move on the ocean floor beneath the wave.

FOR FURTHER ANALYSIS

1. Considering the distribution of all the water on Earth, which presently unavailable category would provide the most freshwater at the least cost for transportation, processing, and storage?
2. Describe a number of ways that you believe would increase the amount of precipitation going into groundwater rather than runoff.
3. Some people believe that constructing a reservoir for water storage is a bad idea because (1) it might change the downstream habitat below the dam and (2) the reservoir will eventually fill with silt and sediments. Write a "letter to the editor" that agrees that a reservoir is a bad idea for the reasons given. Now write a second letter that disagrees with the "bad idea" letter and supports the construction of the reservoir.
4. Explain how the average salinity of seawater has remained relatively constant over the past six hundred million years in spite of the continuous supply of dissolved salts in the river waters of the world.
5. Can ocean waves or ocean currents be used as an energy source? Explain why or why not.
6. What are the significant similarities and differences between a river and an ocean current?

INVITATION TO INQUIRY

Water Use

Investigate the source of water and the amount used by industrial processes, agriculture, and homes in your area. Make pie graphs to compare your area to national averages and develop explanations for any differences. What could be done to increase the supply of water in your area?

19

Organic and Biochemistry

These are ball-and-stick models of organic molecules used in many science classrooms to help students understand the 3-D nature of various organic compounds.

CORE **CONCEPT**

The nature of the carbon atom allows for a great variety of organic compounds, many of which play vital roles in living.

OUTLINE

A great variety of organic molecules can be formed (p. 436).

All living things are comprised of organic molecules (p. 449).

Humans have created many organic compounds (p. 446).

Physics

▷ Organic molecules are involved in energy storage and release (Ch. 3).

Chemistry

▷ The octet rule is followed when bonds are formed (Ch. 9).
▷ Chemistry of the carbon atom is responsible for the traits of organic compounds (Ch. 8).

Earth Science

▷ Fossil fuels were originally the organic components of living things (Ch. 3).
▷ Pollution of air, land, and water may be the result of many types of organic molecules (Ch. 15).

Life Science

▷ The structure of living things is comprised of organic molecules (Ch. 20).
▷ Living things are and utilize organic compounds (Ch. 20).

Astronomy

▷ Extraterrestrial organic compounds are known to exist and may have been the source of the original molecules of life on Earth (Ch. 19).
▷ The exploration for extraterrestrial life centers on the identification of organic compounds (Ch. 19).

OVERVIEW

The impact of ancient Aristotelian ideas on the development of understandings of motion, elements, and matter was discussed in earlier chapters. Historians also trace the "vitalist theory" back to Aristotle. According to Aristotle's idea, all living organisms are composed of the four elements (earth, air, fire, and water) and have in addition an *actuating force*, the life or soul that makes the organism different from nonliving things made of the same four elements. Plants, as well as animals, were considered to have this actuating, or vital, force in the Aristotelian scheme of things.

There were strong proponents of the vitalist theory as recently as the early 1800s. Their basic argument was that organic matter, the materials and chemical compounds recognized as being associated with life, could not be produced in the laboratory. Organic matter could only be produced in a living organism, they argued, because the organism had a vital force that is not present in laboratory chemicals. Then, in 1828, a German chemist named Friedrich Wöhler decomposed a chemical that was *not organic* to produce urea (N_2H_4CO), a known *organic* compound that occurs in urine. Wöhler's production of an organic compound was soon followed by the production of other organic substances by other chemists. The vitalist theory gradually disappeared with each new reaction, and a new field of study, organic chemistry, emerged.

This chapter is an introductory survey of the field of organic chemistry, which is concerned with compounds and reactions of compounds that contain carbon. You will find this an interesting, informative introduction, particularly if you have ever wondered about synthetic materials, natural foods and food products, or any of the thousands of carbon-based chemicals you use every day. The survey begins with the simplest of organic compounds, those consisting of only carbon and hydrogen atoms, compounds known as hydrocarbons. Hydrocarbons are the compounds of crude oil, which is the source of hundreds of petroleum products (figure 19.1).

Most common organic compounds can be considered derivatives of the hydrocarbons, such as alcohols, ethers, fatty acids, and esters. Some of these are the organic compounds that give flavors to foods, and others are used to make hundreds of commercial products, from face cream to margarine. The main groups, or classes, of derivatives will be briefly introduced, along with some interesting examples of each group. A very strong link exists between organic chemistry and the chemistry of living things, which is called biochemistry, or biological chemistry. The organic compounds of life include proteins, carbohydrates, nucleic acids, and fats. Because there are an infinite variety of organic compounds, this chapter is divided into two parts to help in your understanding. Part I focuses on the features that characterize these compounds, the major groupings into which they are placed, and examples of each. Part II centers on the main classes of biologically important compounds in preparation for your study of living things.

FIGURE 19.1 Refinery and tank storage facilities, such as this one in Texas, are needed to change the hydrocarbons of crude oil into many different petroleum products. The classes and properties of hydrocarbons form one topic of study in organic chemistry.

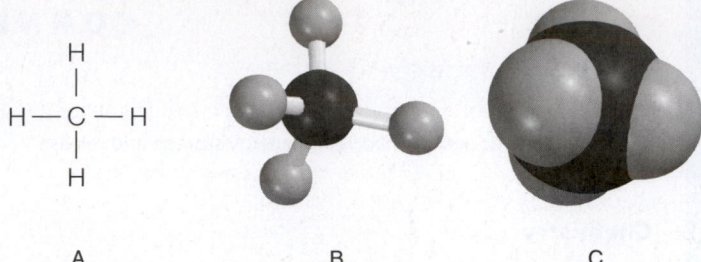

A   B   C

FIGURE 19.2 Models of a methane molecule. Each of the four bondable electrons in a carbon atom inhabits an area as far away from the other three as possible. Carbon can share with four other atoms at each of these sites. For the sake of simplicity, diagrams of molecules such as the gas methane can be (*A*) two-dimensional drawings, although in reality they are three-dimensional molecules and take up space. Recall from chapter 8 that the higher the atomic number, the larger the space taken up by atoms of that element. The model shown in (*B*) above is called a ball-and-stick model. Part (*C*) is a space-filling model. Each time you see the various ways in which molecules are displayed, try to imagine how much space they actually occupy.

PART I
THE NATURE OF ORGANIC COMPOUNDS

19.1 ORGANIC CHEMISTRY

Today, **organic chemistry** is defined as the study of compounds in which carbon is the principal element, whether the compound was formed by living things or not. The study of compounds that do not contain carbon as a central element is called **inorganic chemistry.** An *organic compound* is thus a compound in which carbon is the principal element, and an *inorganic compound* is any other compound.

The majority of known compounds are organic. Several million are known and thousands of new ones are created every year. You use organic compounds every day, including gasoline, plastics, grain alcohol, foods, cosmetics, and many others.

It is the unique properties of carbon that allow it to form so many different compounds. The carbon atom has a valence of four and can combine with one, two, three, or four *other carbon atoms,* in addition to a wide range of other kinds of atoms (figure 19.2). The number of possible molecular combinations is almost limitless, which explains why there are so many organic compounds. Fortunately, there are patterns of groups of carbon atoms and groups of other atoms that lead to similar chemical characteristics, making the study of organic chemistry less difficult.

19.2 HYDROCARBONS

Organic compounds consisting of only two elements, carbon and hydrogen, are called **hydrocarbons.** The simplest hydrocarbon has one carbon atom and four hydrogen atoms (figure 19.3), but since carbon atoms can combine with one another, there are thousands of possible structures and arrangements. The carbon-to-carbon bonds are covalent and can be

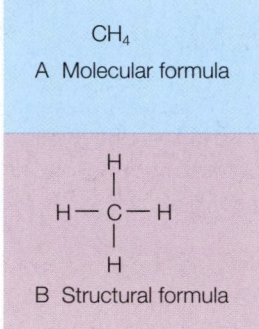

A Molecular formula

B Structural formula

FIGURE 19.3 A molecular formula (*A*) describes the numbers of different kinds of atoms in a molecule, and a structural formula (*B*) represents a two-dimensional model of how the atoms are bonded to each other. Each dash represents a bonding pair of electrons.

single, double, or triple (figure 19.4). Recall that the dash in a structural formula represents a single pair of electrons being shared by two atoms. To satisfy the octet rule, this means that each satisfied carbon atom must have a total of four dashes around it, each attached to another atom—no more and no less. A double covalent bond is one formed when two pairs of electrons are shared by two atoms; and a triple covalent bond is one formed when three pairs of electrons are shared by two atoms.

Note that when the carbon atom has double or triple bonds, fewer hydrogen atoms can be attached. One category of hydrocarbons contains only single covalent bonds and are large enough to form chains of carbon atoms with a straight, a branched, or a ring structure (figure 19.5). Examples of these compounds include petroleum and petroleum products.

Members of the *paraffin series* are not as chemically reactive as the other hydrocarbons. They are called a series because *each higher molecular weight has an additional* CH_2. The simplest is methane, CH_4, and the next highest molecular weight is ethane, C_2H_6. As you can see, C_2H_6 is CH_4 with an additional CH_2.

People Behind the Science

Roy J. Plunkett (1910 –1994)

Roy J. Plunkett was a chemist at Kinetic Chemicals in New Jersey in 1938 when he accidentally invented polytetrafluoroethylene (Teflon). At the time, Plunkett needed some tetrafluoroethylene for his laboratory work, which was kept in a pressurized bottle. He noticed that the valve on the bottle was frozen shut and was concerned that when he tried to open it, gas inside might explode. To protect himself and co-workers, he moved the bottle outside and placed it behind a protective shield. After cutting the bottle open, with no explosion, he found a white powder inside that did not stick to the container. Upon analysis, he found that the tetrafluoroethylene had polymerized into polytetrafluoroethylene, later to become known as Teflon.

Plunkett was born in New Carlisle, Ohio, and attended Newton High School, Manchester College (BA chemistry 1932), and Ohio State University (Ph.D. chemistry 1936). He was the chief chemist involved in the production of Freon and the gasoline additive Tetraethyl lead. He was inducted to the Plastics Hall of Fame in 1973 and the Inventors Hall of Fame in 1985.

Source: Modified from the *Hutchinson Dictionary of Scientific Biography.* © RM, 2011. All rights reserved. Helicon Publishing is a division of RM.

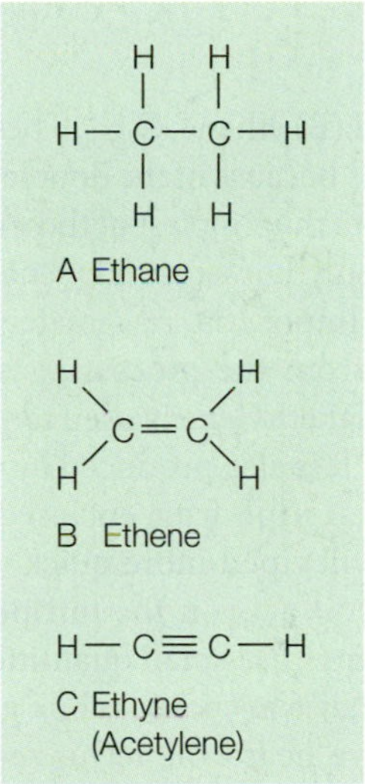

FIGURE 19.4 Carbon-to-carbon bonds can be single (*A*), double (*B*), or triple (*C*). Note that in each example, each carbon atom has four dashes, which represent four bonding pairs of electrons, satisfying the octet rule.

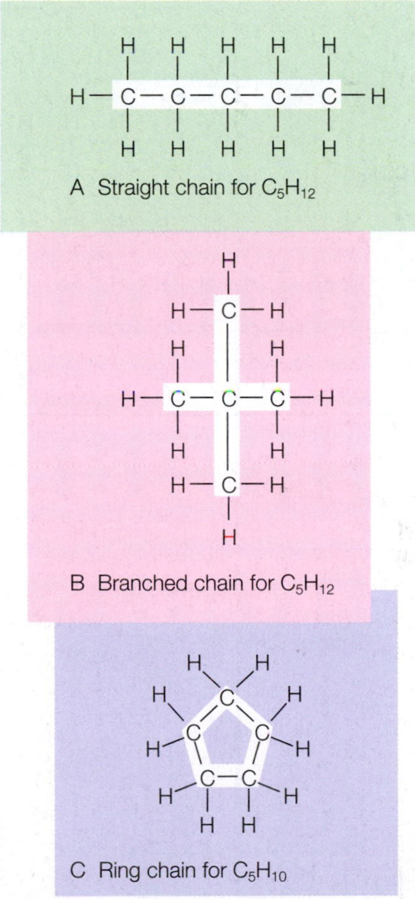

FIGURE 19.5 Carbon-to-carbon chains can be (*A*) straight, (*B*) branched, or (*C*) in a closed ring. (Some carbon bonds are drawn longer but are actually the same length.)

Note the names of the compounds listed in table 19.1. The names have a consistent prefix and suffix pattern. The prefix and suffix pattern is a code that provides a clue about the compound. The Greek prefix tells you the *number of carbon atoms* in the molecule; for example, *oct-* means eight, so *octane* has eight carbon atoms. The suffix *-ane* tells you this hydrocarbon is a member of this series, so it has single bonds only. With the general formula of $C_nH_{2n + 2}$, you can now write the formula when you hear the name. Octane has eight carbon atoms with single bonds and $n = 8$. Two times 8 plus 2($2n + 2$) is 18, so the formula for octane is C_8H_{18}. Most organic chemical names provide clues like this.

The compounds listed in table 19.1 all have straight chains. Figure 19.6A shows butane with a straight chain and a molecular formula of C_4H_{10}. Figure 19.6B shows a different branched structural formula that has the same C_4H_{10} molecular formula. Compounds with the same molecular formulas with different structures are called **isomers.** The branched isomer is called *isobutane.* The isomers of a particular compound, such as butane, have different physical and chemical properties because they have different structures.

Methane, ethane, and propane can have only one structure each, and butane has two isomers. The number of possible isomers for a particular molecular formula increases rapidly as the number of carbon atoms increase. After butane, hexane has five isomers and octane eighteen isomers. Because they have different structures, each isomer has different physical properties.

TABLE 19.1 Examples of Straight-Chain Hydrocarbons

Name	Molecular Formula	Structural Formula
Methane	CH_4	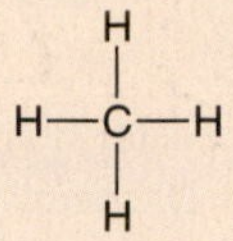
Ethane	C_2H_6	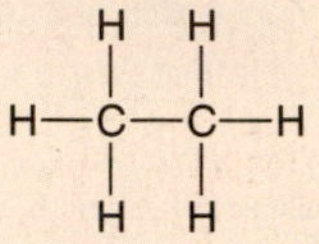
Propane	C_3H_8	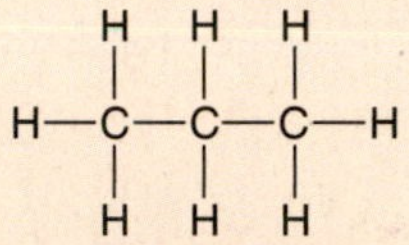
Butane	C_4H_{10}	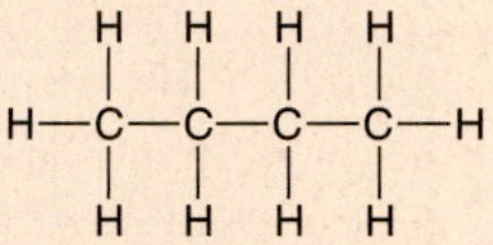
Octane	C_8H_{18}	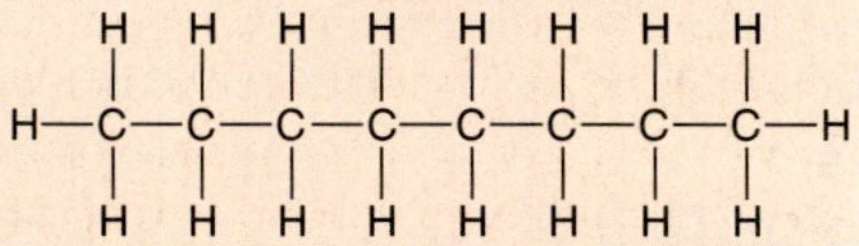

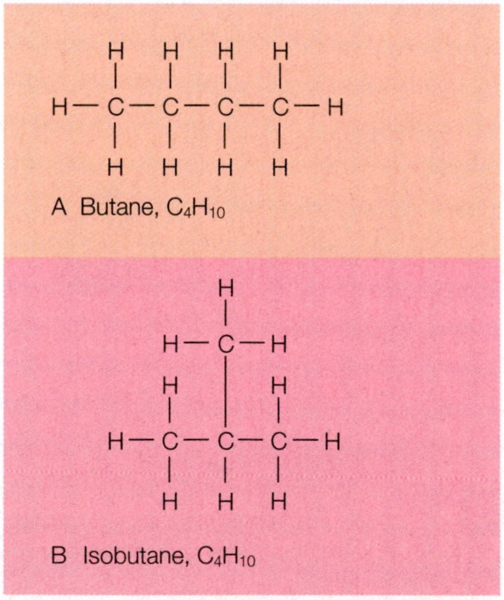

A Butane, C_4H_{10}

B Isobutane, C_4H_{10}

FIGURE 19.6 (A) A straight-chain. (B) A branched-chain isomer.

Hydrocarbons with Double or Triple Bonds

Another important category of *hydrocarbons have double cova-lent carbon-to-carbon bonds*. To denote the presence of a double bond, the suffix *-ane* is changed to *-ene,* as in the compound ethene (table 19.2). Figure 19.4 shows the structural formula for

TABLE 19.2 General Molecular Formulas and Molecular Structures of Different Hydrocarbons

Example Compound	Molecular Structure	General Molecular Formula
Ethane	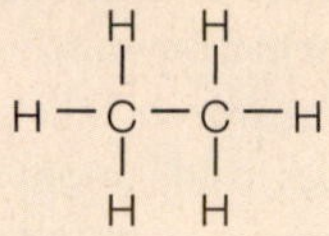	C_nH_{2n+2}
Ethene	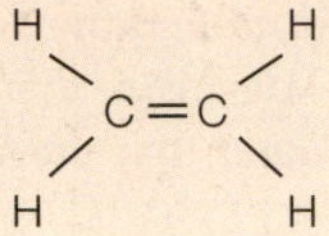	C_nH_{2n}
Ethyne	$H{-}C \equiv C{-}H$	C_nH_{2n-2}

(A) ethane, C_2H_6, and (B) ethene, C_2H_4. These have room for two fewer hydrogen atoms because of the double bond, so the general formula is C_nH_{2n}. Note the simplest of these compounds is called ethene but is commonly known as ethylene.

Ethylene is an important raw material in the chemical industry. Obtained from the processing of petroleum, about half of the commercial ethylene is used to produce the familiar polyethylene plastic. It is also produced by plants to ripen fruit, which explains why unripe fruit enclosed in a sealed plastic bag with ripe fruit will ripen more quickly. The ethylene produced by the ripe fruit acts on the unripe fruit. Commercial fruit packers sometimes use small quantities of ethylene gas to quickly ripen fruit that was picked while green.

Perhaps you have heard the terms *saturated* and *unsaturated* in advertisements for cooking oil or margarine. An organic molecule, such as a hydrocarbon, that does not contain the maximum number of hydrogen atoms is an **unsaturated** hydrocarbon. For example, ethylene can add more hydrogen atoms by reacting with hydrogen gas to form ethane:

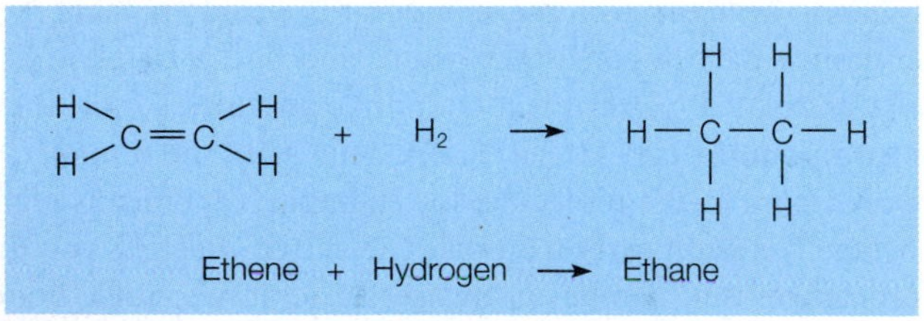

Ethene + Hydrogen → Ethane

The ethane molecule has all the hydrogen atoms possible, so ethane is a **saturated** hydrocarbon. Unsaturated molecules are less stable, which means that they are more chemically reactive than saturated molecules.

Another category of *hydrocarbons has carbon-to-carbon triple bonds* and the general formula of C_nH_{2n-2}. These are highly reactive, and the simplest one, ethyne, has a common name of acetylene. Acetylene is commonly burned with oxygen gas in a welding torch because the flame reaches a temperature of about 3,000°C. Acetylene is also an important raw material in the production of plastics.

Hydrocarbons That Form Rings

The hydrocarbons discussed up until now have been straight or branched open-ended chains of carbon atoms. Carbon atoms can also bond to each other to form a ring, or *cyclic*, structure (figure 19.7). Because of their ring structure, cyclic versions of hydrocarbons have different chemical properties than similar straight chain hydrocarbons. This shows the importance of structural, rather than simply molecular, formulas in describing organic compounds.

The six-carbon ring structure shown in figure 19.8A has three double bonds that do not behave as you would expect. In this six-carbon ring, the double bonds are not in one place but move around, spreading over the whole molecule. Instead of alternating single and double bonds, all the bonds are something in between. This gives the C_6H_6 molecule increased stability. As a result, the molecule does not behave like other unsaturated compounds; that is, it does not readily react in order to add hydrogen to the ring. The C_6H_6 molecule is the organic compound named *benzene*. To denote the six-carbon ring with delocalized electrons, benzene is represented by the symbol shown in figure 19.8B. Notice in this figure that the four molecules differ from one another, depending on the type and location of atoms attached to the benzene ring.

You may have noticed some of the names on labels of paints, paint thinners, and lacquers. Toluene and the xylenes are commonly used in these products as solvents. A benzene ring attached to another molecule is given the name *phenyl*. Phenyl products are good disinfectants and are found in many pine oil products such as Pine Sol.

19.3 PETROLEUM

Petroleum is a mixture containing many of the hydrocarbons previously described. While these hydrocarbons were once the products of photosynthesis, as time passed, these complex carbohydrates were significantly altered. The origin of petroleum is uncertain, but it is believed to have formed from the slow decomposition of buried marine life, primarily plankton and algae in the absence of oxygen (i.e., anaerobic). Time, temperature, pressure, and perhaps bacteria are considered important

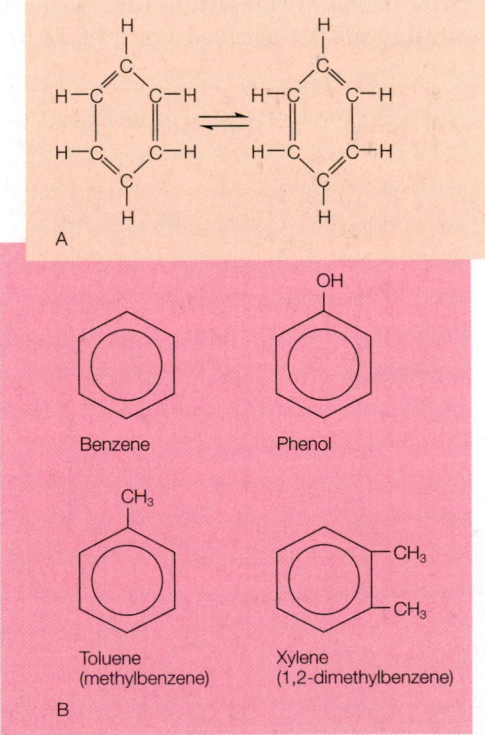

FIGURE 19.7 (A) The "straight" chain has carbon atoms that are able to rotate freely around their single bonds, sometimes linking up in a closed ring. (B) Ring or cyclic compounds.

FIGURE 19.8 (A) The bonds in C_6H_6 are something between single and double, which gives it different chemical properties than double-bonded hydrocarbons. (B) The six-sided symbol with a ring compound known as benzene.

in the formation of petroleum. As the petroleum formed, it was forced through porous rock until it reached a rock type or rock structure that stopped it. Here, it accumulated to saturate the porous rock, forming an accumulation called an **oil field.** The composition of petroleum varies from one oil field to the next. The oil from a given field might be dark or light in color. Some oil fields contain oil with a high quantity of sulfur, referred to as "sour crude." Because of such variations, some fields have oil with more desirable qualities than oil from other fields.

Early settlers found oil seeps in the eastern United States and collected the oil for medicinal purposes. One enterprising oil peddler tried to improve the taste by running the petroleum through a whiskey still. He obtained a clear liquid by distilling

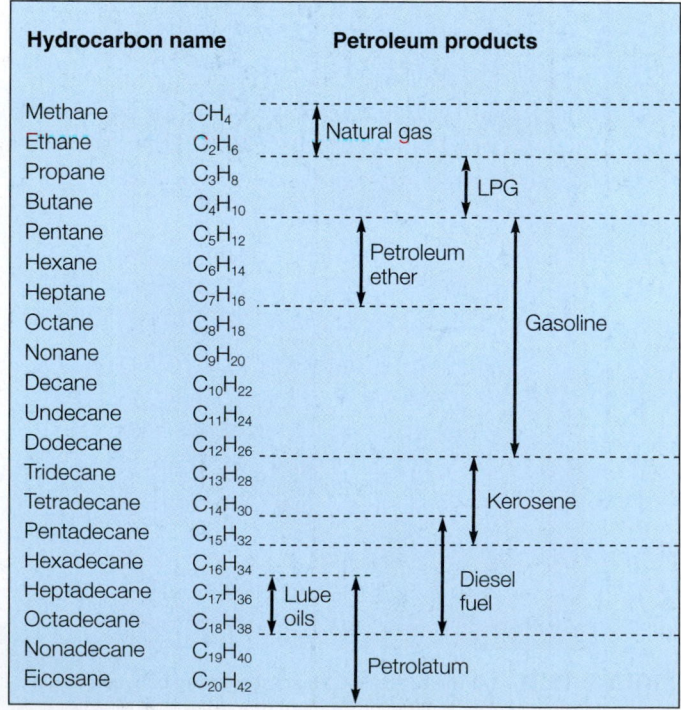

FIGURE 19.9 Petroleum products and the ranges of hydrocarbons in each product.

the petroleum and, by accident, found that the liquid made an excellent lamp oil. This was fortunate timing, for the lamp oil used at that time was whale oil, and whale oil production was declining. This clear liquid obtained by distilling petroleum is today known as *kerosene.*

Wells were drilled, and crude oil refineries were built to produce the newly discovered lamp oil. Gasoline was a by-product and was used primarily as a spot remover. With Henry Ford's automobile production and Thomas Edison's electric light invention, the demand for gasoline increased, and the demand for kerosene decreased. The refineries were converted to produce gasoline, and the petroleum industry grew to become one of the world's largest industries. With the expansion of this industry have come downsides. One of great concern has been climate change associated with the emission of carbon dioxide from a worldwide increase in the number of gasoline-burning engines.

Crude oil is petroleum that is pumped from the ground, a complex and variable mixture of hydrocarbons with an upper limit of about fifty carbon atoms. This thick, smelly black mixture is not usable until it is *refined*, that is, separated into usable groups of hydrocarbons called petroleum products. Petroleum products are separated by distillation—the larger the molecule, the higher the boiling point. Each product has a boiling point range, or "cut," of the distilled vapors. Thus, each product, such as gasoline or heating oil, is made up of hydrocarbons within a range of carbon atoms per molecule (figure 19.9). The products, their boiling ranges, and ranges of carbon atoms per molecule are listed in table 19.3.

The hydrocarbons that have one to four carbon atoms (CH_4 to C_4H_{10}) are gases at room temperature. They can be pumped from certain wells as a gas, but they also can be dissolved in crude oil. *Natural gas* is a mixture of hydrocarbon gases, but it is about 95 percent methane (CH_4). Propane (C_3H_8) and butane (C_4H_{10}) are liquified by compression and cooling and are sold as liquified petroleum gas, or *LPG*. LPG is used where natural gas is not available for cooking or heating and is widely used as a fuel in barbecue grills and camp stoves.

Gasoline is a mixture of hydrocarbons that may have five to twelve carbon atoms per molecule. Gasoline distilled from crude oil consists mostly of straight-chain molecules not suitable for use as an automotive fuel. Straight-chain molecules burn too rapidly in an automobile engine, producing more of

TABLE 19.3   Petroleum Products

Name	Boiling Range (°C)	Carbon Atoms per Molecule
Natural gas	Less than 0	C_1 to C_4
Petroleum ether	35–100	C_5 to C_7
White "gas"	100–350	C_5 to C_9
Gasoline	35–215	C_5 to C_{12}
Kerosene	35–300	C_{12} to C_{15}
Diesel fuel	300–400	C_{15} to C_{18}
Motor oil, grease	350–400	C_{16} to C_{18}
Paraffin	Solid, melts at about 55	C_{20}
Asphalt	Boiler residue	C_{36} or more

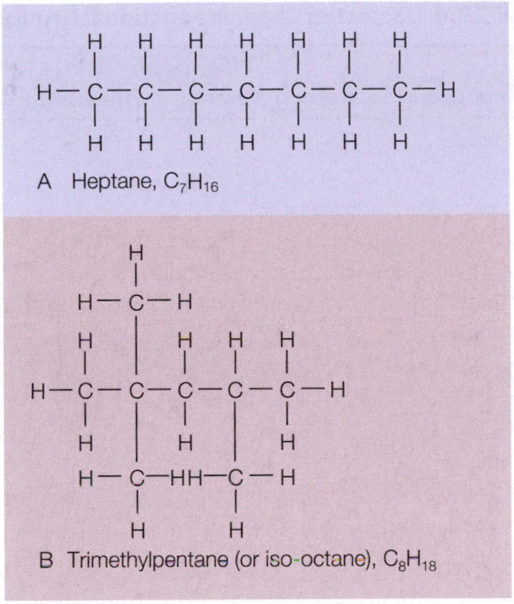

A Heptane, C_7H_{16}

B Trimethylpentane (or iso-octane), C_8H_{18}

FIGURE 19.10 The octane rating scale is a description of how rapidly gasoline burns. It is based on (A) heptane, with an assigned octane number of 0, and (B) trimethylpentane, with an assigned number of 100.

an explosion than a smooth burn. You hear these explosions as a knocking or pinging in the engine, and they indicate poor efficiency and could damage the engine. On the other hand, branched chain molecules burn comparatively slower, without the pinging or knocking explosions. The burning rate of gasoline is described by the *octane number* scale. The scale is based on pure heptane, straight-chain molecules that are assigned an octane number of 0, and a multiple branched isomer of octane, trimethylpentane, which is assigned an octane number of 100 (figure 19.10). Most unleaded gasolines have an octane rating of 87, which are made up of a mixture that is 87 percent trimethylpentane and 13 percent of a heptane complex mixture of other hydrocarbons.

Kerosene is a mixture of hydrocarbons that have from twelve to fifteen carbon atoms. The petroleum product called kerosene is also known by other names, depending on its use. Some of these names are lamp oil (with coloring and fragrance added), jet fuel (with a flash flame retardant added), heating oil, #1 fuel oil, and in some parts of the country, "coal oil."

Diesel fuel is a mixture of a group of hydrocarbons that have from fifteen to eighteen carbon atoms per molecule. Diesel fuel also goes by other names, again depending on its use, that is, distillate fuel oil or #2 fuel oil.

Motor oil and *lubricating oils* have sixteen to eighteen carbon atoms per molecule. Lubricating grease is heavy oil that is thickened with soap. *Petroleum jelly,* also called petrolatum (or Vaseline), is a mixture of hydrocarbons with sixteen to thirty-two carbon atoms per molecule. *Mineral oil* is a light lubricating oil that has been decolorized and purified.

Depending on the source of the crude oil, varying amounts of *paraffin* wax (C_{20} or greater) or *asphalt* (C_{36} or more) may be present. Paraffin is used for candles, waxed paper, and home canning. Asphalt is mixed with gravel and used to surface roads.

19.4 HYDROCARBON DERIVATIVES

The hydrocarbons account for only about 5 percent of the known organic compounds, but the other 95 percent can be considered as hydrocarbon derivatives. **Hydrocarbon derivatives** are formed when *one or more hydrogen atoms on a hydrocarbon have been replaced by some element or group of elements other than hydrogen.* For example, the halogens (Fluorine, F_2, Chlorine, Cl_2, and Bromine, Br_2) react in sunlight or when heated, replacing a hydrogen:

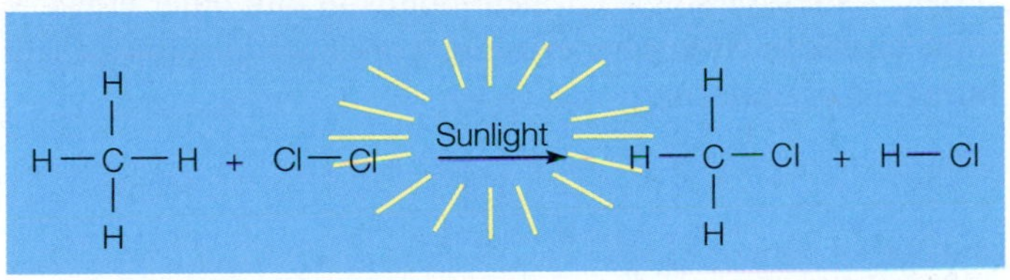

In this particular *substitution reaction,* a hydrogen atom on methane is replaced by a chlorine atom to form methyl chloride. Replacement of any number of hydrogen atoms is possible, and a few *organic halides* are illustrated in figure 19.11. Common organic halides include Freons (chlorofluorocarbons) and Teflon (polytetrafluoroethylene).

If a hydrocarbon molecule is unsaturated (has a double or triple bond), a hydrocarbon derivative can be formed by an *addition reaction:*

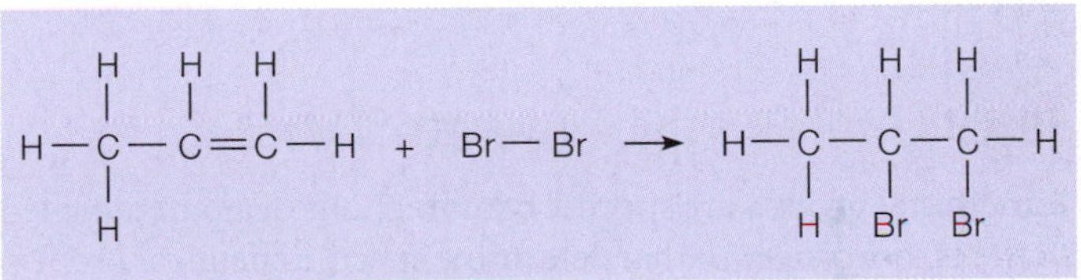

The bromine atoms add to the double bond on propene, forming dibromopropane. This derivative is used in the synthesis of pharmaceuticals and other organic compounds.

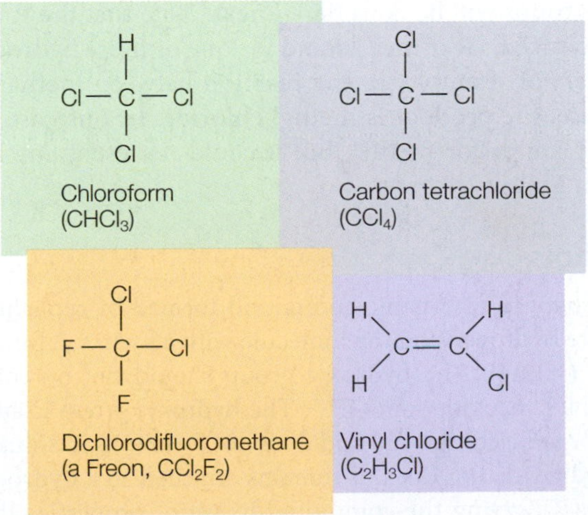

Chloroform ($CHCl_3$)

Carbon tetrachloride (CCl_4)

Dichlorodifluoromethane (a Freon, CCl_2F_2)

Vinyl chloride (C_2H_3Cl)

FIGURE 19.11 Common examples of organic halides.

Certain hydrocarbons containing double bonds can combine with each other in an addition reaction to form a very long chain consisting of hundreds of such subunits. A long chain of repeating units is called a **polymer** (*poly* = many; *mer* = segment), individual pieces are called **monomers** (*mono* = single; *mer* = segment or piece), and the reaction is called *addition polymerization*. Ethylene, for example, is heated under pressure with a catalyst to form *polyethylene,* the most popular plastic in the world. It is used to make such products as grocery bags, shampoo bottles, children's toys, and bullet-proof vests. Heating breaks the double bond between the carbons, which provides sites for single covalent bonds to join other ethylene units together. This polymerization continues forming a chain hundreds of units long.

Ethylene molecules   Polyethylene molecules

Functional Groups Generate Variety

Functional groups are specific combinations of atoms attached to the carbon skeleton that determine specific chemical properties of the organic compound. Functional groups usually have (1) multiple bonds or (2) lone pairs of electrons that cause them to be sites of reactions. Table 19.4 lists some of the common hydrocarbon functional groups. Look over this list and compare the structure of the functional group with the group name. Some of the more interesting examples from a few of these groups will be considered next. Note that the R and R′ (pronounced, "R prime") stand for one or more hydrocarbon groups. For example, in the reaction between methane and chlorine, the product is methyl chloride. In this case, the R in RCl stands for methyl, but it could represent any hydrocarbon group.

Alcohols

An *alcohol* is an organic compound formed by replacing one or more hydrogens on the molecule with a hydroxyl functional group (—OH). The hydroxyl group should not be confused with the hydroxide ion, OH⁻. The hydroxyl group is attached to an organic compound and does not form ions in solution, as the hydroxide ion does. It remains attached to a hydrocarbon group (R), giving the compound its set of properties that are associated with alcohols.

TABLE 19.4   Selected Organic Functional Groups

Name of Functional Group	General Formula	General Structure
Organic halide	RCl	
Alcohol	ROH	
Ether	ROR′	
Aldehyde	RCHO	
Ketone	RCOR′	
Organic acid	RCOOH	
Ester	RCOOR′	
Amino	RNH₂	
Phosphate	RPO₄	

The name of the hydrocarbon group determines the name of the alcohol. If the hydrocarbon group in ROH is methyl, for example, the alcohol is called *methyl alcohol,* or *methanol.* The name of an alcohol has the suffix *-ol* (figure 19.12).

All alcohols have the hydroxyl functional group, and all are chemically similar. Alcohols are toxic to humans except that ethyl alcohol (ethanol) can be consumed in limited quantities. Alcohol has been consumed by people for thousands of years in many cultures. As in the past, there continue to be many health-related problems associated with drinking alcohol, such as fetal alcohol syndrome, liver damage, social disruption, and death. Consumption of small quantities of methanol can result in blindness and death. Ethanol, C_2H_5OH, is produced by the action of yeast or by a chemical reaction of ethylene derived from petroleum refining. Yeast acts on sugars to produce ethanol and CO_2. When beer, wine, and other such beverages are the desired products, the CO_2 escapes during fermentation, and the alcohol remains in solution. In baking, the same reaction

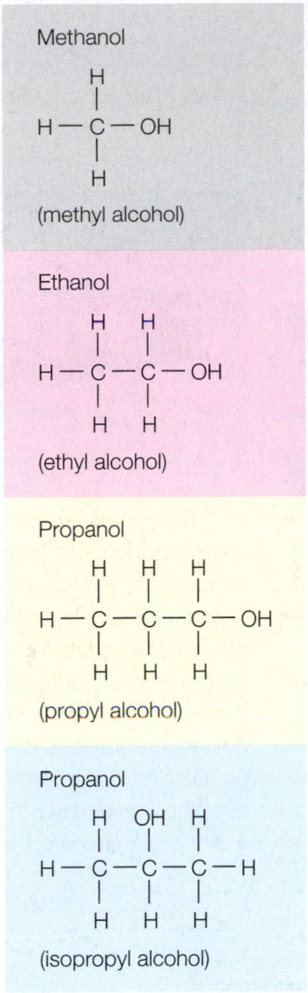

Methanol

$$H-\underset{\underset{H}{|}}{\overset{\overset{H}{|}}{C}}-OH$$

(methyl alcohol)

Ethanol

$$H-\underset{\underset{H}{|}}{\overset{\overset{H}{|}}{C}}-\underset{\underset{H}{|}}{\overset{\overset{H}{|}}{C}}-OH$$

(ethyl alcohol)

Propanol

$$H-\underset{\underset{H}{|}}{\overset{\overset{H}{|}}{C}}-\underset{\underset{H}{|}}{\overset{\overset{H}{|}}{C}}-\underset{\underset{H}{|}}{\overset{\overset{H}{|}}{C}}-OH$$

(propyl alcohol)

Propanol

$$H-\underset{\underset{H}{|}}{\overset{\overset{H}{|}}{C}}-\underset{\underset{H}{|}}{\overset{\overset{OH}{|}}{C}}-\underset{\underset{H}{|}}{\overset{\overset{H}{|}}{C}}-H$$

(isopropyl alcohol)

FIGURE 19.12 Four different alcohols.

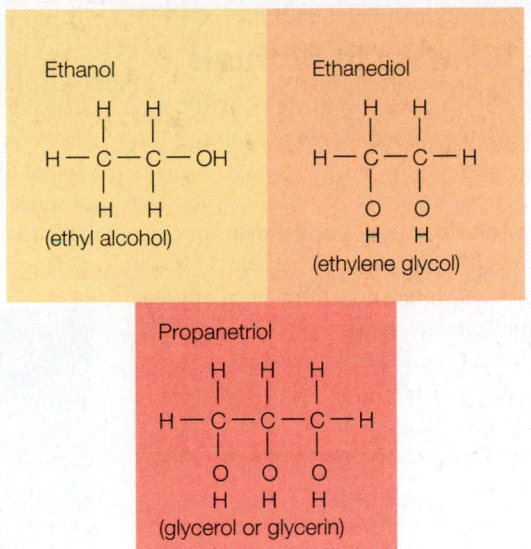

FIGURE 19.13 Common examples of alcohols with one, two, and three hydroxyl groups per molecule.

utilizes the CO_2 to make dough rise (leavened), and the alcohol is evaporated during baking. Most alcoholic beverages are produced by the yeast fermentation reaction:

sugar → carbon dioxide and ethyl alcohol,

but some are made from ethanol derived from petroleum refining.

Alcohols with six or fewer carbon atoms per molecule are soluble in both certain organic solvents and water. A solution of ethanol and gasoline is called *gasohol*. Today, nearly all gasoline contains ethanol to provide oxygen to cause cleaner burning of hydrocarbons ("Connections . . . Biomass to Biofuels"). Gasoline is a mixture of hydrocarbons (C_8H_{18}, for example) that contain no atoms of oxygen. Gasohol contains ethyl alcohol, C_2H_5OH, which does contain oxygen. The addition of alcohol to gasoline, therefore, adds oxygen to the fuel. Because carbon monoxide forms when there is an insufficient supply of oxygen, the addition of alcohol to gasoline helps cut down on carbon monoxide emissions.

Alcoholic beverages are a solution of ethanol and water. The *proof* of such a beverage is double the ethanol concentration by volume. Therefore, a solution of 40 percent ethanol by volume

in water is 80 proof, and wine that is 12 percent alcohol by volume is 24 proof. Distillation alone will produce a 190 proof concentration, but other techniques are necessary to obtain 200 proof absolute alcohol. *Denatured alcohol* is ethanol with acetone, formaldehyde, and other chemicals in solution that are difficult to separate by distillation. Since these denaturants are toxic, they make consumption impossible, so denatured alcohol is sold without the consumption tax.

Methanol, ethanol, and isopropyl alcohol all have one hydroxyl group per molecule. An alcohol with two hydroxyl groups per molecule is called a *glycol*. Ethylene glycol is perhaps the best-known glycol since it is used as an antifreeze. An alcohol with three hydroxyl groups per molecule is called *glycerol* (or *glycerin*). Glycerol is a building block of fat molecules and a by-product in the making of soap. It is added to toothpastes, lotions, cosmetics, and some candies to retain moisture and softness. Ethanol, ethylene glycol, and glycerol are compared in figure 19.13.

Glycerol reacts with nitric acid in the presence of sulfuric acid to produce glyceryl trinitrate, commonly known as *nitroglycerine*. Nitroglycerine is a clear oil that is violently explosive, and when warmed, it is extremely unstable. In 1867, Alfred Nobel (for whom the Nobel Prizes were named) discovered that a mixture of nitroglycerine and siliceous earth was more stable than pure nitroglycerine but was nonetheless explosive. Siliceous earth is also known as diatomaceous earth. The cell walls of diatoms are composed of silicon dioxide (SiO_2). It is composed of the skeletons of algae known as diatoms and provides an enormous surface area upon which the explosive reaction can take place. The mixture is packed in a tube and is called *dynamite*. The name *dynamite* comes from the Greek word meaning "power."

Ethers, Aldehydes, and Ketones

An *ether* has a general formula of ROR′, and the best-known ether is diethylether. In a molecule of diethylether, both the R and the R′ are ethyl groups. Diethylether is a volatile, highly

Connections . . .

Biomass to Biofuels

Biomass is any accumulation of organic material produced by living things. The most commonly used biomass sources are wood, agricultural residue from the harvesting of crops, crops grown for their energy content, and animal waste. These traditional, often noncommercial sources of fuel provide more than 10 percent of the world's energy but are not reported in most statistics about global energy. In many developing countries, these sources of fuel are a large proportion of the energy available.

Biomass conversion is the process of obtaining energy from the chemical energy stored in biomass. It is not a new idea; burning wood is a form of biomass conversion that has been used for thousands of years. Biomass can be burned directly as a source of heat for cooking, burned to produce electricity, converted to alcohol, or used to generate methane. China has an estimated 500,000 small methane digesters in homes and on farms; India and Korea both have

50,000. Brazil is the largest producer of alcohol from biomass. Alcohol provides 50 percent of their automobile fuel. The low price of sugar coupled with the high price of oil has prompted Brazil to use its large crop of sugar cane as a source of energy.

Biomass conversion raises some environmental and economic concerns. Countries that use large amounts of biomass for energy are usually those that have food shortages. Biomass conversion means that fewer nutrients are being returned to the

soil, and this compounds the food shortage. If the price of food rises or the price of oil falls, there could be less biomass conversion.

The energy required to produce usable energy stocks from biomass must be taken into account. Growing corn to produce alcohol requires large energy inputs. The amount of energy present in the alcohol produced from the corn is actually about the same as the amount of energy that went into producing the alcohol. Obviously, this makes no sense from an energy point of view. However, the convenience of a liquid fuel may be worth paying for in economic terms.

---

flammable liquid that was used as an anesthetic in the past. Today, it is used as an industrial and laboratory solvent.

Aldehydes and *ketones* both have a functional group of a carbon atom doubly bonded to an oxygen atom called a *carbonyl group*. The *aldehyde* has a hydrocarbon group, R (or a hydrogen in one case), and a hydrogen attached to the carbonyl group. A *ketone* has a carbonyl group with two hydrocarbon groups attached (figure 19.14).

The simplest aldehyde is *formaldehyde*. Formaldehyde is soluble in water, and a 40 percent concentration called *formalin* has been used as an embalming agent and to preserve biological specimens. Formaldehyde is also a raw material used to make plastics such as Bakelite. All the aldehydes have odors, and the odors of some aromatic hydrocarbons include the odors of almonds, cinnamon, and vanilla. The simplest ketone is *acetone*. Acetone has a fragrant odor and is used as a solvent in paint removers and nail polish removers.

Organic Acids and Esters

Mineral acids, such as hydrochloric acid and sulfuric acid, are made of inorganic materials. Acids that were derived from organisms are called **organic acids.** Because many of these organic acids can be formed from fats, they are sometimes called *fatty acids.* Chemically, they are known as the *carboxylic acids* because they contain the carboxyl functional group, —COOH, and have a general formula of RCOOH.

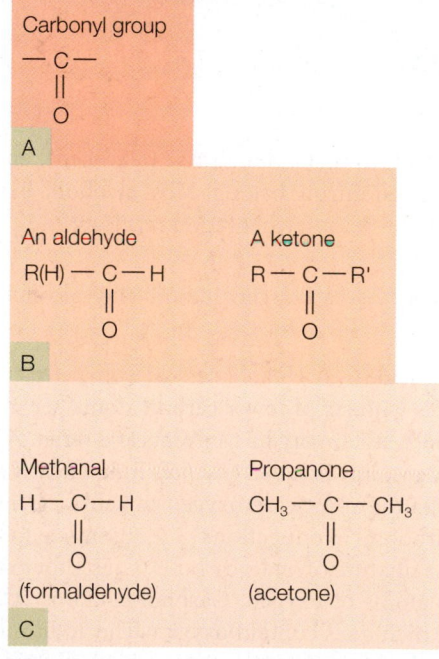

FIGURE 19.14 The carbonyl group (*A*) is present in both aldehydes and ketones, as shown in (*B*). (*C*) The simplest example of each with the more technical name above and the common name below each formula.

The simplest carboxylic acid has been known since the Middle Ages, when it was isolated by the distillation of ants. The Latin word *formica* means "ant," so this acid was given the name *formic acid* (figure 19.15A). Formic acid is

FIGURE 19.15 These African tailer ants (*Oecophylla longinoda*) make the simplest of the organic acids, formic acid. The sting of bees, ants, and some plants contains formic acid, along with some other irritating materials. Formic acid is HCOOH. The plant known as the stinging nettle (*B*) (*Urtica urens*) has small, hollow hairs that contain several irritating substances, including formic acid. These hairs have the ability to scratch the skin and mucous membranes, resulting in almost immediate burning, itching, and irritation.

$$H - C - OH$$
$$\|$$
$$O$$

It is formic acid, along with other irritating materials, that causes the sting of bees, ants, and certain plants, for example the stinging nettle (*Urtica urens*) (figure 19.15B).

Acetic acid, the acid of vinegar, has been known since antiquity. Acetic acid forms from the oxidation of ethanol. An oxidized bottle of wine contains acetic acid in place of the alcohol, which gives the wine a vinegar taste. Before wine is served in a restaurant, the person ordering is customarily handed the bottle cork and a glass with a small amount of wine. You first break the cork in half to make sure it is dry, which tells you

---

CONCEPTS APPLIED

Common Household Compounds

Here is a table listing items commonly found around many U.S. households. In the first blank column, describe the purpose of each compound. For the second column, use the Internet or a reference book such as the *CRC Handbook* or the *Merck Index* and determine whether the material is inorganic or organic. In the third blank column, write the chemical name of the compound; use the last column to write its empirical formula.

CHEMICAL AROUND THE HOUSE

Common Name	Purpose of Compound	Type of Compound (Inorganic or Organic)	Chemical Name	Empirical Formula
1. Acetaminophen				
2. Vinegar				
3. Baking soda				
4. Household bleach				
5. Cetyl palmitate				
6. Super glue				
7. Chocolate				
8. Citric acid				
9. Methyl paraben				
10. Ammonia				
11. Sea salt				
12. Penicillin				

A Very Common Organic Compound—Aspirin

Aspirin was patented in Britain in 1898 and the United States in 1900, and production began when it went on the market as a prescription drug in 1899. However, even primitive humans were familiar with its value as a pain reliever. The bark of the white willow (*Salix alba*) tree was known for its "magical" pain relief power thousands of years ago. Stone tablet writings dated as far back as 3000 B.C. describe the beneficial effects of these trees. People in many cultures stripped and chewed the bark for its medicinal effect. It is estimated that more than 19 billion tablets of aspirin are consumed in the United States daily. It is the most widely used drug in the world. Just what does it do? Aspirin (named by scientists in 1899 at the drug and dye firm Bayer) is really acetylsalicylic acid and is capable of inhibiting the body's production of compounds known as *prostaglandins,* the cause of the pain. In addition to reducing headache pain, there are other conditions aspirin has been found to benefit:

1. In certain people, it lowers the risk of heart attack and stroke.
2. It helps men being treated for prostate or colon cancer to live longer.
3. It controls migraine headaches.
4. It protects against gum disease.
5. It reduces pre-eclampsia (high blood pressure during pregnancy).
6. It reduces the chance of blood clots from forming in veins during long auto/plane trips.

Questions to Discuss

1. What structural parts of aspirin would make it an organic acid?
2. Go to the Internet and find out just how this organic compound acts as a pain reliever.

acid also forms in your muscles as a product of anaerobic carbohydrate metabolism, causing a feeling of fatigue. Anaerobic metabolism of carbohydrates in cells is the breakdown of sugars with the release of usable energy. Citric and lactic acids are small molecules compared to some of the carboxylic acids that are formed from fats. Palmitic acid, for example, is $C_{16}H_{32}O_2$ and comes from palm oil. The structure of palmitic acid is a chain of fourteen CH_2 groups with CH_3- at one end and $-COOH$ at the other. Again, it is the functional carboxyl group, $-COOH$, that gives the molecule its acid properties. Organic acids are also raw materials used in the making of polymers of fabric, film, and paint.

Esters are organic compounds formed from an alcohol and an organic acid by elimination of water. They are common in both plants and animals, giving fruits and flowers their characteristic odor and taste. For example, the ester amyl acetate has the smell of banana. Esters are also used in perfumes and artificial flavorings. A few of the flavors for which particular esters are responsible are listed in table 19.5. These liquid esters can be obtained from natural sources, or they can be chemically synthesized. Natural flavors are complex mixtures of these esters along with other organic compounds. Lower molecular-weight esters are fragrant-smelling liquids, but higher molecular-weight esters are odorless oils and fats.

19.5 SYNTHETIC POLYMERS

Natural polymers are huge, chainlike molecules made of hundreds or thousands of smaller, repeating molecular units called monomers. Cellulose and starch are examples of natural

that the wine has been sealed from oxygen. The small sip is to taste for vinegar before the wine is served. If the wine has been oxidized, the reaction is

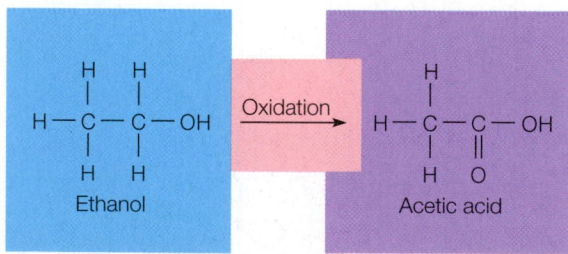

Organic acids are common in many foods. The juice of citrus fruit, for example, contains citric acid, which relieves a thirsty feeling by stimulating the flow of saliva. Lactic acid is found in sour milk, buttermilk, sauerkraut, and pickles. Lactic

TABLE 19.5   Flavors and Esters

Ester Name	Formula	Flavor
Amyl acetate	$CH_3-C-O-C_5H_{11}$, with $\overset{\|}{O}$	Banana
Octyl acetate	$CH_3-C-O-C_8H_{17}$, with $\overset{\|}{O}$	Orange
Ethyl butyrate	$C_3H_7-C-O-C_2H_5$, with $\overset{\|}{O}$	Pineapple
Amyl butyrate	$C_3H_7-C-O-C_5H_{11}$, with $\overset{\|}{O}$	Apricot
Ethyl formate	$H-C-O-C_2H_5$, with $\overset{\|}{O}$	Rum

Name	Chemical unit	Uses
Polyethylene	$\left[\begin{array}{cc} H & H \\ -C-C- \\ H & H \end{array}\right]_n$	Squeeze bottles, containers, laundry and trash bags, packaging
Polypropylene	$\left[\begin{array}{cc} H & H \\ -C-C- \\ H & CH_3 \end{array}\right]_n$	Indoor-outdoor carpet, pipe valves, bottles
Polyvinyl chloride (PVC)	$\left[\begin{array}{cc} H & H \\ -C-C- \\ H & Cl \end{array}\right]_n$	Plumbing pipes, synthetic leather, plastic tablecloths, phonograph records, vinyl tile
Polyvinylidene chloride (Saran)	$\left[\begin{array}{cc} H & Cl \\ -C-C- \\ H & Cl \end{array}\right]_n$	Flexible food wrap
Polystyrene (Styrofoam)	$\left[\begin{array}{cc} H & H \\ -C-C- \\ H & C_6H_5 \end{array}\right]_n$	Coolers, cups, insulating foam, shock-resistant packing material, simulated wood furniture
Polytetrafluoroethylene (Teflon)	$\left[\begin{array}{cc} F & F \\ -C-C- \\ F & F \end{array}\right]_n$	Gears, bearings, coating for nonstick surface of cooking utensils

Name	Chemical unit	Uses
Polyvinyl acetate	$\left[\begin{array}{cc} H & CH_3 \\ -C-C- \\ H & O \\ & | \\ & C-CH_3 \\ & \| \\ & O \end{array}\right]_n$	Mixed with vinyl chloride to make vinylite; used as an adhesive and resin in paint
Styrene-butadiene rubber	$\left[\begin{array}{cccccc} H & H & H & H & H & H \\ -C-C=C-C-C-C- \\ H & & & H & C_6H_5 & H \end{array}\right]_n$	Automobile tires
Polychloroprene (Neoprene)	$\left[\begin{array}{cccc} H & Cl & H & H \\ -C-C=C-C- \\ H & & & H \end{array}\right]_n$	Shoe soles, heels
Polymethyl methacrylate (Plexiglas, Lucite)	$\left[\begin{array}{cc} H & CH_3 \\ -C-C- \\ H & \\ & C-O-CH_3 \\ & \| \\ & O \end{array}\right]_n$	Moldings, transparent surfaces on furniture, lenses, jewelry, transparent plastic "glass"
Polycarbonate (Lexan)	$\left[\begin{array}{c} CH_3 \\ -C_6H_4-C-C_6H_4-O-C-O- \\ CH_3 \quad\quad O \end{array}\right]_n$	Tough, molded articles such as motorcycle helmets
Polyacrylonitrile (Orlon, Acrilan, Creslan)	$\left[\begin{array}{cc} H & H \\ -C-C- \\ H & CN \end{array}\right]_n$	Textile fibers

FIGURE 19.16 Synthetic polymers, the polymer unit, and some uses of each polymer.

polymers made of glucose monomers. *Synthetic polymers* are manufactured from a wide variety of substances. You are familiar with these polymers as synthetic fibers such as nylon and the inexpensive light plastic used for wrappings and containers (figure 19.16).

The first synthetic polymer was a modification of the naturally existing cellulose polymer. Cellulose was chemically modified in 1862 to produce celluloid, the first *plastic*. The term *plastic* means that celluloid could be molded to any desired shape. Celluloid was produced by first reacting cotton with a mixture of nitric and sulfuric acids. The cellulose in the cell walls of the cotton plant reacts with the acids to produce the ester cellulose nitrate. This ester is an explosive compound known as "guncotton," or smokeless gunpowder. When made with ethanol and camphor, the product is less explosive and can be formed and

A Closer Look

Nonpersistent and Persistent Organic Pollutants

A nonpersistent organic pollutant does not remain in the environment for very long. Most nonpersistent pollutants are *biodegradable*. A biodegradable material is chemically changed by living organisms and often serves as a source of food and energy for decomposer organisms, such as bacteria and fungi. Other nonpersistent pollutants decompose as a result of inorganic chemical reactions. Still others quickly disperse to concentrations that are too low to cause harm.

Nonpersistent toxic materials, such as insecticides, are destroyed by sunlight or reaction with oxygen or water in the atmosphere. For example, organophosphates are commonly

used insecticides (ex. parathion) that usually decompose within several weeks. As a result, they do not accumulate in food chains because they are pollutants for only a short period of time.

Persistent pollutants are those that remain in the environment for many years in an unchanged condition. Most of the persistent pollutants are organic, human-made materials. An estimated 30,000 synthetic chemicals are used in the United States. They are mixed in an endless variety of combinations to produce all types of products used in every aspect of daily life. They are part of our food, transportation, clothing, building materials, home appliances, medicine, recreational equipment, and many other items. Our way of life depends heavily on synthetic materials.

An example of a persistent organic pollutant is polychlorinated biphenyls. PCBs are highly stable organic compounds that resist changes from heat, acids, bases, and oxidation. These characteristics make PCBs desirable for industrial use but also make them persistent pollutants when released into the environment. PCBs have been used in inks, plastics, tapes, paints, glues, waxes, and polishes. Although the manufacture of PCBs in the United States stopped in 1977, these persistent chemicals are still present in the soil and sediments and continue to do harm. PCBs are harmful to fish and other aquatic forms of life because they interfere with reproduction. In humans, PCBs produce liver ailments and skin lesions. In high concentration, they can damage the nervous system, and they are suspected carcinogens.

molded into useful articles. This first plastic, celluloid, was used to make dentures, combs, eyeglass frames, and photographic film. Before the discovery of celluloid, many of these articles, including dentures, were made from wood. Today, only Ping-Pong balls are made from cellulose nitrate ("A Closer Look: Nonpersistent and Persistent Organic Pollutants").

Cotton reacted with acetic acid and sulfuric acid produces a cellulose acetate ester. This polymer, through a series of chemical reactions, produces rayon filaments when forced through small holes. The filaments are twisted together to form viscose rayon thread. When forced through a thin slit, a sheet is formed rather than filaments, and the transparent sheet is called *cellophane*. Both rayon and cellophane, as celluloid, are manufactured by modifying the natural polymer of cellulose.

The first truly synthetic polymer was produced in the early 1900s by reacting two chemicals with relatively small molecules rather than modifying a natural polymer. Phenol, an aromatic hydrocarbon, was reacted with formaldehyde, the simplest aldehyde, to produce the polymer named *Bakelite*. Bakelite is a *thermosetting* material that forms cross-links between the polymer chains. Once the links are formed during production, the plastic becomes permanently hardened and cannot be softened or made to flow. Some plastics are *thermoplastic* polymers and soften during heating and harden during cooling because they do not have cross-links.

Polyethylene is a familiar thermoplastic polymer produced by a polymerization reaction of ethylene, which is derived from petroleum. This synthetic was invented just before World War II and was used as an electrical insulating material during the war. Today, there are many variations of polyethylene that are produced by different reaction conditions or by substitution of one or more hydrogen atoms in the ethylene molecule. Polyethylene terephthalate (PET) is used extensively for rigid containers, particularly beverage bottles for carbonated drinks and medicine containers. These containers have the recycle code of number 1. When soft polyethylene near the melting point is rolled in alternating perpendicular directions or expanded and compressed as it is cooled, the polyethylene molecules become ordered in a way that improves the rigidity and tensile strength. One form, high-density polyethylene (HDPE), is used in milk and water jugs, as liners in screw-on jar tops and bottle caps, and as a material for toys. HDPE bottles have the recycling code of number 2 and can be reused to make plastic lumber, motor oil containers, playground equipment, and sheeting. The other form, low-density polyethylene (LDPE), has a recycle code of number 4. Recycled LDPE is used for vegetable, dry cleaning, and grocery bags, and plastic squeeze bottles with a recycle code of number 1.

The properties of polyethylene are also changed by replacing one of the hydrogen atoms in a molecule of ethylene. If the hydrogen is replaced by a chlorine atom, the compound is called vinyl chloride, and the polymer formed from vinyl chloride is

$$\text{H}\diagdown_{\text{H}} \text{C} = \text{C} \diagup^{\text{H}}_{\text{Cl}}$$

polyvinyl chloride (PVC). Polyvinyl chloride is used to make plastic water pipes, synthetic leather, and other vinyl products. It differs from the waxy plastic of polyethylene because of the chlorine atom that replaces hydrogen on each monomer.

Replacement of a hydrogen atom with a benzene ring makes a monomer called *styrene*. Styrene is

$$\text{H}\diagdown_{\text{H}} \text{C} = \text{C} \diagup^{\text{H}}_{\bigcirc}$$

and polymerization of styrene produces *polystyrene*. Polystyrene is puffed full of air bubbles to produce the familiar Styrofoam coolers, cups, and packing materials.

If all hydrogens of an ethylene molecule are replaced with atoms of fluorine, the product is polytetrafluoroethylene, a tough plastic that resists high temperatures and acts more like a metal than a plastic. Since it has a low friction, it is used for bearings, gears, and as a nonstick coating on frying pans. You probably know of this plastic by its trade name of *Teflon*.

There are many different polymers in addition to PVC, Styrofoam, and Teflon, and the monomers of some of these are shown in figure 19.16. There are also polymers of isoprene, or synthetic rubber, in wide use. Fibers and fabrics may be polyamides (such as nylon), polyesters (such as Dacron), or polyacrylonitriles (Orlon, Acrilan, Creslan), which have a CN in place of a hydrogen atom on an ethylene molecule and are called acrylic materials. All of these synthetic polymers have added much to practically every part of your life. It would be impossible to list all of their uses here; however, they present problems since (1) they are manufactured from raw materials obtained from coal and a dwindling petroleum supply, and (2) they do not readily decompose when dumped into rivers, oceans, or other parts of the environment. However, research in the polymer sciences is beginning to reflect new understandings learned from research on biological tissues. This could lead to whole new molecular designs for synthetic polymers that will be more compatible with ecosystems.

PART II
ORGANIC COMPOUNDS OF LIFE

19.6 ORGANISMS AND THEIR MACROMOLECULES

Aristotle and the later proponents of the vitalist theory were *partly* correct in their concept that living organisms are different from inorganic substances made of the same elements. Living organisms, for example, have the ability to (1) exchange matter and energy with their surroundings and (2) transform matter and energy into different forms as they (3) respond to changes in their surroundings. In addition, living organisms can use the transformed matter and energy to (4) grow and (5) reproduce. Living organisms are able to do these things through a great variety of organic reactions that are catalyzed by enzymes, however, and not through some mysterious "vital force." These enzyme-regulated organic reactions take place because living organisms are highly organized and have an incredible number of relationships among many different chemical processes.

Organic molecules in living things have molecular weights of thousands or millions of atomic mass units (u's) and are therefore referred to as **macromolecules.** Many types of macromolecules are polymers comprised of similar repeated monomers. Monomers are usually combined to form a polymer by a chemical reaction (*dehydration synthesis*) that results in a water molecule being removed from between two monomers. The reverse of this reaction (*hydrolysis*) is the process of splitting a larger polymer into its parts by the addition of water. Digestion of food molecules in the stomach and small intestine is an important example of this reaction. There are three main types of polymeric macromolecules: (1) carbohydrates, (2) proteins, and (3) nucleic acids. In addition, there is a fourth category, the lipids, which are not true polymers but are macromolecules.

Carbohydrates

One class of organic molecules, **carbohydrates,** is composed of carbon, hydrogen, and oxygen atoms linked together to form monomers called *simple sugars* or *monosaccharides* (*mono* = single; *saccharine* = sweet, sugar) (see table in "Connections . . . How Sweet It Is!"). The name *carbohydrate* literally means "watered carbon," and the empirical formula (CH_2O) for most carbohydrates indicates one carbon (C) atom for each water (H_2O).

The formula for a simple sugar is easy to recognize because there are equal numbers of carbons and oxygens and twice as many hydrogens—for example, $C_3H_6O_3$ or $C_5H_{10}O_5$. We usually describe simple sugars by the number of carbons in the molecule. The ending *-ose* indicates that you are dealing with a carbohydrate. A tri*ose* has three carbons, a pent*ose* has five, and a hex*ose* has six. If you remember that the number of carbons equals the number of oxygen atoms and that the number of hydrogens is double that number, these names tell you the formula for the simple sugar. Carbohydrates play a number of roles in living things. They serve as an immediate source of energy (sugars), provide shape to certain cells (cellulose in plant cell walls), are components of many antibiotics and coenzymes, and are an essential part of the nucleic acids, DNA and RNA.

Simple sugars can be combined with each other to form polymers of **complex carbohydrates** (figure 19.17). When two simple sugars bond to each other, a *disaccharide* (*di-* = two) is formed; when three bond together, a *trisaccharide* (*tri-* = three) is formed. Generally, we call a complex carbohydrate that is larger than this a *polysaccharide* (many sugar units). In all cases, the complex carbohydrates are formed by the removal of water from between the sugars. For example, when glucose and

Connections . . .

How Sweet It Is!

Simple sugars, such as glucose, fructose, and galactose, provide the chemical energy necessary to keep organisms alive. *Glucose*, $C_6H_{12}O_6$, is the most abundant carbohydrate and serves as a food and a basic building block for other carbohydrates. Glucose (also called dextrose) is found in the sap of plants, and in the human bloodstream, it is called *blood sugar*. Corn syrup, which is often used as a sweetener, is mostly glucose. Fructose, as its name implies, is the sugar that occurs in fruits, and it is sometimes called *fruit sugar*. You also see it on food labels as high fructose corn syrup. Both glucose and fructose have the same molecular formula, but glucose is an aldehyde sugar and fructose is a ketone sugar (box figure 19.1). A mixture of glucose and fructose is found in honey. This mixture also is formed when table sugar (sucrose) is reacted with water in the presence of an acid,

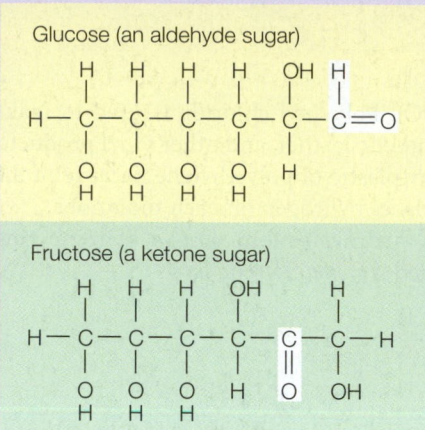

Box Figure 19.1 Glucose (blood sugar) is an aldehyde, and fructose (fruit sugar) is a ketone. Both have a molecular formula of $C_6H_{12}O_6$.

a reaction that takes place in the preparation of canned fruit and candies. The mixture of glucose and fructose is called *invert sugar*.

Thanks to fructose, invert sugar is about twice as sweet to the taste as the same amount of sucrose (refer to figure 19.17).

Relative Sweetness of Various Sugars and Sugar Substitutes	
Type of Sugar or Artificial Sweetener	**Relative Sweetness**
Lactose (milk sugar)	0.16
Maltose (malt sugar)	0.33
Glucose	0.75
Sucrose (table sugar)	**1.00**
Fructose (fruit sugar)	1.75
Cyclamate	30.00
Aspartame	150.00
Stevia	300.00
Saccharin	350.00
Sucralose	600.00

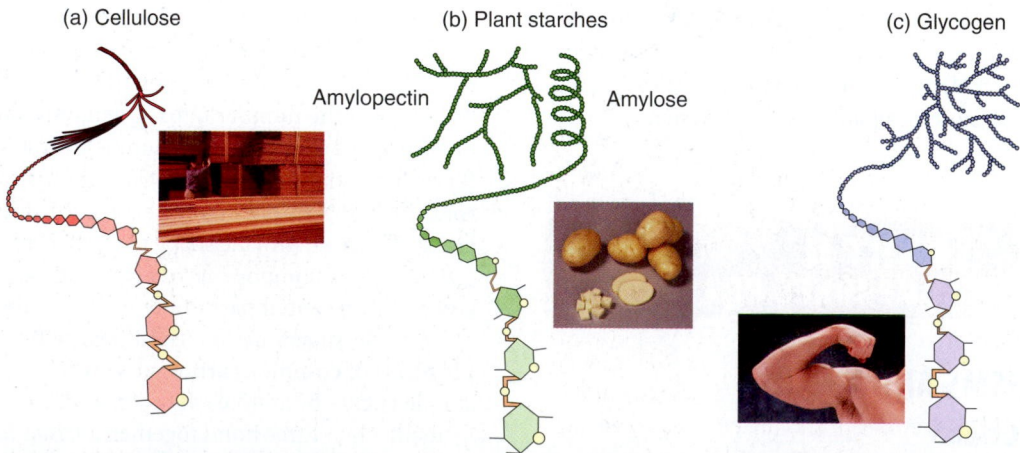

(a) Cellulose   (b) Plant starches   (c) Glycogen

Amylopectin   Amylose

FIGURE 19.17 Simple sugars are attached to each other by the removal of water from between them. Three common complex carbohydrates are the polymers known as (A) cellulose (wood fibers), (B) plant starch (amylopectin and amylose), and (C) glycogen (sometimes called animal starch). Glycogen is found in muscle cells. Notice how they are similar in that they are all polymers of simple sugars, but they differ from one another in how they are joined together. While many organisms are capable of digesting (hydrolyzing) the bonds that are found in glycogen and plant starch molecules, few are able to break those that link together the monosaccharides of cellulose.

So You Don't Eat Meat! How to Stay Healthy

Humans require nine amino acids in their diet: threonine, tryptophan, methionine, lysine, phenylalanine, isoleucine, valine, histidine, and leucine. They are called *essential amino acids* because the body is not able to manufacture them. The body uses these *essential amino acids* in the synthesis of the proteins required for good health. For example, the sulfur-containing amino acid methionine is essential for the absorption and transportation of the elements selenium and potassium. It also prevents excess fat buildup in the liver, and it traps heavy metals, such as lead, cadmium, and mercury, bonding with them so that they can be excreted from the body. Essential amino acids are not readily available in most plant proteins and are most easily acquired through meat, fish, and dairy products.

If this is the case, how do people avoid nutritional deficiency if for economic or personal reasons they do not eat meat, poultry, fish, meat products, dairy products, and honey? People who exclude all animal products from their diet are called *vegans*. Those that include only milk are called lacto-vegetarians, those who include eggs are ovo-vegetarians, and those that include both eggs and milk are called lacto-ovo vegetarians. For anyone but a true vegan, the essential amino acids can be provided in even a small amount of milk and eggs. True vegans can get all their essential amino acids by eating certain combinations of plants or plant products. Even though there are certain plants that contain all these amino acids (soy, lupin, hempseed, chia seed, amaranth, buckwheat, and quinoa), most plants contain one or more of the essential amino acids. However, by eating the right combination of different plants, it is possible to get all the essential amino acids in one meal. These combinations are known as complementary foods.

---

fructose are joined together, they form a *disaccharide (sucrose)* with the loss of a water molecule,

$$C_6H_{12}O_6 + C_6H_{12}O_6 \rightarrow C_{12}H_{22}O_{11} + H_2O$$
$$\text{glucose} \quad \text{fructose} \quad \text{sucrose} \quad \text{water}$$

Sucrose, ordinary table sugar, is the most common disaccharide and occurs in high concentrations in sugar cane and sugar beets. It is extracted by crushing the plant materials, then dissolving the sucrose from the materials with water. The water is evaporated and the crystallized sugar is decolorized with charcoal to produce white sugar. Other common disaccharides include *lactose* (milk sugar) and *maltose* (malt sugar). All three disaccharides have similar properties, but maltose tastes only about one-third as sweet as sucrose. Lactose tastes only about one-sixth as sweet as sucrose. No matter which disaccharide sugar is consumed (sucrose, lactose, or maltose), it is converted into glucose and transported by the bloodstream for use by the body.

Some common examples of polysaccharides are **cellulose, starch,** and **glycogen.** Cellulose is an important polysaccharide used in constructing plant cell walls. Humans cannot digest this complex carbohydrate, so we are not able to use it as an energy source. On the other hand, animals known as ruminants (e.g., cows and sheep) and termites have microorganisms within their digestive tracts that do digest cellulose, making it an energy source for them. Plant cell walls add bulk or fiber to our diet, but no calories. Fiber is an important addition to the diet because it helps control weight and reduces the risks of colon cancer. Its large water-holding molecules also help control constipation and diarrhea. Starch is also a plant product digestible by most other organisms. Once broken down, the monosaccharide can be used as an energy source or building materials. A close but structurally different polysaccharide is glycogen. This macromolecule is found in the muscle cells of many animals as a storage form of polysaccharide.

Many types of sugars can be used by cells as components in other, more complex molecules. Sugar molecules are a part of molecules such as DNA (deoxyribonucleic acid), RNA (ribonucleic acid), or ATP (adenosine triphosphate).

Proteins

Proteins play many important roles. Enzymes are catalysts that speed the rate of chemical reactions in living things. Proteins such as hemoglobin serve as carriers of other molecules such as oxygen. Others such as collagen provide shape and support, and several kinds of protein in muscle cells are responsible for movement. Proteins also act as chemical messengers and are called hormones. **Hormones** are chemical messengers secreted by endocrine glands to regulate other parts of the body. Certain other proteins called antibodies help defend the body against dangerous microbes and chemicals. An **antibody** is a globular protein made by the body in response to the presence of a foreign or harmful molecule called an antigen. Antigens are in many cases proteins, too.

Chemically, proteins are polymers made up of monomers known as *amino acids*. An **amino acid** is a short carbon skeleton that contains an amino group (a nitrogen and two hydrogens) on one end of the skeleton and a carboxylic acid group at the other end (figure 19.18). In addition, the carbon skeleton

Connections . . .

Protein Structure and Sickle-Cell Anemia

Any changes in the arrangement of amino acids within a protein can have far-reaching effects on its function. For example, normal hemoglobin found in red blood cells consists of two kinds of polypeptide chains called the alpha and beta chains. The beta chain is 146 amino acids long. If just one of these amino acids is replaced by a different one, the hemoglobin molecule may not function properly. A classic example of this results in a condition known as *sickle-cell anemia*. In this case, the sixth amino acid in the beta chain, which is normally glutamic acid, is replaced by valine. This minor change causes the hemoglobin to fold differently, and the red blood cells that contain this altered hemoglobin assume a sickle shape when the body is deprived of an adequate supply of oxygen (see figure 26.16).

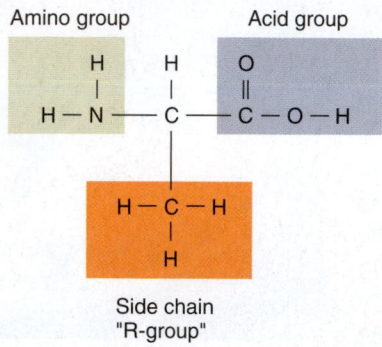

Amino group   Acid group

Side chain
"R-group"

FIGURE 19.18 An amino acid is composed of a short carbon skeleton with three functional groups attached: an amino group, a carboxylic acid group (acid group), and an additional variable group (R-group). It is the variable group that determines which specific amino acid is constructed.

may have one of several different *side chains* on it. These vary in their composition and are generally noted as the amino acid's *R-group*. About twenty common amino acids are important to cells, and each differs from one another in the nature of its R-group (see "A Closer Look: So You Don't Eat Meat! How to Stay Healthy").

Any amino acid can form a bond with any other amino acid. They fit together in a specific way, with the amino group of one bonding to the acid group of the next. You can imagine

Myths, Mistakes, and Misunderstandings

Protein and Your Health

Misconception: The amount of protein per day a person needs for good health is equal to that found by eating a 12-ounce steak.

In fact, health and nutrition experts have determined that most adult men (eighteen years or older) should be eating between 52 and 56 grams of protein per day, while adult women should be consuming about 46 grams per day. Just check out what your daily protein consumption would be if you just ate two eggs (12 grams) for breakfast, a quarter pound hamburger without cheese (28 grams) for lunch, and that 12-ounce steak (84 grams) for dinner! A total of 124 grams of protein! Of course that does not include the protein found in all the other foods on your plate. The next time you are at a restaurant, ask for their nutritional information material to see just what you are eating.

that by using twenty different amino acids as building blocks, you can construct millions of different combinations. There are over three million possible combinations for a molecule five amino acids long. Each of these combinations is termed a **polypeptide chain.** A specific polypeptide is composed of a specific sequence of amino acids bonded end to end.

The specific sequence of amino acids in a polypeptide is controlled by the genetic information of an organism. Genes are specific messages that tell the cell to link particular amino acids in a specific order; that is, they determine a polypeptide's primary structure. Some sequences of amino acids in a polypeptide are likely to twist (as a coil or a pleated sheet), whereas other sequences remain straight (figure 19.19). For example, some proteins (e.g., hair) take the form of a helix, a shape like that of a coiled spring. Other polypeptides form hydrogen bonds that cause them to make several flat folds that resemble a pleated skirt. This is called a pleated sheet. The way a particular protein folds is important to its function. In bovine spongiform encephalitis (mad cow disease) and Creutzfeldt-Jakob diseases, protein structures are not formed correctly, resulting in characteristic nervous system symptoms.

It is also possible for a single polypeptide to contain one or more coils and pleated sheets along its length. As a result, these different portions of the molecule can interact to form an even more complex globular structure. A good example of a protein with highly complex structure is the myoglobin molecule. This is an oxygen-holding protein found in muscle cells.

Frequently, several different polypeptides, each with its own highly complex structure, twist around each other and chemically combine. The larger, globular structure formed by these interacting polypeptides are displayed by the protein molecules called immunoglobulins or *antibodies*, which are involved in fighting diseases such as influenza and tetanus. The protein portion of the hemoglobin molecule (globin is globular in shape) also demonstrates this highest level of complexity.

Energy in the form of heat or light may break the bonds within protein molecules. When this occurs, the chemical and physical properties of the protein are changed and the protein is said to be **denatured.** (Keep in mind, a protein is a molecule, not a living thing, and therefore cannot be "killed.") A common example of this occurs when the gelatinous, clear portion of an egg is cooked and the protein changes to a white solid. Some medications are proteins and must be protected from denaturation so as not to lose their effectiveness. Insulin is an

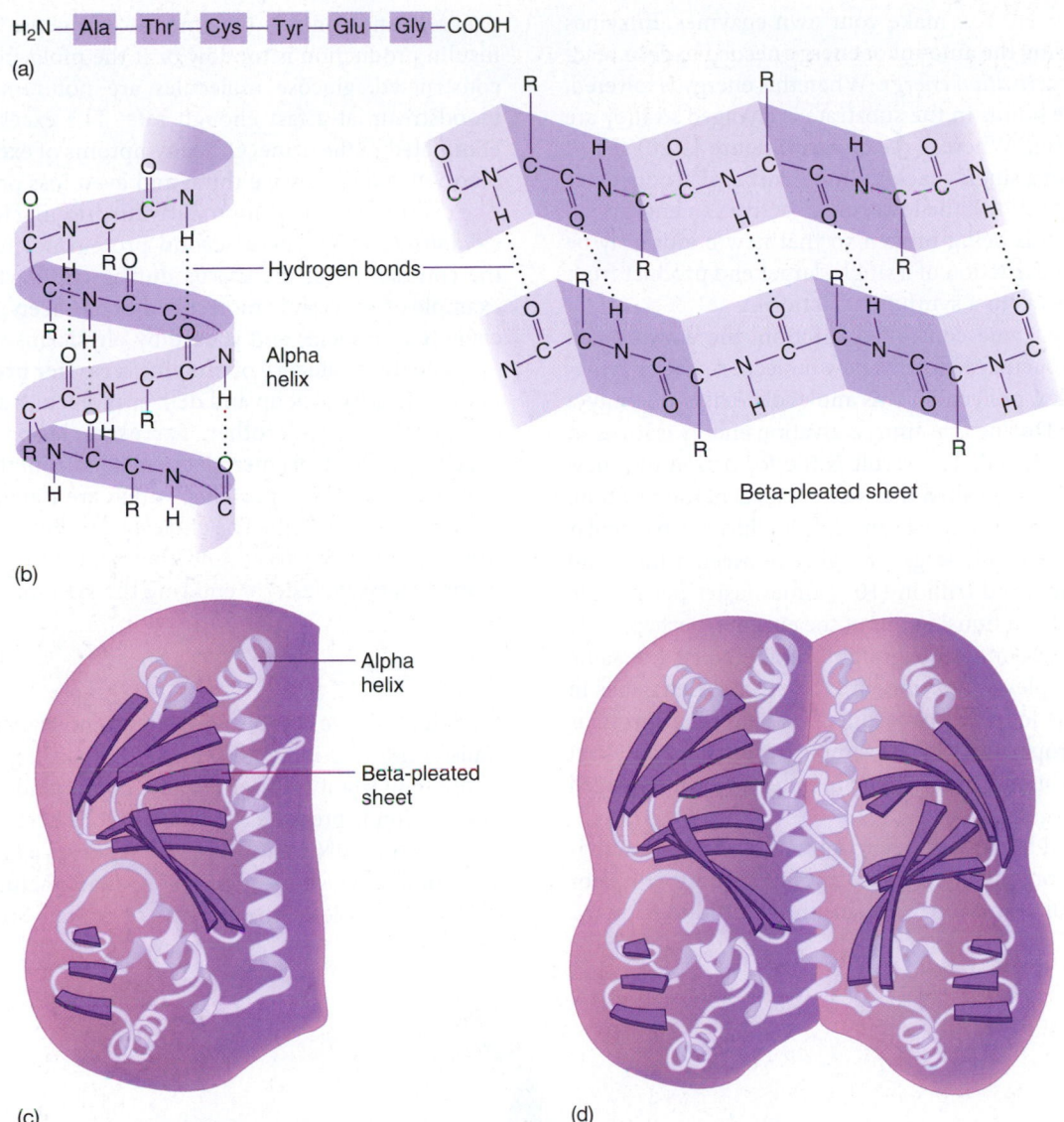

H₂N—$\boxed{\text{Ala}}$—$\boxed{\text{Thr}}$—$\boxed{\text{Cys}}$—$\boxed{\text{Tyr}}$—$\boxed{\text{Glu}}$—$\boxed{\text{Gly}}$—COOH

(a)

Hydrogen bonds

Alpha helix

Beta-pleated sheet

(b)

Alpha helix

Beta-pleated sheet

(c)   (d)

FIGURE 19.19 (*A*) The structure of a protein molecule can be as simple as a list of its amino acids in the order in which they occur. (*B*) This shows the next level of complexity (i.e., how one part of the molecule is attached to another part of the same molecule). (*C*) If already folded parts of a single molecule attach at other places, the molecule displays an even more complex structure. (*D*) The most complex structural arrangements are displayed by protein molecules that are the result of two or more separate molecules combining into one giant macromolecule.

example. For protection, such medications may be stored in brown-colored bottles or kept under refrigeration.

The thousands of kinds of proteins can be placed into three categories. Some proteins are important for maintain-

ing the shape of cells and organisms; they are usually referred to as **structural proteins.** The proteins that make up the cell membrane, muscle cells, tendons, and blood cells are examples of structural proteins. The protein collagen is found throughout the human body and gives tissues shape, support, and strength. The second category of proteins, **regulator proteins,** helps determine what activities will occur in the organism. These regulator proteins include *enzymes, chaperones,* and some *hormones.* These molecules help control the chemical activities of cells and organisms. An enzyme is a protein molecule that acts as a catalyst to speed the rate of a reaction. All *catalysts* are chemicals that speed chemical reactions without increasing the temperature and are not used up in the reaction. Enzymes can be used over and over again until they are worn out or broken. The production of these protein catalysts is under the direct control of an organism's genetic material (DNA). The instructions for the manufacture of all enzymes are found in the genes of the cell. You make your own enzymes. Enzymes operate by lowering the amount of energy needed to get a reaction going—the *activation energy*. When this energy is lowered, the nature of the bonds in the substrate is changed so they are more easily broken. Whereas the cartoon (figure 19.20) shows the breakdown of a single reactant into many end products (as in a hydrolysis reaction), the lowering of activation energy can also result in bonds being broken so that new bonds may be formed in the construction of a single, larger end product from several reactants (as in a synthesis reaction).

During an enzyme-controlled reaction, the enzyme and substrate come together to form a new molecule—the enzyme-substrate complex molecule. This molecule exists for only a very short time. During that time, activation energy is lowered and bonds are changed. The result is the formation of a new molecule or molecules called the end products of the reaction. The number of jobs an enzyme can perform during a particular time period is incredibly large—ranging between a thousand (10^3) and ten thousand trillion (10^{16}) times faster per minute than uncatalyzed reactions! Without the enzyme, perhaps only 50 or 100 substrate molecules might be altered in the same time. Some examples of enzymes are the digestive enzymes in the stomach. The job of a chaperone is to help other proteins fold into their proper shape. For example, some chaperones act as heat shock proteins; that is, they help repair heat damaged proteins.

Enzymes and hormones help control the chemical activities of cells and organisms. Two hormones that are regulator proteins are insulin and oxytocin. Insulin is produced by the pancreas and controls the amount of glucose in the blood. If insulin production is too low or if the molecule is improperly constructed, glucose molecules are not removed from the bloodstream at a fast enough rate. The excess sugar is then eliminated in the urine. Other symptoms of excess sugar in the blood include excessive thirst and even loss of consciousness. The disease caused by improperly functioning insulin is known as *diabetes.* Oxytocin, a second protein hormone, stimulates the contraction of the uterus during childbirth. It is also an example of an organic molecule that has been produced artificially (e.g., pitocin) and is used by physicians to induce labor.

The third category of proteins is **carrier proteins.** Proteins in this category pick up and deliver molecules at one place and transport them to another. For example, proteins regularly attach to cholesterol entering the system from the diet, forming molecules called lipoproteins, which are transported through the circulatory system. The cholesterol is released at a distance from the digestive tract, and the proteins return to pick up more dietary cholesterol entering the system.

Nucleic Acids

Nucleic acids are complex polymeric molecules that store and transfer information within a cell. This section is an overview of this important class of organic compounds. More detailed information is presented in chapter 26. There are two types of nucleic acids, DNA and RNA. DNA serves as genetic material determining which proteins will be manufactured, while RNA plays a vital role in the process of protein manufacture. All

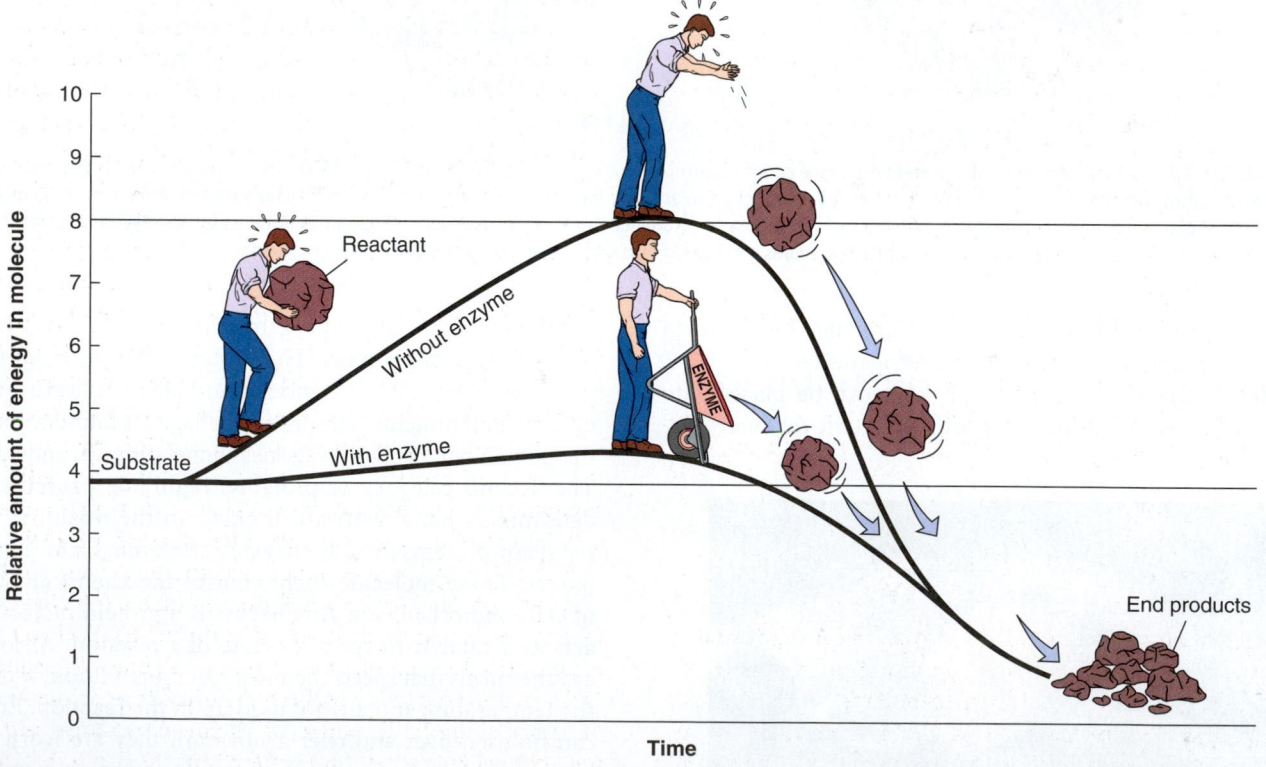

FIGURE 19.20 The enzyme in this figure is represented by the wheelbarrow. Notice how much less energy is required to move the rock (substrate) to its destination (end products). In addition, the wheelbarrow can be reused again and again.

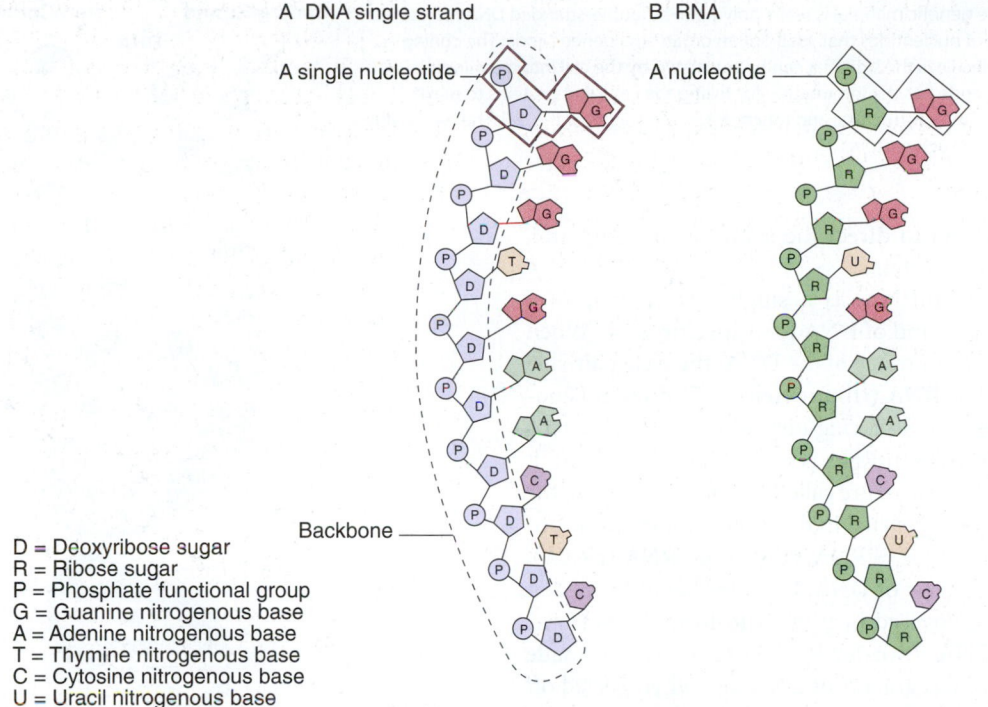

A DNA single strand   B RNA

A single nucleotide

A nucleotide

Backbone

D = Deoxyribose sugar
R = Ribose sugar
P = Phosphate functional group
G = Guanine nitrogenous base
A = Adenine nitrogenous base
T = Thymine nitrogenous base
C = Cytosine nitrogenous base
U = Uracil nitrogenous base

FIGURE 19.21   (*A*) A single strand of DNA is a polymer composed of nucleotides. Each nucleotide (framed at the top of the molecule) consists of deoxyribose sugar, phosphate, and one of four nitrogenous bases: A, T, G, or C. Notice the backbone of blue-highlighted sugar and phosphate. (*B*) RNA is also a polymer, but each nucleotide (framed at the top of this RNA molecule) is composed of ribose sugar, phosphate, and one of four nitrogenous bases: A, U, G, or C. The backbone of ribose and phosphate is highlighted in green.

nucleic acids are constructed of fundamental monomers known as **nucleotides.** Each nucleotide is composed of three parts: (1) a five-carbon simple sugar molecule that may be ribose or deoxyribose, (2) a phosphate group, and (3) a nitrogenous base. There are five types of nitrogenous bases. Two of the bases are the larger, double-ring molecules, *Adenine* and *Guanine*. The smaller bases are the single-ring bases, *Thymine, Cytosine,* and *Uracil.* Nucleotides (monomers) are linked together in long sequences (polymers) so that the sugar and phosphate sequence forms a "backbone" and the nitrogenous bases stick out to the side. DNA has deoxyribose sugar and the bases A, T, G, and C, while RNA has ribose sugar and the bases A, U, G, and C (figure 19.21).

DNA (*deoxyribonucleic acid*) is composed of two strands to form a ladderlike structure thousands of nucleotide bases long. The two strands are attached between their bases according to *the base pair rule;* that is, *Adenine* from one strand always pairs with *Thymine* from the other. *Guanine* always pairs with *Cytosine.*

A T (or A U) and G C

One strand of DNA is called the *coding strand* because it has a meaningful genetic message written using the nitrogenous bases as letters (e.g., the base sequence CATTAGACT) (figure 19.22). This is the basis of the genetic code for all organisms. If these bases are read in groups of three, they make sense to us (i.e., "cat," "tag," and "act"). The opposite strand is called *noncoding* since it makes no "sense" but protects the coding strand from chemical and physical damage. Both strands are twisted into a helix, that is, a molecule turned around a tubular space like a coiled spring.

The information carried by DNA can be compared to the information in a textbook. Books are composed of words (constructed from individual letters) in particular combinations, organized into chapters. In the same way, DNA is composed of tens of thousands of nucleotides in specific sequences (words) organized into genes (a chapter). Each gene carries the information for producing a protein, just as each chapter carries the information relating to one idea. The order of nucleotides in a gene is directly related to the order of amino acids in a protein. Just as chapters in a book are identified by beginning and ending statements, different genes along a DNA strand have beginning and ending signals. They tell when to start and when to stop reading a particular gene. Human body cells contain forty-six strands (books) of helical DNA, each containing thousands of genes (chapters). These strands are called **chromosomes** when they become supercoiled in preparation for cellular reproduction. Before cell reproduction, the DNA makes copies of the coding and noncoding strands, ensuring that the offspring or *daughter cells* will each receive a full complement of the genes required for their survival. Each chromosome is comprised of a sequence of genes. A **gene** is a segment of DNA that is able to (1) replicate by directing the manufacture of copies of itself; (2) mutate, or chemically change, and transmit these changes to future generations; (3) store information that determines the characteristics of cells and organisms; and

FIGURE 19.22 The genetic material is really polymers of double-stranded DNA molecules comprised of sequences of nucleotides that spell out an organism's genetic code. The coding strand of the double molecule is the side that can be translated by the cell into meaningful information. The genetic code has the information for telling the cell what proteins to make, which in turn become the major structural and functional components of the cell. The noncoding strand is unable to code for such proteins.

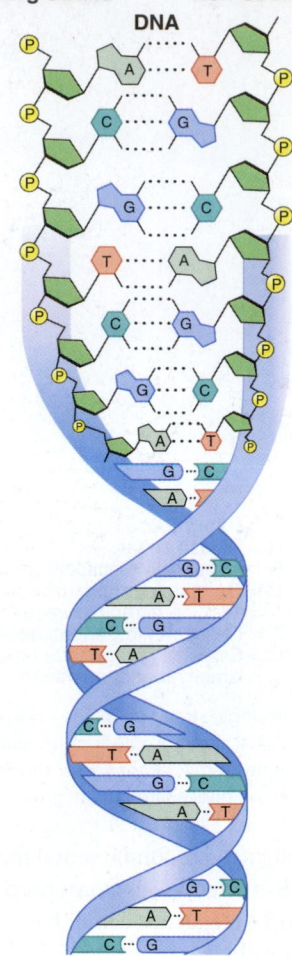

Coding Strand   Non-Coding Strand

DNA

(4) use this information to direct the synthesis of structural, carrier, and regulatory proteins.

Messenger RNA (mRNA) is a single-strand copy of a portion of the coding strand of DNA for a specific gene. When mRNA is formed on the surface of the DNA, the base pair rule applies. However, since **RNA (*ribonucleic acid*)** does not contain thymine, it uses a U-A pairing instead of the T-A pairing between nucleotides. After mRNA is formed and peeled off, it associates with a cellular structure called the *ribosome* where the genetic message can be translated into a protein molecule. Ribosomes contain another type of RNA, **ribosomal RNA (rRNA).** rRNA is also an RNA copy of DNA, but after being formed, it becomes twisted and covered in protein to form a ribosome. The third form of RNA, **transfer RNA (tRNA),** is also made from copies of different segments of DNA, but when peeled off the surface, each takes the form of a cloverleaf. tRNA molecules are responsible for transferring or carrying specific amino acids to the ribosome where all three forms of RNA come together and cooperate in the manufacture of protein molecules.

Lipids

We generally call molecules in this group the **fats.** They are not polymers, as are the previously discussed carbohydrates, proteins, and nucleic acids. However, there are three different types of **lipids:** *true fats* (pork chop fat or olive oil), *phospholipids* (the primary component of cell membranes), and *steroids* (some hormones). In general, lipids are large, nonpolar (do not have a positive end and a negative end), organic molecules that do not easily dissolve in polar solvents such as water. For example, nonpolar vegetable oil molecules do not dissolve in polar water molecules. Lipids are soluble in nonpolar substances such as ether or acetone. Just like carbohydrates, the lipids are composed of carbon, hydrogen, and oxygen. Lipids generally have

---

CONCEPTS APPLIED

You Learned About Monomers And Polymers in Preschool

The concepts of monomers, polymers, and how they interrelate are extremely important and fundamental to understanding the organic compounds important to life. You may have worked with these concepts as early as preschool when you attached colored paper "links" (monomers) together to form a decorative chain (polymer). Return to that experience and create polymeric molecules that better represent proteins, carbohydrates, and nucleic acids as described in this chapter. Label the monomers used to make each type of polymer appropriately.

---

very small amounts of oxygen in comparison to the amounts of carbon and hydrogen.

True (Neutral) Fats

True (neutral) fats are important, complex organic molecules that are used to provide, among other things, energy. The building blocks of a fat are a glycerol molecule and fatty acids. Recall that **glycerol** is a carbon skeleton that has three alcohol groups attached to it. Its chemical formula is $C_3H_5(OH)_3$. At room temperature, glycerol looks like clear, lightweight oil. It is used under the name glycerin as an additive to many cosmetics to make them smooth and easy to spread.

$$
\begin{array}{ccc}
\text{OH} & \text{OH} & \text{OH} \\
| & | & | \\
\text{H}-\text{C}-\text{C}-\text{C}-\text{H} \\
| & | & | \\
\text{H} & \text{H} & \text{H}
\end{array}
$$

Glycerol

The **fatty acids** found in fats are a long-chain carbon skeleton that has a carboxylic acid functional group. If the carbon skeleton has as much hydrogen bonded to it as possible, we call it *saturated*. The saturated fatty acid in figure 19.23A is stearic

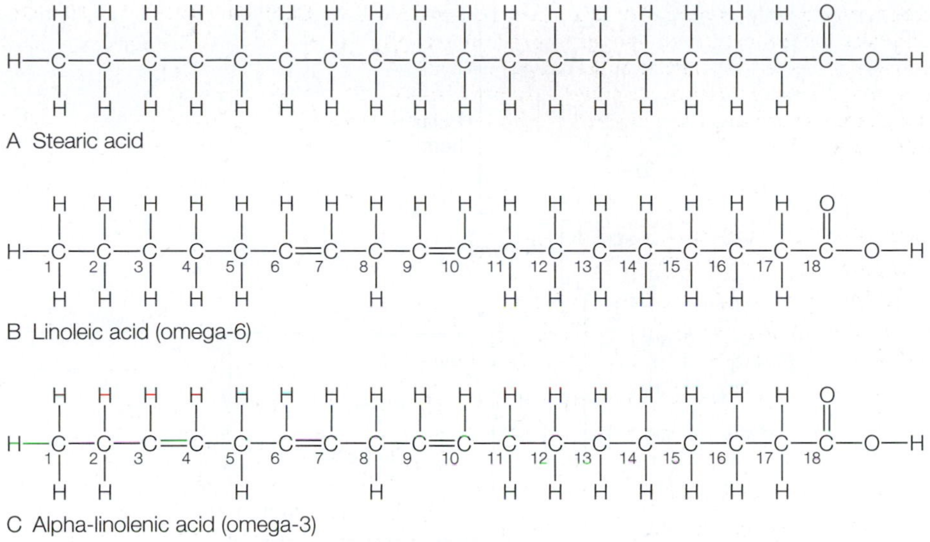

The structures of fatty acids are shown in the top figure.

A Stearic acid

B Linoleic acid (omega-6)

C Alpha-linolenic acid (omega-3)

FIGURE 19.23 Structure of saturated and unsaturated fatty acids. (*A*) Stearic acid is an example of a saturated fatty acid. (*B*) Linoleic acid is an example of an unsaturated fatty acid. It is technically an omega-6 fatty acid because the first double bond occurs at carbon number six. (*C*) An omega-3 fatty acid, linolenic acid. Both linoleic and linolenic acids are essential fatty acids for humans.

acid, a component of solid meat fats such as found in bacon. Notice that at every point in this structure, the carbon has as much hydrogen as it can hold. Saturated fats are generally found in animal tissues and tend to be solids at room temperatures. Some examples of animal products containing saturated fats are butter, whale blubber, suet, and lard.

If the carbons in a fatty acid are double-bonded to each other at one or more points, the fatty acid is said to be *unsaturated*. Unsaturated fats are frequently plant fats or oils, and they are usually liquids at room temperature. Peanut, corn, and olive oil are examples of unsaturated fats that are mixtures of different true fats. When glycerol and three fatty acids are combined, a fat is formed. This reaction is almost exactly the same as the reaction that causes simple sugars to bond together.

Fats are important molecules for storing energy. There is more than twice as much energy in a gram of fat as in a gram of sugar, 9 Calories versus 4 Calories. This is important to an organism because fats can be stored in a relatively small space

and still yield a high amount of energy. Fats in animals also provide protection from heat loss. The thick layer of blubber in whales, walruses, and seals prevents the loss of internal body heat to the cold, watery environment in which they live. This same layer of fat, together with the fat deposits around some internal organs—such as the kidneys and heart—serves as a cushion that protects these organs from physical damage. If a fat is formed from a glycerol molecule and three attached fatty acids, it is called a **triglyceride;** if two, a *diglyceride;* and if one, a *monoglyceride* (figure 19.24). Triglycerides account for about 95 percent of the fat stored in human tissue.

Phospholipids

Phospholipids are a class of complex water-insoluble organic molecules that resemble fats but contain a phosphate group (PO_4) in their structure (figure 19.25). One of the reasons phospholipids are important is that they are a major component of

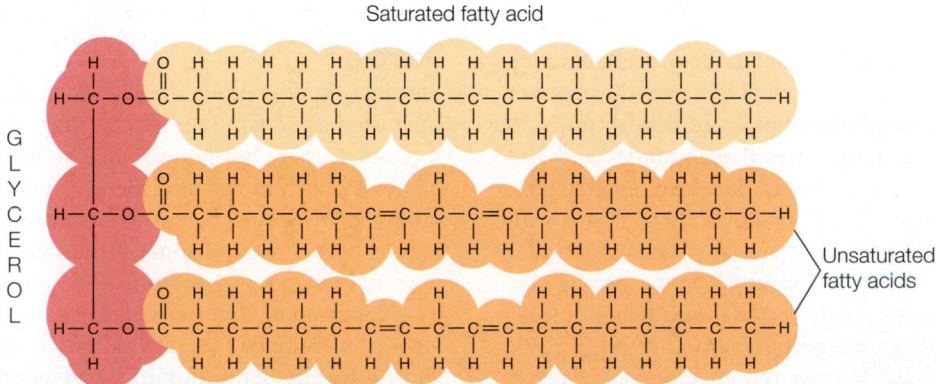

FIGURE 19.24 The arrangement of the three fatty acids attached to a glycerol molecule is typical of the formation of a fat. The structural formula of the fat appears to be very cluttered until you dissect the fatty acids from the glycerol; then it becomes much more manageable. This example of a triglyceride contains a glycerol molecule, two unsaturated fatty acids (linoleic acid), and a third saturated fatty acid (stearic acid).

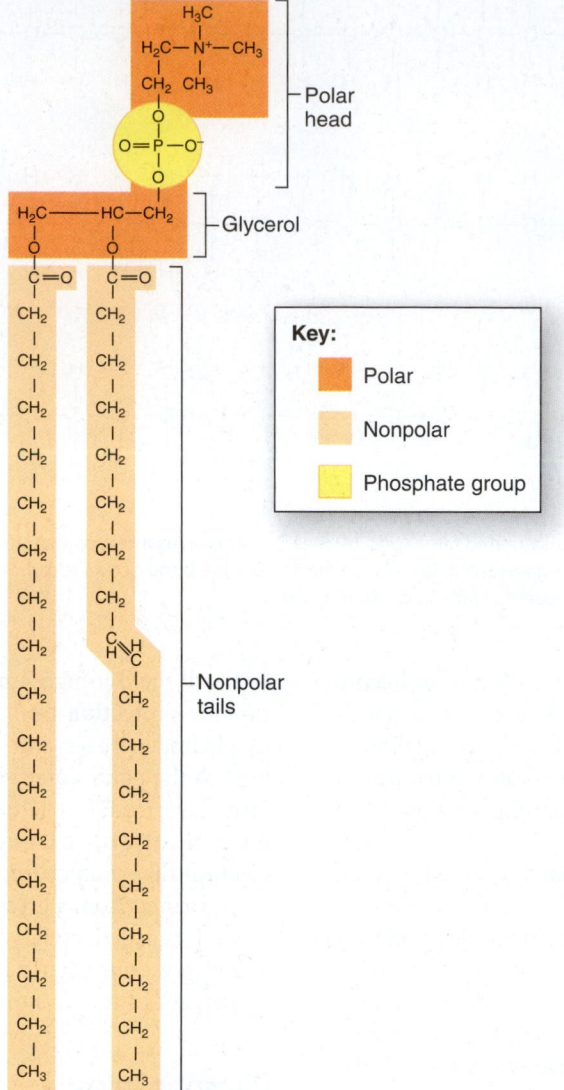

FIGURE 19.25 This molecule is similar to a fat but has a phosphate (yellow) group in its structure. You can think of phospholipid molecules as having a "head" with two strings dangling down. The head portion is the glycerol and phosphate group, which is polar and soluble in water. The strings are the fatty acid segments of the molecule, which are nonpolar.

membranes in cells. Without these lipids in our membranes, the cell contents would not be separated from the exterior environment. Some of the phospholipids are better known as the *lecithins*. Lecithins are found in cell membranes and also help in the emulsification of fats; that is, they help separate large portions of fat into smaller units. This allows the fat to mix with other materials. Lecithins are added to many types of food for this purpose (chocolate bars, for example). Some people take lecithin as nutritional supplements because they believe it leads to healthier hair and better reasoning ability; but once inside your intestines, lecithins are destroyed by enzymes, just like any other phospholipid.

Steroids

Steroids, another group of lipid molecules, are characterized by their arrangement of interlocking rings of carbon. Many steroid molecules are sex hormones. Some of them regulate reproductive processes, such as egg and sperm production (see chapter 21); others regulate such things as salt concentration in the blood. Figure 19.26 illustrates some of the steroid compounds, such as testosterone and progesterone, that are typically manufactured by organisms and in the laboratory as pharmaceuticals. We have already mentioned one steroid molecule: cholesterol. Serum cholesterol (the kind found in your blood and associated with lipoproteins) has been implicated in many cases of athero-

Omega Fatty Acids and Your Diet

The occurrence of a double bond in a fatty acid is indicated by the Greek letter ω (omega) followed by a number indicating the location of the first double bond in the molecule. Oleic acid, one of the fatty acids found in olive oil, is comprised of 18 carbons with a single (1) double bond between carbons 9 and 10. Therefore, it is chemically designated C18:1ω9 and is a monounsaturated fatty acid. This fatty acid is commonly referred to as an omega-9 fatty acid. The unsaturated fatty acid in figure 19.23B is linoleic acid, a component of sunflower and safflower oils. Notice that there are two double bonds between the carbons and fewer hydrogens than in the saturated fatty acid. Linoleic acid is chemically a polyunsaturated fatty acid with two double (2) bonds and is designated C18:2ω6, an omega-6 fatty acid. This indicates that the first double bond of this 18-carbon molecule is between carbons 6 and 7. Since the human body cannot make this fatty acid, it is called an *essential fatty acid* and must be taken in as a part of the diet. The other essential fatty acid, linolenic acid, is C18:3ω3 and has three (3) double bonds. This fatty acid is commonly referred to as an omega-3 fatty acid. One key function of these essential fatty acids is the synthesis of prostaglandin hormones that are necessary in controlling cell growth and specialization.

Sources of Omega-3 Fatty Acids	Sources of Omega-6 Fatty Acids
Certain fish oil (salmon, sardines, herring)	Corn oil
	Peanut oil
Flaxseed oil	Cottonseed oil
Soybeans	Soybean oil
Soybean oil	Sesame oil
Walnuts	Safflower oil
Walnut oil	Sunflower oil

sclerosis. However, your body makes this steroid for use as a component of cell membranes. Cholesterol is necessary for the manufacture of vitamin D, which assists in the proper development of bones and teeth. Vitamin D not only plays an important role in controlling bone disorders, but has also been shown to aid in the control of autoimmune disorders, cancers, and cardiovascular diseases. Cholesterol molecules in your skin react with ultraviolet light to produce vitamin D. The body also uses it to make bile acids. These products of the liver are channeled into your intestine to emulsify fats.

Regulating the amount of cholesterol in the body to prevent its negative effects can be difficult, because the body makes it and it is consumed in the diet. Recall that diets high in saturated fats increase the risk for diseases such as atherosclerosis. By watching your diet, you can reduce the amount of cholesterol in your blood serum by about 20 percent, as much as taking a cholesterol-lowering drug. Therefore, it is best to eat foods that are low in cholesterol. Because many foods that claim to be low- or no-cholesterol have high levels of saturated fats, they should also be avoided in order to control serum cholesterol levels.

CONCEPTS APPLIED

Nutrients in Your Pantry

Pick a food product and write down the number of grams of fats, proteins, and carbohydrates per serving according to the information on the label. Multiply the number of grams of proteins and carbohydrates each by 4 Cal/g and the number of grams of fat by 9 Cal/g. Add the total Calories per serving. Does your total agree with the number of Calories per serving given on the label? Also examine the given serving size. Is this a reasonable amount to be consumed at one time, or would you probably eat two or three times this amount? Write the rest of the nutrition information (vitamins, minerals, sodium content, etc.) and then the list of ingredients. Tell what ingredient you think is providing which nutrient. (Example: vegetable oil—source of fat; milk—provides calcium and vitamin A; MSG—source of sodium.)

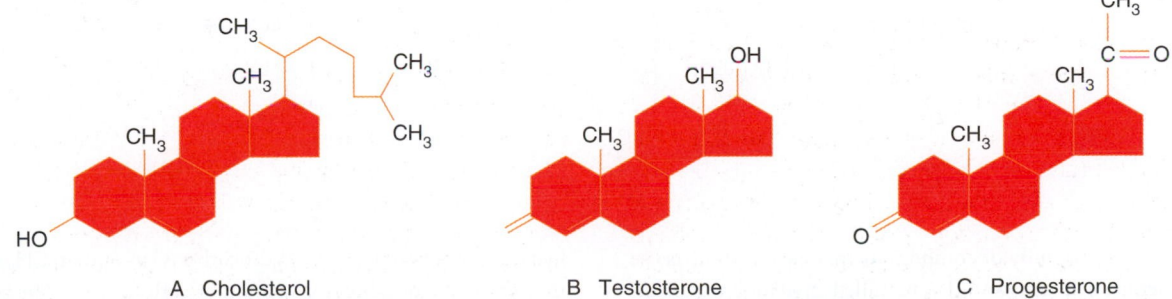

FIGURE 19.26 (*A*) Cholesterol is produced by the human body and is found in your cells' membranes. (*B*) Testosterone increases during puberty, causing the male sex organs to mature. (*C*) Progesterone is a female sex hormone produced by the ovaries and placenta. Notice the slight structural differences among these molecules.

A Closer Look

Fat and Your Diet

When triglycerides are eaten in fat-containing foods, digestive enzymes hydrolyze them into glycerol and fatty acids. These molecules are absorbed by the intestinal tract and coated with protein to form lipoprotein, as shown in the accompanying diagram. The combination allows the fat to dissolve better in the blood so that it can move throughout the body in the circulatory system.

Five types of lipoproteins (box figure 19.2) found in the body are:

1. Chylomicrons
2. Very-low-density lipoproteins (VLDLs)
3. Low-density lipoproteins (LDLs)
4. High-density lipoproteins (HDLs)
5. Lipoprotein a [Lp(a)]

Chylomicrons are very large particles formed in the intestine; they are 80–95 percent triglycerides. As the chylomicrons circulate through the body, cells remove the triglycerides in order to make sex hormones, store energy, and build new cell parts. When most of the triglycerides have been removed, the remaining portions of the chylomicrons are harmlessly destroyed.

The VLDLs and LDLs are formed in the liver. VLDLs contain all types of lipid, protein, and 10–15 percent cholesterol, whereas the LDLs are about 50 percent cholesterol. As with the chylomicrons, the body uses these molecules for the fats they contain. However, in some people, high levels of LDL and lipoprotein a [Lp(a)] in the blood are associated with atherosclerosis, stroke, and heart attack. It appears that saturated fat disrupts the clearance of LDLs from the bloodstream. Thus, while in the blood, LDLs may stick to the insides of the vessels, forming deposits, which restrict blood flow and

contribute to high blood pressure, strokes, and heart attacks. Even though they are 30 percent cholesterol, a high level of HDLs (made in the intestine), compared with LDLs and [Lp(a)], is associated with a lower risk for atherosclerosis. One way to reduce the risk of this disease is to lower your intake of LDLs and [Lp(a)]. This can be done by reducing your consumption of saturated fats. An easy way to remember the association between LDLs and HDLs is "L = Lethal" and "H =Healthy" or "Low = Bad" and "High = Good." The federal government's cholesterol guidelines recommend that all adults get a full lipoprotein profile (total cholesterol, HDL, LDL, and triglycerides) once every five years. They also recommend a sliding scale for desirable LDL levels; however, recent studies suggest that one's LDL level should be as low as possible.

Taking certain drugs is one way to control the level of lipoproteins. Statins are a group of medicines (e.g., simvastatin, atorvastatin) that work by blocking the action of enzymes that control the rate of cholesterol production in the body. Their use can lower cholesterol 20-60 percent. They also increase the liver's ability to remove low-density *lipoproteins*. An additional benefit is a slight increase in high-density lipoproteins and a decrease in triglycerides. There are several ways to increase the level of "good" cholesterol without taking medications. HDL levels will increase if a person

1. Quits smoking.
2. Exercises.
3. Loses weight.
4. Reduces the amount of trans fats in the diet.

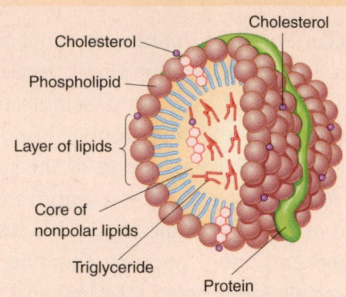

Box Figure 19.2   Diagram of a lipoprotein.

5. Eats more whole grain food instead of "refined" sugars.
6. Eats food rich in niacin (vitamin B$_3$).

Total cholesterol goal values:

- 75–169 mg/dL (milligram per deciliter) for those age twenty and younger
- 100–199 mg/dL for those over age twenty-one

Low-density lipoprotein (LDL) goal values:

- Less than 70 mg/dL for those with heart or blood vessel disease and for other patients at very high risk of heart disease (those with metabolic syndrome)
- Less than 100 mg/dL for high-risk patients (for example, some patients who have diabetes or multiple heart disease risk factors)
- Less than 130 mg/dL otherwise

Very-low-density lipoprotein (VLDL) goal values:

- Less than 40 mg/dL otherwise

High-density lipoprotein (HDL) goal value:

- Greater than 45 mg/dL (the higher the better)

Triglyceride goal value:

- Less than 150 mg/dL

SUMMARY

Organic chemistry is the study of compounds that have carbon as the principal element. Such compounds are called *organic compounds*, and all the rest are *inorganic compounds*. There are millions of organic compounds because a carbon atom can link with other carbon atoms as well as atoms of other elements.

A *hydrocarbon* is an organic compound consisting of hydrogen and carbon atoms. The simplest hydrocarbon is one carbon atom and four

hydrogen atoms, or CH_4. All hydrocarbons larger than CH_4 have one or more carbon atoms bonded to another carbon atom. The bond can be single, double, or triple, and may be found as chains or rings of carbon.

One category of hydrocarbons has single carbon-to-carbon bonds, a second category has a double carbon-to-carbon bond, and the third has a triple carbon-to-carbon bond. These three categories can have straight- or branched-chain molecules. When the number of carbon atoms is greater than three, different arrangements can occur for a particular number of carbon atoms. Molecules with different arrangements with the same molecular formula are called *isomers*. Isomers have different physical properties, so each isomer is given its own name.

Hydrocarbons that have all the hydrogen atoms possible are *saturated* hydrocarbons. Those that can add more hydrogen to the molecule are called *unsaturated* hydrocarbons. Unsaturated hydrocarbons are more chemically reactive than saturated molecules.

Hydrocarbons that occur in a ring or cycle structure are cyclohydrocarbons. A six-carbon cyclohydrocarbon with three double bonds has different properties than the other cyclohydrocarbons because the double bonds are not localized. This six-carbon molecule is *benzene*, the basic unit of the *aromatic hydrocarbons*.

Petroleum is a mixture of many kinds of hydrocarbons that formed from the slow decomposition of buried marine plankton and algae. Petroleum from the ground, or *crude oil*, is distilled into petroleum products of *natural gas, LPG, petroleum ether, gasoline, kerosene, diesel fuel*, and *motor oils*. Each group contains a range of hydrocarbons and is processed according to use.

In addition to oxidation, hydrocarbons react by *substitution, addition*, and *polymerization* reactions. Reactions take place at sites of multiple bonds or lone pairs of electrons on the *functional groups*. The functional group determines the chemical properties of organic compounds. Changes in functional groups result in the *hydrocarbon derivatives* of *alcohols, ethers, aldehydes, ketones, organic acids, esters*, and *amines*.

Polymers occur naturally in plants and animals, and *synthetic polymers* are made today from variations of the ethylene-derived monomers. Among the more widely used synthetic polymers derived from ethylene are polyethylene, polyvinyl chloride, polystyrene, and Teflon. Problems with the synthetic polymers include that (1) they are manufactured from fossil fuels that are also used as the primary energy supply and (2) they do not readily decompose and tend to accumulate in the environment.

Living organisms have an incredible number of highly organized chemical reactions that are catalyzed by *enzymes*, using food and energy to grow and reproduce. These biochemical processes involve building large *macromolecules* such as proteins, carbohydrates, and lipids.

Carbohydrates are a class of organic molecules composed of CHO. The *monosaccharides* are simple sugars such as *glucose* and *fructose*. Glucose is *blood sugar*, a source of energy. The disaccharides are *sucrose* (table sugar), *lactose* (milk sugar), and *maltose* (malt sugar). The polysaccharides are polymers or glucose in straight or branched chains used as a near-term source of stored energy. Plants store the energy in the form of *starch*, and animals store it in the form of *glycogen*. *Cellulose* is a polymer similar to starch that humans cannot digest.

Proteins are macromolecular polymers of *amino acids*. There are twenty amino acids that are used in various polymer combinations to build structural, carrier, and functional proteins.

Nucleic acids are polymers composed of nucleotide units. Two forms are recognized, DNA and RNA. In most organisms, DNA serves as the genetic material, while RNA plays crucial roles in the synthesis of proteins.

Lipids are a fourth important class of biochemicals. *True* or *neutral fats* and oils belong in the category known as *lipids* and are esters formed from three fatty acids and glycerol into a *triglyceride*. True fats are usually solid triglycerides associated with animals, and *oils* are liquid triglycerides associated with plant life, but both represent a high-energy storage material. The other two subgroups are the *phospholipids* used in cell membranes, and *steroids*, which primarily serve as hormones.

KEY TERMS

amino acid (p. **451**)
antibody (p. **451**)
carbohydrates (p. **449**)
carrier proteins (p. **454**)
cellulose (p. **451**)
chromosomes (p. **455**)
complex carbohydrate (p. **449**)
denatured (p. **452**)
deoxyribonucleic acid (DNA) (p. **455**)
fats (p. **456**)
fatty acids (p. **456**)

functional group (p. **442**)
gene (p. **455**)
glycerol (p. **456**)
glycogen (p. **451**)
hormone (p. **451**)
hydrocarbon (p. **436**)
hydrocarbon derivatives (p. **441**)
inorganic chemistry (p. **436**)
isomers (p. **437**)
lipid (p. **456**)
macromolecule (p. **449**)

messenger RNA (mRNA) (p. **456**)
monomer (p. **442**)
nucleic acid (p. **454**)
nucleotide (p. **455**)
oil field (p. **440**)
organic acids (p. **444**)
organic chemistry (p. **436**)
petroleum (p. **439**)
phospholipid (p. **457**)
polymer (p. **442**)
polypeptide chain (p. **452**)

proteins (p. **451**)
regulator proteins (p. **453**)
ribonucleic acid (RNA) (p. **456**)
ribosomal RNA (rRNA) (p. **456**)
saturated (p. **438**)
starch (p. **451**)
steroids (p. **458**)
structural proteins (p. **453**)
transfer RNA (tRNA) (p. **456**)
triglyceride (p. **457**)
true (neutral) fats (p. **456**)
unsaturated (p. **438**)

APPLYING THE CONCEPTS

Answers are located in appendix F.

Part I

1. All organic compounds
   a. contain carbon and were formed only by a living organism.
   b. are natural compounds that have not been synthesized.
   c. contain carbon, no matter if they were formed by a living thing or not.
   d. were formed by a plant.

2. There are millions of organic compounds but only thousands of inorganic compounds because
   a. organic compounds were formed by living things.
   b. there is more carbon on Earth's surface than any other element.
   c. atoms of elements other than carbon never combine with themselves.
   d. carbon atoms can combine with up to four other atoms, including other carbon atoms.

3. A hydrocarbon molecule with 4 carbon atoms would be called saturated if it had how many hydrogen atoms?
   a. 4
   b. 8
   c. 10
   d. 16

4. Isomers are compounds with the same
   a. molecular formula with different structures.
   b. molecular formula with different atomic masses.
   c. atoms but different molecular formulas.
   d. structures but different formulas.

5. The hydrocarbons with a double covalent carbon-carbon bond are called
   a. a polymer.
   b. unsaturated.
   c. petroleum.
   d. None of the above is correct.

6. Petroleum is believed to have formed mostly from the anaerobic decomposition of buried
   a. dinosaurs.
   b. fish.
   c. pine trees.
   d. plankton and algae.

7. Ethylene molecules can add to each other in a reaction to form a long chain called a
   a. monomer.
   b. dimer.
   c. trimer.
   d. polymer.

8. Chemical reactions usually take place on an organic compound at the site of a
   a. double bond.
   b. lone pair of electrons.
   c. functional group.
   d. Any of the above is correct.

9. The R in ROH represents
   a. a functional group.
   b. a hydrocarbon group with a name ending in "-yl."
   c. an atom of an inorganic element.
   d. a polyatomic ion that does not contain carbon.

Part II

10. $-NH_2$ is a specific combination of atoms attached to a carbon skeleton known as
    a. −enes.
    b. hydrocarbon derivative.
    c. alcoholic group.
    d. amino group.

11. A protein is a polymer formed from the linking of many
    a. glucose units.
    b. DNA molecules.
    c. amino acid molecules.
    d. monosaccharides.

12. Which of the following is *not* converted to blood sugar by the human body?
    a. lactose
    b. dextrose
    c. cellulose
    d. glycogen

13. Many synthetic polymers become a problem in the environment because they
    a. decompose to nutrients, which accelerates plant growth.
    b. do not readily decompose and tend to accumulate.
    c. do not contain vitamins as natural materials do.
    d. become a source of food for fish, but ruin the flavor of fish meat.

14. What role is played by tRNA in cells?
    a. telling the cell which protein to make.
    b. transferring a specific amino acid to a ribosome for incorporation into a protein.
    c. transferring a copy of the DNA gene to a ribosome for protein synthesis.
    d. terminating protein synthesis.

15. What is the job of a chaperone protein?
    a. Transferring an amino acid to the ribosome for protein synthesis.
    b. Making sure proteins are synthesized.
    c. Helping other proteins fold into their proper shape.
    d. Keeping mutant proteins out of trouble.

QUESTIONS FOR THOUGHT

Part I

1. What is an organic compound?
2. What features allow organic molecules to form millions of compounds?
3. Is it possible to have an isomer of ethane? Explain.
4. Suggest a reason that ethylene is an important raw material used in the production of plastics but ethane is not.
5. Why are organic molecules three-dimensional?
6. What are (a) natural gas, (b) LPG, and (c) diesel fuel?
7. What does the octane number of gasoline describe? On what is the number based?
8. What is a polymer? Give an example of a naturally occurring plant polymer. Give an example of a synthetic polymer.

Part II

9. Give an example of each of the following classes of organic molecules: carbohydrates, proteins, nucleic acids, lipids.
10. What is a functional group? List four examples.
11. Draw a structural formula for the fatty acid linoleic acid. What feature makes this a saturated fat?
12. A soft drink is advertised to "contain no sugar." The label lists ingredients of carbonated water, dextrose, corn syrup, fructose, and flavorings. Evaluate the advertising.
13. What is the difference between a micromolecule and a macromolecule?
14. What are the key differences between DNA and RNA?
15. List the main roles played by proteins in a cell.

FOR FURTHER ANALYSIS

1. Many people feel that by using a higher-octane gasoline in their vehicles, they will get better performance. What is the difference between high-octane and regular gasoline? Does scientific evidence bear this out? What octane gasoline is recommended for your vehicle?
2. There have been some health concerns about the additives used in gasoline. What are these additives? What purpose do they serve? What do opponents of using such additives propose are their negative impact?
3. The so-called "birth control pill," or "the pill," has been around since the early 1960s. This medication is composed of a variety of organic molecules. What is the nature of these compounds? How do they work to control conception since they do not control birth? What are some of the negative side effects and what are some of the related benefits of taking "the pill"?
4. Many communities throughout the world are involved in recycling. One of the most important classes of materials recycled is plastic. There has been concern that the cost of recycling plastics is higher than the cost of making new plastic products. Go to the Internet and
   a. Identify the kinds of plastics that are commonly recycled.
   b. Explain how a plastic such as HDPE is recycled.
   c. Present an argument for eliminating the recycling of HDPE and for expanding the recycling of HDPE.

INVITATION TO INQUIRY

Alcohols: What Do You Really Know?

Archaeologists, anthropologists, chemists, biologists, and health care professionals agree that the drinking of alcohol dates back thousands of years. Evidence also exists that this practice has occurred in most cultures around the world. Use the Internet to search out answers to the following questions:

1. What is the earliest date for which there is evidence for the production of ethyl alcohol?
2. In which culture did this occur?
3. What is the molecular formula and structure of ethanol?
4. Do alcohol and water mix?
5. How much ethanol is consumed in the form of beverages in the United States each year?
6. What is the legal limit to be considered intoxicated in your state?
7. How is this level measured?
8. Why is there a tax on alcoholic beverages?
9. How do the negative effects of drinking alcohol compare between men and women?
10. Have researchers demonstrated any beneficial effects of drinking alcohol?

Compare what you thought you knew to what is now supported with scientific evidence.

20

The Nature of Living Things

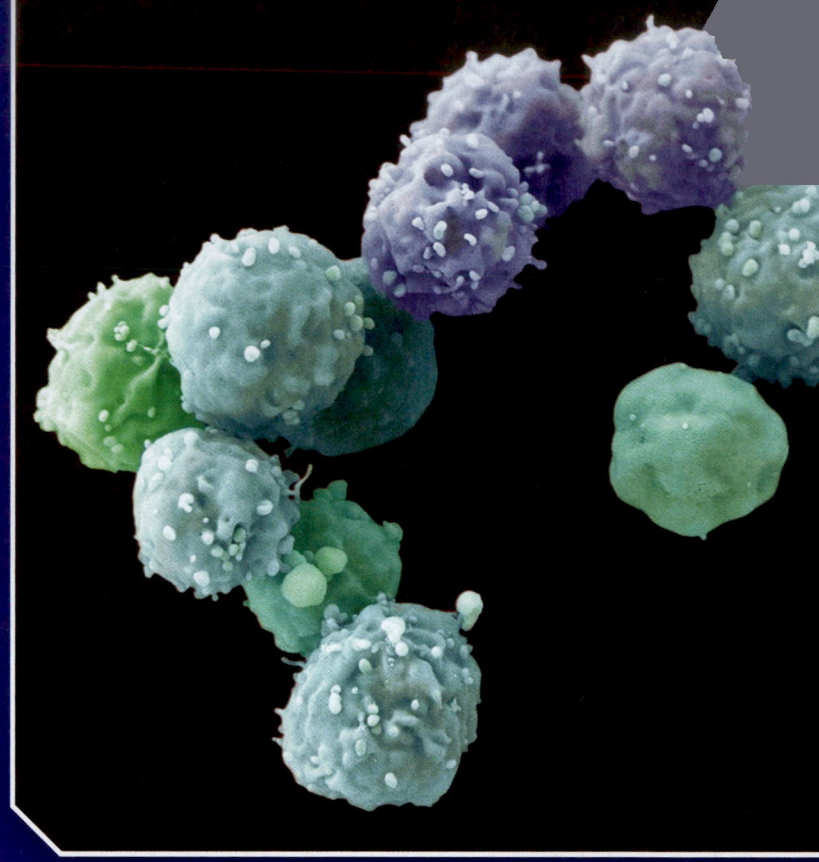

The cell is the simplest structure capable of existing as an individual living unit. Within this unit, many chemical reactions are required for maintaining life. These reactions do not occur at random but are associated with specific parts of the many kinds of cells. These are embryonic stem cells with the potential to repair damaged tissue.

CORE **CONCEPT**

All living things are composed of cells.

OUTLINE

Physics

▷ Energy in a useful form is essential for all life (Ch. 3).

Chemistry

▷ Molecular movement is required for the transport of atoms and molecules into and out of cells (Ch. 4).

▷ Chemical reactions are the essence of all living things (Ch. 9).

Life Science

▷ The structure of living things is basically the same—a cell (Ch. 20).

▷ The physiological processes of living things are common to all life forms (Ch. 20).

▷ Cell division is required for life to continue (Ch. 20).

▷ All living things have genetic material (Ch. 20).

▷ Many types of diseases are the result of one life form living on another (Ch. 20).

OVERVIEW

What does it mean to be alive? You would think that a science textbook could answer this question easily. However, this is more than just a theoretical question because in recent years, it has been necessary to construct legal definitions of what life is and especially of when it begins and ends. The legal definition of death is important because it may determine whether a person will receive life insurance benefits or if body parts may be used in transplants. In the case of heart transplants, the person donating the heart may be legally "dead," but the heart certainly isn't since it can be removed while it still has "life." In other words, there are different kinds of death. There is the death of the whole living unit and the death of each cell within the living unit. A person actually "dies" before every cell has died. Death, then, is the absence of life, but that still doesn't tell us what life is. At this point, we won't try to define life, but we will describe some of the basic characteristics of living things.

PART I
THE CHARACTERISTICS OF LIFE

This chapter is divided into three parts. The first part focuses on the general structural and functional features displayed by all living things. It is these features that separate them from our nonliving world.

20.1 WHAT MAKES SOMETHING ALIVE?

The science of **biology** is, broadly speaking, the study of living things. It draws on chemistry and physics for its foundation and applies these basic physical laws to life processes. Living things have abilities and structures not typically found in things that were never living. The ability to manipulate energy and matter is unique to living things. Developing an understanding of how they modify matter and use energy will help you appreciate how they differ from nonliving objects. Living things show five characteristics that nonliving things do not: (1) metabolic processes, (2) generative processes, (3) responsive processes, (4) control processes, and (5) a unique structural organization (figure 20.1). For something to be "alive," it must display all these characteristics; that is, they must all work together. It is important to recognize that

although these characteristics are typical of all living things, they may not all be present in each organism at every point in time. For example, some individuals may reproduce or grow only at certain times. Should a living organism lose one or more of these features for an extended period, it would "die." Something is living when it displays certain unique chemical processes occurring in association with certain unique structures. This section briefly introduces each of these basic characteristics.

1. **Metabolic processes**   All the chemical reactions involving molecules required for a cell to grow, reproduce, and make repairs are referred to as its **metabolism.** Metabolic properties keep a cell alive. The energy that organisms use is stored in the chemical bonds of complex molecules. Even though different kinds of organisms have different ways of metabolizing **nutrients** or food, we are usually talking about three main activities: *nutrient uptake*, *nutrient processing*, and *waste elimination*. All living things expend energy to take in nutrients (raw materials) from their environment. Many animals take in these materials by eating or swallowing other organisms. Microorganisms and plants absorb raw materials into their cells to maintain their lives. Once inside, raw materials are used in a series of chemical reactions to manufacture new parts, make repairs, reproduce, and provide energy for essential activities. However, not all the raw materials entering a living thing are valuable to it. There may be portions that

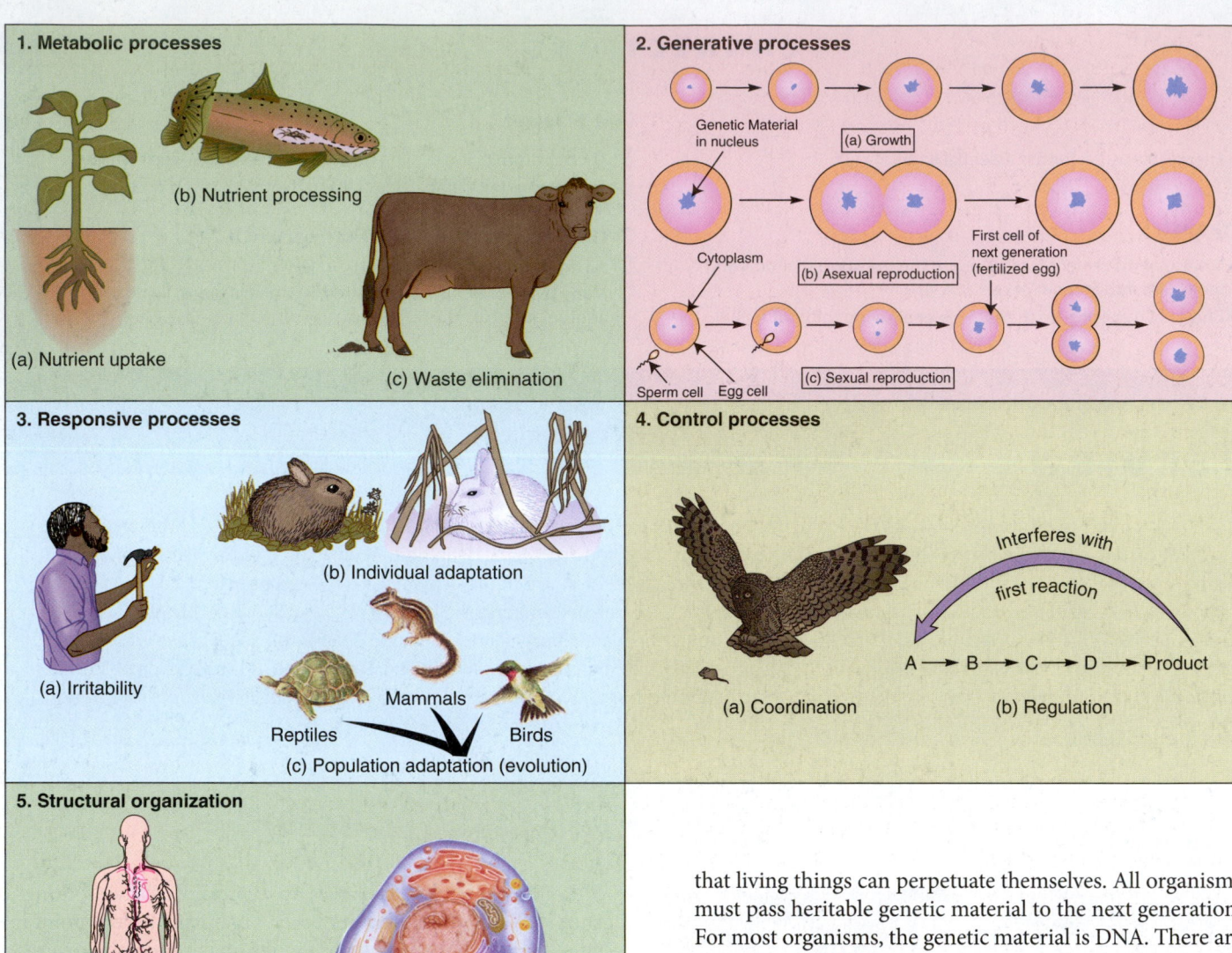

1. Metabolic processes

(a) Nutrient uptake

(b) Nutrient processing

(c) Waste elimination

2. Generative processes

Genetic Material in nucleus

(a) Growth

Cytoplasm

First cell of next generation (fertilized egg)

(b) Asexual reproduction

(c) Sexual reproduction

Sperm cell   Egg cell

3. Responsive processes

(a) Irritability

(b) Individual adaptation

Mammals

Reptiles   Birds

(c) Population adaptation (evolution)

4. Control processes

Interferes with first reaction

A → B → C → D → Product

(a) Coordination

(b) Regulation

5. Structural organization

(a) Organismal organization

(b) Cellular organization

FIGURE 20.1 Living things demonstrate many common characteristics.

are useless or even harmful. Organisms eliminate these portions as waste. Metabolic processes also produce unusable heat energy that may also be considered a waste product.

2. The second group of characteristics of life, **generative processes,** are reactions that result in an increase in the size of an individual organism—*growth*—or an increase in the number of individuals in a population of organisms—*reproduction*. During growth, living things add to their structure, repair parts, and store nutrients for later use. Growth and reproduction are directly related to metabolism, since neither can occur without the acquisition and processing of nutrients.

Reproduction of individuals must occur because each organism eventually dies. Thus, reproduction is the only way

that living things can perpetuate themselves. All organisms must pass heritable genetic material to the next generation. For most organisms, the genetic material is DNA. There are a number of different ways that various kinds of organisms reproduce and guarantee their continued existence. Some reproductive processes known as sexual reproduction involve two organisms contributing to the creation of a unique, new organism. Asexual (without sex) reproduction occurs when organisms make identical copies of themselves.

3. Organisms also respond to changes within their bodies and in their surroundings in a meaningful way. These **responsive processes** have been organized into three categories: *irritability, individual adaptation,* and *population adaptation* or *evolution.* The term *irritability* is not used in the everyday context of an individual being "irritable" or angry. Here, *irritability* means an individual's rapid response to a stimulus, such as your response to a loud noise, beautiful sunset, or bad smell. This type of response occurs only in the individual receiving the stimulus and is rapid because the mechanisms that allow the response to occur (e.g., muscles, bones, and nerves) are already in place. Individual adaptation is also an individual response but is slower since it requires growth or some other fundamental change in an organism. For example, a weasel's fur color will change from its brown summer coat to its white winter coat when genes responsible for the production of brown pigment are "turned off." Or the response of our body to disease organisms requires a

change in the way cells work that eventually gets control of the organism causing the disease. Population adaptation involves changes in the kinds of characteristics displayed by individuals within the population. It is also known as *evolution*, which is a change in the genetic makeup of a *population* of organisms. This process occurs over long periods of time and enables a species (specific kind of organism) to adapt and better survive long-term changes in its environment over many generations. For example, the structures that give birds the ability to fly long distances allow them to respond to a world in which the winter season presents severe conditions that would threaten survival. Similarly, the ability of humans to think and use tools allows them to survive and be successful in a great variety of environmental conditions.

4. The **control processes** of *coordination* and *regulation* constitute the fourth characteristic of life. Control processes are mechanisms that ensure that an organism will carry out all metabolic activities in the proper sequence (coordination) and at the proper rate (regulation). All the chemical reactions of an organism are coordinated and linked together in specific biochemical pathways. The orchestration of all the reactions ensures that there will be specific stepwise handling of the nutrients needed to maintain life. The molecules responsible for coordinating these reactions are known as *enzymes*. **Enzymes** are molecules produced by organisms, molecules that are able to increase and control the rate at which life's chemical reactions occur. Enzymes also regulate the amount of nutrients processed into other forms.

   Many of the internal activities of organisms are interrelated and coordinated to maintain a constant internal environment, a process called **homeostasis.** For example, when we begin to exercise, we use up oxygen more rapidly so the amount of oxygen in the blood falls. To maintain a "constant internal environment," the body must obtain more oxygen. This involves more rapid contractions of the muscles that cause breathing and a more rapid and forceful pumping of the heart to get blood to the lungs. These activities must occur together at the right time and at the correct rate, and when they do, the level of oxygen in the blood will remain normal while supporting the additional muscular activity.

5. In addition to these four basic processes that are typical of living things, living things also share some basic **structural similarities.** All living things are organized into complex structural units called *cells.* The cells are units that have an outer limiting membrane and internal structural units that have specific functions. Some living things, such as you, consist of trillions of cells with specialized abilities that interact to provide the independently functioning unit called an *organism* (figure 20.2). Typically, in such large, multicellular organisms as humans, cells cooperate with one another in units called *tissues* (e.g., muscle, nerves). Groups of tissues are organized into larger units known as *organs* (e.g., heart) and, in turn, into *organ systems* (e.g., circulatory system). Other organisms, such as bacteria or yeast, carry out all four of the life processes within a single cell. Nonliving materials, such as rocks, water, or gases, do not share a structurally complex common subunit.

Biologists and other scientists like to organize vast amounts of information into conceptual chunks that are easier to relate to one another. One important concept in biology is that all living things share the structural and functional characteristics we have just discussed. Another important organizing concept is that organisms are special kinds of matter that interact with their surroundings at several different levels (table 20.1). When biologists seek answers to a particular problem, they may attack it at several different levels simultaneously. They must understand the molecules that make up living things; how the molecules are incorporated into cells; how tissues, organs, or systems within an organism function; and how populations and ecosystems are affected by changes in individual organisms.

20.2 THE CELL THEORY

One of the characteristics of life, the *cell,* is one of the most important ideas in biology because it applies to all living things. This

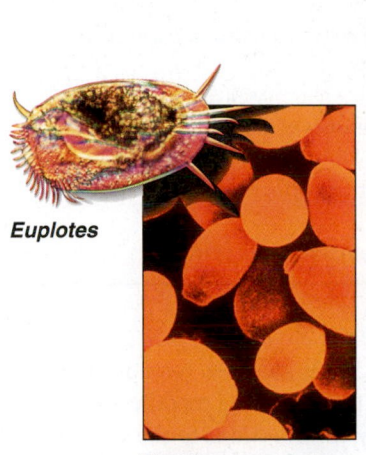
Euplotes

DNA helix

Yeast

Orchid

Humans

FIGURE 20.2   Each organism, whether it is simple or complex, independently carries on metabolic, generative, responsive, and control processes. It also contains special molecules, a cellular structure, and other structural components. DNA is a molecule unique to living things. Some organisms, such as yeast or the protozoan *Euplotes,* consist of single cells, whereas others, such as orchids and humans, consist of many cells organized into complex structures.

TABLE 20.1 Levels of Organization for Living Things

Category	Characteristics/Explanation	Example/Application
Biosphere	Worldwide ecosystem; human activity affects the climate of Earth.	Global climate change, hole in ozone layer
Ecosystem	Communities (groups of populations) that interact with the physical world in a particular place	The Everglades ecosystem involves many kinds of organisms, the climate, and the flow of water to south Florida.
Community	Populations of different kinds of organisms that interact with one another in a particular place	The populations of trees, insects, birds, mammals, fungi, bacteria, and many other organisms that interact in any location
Population	A group of individual organisms of a particular kind	The human population currently consists of over 6 billion individual organisms. The current population of the California condor is about 220 individuals.
Individual organism	An independent living unit	A single organism Some organisms consist of many cells—you, a morel mushroom, a rose bush. Others are single cells—yeast, pneumonia bacterium, *Amoeba*.
Organ system	Groups of organs that perform particular functions	The circulatory system consists of a heart, arteries, veins, and capillaries, all of which are involved in moving blood from place to place.
Organ	Groups of tissues that perform particular functions	An eye contains nervous tissue, connective tissue, blood vessels, and pigmented tissues, all of which are involved in sight.
Tissue	Groups of cells that perform particular functions	Blood, groups of muscle cells, and the layers of the skin are all groups of cells that perform a particular function.
Cell	The smallest unit that displays the characteristics of life	Some organisms are single cells. Within multicellular organisms are several kinds of cells—heart muscle cells, nerve cells, white blood cells.
Molecules	Specific arrangements of atoms	Living things consist of special kinds of molecules, such as proteins, carbohydrates, and DNA.
Atoms	The fundamental units of matter	Hydrogen, oxygen, nitrogen and about one hundred others

concept did not emerge all at once but has been developed and modified over hundreds of years. It is still being modified today.

Several individuals made key contributions to the cell concept. Anton van Leeuwenhoek (1632–1723) was one of the first to make use of a *microscope* to examine biological specimens. When van Leeuwenhoek discovered that he could see things moving in pond water using his microscope, his curiosity stimulated him to look at a variety of other things. He studied blood, semen, feces, pepper, and scrapings from the surface of teeth, for example. He was the first to see individual cells and recognize them as living units, but he did not call them cells. The name he gave to these "little animals" that he saw moving around in the pond water was *animalcules.*

The first person to use the term *cell* was Robert Hooke (1635–1703) of England, who was also interested in how things looked when magnified (figure 20.3). He chose to study thin slices of cork from the bark of a cork oak tree. He saw a mass of cubicles fitting neatly together, which reminded him of the barren dormitory rooms in a monastery. Hence, he called them *cells.* As it is currently used, the term **cell** refers to the basic functional and structural unit that makes up all living things. When Hooke looked at cork, the tiny boxes he saw were, in fact, only the cell walls that surrounded the living portions of plant cells. We now know that the **cell wall** is composed of the complex carbohydrate

cellulose, which provides strength and protection to the living contents of the cell. The cell wall appears to be a rigid, solid layer of material, but in reality, it is composed of many interwoven strands of cellulose molecules. While the loose weave of the cell wall provides a rigid outer framework for the cell, it is porous to many types of molecules.

Hooke's use of the term *cell* in 1665 in his publication *Micrographia* was only the beginning, for nearly two hundred years passed before it was generally recognized that all living things are made of cells and that these cells can reproduce themselves; that is, they come from preexisting cells. However, many others were involved in understanding how this concept applied to all living things (see "People Behind the Science" in this chapter).

Soon after the term *cell* caught on, it was recognized that the cell's vitally important portion is inside the cell wall. This living material was termed **protoplasm,** which means *first-formed substance.* The term *protoplasm* allowed scientists to distinguish between the living portion of the cell and the nonliving cell wall. Very soon microscopists were able to distinguish two different regions of protoplasm. One type of protoplasm was more viscous and darker than the other. This core region, called the **cell nucleus,** is a central body within a more fluid material surrounding it. **Cytoplasm** is the name given to the colloidal fluid portion of the protoplasm (protoplasm = cyto-

FIGURE 20.3 The cell concept has changed considerably over the last three hundred years. Hooke invented the compound microscope and illumination system shown in (A), one of the best such microscopes of his time. Anton van Leeuwenhoek made many observations of various materials and made drawings of these "animalcules" (B). One of the first subcellular differentiations was to divide the protoplasm into cytoplasm and nucleus. We now know that cells are much more complex than this; they are composed of many kinds of subcellular structures, some components numbering in the thousands.

A

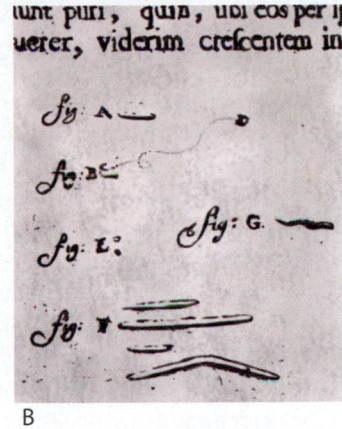

B

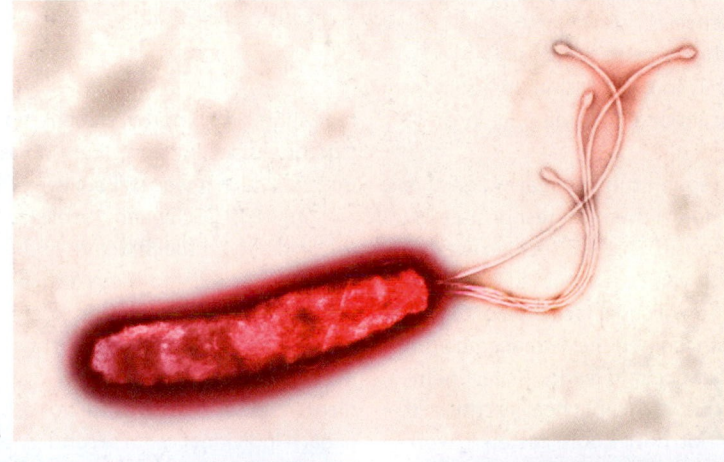

A

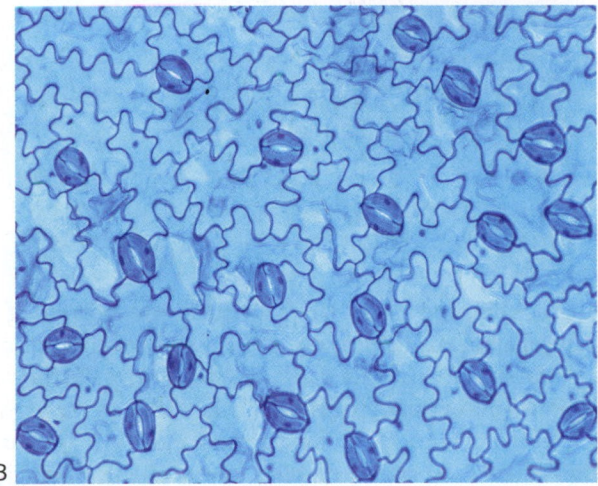

B

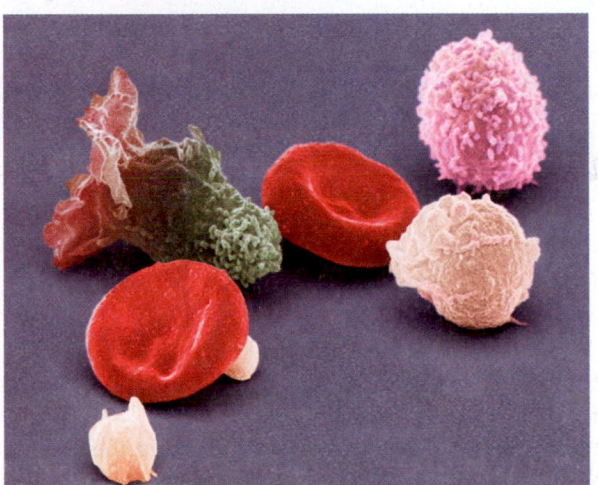

C

FIGURE 20.4 There are two major types of cells, the prokaryotes and the eukaryotes. Prokaryotic cells are represented by the (A) bacteria (Helicobacter pylori), and eukaryotic cells by (B) plant (epidermis of a leaf) and (C) animal cells (various human blood cells—stained).

plasm + nucleus). *Colloids* are mixtures that contain suspended particles larger than those in true solutions but smaller than those in a coarse suspension. The solutes in a colloid cannot be filtered out and do not settle out by gravity. While the term *protoplasm* is seldom used today, the term *cytoplasm* is still very common in the vocabulary of cell biologists.

The development of better microscopes and better staining techniques revealed that protoplasm contains many tiny structures called **organelles.** It has been determined that certain functions characteristic of life are performed in certain

organelles. The essential job an organelle does is related to its structure. Each organelle is dynamic in its operation, changing shape and size as it works. Organelles move throughout the cell, and some even self-duplicate. The structure and function of some of these organelles are compared in table 20.3 at the end of the chapter.

All living things are cells or composed of cells. To date, most biologists recognize two major cell types, *prokaryotes* and *eukaryotes* (figure 20.4). Whether they are **prokaryotic cells** or **eukaryotic cells,** all have certain things in common: (1) cell

Matthias Jakob Schleiden (1804–1881) and Theodor Schwann (1810–1882)

Matthias Schleiden was a German botanist who, with Theodor Schwann, is best known for establishing the cell theory.

The existence of cells had been known since the seventeenth century (Robert Hooke is generally credited with their discovery in 1665). However, Schleiden was the first to recognize their importance as the fundamental units of living organisms. In 1838, he announced that the various parts of plants consist of cells or derivatives of cells. In the following year, Schwann published a paper in which he confirmed for animals Schleiden's idea of the basic importance of cells in the organization of organisms. Thus, Schleiden and Schwann established the cell theory, a concept that is common knowledge today and is as fundamental to biology as atomic theory is to the physical sciences.

Schleiden also researched other aspects of cells. He recognized the importance of the nucleus in cell division. In addition, he noted the active movement of intracellular material in plant tissues.

In 1834, Theodor Schwann began to investigate digestive processes and two years later isolated a chemical responsible for protein digestion, which he called pepsin. This was the first *enzyme* to be isolated from animal tissue. Schwann then studied fermentation and showed that the fermentation of sugar is a result of the life processes of living yeast cells. He later coined the term *metabolism* to denote the chemical changes that occur in living tissue. Chemists such as Friedrich Wöhler and Justus von Liebig heavily criticized his work on fermentation. It was not until Louis Pasteur's work on fermentation in the 1850s that Schwann was proved correct. Meanwhile, however, Schwann investigated putrefaction (rotting) in an attempt to disprove the theory of spontaneous generation,

Matthias Schleiden   Theodor Schwann

repeating, with improved techniques, Lazzaro Spallanzani's earlier experiments. Schwann found no evidence to support the theory, despite which it was still believed by some scientists.

In 1839, Schwann published *Microscopical Researches on the Similarity in the Structure and Growth of Animals and Plants* in which he formulated the cell theory. In the previous year, Matthias Schleiden—whom Schwann knew well—had stated the theory in connection with plants, but Schwann extended the theory to animals. Schwann concluded that all organisms (both animals and plants) consist entirely of cells or of products of cells and that the life of each individual cell is secondary to that of the whole organism.

Source: Modified from the *Hutchinson Dictionary of Scientific Biography*. © RM, 2011. All rights reserved. Helicon Publishing is a division of RM.

---

membranes, (2) cytoplasm, (3) genetic material, (4) energy transfer molecules, (5) enzymes and coenzymes. These are all necessary to carry out life's functions. Should any of these not function properly, a cell would die.

The differences among cell types are found in the details of their structure. While prokaryotic cells lack most of the complex internal organelles typical of eukaryotes, they are cells and can carry out life's functions.

Most single-celled organisms that we commonly refer to as *bacteria* are prokaryotic cells (figure 20.4A). Algae, protozoa, fungi, plants, and animals are all comprised of eukaryotic cells (figure 20.4 B, C).

20.3 CELL MEMBRANES

One feature common to all cells and many of the organelles they contain is a thin sheet of material called *membrane*. Sheets of membrane can be folded and twisted into many different struc-

tures, shapes, and forms. The particular arrangement of membrane of an organelle is related to the functions that it is capable of performing. This is similar to the way a piece of fabric can be fashioned into a pair of pants, shirt, pillow case, or chair cover. All cellular membranes have a fundamental molecular structure that allows them to be fashioned into a variety of different organelles.

Cell membranes are thin sheets composed primarily of phospholipids (refer to figure 19.25) and proteins. The current hypothesis of how membranes are constructed is known as the *fluid-mosaic model,* which proposes that the various molecules of the membrane are able to flow and move about (figure 20.5). The membrane maintains its form because of the physical interaction of its molecules with its surroundings. The phospholipid molecules of the membrane are polar molecules since they have one end (the glycerol portion) that is soluble in water and is therefore called *hydrophilic* (water-loving). The other end that is not water soluble, called *hydrophobic* (water-hating), is comprised of fatty acid. When phospholipid molecules are placed in water, they form a

To view very small objects, we use a magnifying glass as a way of extending our observational powers. A magnifying glass is a lens that bends light in such a way that the object appears larger than it really is. Such a lens might magnify objects ten or even fifty times. Anton van Leeuwenhoek, a Dutch drape and clothing maker, was one of the first individuals to carefully study magnified cells. He made very detailed sketches of the things he viewed with his simple microscopes and communicated his findings to Robert Hooke and the Royal Society of London. His work stimulated further investigation of magnification techniques and descriptions of cell structures. These first microscopes were developed in the early 1600s.

Compound microscopes (box figure 20.1A), developed soon after the simple microscopes, are able to increase magnification by bending light through a series of lenses. One lens, the *objective lens,* magnifies a specimen that is further magnified by the second lens, known as the *ocular lens* or eye piece.

There are many types of microscopes in use today, including dissecting (stereo), fluorescence, dark field, confocal, scanning acoustic, digital image, tunneling, atomic force, and electron (transmission and scanning). An *electron microscope* (box figure 20.1B) is able to magnify two-hundred thousand times and still resolve individual structures. The difficulty is, of course, that you are unable to see electrons with your eyes. Therefore, in an electron microscope, the electrons strike a photographic film or video monitor, and this "picture" shows the individual structures. The techniques for preparing the material to be viewed—slicing the specimen very thinly and focusing the electron beam on the specimen—make electron microscopy an art as well as a science.

More recently, the scanning *tunneling microscope* was developed. The developers of this microscope, Gerd Binning and Heinrich Rohrer, were awarded the Nobel Prize in 1985 for their work. The invention of the scanning *tunneling microscope followed by the atomic force microscope* enabled researchers to visualize previously unseen molecules and even the surface of atoms such as chlorine and sodium. A scanning tunneling microscope utilizes a thin platinum and viridium wire to trace the surface of the specimen. Electrons on the surface of the probe interact with those on the specimen's surface to produce a "tunnel" that is seen as a current. The stronger the current, the closer the probe is to the surface. The *atomic force microscope* enables researchers to see three-dimensional objects even as small as molecules and the surface of atoms such as chlorine and sodium.

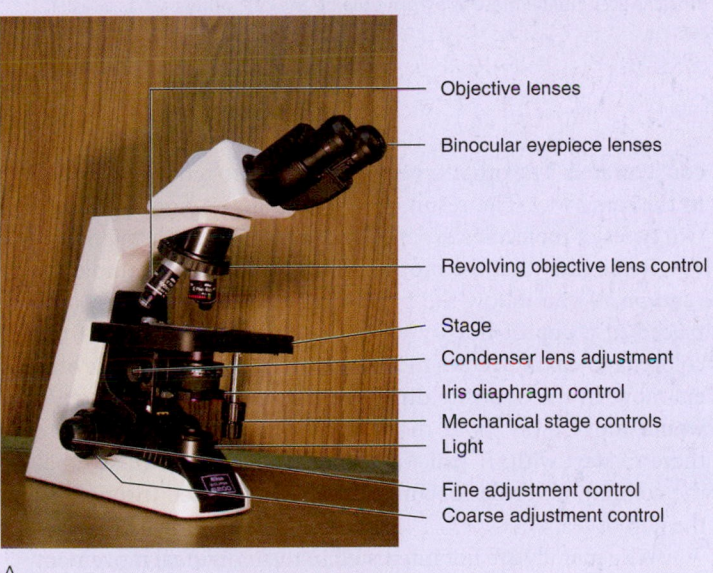

Objective lenses

Binocular eyepiece lenses

Revolving objective lens control

Stage
Condenser lens adjustment
Iris diaphragm control
Mechanical stage controls
Light

Fine adjustment control
Coarse adjustment control

A

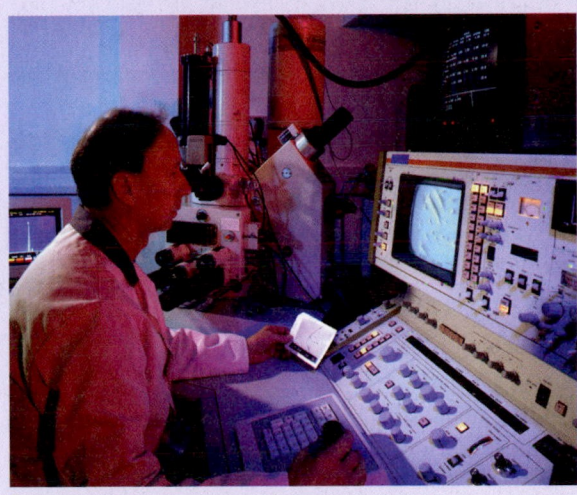

B

Box Figure 20.1   (*A*) Compound microscope. (*B*) Electron microscope.

double-layered sheet, with the water-soluble (hydrophilic) portions of the molecules facing away from each other. This is commonly referred to as a *phospholipid bilayer.*

The protein component of cellular membranes can be found on either surface of the membrane or in the membrane, among the phospholipid molecules. Many of the membrane proteins are capable of moving from one side to the other, while others aid in the movement of molecules across the membrane by forming channels through which substances may travel or by acting as transport molecules. Some protein molecules found on the outside surfaces of cellular membranes have carbohydrates or fats attached to them. These combination molecules are important in determining the "sidedness" (inside-outside) of the membrane and also help organisms recognize differences between types of cells. Your body can recognize disease-causing organisms because their surface proteins are different from those of its own cellular membranes. Some of these molecules also serve as attachment sites for specific chemicals, bacteria, protozoa, white blood cells, and viruses. Many dangerous agents cannot stick to the surface of cells and therefore cannot cause harm. For this reason, cell biologists explore the exact structure and function of these molecules. They are also attempting to identify molecules that can interfere with the binding of such agents as viruses and bacteria in the hope of controlling infections.

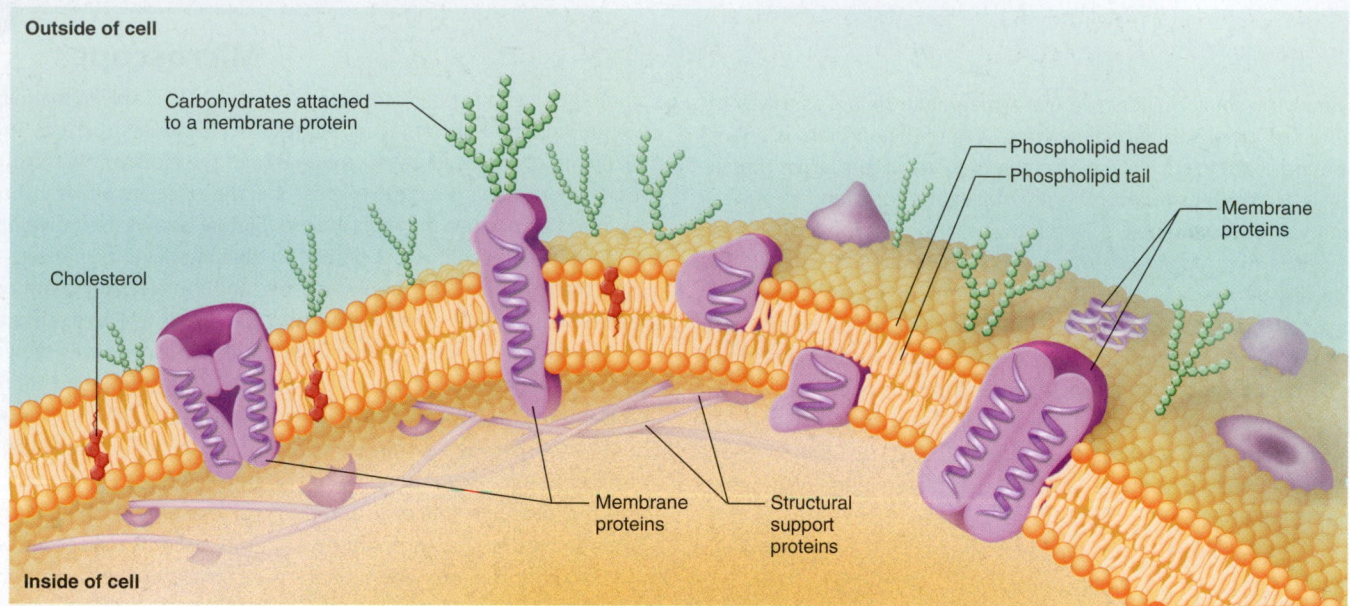

FIGURE 20.5 Notice in this section of a generalized human cell that there is no surrounding cell wall as described by Hooke. Membranes in all cells are composed of protein and phospholipids. Two layers of phospholipid are oriented so that the hydrophobic fatty tails extend toward each other and the hydrophilic glycerol heads are on the outside. The phosphate-containing chain of the phospholipid is coiled near the glycerol portion. Buried within the phospholipid layer and/or floating on it are the globular proteins. Some of these proteins accumulate materials from outside the cell; others act as sites of chemical activity. Carbohydrates are often attached to one surface of the membrane.

Other types of molecules found in cell membranes are cholesterol and carbohydrates. Cholesterol appears to play a role in stabilizing the membrane and keeping it flexible. Carbohydrates are usually found on the outside of the membrane, where they are bound to proteins or lipids. These carbohydrates appear to play a role in cell-to-cell interactions and are involved in binding with regulatory molecules.

20.4 GETTING THROUGH MEMBRANES

If a cell is to stay alive, it must meet the characteristics of life outlined earlier. This includes taking in nutrients and eliminating wastes and other by-products of metabolism. Several mechanisms allow cells to carry out life's process including diffusion, osmosis, facilitated diffusion, active transport, and phagocytosis. There are two categories by which cells move materials through the membrane: passive transport and active transport. Passive transport methods include diffusion, osmosis, and facilitated diffusion. Active transport mechanisms include endocytosis, exocytosis, phagocytosis, and pinocytosis.

Diffusion

There is a natural tendency in gases and liquids for molecules of different types to completely mix with each other (refer to chapter 10, Water and Solutions). This is because they are constantly moving about with various levels of kinetic energy. Consider two types of molecules. As the molecules of one type move about, they tend to scatter from the place where they are most concentrated. The other type of molecule also tends to disperse in the same way. The result of this random motion is that the two types of molecules eventually become mixed throughout.

Remember that the motion of the molecules is completely random. If you follow the paths of molecules from tea leaves placed in a cup of hot water, you will find that some of the colored tea leaf molecules move away from the leaves while others move in the opposite direction. However, more molecules would move away from the leaves because there were more there to start with. If you wait long enough, you will see that the colored molecules become equally distributed throughout the cup of tea.

We generally are not interested in the individual movement but rather in the overall movement, which is called the *net movement*. The direction of greatest movement (net movement) is determined by the relative concentration of the molecules, for example, the amount in one area in comparison to the amount in another area. **Diffusion** is the net movement of a kind of molecule from a place where that molecule is in higher concentration to a place where that molecule is more rare. When a kind of molecule is completely dispersed and movement is equal in all directions, we say that the system has reached a state of *dynamic equilibrium*. There is no longer a net movement because movement in one direction equals movement in the other. It is dynamic, however, because the system still has energy, and the molecules are still moving.

Because the cell membrane is composed of phospholipid and protein molecules that are in constant motion, temporary openings are formed that allow small molecules to cross from one side of the membrane to the other. Molecules close to the membrane

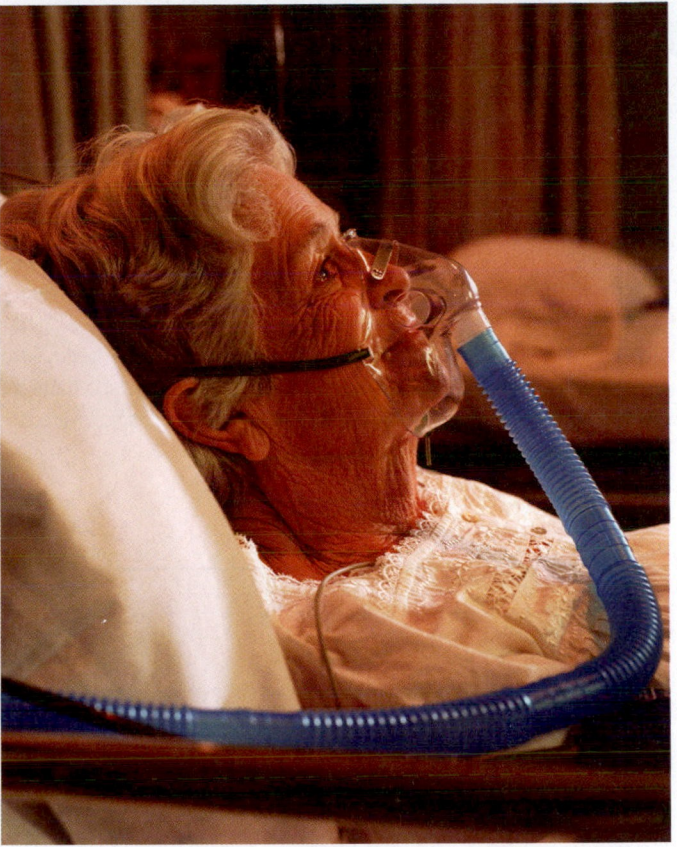

FIGURE 20.6   As a result of molecular motion, molecules move from areas where they are concentrated to areas where they are less concentrated. Chronic obstructive pulmonary disease (COPD) is an illness in which the lungs become inflamed or damaged, restricting air flow into and out of the lungs. This makes it hard to breathe and take in the amount of oxygen needed for good health. Therefore, such people are often "put on oxygen" as seen with this woman.

are in constant motion as well. They are able to move into and out of a cell by passing through these openings in the membrane.

The rate of diffusion is related to the kinetic energy and size of the molecules. Since diffusion occurs only when molecules are unevenly distributed, the relative concentration of the molecules is important in determining how fast diffusion occurs. The difference in concentration of the molecules is known as a *concentration gradient* or *diffusion gradient.* When the molecules are equally distributed, no such gradient exists.

Diffusion can take place only as long as there are no barriers to the free movement of molecules. In the case of a cell, the membrane permits some molecules to pass through, while others are not allowed to pass or are allowed to pass more slowly. This permeability is based on size, ionic charge, and solubility of the molecules involved. The membrane does not, however, distinguish direction of movement of molecules; therefore, the membrane does not influence the direction of diffusion. The direction of diffusion is determined by the relative concentration of specific molecules on the two sides of the membrane, and the energy that causes diffusion to occur is supplied by the kinetic energy of the molecules themselves. The health of persons who have difficulty getting enough oxygen to their cells can be improved by increasing the concentration gradient. Oxygen makes up about 20 percent of the air. If this concentration is artificially raised by supplying a special source of oxygen, diffusion from the lungs to the blood will take place more rapidly. This will help assure that oxygen reaches the body cells that need it, and some of the person's symptoms can be controlled (figure 20.6).

Diffusion is an important means by which materials are exchanged between a cell and its environment. Since the movement of the molecules is random, the cell has little control over the process; thus, diffusion is considered a passive process, that is, chemical bond energy does not have to be expended. For example, animals are constantly using oxygen in various chemical reactions. Consequently, the oxygen concentration in cells always remains low. The cells, then, contain a lower concentration of oxygen than the oxygen level outside of the cells. This creates a diffusion gradient, and the oxygen molecules diffuse from the outside of the cell to the inside of the cell.

Osmosis

An important characteristic of all membranes is that they are selectively permeable. **Selectively permeable** means that a membrane will allow certain molecules to pass across it and will prevent others from doing so. Molecules that are able to dissolve in phospholipids, such as vitamins A and D, can pass through the membrane rather easily; however, many molecules cannot pass through at all. In certain cases, the membrane differentiates on the basis of molecular size; that is, the membrane allows small molecules, such as water, to pass through and prevents the passage of larger molecules. The membrane may

Connections . . .

The Other Outer Layer—The Cell Wall

The *cell walls* of microorganisms, plants, and fungi appear to be rigid, solid layers of material but are really loosely woven layers. And like water pouring through a leaky straw basket, many types of molecules easily pass through a cell wall. The cell wall lends strength and protection to the contents of the cell but hampers flexibility and movement. There are three kinds of materials typically used for cell walls. All the higher plants and most algae have *cellulose* as their wall material. When found in large amounts, it is known as *wood*. Another cell wall material, *chitin*, is found in the fungi. Chitin also constitutes the exoskeleton material in insects. The outer shell of a beetle or a shrimp is chitin. But in those animals, the chitin surrounds masses of tissue instead of individual cells. The fungi are not crunchy or brittle like shrimp skeletons because the chitin is very thin. Many bacteria have their walls composed of *peptidoglycan*. It is composed of both amino acids and carbohydrates. The lengths of peptidoglycan chains and how they are interlinked may determine the shape of a bacterium. *Mycobacterium tuberculosis*, the bacterium that causes tuberculosis (TB) contains a special type of peptidoglycan that makes it particularly difficult for the body to fight and more difficult for antibiotics to get into the cell. If antibiotics cannot get into the cells, they cannot disrupt cellular processes and destroy the microbe.

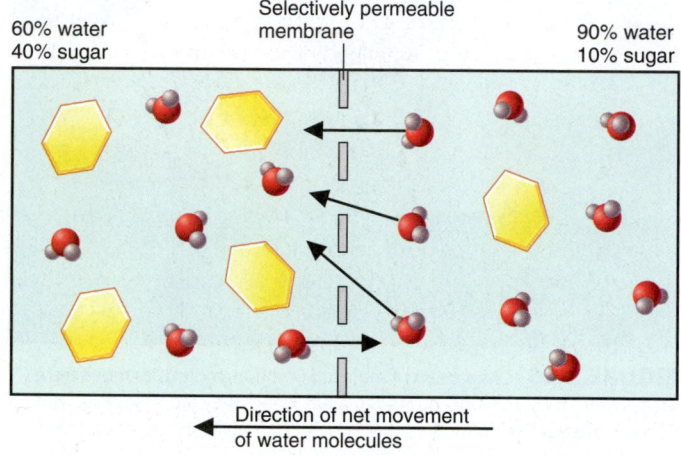

FIGURE 20.7 When two solutions with different percentages of water are separated by a selectively permeable membrane, there will be a net movement of water from the solution with the highest percentage of water to the one with the lowest percentage of water.

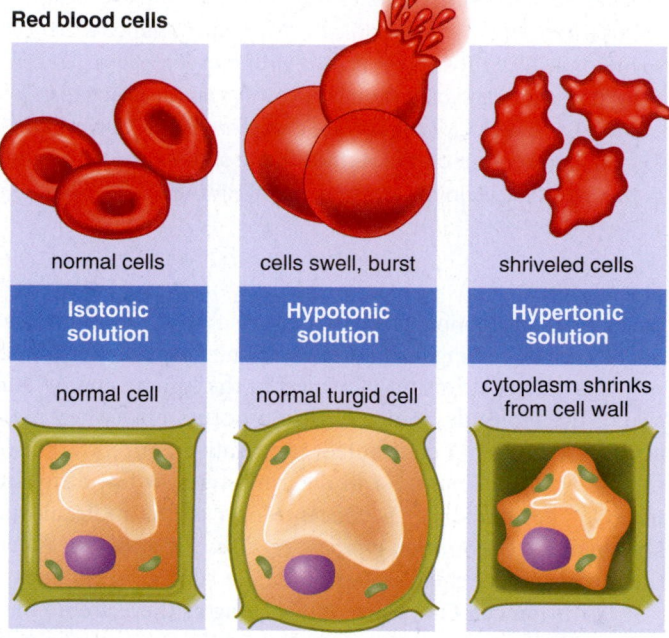

FIGURE 20.8 Cells are affected by the amount of dissolved materials in the water that surrounds them. When in an isotonic situation, the cells neither gain nor lose water. In a hypotonic solution water diffuses from the surroundings into the cell. Animal cells will swell and burst, but plant cells have a tough cell wall surrounding the cell contents, and the pressure generated on the inside of the cell causes it to become rigid. Both plant and animal cells shrink when in a hypertonic solution because water moves from the cells, which have the higher water concentration, to the surroundings.

also regulate the passage of ions. If a particular portion of the membrane has a large number of positive ions on its surface, positively charged ions in the environment will be repelled and prevented from crossing the membrane.

Water molecules diffuse through cell membranes. The net movement (diffusion) of water molecules (the solvent) through a selectively permeable membrane is known as **osmosis.** In any osmotic situation, there must be a selectively permeable membrane separating two solutions. For example, a solution of 90 percent water and 10 percent sugar separated by a selectively permeable membrane from a different sugar solution, such as one of 80 percent water and 20 percent sugar, demonstrates osmosis (figure 20.7). The membrane allows water molecules to pass freely but prevents the larger sugar molecules from crossing. There is a higher concentration of water molecules in one solution compared to the concentration of water molecules in the other. Therefore, more of the water molecules move from the solution with 90 percent water to the other solution with 80 percent water. Be sure that you recognize that osmosis is really diffusion in which the diffusing substance is water and that the regions of different concentrations are separated by a membrane that is more permeable to water.

A proper amount of water is required if a cell is to function efficiently. Too much water in a cell may dilute the cell contents and interfere with the chemical reactions necessary to keep the cell alive. Too little water in the cell may result in a buildup of poisonous waste products. If a cell contains a concentration of water and dissolved materials that is equal to that of its surroundings, the cell is said to be *isotonic* to its surroundings. Cells within the human body must have a concentration of water and dissolved materials within the cells that is equal to the concentration outside the cells. For example, red blood cells are isotonic

Connections . . .

Kidney Machines in Action

Blood inlet from patient

Dialysis solution outlet

Tubular cellulose membranes

Dialysis solution inlet

Blood outlet return to patient

A

Box Figure 20.2 (*A*) Dialysis Process. (*B*) Dialysis in a hospital.

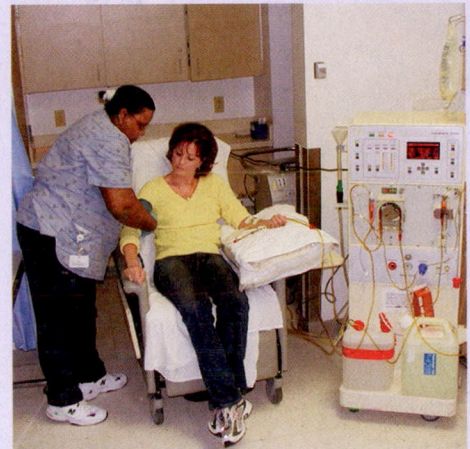

B

When the kidneys do not function properly, the toxic waste urea builds up in the blood because the ailing kidneys cannot filter it out. This eventually leads to a condition known as *uremia*, "urine in the blood." Unless the urea is removed from the body, the concentration of urea increases and will eventually lead to death. Kidney or dialysis machines use tubular cellulose membranes to separate the urea from the blood.

Cellulose is a polymer of glucose molecules and contains openings through which urea molecules pass. In a dialysis machine, cellulose membrane tubes are immersed in a large volume of "cleansing" or dialysis solution. The blood is pumped from the vein of a person with uremia through this tubing and then returned to the patient. While in the machine, the membrane allows most of the urea in the blood to pass through the tubing walls, but the larger blood proteins and blood cells do not. Therefore, the patient does not lose vital proteins or blood cells. This system works because the dialysis solution in which the tubing is immersed has a lower amount of urea than does the blood. Because of this concentration gradient, urea (the solute) in the blood passes from the high concentration in the blood through the cellulose membrane tubing pores into the surrounding dialysis solution. This results in a decrease in urea concentration in the blood. As the concentration of the urea increases in the dialysis solution, the solution is flushed out. This enables the dialysis solution to maintain a constant, low concentration of urea and maintain the movement of urea molecules out of the person's body. To maintain the blood's concentration of other vital ions, the solution is made with the same concentration of those ions as found in the blood. This means that these other ions are in dynamic equilibrium, and so their concentration does not change during kidney dialysis (box figure 20.2).

when they have the same percentage of water and dissolved materials as the surrounding plasma. Many products are isotonic. For example, over-the-counter eyewashes are labeled *isotonic,* as are many medically important solutions. *Physiological* or *"normal" saline* solution is isotonic, as are other intravenous (IV) solutions (0.9 percent dissolved salts or 5 percent glucose).

However, if cells or tissues are going to survive in an environment that has a different concentration of water, they must expend energy to maintain this difference. Red blood cells having a lower concentration of water (higher concentration of dissolved materials, solutes) than their surroundings tend to gain water by osmosis very rapidly. They are said to be *hypertonic* to their surroundings and the surroundings are *hypotonic.* These two terms are always used to compare two different solutions. The hypertonic solution is the one with more solutes (dissolved material) and less solvent (water); the hypotonic solution has less solutes (dissolved material) and more solvent (water). It may help to remember that the water goes where the salt is (figure 20.8).

As with the diffusion of other molecules, osmosis is a passive process because the cell has no control over the diffusion of water molecules. This means that the cell can remain in balance with an environment only if that environment does not cause the cell to lose or gain too much water.

Controlled Methods of Transporting Molecules

Some molecules move across the membrane by interacting with specific membrane proteins. When the rate of diffusion of a substance is increased in the presence of such a protein, it is called **facilitated diffusion.** Because this movement is still diffusion, the net direction of movement is from high to low concentration. The action of the carrier does not require an input of energy other than the molecules' kinetic energy. Therefore, this is considered a *passive transport* method, although it can occur only in living organisms with the necessary proteins. There are two

Connections . . .

Cell Membrane Channels and the Flu

Understanding the structure of cell membranes can lead to the control of disease. In 2010, researchers at Brigham Young University developed a more clear understanding of how flu viruses enter human cells through so-called M2 protein cell membrane channels. If their work is successful, this research could lead to the development of drugs that would interfere with the M2 protein channel, preventing the flu virus from entering human cells. If the virus does not enter cells, it cannot reproduce and cause symptoms of the flu. It is hoped that drugs can be developed that would stop all types of the flu virus from entering through this M2 channel.

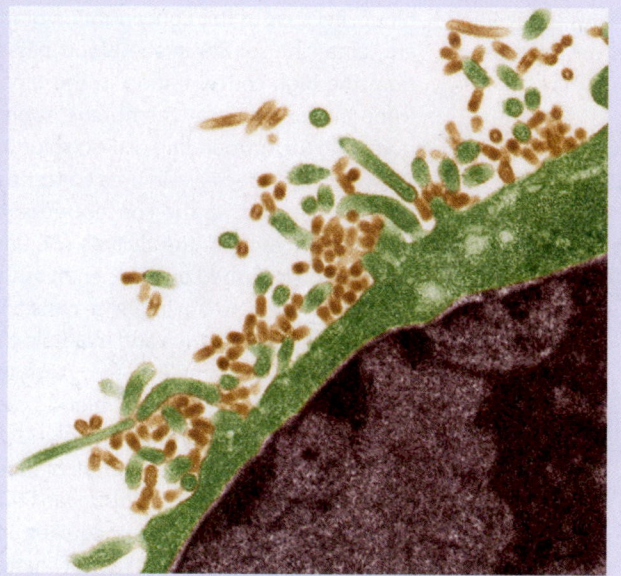

Box Figure 20.3 Influenza viruses. Colored transmission electron micrograph (TEM) of influenza (flu) viruses (yellow) budding from a host cell. The host cell nucleus is black. Magnification: x83,300 when printed at 10 centimeters wide.

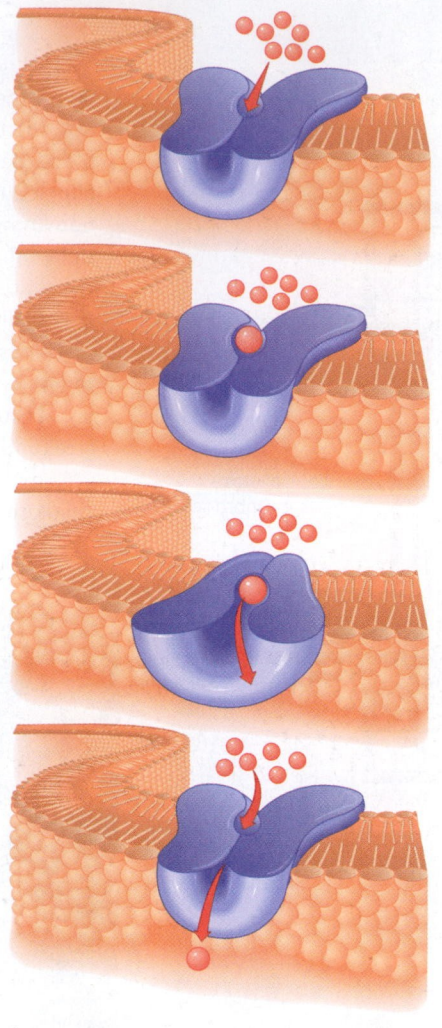

A

FIGURE 20.9 (A) The molecules being moved through the membrane attach to a specific transport carrier protein in the membrane. This causes a change in its shape that propels the molecule or ion through to the other side. *(continued on next page)*

groups of membrane proteins involved in facilitated diffusion: (a) *carrier proteins* and (b) *ion channels*. When a carrier protein attaches to the molecule to be moved across the membrane, the combination molecule changes shape. This shape change enables the molecule to be shifted from one side of the membrane to the other. The carrier then releases the molecule and returns to its original shape (figure 20.9a). Ion channels do not really attach to the molecule being transported through the membrane, but operate like gates. The opening and closing of a channel is controlled by changes in electrical charge at the pore or "gate-keeping" signal molecules (figure 20.9b).

When molecules are moved across the membrane from an area of *low* concentration to an area of *high* concentration, the cell must expend energy. The process of using a carrier protein to move molecules up a concentration gradient is called

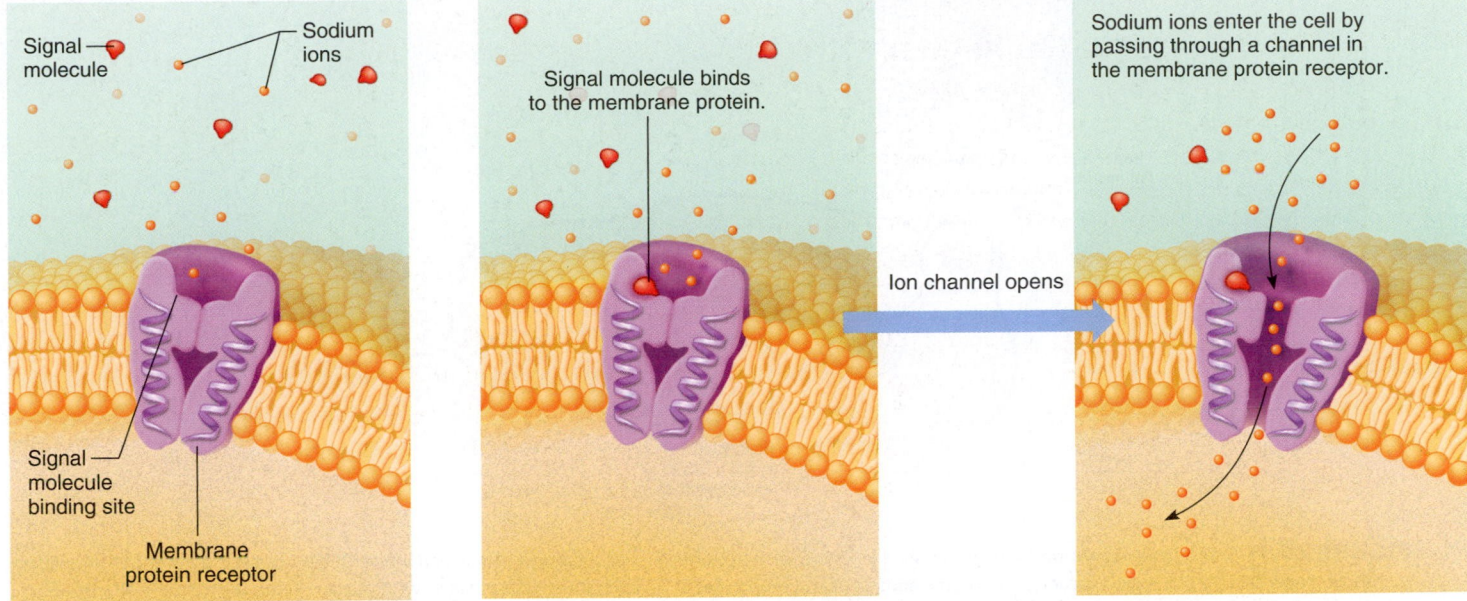

B

FIGURE 20.9   (*B*) Ion channels can be opened or closed to allow these sodium ions to be transported to the other side of the membrane. When the signal molecule binds to the ion channel protein, the gate is opened.

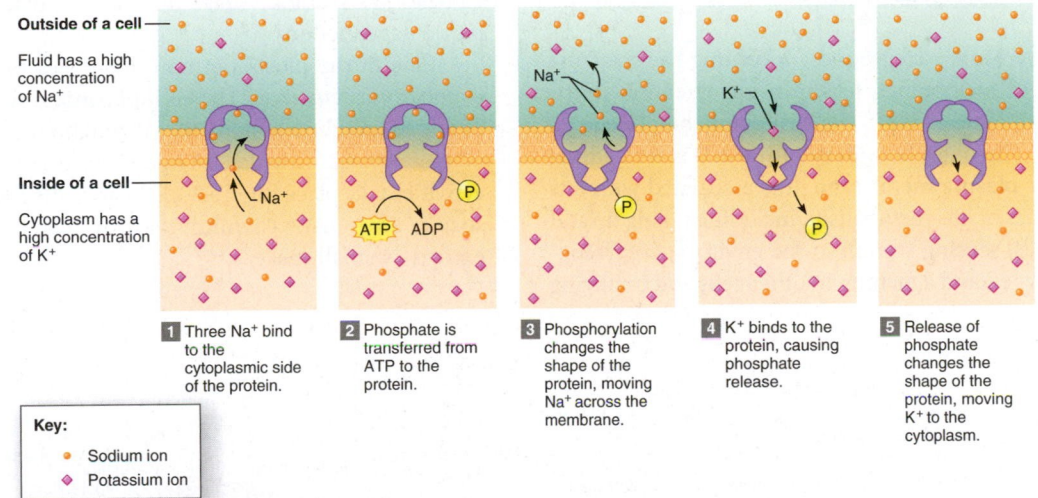

Outside of a cell ——

Fluid has a high concentration of Na⁺

Inside of a cell ——

Cytoplasm has a high concentration of K⁺

1 Three Na⁺ bind to the cytoplasmic side of the protein.

2 Phosphate is transferred from ATP to the protein.

3 Phosphorylation changes the shape of the protein, moving Na⁺ across the membrane.

4 K⁺ binds to the protein, causing phosphate release.

5 Release of phosphate changes the shape of the protein, moving K⁺ to the cytoplasm.

Key:
• Sodium ion
◆ Potassium ion

FIGURE 20.10   The action of the carrier protein requires an input of energy (the compound ATP) other than the kinetic energy of the molecules; therefore, this is an *active* (not passive) process termed *active* transport. Active transport mechanisms can transport molecules or ions up a concentration gradient from a low concentration to a higher concentration.

active transport (figure 20.10). Active transport is very specific: Only certain molecules or ions can be moved in this way, and specific proteins in the membrane must carry them. The action of the carrier requires an input of energy other than the kinetic energy of the molecules; therefore, this process is termed *active* transport. For example, some ions, such as sodium and potassium, are actively pumped across cell membranes. Sodium ions are pumped *out of cells* up a concentration gradient. Potassium ions are pumped *into cells* up a concentration gradient.

In addition to active transport, materials can be transported into a cell by *endocytosis* and out by *exocytosis*. **Phagocytosis** is another name for one kind of endocytosis that is the process cells use to wrap membrane around a particle (usually food) and engulf it (figure 20.11). This is the process leukocytes (white blood cells) use to surround invading bacteria, viruses, and other foreign materials. Because of this, these kinds of cells are called *phagocytes*. When phagocytosis occurs, the material to be engulfed touches the surface of the phagocyte and causes a portion of the outer cell membrane to be indented. The indented cell membrane is pinched off inside the cell to form a sac containing the engulfed material. This sac, composed of a single membrane, is called a **vacuole.** Once inside the cell, the membrane of the vacuole is broken down, releasing its contents inside the cell, or it may combine with another vacuole containing destructive enzymes.

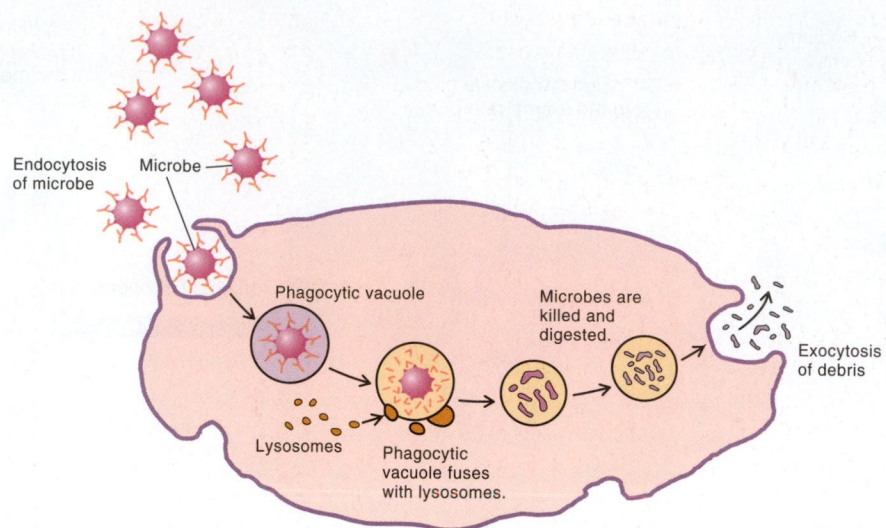

FIGURE 20.11 The sequence illustrates a cell engulfing (endocytosis) a microbe and surrounding it with a membrane. Once encased in a portion of the cell membrane (now called a phagocytic vacuole), a lysosome adds its digestive enzymes to it, which speeds the breakdown of the dangerous microbes. Finally, the hydrolyzed (digested) material moves from the vacuole to the inner surface of the cell membrane, where the contents are discharged (exocytosis).

20.5 ORGANELLES COMPOSED OF MEMBRANES

Now that you have some background concerning the structure and the function of membranes, let's turn our attention to the way cells use membranes to build the structural components of their protoplasm. The outer boundary of the cell is termed the cell membrane, or **plasma membrane.** It is associated with one of the characteristics of life, metabolism, including taking up and releasing molecules, sensing stimuli in the environment, recognizing other cell types, and attaching to other cells and nonliving

objects. In addition to the cell membrane, many other organelles are composed of membranes. Each of these membranous organelles has a unique shape or structure that is associated with particular functions.

One of the most common organelles found in cells is the *endoplasmic reticulum*. The **endoplasmic reticulum** (ER) is a set of folded membranes and tubes throughout the cell. This system of membranes provides a large surface upon which chemical activities take place (figure 20.12). Since the ER has an enormous surface area, many chemical reactions can be carried out in an extremely small space.

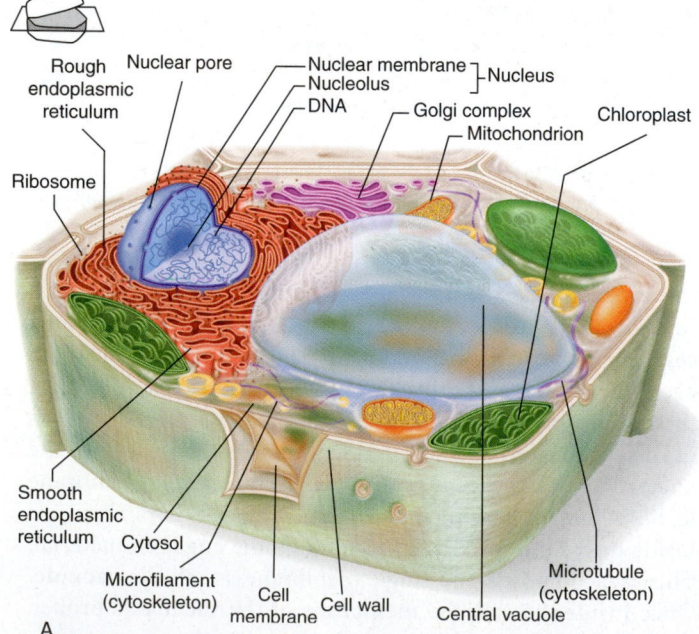

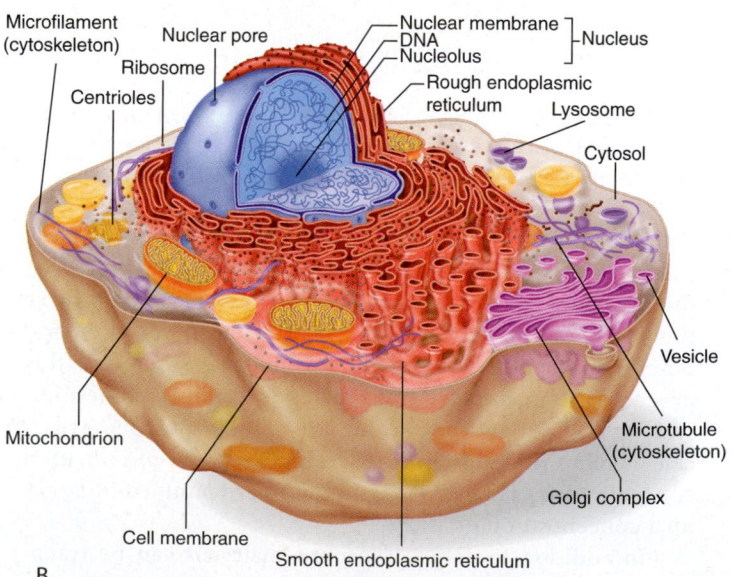

FIGURE 20.12 Certain structures in the cytoplasm are constructed of membranes. Membranes are composed of protein and phospholipids. The membranous cytoplasmic organelles shown are in typical (*A*) plant and (*B*) animal cells.

The **Golgi apparatus** is another organelle composed of membrane. Even though this organelle is also composed of membrane, the way in which it is structured enables it to perform jobs different from those performed by the ER. The typical Golgi is composed of from five to twenty flattened, smooth, membranous sacs, which resemble a stack of pancakes. The Golgi apparatus is the site of the synthesis and packaging of certain molecules produced in the cell. It is also the place where particular chemicals are concentrated prior to their release from the cell or distribution within the cell.

An important group of molecules that is necessary to the cell includes the hydrolytic enzymes. This group of enzymes is capable of destroying carbohydrates, nucleic acids, proteins, and lipids. Since cells contain large amounts of these molecules, these enzymes must be controlled in order to prevent the destruction of the cell. The Golgi apparatus is the site where these enzymes are converted from their inactive to their active forms and packaged in protective membranous sacs. These vesicles are pinched off from the outside surfaces of the Golgi sacs and given the special name **lysosomes,** or "bursting body." Cells use the lysosomes in four major ways:

1. Decompose dead and dying cells.
2. Selectively destroy cells to fashion a developing organism from its immature to mature state (figure 20.13).
3. Digest macromolecules that have been ingested into the cell.
4. Kill dangerous microorganisms that have been taken into the cell by phagocytosis.

A nucleus is a place in a cell—not a solid mass. Just as a room is a place created by walls, a floor, and a ceiling, the nucleus is a place in the cell created by the **nuclear membrane.** This membrane separates the *nucleoplasm,* liquid material in

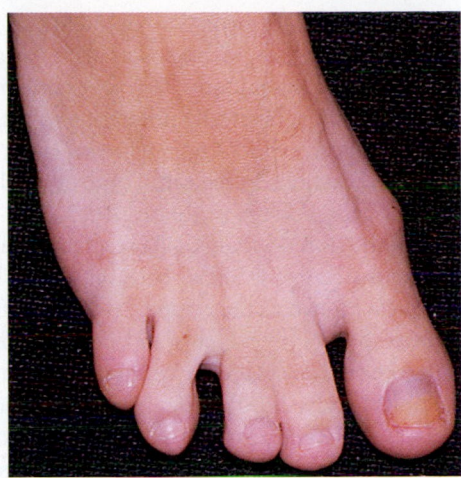

FIGURE 20.13   This person displays the trait known as syndactylism (*syn* = connected; *dactyl* = finger or toe). In most people, lysosomal enzymes break down the cells between the toes, allowing the toes to separate. In this genetic abnormality, these enzymes fail to do their job.

the nucleus, from the cytoplasm. Because they are separated, the cytoplasm and nucleoplasm can maintain different chemical compositions. If the membrane were not formed around the genetic material, the organelle we call the nucleus would not exist. The nuclear membrane is formed from many flattened sacs fashioned into a hollow sphere around the genetic material, DNA (*d*eoxyribo*n*ucleic *a*cid). It also has large openings called nuclear pores that allow thousands of relatively large molecules to pass into and out of the nucleus each minute. These pores are held open by donut-shaped molecules that resemble the "eyes" in shoes through which the shoelace is strung.

All of the membranous organelles just described can be converted from one form to another (figure 20.14). For example,

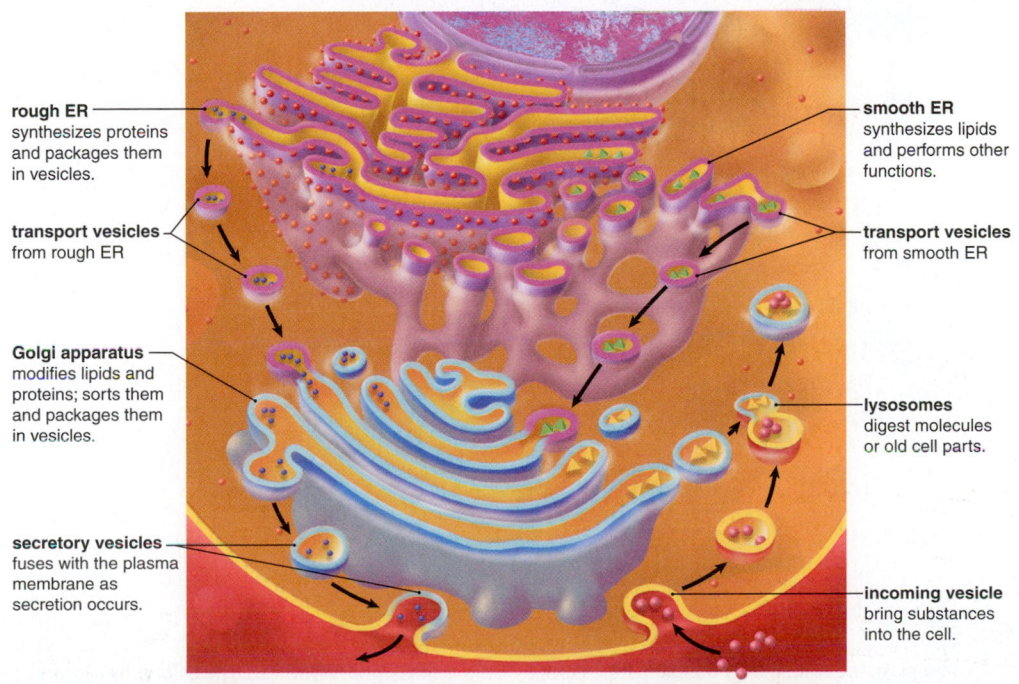

rough ER
synthesizes proteins and packages them in vesicles.

transport vesicles
from rough ER

Golgi apparatus
modifies lipids and proteins; sorts them and packages them in vesicles.

secretory vesicles
fuses with the plasma membrane as secretion occurs.

smooth ER
synthesizes lipids and performs other functions.

transport vesicles
from smooth ER

lysosomes
digest molecules or old cell parts.

incoming vesicle
bring substances into the cell.

FIGURE 20.14   Eukaryotic cells contain a variety of organelles composed of membranes that consist of two layers of phospholipids and associated proteins. Each organelle has a unique shape and function. Many of these organelles are interconverted from one to another as they perform their essential functions.

phagocytosis results in the formation of vacuolar membrane from cell membrane that fuses with lysosomal membrane, which in turn came from Golgi membrane. Two other organelles composed of membranes are chemically different and are incapable of interconversion. Both types of organelles are associated with energy conversion reactions in the cell. These organelles are the *mitochondrion* and the *chloroplast* (figure 20.15).

The **mitochondrion** is an organelle resembling a small bag with a larger bag inside that is folded back on itself. These inner folded surfaces are known as the *cristae*. Located on the surface of the cristae are particular proteins and enzymes involved in *aerobic cellular respiration*. **Aerobic cellular respiration** is the series of reactions involved in the release of usable energy from food molecules, which requires the participation of oxygen molecules. The average human cell contains upwards of ten thousand mitochondria. Cells that are involved in activities that require large amounts of energy, such as muscle cells, contain many more mitochondria. When properly stained, they can be seen with a compound light microscope. In cells that are functioning aerobically, the mitochondria swell with activity. But when this activity diminishes, they shrink and appear as thread-like structures.

A second energy-converting organelle is the **chloroplast.** Some cells contain only one large chloroplast, while others contain hundreds of smaller chloroplasts. A study of the ultrastructure—that is, the structures seen with an electron microscope—of a chloroplast shows that the entire organelle is enclosed by a membrane, while other membranes are folded and interwoven throughout. As shown in figure 20.15A, in some areas, concentrations of these membranes are stacked up or folded back on themselves. Chlorophyll and other photosynthetic molecules are attached to these membranes. These areas of concentrated chlorophyll are called the *grana* of the chloroplast. The space between the grana, which has no chlorophyll, is known as the *stroma*.

These membranous, saclike organelles are found only in plants and algae. In this organelle, light energy is converted to chemical-bond energy in a process known as *photosynthesis*. **Photosynthesis** is the *trapping of radiant energy and its conversion into the energy of chemical bonds*. Chloroplasts contain a variety of photosynthetic pigments including green *chlorophyll*.

Mitochondria and chloroplasts are different from other kinds of membranous structures in several ways. First, their

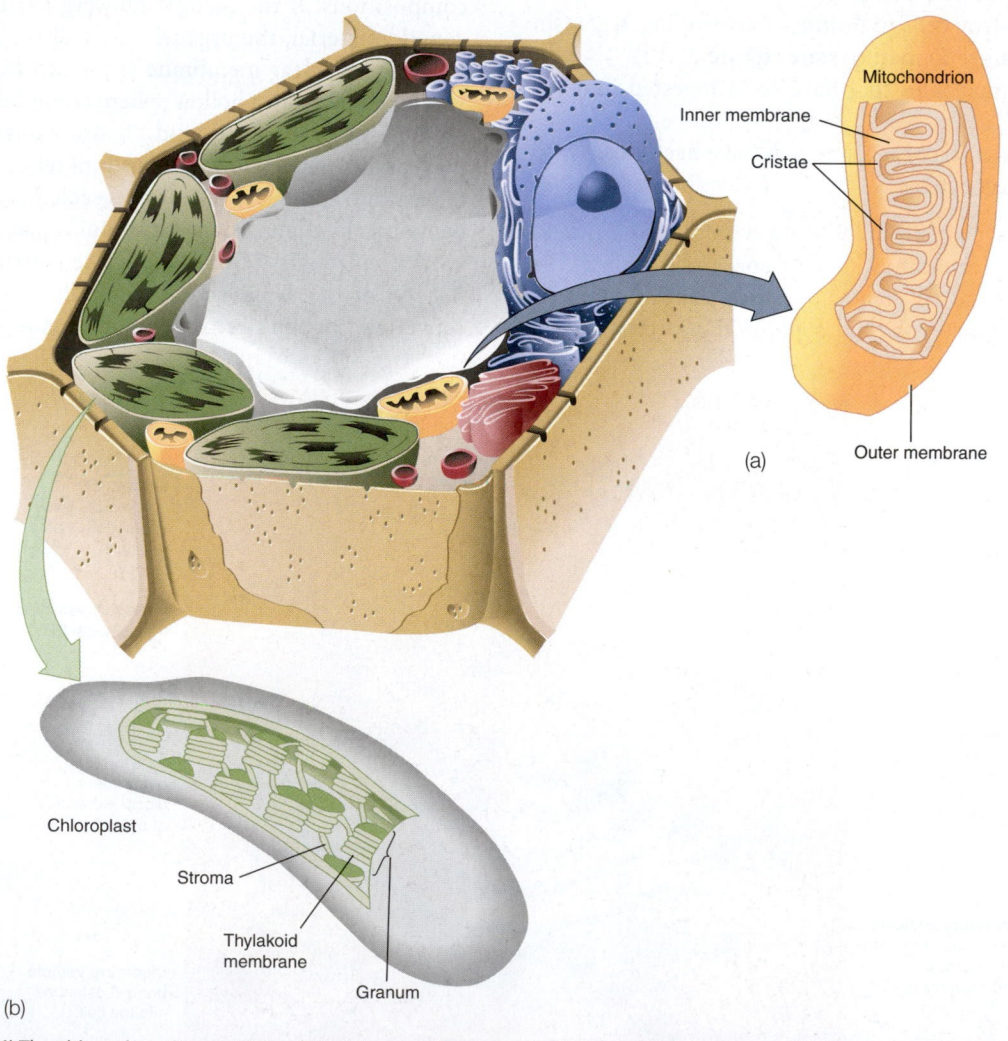

Mitochondrion

Inner membrane

Cristae

(a)

Outer membrane

Chloroplast

Stroma

Thylakoid membrane

Granum

(b)

FIGURE 20.15   (*A*) The chloroplast, the container of the pigment chlorophyll, is the site of photosynthesis. The chlorophyll, located in the grana, captures light energy that is used to construct organic molecules in the stroma. (*B*) The mitochondria with their inner folds, called cristae, are the site of aerobic cellular respiration, where food energy is converted to usable cellular energy. Both organelles are composed of phospholipid and protein membranes.

membranes are chemically different from those of other membranous organelles. Second, they are composed of double layers of membrane—an inner and an outer membrane. Third, both of these structures have ribosomes and DNA that are similar to those of bacteria. Finally, these two structures have a certain degree of independence from the rest of the cell—they have a limited ability to reproduce themselves but must rely on DNA from the cell's nucleus for assistance.

The biochemical pathways associated with mitochondria and chloroplasts are respiration and photosynthesis, both oxidation-reduction reactions.

20.6 NONMEMBRANOUS ORGANELLES

Suspended in the cytoplasm and associated with the membranous organelles are various kinds of structures that are not composed of phospholipids and proteins arranged in sheets. These are nonmembranous organelles.

In the cytoplasm are many very small structures called **ribosomes** that are composed of ribonucleic acid (RNA) and protein. Ribosomes function in the manufacture of protein. Many ribosomes are found floating freely in the cytoplasm and attached to the endoplasmic reticulum (figure 20.16). Cells that are actively producing protein (e.g., liver cells) have great numbers of free and attached ribosomes.

Among the many types of nonmembranous organelles found there are elongated protein structures that are known as *microtubules, microfilaments,* and *intermediate filaments.* Their various functions are as complex as those provided by the structural framework and cables of a high-rise office building, geodesic dome, or skeletal and muscular systems of a large animal. All three types of organelles interconnect, and some

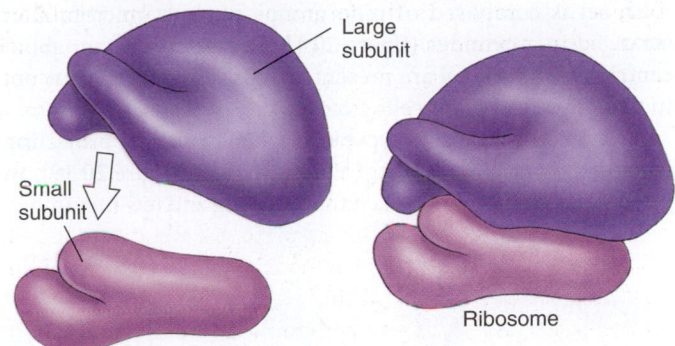

FIGURE 20.16 Each ribosome is constructed of two subunits. Each of the subunits is composed of protein and RNA. These globular organelles are associated with the construction of protein molecules from individual amino acids. The Nobel Prize in Chemistry 2009 was awarded to Drs. Venkatraman Ramakrishan, Thomas A. Steitz, and Ada E. Yonath for determining the structure and function of the ribosomes.

are attached to the inside of the cell membrane, forming what is known as the *cytoskeleton* of the cell (figure 20.17). These cellular components provide the cell with shape, support, and ability to move about the environment. In certain diseases such as ALS (amyotrophic lateral sclerosis, or Lou Gehrig's disease), Alzheimer's, and Down syndrome, there is a buildup of intermediate-type filaments. As a result, researchers are investigating control methods that prevent such buildup.

An arrangement of two sets of microtubules at right angles to each other makes up the structures known as the *centrioles,* and they operate by organizing microtubules into a complex of strings called *spindle fibers.* The *spindle* is the structure upon which chromosomes are attached in order that they may be properly separated during cell division (refer to figure 20.20).

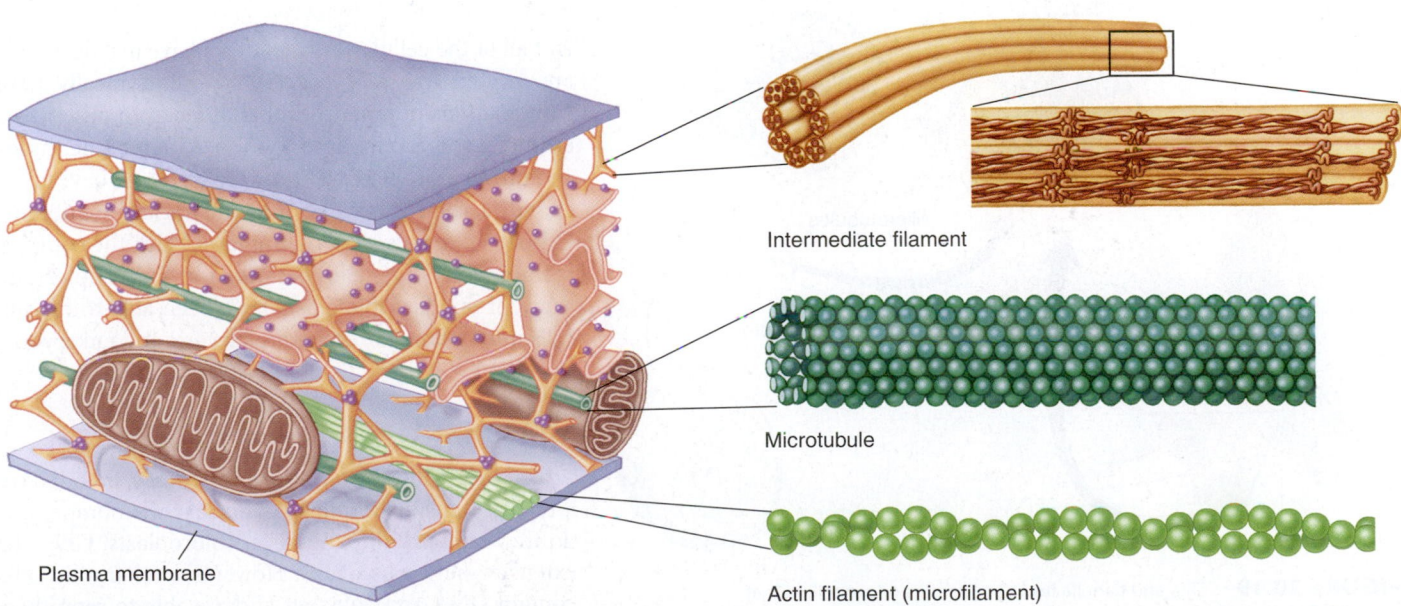

FIGURE 20.17 Microtubules, actin filaments, and intermediate filaments are all interconnected within the cytoplasm of the cell. These structures, along with connections to other cellular organelles, form a cytoskeleton for the cell. The cellular skeleton is not a rigid, fixed-in-place structure but rather changes as the actin and intermediate filaments and microtubule component parts are assembled and disassembled.

Each set is composed of nine groups of short microtubules arranged in a cylinder (figure 20.18). One curious fact about centrioles is that they are present in most animal cells but not in many types of plant cells.

Many cells have microscopic, hairlike structures projecting from their surfaces; these are *cilia* or *flagella* (figure 20.19). In general, we call them flagella if they are long and few in number,

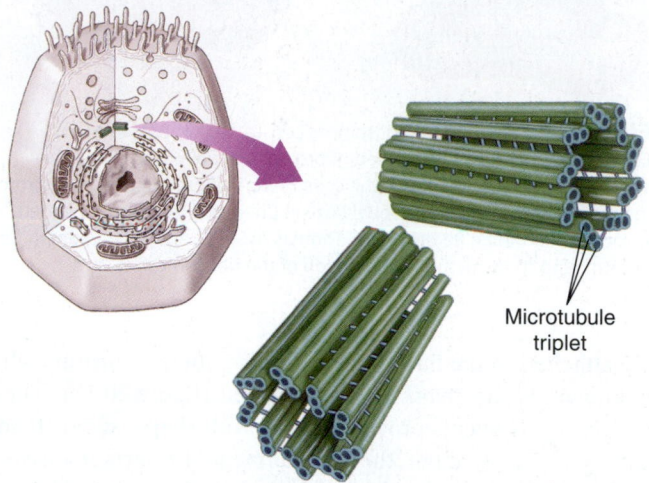

and cilia if they are short and more numerous. The cell has the ability to control the action of these microtubular structures, enabling them to be moved in a variety of different ways. Their coordinated actions either propel the cell through the environment or the environment past the cell surface. The flagella of sperm propel them through the reproductive tract toward the egg. The protozoan *Paramecium* is covered with thousands of cilia that actively beat in a rhythmic motion to move the cell through the water. The cilia on the cells that line your trachea move mucus-containing particles from deep within your lungs.

20.7 NUCLEAR COMPONENTS

When nuclear structures were first identified, it was noted that certain dyes stained some parts more than others. The parts that stained more heavily were called **chromatin,** which means "colored material." Chromatin is composed of long molecules of deoxyribonucleic acid (DNA) in association with proteins. Chromatin is loosely organized DNA/protein strands in the nucleus. When the chromatin is tightly coiled into shorter, denser structures, we call them **chromosomes.** Chromatin and chromosomes are really the same molecules but differ in structural arrangement. In addition to chromosomes, the nucleus may also contain one, two, or several *nucleoli.* A **nucleolus** is composed of granules and fibers in association with the cell's DNA used in the manufacture of ribosomes.

The final component of the nucleus is its liquid matrix called the *nucleoplasm.* It is a colloidal mixture composed of water and the molecules used in the construction of ribosomes, nucleic acids, and other nuclear material.

FIGURE 20.18 These two sets of short microtubules are located just outside the nuclear membrane in many types of cells.

20.8 MAJOR CELL TYPES

Not all of the cellular organelles we have just described are located in every cell. Some cells typically have combinations of organelles that differ from others. For example, some cells have a nuclear membrane, mitochondria, chloroplasts, ER, and Golgi; others have mitochondria, centrioles, Golgi, ER, and nuclear membrane. Other cells are even more simple and lack the complex membranous organelles described in this chapter. Because of this fact, biologists have been able to classify cells into two major types: prokaryotic and eukaryotic.

The Prokaryotic Cell Structure

Prokaryotic cells, the *bacteria* and *archaea,* do not have a typical nucleus bound by a nuclear membrane, nor do they contain mitochondria, chloroplasts, Golgi, or extensive networks of ER. However, prokaryotic cells contain DNA and enzymes and are able to reproduce and engage in metabolism. They perform all of the basic functions of living things with fewer and more simple organelles. Members of the Archaea are of

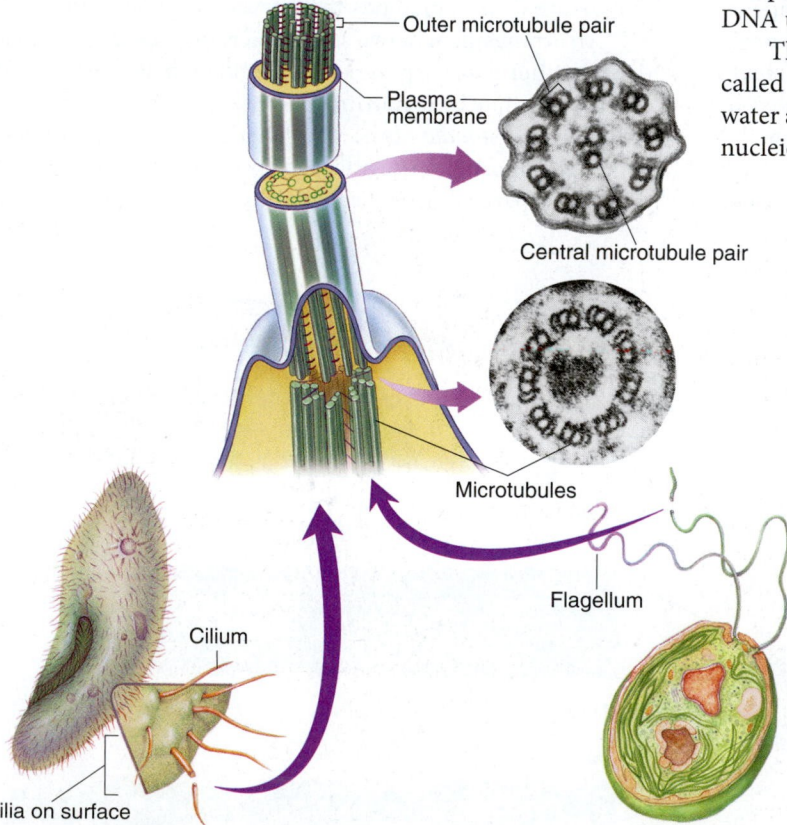

FIGURE 20.19 Cilia and flagella function like oars or propellers that move the cell through its environment or move the environment past the cell. Cilia and flagella are constructed of groups of microtubules, as in the ciliated protozoan shown on the left and the flagellated alga on the right. Flagella are usually less numerous and longer than cilia.

How We Are Related

Now that you have an idea of how cells are constructed, we can look at the great diversity of the kinds of cells that exist. You already know that there are significant differences between prokaryotic and eukaryotic cells.

Because prokaryotic (noneukaryotic) and eukaryotic cells are so different and prokaryotic cells show up in the fossil records much earlier, the differences between the two kinds of cells are used to classify organisms. Thus, biologists have classified organisms into three large categories, called **domains**. The following diagram illustrates how living things are classified:

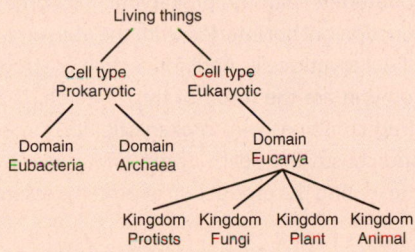

The Domain Bacteria contains most of the microorganisms and can be found in a wide variety of environments. The Domain Archaea contains many kinds of microorganisms that have significant biochemical differences from the Bacteria. Many of the Archaea have special metabolic abilities and live in extreme environments of high temperature or extreme saltiness. Although only a few thousand Bacteria and only about 200 Archaea have been described, recent DNA studies of seawater and soil suggest that there are millions of undescribed species. In all likelihood, these noneukaryotic organisms far outnumber all the species of eukaryotic organisms combined. All other living things are comprised of eukaryotic cells.

little concern to the medical profession because none has been identified as disease-causing. They are typically found growing in extreme environments where the pH, salt concentration, or temperatures make it impossible for most other organisms to survive. The other prokaryotic cells are called bacteria, and about 5 percent cause diseases such as tuberculosis, strep throat, gonorrhea, and acne. Other bacteria are responsible for the decay and decomposition of dead organisms. Although some bacteria have a type of green photosynthetic pigment and carry on photosynthesis, they do so without chloroplasts and use different chemical reactions.

Most prokaryotic cells are surrounded by a *capsule,* slime layer, or spore coat that can be composed of a variety of compounds (figure 20.20). In certain bacteria, this layer is responsible for their ability to stick to surfaces (including host cells) and to resist phagocytosis. Many bacteria also have fimbriae, hairlike protein structures, which help the cell stick to objects. Those with flagella are capable of propelling themselves through the environment. Below the capsule is the rigid cell wall composed of a unique protein-carbohydrate complex called peptidoglycan. This complex provides the cell with the strength to resist osmotic pressure changes and gives the cell shape. Just beneath the wall is the cell membrane. Thinner and with a slightly different chemical composition from eukaryotes, it carries out the same functions as the cell membranes in eukaryotes. Most bacteria are either rods (bacilli), spherical (cocci), or curved (spirilla). The genetic material within the cytoplasm is DNA in the form of a loop.

The Eukaryotic Cell Structure

Eukaryotic cells contain a true nucleus and most of the membranous organelles described earlier. Eukaryotic organisms can be further divided into several categories based on the specific combination of organelles they contain. The cells of plants, fungi, protozoa and algae, and animals are all eukaryotic. The most obvious characteristic that sets the plants and algae apart

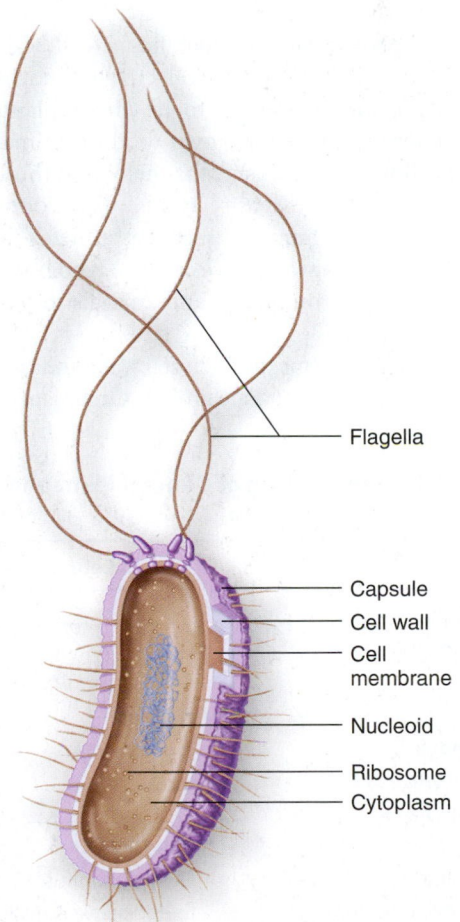

FIGURE 20.20 All bacteria are prokaryotic cells. While smaller and less complex than eukaryotic cells, each is capable of surviving on its own. Most bacteria are involved with decay and decomposition, but a small percentage are pathogens responsible for disease such as strep throat, TB, syphilis, and gas gangrene. The cell illustrated here is a bacillus because it has a rod shape.

A Closer Look

Antibiotics and Cell Structural Differences

One significant difference between prokaryotic and eukaryotic cells is in the chemical makeup of their ribosomes. The ribosomes of prokaryotic cells contain different proteins from those found in eukaryotic cells. Prokaryotic ribosomes are also smaller. This discovery was important to medicine because many cellular forms of life that cause common diseases are bacterial. As soon as differences in the ribosomes were noted, researchers began to look for ways in which to interfere with the prokaryotic ribosome's function but *not* interfere with the ribosomes of eukaryotic cells. Antibiotics such as streptomycin are the result of this research. This drug combines with prokaryotic ribosomes and causes the death of the prokaryote by preventing the production of proteins essential to its survival. Because eukaryotic ribosomes differ from prokaryotic ribosomes, streptomycin does not interfere with the normal function of ribosomes in human cells. Keep in mind that viruses have no cellular structures including ribosomes. Therefore, antibacterial antibiotics are unable to control virus infections such as the common cold, virally caused flu, or herpes.

---

from other organisms is their green color, which indicates that the cells contain chlorophyll. Chlorophyll is necessary for the process of photosynthesis—the conversion of light energy into chemical-bond energy in food molecules. These cells, then, are different from the other cells in that they contain chloroplasts in their cytoplasm. Another distinguishing characteristic of plants and algae is the presence of cellulose in their cell walls (table 20.2).

The group of organisms that has a cell wall but lacks chlorophyll in chloroplasts is collectively known as *fungi*. They were previously thought to be either plants that had lost their ability to make their own food or animals that had developed cell walls. Organisms that belong in this category of eukaryotic cells include yeasts, molds, mushrooms, and the fungi that cause such human diseases as athlete's foot, "jungle rot," and ringworm. Now we have come to recognize this group as different enough from plants and animals to place them in a separate kingdom.

Eukaryotic organisms that lack cell walls and cannot photosynthesize are placed in separate groups. Organisms that consist of only one cell are called protozoans—examples are *Amoeba* and *Paramecium*. They have all the cellular organelles described in this chapter except the chloroplast; therefore, protozoans must consume food as do the fungi and the multicellular animals.

Although the differences in these groups of organisms may seem to set them worlds apart, their similarity in cellular structure is one of the central themes unifying the field of biology. You can obtain a better understanding of how cells operate in general by studying specific examples. Because the organelles have the same general structure and function regardless of the kind of cell in which they are found, you can learn more about how mitochondria function in plants by studying how mitochondria function in animals. There is a commonality among all living things with regard to their cellular structure and function.

PART II
ENERGY TRANSFORMATIONS IN CELLS

This part of the chapter explains two of the most fundamental biochemical processes performed by organisms. These processes are the energy transformation functions that keep all living things alive.

20.9 RESPIRATION AND PHOTOSYNTHESIS

In living things, oxidation-reduction reactions are metabolic processes that do not take place in one single step but occur in a series of small steps. This allows the potential energy in the oxidized molecule to be released in smaller, more useful amounts. A unique protein catalyst (enzyme) controls each of the steps. Each step begins with a molecule or molecules that serve as the reactant and are called the *substrate*. When the reaction is complete, the substrate is converted to a product, which in turn becomes the new substrate for the next enzyme-controlled reaction. Such a series of enzyme-controlled reactions is often called a *biochemical pathway* or a *metabolic pathway*:

| Enzyme 1 | Enzyme 2 | Enzyme 3 | Enzyme 4 |

Substrate A → Substrate B → Substrate C → Substrate D → End product

TABLE 20.2 Comparison of General Plant and Animal Cell Structure

Plant Cells	Animal Cells
CELL WALL	_____
Cell membrane	Cell membrane
Cytoplasm	Cytoplasm
Nucleus	Nucleus
Mitochondria	Mitochondria
CENTRAL VACUOLE	_____
CHLOROPLASTS	
_____	**CENTRIOLE**
Golgi apparatus	Golgi apparatus
Endoplasmic reticulum	Endoplasmic reticulum
Lysosomes	Lysosomes
Vacuoles/vesicles	Vacuoles/vesicles
Ribosomes	Ribosomes
Nucleolus	Nucleolus
Inclusions	Inclusions
Cytoskeleton	Cytoskeleton

Such pathways can be used to create molecules (e.g., photosynthesis), release energy (e.g., aerobic cellular respiration), and perform many other actions. One of the amazing facts of nature is that most organisms use the same basic biochemical or metabolic pathways. So, if you study aerobic cellular respiration in an elephant, it will be essentially the same as that in a petunia, shark, or earthworm. However, since the kinds of enzymes an organism is able to produce depend on the genes that it has, we should expect some variation in the details of biochemical pathways in different organisms. The fact that so many organisms use essentially the same biochemical processes is a strong argument for the idea of evolution from common ancestors. Once a successful biochemical strategy evolved, the genes and the pathway were retained by evolutionary descendants with slight modifications of the scheme. Two such pathways of importance are aerobic cellular respiration and photosynthesis, both of which involve the transfer of chemical bond energy in the form of ATP.

The Energy Transfer Molecules of Living Things—ATP

To transfer the right amount of chemical-bond energy from energy-releasing to energy-requiring reactions, cells use the molecule ATP. **Adenosine triphosphate (ATP)** is a handy source of the right amount of usable chemical-bond energy. Each ATP molecule used in the cell is like a rechargeable AAA battery used to power small toys and electronic equipment. Each contains just the right amount of energy to power the job. When the power has been drained, it can be recharged numerous times before it must be recycled.

Recharging the AAA battery requires getting a small amount of energy from a source of high energy such as a hydroelectric power plant (figure 20.21). Energy from the electric plant is too powerful to directly run a small flashlight

or cell phone. If you plug your cell phone directly into the power plant, the cell phone would be destroyed. However, the recharged AAA battery delivers just the right amount of energy at the right time and place. ATP functions in much the same manner. After the chemical-bond energy has been drained by breaking one of its bonds:

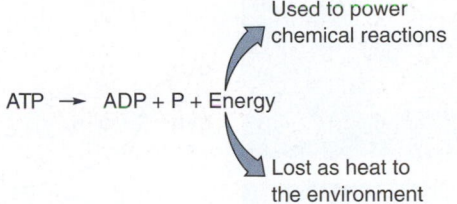

the discharged molecule (ADP) is recharged by "plugging it in" to a high-powered energy source. This source may be (1) sunlight (photosynthesis) or (2) chemical-bond energy (released from cellular respiration):

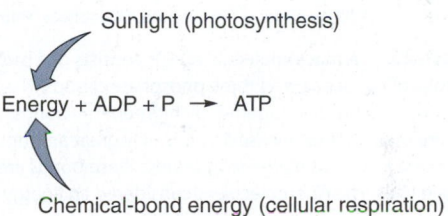

An ATP molecule is formed from adenine (nitrogenous base), ribose (sugar), and phosphates (figure 20.22). These three are chemically bonded to form AMP, *adenosine monophosphate* (one phosphate). When a second phosphate group is added to the AMP, a molecule of ADP (diphosphate) is formed. The ADP, with the addition of more energy, is able to bond to a third phosphate group and form ATP. The covalent bond that attaches the second phosphate to the AMP molecule is easily broken to release energy for energy-requiring cell processes. Because the energy in this bond is so easy for a cell to use, it is called a *high-energy phosphate*

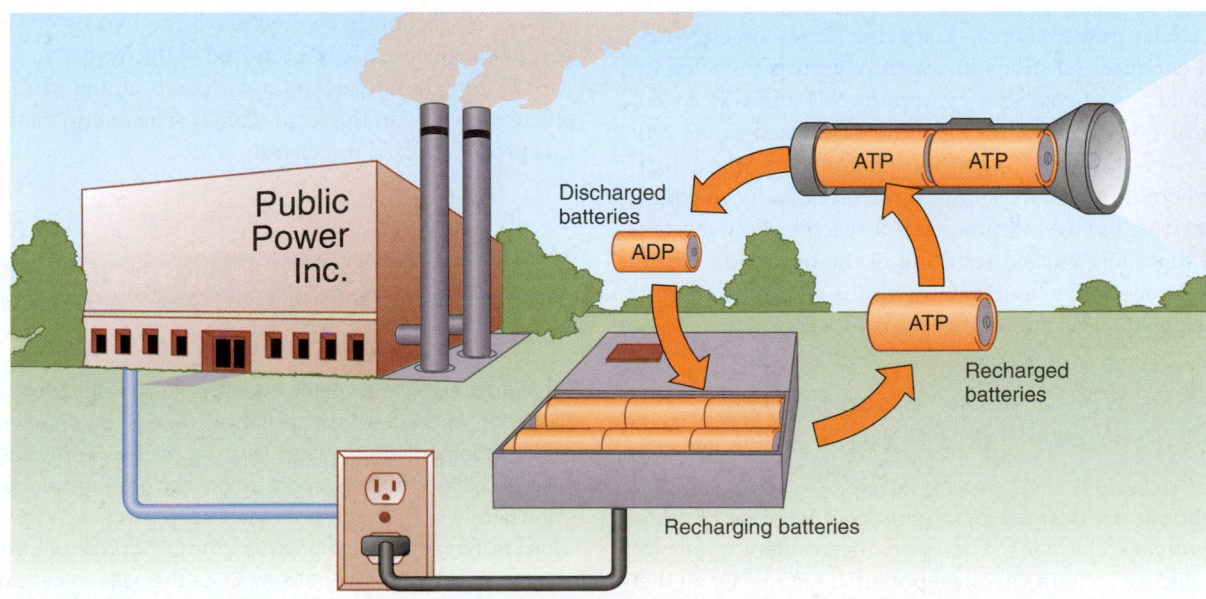

FIGURE 20.21 When rechargeable batteries in a flashlight have been drained of their power, they can be recharged by placing them in a specially designed battery charger. This enables the right amount of power from a power plant to be packed into the batteries for reuse. Cells operate in much the same manner. When the cell's "batteries," ATPs, become drained as a result of powering a job such as muscle contraction, these discharged "batteries," ADPs, can be recharged back to full ATP power.

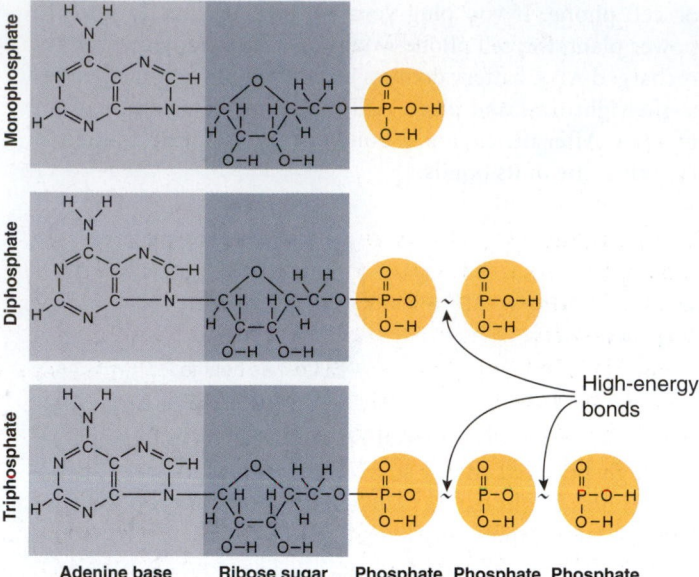

Monophosphate

Diphosphate

High-energy bonds

Triphosphate

Adenine base   Ribose sugar   Phosphate   Phosphate   Phosphate

FIGURE 20.22 A macromolecule of ATP consists of a molecule of adenine, a molecule of ribose, and three phosphate groups. The two end phosphate groups are bonded together by high-energy bonds. When these bonds are broken, they release an unusually great amount of energy; therefore, they are known as *high-energy bonds*. These bonds are represented by curved, solid lines. The ATP molecule is considered an energy carrier.

bond. ATP has two high-energy phosphate bonds represented by curved solid lines. Both ADP and ATP, because they contain high-energy bonds, are very unstable molecules and readily lose their phosphates. When this occurs, the energy held in the high-energy bonds of the phosphate can be transferred to another molecule or released to the environment. It is this ATP energy that is used by cells to perform all the characteristics that keep them alive.

Aerobic Cellular Respiration

Aerobic cellular respiration is a specific series of enzyme-controlled chemical reactions in which oxygen is involved in the breakdown of glucose into carbon dioxide and water, and the chemical-bond energy from glucose is released to the cell in the form of ATP (adenosine triphosphate). While the actual process of aerobic cellular respiration involves many enzyme-controlled steps, the overall process is a reaction between the substrates sugar and oxygen resulting in the formation of the products carbon dioxide and water with the release of energy. The following equation summarizes this process:

$$\text{glucose} + \text{oxygen} \rightarrow \overset{\text{carbon}}{\text{dioxide}} + \text{water} + \text{energy}$$
$$C_6H_{12}O_6 + O_2 \rightarrow CO_2 + H_2O + ATP + \text{heat}$$

Covalent bonds are formed by atoms sharing pairs of fast-moving, energetic electrons. Therefore, the covalent bonds in the sugar glucose contain chemical potential energy. Of all the covalent bonds in glucose (O—H, C—H, C—C), those easiest to get at are the C—H and O—H bonds on the outside of the molecule. The chemical activities that remove electrons from glucose result in the glucose being oxidized. When these bonds are broken, several things happen:

1. Some ATP is produced;
2. Hydrogen ions (H^+ or protons) are produced; and,
3. Electrons are released and taken up by special carrier molecules.

The ATP is used to power the metabolic activities of the cell. The chemical activities that remove electrons from glucose result in the glucose being oxidized.

The electrons produced during aerobic cellular respiration must be controlled. If they were allowed to fly about at random, they would quickly combine with other molecules, causing cell death. Electron transfer molecules temporarily hold the electrons and transfer them to other electron transfer molecules. ATP is formed when these transfers take place. Once energy has been removed from electrons for ATP production, the electrons must be placed in a safe location. In *aerobic* cellular respiration, these electrons are ultimately attached to oxygen. Oxygen serves as the final resting place of the less-energetic electrons. When the electrons are added to oxygen, it becomes a negatively charged ion, O=. Since the oxygen has gained electrons, it has been *reduced*. So, in the aerobic cellular respiration of glucose, glucose is oxidized and oxygen is reduced. If something is oxidized (loses electrons), something else must be reduced (gains electrons). A simple way to help identify an oxidation-reduction reaction is to use the mnemonic device "LEO the lion says GER." LEO stands for "Loss of Electrons is Oxidation" and GER stands for "Gain of Electrons is Reduction." A molecule cannot simply lose its electrons—they have to go someplace! Eventually, the positively charged hydrogen ions that were released from the glucose molecule combine with the negatively charged oxygen ion to form water.

Once all the hydrogens have been stripped off the glucose molecule, the remaining carbon and oxygen atoms are rearranged to form individual molecules of CO_2. The oxidation-reduction reaction is complete. All the hydrogen originally a part of the glucose has been moved to the oxygen to form water. All the remaining carbon and oxygen atoms of the original glucose are now in the form of CO_2. The energy released from this process is used to generate ATP.

CONCEPTS APPLIED

Activation of Dried Yeast

A simple exercise to demonstrate cellular respiration is to add a packet of dried yeast to a narrow-mouthed bottle with a few tablespoons of warm water and a tablespoon of table sugar. Place a balloon over the top, tie it on with a string or rubber band, and gently swirl. As the yeast metabolizes the sugar through cellular respiration, it generates CO_2. The CO_2 that results from respiration will blow up the balloon. Once the balloon has expanded, wait a couple of days before you tear down this demonstration. What happens to the balloon? How can you explain this?

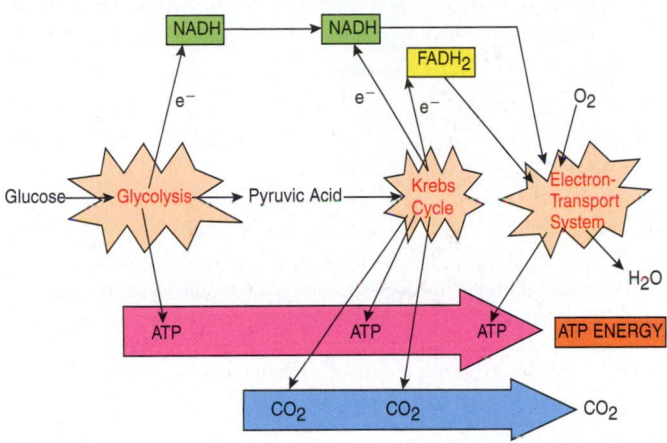

Glucose + specific sequence of reactions controlled by enzymes + O_2 ⟶ H_2O + CO_2 + ATP

A

B

FIGURE 20.23 (*A*) This sequence of reactions in the aerobic oxidation of glucose is an overview of the energy-yielding reactions of a cell. (*B*) Glycolysis, the Krebs cycle, and the electron-transport system are each a series of enzyme-controlled reactions that extract energy from the chemical bonds in a glucose molecule. During glycolysis, glucose is split into pyruvic acid and ATP and electrons are released. During the Krebs cycle, pyruvic acid is further broken down to carbon dioxide with the release of ATP and electrons. During the electron-transport system, oxygen is used to accept electrons and water is produced. A very large amount of ATP is also produced. Glycolysis takes place in the cytoplasm of the cell. Pyruvic acid enters mitochondria, where the Krebs cycle and the electron-transport system take place.

In cells, these oxidation reactions take place in a particular order and in particular places within the cell. In eukaryotic cells, the process of releasing energy from food molecules begins in the cytoplasm and is completed in the mitochondrion. Three distinct enzymatic pathways are involved (figure 20.23):

1. *Glycolysis* (*glyco* = sugar; *lysis* = to split) is a series of enzyme-controlled reactions that takes place in the cytoplasm of cells and results in the breakdown of glucose with the release of electrons and the formation of ATP. The electrons are sent to the electron-transport system (ETS) for processing.
2. The *Krebs cycle* is a series of enzyme-controlled reactions that takes place inside the mitochondrion and completes the breakdown of the remaining fragments of the original glucose with the release of carbon dioxide, more ATP, and more electrons are sent to the electron-transport system (ETS) for processing.
3. The *electron-transport system (ETS)* occurs in the mitochondria. It is a series of enzyme-controlled reactions that converts the kinetic energy of electrons it receives from glycolysis and the Krebs cycle to ATP. In fact, the majority of ATP produced during aerobic cellular respiration comes from the ETS. The electrons are transferred through a series of oxidation-reduction reactions involving enzymes until eventually, the electrons are accepted by oxygen atoms to form oxygen ions (O=). The negatively charged oxygen atoms attract two positively charged hydrogen ions to form water (H_2O).

Photosynthesis

Ultimately, the energy to power all organisms comes from the Sun's radiant energy. Chlorophyll is a green pigment that absorbs light energy for the process of photosynthesis. Through the process of photosynthesis, plants, algae, and certain bacteria transform light energy to chemical-bond energy in the form of ATP and then use ATP to produce complex organic molecules such as glucose. In algae and the leaves of green plants, photosynthesis occurs in cells that contain chloroplasts.

The following equation summarizes the chemical reactions green plants and many other photosynthetic organisms use to make ATP and organic molecules:

$$\text{light energy} + \text{carbon dioxide} + \text{water} \rightarrow \text{glucose} + \text{oxygen}$$
$$\text{light energy} + CO_2 + H_2O \rightarrow C_6H_{12}O_6 + O_2$$

There are three distinct events in the photosynthetic pathway:

1. *Light-capturing events.* In eukaryotic cells, the process of photosynthesis takes place within chloroplasts. Each chloroplast is surrounded by membranes and contains the green pigment, chlorophyll, along with other pigments. Chlorophyll and other *accessory* pigments (yellow, red, and orange) absorb specific wavelengths of light (figure 20.24). After specific amounts of light are absorbed by photosynthetic pigments, their electrons become "excited." With this added energy, these excited electrons are capable of entering into the chemical reactions responsible for the production of ATP.

FIGURE 20.24 In the fall, a layer of waterproof tissue forms at the base of each leaf, cutting off the flow of water and other nutrients. The cells of the leaf die and their chlorophyll disintegrates. The color change seen in leaves in the fall in certain parts of the world is the result of the breakdown of the green chlorophyll. Other pigments (red, yellow, orange, brown) are always present but are masked by the green chlorophyll pigments. When the chlorophyll disintegrates, the reds, oranges, yellows, and browns are revealed.

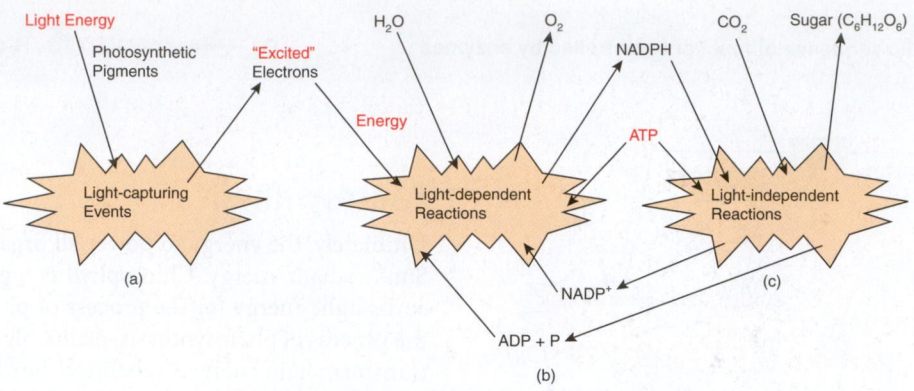

FIGURE 20.25 Photosynthesis is a complex biochemical pathway in plants, algae, and certain bacteria. The upper portion of this figure shows the overall process. Sunlight, along with CO_2 and H_2O, is used to make organic molecules such as sugar. The lower portion illustrates the three parts of the process: (A) the light-capturing events, (B) the light-dependent reactions, and (C) the light-independent reactions. Notice that the end products of the light-dependent reactions, electron carriers and ATP, are necessary to run the light-independent reactions, while the water and carbon dioxide are supplied from the environment.

2. *Light-dependent reactions.* The light-dependent reactions utilize the excited electrons produced by the light-capturing activities. The light-dependent reactions are also known as the *light reactions.* During these reactions, "excited" electrons from the light-capturing reactions are used to do two different things. Some of the "excited" electrons are used to make ATP, while others are used to split water into hydrogen and oxygen. The oxygen from the water is released to the environment as O_2 molecules, and the hydrogens are transferred to an electron carrier.

3. *Light-independent reactions.* These reactions are also known as the *dark reactions,* since light is not needed for the reactions to take place. During these reactions, ATP and electron-carrying molecules from the light-dependent reactions are used to attach CO_2 to a five-carbon molecule, already present in the cell, to manufacture new, larger organic molecules. Ultimately, glucose ($C_6H_{12}O_6$) is produced. The ADP and the electron carriers produced during the light-independent reactions are recycled back to the light-dependent reactions to be used over again (figure 20.25).

Because prokaryotic cells lack mitochondria and chloroplasts, they carry out photosynthesis and cellular respiration within the cytoplasm, on the inner surfaces of the cell membrane, or on other special membranes. Photosynthesis and cellular respiration both involve a series of chemical reactions that control the flow of energy.

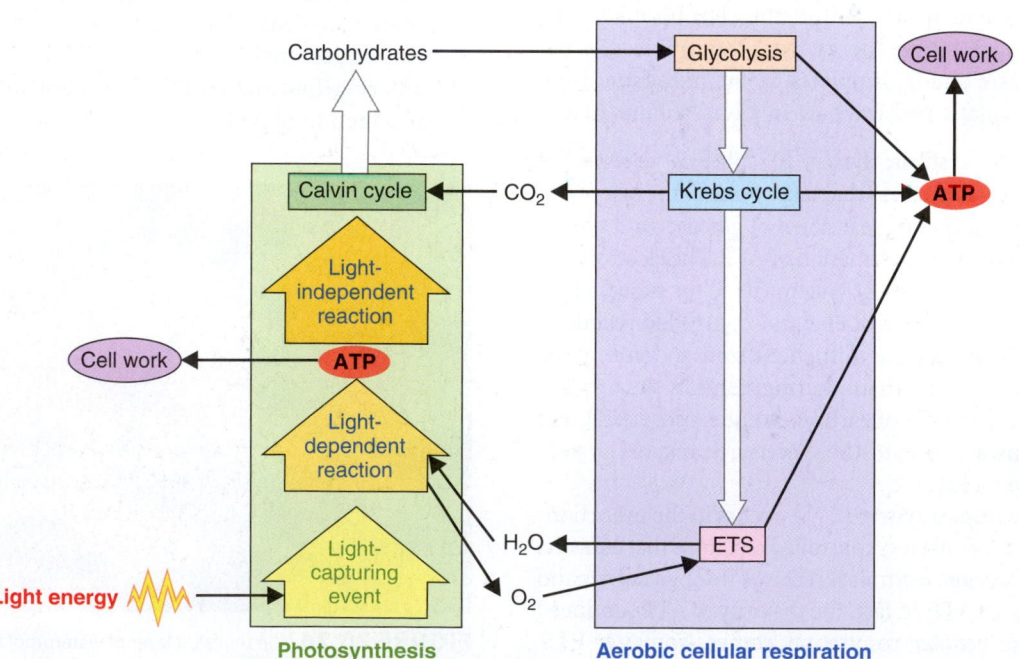

FIGURE 20.26 Although both autotrophs and heterotrophs carry out cellular respiration, the photosynthetic process that is unique to photosynthetic autotrophs provides essential nutrients for both processes. Photosynthesis captures light energy, which is ultimately transferred to heterotrophs in the form of carbohydrates and other organic compounds. Photosynthesis also generates O_2, which is used in aerobic cellular respiration. The ATP generated by cellular respiration in both heterotrophs, (e.g., animals) and autotrophs (e.g., plants) is used to power their many metabolic processes. In return, cellular respiration supplies two of the most important basic ingredients of photosynthesis, CO_2 and H_2O.

Stem cells have been a hot topic in the news. Controversy surrounds their use in research because of their sources and how they are viewed in the debate over the definition of life. Just what are stem cells, how are they acquired, and of what value might they be?

Stem cells are primitive, undifferentiated cells that have the ability to become many other types of cells, including liver, skin, and brain cells. Theoretically, they can divide for an indefinite period in laboratory culture. They are found in many tissues, such as circulating blood and the red bone marrow of certain bones, including the pelvis, and in embryos, or they can be generated from embryos in the lab. Adult stem cells can differentiate to produce all the specialized cell types of the tissue from which they originated. Embryonic stem cells are undifferentiated cells from the embryo. Their ability to differentiate into specialized cells may make them extremely valuable as replacement cells for individuals who have lost cells as a result of illness or injury.

There are three sources of stem cells: lab cultures, isolation from embryos, and adult donors. One method of obtaining embryonic stem cells is to use donated sperm and an egg and fertilize them in a Petri dish. The fertilized egg or zygote is then allowed to undergo mitosis for several days until it becomes a ball of cells. Some of these cells are then removed and grown in a separate dish, serving as a source of embryonic stem cells.

Since embryonic stem cells can differentiate into many other types of human cells, researchers are hoping that they could potentially be provided to people to resolve many problems. Such problems might include: the repair of spinal injuries to restore function to paralyzed limbs, the replacement of useless scar tissue after a heart attack, or adding cells to have them produce chemicals not normally produced as a result of genetic or other disease (e.g., dopamine in Parkinson's patients, insulin in diabetics, specific receptor sites that help people resist HIV or bubonic plague infections). Embryonic stem cells might also be used to provide sickle-cell disease patients with the healthy cells they need to produce normal red blood cells. This is still in the research stage, and there is still much to be accomplished before such therapies are available to the public.

Adults can also be a source of stem cells. People with certain diseases (e.g., chronic lymphocytic leukemia, CLL) or those receiving chemotherapy or radiation for cancer, lose stem cells and therefore have lower numbers of vital red and white blood cells. However, it is possible to replace these cells with healthy, blood-forming stem cells from an adult donor, provided there is a tissue match. There are two sources of adult stem cells: bone marrow and circulating blood.

Adult stem cells can be collected during an outpatient procedure by using a hypodermic needle and syringe. The sterile needle is inserted into the pelvis and marrow is withdrawn. Stem cells can also be taken in a manner similar to whole blood donation called apheresis. During apheresis, the donor's blood is passed through a machine that separates the cells from their plasma and their stem cells are separated out from other types of blood cells. Once this process is completed, the unneeded cells and plasma are returned to the donor. This may take several hours. To improve the harvest, a stem cell growth factor is given to the donor before apheresis. This stimulates mitosis of the stem cells and increases their numbers. Once harvested, the stem cells are isolated and concentrated for injection into the recipient. If successful, the donated stem cells will repopulate the recipient's marrow, continue to undergo mitosis, and become a source of the red and white blood cells needed by the recovering patient.

Under development is a technique that creates stem cells from patients with certain genetic diseases, such as Amyotrophic Lateral Sclerosis, or ALS (Lou Gehrig's disease), for use in medical research. By adding foreign genes to adult skin cells called fibroblasts of patients with such a disease, the cells are reprogrammed to become embryonic. They are then exposed to signaling molecules that cause them to reproduce and differentiate into cells that display the disease characteristics of that patient such as neurons. These customized cells can be used to screen the potential effectiveness of new drugs.

Many people believe that plants only give off oxygen and never require it. This is incorrect! Plants do give off oxygen in the light-dependent reactions of photosynthesis, but in aerobic cellular respiration, they use oxygen as does any other organism. During their life spans, green plants give off more oxygen to the atmosphere than they take in for use in respiration. The surplus oxygen given off is the source of oxygen for aerobic cellular respiration in both plants and animals. Animals are not only dependent on plants for oxygen but are ultimately dependent on plants for the organic molecules necessary to construct their bodies and maintain their metabolism (figure 20.26).

All of the organelles just described are composed of membranes. Many of these membranes are modified for particular functions. Each membrane is composed of the double phospholipid layer with protein molecules associated with it.

PART III
CELLULAR REPRODUCTION

The final portion of the chapter reviews the basic reproductive activities that occur in living cells. Without cellular reproduction, all life on our planet would become extinct.

20.10 THE IMPORTANCE OF CELL DIVISION

The process of cell division is a generative process, one of the characteristics of life. It replaces dead cells with new ones, repairs damaged tissues, and allows living organisms to grow. For example, you began as a single cell that resulted from the union of a sperm and an egg. One of the *first* activities of this single cell was to divide. As this process continued, the number of cells in your body increased, so that as an adult your body consists of several trillion cells.

The *second* function of cell division is to maintain the body. Certain cells in your body, such as red blood cells and cells of the gut lining and skin, wear out. As they do, they must be replaced with new cells. Altogether, you lose about fifty million cells per second; this means that millions of cells are dividing in your body at any given time.

A *third* purpose of cell division is repair. When a bone is broken, the break heals because cells divide, increasing the number of cells available to knit together the broken pieces. If some skin cells are destroyed by a cut or abrasion, cell division produces new cells to repair the damage.

During eukaryotic cell division, two events occur. The replicated genetic information of a cell is equally distributed to two daughter nuclei in a process called **mitosis** (Latin, from Greek *mitos,* meaning "thread"). As the nucleus goes through its division, the cytoplasm also divides into two new cells. This division of the cell's cytoplasm is called **cytokinesis**—cell splitting. Each new cell gets one of the two daughter nuclei so that both have a complete set of genetic information.

20.11 THE CELL CYCLE

All eukaryotic cells go through the same basic life cycle, but they vary in the amount of time they spend in the different stages. A generalized picture of a cell's life cycle may help you understand it better (figure 20.27). Once begun, cell division is a continuous process without a beginning or an end. It is a cycle in which cells continue to grow and divide. There are five stages to the life cycle of a eukaryotic cell: (1) G_1, gap (growth)—phase one; (2) S, synthesis; (3) G_2, gap (growth)—phase two; (4) cell division (mitosis and cytokinesis); and (5) G_0, gap (growth)—mitotic dormancy or differentiation.

During the G_0 phase, cells are not considered to be in the cycle of division but become *differentiated* or specialized in their function. It is at this time that they "mature" to play the role specified by their genetic makeup. Whereas some cells entering the G_0 phase remain there more or less permanently (e.g., nerve cells), others have the ability to move back into the cell cycle of mitosis—G_1, S, and G_2—with ease (e.g., skin cells).

The first three phases of the cell cycle—G_1, S, and G_2—occur during a period of time known as interphase. **Interphase** is the stage between cell divisions. During the G_1 stage, the cell grows in volume as it produces tRNA, mRNA, ribosomes, enzymes, and other cell components. During the S stage, DNA replication occurs in preparation for the distribution of genes to daughter cells. During the G_2 stage that follows, final preparations are made for mitosis with the synthesis of spindle-fiber proteins.

During interphase, the cell is not dividing but is engaged in metabolic activities such as photosynthesis or glandular-cell secretion. During interphase, the nuclear membrane is intact and the individual chromosomes are not visible. The individual loosely coiled *chromatin* threads are too thin and tangled to be seen. Remember that chromosomes are highly coiled strands of DNA and contain coded information that tells the cell such

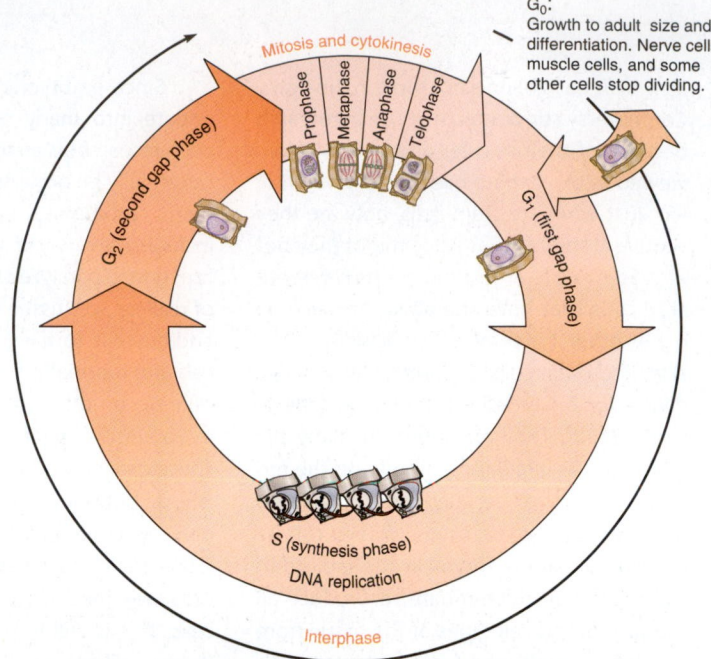

FIGURE 20.27 During the cell cycle, tRNA, mRNA, ribosomes, and enzymes are produced in the G_1 stage. DNA replication occurs in the S stage. Proteins required for the spindles are synthesized in the G_2 stage. The nucleus is replicated in mitosis and two cells are formed by cytokinesis. Once some organs, such as the brain, have completely developed, certain types of cells, such as nerve cells, enter the G_0 stage. The time periods indicated are relative and vary depending on the type of cell and the age of the organism.

things as (1) how to produce enzymes required for the digestion of nutrients, (2) how to manufacture enzymes that will metabolize the nutrients and eliminate harmful wastes, (3) how to repair and assemble cell parts, (4) how to reproduce healthy offspring, (5) when and how to react to favorable and unfavorable changes in the environment, and (6) how to coordinate and regulate all of life's essential functions. The double helix of DNA and the nucleosomes are arranged as a **chromatid,** and there are two attached chromatids for each replicated chromosome after the S stage (figure 20.28). These chromatids (chromosomes) are what will be distributed during mitosis.

20.12 THE STAGES OF MITOSIS

All stages in the life cycle of a cell are continuous; there is no precise point when the G_1 stage ends and the S stage begins or when the interphase period ends and mitosis begins. Likewise, in the individual stages of mitosis, there is a gradual transition from one stage to the next. However, for purposes of study and communication, scientists have divided the process of mitosis into four stages based on recognizable events. These four phases are prophase, metaphase, anaphase, and telophase (figure 20.29).

Prophase

During this phase, the replicated chromatin strands begin to coil into recognizable chromosomes. The nuclear membrane frag-

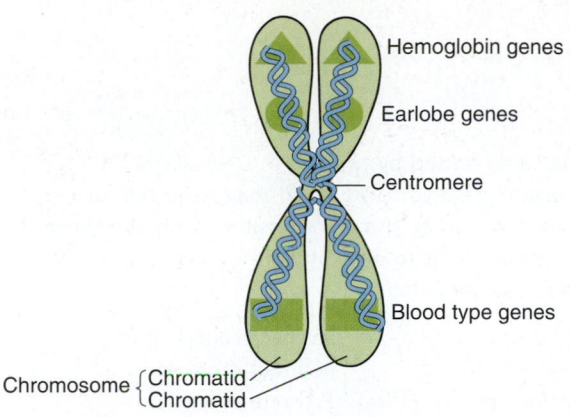

FIGURE 20.28 During interphase, when chromosome replication occurs, the original double-stranded DNA unzips to form two identical double strands that remain attached at the centromere. Each of these double strands is a chromatid. The two identical chromatids of the chromosome are sometimes termed a dyad, to reflect that there are two double-stranded DNA molecules, one in each chromatid. The DNA contains the genetic data. (The examples presented here are for illustrative purposes only. Do not assume that the traits listed are actually located in the positions shown on these hypothetical chromosomes.)

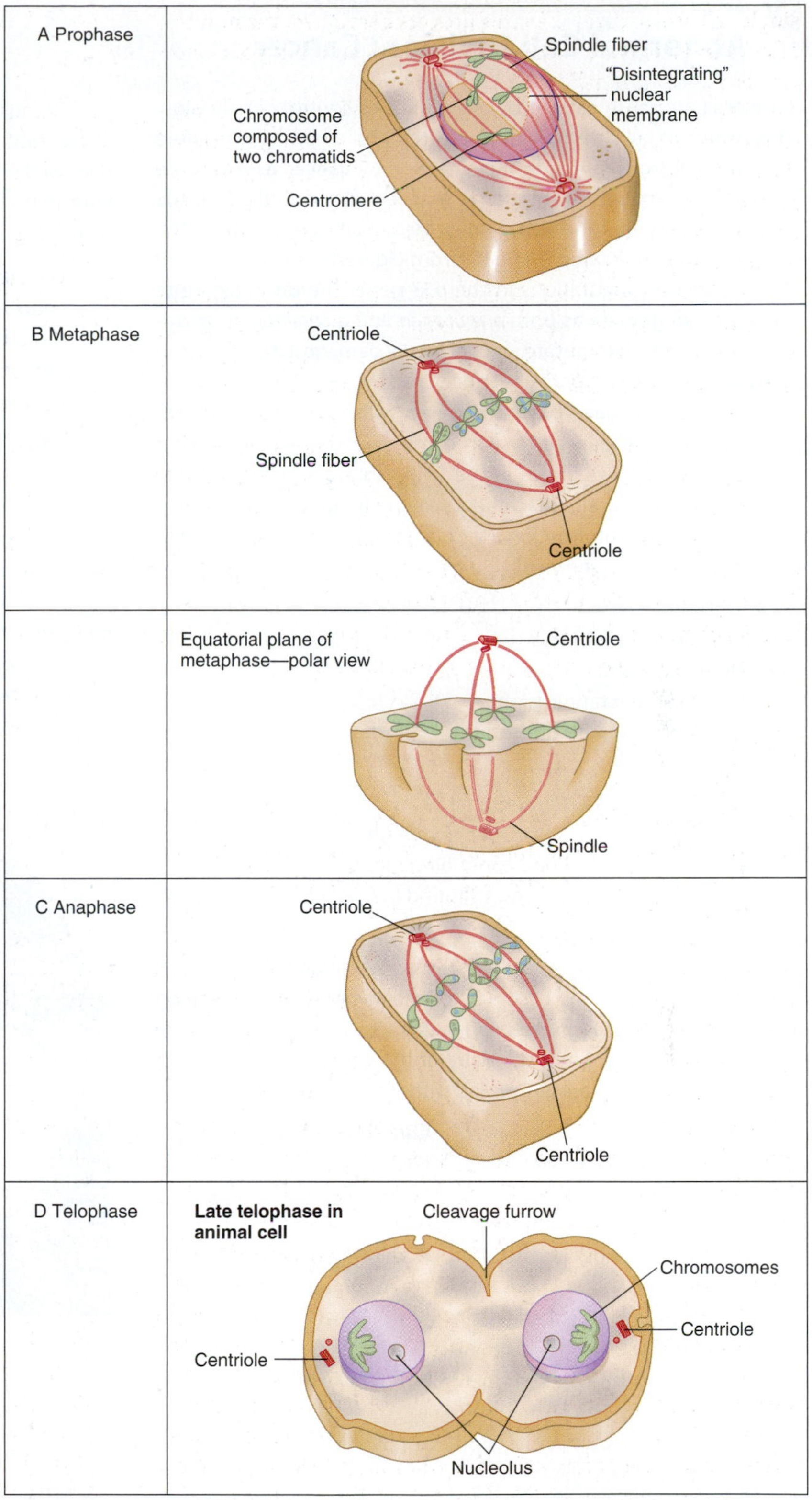

ments and the pieces become parts of other membranous organelles. The centrioles move away from one another to opposite ends of the cell, forming the cell's poles. Spindle fibers form and grow from one pole to another. As prophase proceeds and as the chromosomes become more visible, we recognize that each chromosome is made of two parallel, threadlike parts lying side by side, the chromatids.

Metaphase

This is the phase when the condensed chromosomes move to the equator of the cell. The phrase "equator of the cell" is used in the same manner as "equator of Earth." The equator is the imaginary circle around Earth's (cell's) surface. When they arrive, the chromosomes attach to the spindle fibers at a point on the chromosomes known as the centromeres. At this stage in mitosis, each chromosome still consists of two chromatids attached at a centromere. In a human cell, there are forty-six chromosomes, or ninety-two chromatids, aligned at the cell's equatorial plane during metaphase. In late metaphase, each chromosome splits as the DNA of each centromere replicates, and the cell enters the next phase, anaphase.

Anaphase

Centromeres complete DNA replication during anaphase, allowing the chromatids to separate and move toward the

FIGURE 20.29 The stages of mitosis.

poles. As this separation of chromatids occurs, the chromatids are now called daughter chromosomes. Daughter chromosomes contain identical genetic information. It is later during anaphase that a second important event occurs, cytokinesis. When cytokinesis (cytoplasm splitting) is completed during the last phase, telophase, the cytoplasm

Abnormal Cell Division: Cancer

Cancer is a disease caused by the failure to control cell division. This results in cells that divide too often and eventually interfere with normal body function. Scientists view cancer as a disease caused by mutations in the genes that regulate cell division. The mutations may be inherited or may be caused by agents in the environment. For example, the tar from cigarette smoke has been directly linked to mutations in the *p53* gene. The tar in cigarette smoke is categorized as both a *mutagen* and a *carcinogen*. **Mutagens** are agents that mutate, or chemically damage, DNA. **Carcinogens** are mutagens that cause cancer.

Many agents have been associated with higher rates of cancer. The one thing they all have in common is their ability to alter the sequence of nucleotides in the DNA molecule. When damage occurs to DNA, the cell's machinery may no longer be able to read the DNA's genetic information. This is a partial list of mutagens that are found in our environment.

Radiation
X rays and gamma rays
Ultraviolet light:
  UV-A from tanning lamps
  UV-B, the cause of sunburn

Chemicals

Arsenic	Cigarette tar
Asbestos	Polyvinyl chloride (PVC)
Benzene	Food containing nitrates (e.g., bacon)
Alcohol	Chemicals found in smoked meats and fish
Dioxin	

Some viruses insert a copy of their genetic material into a cell's DNA. When this insertion occurs in a gene involved with regulating the cell cycle, it creates an insertion mutation, which may disrupt the cell's ability to control mitosis. Many of the viruses associated with higher rates of cancer are associated with a particular type of cancer.

Viruses	*Cancer*
Hepatitis B virus (HBV)	Liver cancer
Herpes simplex virus (HSV) type II	Uterine cancer
Epstein-Barr virus	Burkitt's lymphoma
Human T-cell lymphotropic virus (HTLV-1)	Lymphomas and leukemias
Papillomavirus	Several cancers

Because cancer is caused by *changes* in DNA, scientists have found that a person's genetic background may leave him or her disposed to developing cancer. This predisposition is inherited from your parents. Predispositions to developing the following cancers have been shown to be inherited:

Leukemia	Stomach cancer
Lung cancer	Retinoblastomas
Certain Skin cancers	Prostate cancer
Endometrial cancer	Breast cancer
Colorectal cancer	

When uncontrolled mitotic division occurs, a group of cells forms a *tumor*. A **tumor** is a mass of cells not normally found in a certain portion of the body. A **benign tumor** is a cell mass that does not fragment and spread beyond its original area of growth. A benign tumor can become harmful, however, by growing large enough to interfere with normal body functions. Some tumors are *malignant*. **Malignant tumors** are harmful because they may spread or invade other parts of the body. Cells of these tumors **metastasize**, or move from the original site and begin to grow new tumors in other regions of the body.

of the original cell is divided so that two smaller, separate daughter cells result.

Telophase

Two daughter cells are formed from the divided cells as cytokinesis (started during anaphase) is completed. The nuclear membranes and nucleoli re-form. Spindle fibers fragment, and the chromosomes unwind and change from chromosomes to chromatin. Depending on the cell type, these newly formed daughter cells can differentiate and become specialized in their functions, or they can reenter the mitotic portion of the cell cycle.

TABLE 20.3 Summary of the Structure and Function of the Cellular Organelles

Organelle	Type of Cell in Which Located	Structure	Function
Plasma membrane	Prokaryotic and eukaryotic	Membranous; typical membrane structure; phospholipid and protein present	Controls passage of some materials to and from the environment of the cell
Inclusions (granules)	Prokaryotic and eukaryotic	Nonmembranous; variable	May have a variety of functions
Chromatin material	Prokaryotic and eukaryotic	Nonmembranous; composed of DNA and proteins (histones in eukaryotes and HU proteins in prokaryotes)	Contains the hereditary information that the cell uses in its day-to-day life and passes it on to the next generation of cells
Ribosomes	Prokaryotic and eukaryotic	Nonmembranous; protein and RNA structure	Site of protein synthesis
Microtubules, microfilaments, and intermediate filaments	Eukaryotic	Nonmembranous; strands composed of protein	Provide structural support and allow for movement
Nuclear membrane	Eukaryotic	Membranous; double membrane formed into a single container of nucleoplasm and nucleic acids	Separates the nucleus from the cytoplasm
Nucleolus	Eukaryotic	Nonmembranous; group of RNA molecules and DNA located in the nucleus	Site of ribosome manufacture and storage
Endoplasmic reticulum	Eukaryotic	Membranous; folds of membrane forming sheets and canals	Surface for chemical reactions and intracellular transport system
Golgi apparatus	Eukaryotic	Membranous; stack of single-membrane sacs	Associated with the production of secretions and enzyme activation
Vacuoles and vesicles	Eukaryotic	Membranous; microscopic single-membranous sacs	Containers of materials
Peroxisomes	Eukaryotic	Membranous; submicroscopic membrane-enclosed vesicle	Peroxisomes contain several enzymes that play important roles in fat metabolism.
Lysosomes	Eukaryotic	Membranous; submicroscopic membrane-enclosed vesicle	Isolate very strong enzymes from the rest of the cell
Mitochondria	Eukaryotic	Membranous; double-membranous organelle: large membrane folded inside a smaller membrane	Associated with the release of energy from food; site of aerobic cellular respiration
Chloroplasts	Eukaryotic	Membranous; double-membranous organelle: large membrane folded inside a smaller membrane (grana)	Associated with the capture of light of energy and synthesis of carbohydrate molecules: site of photosynthesis
Centriole	Eukaryotic	Two clusters of 9 microtubules	Associated with cell division
Contractile vacuole	Eukaryotic	Membranous; single-membrane container	Expels excess water
Cilia and flagella	Eukaryotic and prokaryotic	Nonmembranous; prokaryotes composed of single type of protein arranged in a fiber that is anchored into the cell wall and membrane; 9 + 2 tubulin protein in eukaryotes	Flagellar movement in prokaryotic type rotate; ciliary and flagellar movement in eukaryotic type seen as waving or twisting

SUMMARY

Living things show the *characteristics* of (1) *metabolic processes*, (2) *generative processes*, (3) *responsive processes*, (4) *control processes*, and (5) *a unique structural organization*. The concept of the *cell* has developed over a number of years. Initially, only two regions, the *cytoplasm* and the *nucleus*, could be identified. At present, numerous *organelles* are recognized as essential components of both major cell types, *prokaryotic* and *eukaryotic*. The structure and function of some of these organelles are compared in table 20.3. This table also indicates whether the organelle is unique to prokaryotic or eukaryotic cells or found in both.

The *cell* is the common unit of life. We study individual cells and their structures to understand how they function as *individual living organisms* and as parts of *many-celled beings*. Knowing how *prokaryotic* and *eukaryotic cell* types resemble or differ from each other helps physicians control some organisms dangerous to humans.

In the process of respiration, organisms convert foods into energy (ATP) and waste materials (carbon dioxide and water). Aerobic cellular respiration uses oxygen (O_2) in this biochemical pathway. This energy-releasing process is composed of three stages: (1) *glycolysis,*

(2) *Krebs cycle*, and (3) the *electron-transport system*. Plants use the products of respiration in the *photosynthesis* pathway. Photosynthesis is comprised of three stages: (1) *light-capturing events*, (2) *light-dependent reactions*, and (3) *light-independent reactions*. Photosynthetic organisms carry out both biochemical pathways. There is also a constant cycling of materials between plants and animals. Sunlight supplies the essential initial energy for making the large organic molecules necessary to maintain the forms of life we know.

All cells come from preexisting cells as a result of cell division. This process is necessary for growth, repair, and reproduction. Eukaryotic cells go through a *cell cycle* that includes *cell division* (*mitosis and cytokinesis*) and *interphase. Interphase* is the period of growth and preparation for division. Mitosis is divided into four stages: *prophase, metaphase, anaphase,* and *telophase.* During mitosis, *two daughter nuclei* are formed from *one parent nucleus.* These nuclei have *identical* sets of *chromosomes* and *genes* that are *exact copies* of those of the parent. Although the process of *mitosis* has been presented as a series of phases, you should realize that it is a *continuous, flowing process* from *prophase* through *telophase.* Following mitosis, *cytokinesis* divides the *cytoplasm,* and the cell returns to *interphase.*

KEY TERMS

active transport (p. **477**)
adenosine triphosphate (ATP) (p. **485**)
aerobic cellular respiration (p. **480**)
benign tumor (p. **492**)
biology (p. **465**)
cancer (p. **491**)
carcinogens (p. **492**)
cell (p. **468**)
cell membrane (p. **470**)
cell nucleus (p. **468**)
cell wall (p. **468**)
chloroplast (p. **480**)

chromatid (p. **490**)
chromatin (p. **482**)
chromosomes (p. **482**)
control processes (p. **467**)
cytokinesis (p. **490**)
cytoplasm (p. **468**)
diffusion (p. **472**)
endoplasmic reticulum (p. **478**)
enzymes (p. **467**)
eukaryotic cells (p. **469**)
facilitated diffusion (p. **475**)
generative processes (p. **466**)
Golgi apparatus (p. **479**)

homeostasis (p. **467**)
interphase (p. **490**)
lysosomes (p. **479**)
malignant tumor (p. **492**)
metabolic processes (p. **465**)
metabolism (p. **465**)
metastasize (p. **492**)
mitochondrion (p. **480**)
mitosis (p. **490**)
mutagens (p. **492**)
nuclear membrane (p. **479**)
nucleolus (p. **482**)
nutrients (p. **465**)
organelles (p. **469**)

osmosis (p. **474**)
phagocytosis (p. **477**)
photosynthesis (p. **480**)
plasma membrane (p. **478**)
prokaryotic cells (p. **469**)
protoplasm (p. **468**)
responsive processes (p. **466**)
ribosomes (p. **481**)
selectively permeable (p. **473**)
structural similarities (p. **467**)
tumor (p. **492**)
vacuole (p. **477**)

APPLYING THE CONCEPTS

Answers are located in appendix F.

Part I

1. Metabolic processes include
   a. nutrient processing.
   b. aerobic cellular respiration.
   c. waste elimination.
   d. All of these are correct.

2. Tanning as a result of exposure to the sun is an example of which characteristic of life?
   a. metabolic processes
   b. responsive processes
   c. generative processes
   d. control processes

3. Which is not a passive form of transport through cell membranes?
   a. facilitated diffusion
   b. diffusion
   c. phagocytosis
   d. ion channel operation

4. Which is a role played by lysosomes?
   a. It is the location for the production of ATP.
   b. It kills dangerous microorganisms that have been taken into the cell by phagocytosis.
   c. It provides for the storage of starch.
   d. It is the surface upon which protein synthesis occurs.

5. Which is an example of an organism composed of eukaryotic cells?
   a. bacterium
   b. virus
   c. maple tree
   d. limestone

Part II

6. A useful chemical bond form of energy used in all cells is
   a. DNA.
   b. ATP
   c. protein.
   d. centrioles.

7. A series of enzyme-controlled reactions operating in a cell is
   a. photosynthesis.
   b. ATP formation.
   c. biochemical pathway.
   d. All the above are correct.

8. The *ultimate* energy source for all life is
   a. ATP.
   b. radiant energy from the Sun.
   c. biochemicals.
   d. DNA.

9. Associated with the release of energy from food, these are the site of aerobic cellular respiration.
   a. chloroplasts
   b. mitochondria
   c. ribosomes
   d. centrioles

10. The biochemical pathways of cellular respiration and photosynthesis are controlled by
    a. mitochondria.
    b. lysosomes.
    c. enzymes.
    d. sunlight.

Part III

11. During which stage of the cell cycle does DNA replication occur?
    a. the S stage of interphase
    b. anaphase of mitosis
    c. G 2 stage of metaphase
    d. prophase

12. Chromosomes move to the equator of the cell and attach to the spindle fibers at a point on the chromosomes known as the centromeres at the ____ stage of mitosis.
   a. interphase
   b. prophase
   c. telophase
   d. metaphase

13. The centromeres split during
   a. anaphase.
   b. prophase.
   c. interphase.
   d. metaphase.

14. Which is not a portion of the cell cycle?
   a. G1, gap (growth)—phase one
   b. Interphase, synthesis
   c. G2, gap (growth)—phase two
   d. Cell division (mitosis and cytokinesis)

15. Individual, loosely coiled threads of DNA that are too thin and tangled to be seen with the naked eye are called
   a. chromatin.
   b. chromosomes.
   c. chromatids.
   d. centromeres.

QUESTIONS FOR THOUGHT

Part I

1. Make a list of the membranous organelles of a eukaryotic cell and describe the function of each.

2. How do diffusion, facilitated diffusion, osmosis, and active transport differ?

3. What are the differences between the cell wall and the cell membrane?

4. Diagram a cell and show where proteins, nucleic acids, carbohydrates, and lipids are located.

5. There are several over-the-counter eyewash products on the market. Describe what would happen if they were not isotonic solutions.

Part II

6. What is it about ATP that allows it to be compared to a rechargeable battery?

7. Briefly describe what happens during the glycolytic pathway of cellular respiration.

8. When does O_2 come into the picture of aerobic cellular respiration?

9. Where in a eukaryotic cell do the following reactions take place: Glycolysis, Krebs cycle, and electron transport system?

10. When there is no light, which portion(s) of the photosynthetic pathway will cease to operate?

Part III

11. Name the four stages of mitosis and describe what occurs in each stage.

12. What is meant by cell cycle?

13. During which stage of a cell's cycle does DNA replication occur?

14. At what phase of mitosis does the DNA become most visible?

15. How might spindle fibers aid in the separation of chromatids during mitosis?

FOR FURTHER ANALYSIS

1. Some people believe that the common cold virus can be controlled by using the same antibiotics that are used to control infections such as strep throat. What is the true story?

2. Local community blood programs are always seeking donors. They want to collect whole blood, red blood cells, and platelets from donors. How do these differ from one another? What is the value and application for each in a medical situation?

3. The sources of red and white blood cells are undifferentiated cells called stem cells. Currently, a great controversy exists over the source and use of these cells for research and medical purposes. What is the nature of the controversy? What kinds of abnormalities are associated with stem cells? To what medical purposes might these cells be used?

4. Cancer often results from mutations in genes like p53. However, some cancers, like cervical cancer, can be initiated by viruses (human papilloma virus, HPV). How are these two types of cancers different? Can either be prevented by vaccination? What do the vaccines do?

INVITATION TO INQUIRY

Athletes and Enzymes

We all know that as we exercise, we sweat and as a result, lose water and salt. These materials must be replaced. Athletes who participate in extremely long events of several hours have a special concern. They need to replace the water on a regular basis during the event. If they drink large quantities of water at one time at the end of the event, they may dilute their blood to the point that they develop hyponatremia. This condition can result in swelling of the cells of the brain and lead to mental confusion and in extreme cases collapse and death.

1. What is hyponatremia?

2. What is a "sports drink"?

3. How is one of these drinks supposed to help an athlete?

4. What is the point of the various colors and flavors?

5. How could the kinds of liquids you drink affect your cell's osmotic balance?

6. Why can drinking electrolyte-free water at the end of an endurance athletic event cause the brain to swell?

7. Should sports drinks be available to children in school cafeterias?

21

The Origin and Evolution of Life

Earth is the only planet in our solar system with liquid water and the only planet with living things. When seen from space, Earth is blue because of its vast oceans. Early in Earth's history, inorganic materials were converted to organic materials that are thought to have accumulated in the oceans. These organic materials have become combined into living units called cells. Cells are the basic unit of life. While we know a great deal about cells and the organisms they make up, scientists still ask these two questions: What is the nature of life? And how did life originate?

CORE **CONCEPT**

Earth and the life on it have changed over billions of years.

OUTLINE

Species are genetically isolated from each other (p. 515).

CONNECTIONS

Physics

▷ Newton's laws of motion provide the background for understanding ingestion and elimination at the cellular level (Ch. 2).

▷ Measuring certain radioactive isotopes provides a clock to measure the age of Earth (Ch. 11 and 22).

Chemistry

▷ Biochemical reactions evolved with living things (Ch. 19).

▷ Carbon chemistry is the chemistry of living things (Ch. 19).

Astronomy

▷ Understanding the origin of the universe is critical for the development of hypotheses related to the origin of life on Earth (Ch. 12).

▷ Conditions on other planets help us understand what the ancient Earth may have been like (Ch. 13).

Earth Science

▷ Atmospheric gases have changed over time and influence the nature of life (Ch. 17).

Life Science

▷ Photosynthesis uses sunlight energy to make organic matter and release oxygen (Ch. 20).

▷ Respiration releases usable energy from organic molecules (Ch. 20).

▷ DNA stores genetic information that determines the kinds of biochemical reactions an organism can perform. (Ch. 19 and 26).

OVERVIEW

Understanding how Earth came to have such a complex and diverse combination of living things has challenged thinkers for thousands of years. Has Earth always been filled with living things? Were there kinds of living things that have ceased to exist? Did living things originate on Earth? Do new kinds of living things originate today? As our understanding of the laws of nature and the nature of living things has developed, the way we approach these questions has changed.

Although at one time these questions were just interesting intellectual exercises, today they have many practical applications and affect the decision making of governments, corporations, and individual citizens. For example, we are concerned about the extinction of various species of organisms, and governments spend considerable time and money to prevent extinctions. Understanding how evolutionary processes occur in populations is important for controlling populations of disease organisms that continually develop strains resistant to commonly used antibiotics. People are concerned about the manipulation of the DNA of organisms, which can cause major changes in the nature of living things. And governments are spending vast amounts of money to determine if there is or was life on other planets. However, even from a modern scientific perspective, we are still asking the same basic questions. How did life originate? And how and why do living things change?

All the ideas in this chapter are related to the origin and modification of living things. However, it may be easier to think about the many aspects of this topic if the chapter is divided into separate parts. Therefore, you will find three parts (Part I, How Did Life Originate?, Part II, The Process of Evolution, and Part III, Speciation), each focusing on one aspect of the general topic of understanding the origin and evolution of life. One of the fundamental questions most ancient societies have sought to answer is, Where did we come from? An extension of this question is, How did life originate? In this part we will look at the various kinds of thinking that apply to this question.

21.1 EARLY ATTEMPTS TO UNDERSTAND THE ORIGIN OF LIFE

In earlier times, no one doubted that life originated from nonliving things. The concept that life could be generated from nonliving matter is known as **spontaneous generation.** The Greeks, Romans, Chinese, and many other ancient peoples believed that maggots arose from decaying meat; mice developed from wheat stored in dark, damp places; lice formed from sweat; and frogs originated from damp mud. Although these ideas were widely accepted until the seventeenth century, there were some who doubted that spontaneous generation occurred. These people subscribed to an opposing concept that life originates only from preexisting life. The concept is known as **biogenesis.**

One of the earliest challenges to the idea that life can be generated from nonliving matter came in 1668. Francesco Redi, an Italian physician, set up a controlled experiment (figure 21.1). He used two sets of jars that were identical except

for one aspect. Both sets of jars contained decaying meat, and both were exposed to the atmosphere; however, one set of jars was covered by gauze, and the other was uncovered. Redi observed that flies settled on the meat in the open jar, but the gauze blocked their access to the meat in the covered jars. When maggots appeared on the meat in the uncovered jars but not on the meat in the covered ones, Redi concluded that the maggots arose from the eggs of the flies and were not produced from the decaying meat.

In 1861, the French chemist Louis Pasteur convinced most scientists that spontaneous generation could not occur. He placed a fermentable sugar solution in a flask that had a long swan neck. The mixture and the flask were boiled for a long time, which would have killed any organisms already in the solution. The long swan neck of the flask was left open to allow oxygen to enter. This was an important experiment because at the time, many thought that oxygen was the "vital element" necessary for the spontaneous generation of living things. Pasteur postulated that it was not the oxygen that caused life to

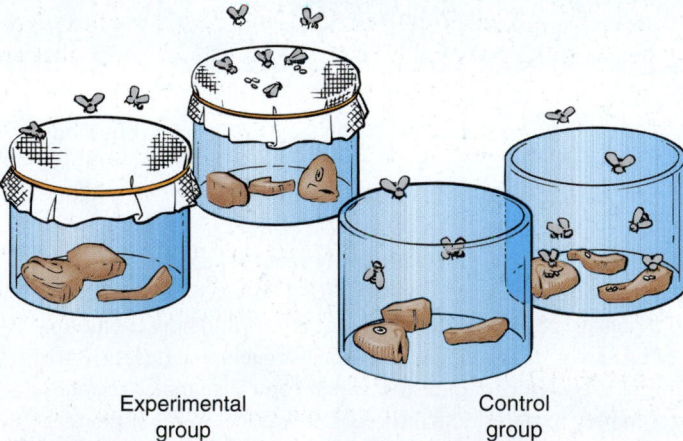

FIGURE 21.1 Francesco Redi performed an experiment in which he prepared two sets of jars that were identical in every way except one. One set of jars had a gauze covering. The uncovered set was the control group; the covered set was the experimental group. Any differences seen between the control and the experimental groups were the result of a single variable—being covered by gauze. In this manner, Redi concluded that the presence of maggots in meat was due to flies laying their eggs on the meat and not spontaneous generation.

Experimental group | Control group

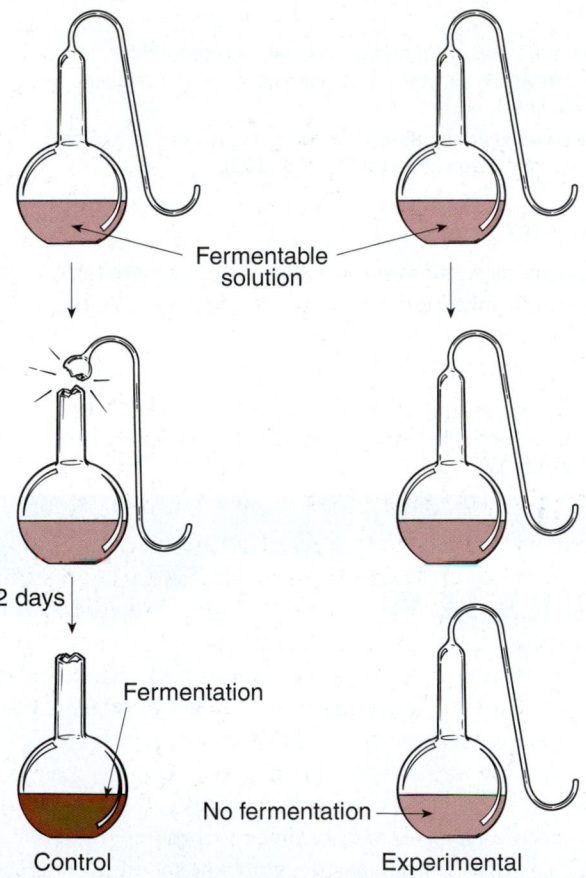

Fermentable solution

2 days

Fermentation

No fermentation

Control | Experimental

FIGURE 21.2 Louis Pasteur conducted an experiment to test the idea that oxygen was necessary for spontaneous generation to take place. He used swan-neck flasks that allowed oxygen but not airborne organisms to enter the flask. The flasks contained a fermentable solution that had been boiled. One flask was kept intact, but he broke the neck off another flask. The intact flask was the experimental flask and the flask with the broken neck was the control. Within two days, there was growth in the flask with the broken neck but none in the intact flask. Thus, Pasteur demonstrated that it was not oxygen in the air that caused growth in the flasks but living things, which were prevented from entering the flask with the unbroken swan neck. This provided additional evidence against the theory of spontaneous generation.

"originate" in such mixtures but that there were organisms in the air that entered nutrient mixtures and caused fermentation. He further postulated that airborne organisms would settle on the bottom of the curved portion of the neck and be unable to reach the sugar-water mixture and thus the sugar solution should not ferment. When he performed the experiment, the solution in the swan-neck flask did not ferment. As a control he cut off the swan neck of other flasks (figure 21.2). This allowed microorganisms from the air to fall into the flask, and within two days, the fermentable solution was supporting a population of microorganisms. In his address to the French Academy, Pasteur stated, "Never will the doctrine of spontaneous generation recover from the mortal blow of this simple experiment. No, there is now no circumstance known in which it can be affirmed that microscopic beings came into the world without germs, without parents similar to themselves."

21.2 CURRENT THINKING ABOUT THE ORIGIN OF LIFE

Extraterrestrial or Earth Origin?

Today, when scientists look at the question of how life on Earth originated, we still have basically the same two theories. One holds that life arrived on Earth from some extraterrestrial source. This is essentially a variation of the biogenesis argument. The other maintains that life was created on Earth from nonliving material through a process of chemical evolution, which is a variation of the spontaneous generation argument. It is important to recognize that we will probably never know for sure how life on Earth came to be, but it is interesting to speculate and examine the evidence related to this fundamental question.

Extraterrestrial Origin for Life on Earth

Early in the 1900s, Swedish scientist Svante Arrhenius popularized the idea of *panspermia*. Panspermia is the concept that life arose outside Earth and that living things were transported to Earth to seed the planet with life. However, this idea does not explain *how* life arose originally. It sees Earth as similar to one of Spallanzani's or Pasteur's open flasks. Although Arrhenius's

CONCEPTS APPLIED

Spontaneous Generation?

At one time, people believed that life could be generated spontaneously from nonliving matter. To get an idea of why they thought this was possible, you can do the following: Take a clear bottle or jar and place a small amount of soil in it. Fill it with water. Place the jar in a warm, well-lighted place. A sunny window is ideal. Observe the changes in it over the next two to three weeks. What evidence do you have that there are living things in the jar? Examine the water under a microscope if you have access to one.

ideas had little scientific support at that time, his basic concept has since been revived and modified as a result of new evidence gained from examinations of meteorites and space explorations.

Because all life forms that we know about are based on organic molecules, the presence of organic molecules in space and in extraterrestrial objects such as meteorites suggest that life or the conditions necessary for life may have existed in other worlds.

In 1996 a meteorite found in Antarctica generated excitement about the possibility of life on other planets. Its chemical makeup suggested it had been a portion of the planet Mars that had been ejected from that planet as a result of a collision with an asteroid. Analysis of the meteorite showed the presence of complex organic molecules and tiny, microscopic objects that were thought to be ancient microorganisms. While scientists no longer think these objects are microorganisms or were formed by living things, many still think conditions on Mars may have been able to support life in the past.

In June and July 2003, two spacecrafts were launched by the National Aeronautics and Space Administration (NASA) to explore the surface of Mars. One of the important goals of these missions is to search for signs of present or past life. The robotic rover vehicles from these two spacecrafts have gathered much information about the surface of Mars. One important piece of information is that it is highly likely that in the past, liquid water existed in large enough quantities to form rivers, lakes, and perhaps salty oceans. In 2009, it was discovered that methane (a simple organic molecule) exists on Mars. However, methane can be produced by either geochemical or biological means, and at this point, it is not clear which processes are responsible.

In recent years, astronomers have been able to detect the presence of planets in other solar systems. By 2010, more than 400 such planets have been identified as orbiting stars in other solar systems. Although many are large gas planets like Jupiter, a few appear to be smaller rocky planets like Earth or Mars. Astronomers have named planets that are close to the size of Earth, fall within a star's "habitable zone," and might support life, *Goldilocks planets*. That is, they have conditions that are "just right" for the existence of life as we define it on Earth. The *habitable zone* is the distance from a star where an Earth-like planet can maintain liquid water on its surface and Earth-like life. Scientists estimate that there are tens of billions of such planets in the Milky Way galaxy alone.

Although none of these discoveries proves that life exists or existed elsewhere in the universe, they keep open the possibility that life may have originated elsewhere and arrived on Earth.

Earth Origin for Life on Earth

There has been much research and speculation about the conditions that would have been necessary for life to originate on Earth and the kinds of problems that early living things would have had to solve. Several different kinds of information are important to this discussion.

1. **Water is necessary for life.** Earth is currently the only planet in our solar system that has a temperature range that

allows for water to exist as a liquid on its surface, and water is the most common compound in most kinds of living things.

2. **The atmosphere of the early Earth lacked molecular oxygen.** Analysis of the atmospheres of other planets shows that they all lack oxygen. The oxygen in Earth's current atmosphere is the result of photosynthetic activity. Therefore, before there was life on Earth, the atmosphere probably lacked oxygen.

3. **Organic molecules can form spontaneously in the absence of oxygen.** Experiments demonstrate that organic molecules can be generated in an atmosphere that lacks oxygen. Furthermore, these organic molecules could have accumulated in the oceans, since there would have been no oxygen to bring about oxidation to cause their breakdown. (See figure 21.3.)

4. **The early Earth was hot.** Since it is assumed that all of the planets have been cooling off as they age, it is very likely that Earth was much hotter in the past. The large portions of Earth's surface that are of volcanic origin strongly support this idea.

5. **Sources of energy on the early Earth were different from today.** The surface of the earth was hotter. A hotter Earth would have contributed to more water vapor in the atmosphere resulting in more storms and lightning. Without oxygen in the atmosphere, much more ultraviolet light energy would have reached Earth's surface.

6. **Simple organisms today often live in conditions similar to those thought to have existed on a primitive Earth.** Prokaryotic organisms (Bacteria and Archaea) are relatively simple and extremely common. Today, certain prokaryotic organisms live in extreme environments of high temperature, high salinity, low pH, or the absence of oxygen. This suggests that they may have been adapted to life in a world that is very different from today's Earth. These specialized organisms are found today in unusual locations such as hot springs and around thermal vents in the ocean floor and may be descendants of the first organisms formed on the primitive, more hostile Earth.

7. **Putting it all together.** If we use all these ideas, we can create the following scenario. A hot planet Earth with much ultraviolet light, lightning, and heat to serve as sources of energy and an atmosphere that lacked oxygen could have allowed organic molecules to form from inorganic molecules. These organic molecules could have been washed from the atmosphere by rain and carried by rivers to the oceans and produced a dilute organic "soup" because oxygen was not present to cause them to break down. They could have served as "building blocks" for the development of simple organisms. They could also have served as a source of energy for newly formed living things. The earliest forms of life would have been adapted to conditions that are very different from those that exist on Earth today.

Meeting Metabolic Needs

Fossil evidence indicates that there were primitive, bacterialike forms of life on Earth at least 3.5 billion years ago. Regardless

FIGURE 21.3 The environment of the primitive Earth was harsh and lifeless. But many scientists believe that it contained the necessary molecules and sources of energy to fashion the first living cell. The energy furnished by volcanoes, lightning, and ultraviolet light could have broken the bonds in the simple inorganic molecules such as carbon dioxide (CO_2), ammonia (NH_3), and methane (CH_4) in the atmosphere. New bonds could have formed as the atoms from the smaller molecules were rearranged and bonded to form simple organic compounds in the atmosphere. Rain and runoff from the land would have carried these chemicals into the oceans. Here, they could have reacted with each other to form more complex organic molecules.

Carbon
Nitrogen
Oxygen
Hydrogen

of how they developed, these first primitive cells would have needed a way to obtain energy and to add new organic molecules to their structure as the organisms grew or as previously existing molecules were lost or destroyed. There are two ways to accomplish this.

Heterotrophs consume organic molecules from their surroundings, which they use to make new molecules and to provide themselves with a source of energy. Today, we recognize all animals, fungi, and most protozoa and Bacteria as being heterotrophs.

Autotrophs use an external energy source such as the energy from inorganic chemical reactions (chemoautotrophs) or sunlight (photoautotrophs) to combine simple inorganic molecules such as water and carbon dioxide to make new organic molecules. These new organic molecules can then be used as building materials for new cells or can be broken down at a later date to provide a source of energy. Today, we recognize eukaryotic plants and algae as being photoautotrophs. Among the Bacteria and Archaea, there are many photoautotrophs and chemoautotrophs.

The Heterotroph Hypothesis

Many scientists support the idea that the first living things produced on Earth were primitive, bacterialike heterotrophs that lived off the organic molecules that would have accumulated in the oceans. Evidence exists to suggest that a wide variety of compounds were present in the early oceans. Since the early heterotrophs are thought to have developed in an atmosphere that lacked oxygen, they would have been, of necessity, anaerobic organisms. Therefore, they did not obtain the maximum amount of energy from the organic molecules they obtained from their environment. At first, this would not have been a problem. The organic molecules that had been accumulating in the ocean for hundreds of millions of years served as an ample source of organic material for the heterotrophs.

Initially these primitive heterotrophs did not have to modify the compounds to meet their needs, and they probably carried out a minimum of enzyme-controlled, metabolic reactions. Those compounds that could be used easily by heterotrophs would have been the first to become depleted from the early environment.

As easily used organic molecules became scarce, it is possible that some of the primitive cells may have had altered (mutated) genetic material that allowed them to convert material that was not directly usable into a compound that could be used. Genetic mutations may have been common because the amount of ultraviolet light, which is known to cause mutations, would have been high. Today, ozone is formed from oxygen in the atmosphere, and ozone screens out much of the ultraviolet light. Before there was oxygen in the atmosphere, there would have been no ozone and much more ultraviolet light would have reached Earth. Heterotrophs with such mutations could have survived, while those without it would have become extinct as the compounds they used for food became scarce. It has been suggested that through a series of mutations in the early heterotrophs, a more complex series of biochemical reactions originated within some of the cells. Such cells could use controlled reactions to modify chemicals from their surroundings and convert them into usable organic compounds.

FIGURE 21.4 Thermal vents like this "black smoker" in the Pacific Ocean are places where hot mineral-rich water enters the ocean water from holes in the ocean bottom. Surrounding these vents are collections of many kinds of animals. Chemoautotrophic members of the Archaea use the hot mineral-rich water as a source of energy with which they synthesize organic molecules. These organisms and the organic molecules they produce are the base of the food chain for a variety of kinds of animals found only in these locations.

The Autotroph Hypothesis

As with many areas of science, there are often differences of opinion. Although the heterotroph hypothesis for the origin of living things was the prevailing theory for many years, recent discoveries have caused many scientists to consider an alternative. They propose that the first organism was an autotroph that used the energy released from inorganic chemical reactions to synthesize organic molecules. Such organisms are known as chemoautotrophs. Several kinds of information support this theory. Many kinds of Bacteria and Archaea are chemoautotrophs and live in extremely hostile environments around the world. These organisms are found in hot springs such as those in Yellowstone National Park or near hot thermal vents—areas where hot, mineral-rich water enters seawater from the ocean floor (figure 21.4). They use inorganic chemical reactions as a source of energy to allow them to synthesize organic molecules from inorganic components. The fact that many of these organisms live in very hot environments suggests that they may have originated on an Earth that was much hotter than it is currently. If the first organism was an autotroph, there could have been subsequent evolution of a variety of kinds of cells, both autotrophic and heterotrophic, that could have led to the variety of different prokaryotic cells seen today.

Summary of Ideas About the Origin of Life

As a result of this discussion, you should understand that we do not know how life on Earth originated. Scientists look at many kinds of evidence and continue to explore new avenues of research. So we currently have three competing scientific theories for the origin of life on Earth:

1. Life originated elsewhere in the universe and arrived from some extraterrestrial source.
2. Life originated on Earth as a heterotroph.
3. Life originated on Earth as an autotroph.

From our modern perspective, we can see that today, all life comes into being as a result of reproduction. Life is generated from other living things, the process of biogenesis. However, reproduction does not answer the question: Where did life come from in the first place? We can speculate, test hypotheses, and discuss various possibilities, but we will probably never know for sure. Life either always was, or it started at some point in the past. If it started, then spontaneous generation of some type had to occur at least once, but it is not happening today.

21.3 MAJOR EVENTS IN THE EARLY EVOLUTION OF LIVING THINGS

Regardless of what the first living things were like, once they existed, a process began that resulted in changes in their abilities and structures. Furthermore, the nature of the planet Earth also changed. When we compare the primitive nature of the first living things with the characteristics of current organisms, we recognize that there were several major steps necessary to go from those first organisms to the great diversity and complexity we see today.

Reproduction and the Origin of Genetic Material

In the previous discussion about possible metabolic activities of early life-forms, we made the assumption that there was some kind of genetic material to direct their metabolic activities. There has been considerable thought about the nature of this genetic material. Today, we know that two important *nucleic acid* molecules, DNA and RNA, store and transfer information within a cell. RNA is a simpler molecule than DNA. In most current cells, DNA stores genetic information and RNA assists the DNA in carrying out its instructions. The details of how DNA and RNA function are discussed in chapter 26.

However, it has been discovered that in some viruses, RNA rather than DNA can serve as the genetic material. Other research about the nature of RNA provides interesting information. RNA can be assembled from simpler subunits that could have been present on the early Earth. Scientists have also shown that RNA molecules are able to make copies of themselves without the need for enzymes, and they can do so without being inside cells.

These pieces of evidence suggest that RNA may have been the first genetic material, which helps to solve one of the problems associated with the origin of life: how genetic information was stored in these primitive life-forms. Once a primitive life-form had the ability to copy its genetic material, it would be able to reproduce. Reproduction is one of the most fundamental characteristics of living things.

Once living things existed and had a molecule that stored information, the stage was set for the evolution of new forms of life. Our current knowledge about how DNA and RNA store and use genetic information includes the idea that these molecules are changeable. In other words, *mutations* can occur. Thus, the

stage was set for living things to proliferate into a variety of kinds that were adapted to specific environmental conditions.

The Development of an Oxidizing Atmosphere

Ever since its formation, Earth has undergone constant change. In the beginning, it was too hot to support an atmosphere. Later, as it cooled and as gases escaped from volcanoes, a reducing atmosphere (lacking oxygen) was likely to have been formed. The early life-forms would have lived in this reducing atmosphere. However, today we have an oxidizing atmosphere and most organisms use this oxygen as a way to extract energy from organic molecules through a process of aerobic respiration. But what caused the atmosphere to change?

Atmospheric Oxygen Produced by Photosynthesis

Today, it is clear that the oxygen in our atmosphere is the result of the process of photosynthesis. Prokaryotic cyanobacteria are the simplest organisms that are able to photosynthesize. The first fossils of primitive living things show up at about 3.5 billion years ago, and it appears that an oxidizing atmosphere began to develop about 2 billion years ago. Because the fossil record shows that living things existed before there was oxygen in the atmosphere and that cyanobacteria were present early in the evolution of life, it seemed logical that the first organisms could have accumulated many mutations over time that could have resulted in photosynthetic autotrophs (photoautotrophs). This would have been a significant change because eventually the release of oxygen from photosynthesis would have resulted in an accumulation of oxygen in the atmosphere—an **oxidizing atmosphere.** Once oxygen was present in the atmosphere it would not have been possible to accumulate organic molecules in the oceans as was described earlier in this chapter.

The Significance of Ozone

The presence of oxygen in the atmosphere had an additional significance. Oxygen molecules react with one another to form ozone (O_3). Ozone collects in the upper atmosphere and acts as a screen to prevent most of the ultraviolet light from reaching Earth's surface. The reduction of ultraviolet light reduced the number of mutations in cells.

Evolution of Aerobic Respiration Possible

The presence of oxygen in the atmosphere also allowed for the evolution of aerobic respiration. Since any early heterotrophs were, of necessity, anaerobic organisms, they did not derive large amounts of energy (ATP) from the organic materials available as food. With the evolution of aerobic heterotrophs, there could be a much more efficient conversion of food into usable energy. They could use the oxygen for aerobic respiration and, therefore, generate many more energy-rich ATP molecules from the food molecules they consumed. Because of this, aerobic organisms would have a significant advantage over anaerobic organisms.

The Establishment of Three Major Domains of Life

Although biologists have traditionally divided organisms into kingdoms based on their structure and function, it was very difficult to do this with microscopic organisms. In 1977, Carl Woese published the idea that the "bacteria" (organisms that lack a nucleus), which had been considered a group of similar organisms, were really made up of two very different kinds of organisms, the Bacteria and Archaea. With the newly developed ability to decode the sequence of nucleic acids, it became possible to look at the genetic nature of organisms without being confused by their external structures. Woese studied the sequences of ribosomal RNA and compared similarities and differences. As a result of his studies and those of many others, a new concept of the relationships between various kinds of organisms has emerged. There are three main categories of living things, Bacteria, Archaea, and Eucarya, that have been labeled **"domains."**

This new picture of living things requires us to reorganize our thinking. It appears that the oldest organisms were members of the Domain Bacteria and that the Domains Archaea and Eucarya are both derived from the Bacteria. Perhaps most interesting is the idea that the Archaea and Eucarya share many characteristics, suggesting that they are more closely related to each other than either is to the Bacteria. It appears that each domain developed specific abilities.

The Bacteria developed many different metabolic capabilities. Today, many are able to use organic molecules as a source of energy (heterotrophs). Some of these heterotrophs use anaerobic respiration, whereas others use aerobic respiration. Other Bacteria are autotrophic. Some, such as the cyanobacteria, carry on photosynthesis, whereas others are chemosynthetic and get energy from inorganic chemical reactions.

The Archaea have very diverse metabolic abilities. Some are chemoautotrophic and use inorganic chemical reactions to generate the energy they need to make organic matter. The reactions that provide them with energy often produce methane (CH_4) or hydrogen sulfide (H_2S). In addition, most of these organisms are found in extreme environments such as hot springs or in extremely salty or acidic environments. However, it is becoming clear that they also inhabit soil, and the guts of animals, and are particularly abundant in the ocean.

The Eucarya (animals, plants, fungi, protozoa, and algae) are familiar to most people. The cells of eukaryotic organisms are much larger than those of prokaryotic organisms and appear to have incorporated entire cells of other organisms within their own cellular structure. Chloroplasts and mitochondria are both bacterialike structures found inside eukaryotic cells.

The Endosymbiotic Theory and the Origin of Eukaryotic Cells

The earliest fossils occur at about 3.5 billion years ago and appear to be similar in structure to that of present-day Bacteria and Archaea. Therefore, it is likely that the early heterotrophs and autotrophs were probably simple one-celled prokaryotic

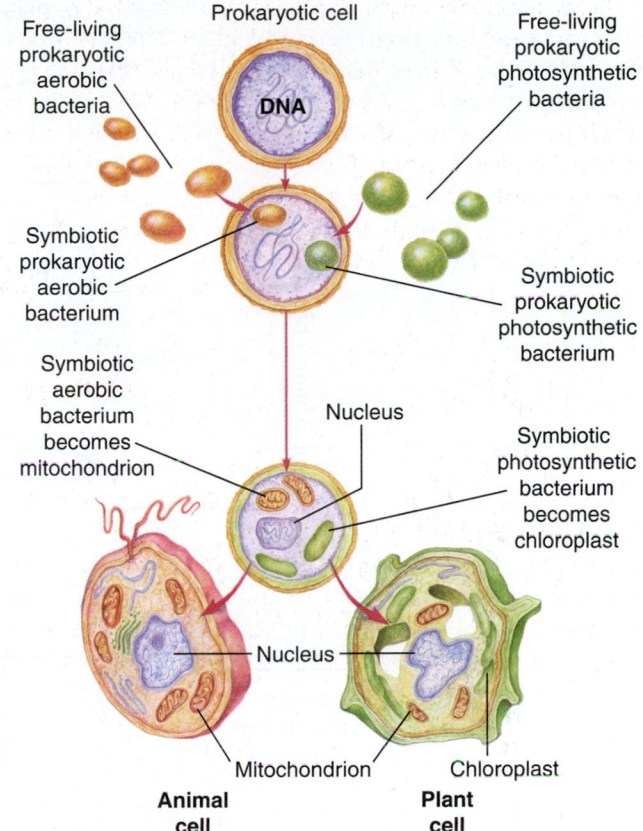

FIGURE 21.5 The endosymbiotic theory proposes that some free-living prokaryotic Bacteria developed symbiotic relationships with a host cell. Some aerobic Bacteria appear to have developed into mitochondria and photosynthetic Bacteria appear to have developed into chloroplasts.

organisms. The earliest fossils of eukaryotic cells appear at about 1.8 billion years ago.

The **endosymbiotic theory** attempts to explain this evolution. This theory states that eukaryotic cells came into being as a result of the combining of several different types of primitive prokaryotic cells. It is thought that some organelles found in eukaryotic cells originated as free-living prokaryotes. For example, since mitochondria and chloroplasts contain bacterialike DNA and ribosomes, control their own reproduction, and synthesize their own enzymes, it has been suggested that they originated as free-living prokaryotic Bacteria. These bacterial cells could have established a symbiotic relationship with another primitive nuclear-membrane-containing cell type (figure 21.5). When this theory was first suggested, it met with a great deal of criticism and even ridicule. However, continuing research has uncovered several other instances of the probable joining of two different prokaryotic cells to form one. If these cells adapted to one another and were able to survive and reproduce better as a team, it is possible that this relationship may have evolved into present-day eukaryotic cells.

If this relationship had included only a nuclear-membrane-containing cell and aerobic Bacteria, the newly evolved cell would have been similar to present-day heterotrophic protozoa, fungi, and animal cells. If this relationship had included both aerobic Bacteria and photosynthetic Bacteria, the newly formed cell would have been similar to present-day autotrophic algae and plant cells. In addition, it is likely that endosymbiosis occurred

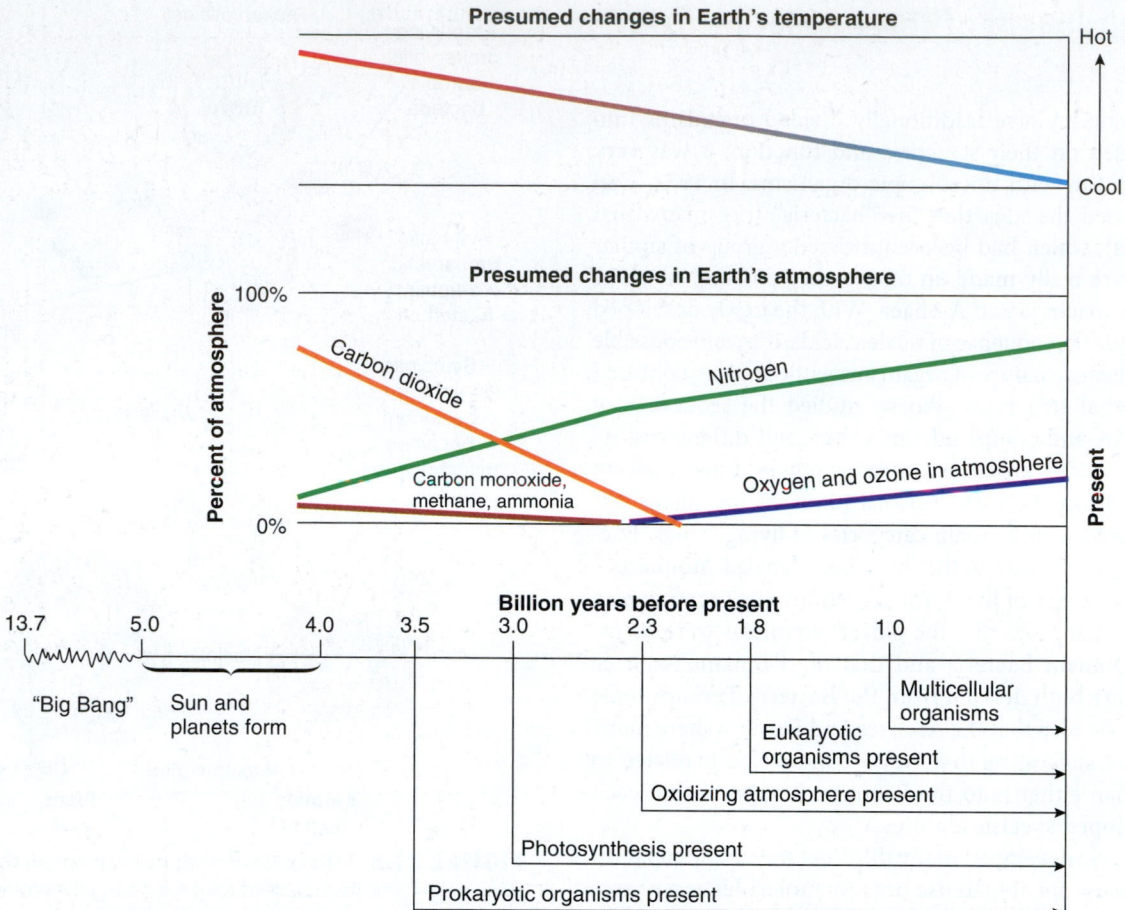

FIGURE 21.6 Current thinking suggests that several factors shaped the nature of Earth and its atmosphere. Conceivably, the initial atmosphere was primarily carbon dioxide (CO_2) with smaller amounts of nitrogen (N_2), ammonia (NH_3), methane (CH_4), and carbon monoxide (CO). Carbon dioxide decreased as a result of chemical reactions at the surface of Earth. Living organisms have had a significant effect on the atmosphere because once photosynthesis began, carbon dioxide was further reduced and oxygen (O_2) began to increase. The presence of oxygen allowed for the increase in ozone (O_3), which reduces the amount of ultraviolet light penetrating the atmosphere to reach the surface of Earth. Because ultraviolet light causes mutations, a protective layer of ozone probably has reduced the frequency of mutations from that which was present on primitive Earth.

among eukaryotic organisms as well. Several kinds of eukaryotic red and brown algae contain chloroplastlike structures that appear to have originated as free-living eukaryotic cells.

A Summary of the Early Evolution of Life

If we incorporate information about endosymbiosis along with information about the nature of genetic material, we can develop a diagram that helps us see how these various kinds of organisms are related to one another.

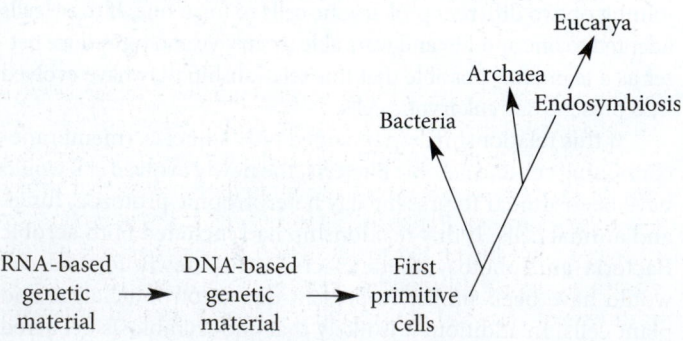

Table 21.1 summarizes the relationships among these three major kinds of organisms. Figure 21.6 summarizes several of the major events in the formation of the planet Earth and the development of life on it.

Although we have introduced the idea of evolutionary change while discussing the origin of life, it is important to have a clear understanding of what is meant by the evolution of organisms and the mechanisms that drive this important process. This part focuses on the mechanisms that cause evolutionary change.

21.4 THE DEVELOPMENT OF EVOLUTIONARY THOUGHT

For centuries, people believed that the various species of plants and animals were fixed and unchanging; that is, they were

TABLE 21.1 Summary of Characteristics of the Three Major Domains of Life

Characteristics	DOMAIN		
	Bacteria	Archaea	Eucarya
Cell Structure	Few membranous structures	Few membranous structures	Many kinds of membranous organelles are present in cells.
	There is no nuclear membrane.	There is no nuclear membrane.	A nuclear membrane is present.
			Chloroplasts are probably derived from cyanobacteria and entered cells through endosymbiosis.
			Mitochondria are probably derived from certain aerobic bacteria and entered cells through endosymbiosis.
Metabolic Activity	Some bacteria are chemoautotrophs that use energy from inorganic chemical reactions to produce organic molecules.	Most Archaea are chemoautotrophs that obtain energy from inorganic reactions to make organic matter.	
	Most Bacteria are anaerobic heterotrophs.	There are few heterotrophs.	A few Eucarya use only anaerobic respiration—fungi, some protozoa.
	Some Bacteria are aerobic heterotrophs.		Many Eucarya have tissues that use anaerobic respiration—muscle.
	Chlorophyll-based, oxygen-generating photosynthesis was an invention of the cyanobacteria.		Nearly all Eucarya have mitochondria and use aerobic respiration.
			Plants and algae have choloroplasts and use photosynthesis in addition to aerobic respiration.
Evolutionary Status	Probably related to the first living thing		
	Some live at high temperatures and are probably ancestral to Archaea.	Probably derived from Bacteria	
		Archaea probably have a common ancestor with Eucarya.	Eucarya probably have a common ancestor with Archaea. The common evolutionary theme is the development of complex cells through endosymbiosis of other organisms.
Ecological Status	Major role as photosynthesizers in aquatic environments		Major role as photosynthesizers on terrestrial and aquatic environments
	Major category of decomposers	Archaea are typically found in extreme environments.	Dominant form of life today
	Some are pathogenic.		
		None have been identified as pathogenic.	Various eukaryotes fill ecological roles of producer, consumer, pathogen, and decomposers.
Example	*Streptococcus pneumoniae*, one cause of pneumonia	*Pyrolobus fumarii*, deep-sea hydrothermal vents, hot springs, volcanic areas, growth to 113°C (235°F)	*Spirogyra*, a filamentous alga found in freshwater

thought to have remained unchanged from the time of their creation. This was a reasonable assumption because people knew nothing about DNA, how sperm and eggs were formed, or population genetics. Furthermore, the process of evolution is so slow that the results of evolution were usually not evident during a human lifetime. It is even difficult for modern scientists to recognize this slow change in many kinds of organisms. It has taken the development of language, writing, science, and tech-

nology to be able to gain the long-term perspective necessary to understand the evolutionary process of living things.

Charles Darwin (1809–1882) is famous for his contribution of the idea of natural selection as the process that directs evolution. However, he probably would not have recognized the importance of his observations of nature if he had not had the advantage of reading about the ideas of or engaging in discussions with others who had considered the possibility that organisms change over time. The following discussion lists some of the people known to have influenced Darwin's thinking.

Georges-Louis Leclerc, Comte de Buffon (1707–1788)— In his writings, Buffon questioned the established church doctrines of the day that all organisms were created as we currently see them and that they are unchanging. In addition, he questioned the doctrine that Earth was only about six thousand years old. However, he did not suggest any mechanism for evolution. Buffon produced a thirty-six–volume set of books that covered essentially everything known about the natural world. Darwin referred to Buffon as the first person to suggest the idea of natural selection.

Erasmus Darwin (1731–1802)—Erasmus Darwin was Charles Darwin's grandfather. He was a major thinker of his day and had many thoughts about the processes by which organisms could change over time. Although he died before Charles was born, in many ways his thinking anticipated what his grandson would make a basic tenet of biological thought. Many of these ideas are in evidence in the following poem.

Organic life beneath the shoreless waves
Was born and nurs'd in ocean's pearly caves;
First forms minute, unseen by spheric glass,
Move on the mud, or pierce the watery mass;
These, as successive generations bloom,
New powers acquire and larger limbs assume;
Whence countless groups of vegetation spring,
And breathing realms of fin and feet and wing.

Erasmus Darwin, *The Temple of Nature* (1802)

Jean-Baptiste de Lamarck (1744–1829)—In 1809, Lamarck, a student of Buffon's, suggested a process by which evolution could occur. He proposed that acquired characteristics could be transmitted to offspring. For example, he postulated that giraffes originally had short necks. Since giraffes constantly stretched their necks to obtain food, their necks got slightly longer. He thought that the slightly longer neck acquired through stretching could be passed to the offspring, who were themselves stretching their necks, and over time, the necks of giraffes would get longer and longer. Although we now know Lamarck's theory was wrong (because acquired characteristics are not inherited), it stimulated further thought as to how evolution could occur. All during this period, from the mid-1700s to the mid-1800s, lively arguments continued about the possibility of evolutionary change. Darwin was aware of Lamarck's writings and considered him to be an important thinker of the day.

Thomas Malthus (1766–1834)—Malthus' *Essay on the Principle of Population* put forward the concept that people reproduce faster than do resources needed to sustain the population. Therefore, there would be a constant struggle among the members of the population for the limited resources available. Thus, the human population could look forward to poverty and famine unless human repro-

duction was controlled. Darwin's reading of this essay and its implication of a struggle for limited resources was instrumental in his formulating the theory of natural selection as a mechanism for evolution.

Charles Lyell (1797–1875)—Lyell was an eminent geologist of his day. He is most famous for his elucidation of the idea of uniformitarianism. The central idea of uniformitarianism is that we can interpret evidence of past geologic events by understanding how things are working today. He felt that processes such as sedimentation, volcanic activity, and erosion did not operate differently today from the past. One of the obvious outcomes of this thinking is that the age of Earth must be much greater than a few thousand years. Darwin knew Lyell and Lyell's ideas greatly influenced Darwin's thinking about the time available for the mechanisms of evolution to operate.

Alfred Russel Wallace (1823–1913)—Like Darwin, Wallace traveled extensively and, as a result of his travels, came to the same conclusions about natural selection as did Darwin. He was also influenced by Malthus's ideas, as was Darwin. Although today Wallace is often mentioned as an afterthought, it was his essay concerning his ideas about natural selection that prompted Darwin to publish *On the Origin of Species by Means of Natural Selection, or the Preservation of Favoured Races in the Struggle for Life.*

21.5 EVOLUTION AND NATURAL SELECTION

In many cultural contexts, the word *evolution* means progressive change. We talk about the evolution of economies, fashion, or musical tastes. From a biological perspective, the word has a more specific meaning. One of the key characteristics of life is that organisms are able to respond to their surroundings. Evolution is a responsive process that takes place at the population level.

Defining Evolution

Evolution is the change in frequency of genetically determined characteristics within a population over time. There are three key points to this definition. First of all, evolution only occurs in populations. *Populations* are groups of organisms of the same species that are able to interbreed and are thus genetically similar. The second point is that *genes* (specific pieces of DNA) determine the characteristics displayed by organisms. The third point is that the mix of genes (DNA) within populations can change. Thus, evolution involves changes in the genes that are present in a population. By definition, individual organisms are not able to evolve—only populations can.

The Role of the Environment in Evolution

The organism's surroundings determine which characteristics favor survival and reproduction (i.e., which characteristics best fit the organism to its environment). A primary mechanism by which evolution occurs involves the selective passage of genes from one generation to the next through sexual reproduction. This adaptation can take place at different levels and over different time periods. At the smallest level, populations show changes in the genes within a local population. For example, some populations of potato beetles became resistant to certain insecticides and some populations of crabgrass became resistant to specific herbicides when these chemicals became a constant part of their environment. The susceptible members died and the resistant ones lived, leading to populations that consisted primarily of resistant individuals. These populations became genetically adapted to an altered environment. The genes that allowed the organisms to resist the effects of the poisons became more common in these populations. When we look at evolutionary change over millions or billions of years, we recognize that Earth has gone through major changes of climate, sea level, and other conditions. These changes have resulted in changes in the kinds of organisms present as well. Some species went extinct as new species came into being.

Natural Selection Leads to Evolution

The various processes that encourage the passage of beneficial genes to future generations and discourage the passage of harmful or less valuable genes are collectively known as **natural selection.** The idea that some individuals whose gene combinations favor life in their surroundings will be most likely to survive, reproduce, and pass their genes on to the next generation is known as the **theory of natural selection.** The *theory of evolution* states that populations of organisms experience changes in the frequency of genetically determined characteristics over time. Thus, natural selection is a process that brings about evolution by "selecting" which genes will be passed to the next generation. Although natural selection is the major mechanism that drives evolution, there are other processes that can cause changes in genes from generation to generation. "A Closer Look: Other Mechanisms that Cause Evolution" discusses some of these mechanisms.

A theory is a well-established generalization supported by many different kinds of evidence. At one time, the idea that populations of species changed over time was revolutionary. Today, the concepts of evolution and natural selection are central to the study of all of the biological sciences.

The theory of natural selection was first proposed by Charles Darwin and Alfred Wallace and was clearly set forth in 1859 by Darwin in his book *On the Origin of Species by Means of Natural Selection, or the Preservation of Favoured Races in the Struggle for Life.* Since the time it was first proposed, the theory of natural selection has been subjected to countless tests and remains the core concept for explaining how evolution occurs.

There are two common misinterpretations associated with the process of natural selection. The first involves the phrase "survival of the fittest," commonly associated with the theory of natural selection. Individual survival is certainly important because those that do not survive will not reproduce. But the more important factor is the number of descendants an organism leaves. An organism that has survived for hundreds of years but has not reproduced has not contributed any of its genes to the next generation and so

FIGURE 21.7 Many kinds of birds, such as this shafted or northern flicker, nest in holes in trees. If old and dead trees are not available, they may not be able to breed.

has been selected against. The key, therefore, is not survival alone but survival *and reproduction* of the more fit organisms.

Second, the phrase "struggle for life" does not necessarily refer to open conflict and fighting. It is usually much more subtle than that. When a resource such as nesting material, water, sunlight, or food is in short supply, some individuals survive and reproduce more effectively than others. For example, many kinds of birds require holes in trees as nesting places (figure 21.7). If these are in short supply, some birds will be fortunate and find a top-quality nesting site, others will occupy less suitable holes, and some may not find any. There may or may not be fighting for possession of a site. If a site is already occupied, a bird may not necessarily try to dislodge its occupant but may just continue to search for suitable but less valuable sites. Those that successfully occupy good nesting sites will be much more successful in raising young than will those that must occupy poor sites or those that do not find any.

Similarly, on a forest floor where there is little sunlight, some small plants may grow fast and obtain light while shading out plants that grow more slowly. The struggle for life in this instance involves a subtle difference in the rate at which the plants grow. But the plants are indeed engaged in a struggle, and a superior growth rate is the weapon for survival.

21.6 GENETIC DIVERSITY IS IMPORTANT FOR NATURAL SELECTION

Now that we have a basic understanding of how natural selection works, we can look in more detail at factors that influence it. For natural selection to occur there must be genetic differences among the many individuals of an interbreeding population of organisms. If all individuals are identical genetically, it does not matter which ones reproduce—the same genes will

A Closer Look

The Voyage of HMS *Beagle,* 1831–1836

Probably the most significant event in Charles Darwin's life was his opportunity to sail on the British survey ship HMS *Beagle* (box figure 21.1). Surveys were common at this time; they helped refine maps and chart hazards to shipping. Darwin was twenty-two years old and probably would not have had the opportunity had his uncle not persuaded Darwin's father to allow him to take the voyage. Darwin was to be a gentleman naturalist and companion to the ship's captain, Robert Fitzroy. When the official naturalist left the ship and returned to England, Darwin became the official naturalist for the voyage. The appointment was not a paid position.

The voyage of the *Beagle* lasted nearly five years. During the trip, the ship visited South America, the Galápagos Islands, Australia, and many Pacific Islands (box figure 21.2). The *Beagle's* entire route is shown on the accompanying map. Darwin suffered greatly from seasickness, and perhaps because of it, he made extensive journeys by mule and on foot some distance inland from wherever the *Beagle* happened to be at anchor. These inland trips provided Darwin the opportunity to make many of his observations. His experience was unique for a man so young and very difficult to duplicate because of the slow methods of travel used at that time.

Although many people had seen the places that Darwin visited, never before had a student of nature collected volumes of information on them. Also, most other people who had visited these faraway places were military men or adventurers who did not recognize the significance of what they saw. Darwin's notebooks included information on plants, animals, rocks, geography, climate, and the native peoples he encountered. The natural history notes he took during the voyage served as a vast storehouse of information that he used in his writings for the rest of his life. Because Darwin was wealthy, he did not need to work to earn a living and could devote a good deal of his time to the further study of natural history and the analysis of his notes. He was a semi-invalid during much of his later life. Many people think his ill health was caused by a tropical disease he contracted during the voyage of the *Beagle*. As a result of his experiences, he wrote several volumes detailing the events of the voyage, which were first published in 1839 in conjunction with other information related to the voyage of the *Beagle*. His volumes were revised several times and eventually were entitled *The Voyage of the Beagle*. He also wrote books on barnacles, the formation of coral reefs, how volcanos might have been involved in reef formation, and, finally, *On the Origin of Species by Means of Natural Selection, or the Preservation of Favoured Races in the Struggle for Life*. This last book, written twenty-three years after his return from the voyage, changed biological thinking for all time.

Box Figure 21.1 Portrait of Charles Darwin

Box Figure 21.2 The Galápagos Islands

be passed to the next generation and natural selection cannot occur. There are two primary ways in which genetic diversity is generated within the organisms of a species: mutation and genetic recombination during sexual reproduction.

Genetic Diversity Resulting from Mutation

A **mutation** is any change in the genetic information (DNA) of an organism. **Spontaneous mutations** are changes in DNA that cannot be tied to a particular causative agent. It is suspected that cosmic radiation or naturally occurring mutagenic chemicals might be the cause of many of these mutations. It is known that subjecting organisms to high levels of radiation or to certain chemicals increases the rate at which mutations occur. It is for this reason that people who work with radioactive materials or other mutagenic agents take special safety precautions (figure 21.8). It is also known that when cells make copies of DNA, the DNA may not be copied perfectly and small errors can be introduced.

Naturally occurring mutation rates are low (perhaps one chance in 100,000 that a gene will be altered). There are

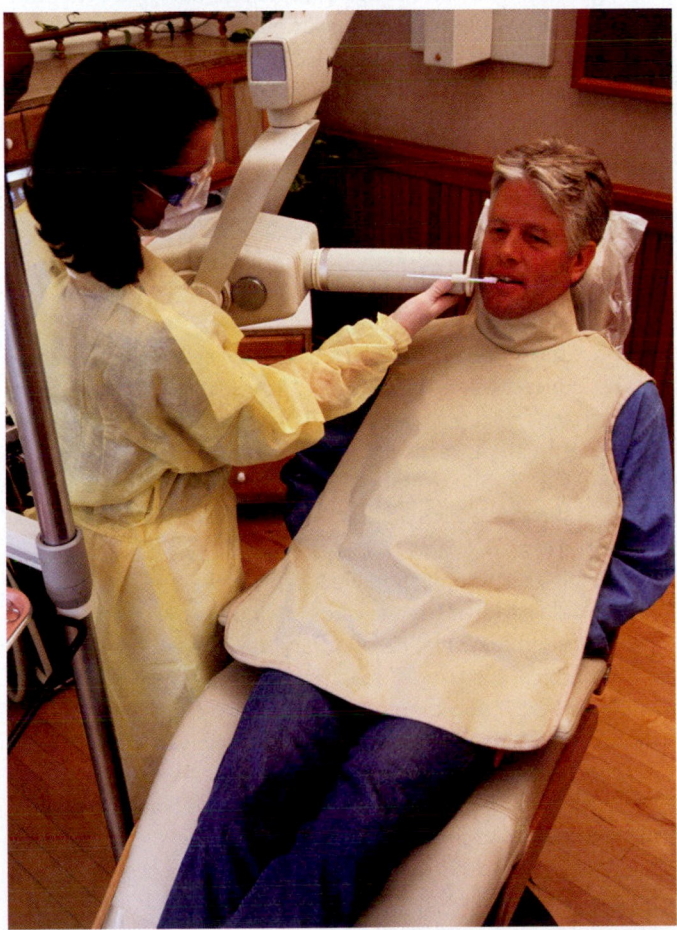

FIGURE 21.8 Because radiation and certain chemicals increase the likelihood of mutations, people who work in hazardous environments receive special training and use protective measures to reduce their exposure to mutagenic agents. It is particularly important to protect the ovaries and testes, since mutations that alter the DNA of eggs and sperm can be passed on to children. This dental patient is being protected with a lead-containing blanket.

Genetic Diversity and Health Care

People turn to their health care providers when they experience a medical problem, whether it is the result of an accident, infection, or some abnormality. In many large cities, the emergency rooms (ERs) of large hospitals have become a substitute for a visit to a physician's office or a neighborhood clinic. Medical facilities are thought of as places where everyone always gets better and no one gets sick. However, 2 million people a year get bacterial infections while they are being treated in hospitals as patients. An estimated 90,000 people die from these infections each year.

Many people do not realize that the hospital is a place where patients who have not been able to have their infections resolved by home care bring all of those nasty microbes. Studies have shown that the farther away you are from the hospital, the less dangerous the microbes. What makes this situation worse is the fact that these bacteria are undergoing genetic changes and becoming more unbeatable. Populations of hospital microbes contain mutations that protect them from specific antibiotics— that is, they are antibiotic resistant. If resistant microbes are transmitted, the infected person will find the infection even harder to control. For example, methicillin-resistant *Staphylococcus aureus* (MRSA) was responsible for more than 94,000 potentially fatal infections and nearly 19,000 deaths in the United States in 2005. Eighty-five percent of these deaths were associated with health care settings.

three possible outcomes when a gene is altered. The mutation may be so minor that it has no effect, the mutation may be harmful, or the mutation may be beneficial. Although it is likely that most mutations are neutral or harmful, beneficial mutations do occur. In populations of millions of individuals, each of whom has thousands of genes, over thousands of generations it is highly likely that a new beneficial piece of genetic information could come about as a result of mutation. When we look at the various genes that exist in humans or in any other organism, we should remember that every gene originated as a modification of a previously existing gene. For example, the gene for blue eyes may be a mutated brown-eye gene, or blond hair may have originated as a mutated brown-hair gene. Thus, mutations have been very important for introducing new genetic material into species over time.

In order for mutations to be important in the evolution of organisms, they must be in cells that will become sex cells. Mutations to the cells of the skin or liver will affect only those specific cells and will not be passed to the next generation.

Genetic Diversity Resulting from Sexual Reproduction

A second very important process involved in generating genetic diversity is sexual reproduction. While sexual reproduction does not generate new genetic information the way mutation does, it allows for the recombination of genes into mixtures that did not occur previously. Each individual entering a population by sexual reproduction carries a unique combination of genes; half are donated by the mother and half are donated by the father. During the formation of sex cells, the activities of chromosomes results in new combinations of genes. This means that there are millions of possible combinations of genes in the sex cells of any individual. When fertilization occurs, one of the millions of possible sperm unites with one of the millions of possible eggs, resulting in a genetically unique individual. The gene mixing that occurs during sexual reproduction is known as **genetic recombination.** The newly conceived individual has a complete set of genes that is different from that of any other organism that ever existed.

The power of genetic recombination is easily seen when you look at members of your family. Compare yourself with your siblings, parents, and grandparents. Although your genes are traceable back to your grandparents, your specific combination of genes is unique, and this uniqueness shows up in differences in physiology, physical structure, and behavior.

21.7 PROCESSES THAT DRIVE NATURAL SELECTION

Several mechanisms allow for selection of certain individuals for survival and successful reproduction. The specific environmental factors that favor certain characteristics are called **selecting agents.** If predators must pursue swift prey organisms, then the faster predators will be selected for, and the selecting agent is the swiftness of available prey. If predators must find prey that are slow but hard to see, then the selecting agent is the camouflage coloration of the prey, and keen eyesight is selected for. If plants are eaten by insects, then the production of toxic materials in the leaves is selected for. All selecting agents influence the likelihood that certain characteristics will be passed to subsequent generations. Three kinds of mechanisms influence which genes are passed to future generations: differential survival, differential reproductive rates, and differential mate selection.

Differential Survival

As stated previously, the phrase "survival of the fittest" is often associated with the theory of natural selection. Although this is recognized as an oversimplification of the concept, survival is an important factor in influencing the flow of genes to subsequent generations. If a population consists of a large number of genetically and structurally different individuals, it is likely that some of them will possess characteristics that make their survival difficult. Therefore, they are likely to die early in life and not have an opportunity to pass their genes on to the next generation.

Charles Darwin described several species of finches on the Galápagos Islands, and scientists have often used these birds in scientific studies of evolution. On one of the islands, scientists studied one of the species of seed-eating ground finches, *Geospiza fortis*. They measured the size of the animals and the size of their bills and related these characteristics to their survival. They found the following: During a drought, the birds ate the smaller, softer seeds more readily than the larger, harder seeds. As the birds consumed the more easily eaten seeds, only the larger, harder seeds remained. During the drought, mortality was extremely high. When scientists looked at ground finch mortality, they found that the larger birds that had stronger, deeper bills survived better than smaller birds with weaker, narrower bills. They also showed that the offspring of the survivors tended to show larger body and bill size as well. The lack of small, easily eaten seeds due to the drought resulted in selection for larger birds with stronger bills that could crack open larger, tougher seeds. Table 21.2 shows data on two of the parameters measured in this study.

As another example of how differential survival can lead to changes in the genes within a population, consider what has happened to many insect populations as we have subjected them to a variety of insecticides. Since there is genetic diversity within all species of insects, an insecticide that is used for the first time on a particular species kills all those that are genetically susceptible. However, the insecticide may not kill individuals with slightly different genetic compositions that give them resistance to the insecticide.

Suppose that in a population of a particular species of insect, 5 percent of the individuals have genes that make them resistant to a specific insecticide. The first application of the insecticide could, therefore, kill 95 percent of the population. However, tolerant individuals would then constitute the majority of the breeding population that survived. This would mean that a larger proportion of the insects in the second generation would be tolerant. The second use of the insecticide on this population would not be as effective as the first. With continued use of the same insecticide, each generation would become more tolerant. Please note in this example that the spraying did not cause mutations that made the population resistant. The use of insecticide killed susceptible individuals, thus leaving resistant individuals to reproduce and pass their resistance genes to the next generation.

Many species of insects produce a new generation each month. In organisms with a short generation time, 99 percent of the population could become resistant to the insecticide in just five years. As a result, the insecticide would no longer be useful in controlling the species. As a new factor (the insecticide) was introduced into the environment of the insect, natural selection resulted in a population that was tolerant of the insecticide.

The same kind of selection process has occurred with herbicides. Within the past 50 years, many kinds of herbicides have been developed to control weeds in agricultural fields. After several years of use, a familiar pattern develops as more and more species of weeds show resistance to the herbicide. Figure 21.9 shows several kinds of herbicides and the number of weed

TABLE 21.2 Changes in the Body Structure of *Geospiza fortis*

	Before Drought	After Drought
Average Body Weight	16.06 g	17.13 g
Average Bill Depth	9.21 mm	9.70 mm

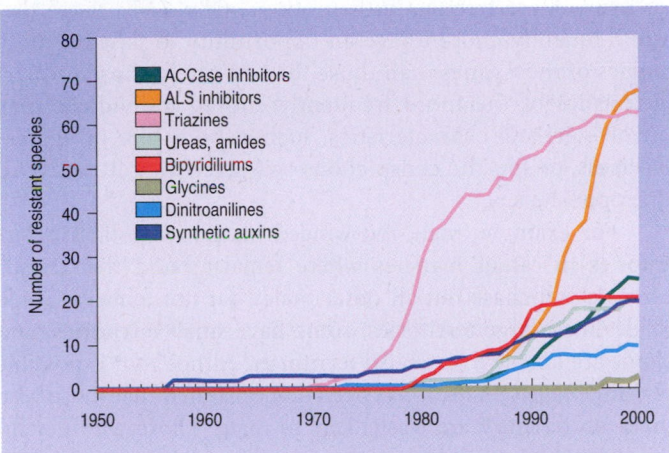

FIGURE 21.9 Populations of weed plants that have been subjected repeatedly to herbicides often develop resistant populations. The individual weed plants that have been able to resist the effects of the herbicide have lived to reproduce and pass on their genes for resistance to their offspring; thus, resistant populations of weeds have developed.

species that have become resistant over time. In each weed species, there has been selection for the individuals that have the genetic information that allows them to tolerate the presence of the herbicide.

Differential Reproductive Rates

Survival alone does not always ensure reproductive success. For a variety of reasons, some organisms may be better able to utilize available resources to produce offspring. If one individual leaves 100 offspring and another leaves only 2, the first organism has passed more copies of its genetic information to the next generation than has the second and the mix of genes in the population of offspring is different from that of their parents.

Scientists have conducted studies of the frequencies of genes for the height of clover plants. Two identical fields of clover were planted and cows were allowed to graze in one of them. Cows acted as a selecting agent by eating the taller plants first. These tall plants rarely got a chance to reproduce. Only the shorter plants flowered and produced seeds. After some time, seeds were collected from both the grazed and ungrazed fields and grown in a greenhouse under identical conditions. The average height of the plants from the ungrazed field was compared to that of the plants from the grazed field. The seeds from the ungrazed field produced some tall, some short, but mostly medium-sized plants. However, the seeds from the grazed field produced many more shorter plants than medium or tall ones. The cows had selectively eaten the plants that had the genes for tallness. Since the flowers are at the tip of the plant, tall plants were less likely to successfully reproduce, even though they might have been able to survive grazing by cows.

Differential Mate Selection

Within animal populations, some individuals may be chosen as mates more frequently than others. Obviously, those that are frequently chosen have an opportunity to pass on more copies of their genes than those that are rarely chosen. Characteristics of the more frequently chosen individuals may involve general characteristics, such as body size or aggressiveness, or specific conspicuous characteristics attractive to the opposite sex.

For example, male red-winged blackbirds establish territories in cattail marshes where females build their nests. A male will chase out all other males but not females. Some males have large territories, some have small territories, and some are unable to establish territories. Although it is possible for any male to mate, it has been demonstrated that those who have no territory are least likely to mate. Those who defend large territories may have two or more females nesting in their territories and are very likely to mate with those females. It is unclear exactly why females choose one male's territory over another, but the fact is that some males are chosen as mates and others are not.

In other cases, it appears that the females select males that display conspicuous characteristics. Male peacocks have very conspicuous tail feathers. Those with spectacular tails are more likely to mate and have offspring (figure 21.10). Darwin was puzzled by such cases as the peacock, in which the large and conspicuous tail should have been a disadvantage to the bird.

FIGURE 21.10 In many animal species, the males display very conspicuous characteristics that are attractive to females. Because the females choose the males they will mate with, those males with the most attractive characteristics will have more offspring, and in future generations, there will be a tendency to enhance the characteristic. With peacocks *(Pavo cristatus)*, those individuals with large colorful displays are more likely to mate. Female peacocks are not as highly colored and have short tails.

Long tails require energy to produce, make it more difficult to fly, and make it more likely that predators will capture the individual. The current theory that seeks to explain this paradox involves female choice. If the females have an innate (genetic) tendency to choose the most elaborately decorated males, genes that favor such plumage will be regularly passed to the next generation because females choose the gaudy males as mates. Such special cases in which females choose males with specific characteristics have been called sexual selection.

21.8 ACQUIRED CHARACTERISTICS DO NOT INFLUENCE NATURAL SELECTION

Many individual organisms acquire characteristics during their lifetime that are not the result of the genes that they have. For example, a squirrel may learn where a bird feeder is and visit it frequently. It may even learn to defeat the mechanics of special bird feeders designed to prevent it from getting to the food. Such an ability can be very important for survival if food becomes scarce. Such **acquired characteristics** are gained during the life of the organism; they are not genetically determined and, therefore, cannot be passed on to future generations through sexual reproduction.

For example, we often desire a specific set of characteristics in our domesticated animals. The breed of dog known as boxers, for example, is "supposed" to have short tails. However, the genes for short tails are rare in this breed. Consequently, the tails of these dogs are amputated—a procedure called docking. Similarly, the tails of lambs are also usually amputated. These acquired characteristics are not passed on to the next generation. Removing the tails of these animals does not remove the genes for tail production from their genetic information, and each generation of puppies or sheep is born with tails.

However, before the nature of the gene and mechanisms of heredity were understood it was commonly thought that acquired characteristics could be passed from generation to generation. As mentioned earlier, Lamarck postulated that giraffes could have developed long necks as a result of stretching them to reach food and then these acquired characteristics could be passed to offspring. Darwin's theory of natural selection provides an alternative explanation for long necks in giraffes.

1. In each generation, more giraffes would be born than the food supply could support.
2. In each generation, some giraffes would inherit longer necks, and some would inherit shorter necks.
3. All giraffes would compete for the same food sources.
4. Giraffes with longer necks would obtain more food, have a higher survival rate, and produce more offspring.
5. As a result, succeeding generations would show an increase in the neck length of the giraffe species.

Figure 21.11 contrasts how Lamarck and Darwin provide different explanations for the phenomenon of long necks in giraffes.

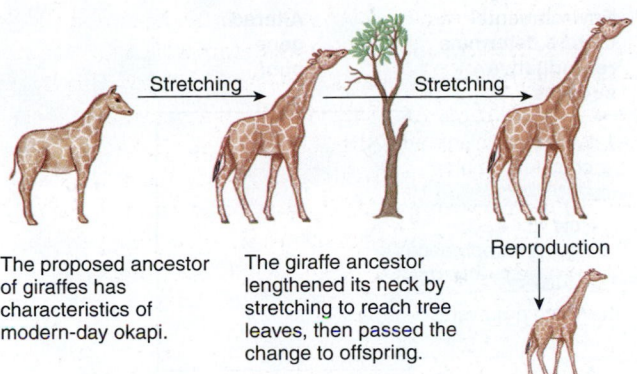

The proposed ancestor of giraffes has characteristics of modern-day okapi.

The giraffe ancestor lengthened its neck by stretching to reach tree leaves, then passed the change to offspring.

Stretching → Stretching →

Reproduction

A Lamarck's theory: variation is acquired.

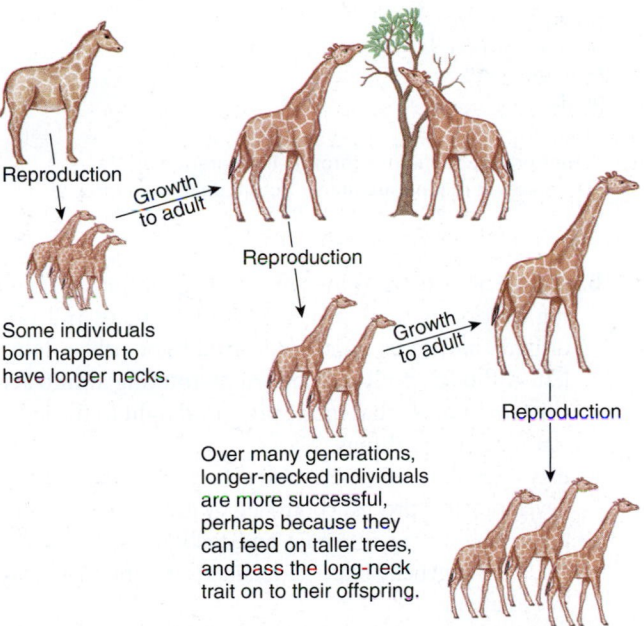

Reproduction

Growth to adult →

Reproduction

Some individuals born happen to have longer necks.

Growth to adult →

Reproduction

Over many generations, longer-necked individuals are more successful, perhaps because they can feed on taller trees, and pass the long-neck trait on to their offspring.

B Darwin-Wallace theory: variation is inherited.

FIGURE 21.11 (*A*) Lamarck thought that acquired characteristics could be passed on to the next generation. Therefore, he postulated that as giraffes stretched their necks to get food, their necks got slightly longer. This characteristic was passed on to the next generation, which would have longer necks. (*B*) The Darwin-Wallace theory states that there is variation within the population and that those with longer necks would be more likely to survive, reproduce, and pass on their genes for long necks to the next generation.

21.9 THE HARDY-WEINBERG CONCEPT

In the section on evolution and natural selection earlier in this chapter, evolution was described as the change in the genetic makeup of a population over time. We can think of all the genes of all the individuals in a population as a **gene pool** and that a change in the gene pool indicates that evolution is taking place.

In the early 1900s, an English mathematician, G. H. Hardy, and a German physician, Wilhelm Weinberg, recognized that it was possible to apply a simple mathematical relationship to the study of genes in populations if certain conditions were met. An unchanging gene pool over several generations would

imply that evolution is *not* taking place. A changing gene pool would indicate that evolution is taking place.

The conditions they felt were necessary for the genetic makeup to *remain constant* (for evolution *not* to occur) are:

1. Mating must be completely random.
2. Mutations must not occur.
3. Migration of individual organisms into and out of the population must not occur.
4. The population must be very large.
5. All genes must have an equal chance of being passed on to the next generation. (Natural selection is not occurring.)

The concept that the genetic makeup will remain constant if these five conditions are met has become known as the **Hardy-Weinberg concept.** The Hardy-Weinberg concept is important because it allows a simple comparison of genes in populations to indicate if genetic changes are occurring within the population. Two different populations of the same species can be compared to see if they have the same genetic makeup, or populations can be examined at different times to see if the genetic makeup is changing.

Although Hardy and Weinberg described the conditions necessary to prevent evolution, an examination of these conditions becomes a powerful argument for why evolution must occur.

Random mating rarely occurs. For example, mating between individuals is more likely between those that are nearby than those that are distant. In addition, in animal populations, the possibility exists that females may choose specific males with which to mate.

Mutations occur and alter genetic diversity by either changing one genetic message into another or introducing an entirely new piece of genetic information into the population.

Migration alters the gene pool, since organisms carry their genes with them when they migrate from one place to another. Even plants and fungi migrate when their reproductive stages are carried to new places by wind, water, or animals. Immigration introduces new genetic information, and emigration removes genes from the gene pool.

Population size influences the gene pool. Small populations typically have less genetic diversity than large populations.

Finally, *natural selection* systematically filters some genes from the population, allowing other genes to remain and become more common. The primary mechanisms involved in natural selection are differences in death rates, reproductive rates, and the rate at which individuals are selected as mates. The diagram in figure 21.12 summarizes these ideas.

21.10 ACCUMULATING EVIDENCE OF EVOLUTION

The theory of evolution has become the major unifying theory of the biological sciences. Medicine recognizes the dangers of mutations, the similarity in function of the same organ in related

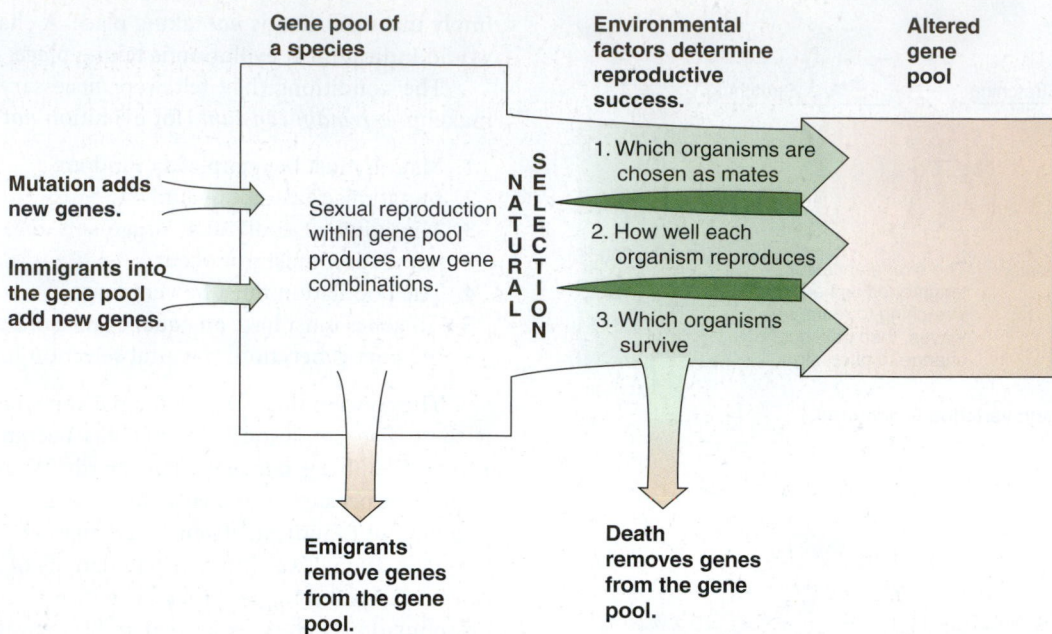

FIGURE 21.12 Several different processes cause gene frequencies to change. Genes enter populations through immigration and mutation. Genes leave populations through death and emigration. Natural selection operates within populations through death and rates of reproduction.

species, and the way in which the environment can interfere with the preprogrammed process of embryological development. Agricultural science recognizes the importance of selecting specific genes for passage into new varieties of crop plants and animals. The concepts of mutation, selection, and evolution are so fundamental to understanding what happens in biology that we often forget to take note of the many kinds of observations (facts) that support the theory of evolution. The following list describes some of the more important pieces of evidence that support the idea that evolution has been and continues to be a major force in shaping the nature of living things.

1. Species and populations are not genetically fixed. Change occurs in individuals and populations.
   a. Mutations cause slight changes in the genetic makeup of an individual organism.
   b. Different populations of the same species show adaptations suitable for their local conditions and have different combinations of genes.
   c. Changes in the characteristics displayed by species can be linked to environmental changes.
   d. Selective breeding of domesticated plants and animals indicates that the shape, color, behavior, metabolism, and many other characteristics of organisms can be selected for.
   e. Extinction of poorly adapted species is common.
2. Evolution occurs by small, incremental steps, not by major changes. All evidence suggests that, once embarked on a particular evolutionary road, the system is not abandoned, only modified. The following list supports the concept that evolution proceeds by modification of previously existing structures and processes rather than by catastrophic change.
   a. All species use the same DNA code.

   b. All species use the same left-handed, amino acid building blocks in their proteins. Amino acids and other organic molecules rotate polarized light either right or left. Although both left- and right-rotating amino acids exist, only those that rotate polarized light to the left are found in living things.
   c. It is difficult to eliminate a structure when it is part of a developmental process controlled by genes. Vestigial structures such as the appendix and tailbone in humans are evidence of genetic material retained from previous stages in evolution.
   d. Embryological development of related animals is similar, regardless of the peculiarities of adult anatomy. The embryos of all vertebrates have an early stage that contains structures that resemble gill slits.
   e. Species of organisms that are known to be closely related show greater similarity in their DNA than those that are distantly related.
3. The fossil record supports the concept of evolution.
   a. The nature of Earth has changed significantly over time.
   b. The fossil record shows vast changes in the kinds of organisms present on Earth. New species appear and most go extinct. This is evidence that living things change in response to changes in their environment.
   c. The fossils found in old rocks do not reappear in younger rocks. Once an organism goes extinct, it does not reappear, but new organisms arise that are modifications of previous organisms.
4. New techniques and discoveries invariably support the theory of evolution.
   a. The recognition that Earth was formed billions of years ago supports the slow development of new kinds of organisms.

The Reemerging of Infectious Diseases

Infectious diseases caused by bacteria, viruses, fungi, and parasitic worms continue to be a major cause of suffering and death throughout the world. They are the third leading cause of death in the United States. *Reemerging infectious diseases* (for example, diphtheria, malaria, whooping cough) are diseases that were once major health concerns but then declined significantly. However, they are beginning to increase in frequency. The reemergence of many kinds of infectious diseases is the result of two primary factors: our failure to immunize against these diseases and evolutionary changes in the microbes. People who are not being immunized against diseases are susceptible and may become ill with the disease or become asymptomatic carriers of the microbe. A further contributing factor to the reemergence of old diseases is the increased number of people with poorly functioning immune systems. HIV/AIDS has created a huge population of people with compromised immune systems. Famine and malnutrition also impair the immune system. War and the crowding that occurs in refugee camps and prisons enhance the easy spread of disease.

The reemergence of some diseases is also the result of evolution. Mutations are necessary if evolution is to take place. As parasites and their hosts interact, they constantly react to each other in an evolutionary fashion. Hosts develop new mechanisms to combat parasites, and parasites develop new mechanisms to overcome the hosts' defenses (for example, antibiotic resistance).

One of the mechanisms viruses use is a high rate of mutation. This ability to mutate has resulted in many new, serious human diseases. In addition, many new diseases arise when viruses that cause disease in another animal are able to establish themselves in humans. Many kinds of influenza originated in pigs, ducks, or chickens and were passed to humans through close contact with infected animals or by eating meat from infected animals. In many parts of the world, these domesticated animals live in close contact with humans (often in the same building), making conditions favorable for the transmission of animal viruses to humans.

Each year, mutations result in new varieties of influenza and colds, which pass through the human population. Occasionally, the new varieties are deadly. In 1918, a new variety of influenza virus originated in pigs in the United States and spread throughout the world. During the 1918–1919 influenza pandemic that followed, 20 to 40 million people died. In 1997 in Hong Kong, a new kind of influenza was identified that killed six of the eighteen people infected. When public health officials discovered the virus had come from chickens, they ordered the slaughter of all the live chickens in Hong Kong, which stopped the spread of the disease.

In early 2003, an outbreak of a new viral disease, known as severe acute respiratory syndrome (SARS), originated in China. SARS is a variation of a coronavirus—a class of virus commonly associated with the common cold—but it causes severe symptoms and, if untreated, can result in death. In June 2003, the SARS virus was isolated from an animal known as the masked palm civet (*Paguma larvata*). This animal is used for food in China and is a possible source of the virus that caused SARS in humans. However, other animals have also tested positive for the virus. The disease spread rapidly to several countries as people traveled by airplane from China to other parts of the world. A recognition of the seriousness of the disease and the isolation of infected persons prevented further spread, and this new disease was brought under control. However, if the virus still exists in some unknown wild animal host, it could reappear in the future.

In addition to cold and influenza, other kinds of diseases often make the leap from nonhuman to human hosts. The swine flu virus outbreaks of 2009 were traced to a population of pigs in Mexico. It is likely that genetic mixing occurred in the pigs' cells, resulting in a new kind of virus containing bird, human, and swine genes. The emergence of these new kinds of viruses enables them to more easily move from one species to another.

---

b. The recognition that the continents of Earth have separated and drifted apart helps explain why organisms on Australia are so different from those found elsewhere.

c. The discovery of DNA and how it works helps explain mutation and allows us to demonstrate the genetic similarity of closely related species.

PART III
SPECIATION

Natural selection provides a mechanism for evolutionary change. However, it requires some additional thinking to visualize how new species come into being. This section focuses on the nature of species and how new species originate.

21.11 SPECIES: A WORKING DEFINITION

The smallest irreversible step in the evolutionary process is the development of a new species. Before we consider how new species are produced, let's establish how one species is distinguished from another. A **species** is commonly defined as a population of organisms whose members have the potential to interbreed naturally to produce fertile offspring but do not interbreed with other groups. This is a working definition since it applies in most cases but must be interpreted to encompass some exceptions.

There are three key ideas within this definition. First, a species is a population of organisms. An individual—you, for example—is not a species. You can only be a member of a group that is recognized as a species. The human species, *Homo*

Science and Society

Antibiotic Resistance and Human Behavior

Antibiotics have been a major factor in improving the health of people and other animals throughout the world. We have become very dependent on these molecules to combat various kinds of disease-causing organisms—particularly Bacteria. However, when a specific antibiotic is used widely, it eventually becomes ineffective against many kinds of Bacteria. This occurs because exposure of Bacteria to the antibiotic acts as a selecting agent on the bacterial population. Those individual bacterial cells that are susceptible to the antibiotic die, while those that are resistant live and reproduce. When the surviving Bacteria reproduce, they pass their antibiotic resistance genes on to their offspring, and eventually, the majority of that particular population of bacterium becomes resistant. The gene pool of the population has changed; thus, we can say that evolution has occurred. Some species of disease-causing Bacteria have populations (often called strains) that are resistant to several antibiotics.

Recognition that antibiotic resistance is a problem has led to requests to change the way we use antibiotics. Widespread use of antibiotics in animal feed has the high probability of causing the development of resistant strains of the bacterium that causes tuberculosis in both cattle and humans. Indiscriminate prescription of antibiotics for healthy patients with minor bacterial infections has led to resistant strains. Thus, public health officials are calling on veterinarians and physicians to change their behavior and use antibiotics only when needed to protect the health of the patient.

Even the behavior of the patient can encourage the development of resistant strains. Often people stop taking an antibiotic when they start to feel better. If someone they know has what they consider to be similar symptoms, they might give their leftover or extra pills to that person. However, they should not do this because they may still have the disease organisms present in their own body, and the amount of drug they are giving will probably not be enough to control the infection in the person receiving the antibiotic. Taking a lower dose of an antibiotic or stopping the medication before the disease is under control can result in the development of resistant strains. Tuberculosis is a particularly difficult disease, since it is a chronic disease that requires long periods of antibiotic drug therapy and is likely to be caused by organisms that are already resistant to several antibiotics. Resistant strains of tuberculosis have become so common that one of the standard methods of administering the drug is to have a health care professional directly observe the patient taking the medication as prescribed. Some particularly difficult or stubborn patients have been institutionalized until their course of treatment is complete.

Questions to Discuss

1. Should persons be deprived of their liberty so that a disease can be controlled?
2. Should there be laws about when physicians and veterinarians can prescribe antibiotics?
3. Should physicians withhold antibiotics from patients who are very likely to recover without the antibiotic?

Connections . . .

Evolution and Domesticated Cats

The evolutionary history of the domestic cat (*Felis silvestris lybica*) has been unclear because of incomplete fossil records. However, an international team sampling both mitochondrial DNA and the DNA from both X and Y sex chromosomes has finally come up with an evolutionary tree for felines. The group proposes that about 11 million years ago, a single, ancestral feline-like species migrated from Asia throughout the world, except Australia.

Researchers believe that 3 to 10 million years ago (MYA) land bridges between continents were created when sea levels fell. The common ancestor to all of today's cats probably migrated south to Africa from Asia. The cats also moved north, crossing the Bering land bridge (as wide as 1,000 miles) to North America, and migrated to South America by the Panamanian land bridge. When sea levels rose, they covered the land bridges and cut off cat species from their original groups. These isolated subpopulations genetically drifted apart, each adapting to its unique environment. When the subpopulations had the chance of coming back together, they were no longer able to interbreed and, at that point, found themselves to be different species. Ancestral felines, originally a Eurasian genus, successfully migrated throughout the globe because they encountered little or no competition from other carnivores. They continue to be one of the most successful of carnivore families.

Traditionally, domestication was thought to have occurred about 3,600 years ago in Egypt. Archeological evidence in Egyptian hieroglyphics portrays cats, and bones of cats have been found buried with humans in tombs. However, more recent archeological and genetic evidence strongly suggests that cats were domesticated about 9,500 years ago in the Fertile Crescent of the Middle East. Today, this region includes Egypt, Israel, the West Bank, Gaza strip, and Lebanon and parts of Jordan, Syria, Iraq, southeastern Turkey, and southwestern Iran and Kuwait.

A Horse   B Donkey   C Mule

FIGURE 21.13 Even though they do not do so in nature, (*A*) horses (*Equus caballus*) and (*B*) donkeys (*Equus asinus*) can be mated. The offspring produced by mating a female horse with a male donkey is called a (*C*) mule (*Equus asinus x caballus*) and is sterile. Because the mule is sterile, the horse and the donkey are considered to be of different species.

sapiens, consists of more than six billion individuals, whereas the endangered California condor species, *Gymnogyps californianus,* consists of about three hundred twenty-five individuals. Second, the definition involves the ability of individuals within the group to produce fertile offspring. Obviously, we cannot check every individual to see if it is capable of mating with any other individual that is similar to it, so we must make some judgment calls. Do most individuals within the group potentially have the capability of interbreeding to produce fertile offspring? In the case of humans, we know that some individuals are sterile and cannot reproduce, but we don't exclude them from the human species because of this. If they were not sterile, they would have the potential to interbreed. We recognize that humans from all parts of the world are potentially capable of interbreeding. We know this to be true because of the large number of instances of reproduction involving people of different ethnic and racial backgrounds. The same is true for many other species that have local subpopulations but have a wide geographic distribution. Third, the species concept also takes into account an organism's evolutionary history. A species is a group of organisms that shares a common ancestor with other species but is set off from those others by having newer, genetically unique traits.

One way to find out if two populations belong to the same species is investigate *gene flow.* **Gene flow** is the movement of genes from one generation to the next or from one region to another. Two or more populations that demonstrate gene flow between them constitute a single species. Conversely, two or more populations that do not demonstrate gene flow between them are generally considered to be different species. Some examples will clarify this working definition.

The mating of a male donkey and a female horse produces young that grow to be adult mules, incapable of reproduction (figure 21.13). Since mules are nearly always sterile, there can be no gene flow between horses and donkeys and they are considered to be separate species. Similarly, lions and tigers can be mated in zoos to produce offspring. However, this does not happen in nature, so gene flow does not occur naturally; thus, they are considered to be two separate species.

Still another way to try to determine if two organisms belong to different species is to determine their genetic similarity. The

Myths, Mistakes, and Misunderstandings

Common Misconceptions About the Theory of Evolution

1. *Evolution happened only in the past and is not occurring today.* In fact, we see lots of evidence that genetic changes are occurring in the populations of current species (antibiotic resistance, pesticide resistance, and domestication).

2. *Evolution has a specific goal.* Evolution does not move toward a specific goal. Natural selection selects those organisms that best fit the current environment. As the environment changes, so do the characteristics that have value. Random events such as changes in sea level, major changes in climate such as ice ages, or collisions with asteroids have had major influences on subsequent natural selection and evolution. Evolution results in organisms that "fit" the current environment.

3. *Changes in the environment cause mutations that are needed by an organism to survive under the new environmental conditions.* Mutations are random events and are not necessarily adaptive. However, when the environment changes, mutations that were originally detrimental or neutral may have greater value. The gene did not change but the environmental conditions did. In some cases, the mutation rate may increase or there may be more frequent exchanges of genes between individuals when the environment changes, but the mutations are still random. They are not directed to a particular goal.

4. *Individual organisms evolve.* Individuals are stuck with the genes they inherited from their parents. Although individuals may adapt by changing their behavior or physiology, they cannot evolve; only populations can show changes in the gene pool.

5. *Today's species frequently can be shown to be derived from other present-day species. For example, it is commonly stated that apes gave rise to humans.* There are few examples in which it can be demonstrated that one current species gave rise to another. Apes did not become humans, but apes and humans had a common ancestor several million years ago.

Human-Designed Organisms

Humans have designed several kinds of plants and animals for their own purposes through the process of domestication. Most modern cereal grains are special plants that rely on human activity for their survival; most would not live without fertilizer, cultivation, and other helps. These grains are the descendants of wild plants. Initially, the process of domestication of plants by humans probably was not a conscious effort to develop crops but the unconscious selection of individual plants that had particularly useful characteristics from among all the plants of that species. For example, the seeds may have been larger than those of other plants of the same species or they may have tasted better or they may have been easier to chew. The unconscious selection of these seeds could have led to the dispersal of these particular seeds along traditional travel routes and eventually through trading with other groups of people. Wheat, barley, rice, and corn are all plants that have been domesticated and currently supply a major part of the food for the world's population. However, modern genetic techniques have resulted in domesticated plants that only superficially resemble their wild ancestors.

Most domesticated animals are mammals or birds that are herbivores that grow rapidly, are reasonably docile, and will reproduce in captivity. Examples are cattle, sheep, pigs, goats, horses, camels, chickens, turkeys, ducks, and geese. (Carnivorous cats and dogs are obvious exceptions to the general herbivore rule.) Obviously, animals that do not reproduce in captivity would not be good candidates for domestication, and those that had behaviors that made them dangerous to be around would not be chosen to be domesticated. Again, the choices by humans would have been "unconscious." They would not have planned to domesticate an animal for a particular purpose (milk, eggs, meat, power), and modern science has greatly modified domesticated animals to something that is quite different from their wild ancestors.

In fact, many of the wild ancestors of modern domesticated plants and animals are extinct. Thousands of generations of selection have, in effect, caused the development of species that are known only by their domesticated remnants.

In recent years, the genetic manipulation of organisms has expanded into a new arena. With the development of biotechnology techniques, genes can be moved from one species to another. The transfer of genes can be between plants and animals, Bacteria and plants, or Bacteria and animals. Genes from certain kinds of Bacteria produce a compound that kills insects. This gene has been inserted into several kinds of plants so that the plants produce the natural insecticide and are protected from some of their insect pests. In addition, animal genes have been inserted into plants and human genes have been inserted into Bacteria. It appears that gene transfer between species occurs naturally as well. This knowledge requires us to rethink our concept of what a species is and how evolution occurs.

---

recent advances in molecular genetics allow scientists to examine the structure of the genes present in individuals from a variety of different populations. Those that have a great deal of similarity are assumed to have resulted from populations that have exchanged genes through sexual reproduction in the recent past. If there are significant differences in the genes present in individuals from two populations, they have not exchanged genes recently and are more likely to be members of separate species. Interpretation of the results obtained by examining genetic differences still requires the judgment of experts. It will not unequivocally settle every dispute related to the identification of species, but it is another tool that helps to clarify troublesome situations.

This concept that species can be distinguished from one another by their inability to interbreed or produce fertile offspring is often called the **biological species concept.** Although this concept of a species is useful from a theoretical point of view, it is often not a practical way to distinguish species. Thus, biologists often use specific observable physical, chemical, or behavioral characteristics as guides to distinguishing species. This method of using physical characteristics to identify species is called the **morphological species concept.** Structural differences are useful but not foolproof ways to distinguish species. However, we must rely on such indirect ways to identify species because we cannot possibly test every individual by breeding it with another to see if they will have fertile offspring.

Furthermore, many kinds of organisms reproduce primarily by asexual means. Because organisms that reproduce exclusively by asexual methods do not exchange genes with any other individuals, they do not fit our *biological species* definition very well. In addition, the study of fossil species must rely on structural characteristics to make species distinctions since it is impossible to breed extinct organisms.

21.12 HOW NEW SPECIES ORIGINATE

The geographic area over which a species can be found is known as its **range.** The range of the human species is the entire world, while that of a bird known as a snail kite is a small region of southern Florida. As a species expands its range or environmental conditions change in some parts of the range, portions of the population can become separated from the rest. This means that new colonies or isolated populations have infrequent gene exchange with their geographically distant relatives. Thus, many species consist of partially isolated populations that display characteristics that differ significantly from other local populations. Many of the differences observed may be directly related to adaptations to local environmental conditions. These genetically distinct populations are known as subspecies (or breeds, varieties, strains, races, and types).

Other Mechanisms That Cause Evolution

Although natural selection is the primary mechanism for evolutionary change, several special cases are alternative causes of evolution.

Artificial Selection

Artificial selection was recognized by Darwin as a special case of selection. With domesticated plants and animals, humans either consciously or unconsciously choose specific individual organisms for reproduction. This results in certain characteristics becoming common in the populations of domesticated organisms while others become rare. For example, populations of dogs, horses, chickens, pigs, and cattle have been specifically bred to show particular characteristics. Selective breeding in chickens has resulted in some populations with enhanced egg production. Other populations were selected for enhanced meat production. Others were selected to produce interesting colors or shapes. Although artificial selection is considered a special case, it could be argued that humans simply become an important part of the environment of domesticated organisms; thus, artificial selection is a special case of natural selection.

Horizontal Gene Transfer

One of the most interesting discoveries of the recent past is that there are many cases of genetic exchanges between organisms that are not related. Among Bacteria it is common for pieces of DNA to be transferred from one species to another. This means that entire collections of genes and the chemical processes the genes control can be added to an organism. Another example of horizontal gene transfer is endosymbiosis (discussed earlier in this chapter) that resulted in the development of such structures as chloroplasts and mitochondria. When one organism incorporates a different organism into its cells, it also accepts all the genes that were part of the captured cell. With the development of techniques to rapidly decipher the genes in DNA, it is being recognized that certain collections of genes appear to have been transferred in all kinds of organisms. In other words, horizontal gene transfer has been a common way in which new combinations of genes have been developed in organisms. Once collections of genes have been incorporated into an organism, they are subject to the process of natural selection as are all other genes.

Genetic Drift

Sometimes the mix of genes present in a population can change for reasons unrelated to natural selection. This is called *genetic drift*. Genetic drift is most likely to occur in small populations where accidental events may greatly affect the frequency of a gene in a population.

In large populations any unusual shifts in gene frequency in one part of the population usually would be counteracted by reciprocal changes in other parts of the population. However, in small populations the random distribution of genes to gametes may not reflect the percentages present in the population. For example, consider a situation in which there are

Box Figure 21.3 (A) Rhode Island Red chickens were selectively bred, resulting in the (B) leghorn variety, a breed that lays more eggs.

one hundred plants in a population and ten have dominant alleles for patches of red color, whereas the others do not. If in those ten plants the random formation of gametes resulted in no red alleles present in the gametes that were fertilized, the allele could be eliminated. Similarly, if all those plants with the red allele happened to be in a hollow that was subjected to low temperatures, they could be killed by a late frost and would not pass on their alleles to the next generation. Therefore, the allele would be lost, but the loss would not be the result of natural selection.

Speciation is the process of generating new species. The process can occur in several ways. One common way that speciation occurs is through geographic isolation.

The Role of Geographic Isolation in Speciation

A portion of a species can become totally isolated from the rest of the gene pool by some geographic change, such as the formation of a mountain range, river valley, desert, or ocean. When this happens, the portion of the species is said to be in **geo**-

graphic isolation from the rest of the species. If two populations of a species are geographically isolated, they are also reproductively isolated, and gene exchange is not occurring between them. The geographic features that keep the different portions of the species from exchanging genes are called **geographic barriers.** The uplifting of mountains, the rerouting of rivers, and the formation of deserts all may separate one portion of a gene pool from another. For example, two kinds of squirrels are found on opposite sides of the Grand Canyon. Some people consider them to be separate species, while others consider them to be different isolated subpopulations of the same species (figure 21.14). Even small changes may cause geographic isolation in species

People Behind the Science

Ernst Mayr (1904–2005)

In March 1923, just after completing high school, Ernst Mayr visited a lake about 15 km (9 mi) north of Dresden, Germany, and observed a pair of red-crested pochard ducks, which are rare in Germany. He was encouraged to stop in Berlin on his way to the University of Griefswald, where he was to study medicine, to tell noted ornithologist Edwin Stresemann about his discovery. This was to change the direction of his life. After Mayr had completed his preclinical studies, Stresemann invited him to complete a Ph.D. with him at the University of Berlin. He completed his Ph.D. at age twenty-one. Stresemann introduced Mayr to Lord Walter Rothschild of England, an avid collector of birds. Rothschild sponsored Mayr to visit the island of New Guinea to study and collect birds.

While in New Guinea, Mayr met researchers from the American Museum of Natural History who invited him to accompany them on a trip to the Solomon Islands. Shortly after returning to Germany, Mayr was invited to join the staff of the American Museum of Natural History (1931), and he joined the faculty of Harvard University in 1953.

Mayr was one of the twentieth century's leading evolutionary biologists. In 1942, he published *Systematics and the Origin of Species*, in which he introduced the biological species concept—the idea that a species is a population of organisms potentially capable of interbreeding, and, furthermore, if individuals of two populations are not able to interbreed, they are of different species. Previously, scientists had sought to identify individuals as belonging to distinct species by the physical characteristics they possessed.

He also described the process of allopatric speciation, which is the idea that an isolated population is subjected to unique environmental conditions and may have different mutations from the parent population from which it was split. This can ultimately lead to the isolated population becoming a distinct species. These two concepts are central ideas in modern evolutionary thought.

Mayr received numerous awards and honors throughout his life. He retired in 1975 but continued to write about evolution and the philosophy of biology until his death at age one hundred in 2005.

Source: Modified from the *Hutchinson Dictionary of Scientific Biography*. © RM, 2011. All rights reserved. Helicon Publishing is a division of RM.

that have little ability to move. A fallen tree, a plowed field, or even a new freeway may effectively isolate populations within such species. Snails in two valleys separated by a high ridge have been found to be closely related but different species. The snails cannot get from one valley to the next because of the height and climatic differences presented by the ridge (figure 21.15).

The Role of Natural Selection in Speciation

The separation of a species into two or more isolated sub-populations is not enough to generate new species. Even after many generations of geographic isolation, these separate groups may still be able to exchange genes (mate and produce fertile offspring) if they overcome the geographic barrier, because they have not accumulated enough genetic differences to prevent reproductive success. Differences in environments and natural selection play very important roles in the process of forming new species.

 Kaibab squirrel

 Aberts squirrel

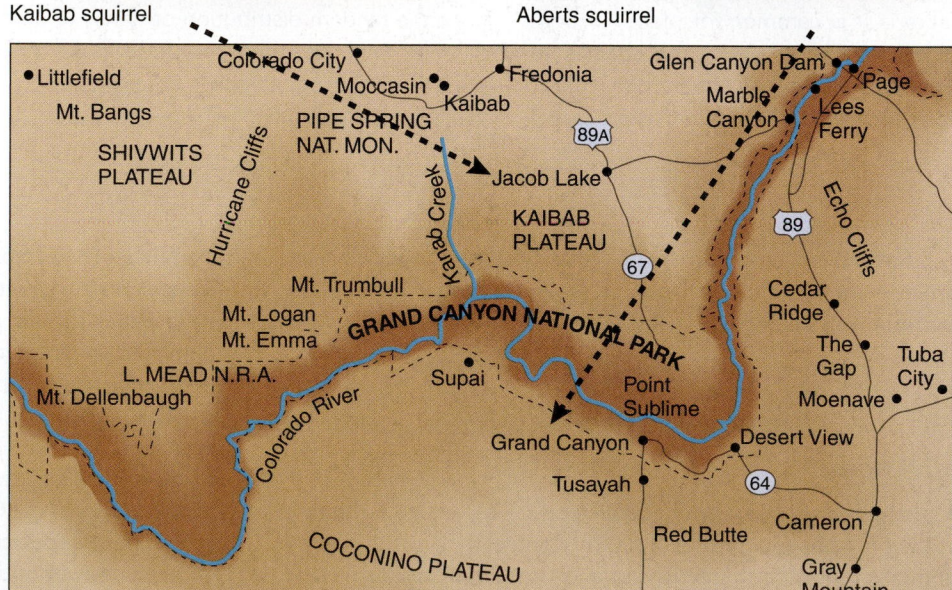

FIGURE 21.14 These two squirrels are found on opposite sides of the Grand Canyon. Some people consider them to be different species; others consider them to be distinct populations of the same species.

520 Chapter 21 *The Origin and Evolution of Life*

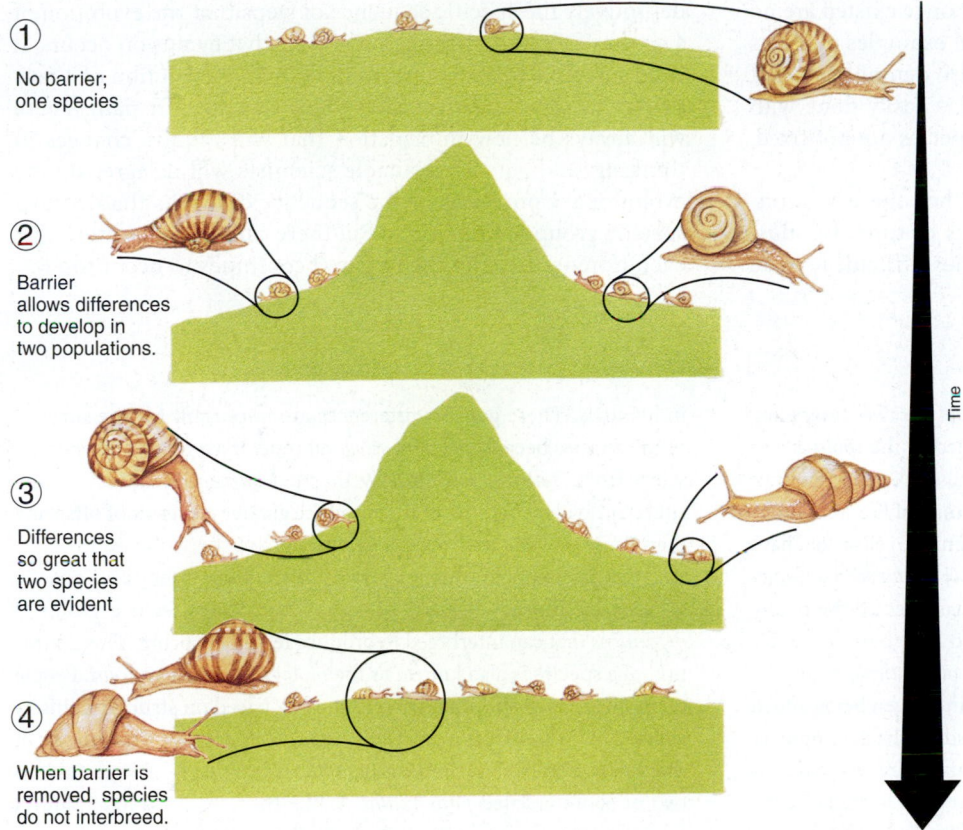

① No barrier; one species

② Barrier allows differences to develop in two populations.

③ Differences so great that two species are evident

④ When barrier is removed, species do not interbreed.

Time

FIGURE 21.15 If a single species of snail were to be divided into two different populations by the development of a ridge between them, the two populations could be subjected to different environmental conditions. This could result in a slow accumulation of changes that could ultimately result in two populations that would not be able to interbreed even if the ridge between them were to erode. They would be different species.

Following separation from the main portion of the gene pool by geographic isolation, the organisms within the small, local population are likely to experience different environmental conditions. If, for example, a mountain range has separated a species into two populations, one of them may receive more rain or more sunlight than the other. These environmental differences act as natural selecting agents on the two gene pools and, acting over a long period of time, account for different genetic combinations in the two places. Furthermore, different mutations may occur in the two isolated populations, and each may generate different random combinations of genes as a result of sexual reproduction. This would be particularly true if one of the populations was very small. As a result, the two populations may show differences in color, height, enzyme production, time of seed germination, or many other characteristics.

Reproductive Isolation

Over a long period of time, the genetic differences that accumulate may result in regional populations called **subspecies** that are significantly modified structurally, physiologically, or behaviorally. The differences among some subspecies may be so great that they have reduced reproductive success when the subspecies mate. If enough genetic differences accumulate that two populations are not able to interbreed, then speciation has

occurred. Speciation has occurred only if gene flow between isolated populations does not occur even after barriers are removed. In other words, the process of speciation can begin with the geographic isolation of a portion of the species, but new species are generated only if isolated populations become separate from one another genetically. Speciation by this method is really a three-step process. It begins with geographic isolation, is followed by the action of selective agents that choose specific genetic combinations as being valuable, and ends with the genetic differences becoming so great that reproduction between the two groups is impossible.

Speciation Without Isolation

It is also possible to envision ways in which speciation could occur without geographic isolation being necessary. Any process that could result in the reproductive isolation of a portion of a species could lead to the possibility of speciation. For example, within populations, some individuals may breed or flower at a somewhat different time of the year. If the difference in reproductive time is genetically based, different breeding populations could be established, which could eventually lead to speciation. Among animals, variations in the genetically determined behaviors related to courtship and mating could effectively separate one species into two or more separate breeding populations. In plants, genetically determined incompatibility of the pollen of one population of flowering plants with the flowers of other populations of the same species could lead to separate species.

21.13 THE TENTATIVE NATURE OF THE EVOLUTIONARY HISTORY OF ORGANISMS

It is important to understand that thinking about the concept of evolution can take us in several different directions. First, it is clear that genetic changes do occur. Mutations introduce new genes into a species. This has been demonstrated repeatedly with chemicals and radiation. Our recognition of this danger is evident by the ways we protect ourselves against excessive exposure to mutagenic agents. We also recognize that species can change. We purposely manipulate the genetic constitution of our domesticated plants and animals and change their characteristics to suit our needs. We also recognize that different populations of the same species show genetic differences. Examination of fossils shows that species of organisms that once existed are no longer in existence. We even have historical examples of plants and animals that are now extinct. We can also demonstrate that new species come into existence, and this is easily done with plants. It is clear from this evidence that species are not fixed, unchanging entities.

However, when we try to piece together the evolutionary history of organisms over long periods of time, we must use much indirect evidence, and it becomes difficult to state definitively the specific sequence of steps that the evolution of a species followed. Although it is clear that evolution occurs, it is not possible to state unconditionally that evolution of a particular group of organisms has followed a specific path. There will always be new information that will require changes in thinking, and equally reputable scientists will disagree on the evolutionary processes or the sequence of events that led to a specific group of organisms. But there can be no question that evolution occurred in the past and continues to occur today.

SUMMARY

Current theories about the *origin of life* include the two competing ideas (1) that the primitive Earth environment led to the *spontaneous organization of organic chemicals* into primitive cells or (2) that primitive forms of life *arrived on Earth from space*. Basic units of life were probably similar to present-day *prokaryotes*. These primitive cells could have *changed through time* as a result of *mutation* and *in response to a changing environment*. The presence of living things has affected the nature of Earth's atmosphere. The first organisms would have been *anaerobic* since there was no oxygen in the atmosphere. *Photosynthesis* by organisms such as cyanobacteria would have resulted in oxygen being added to the atmosphere. The presence of oxygen allowed for the development of *aerobic respiration*. Three major kinds of organisms are referred to as *domains: Bacteria, Archaea,* and *Eucarya*. It appears that Bacteria are the oldest group, followed by Archaea and Eucarya. The Eucarya appear to have arisen as a result of the process of *endosymbiosis*.

All sexually reproducing organisms naturally exhibit genetic diversity among the individuals in the population as a result of *mutations* and the *genetic recombination* resulting from sexual reproduction. The *genetic differences* are reflected in physical differences among individuals. These genetic differences are important for the survival of the species because *natural selection* must have *genetic diversity* to select from. Natural selection by the *environment* results in better-suited individual organisms that produce greater numbers of offspring than those that are less well-off genetically.

Organisms with wide *geographic distribution* often show genetic differences in separate parts of their *range*. A *species* is a group of *organisms* that can interbreed to produce fertile offspring. This definition of a species is also known as the *biological species concept*. People often distinguish species from one another based on structural differences. This is known as a *morphological species concept*. The process of *speciation* usually involves the *geographic separation* of the species into two or more isolated *populations*. While they are separated, natural selection operates to *adapt each population to its environment*.

At one time, people thought that all organisms had remained *unchanged* from the time of their creation. Lamarck suggested that *change did occur* and thought that *acquired characteristics could be passed from generation to generation*. Darwin and Wallace proposed the *theory of natural selection* as the mechanism that drives *evolution*.

KEY TERMS

acquired characteristics (p. **512**)
autotroph (p. **501**)
biogenesis (p. **498**)
biological species concept (p. **518**)
domains (p. **503**)
endosymbiotic theory (p. **503**)
evolution (p. **506**)

gene flow (p. **517**)
gene pool (p. **513**)
genetic recombination (p. **510**)
geographic barriers (p. **519**)
geographic isolation (p. **519**)
Hardy-Weinberg concept
  (p. **513**)

heterotroph (p. **501**)
morphological species concept
  (p. **518**)
mutation (p. **509**)
natural selection (p. **507**)
oxidizing atmosphere (p. **502**)
range (p. **518**)

selecting agents (p. **510**)
speciation (p. **519**)
species (p. **515**)
spontaneous generation (p. **498**)
spontaneous mutation (p. **509**)
subspecies (p. **521**)
theory of natural selection (p. **507**)

APPLYING THE CONCEPTS

Answers are located in appendix F.
Part I

1. Which was *not* a major component of the early Earth's reducing atmosphere?
   a. H_2
   b. CO_2
   c. NH_3
   d. O_2

2. Evidence from fossils shows that prokaryotic cells came into existence approximately _____ years ago.
   a. 20 billion
   b. 3.5 billion
   c. 4–5 billion
   d. 1.5 billion

3. Oxygen present in our atmosphere is the result of
   a. organisms carrying on respiration.
   b. gases from volcanoes.
   c. the breakdown of carbon dioxide into carbon and oxygen.
   d. organisms carrying on photosynthesis

4. The Domain Eucarya
   a. has the simplest cell structure.
   b. is probably the result of endosymbiosis.
   c. was the first to be produced.
   d. began about 3.5 billion years ago.

5. As a control, Pasteur cut off the tops of swan-necked flasks and allowed microorganisms from the air to fall into the flask. His experimental swan-necked flasks were
   a. not broken.
   b. covered with gauze.
   c. deliberately seeded with microbes.
   d. also heat-sealed.

Part II

6. Which of the following is involved in generating new genetic combinations in organisms?
   a. acquired characteristics
   b. sexual reproduction
   c. Hardy-Weinberg
   d. natural selection

7. Which of the following could lead to a changed gene frequency in a population?
   a. Individuals with certain characteristics die.
   b. Individuals with certain characteristics are attractive to the opposite sex.
   c. Individuals with certain characteristics have more offspring than others.
   d. All of the above are correct.

8. The Darwin-Wallace theory of natural selection differs from Lamarck's ideas in that
   a. Lamarck understood the role of genes, and Darwin and Wallace did not.
   b. Lamarck assumed that characteristics obtained during an organism's lifetime could be passed to the next generation; Darwin and Wallace did not.
   c. Lamarck did not think that evolution took place; Darwin and Wallace did.
   d. Lamarck developed the basic ideas of speciation, which Darwin and Wallace refined.

9. The Hardy-Weinberg concept requires that
   a. mutations occur
   b. mating be random
   c. migration occur
   d. natural selection take place

10. The various processes that encourage the passage of beneficial genes to future generations and discourage the passage of harmful or less valuable genes are collectively known as
    a. natural selection.   c. sex.
    b. evolution.   d. adaptation of the fittest.

Part III

11. Hybrid animals such as mules are not considered to be a species because they
    a. do not reproduce through many generations.
    b. are not common enough.
    c. can be maintained only by humans.
    d. look different from both parents.

12. Two closely related organisms are not considered to be separate species unless they
    a. look different.
    b. are reproductively isolated.
    c. are able to interbreed.
    d. are in different geographic parts of the world.

13. Which of the following is required for speciation to occur?
    a. genetic isolation from other species
    b. genetic diversity within a species
    c. hundreds of millions of years
    d. reproduction

14. Two groups of organisms belong to different species if
    a. gene flow between the two groups is not possible, even in the absence of physical barriers.
    b. physical barriers separate the two groups, thereby preventing cross matings.
    c. the two groups of organisms have a different physical appearance.
    d. individuals from the two groups, when mated, produce fertile offspring.

15. The movement of genes from one generation to the next or from one region to another is known as
    a. mitosis.
    b. meiosis.
    c. gene flow.
    d. gene transfer.

QUESTIONS FOR THOUGHT

Part I

1. In what sequence did the following things happen: living cell, oxidizing atmosphere, eukaryotes developed, reducing atmosphere, first organic molecule?

2. What is meant by *spontaneous generation?* What is meant by *biogenesis?*

3. List two important effects caused by the increase of oxygen in the atmosphere.

4. What evidence supports the theory that eukaryotic cells arose from the development of a symbiotic relationship between primitive prokaryotic cells?

5. Describe Francesco Redi's experiment to disprove spontaneous generation.

Part II

6. Why are acquired characteristics of little interest to evolutionary biologists?

7. What is natural selection? How does it work?

8. Give two examples of selecting agents and explain how they operate.

9. Why has Lamarck's theory been rejected?

10. What role did Alfred Russel Wallace play in developing the theory of natural selection?

11. The concept of evolution involves three important items. Describe them.

12. What are the two most common misinterpretations associated with the process of natural selection?

13. What does sexual reproduction have to do with genetic diversity, evolution, and natural selection?

Part III

14. "Evolution is a fact." "Evolution is a theory." Explain how both of these statements can be true.

15. Why is geographic isolation important in the process of speciation?

16. Why aren't mules considered a species?

17. List the series of events necessary for speciation to occur.

18. Explain the morphological species concept.

FOR FURTHER ANALYSIS

Part I

1. In earlier times, people were certain that horse hairs could turn into worms. There is even a group of simple kinds of worms known as horse-hair worms. Consider the following evidence:

   - Watering troughs were provided for horses to drink from.
   - Hairs were observed to fall from the horse's mane into the water of the watering trough.
   - Long, thin worms were seen to be swimming in the watering troughs.

   Since hair is not alive, if this were to occur, it would be a case of spontaneous generation. Devise an experiment that would determine if nonliving horse hairs become living worms.

Part II

2. People who are hospitalized often develop infections after they have entered the hospital. Often these infections are very difficult to treat with conventional antibiotics. Why do you think this is so?

3. How much diversity is there in hair color in the students of your class? Assume all colors seen are natural. Have the students stand in groups based on distinguishable differences in hair color. How many different categories do you have? What does this tell you about genetic diversity within our species?

4. During the early part of the twentieth century, a major social movement known as the eugenics movement occurred. The purpose of eugenics was to improve the quality of the human gene pool by eliminating "bad genes" from the population. Universities taught courses in eugenics, states passed laws that allowed certain persons to be sterilized against their will, and judging contests were held at state fairs to determine the family with the best characteristics. Hitler's concept of developing a "master race" was an extension of this thinking. Scientists who study evolutionary processes have thoroughly discredited eugenics. However, there are still vestiges of this thinking in society. Describe several examples.

Part III

5. Scientist regularly reorganize how they think specific species of organisms evolved. Some people argue that this is evidence that evolution is "only a theory" and that we can't ever know how change takes place. How would you refute this criticism?

INVITATION TO INQUIRY

The Evolution of Technology

A circular computer disc stores information on its surface in the form of magnetic "spots." This disc spins and is read by a device that detects the different magnetic spots. Assume you are trying to determine the evolutionary process that led to this current level of technology. How would you fit the following "fossils" into your evolutionary scheme: Edison's cylinder phonograph, a magnetic tape machine, an LP record, and a CD. Reread "A Closer Look: The Compact Disc" on page 171 to review how a CD stores information. This is similar to how Edison's cylinder and an LP record store information. Determine the approximate time each technology was developed and describe a logical evolutionary relationship among these technologies. Can you see a direct development from one technology to another? Were there instances of "endosymbiosis" in which one technology was applied within a new system?

22

The History of Life on Earth

These are modern-day stromatolites from Hamlin Pool in Western Australia. The dome-shaped structures shown in the photograph are composed of layers of cyanobacteria and materials they secrete. They grow up to 60 cm (about 2 ft) tall. Some of the oldest fossils are of ancient stromatolites that developed in shallow marine environments about 3.5 billion years ago. When samples from fossil stromatolites are cut into slices, microscopic images can be produced that show the fossil remains of some of the world's oldest cells.

CORE CONCEPT

Earth and its kinds of living things changed greatly over billions of years.

OUTLINE

525

OVERVIEW

At one time, people assumed that Earth and its inhabitants were fixed and unchanging. However, as they began to understand the nature of events such as volcanic eruptions, earthquakes, and erosion, it became obvious that Earth was changeable. Furthermore, the presence of the fossilized remains of extinct organisms showed that the nature of the kinds of things that inhabited Earth had changed as well. These ideas challenged core beliefs and began a long process of modifying the way we look at the history of Earth and life on it. In particular, people needed to develop an understanding that the history of Earth and its life was very long and should be measured in billions of years rather than hundreds or thousands of years.

PART I
KINDS OF ORGANISMS

This chapter is divided into two parts. Part I is a brief description of the kinds of organisms that live on Earth and the processes involved in organizing this vast biodiversity into logical arrangements.

22.1 THE CLASSIFICATION OF ORGANISMS

To talk about items in our world, we must have names for them. As new items come into being or are discovered, new words are devised to describe them. For example, the words *laptop, smartphone,* and *text message* describe new technology that did not exist thirty years ago. Similarly, in the biological world, people have given names to newly discovered organisms so they can communicate with others about the organism.

The Problem with Common Names

The common names used by people of distinct cultures are usually different. A *dog* in English is *chien* in French, *perro* in Spanish, and *cane* in Italian. It is even possible that different names can be used to identify the same organism in different regions within a country. For example, the common garter snake may be called a garden snake or gardner snake depending on where you live (figure 22.1). Actually, there are several different species of "garter snakes" that have been identified as distinct from one another. Thus, common names can be confusing, and scientists sought a more acceptable way to give organisms names that would eliminate confusion and would be used by all scientists.

FIGURE 22.1   Depending on where you live you may call this organism a garter snake, a garden snake, or a gardner snake. These common names can lead to confusion about what kind of snake a person might be talking about. However, the scientific name (*Thamnophis sirtalis*) is recognized worldwide by the scientific community.

FIGURE 22.2 Carolus Linnaeus (1707–1778), a Swedish doctor and botanist, originated the modern system of taxonomy.

The naming of organisms is a technical process, but it is extremely important. When biologists are describing their research, common names such as robin, maple tree, or garter snake are not good enough. They must be able to accurately identify the organisms involved so that everyone who reads the report, wherever they live in the world, will be able to know what organism is being discussed. The scientific identification of organisms really involves two different but related activities. One, *taxonomy,* involves the naming of organisms, and the other, *phylogeny,* involves showing how organisms are related evolutionarily. In reality no taxonomic decisions are made without considering the evolutionary history of the organism.

Taxonomy

Taxonomy is the science of naming organisms and grouping them into logical categories. Various approaches have been used to classify organisms. The Greek philosopher Aristotle (384–322 B.C.) had an interest in nature and was the first person to attempt a logical classification system. The root word for *taxonomy* is the Greek word *taxis,* which means "arrangement." Aristotle used the size of plants to divide them into the categories of trees, shrubs, and herbs.

During the Middle Ages, Latin was widely used as the scientific language. As new species were identified, they were given Latin names, often using as many as fifteen words to describe a single organism. Although using Latin meant that most biologists, regardless of their native language, could understand a species name, it did not completely do away with duplicate names. Because many of the organisms could be found over wide geographic areas and communication was slow, there could still be two or more Latin names for a species. To make the situation even more confusing, ordinary people used common local names.

Binomial System of Nomenclature

The modern system of classification began in 1758, when Carolus Linnaeus (1707–1778), a Swedish doctor and botanist, published his tenth edition of *Systema Naturae* (figure 22.2). (Linnaeus's original name was Karl von Linne, which he "latinized" to Carolus Linnaeus.) In the previous editions, Linnaeus had used the polynomial (many-names) Latin system. However, in the tenth edition, he introduced the *binomial system of nomenclature.* The **binomial system of nomenclature** uses only two Latin names—the genus name and the specific epithet (*epithet* = descriptive word)—for each species of organism. Recall that a species is a population of organisms capable of interbreeding and producing fertile offspring. Individual organisms are members of a species. A **genus** (plural, *genera*) is a group of closely related species of organisms; the specific epithet is a word added to the genus name to identify which one of several species within the genus we are discussing. This is similar to the naming system we use with people. When you look in the phone book, you look for the last name (surname), which gets you in the correct general category. Then you look for the first name (given name) to identify the individual you wish to call. The unique name given to a particular type of organism is called its species name or scientific name. To clearly identify the scientific name from other words, binomial names are either *italicized* or <u>underlined</u>. The first letter of the genus name is capitalized. The specific epithet is always written in lowercase. For example, *Thamnophis sirtalis* is the binomial name for the common garter snake.

In addition to assigning a specific name to each species, Linnaeus recognized a need for placing organisms into meaningful groups. In his system, he divided all forms of life into two *kingdoms,* Plantae and Animalia, and then further subdivided the kingdoms into other smaller units.

Organizing Species into Logical Groups

Since Linnaeus's initial attempts to place all organisms into categories, there have been many changes. One of the most fundamental is the recent recognition that there are three major categories of organisms that have been called *domains.*

A **domain** is the largest category into which organisms are classified. There are three domains: Bacteria, Archaea, and Eucarya (figure 22.3). Organisms are separated into these three domains based on specific structural and biochemical features of their cells. The Bacteria and Archaea are prokaryotic, and the Eucarya are eukaryotic.

A **kingdom** is a subdivision of a domain. There are several kingdoms within the Bacteria and Archaea that are based primarily on differences in the metabolism and genetic composition of the organisms. Within the domain Eucarya, there

Connections . . .

The Current Status of Taxonomy

The Tree of Life website is an international effort to provide easy access to current information on the taxonomic and phylogenetic relationships of living things. People from all over the world contribute information about organisms and where information about particular kinds of organisms can be found. It is designed for professional biologists, educators, and students. Like the fields of taxonomy and phylogeny, it will continue to be a work in progress. As new information is submitted, it is being evaluated by professional biologists to ensure that it is as up-to-date and accurate as possible. To see its current status, type "tree of life" into your search engine or go to http://tolweb.org/tree/.

are four kingdoms: Plantae, Animalia, Fungi, and Protista (protozoa and algae). Figure 22.4 provides examples of the three domains. However, no clear consensus exists at this time about how many kingdoms there are in the Bacteria and Archaea.

A **phylum** is a subdivision of a kingdom. However, microbiologists and botanists often use the term *division* in place of the term *phylum*.

All kingdoms have more than one phylum. For example, the kingdom Plantae contains several phyla, including flowering plants, conifer trees, mosses, ferns, and several other groups. Organisms are placed in phyla based on careful investigation of the specific nature of their structure, metabolism, and biochemistry. An attempt is made to identify natural groups rather than artificial or haphazard arrangements. For example, while nearly all plants are green and carry on photosynthesis, only flowering plants have flowers and produce seeds; conifers lack flowers but have seeds in cones; ferns lack flowers, cones, and seeds; and mosses are so simple in structure that they even lack tissues for transporting water.

A **class** is a subdivision within a phylum. For example, within the phylum Chordata, there are seven classes: mammals, birds, reptiles, amphibians, and three classes of fishes.

An **order** is a category within a class. Carnivora is an order of meat-eating animals within the class Mammalia. There are several other orders of mammals, including horses and their relatives, cattle and their relatives, rodents, rabbits, bats, seals, whales, humans, and many others.

A **family** is a subdivision of an order and consists of a group of closely related genera, which in turn are composed of groups of closely related species. The cat family, Felidae, is a subgrouping of the order Carnivora and includes many species in several genera, including the Canada lynx and bobcat (genus *Lynx*), the cougar (genus *Puma*), the leopard, tiger, jaguar, and lion (genus *Panthera*), the house cat (genus *Felis*), and several others. Thus, in the present-day science of taxonomy, each organism that has been classified has its own unique binomial name. In turn, it is assigned to larger groupings that are thought to have a common evolutionary history. Table 22.1 uses the classification of humans to show how the various categories are used.

Phylogeny

Phylogeny is the science that explores the evolutionary relationships among organisms and seeks to reconstruct evolutionary history. Taxonomists and phylogenists work together so that the products of their work are compatible. A taxonomic ranking should reflect the phylogenetic (evolutionary) relationships among the organisms being classified. Although taxonomy and phylogeny are sciences, there is no complete agreement as to how organisms are classified or how they are related. New organisms and new information about organisms are discovered constantly. Therefore, taxonomic and phylogenetic relationships are constantly being revised. During this revision

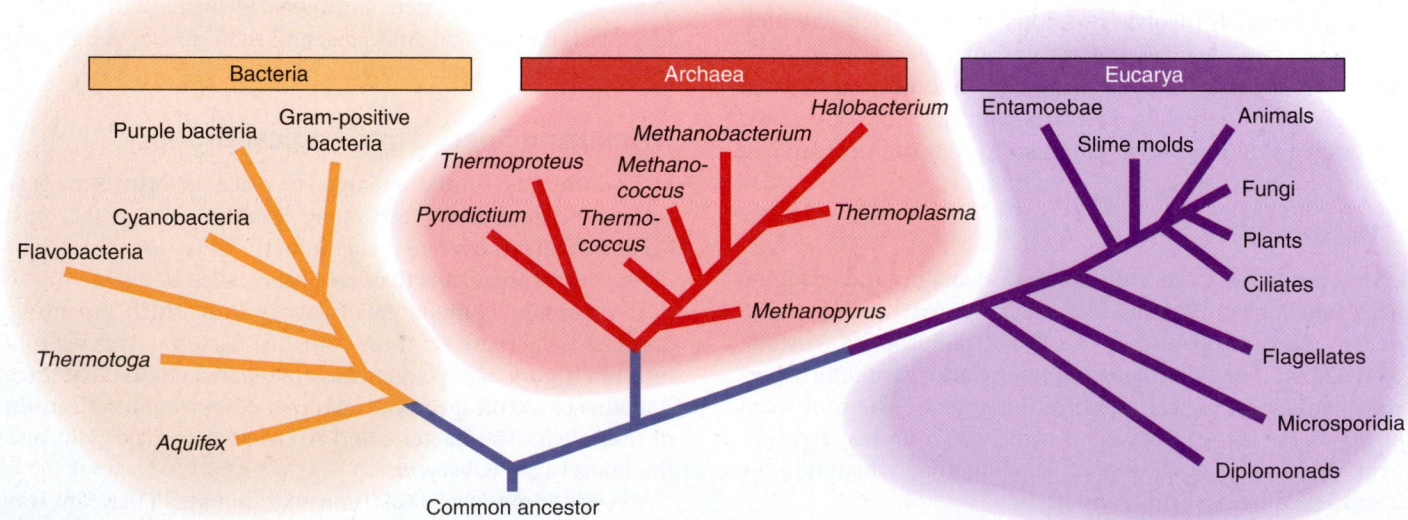

FIGURE 22.3 This diagram shows the three domains of living things and the way they are related to one another evolutionarily. The domain Bacteria is the oldest group. The domains Archaea and Eucarya are derived from the Bacteria.

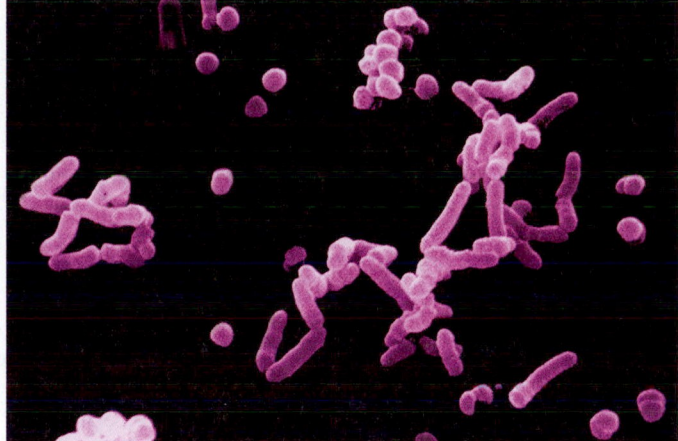

A *Streptococcus pyogenes*

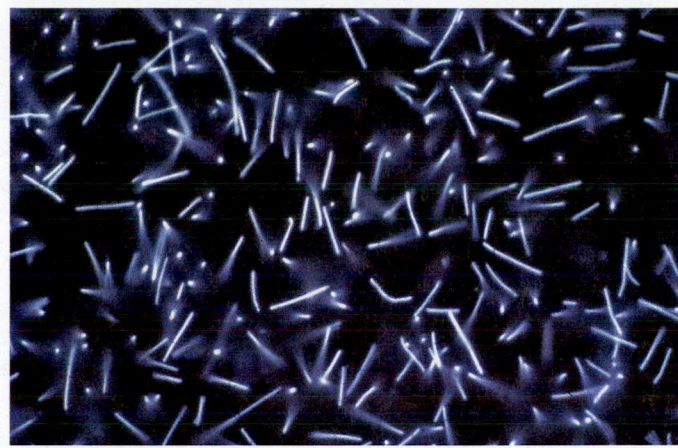

B *Methanopyrus*

D *Amoeba proteus*

C *Morchella esculenta*

E *Homo sapien*

F *Acer saccharum*

FIGURE 22.4 (*A*) The domain Bacteria is represented by the bacterium *Streptococcus pyogenes*. (*B*) The domain Archaea is represented by the methane-producing *Methanopyrus*. The domain Eucarya is represented by members of the following kingdoms: (*C*) Fungi, by the ascomycete *Morchella esculenta*; (*D*) Protista, by the one-celled *Amoeba proteus*; (*E*) Animalia, by the animal *Homo sapiens sapiens*; and (*F*) Plantae, by the tree *Acer saccharum*.

process, scientists often have differences of opinion about the significance of new information.

Evidence Used to Establish Phylogenetic Relationships

Phylogenists use several lines of evidence to develop evolutionary histories: fossils, comparative anatomy, life-cycle information, and biochemical/molecular evidence.

1. *Fossils* are evidence of previously existing life. Evidence obtained from the discovery and study of fossils allows

biologists to place organisms in a time sequence. It is also possible to compare subtle changes in particular kinds of fossils over time. For example, the size of the leaf of a specific fossil plant has been found to change extensively through long geological periods. A comparison of the extremes, the oldest with the newest, would lead to their classification into different categories. However, the fossil links between the extremes clearly show that the younger plant is a descendant of the older.

2. *Comparative anatomy studies* of fossils or currently living organisms can be very useful in developing a phylogeny.

TABLE 22.1 Classification of Humans

Taxonomic Category	Taxonomic Name	Characteristics	Other Representatives
Domain	Eucarya	Cells contain a nucleus and many other kinds of organelles.	Plants, animals, fungi, protozoa, algae
Kingdom	Animalia	Eukaryotic, heterotrophs that are usually motile and have specialized tissues	Sponges, jellyfish, worms, clams, insects, snakes, cats
Phylum	Chordata	Animals that have a stiffening rod down their back	Fish, amphibians, reptiles, birds, mammals
Class	Mammalia	Animals with hair and mammary glands	Platypus, kangaroos, mice, whales, skunks, monkeys
Order	Primates	Mammals with relatively large brains and opposable thumbs	Monkeys, gorillas, chimpanzees, baboons
Family	Hominidae	Primates that lack a tail and have upright posture	Humans and extinct relatives in several genera (*Australopithecus, Paranthropus,* and *Homo*)
Genus	*Homo*	Hominids with large brains	Humans and extinct relatives such as *Homo erectus* and *Homo neanderthalensis*.
Species	*Homo sapiens*	Humans	
Subspecies	*Sapiens*		

Since the structures of an organism are determined by its genes and developmental processes, those organisms having similar structures are thought to be related. For example, plants can be divided into several categories: All plants that have flowers are thought to be more closely related to one another than to plants like ferns, which do not have flowers. In the animal kingdom, all organisms that have hair and mammary glands are grouped together, and all animals in the bird category have feathers, wings, and beaks.

3. *Life-cycle information* is another line of evidence useful to phylogenists and taxonomists. Many organisms have complex life cycles that include many completely different stages. After fertilization, the fertilized eggs of some organisms grow into free-living developmental stages that do not resemble the adults of their species. These are called *larvae* (singular, *larva*). Larval stages often provide clues to the relatedness of organisms. For example, barnacles live attached to rocks and other solid marine objects and look like small, hard cones. Their outward appearance does not suggest that they are related to shrimp; however, the larval stages of barnacles and shrimp are very similar. Detailed anatomical studies of barnacles confirm that they share many structures with shrimp; their outward appearance tends to be misleading (figure 22.5).

Both birds and reptiles lay eggs with shells. However, reptiles lack feathers and have scales covering their bodies. The fact that these two groups share this fundamental eggshell characteristic implies that they are more closely related to each other than they are to other groups, but they can be divided into two groups based on their anatomical differences.

This same kind of evidence is available in the plant kingdom. Many kinds of plants, such as peas, peanuts, and lima beans, produce large, two-parted seeds in pods (you

can easily split the seeds into two parts). Even though peas grow as vines, lima beans grow as bushes, and peanuts have their seeds underground, all these plants are considered to be related.

4. *Biochemical and molecular studies* are recent additions to the toolbox of phylogenists. Like all aspects of biology, the science of phylogeny is constantly changing as new techniques develop. Recent advances in DNA analysis are being used to determine genetic similarities among species. In the field of ornithology, which is the study of birds, there are those who have evidence that storks and flamingos are closely related; others have evidence that flamingos are more closely related to geese. An analysis of the DNA points to a higher degree of affinity between flamingos and storks than between flamingos and geese. This is interpreted to mean that the closest relationship is between flamingos and storks.

Algae and plants have several different kinds of chlorophyll: chlorophyll *a*, *b*, *c*, *d*, and *e*. Most photosynthetic organisms contain a combination of two of these chlorophyll molecules. Members of the kingdom Plantae have chlorophyll *a* and *b*. The large seaweeds, such as kelp, superficially resemble terrestrial plants like trees and shrubs. However, a comparison of the chlorophylls present shows that kelp has chlorophyll *a* and *d*. When another group of algae, called the *green algae*, are examined, they are found to have chlorophyll *a* and *b*, as do plants. Along with other anatomical and developmental evidence, this biochemical information has helped to establish an evolutionary link between the green algae and plants. All of these kinds of evidence (fossils, comparative anatomy, developmental stages, and biochemical evidence) have been used to develop the various taxonomic categories, including kingdoms.

Given all these sources of evidence, biologists have developed an idea of how all organisms are related evolutionarily

A Barnacle

C Nauplius larva of a barnacle

B Shrimp

D Nauplius larva of a shrimp

FIGURE 22.5 The adult barnacle (*A*) and shrimp (*B*) are very different from each other, but their early larval stages (*C* and *D*) look very much alike.

(figure 22.6). At the base of this evolutionary scheme is the biochemical evolution of cells. These first cells are thought to be the origin of all organisms. While these first cells no longer exist, their descendants have diversified over millions of years. Of these groups, the members of the domains Bacteria and Archaea have the simplest structure and are probably most similar to some of the first cellular organisms on Earth. Members of the domain Eucarya evolved later and have greater structural and functional complexity.

22.2 A BRIEF SURVEY OF BIODIVERSITY

There is great variety in the kinds of living things on Earth. Some groups such as the Bacteria and Archaea have been in existence for more than 3 billion years. Other groups such as plants and animals have been in existence for less than a billion years. In this section, we will look at the three domains of living things and briefly describe some of their more distinctive characteristics.

Domains Bacteria and Archaea

Because both the Bacteria and Archaea are prokaryotic, they were previously lumped together as one taxonomic unit and are still commonly referred to as "bacteria." However, members of the two domains are very different from one another and now are assigned separate positions on the evolutionary tree. Evidence gained from studying DNA and RNA nucleotide sequences and a comparison of the amino acid sequences of proteins indicates that the Bacteria are older evolutionarily than the Archaea.

Bacteria

The Bacteria are small, prokaryotic, single-celled organisms. Their cell walls contain a unique, complex organic molecule known as peptidoglycan. Peptidoglycan is only found in the Bacteria and is composed of two kinds of sugars linked together by amino acids. One of these sugars, muramic acid, is found only in the Bacteria. This characteristic along with differences in the structure of the cell membrane and the nature of their DNA is used to distinguish the Bacteria from the Archaea and Eucarya.

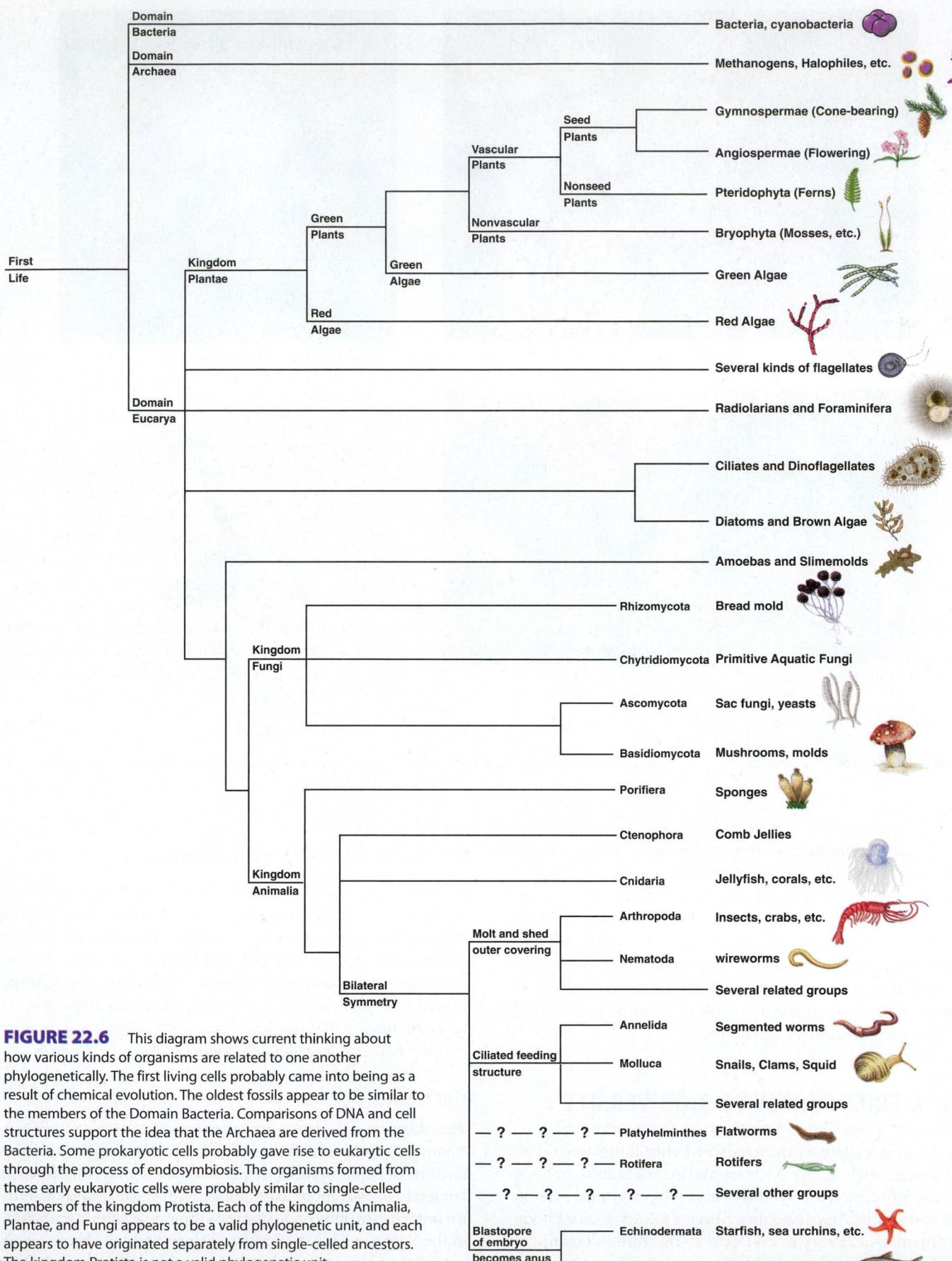

FIGURE 22.6 This diagram shows current thinking about how various kinds of organisms are related to one another phylogenetically. The first living cells probably came into being as a result of chemical evolution. The oldest fossils appear to be similar to the members of the Domain Bacteria. Comparisons of DNA and cell structures support the idea that the Archaea are derived from the Bacteria. Some prokaryotic cells probably gave rise to eukaryotic cells through the process of endosymbiosis. The organisms formed from these early eukaryotic cells were probably similar to single-celled members of the kingdom Protista. Each of the kingdoms Animalia, Plantae, and Fungi appears to be a valid phylogenetic unit, and each appears to have originated separately from single-celled ancestors. The kingdom Protista is not a valid phylogenetic unit.

A Closer Look

Cladistics—A Tool for Taxonomy and Phylogeny

Classification, or taxonomy, is one part of the much larger field of phylogenetic systematics. Classification involves placing organisms into logical categories and assigning names to those categories. Phylogeny, or *systematics,* is an effort to understand the evolutionary relationships of living things in order to interpret the way in which life has diversified and changed over billions of years of biological history. Phylogeny attempts to understand how organisms have changed over time. *Cladistics* (*klados* = branch) is a method biologists use to evaluate the degree of relatedness among organisms within a group, based on how similar they are genetically. The basic assumptions behind cladistics are that

1. Groups of organisms are related by descent from a common ancestor.
2. The relationships among groups can be represented by a branching pattern, with new evolutionary groups arising from a common ancestor.
3. Changes in characteristics occur in organisms over time.

Several steps are involved in applying cladistics to a particular group of organisms. First, you must select characteristics that vary and collect information on the characteristics displayed by the group of organisms you are studying. The second step is to determine which expression of a characteristic is ancestral and which is more recently derived. Usually, this involves comparing the group in which you are interested with an *outgroup* that is related to but not a part of the group you are studying. The characteristics of the outgroup are then considered to be ancestral. Finally, you must compare the characteristics displayed by the group you are studying and construct a diagram known as a *cladogram* (box figure 22.1). For

Characteristic Organism	Lungs Present	Skin Dry	Warm-Blooded	Hair Present
Shark	0	0	0	0
Frog	+	0	0	0
Lizard	+	+	0	0
Crow	+	+	+	0
Bat	+	+	+	+

Box Table 22.1

example, if you were interested in studying how various kinds of terrestrial vertebrates are related, you could look at the characteristics shown in box table 22.1.

In this example, the shark is the outgroup, and the ancestral conditions are lungs absent, skin not dry, cold-blooded, and hair absent. Using this information, you could construct the cladogram below.

All of the organisms, except sharks, have lungs. Lizards, crows, and bats have dry skin, as well as lungs and so on. Crows and bats share the following characteristics; they have lungs, they have dry skin, and they are warm-blooded. Because they share more characteristics with each other than with the other groups, they are considered to be more closely related.

The choice of characteristics with which to make comparisons is important. Two organisms may share many characteristics but not be members of the same evolutionary group if the characteristics being compared are not from the same genetic background. For example, if you were to compare butterflies, birds, and squirrels using the presence or absence of wings and the presence or absence of bright colors as your characteristics, you would conclude that butterflies and birds are more closely related than birds and squirrels. However, this is not a valid comparison, because the wings of birds and butterflies are not of the same evolutionary origin.

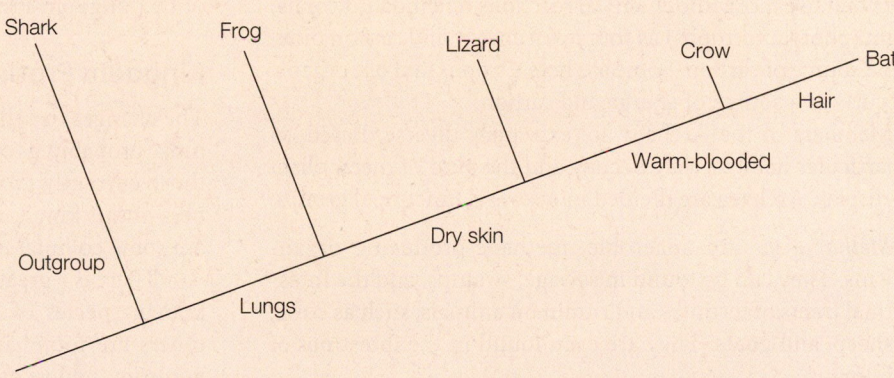

Box Figure 22.1 Cladogram

There are thousands of species of Bacteria that can be divided into groups based on their metabolic abilities. Most Bacteria are heterotrophs that require organic molecules as a source of food. Many are important decomposers of dead organic matter. Sewage treatment plants rely on Bacteria along with other organisms to break down organic wastes. A few kinds are **parasites** that live on the tissues of other organisms and cause disease. Because parasitic Bacteria cause us problems, they have

been intensively studied. Pneumonia, tuberculosis, syphilis, gonorrhea, strep throat, and staphylococcus infections are just a few Bacteria we know all too well. In addition, heterotrophic Bacteria can be differentiated based on their ability to use oxygen. Some are aerobic (use oxygen to break down organic food molecules), while others are anaerobic (do not use oxygen). Since Earth's early atmosphere is thought to have been a reducing atmosphere, the first Bacteria were probably anaerobic organisms.

Several kinds of autotrophic Bacteria exist. The cyanobacteria (blue-green bacteria) were probably the first organisms to carry on photosynthesis. They contain a blue-green pigment that allows them to capture sunlight and carry on a kind of photosynthesis. Cyanobacteria are extremely common in marine and freshwater, and contribute significantly to the production of oxygen in the atmosphere. Other Bacteria are able to use inorganic chemical reactions to provide the energy needed to build new organic molecules. One that has important environmental implications is *Thiobacillus ferrooxidans,* which derives energy from the oxidation of iron and sulfur, and is at least partly responsible for the development of acid mine drainage.

Archaea

The term *Archaea* comes from the Greek term *archaios,* meaning "ancient." This is a little misleading since the Bacteria preceded them and the Archaea are thought to have branched off from the Bacteria somewhere between two to three billion years ago. The Archaea differ from Bacteria in several fundamental ways. The Archaea do not have peptidoglycan in their cell walls but do have a unique chemical structure for their cell membranes that is not found in either the Bacteria or Eucarya. The DNA of Archaea appears to have a large proportion of genes that are different from those of either the Bacteria or Eucarya. However, the structure of the DNA is similar in many ways to that found in the Eucarya.

Because many members of the Archaea are found in extreme environments, they have become known as *extremophiles.* However, as more species are discovered and organisms that were once thought to be Bacteria are reclassified as Archaea, it is becoming clear there are many that do not live in extreme environments. Archaea use a variety of ways of obtaining energy. Many are autotrophs that use inorganic chemical reactions (chemoautotrophs) or light (photoautotrophs) as sources of energy and carbon dioxide as a source of carbon. Some are heterotrophs and use organic molecules as a source of energy and carbon.

Members of the Archaea are extremely diverse. Based on the particular habitats they occupy and the kind of metabolism they display, Archaea are divided into several functional groups:

1. *Methanogens* are anaerobic, methane-producing organisms. They can be found in sewage, swamps, and the intestinal tracts of termites and ruminant animals, such as cows, sheep, and goats. They are even found in the intestines of humans.
2. *Halobacteria* (*halo* = salt) live in very salty environments such as the Great Salt Lake (Utah), salt ponds, and brine solutions. Many have a reddish pigment and can be present in such high numbers that they color the water red. Some contain a special kind of chlorophyll and are, therefore, capable of generating their ATP by a kind of photosynthesis but they do not release oxygen.
3. The *thermophilic* Archaea live in environments that normally have very high temperatures and high concentrations of sulfur (e.g., hot sulfur springs or around deep-sea hydrothermal vents). Over five hundred species of these thermophiles have been identified at the openings of hydrothermal vents in the ocean. Some of these organisms are heterotrophs, while others are chemoautotrophs and use reactions with sulfur to provide energy to synthesize organic molecules. One such thermophile, *Pyrolobus fumarii,* grows in a hot spring in Yellowstone National Park. Its maximum growth temperature is 113°C (235°F), its optimum is 106°C (223°F), and its minimum is 90°C (194°F).
4. *Marine, freshwater, and soil Archaea* have recently been discovered to be extremely abundant, but little is yet known about their role in these habitats.
5. Recently an archeon has been discovered that appears to be parasitic on another archeon.

Domain Eucarya

Most biologists accept the evidence that eukaryotic cells evolved from prokaryotic cells by a process of endosymbiosis. (See "The Endosymbiotic Theory and the Origin of Eukaryotic Cells" section in chapter 21.) This hypothesis proposes that structures such as mitochondria, chloroplasts, and other membranous organelles originated from separate cells that were ingested by larger, more primitive cells. Once inside, these structures and their functions became integrated with the host cell and ultimately became essential to its survival. This new type of cell was the forerunner of present-day eukaryotic cells, which are usually much larger than the prokaryotes, typically having more than a thousand times the volume of prokaryotic cells. Their larger size was made possible by the presence of specialized membranous organelles, such as mitochondria, the endoplasmic reticulum, chloroplasts, and nuclei. Members of the kingdoms Protista, Fungi, Plantae, and Animalia are eukaryotic. Single-celled eukaryotic organisms are members of the kingdom Protista.

Kingdom Protista

The changes in cell structure that led to eukaryotic organisms most probably gave rise to single-celled organisms similar to those currently grouped in the kingdom Protista. Most members of this kingdom are one-celled organisms, although there are some colonial forms.

There is a great deal of diversity among the sixty-thousand known species of Protista. Many species live in freshwater; others are found in marine or terrestrial habitats. Some are *parasitic* and live on the tissues of other living things. Some are *commensal* organisms that live in another organism without causing harm, and some are *mutualistic* organisms that live in partnership with another organism in a relationship in which both organisms benefit. All species can undergo mitosis, resulting in asexual reproduction. Most species can also reproduce sexually. Many contain chlorophyll in chloroplasts and are autotrophic; others require organic molecules as a source of energy and are heterotrophic. Both autotrophs and heterotrophs have mitochondria and respire aerobically. Some specialized parasitic types lack mitochondria and other cellular structures.

Because members of this kingdom are so diverse with respect to form, metabolism, and reproductive methods, most

Green Algae

Kelp

Slime mold

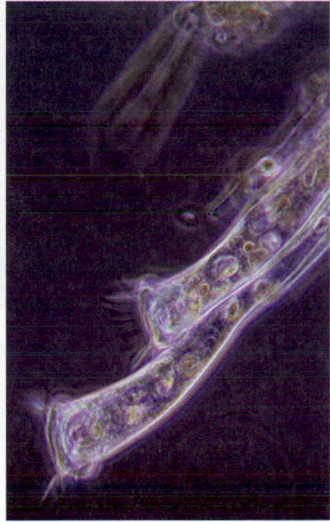
Protozoan

FIGURE 22.7 The kingdom Protista includes a wide variety of organisms that are simple in structure. They are not a phylogenetic group. These are some examples of this diverse group.

CONCEPTS APPLIED

Diversity of Life

Find a place where you can measure a square meter in your yard or a park. Get down on your belly and differentiate the kinds of plants, animals, and fungi you see. It does not matter if you know the specific names for each. Make three separate lists (Plants, Animals, Fungi) of the different organisms you identify and the characteristics you use to identify them. How many different organisms did you find? Which was the most numerous?

heterotrophs probably gave rise to the kingdom Animalia, and the funguslike heterotrophs were probably the forerunners of the kingdom Fungi.

Kingdom Fungi

Fungus is the common name for members of the kingdom Fungi. The majority of fungi are not able to move about. They have a rigid, thin cell wall, which contains chitin, a complex carbohydrate containing nitrogen. This is an important diagnostic feature, since plants have cell walls of cellulose. Members of the kingdom Fungi are nonphotosynthetic, eukaryotic organisms. The majority (mushrooms and molds) are multicellular, but a few, such as yeasts, are single-celled. In the multicellular fungi, the basic structural unit is a network of multicellular filaments.

Because all of these organisms are heterotrophs, they must obtain nutrients from organic sources. Most secrete enzymes outside their cells that digest large molecules into smaller units that can be absorbed. (This method of obtaining nutrients is called external digestion.) They are very important as decomposers in all ecosystems. They feed on a variety of nutrients ranging from dead organisms to such products as shoes, foodstuffs, and clothing. Most synthetic organic molecules are not attacked as readily by fungi; this is one reason plastic bags, foam cups, and organic pesticides are slow to decompose.

Some fungi are parasitic and others are mutualistic. Many of the parasitic fungi are important plant pests. Some attack and kill plants (chestnut blight, Dutch elm disease); others injure the fruit, leaves, roots, or stems and reduce yields. The fungi that are human parasites are responsible for athlete's foot, vaginal yeast infections, valley fever, "ringworm," and other diseases. Many kinds of fungi form mutualistic relationships with other kinds of organisms. Mutualistic fungi are important in lichens and in combination with the roots of certain kinds of plants, where they assist the plant in obtaining nutrients from the soil. Figure 22.8 shows some examples of this group of organisms.

Kingdom Plantae

Another major group thought to be derived from the kingdom Protista is the green, photosynthetic plant kingdom. The ancestors of plants were most likely specific kinds of algae commonly called *green algae*. Members of the kingdom Plantae are non-

biologists do not feel that the Protista form a valid phylogenetic unit. However, it is still a convenient taxonomic grouping. By placing these organisms together in this group, it is possible to gain a useful perspective on how they relate to other kinds of organisms. After the origin of eukaryotic organisms, evolution proceeded along several different pathways. Three major lines of evolution can be seen today in the plantlike autotrophs (algae), animal-like heterotrophs (protozoa), and the funguslike heterotrophs (slime molds). *Amoeba* and *Paramecium* are commonly encountered examples of protozoa. Many seaweeds and pond scums are collections of large numbers of algal cells. Slime molds are less frequently seen because they live in and on the soil in moist habitats; they are most often encountered as slimy masses on decaying logs. Figure 22.7 shows some examples of this diverse group of organisms.

Through the process of evolution, the plantlike autotrophs probably gave rise to the kingdom Plantae, the animal-like

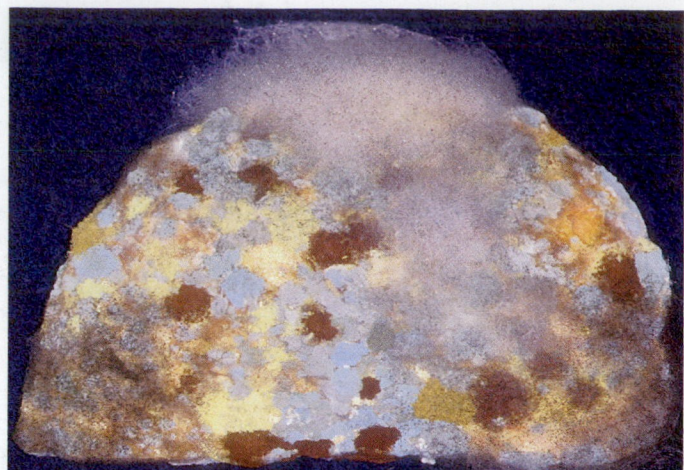
Several kinds of Mold

Puffball

Mushroom

FIGURE 22.8 Molds, mushrooms, and puffballs are commonly seen examples of the kingdom Fungi.

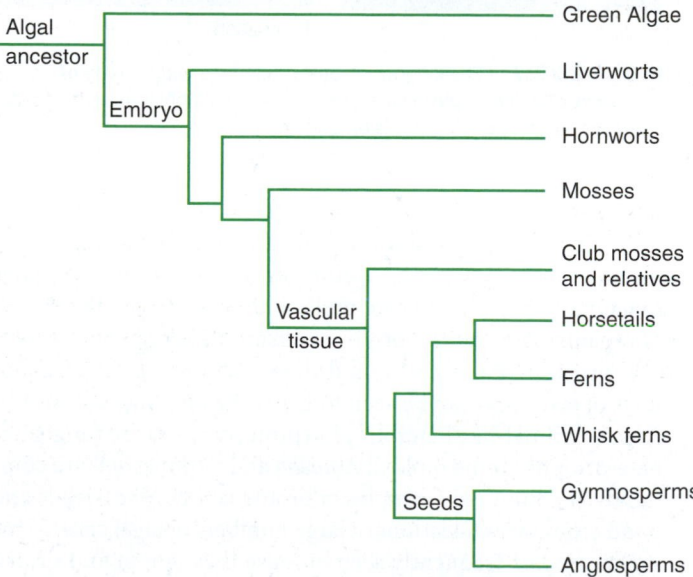

FIGURE 22.9 The evolution of plants involves specializations for living on land and changes in the way they reproduce. Plants differ from their algal ancestors in that they have a multicellular immature stage known as an embryo. Vascular tissue and seeds were important evolutionary steps that allowed plants to be successful away from moist habitats.

motile, terrestrial, multicellular organisms that contain chlorophyll and produce their own organic compounds through photosynthesis. All plant cells have a cellulose cell wall. Over three hundred thousand species of plants have been classified; about 85 percent are flowering plants, 14 percent are mosses and ferns, and the remaining 1 percent are cone-bearers and several other small groups within the kingdom.

A wide variety of plants exist on Earth today. Members of the kingdom Plantae range from simple mosses to vascular plants with stems, roots, leaves, and flowers. Most biologists feel that the evolution of this kingdom began nearly five hundred million years ago when the green algae of the kingdom Protista gave rise to nonvascular plants such as the mosses. Early in the evolution of plants, the development of vascular tissue led to a second line of evolution that includes the ferns, cone-bearing plants, and flowering plants (figure 22.9). Some of the vascular plants evolved into seed-producing plants, which today are the cone-bearing and flowering plants, while the ferns lack seeds. The development of vascular plants was a major step in the evolution of plants from an aquatic to a terrestrial environment.

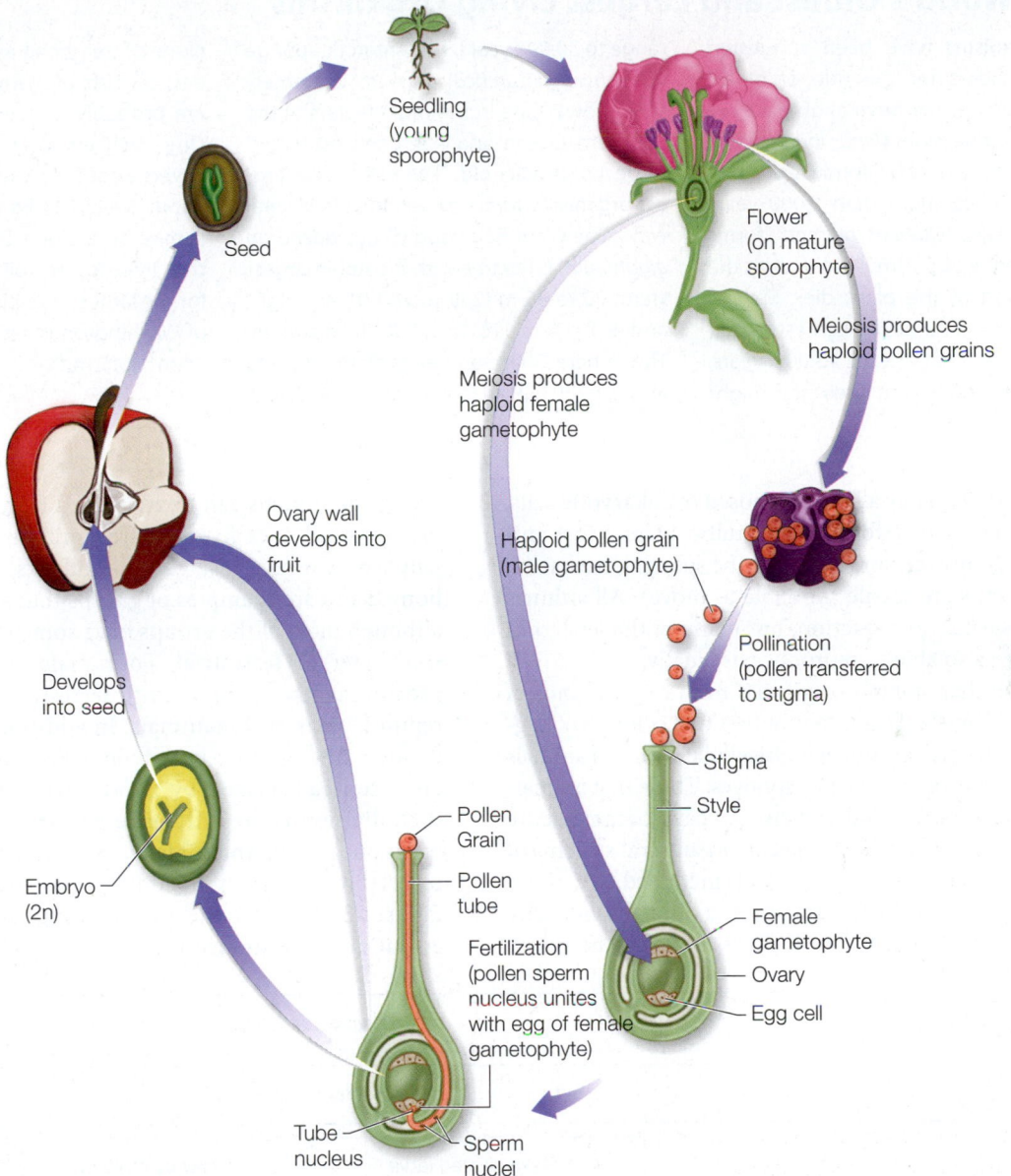

FIGURE 22.10 In flowering plants, the sporophyte generation consists of roots, stems, leaves, and flowers. The flower produces two kinds of gametophyte plants. The pollen grain is the male gametophyte, and the female gametophyte is produced within the ovary of the flower. The male gametophyte produces the equivalent of sperm and the female gametophyte produces an egg. Pollination involves the transfer of pollen to the stigma of a flower. Fertilization occurs following pollination, when the pollen tube releases a sperm nucleus to fertilize the egg of the female gametophyte inside the ovary of the flower. The fertilized egg develops into the embryo plant, and other cells of the female gametophyte produce stored food. A tough coat develops around the embryo and food. This package is known as a seed. The seed germinates to produce a new sporophyte plant.

Plants have a unique life cycle that has two distinctly different stages. This is known as alternation of generations. The presence of this kind of life cycle is a unifying theme that ties together all members of this kingdom. There is a gametophyte stage that produces sex cells by mitosis. The sex cells unite and give rise to a sporophyte stage that produces spores by meiosis. The spores give rise to a new gametophyte stage. Mosses and ferns have life cycles with relatively clearly defined gametophyte and sporophyte generations. Figure 22.10 illustrates the life cycle of a flowering plant. In addition to sexual reproduction, plants are able to reproduce asexually.

Kingdom Animalia

Like fungi and plants, animals are thought to have evolved from the Protista. More than a million species of animals have been classified. These range from microscopic types, such as mites or aquatic larvae of marine animals, to huge animals such as elephants or whales. Regardless of their type, all animals have

A Closer Look

The World's Oldest and Largest Living Organisms

Several organisms have been suggested as record holders for the title of oldest and largest organisms. Several of these are plants. Bristlecone pines (*Pinus longaeva*) in the White Mountains of California have been determined to be more than 5,000 years old. Jurupa Oak (*Quercus palmeri*) forms clones that grow out from the center as the central portion of the plant dies. Several clones of Jurupa Oak have been estimated by some to have an age of 13,000 years, but others speculate that their age might range to 30,000 years. The Antarctic sponge (*Cinachyra antarctica*) has an extremely slow growth rate in the frigid waters of the Southern Ocean and has been estimated to be 1,550 years old. The title of the largest organism can be determined in several ways. The Giant Redwood (*Sequoiadendron giganteum*) is the tree with the single largest stem, 70 to 85 m (230 to 280 ft) in height and 5 to 7 m (16 to 23 ft) in diameter. The General Sherman tree probably weighs about 1,270,058 kg (1,400 tons). However, a clone of trembling aspen (*Populus tremuloides*) consists of many individual stems that are probably all joined together by roots. One such clone in the Rocky Mountains covers about 0.4 km^2 and probably weighs about 5,443,108 kg (6,000 tons). However, it may be a clone of a fungus (*Armillaria*) that lives in the soil that holds the record for the largest organism. A clone in the state of Washington is estimated to cover about 3 km^2 (1.15 mi^2).

some common traits. They all are composed of eukaryotic cells, and all species are heterotrophic and multicellular. Most animals are motile; however, some, such as the sponges, barnacles, mussels, and corals are sessile (not able to move). All animals are capable of sexual reproduction, but many of the less complex animals are also able to reproduce asexually.

It is thought that animals originated from certain kinds of Protista that had flagella. This idea proposes that colonies of flagellated Protista gave rise to simple multicellular forms of animals such as the ancestors of present-day sponges. These first animals lacked specialized tissues and organs. As cells became more specialized, organisms developed special organs and systems of organs and the variety of kinds of animals increased.

Animals originated in the ancient sea, and the majority of kinds of animals are still found there. Most of the major groups of animals can be identified from fossils of the Cambrian period over five hundred million years ago. Sponges, jellyfish, worms, crustaceans, mollusks, starfish, sharks, and bony fishes are examples of groups that are primarily marine, although most of the groups have some species that are either freshwater or terrestrial. Four major groups have become predominantly air-breathing, terrestrial organisms: insects, reptiles, birds, and mammals. In addition to being artificially divided into groups based on where they live, organisms are often categorized by major structural differences. One such division is the difference between vertebrates that have backbones (fish, amphibians, reptiles, birds, mammals) and invertebrates, which include all the other animal groups. Figure 22.11 shows the major groups of animals and their evolutionary relationships.

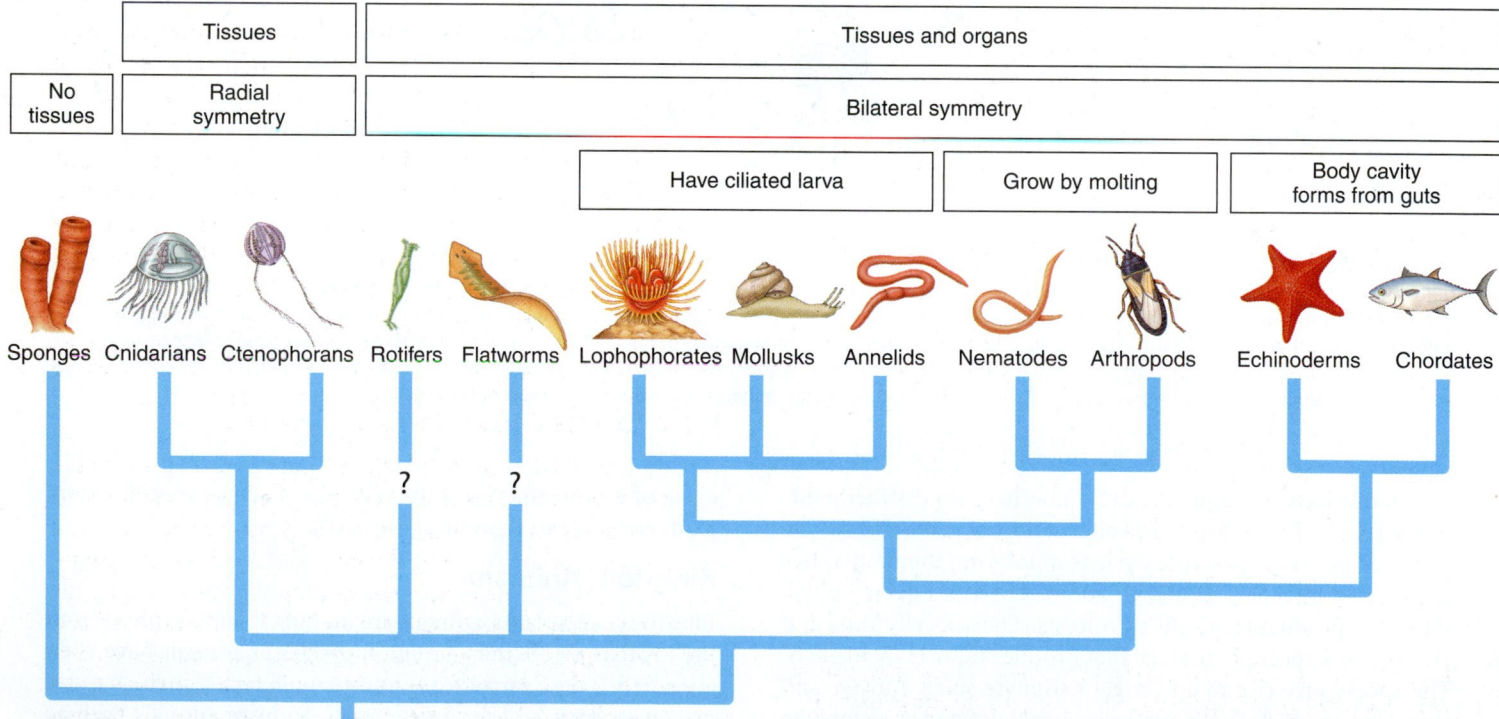

FIGURE 22.11 The classification of animals reflects complexity of body form and developmental features.

Lynn (Alexander) Margulis (1938–2011)

Lynn (Alexander) Margulis taught at Brandeis University and Boston College, and was a professor in the geology department at the University of Massachusetts Amherst.

Margulis had a major impact in two areas related to taxonomy and phylogeny. She championed the idea that eukaryotic cells came about as a result of the combining of prokaryotic cells (the endosymbiotic theory) and argued for changes in taxonomy that better reflect the true nature of living things. Her study of the organisms in the kingdom Protista (algae, protozoa, related

organisms) led her to recognize that they were distinct from plants and animals. Thus, she argued that the traditional way of dividing living things into two kingdoms (plants and animals) was inadequate. At least in part due to her efforts, today the former plant and animal kingdoms have been divided into Protists, Fungi, Plants, and Animals.

Her support for the Gaia hypothesis, which suggests that Earth and all the living things on it are in a symbiotic relationship, led her to be a champion for various environmental concerns.

Although all animals are heterotrophs, they use various methods for obtaining nutrients. Most collect and consume food. However, some are parasites and live within or upon other organisms and use them as food. Many kinds of insects, worms, and ticks, and even some mollusks and fish are involved in parasitic relationships. Other animals are mutualistic and cooperate with other organisms to obtain food. For example, coral organisms have a mutualistic relationship with algae, which allows the coral to obtain some of its nutrients from the photosynthesis of the algae. (See chapter 23 for a more complete discussion of how animals interact.)

22.3 ACELLULAR INFECTIOUS PARTICLES

All of the groups discussed so far fall under the category of cellular forms of life. They all have at least the following features in common: They have (a) cell membranes, (b) nucleic acids as their genetic material, (c) cytoplasm, (d) enzymes and coenzymes, (e) ribosomes, and (f) use ATP as their source of chemical-bond energy. However, there are some particles that show some characteristics of life and cause disease but do not have a cell structure. Because they lack a cell structure, they are referred to as *acellular* (the prefix *a* = lacking). Because they enter cells and cause disease and can be passed from one organism to another, they are often called infectious particles. In the

process of causing disease, they make copies of themselves. There is no clear explanation for how these particles came to be. Therefore, they are not included in the classification system used for cellular organisms. There are three kinds of these acellular infectious particles: *viruses, viroids,* and *prions*.

Viruses

A **virus** is an acellular infectious particle that consists of a nucleic acid core surrounded by a coat of protein (figure 22.12). Viruses are often called **obligate intracellular parasites,** which means they are infectious particles that can function only when inside a living cell. Viruses are not considered to be living because they are not capable of living and reproducing by themselves and show some characteristics of life only when inside living cells.

How Did Viruses Originate?

Soon after viruses were discovered in the late part of the nineteenth century, biologists began to speculate on how they originated. One early hypothesis was that they were descendents of the first precells that did not evolve into cells. This idea was discarded as biologists learned more about the complex relationship between viruses and host cells. A second hypothesis was that viruses developed from intracellular parasites that became so specialized that they needed only the nucleic acid to continue their existence. A third hypothesis is that viruses are runaway genes that have escaped from cells and must return to a host cell to replicate. Regardless of how the viruses came into being, they are important today as parasites in all forms of life.

How Viruses Cause Disease

Viruses are typically host-specific, which means that they usually attack only one kind of host cell that provides what the virus needs to function. To enter cells, viruses must attach to a cell surface. Viruses can infect only those cells that have the proper receptor sites to which the virus can attach. For example, the virus responsible for measles attaches to the membranes of skin cells, hepatitis viruses attach to liver cells, and mumps viruses attach to cells in the salivary glands. Host cells for the human immunodeficiency virus (HIV) include some types of human brain cells and several types belonging to the immune system.

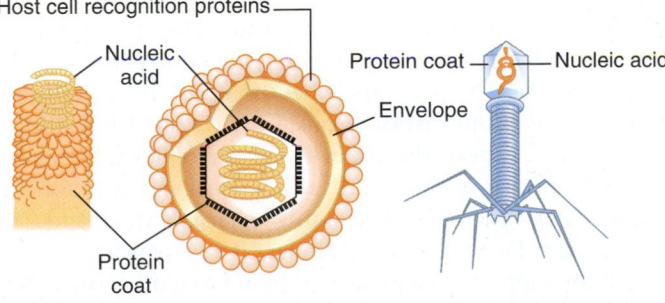

FIGURE 22.12   Viruses consist of a core of nucleic acid, either DNA or RNA, depending on the kind of virus. The nucleic acid is surrounded by a protein coat. In addition some have an envelope surrounding the protein coat.

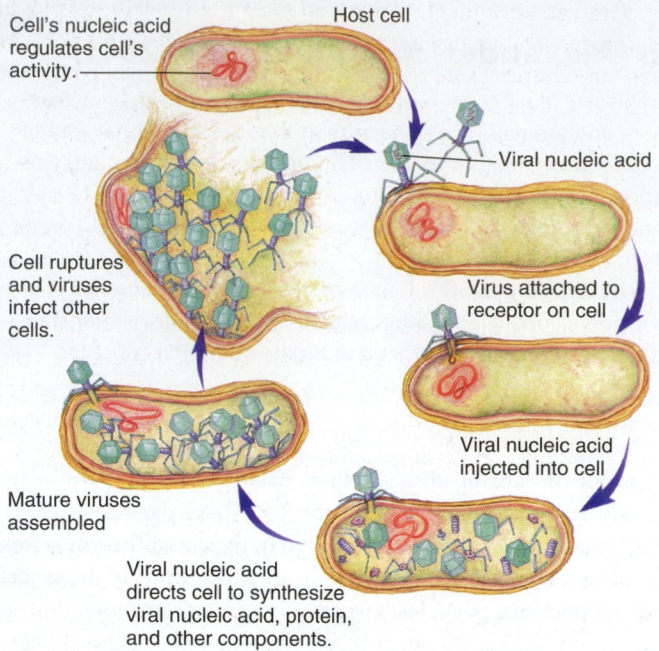

Cell's nucleic acid regulates cell's activity.

Host cell

Viral nucleic acid

Virus attached to receptor on cell

Viral nucleic acid injected into cell

Cell ruptures and viruses infect other cells.

Mature viruses assembled

Viral nucleic acid directs cell to synthesize viral nucleic acid, protein, and other components.

FIGURE 22.13 The viral nucleic acid takes control of the activities of the host cell. Because the virus has no functional organelles of its own, it can become metabolically active only while it is within a host cell.

Once it has attached to the host cell, the virus either enters the cell or injects its nucleic acid into the cell. If it enters the cell, the virus loses its protein coat, releasing the nucleic acid. Once released into the cell, the nucleic acid of the virus may remain free in the cell or it may link with the host's genetic material. Some viruses contain as few as three genes, others contain as many as five hundred. A typical eukaryotic cell contains tens of thousands of genes.

Viral genes are able to take command of the host's metabolic pathways and direct it to carry out the work of making new copies of the original virus. The virus makes use of the host's available enzymes and ATP for this purpose. When enough new viral nucleic acid and protein coat are produced, complete virus particles are assembled and released from the host (figure 22.13). In many cases, this process results in the death of the host cell. When the virus particles are released, they can infect adjacent cells and the infection spreads. The

TABLE 22.2 Viral Diseases

Type of Virus	Disease
Papovaviruses	Warts in humans
Paramyxoviruses	Mumps and measles in humans; distemper in dogs
Adenoviruses	Respiratory infections in most mammals
Poxviruses	Smallpox
Wound-tumor viruses	Diseases in corn and rice
Potexviruses	Potato diseases
Bacteriophage	Infections in many types of bacteria

number of viruses released ranges from ten to thousands. The virus that causes polio affects nerve cells and releases about ten thousand new virus particles from each infected human host cell. Some viruses remain in cells and are only occasionally triggered to replicate, causing symptoms of disease. Herpes viruses, which cause cold sores, genital herpes, and shingles, are such viruses that reside in nerve cells and occasionally become active.

We know much about the viruses that cause human disease and disease of other organisms we value, but we know very little about viruses that may infect other kinds of organisms. It is likely that most species serve as hosts to some form of virus (table 22.2).

Viroids: Infectious RNA

The term **viroid** refers to infectious particles that are composed solely of a small, single strand of RNA in the form of a loop. To date, no viroids have been found that parasitize animals. The hosts in which they have been found are cultivated crop plants such as potatoes, tomatoes, and cucumbers. Viroid infections result in stunted or distorted growth and may sometimes cause the plant to die. Pollen, seeds, or farm machinery can transmit viroids from one plant to another. Some scientists believe that viroids may be parts of normal RNA that have gone wrong.

Prions: Infectious Proteins

Prions are proteins that can be passed from one organism to another and cause disease. All the diseases of this type currently known cause changes in the brain that result in a spongy appearance called spongiform encephalopathies. Because the disease can be transmitted from one animal to another, the disease is often called a transmissible spongiform encephalopathy. The symptoms typically involve abnormal behavior and eventually death. In animals, the most common examples are scrapie in sheep and goats, mad cow disease in cattle, and chronic wasting diseas in deer and elk. Scrapie got its name because one of the symptoms of the disease is an itching of the skin associated with nerve damage that causes the animals to rub against objects and scrape off their hair.

The occurrence of mad cow disease (bovine spongiform encephalopathy—BSE) in Great Britain was apparently caused by the spread of prions from sheep to cattle. This occurred because of the practice of processing unusable parts of sheep carcasses into a protein supplement that was fed to cattle. A human form of this disease is called Creutzfeldt-Jakob disease (CJD). It now appears that the original form of BSE has changed to a variety that is able to infect humans. This new form is called vCJD, which makes scientists believe that BSE and CJD (Creutzfeldt-Jakob disease) are in fact the same prion.

Prions appear to cause disease by altering the form of normal proteins to that of the abnormal prion shape. However, in cases of human prion disease, the genes a person has appears to play a role in determining if an individual is susceptible to the disease.

Part II deals with how to interpret the geologic history of Earth and how evidence from geology is used to gain an understanding of how organisms have changed through time.

22.4 GEOLOGIC TIME

Reading history from the rocks of Earth's crust requires both a feel for the immensity of geologic time and an understanding of geologic processes. By "geologic time," we mean the age of Earth; the very long span of Earth's history. This span of time is difficult for most of us to comprehend. Human history is measured in units of hundreds and thousands of years, and even the events that can take place in a thousand years are hard to imagine. Geologic time, on the other hand, is measured in units of millions and billions of years.

The understanding of geologic processes has been made possible through the development of various means of measuring ages and time spans in geologic systems. An understanding of geologic time leads to an understanding of geologic processes, which then leads to an understanding of the environmental conditions that must have existed in the past. Thus, the mineral composition, texture, and sedimentary structure of rocks are clues about past events, events that make up the geological and biological history of Earth.

Early Attempts at Earth Dating

Most human activities are organized by time intervals of minutes, hours, days, months, and years. These time intervals are based on the movements of Earth and the Moon. Short time intervals are measured in minutes and hours, which are typically tracked by watches and clocks. Longer time intervals are measured in days, months, and years, which are typically tracked by calendars. How do you measure and track time intervals for something as old as Earth? First, you would need to know the age of Earth; then you would need some consistent, measurable events to divide the overall age into intervals. Questions about the age of Earth have puzzled people for thousands of years, dating back at least to the time of the ancient Greek philosophers. Many people have attempted to answer this question and understand geologic time but with little success until the last few decades.

Early Scientific Approaches

Considered the father of modern geology, James Hutton (1726–1797) in Scotland reasoned out the concept that the geologic processes we see today have operated since Earth began. This is known as the *principle of uniformity,* and since many geologic processes such as erosion and sedimentation are slow, people began to assume a much older Earth. Because the principle of uniformity assumed a slow pace for geologic changes, this view of geologic processes became known as gradualism to contrast it with catastrophism, which had been the prevailing view for centuries. (Many of Hutton's ideas were restated and popularized by Charles Lyell, whose writings greatly influenced Charles Darwin.) The problem then became one of finding some uniform change or process that could serve as a geologic clock to measure the age of Earth. To serve as a geologic clock, a process or change would need to meet three criteria: (1) the process must have been operating since Earth began, (2) the process must be uniform or at least subject to averaging, and (3) the process must be measurable.

During the nineteenth century (1801–1900), many attempts were made to find earth processes that would meet the criteria to serve as a geologic clock. Among others, the processes explored were (1) the rate that salt is being added to the ocean, (2) the rate that sediments are being deposited, and (3) the rate that Earth is cooling. Comparing the load of salts being delivered to the ocean by all the rivers, and assuming the ocean was initially pure water, it was calculated that about 100 million years would be required for the present salinity to be reached. The calculations did not consider the amount of materials being removed from the ocean by organisms and by chemical sedimentation, however, so this technique was considered to be unacceptable. Even if the amount of materials removed were known, it would actually result in the age of the ocean, not the age of Earth.

A number of separate and independent attempts were made to measure the rate of sediment deposition, then compare that rate to the thickness of sedimentary rocks found on Earth. Dividing the total thickness by the rate of deposition resulted in estimates of an Earth age that ranged from about 20 million to 1,500 million years. The wide differences occurred because there are gaps in many sedimentary rock sequences, periods when sedimentary rocks were being eroded away to be deposited again elsewhere as sediments. There were just too many unknowns for this technique to be considered as acceptable.

The idea of measuring the rate that Earth is cooling for use as a geologic clock assumed that Earth was initially a molten mass that has been cooling ever since. Calculations estimating the temperature that Earth must have been to be molten were compared to Earth's present rate of cooling. This resulted in an estimated age of 20 to 40 million years. These calculations were made back in the nineteenth century before it was understood that natural radioactivity is adding heat to Earth's interior, so it has required much longer to cool down to its present temperature.

Modern Techniques for Determining the Age of Earth

Soon after the beginning of the twentieth century (1901–2000), the discovery of the radioactive decay process in the elements of minerals and rocks led to the development of a new, accurate geologic clock. This clock finds the *radiometric age* of rocks in years by measuring how much of an unstable radioactive isotope within the crystals of certain minerals has decayed. Since radioactive decay for each kind of radioactive isotope occurs

TABLE 22.3 Radioactive Isotopes and Half-Lives

Radioactive Isotope	Stable Daughter Product	Half-Life
Samarium-147	Neodymium-143	106 billion years
Rubidium-87	Strontium-87	48.8 billion years
Thorium-232	Lead-208	14.0 billion years
Uranium-238	Lead-206	4.5 billion years
Potassium-40	Argon-40	1.25 billion years
Uranium-235	Lead-207	704 million years
Carbon-14	Nitrogen-14	5,730 years

at a constant, known rate, the ratio of the remaining amount of a radioactive element to the amount of decay products present can be used to calculate the time that the radioactive element has been a part of that crystal (see chapter 11). Certain radioactive isotopes of potassium, uranium, and thorium are often included in the minerals of rocks, so they are often used as "radioactive clocks." By using radiometric aging techniques along with other information, we arrive at a generally accepted age for Earth of about 4.5 billion years. It should be noted that radiometric aging is only useful in aging igneous rocks, since sedimentary rocks are the result of weathering and deposition of other rocky materials.

Table 22.3 lists several radioactive isotopes, their decay products, and half-lives. Often two or more isotopes are used to determine an age for a rock. Agreement between them increases the scientist's confidence in the estimates of the age of the rock. Because there are great differences in the half-lives, some are useful for dating things back to several billion years, while others, such as carbon-14, are only useful for dating things to perhaps 50,000 years. Carbon-14 is not used to age rock but is very useful in aging materials that are of relatively recent biological origin, since carbon is an important part of all living things. Also, a slightly different method is used to determine carbon-14 dating (see "A Closer Look: What Is Carbon-14 Dating?" on p. 547).

A recently developed geologic clock is based on the magnetic orientation of magnetic minerals. These minerals become aligned with Earth's magnetic field when the igneous rock crystallizes, making a record of the magnetic field at that time. Earth's magnetic field is global and has undergone a number of reversals in the past. A *geomagnetic time scale* has been established from the number and duration of magnetic field reversals occurring during the past six million years. Combined with radiometric age dating, the geomagnetic time scale is making possible a worldwide geologic clock that can be used to determine local chronologies.

Interpreting the Geologic Record

Features of the geology of an area provide many clues that help us interpret what has happened in the past. However, it is important to keep in mind several basic principles of geology. The following is a summary of these basic guiding principles that are used to read a story of geologic events from the rocks.

The *principle of uniformity* is the cornerstone of the logic used to guide thinking about geologic time. As described in chapter 16, this principle is sometimes stated as "the present is the key to the past." This means that the geologic features that you see today were formed in the past by the same processes of crustal movement, erosion, and deposition that are observed today. By studying the processes now shaping Earth, you can understand how it has evolved through time. This principle establishes the understanding that the surface of Earth has been continuously and gradually modified over the immense span of geologic time.

The *principle of original horizontality* is a basic principle that is applied to sedimentary rocks. (See chapter 15 for a discussion of the nature of igneous, sedimentary, and metamorphic rocks.) It is based on the observation that, on a large scale, sediments are commonly deposited in flat-lying layers. Old rocks are continually being changed to new ones in the continuous processes of crustal movement, erosion, and deposition. As sediments are deposited in a basin of deposition, such as a lake or ocean, they accumulate in essentially flat-lying, approximately horizontal layers (figure 22.14). Thus, any layer of sedimentary rocks that is not horizontal has been subjected to forces that have deformed Earth's surface.

The *principle of superposition* is another logical and obvious principle that is applied to sedimentary rocks. Layers of sediments are usually deposited in succession in horizontal layers, which later are compacted and cemented into layers of sedimentary rock. An undisturbed sequence of horizontal layers is thus arranged in chronological order with the oldest layers at the bottom. Each consecutive layer will be younger than the one below it (figure 22.15). This is true, of course, only if the layers have not been turned over by deforming forces.

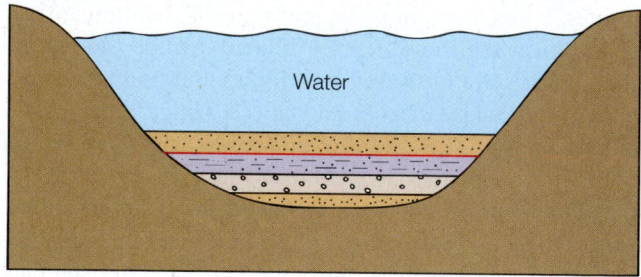

A

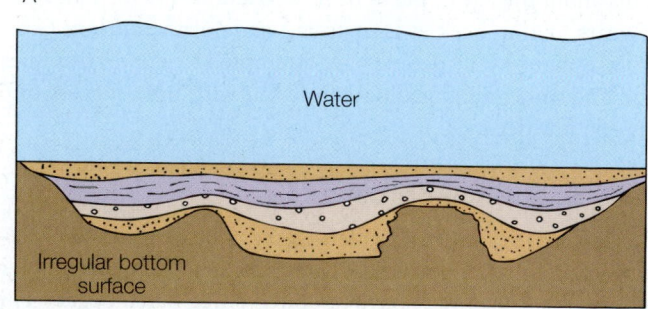

B

FIGURE 22.14 (*A*) The principle of original horizontality states that sediments tend to be deposited in horizontal layers. (*B*) Even where the sediments are draped over an irregular surface, they tend toward the horizontal.

FIGURE 22.15 The Grand Canyon, Arizona, provides a majestic cross section of horizontal sedimentary rocks. According to the principle of superposition, traveling deeper and deeper into the Grand Canyon means that you are moving into older and older rocks.

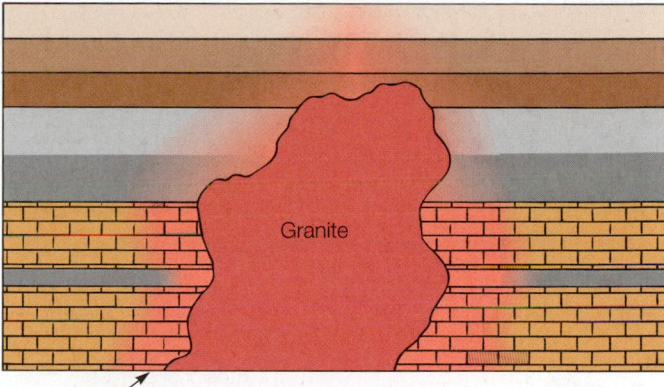

Rocks adjacent to intruding magma may also be metamorphosed by its heat.

A

B

FIGURE 22.16 (*A*) According to the principle of crosscutting relationships, since the granite intrusion is cutting across the sedimentary rock, the granite intrusion is younger than the surrounding sedimentary rock. (*B*) The photo shows two examples of crosscutting relationships. The darker rock was the original rock formed first. Later, the white layers were intruded into the darker rock. Finally, a fault line cuts acoss the white intrusions. So it must have occurred later than the intrusions.

The *principle of crosscutting relationships* is concerned with igneous and metamorphic rock, in addition to sedimentary rock layers. Any geologic feature that cuts across or is intruded (forced) into a rock mass must be younger than the rock mass. Thus, if a fault cuts across a layer of sedimentary rock, the fault is the youngest feature. Faults, folds, and igneous intrusions are always younger than the rocks in which they occur. Often the presence of metamorphic rock provides a further clue to the correct sequence. When hot igneous rock is intruded into preexisting rock, the rock surrounding the hot igneous rock can be "baked," or metamorphosed, into a different form. Thus, the rock from which the metamorphic rock formed must have preceded the igneous rock (figure 22.16).

Shifting sites of erosion and deposition: The principle of uniformity states that the earth processes going on today have always been occurring. This does not mean, however, that they always occur in the same place. As erosion wears away the rock layers at a site, the sediments produced are deposited someplace else. Later, the sites of erosion and deposition may shift, and the sediments are deposited on top of the eroded area. When the new sediments later are formed into new sedimentary rocks, there will be a time lapse between the top of the eroded layer and the new layers. A time break in the rock record is called an **unconformity.** The unconformity is usually shown by a surface within a sedimentary sequence on which there was a lack of sediment deposition or where active erosion may even have occurred for some period of time. When the rocks are later examined, that time span will not be represented in the record, and if the unconformity is erosional, some of the record once present will have been lost. An unconformity may occur within a sedimentary sequence of the same kind or between different kinds of rocks. The most obvious kind of unconformity to spot is an *angular unconformity.* An angular unconformity, as illustrated in figure 22.17, is one in which the bedding planes above and below the unconformity are not parallel. An angular unconformity usually implies some kind of tilting or folding, followed by a significant period of erosion, which in turn was followed by a period of deposition (figure 22.18).

The principle of superposition, the principle of crosscutting relationships, and the presence of an unconformity help us interpret the order in which geologic events occurred. This order can be used to unravel a complex sequence of events, such as the one shown in figure 22.19.

22.5 GEOLOGIC TIME AND THE FOSSIL RECORD

A **fossil** is *any* evidence of former life, so the term means more than fossilized remains such as those pictured in figure 22.20. Evidence can include actual or altered remains of plants, animals, fungi, algae, or even bacteria. It could also be just simple evidence of former life such as the imprint of a leaf, the footprint of a dinosaur, or droppings from bats in a cave.

Early Ideas About Fossils

Our current understanding of fossils requires a great deal of information about geologic time and geologic processes. Without this information, it was difficult for people to appreciate how fossils formed and how to interpret them. The ancient

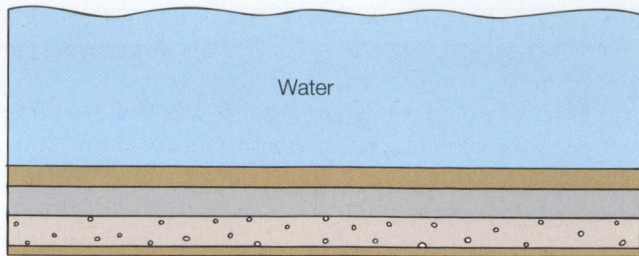

1. Deposition

Erosional surface

2. Rocks tilted, eroded

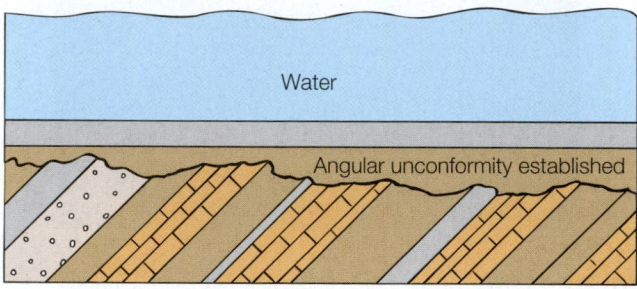

Water

Angular unconformity established

3. Subsequent deposition

FIGURE 22.17 An unconformity is a gap in the geologic record caused when the original deeper rocks were subject to erosion and then a subsequent layer of sediments was deposited on top of them. An angular unconformity is easy to identify because the layers of the deeper rocks are not parallel to the horizontal surface rocks.

Greek historian Herodotus was among the first to realize that fossil shells found in rocks far from any ocean were remnants of organisms left by a bygone sea. However, one hundred years later, Aristotle still believed that fossils had no relation to living things and were formed inside the rocks. Note that he also believed that living organisms could arise by spontaneous generation from mud. A belief that the fossils must have "grown" in place in rocks would be consistent with a belief in spontaneous generation.

Even when people recognized some similarities between living organisms and certain fossils, they did not make a connection between the fossil and previously living things. Fossils were considered to be the same as quartz crystals or any other mineral crystals that either formed with Earth or grew there later (depending on the philosophical view of the interpreter). With the development of Judeo-Christian beliefs, many people came to regard well-preserved fossils that were very similar to living organisms as remains of once-living organisms—that were buried in Noah's flood. Leonardo da Vinci, like other Renaissance scholars, argued that fossils were the remains of organisms that had lived in the past.

FIGURE 22.18 This photograph shows an angular unconformity. The horizontal sedimentary rock layers overlie almost vertically oriented metamorphic rocks. Metamorphic rocks form deep in Earth, so they must have been uplifted, and the overlying material eroded away before being buried again.

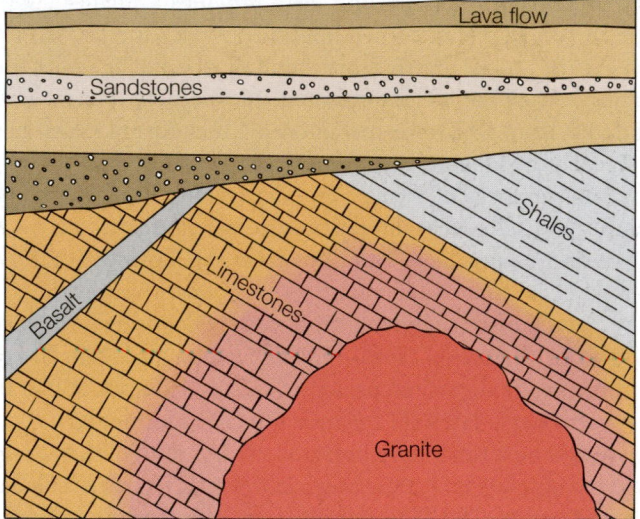

Lava flow

Sandstones

Shales

Basalt

Limestones

Granite

FIGURE 22.19 A complex rock sequence can be deciphered by using geologic principles. The limestones must be oldest (law of superposition), followed by the shales. The granite and basalt must both be younger than the limestone they crosscut (note the metamorphosed zone around the granite). It is not possible to tell whether the igneous rocks (granite and basalt) predate or postdate the shales or to determine whether the sedimentary rocks were tilted before or after the igneous rocks were emplaced. After the limestones and shales were tilted, they were eroded, and then the sandstones were deposited on top. Finally, the lava flow covered the entire sequence.

By the early 1800s, the true nature of fossils was becoming widely accepted. William "Strata" Smith, an English surveyor, discovered at this time that sedimentary rock strata could be identified by the fossils they contain. Smith grew up in a region

FIGURE 22.20 The hard parts of this fish fossil, *Diplomystus dentatus,* from the Eocene period, are beautifully preserved, along with a carbon film, showing a detailed outline of the fish and some of its internal structure.

of England where fossils were particularly plentiful. He became a collector, keeping careful notes on where he found each fossil and in which type of sedimentary rock layers. During his travels, he discovered that the succession of rock layers on the south coast of England was the same as the succession of rock layers on the east coast. Through his keen observations, Smith found that each kind of sedimentary rock had a distinctive group of fossils that was unlike the group of fossils in other rock layers. Smith amazed his friends by telling them where and in what type of rock they had found their fossils.

Today, the science of discovering fossils, studying the fossil record, and deciphering the history of life from fossils is known as **paleontology.** The word *paleontology* was invented in 1838 by the Scottish geologist Charles Lyell to describe this newly established branch of geology. It is derived from classic Greek roots and means "study of ancient life," and this means a study of fossils.

People sometimes blur the distinction between *paleontology* and *archaeology.* **Archaeology** is the study of past human life and culture from material evidence of artifacts, such as graves, buildings, tools, pottery, landfills, and so on (*artifact* literally means "something made"). The artifacts studied in archaeology can be of any age, from the garbage added to the city landfill yesterday to the fragments of pottery of an ancient tribe that disappeared hundreds of years ago. The word *fossil* originally meant "anything dug up" but today carries the meaning of any *evidence of ancient organisms in the history of life.* Artifacts are therefore not fossils, as you can see from the definitions.

Types of Fossils

Considering all the different things that can happen to the remains of an organism, and considering the conditions needed to form a fossil, it seems amazing that any fossils are formed and then found. For example, the animals you see killed beside the road or the dead trees that fall over in a forest rarely become fossils because scavengers eat the remains of dead animals, and

decay organisms break down the organic remains of plants and animals. As a result, very little digestible organic matter escapes destruction.

Thus, a fossil is not likely to form unless there is rapid burial of a recently deceased organism. The presence of hard parts, such as a shell or a skeleton, will also favor the formation of a fossil if there is rapid burial. Several different kinds of fossils can be differentiated based on the kind of material that is preserved and the processes of preservation.

Preserved Whole Organisms

Occasionally, entire organisms are preserved, and in rare cases, even the unaltered remains of an organism's *soft parts* are found. For this to occur, the organism must be quickly protected from scavengers and decomposers following death. Several conditions allow this to occur. The best examples of this uncommon method of fossilization include protection by freezing or entombing in tree resin. Mammoths, for example, have been found frozen and preserved by natural refrigeration in the ices of Alaska and Siberia. The bodies of ice age humans have also been discovered in glaciers, and early inhabitants of South America have been found in ice caves high in mountains. In 1991 the body of a man was discovered in a glacier on the border between Austria and Italy. Subsequent studies of the body showed that he dated from about 3300 B.C. Furthermore, study of the body determined much information about him including what he had recently eaten, that he had been wounded by an arrow, and that he had tattoos. Insects and spiders, complete with delicate appendages, have been found preserved in amber, which is fossilized tree resin (figure 22.21). In each case—ice or resin—the soft parts were protected from scavengers, insects, and bacteria. Additional examples of organisms that have been preserved in a relatively unaltered state include organisms buried in acid bogs or those quickly desiccated following death.

FIGURE 22.21 This mosquito was stuck in and covered over with plant pitch. When the pitch fossilized to form amber, the mosquito was preserved as well. The entire mosquito is preserved in the amber.

FIGURE 22.22 Skeletons of recent organisms are often preserved as fossils in sediments. These are the bones of a Cretaceous turtle fossil found at Las Hoyas, Madrid, Spain.

Preserved Hard Parts

Although fossils of soft parts are occasionally formed, the *remains of hard parts* such as shells, bones, and teeth of animals or the pollen and spores of plants, which have hard coverings, are much more common. The turtle bones shown in figure 22.22 are a good example of fossils of hard parts of organisms. The bones were covered with sediment and preserved.

Because hard parts are more easily preserved, organisms that have such parts are more often found as fossils than those that lack hard parts. Worms of various kinds are hard to find as fossils for this reason. The hard parts of various organisms are composed of different kinds of materials. Calcium carbonate is present in the shells of most mollusks and certain kinds of protozoans, and the skeletal structures of corals and some sponges. Some thick deposits of limestone are composed of the calcium carbonate shells of tiny marine protozoans known as foraminiferans. Calcium phosphate was found in the shells of some kinds of extinct brachiopods. Silica is found in the skeletal structures of certain protozoans and sponges. The silica shells of the marine microorganisms known as diatoms have accumulated to such a degree that ancient deposits of the shells are mined as diatomaceous earth and used for filters and certain buffing compounds. Chitin is the tough organic material that makes up the exoskeletons of insects, spiders, crabs, lobsters, and similar organisms. Since chitin resists decomposition, organisms that have exoskeletons of chitin are common in the fossil record.

Traces, Molds, and Casts

Many kinds of fossils do not preserve the organisms but give a general indication of shape and size. A carbon trace is an example of another kind of fossil, one that does not preserve the detail of internal structures but gives a good idea of the general shape of the organism. Plant fossils are often found as carbon traces, sometimes looking like a photograph of a leaf on a slab of shale or limestone. Other fossils that give a good idea of the general shape of an organism are molds and casts. Calcium

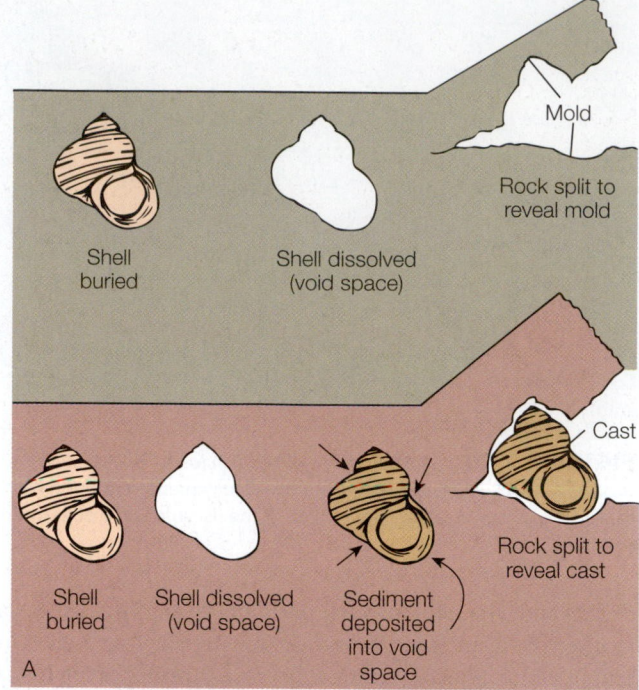

FIGURE 22.23 (*A*) How molds and casts are formed. (*B*) A trace fossil of a leaf. (*C*) Mold fossils of clams and snails. (*D*) Cast fossils of snails.

carbonate shell material may be dissolved by groundwater in certain buried environments, leaving an empty *mold* in the rock. Sediment or groundwater deposits may fill the mold and make a *cast* of the organism (figure 22.23).

Petrified Fossils

Some fossils are formed by the modification of the original chemical nature of the organism. Figure 22.24 is a photograph of part of the Petrified Forest National Park in Arizona. Petrified trees are not trees that have been "turned to stone." They are stony *replicas* that were formed as the wood was altered by circulating groundwater carrying elements in solution. There are two processes involved in the making of petrified fossils, and they are not restricted to wood. The processes involve (1) *mineralization,* which is the filling of pore spaces within the structure of the buried organism with deposits of calcium carbonate, silica, or pyrite, and/or (2) *replacement,* which is the

What Is Carbon-14 Dating?

Carbon is an element that occurs naturally as several isotopes. The most common isotope is carbon-12. A second, heavier radioactive isotope, carbon-14, is constantly being produced in the atmosphere by cosmic rays. Radioactive elements are unstable and break down into other forms of matter. Hence, radioactive carbon-14 naturally decays. The rate at which carbon-14 is formed and the rate at which it decays are about the same; therefore, the concentration of carbon-14 on Earth stays relatively constant. All living things contain large quantities of the element carbon. Plants take in carbon in the form of carbon dioxide from the atmosphere, and animals obtain carbon from the

food they eat. While an organism is alive, the proportion of carbon-14 to carbon-12 within its body is equal to its surroundings. When an organism dies, it is no longer adding carbon-12 or carbon-14 to its tissues. As the carbon-14 within its tissues undergoes nuclear decay, it is not replaced, and the relative percentages of carbon-12 and carbon-14 change. Therefore, the age of plant and animal remains can be determined by the ratio of carbon-14 to carbon-12 in the tissues. The older the specimen, the less carbon-14 present. Radioactive decay rates are measured in half-life. One half-life is the amount of time it takes for one-half of a radioactive sample to decay.

The half-life of carbon-14 is 5,730 years. Therefore, a bone containing one-half the normal proportion of carbon-14 is 5,730 years old. If the bone contains one-quarter of the normal proportion of carbon-14, it is $2 \times 5,730 = 11,460$ years old, and if it contains one-eighth of the naturally occurring proportion of carbon-14, it is $3 \times 5,730 = 17,190$ years old. As the amount of carbon-14 in a sample becomes smaller, it becomes more difficult to measure the amount remaining. Therefore, carbon-14 dating is generally useful only for dating things that are less than 50,000 years old.

dissolving of the original material and depositing of new material an ion at a time. Petrified wood is formed by both processes over a long period of time. As it decayed, the original wood was replaced by mineral matter an ion at a time. Over time, the "mix" of minerals being deposited changed, resulting in various colors. Since the mineralization or replacement processes take place an ion at a time, a great deal of detail can be preserved in such fossils. The size and shape of the cells and growth rings in petrified wood are preserved well enough that they can be compared to modern plants. The skeletons or shells of many extinct organisms are also typically preserved in this way. The "fossilized bones" of dinosaurs or the "fossilized shells" of many invertebrates are not actually bones or shells but are examples of petrified replicas.

Fossils That Are Not Remains

Finally, there are many kinds of fossils that are not the preserved remains of an organism but are preserved indications of the activities of organisms. Some of the most interesting of such fossils are the footprints of various kinds of animals. There are examples of footprints of dinosaurs and various kinds of extinct mammals, including ancestors of humans. If an animal walked through mud or other soft substrates and the substrates were covered with silt or volcanic ash, the pattern of the footprints can be preserved (figure 22.25). The tunnels of burrowing animals, the indentations left by crawling worms, and the nests of dinosaurs have been preserved in a similar manner. Even the dung of dinosaurs and other animals has been preserved in a mineralized form known as coprolites.

FIGURE 22.24 Petrified fossils are the result of an ion-for-ion replacement of the buried organisms with mineral material. This is a section of a fossil tree (*Araucarioxylon arizonicum*) on the ground at the Petrified Forest, Arizona.

FIGURE 22.25 This dinosaur footprint is in Moenave sandstone, Utah. It tells you something about the relative age of the sandstone, since it must have been soft mud when the dinosaur stepped here.

Connections . . .

Biomechanics and Fossil Animals

Many books have illustrations of various kinds of extinct animals. Have you ever wondered how anyone can know what these animals looked like? Certainly the general size and shape can be determined by assembling fossils of individual skeletal parts, but most of the shape of an animal is determined by the size and distribution of its muscles. An understanding of basic principles of physics provides part of the answer. The skeleton of an animal is basically arranged like a large number of levers. For example, a leg and foot consists of several bones linked end to end. The size of a bone gives a clue to the amount of mass that particular bone had to support. Various bumps on the bones are places where muscles and tendons attach. By knowing where the muscles are attached and analyzing the kinds of levers they represent, the size of the muscle can be calculated. Thus, the general shape of the animal can be estimated. However, this doesn't provide any information about the color or surface texture of the animal. That information must come from other sources.

CONCEPTS APPLIED

Examining Sedimentary Rock

Visit a place where layers of sedimentary rock are exposed. In many parts of the country, this may require you to travel to a place where a hill has been cut through to allow for the building of a road. Make a drawing of the layers and measure their thickness. If it took a hundred years to form a millimeter of sediment, how many years are represented by each different layer? Look for fossils in each of the layers.

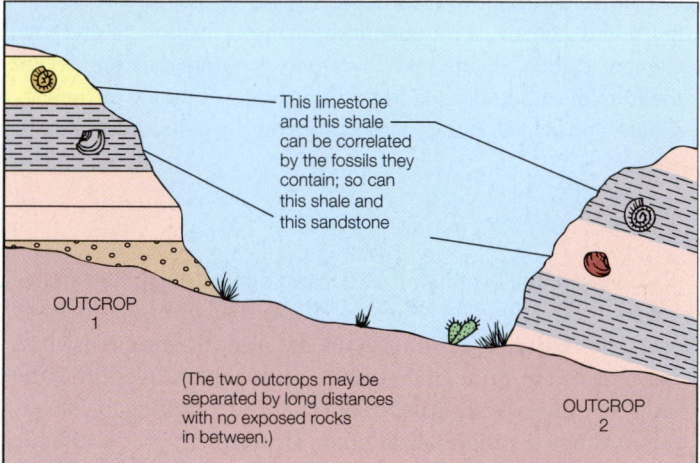

FIGURE 22.26   Index fossils can be used to help to compare the ages of rocks in different parts of the world. Similarity of fossils suggests similarity of ages, even in different kinds of rocks widely separated in space.

Using Fossils to Determine the Order of Geologic Events

As you can see, there are many different ways in which a fossil can be formed, but it must be *found and studied* if it is to reveal its part in the history of life. This means the rocks in which the fossil formed must somehow make it back to the surface of Earth. This usually involves movement and uplift of the rock and weathering and erosion of the surrounding rock to release or reveal the fossil. Most fossils are found in recently eroded sedimentary rocks—sometimes atop mountains that were under the ocean a long, long time ago. The complete record of what has happened in the past is not found in the fossil alone but requires an understanding of the layers of rocks present, the relationship of the layers to each other, and their age.

The *principle of faunal succession* recognizes that life-forms have changed through time. Old life-forms disappear from the fossil record and new ones appear, but the same form is never exactly duplicated independently at two different times in history. This principle implies that the same type of fossil organisms that lived only a brief geologic time should occur only in rocks that are the same age. According to the principle of faunal succession, then, once the basic sequence of fossil forms in the rock record is determined, rocks can be placed in their correct relative chronological position on the basis of the fossils contained in them. The principle also means that if the same type of fossil organism is preserved in two different rocks, the rocks should be the same age. This is logical even if the two rocks have very different compositions and are from places far, far apart (figure 22.26).

Distinctive fossils of plant or animal species that were distributed widely over Earth but lived only a brief time are called **index fossils.** Index fossils, together with the other principles used in reading rocks, make it possible to compare the ages of rocks exposed in two different locations. This is called *age correlation* between rock units. Correlations of exposed rock units separated by a few kilometers are easier to do, but correlations have been done with exposed rock units that are separated by an ocean. Correlation allows the ordering of geologic events according to age. Since this process is only able to determine the age of a rock unit or geologic event relative to some other unit or event, it is called *relative dating*.

The usefulness of correlation and relative dating through the concept of faunal succession is limited because the principles can be applied only to rocks in which fossils are well preserved, which are almost exclusively sedimentary rocks. Correlation also can be based on the occurrence of unusual rock types, distinctive rock sequences, or other geologic similarities. All this is useful in clarifying relative age relationships among rock units. It is not useful in answering questions about the age of rocks or the time required for certain events, such as the eruption of a volcano, to occur. Questions requiring numerical answers went unanswered until the twentieth century, when such tools as radiometric and geomagnetic dating techniques became available.

A Trilobite

B Ammonites

FIGURE 22.27 Since both trilobites (*A*) and ammonites (*B*) were common in the oceans throughout the world at specific times in the past, they have been used as index fossils. Trilobites are always found in older rocks than are ammonites. Trilobites were common in the Paleozoic era, and ammonites were common in the earlier parts of the Mesozoic era.

The Geologic Time Scale

A yearly calendar helps you keep track of events over periods of time by dividing the year into months, weeks, and days. In a similar way, the **geologic time scale** helps you keep track of events that have occurred in Earth's geologic history (table 22.4). The first development of this scale came from the work of William "Strata" Smith, the English surveyor described in the section on fossils earlier in this chapter. Recall that Smith discovered that certain sedimentary rock layers in England occurred in the same order, top to bottom, wherever they were located. He also found that he could correlate and identify each layer by the kinds of fossils in the rocks of the layers. In 1815, he published a geologic map of England, identifying the rock layers in a sequence from oldest to youngest. Smith's work was followed by extensive geological studies of the rock layers in other countries. Soon it was realized that similar, distinctive index fossils appeared in rocks of the same age when the principle of superposition was applied. For example, the layers at the bottom contained fossils of trilobites (figure 22.27A), but trilobites

were not found in the upper levels. (Trilobites are extinct marine arthropods that may be closely related to living crustaceans, spiders, and horseshoe crabs.) On the other hand, fossil shells of ammonites (figure 22.27B) appeared in the middle levels but not the lower nor upper levels of the rocks. The topmost layer was found to contain the fossils of animals identified as still living today. The early appearance and later disappearance of fossils in progressively younger rocks are explained by organic evolution and extinction, events that could be used to mark the time boundaries of Earth's geologic history.

The geologic time scale developed in the 1800s was without dates because geologists did not yet have a way to measure the length of the eras, periods, or epochs. Their time scale was a *relative time scale,* based on the superposition of rock layers and fossil records of organic evolution. With the development of radiometric dating, it became possible to attach numbers to the time scale. When this was accomplished, it became apparent that geologic history spanned very long periods of time. In addition, the part of the time scale with evidence of life makes up about 75 percent of all of Earth's history.

Names for Units of Geologic Time

The major blocks of time in Earth's geologic history are divided into several different categories. The largest are called **eons.** The *Phanerozoic eon* spans the period of time from about 540 million years ago, when the various groups of animals we currently see on Earth developed, until the present. The *Proterozoic eon* encompasses the period of time (2,500 million to 540 million years ago) when most living things were simple, marine organisms. The first eukaryotic organisms developed during this time. The *Archean eon* (3,800 million to 2,500 million years ago) constitutes a time when all organisms were prokaryotic and marine. The *Hadean eon* is the earlier period before life and before Earth solidified.

The Phanerozoic eon is divided into subunits of time called **eras,** and each era is identified by the appearance and disappearance of particular fossils in the sedimentary rock record. The eras are: (1) *Cenozoic,* which refers to the time of recent life. Recent life means that the fossils for this time period are similar to the life found on Earth today. (2) *Mesozoic,* which refers to the time of middle life. Middle life means that some of the fossils for this time period are similar to the life found on Earth today, but many are different from anything living today. (3) *Paleozoic,* which refers to the time of ancient life. Ancient life means that the fossils for this time period are very different from anything living on Earth today. (4) Sometimes the Proterozoic, Archean, and Hadean eons are referred to as the *Precambrian,* which refers to the time before the time of ancient life. This means that the rocks for this time period contain very few fossils.

The eras were divided into blocks of time called **periods,** and the periods were further subdivided into smaller blocks of time called **epochs** (see table 22.4).

Geologic Time Periods and Typical Fossils

1. Precambrian rocks contain the earliest fossils. The Precambrian fossils that have been found are chiefly those of deposits from bacteria, algae, a few fungi, unusual

TABLE 22.4   Geologic Time and the History of Life (To trace the history of life, start with the oldest time period, the Hadean.)

Eon	Era	Period	Epoch	Million Years Ago*	Major Physical Events	Major Biological Events
Phanerozoic	Cenozoic	Quaternary	Holocene (Recent)	0.01 to present	—Much of Earth has been modified by human activity	—Dominance of modern humans
			Pleistocene	0.01 ← 1.8	—Extinction of many kinds of large mammals in late Pleistocene, perhaps due to human activity —Most recent periods of glaciation	—Origin of modern humans —Many kinds of large mammals present in early Pleistocene
		Tertiary	Pliocene	1.8 ← 5.3	—Cool, dry period —North and South America join —The Indian Plate collides with Asia to form the Himalaya Mountains	—Origin of ancestors of humans —Grasslands and grazing mammals widespread —All modern groups of mammals present
			Miocene	5.3 ← 23.8	—Warm, moist period —Current arrangement of continents present	—Grasslands and grazers common —Mammals widespread —Marine kelp forests common
			Oligocene	23.8 ← 33.7	—Cool period —India, South America, and Australia are separate continents	—Tropics diminish —Woodlands and grasslands expand —Many new grazing mammals
			Eocene	33.7 ← 54.8	—Australia and South America separate from Antarctica —North America separates from Europe	—All modern forms of flowering plants present —Forests common —All major groups of mammals present —Grasses found for the first time —First primates
			Paleocene	54.8 ← 65.0	—Warm period	—Many new kinds of mammals —Many new kinds of birds
	Mesozoic	Cretaceous		65 ← 144	—Major extinction at the end of Cretaceous affected both land and ocean organisms, probably because of meteorite impact —Warm, moist period —Continents continue to separate	—Dinosaurs dominant —Evolution of birds —Coevolution of flowering plants and their insect pollinators —The three modern mammal groups found —First flowering plants
		Jurassic		144 ← 206	—Pangaea begins to split up —North and South America split —South America and Africa split	—Dinosaurs dominant —First birds —Many small primitive mammals
		Triassic		206 ← 248	—Climate cooler and wetter as Pangaea splits —Major extinction of about 50% of species —No clear reason for extinction —Supercontinent, Pangaea, near equator —Warm, dry climate	—Many modern insects —Explosion in reptile diversity —Evolution of modern gymnosperms —Evolution of modern corals —First dinosaurs —First mammals —Mollusks are dominant aquatic invertebrates
	Paleozoic	Permian		248 ← 290	—Largest of all extinctions occurred at the end of the Permian (90% of all species) but no clear cause for this extinction —New giant continent of Pangaea forms —Much of Pangaea was probably desert —Modern levels of oxygen in atmosphere	—Gymnosperms dominant —Many new species of insects —Amphibians and reptiles dominant

Eon	Era	Period	Epoch	Years Ago*	Major Physical Events	Major Biological Events
Phanerozoic (continued)	Paleozoic (continued)	Carboniferous		290 ↑ 354	—Major coal deposits formed —Two major continents moving toward each other to form Pangaea —Warm climate	—Gymnosperms (cone-bearing plants) present by the end of the period —Reptiles present by the end of the period —Vast swamps of club mosses, horsetails, and ferns common —Many kinds of sharks —First winged insects, some very large —Amphibians common —Armored fish extinct
		Devonian		354 ↓ 417	—Glaciation probable cause of mass extinction at end of Devonian —About 60% of genera extinct —Extinction primarily affected warm-water, marine life —Warm, shallow seas common for much of this period —Two large continents present	—First seed plants (seed ferns) —Both jawless and jawed fish abundant —Coral reefs abundant —Land plants abundant (first forests) —Many new kinds of wingless insects —Amphibians on land
		Silurian		417 ↓ 443	—Melting of glaciers led to higher sea level —A large continent, Gondwana, is present at South Pole, but there are many smaller equatorial continents	—Numerous coral reefs —Fungi present —First vascular plants —Widespread nonvascular plants —First land animals—Arthropods (spiders and centipedes) —First jawed fish —Jawless armored fish abundant —Some jawless fish found in freshwater
		Ordovician		443 ↓ 490	—Giant continent, Gondwana, located at South Pole resulted in formation of glaciers —Formation of glaciers led to drop in sea level, which is the probable cause of mass extinction at the end of Ordovician (60% of genera go extinct)	—Primitive, nonvascular land plants present —Jawless, armored fish common —Diverse marine invertebrates
		Cambrian		490 ↓ 543	—Rodinia breaks up into pieces —Glaciation probable cause of mass extinction at the end of Cambrian	—All major groups of animals present —Mollusks, brachiopods, echinoderms, trilobites, and nautiloids common
Proterozoic	This period of time is also known as the Precambrian			543 ↓ 2,500	—Single large supercontinent (Rodinia) existed about 1.1 billion years ago —Oxidation of terrestrial iron deposits forms red beds that indicate presence of atmospheric oxygen (oxidized iron is red) at about 2.5 billion years ago —About 1.8 billion years ago, the atmospheric oxygen concentration was probably about 15% of current level	—Primitive multicellular, soft-bodied marine animals present about 1 billion years ago —Algae present about 1 billion years ago —Fossil record poorly preserved —First eukaryotic cells present about 1.8 billion years ago
Archean				2,500 ↓ 3,800	—Formation of continents —Atmosphere lacked oxygen —Oldest rocks 3.8 billion years ago	—Fossil stromatolites (layers of cyanobacteria) common —Cyanobacteria carried on photosynthesis, releasing oxygen —Fossil cyanobacteria present 3.5 billion years ago —Origin of life about 3.8 billion years ago
Hadean				3,800 ↓ 4,500	—Crust of Earth in process of solidifying —No rocks of this age —Origin of Earth	

* Time scale is based on *1999 Geologic Time Scale* of the Geological Society of America.

soft-bodied animals, and the burrow holes of worms. It appears that there were no animals with hard parts; thus, the fossil record is incomplete since it is the hard parts of animals or plants that form fossils, usually after rapid burial. Another problem associated with finding fossils in these extremely old rocks is that heat and pressure have altered many of the ancient rocks over time, destroying any fossil evidence that may have been present.

2. The Paleozoic era was a time when there was great change in the kinds of plants and animals present. In general, the earliest abundant fossils are found in rocks from the Cambrian period at the beginning of the Paleozoic era. They show an abundance of oceanic life that represents all the major groups of marine animals found today. There is no fossil evidence of life of any kind living on the land during the Cambrian. The dominant life-forms of the Cambrian ocean were echinoderms, mollusks, trilobites, and brachiopods. The trilobites, now extinct, made up more than half of the kinds of living things during the Cambrian.

   During the Ordovician and Silurian periods of the Paleozoic era, most living things were still marine organisms with various kinds of jawless fish becoming prominent. Also, by the end of the Silurian, some primitive plants were found on land. The Devonian period saw the further development of different kinds of fish, including those that had jaws, and many kinds of land plants and animals. Coral reefs were also common in the Devonian. The Carboniferous period was a time of vast swamps of ferns, horsetails, and other primitive nonseed plants that would form great coal deposits. Fossils of the first reptiles and the first winged insects are found in rocks from this age. The Paleozoic era closed with the extinction at the end of the Permian period of about 90 percent of plant and animal life of that time.

3. The Mesozoic era was a time when the development of life on land flourished. The dinosaurs first appeared in the Triassic period, outnumbering all the other reptiles until the close of the Mesozoic. Fossils of the first mammals and modern forms of gymnosperms (cone-bearing plants) developed in the Triassic. The first flowering plants, the first deciduous trees, and the first birds appeared in the Cretaceous period. The Cretaceous is the final period of the Mesozoic era and is characterized by the dominance of dinosaurs and extensive evolution of flowering plants and the insects that pollinate them. Birds and mammals also increased in variety. Like the close of the Paleozoic, the Mesozoic era ended with a great dying of land and marine life that resulted in the extinction of many species, including the dinosaurs.

4. As the Cenozoic era opened, the dinosaurs were extinct and the mammals became the dominant vertebrate life-form. The Cenozoic is thus called the Age of the Mammals. However, there were also major increases in the kinds of insects, flowering plants (particularly grasses), and birds. Finally, toward the end of this period of time, humans arrived on the scene, and many other kinds of large mammals such as mammoths, mastodons, giant ground sloths, and saber-toothed tigers went extinct.

Mass Extinctions

When we look at the fossil record, there is evidence of several mass extinctions. Five are recognized for causing the extinction of 50 percent or more of the species present. It is important to understand that although these extinctions were "sudden" in geologic time, they occurred over millions of years. Each resulted in a change in the kinds of organisms present with major groups going extinct and the evolution of new kinds of organisms. The boundaries between many of the geologic time designations are defined by major extinction events. Geologists have developed theories about the causes of each of these mass extinctions. Many of these theories involve changes in the size and location of continents as a result of plate tectonics.

1. The mass extinction at the end of the Ordovician period resulted in the extinction of 60 percent of genera of organisms. At that time, most organisms lived in the oceans. It is thought that the large continent of Gondwana migrated to the South Pole and this resulted in the development of large glaciers and a drastic drop in sea level along with a cooling of the waters.

2. At the end of the Devonian period, there was a mass extinction that affected primarily marine organisms. Approximately 60 percent of genera went extinct. Since many of the organisms that went extinct were warm-water, marine organisms, glaciation along with a cooling of the oceans is a widely held theory for the cause of this extinction.

3. The mass extinction at the end of the Permian period is unusual in several ways. It resulted in the extinction of about 90 percent of organisms and took place over a very short time—less than a million years. Because of this, it is often referred to as the "Great Dying." Both marine and terrestrial organisms were affected. Because this extinction event occurred over a short time and affected all species, it is assumed that a major, worldwide event was responsible. However, at this time, there is no clearly identifiable cause. Suggestions include a meteorite impact, a supernova, massive volcanic activity, or a combination of factors.

4. The extinction at the end of the Triassic period was relatively mild compared to others. About 50 percent of species appear to have gone extinct. There is no clear cause for this extinction.

5. The mass extinction at the end of the Cretaceous period resulted in the extinction of about 60 percent of species. Based on evidence of a thin clay layer marking the boundary between the Cretaceous and Tertiary periods, one theory proposes that a huge (16 km diameter and 10^{15} kg mass) meteorite struck Earth (figure 22.28). The impact would have thrown a tremendous amount of dust into the atmosphere, obscuring the Sun and significantly changing the climate and thus the conditions of life on Earth. The resulting colder climate may have led to the extinction of many plant and animal species, including the dinosaurs. This theory is based on the clay layer, which theoretically formed as the dust settled, and its location in the rock record at the time of the extinctions. The layer is enriched with a rare metal, iridium, which is not found on Earth

FIGURE 22.28 A widely held theory is that a large meteorite was responsible for the mass extinction at the end of the Cretaceous period. Several of the other mass extinctions may have also been caused by meteorite impacts. This is an artist's depiction of the kinds of impact a large meteorite could have had on Earth.

in abundance but occurs in certain meteorites in greater abundance.

6. During the Quaternary period at the end of the last ice age, there was a relatively minor extinction of the many species of large mammals. This is thought to be due to either a major change in climate at the end of the last ice age or hunting by humans as they expanded their range from Africa to Europe, Asia, and the Americas.

7. Based on current scientific evidence, many biologists are convinced that we are currently experiencing a human-caused mass extinction because of our ability to alter the face of Earth and destroy the habitats needed by plants and animals.

Interpreting Geologic History—A Summary

When interpreting the fossil record in any part of the world, there are several things to keep in mind:

1. *We are dealing with long periods of time.* The history of life on Earth goes back to 3.5 billion years ago, and the evolution of humans took place over a period of several million years. It is important to understand that many of the processes of sedimentation, continental drift, and climate change took place slowly over many millions of years.

2. *Earth has changed greatly over its history.* There have been repeated periods of warming and cooling. During periods of cooling, glaciers formed, which resulted in lower sea level, which in turn exposed more land and changed the climate of any particular continent. In addition, the continents were not fixed in position. Changes in position affected the climate that the continents experienced. For example, at one time, what is now North America was attached to Antarctica near the South Pole.

3. *There have been many periods in the history of Earth when most of the organisms went extinct.* Cooling climates, changes in sea level, and meteorite impacts are all suspected of caus-

ing mass extinctions. However, it should not be thought that these were sudden extinctions. Most took place over millions of years. Some of the extinctions affected as much as 90 percent of the things living at the time.

4. *New forms of life evolved that replaced those that went extinct.* The earliest organisms we see in the fossil record were prokaryotic marine organisms similar to present-day bacteria. The oldest fossils of these organisms date to about 3.5 billion years ago. The next major step was the development of eukaryotic marine organisms about 1.8 billion years ago. The first eukaryotic multicellular organisms were present by about 1 billion years ago. The development of multicellular organisms ultimately led to the colonization of land by plants and animals, with plants colonizing about 500 million years ago and animals at about 450 million years ago.

5. *Although there were massive extinctions, there are many examples of the descendants of early life-forms present today.* Bacteria and many kinds of simple eukaryotic organisms are extremely common today, as are various kinds of algae and primitive forms of plants. In the oceans, many kinds of marine animals such as starfish, jelly fish, and clams are descendants of earlier forms.

6. *The kinds of organisms present have changed the nature of Earth.* The oxygen in the atmosphere is the result of the process of photosynthesis. Its presence has altered the amount of ultraviolet radiation reaching Earth. Plants tend to reduce the erosive effects of running water and humans have significantly changed the surface of Earth.

22.6 PALEONTOLOGY, ARCHAEOLOGY, AND HUMAN EVOLUTION

There is intense curiosity about how our species (*Homo sapiens sapiens*) came to be, and the evolution of the human species remains an interesting and hot topic. Human beings are classified as mammals belonging to a group known as primates. Primates are thought to have come into existence approximately

Myths, Mistakes, and Misunderstandings

The Best Species on Earth

Misconception: Humans are the top of the evolutionary pyramid, and therefore, the "best" organisms are humans.

In fact, there is no such thing as an evolutionary pyramid; it is not a fixed condition. Evolution is a process and all living things are constantly involved in this process. If any group of organisms stops evolving, they will become extinct. There is no one "best" type of organism. If the term *best* means that a species has reached the end of the process, the species is in the process of becoming extinct. However, it is possible to refer to a species that "best suits" its environment at a designated time. If this is the meaning of the term *best*, then any species of organism can be its "best" at a designated time.

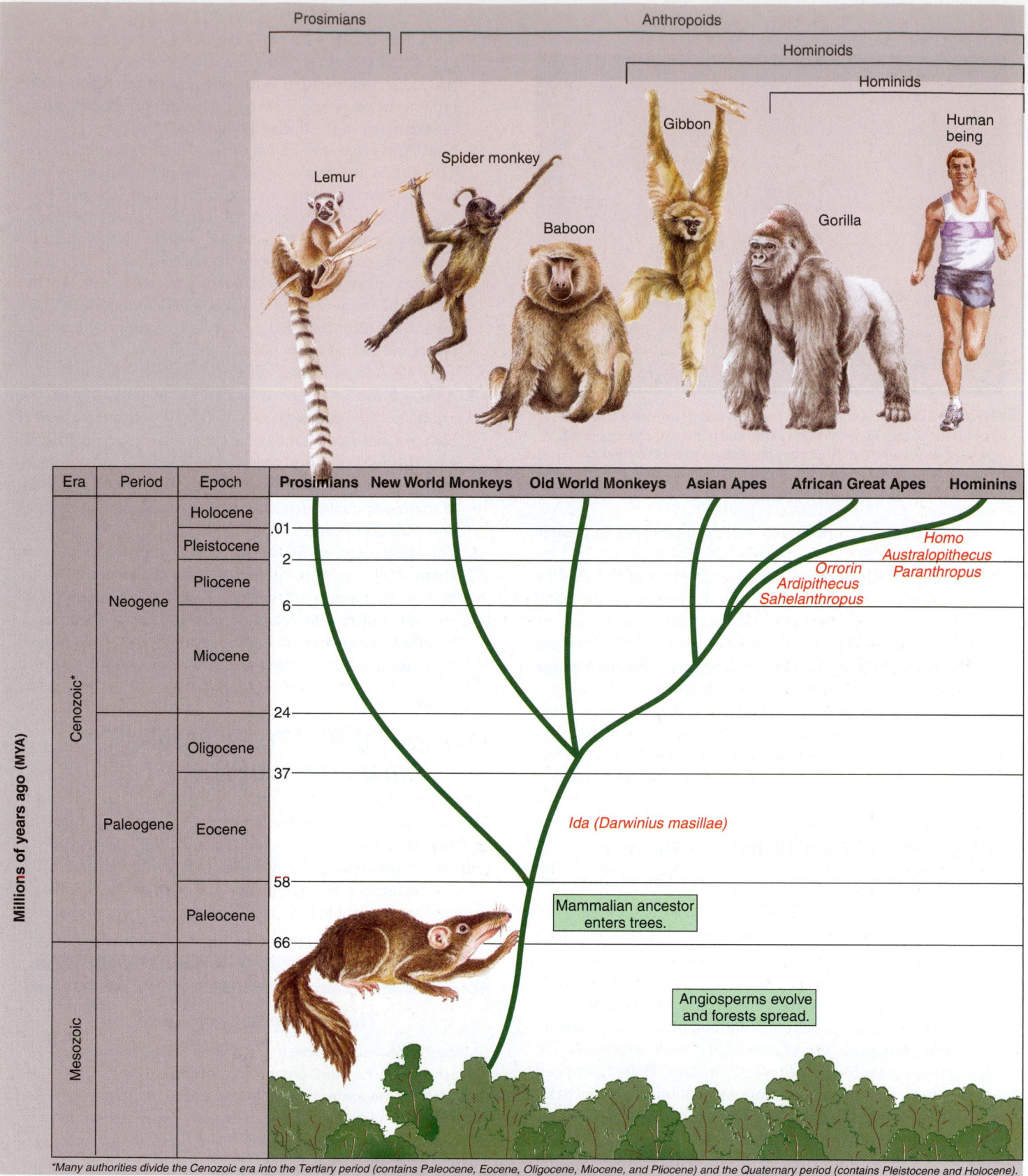

Era	Period	Epoch		Prosimians	New World Monkeys	Old World Monkeys	Asian Apes	African Great Apes	Hominins

Prosimians | Anthropoids

Hominoids

Hominids

Lemur | Spider monkey | Baboon | Gibbon | Gorilla | Human being

.01

2

6

24

37

58

66

Homo
Australopithecus
Paranthropus

Orrorin
Ardipithecus
Sahelanthropus

Ida (Darwinius masillae)

Mammalian ancestor enters trees.

Angiosperms evolve and forests spread.

Millions of years ago (MYA)

Cenozoic*

Mesozoic

Neogene

Paleogene

Holocene
Pleistocene
Pliocene
Miocene
Oligocene
Eocene
Paleocene

*Many authorities divide the Cenozoic era into the Tertiary period (contains Paleocene, Eocene, Oligocene, Miocene, and Pliocene) and the Quaternary period (contains Pleistocene and Holocene).

FIGURE 22.29 As new evidence is uncovered, scientists reconsider how that information best fits into the hypothesis on the evolution of humans. As the pieces of the puzzle are reexamined and rearranged, we see a clearer picture of where human beings fit in the scheme of life.

66 million years ago. They include animals with enlarged, complex brains; five digits, with nails, on the hands and feet; and hands and feet adapted for grasping. Their bodies, except those of humans, are covered with hair. There are two groups of primates, (1) the prosimians—lemurs and tarsiers—and (2) the anthropoids—monkeys, apes, and humans (figure 22.29).

You cannot shake hands with any other species belonging to the genus *Homo*. All other versions of our close evolution-

A Closer Look

Another Piece of the Human Evolution Puzzle Unearthed?

Just where would you expect to find a 47-million-year-old primate fossil? Africa, of course! But not this time. "Ida" (*Darwinius masillae*) was found in Messel Pit, a pit created by an oil shale mining operation in Germany in 1983, and not by a professional paleontologist, but by an amateur collector. Fossil exploration began after the mining operation was completed and the pit was authorized to become a garbage dump. Ida was kept in a private collection for twenty-five years before she was acquired by the Natural History Museum of the University of Oslo for scientific study.

Ida is the most complete primate skeleton known in the fossil record. She has a complete skeleton, a soft body outline, and food in her digestive tract. Preliminary evidence reveals that she lived during the Eocene Epoch, after the extinction of dinosaurs and when primates split into two major groups: prosimians and anthropoids. The region was experiencing continental drift and just beginning to take on features we would recognize as Germany's landscape today. During the Eocene, many modern plants and animals were evolving in a subtropical, junglelike environment. Evolutionarily, Ida and her relatives are thought to have been the evolutionary base of the anthropoid branch that led to monkeys, apes, and humans.

Ida lacks traits found in lemurs, such as a grooming claw on the second toe of the foot, a fused row of teeth in the middle of her lower jaw (known as a toothcomb), and claws. Her more advanced traits include the presence of fingernails, forward-facing eyes (allowing her to have 3-D vision and the ability to judge distance), and teeth similar to those of monkeys. Ida also has a talus bone in her feet. This bone allows her entire weight to be transmitted to the foot, an important feature in bipedal animals.

ary relatives are extinct. This makes it difficult to visualize our evolutionary development. Therefore, we tend to think we are not subject to the laws of nature. However, humans show genetic diversity, experience mutations, and are subject to the same evolutionary forces as other organisms.

Scientists use several kinds of evidence to try to sort out our evolutionary history. Fossils of various kinds of prehuman and ancient human ancestors have been found, but many of these are only fragments of skeletons, which are difficult to interpret and are hard to date. Stone tools of various kinds have also been found that are associated with prehuman and early human sites. Finally, other aspects of the culture of our ancestors have been found in burial sites, including cave paintings and ceremonial objects. Various methods have been used to date these findings. When fossils are examined, anthropologists can identify differences in the structures of bones that are consistent with changes in species. Based on the amount of change they see and the ages of the fossils, scientists make judgments about the species to which the fossil belongs.

As new discoveries are made, experts' opinions will change, and our evolutionary history may become clearer as old ideas are replaced.

Scientists must also review and make changes in the terminology they use to refer to our ancestors. The term *hominin* now refers to humans and their humanlike ancestors, whereas previously the term *hominid* was used. The term *hominid* now refers to the broader group that includes all humanlike organisms plus the great apes—gorillas, orangutans, chimpanzees, and bonobos. When you read material about this topic, you will need to determine how the terms are being used. Although there is no clear picture of how humans evolved, the fossil record shows that humans are a relatively recent addition to the forms of life. Members of the genus *Homo* are believed to have evolved at least 2.2 million to 2.5 million years ago (table 22.5).

When scientists put all the bits of information together, they constructed the following scenario for the likely evolu-

tion of our species. Rather than humans evolving from a chimpanzeelike ancestor, chimps and humans evolved from a common primate ancestor. Early primates, ancestors to modern monkeys, chimpanzees, and apes, were adapted to living in forested areas, where their grasping hands, opposable thumbs, big toes, and wide range of shoulder movement allow them to move freely in the trees. As the climate became drier, grasslands replaced the forests. Early hominins were adapted to drier conditions. Walking upright was probably an adaptation to these conditions. Most later hominins had large brains and used tools. A recent find, however, illustrates just how tentative the understanding of human evolution really is.

TABLE 22.5   Primate Classification—Order to Subfamily

Classification Category	Name	Members
Order	Primates (*Primates*)	Prosimians, tarsiers, monkeys, gibbons, orangutans, gorillas, chimps, humans
Suborder	Anthropoidea (*Anthropoid*)	Tarsiers, monkeys, gibbons, orangutans, gorillas, chimpanzee, humans
Superfamily	Hominoidea (*Hominoid*)	Gibbons, orangutans, gorillas, chimps, humans
Family	Hominidae (*Hominid*)	Orangutans, gorillas, chimps, humans
Subfamily	Homininae (*Hominin*)	Humans and direct ancestors

		Hominoidea			*superfamily*
	Hominidae			Hylobatidae	*family*
Homininae		Ponginae			*subfamily*
Hominin	Gorillini				*tribe*
Homo	Pan	Gorilla	Pongo	4 genera	*genus*
Humans	Chimps	Gorillas	Orangutans	Gibbons*	

*Based on DNA evidence

Science and Society

Neandertals, Denisovans, and *Homo sapiens*

An ongoing controversy surrounds the relationship between the Neandertals of western Europe, Denisovans of Asia, and modern humans. The names of these ancient people typically are derived from the place where the fossils were first discovered. The Neandertals were first found in the Neander Valley of Germany, and the Cro-Magnons, considered to be modern *Homo sapiens sapiens*, were initially found in the Cro-Magnon caves in France. The Denisovans got their name from the discovery of a 41,000-year-old female skeleton found in Denisova Cave, Siberia, in 2008.

It has been proposed that the Neandertals and Denisovans branched off from the human evolutionary tree about 400,000 years ago, and that these two groups split from each other about 200,000 years ago. Some scientists believe that they were small, separate races or subspecies of humans.

The use of molecular genetic technology has shed some light on the relationship of Neandertals and Denisovans to other kinds of humans. Examination of the

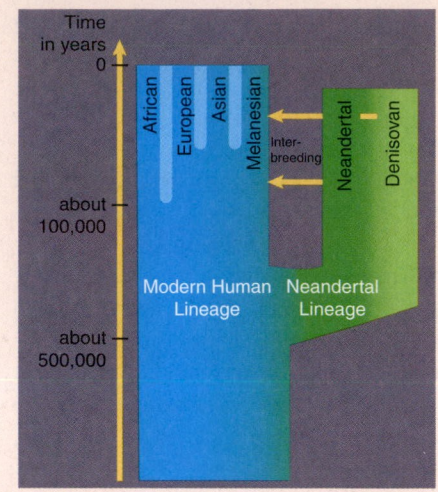

nuclear and mitochondrial DNA obtained from the bones and teeth of Neandertals and Denisovans reveals that there are significant differences between them and other kinds of early humans. This strengthens the argument that the Neandertals and Denisovans separated from the lineage of modern humans some 400,000 years ago. Using these biotech techniques, scientists

have found that Neandertal and human genomes are between 99.5 percent and 99.9 percent identical. The Denisovan girl's DNA was most like that of a Neandertal, but her people appear to have been a distinct group that had long been separated from Neandertals.

Genetic evidence now suggests that "pure bred" *Homo sapiens sapiens* spread from Africa through a bottleneck created by a strong glacial phase and moved into Europe and Asia. During the time Neandertals and *H. sapiens* were in contact, and Denisovans and *H. sapiens* were in contact, some interbred. The result is that all modern humans descended from this interbreeding European and Asian ancestry are not "pure," but display genetic ties to the Neandertals and Denisovans. Evidence also suggests that extinction of "pure bred" Neandertals may have been accelerated by volcanic eruptions, climate change, and genetic differences that, at the same time, paved the way for the dominance of modern European *Homo sapiens sapiens*.

Also, Neandertals and Denisovans may have disappeared because their subspecies were eliminated by interbreeding with more populous, more successful groups. (Many small, remote tribes have been lost as distinct cultural and genetic entities in the same way in recent history.) Others maintain that they showed such great differences from other early humans that they must have been different species and became extinct because they could not compete with the more successful modern human immigrants from Africa.

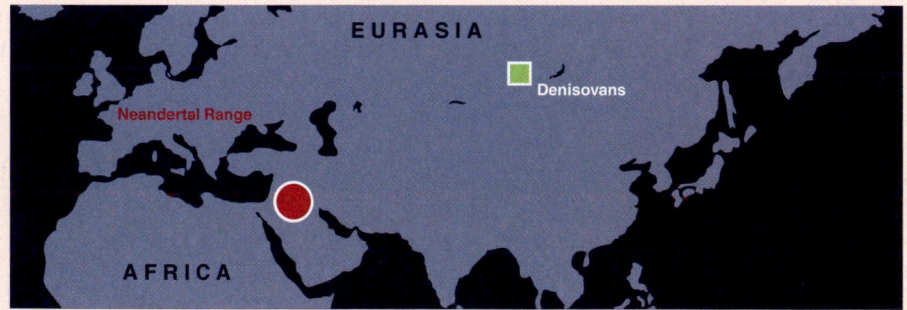

The discovery of fossils of a small human—known as the "hobbit"—in the sediments of a cave on an Indonesian island in 2003 has led to much speculation about a possible new species of *Homo*: *Homo floresiensis*. The fossils were dated to about 18,000 years ago. There has been much discussion among scientists about the significance of this discovery. Some feel that the small size indicates a developmental abnormality and that the remains could be that of an abnormally small *H. sapiens sapiens*. Others suggest that there are many examples of island animals that show small size and that these fossils were adapted to island conditions. Research continues and a more complete understanding will develop as more information is gathered. Like other human relatives, *H. floresiensis* is extinct. Therefore, the *sole surviving member of our evolution-*

ary line is Homo sapiens sapiens. We are now going to look at several organisms important to understanding the evolution of human beings.

Ardipithecus and Other Early *Hominins*

Ardipithecus ramidus was discovered in Ethiopia and lived in woodlands. The 125-piece skeletal remains revealed the first anatomical changes that laid the groundwork for walking upright. "Ardi" did not look like an ape and was able to walk upright when on the ground as well as navigate on branches of trees. The fossils from the best-studied site have been aged at about 4.4 million years ago. Studies of these fossils indicate that *Ardipithecus ramidus* was about 120 cm (4 ft) tall when standing, weighed about 50 kg (110 lb), and had a brain capacity of less than 400 cm^3. It was able to walk and run on its hind limbs but had an opposable big toe and was able to climb and walk on the tops of branches on all fours. Males and females were about the same size. They ate a variety of plant and animal materials in their woodland habitat. The hominins *Orrorin* and *Sahelanthropus* are older than *Ardipithecus* and are poorly understood. Fossils of all three are found only in Africa.

The Genera *Australopithecus* and *Paranthropus*

Both *Australopithecus* and *Paranthropus* are important evolutionary links in understanding our evolutionary past. Various species of *Australopithecus* and *Paranthropus* were present in Africa from about 4.4 million years ago until about 1 million years ago. Various members of these two extinct genera are often referred to as australopiths. Earlier fossils, such as *Ardipithecus* (which could have included a number of species), *Sahelanthropus* (7 to 6 mya), and *Orrorin* (6 to 5.2 mya), are thought to be ancestral to the genus *Australopithecus*. *Ardipithecus* would most likely be a "distant cousin" to human beings rather than a direct ancestor. *Australopithecus* and *Paranthropus* were herbivores and walked upright, the males being much larger than females.

Australopithecus is the genus to which the famed 3.2-million-year-old "Lucy" belongs, together with a 3.3-million-year-old *A. afarensis*, dubbed "Lucy's baby." Lucy's baby, Selam, is estimated to have been about three years of age when she died and is the most complete hominin skeleton ever found. There are few fossils of these early humanlike organisms, however, and many fossils are fragments of organisms. This has led to much speculation and argument among experts about the specific position each fossil has in the evolutionary history of humans. However, from examinations of the fossil bones of the leg, pelvis, and foot, it is apparent that the australopiths were relatively short (males, 1.5 meters or less; females, about 1.1 meters) and stocky and walked upright, as humans do. They had relatively small brains (cranial capacity 530 cm^3 or less—about the size of a standard softball).

An upright posture had several advantages in a world that was becoming drier. It allowed for more rapid movement over long distances and the ability to see longer distances, and it reduced the amount of heat gained from the sun. In addition, upright posture freed the arms for other uses, such as carrying and manipulating objects and using tools. The various species of *Australopithecus* and *Paranthropus* shared these characteristics and, based on the structure of their skulls, jaws, and teeth, appear to have been herbivores with relatively small brains.

The Genus *Homo*

Based on their findings, scientists understand that the evolutionary path leading to *Homo sapiens sapiens* was not a straight line, but included a number of diverse ancestors. They estimate that the animals that would eventually evolve to humans branched off from a common ancestor with the chimpanzees 5 to 7 million years ago. About 2.5 million years ago, the first members of the genus *Homo* appeared on the scene. There is disagreement about how many species and subspecies there were, but we know that all are now extinct. Because the fossils of *Homo* species found in Asia and Europe are generally younger than the early *Homo* species found in Africa, it is assumed that species of *Homo* migrated from Africa to Europe and Asia from Africa. *Homo habilis* is one of the earliest. *H. habilis* had a larger brain (650 cm^3) and smaller teeth than australopiths and made much more use of stone tools. Some scientists feel that it was a direct descendant of *Australopithecus africanus*. They feel that *H. habilis* was a scavenger that made use of group activities, tools, and higher intelligence to take over the kills made by other carnivores. The higher-quality diet would have supported the metabolic needs of the larger brain.

About 1.8 million years ago, *Homo ergaster* appeared on the scene. It was much larger, up to 1.6 m (6 ft) tall, than *H. habilis*, who was about 1.3 m (4 ft) tall, and also had a much larger brain (cranial capacity of 850 cm^3). A little later, a similar species primarily found in Asia (*Homo erectus*) appears in the fossil record. Some people consider *H. ergaster* and *H. erectus* to be variations of the same species. Along with their larger brain and body size, *H. ergaster* and *H. erectus* are distinguished from earlier species by their extensive use of stone tools. Hand axes were manufactured and used to cut the flesh of prey and crush the bones to obtain the fatty marrow. These organisms appear to have been predators, whereas *H. habilis* was a scavenger. The use of meat as food allows animals to move about more freely, because appropriate food is available almost everywhere. By contrast, herbivores are often confined to places that have foods appropriate to their use: fruits for fruit eaters, grass for grazers, forests for browsers, etc. In fact, fossils of *H. erectus* have been found in the Middle East and Asia as well as Africa. Most experts feel that *H. erectus* originated in Africa and migrated through the Middle East to Asia.

At about 800,000 years ago, another hominin classified as *Homo heidelbergensis* appears in the fossil record. Since fossils of this species are found in Africa, Europe, and Asia, it appears that they constitute a second wave of migration of early *Homo* from Africa to other parts of the world. Both *H. erectus* and *H. heidelbergensis* disappear from the fossil record as new species (*Homo sapiens neanderthalensis* and *Homo sapiens sapiens* and possibly *Denisovan*) become common.

The Neandertals and Denisovans were primarily found in Europe and adjoining parts of Asia and were not found in Africa. Because Neandertals were common in Europe, many scientists feel Neandertals are descendants of *H. heidelbergensis*, who also were common in Europe and preceded Neandertals.

Where Did It All Start?

It is thought that *Homo sapiens* evolved between 400,000 and 250,000 years ago. *Homo sapiens sapiens*, as we are classified today, came about approximately 130,000 years ago and is now found throughout the world. We are the only hominin species remaining of a long and branched history of ancestors. But

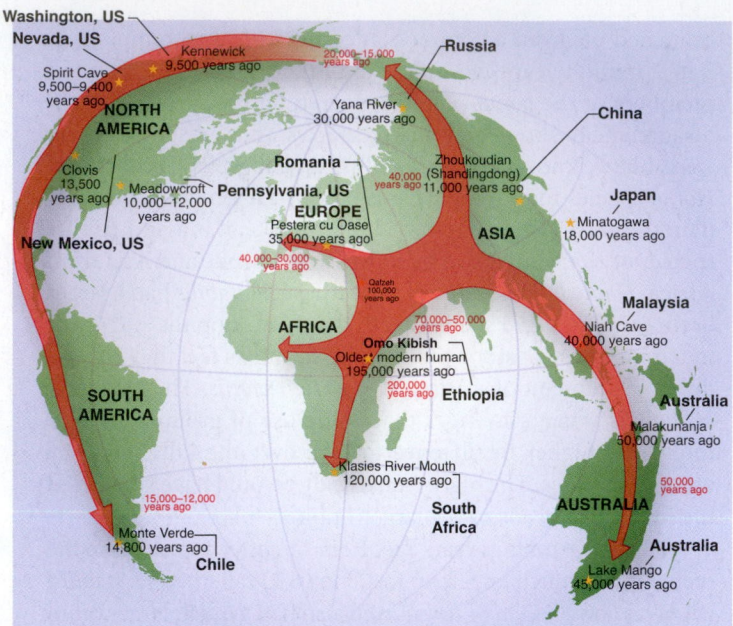

FIGURE 22.30 Most scientists favor this explanation on the origin and dispersal of *Homo sapiens sapiens* that arose in Africa about 200,000 years ago.

years ago. Over these tens of thousands of years, *Homo sapiens sapiens* replaced earlier human populations such as *Homo sapiens neanderthalensis*, *Denisovans*, and *Homo erectus*. Speculation is that they moved into the Americas about 20,000 to 15,000 years ago, taking advantage of low sea levels and a land bridge that connect Siberia to Alaska (figure 22.30).

Large numbers of fossils of prehistoric humans have been found in all parts of the world. Many of these show evidence of a collective group memory we call *culture*. Cave paintings, carvings in wood and bone, tools of various kinds, and burials are examples. These are also evidence of a capacity to think and invent, and "free time" to devote to things other than gathering food and other necessities of life. We may never know how we came to be, but we will always be curious and will continue to search and speculate about our beginnings. Figure 22.31 summarizes the current knowledge of the historical record of humans and their relatives.

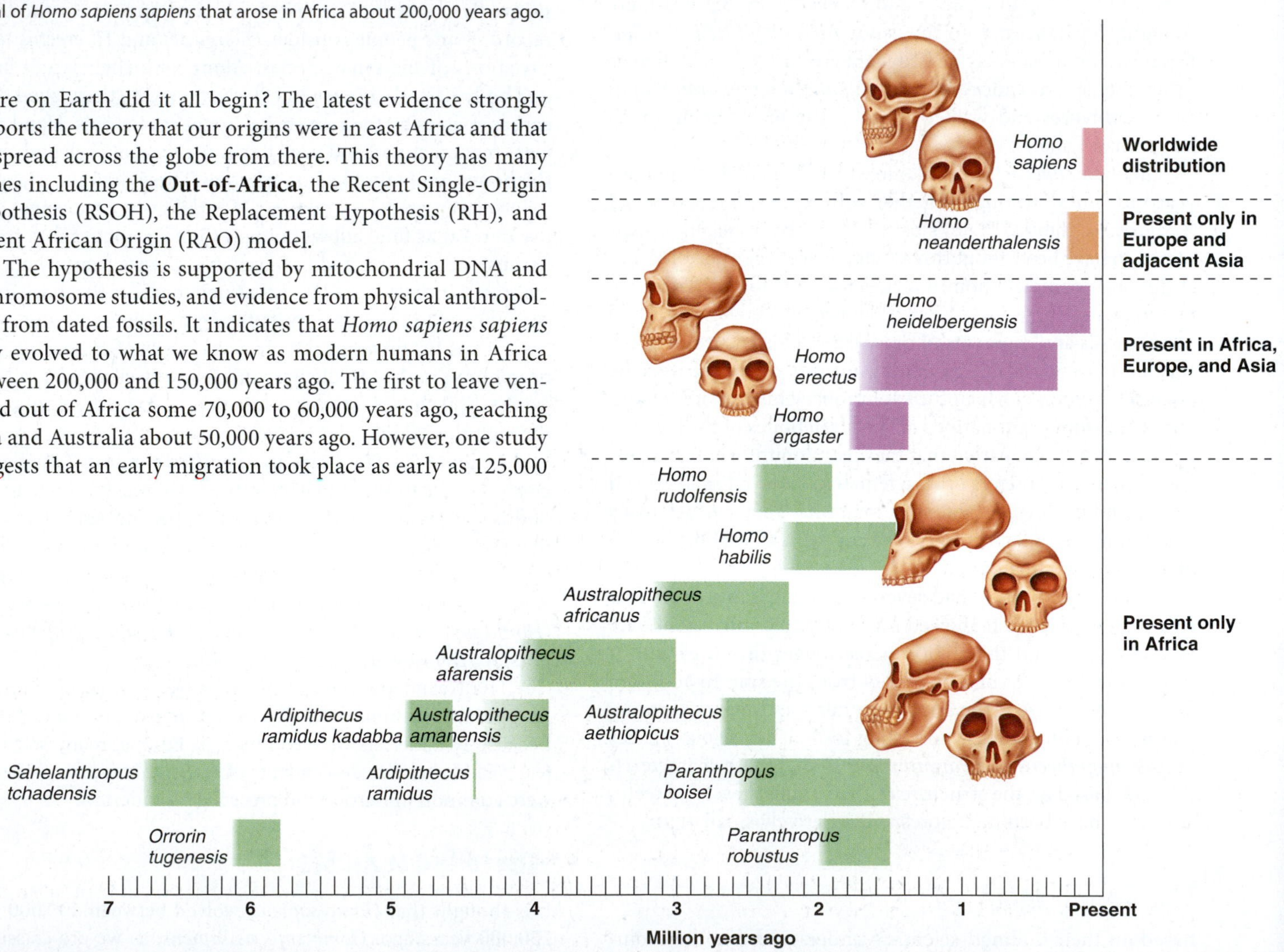

where on Earth did it all begin? The latest evidence strongly supports the theory that our origins were in east Africa and that we spread across the globe from there. This theory has many names including the **Out-of-Africa**, the Recent Single-Origin Hypothesis (RSOH), the Replacement Hypothesis (RH), and Recent African Origin (RAO) model.

The hypothesis is supported by mitochondrial DNA and Y-chromosome studies, and evidence from physical anthropology from dated fossils. It indicates that *Homo sapiens sapiens* only evolved to what we know as modern humans in Africa between 200,000 and 150,000 years ago. The first to leave ventured out of Africa some 70,000 to 60,000 years ago, reaching Asia and Australia about 50,000 years ago. However, one study suggests that an early migration took place as early as 125,000

Homo sapiens	**Worldwide distribution**
Homo neanderthalensis	**Present only in Europe and adjacent Asia**
Homo heidelbergensis *Homo erectus* *Homo ergaster*	**Present in Africa, Europe, and Asia**
Homo rudolfensis *Homo habilis* *Australopithecus africanus* *Australopithecus afarensis* *Ardipithecus ramidus kadabba* *Australopithecus amanensis* *Australopithecus aethiopicus* *Sahelanthropus tchadensis* *Ardipithecus ramidus* *Paranthropus boisei* *Orrorin tugenesis* *Paranthropus robustus*	**Present only in Africa**

7   6   5   4   3   2   1   Present

Million years ago

FIGURE 22.31 This diagram shows the various organisms thought to be relatives of humans. The bars represent approximate times the species are thought to have existed. Notice that (1) all species are extinct today except for modern humans; (2) several different species of organisms coexisted for extensive periods; (3) there are no evolutionary lines connecting these organisms because scientists are still investigating just how they might have been related.

SUMMARY

To facilitate accurate *communication,* biologists assign a *specific name* to each *species* that is cataloged. The various species are cataloged into larger groups on the basis of similar traits. *Taxonomy* is the science of classifying and naming organisms. *Phylogeny* is the science of trying to figure out the evolutionary history of a particular organism. The *taxonomic ranking of organisms* reflects their *evolutionary relationships.* Fossil evidence, comparative anatomy, developmental stages, and biochemical evidence are employed in the sciences of taxonomy and phylogeny.

The first organisms thought to have evolved were single-celled, prokaryotic organisms. These simple organisms gave rise to two different prokaryotic domains, Bacteria and Archaea. The Bacteria are a very diverse group of organisms. Some are autotrophs, while others are heterotrophs. The cyanobacteria were probably the first organisms to carry on photosynthesis. Many kinds of heterotrophic Bacteria are decomposers, while some are parasites, mutualistic, or commensal organisms. The Archaea are distinguished from Bacteria by differences in their cell walls, cell membranes, and the nature of their DNA. Many of the Archaea live in very hostile environments. Eukaryotic organisms were probably derived from the Bacteria and Archaea and constitute the domain Eucarya. There are four kingdoms in the Eucarya: *Protista, Fungi, Plantae,* and *Animalia.* The Protista includes organisms such as protozoa and algae that consist of single cells or small colonies of cells. The Fungi kingdom includes multicellular, heterotrophic organisms with cell walls made of chitin. The Plantae are complex, multicellular, autotrophic organisms that have cell walls made of cellulose and are primarily terrestrial. The Animalia are complex, multicellular, heterotrophic organisms that lack cell walls and typically have the ability to move.

Geologic time is measured through the radioactive decay process, determining the *radiometric age* of rocks in years. A *geomagnetic time scale* has been established from the number and duration of reversals in the magnetic field of Earth's past.

Correlation of the ages of rocks with physical and biological events has led to the development of a *geologic time scale.* The major blocks of time on this calendar are called *eons.* The Hadean eon, 3.8

to 4.5 billion years ago, is the period of time before Earth's surface solidified. The Archean eon, 2.5 to 3.8 billion years ago, is a time when the atmosphere lacked oxygen and only certain kinds of prokaryotic organisms existed. The Proterozoic eon, 2.5 billion to 540 million years ago, saw the development of an oxidizing atmosphere, the first eukaryotic cells, and the first multicellular organisms. The time encompassed by the Hadean, Archaean, and Proterozoic is also referred to as the Precambrian. The Phanerozoic eon, 540 million years ago to the present, is a period of time that has seen the elaboration of the various kinds of living things we see today as well as many groups that have gone extinct. The Phanerozoic eon is divided into smaller units of time known as *eras.* The eras are the (1) *Cenozoic,* the time of recent life; (2) *Mesozoic,* the time of middle life; and (3) *Paleozoic,* the time of ancient life. The eras are divided into smaller blocks of time called *periods,* and the periods are further subdivided into *epochs.*

There are many different kinds of *fossils*—evidence of former living things. They are formed when organisms are covered over and prevented from being destroyed. Most are modified chemically but still give information about past living things. The fossil record is seen to change over geologic time, with certain kinds of fossils being associated with certain periods of time. There have been several *great extinctions* of living things. Some of the extinctions eliminated more than 80 percent of the organisms alive at the time. These extinctions appear to be related to changes in climate that may have been initiated by changes in the location and arrangement of continents, volcanic activity, or meteorite impact.

The early evolution of humans has been difficult to piece together because of the fragmentary evidence. Beginning about 4.4 million years ago, the earliest forms of *Australopithecus* and *Paranthropus* showed upright posture and other humanlike characteristics. The structure of the jaw and teeth indicates that the various kinds of australopiths were herbivores. *Homo habilis* had a larger brain and appears to have been a scavenger. Several other species of the genus *Homo* arose in Africa. These forms appear to have been carnivores. Some of these migrated to Europe and Asia.

KEY TERMS

archaeology (p. **545**)
binomial system of nomenclature
  (p. **527**)
class (p. **528**)
domain (p. **527**)
eon (p. **549**)
epoch (p. **549**)
era (p. **549**)

family (p. **528**)
fossil (p. **543**)
fungus (p. **535**)
genus (p. **527**)
geologic time scale (p. **549**)
index fossils (p. **548**)
kingdom (p. **527**)

obligate intracellular parasite
  (p. **539**)
order (p. **528**)
out-of-Africa hypothesis (p. **558**)
paleontology (p. **545**)
parasite (p. **533**)
period (p. **549**)

phylogeny (p. **528**)
phylum (p. **528**)
prions (p. **540**)
taxonomy (p. **527**)
unconformity (p. **543**)
viroid (p. **540**)
virus (p. **539**)

APPLYING THE CONCEPTS

Answers are located in appendix F.
Part I

1. The three main kinds of living things, Bacteria, Archaea, and Eucarya, have been labeled as
  a. species.
  b. kingdoms.
  c. domains.
  d. families.

2. Which is commonly used as sources of information when developing a phylogeny?
  a. fossil evidence
  b. biochemical information
  c. embryological development
  d. All of the above are correct.

3. Which of the following kingdoms contain members that are autotrophs?
   a. Protista
   b. Plantae
   c. Bacteria
   d. All of the above are correct.
4. Plants and fungi differ in that
   a. plants are autotrophic and fungi are heterotrophic.
   b. plants have cell walls of cellulose and fungi have cell walls of chitin.
   c. plants carry on photosynthesis and fungi do not.
   d. All of the above are correct.
5. Viruses
   a. can be free-living or parasitic.
   b. belong to the domain Archaea.
   c. can function and reproduce only inside a host cell.
   d. All of the above are correct.
6. The Archaea
   a. are ancestors of Bacteria.
   b. are eukaryotic.
   c. lack a nucleus.
   d. None of the above is correct.

Part II

7. Some of the oldest fossils are about how many years old?
   a. 4.55 billion
   b. 3.5 billion
   c. 250 million
   d. 10,000
8. A fossil of a jellyfish, if found, would most likely be
   a. a preserved fossil.
   b. one formed by mineralization.
   c. a carbon film.
   d. a cast or a mold.
9. In any sequence of sedimentary rock layers that has not been subjected to stresses, you would expect to find
   a. essentially horizontal stratified layers.
   b. the oldest layers at the bottom and the youngest at the top.
   c. younger faults, folds, and intrusions in the rock layers.
   d. All of the above are correct.

10. You would expect to find the least number of fossils in rocks from which geologic time?
    a. Cenozoic
    b. Mesozoic
    c. Paleozoic
    d. Precambrian
11. The numerical dates associated with events on the geologic time scale were determined by
    a. relative dating of the rate of sediment deposition.
    b. radiometric dating using radioactive decay.
    c. the temperature of Earth.
    d. the rate that salt is being added to the ocean.
12. In the evolution of humans, the fossil evidence suggests that
    a. the earliest humanlike primates were present in Africa, Europe, and North America.
    b. *Homo sapiens* has been around for about two million years.
    c. there were several different species of human ancestors in Africa.
    d. large numbers of fossils clearly show the way humans evolved.
13. Over these tens of thousands of years, *Homo sapiens sapiens* replaced earlier human populations such as
    a. *Homo neanderthalensis.*
    b. *Homo erectus.*
    c. *Homo ergaster.*
    d. All of the above.
14. This term now refers to humans and their humanlike ancestors, whereas previously the term *hominid* was used.
    a. *Hominin*
    b. *Australopithecus*
    c. *Homo*
    d. *Neanderthalensis*
15. Which is not an era in geologic time?
    a. Paleozoic
    b. Mesozoic
    c. Proterozoic
    d. Cenozoic

QUESTIONS FOR THOUGHT

Part I

1. How do the domains Bacteria and Archaea differ?
2. What is the value of taxonomy?
3. How do viruses reproduce?
4. Why are Latin names used for genus and species?
5. List two characteristics typical for each of the following kingdoms: Protista, Fungi, Plantae, Animalia.
6. Describe four kinds of evidence used to establish phylogenetic relationships among organisms.

Part II

7. Why does the rock record go back only 3.8 billion years? If this missing record were available, what do you think it would show? Explain.
8. What major event marked the end of the Paleozoic and Mesozoic eras according to the fossil record? Describe one theory that proposes to account for this.
9. Describe how the principles of superposition, horizontality, and faunal succession are used in the relative dating of sedimentary rock layers.
10. Describe four different kinds of fossils and how they were formed.
11. Describe four things that fossils can tell you about Earth's history.
12. What are some of the major steps thought to have been involved in the evolution of humans?

FOR FURTHER ANALYSIS

Part I

1. Develop a "phylogeny" for motor vehicles and draw a diagram of your phylogeny. Consider the following questions among those you include in determining your phylogeny:

   How many "kinds" currently exist?

   Are there "fossils" that may have been precursors to motor vehicles?

   What was the ancestor of motor vehicles?

   What was the first motor vehicle?

   What motor vehicles have gone extinct?

   What major evolutionary changes have occurred?

   What "environmental factors" shaped the evolution of motor vehicles?

   Present your phylogeny to your class and have them critique your effort.

Part II

2. Take a long, narrow strip of paper. Tear it in half crossways. Tear one of the remaining pieces in half. Continue until you have made five tears. Measure the length of the piece you have remaining. Use this information and the number of times you "halved" the paper to determine the length of the original strip of paper. This is analogous to the way in which the measurement of radioactive isotopes can be used to determine the age of rocks.

3. Consider the following questions:

   Could our early hominin ancestors have seen living dinosaurs?

   What did the animals in the Cambrian period eat?

   What organisms have dominated most of the history of Earth?

   How might viruses and viroids be related?

INVITATION TO INQUIRY

Understanding Geological Time

It is often difficult to appreciate the extremely long period of time involved in the history of Earth and life. It is also often difficult to appreciate that the kinds of living things on Earth have changed significantly. From table 22.4, obtain the approximate dates for the following significant events and place them on the time line.

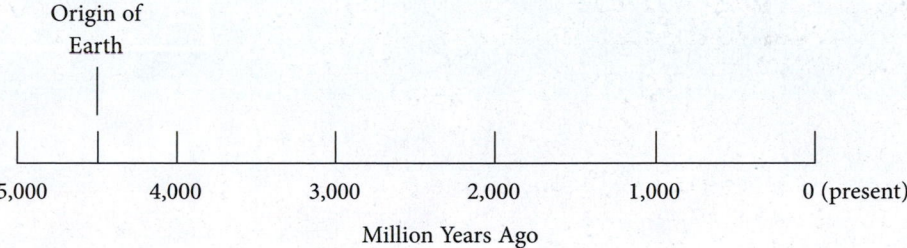

Origin of Earth

5,000   4,000   3,000   2,000   1,000   0 (present)

Million Years Ago

1. First fossil evidence of life
2. First evidence of eukaryotic cells
3. First animals
4. First terrestrial plants
5. First dinosaurs
6. First mammals
7. Dinosaurs go extinct
8. First humans

Now do the following:

1. Circle the period of time in which only prokaryotic organisms were present.
2. Circle the period of time during which dinosaurs were present.
3. Circle the period of time in which humans have been present.
4. Circle the period of time in which only marine organisms were present.

23

Ecology and Environment

These are two White Storks (*Ciconia ciconia*) perched on their chimney top nest near a palm tree. The White Stork, although declining in numbers, has a wide range, breeding in the warmer parts of Europe and wintering in tropical Africa.

CORE **CONCEPT**

Everything is interconnected.

OUTLINE

Populations are limited by environmental factors (p. 590).

Human activities alter natural systems and processes (p. 592).

CONNECTIONS

Physics
▷ Second law of thermodynamics applies to all ecosystems (Ch. 4).
▷ Energy flows through ecosystems (Ch. 3).

Astronomy
▷ Energy comes from the Sun (Ch. 12 and 13).
▷ Seasons occur because Earth orbits the Sun (Ch. 17).

Chemistry
▷ Biochemistry is the chemistry of living things (Ch. 19).

▷ Chemical reactions cause material and energy changes (Ch. 9).

Earth Science
▷ Atmospheric gases affect living things (Ch. 17).
▷ Climate is different at different places on Earth (Ch. 17).
▷ The greenhouse effect warms Earth (Ch. 17).

Life Science
▷ Evolutionary processes lead to adaptation (Ch. 21).
▷ Photosynthesis traps sunlight energy as chemical energy (Ch. 20).
▷ Respiration releases chemical bond energy for organisms (Ch. 20).

OVERVIEW

Today, we recognize that environmental problems are a worldwide concern. Poor agricultural practices in Africa and China result in dust storms that affect the local people and are also carried to the rest of the world. Concern about climate change has led to the convening of international conferences and the drafting of treaties whose goal is to reduce the impact of energy use. Many are concerned about protecting endangered species and the forests, grasslands, and oceans that provide habitats for them. In many ways, these problems are the result of an incredible increase in the size of the human population. As the human population has increased, humans have sought to use land for agriculture and other purposes. Converting land to agricultural use results in its loss for other purposes such as habitat and watershed protection. To understand the nature of environmental problems and steps we can take to solve them, we need to be familiar with some of the central ideas of the science of ecology.

23.1 A DEFINITION OF ENVIRONMENT

Ecology is the branch of biology that studies the relationships between organisms and their environments. This is a very simple definition for a very complex branch of science. Most ecologists define the word **environment** very broadly as anything that affects an organism during its lifetime. Environmental influences can be divided into two categories: **biotic factors** are other living things that affect an organism, and **abiotic factors** are nonliving influences that affect an organism (figure 23.1). If we consider a fish in a stream, we can identify many environmental factors that are important to its life. The temperature of the water is extremely important as an abiotic factor that may be influenced by the presence of trees (biotic factor) along the stream bank that shade the stream and prevent the Sun from heating it. Obviously, the kind and number of food organisms in the stream are important biotic factors as well. The type of material that makes up the stream bottom and the amount of oxygen dissolved in the water are

other important abiotic factors, both of which are related to how rapidly the water is flowing.

Characterizing the environment of an organism is a complex and challenging process; everything seems to be influenced or modified by other factors. A plant is influenced by many different factors during its lifetime: the types and amounts of minerals in the soil, the amount of sunlight hitting the plant, the animals that eat the plant, and the wind, water, and temperature. Each item on this list can be further subdivided into other areas of study. For instance, water is important in the life of plants, so rainfall is studied in plant ecology. But even the study of rainfall is not simple. In some areas of the world, it rains only during one part of the year, while in others, rain is evenly distributed throughout the year. The rainfall could be hard and driving, or it could come as gentle, misty showers of long duration. The water could soak into the soil for later use, or it could run off into streams and be carried away.

The animals in an area are influenced as much by environmental factors as are the plants. If environmental factors do not

FIGURE 23.1 (*A*) The sticks and branches Frigate birds (*Fregata magnificens*) use to build their nest are part of their biotic environment. The tree in which this nest is built is also part of the Frigate bird's biotic environment. (*B*) The irregular shape of the trees is the result of wind and snow, both abiotic factors. Snow driven by these prevailing winds tends to "sandblast" one side of the tree and prevent limb growth.

A Biotic Factor (nesting material and tree)   B Abiotic Factor (wind-driven snow)

favor the growth of plants, there will be little food and few hiding places for animal life. Two types of areas that support only small numbers of living animals are deserts and polar regions. Near the poles, the low temperature and short growing season inhibit growth; therefore, there are relatively few species of animals with relatively small numbers of individuals. Deserts receive little rainfall and therefore have poor plant growth and low concentrations of animals. On the other hand, tropical rainforests have high rates of plant growth and large numbers of animals of many kinds.

23.2 THE ORGANIZATION OF ECOLOGICAL SYSTEMS

Ecologists can study ecological relationships at several different levels. The smallest living unit is the individual **organism,** which is composed of atoms and molecules arranged in a highly organized manner. Each kind of organism has a particular functional role in the community, which is known as its niche. Groups of organisms of the same species are called populations. Interacting groups of populations of different species are called **communities.** An **ecosystem** consists of all the interacting organisms in an area and their interactions with their abiotic surroundings. Figure 23.2 shows how these different levels of organization are related to one another.

All living things require a continuous supply of energy to maintain life. Therefore, many people like to organize living systems by the energy relationships that exist among the different kinds of organisms present. An ecosystem contains several different kinds of organisms. Those that trap sunlight for photosynthesis, resulting in the production of organic material from inorganic material, are called **producers.** Green plants and other photosynthetic organisms such as algae and cyanobacteria are, in effect, converting sunlight energy

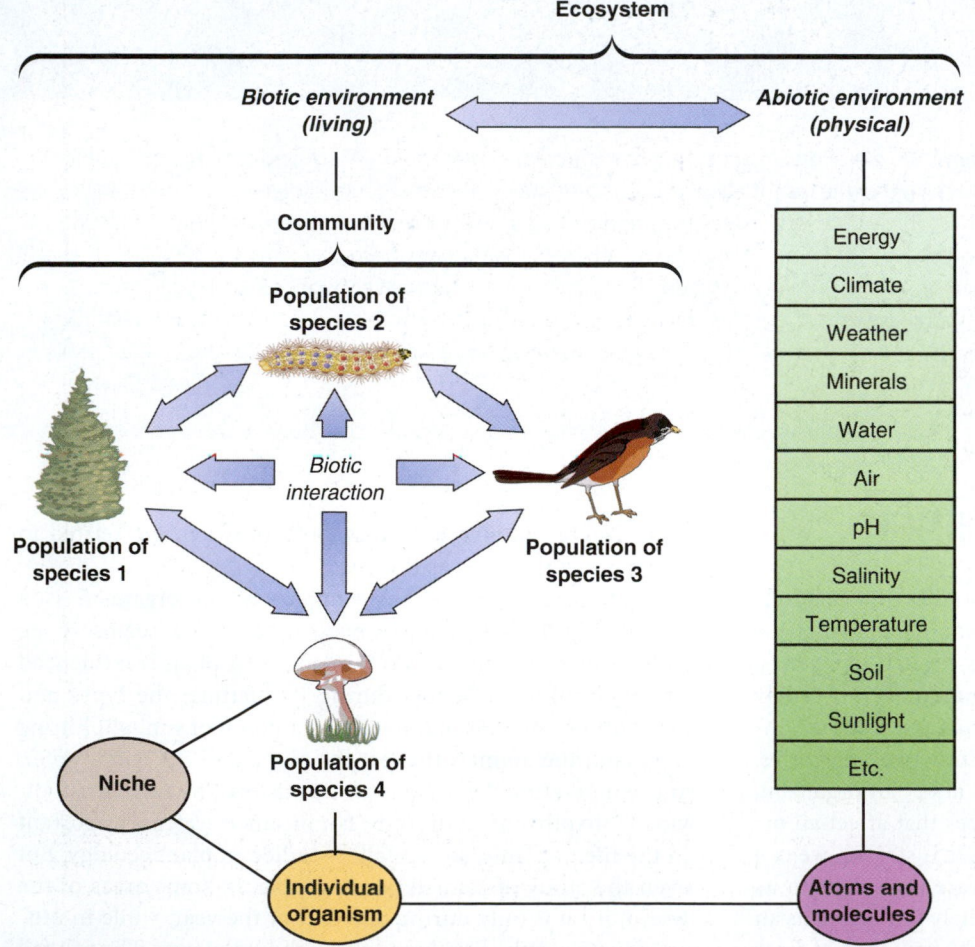

FIGURE 23.2 Ecology is the science that deals with the interactions between organisms and their environment. This study can take place at several different levels, from the broad ecosystem level through community interactions and population studies to the study of the niche of individual organisms. Ecology also involves study of the physical environment and the atoms and molecules that make up both the living and nonliving parts of an ecosystem. The same organism can be viewed in many ways. We can study it as an individual, as a member of a population, or as a participant in a community or ecosystem.

People Behind the Science

Dr. Jane Lubchenco, (December 4, 1947–)

Dr. Jane Lubchenco was confirmed by the U.S. Senate in 2009 as the first woman to serve as the head of the National Oceanic and Atmospheric Administration (NOAA). A Denver native, she attended Colorado College, earned master's and doctoral degrees in marine ecology from the University of Washington and Harvard, and became a professor at Oregon State University in 1977. At Oregon State, Lubchenco spent 30 years building a database about the state's coast. Her studies include the ecology of algae and seaweed, the deciphering of biological interactions along rocky shorelines, and assessments of environmental sustainability.

On the way to her new post, she founded the Aldo Leopold Leadership Program, which trains environmental researchers in communication and policy making. In 1999, she also organized the Communication Partnership for Science and the Sea, which provides information on marine conservation science to the public and policymakers. Dr. Lubchenco is the founding director of Climate Central, a website that went online with what she calls "credible and nonadvocacy" information on climate change. She has held the posts of president of the American Association for the Advancement of Science and president of the Ecological Society of America and is a continuing member of the National Academy of Sciences.

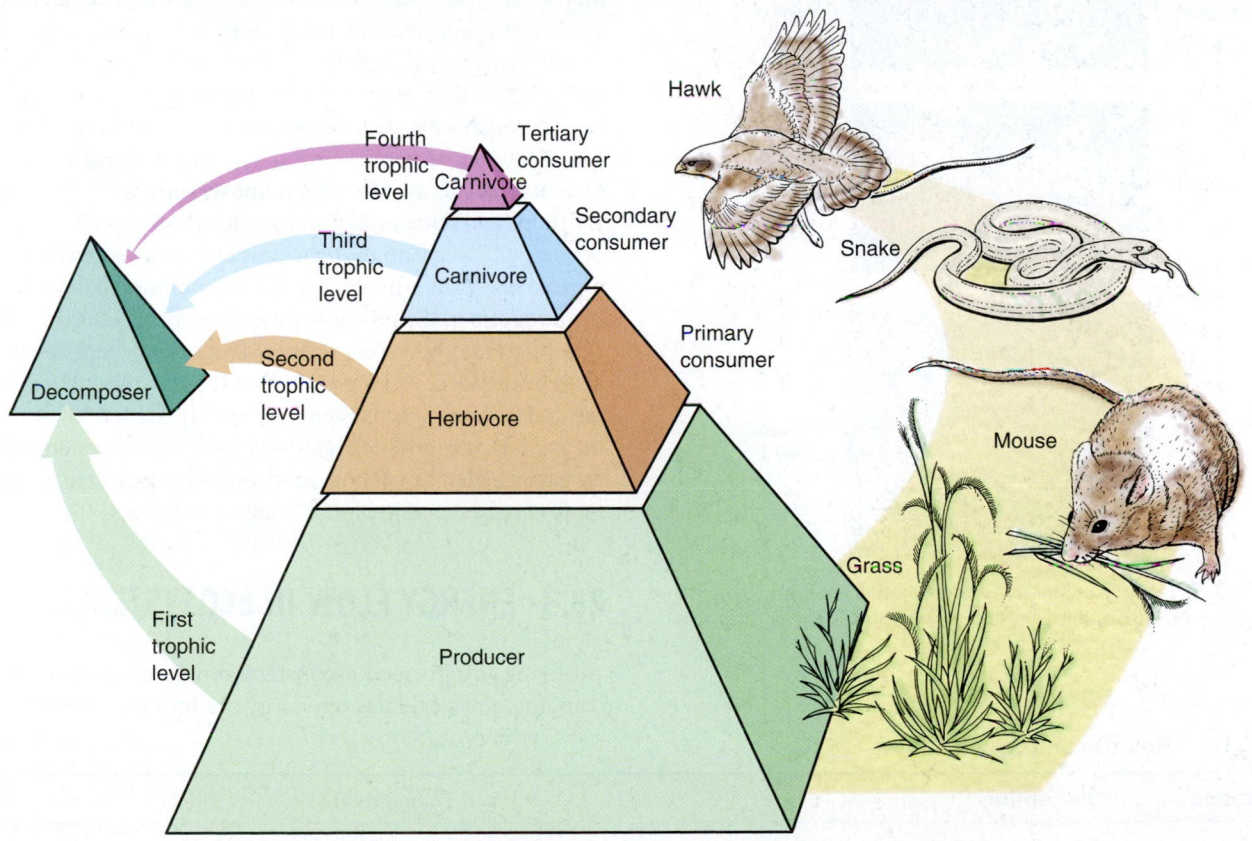

FIGURE 23.3 Organisms within ecosystems can be divided into various trophic levels on the basis of how they obtain energy. Several different sets of terminology are used to identify these different roles. This illustration shows how the different sets of terminology are related to one another. Each colored block in this diagram represents the amount of energy or living material at each trophic level. The amount of energy and the amount of living matter decreases sharply as they are passed to higher trophic levels.

into the energy contained within the chemical bonds of organic compounds. There is a flow of energy from the Sun into the living matter of plants.

The energy that plants trap can be transferred through a number of other organisms in the ecosystem. Since all organisms, other than producers, must obtain energy in the form of organic matter, they are called **consumers.** Consumers cannot capture energy from the Sun as plants do. All animals are consumers. They either eat plants directly or eat other sources of organic matter derived from plants. Each time the energy enters a different organism, it is said to enter a different **trophic level,** which is a step, or stage, in the flow of energy through an ecosystem (figure 23.3).

Quaternary carnivore — Trophic level 6

Tertiary carnivore — Trophic level 5

Secondary carnivore — Trophic level 4

Primary carnivore — Trophic level 3

Herbivore — Trophic level 2

Producer — Trophic level 1

FIGURE 23.4 As one organism feeds on another organism, there is a flow of energy from one trophic level to the next. This illustration shows six trophic levels.

The plants (producers) receive their energy directly from the Sun and are said to occupy the *first trophic level.*

Various kinds of consumers can be divided into several categories, depending on how they fit into the flow of energy through an ecosystem. Animals that feed directly on plants are called **herbivores,** or **primary consumers,** and occupy the *second trophic level.* Animals that eat other animals are called **carnivores,** or **secondary consumers,** and can be subdivided into different trophic levels depending on what animals they eat. Animals that feed on herbivores occupy the *third trophic level* and are known as **primary carnivores.** Animals that feed on the primary carnivores are known as **secondary carnivores** and occupy the *fourth trophic level.*

This sequence of organisms feeding on one another is known as a **food chain.** Often food chains are very complex and can involve many trophic levels. For example, a human may eat a fish that ate a frog that ate a spider that ate an insect that consumed plants for food. Figure 23.4 shows the six different trophic levels in this food chain. Obviously, there can be higher categories, and some organisms don't fit neatly into this theoretical scheme. Some animals are carnivores at some times and herbivores at others; they are called **omnivores.** They are classified into different trophic levels depending on what they happen to be eating at the moment.

If an organism dies, the energy contained within the organic compounds of its body is finally released to the environment as heat by organisms that decompose the dead body into carbon dioxide, water, ammonia, and other simple inorganic molecules. Organisms of decay, called **decomposers,** are things such as bacteria, fungi, and other organisms that use dead organisms as sources of energy. This group of organisms efficiently converts nonliving organic matter into simple inorganic molecules that can be reused by producers in the process of trapping energy. Decomposers are thus very important components of ecosystems that cause materials to be recycled. As long as the Sun supplies the energy, elements are cycled through ecosystems repeatedly. Table 23.1 summarizes the various categories of organisms within an ecosystem. Now that we have a better idea of how ecosystems are organized, we can look more closely at energy flow through ecosystems.

23.3 ENERGY FLOW IN ECOSYSTEMS

All living things need a constant source of energy. Two fundamental physical laws of energy are important when looking

TABLE 23.1   **Roles in an Ecosystem**

Classification	Description	Examples
Producers	Organisms that convert simple inorganic compounds into complex organic compounds by photosynthesis	Trees, flowers, grasses, ferns, mosses, algae, cyanobacteria
Consumers	Organisms that rely on other organisms as food; animals that eat plants or other animals	
Herbivore	Eats plants directly	Deer, goose, cricket, vegetarian human, many snails
Carnivore	Eats meat	Wolf, pike, dragonfly
Omnivore	Eats plants and meat	Rat, most humans
Scavenger	Eats food left by others	Coyote, skunk, vulture, crayfish
Parasite	Lives in or on another organism, using it for food	Tick, tapeworm, many insects
Decomposers	Organisms that return organic compounds to inorganic compounds; important components in recycling elements	Bacteria, fungi

at ecological systems from an energy point of view. The *first law of thermodynamics*—also known as the *law of conservation of energy*—states that energy is neither created nor destroyed. This means that we should be able to describe the amounts of energy in each trophic level and follow energy as it flows through successive trophic levels. The *second law of thermodynamics* states that natural processes proceed toward a state of greater disorder. This disordered condition is known as entropy. Another way to state the second law of thermodynamics is to say that when energy is converted from one form to another, useful energy is lost, which is to say there is an increase in entropy. This means that as energy passes from one trophic level to the next, there will be a reduction in the amount of available energy in living things and an increase in the amount of heat in their surroundings.

The total energy of an ecosystem can be measured in several ways. An indirect way to get an idea of the amount of energy at a trophic level is to determine its **biomass,** which is the weight of the organisms at that trophic

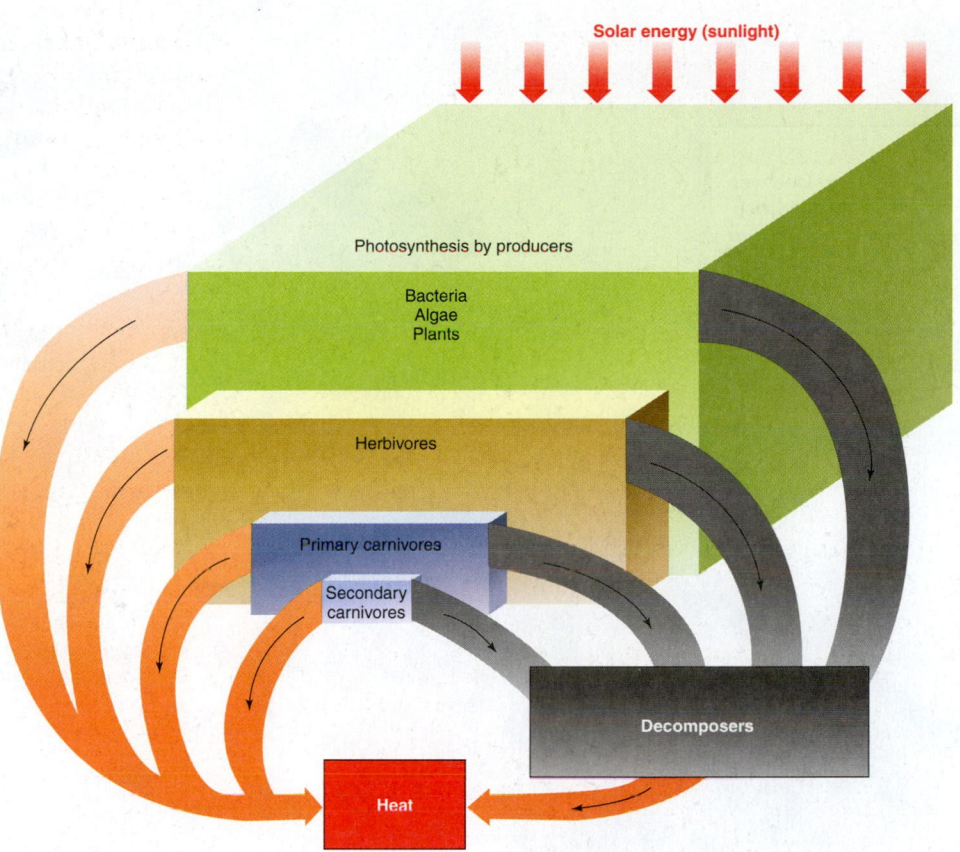

FIGURE 23.5   As energy flows from one trophic level to the next, approximately 90 percent of it is lost. This means that the amount of energy at the producer level is ten times larger than the amount of energy at the herbivore level.

level. Since organic matter will burn, it is possible to convert the energy in biomass into heat and light. For example, the total producer trophic level can be harvested and burned. The number of kilocalories of heat energy produced by burning is equivalent to the energy content of the organic material of the plants. This is what you do when you burn wood at a campfire. Another way of determining the energy present is to measure the rate of photosynthesis and respiration, and calculate the amount of energy being trapped in the living material of the plants. However, this is technically difficult. It requires careful measurements of tiny changes in the amount of carbon dioxide and oxygen in the atmosphere as a result of photosynthesis and respiration or the use of techniques to follow traceable atoms, such as radioisotopes, from one organism to another.

Since only the plants, algae, and cyanobacteria in the producer trophic level are capable of capturing energy from the Sun, all other organisms are directly or indirectly dependent

on the producer trophic level. The second trophic level consists of herbivores that eat the producers. This trophic level has significantly less energy and biomass in it for several reasons. *In general, there is about a 90 percent loss of available energy as we proceed from one trophic level to the next higher level.* Actual measurements will vary from one ecosystem to another, but 90 percent is a good rule of thumb. This loss in energy content at the second and subsequent trophic levels is predicted by the second law of thermodynamics. When the energy in producers is converted to the energy of herbivores, much of the energy is lost as heat to the surroundings.

In addition to this loss of available energy, there is an additional loss involved in the capture and processing of food material by herbivores. Although herbivores don't need to chase their food, they do need to travel to where food is available, then gather, chew, digest, and metabolize it. All these processes require energy.

Just as the herbivore trophic level experiences a 90 percent loss in energy content, the higher trophic levels of primary carnivores, secondary carnivores, and tertiary carnivores also experience a reduction in the energy available to them. This energy loss is reflected in the number of organisms at each trophic level. A field may have millions of producers (grass and other plants), thousands of herbivores that eat grass (grasshoppers, prairie dogs, etc.), and a few carnivores (birds, weasels, etc.) that eat the herbivores. Finally, decomposers act on dead organisms with this same kind of available energy loss. Figure 23.5 shows a diagram in which the energy content decreases by 90 percent as we pass from one trophic level to the next.

CONCEPTS APPLIED

Ecological Categories

Place each organism from the following list into one of the four categories below: honeybee, fungus, moss, squirrel, spider, shark, corn.

Producer   **Herbivore**   **Carnivore**   **Decomposer**

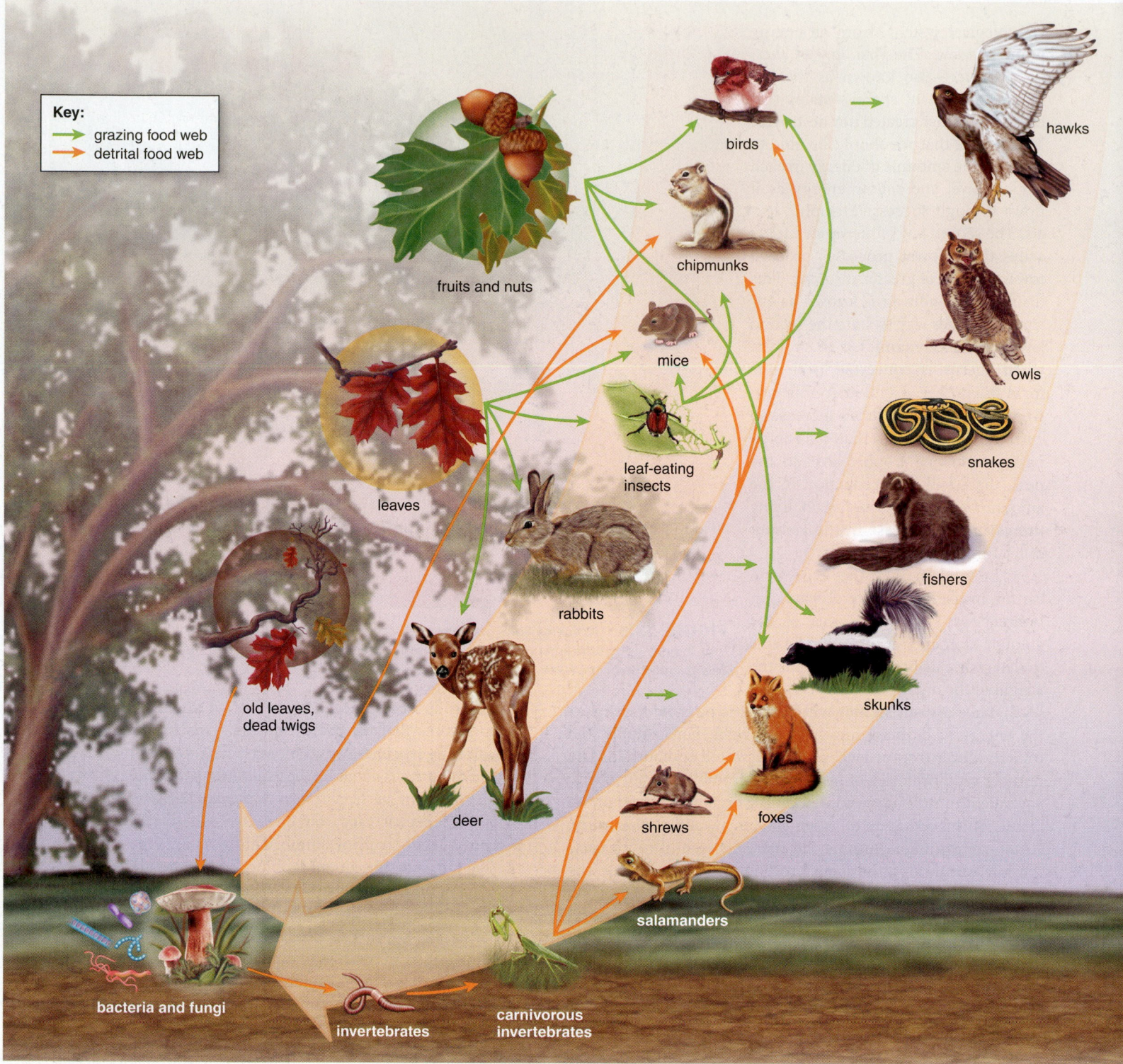

fruits and nuts

birds

hawks

chipmunks

owls

mice

leaves

leaf-eating insects

snakes

rabbits

fishers

old leaves, dead twigs

skunks

deer

foxes

shrews

salamanders

bacteria and fungi

invertebrates

carnivorous invertebrates

FIGURE 23.6 As organisms feed on one another they establish a web of relationships known as a food web. This illustration shows the interactions between grazing and detrital food webs. In grazing food webs, photosynthesis by plants provides the energy for grazing animals that eat plants, which in turn provide energy for carnivores. Since all organisms die they ultimately become part of a detrital food web in which dead organic matter and waste products supply the energy for a series of bacteria, fungi, and animals.

23.4 COMMUNITY INTERACTIONS

In the section on energy flow in ecosystems, we looked at ecological relationships from the point of view of ecosystems and the way energy flows through them. But we can also study relationships at the community level and focus on the kinds of interactions that take place among organisms.

As you know, one of the ways that organisms interact is by feeding on one another. A community includes many different food chains, and many organisms may be involved in several of the food chains at the same time, so the food chains become interwoven into a *food web* (figure 23.6). In general, communities are relatively stable in terms of the kinds of organisms present and how they interact. Some communities, such as tropical rainforests, have large numbers of different kinds of organisms present. Such communities have high **biodiversity.** Others, such as tundra communities, have low biodiversity. People also talk about a loss of biodiversity when a specific kind of organism is eliminated from a region. For example, when wolves were eliminated from much of North America, there was a loss of biodiversity. Although communities are relatively stable, we need to recognize that they are also dynamic collections of organisms. Although the kinds of organisms present may not change, the numbers of each kind may change significantly throughout the year or over several years. For example, a drought will reduce the survival of plants, and an epidemic of disease will reduce the survival of many birds.

If numbers of a particular kind of organism in a community increase or decrease significantly, there will be a ripple effect through the community. For example, the populations of many kinds of small mammals fluctuate from year to year. This results in changes in the numbers of their predators, and often the predators switch to other prey species, which impacts other parts of the community.

Not all organisms within a community have the same level of importance. Obviously, plants are key organisms that supply the energy for all other organisms. However, even certain animal species can have a very high impact on the nature of a community. For example, sea otters feed on sea urchins that feed on a seaweed called kelp. When sea otter populations were reduced significantly, sea urchin populations increased, they ate much of the kelp, and other species of animals had fewer places to hide. Thus, the sea otters played a key role in shaping the nature of the community. Such organisms are often referred to as **keystone species.**

23.5 TYPES OF TERRESTRIAL COMMUNITIES

Although each community of organisms is unique, similar communities can be combined into broad categories based on specific characteristics of the physical environment and the kinds of organisms that live in the area. The kind of terrestrial community that develops in an area is determined primarily by climatic factors of precipitation patterns and temperature ranges. These large regional communities are known as **biomes.** The map in figure 23.7 shows the distribution of the major terrestrial biomes of the world.

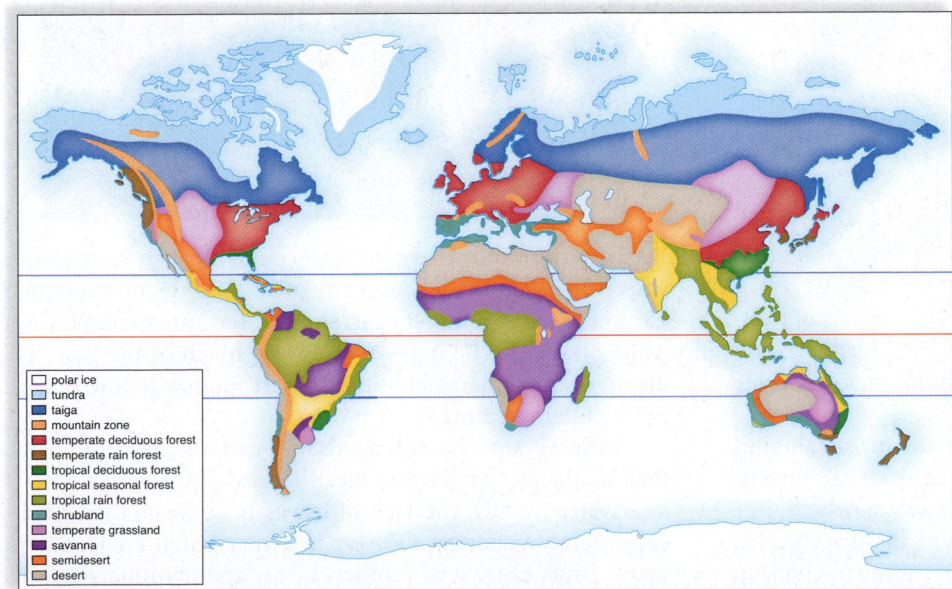

- polar ice
- tundra
- taiga
- mountain zone
- temperate deciduous forest
- temperate rain forest
- tropical deciduous forest
- tropical seasonal forest
- tropical rain forest
- shrubland
- temperate grassland
- savanna
- semidesert
- desert

FIGURE 23.7 Major climatic differences determine the kind of vegetation that can live in a region of the world. Associated with specialized groups of plants are particular kinds of animals. These regional ecosystems are called biomes.

Temperate Deciduous Forest

The *temperate deciduous forest* biome is found in parts of the world (primarily eastern North America, parts of Europe, Japan, Korea, and parts of China) that have 75 to 100 cm (30 to 40 in) of rainfall and cold weather for a significant part of the year. Precipitation is spread throughout the year. The predominant plants are large trees that lose their leaves more or less completely during the fall of the year and are therefore called *deciduous*. Aspen, birch, cottonwood, oak, hickory, beech, and maple are typical trees found in this geographic region. Typical animals of this biome are skunks, porcupines, deer, frogs, opossums, owls, mosquitoes, and beetles.

In much of this region, the natural vegetation has been removed to allow for agriculture, so the original character of the biome is gone except where farming is not practical or the original forest has been preserved.

Temperate Grassland or Prairie

Temperate grasslands are common in temperate regions of western North America and parts of Eurasia, Africa, Australia, and South America. The dominant vegetation in this region is made up of various species of grasses.

The annual rainfall is 25 to 75 cm (10 to 30 in), an amount that is not adequate to support the growth of dense forests. Trees in grasslands grow primarily along streams where they can obtain sufficient water. Animals found in this area include the prairie dog, pronghorn antelope, prairie chicken, grasshopper, rattlesnake, and meadowlark. Most of the original

grasslands, like the temperate deciduous forest, have been converted to agricultural uses.

Savanna

Savannas are found in tropical regions of central Africa, northern Australia, and parts of South America that have pronounced rainy and dry seasons. Although these regions may receive 100 cm (40 in) of rain per year, there is an extended dry season of three months or more.

Therefore, savannas consist of grasses with scattered trees. Grazing animals such as antelope, zebra, and buffalo are common, as are termites and many other insects, which serve as food for many birds and reptiles.

Desert

Very dry areas known as *deserts* are found throughout the world wherever rainfall is low and irregular. They receive less than 25 cm (10 in) of rain per year. Some deserts are extremely hot, while others can be quite cold during much of the year. The distinguishing characteristic of desert biomes is low rainfall, not high temperature.

Deserts are characterized by scattered, thorny plants that lack leaves or have reduced leaves. Since leaves tend to lose water rapidly, the lack of leaves is an adaptation to dry conditions. Although a desert is a very harsh environment, many kinds of insects, reptiles, birds, and mammals can live in this biome. The animals usually avoid the hottest part of the day by staying in burrows or other shaded, cooler areas. Staying underground or in the shade also allows the animals to conserve water.

Boreal Coniferous Forest

The *boreal coniferous forest* (also known as the *northern coniferous forest,* or *taiga*) is found through parts of southern Canada, extending southward along the mountains of the United States, and in much of northern Europe and Asia. This region has short, cool summers and long, cold winters with abundant snowfall. Precipitation ranges from 25 to 100 cm (10 to 40 in) per year. However, even if precipitation is low, the extensive snow-melt in spring and low evaporation rate result in humid summers. The dominant vegetation consists of evergreen trees that are especially adapted to withstand long, cold winters with abundant snowfall. Spruces and firs are common tree species that are intermingled with many other kinds of vegetation. Small lakes and bogs are common. Characteristic animals in this biome include mice, bears, wolves, squirrels, moose, midges, and flies.

Mediterranean Shrublands (Chaparral)

Mediterranean shrublands are located near an ocean and are dominated by shrubby plants. As the name implies, this biome is typical of the Mediterranean coast and is also found in coastal southern California, the southern tip of Africa, a portion of the west coast of Chile, and southern Australia. The climate has wet, cool winters and hot, dry summers. Rainfall is

40 to 100 cm (15 to 40 in) per year. The vegetation is dominated by woody shrubs that are adapted to withstand the hot, dry summer. Often the plants are dormant during the summer. Fire is a common feature of this biome, and the shrubs are adapted to withstand occasional fires. The kinds of animals vary widely in the different regions of the world with this biome. Many kinds of insects, reptiles, birds, and mammals are found in these areas. In the chaparral of California, rattlesnakes, spiders, coyotes, lizards, and rodents are typical inhabitants.

Temperate Rainforest

The coastal areas of northern California, Oregon, Washington, British Columbia, and southern Alaska contain an unusual set of environmental conditions that supports a *temperate rainforest.* There are also small areas of temperate rainforest in southern Chile and on the west coast of New Zealand. The temperate rainforests occur along the coast where the prevailing winds from the west bring moisture-laden air. As the air meets the coastal mountains and is forced to rise, it cools and the moisture falls as rain or snow. Most of these areas receive 200 cm (80 in) or more of precipitation per year. This abundance of water, along with fertile soil and mild temperatures, results in a lush growth of plants.

Sitka spruce, Douglas fir, and western hemlock are typical evergreen coniferous trees in the temperate rainforest of North America. Undisturbed (old growth) forests of this region have trees as old as eight hundred years that are nearly 100 m (about 300 ft) tall. Deciduous trees of various kinds (red alder, big leaf maple, black cottonwood) also exist in open areas where they can get enough light. All trees are covered with mosses, ferns, and other plants that grow on the surface of the trees. The kinds of animals present are similar to those of a temperate deciduous forest.

Tundra

North of the coniferous forest biome in North America and northern Europe and Asia is an area known as the *tundra*. It is characterized by extremely long, severe winters and short, cool summers. The deeper layers of the soil remain permanently frozen, forming a layer called the *permafrost*. Although precipitation is typically less than 25 cm (10 in) per year, the melting of snow during the brief summer (about one hundred days) results in moist conditions. Summer temperatures may be near freezing at night and rise to about 10°C (50°F) during the day. Because the deeper layers of the soil are frozen, when the surface thaws, the water forms puddles.

Under these conditions of low temperature and short growing season, very few kinds of animals and plants can survive. No trees can live in this region. Typical plants and animals of the area are dwarf willow and some other shrubs, reindeer moss (actually a lichen), some flowering plants, caribou, wolves, musk oxen, fox, snowy owls, mice, and many kinds of insects. Many kinds of birds are summer residents only.

Tropical Rainforest

Tropical rainforests are found primarily near the equator in Central and South America, Africa, parts of southern Asia, and some Pacific Islands. The temperature is quite warm and constant, and

rain falls nearly every day. Most areas receive over 200 cm (80 in) of rain per year and some exceed 500 cm (200 in).

These areas have high biodiversity. Balsa (a very light wood), teak (used in furniture), and ferns the size of trees are examples of plants from the tropical rainforest. Typically, every plant has other plants growing on it. Tree trunks are likely to be covered with orchids, many kinds of vines, and mosses. Tree frogs, bats, lizards, birds, monkeys, and an almost infinite variety of insects inhabit the rainforest. These forests are very dense and little sunlight reaches the forest floor. When the forest is opened up (by a hurricane or the death of a large tree) and sunlight reaches the forest floor, the opened area is rapidly overgrown with vegetation.

Tropical Dry Forest

Tropical dry forests are found in parts of Central and South America, Australia, Africa, and Asia (particularly India and Myanmar). A major characteristic of this biome is seasonal rainfall. Many of the tropical dry forests have a monsoon climate in which several months of heavy rainfall are followed by extensive periods without rain. Some tropical dry forests have as much as eight months without rain. The rainfall may be as low as 50 cm (20 in) or as high as 200 cm (80 in). Since the rainfall is highly seasonal, many of the plants have special adaptations for enduring drought. In many of the regions that have extensive dry periods, many of the trees drop their leaves during the dry period. Many of the species of animals found here are also found in more moist tropical forests of the region. However, there are fewer kinds in dry forests than in rainforests.

23.6 TYPES OF AQUATIC COMMUNITIES

Terrestrial biomes are determined by the amount and kind of precipitation and by temperature ranges. Other factors, such as soil type and wind, also play a part. Aquatic ecosystems also are shaped by key environmental factors. Several important factors are the ability of the Sun's rays to penetrate the water, the depth of the water, the nature of the bottom substrate, the water temperature, and the amount of dissolved salts. Aquatic systems are often divided into **marine communities,** which have a high salt content, and **freshwater communities,** which have low salt concentrations.

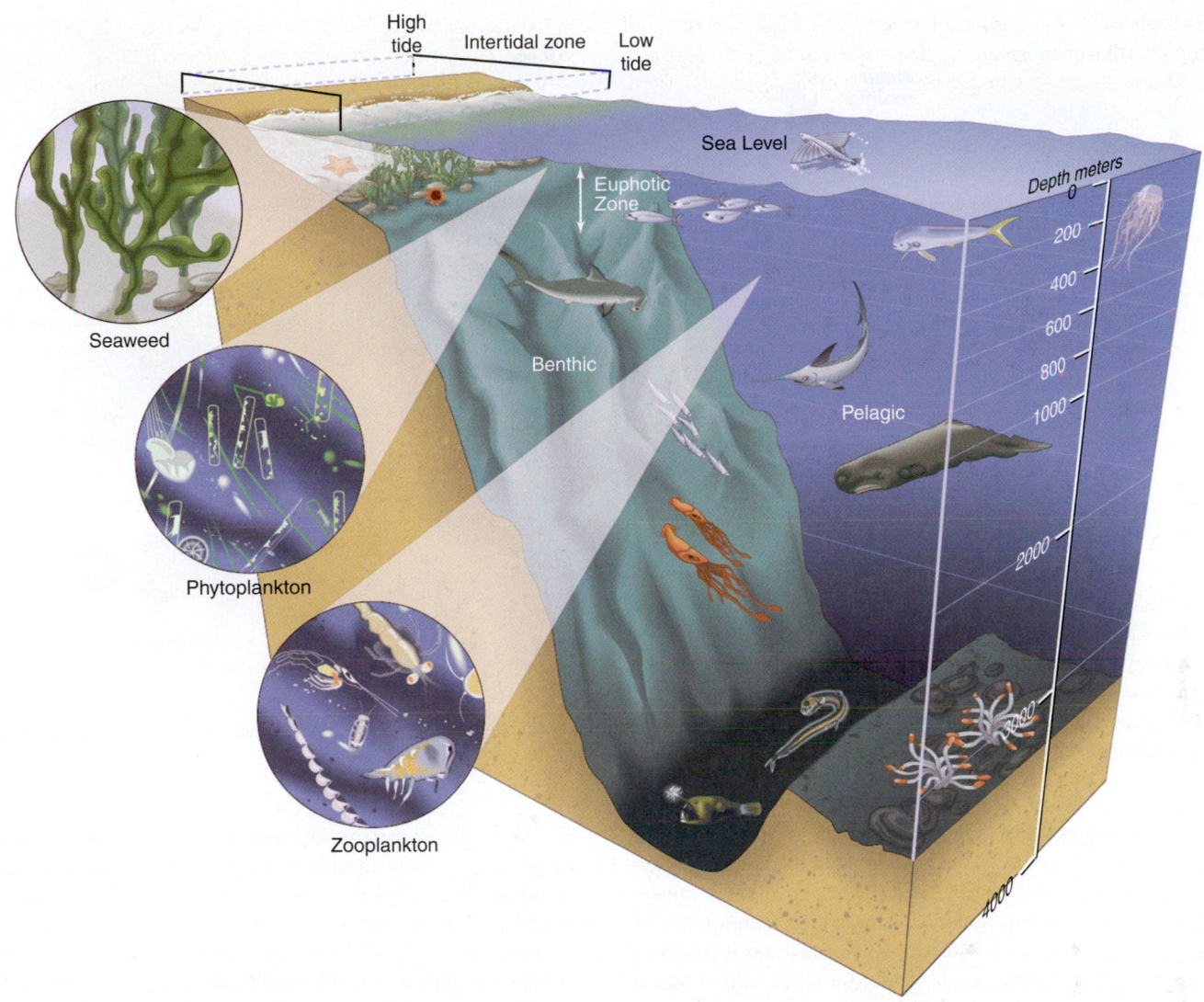

High tide · Intertidal zone · Low tide

Sea Level

Euphotic Zone

Benthic

Pelagic

Seaweed

Phytoplankton

Zooplankton

Depth meters

0
200
400
600
800
1000
2000
4000

FIGURE 23.8 All of the photosynthetic activity of the ocean occurs in shallow water called the euphotic zone, either by attached algae near the shore or by minute phytoplankton in the upper levels of the open ocean. Consumers are either free-swimming pelagic organisms or benthic organisms that live on the bottom. Small animals that feed on phytoplankton are known as zooplankton.

Marine Communities

In the open ocean, many kinds of organisms float or swim actively and are called **pelagic.** Among the pelagic organisms are two types: larger organisms that actively swim and smaller, weakly swimming organisms. The term **plankton** is used to describe aquatic organisms that are so small and weakly swimming that they are simply carried by currents. The planktonic organisms that carry on photosynthesis are called *phytoplankton.* In the open ocean, a majority of these organisms are microscopic floating algae and cyanobacteria. The upper layer of the ocean, where the Sun's rays penetrate, is known as the *euphotic zone.* It is in this euphotic zone where phytoplankton are most common. Small, weakly swimming animals of many kinds, known as *zooplankton,* feed on the phytoplankton. The zooplankton are in turn eaten by larger animals such as fish and larger shrimp, which are eaten by larger fish such as salmon, tuna, sharks, and mackerel (figure 23.8).

Those marine organisms that live on the bottom are called **benthic.** Many seaweeds, some fish, clams, oysters, various crustaceans, sponges, sea anemones, and many other kinds of organisms live on the bottom. In shallow water, sunlight can penetrate to the bottom, and a variety of attached seaweeds trap this energy as they carry on photosynthesis. Many of the benthic animals graze on the seaweeds and are in turn eaten by other larger animals. Some benthic animals, such as clams, filter water to obtain plankton for food.

As with terrestrial communities, environmental conditions influence the kinds of marine communities that exist. Coral reefs are found in warm, tropical, shallow seas, and "forests" of large seaweeds are found along cool, rocky shores. Sandy shores with lots of wave action typically have few plants, while shallow, protected, muddy areas may support mangrove swamps or salt marshes.

Freshwater Communities

Freshwater ecosystems differ from marine ecosystems in several ways. In comparison to marine communities, the amount of salt

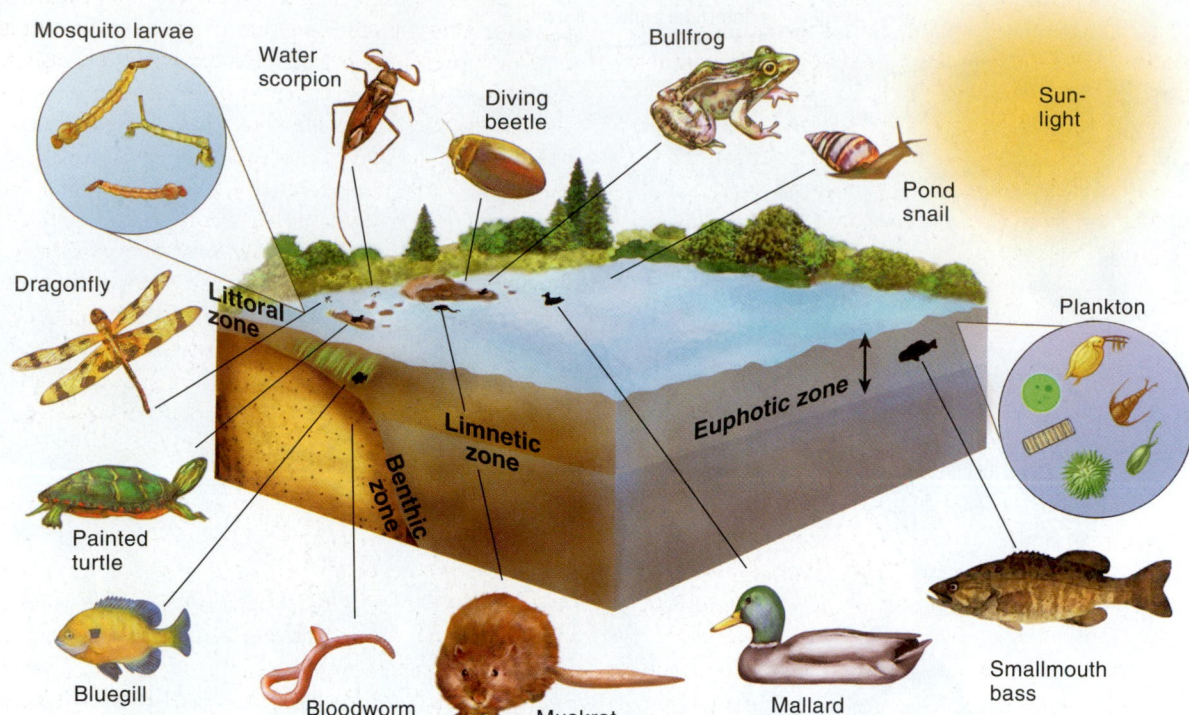

Mosquito larvae

Water scorpion

Diving beetle

Bullfrog

Sun-light

Pond snail

Dragonfly

Littoral zone

Plankton

Limnetic zone

Euphotic zone

Benthic zone

Painted turtle

Bluegill

Bloodworm

Muskrat

Mallard

Smallmouth bass

FIGURE 23.9 Lakes are similar in structure to oceans except that the species are different because most marine organisms cannot live in freshwater. Insects are common organisms in freshwater lakes, as are many kinds of fish, zooplankton, and phytoplankton.

present is much less, the temperature of the water can change greatly, the water is in the process of moving to the ocean, oxygen can often be in short supply, and the organisms that inhabit freshwater systems are different. Freshwater ecosystems can be divided into two categories: those in which the water is relatively stationary, such as lakes, ponds, and reservoirs, and those in which the water is running downhill, such as streams and rivers.

Large lakes have many of the same characteristics as the ocean. If the lake is deep, there is a euphotic zone at the top, with many kinds of phytoplankton, and zooplankton that feed on the phytoplankton. Small fish feed on the zooplankton, which are in turn eaten by larger fish. The species of organisms found in freshwater lakes are different from those found in the ocean, but the roles played are similar, so the same terminology is used.

Along the shore and in the shallower parts of lakes, many kinds of flowering plants are rooted in the bottom. Some are completely submerged, while others have leaves that float on the surface or protrude above the water. Cattails, bulrushes, arrowhead plants, and water lilies are examples. In addition, many kinds of freshwater algae are present.

Associated with the plants and algae are a large number of different kinds of animals. Fish, crayfish, clams, and many kinds of aquatic insects are common inhabitants of this mixture of plants and algae. This region, with rooted vegetation, is known as the *littoral zone,* and the portion of the lake that does not have rooted vegetation is called the *limnetic zone* (figure 23.9).

Streams and rivers present a different set of conditions. Since the water is moving, planktonic organisms are less important than are attached organisms. Most algae grow attached to rocks and other objects on the bottom. Since the water is

shallow, light can penetrate easily to the bottom (except for large or extremely muddy rivers). Even so, it is difficult for photosynthetic organisms to accumulate the nutrients necessary for growth, and most streams are not very productive. As a matter of fact, the major input of nutrients is from organic matter that falls into the stream from terrestrial sources. These are primarily the leaves from trees and other vegetation, as well as the bodies of living and dead insects.

Within the stream is a community of organisms specifically adapted to use the debris as a source of food. Bacteria and fungi colonize the organic matter, and many kinds of insects shred and eat this organic matter as well as the fungi and bacteria living on it. The feces (intestinal wastes) of these insects and the tiny particles produced during the eating process become food for other insects that build nets to capture the tiny bits of organic matter that drift their way. These herbivorous insects are in turn eaten by carnivorous insects and fish.

Estuaries

An **estuary** is a special category of aquatic ecosystem that consists of shallow, partially enclosed areas where freshwater enters the ocean. The saltiness of the water in the estuary changes with tides and the flow of water from rivers. The organisms that live here are specially adapted to this set of physical conditions, and the number of species is less than in the ocean or in freshwater. Estuaries are especially productive ecosystems because of the large quantity of nutrients introduced into the basin from the rivers that run into them. This productivity is further enhanced by the fact that the shallow water allows light to penetrate to most of the water in the basin. Phytoplankton and attached algae and plants are able to use the sunlight and the nutrients for rapid growth. This photosynthetic activity supports many kinds of organisms in the estuary. Estuaries are especially important as nursery sites for fish and crustaceans such as flounder and shrimp. The adults enter these productive, sheltered areas to reproduce and then return to the ocean. The young spend their early life in the estuary and eventually leave as they get larger and are more able to survive in the ocean.

23.7 INDIVIDUAL SPECIES REQUIREMENTS: HABITAT AND NICHE

People approach the study of organism interactions in two major ways. Many people look at interrelationships from the broad ecosystem point of view, while others focus on individual organisms and the specific things that affect them in their daily lives. The first approach involves the study of all the organisms that interact with one another—the community—and usually looks at general relationships among them. For example, organisms are lumped into categories such as producers, consumers, and decomposers because they perform different functions in a community. These ideas were described earlier in this chapter. The second way of looking at interrelationships is to study in detail the ecological relationships of a certain species of organism. Two important concepts related to the ecological study of individual species are the concepts of habitat and niche.

Habitat

Each organism has particular requirements for life and lives where the environment provides what it needs. The environmental requirements of a whale include large expanses of ocean but with seasonally important feeding areas and protected locations used for giving birth. The kind of place, or part of an ecosystem, occupied by an organism, is known as its **habitat.** Habitats are usually described in terms of conspicuous or particularly significant features in the area where the organism lives. For example, the habitat of a prairie dog is usually described as a grassland, while the habitat of a tuna is described as the open ocean. The key thing to keep in mind when you think of habitat is that it is the *place* a particular kind of organism lives. When describing the habitats of organisms, we sometimes use the terminology of the major biomes of the world, such as desert, grassland, or savanna. It is also possible to describe the habitat of the bacterium *Escherichia coli* as the gut of humans and other mammals or the habitat of a fungus as a rotting log.

Niche

Each species has particular requirements for life and places specific demands on the habitat in which it lives. The specific functional role of an organism is its **niche.** Its niche is the way it goes about living its life. Just as the word *place* is the key to understanding the concept of habitat, the word *function* is the key to understanding the concept of a niche. Understanding the niche of an organism involves a detailed understanding of the impacts an organism has on its biotic and abiotic surroundings as well as all of the factors that affect the organism. For example, the niche of an earthworm includes abiotic items such as soil particle size, soil texture, and the moisture, pH, and temperature of the soil. The earthworm's niche also includes biotic impacts such as serving as food for birds, moles, and shrews; as bait for anglers; or as a consumer of dead plant organic matter (figure 23.10). In addition, an earthworm serves as a host for a variety of parasites, transports minerals and nutrients from deeper soil layers to the surface, incorporates organic matter into the soil, and creates burrows that allow air and water to

A Closer Look

The Importance of Habitat Size

Many people interested in songbird populations have documented a significant decrease in the numbers of certain songbird species. Species particularly affected are those that migrate between North America and South America and require relatively large areas of undisturbed forest in both their northern and southern homes. Many of these species are being hurt by human activities that fragment large patches of forest into many smaller patches, creating more edges between different habitat types.

Bird and other animal species that thrive in edge habitats replace the songbirds, which require large patches of undisturbed forest. Cowbirds that normally live in open areas are a particular problem. Cowbirds do not build nests but lay their eggs in the nests of other birds after removing the eggs of the host species. When forests are broken into small patches, cowbirds reach a larger percentage of forest-nesting birds because the forest birds must nest closer to the edge.

The species most severely affected are those that have both their northern and southern habitats disturbed. A study of migrating sharp-shinned hawks indicated that their numbers have also been greatly reduced in recent years. It is thought that since sharp-shinned hawks use small songbirds as their primary source of food, the reduction in hawks is directly related to the reduction in migratory songbirds.

A Closer Look

Alien Invasion

Yard and garden centers often sell plant species that are not native to the area in which you live. Furthermore, homeowners often want unusual plants that are particularly colorful or have other striking characteristics. Some of these exotic plants are invasive. They have characteristics such as fruits or seeds that are easily spread from place to place. When this occurs, the exotic plant may become a pest because it competes with local native plants and replaces them, causing local extinctions of native species.

In the United States, there are many examples of exotic invasive species. Glossy buckthorn and autumn olive have replaced understory species in forests of the Northeast. Tamarisk (salt cedar) has become a dominant species along rivers in the Southwest. Brazilian pepper and *Melaleuca* have become major problems in south Florida. Kudzu (a vine) and water hyacinth have become significant problems in areas of the South. Purple loosestrife has taken over wetlands in many areas of the northern parts of the United States and southern parts of Canada.

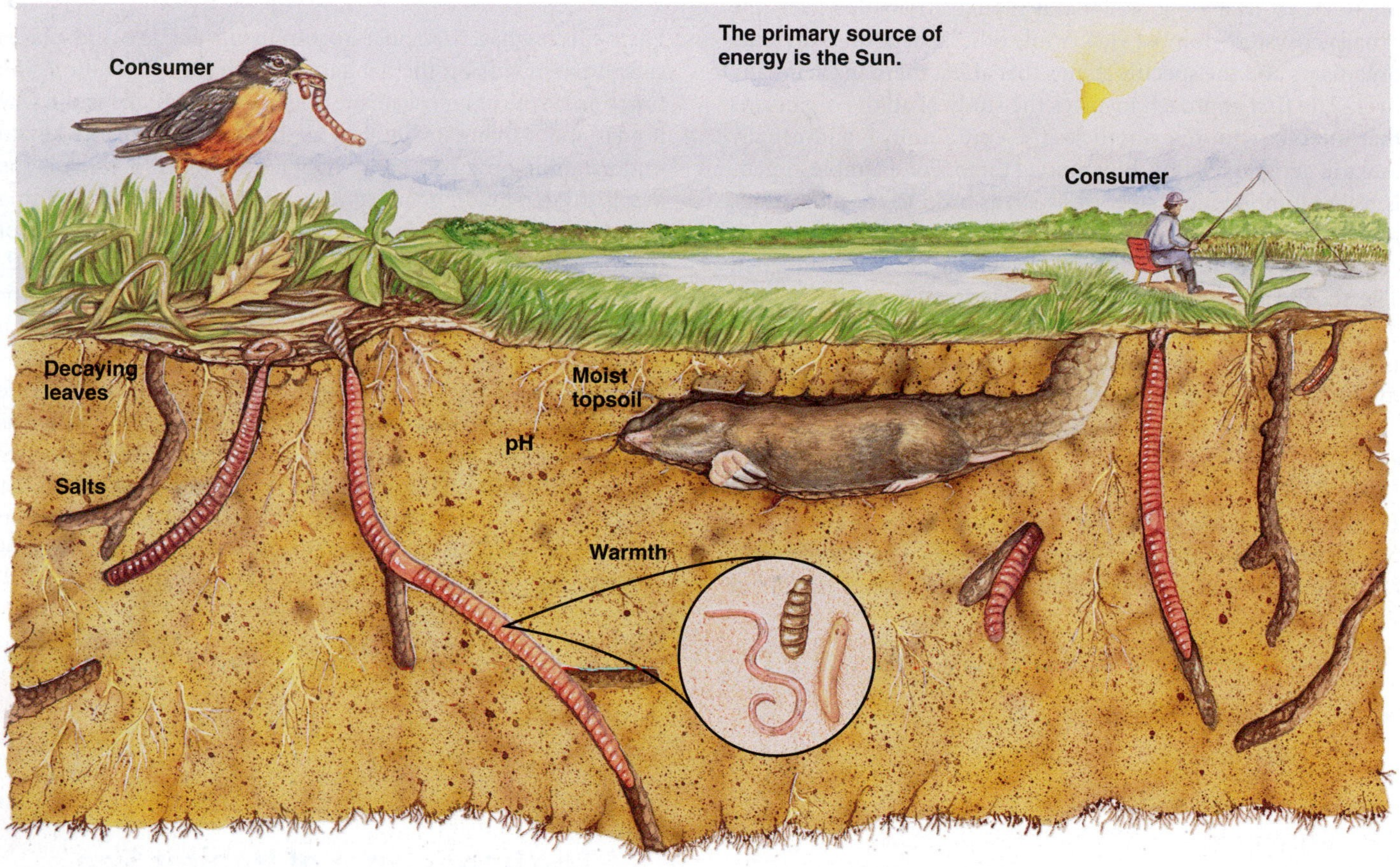

Consumer

The primary source of energy is the Sun.

Consumer

Decaying leaves

Moist topsoil

pH

Salts

Warmth

FIGURE 23.10 The niche of an earthworm involves a great many factors. It includes the fact that the earthworm is a consumer of dead organic matter, a source of food for other animals, host to parasites, and bait for an angler. Furthermore, the earthworm loosens the soil by its burrowing and "plows" the soil when it deposits materials on the surface. In addition, the pH, texture, and moisture content of the soil have an impact on the earthworm. Keep in mind that this is but a small part of what the niche of the earthworm includes.

penetrate the soil more easily. And this is only a limited sample of all of the aspects of its niche.

Some organisms have rather broad niches; others, with very specialized requirements and limited roles to play, have niches that are quite narrow. The opossum (figure 23.11A) is an animal with a very broad niche. It eats a wide variety of plant and animal foods, can adjust to a wide variety of climates, is used as food by many kinds of carnivores (including humans), and produces large numbers of offspring. By contrast, the koala of Australia (figure 23.11B) has a very narrow niche. It can live only in areas of Australia with specific species of *Eucalyptus* trees, because it eats the leaves of only a few kinds of these trees. Furthermore, it cannot tolerate low temperatures and does not produce large numbers of offspring. As you might guess, the opossum is expanding its range and the koala is endangered in much of its range.

A The oppossum has a broad niche.

FIGURE 23.11   (*A*) The oppossum has a very broad niche. It eats a variety of foods, is able to live in a variety of habitats, and has a large reproductive capacity. It is generally extending its range in the United States. (*B*) The koala (*Phascolarctos cinereus*) has a narrow niche. It feeds on the leaves of only a few species of *Eucalyptus* trees, is restricted to relatively warm, forested areas of Australia, and is generally endangered in much of its habitat.

B The koala has a narrow niche.

It is often easy to overlook important roles played by some organisms. For example, when Europeans introduced cattle into Australia—a continent where there had previously been no large, hoofed mammals—they did not think about the impact of cow manure or the significance of the niche of a group of beetles called *dung beetles*. These beetles rapidly colonize fresh dung and cause it to be broken down. No such beetles existed in Australia; therefore, in areas where cattle were raised, a significant amount of land became covered with accumulated cow dung. This reduced the area where grass could grow and reduced productivity. The problem was eventually solved by the importation of several species of dung beetles from Africa, where large, hoofed mammals are common. The dung beetles made use of what the cattle did not digest, returning it to a form that plants could more easily recycle into plant biomass.

23.8   KINDS OF ORGANISM INTERACTIONS

When organisms encounter one another, they can influence one another in numerous ways. Some interactions are harmful to one or both of the organisms. Others are beneficial. Ecologists have classified kinds of interactions between organisms into several broad categories.

Predation

Predation occurs when one animal captures, kills, and eats another animal. The organism that is killed is called the *prey*, and the one that does the killing is called the *predator*. The predator obviously benefits from the relationship, while the prey organism is harmed. Most predators are relatively large compared to their prey and have specific adaptations that aid them in catching prey. Many spiders build webs that serve as nets to catch flying insects. The prey are quickly paralyzed by the spider's bite and wrapped in a tangle of silk threads. Other rapidly moving spiders, such as wolf spiders and jumping spiders, have large eyes that help them find prey without using webs. Dragonflies patrol areas where they can capture flying insects. Hawks and owls have excellent eyesight that allows them to find their prey. Many predators, such as leopards, lions, and cheetahs, use speed to run down their prey, while others such as frogs, toads, and many kinds of lizards blend in with their surroundings and strike quickly when a prey organism happens by (figure 23.12).

Parasitism

Another kind of interaction in which one organism is harmed and the other aided is the relationship of parasitism. In fact, there are more species of parasites in the world than there are nonparasites, making this a very common kind of relationship. **Parasitism** involves one organism living in or on another living organism from which it derives nourishment. The *parasite* derives the benefit and harms the *host*, the organism it lives in or on (figure 23.13). Many kinds of fungi live on trees and other kinds of plants, including those that are commercially valuable. Dutch elm disease is caused by a fungus that infects the living, sap-carrying parts of the tree. Mistletoe is a common plant that is a

A Cheetahs use strength and speed to capture prey.

B Spiders use nets to trap prey.

FIGURE 23.12 (*A*) Many predators, such as a cheetah (*Acinonyx jubatus*), capture prey by making use of speed and strength. (*B*) Other predators, such as many kinds of spiders, use nets to trap prey which they then bite and kill. Obviously, predators benefit from the food they obtain to the detriment of the prey organism.

A Roundworm—an internal parasite

B Tick—an external parasite

C Indian Pipe—a parasitic plant

FIGURE 23.13 Parasites benefit from the relationship because they obtain nourishment from the host. Roundworms (*Ascaris lumbricoides*) (*A*) are internal parasites in the guts of their host, where they absorb food from the host's gut. The tick (*B*) is an external parasite that sucks blood from its host. Indian pipe (*Monotropa uniflora*) (*C*) is a flowering plant that lacks chlorophyll and is parasitic on fungi that have a mutualistic relationship with tree roots.

A A remora on a shark.

FIGURE 23.14   In the relationship called commensalism, one organism benefits and the other is not affected. (*A*) The remora fish shown here hitchhikes a ride on the shark. It eats scraps of food left over by the messy eating habits of the shark. The shark does not seem to be hindered in any way. (*B*) The grey Spanish moss hanging on this oak tree is a good example of an epiphyte. The Spanish Moss does not harm the tree but benefits from using the tree surface as a place to grow.

B Spanish moss (*Tillandsia usneoides*), a flowering plant, is an epiphyte.

parasite on other plants. The mistletoe plant invades the tissues of the tree it is living on and derives nourishment from the tree. There are even parasitic plants that have lost their ability to produce chlorophyll.

Many kinds of worms, protozoa, bacteria, and viruses are important parasites. Parasites that live on the outside of their hosts are called *external parasites.* For example, ticks live on the outside of the bodies of mammals such as rats, dogs, cats, and humans, where they suck blood and do harm to their host. At the same time, the host could also have a tapeworm in its intestine. Since the tapeworm lives inside the host, it is called an *internal parasite.* Bacterial and viral parasites are common causes of disease in humans and other organisms.

Commensalism

Predation and parasitism both involve one organism benefiting while the other is harmed. Another common relationship is one in which one organism benefits while the other is not affected. This is known as **commensalism.** For example, sharks often have another fish, the remora, attached to them. The remora has a sucker on the top side of its head that allows it to attach to the shark and get a free ride (figure 23.14A). While the remora benefits from the free ride and by eating leftovers from the shark's meals, the shark does not appear to be troubled by this uninvited guest, nor does it benefit from the presence of the remora.

Another example of commensalism is the relationship between trees and epiphytic plants. Epiphytes are plants that live on the surface of other plants but do not derive nourishment from them (figure 23.14B). Many kinds of plants (e.g., orchids, ferns, Spanish moss, and mosses) use the surface of trees as a place to live. These kinds of organisms are particularly common in tropical rainforests. Many epiphytes derive benefit from the relationship because they are able to be located in the top of the

tree, where they receive more sunlight and moisture. The trees derive no benefit from the relationship, nor are they harmed; they simply serve as a support surface for epiphytes.

Mutualism

So far in our examples, only one species has benefited from the association of two species. There are also many situations in which two species live in close association with one another, and both benefit. This is called **mutualism.** One interesting example of mutualism involves digestion in animals such as rabbits, cows, and termites. These animals eat plant material that is high in cellulose, even though they do not produce the enzymes capable of breaking down cellulose molecules into simple sugars. They manage to get energy out of these cellulose molecules with the help of special microorganisms living in their digestive tracts. The microorganisms produce cellulose-digesting enzymes, called *cellulases,* which break down cellulose into smaller carbohydrate molecules that the host's digestive enzymes can break down into smaller glucose molecules. The microorganisms benefit because the host's gut provides them with a moist, warm, nourishing environment in which to live. The hosts benefit because the microorganisms provide them with a source of food.

Another kind of mutualistic relationship exists between flowering plants and bees. Undoubtedly, you have observed bees and other insects visiting flowers to obtain nectar from the blossoms. Usually the flowers are constructed in such a manner that the bees pick up pollen (sperm-containing packages) on their hairy bodies and transfer it to the female part of the next flower they visit (figure 23.15). Because bees normally visit many individual flowers of the same species for several minutes and ignore other species of flowers, they can serve as pollen carriers between two flowers of the same species. Plants pollinated in this manner produce less pollen than do plants that rely on the wind to transfer pollen. This saves the plant energy because it doesn't need to produce huge quantities of pollen. It does, however, need to transfer some of its energy savings into the production of showy flowers and nectar to attract the bees. The bees benefit from both the nectar and pollen; they use both for food.

Symbiosis is a close physical relationship between two kinds of organisms. This term is sometimes used to mean the same thing as mutualism. At other times, it is used to include mutualism, parasitism, and commensalism as different kinds of symbiotic relationships. So you will need to determine the meaning from the context in which the word is used.

Competition

So far in our discussion of organism interactions we have left out the most common one. It is reasonable to envision every organism on the face of Earth being involved in *competitive* interactions. **Competition** is a kind of interaction between organisms in which both organisms are harmed to some extent. Competition occurs whenever two organisms both need a vital resource that is in short supply (figure 23.16). The vital resource could be food, shelter, nesting sites, water, mates, or space. It may involve a snarling tug-of-war between two dogs over a scrap of food, or it can be a silent struggle between plants for access to available light. If you have ever started tomato seeds (or other garden plants) in a garden and failed to eliminate the weeds, you have witnessed competition. If the weeds are not removed, they compete with the garden plants for available sunlight, water, and nutrients, resulting in poor growth of both the garden plants and the weeds.

FIGURE 23.15 Mutualism is an interaction between two organisms in which both benefit. The plant benefits because cross-fertilization (exchange of gametes from a different plant) is more probable; the bee benefits by acquiring nectar and pollen for food.

Competition and Natural Selection

It is important to recognize that although competition results in harm to both organisms, there can still be winners and losers. The two organisms may not be harmed to the same extent, which results in one having greater access to the limited resource. Furthermore, even the loser can continue to survive if it migrates to an area where competition is less intense. Over many genera-

B Interspecific competition for food.

FIGURE 23.16 Whenever a needed resource such as sunlight is in limited supply, organisms compete for it. When competition occurs between members of the same species it is called *intraspecific competition*. (*A*) Intraspecific competition for sunlight among these pine trees in the Irati forest, Navarre, Spain, has resulted in the tall, straight trunks. Those trees that did not grow fast enough died. Competition between different species is called *interspecific competition*. (*B*) The hyena and vultures are competing for the hyena's kill in Kenya.

A Intraspecific competition for light.

tions, when individuals of two different species compete, it is possible that the species may evolve through natural selection to exploit different niches. Thus, competition provides a major mechanism for natural selection. With the development of slight differences between niches, the intensity of competition is reduced. For example, many birds catch flying insects as food. However, they do not compete directly with each other because some feed at night, some feed high in the air, some feed only near the ground, and still others perch on branches and wait for insects to fly past. The insect-eating niche can be further subdivided by specializing on particular sizes or kinds of insects.

Many of the relationships just described involve the transfer of nutrients from one organism to another (predation, parasitism, mutualism). Another important way scientists look at ecosystems is to look at how materials are cycled from organism to organism.

23.9 THE CYCLING OF MATERIALS IN ECOSYSTEMS

Although some new atoms are being added to Earth from cosmic dust and meteorites, this amount is not significant in relation to the entire mass of Earth. Therefore, Earth can be considered to be a closed ecosystem as far as matter is concerned. Only sunlight energy comes to Earth in a continuous stream, and even this is ultimately returned to space as radiant energy. However, it is this flow of energy that drives all biological processes. Living systems have evolved ways of using this energy to continue life through growth and reproduction and the continual reuse of existing atoms. In this recycling process, inorganic molecules are combined to form the organic compounds of living things. If there were no way of recycling this organic matter back into its inorganic forms, organic material would build up as the bodies of dead organisms. This is thought to have occurred millions of years ago when the present deposits of coal, oil, and natural gas were formed. However, under most conditions, decomposers are available to break down organic material to inorganic material that can then be reused by other organisms to rebuild organic material.

Carbon, hydrogen, phosphorus, potassium, nitrogen, sulfur, oxygen, calcium, and many other kinds of atoms are involved in the structure of living things. These atoms are constantly recycled. One way to get an appreciation of how various kinds of organisms interact to cycle materials is to look at a specific kind of atom and follow its progress through an ecosystem.

The Carbon Cycle

The **carbon cycle** is the series of activities of organisms that result in the cycling of carbon atoms between the atmosphere and living things. All living things contain organic molecules that are composed of long chains of carbon atoms. Carbon and oxygen atoms combine to form the molecule carbon dioxide (CO_2), which is a gas found in small quantities (about 0.04 percent) in the atmosphere. Producers combine carbon dioxide (CO_2) and

water (H_2O) to form complex organic molecules such as sugar ($C_6H_{12}O_6$) during photosynthesis. At the same time, oxygen molecules (O_2) are released. Terrestrial producers obtain CO_2 from and release O_2 to the atmosphere. Aquatic producers obtain CO_2 from and release O_2 to the water in which they live.

The organic matter in the bodies of plants may be used by herbivores as food. When an herbivore eats a plant, it breaks down the complex organic molecules into simpler organic molecules, such as simple sugars, amino acids, glycerol, and fatty acids. These can be used as building blocks in the construction of its own body. Thus, the atoms in the body of the herbivore can be traced back to the plants that were eaten. Similarly, when carnivores eat herbivores, these same atoms are transferred to them. Finally, the waste products of plants and animals and the remains of dead organisms are used by decomposer organisms as sources of carbon and other atoms they need for survival.

In addition, all the organisms in this cycle—plants, herbivores, carnivores, and decomposers—obtain energy (ATP) from the process of respiration, in which oxygen (O_2) is used to break down organic compounds into carbon dioxide (CO_2) and water (H_2O). Thus, the carbon atoms that started out as components of carbon dioxide (CO_2) molecules have passed through the bodies of living organisms as parts of their organic molecules and returned as carbon dioxide as a result of respiration, ready to be cycled again. Similarly, the oxygen atoms (O) released as oxygen molecules (O_2) during photosynthesis have been used during the process of respiration (figure 23.17).

The Nitrogen Cycle

Another important element for living things is nitrogen (N). Nitrogen is essential in the formation of amino acids, which are needed to form proteins, and in the formation of nitrogenous bases, which are a part of ATP and the nucleic acids DNA and RNA. Nitrogen (N) is found as molecules of nitrogen gas (N_2) in the atmosphere. Although nitrogen gas (N_2) makes up approximately 80 percent of Earth's atmosphere, only a few kinds of bacteria are able to convert it into nitrogen compounds that other organisms can use. Therefore, in most terrestrial ecosystems, the amount of nitrogen available limits the amount of plant biomass that can be produced. (Most aquatic ecosystems are limited by the amount of phosphorus rather than the amount of nitrogen.) Plants utilize several different nitrogen-containing compounds to obtain the nitrogen atoms they need to make amino acids and other compounds (figure 23.18).

Symbiotic nitrogen-fixing bacteria live in the roots of certain kinds of plants, where they convert nitrogen gas molecules into compounds that the plants can use to make amino acids and nucleic acids. The most common plants that enter into this mutualistic relationship with bacteria are the legumes, such as beans, clover, peas, alfalfa, and locust trees. Some other organisms, such as alder trees and even a kind of aquatic fern, can also participate in this relationship. There are also **free-living nitrogen-fixing bacteria** in the soil that provide nitrogen compounds that can be taken up through the roots, but the bacteria do not live in a close physical union with plants. In

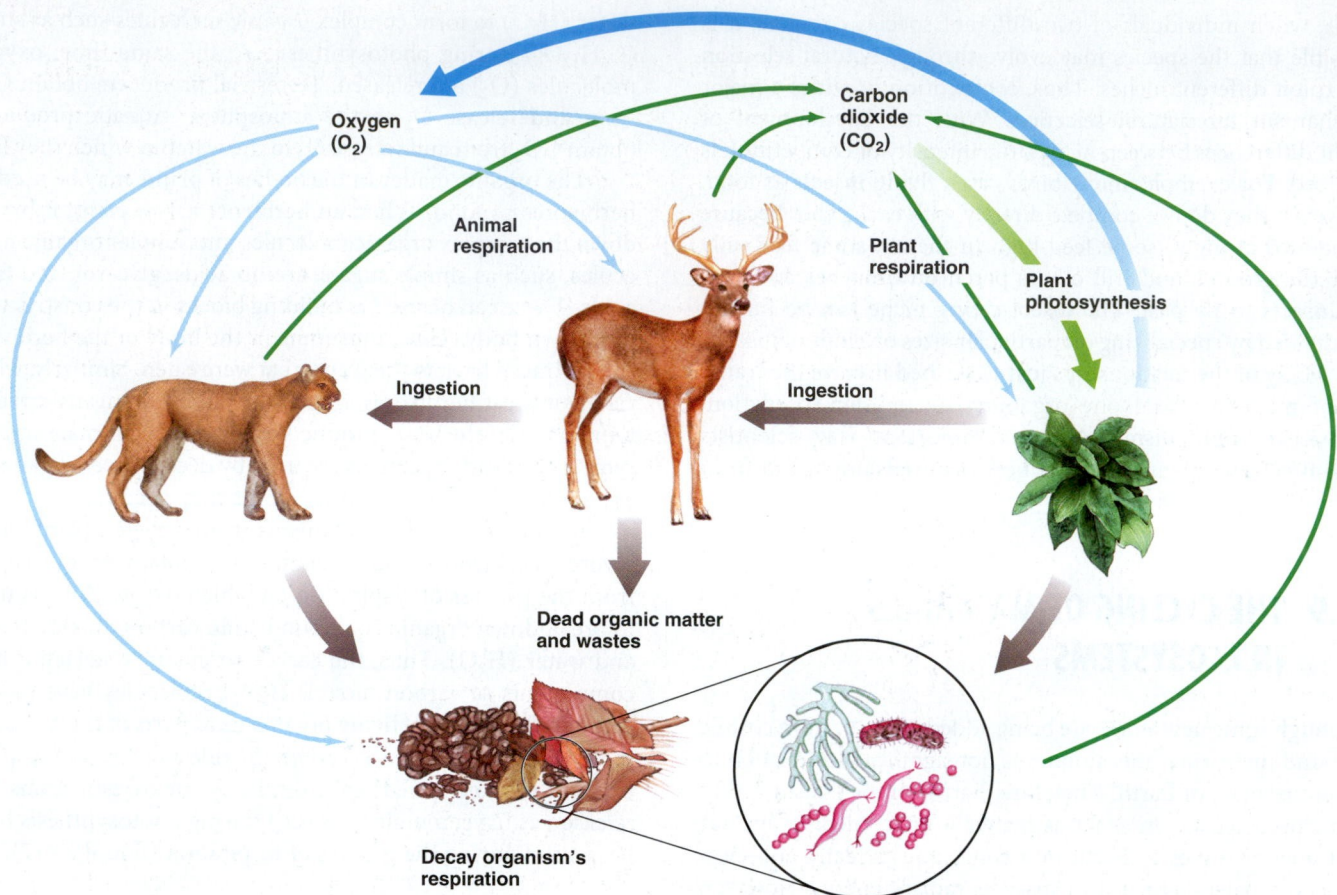

FIGURE 23.17 Carbon atoms are cycled through ecosystems. Carbon dioxide (green arrows) produced by respiration is the source of carbon that plants incorporate into organic molecules when they carry on photosynthesis. These carbon-containing organic molecules (black arrows) are passed to animals when they eat plants and other animals. Organic molecules in waste or dead organisms are consumed by decomposer organisms in the soil when they break down organic molecules into carbon dioxide and water. All organisms (plants, animals, and decomposers) return carbon atoms to the atmosphere or water as carbon dioxide when they carry on cellular respiration. Oxygen (blue arrows) is being cycled at the same time that carbon is. The oxygen is released into the atmosphere and into the water during photosynthesis and taken up during cellular respiration.

addition to these soil nitrogen-fixing bacteria, cyanobacteria are important in providing nitrogen in aquatic environments, and recent studies have shown that several kinds of Archaea may also be involved in the nitrogen cycle as nitrogen-fixers.

Another way plants get usable nitrogen compounds involves a series of different bacteria. *Decomposer bacteria* convert organic nitrogen-containing compounds into ammonia (NH_3) and then to ammonium ions (NH_4^+). *Nitrifying bacteria* can convert ammonium ions (NH_4^+) into nitrite-containing (NO_2) compounds, which in turn can be converted into nitrate-containing (NO_3) compounds. Many kinds of plants can use either ammonium ions (NH_4^+) or nitrate (NO_3) from the soil as building blocks for amino acids and nucleic acids.

All animals obtain their nitrogen from the food they eat. The ingested proteins are broken down into their component amino acids during digestion. These amino acids can then be reassembled into new proteins characteristic of the animal. All dead organic matter and waste products of plants and animals are acted upon by decomposer organisms, and the nitrogen is released as ammonia (NH_3), which can be taken up by plants or acted upon by nitrifying bacteria to make nitrate (NO_3^-).

Finally, other kinds of bacteria called *denitrifying bacteria* are capable of converting nitrite (NO_2^-) to nitrogen gas (N_2), which is released into the atmosphere. Thus, in the **nitrogen cycle,** nitrogen from the atmosphere is passed through a series of organisms, many of which are bacteria, and ultimately returns to the atmosphere to be cycled again. However, the nitrogen cycle differs from the carbon cycle in a fundamental way. Inorganic nitrogen compounds are not necessarily returned to the atmosphere but are returned to plants by way of decomposer, nitrate, and nitrite bacteria.

The Phosphorus Cycle

Phosphorus is another atom common in the structure of living things. It is present in many important biological molecules, such as DNA, and in the membrane structure of cells. In addition, the bones and teeth of vertebrate animals contain significant quantities of phosphorus. The ultimate source of phosphorus atoms is rock. In nature, new phosphorus compounds are released by the erosion of rock and dissolved in water. Plants use the dissolved phosphorus compounds to

Connections . . .

Chemical Fertilizers and the Nitrogen Cycle

Modern agriculture depends on the use of fertilizers to replace the elements removed from the soil by the plants that are harvested. The three most important elements in fertilizers are nitrogen, phosphorus, and potassium. The numbers on bags of fertilizer tell you how much of each of these three components is present. For example, a 6-12-12 fertilizer has 6 percent nitrogen, 12 percent phosphorus, and 12 percent potassium compounds. Although molecular nitrogen makes up 80 percent of the atmosphere, it is not available to plants. From a chemical point of view, it is very difficult to get molecular nitrogen, which consists of two nitrogen atoms bonded to each other, to react with other atoms. Although some bacteria are able to attach hydrogen to nitrogen to form ammonia, a chemical method for producing ammonia was not determined until 1909, when Fritz Haber determined the conditions under which nitrogen and hydrogen would combine to form ammonia. The process of attaching hydrogen to ammonia requires a great deal of energy. This was an important discovery because it led to a practical way to produce a variety of nitrogen-containing compounds that can be used for fertilizer. Farmers apply ammonia directly into the soil to provide nitrogen for many kinds of plants. Ammonia can be reacted with other materials to produce a variety of compounds that are used as nitrogen fertilizers, including urea, ammonium nitrate, and ammonium sulfate.

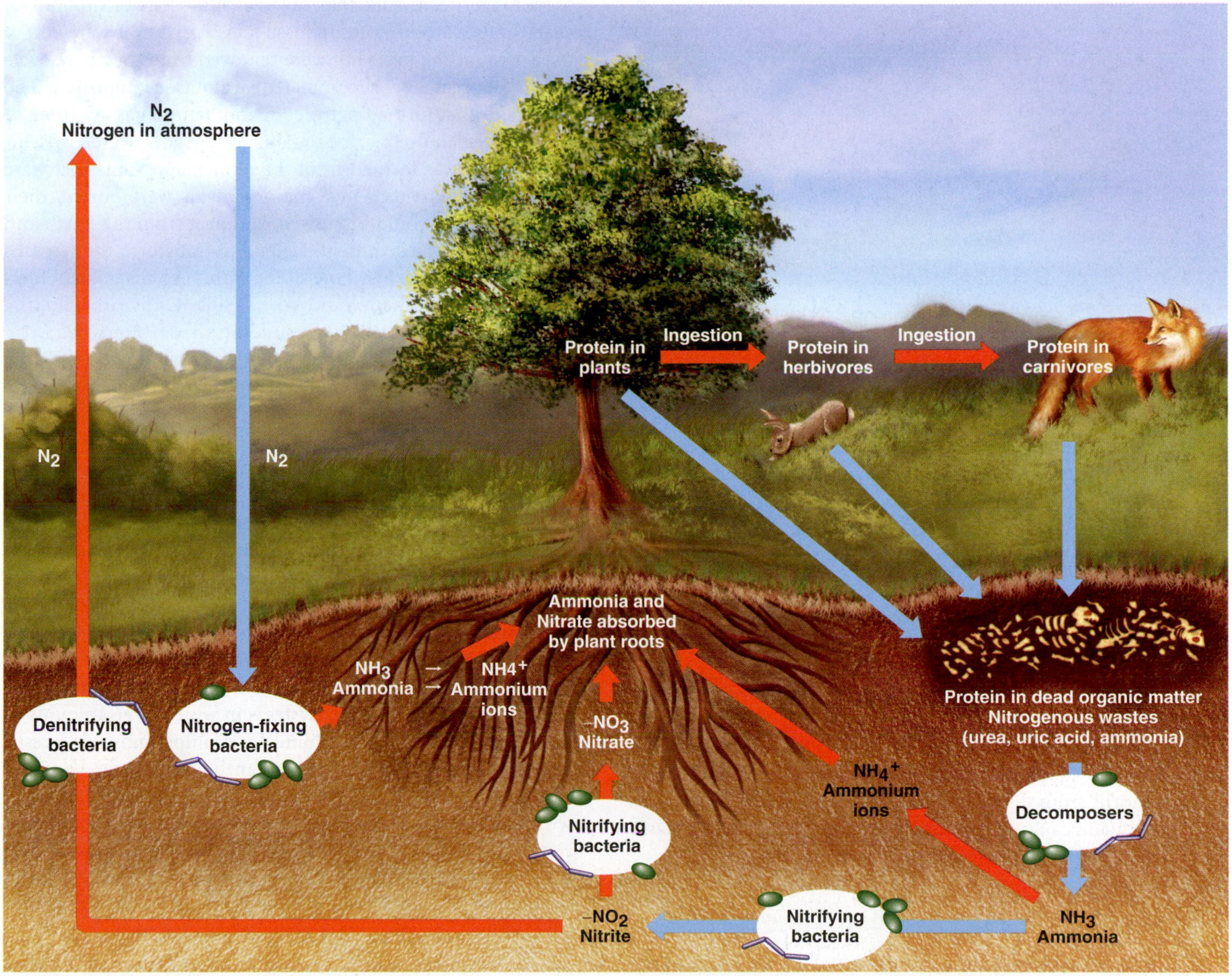

FIGURE 23.18 Nitrogen atoms are cycled through ecosystems. Atmospheric nitrogen is converted by nitrogen-fixing bacteria to nitrogen-containing compounds that plants can use to make proteins and other compounds. Proteins are passed to other organisms when one organism is eaten by another. Dead organisms and their waste products are acted upon by decomposer organisms to form ammonia, which may be reused by plants and converted to other nitrogen compounds by nitrifying bacteria. Denitrifying bacteria return nitrogen as a gas to the atmosphere.

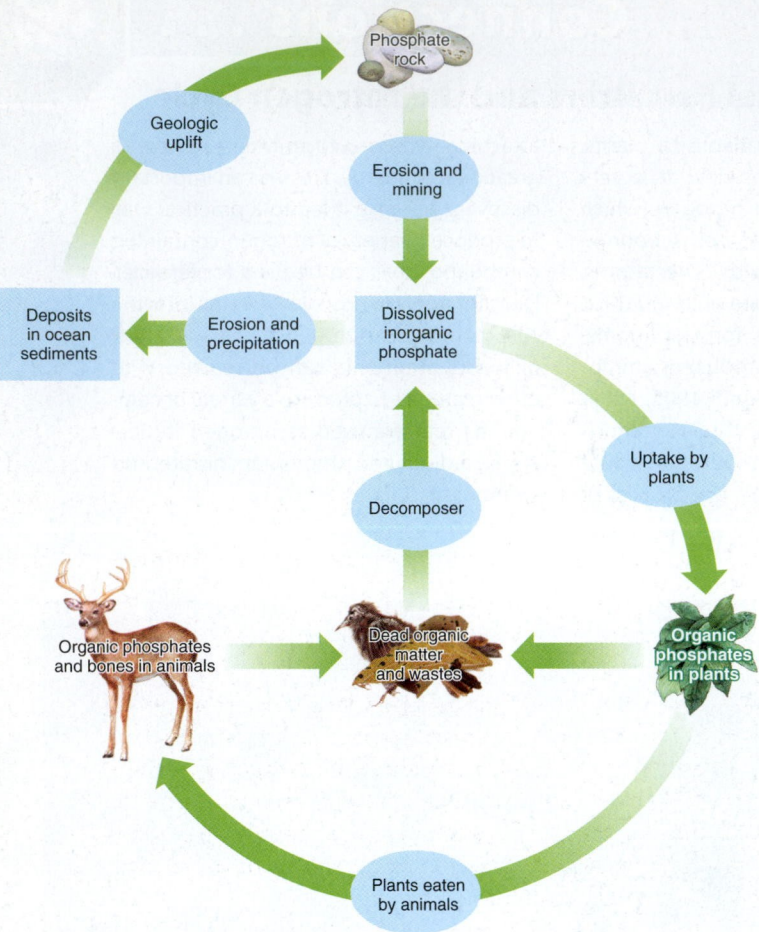

FIGURE 23.19 The source of phosphorus is rock that, when dissolved, provides phosphorus used by plants and animals.

construct the molecules they need. Animals obtain the phosphorus they need when they consume plants or other animals. When an organism dies or excretes waste products, decomposer organisms recycle the phosphorus compounds back into the soil. Phosphorus compounds that are dissolved in water are ultimately precipitated as deposits. Geologic processes elevate these deposits and expose them to erosion, thus making these deposits available to organisms (figure 23.19). Waste products of animals often have significant amounts of phosphorus. In places where large numbers of seabirds or bats have congregated for hundreds of years, the thickness of their droppings (called *guano*) can be a significant source of phosphorus for fertilizer. In many soils, phosphorus is in short supply and must be provided to crop plants to get maximum yields. Phosphorus is also in short supply in aquatic ecosystems, and the addition of phosphorus from detergents in sewage and runoff from lawns and farmland can cause undesirable growth of algae and plants in lakes, rivers, estuaries, and ocean waters near land.

Nutrient Cycles and Geologic Time

The nutrient cycles we have just discussed act on a short-term basis in which elements are continually being reused among organisms and on a long-term basis in which certain elements are tied up for long time periods and are not part of the active nutrient cycle. In our discussion of the phosphorus cycle, it was mentioned

that the source of phosphorus is rock. While phosphorus moves rapidly through organisms in food chains, phosphorus ions are not very soluble in water and tend to precipitate in the oceans to form sediments that eventually become rock on the ocean floor. Once this has occurred it takes the process of geologic uplift followed by erosion to make phosphorus ions available to terrestrial ecosystems. Thus we can think of the ocean as a place where phosphorus is removed from the active nutrient cycle (this situation is known as a *sink*).

There are also long-term aspects to the carbon cycle. Organic matter in soil and sediments is the remains of once living organisms. Thus, these compounds constitute a sink for carbon, particularly in ecosystems in which decomposition is slow (tundra, northern forests, grasslands, swamps, marine sediments). These materials can tie up carbon for hundreds to thousands of years. Fossil fuels (coal, petroleum, and natural gas), which were also formed from the remains of organisms, are a longer-term sink that involves hundreds of millions of years. These carbon atoms in fossil fuels at one time were part of the active carbon cycle but were removed from the active cycle when the organisms accumulated without decomposing. The organisms that formed petroleum and natural gas are thought to be the remains of marine organisms that got covered by sediments. Coal was formed from the remains of plants that were buried by sediments. Once the organisms were buried, their decomposition would be slowed, and heat from the Earth and pressure from the sediments helped to transform the remains of living things into fossil fuels. The carbon atoms in fossil fuels have been locked up for hundreds of millions of years. Thus the formation of fossil fuels was a sink for carbon atoms.

Oceans are a major carbon sink. Carbon dioxide is highly soluble in water. Many kinds of carbonate sedimentary rock are formed from the precipitation of carbonates from solution in oceans. In addition, many marine organisms form skeletons or shells of calcium carbonate. These materials accumulate on the ocean floor as sediments that over time can be converted to limestone. Limestone typically contains large numbers of fossils. The huge amount of carbonate rock is an indication that there must have been higher amounts of carbon dioxide in the Earth's atmosphere in the past.

Since fossil fuels are the remains of once living things and living things have nitrogen as a part of protein, nitrogen that was once part of the active nitrogen cycle was removed when the fossil fuels were formed. In ecosystems in which large amounts of nonliving organic matter accumulate (swamps, humus in forests, and marine sediments) nitrogen can be tied up for relatively long time periods. In addition, some nitrogen may be tied up in sedimentary rock and in some cases is released with weathering. However, it appears that the major sink for nitrogen is as nitrogen in the atmosphere. Nitrogen compounds are very soluble in water so when sedimentary rock is exposed to water, these materials are dissolved and reenter the active nitrogen cycle.

Bioaccumulation and Biomagnification

Bioaccumulation occurs when organisms capture and store materials. Carbon, nitrogen, and phosphorus are examples of

A Closer Look

Scientists Accumulate Knowledge About Climate Change

Humans have significantly altered the carbon cycle. As we burn fossil fuels, the amount of carbon dioxide in the atmosphere continually increases. Carbon dioxide allows light to enter the atmosphere but does not allow heat to exit. Because this is similar to what happens in a greenhouse, carbon dioxide and the other gases that have similar effects are called greenhouse gases. Therefore, many scientists are concerned that increased carbon dioxide levels are leading to a warming of the planet, which will cause major changes in our weather and climate. In science, when a new discovery is made or a new issue is raised, it stimulates a large number of observations and experiments that add to the body of knowledge about the topic (box figure 23.1). Concerns about global climate change and the role that carbon dioxide plays in causing climate change have resulted in scientists studying many aspects of the problem. This has been a worldwide effort and has in-volved many different branches of science. This effort has resulted in critical examination of several basic assumptions about climate change, the collection of much new information, and new predictions about the consequences of global climate change.

Several significant studies include:

- Examination of gas bubbles trapped in the ice of glaciers has allowed scientists to measure the amount of carbon dioxide in the atmosphere at the time the ice formed. This provides information about carbon dioxide concentrations prior to human-caused carbon dioxide releases and allows scientists to track the rate of change.
- Long-term studies of the atmosphere at various locations throughout the world show that carbon dioxide levels are increasing.

- Measurements show that sea level is rising almost 2 mm per year.
- Measurements of the temperature of the Earth's atmosphere have allowed tracking of temperature. According to NASA, 10 of the warmest years on record occurred in the 12-year period between 1998 and 2009.
- Satellite images of the Arctic Ocean show reduced ice cover.
- Observations of bird migration in Europe document that birds that migrate long distances are arriving in Europe earlier in the spring.
- Many studies of the rate at which different ecosystems take up carbon dioxide have been done to determine if assumptions about the carbon dioxide–trapping role of natural ecosystems are correct.
- Warming of the Arctic has resulted in less permafrost.
- Increased water temperatures have been linked to increases in the number

and extent of blooms of cyanobacteria in lakes and oceans.

- Studies suggest that an increase in the level of carbon dioxide in the atmosphere could result in increased amounts of dissolved carbon dioxide in the ocean. Increased carbon dioxide will lower the pH of the ocean, which could have a negative effect on animals that make shells.
- Warming of the oceans is linked to more intense hurricanes.
- Earlier arrival of spring is linked to increased numbers and intensity of forest fires in the western United States.

The United Nations established the Intergovernmental Panel on Climate Change (IPCC)—a panel of scientists, political leaders, and economists—to analyze the large amount of information generated on the topic of climate changes. The IPCC has issued several reports about the nature, causes, and the impacts of climate change on ecosystems and culture.

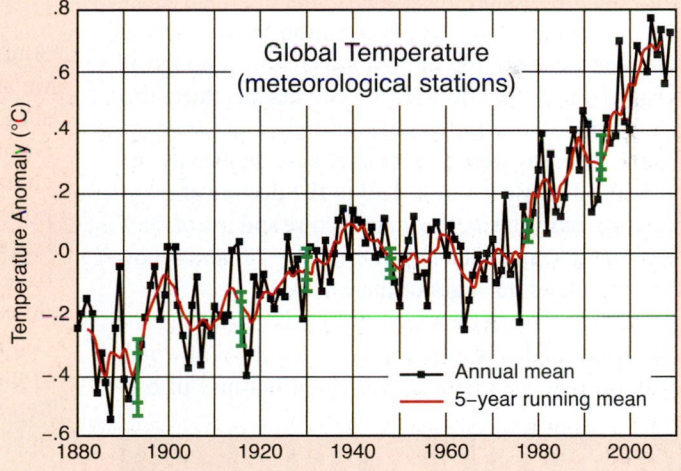

Box Figure 23.1

elements that are temporarily bioaccumulated by living things. These materials are released from organisms as part of their metabolism or when they die. However, it is also possible for organisms to bioaccumulate toxins that are taken in by organisms but not metabolized or excreted by them. Two kinds of materials that have received attention because they bioaccumulate are methylmercury (CH_3Hg^+) and persistent organic pollutants.

Methylmercury is formed by certain kinds of bacteria that live in low-oxygen environments. Although elemental mercury is present in rocks, the actions of humans have increased the amount of mercury available for action by these bacteria. Mining activity and the burning of coal are major sources of mercury. Persistent organic pollutants result from the manufacture of certain plastics, fire retardants, and pesticides. Since these chemicals are not a product of nature, most organisms, including decomposers, do not have the necessary metabolic pathways to cause their decomposition. Therefore, some of these materials are taken up by organisms and bioaccumulate.

The more methylmercury or persistent organic compounds an organism is exposed to, the more they accumulate, and the more likely they will affect the health of the organism.

Compounding the problem is the phenomenon of biomagnification in food chains. When an herbivore eats plants, it accumulates the persistent materials already accumulated by the plant. Similarly, a carnivore further concentrates the materials accumulated by the herbivores it eats. Thus, as one proceeds to higher trophic levels, the concentration found in animals increases—it is biomagnified.

The well-documented case of DDT is an example of how biomagnification occurs. At one time DDT was a commonly used insecticide. DDT is not very soluble in water but dissolves in oil or fatty compounds. When DDT falls on an insect or is consumed by the insect, the DDT is accumulated in the insect's fatty tissue. Large doses kill insects but small doses do not, and their bodies may contain as much as one part per billion of DDT. This is not very much, but it can have a tremendous effect on the animals that feed on the insects.

When an aquatic habitat was sprayed with a small concentration of DDT or received DDT from runoff, small aquatic organisms accumulated a concentration that was up to 250 times greater than the concentration of DDT in the surrounding water. These organisms were eaten by shrimp, clams, and small fish, which were in turn eaten by larger fish. DDT concentrations of large fish were as much as 2000 times the original concentration sprayed on the area. What was a very small initial concentration had become so high that it could be fatal to animals at higher trophic levels. This was of particular concern for birds, since DDT interferes with the production of eggshells, making them much more fragile. This problem was more common in carnivorous birds because they are at the top of the food chain. Although all birds of prey probably were affected to some degree, those that relied on fish for food were affected most severely. Eagles, osprey, cormorants, and pelicans were particularly susceptible species. (See figure 23.20.) Because it was linked to the decline of many populations of fish-eating birds, the manufacture and use of DDT was prohibited in the United States in the early 1970s and is now prohibited for most uses throughout the world.

The populations of several species of birds, including the brown pelican, bald eagle, osprey, and cormorant, all of which feed primarily on fish, were severely affected because of bio-

magnification of persistent organic chemicals. With the control or restriction of most persistent pesticides and several other chemicals, the levels of these chemicals in their body tissues have declined, and their populations have rebounded. The bald eagle has been removed from the endangered species list, and the pelican has been removed from the endangered species list in part of its range.

Today we still see evidence of the problem of biomagnifications in advisories that caution people (particularly children and women of childbearing age) from eating certain kinds of fish. This is primarily because of the level of methylmercury in predator fish that are at the top of the food chain.

23.10 POPULATION CHARACTERISTICS

A **population** is a group of organisms of the same species located in the same place at the same time. Examples are the number of dandelions in your front yard, the rat population in the sewers of your city, or the number of people in New York City. On a larger scale, all the people of the world constitute the human population. The terms *species* and *population* are interrelated because a species is a population—the largest possible population of a particular kind of organism. The term *population,* however, is often used to refer to portions of a species by specifying a space and time. For example, the size of the human population in a city changes from hour to hour during the day and varies according to where you set the boundaries of the city.

Genetic Differences

Since each local population is a small portion of a species, we should expect distinct populations of the same species to show differences. Some of these differences will be genetic. *Gene frequency* is a measure of how often a specific gene for a characteristic shows up in the individuals of a population. Two populations of the same species often have quite different gene frequencies. For example, many populations of mosquitoes have high frequencies of insecticide-resistant genes, whereas others do not. The frequency of the genes for tallness in humans is greater in certain African tribes than in any other human population. Figure 23.21 shows that the frequency of the genetic information for type B blood differs significantly from one human population to another.

Age Structure

Another feature of a population is its *age distribution,* which is the number of organisms of each age in the population. In addition, organisms are often grouped into the following categories:

- prereproductive juveniles—insect larvae, plant seedlings, or babies;
- reproductive adults—mature insects, plants producing seeds, or humans in early adulthood; or
- postreproductive adults no longer capable of reproduction—annual plants that have shed their seeds, salmon that have spawned, and many elderly humans.

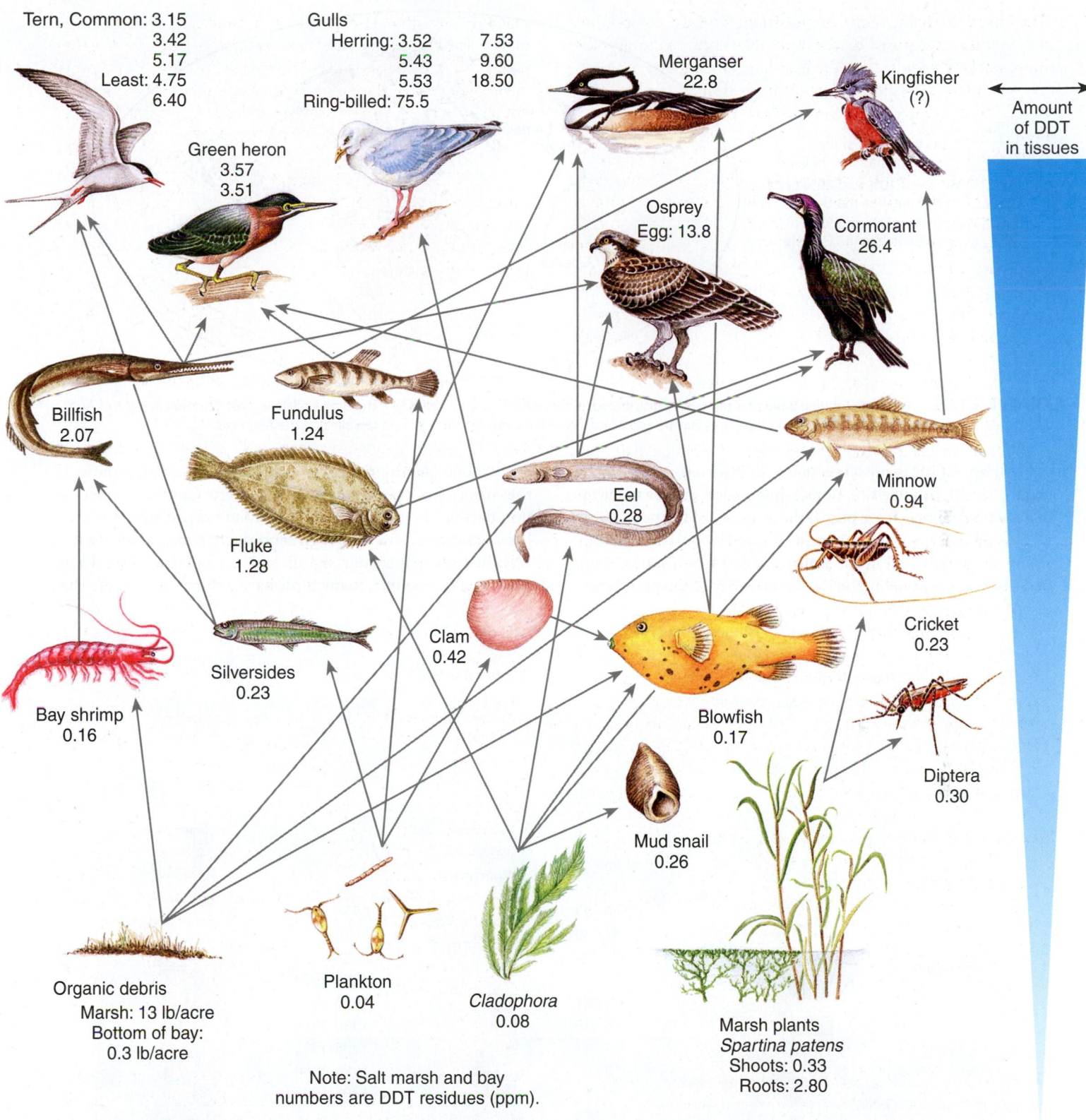

Tern, Common: 3.15
3.42
5.17
Least: 4.75
6.40

Gulls
Herring: 3.52   7.53
5.43   9.60
5.53   18.50
Ring-billed: 75.5

Merganser
22.8

Kingfisher
(?)

Amount
of DDT
in tissues

Green heron
3.57
3.51

Osprey
Egg: 13.8

Cormorant
26.4

Billfish
2.07

Fundulus
1.24

Minnow
0.94

Fluke
1.28

Eel
0.28

Cricket
0.23

Silversides
0.23

Clam
0.42

Blowfish
0.17

Diptera
0.30

Bay shrimp
0.16

Mud snail
0.26

Organic debris
Marsh: 13 lb/acre
Bottom of bay:
0.3 lb/acre

Plankton
0.04

Cladophora
0.08

Marsh plants
Spartina patens
Shoots: 0.33
Roots: 2.80

Note: Salt marsh and bay
numbers are DDT residues (ppm).

FIGURE 23.20 This illustration shows the biomagnification of DDT in an aquatic food chain. All the numbers shown are in parts per million (ppm). A concentration of one part per million means that in a million equal parts of the organism, one of the parts would be DDT. Notice how the amount of DDT in the bodies of the organisms increases as we go from producers to herbivores to carnivores. Because DDT is persistent, it builds up in the top trophic levels of the food chain.

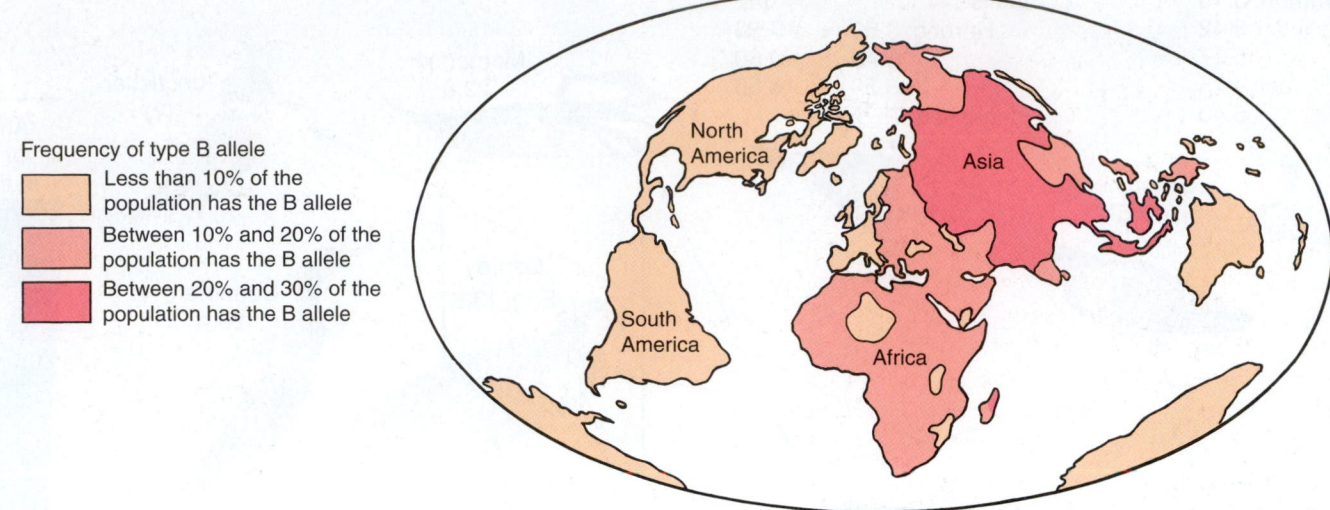

Frequency of type B allele

Less than 10% of the population has the B allele

Between 10% and 20% of the population has the B allele

Between 20% and 30% of the population has the B allele

FIGURE 23.21 The allele for type B blood is not evenly distributed in the world. This map shows that the type B allele is most common in parts of Asia and has been dispersed to the Middle East and parts of Europe and Africa. There has been very little flow of the allele to the Americas.

A population is not necessarily divided into equal thirds (figure 23.22). In some situations, a population may be made up of a majority of one age group. If the majority of the population is prereproductive, then a period of rapid population growth should be anticipated in the future as the prereproductive individuals reach sexual maturity. If a majority of the population is reproductive, the population should be growing rapidly. If the majority of the population is postreproductive, a population decline should be anticipated. Many organisms that only live a short time and have high reproductive rates can have age distributions that change significantly in a matter of weeks or months. For example, many birds have a flurry of reproductive

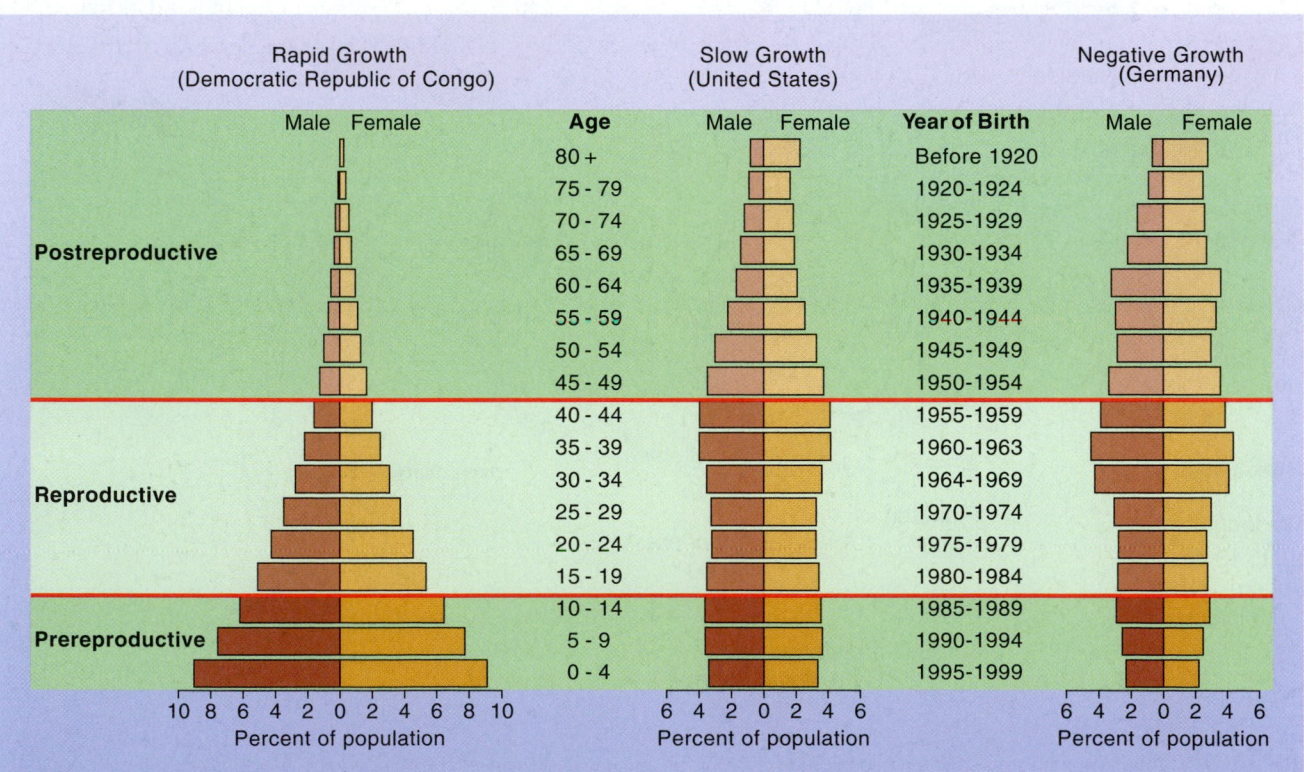

FIGURE 23.22 The relative numbers of individuals in each of the three categories (prereproductive, reproductive, and postreproductive) are good clues to the future growth of a population. The Democratic Republic of Congo has a large number of young individuals who will become reproducing adults. Therefore, this population is likely to grow rapidly. The United States has a large proportion of reproductive individuals and a moderate number of prereproductive individuals. Therefore, this population is likely to grow slowly. Germany has a declining number of reproductive individuals and a very small number of prereproductive individuals. Therefore, its population has begun to decline.

Source: Data from Population Reference Bureau, Inc.

ity for squirrels, which need mature, fruit-producing trees as a source of food and old, hollow trees for shelter.

23.13 LIMITING FACTORS TO HUMAN POPULATION GROWTH

Today, we hear differing opinions about the state of the world's human population. On one hand, we hear that the population is growing rapidly. In contrast, we hear that some European countries have shrinking populations. Other countries are concerned about the aging of their populations because birthrates and death rates are low. In magazines and on television, we see that there are starving people in the world. At the same time, we hear discussions about the problem of food surpluses and obesity in many countries. Some have even said that the most important problem in the world today is the rate at which the human population is growing; others maintain that the growing population will provide markets for goods and be an economic boon. How do we reconcile this mass of conflicting information?

The world population was estimated to be 6.8 billion at mid-year 2010, and is expected to grow to 7 billion by 2012 and to 9 billion by 2044. The distribution of this growth, however, will not be even. The human population is divided into two rather distinct segments: the more economically developed countries of the world (Europe, North America, Australia, New Zealand, and Japan) and the less developed nations. Sharp contrasts exist between the populations of these two groups of nations. Table 23.2 highlights some of these differences.

Regardless of whether we are considering the developed world, the less developed world, or the total world population, it is important to realize that human populations follow

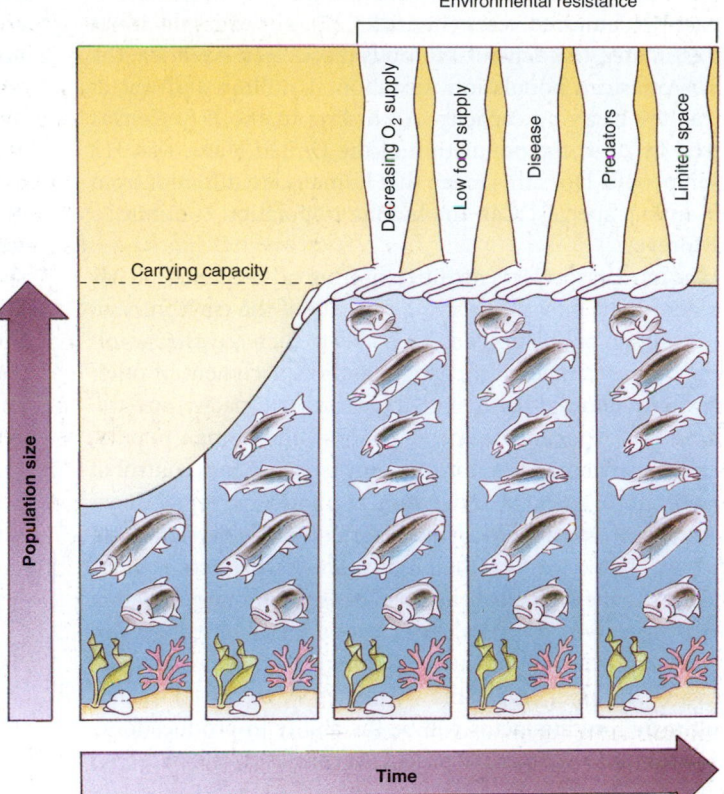

FIGURE 23.26 A number of factors in the environment, such as food, oxygen supply, diseases, predators, and space, determine the number of organisms that can survive in a given area—the carrying capacity of that area. The environmental factors that limit populations are collectively known as environmental resistance.

the same patterns of growth and are acted upon by the same kinds of limiting factors as are populations of other organisms. When we look at the curve of population growth over the past several thousand years, estimates are that the human population remained low and constant for thousands of years

TABLE 23.2 Comparison of Population Characteristics of More and Less Developed Nations (2010)

Population Characteristic	More Developed Nations	Less Developed Nations	Significance
Total population (2010)	1.237 billion	5.656 billion	Less developed nations constitute nearly 82 percent of world's population.
Expected 2025 population	1.290 billion	6.819 billion	Less developed nations are growing much faster.
Birthrate	11 births per 1,000 in population	22 births per 1,000 in population	Less developed nations have high birthrates.
Death rate	10 deaths per 1,000 in population	8 deaths per 1,000 in population	More developed nations have higher death rates because their populations are older.
Total fertility rate	1.7 births per woman per lifetime	2.7 births per woman per lifetime	Women in less developed nations have nearly twice as many children as those in more developed nations.
Infant mortality rate	6 deaths per 1,000 births	50 deaths per 1,000 births	Less developed nations have much higher infant mortality rates.
Percent under 15 years of age	17	30	Less developed nations have young populations.
Percent over 65 years of age	16	6	More developed nations have aging populations.
Rate of increase (%)	0.2	1.4	India will become the world's most populous country and will exceed China's population by 2025.

Source: Data from the Population Reference Bureau, *2010 World Population Data Sheet.*

but has increased explosively (from 1 billion to 6.7 billion) in the past two hundred years (figure 23.27). For example, it has been estimated that when Columbus discovered America, the Native American population was about 1 million and was at or near the carrying capacity. According to the U.S. Census released in 2011, the population of the United States was 312 million people. Does this mean that humans are different from other animal species? Can the human population continue to grow forever?

The human species is no different from other animals. It has an upper limit set by the carrying capacity of the environment, but the human population has been able to increase astronomically because technological changes and displacement of other species have allowed us to shift the carrying capacity upward. Much of the exponential growth phase of the human population can be attributed to improvements in sanitation, control of diseases, improvement in agricultural methods, and replacement of natural ecosystems with artificial agricultural ecosystems. But even these conditions have their limits. There will be some limiting factors that eventually cause a leveling off of our population growth curve. We cannot increase beyond our ability to get raw materials and energy, nor can we ignore the waste products we produce or the other organisms with which we interact. Probably the ultimate limiting factor will be the ability to produce food. Improvements in agricultural production have not kept pace with population growth. Indeed, in many parts of the world, food is in short supply despite surpluses in much of the developed world.

Humans are different from most other organisms in a fundamental way: We are able to predict the outcome of a specific course of action. Current technology and medical knowledge are available to control human population and improve the health and well-being of the people of the world. Why then does the human population continue to grow, resulting in human suffering and stressing the environment in which we live? Because we are social animals that have freedom of choice, we frequently do not do what is considered "best" from an unemotional, unselfish, biological point of view. People make decisions based on historical, social, cultural, ethical, and personal considerations. What is best for the population as a whole may be bad for you as an individual. The biggest problems associated with control of the human population are not biological problems but rather require the efforts of philosophers, theologians, politicians, and sociologists. If our population increases, so will political, social, and biological problems; there will be less individual freedom, there will be intense competition for resources, and famine and starvation will become even more common. The knowledge and technology necessary to control the human population are available, but the will is not. What will eventually limit the size of our population? Will it be lack of resources, lack of energy, accumulated waste products, competition among ourselves, or rational planning of family size?

23.14 HUMAN POPULATION GROWTH AND THE GLOBAL ECOSYSTEM

The growth of the human population has had major consequences for the global ecosystem. As the size of the human population has increased, more and more resources have been diverted to human use and fewer are available for other organisms. Nearly Earth's entire surface has been affected by human activity. More than 80 percent of the land surface has been modified for human uses such as agricultural and grazing land, managed forests, urban space, and transportation corridors. Essentially all of the land that can support agriculture has been converted from forests or grasslands to raising crops. Oceans

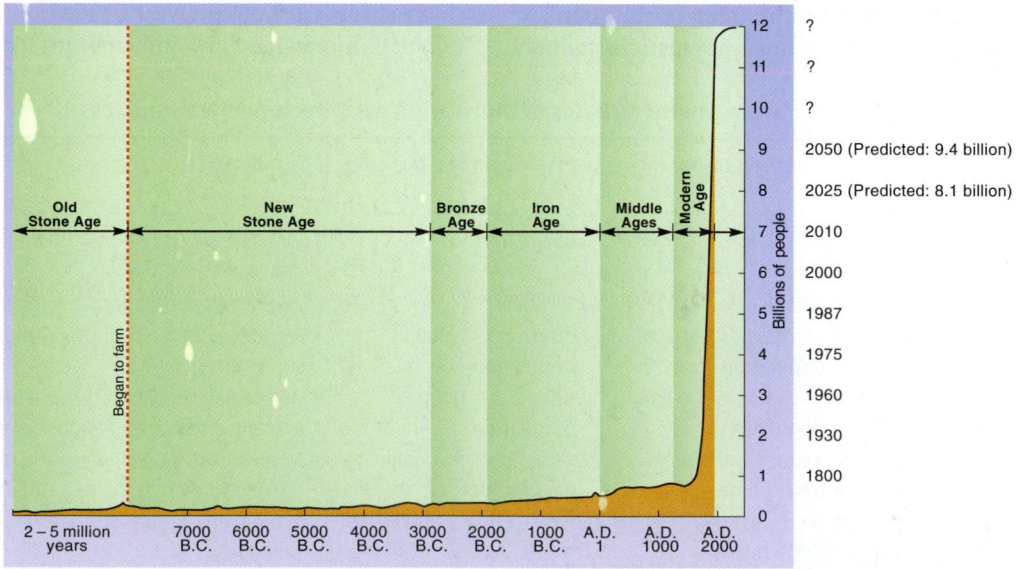

FIGURE 23.27   The number of humans doubled from A.D. 1800 to 1930 (from 1 billion to 2 billion), doubled again by 1975 (4 billion), and is projected to double again by the year 2025. How long can the human population continue to double before Earth's ultimate carrying capacity is reached?

Source: Data from Jean Van Der Tak, et al., "Our Population Predicament: A New Look," in *Population Bulletin,* vol. 34, no. 5, December 1979, Population Reference Bureau, Washington, D.C., and more recent data from the Population Reference Bureau. Updated: 2011 U.S. Census Bureau.

Government Policy and Population Control

The actions of government can have a significant impact on the population growth patterns of nations. Some countries have policies that encourage couples to have children. The U.S. tax code indirectly encourages births by providing tax advantages to the parents of children. Some countries in Europe are concerned about the lack of working-age people in the future and are considering ways to encourage births.

China and India are the two most populous countries in the world with about 41 percent of the world's population. Both have over 1 billion people. However, China has taken steps to control its population and now has a total fertility rate of 1.6 children per woman (below replacement levels), while India has a total fertility rate of 2.6. The total fertility rate is the average number of children born to a woman during her lifetime. The difference in the total fertility rate between these two countries is the result of different policy decisions over the last fifty years. The history of China's population policy is an interesting study of how government policy affects reproductive activity among its citizens. When the People's Republic of China was established in 1949, the official policy of the government was to encourage births because more Chinese would be able to produce more goods and services, and production was the key to economic prosperity. The population grew from 540 million to 614 million between 1949 and 1955 while economic progress was slow. Consequently, the government changed its policy and began to promote population control.

The first family-planning program began in 1955, as a means of improving maternal and child health. Birthrates fell, but other social changes resulted in widespread famine and increased death rates and low birthrates in the late 1950s and early 1960s.

The present family-planning policy began in 1971 with the launching of the *wan xi shao* campaign. Translated, this phrase means "later" (marriages), "longer" (intervals between births), and "fewer" (children). As part of this program, the legal ages for marriage were raised. For women and men in rural areas, the ages were raised to twenty-three and twenty-five, respectively; for women and men in urban areas, the ages were raised to twenty-five and twenty-eight, respectively. These policies resulted in a reduction of birthrates by nearly 50 percent between 1970 and 1979.

An even more restrictive "one child" campaign was begun in 1978–1979. The program offered incentives for couples to restrict their family size to one child. Couples enrolled in the program would receive free medical care, cash bonuses for their work, special housing treatment, and extra old-age benefits. Those who broke their pledge were penalized by the loss of these benefits as well as other economic penalties. By the mid-1980s, fewer than 20 percent of the eligible couples were signing up for the program. Rural couples, particularly, desired more than one child. In fact, in a country where about 60 percent of the population is rural, the rural total fertility rate was 2.5 children per woman. In 1988, a second child was sanctioned for rural couples if their first child was a girl, which legalized what had been happening anyway.

The programs appear to have had an effect because the current total fertility rate has fallen to 1.6 children per woman. Replacement fertility, the total fertility rate at which the population would eventually stabilize, is 2.1 children per woman per lifetime. Furthermore, more than 90 percent of couples use contraception. Abortion is also an important aspect of this program, with a ratio of more than six hundred abortions per thousand live births.

Although the program has been effective in reducing the birthrate, there has been one unexpected outcome: the sex ratio shifting in favor of males. (Among children the ratio is about 120 boys to 100 girls.) It is not clear what is causing the change; however, as in many societies, there is a strong social desire for sons. Because of this desire for sons, many speculate that one or more of the following may be the cause: female fetuses may be aborted more frequently than males; there may be a higher infant mortality rate among female children due to neglect; or female offspring may be given up for adoption more frequently than males. Recently the Chinese government has responded to this inequality with policy changes. Public information sources, advice from health care workers, and small financial incentives are all aimed at making the raising of girl children socially acceptable.

By contrast, during the same fifty years, India has had little success in controlling its population. In 2000, a new plan was unveiled that includes the goal of bringing the total fertility rate from 3.2 children per woman to 2 (replacement level) by 2010. In the past, the emphasis of government programs was on meeting goals of sterilization and contraceptive use, but this has not been successful. Today, the percentage of couples in various states using contraceptives ranges between 52.6 and 72.6 percent. This new plan emphasizes improvements in the quality of life of the people. The major thrusts are to reduce infant and maternal death, immunize children against preventable disease, and encourage girls to attend school. It is hoped that improved health will remove the perceived need for large numbers of births. According to the census held in 2001, the percentage of women in India who could read and write was 54.16 percent. The emphasis on improving the educational status of women is related to the experiences of other developing countries. In many other countries, it has been shown that an increase in the education level of women has been linked to lower fertility rates.

It seems overly optimistic to think that the total fertility rate can be reduced from 3.2 to 2 in just ten years, but it had declined to 2.72 in 2010. This is an encouraging sign and programs that emphasize improvements in maternal and child health and increasing the educational level of women have been very effective in reducing total fertility in other countries.

have been significantly impacted as well. Most fish populations are either overfished or are being fished at capacity.

When humans convert terrestrial or aquatic ecosystems to their use, these ecosystems are unavailable for their original inhabitants. This habitat destruction has resulted in a loss of biodiversity due to extinction and endangerment of species that compete with humans for resources. For example, the bison population of North America is restricted to a few protected areas, since nearly all the grasslands have been converted to agriculture and grazing. Habitat destruction in the ocean also occurs. The use of trawls—weighted nets that are pulled along the ocean bottom—to catch fish has greatly modified the ocean bottom in areas of high fishing activity.

Other human activities affect Earth globally. Global climate change is linked to increased amounts of carbon dioxide, methane, and other greenhouse gases in the atmosphere. Increased greenhouse gases lead to a general warming of Earth that results in the melting of glaciers, affects precipitation patterns, and causes a rise in sea level. These changes have the potential to cause significant change in biomes. The increase in atmospheric carbon dioxide is a direct result of the use of fossil fuels and the destruction of forests. The increases in methane can be traced to, in large part, domesticated livestock

that release methane from their guts and decomposition that occurs in flooded rice paddies.

In addition, persistent organic pollutants such as DDT, polychlorinated biphenyls (PCBs), and many other compounds from industry and mercury from the burning of coal in electric power plants have a global circulation and are found in humans and other animals throughout the world. Many kinds of ocean fish have elevated levels of mercury.

Lifestyles affect the degree of impact. Consumption of energy and other resources is much higher in the nations of the developed world than in those of the less developed world. The developed world, with less than 20 percent of the world's human population, consumes about 60 percent of the world's energy. The United States, with less than 5 percent of the world's population, consumes about 25 percent of the world's energy. The consumption of goods and services by the people of the developed world has a global impact. Even the food consumption habits of people have wide-ranging effects. Rainforests in Brazil are burned to create grazing land to raise beef to sell to developed nations.

As the human population continues to increase, so will the pressure to raise more food, produce more consumer products, and supply more energy. These needs will place increased pressure on the natural resources of Earth.

SUMMARY

Ecology is the study of how organisms interact with their environment. The *environment* consists of *biotic* and *abiotic* components that are interrelated in an *ecosystem*. All *ecosystems* must have a constant *input of energy* from the Sun. *Producer organisms* are capable of trapping the Sun's energy and converting it into *biomass*. *Herbivores* feed on producers and are in turn eaten by *carnivores*, which may be eaten by *other carnivores*. Each level in the *food chain* is known as a *trophic level*. Other kinds of organisms involved in food chains are *omnivores*, which eat both plant and animal food, and *decomposers*, which break down dead organic matter and waste products. All *ecosystems* have a large *producer* base with successively smaller amounts of energy at the *herbivore, primary carnivore,* and *secondary carnivore trophic levels*. This is because each time *energy* passes from *one trophic level* to the next, about 90 percent of the energy is lost from the *ecosystem*. A *community* consists of the interacting *populations of organisms* in an area. The organisms are interrelated in many ways in food chains that interlock to create *food webs*. Because of this interlocking, changes in one part of the *community* can have effects elsewhere.

Major land-based regional *ecosystems* are known as *biomes*. The *temperate deciduous forest, boreal coniferous forest, temperate rainforest, tropical rainforest, tropical dry forest, Mediterranean shrublands, temperate grasslands, desert, savanna,* and *tundra* are examples of biomes. Aquatic communities can be divided into marine, freshwater, and estuarine communities.

Each organism in a community occupies a specific space known as its *habitat* and has a specific functional role to play known as its *niche*. An *organism's habitat* is usually described in terms of some conspicuous element of its surroundings. The *niche* is difficult to describe because it involves so many interactions with the physical environment and other living things.

Interactions between organisms fit into several categories. *Predation* is one organism benefiting (*predator*) at the expense of the organism killed and eaten (*prey*). *Parasitism* is one organism benefiting (*parasite*) by living in or on another organism (*host*) and deriving nourishment from it. *Commensal* relationships exist when one organism is helped but the other is not affected. *Mutualistic* relationships benefit both organisms. *Competition* causes harm to both of the organisms involved, although one may be harmed more than the other and may become extinct, evolve into a different niche, or be forced to migrate.

Many atoms are cycled through ecosystems. The *carbon atoms* of living things are trapped by *photosynthesis,* passed from organism to organism as food, and released to the atmosphere by *respiration. Nitrogen* originates in the atmosphere, is trapped by *nitrogen-fixing bacteria,* passes through a series of organisms, and is ultimately released to the atmosphere by *denitrifying bacteria. Phosphorus* originates in rock and is used by organisms and eventually deposited as sediments.

A *population* is a group of organisms of the same species in a particular place at a particular time. *Populations* differ from one another in *gene frequency, age distribution, sex ratio,* and *population density*. A typical population growth curve consists of a *lag phase* in which population rises very slowly, followed by an *exponential growth phase* in which the population increases at an accelerating rate, followed by a leveling off during the *deceleration phase* until population growth stops in the *stable equilibrium phase* as the *carrying capacity* of the environment is reached.

Humans as a species have the same limits and influences that other organisms do. However, humans can reason and predict, thus offering the possibility of population control through *conscious population limitation*. Human activities have had many impacts on the global ecosystem. These include: habitat destruction, release of pollutants, climate change, and reduced biodiversity.

KEY TERMS

abiotic factors (p. **563**)
benthic (p. **573**)
biodiversity (p. **569**)
biomass (p. **567**)
biomes (p. **569**)
biotic factors (p. **563**)
carbon cycle (p. **581**)
carnivore (p. **566**)
carrying capacity (p. **590**)
commensalism (p. **579**)
community (p. **564**)
competition (p. **580**)
consumer (p. **565**)
deceleration phase (p. **590**)

decomposers (p. **566**)
ecology (p. **563**)
ecosystem (p. **564**)
environment (p. **563**)
environmental resistance (p. **590**)
estuary (p. **574**)
exponential growth phase (p. **590**)
food chain (p. **566**)
free-living nitrogen-fixing
    bacteria (p. **581**)
freshwater community (p. **572**)
habitat (p. **575**)
herbivore (p. **566**)
keystone species (p. **569**)

lag phase (p. **590**)
limiting factors (p. **590**)
marine community (p. **572**)
mortality (p. **590**)
mutualism (p. **579**)
natality (p. **590**)
niche (p. **575**)
nitrogen cycle (p. **582**)
omnivore (p. **566**)
organism (p. **564**)
parasitism (p. **577**)
pelagic (p. **573**)
plankton (p. **573**)
population (p. **586**)

population growth curve (p. **590**)
predation (p. **577**)
primary carnivore (p. **566**)
primary consumer (p. **566**)
producer (p. **564**)
secondary carnivore (p. **566**)
secondary consumer (p. **566**)
stable equilibrium phase
    (p. **590**)
symbiosis (p. **580**)
symbiotic nitrogen-fixing
    bacteria (p. **581**)
trophic level (p. **565**)

APPLYING THE CONCEPTS

Answers are located in appendix F.

1. As energy is passed from one trophic level to the next in a food chain, about _____ percent is lost at each transfer.
   a. 10
   b. 50
   c. 75
   d. 90

2. The primary factor that determines whether a geographic area will support temperate deciduous forest or prairie is
   a. the amount of rainfall.
   b. the severity of the winters.
   c. the depth of the soil.
   d. the kinds of animals present.

3. Which one of the following organisms is most likely to be at the second trophic level?
   a. a maple tree
   b. a snake
   c. a fungus
   d. an elephant

4. Which one of the following processes is necessary for the production of organic molecules from inorganic molecules?
   a. respiration by animals
   b. decomposer organisms releasing carbon dioxide
   c. carbon dioxide uptake by plants
   d. herbivores consuming plants

5. If two species of organisms occupy the same niche,
   a. mutualism will result.
   b. competition will be very intense.
   c. both organisms will become extinct.
   d. both will need to enlarge their habitat.

6. If nitrogen-fixing bacteria were to become extinct,
   a. life on Earth would stop immediately since there would be no source of nitrogen.
   b. life on Earth would continue indefinitely.
   c. life on Earth would slowly dwindle as useful nitrogen became less available.
   d. life on Earth would be unchanged except that proteins would be less common.

7. You obtain nitrogen for the organic molecules in your body from
   a. the air you breathe.
   b. the water you drink.
   c. carbohydrates in the food you eat.
   d. proteins in the food you eat.

8. Many plants (flowers) provide nectar for insects. The insects, in turn, pollinate the flower. This relationship between the insect and plant represents
   a. parasitism.
   b. commensalism.
   c. mutualism.
   d. predation.

9. As population density increases, which one of the following is likely to occur?
   a. Natality will increase.
   b. Mortality will decrease.
   c. The population will experience exponential growth.
   d. Individuals will migrate from the area of highest density.

10. A population made up primarily of prereproductive individuals will
    a. increase rapidly in the future.
    b. become extinct.
    c. rarely occur.
    d. remain stable for several generations.

11. Listed below are the sex ratios for four populations. All other things being equal (including current population size), which population should experience the greatest future growth?
    a. 1 male:1 female   c. 1 male: 2 female
    b. 2 male:1 female   d. 3 male: 1 female

12. The current human population of the world is experiencing
    a. a population decline.   c. stable equilibrium.
    b. slow steady growth.   d. rapid growth.

13. The kind and number of food organisms in an ecosystem are best characterized as
    a. biotic factors.   c. energy factors.
    b. abiotic factors.   d. ecological essentials.

14. Which is the correct hierarchy?
    a. ecosystem, community, individual, population
    b. ecosystem, community, population, individual
    c. community, ecosystem, individual, population
    d. individual, population. ecosystem, community

15. Considering various trophic levels on the basis of energy, which of the first in each pair is above the other?
    a. producer, carnivore
    b. herbivore, carnivore
    c. primary consumer, producer
    d. herbivore, producer

QUESTIONS FOR THOUGHT

1. Describe the flow of energy through an ecosystem.
2. What role does each of the following play in an ecosystem: sunlight, plants, the second law of thermodynamics, consumers, decomposers, herbivores, carnivores, and omnivores?
3. Why is there usually a larger herbivore biomass than a carnivore biomass?
4. List several biotic and abiotic factors that interact to give rise to the following biomes: temperate deciduous forest, boreal coniferous forest, desert, tundra, tropical rainforest, and savanna.
5. Can energy be recycled through an ecosystem? Explain why or why not.
6. Describe your niche.
7. What do parasites, commensal organisms, and mutualistic organisms have in common? How are they different?
8. Trace the flow of carbon atoms through a community that contains plants, herbivores, decomposers, and parasites.
9. Describe four different roles played by bacteria in the nitrogen cycle.
10. What kinds of organisms have their population size limited by the carrying capacity of their environment?
11. If competition is intense, how can both kinds of competing organisms continue to exist?
12. How does the population growth curve of humans compare with that of other kinds of animals?
13. List three environmental factors that would change as a result of building a new bridge over a river.
14. If variety in the number and kinds of producers is reduced in a field as the result of farming, what might happen to the variety of herbivores in that same field?
15. Look around your neighborhood and identify the various kinds of plants and animals you see. Create a food web based on how these organisms interact.

FOR FURTHER ANALYSIS

1. Many cities have serious problems with large populations of deer. They eat the plants in people's yards and cause serious traffic accidents. You are a consultant hired to help a city solve its deer population problem. Using what you have learned about the principles of population growth, describe two alternative plans that would work.
2. Sulfur is present in all organisms, particularly in proteins. Based on what you know about the carbon and nitrogen cycles, develop an outline of what you think the sulfur cycle would look like.
3. Many people advocate the use of biofuels (biodiesel, ethanol, etc.) as a substitute for fossil fuels. They suggest that these fuels cause less environmental damage and are renewable. Based on what you know about energy flow through ecosystems and the effect of agriculture on biodiversity, what kinds of questions would you like to ask these advocates of biofuels production?
4. Describe how one might go about the process of conserving an endangered species.
5. Compare your social interactions with other people to the kinds of ecological interactions that occur among organisms. List examples of social interactions that could be considered to be:
   mutualism
   parasitism
   commensalism
   competition

INVITATION TO INQUIRY

World Population Characteristics

The Population Reference Bureau maintains a website that contains the World Population Data Sheet. It produces a new data sheet each year. Go to that website (http://www.prb.org) and access the World Population Data Sheet. Determine which countries have each of the following:

1. The highest infant mortality rate
2. The highest total fertility rate
3. The lowest life expectancy

Compare the countries you have listed. In what ways are they similar? Now determine which countries have each of the following:

1. The lowest infant mortality rate
2. The lowest total fertility rate
3. The highest life expectancy

Compare the countries you have listed. In what ways are they similar?

24

Human Biology: Materials Exchange and Control Mechanisms

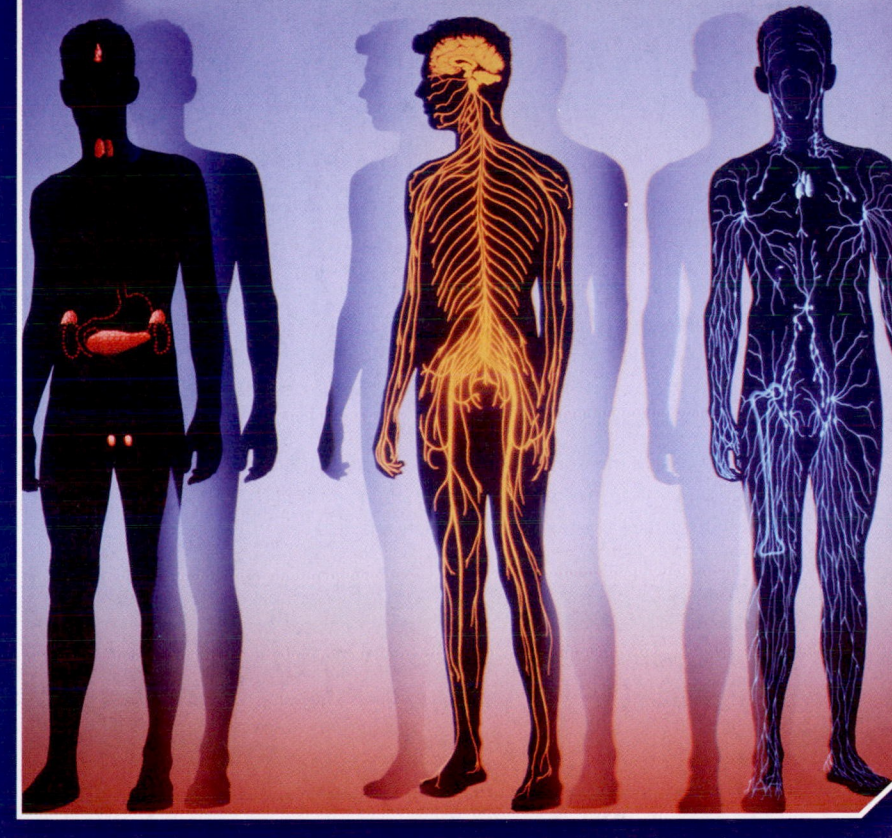

Because many different systems each perform a specific set of tasks in the human body, it is necessary to coordinate these various activities with one another

CORE CONCEPT

The human body consists of many interacting systems.

OUTLINE

597

Physics

▷ The kinetic molecular theory is key to understanding basic physiology (Ch. 4).

▷ Pressure is a force (Ch. 4).

▷ Gas laws explain certain physiological processes (Ch. 4).

Chemistry

▷ pH is a measure of hydrogen ion concentration (Ch. 10).

▷ Many large organic molecules are polymers (Ch. 19).

▷ Ions are positively or negatively charged (Ch. 10).

Earth Science

▷ Minerals are inorganic materials (Ch. 15).

Life Science

▷ Enzymes control biological reactions (Ch. 20).

▷ Aerobic cellular respiration requires oxygen (Ch. 20).

▷ Cell membranes are selectively permeable (Ch. 20).

OVERVIEW

All cells must have nutrients, waste products, gases, and other molecules pass through their cell membranes. In single-celled organisms, the cell membrane is in direct contact with the cell's surroundings so diffusion and other processes accomplish this exchange. However, large, multicellular organisms like humans consist of trillions of individual cells, most of which are not located near the surface of the organism yet must exchange materials in the same manner as in single-celled creatures. Therefore, large organisms consist of many different systems that assist in the exchange of materials such as food, wastes, and gases between the cells of the organisms and the external environment. The primary systems involved are the circulatory, respiratory, digestive, and excretory systems.

These systems must be coordinated and regulated in such a way that they supply the needs of the individual cells. Two other systems, the nervous and endocrine systems, are involved in managing these functions. This chapter will look at the systems involved in the exchange of materials and coordination of activities within the human body.

24.1 HOMEOSTASIS

Living things are like complex machines with many parts that must function in a coordinated fashion. All of their systems are integrated and affect one another in many ways. For example, when you run up a hill, your leg and arm muscles move in a coordinated way to provide power. They burn fuel (glucose) for energy and produce carbon dioxide and lactic acid as waste products, which tend to lower the pH of the blood. Your heart beats faster to provide oxygen and nutrients to the muscles, you breathe faster to supply the muscles with oxygen and get rid of carbon dioxide, and the blood vessels in the muscles dilate to allow more blood to flow to them. As you run, you generate excess heat. As a result, more blood flows to the skin to get rid of the heat and sweat glands begin to secrete, thus cooling the skin. All of these automatic internal adjustments help to regulate the pH of your blood; the concentration of oxygen, carbon dioxide, and glucose in the blood; and body temperature within narrow ranges. These processes can be summed up in the concept of *homeostasis*. **Homeostasis** is the process of maintaining a constant internal environment as a result of monitoring and modifying the functioning of various systems.

24.2 EXCHANGING MATERIALS: BASIC PRINCIPLES

In order to maintain homeostasis, all organisms must exchange materials with their surroundings. For materials to enter or leave an organism, they must pass through a membrane. The ability to transport materials into or out of an organism is determined by the surface area where exchange takes place. As you increase the size of an object, the volume increases faster than the surface area. (See chapter 1, page 8, for a discussion of surface-area-to-volume ratio.) So, the larger an organism or cell becomes, the more difficult it is to satisfy its needs.

A major mechanism for exchange of materials between organisms and their environment is diffusion. In addition to the limitation that surface area presents to the transport of materials, cells also have a problem with diffusion of materials within the cell. The molecular process of diffusion is quite rapid over short distances but becomes very slow over longer distances. Diffusion is generally insufficient to handle the needs of cells if it must take place over a distance of more than 1 mm. The center of the cell would die before it received the molecules it needed if the distance

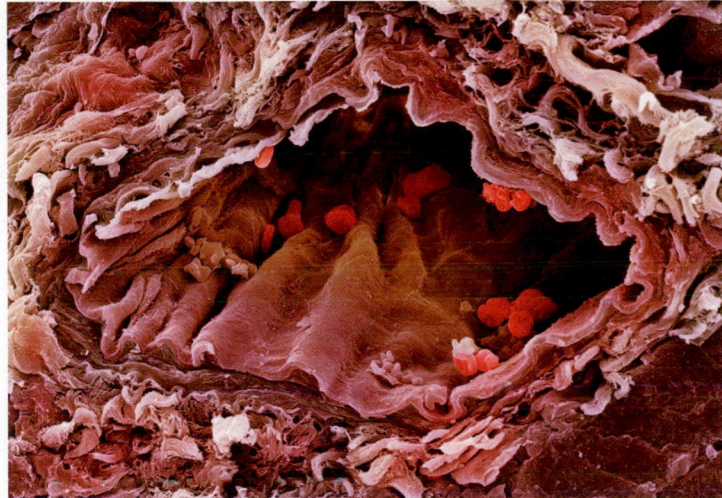

FIGURE 24.1 This color-highlighted photo was taken with a scanning electron microscope. It shows how a blood vessel (artery) brings oxygen-carrying red blood cells into close contact with oxygen-requiring cells found deep inside the body.

were greater. Because of this, it is understandable that the basic unit of life, the cell, must remain small.

When we look at the evolution of organisms, we find that the first organisms were small, single-celled, and marine. Under these conditions, most exchanges could simply occur through their cell membranes. As more complex, multicellular organisms evolved, they had many cells that were not in contact with their surroundings and developed other ways to allow for efficient exchange of materials (figure 24.1). The evolution of organs and organ systems allowed them to exchange and distribute materials to all cells even if many cells were not exposed to the sea. Often it is possible to see a pattern of evolutionary development of the structures involved in the exchange of materials. For example, many tiny multicellular marine animals still can get sufficient oxygen through their surfaces without special structures. Larger marine animals have some sort of gill (a folding of the body surface) to assist gas exchange.

24.3 TRANSPORTING MATERIALS: THE CIRCULATORY SYSTEM

Large, multicellular organisms such as humans consist of trillions of cells. Since many of these cells are buried within you far from your body's surface, there must be some sort of distribution system to assist them in solving their materials exchange problems. The primary mechanism used is your circulatory system.

The human circulatory system consists of several fundamental parts. **Blood** is the fluid that assists in the transport of materials and heat. Your **heart** is a pump that forces the fluid blood from one part of your body to another. The heart pumps blood into **arteries,** which distribute blood to organs. It flows into successively smaller arteries until it reaches tiny vessels called **capillaries,** where materials are exchanged between the blood and tissues through the walls of the capillaries. The blood flows from the capillaries into **veins** that combine into larger veins that ultimately return the blood to the heart from the tissues.

The Nature of Blood

Your blood is a fluid that consists of several kinds of cells suspended in a watery matrix called **plasma.** Table 24.1 lists the chemical composition of the plasma and the variety of cells found in blood. The plasma contains many kinds of dissolved molecules. The primary function of the blood is to transport molecules, cells, and heat from one part of the body to another. The major kinds of molecules that are distributed by the blood are respiratory gases (oxygen and carbon dioxide), nutrients, waste products, disease-fighting antibodies, and chemical messengers (hormones). Blood has special characteristics that allow it to distribute respiratory gases very efficiently. Although little oxygen is carried as free, dissolved oxygen molecules in the plasma, *red blood cells* (*RBCs*)—also known as *erythrocytes*—contain **hemoglobin,** an iron-containing molecule, to which oxygen molecules readily bind. This allows for much more oxygen to be carried than could be possible if it were simply dissolved in your blood. Thus, the circulating blood is constantly picking up and delivering molecules to where they are needed. This is important for maintaining homeostasis.

Another important regulatory function of your blood involves the various kinds of cells involved in immunity. The *white blood cells* (*WBCs*) carried in your blood help the body resist many diseases. They constitute the core of the *immune system.* The various WBCs participate in providing immunity in several ways. First of all, immune system cells are able to recognize cells and molecules that are harmful to your body. If a molecule is recognized as harmful, certain white blood cells called *lymphocytes* produce *antibodies* (*immunoglobulins*) that attach to the dangerous materials. The dangerous molecules that stimulate the production of antibodies are called *antigens* (*immunogens*). When harmful microorganisms (e.g., bacteria, viruses, fungi) or toxic molecules (e.g., botulism poison, tetanus toxin, diphtheria toxin) enter the body or cancer cells are produced, a combination of several kinds of lymphocytes (1) recognize, (2) boost their abilities to respond to, (3) move toward, and (4) destroy the problem causers. *Neutrophils* and *monocytes* are specific kinds of WBCs capable of engulfing harmful material, a process called phagocytosis. Thus, they are often called *phagocytes.* While most can move from the bloodstream into the surrounding tissue, monocytes undergo such a striking increase in size that they are given a different name—*macrophages.* Macrophages can be found throughout your body and are the most active of the phagocytes. *Eosinophils* are primarily involved in destroying multicellular parasites such as worms, and *basophils* appear to be primarily involved in the inflammation response, which results in increased blood flow to an injured area, causing swelling and reddening.

Another kind of cellular particle in your blood is the *platelet.* These are fragments of specific kinds of white blood cells and are important in blood clotting. They collect at the site of a wound, where they break down, releasing specific molecules, which begin a series of reactions that results in the formation of fibers that trap blood cells and form a plug in the opening of the wound. This mixture of fibers and trapped blood cells dries and hardens to form a scab.

TABLE 24.1   The Composition of Blood

Component	Quantity Present	Function
Plasma	55%	Liquid that acts as the transportation, pH and osmosis regulator, and "carrier." Of various ions (clotting factors, nutrients, CO_2, O_2, N_2, hormones, cholesterol, lipids, fatty acids, waste products, heat, cellular materials, platelets); protein is the regulation of osmotic pressure.
Water	91.5%	
Protein	7.0%	
Other materials	1.5%	
Cellular material	45%	
Red blood cells (erythrocytes)	4.3–5.8 million/mm³	Carry oxygen and carbon dioxide
White blood cells (leukocytes)	5–9 thousand/mm³	Immunity, phagocytosis, pus formation, produce antibodies, histamines, and cytokines; associated with some allergies; attack infecting microbes, parasites, and cancer cells.
Lymphocytes	25%–30% of white cells present	
Monocytes	3%–7% of white cells present	
Neutrophils	57%–67% of white cells present	
Eosinophils	1%–3% of white cells present	
Basophils	Less than 1% of white cells present	
Platelets	250–400 million/ml	Clotting is source of growth factors to aid healing, stimulate inflammation, transport hormones, aid in new blood vessel formation; involved in cancer.

Neutrophils   Eosinophils   Basophils

Lymphocytes   Monocytes   Platelets   Erythrocytes

The Heart

Your blood can perform its transportation function only if it moves. The organ responsible for providing the energy to pump the blood is your heart. The heart is a muscular pump that provides the pressure necessary to propel the blood throughout your body. It must continue its cycle of contraction and relax-ation, or blood stops flowing and body cells are unable to obtain nutrients or get rid of wastes. Some cells, such as brain cells, are extremely sensitive to having their flow of blood interrupted because they require a constant supply of glucose and oxygen. Others, such as muscle cells or skin cells, are much better able to withstand temporary interruptions of blood flow.

The use of performance-enhancing techniques and drugs in sports is not new. For example, one way to increase the amount of oxygen available to muscles is to inhale higher concentrations of the gas. You may have seen athletes breathing from an oxygen tank while on the sidelines during a sporting event. Another approach to increasing the amount of available oxygen is to increase the number of oxygen-carrying red blood cells. This can be accomplished in several ways. An athlete can have a pint of his or her blood removed and put into storage about three weeks before an event. During that time, the athlete's body replaces the removed red blood cells. Just before the event, the "donated" blood is transfused back into the athlete, increasing the total number of red blood cells. This method of enhancing an athlete's performance has been banned because it is considered to give an unfair advantage and because there can be dangerous side effects.

Another technique involves artificially stimulating the body to produce more red blood cells by injecting the protein hormone erythropoietin, commonly referred as EPO. The kidney produces EPO; after it is released into the bloodstream, it binds to receptors on stem cells in the red bone marrow, where it stimulates the production of new red blood cells. The increased red blood cell production in turn increases the oxygen-carrying capacity of the blood. This fact has resulted in some members of the athletic community using EPO as a performance-enhancing drug to artificially increase the number of red blood cells. Still another method of increasing available oxygen is the use of synthetic oxygen carriers, such as hemoglobin-based oxygen carriers (HBOCs) or perfluorocarbons (PFCs). These are purified proteins or chemicals that have the ability to carry oxygen. They can be useful in emergencies when human blood is not available, when the risk of blood infection is high, or when there is not enough time to properly cross-match donated blood with a recipient. However, their misuse for doping carries the risk of cardiovascular disease, in addition to other side effects such as stroke, heart attack, or blood clot.

The use of blood transfusions, EPO, and synthetic oxygen carriers is prohibited under the World Anti-Doping Agency's (WADA) List of Prohibited Substances and Methods.

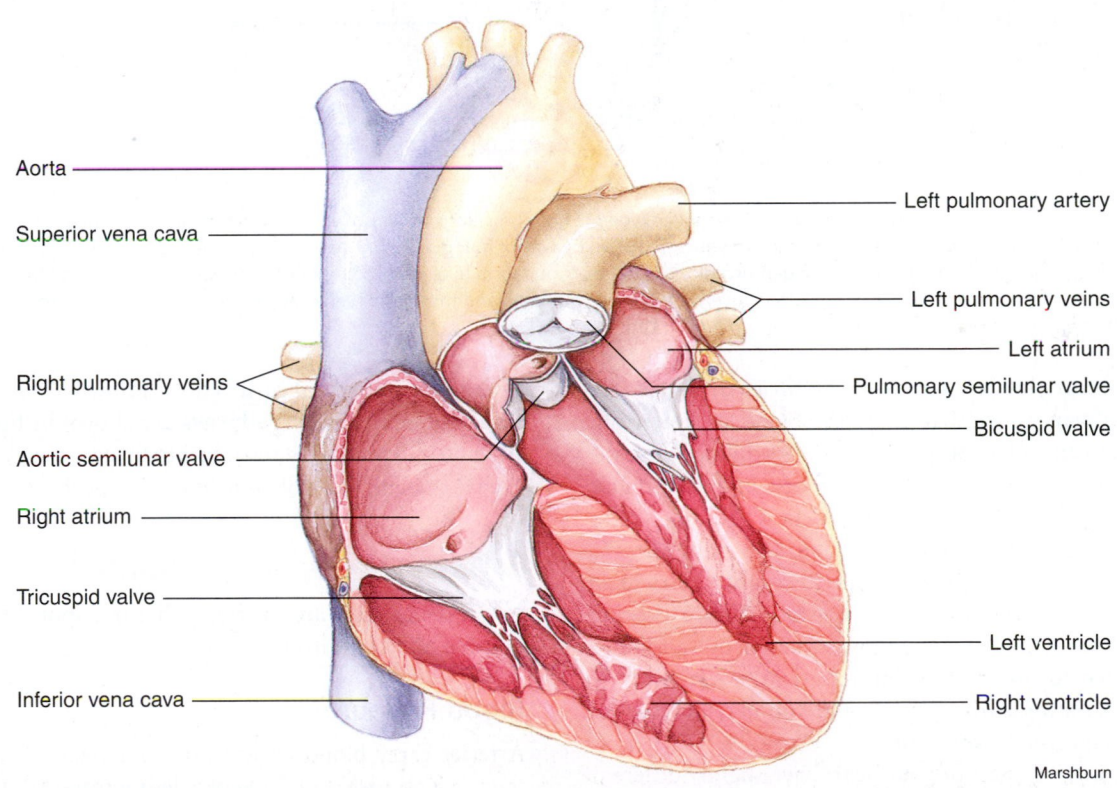

Aorta

Superior vena cava

Right pulmonary veins

Aortic semilunar valve

Right atrium

Tricuspid valve

Inferior vena cava

Left pulmonary artery

Left pulmonary veins

Left atrium

Pulmonary semilunar valve

Bicuspid valve

Left ventricle

Right ventricle

Marshburn

FIGURE 24.2 The heart consists of two thin-walled chambers called atria that contract to force blood into the two ventricles. When the ventricles contract, the atrioventricular valves (bicuspid and tricuspid) close, and blood is forced into the aorta and pulmonary artery. Semilunar valves in the aorta and pulmonary artery prevent the blood from flowing back into the ventricles when they relax.

The hearts of humans, other mammals, and birds consist of four chambers and four sets of valves that work together to ensure that blood flows in one direction only. Two of these chambers, the right and left *atria* (singular, *atrium*), are relatively thin-walled structures that collect blood from the major veins and empty it into the larger, more muscular ventricles (figure 24.2).

The right and left *ventricles* are chambers that have powerful muscular walls whose contraction forces blood to flow through the arteries to all parts of the body. The valves between

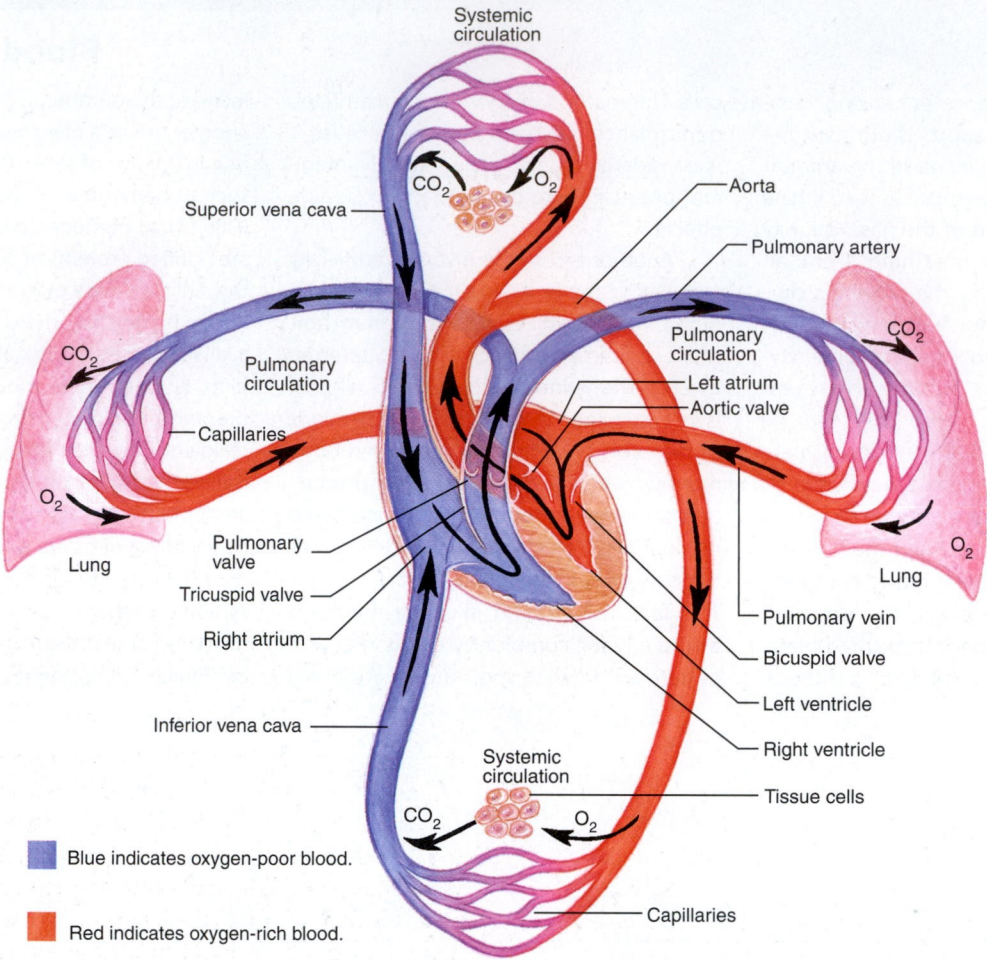

Systemic circulation

Superior vena cava

Pulmonary circulation

Capillaries

Pulmonary valve

Tricuspid valve

Right atrium

Inferior vena cava

Systemic circulation

Lung

O_2

O_2

CO_2

CO_2

Aorta

Pulmonary artery

Pulmonary circulation

Left atrium

Aortic valve

Lung

CO_2

O_2

Pulmonary vein

Bicuspid valve

Left ventricle

Right ventricle

Tissue cells

Capillaries

Blue indicates oxygen-poor blood.

Red indicates oxygen-rich blood.

FIGURE 24.3 The right ventricle pumps blood that is poor in oxygen to your two lungs by way of your pulmonary arteries, where the blood receives oxygen and turns bright red. Your blood is then returned to your left atrium by way of four pulmonary veins. This part of the circulatory system is known as pulmonary circulation. Your left ventricle pumps oxygen-rich blood by way of your aorta to all parts of your body except the lungs. This blood returns to your right atrium, depleted of its oxygen, by way of the superior vena cava from your head and the inferior vena cava from the rest of your body. This portion of the circulatory system is known as systemic circulation.

the atria and ventricles, known as *atrioventricular valves,* are important one-way valves that allow the blood to flow from the atria to the ventricles but prevent flow in the opposite direction. Similarly, there are valves in the aorta and pulmonary artery, known as *semilunar valves.* The **aorta** is the large artery that carries blood from the left ventricle to the body, and the **pulmonary artery** carries blood from the right ventricle to the lungs. The semilunar valves prevent blood from flowing back into the ventricles. If the atrioventricular or semilunar valves are damaged or function improperly, the efficiency of your heart as a pump is diminished, and you may develop an enlarged heart or other symptoms.

The right and left sides of your heart have slightly different jobs, because they pump blood to different parts of the body. The right side of your heart receives blood from the general body and pumps it through the pulmonary arteries to your lungs, where exchange of oxygen and carbon dioxide takes place, and the blood returns from the lungs to the left atrium. This is called **pulmonary circulation.** The larger, more powerful left side of your heart receives oxygenated blood from your lungs, delivers it through the aorta to all parts of your body, and returns it to the right atrium by way of large veins known as the *superior vena*

cava and *inferior vena cava.* This is known as **systemic circulation.** Both circulatory pathways are shown in figure 24.3. The systemic circulation is responsible for gas, nutrient, and waste exchange in all parts of your body except the lungs.

Arteries, Veins, and Capillaries

Arteries and veins are the tubes that transport blood from one place to another within the body.

Blood Pressure

Arteries carry blood away from your heart because the contraction of the walls of the ventricles increases the pressure in the

CONCEPTS APPLIED

Go with the Flow

Look at figure 24.3. Start with blood in the left lung and trace the flow of blood until it arrives back at the left lung. Make a list of the structures encountered.

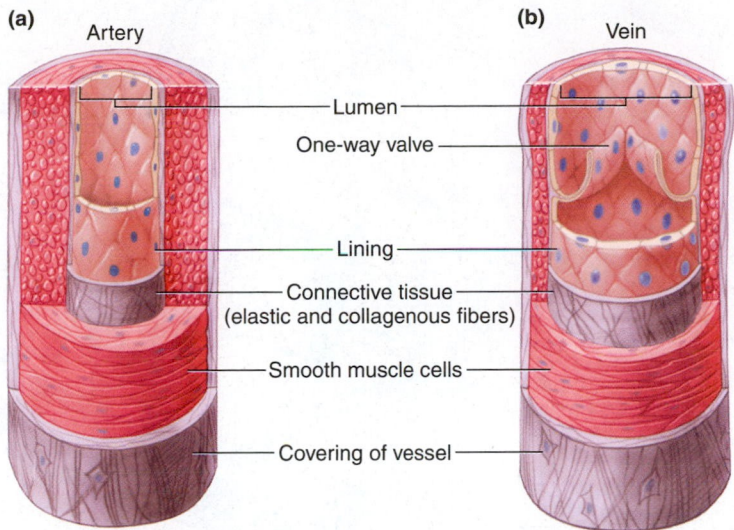

(a) Artery

(b) Vein

Lumen

One-way valve

Lining

Connective tissue (elastic and collagenous fibers)

Smooth muscle cells

Covering of vessel

FIGURE 24.4   The walls of arteries are much thicker than the walls of veins. (The pressure in arteries is much higher than the pressure in veins.) The pressure generated by the ventricles of the heart forces blood through the arteries. Veins often have very low pressure. The valves in the veins prevent the blood from flowing backward, away from the heart.

arteries compared to that in the capillaries and veins downstream of the heart. A typical pressure recorded in a large artery while the heart is contracting is about 120 millimeters of mercury. This is known as your **systolic blood pressure.** The pressure recorded while the heart is relaxing is about 80 millimeters of mercury. This is known as the **diastolic blood pressure.** A blood pressure reading includes both of these numbers and is recorded as 120/80. Most physicians consider a systolic pressure above 140 or a diastolic pressure above 90 to be cause for concern and will recommend treatment for high blood pressure.

Structure and Function of Arteries and Veins

The walls of arteries are relatively thick and muscular yet elastic. Healthy arteries have the ability to expand as blood is pumped into them and return to normal as the pressure drops. This ability to expand absorbs some of the pressure and reduces the peak pressure within the arteries, thus reducing the likelihood that they will burst. If arteries become hardened and less resilient, the peak blood pressure rises and they are more likely to rupture. The elastic nature of the arteries is also responsible for assisting the flow of blood. When they return to normal from their stretched condition, they give a little push to the blood that is flowing through them. Depending on their size, a cut artery will release blood in spurts.

Blood is distributed from the large aorta through smaller and smaller blood vessels to millions of tiny capillaries. Some of the smaller arteries, called **arterioles,** may contract or relax to regulate the flow of blood to specific parts of the body. Major parts of the body that receive differing amounts of blood, depending on need, are the digestive system, muscles, and skin.

Veins collect blood from the capillaries and return it to the heart. The pressure in these blood vessels is very low. Some of the largest veins may have a blood pressure of 0.0 mmHg for brief periods. Since pressure in veins is so low, muscular movements

of the body are important in helping to return blood to the heart. When muscles of the body contract, they compress nearby veins, and this pressure pushes blood along in the veins. Because valves in the veins allow blood to flow only toward the heart, this activity acts as an additional pump to help return blood to your heart. Depending on their size, a cut vein will *not* release blood in spurts; it will ooze out in a slow, steady flow. Figure 24.4 compares the structure and function of arteries and veins.

Structure and Function of Capillaries

Although the arteries are responsible for distributing blood to various parts of your body and arterioles regulate where blood goes, it is the function of capillaries to assist in the exchange of materials between the blood and cells. Capillaries are tiny, thin-walled tubes that receive blood from arterioles. They are so small that red blood cells must go through them in single file. They are so numerous that each cell in the body has a capillary located near it. It is estimated that there are about 1,000 m^2 of surface area represented by your capillary surfaces. Each capillary wall consists of a single layer of cells and therefore presents only a thin barrier to the diffusion of materials between your blood and the cells that surround the capillaries. Figure 24.5 shows the close physical relationship between capillaries and body cells. The flow of blood through these smallest blood vessels is relatively slow. This allows time for the diffusion of such materials as oxygen, glucose, and water from your blood to surrounding cells and for the movement of such materials as carbon dioxide, lactic acid, and ammonia from the cells into your blood.

In addition to molecular exchange by diffusion, considerable amounts of water and dissolved materials leak through the small holes in the capillaries. This liquid is known as **lymph.** Lymph is produced when the blood pressure forces water and some small dissolved molecules through the walls of the capillaries. Lymph bathes your cells but must eventually be returned

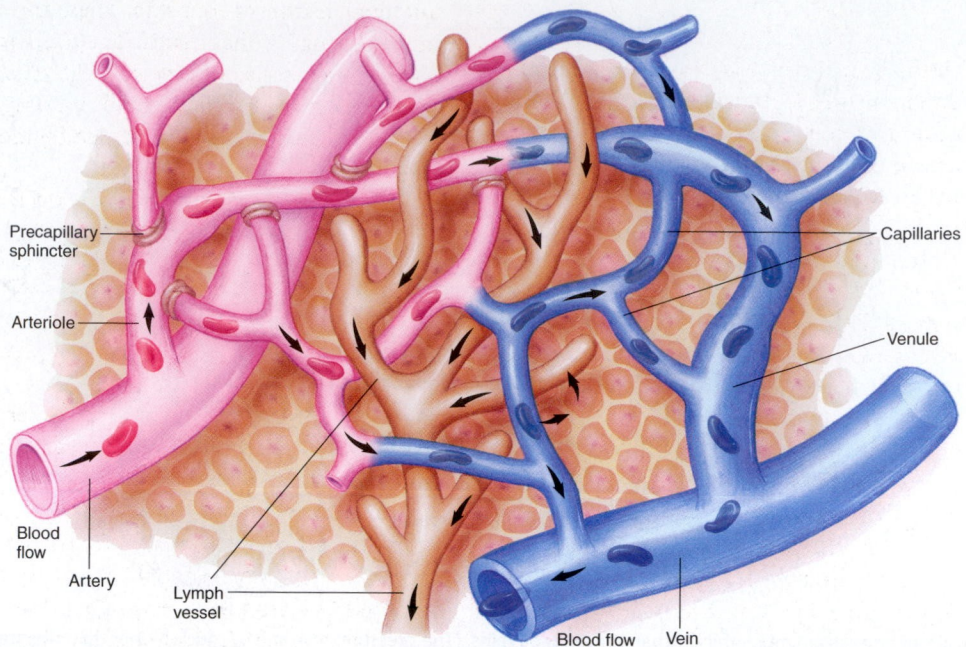

FIGURE 24.5 Capillaries are tiny blood vessels. Exchange of cells and molecules can occur between blood and tissues through their thin walls. Molecules diffuse in and out of your blood, and cells such as monocytes can move from the blood through the thin walls into the surrounding tissue. There is also a flow of liquid through holes in the capillary walls. This liquid, called lymph, bathes the cells and eventually enters small lymph vessels that return lymph to the circulatory system near your heart.

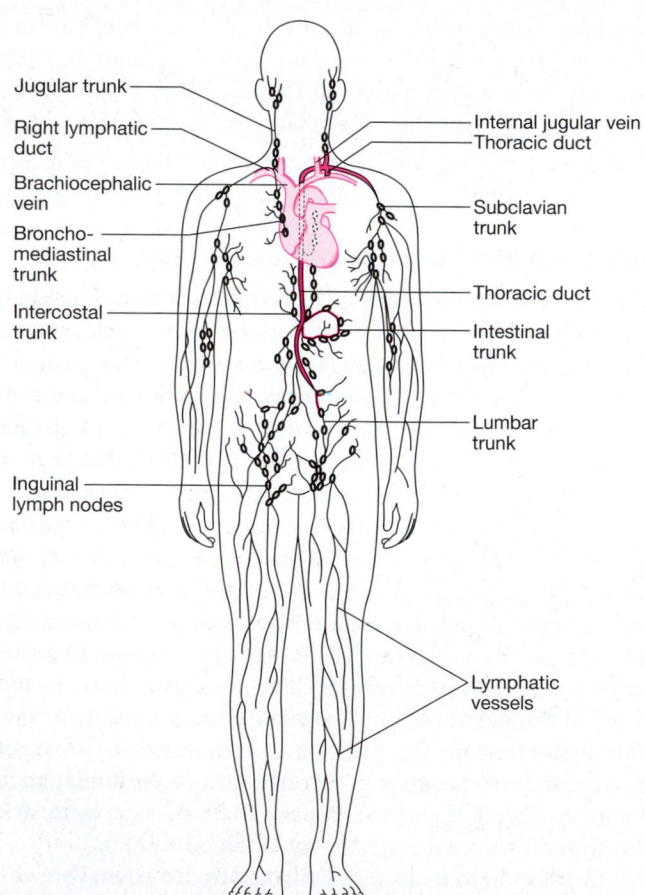

FIGURE 24.6 The lymphatic system consists of many ducts that transport lymph fluid back toward the heart. Along the way the lymph is filtered in the lymph nodes, and bacteria and other foreign materials are removed before the lymph is returned to the circulatory system near the heart.

to your circulatory system by lymph vessels or swelling will occur. Return is accomplished by the *lymphatic system,* a collection of thin-walled tubes that branch throughout your body. These tubes collect lymph that left the circulatory system and ultimately empty it into major blood vessels near your heart. Figure 24.6 shows the structure of the lymphatic system.

24.4 SKIN: THE BODY'S CONTAINER

Your skin is the outer covering of your body. It is the largest organ in the body, has a relatively complex structure, and serves several important functions.

Primary Functions of the Skin

Although it serves several important functions, the importance of your skin is often overlooked. It acts as an impermeable barrier between you and your surroundings. While aquatic animals such as fish and amphibians can get by with a water-permeable skin, terrestrial animals such as reptiles, birds, and mammals must prevent the loss of water from their body surface. An impermeable skin also is a barrier to the entry of microorganisms and chemicals.

Your skin constitutes about 7,600 cm² (3,000 in²). This is about the surface area of a 4-by-5-ft rug. The large surface area of your skin is important in getting rid of heat. Shunting blood to the skin results in heat loss (as long as the environmental temperature is less than the body temperature), and the secretions of sweat glands spread over this large surface provide evaporative cooling.

The Structure of the Skin

The skin is divided into three distinct layers: the epidermis, the dermis, and the subcutaneous layers. The epidermis is the outer surface of your skin that you can see. It consists of a layer of cells at its base, next to the dermis, that undergoes mitosis to produce new cells. This layer of cells is called the germinating layer. The cells produced by the germinating layer migrate toward the surface. As they migrate to the surface, they produce a protein called keratin, lose their nuclei and other structures, and become flattened disks. This outer layer of the epidermis is called the cornified layer and consists of several layers of these flattened, scalelike, keratin-filled cell remnants. The cornified layer provides toughness to the epidermis and is impermeable to most chemicals and microorganisms. Thus, the cornified layer prevents the loss of water from and the entry of microorganisms to the underlying tissues of the body.

Since the cornified layer is in direct contact with the environment, it is constantly worn away by friction with surroundings. The constant replenishment of this layer from below, as a result of the germinating layer, is essential to maintain this function of the skin. If the cornified layer is destroyed, people lose fluids very rapidly and are subject to invasion by various kinds of disease microorganisms. This is probably most dramatically illustrated by the problems experienced by victims of heat or chemical burns in which the cornified layer is destroyed. They must receive intravenous fluids to replace the liquids lost and often succumb to generalized infections because they have no barrier to microorganisms.

The dermis is immediately below the epidermis. It contains a great many connective tissue fibers that provide the flexible but tough texture of your skin. There are many blood vessels and nerve endings in the dermis. The overlying epidermis does not have blood vessels or nerve endings. The cells of the epidermis receive nutrition from the blood vessels in the dermis, and pain, touch, and temperature are sensed by the nerve endings in the dermis.

The subcutaneous layer lies below the dermis and contains a great many connective tissue fibers and fat cells. The fibers in this layer bind the skin to underlying tissues, primarily muscle. There are also blood vessels and nerves in this layer.

Other Features of the Skin

Several other features of your skin deserve comment. Melanin is a brown-black pigment produced by special cells in the germinating layer. Dark-skinned people have melanin-producing cells that are more active than those of light-skinned people. Melanin is a protection against the damaging effects of ultraviolet light. When people are exposed to ultraviolet light, the melanin-producing cells are stimulated to produce more melanin, they spread out, and tanning occurs. There are two types of melanin: eumelanin and phaeomelanin. When UV radiation strikes eumelanin, the color of the skin becomes golden-brown. When UV radiation strikes phaeomelanin, the color of the skin becomes redder in color. People with red hair and blondes produce more phaeomelanin and less eumelanin and do not tan as well.

Hair and nails (and hooves, claws, and horns in other mammals) are composed of keratin. Therefore, they are specialized accumulations produced by the epidermis of the skin.

Several kinds of glands are derived from the epidermis but project down into the underlying dermis. Oil glands associated with hair follicles secrete oily material that is released onto your skin surface where it lubricates the skin and hair. Scent glands are also associated with hair follicles and produce a watery secretion that contains particular odor molecules. In humans, scent glands are found primarily in your armpit and pubic area. Other mammals may have scent glands near their eyes, around the anus, on their feet, or on their belly. Humans and a few other mammals have sweat glands over much of their skin surfaces. As mentioned earlier, evaporation of sweat from the skin surface cools your body. Finally, mammary glands, which are responsible for milk production, are specialized sweat glands. Figure 24.7 illustrates the various parts of the skin.

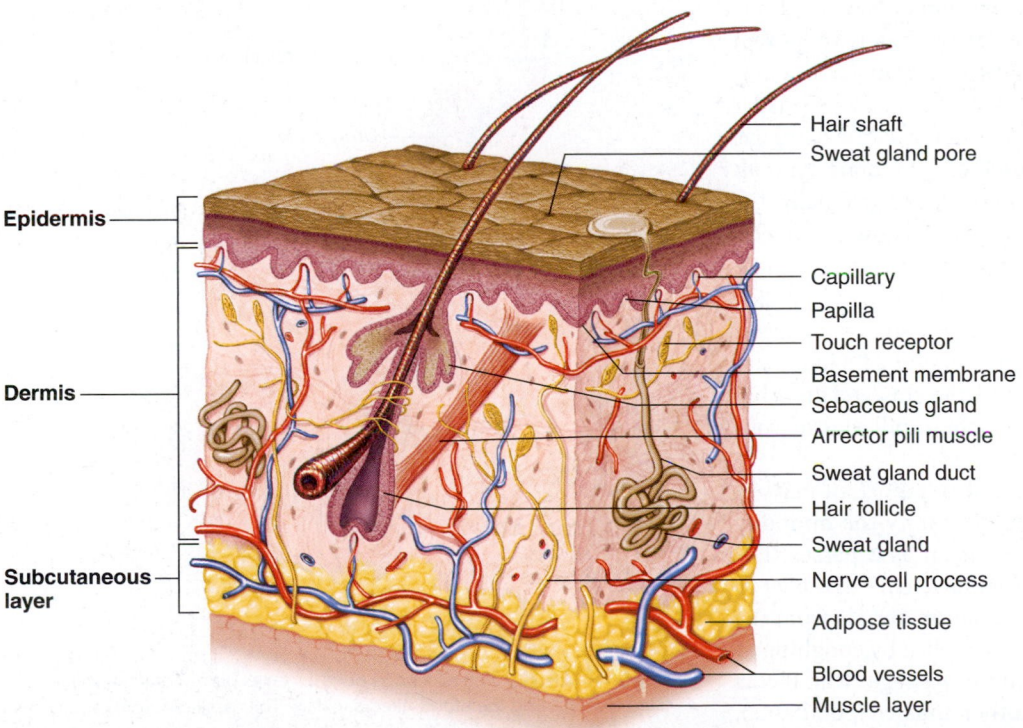

Epidermis

Dermis

Subcutaneous layer

Hair shaft
Sweat gland pore

Capillary
Papilla
Touch receptor
Basement membrane
Sebaceous gland
Arrector pili muscle
Sweat gland duct
Hair follicle
Sweat gland
Nerve cell process
Adipose tissue
Blood vessels
Muscle layer

FIGURE 24.7 The skin is the largest organ of the body and consists of several layers and structures that prevent water loss, protect against infection, provide for cooling, and serve as sense organs.

24.5 EXCHANGING GASES: THE RESPIRATORY SYSTEM

In terrestrial vertebrates, the primary structures involved in gas exchange are the lungs. Since the lungs are internal structures, the amount of water lost is much less than if these moist surfaces were on the outer surface of the body, as are the gills of fish.

Structure and Function of Lungs

The **lungs** are organs of your body that allow gas exchange to take place between the air and blood. Associated with the lungs is a set of tubes that conducts air from outside the body to your lungs. The single large-diameter **trachea** is supported by rings of cartilage that prevent its collapse. It branches into two major **bronchi** (singular, *bronchus*) that deliver air to smaller and smaller branches. Bronchi are also supported by cartilage. The smallest tubes, known as **bronchioles,** contain smooth muscle and are therefore capable of constricting. Finally, the bronchioles deliver air to clusters of tiny sacs, known as **alveoli** (singular, *alveolus*), where the exchange of gases takes place between the air and blood. The alveoli are very thin-walled and surrounded by capillaries. This close association makes gas exchange by diffusion relatively easy. Figure 24.8 illustrates the parts of the respiratory system and the close relationship between the alveoli of your lungs and the capillaries of your circulatory system.

Your nose, mouth, and throat are also important parts of the air-transport pathway because they modify the humidity and temperature of the air and clean the air as it passes. The lining of the trachea contains cells that have cilia that beat in a direction that moves mucus and foreign materials from the lungs. The foreign matter may then be expelled by coughing or "clearing your throat," thus keeping the air passage clear. If cilia are inhibited (e.g., by nicotine or other inhaled pollutants), bacteria, viruses, and dangerous chemicals can move into the lungs, leading to diseases such as bronchitis, pneumonia, and emphysema.

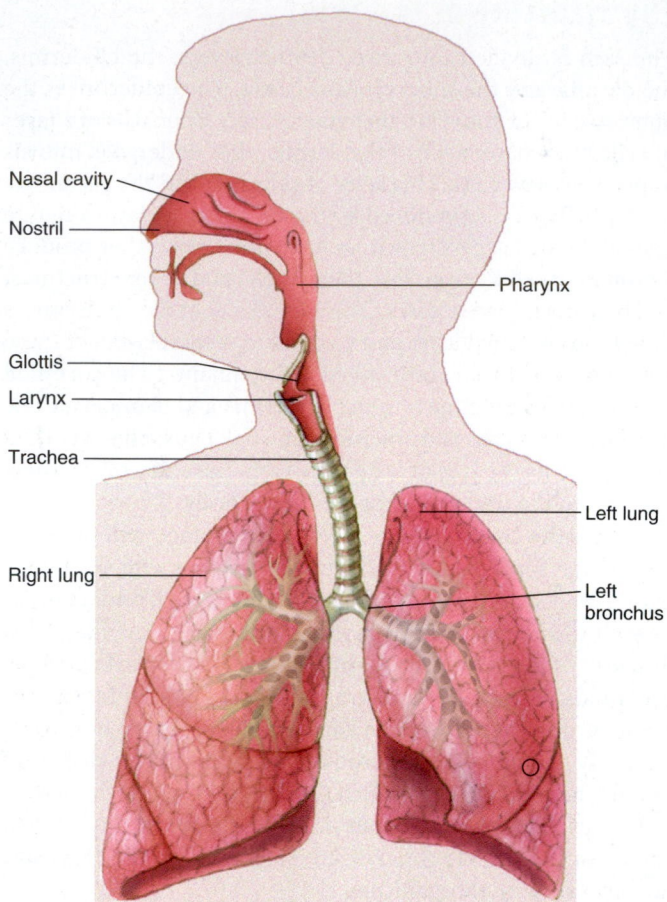

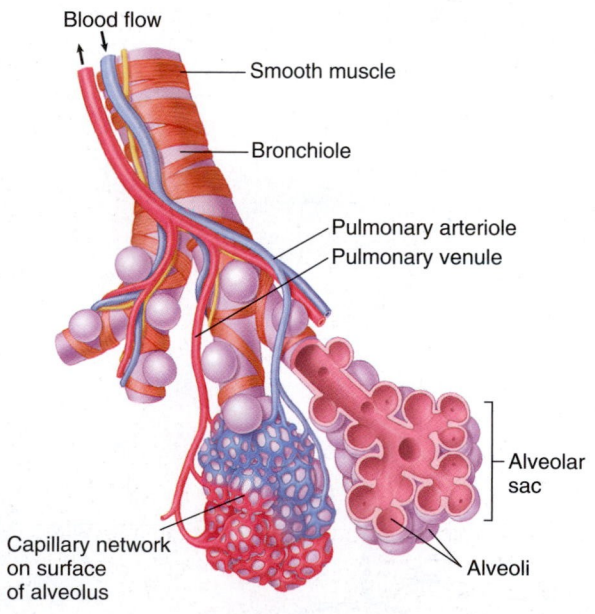

FIGURE 24.8   The exchange of gases takes place between the air-filled alveolus and the blood-filled capillary. The capillaries form a network around the saclike alveoli. The thin walls of the alveolus and capillary are in direct contact with one another; their combined thickness is usually less than 1 micron (a thousandth of a millimeter). This close relationship and the thinness of the walls allow for the easy diffusion of gases.

Cigarette Smoking and Your Health

Cigarette smoking is becoming less and less acceptable in our society. The banning of smoking in public buildings and on domestic air flights attests to this fact. Yet in spite of social pressure to quit smoking, research linking smoking with lung and heart disease, and evidence that even secondhand smoke can be harmful, about 23.1 percent of men and 18.3 percent of women are smokers. The same is true for high school students—about 20 percent are smokers.

Hazards of Cigarette Smoking

Bronchitis. Cigarette smoking is the leading cause of chronic bronchitis, which involves the inflammation of the bronchi. A common symptom of bronchitis is a harsh cough that expels a greenish-yellow mucus.

Emphysema. Emphysema is a progressive disease in which some of the alveoli are lost. People afflicted with this disease have less and less respiratory surface area and experience greater difficulty getting adequate oxygen, even though they may be breathing more rapidly. It may be caused by cigarette smoke and other air pollutants that damage alveoli. This damage reduces the capacity of the lungs to exchange gases with the bloodstream. A common symptom of emphysema is difficulty exhaling. Several years of an emphysema sufferer's forced breathing can increase the size of the chest and give it a barrel appearance.

Asthma. Cigarette smoke is one of many environmental factors that may trigger an asthma attack. Asthma is an allergic reaction that results in the narrowing of the bronchioles of the lungs and the excess production of fluids that limit the amount of air that can enter the lungs. Symptoms of asthma include coughing, wheezing, and difficulty breathing.

Lung cancer. Lung cancer develops twenty times more frequently in heavy smokers than in nonsmokers. Typically, lung cancer starts in the bronchi. Cigarette smoke and other pollutants cause cells below the surface of the bronchi to divide at an abnormally high rate. This malignant growth may spread through the lung and move into other parts of the body. Occurrence of cancers of the mouth, larynx, esophagus, pancreas, and bladder is also significantly greater in smokers than in nonsmokers.

Pneumonia. Cigarette smokers have an increased risk of developing pneumonia. Pneumonia involves the infection or inflammation of alveoli, which leads to fluid filling the alveolar sacs. Pneumonia is typically caused by the bacterium *Streptococcus pneumoniae* but in some cases is caused by other bacteria, fungi, protists, or viruses.

Pregnancy complications. Cigarette smoking during pregnancy has been linked to low birth weight and higher rates of fetal and infant death. Children of mothers who smoke during pregnancy have a higher incidence of heart abnormalities, cleft palate, and sudden infant death syndrome (SIDS). Nursing infants of smoking mothers have higher than normal rates of intestinal problems, and infants exposed to secondhand smoke have an increased incidence of respiratory disorders.

Heart disease. Smoking is a major contributor to heart disease. The action of nicotine from cigarette smoke results in constriction of blood vessels and the reduction of blood flow.

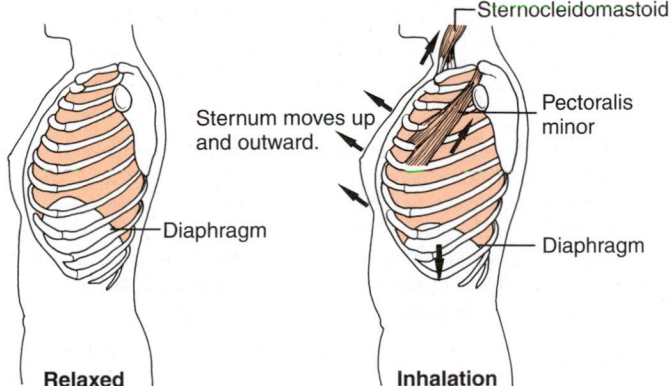

FIGURE 24.9 During inhalation, the diaphragm and external intercostal muscles between the ribs contract, causing the volume of the chest cavity to increase. During a normal exhalation, these muscles relax, and the chest volume returns to normal.

The Mechanism of Breathing

Breathing is the process of moving air in and out of the lungs. It is accomplished by the movement of a muscular organ known as the **diaphragm,** which separates your chest cavity, which contains the lungs, from your abdominal cavity, which contains

your stomach, liver, and other organs. In addition, muscles located between your ribs (*intercostal* muscles) assist in breathing. During inhalation, the diaphragm moves downward, and the external intercostal muscles of the chest wall contract, causing the chest wall to move outward and upward. Both of these muscular contractions cause the volume of your chest cavity to increase. This increase in volume results in a lower pressure in the chest cavity compared to the outside air pressure. Consequently, air flows from the outside, higher-pressure area through the trachea, bronchi, and bronchioles to the alveoli (figure 24.9).

During normal relaxed breathing, exhalation is accomplished by the chest wall and diaphragm simply returning to their normal, relaxed position. Muscular contraction is not involved. During vigorous inhalation, additional muscles of your chest wall are involved, making the chest cavity larger. In vigorous exhalation, internal intercostal muscles of the chest wall and muscles of the abdominal wall contract to compress the chest cavity and more completely empty the lungs of air.

Homeostasis and Breathing

Exercising causes an increase in the amount of carbon dioxide in your blood because muscles are oxidizing glucose more rapidly. When carbon dioxide dissolves in water, it forms a weak acid (carbonic acid). This lowers the pH of the blood. Certain brain cells and specialized cells in the aortic arch and carotid arteries are sensitive to changes in blood pH. When they sense a lower pH, nerve impulses are sent more frequently to the diaphragm and intercostal muscles. These muscles contract more rapidly and more forcefully, resulting in more rapid, deeper breathing. Because more air is being exchanged per minute, carbon dioxide is lost from the lungs more rapidly. When exercise stops, the carbon dioxide level drops, blood pH rises, and breathing eventually returns to normal.

24.6 OBTAINING NUTRIENTS: THE DIGESTIVE SYSTEM

All cells must have a continuous supply of nutrients to provide the energy they require and the building blocks needed to construct the macromolecules typical of living things. This section will deal with the processing and distribution of different kinds of nutrients.

Your digestive system consists of a muscular tube with several specialized segments. In addition, there are glands that secrete digestive juices into the tube. Several different kinds of activities are involved in getting nutrients to the cells that need them.

Processing Food

The digestive system is designed as a disassembly system. Its purpose is to take large chunks of food and break them down to small molecules that can be taken up by the circulatory system and distributed to your cells. The first steps in this process take place in the mouth.

Digestive Activities in the Mouth

It is important to grind large particles into small pieces by chewing in order to increase their surface area and allow for more efficient chemical reactions. It is also important to add water to the food, which further disperses the particles and provides the watery environment needed for these chemical reactions. Materials must also be mixed, so that all the molecules that need to interact with one another have a good chance of

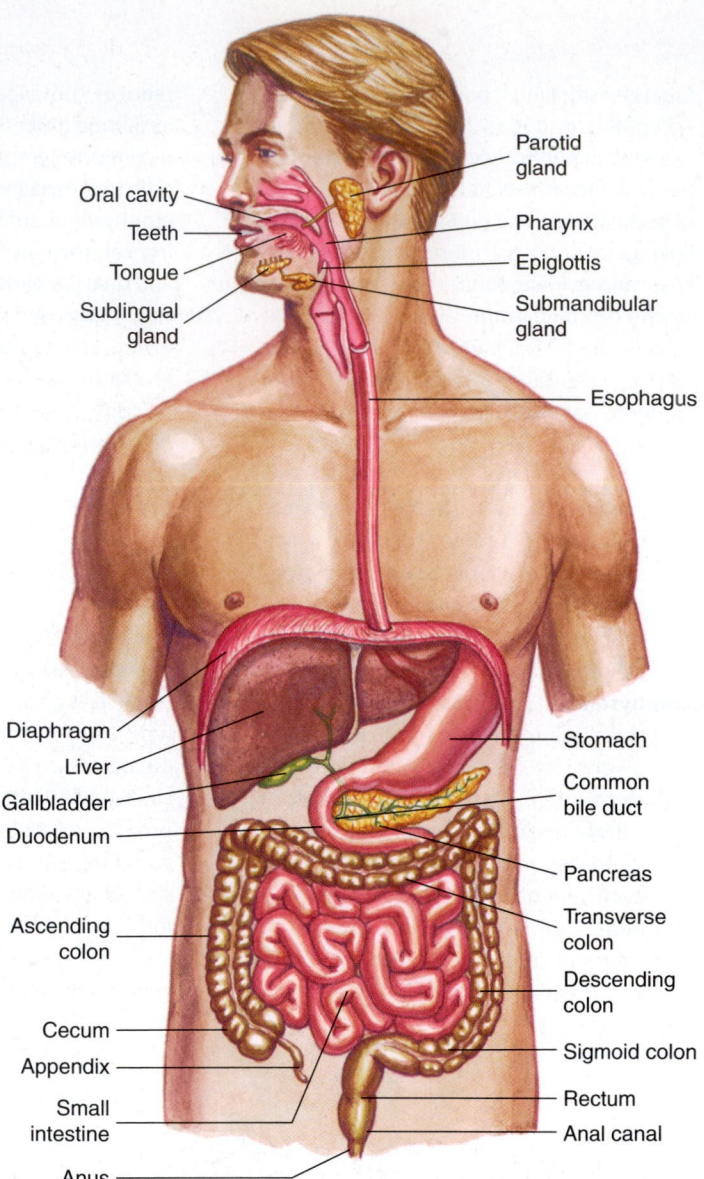

FIGURE 24.10   In your digestive system, the teeth, tongue, and enzymes from the salivary glands modify the food before it is swallowed. The stomach adds acid and enzymes and further changes the texture of the food. The food is eventually emptied into the duodenum, where the liver and pancreas add their secretions. The small intestine also adds enzymes and is involved in absorbing nutrients. The large intestine is primarily involved in removing water.

doing so. Your mouth and stomach are the major body regions involved in reducing the size of food particles. Your teeth are involved in cutting and grinding food to increase its surface area. The watery mixture that is added to the food in the mouth is known as *saliva*, and the three pairs of glands that produce saliva are known as *salivary glands*. Saliva contains the enzyme *salivary amylase*, which initiates the chemical breakdown of starch. Saliva also lubricates the mouth and helps to bind food before swallowing. Figure 24.10 shows the structures of the digestive system.

Henry Molaison (1926–2008) and William Beecher Scoville (1906–1984)

Henry Molaison (also known as HM) suffered from epilepsy from the age of nine, experiencing both partial and tonic-clonic seizures. The investigation into controlling his condition began in the 1950s. These studies took place while he was institutionalized in a Connecticut facility, where he socialized with other patients and staff.

In 1953, while exploring possible treatments for epilepsy, the neurosurgeon William Beecher Scoville determined that HM's epileptic abnormalities were located in the left and right medial temporal lobes of his brain. To control the epilepsy, Scoville surgically removed portions of HM's brain. Even though the surgery was successful, HM developed a new problem: anterograde amnesia. This condition results in an inability to commit new life events to long-term memory. In addition, HM could no longer remember events that occurred one to two years prior to his surgery nor recall some events up to eleven years before. However, his long-term procedural memories were just fine; that is, he could learn new motor skills, even though he could not remember learning them. The knowledge gained from this work resulted in a better understanding of the nature of amnesia and how particular areas of the brain are connected to specific processes associated with memory impairment and formation.

After his death in 2008, Henry's brain was removed and sliced into histological sections for further study. It is now at the University of California, San Diego.

Connections . . .

Newborn Jaundice

Albumin transports bilirubin, a yellowish molecule that is a breakdown product of hemoglobin from destroyed RBCs. Normally, bilirubin is transported to the liver, where it is destroyed and the broken-down products are released with the bile into the intestine. However, some newborns become jaundiced (a yellowing of the skin and whites of the eyes) because their livers have not matured enough to do this important job. If bilirubin is allowed to accumulate in their blood, jaundiced babies can develop deafness, cerebral palsy, or brain damage. Placing babies under a special ultraviolet ("bili") light helps relieve the problem. The light has a wavelength that helps break down the bilirubin to a form that can be eliminated through the kidneys until the baby's liver matures and can adequately handle the bilirubin.

Swallowing

Once the food has been chewed, it is swallowed and passes down your esophagus to your stomach. The process of swallowing involves a complex series of events. First, a ball of food, known as a *bolus*, is formed by your tongue and moved to the back of your mouth cavity. Here it stimulates the walls of your throat, also known as the *pharynx*. Nerve endings in the lining of the pharynx are stimulated, causing a reflex contraction of the walls of the esophagus, which transports the bolus to your stomach.

Since both food and air pass through the pharynx, it is important to prevent food from getting into your lungs. During swallowing, the larynx is pulled upward. This causes a flap of tissue called the *epiglottis* to cover the opening to your trachea and prevent food from entering your trachea.

Digestive Activities in the Stomach

In your stomach, additional liquid, called *gastric juice*, is added to the food. Gastric juice contains enzymes and hydrochloric acid. The major enzyme of the stomach is *pepsin*, which initiates the chemical breakdown of protein. The pH of gastric juice is very low, generally around pH 2. The entire mixture is churned by the contractions of the three layers of muscle in your stomach wall. The combined activities of enzymatic breakdown, chemical breakdown by hydrochloric acid, and mechanical processing by muscular movement result in a thoroughly mixed liquid called *chyme*. Chyme eventually leaves the stomach through a valve known as the pyloric sphincter and enters the small intestine. If there is a backward flow of acid-containing stomach contents into your esophagus, a condition is created called esophageal or gastroesophageal reflux disease (GERD). The acid causes damage to the mucosal lining of the esophagus, coughing, pain, sore throat, and vomiting.

Digestive Activities in the Small Intestine

The first part of your **small intestine** is known as the **duodenum.** In addition to producing enzymes, the duodenum secretes several kinds of hormones that regulate the release of food from your stomach and the release of secretions from your pancreas and liver. The **pancreas** produces a number of different digestive enzymes and also secretes large amounts of bicarbonate ions, which neutralize stomach acid so that the pH of the duodenum is about pH 8. The **liver** is a large organ in the upper abdomen that performs several functions. One of its functions is the secretion of *bile*. When bile leaves the liver, it is stored in the *gallbladder* prior to being released into the duodenum. When bile is released from the gallbladder, it assists mechanical mixing by breaking large fat globules into smaller particles. This process is called *emulsification*.

Digestive Activities in the Large Intestine

Along the length of the small intestine, additional watery juices are added until the mixture reaches the **large intestine.** Your large intestine is primarily involved in reabsorbing the water that has been added to the food tube along with saliva, gastric juice, bile, pancreatic secretions, and intestinal juices. Your large intestine is also home to a variety of different kinds of bacteria. Most live on the undigested food that makes it through the small intestine. Some provide additional benefit by producing vitamins that can be absorbed from the large intestine. A few kinds may cause disease. If the "good bacteria" are lost as a result of infection or severe diarrhea, you can encourage their reestablishment by being sure that the bacteria are well fed. Because one of the primary bacterial species is *Lactobacillus acidophilus*, eating fermented dairy products such as yogurt can be a great benefit. Also, probiotics foods (i.e., foods containing these living organisms) will more likely increase their population in your gut and enable them to out-compete pathogens that may be present.

Nutrient Uptake

The process of digestion has resulted in a variety of simple organic molecules that are available for absorption from the tube of the gut into your circulatory system. Nearly all nutrients are absorbed in your small intestine. Many features of the small intestine provide a large surface area for the absorption of nutrients. First of all, your small intestine is a very long tube; the longer the tube, the greater the internal surface area. In a typical adult human, it is about 6 m (about 20 ft) long. In addition to length, the lining of your intestine consists of millions of tiny projections called *villi,* which increase the surface area. When we examine the cells that make up the villi, we find that they also have folds in their surface membranes. All of these characteristics increase the surface area available for the transport of materials from the gut into your circulatory system (figure 24.11).

Scientists estimate that the cumulative effect of all of these features produces a total intestinal surface area of about 250 m^2 (about 2,600 ft^2). That is equivalent to about the area of a basketball court.

The surface area by itself would be of little value if it were not for the intimate contact of the circulatory system with this lining. Each villus contains several capillaries and a branch of the lymphatic system called a *lacteal.* The close association between the intestinal surface and the circulatory and lymphatic systems allows for the efficient uptake of nutrients from the cavity of the gut into your circulatory system.

Several different kinds of processes are involved in the transport of materials from the intestine to your circulatory system. Some molecules, such as water and many ions, simply diffuse through the wall of the intestine into the circulatory system. Other materials, such as amino acids and simple sugars, are assisted across the membrane by carrier molecules. Fatty acids and glycerol are absorbed into the intestinal lining cells where they are resynthesized into fats and enter lacteals in the villi. Since the lacteals are part of the lymphatic system, which eventually empties its contents into your circulatory system, fats are also transported by the blood. They just reach the blood by a different route.

24.7 NUTRITION

Organisms maintain themselves by constantly processing molecules to provide building blocks for new living material and energy to sustain themselves. The word **nutrition** is used in two related contexts. First of all, nutrition is a branch of science that seeks to understand food, its nutrients, how the nutrients are used by the body, and how inappropriate combinations or quantities of nutrients lead to ill health. The word *nutrition* is also used in a slightly different context to refer to all the processes by which we take in food and utilize it, including *ingestion, digestion, absorption,* and *assimilation. Ingestion* involves the process of taking food into the body through eating. *Digestion* involves the breakdown of complex food molecules to simpler molecules. *Absorption* involves the movement of simple molecules from the digestive system to the circulatory system for dispersal throughout the body. *Assimilation* involves the modification and incorporation of absorbed molecules into the structure of the organism.

Kinds of Nutrients

Nutrients are the kinds of molecules that must be consumed to maintain your body. They can be organized into several categories based on their chemical structure and the functions they serve in the body. Commonly accepted categories are: carbohydrates, fats, proteins, vitamins, minerals, and water. The chemical nature of organic molecules is discussed in chapter 19.

1. *Carbohydrates*: Common carbohydrates that serve as nutrients are various kinds of sugars and starches and are a quick source of energy. Each gram of carbohydrate provides about 4 kilocalories of energy. The sugar glucose undergoes aerobic cellular respiration to provide ATP. Starch and some other kinds of sugars can be converted to glucose and thus are able to undergo aerobic cellular respiration.
2. *Fats*: Fats are excellent sources of energy since they can be broken down into smaller units that enter the Krebs cycle to provide ATP. Fats provide about 9 kilocalories of energy per gram—more than twice the energy of carbohydrates.
3. *Proteins*: Proteins are important in the structure of cells and the body. All cells and parts of the body contain protein. Thus, protein is needed as a nutrient to provide for growth

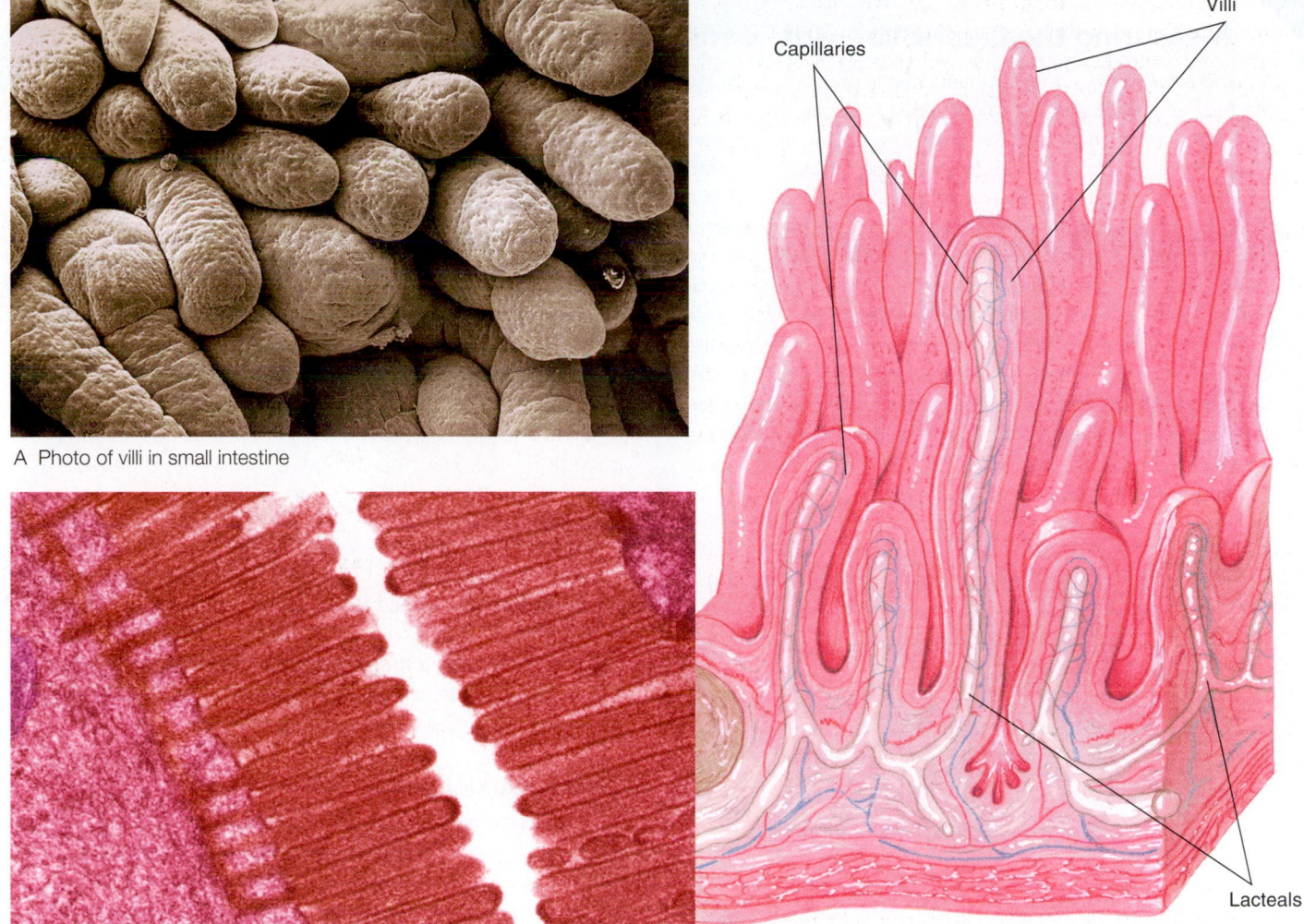

A Photo of villi in small intestine

B Photo of microvilli in intestinal cell

C Relationship of villi, capillaries, and lacteals

FIGURE 24.11 (*A*) The surface area of the intestinal lining is increased by the many fingerlike projections known as villi. (*B*) The cells that make up the surface of the villi have many tiny projections on them known as microvilli. Both the villi and microvilli provide a large surface area for the exchange of materials between the gut and the circulatory system. (*C*) Within each villus are capillaries and lacteals. Most kinds of materials enter the capillaries, but most fat-soluble substances enter the lacteals, giving them a milky appearance. Lacteals are part of the lymphatic system. Because the lymphatic system empties into the circulatory system, fat-soluble materials also eventually enter the circulatory system. The close relationship between these vessels and the epithelial lining of the villus allows for efficient exchange of materials from the intestinal cavity to the circulatory system.

and repair of cells, tissues, and organs. Proteins also contain energy (about 4 kilocalories per gram) but yield less energy than fats or carbohydrates because they are more difficult to break down in aerobic cellular respiration.

4. *Vitamins*: Organic molecules that cannot be manufactured by the body but are needed in small amounts to assist enzymes in their functions.

5. *Minerals*: Inorganic molecules needed for a variety of purposes. Some such as calcium are needed as a component of bone. Others are needed as components of enzymes or other important molecules. Most minerals are needed in relatively small amounts.

6. *Water*: Often not even classified as a nutrient but is vitally important to sustain life. Water is the most common molecule in the body. Since it is constantly lost in urine, feces, and by evaporation, it must be replaced.

All of these nutrients must be obtained in adequate amounts on a daily basis or ill health results.

Counting Calories

One of the primary pieces of information on the nutritional facts label on foods you purchase is the energy content of the food item. The unit used to measure the amount of energy in foods is the **kilocalorie (kcal)**. One kilocalorie is the amount of heat needed to increase the temperature of one *kilogram* of water one degree Celsius. Remember that the prefix *kilo-* means "1,000 times" the value listed. Therefore, a kilocalorie is 1,000 times more heat energy than a **calorie**, which is the amount of heat needed to raise the temperature of one *gram* of water one degree Celsius. Although the amount of energy in food is measured in kilocalories, for nutritional purposes a kilocalorie is usually designated as Calorie with a capital C. This is unfor-

Connections . . .

Measuring the Caloric Value of Foods

A *bomb calorimeter* is an instrument that is used to determine the heat of combustion or caloric value of foods. To operate the instrument, a small food sample is formed into a pellet and sealed inside a strong container called the bomb. The bomb is filled with 30 atmospheres of oxygen and then placed in a surrounding jacket that is filled with water. The sample is electrically ignited, and as it aerobically reacts with the oxygen, the bomb releases heat to the surrounding water jacket. As the heat is released, the temperature of the water rises. Recall that a calorie is the amount of energy (or heat) needed to increase the temperature of one gram of water one degree Celsius. These are the same types of calories that we digest, even if our bodies use enzymes and cells to do so. Therefore, if one gram of food results in a water temperature increase of 13°C for each gram of water, that food has the equivalent of 13 calories.

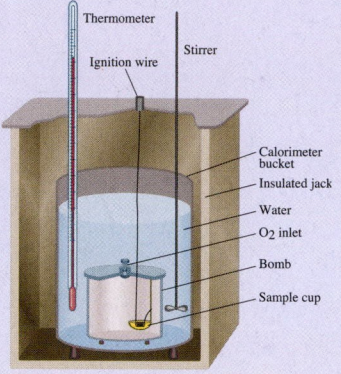

Thermometer
Ignition wire
Stirrer
Calorimeter bucket
Insulated jack
Water
O_2 inlet
Bomb
Sample cup

tunate because it is easy to confuse a Calorie, which is really a kilocalorie, with a calorie.

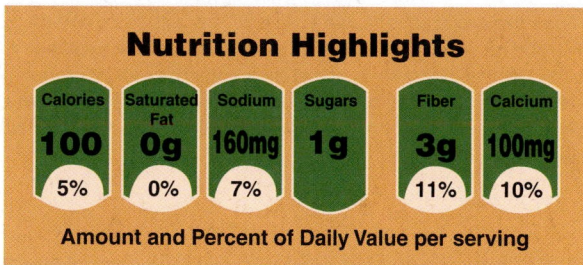

Nutrition Highlights

Calories	Saturated Fat	Sodium	Sugars	Fiber	Calcium
100	0g	160mg	1g	3g	100mg
5%	0%	7%		11%	10%

Amount and Percent of Daily Value per serving

Most books on nutrition and dieting use the term *Calorie* to refer to *food Calories*. Thus, the information on a nutrition facts label can be used to help a person determine the number of Calories they are consuming per day. A commonly used standard for a daily diet is 2,000 Calories. However, the number of Calories needed by an individual is highly variable and depends on a person's specific metabolism and level of physical activity. Research has also made it clear that dietary restriction without malnutrition can increase life span and reduce age-related diseases.

Guidelines for Obtaining Adequate Nutrients

The food and drink consumed from day to day constitutes a person's **diet**. A good understanding of nutrition can promote

good health and involves an understanding of the energy and nutrient content in various foods. Nutritional scientists and government officials in the United States and many other countries have developed their own guidelines (e.g. University of Michigan's *Healing Foods Pyramid™ 2010*, and Harvard University's *Healthy Eating Plate and Healthy Eating Pyramid*); this discussion focuses on the nutritional guides published in 2011 by the U.S. Department of Agriculture.

Dietary Guidelines for Americans

Over the last 30 years, the U.S. Department of Agriculture (USDA) and Department of Health and Humans Services (HHS) have published various guidelines as aids to maintaining good nutritional health. These include Dietary Guidelines for Americans, Dietary Reference Intakes (Table 24.2), and MyPlate. The Dietary Guidelines for Americans was first published in 1980 and was revised in 2010. These strategies provide scientifically based dietary recommendations for Americans older than 2 years that can promote good health and reduce the risk for major chronic diseases. The majority of Americans are overweight or obese and yet under-nourished in several key nutrients; this is of great concern since the number of deaths related to poor diet and physical inactivity is increasing and may soon overtake tobacco as the leading cause of death in the country. A poor diet is also linked to numerous diseases such as cardiovascular disease and type 2 diabetes and their related risk factors.

The Guidelines and recommendations focus on halting and reversing the obesity problem through prevention and changes in behavior, the environment, and the food supply. The USDA and HHS address eight items, each of which is summarized here.

1. Energy balance and weight management
Conclusions:
Eating too few vegetables and fruits is associated with higher body weights. Neighborhoods with more supermarkets stocked with vegetables and fruits are associated with lower body mass index (BMI) levels, while those with more fast food restaurants and convenience stores are associated with higher BMIs. Adults

TABLE 24.2 Dietary Reference Intakes for Some Common Nutrients

These are recommendations from the U.S. Department of Agriculture on the amounts of different kinds of nutrients people should receive. These guidelines are very detailed and include specific guidelines for males and females of various ages and for pregnant and nursing mothers. They also provide information about the maximum amount of certain nutrients that people should not exceed. In other words, you can have too much of a good thing.

Nutrient	Women 19–30 Years Old	Men 19–30 Years Old	Maximum for Persons 19–30 Years Old	Value of Nutrient
Carbohydrates	130 g/day (45–65% of kilocalories)	130 g/day (45–65% of kilocalories)	No maximum set, but refined sugars should not exceed 25% of total kilocalories.	A source of energy
Proteins	46 g/day (10–35% of kilocalories)	56 g/day (10–35% of kilocalories)	No maximum set, but high protein diets stress kidneys.	Proteins are structural components of all cells. There are ten essential amino acids that must be obtained in the diet.
Fats	20–35% of kilocalories	20–35% of kilocalories	Up to 35% of total kilocalories	Energy source and building blocks needed for many molecules
Saturated and trans fatty acids	No reference amount determined	No reference amount determined	As low as possible	
Linoleic acid	12 g/day	17 g/day	No maximum set	Essential fatty acid needed for enzyme function and to maintain epithelial cells
Linolenic acid	1.1 g/day	1.6 g/day	No maximum set	Essential fatty acid needed to reduce risk of coronary heart disease
Cholesterol	No reference amount determined	No reference amount determined	As low as possible	None needed, since the liver makes cholesterol
Total fiber	25 g/day	38 g/day	No maximum set	Improve gut function
Water	2.7 L/day	3.7 L/Day	No maximum set	Improve gut and kidney function
Calcium	1 g/day	1 g/day	2.5 g/day	Needed for the structure of bones and many other functions
Iron	18 mg/day	8 mg/day	45 mg/day	Needed to build hemoglobin of red blood cells
Sodium	1.5 g/day	1.5 g/day	2.3 g/day (most people exceed this limit)	Needed for normal cell function
Vitamin A	700 ug/day	900 ug/day	3,000 ug/day	Maintains skin and intestinal lining
Vitamin C	75 mg/day	90 mg/day	2,000 mg/day	Maintains connective tissue and skin
Vitamin D	5 ug/day	5 ug/day	50 ug/day	Needed to absorb calcium for bones

and children whose diets consist of one or more fast food meals per week report higher BMIs. Both adults and children show a positive correlation between television and computer screen time and being obese or overweight. Not eating breakfast also puts adults and children at risk. Physically active people have a reduced risk of becoming overweight or obese.

Women who gain excessive weight during pregnancy are less likely to lose that weight after delivery and more likely to undergo caesarean section. They are also more likely to deliver larger than gestational age babies who are in turn more likely to become obese later in life. Sugar-sweetened beverages (soft drinks and juices) are associated with increased fat tissue in children.

Recommendations:
1. Consume smaller portions.
2. Eat a healthy breakfast (while staying within energy needs) of minimally processed foods, i.e., grains, fruits.
3. Limit screen time for children and adolescents to no more than 1 or 2 hours per day, and discourage eating while watching TV or using the computer. Currently, adults average 35.5 hours of screen time per week.

4. Become actively involved in self-monitoring your body weight, food intake, and physical activity.
5. During pregnancy, keep your weight within ranges recommended by your physician.
6. Reduce the amount of sugar-sweetened beverages in your diet.
7. Diets that are reduced in calories and have macronutrients within these ranges (protein: 10%–35%; carbohydrate: 45%–65%; fat: 20%–35%) will help those who want to lose or maintain their weight.
8. Engage in physical activities on a daily basis.

2. Nutrient adequacy
Conclusions:

Americans eat an excess of solid fats, refined sugars and sodium. We eat too few vegetables, fruits, whole grains, fluid milk and milk products, and oils. Neural tube defects in newborns are associated with folic acid deficiency during pregnancy.

Recommendations:
1. Keep consumption of solid fats, refined sugars, and sodium as low as possible.

A Closer Look

Body Mass Index

The **body mass index (BMI)** is an indirect measurement of the amount of fat a person has stored. BMI is a ratio of height to weight and for most people, a reasonably good tool for determining how close you are to a "healthy weight."

$$BMI = \frac{weight\ in\ kilograms}{height\ in\ meters^2}$$

Thin people have a low BMI, while overweight and obese people have a high BMI. However, because a given volume of muscle tissue weighs more than an equal volume of fat, extremely muscular people may have a high BMI and actually have a low amount of fat. This is only likely to be true in people who engage in heavy muscular exercise. Box Figure 24.1 shows the body mass index for various heights and weights. Although the height and weight in the chart are in feet and pounds, the BMI is given in kg/m^2.

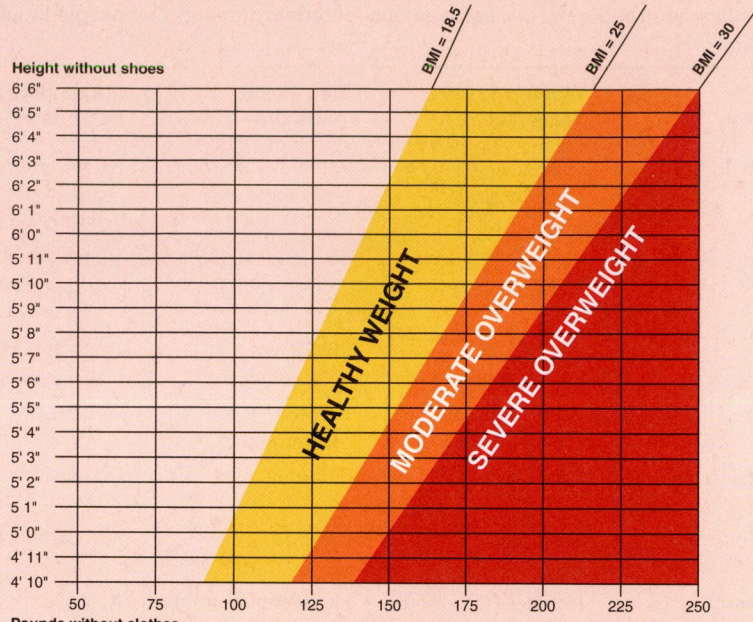

Box Figure 24.1 Find your height and weight and determine your approximate BMI.

2. Increase the amount of vegetables (especially dark-green and red-orange vegetables; cooked dry beans and peas), fruits, whole grains, fat-free or low-fat milk (or vitamin D fortified soymilk) and milk products to that recommended by the Dietary Reference Intakes (Table 24.2).
3. Substitute oils for fats in your diet; do not add them to your diet.
4. Be sure that your diet contains the amount of vitamin D, calcium, potassium, and dietary fiber recommended by the USDA and HHS.
5. Folic acid and iron supplements might be necessary during pregnancy.
6. People over 50 years of age should be encouraged to eat food fortified with vitamin B$_{12}$.
7. Certain people should be encouraged to take a variety of micronutrients (multivitamin/mineral supplements).

3. Fatty acids and cholesterol
Conclusions:

The consumption of saturated fatty acids is associated with increased levels of low-density lipoproteins (LDLs), serum cholesterol, and increased risk of cardiovascular disease and type 2 diabetes. The amount of *trans* fatty acids in the American food supply has decreased since mandatory *trans* fatty acids labeling regulation went into effect in 2006. The amount of omega-3 fatty acids in two servings of seafood per week (about 4 ounces per serving) will reduce the incidence of death from cardiac heart and cardiovascular diseases. The consumption of dark chocolate, unsalted peanuts, and tree nuts (i.e., walnuts, almonds, and pistachios) in a nutritionally adequate diet (while

staying within energy needs), can reduce cardiovascular disease risk factors.

Recommendations:
1. Reduce consumption of saturated fatty acids to less than 7 percent of intake of total fats and replace with monounsaturated fatty acids or polyunsaturated fatty acids.
2. Eat fewer than 7 eggs per week
3. Limit dietary cholesterol to less than 300mg per day; for those with cardiovascular disease or type 2 diabetes, the recommendation is 200mg per day.
4. Limit consumption of cholesterol-raising fats (C12, C14, C16 fatty acids and *trans* fatty acids) to less than 5–7 percent of your diet to maintain blood cholesterol at desirable levels.
5. Add small portions of dark chocolate and nuts to your diet, but be sure to stay within energy needs, as these foods are also high in calories.

4. Protein
Conclusions:

Animal protein products (i.e., beef, chicken, pork) contain fat and a high calorie load per serving. Soy protein has a small to moderate effect on total cholesterol and low density lipoproteins, and no beneficial effect in controlling body weight or blood pressure. While vegetarian diets show no special benefit in controlling cancer, they are helpful in lowering BMI and blood pressure. Many children do not consume the recommended amounts of milk and milk products, increasing their risk for poor bone health, cardiovascular disease and type 2 diabetes. Evidence suggests that cooked dry beans and peas lower serum lipids.

Recommendations:
1. Serving sizes and total protein intake should be within limits proposed by the USDA and HHS.
2. For those deficient in protein, it is advisable to have complete vegetable protein products, especially during pregnancy, lactation, and childhood.
3. A vegetarian diet should include adequate amounts of complete protein, calcium, iron, B12, zinc, and omega-3 fatty acids to obtain the essential amino acids.
4. Include daily amounts of low-fat or fat-free milk and milk products in your diet (2 cups for children ages 2 to 8 years, 3 cups for those ages 9 years and older).
5. Eat the recommended amount of legumes to increase total vegetable consumption and dietary fiber intake.

5. Carbohydrates
Conclusions:
Most Americans do not consume adequate amounts of whole grains. The fiber found in whole grain foods protects against cardiovascular disease, type 2 diabetes, and obesity, and is necessary for good digestive system health. Whole foods such as apples and carrots provide a feeling of fullness, and can decrease the amount of food eaten at later meals. Fiber added to a drink is not effective in reducing food intake at later meals. Proteins and carbohydrates may be more effective at satisfying the appetite than fats. Probiotic foods provide health benefits to the intestinal tract.

Recommendations:
1. A healthy and balanced diet should include whole grain foods, cooked dry beans and peas, vegetables, fruits, and nuts.
2. Probiotic foods are a valuable addition to the diet.
3. Consume the amounts of carbohydrate described by the USDA and HHS, while staying within total energy needs.

6. Sodium, potassium, and water
Conclusions:
In adults, blood pressure decreases with a decrease of sodium intake, as will the risks of stroke and cardiovascular disease. The goal for sodium intake is 1,500 mg per day. A high intake of potassium is associated with lower blood pressure in adults and reduces the negative effects of sodium on blood pressure. Potassium also reduces the risk of developing kidney stones and bone loss. For adults, 4,700 mg per day is recommended. The consumption of fluids with means is sufficient to maintain normal hydration. There is no evidence indicating a need to drink 8 glasses of water per day.

Recommendations:
1. Reduce consumption of high sodium foods, e.g., yeast breads, pizza, pasta dishes, cold cuts, franks, bacon.
2. Milk and dairy products are a good source of potassium.
3. Exposure to heat stress or involvement in strenuous activities requires extra fluids to remain hydrated.

7. Alcohol
Conclusions:
Heavy consumption (more than two drinks per day) of alcohol is associated with weight gain and is not a good source of nutrients. Heavy or binge drinking is detrimental to age-related cognitive (mental processes of perception, memory, judgment, and reasoning) decline and increases the risk of coronary heart disease and bone health. Evidence shows that when a lactating mother consumes alcohol, alcohol enters the breast milk, and the quantity of milk produced is reduced; this leads to reduced milk consumption by the infant. It is also associated with modified post-natal growth, sleep patterns, and psychomotor patterns of the baby. Heavy or binge drinking can result in fetal alcohol syndrome and appears to have the most harmful effects when consumed during the first 3 months of pregnancy.

Recommendations:
1. Drink only the recommended amount of alcoholic beverages per day (14–28g/day): two for adult males and one for females.
2. Avoid heavy or binge drinking.
3. Nursing mothers should wait 3 to 4 hours after a single drink before breastfeeding.

8. Food safety and technology
Conclusions:
The easiest ways to prevent food safety problems include proper hand sanitation techniques, proper washing of vegetables and fruit, prevention of cross-contamination, and appropriate cooking and storage of foods in the home kitchen. Evidence shows that pregnant women, college students, and older adults commonly practice unsafe food handling and consumption behaviors.

Recommendations:
1. Wash vegetables and fruit by running water over them.
2. Sinks, refrigerators, cooking utensils, cutting boards, cutlery and counter tops need regular cleaning with soap and/or a disinfectant.
3. Refrigerators (40°F or 5°C, or lower) and freezers (0°F or minus 18°C) must be maintained at the appropriate temperatures.
4. Regularly use accurate and clean food thermometers.
5. Do not consume raw or undercooked animal-source food products such as eggs and egg-containing products, and ground beef products.

MyPlate

MyPlate 2011 is the most recent nutritional guide published by the USDA. The MyPlate icon shows a plate and glass, and illustrates the relative proportions of the five food groups required for good health. The icon replaces the USDA's MyPyramid nutritional guide (Figure 24.12). When at the web site, MyPlate allows you to learn about:

1. Eating healthy on a budget
2. Getting information en español
3. Looking up a food

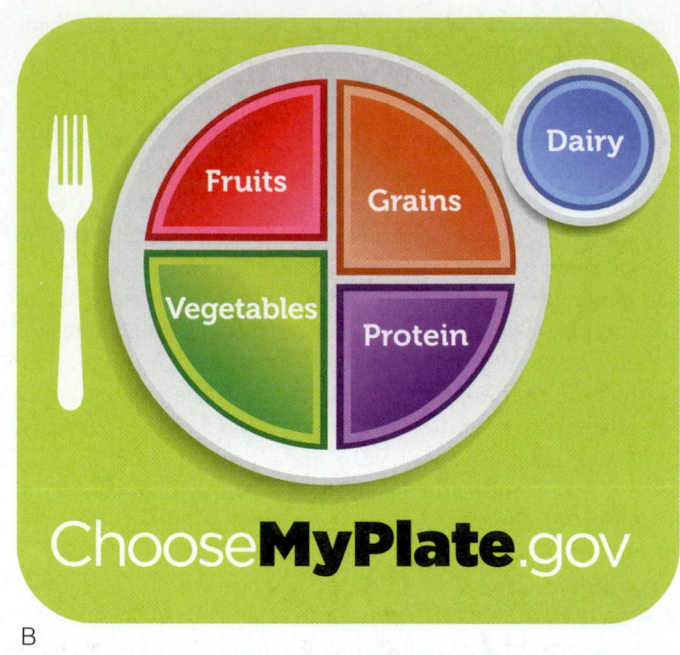

A

B

FIGURE 24.12 Over the years, as new nutritional information became available, nutritional suggestions have been modified. (A) The Food Guide Pyramid, released in 2005, has been modified and is now displayed as (B) the MyPlate icon. For more information see the website http://www.choosemyplate .gov/ *Source*: U.S. Department of Agriculture, 2011

4. Learning about food groups
5. Getting a personalized plan
6. Learning healthy eating tips
7. Getting weight loss information
8. Planning a healthy menu
9. Analyzing my diet
10. Getting MyPlate updates
11. Asking a questions

By clicking on each food category shown on the plate you can find out such things as:

1. The nature of the food in that group
2. How much is required by age and sex
3. The health benefits of the group
4. Tips on how to get food within that group
5. Recipes for preparing food within the group.

MyPlate designates five food groups (1) grains, (2) vegetables) (3) fruits, (4) dairy, and (5) protein foods.

1. Grains Group Grains products are divided into two groups. *Whole grains* contain the entire grain kernel, that is, the bran, germ, and endosperm. *Refined grains* have been processed to remove these parts. In many cases, they are enzymatically treated to break down the complex carbohydrates. This gives grain products a softer texture and improves shelf life. To compensate for the loss of dietary fiber, iron, and many B vitamins, products such as breads, bagels, buns, crackers, dry and cooked cereals, pancakes, pasta, and tortillas are enriched. Because whole-grain foods contain more nutrients than those that are not whole-grain, it is recommended that half of all grains consumed be whole-grain products. MyPlate recom-

mends that people consume 3–8 ounces of grain products per day, depending on their age, sex, and level of physical activity. This is relatively easy to meet, because a slice of bread, tortilla, half a bagel, or cup of dry cereal is equivalent to an ounce.

Significant nutritional components: carbohydrate, fiber, several B vitamins, vitamin E, and the minerals iron, selenium, and magnesium

2. Vegetables Group Vegetables include either raw or cooked nonsweet plant materials, such as broccoli, carrots, cabbage, corn, green beans, tomatoes, potatoes, lettuce, and spinach. MyPlate recommends 1–3 cups of items from this food group each day depending on a person's age, sex, and level of physical activity. Since different vegetables contain different amounts of vitamins, it is wise to include as much variety as possible in this group. Therefore, vegetables are divided into five subgroups: dark green vegetables, orange vegetables, dry beans and peas, starchy vegetables, and other vegetables. On a weekly basis, people should be sure to consume some vegetables from each of the following five groups: dark green, red and orange, beans and peas, starchy vegetables, and "other" vegetables.

Significant nutritional components: carbohydrate, fiber, several B vitamins, vitamins A, C, E, and K, and the minerals iron, potassium, and magnesium.

3. Fruits Group Fruits include such sweet plant products as melons, berries, apples, oranges, bananas, or 100 percent fruit juices. MyPlate recommends 1–2 cups of fruit per day depending on a person's age, sex, and level of physical activity. Fruit may be fresh or preserved or 100% fruit juice. Many individual pieces of fruit (apple, orange, banana) are about 1 cup.

Most people think of the skeleton as a nonliving framework that supports the more "soft" body parts. However, the skeleton is anything but inert and serves functions in addition to that of support. Box figure 24.2 shows the primary parts of the human skeleton.

A primary function of the skeleton is to provide support while allowing motion. At joints, tough, fibrous cords called ligaments hold bones together, and muscles are attached to the bones by tough, fibrous cords called *tendons.* Muscles are typically attached to two different bones, and muscular contraction results in one bone moving with respect to another, causing a bending or rotation at the joint. The actions of muscles are also important in providing support because they generate tension that provides rigidity to the skeletal framework, like the cables that hold up the mast of a ship.

Protection is another major function of the skeleton. The brain, spinal cord, and organs of the chest are enclosed by and protected by the bony structures of the skull, vertebrae, and ribs and associated structures, respectively.

Chemically, the skeleton is a combination of an inorganic mineral, hydroxyapatite, and organic fibers of various kinds of protein. Hydroxyapatite contains calcium and phosphorus and has the chemical formula $Ca_{10}(PO_4)_6(OH)_2$. There is constant turnover of both hydroxyapatite and protein in the bones of the skeleton. For example, calcium and phosphorus have several important functions in the body and are distributed throughout the body by the circulatory system. If the blood does not contain adequate calcium and phosphorus from the diet, these minerals are taken from the skeleton to supply immediate needs. These substances can be replaced later, when they are present in the diet. This function is particularly important in pregnant and nursing mothers, who must provide calcium and phosphorus for bone development of embryos and as a source of calcium in the mother's

milk for nursing infants. Without adequate calcium and phosphorus in the diet, the mother can experience significant bone loss and skeletal development in the infant is impaired.

Nowhere is the dynamic nature of the skeleton more evident than in the process of growth. Because long bones are typically connected to other bones at a joint, it is important that the joint continue to function while the bone is growing. Long bones of the body have special growth areas near their ends but not on the very end. Thus, new bone is added near the ends so that the bones become longer without interfering with the function of the joint. Bones of the skull and certain other bones grow directly from their edges, eventually growing together and fusing into a solid structure. The bones of the skull of newborn babies are not joined and are capable of being moved around during the birth process. As the individual bones of the skull grow, the bones eventually knit together to form a solid skull, but a "soft spot" persists on the top of the head for a time after birth.

One of the activities that stimulates change in the nature of the skeleton is the stress and strain placed on it. When the skeleton is "challenged," it gets stronger by adding material. When it is not challenged, material is lost and the bones get weaker. For example, astronauts in weightless conditions lose bone mass. Therefore, exercise programs are very important parts of the schedule of astronauts. Since exercise strengthens bones, it is particularly important in elderly persons who tend to

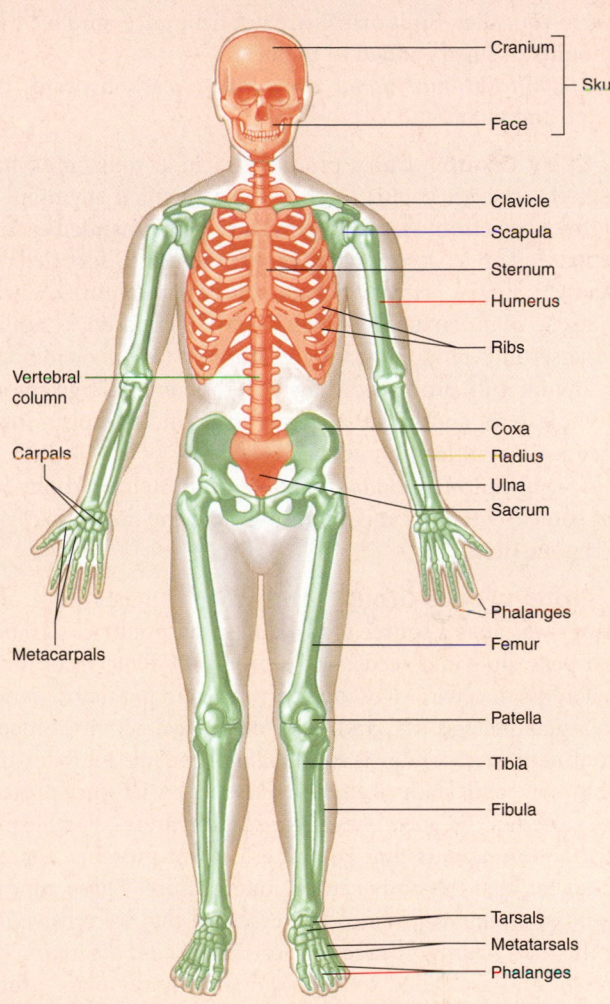

Box Figure 24.2 Major bones of the skeleton.

have reduced bone mass for age-related reasons.

The skeleton is even involved in the production of blood. The red bone marrow, found in the core of certain bones of the skeleton (vertebrae, ribs, breast bone, and upper arm and leg bones), is the site of production of red and white blood cells and platelets. The lymph nodes, tonsils, spleen, and thymus also produce some kinds of white blood cells. Bone marrow transplants are often provided for people whose blood cell production capacity has been destroyed by chemotherapy or radiation.

A half-cup of dried fruit is equivalent to 1 cup. However, since these foods tend to be high in natural sugars, consumption of large amounts of fruits can add significant amounts of kilocalories to your diet. Fruit drinks are not fruit juices and have high amounts of sugars added to them.

Significant nutritional components: carbohydrate, fiber, water, and vitamin C.

4. Dairy Group Dairy products include milk, milk-based desserts, cheese, yogurt, and calcium-fortified soymilk. Two to three cups from this food group are recommended each day depending on a person's age and sex. Vitamin D fortified dairy products are the primary dietary source of vitamin D. Vitamin D is also manufactured in the skin when exposed to sunlight. Because milk products contain saturated fats, choose non-fat or low-fat milk products. Many cheeses contain large amounts of cholesterol and fat for each serving. Cholesterol in low-fat cheeses is significantly reduced.

Significant nutritional components: protein, carbohydrate, fat, some B vitamins and vitamin D, and the minerals calcium and potassium.

5. Protein Foods Group This group contains most of the things we eat as a source of protein: meat, poultry, eggs, beans and peas, nuts and seeds, and seafood. MyPlate recommends that adults receive 5–61/2 ounces of protein per day depending on a person's age, sex, and level of physical activity. Since the amount needed per day is so small, most people eat many times what is needed. Daily protein intake is essential since protein is not stored in the body like fats or carbohydrates. Furthermore, of the amino acids that combine to form proteins, there are about ten that the body cannot manufacture. These are called *essential amino acids* since it is essential that we receive them in the diet. Animal sources of protein have *all the amino acids present* and are called *complete proteins*. Many plant proteins may lack specific amino acids and are called *incomplete proteins*. Vegetarians must pay particular attention to acquiring adequate sources of protein because they have eliminated animal flesh from their diet. They can get all the essential amino acids if they eat proper combinations of plant materials. Since many sources of protein (particularly meat and nuts) also include significant fat and health recommendations suggest reduced fat, choosing low-fat sources of protein is important. Selecting foods that have less fat, broiling rather than frying, or removing the fat before cooking, reduces fat.

Significant nutritional components: protein, fat, several B vitamins and vitamin E from seeds and nuts, and the minerals iron, zinc, and magnesium.

Not Displayed but Important
While not displayed on the plate, two other items need to be considered: *empty calories* and physical activity. Empty calories are a measure of the energy in high-energy foods with poor nutritional values including processed oils ad fats, carbohydrates, and alcoholic beverages.

Oils are fats that are liquids at room temperature. MyPlate recommends 3–7 teaspoons per day depending on age, sex, and level of physical activity. Small amounts of fats and oils are important in the diet because certain essential fatty acids cannot be manufactured by the body and must be obtained in the diet. The significant nutritional components are the omega-3 and omega-6 fatty acids, essential fatty acids, and vitamin E. Oils and fats are found in such foods as meats, nuts, fish, cooking oil, and salad dressings. Most people get more than they need from salad dressing and oils used in cooking and baking. However, since they have a high empty caloric content, eating in excess of the recommended amount can lead to a variety of health problems. Some empty calorie foods include candy, ice cream, margarine, butter, and lard.

Since many "snack foods" such as candy, sweets, and soda also contain empty calories, the consumption of an excess amount can lead to diet-related health problems such as diabetes, obesity, tooth decay, and cardiovascular diseases. The empty calories in such foods stem from their containing large amount of refined carbohydrates such as sucrose, fructose, and lactose. The alcohol in certain beverages (ethanol) is also considered to contain empty calories and their excess consumption can also result in a variety of health problems (refer to Science and Society – What Happens When you Drink Alcohol).

Since more than 65 percent of adult Americans are overweight and more than 30 percent are obese, exercise is important to improving nutritional health. Although it is not directly related to nutrition, the amount of exercise people get affects the number of food calories they can consume on a daily basis without gaining weight. Exercise has other health benefits (refer to A Closer Look – Exercise: More than Just Maintaining Your Weight). MyPlate recommends at least thirty minutes of moderate exercise per day, and longer periods or more vigorous exercise has additional health benefits. Moderate exercise elevates the heart rate significantly. There are also exercises that strengthen the core of the body, resulting in better balance and fewer falls. Activities such as doing household chores or walking while shopping do not elevate heart rate and therefore do not count as moderate exercise.

Your Health and Body Weight

Not maintaining a healthy weight can adversely affect a person's health, quality of life, and life span. The reasons for an inability to maintain a healthy weight are complex and involve such factors as (1) metabolism, (2) prevailing cultural perceptions and values, and (3) psychological components. In general, we classify people as being ideal, under-, or overweight. By "ideal" we mean that their weight is within a range that fosters good health and contributes to a high quality of life and life span. When we use the terms *underweight* or *overweight*, we mean that their weight is outside their healthy range and that by either gaining or losing weight they will become healthier. Being under- or overweight may be the result of hormonal imbalances, exceptionally high or low metabolic rates, depression, diabetes, parasitic worm infection, eating disorders, medications, or other issues that have nothing to do with food intake.

What Happens When You Drink Alcohol

Ethyl alcohol (CH_3CH_2OH) is a 2-carbon organic compound with a single alcoholic functional group. Because it is soluble in water, it is easily absorbed into the bloodstream. After an alcoholic beverage enters the body, it is spread by the circulatory system rapidly throughout the body and enters the brain. The majority of the alcohol is absorbed from the stomach (20 percent) and small intestine (80 percent). The more a person drinks, the higher the blood alcohol level. How fast alcohol is absorbed depends on several factors:

1. Food in the stomach slows absorption.
2. Strenuous physical exercise decreases absorption.
3. Drugs (e.g., nicotine, marijuana, and ginseng) increase absorption.

Ninety percent of ethyl alcohol is oxidized in mitochondria to acetate ($CH_3CH_2OH + NAD \rightarrow CH_3CHO + NADH + H$). The acetate is then converted to acetyl-CoA that enters the Krebs cycle, where ATP is produced. Alcohol is high in calories (1 g = 7,000 calories, or 7 *food* calories). A standard glass of wine has about 15 g of alcohol and about 100 kilocalories. The 10 percent not metabolized is eliminated in sweat or urine, or given off in breath. It takes the liver one hour to deal with one unit of alcohol. A unit of alcohol is:

- 250 ml (1/2 pint) of ordinary-strength beer/lager.
- One glass (125 ml/4 fl oz) of wine.
- 47 ml/1.5 oz of sherry/vermouth.
- 47 ml/1.5 oz of liquor.

If alcohol is consumed at a rate faster than the liver can break it down, the blood alcohol level rises. This causes an initial feeling of warmth and light-headedness. However, alcohol is a depressant; that is, it decreases the activity of the nervous system. At first, it may inhibit circuits in the brain that normally inhibit a person's actions. This usually results in a person becoming more talkative and active—uninhibited. However, as the alcohol's effect continues, other changes can take place. These include increased aggression, loss of memory, and loss of motor control.

According to the World Health Organization, alcohol is responsible for nearly 4 percent of deaths worldwide, more than AIDS, tuberculosis, or violence. It is also the causal factor in sixty types of diseases, such as cirrhosis of the liver, epilepsy, poisonings, diseases or environmental factors such as cirrhosis, and cancers of the colon, breast, larynx, and liver; and many types of injuries. Long-term, excessive use of alcohol can cause damage to the liver, resulting in the development of a fatty liver, alcoholic hepatitis, and alcoholic cirrhosis. It can also interfere with the kidneys' regulation of water, sodium, potassium, calcium, and phosphate and with the kidneys' ability to maintain a proper acid-base balance, and produce hormones. It also causes low blood sugar levels, dehydration, high blood pressure, strokes, heart disease, birth defects, osteoporosis, and certain cancers. Drinking alcohol in moderation does have some health benefits if the beverage contains antioxidants (e.g., red wines and dark beers). The antioxidants in red wine (polyphenols) appear to counteract the negative effect of chemicals, called free radicals, released during metabolism. Free radicals are known to destroy cell components and cause mutations or damage, which can lead to heart disease and cancers. Antioxidants protect against this kind of harm by capturing free radicals.

Being Overweight, Obese, and Underweight

Technically the term **overweight** is defined as a BMI of 25 or more. The term **obese** describes a person with a BMI over 30 and is overweight to the extent that their health and life span are adversely affected. Being overweight or obese occurs when people consistently take in more food energy than is necessary to meet their daily requirements. About 64 percent of the U.S. population is considered to be overweight or obese, and rates worldwide have doubled since 1980, even as blood pressure and cholesterol levels have dropped. The term **underweight** describes a person who has a BMI under 18.5 and occurs when people consistently take in less food energy than is necessary to meet their daily requirements. About 3.3 percent of children and teens ages two to nineteen and 1.8 percent of adults ages twenty to seventy-four in the United States are considered to be underweight.

At first glance, it would appear that an underweight or overweight problem is a simple problem to solve. To gain weight or lose weight, all that people must do is consume more or fewer kilocalories, exercise accordingly, or do both. While all people outside their ideal body weights have an imbalance between their need for kilocalories and the amount of food they eat, the reasons for this imbalance are complex and varied.

1. *Metabolism* Even at rest, energy is required to maintain breathing, heart rate, and other normal body functions. The **basal metabolic rate** is the rate at which a person uses energy when at rest. Some people have a low basal metabolic rate and, therefore, use fewer kilocalories for maintenance than other people, while others have a higher basal metabolic rate and, therefore, use more kilocalories for maintenance. There are about twenty genes that are

A Closer Look

Exercise: More Than Just Maintaining Your Weight

The Food Guide Pyramid recommends 30 minutes of moderate exercise above normal daily activities. This might be a brisk walk at 3.5 miles/hour, golfing, bicycling (10 miles/hour), and hiking. Workouts such as weight lifting—or riding a cart while golfing—do not fall into this category. In addition to planned exercise, there are other ways to be active, such as taking the stairs instead of the elevator or escalator, parking at the far end of the lot when shopping, walking to the corner store instead of driving, or cutting the grass with a push mower instead of a riding mower.

When most people talk about exercise, they often focus on weight control. However, research in many diverse areas has revealed benefits that influence many aspects of a person's health. In addition to helping control weight, exercise:

1. Increases the strength of muscles and general muscle tone.
2. Reduces the likelihood of injuries because of improved strength and balance.
3. Strengthens bones and joints; bones respond to the stress placed on them by exercise by adding bone mass.
4. Improves flexibility.
5. Increases efficiency of the respiratory system.
6. Increases the efficiency of aerobic respiration in mitochondria.
7. Heightens the immune response to better protect against infection.
8. Increases endorphins in the brain to reduce pain threshold
9. Increases pleasure sensation.
10. Improves self-esteem and feelings of well-being.
11. Reduces feelings of depression and anxiety.
12. Helps control diabetes.
13. Strengthens heart muscle.
14. Improves cardiovascular health.
15. Lowers serum cholesterol.
16. Lowers blood pressure.
17. Improves sex life.
18. Delays onset of dementia.
19. Increases muscle stem cell formation.
20. Blunts the aging process by decreasing the number of senescent (aging) cells in the body.

---

associated with weight gain or loss; however, it appears they are not the primary reason for a person being under- or overweight. Research has demonstrated that the brain (hypothalamus and prefrontal cortex) plays an important role in regulating the feeling of fullness, or satiety. In fact, chronic overeaters and drug addicts demonstrate similar brain biochemistry. Susceptibility to obesity is also influenced by genes and the interaction between intestinal microbes and the immune system.

2. *Physical activity*   Often, exercise is all that is needed to control an overweight problem. In addition to increasing metabolism during the exercise itself, exercise tends to raise the basal metabolic rate for a period of time following exercise. For underweight people, increased inactivity would no longer provide the benefits of regular exercise. Table 24.3 lists several common activities and the amount of energy needed to sustain the activity.

3. *Cultural influences*   Our culture encourages food consumption. Social occasions and business meetings frequently involve eating, and restaurants and other food preparers have increased the size of portions significantly over the past fifty years. For example, in the 1950s, a fast-food serving of French fries was 2.4 ounces. Today, that size is still available, but it is the small size, and medium and large sizes contain two to three times the quantity of the small size.

The abundance of snack foods high in sugars and fats is another important cultural influence. Consumption of these foods at other than meal times increases the total number of calories consumed during the day. Furthermore, many of these foods are said to provide "empty calories"—calories are provided by sugar or fat but there is little or no other nutritional benefit (protein, vitamins, minerals, or fiber).

4. *Psychological factors*   Undereating and overeating are associated with a variety of psychological factors. Eating is a pleasurable activity, and with all pleasurable activities it is sometimes difficult to determine when to stop. Conversely, overeating is also associated with depression or medica-

---

TABLE 24.3   Typical Energy Requirements for Common Activities

Light Activities (120–150 kcal/h)	Light to Moderate Activities (150–300 kcal/h)	Moderate Activities (300–400 kcal/h)	Heavy Activities (420–600 kcal/h)
Dressing	Sweeping floors	Pulling weeds	Chopping wood
Typing	Painting	Walking behind a lawnmower	Shoveling snow
Slow walking	Walking 2–3 mi/h	Walking 3.5–4 mi/h on level surface	Walking or jogging 5 mi/h
Standing	Bowling	Golf (no cart)	Walking up hills
Studying	Store clerking	Doubles tennis	Cross-country skiing
Sitting activities	Canoeing 2.5–3 mi/h	Canoeing 4 mi/h	Swimming
	Bicycling on level surface at 5.5 mi/h	Volleyball	Bicycling 11–12 mi/h or up and down hills

Myths, Mistakes, and Misunderstandings

Diet and Nutrition

There is probably no area of modern life in which there is more misinformation, pseudoscience, and myth than nutrition. Diet books abound, each purporting to provide the answer to nutritional and weight control problems. It is helpful to look at the science behind these various statements. The list below provides several examples of the mismatch between claims and scientific reality.

Myth or Misunderstanding	Scientific Basis
1. Exercise burns calories.	Calories are not molecules, so they cannot be burned in the physical sense, but you do oxidize (burn) the fuels to provide yourself with the energy needed to perform various activities.
2. Active people who are increasing their fitness need more protein.	The amount of protein needed per day is very small—about 50 g. Most people get many times the amount of protein required.
3. Vitamins give you energy.	Most vitamins assist enzymes in bringing about chemical reactions, some of which may be energy-yielding, but vitamins are not sources of energy.
4. Large amounts of protein are needed to build muscle.	A person can build only a few grams of new muscle per day. Therefore, consuming large amounts of protein will not increase the rate of muscle growth.
5. Large quantities (mega doses) of vitamins will fight disease, build strength, and increase length of life.	Quantities of vitamins that greatly exceed recommendations have not been shown to be beneficial. Large doses of some vitamins (vitamins A, D, B_3) are toxic.
6. Special protein supplements are more quickly absorbed and can build muscle faster.	Adequate protein is obtained in nearly all diets. The supplements may be absorbed faster, but that does not mean that they will be incorporated into muscle mass faster.
7. Vitamins prevent cancer, heart disease, and other health problems.	Vitamins are important to health. However, it is a gross oversimplification to suggest that consumption of excess amounts of specific vitamins will prevent certain diseases. Many factors contribute to the cause of disease.

tions. Others are strongly influenced by prevailing cultural perceptions and values and pursue a body weight that is "skinny," muscular, and underweight. Two conditions considered to be eating disorders are bulimia and anorexia nervosa.

Bulimia ("hunger of an ox" in Greek) is an eating disorder in which the person engages in a cycle of eating large amounts of food (binges) followed by purging the body of the food by inducing vomiting or using laxatives. Many bulimics also use diuretics that cause the body to lose water and therefore reduce weight. It is often called the silent killer because it is difficult to detect. Bulimics are usually of normal body weight or overweight. The behavior has a strong psychological component and is often associated with depression or other psychological problems.

Vomiting may be induced physically or by the use of some nonprescription drugs. Case studies have shown that bulimics may take forty to sixty laxatives a day to rid themselves of food. For some, the laxative becomes addictive. The binge-purge cycle and associated use of diuretics results in a variety of symptoms that can be deadly.

Anorexia nervosa is an eating disorder characterized by severe, prolonged weight loss as a result of a voluntary severe restriction in food intake. It is most common among adolescent and preadolescent females. An anorexic person's fear of becoming overweight is so intense that even though weight loss occurs, it does not lessen the fear of obesity, and the person continues to eat less, often even refusing to maintain the optimum

body weight for his or her age, sex, and height. Persons who have this disorder have a distorted perception of their bodies. They see themselves as fat when in fact they are starving to death. Society's preoccupation with body image, particularly among young people, may contribute to the incidence of this disorder.

24.8 WASTE DISPOSAL: THE EXCRETORY SYSTEM

Because cells are modifying molecules during metabolic processes, harmful waste products are constantly being formed. Urea is a common waste product formed during the metabolic breakdown of amino acids, and hydrogen ions are produced by many metabolic reactions. Other molecules, such as water and salts, may be consumed in excessive amounts and must be removed. The primary organs involved in regulating the level of toxic or unnecessary molecules are the **kidneys.** The urine they produce flows to the urinary bladder through tubes known as *ureters.* The urine leaves the bladder through a tube known as the urethra (figure 24.13).

The kidneys consist of about 2.4 million tiny units called **nephrons.** At one end of a nephron is a cup-shaped structure called *Bowman's capsule,* which surrounds a knot of capillaries known as a *glomerulus.* In addition to Bowman's capsule, a nephron consists of three distinctly different regions: the *proximal convoluted tubule,* the *loop of Henle,* and the *distal convoluted tubule.* The distal convoluted tubule of a nephron is connected to a collecting duct that transports fluid to the ureters and ultimately to the urinary bladder, where it is stored until it can be eliminated (figure 24.14).

As in the other systems, your excretory system involves a close connection between your circulatory system and a surface. In this case, the large surface is provided by the walls of the millions of nephrons, which are surrounded by capillaries. Three major activities occur at these surfaces: filtration, reabsorption, and secretion.

Filtration occurs through the glomerulus. The glomerulus presents a large surface for the filtering of material from the blood to Bowman's capsule. Blood that enters the glomerulus is under pressure from the muscular contraction of your heart. The capillaries of the glomerulus are quite porous and provide a large surface area for the movement of water and small dissolved molecules from the blood into Bowman's capsule. Normally, only the smaller molecules, such as glucose, amino acids, and ions, are able to pass through the glomerulus into Bowman's capsule at the end of the nephron. The various kinds of blood cells and larger molecules such as proteins do not pass out of the blood into the nephron. This physical filtration process allows many kinds of molecules to leave your blood and enter your nephron. The volume of material filtered in this way is about 7.5 liters per hour. Since your entire blood supply is about 5 to 6 liters, there must be some method of recovering much of this fluid.

Reabsorption occurs in several parts of your nephron. The proximal convoluted tubule reabsorbs most of the useful molecules (water, phosphate, salts, sugars, amino acids) as the mate-

rial filtered from your glomerulus passes through your nephron. The loop of Henle is involved in the reabsorption of water.

The distal convoluted tubule is involved in both reabsorption and secretion and is involved in the regulation of ions, such as hydrogen, sodium, potassium, and calcium ions.

Some molecules that pass through the nephron are relatively unaffected by the various activities going on in the kidney. One of these is urea, which is filtered through your glomerulus into Bowman's capsule. As it passes through the nephron, much of it stays in the tubule and is eliminated in your urine. Many other kinds of molecules, such as minor metabolic waste products and some drugs, are also lost in the urine.

24.9 CONTROL MECHANISMS

Your nervous and endocrine systems are the major systems of the body that integrate stimuli and generate appropriate responses necessary to maintain homeostasis. There are many

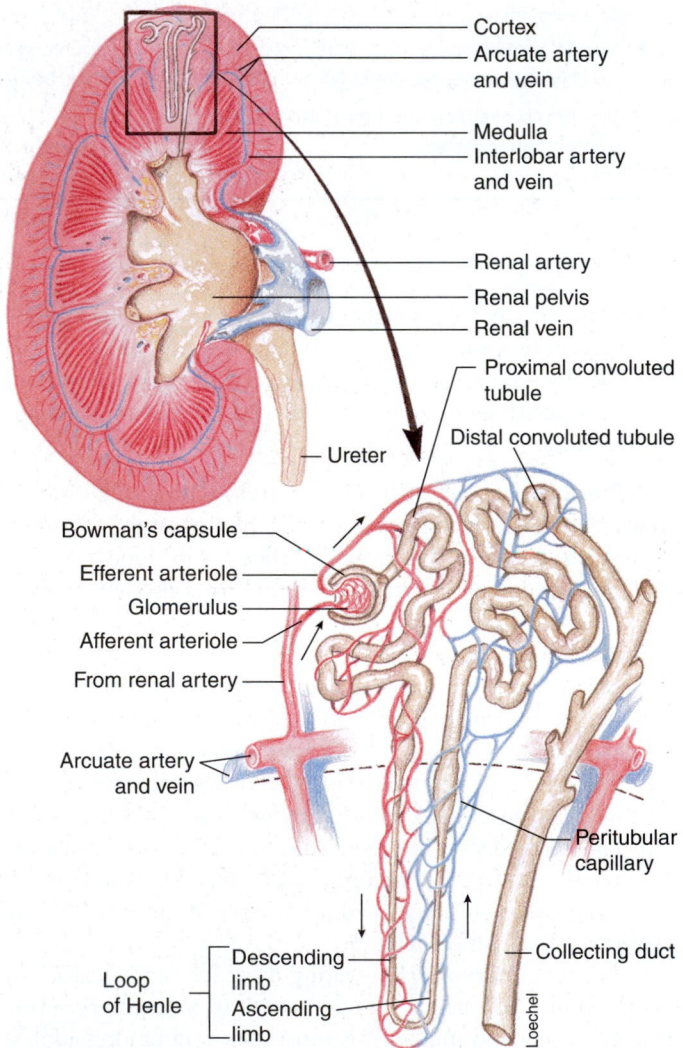

FIGURE 24.14 The nephron and the closely associated blood vessels create a system that allows for the passage of materials from the circulatory system to the nephron by way of the glomerulus and Bowman's capsule. Materials are added to and removed from the fluid in the nephron via the tubular portions of the nephron.

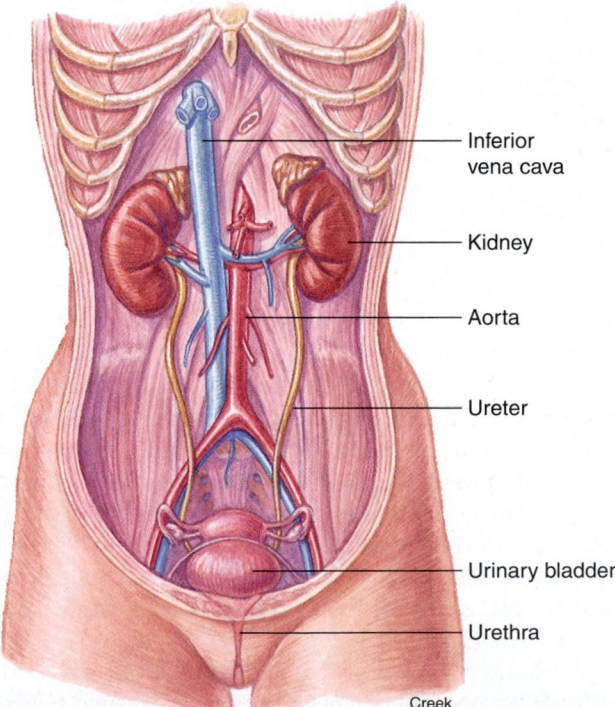

FIGURE 24.13 The primary organs involved in removing materials from the blood are the kidneys. The urine produced by the kidneys is transported by the ureters to the urinary bladder. From the bladder, the urine is emptied to the outside of the body by way of the urethra.

kinds of sense organs located within organs and on the surface of your body that respond to specific kinds of stimuli. A **stimulus** is any change in the environment that the organism can detect. Some stimuli, such as light or sound, are typically external to the organism; others, such as the pain generated by an infection, are internal. The reaction of the organism to a stimulus is known as a **response**.

Your *nervous system* consists of a network of cells with fibrous extensions that carry information throughout the body. Your *endocrine system* consists of a number of glands that communicate with one another and with other tissues through chemicals distributed throughout the organism. **Glands** are organs that manufacture molecules that are secreted either through ducts or into surrounding tissue, where they are picked up by your circulatory system. *Endocrine glands* have no ducts and secrete their products, called **hormones,** into the circulatory system; other glands, such as your digestive glands and sweat glands, empty their contents through ducts. These kinds of glands are called *exocrine glands*.

Although the functions of your nervous and endocrine systems can overlap and be interrelated, these two systems have quite different methods of action. The nervous system functions very much like a computer. A message is sent along established pathways from a specific initiating point to a specific end point, and the transmission is very rapid. Your endocrine system functions in a manner analogous to a radio broadcast system. Radio stations send their signals in all directions, but only those radio receivers that are tuned to the correct frequency can receive the message. Messenger molecules (hormones) are typically distributed throughout the body by the circulatory system, but only those cells that have the proper receptor sites can receive and respond to the molecules.

The Structure of the Nervous System

The basic unit of your nervous system is a specialized cell called a *neuron,* or *nerve cell.* A typical neuron consists of a central body called the *soma,* or *cell body,* that contains the nucleus, and several long parts of the neural cell that stretch outward, called nerve fibers. There are two kinds of fibers: *axons,* which carry information away from the cell body, and *dendrites,* which carry information toward the cell body. Most nerve cells have one axon and several dendrites.

Neurons are arranged into two major systems. Your *central nervous system,* which consists of the brain and spinal cord, is surrounded by your skull and the vertebrae of your spinal column. It receives input from sense organs, interprets information, and generates responses. The *peripheral nervous system* is located outside your skull and spinal column and consists of bundles of long axons and dendrites called nerves. There are two different sets of neurons in the peripheral nervous system. *Motor neurons* carry messages from your central nervous system to muscles and glands, and *sensory neurons* carry input from sense organs to your central nervous system. Motor neurons typically have one long axon that runs from the spinal cord to a muscle or gland, while sensory neurons have long dendrites that carry input from the sense organs to the central nervous system (figure 24.15).

The Nature of the Nerve Impulse

Since most nerve cells have long fibrous extensions, it is possible for information to be passed along the nerve cell from one end to the other. The message that travels along a neuron is known as a **nerve impulse.** A nerve impulse is not a simple electric current but involves a specific sequence of chemical events involving activities at the cell membrane.

Since all cell membranes are differentially permeable, it is difficult for some ions to pass through

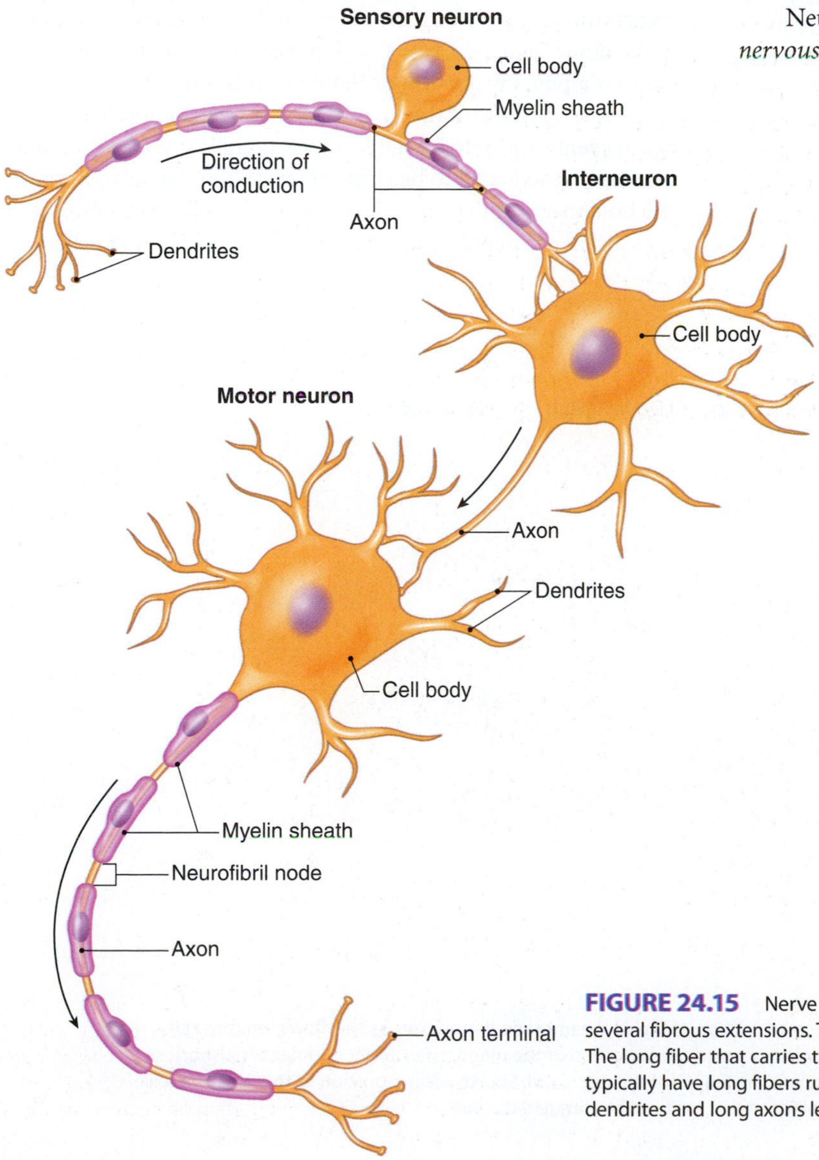

FIGURE 24.15 Nerve cells consist of a nerve cell body that contains the nucleus and several fibrous extensions. The fibers that carry impulses to the nerve cell body are dendrites. The long fiber that carries the impulse away from the cell body is the axon. Sensory neurons typically have long fibers running to and from the cell body. Motor neurons have short dendrites and long axons leaving the cell body.

Connections . . .

Because a normal resting cell has more positively charged Na^+ ions on the outside of the cell than on the inside, a small but measurable voltage exists across the membrane of the cell. The voltage difference between the inside and outside of a cell membrane is about 70 millivolts (0.07 volt). The two sides of the cell membrane are, therefore, polarized in the same sense that a battery is polarized, with a positive and negative pole. A resting neuron has its positive pole on the outside of the cell membrane and its negative pole on the inside of the membrane.

When a cell is stimulated at a specific point on the cell membrane, the cell membrane changes its permeability and lets sodium ions (Na^+) pass through it from the outside to the inside and potassium ions (K^+) diffuse from the inside to the outside. The membrane is thus *depolarized*; it loses its difference in charge as sodium ions (Na^+) diffuse into the cell from the outside. Sodium ions diffuse into the cell because initially they are in greater concentration outside the cell than inside. When the membrane becomes more permeable, they are able to diffuse into the cell, toward the area of lower concentration.

The depolarization of one point on the cell membrane causes the adjacent portion of the cell membrane to change its permeability as well, and it also depolarizes. Thus, a wave of depolarization passes along the length of the neuron from one end to the other (figure 24.16). The depolarization and passage of an impulse along any portion of the neuron is a momentary event. As soon as a section of the membrane has been depolarized, a series of events begins to reestablish the original polarized state, and the membrane is said to be *repolarized*. When the nerve impulse reaches the end of the axon, it stimulates the release of a molecule that stimulates depolarization of the next neuron in the chain.

the membrane, and the combination of ions inside the membrane is different from that on the outside. Cell membranes also contain proteins that actively transport specific ions from one side of the membrane to the other. Active transport involves the cell's use of ATP to move materials from one side of the cell membrane to the other. One of the ions that is actively transported from cells is the sodium ion (Na^+). At the same time sodium ions (Na^+) are being transported out of cells, potassium ions (K^+) are being transported into the normal resting cells. However, there are more sodium ions (Na^+) transported out than potassium ions (K^+) transported in.

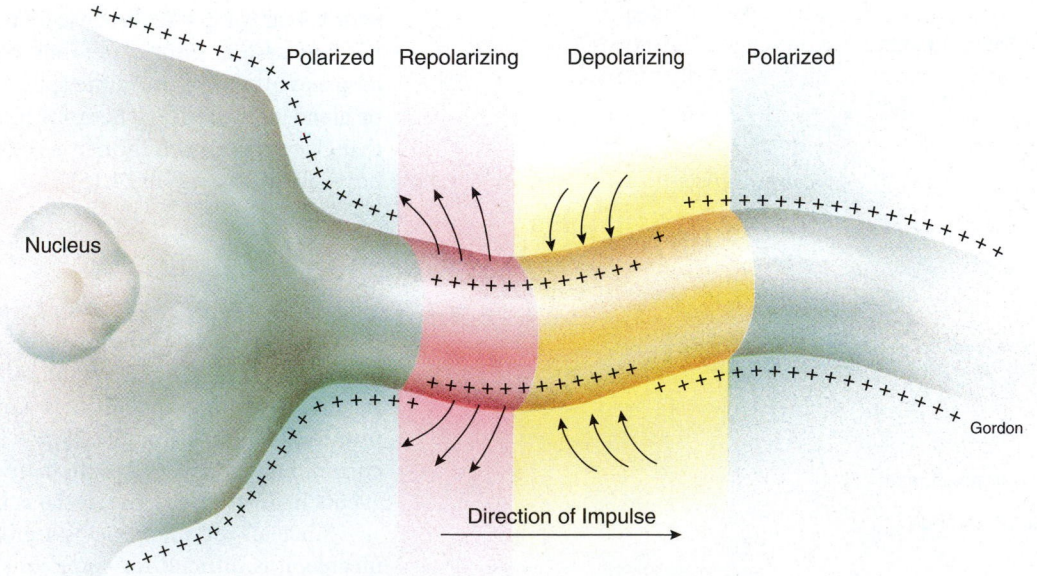

FIGURE 24.16   When a nerve cell is stimulated, a small portion of the cell membrane depolarizes as Na^+ flows into the cell through the membrane. This encourages the depolarization of an adjacent portion of the membrane, and it depolarizes a short time later. In this way a wave of depolarization passes down the length of the nerve cell. Shortly after a portion of the membrane is depolarized, the ionic balance is reestablished. It is repolarized and ready to be stimulated again.

Activities at the Synapse

Neurons are arranged end-to-end to form long chains of cells. Between the fibers of adjacent neurons is a space called the **synapse.** Many chemical events occur in the synapse that are important in the function of your nervous system. When a neuron is stimulated, an impulse passes along its length from one end to the other. When the impulse reaches a synapse, a molecule called a **neurotransmitter** is released into the synapse from the axon. It diffuses across the synapse and binds to specific receptor sites on the dendrite of the next neuron. When enough neurotransmitter molecules have bound to the second neuron, an impulse is initiated in it as well. Several kinds of neurotransmitters are produced by specific neurons. These include dopamine, epinephrine, acetylcholine, and several other molecules. The first neurotransmitter identified was *acetylcholine.*

As long as a neurotransmitter is bound to its receptor, it continues to stimulate the nerve cell. Thus, if acetylcholine continues to occupy receptors, the neuron continues to be stimulated again and again. An enzyme called *acetylcholinesterase* destroys acetylcholine and prevents this from happening. The breakdown products of the acetylcholine can be used to remanufacture new acetylcholine molecules at the end of the axon. The destruction of acetylcholine allows the second neuron in the chain to return to normal. Thus, it will be ready to accept another burst of acetylcholine from the first neuron a short time later. Neurons must also constantly manufacture new acetylcholine molecules, or they will exhaust their supply and be unable to conduct an impulse across a synapse (figure 24.17).

Caffeine and nicotine are considered stimulants because they make it easier for impulses to pass through the synapse. The nerve cells are sensitized and will respond to smaller amounts of acetylcholine than normal. Drinking coffee, taking caffeine pills, or smoking tobacco will increase the sensitivity of your nervous system to stimulation. However, as with most kinds of drugs, continual use or abuse tends to lead to a loss of the effect as the nervous system adapts to the constant presence of the drugs.

Because of the way the synapse works, impulses can go in only one direction: Only axons secrete acetylcholine, and only dendrites have receptors. This explains why there are separate sensory and motor neurons to carry messages to and from your central nervous system.

Endocrine System Function

As mentioned previously, your endocrine system is basically a broadcasting system in which glands secrete messenger molecules, called hormones, which are distributed throughout your body by the circulatory system (figure 24.18). However, each kind of hormone attaches only to appropriate receptor molecules on the surfaces of certain cells.

How Cells Respond to Hormones

The cells that receive the message typically respond in one of three ways: (1) Some cells release products that have been previously manufactured, (2) other cells are stimulated to synthesize molecules or to begin metabolic activities, and (3) some are stimulated to divide and grow.

These different kinds of responses mean that some endocrine responses are relatively rapid, while others are very slow. For example, the release of the hormones *epinephrine* and *norepinephrine* (also known as adrenaline and noradrenaline) from the adrenal medulla, located near your kidney or certain nerve cells, causes a rapid change in the behavior of an organism. The heart rate increases, blood pressure rises, blood is shunted to muscles, and the breathing rate increases. You have certainly experienced this reaction many times in your lifetime as when you nearly have an automobile accident or slip and nearly fall.

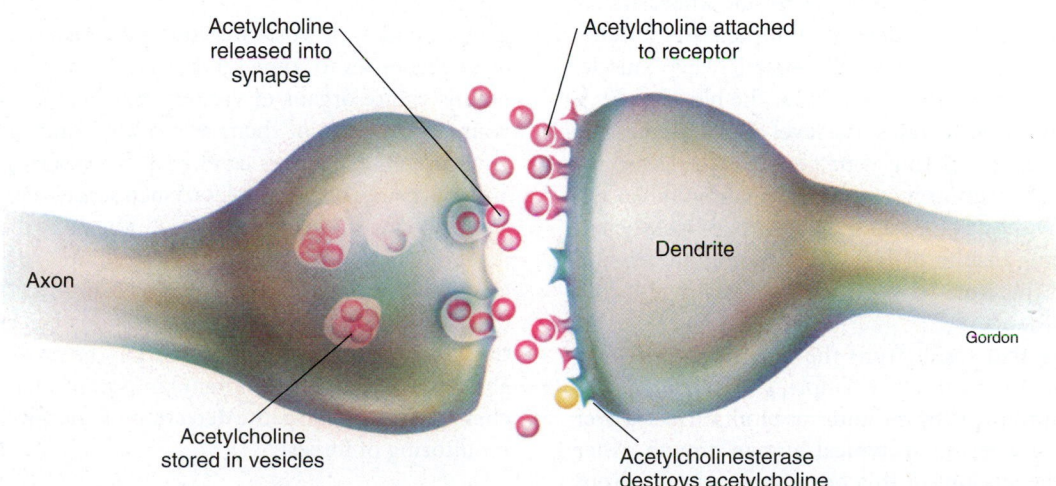

FIGURE 24.17 When a nerve impulse reaches the end of an axon, it releases a neurotransmitter into the synapse. In this illustration, the neurotransmitter is acetylcholine. When acetylcholine is released into the synapse, acetylcholine molecules diffuse across the synapse and bind to receptors on the dendrite, initiating an impulse in the next neuron. Acetylcholinesterase is an enzyme that destroys acetylcholine, preventing continuous stimulation of the dendrite.

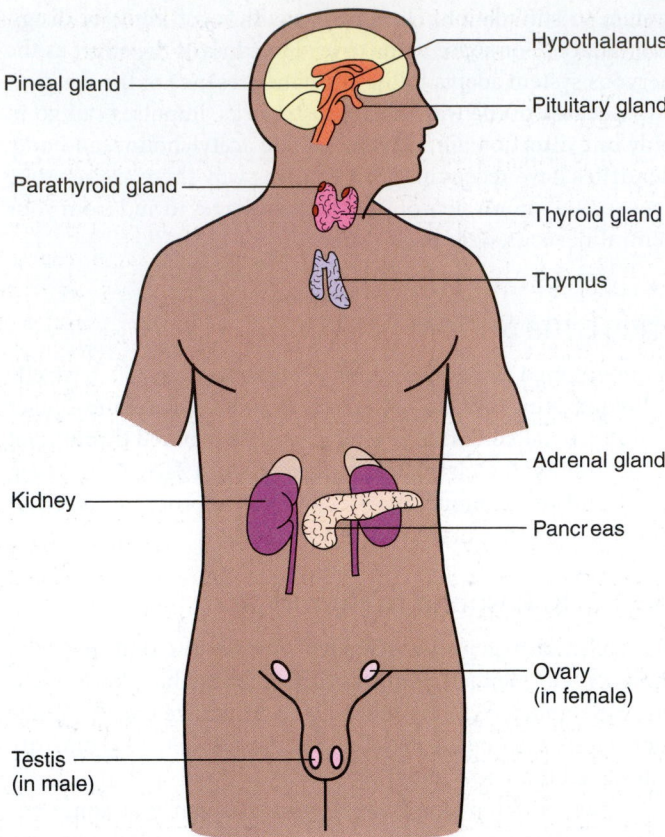

Pineal gland

Hypothalamus

Pituitary gland

Parathyroid gland

Thyroid gland

Thymus

Adrenal gland

Kidney

Pancreas

Ovary
(in female)

Testis
(in male)

FIGURE 24.18   The endocrine glands are located at various places within the body and cause their effects by secreting hormones.

Another hormone, called *antidiuretic hormone (ADH),* acts more slowly. It is released from the posterior pituitary gland at the base of the brain and regulates the rate at which your body loses water through the kidneys. It does this by allowing the reabsorption of water from the kidney back into your bloodstream. The effects of this hormone can be noticed in a matter of minutes to hours. Insulin is another hormone whose effects are quite rapid. Insulin is produced by the pancreas, located near the stomach, and stimulates cells—particularly muscle, liver, and fat cells—to take up glucose from the blood. After a meal that is high in carbohydrates, the level of glucose in your blood begins to rise, stimulating your pancreas to release insulin. The increased insulin causes glucose levels to fall as the sugar is taken up by cells. People with diabetes have insufficient or improperly acting insulin or lack the receptors to respond to the insulin and therefore have difficulty regulating glucose levels in their blood.

The responses that result from the growth of cells may take weeks or years to occur. For example, *growth-stimulating hormone (GSH)* is produced by the anterior pituitary gland over a period of years and results in typical human growth. After sexual maturity, the amount of this hormone generally drops to very low levels and body growth stops. Sexual development is also largely the result of the growth of specific tissues and organs. The male sex hormone *testosterone,* produced by the testes, causes the growth of male sex organs and a change to

the adult body form. The female counterpart, *estrogen,* results in the development of female sex organs and the adult body form. In all of these cases, it is the release of hormones over long periods, continually stimulating the growth of tissues sensitive to the hormones, that results in the normal developmental pattern. The absence or inhibition of any of these hormones early in life changes the normal growth process.

Interactions of the Nervous and Endocrine Systems

Because the pituitary is constantly receiving information from the hypothalamus of the brain, many kinds of sensory stimuli to the body can affect the functioning of your endocrine system. One example is the way in which the nervous system and endocrine system interact to influence the menstrual cycle. At least three different hormones are involved in the cycle of changes that affect the ovary and the lining of the uterus. It is well documented that stress caused by tension or worry can interfere with the normal cycle of hormones and delay or stop menstrual cycles. In addition, young women living in groups, such as in college dormitories, often find that their menstrual cycles become synchronized. Although the exact mechanism involved in this phenomenon is unknown, it is suspected that input from the nervous system causes this synchronization. (Odors and sympathetic feelings have been suggested as causes.)

Although we still tend to think of the nervous and endocrine systems as being separate and different, it is becoming clear that they are interconnected. These two systems cooperate to bring about appropriate responses to environmental challenges. The nervous system is specialized for receiving and sending short-term messages, whereas activities that require long-term, growth-related actions are handled by the endocrine system.

24.10 SENSORY INPUT

The activities of your nervous and endocrine systems are often responses to some kind of input received from the sense organs. Sense organs of various types are located throughout your body. Many of them are located on the surface, where environmental changes can be easily detected. Hearing, sight, and touch are good examples of such senses. Other sense organs are located within your body and indicate to the organism how its various parts are changing. For example, pain and pressure are often used to monitor internal conditions. The sense organs detect changes, but the brain is responsible for perception—the recognition that a stimulus has been received. Sensory abilities involve many different kinds of mechanisms, including chemical recognition, the detection of energy changes, and the monitoring of forces.

Chemical Detection

All cells have receptors on their surfaces that can bind selectively to molecules they encounter. This binding process can cause changes in the cell in several ways. In some cells, it causes depolarization. When this happens, the cells can stimulate neurons and cause messages to be sent to the central nervous system, informing it of some change in your surroundings. In other cases, a molecule binding to the cell surface may cause certain genes to be expressed, and the cell responds by changing the molecules it produces. This is typical of the way the endocrine system receives and delivers messages.

Taste

Most cells have specific binding sites for particular molecules. Others, such as the taste buds on your tongue, appear to respond to classes of molecules. Traditionally we have distinguished four kinds of tastes: sweet, sour, salt, and bitter. However, recently, a fifth kind of taste, *umami* (meaty), has been identified that responds to the amino acid glutamate, which is present in many kinds of foods and can be added as a flavor enhancer (monosodium glutamate). Sourness appears to result from taste buds that respond to the presence of hydrogen ions (H^+), because acids taste sour. Saltiness is sensed when sodium chloride stimulates taste buds. However, the sensations of sweetness, bitterness, and umami occur when molecules bind to specific surface receptors on the cell. Sweetness can be stimulated by many kinds of organic molecules, including sugars and artificial sweeteners, and also by inorganic lead compounds. When a molecule binds to a sweetness receptor, a molecule is split, and its splitting stimulates an enzyme that leads to the depolarization of the cell. The sweet taste of lead salts in old paints partly explains why children sometimes eat paint chips. Because the lead interferes with normal brain development, this behavior can have disastrous results. Many kinds of compounds of diverse structures give the bitter sensation. The cells that respond to bitter sensations have a variety of receptor molecules on their surface. When a substance binds to one of the receptors, the cell depolarizes. In the case of umami, it is the glutamate molecule that binds to receptors on the cells of the taste buds.

It is also important to understand that much of what we often refer to as taste involves such inputs as temperature, texture, and smell. Cold coffee has a different taste than hot coffee even though they are chemically the same. Lumpy, cooked cereal and smooth cereal have different tastes. If you are unable to smell food, it doesn't taste as it should, which is why you sometimes lose your appetite when you have a stuffy nose. We still have much to learn about how the tongue detects chemicals and the role that other associated senses play in modifying taste.

Smell

The other major chemical sense, the sense of smell, is much more versatile; it can detect thousands of different molecules at very low concentrations. The cells that make up the *olfactory epithelium,* the cells that line the nasal cavity and respond to smells, apparently bind molecules to receptors on their surface. Exactly how this can account for the large number of recognizably different odors is unknown, but the receptor cells are extremely sensitive. In some cases, a single molecule of a substance is sufficient to cause a receptor cell to send a message to your brain, where the sensation of odor is perceived. These sensory cells for smell also fatigue rapidly. You have probably noticed that when you first walk into a room, specific odors are readily detected, but after a few minutes, you are unable to detect them. Most perfumes and aftershaves are undetectable after fifteen minutes of continuous stimulation.

Other Chemical Senses

Many internal sense organs also respond to specific molecules. For example, your brain and aorta contain cells that respond to concentrations of hydrogen ions, carbon dioxide, and oxygen in the blood. Remember, too, that your endocrine system relies on the detection of specific messenger molecules to trigger its activities.

Light Detection

The eyes primarily respond to changes in the flow of light energy. The structure of the eye is designed to focus light on a light-sensitive layer of the back of the eye known as the **retina** (figure 24.19). There are two kinds of receptors in the retina of your eye. The cells called *rods* respond to a broad range of wavelengths of light and are responsible for black-and-white vision. Since rods are very sensitive to light, they are particularly useful in dim light. Rods are located over most of the retinal surface except for the area of most acute vision known as the *fovea centralis.* The other receptor cells, called *cones,* are found throughout the retina but are particularly concentrated in the fovea centralis. Cones are not as sensitive to light, but they can detect light of different wavelengths (color). This combination of receptors gives us the ability to detect color when light levels are high, but we rely on black-and-white vision at night. There are three different varieties of cones: One type responds best to red light, another responds best to green light, and the third responds best to blue light. Stimulation of various combinations of these three kinds of cones allows us to detect different shades of color.

Rods and the three different kinds of cones each contain a pigment that decomposes when struck by light of the proper wavelength and sufficient strength. The pigment found in rods is called *rhodopsin.* This change in the structure of rhodopsin causes the rod to stimulate the nerve cell connected to it and send messages to the brain. Cone cells have a similar mecha-

---

CONCEPTS APPLIED

Taste versus Smell

Distinguish between the taste and smell (olfaction) components of your sensory system. First taste common sweet candy; then rinse your mouth with water. Take note of your responses. Next taste a common sour or bitter candy. Again take note of your responses. Second, pinch your nose shut and repeat the tasting. Do you note any differences? Which flavors are "tasted" and which are "smelled"?

Connections . . .

Negative Feedback Control

Many common mechanical and electronic systems are controlled by negative feedback. In negative feedback control, some output of a process influences the process so that it produces less of the product. For example, a thermostat is used to control the temperature in a building or room. When the temperature becomes too low, an electronic circuit is completed and a heating device is turned on, which raises the temperature. When the temperature reaches the set point, the circuit is disconnected and the heating device is turned off.

A large number of physiological processes are controlled in the same way. For example, for most people the amount of the sugar, glucose, is regulated within very narrow limits. However, when we eat a source of glucose, the amount in the blood rises. The presence of elevated glucose stimulates the pancreas to release insulin. Insulin stimulates muscle and liver cells to take glucose out of the blood and store it. When the glucose level falls, the pancreas is no longer stimulated and the amount of insulin declines.

nism of action, and each of the three kinds of cones has a different pigment. Thus, the pattern of color and light intensity recorded on the retina is detected by rods and cones and converted into a series of nerve impulses that are received and interpreted by the brain.

Sound Detection

Your ears respond to changes in sound waves. Sound is produced by the vibration of molecules. Consequently, your ears are detecting changes in the quantity of energy and the quality of sound waves. Sound has several characteristics. Loudness, or volume, is a measure of the intensity of sound energy that arrives at your ear. Very loud sounds will literally vibrate your body and can cause hearing loss if they are too intense. Pitch is a quality of sound that is determined by the frequency of the sound vibrations. High-pitched sounds have short wavelengths; low-pitched sounds have long wavelengths.

Figure 24.20 shows the anatomy of your ear. The sound that arrives at the ear is first funneled by the external ear to the **tympanum,** also known as the *eardrum.* The cone-shaped nature of the external ear focuses sound on the tympanum and causes it to vibrate at the same frequency as the sound waves reaching it. Attached to your tympanum are three tiny bones known as the *malleus* (hammer), *incus* (anvil), and *stapes* (stirrup). The malleus is attached to the tympanum, the incus is attached to the malleus and stapes, and the stapes is attached to a small, membrane-covered opening called the oval window in a snail-shaped structure known as the cochlea. The vibration of the tympanum causes the tiny bones (malleus, incus, and stapes) to vibrate, and they in turn cause a corresponding vibration in the membrane of the *oval window.*

The *cochlea* of the ear is the structure that detects sound and consists of a snail-shaped set of fluid-filled tubes. When your oval window vibrates, the fluid in the cochlea begins to move, causing a membrane in the cochlea, called the *basilar membrane,* to vibrate. High-pitched, short-wavelength sounds cause the basilar membrane to vibrate at the base of the cochlea near the oval window. Low-pitched, long-wavelength sounds vibrate the basilar membrane far from the oval window. Loud sounds cause the basilar membrane to vibrate more vigorously than do faint sounds. Cells on this membrane depolarize when they are stimulated by its vibrations. Because they synapse with neurons, messages can be sent to your brain.

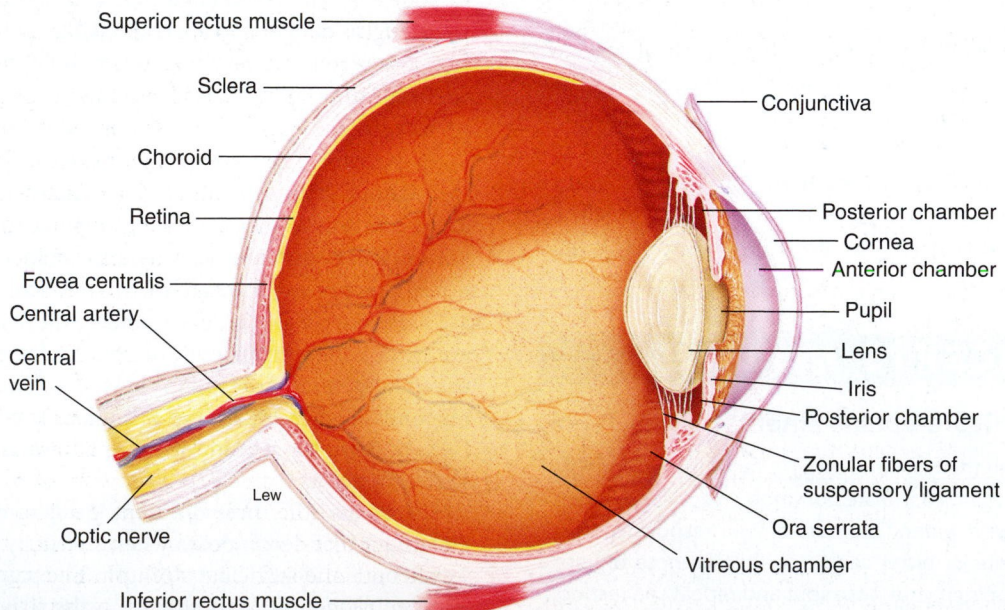

FIGURE 24.19 The eye contains a cornea and lens that focus the light on the retina of the eye. The light causes pigments in the rods and cones of the retina to decompose. This leads to the stimulation of neurons that send messages to the brain.

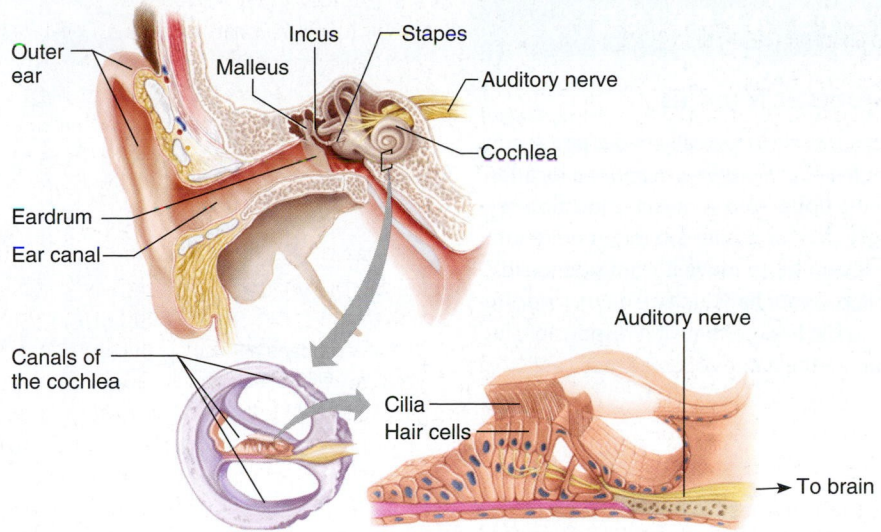

FIGURE 24.20 The ear consists of an external cone that directs sound waves to the tympanum (eardrum). Vibrations of the tympanum move the ear bones and vibrate the oval window of the cochlea, where the sound is detected. The semicircular canals monitor changes in the position of the head, helping us maintain balance.

Because sounds of different wavelengths stimulate different portions of your cochlea, the brain is able to determine the pitch of a sound. Most sounds consist of a mixture of pitches that are heard. Louder sounds stimulate the membrane more forcefully, causing the sensory cells in the cochlea to send more nerve impulses per second. Thus, your brain is able to perceive the loudness of various sounds, as well as the pitch.

Associated with the cochlea are two fluid-filled chambers and a set of fluid-filled tubes called the *semicircular canals.* These structures are not involved in hearing but are involved in maintaining balance and posture. In the walls of these canals and chambers are cells similar to those found on the basilar membrane. These cells are stimulated by movements of your head and by the position of the head with respect to the force of gravity. The constantly changing position of the head results in sensory input that is important in maintaining balance.

Touch

What we normally call the sense of *touch* consists of a variety of different kinds of input. Some receptors respond to pressure, others to temperature, and others, which we call *pain receptors,* usually respond to cell damage. When these receptors are appropriately stimulated, they send a message to your brain. Because receptors are stimulated in particular parts of the body, your brain is able to localize the sensation. However, not all parts of the body are equally supplied with these receptors. The tips of your fingers, lips, and external genitals have the highest density of these nerve endings, while the back, legs, and arms have far fewer receptors.

Some internal receptors, such as pain and pressure receptors, are important in allowing us to monitor our internal activities. Many pains generated by the internal organs are often perceived as if they were somewhere else. For example,

the pain associated with heart attack is often perceived to be in the left arm.

Pressure receptors in joints and muscles are important in providing information about the degree of stress being placed on a portion of your body. This is also important information to send back to the brain so that adjustments can be made in movements to maintain posture. If you have ever had your foot "go to sleep" because the nerve stopped functioning, you have experienced what it is like to lose this constant input of nerve messages from the pressure sensors to assist in guiding the movements you make. Your movements become uncoordinated until the nerve function returns to normal.

24.11 OUTPUT MECHANISMS

Your nervous and endocrine systems cause changes in several ways. Both systems can stimulate muscles to contract and glands to secrete. The endocrine system is also able to alter the metabolic activities of cells and regulate the growth of tissues. The nervous system acts upon two kinds of organs: muscles and glands. The actions of muscles and glands are simple and direct: muscles contract and glands secrete.

Muscles

The ability to move is one of the fundamental characteristics of animals. Through the coordinated contraction of many muscles, the intricate, precise movements of a dancer, basketball player, or writer are accomplished. It is important to recognize that muscles can pull only by contracting; they are unable to push by lengthening. The work of any muscle is done during its contraction. Relaxation is the passive state of the muscle. There must always be some force available that will stretch a muscle after it has stopped contracting and relaxes. Therefore,

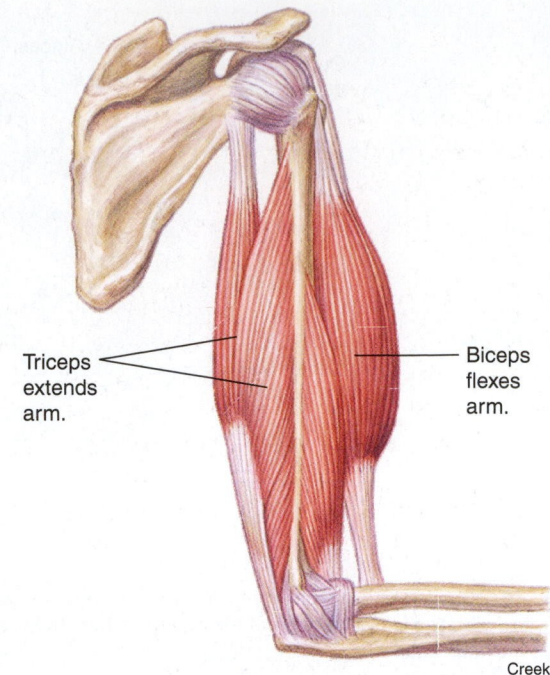

Triceps extends arm.

Biceps flexes arm.

Creek

FIGURE 24.21 Because muscles cannot actively lengthen, it is necessary to have sets of muscles that oppose one another. The contraction and shortening of one muscle cause the stretching of a relaxed muscle.

the muscles that control the movements of your skeleton are present in antagonistic sets—for every muscle's action, there is another muscle that has the opposite action. For example, your biceps muscle causes your arm to flex (bend) as the muscle shortens. The contraction of its antagonist, the triceps muscle, causes the arm to extend (straighten) and at the same time stretches the relaxed biceps muscle (figure 24.21).

There are three major types of muscle: *skeletal, smooth,* and *cardiac.* These differ from one another in several ways. *Skeletal muscle* is voluntary muscle; it is under the control of the nervous system. Your brain or spinal cord sends a message to skeletal muscles, and they contract to move the legs, fingers, and other parts of the body. This does not mean that you must make a conscious decision every time you want to move a muscle. Many of the movements we make are learned initially but become automatic as a result of practice. For example, walking, swimming, or riding a bicycle required a great amount of practice originally, but now you probably perform these movements without thinking about them. They are, however, still considered voluntary actions.

Smooth muscles make up the walls of muscular internal organs, such as the gut, blood vessels, and reproductive organs. They have the property of contracting as a response to being stretched. Since much of the digestive system is being stretched constantly, the responsive contractions contribute to the normal rhythmic movements associated with the digestive system. These are involuntary muscles; they can contract on their own without receiving direct messages from your nervous system. Although they are involuntary, they can be stimulated to contract by the presence of certain hormones (the hormone oxytocin causes the smooth muscle of the uterus to contract), or their degree of contraction can be modified by nervous stimulation.

Cardiac muscle is the muscle that makes up the heart. It has the ability to contract rapidly like skeletal muscle but does not require nervous stimulation to do so. Nervous stimulation can, however, cause the heart to speed or slow its rate of contraction. Hormones, such as epinephrine and norepinephrine, also influence the heart by increasing its rate and strength of contraction. Cardiac muscle also has the characteristic of being unable to

stay contracted. It will contract quickly but must have a short period of relaxation before it will be able to contract a second time. This makes sense in light of its continuous, rhythmic, pumping function.

Glands

The glands of your body are of two different kinds. Those that secrete into the bloodstream are called **endocrine glands.** We have already talked about several of these: the pituitary, thyroid, ovary, and testis are examples. The **exocrine glands** are those that secrete to the surface of the body or into one of the tubular organs of the body, such as the gut or reproductive tract. Examples are the salivary glands, intestinal mucous glands, and sweat glands. Some of these glands, such as salivary glands and sweat glands, are under nervous control. When stimulated by the nervous system, they secrete their contents.

Many other exocrine glands are under hormonal control. Many of the digestive enzymes of the stomach and intestine are secreted in response to local hormones produced in the gut. These are circulated through the blood to the digestive glands, which respond by secreting the appropriate digestive enzymes and other molecules. Endocrine glands also respond to nervous stimulation or stimulation by hormones.

Growth Responses

The hormones produced by the endocrine system can have a variety of effects. As mentioned earlier, hormones can stimulate

Which Type of Exercise Do You Do?

Aerobic exercise occurs when the muscles being contracted are supplied with sufficient oxygen to continue aerobic cellular respiration:

$$C_6H_{12}O_6 + 6O_2 \rightarrow 6CO_2 + 6H_2O + energy$$

This type of exercise involves long periods of activity with elevated breathing and heart rate. It results in strengthened chest muscles, which enable more complete exchange of air during breathing. It also improves the strength of the heart, enabling it to pump more efficiently. Flow of blood to the muscles also improves. All these changes increase endurance.

Anaerobic exercise takes place when insufficient amounts of oxygen reach the contracting muscle cells and they shift to anaerobic cellular respiration:

$$C_6H_{12}O_6 \rightarrow lactic\ acid + less\ energy$$

The buildup of lactic acid results in muscle pain and eventually prevents further contraction. Anaerobic exercise involves explosive bouts of activity, as in sprints or jumping. This kind of exercise increases muscle strength but does little to improve endurance.

Resistance exercise occurs when muscles contract against an object that does not allow the muscles to move that object. This type of exercise does not improve the ability of your body to deliver oxygen to your muscles, nor does it increase your endurance. However, it does stimulate your muscle cells to manufacture more contractile protein fibers.

Make a list of your activities that would be considered aerobic, anaerobic, or resistance exercise. Which activities would result in weight reduction? Weight gain? Improved cardiovascular fitness?

---

smooth muscle to contract and can influence the contraction of cardiac muscle as well. Many kinds of glands, both endocrine and exocrine, are caused to secrete as a result of a hormonal stimulus. However, the endocrine system has one major effect that is not equaled by the nervous system: Hormones regulate growth. *Growth-stimulating hormone (GSH)* is produced over a period of years to bring about the increase in size of most of the structures of the body. A low level of this hormone results in a person with small body size. It is important to recognize that the amount of growth-stimulating hormone (GSH) present varies from time to time. It is present in fairly high amounts throughout childhood and results in steady growth. It also appears to be present at higher levels at certain times, resulting in growth spurts. Finally, as adulthood is reached, the level of this hormone falls, and growth stops.

Similarly, testosterone produced during adolescence influences the growth of bone and muscle to provide men with larger, more muscular bodies than those of women. In addition, there is growth of the penis, growth of the larynx, and increased growth of hair on the face and body. The primary female hormone, estrogen, causes growth of reproductive organs and development of breast tissue. It is also involved, along with other hormones, in the cyclic growth and sloughing of the wall of the uterus.

SUMMARY

The body's various *systems* must be integrated in such a way that the *internal environment* stays relatively constant. This concept is called *homeostasis.*

The *circulatory, respiratory, digestive,* and *excretory systems* are involved in the exchange of materials across cell membranes. All of these systems have special features that provide *large surface* areas to allow for necessary *exchanges.*

The *circulatory system* consists of a pump, the *heart,* and *blood vessels* that distribute the blood to all parts of the body. The *blood* is a carrier fluid that transports molecules and heat. The exchange of materials between the *blood* and body *cells* takes place through the walls of the *capillaries. Hemoglobin* in *red blood cells* is very important in the transport of *oxygen.*

The *skin* is the outer covering of the body and performs several functions, including protecting underlying tissues, preventing water loss, and regulating temperature.

The *respiratory system* consists of the *lungs* and associated tubes that allow air to enter and leave the lungs. The *diaphragm* and *muscles* of the chest wall are important in the process of *breathing.* In the lungs, tiny sacs called *alveoli* provide a large surface area in association with capillaries, which allows for rapid exchange of oxygen and carbon dioxide.

The *digestive system* is involved in disassembling food molecules. This involves several processes: *grinding* by the teeth and stomach, *emulsification* of fats by bile from the liver, *addition of water* to dissolve molecules, and *enzymatic action* to break complex molecules into simpler molecules for *absorption.* The intestine provides a *large surface area* for the *absorption of nutrients* because it is long and its wall contains many tiny projections that increase surface area.

To maintain good health, people must receive *nutrient molecules* that can enter the cells and function in the metabolic processes. The *proper quantity and quality of nutrients* are essential to good health.

An important measure of the amount of energy required to sustain a human at rest is the *basal metabolic rate*. To meet this and all additional requirements, the United States has established recommendations for each nutrient. Should there be metabolic or psychological problems associated with a person's *normal metabolism,* a variety of disorders may occur, including *obesity, anorexia nervosa,* and *bulimia.*

The *excretory system* is a filtering system of the body. The *kidneys* consist of *nephrons* into which the *circulatory system* filters fluid. Most of this fluid is useful and is reclaimed by the cells that make up the walls of these *tubules.* Materials that are present in excess or those that are harmful are allowed to escape.

A *nerve impulse* is caused by sodium ions entering the cell as a result of a change in the permeability of the cell membrane. Thus, a *wave of depolarization* passes down the length of a *neuron* to the *synapse.* The *axon* of a neuron secretes a *neurotransmitter,* such as *acetylcholine,* into the *synapse,* where these molecules bind to the dendrite of the next cell in the chain, resulting in an impulse in it as well. The *acetylcholinesterase* present in the synapse destroys acetylcholine so that it does not repeatedly stimulate the dendrite.

Several kinds of *sensory inputs* are possible. Many kinds of chemicals can bind to cell surfaces and be recognized. This is probably how the *sense of taste* and the *sense of smell* function. Light energy can be detected because light causes certain molecules in the *retina* of the eye to decompose and stimulate neurons. *Sound* can be detected because fluid in the *cochlea* of the *ear* is caused to vibrate, and special cells detect this movement and stimulate neurons. The sense of *touch* consists of a variety of receptors that respond to *pressure, cell damage,* and *temperature.*

There are two major kinds of organs that respond to nervous and hormonal stimulation: *muscles* and *glands.* There are three kinds of muscle: *skeletal muscle,* which moves parts of the skeleton; *cardiac muscle*—the heart; and *smooth muscle,* which makes up the muscular walls of internal organs.

Glands are of two types: *exocrine glands,* which secrete through ducts into the cavity of an organ or to the surface of the skin, and *endocrine glands,* which release their secretions into the circulatory system. It is becoming clear that the endocrine system and the nervous system are interrelated. Actions of the endocrine system can change how the nervous system functions, and the reverse is also true.

KEY TERMS

alveoli (p. **606**)
anorexia nervosa (p. **621**)
aorta (p. **602**)
arteries (p. **599**)
arterioles (p. **603**)
basal metabolic rate (p. **620**)
blood (p. **599**)
body mass index (BMI) (p. **616**)
bronchi (p. **606**)
bronchioles (p. **606**)
bulimia (p. **621**)
calorie (p. **611**)
capillaries (p. **599**)
diaphragm (p. **607**)

diastolic blood pressure (p. **603**)
diet (p. **612**)
Dietary Reference Intakes (p. **612**)
duodenum (p. **609**)
endocrine glands (p. **630**)
exocrine glands (p. **630**)
Food Guide Pyramid (p. **616**)
glands (p. **623**)
heart (p. **599**)
hemoglobin (p. **599**)
homeostasis (p. **598**)
hormones (p. **623**)
kidney (p. **622**)

kilocalorie (kcal) (p. **611**)
large intestine (p. **610**)
liver (p. **610**)
lung (p. **606**)
lymph (p. **603**)
nephrons (p. **622**)
nerve impulse (p. **624**)
neurotransmitter (p. **625**)
nutrients (p. **610**)
nutrition (p. **610**)
obese (p. **619**)
overweight (p. **619**)
pancreas (p. **609**)
plasma (p. **599**)

pulmonary artery (p. **602**)
pulmonary circulation (p. **602**)
response (p. **623**)
retina (p. **627**)
small intestine (p. **609**)
stimulus (p. **623**)
synapse (p. **625**)
systemic circulation (p. **602**)
systolic blood pressure (p. **603**)
trachea (p. **606**)
tympanum (p. **628**)
underweight (p. **619**)
veins (p. **599**)

APPLYING THE CONCEPTS

Answers located in appendix F.

1. The primary structures involved in pumping blood are the
   a. veins.
   b. atria.
   c. capillaries.
   d. ventricles.

2. The fluid portion of the blood that leaves the capillaries and surrounds the cells is
   a. hemoglobin.
   b. edema.
   c. lymph.
   d. lacteal.

3. Blood is carried through vessels to all parts of the body except the lungs by
   a. pulmonary circulation.
   b. the pulmonary artery.
   c. systemic circulation.
   d. the lymphatic system.

4. As air passes through the lungs, it follows the path:
   a. trachea → bronchi → bronchioles → alveoli
   b. trachea → bronchioles → bronchi → alveoli
   c. bronchi → trachea → alveoli → bronchioles
   d. bronchioles → alveoli → bronchi → trachea

5. The levels of water, hydrogen ions, salts, and urea in the blood are regulated by the
   a. liver.
   b. kidneys.
   c. bladder.
   d. rectum.

6. An organism's reaction to a change in the environment is a(n)
   a. stimulus.
   b. impulse.
   c. response.
   d. perception.

7. A light stimulus is received by the nervous system, which results in growth. The growth is the direct result of
   a. hormones stimulating cells.
   b. activating muscles.
   c. the endocrine system stimulating the nervous system.
   d. increasing nervous activity.

8. When a nerve cell is stimulated,
   a. acetylcholine is destroyed.
   b. potassium ions enter the neuron.
   c. sodium ions enter the neuron.
   d. calcium attaches to the dendrites.

9. The central nervous system consists of the
   a. brain only.
   b. brain and spinal cord.
   c. brain, spinal cord, and nerves.
   d. motor neurons and sensory neurons.

10. Olfactory senses detect
    a. light.   c. chemicals.
    b. sound.   d. pain.

11. Strict vegetarians do not eat animals or their products (milk, eggs). Which one of the following nutritional needs would they have the greatest difficulty meeting from food alone?
    a. carbohydrates
    b. essential amino acids
    c. essential fatty acids
    d. vitamin A

12. Large food molecules are chemically broken down to smaller molecules by
    a. bile salts.   c. dissolving.
    b. enzymes.   d. chewing.

13. The maintenance of a constant blood pH is an example of
    a. homeostasis.
    b. the sole activities of the nervous system.
    c. an evolutionary response.
    d. kinetic molecular theory.

14. The watery matrix of blood is known as
    a. hemoglobin.   c. plasma.
    b. WBC.   d. water.

15. Which is an example of a cell type involved in the immune response?
    a. phagocyte   c. nephon
    b. tumor cell   d. axon

QUESTIONS FOR THOUGHT

1. What are the functions of the heart, arteries, veins, arterioles, the blood, and capillaries?

2. Describe three ways in which the digestive system increases its ability to absorb nutrients.

3. Describe the mechanics of breathing.

4. Describe how changing permeability of the cell membrane and the movement of sodium ions cause a nerve impulse.

5. What is the role of acetylcholine in a synapse?

6. What is actually detected by the nasal epithelium, taste buds, cochlea of the ear, and retina of the eye?

7. List the differences between the central and peripheral nervous systems.

8. Describe how a nerve impulse travels along a nerve fiber.

9. Why do large, multicellular organisms need a circulatory system?

10. List the six groups of MyPlate, along with a major contribution of each to the diet.

11. Describe how villi and the length of the small intestine are related to the ability to absorb nutrients.

12. Describe two ways in which smooth muscle and skeletal muscle differ in the way they function.

13. Give an example of homeostasis.

14. List the components of blood.

15. Describe the flow of blood in systemic circulation.

FOR FURTHER ANALYSIS

1. The heart can be replaced with a mechanical pump. Major blood vessels can be replaced with plastic tubing. Can you envision artificial capillaries? What technical problems would need to be overcome before these would be possible?

2. The skin is an excellent organ for getting rid of heat. What role does surface area play in this function?

3. If you were allergic to dairy products, what are the particular nutrients for which you would need to find alternative sources?

4. Describe the ways in which the functions of the nervous and endocrine systems are similar. In what ways are they different?

5. How is the structure of each of the following organs able to provide a large surface area for the exchange of molecules: lungs, small intestine, and kidney?

INVITATION TO INQUIRY

Monitor Your Diet

Many people have a poor idea of their normal dietary intake. For twenty-four hours, keep track of what you eat. Record the food and quantity eaten. Include all food and drink (including alcoholic beverages) consumed, not just those foods eaten at mealtime. If you are eating packaged foods, save the package because it has the quantity contained and the nutritional facts about the food. For other items, estimate the size. You can use the following chart to record your food intake.

Meal	Item	Quantity	Nutritional Information
Breakfast	1. 2. 3. 4.		
Snacks	1. 2. 3. 4.		
Lunch	1. 2. 3. 4.		
Snacks	1. 2. 3. 4.		
Dinner	1. 2. 3. 4.		
Snacks	1. 2. 3. 4.		

Now compare your daily diet with MyPlate. What categories were consumed in excess? What categories do you need to consume more of? Did you get a good mixture of vitamins and minerals?

25

Human Biology: Reproduction

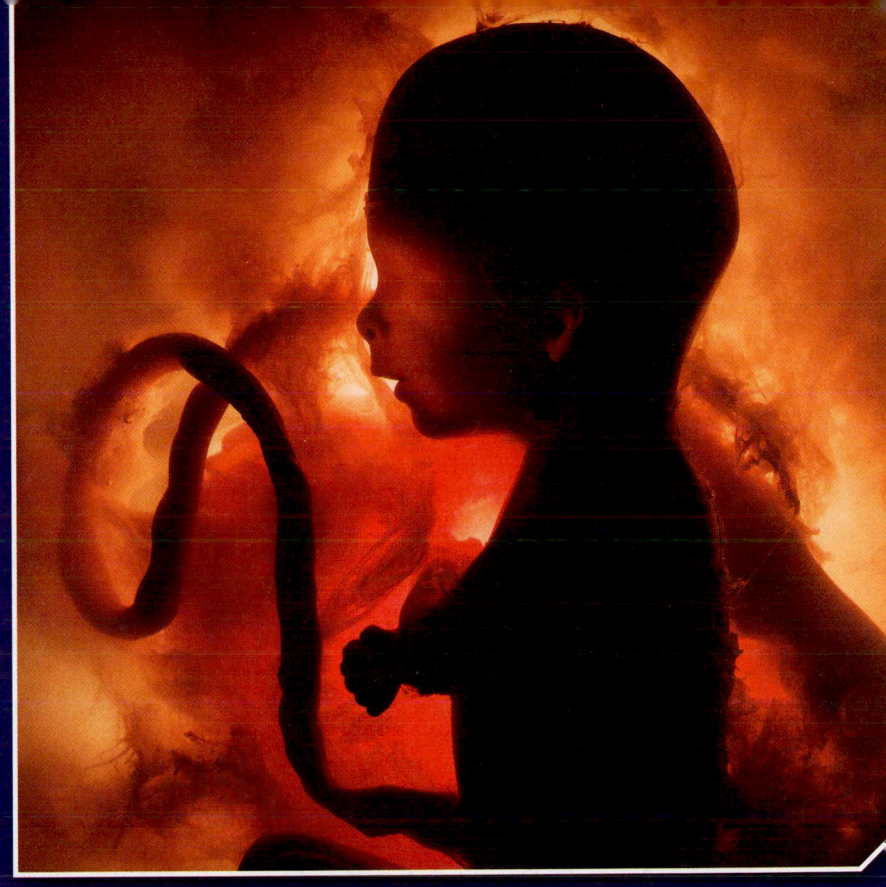

This is a model of an embryo in womb, silhouetted by light from outside the uterus.

CORE CONCEPT

Humans are sexual beings and possess the structural, functional, and behavioral capabilities of transmitting necessary genetic material to the next generation.

OUTLINE

The process of meiosis reduces the number of chromosomes during sex cell formation and introduces genetic variation to the offspring (p. 636).

Sexuality involves distinct hereditary, anatomical, and behavioral aspects (p. 639).

Hormones play important roles in sexual development and function (p. 644).

Chemistry

▷ Physiological processes involve inorganic and organic compounds (Ch. 19).

▷ Understanding the nature of biochemical reactions is essential to the study of physiological processes (Ch. 19).

Life Science

▷ A specialized form of cell division plays a vital role in sexual reproduction (Ch. 25).

▷ One's sexuality is a complex of anatomy, physiology, behavior, and culture (Ch. 25).

▷ Males and females have specialized anatomical features for reproduction (Ch. 25).

▷ Gamete formation in males and females differs (Ch. 25).

▷ Reproduction is under the control of many different hormones (Ch. 25).

▷ Understanding sexuality requires knowledge of the process from gamete production to old age (Ch. 25)

OVERVIEW

If both parents contribute equally to the genetic information of the child, how can the chromosome number in humans remain constant for generation after generation? There is a specialized kind of cell division that results in the formation of sex cells. In this chapter, we will discuss the mechanics of this process. Knowing the mechanics is essential to understanding how genetic variety can occur in sex cells. This variety ultimately shows up as differences in offspring as a result of sexual reproduction. Sex and your sexuality influence you in many ways throughout life. Before birth, sex-determining chromosomes direct the formation of hormones that control the development of sex organs, after which the effects of these hormones diminish. With the start of puberty, renewed hormonal activity causes major structural and behavioral changes that influence you for the remainder of your life.

25.1 SEXUAL REPRODUCTION

Nearly all organisms have a method of shuffling and exchanging genetic information. Humans and many other organisms have two sets of genetic data, one inherited from each of their parents. **Sexual reproduction** is the formation of a new individual by the union of two sex cells, one donated by each parent. The sex cells produced by male organisms are called **sperm,** and those produced by females are called **eggs.** A general term sometimes used to refer to either eggs or sperm is **gamete** (sex cell). During this process, the two sets of genetic information found in each body cell must be reduced to one set, that is, one of each kind of chromosome. The cellular process that is responsible for forming gametes is called **gametogenesis.** This is somewhat similar to shuffling a deck of cards and dealing out hands; the shuffling and dealing ensure that each hand will be different. An organism with two sets of chromosomes can produce many combinations of chromosomes when it produces sex cells, just as many different hands can be dealt from one pack of cards. When a sperm unites with an egg (*fertilization*), a new organism containing two sets of genetic information is formed. This new organism's genetic information might have survival advantages over the information found in either parent; this is the value of sexual reproduction.

The life cycle of a sexually reproducing organism begins with the fertilization of an egg by a sperm. This new cell, called a **zygote,** divides by mitosis, as do the resulting cells, and growth occurs. When the organism reaches sexual maturity, it

is able to produce gametes by another form of cell division, *meiosis* and the life cycle is complete. Notice in figure 25.1 that the zygote and its descendants have two sets of chromosomes. However, male gametes and female gametes each contain only one set of chromosomes. These sex cells are said to be **haploid.** The haploid number of chromosomes is noted as *n*. A zygote contains two sets and is said to be **diploid.**

Meiosis

Sex cell formation

Single adult cell → Single gamete (sperm or egg)

Contains 2 sets of chromosomes → Contains 1 set of chromosomes

$$2n \rightarrow n$$

The diploid number of chromosomes is noted as $2n$ ($n + n = 2n$). Diploid cells have two sets of chromosomes, one set from each parent. Remember, a chromosome is composed of two chromatids, each containing double-stranded DNA. These two chromatids are attached to each other at a point called the *centromere*. In a diploid nucleus, the chromosomes occur as *homologous chromosomes*—a pair of chromosomes in a diploid cell that contain similar (not necessarily identical) genes throughout their length. One of the chromosomes of a homologous pair was donated by the father, the other by the mother (figure 25.2).

It is necessary for organisms that reproduce sexually to form gametes having only one set of chromosomes. If gametes contained two sets of chromosomes, the zygote result-

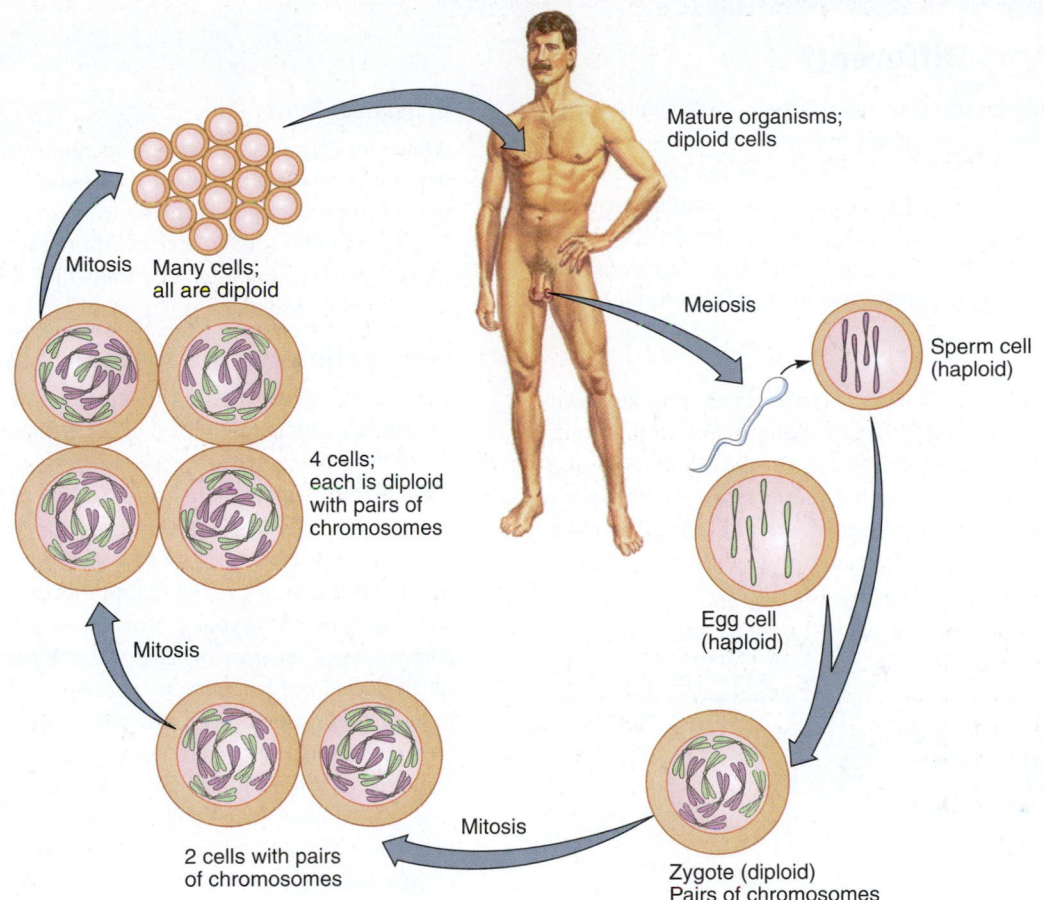

FIGURE 25.1 The cells of this adult human actually have forty-six chromosomes in their nuclei, not the eight shown here. In preparation for sexual reproduction, the number of chromosomes must be reduced by half so that fertilization will result in the original number of forty-six chromosomes in the new individual. The offspring will grow and produce new cells by mitosis, completing the life cycle.

Labels in figure: Mature organisms; diploid cells — Mitosis — Many cells; all are diploid — 4 cells; each is diploid with pairs of chromosomes — Meiosis — Sperm cell (haploid) — Egg cell (haploid) — Mitosis — Mitosis — 2 cells with pairs of chromosomes — Zygote (diploid) Pairs of chromosomes

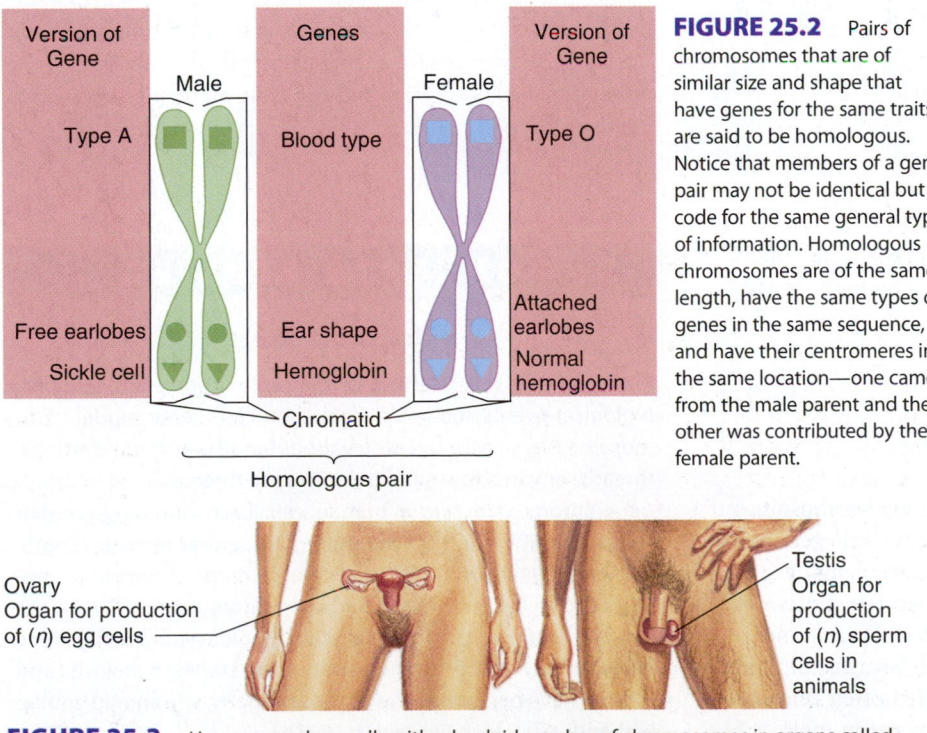

FIGURE 25.2 Pairs of chromosomes that are of similar size and shape that have genes for the same traits are said to be homologous. Notice that members of a gene pair may not be identical but code for the same general type of information. Homologous chromosomes are of the same length, have the same types of genes in the same sequence, and have their centromeres in the same location—one came from the male parent and the other was contributed by the female parent.

Labels in figure 25.2: Version of Gene — Male — Type A — Free earlobes — Sickle cell — Genes — Blood type — Ear shape — Hemoglobin — Chromatid — Female — Version of Gene — Type O — Attached earlobes — Normal hemoglobin — Homologous pair

FIGURE 25.3 Humans produce cells with a haploid number of chromosomes in organs called gonads, the ovaries and the testes.

Labels in figure 25.3: Ovary Organ for production of (*n*) egg cells — Testis Organ for production of (*n*) sperm cells in animals

ing from their union would have four sets of chromosomes. The number of chromosomes would continue to double with each new generation. However, this does not usually happen; the number of chromosomes remains constant generation after generation. How are sperm and egg cells formed so that they get only half the chromosomes of the diploid cell? The answer lies in the process of **meiosis,** the specialized pair of cell divisions that reduce the chromosome number from diploid (2*n*) to haploid (*n*). One of the major functions of meiosis is to produce cells that have one set of genetic information. Therefore, when fertilization occurs between a haploid sperm (*n*) and a haploid egg (*n*), the resulting zygote will have two sets of chromosomes (*n* + *n* = 2*n*), as did each parent.

Not every cell in the body goes through the process of meiosis. Only specialized organs are capable of producing haploid cells (figure 25.3). In humans, the

Connections . . .

With the possible exception of identical twins, every human who has ever been born is genetically unique. The formation of haploid cells by meiosis and the combination of sex cells from two parents to form a diploid offspring by sexual reproduction result in genetic diversity in the offspring. The five factors that influence genetic diversity in offspring are mutations, crossing-over, segregation and independent assortment, and fertilization.

Mutations

Two types of mutations will be discussed in chapter 26: point mutations and chromosomal aberrations. In point mutations, a change in a DNA nucleotide results in the production of a different protein. In chromosomal aberrations, genes are rearranged when parts of chromosomes are duplicated, deleted, or inverted. By causing the production of different proteins, both types of mutations increase diversity.

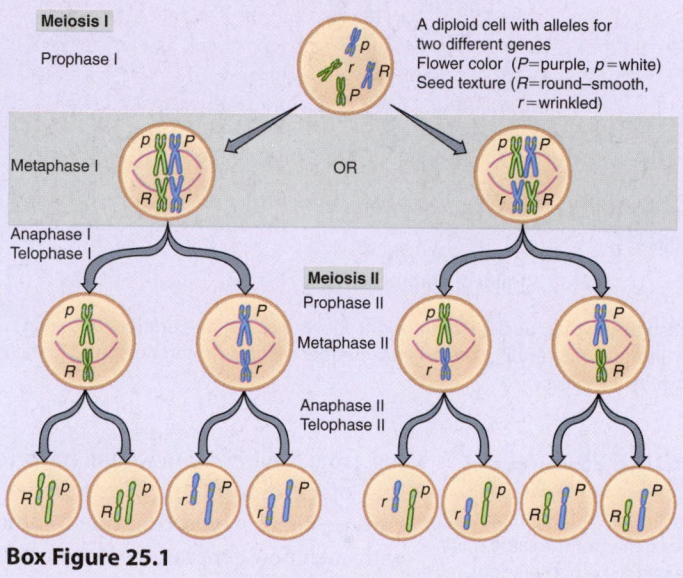

Box Figure 25.1

Crossing-Over

When crossing-over occurs during gametogenesis in your ovaries or testes, pieces of genetic material are exchanged between the chromosomes you received from your mother and father. This exchange results in a chromosome with a new combination of genes, *combinations* of traits that were probably never before found in the family.

Segregation and Independent Assortment

The process of segregation relates to how the chromosomes donated by your father line up or assort themselves in pairs with those donated by your mother. How one pair of chromosomes lines up during metaphase I does not affect how the other pair lines up; i.e., they assort themselves independently of the others (box figure 25.1). If all those donated by your father line up on the right side of the cell during metaphase I of meiosis and all those donated by your mother line up on the left, when the chromosomes separate or *segregate,* the resulting gametes will be of two kinds: half of your gametes will contain only your father's chromosomes and the other half of the gametes you form will contain only your mother's chromosomes. If, however, some of the chromosomes from your father line up on the right side and some of your mother's line up on the left, the gametes you form will contain some of your mother's and some of your father's chromosomes. There are millions of ways (2^{23}) in which these chromosomes can line up and divide.

Fertilization

Because of the large number of possible gametes resulting from independent assortment, segregation, mutations, and crossing-over, an incredibly large number of types of offspring can result. Because human males and females can produce millions of genetically unique sperm and eggs, the number of kinds of offspring possible is infinite.

---

organs in which meiosis occurs are called **gonads.** The female gonads that produce eggs are called **ovaries.** The male gonads that produce sperm are called **testes.**

25.2 THE MECHANICS OF MEIOSIS

The whole point of meiosis is to distribute the chromosomes and the genes they carry so that each daughter cell gets one member of each homologous pair. In this way, each daughter cell gets one complete set of genetic information. Meiosis consists of two divisions known as meiosis I and meiosis II. During meiosis I, the number of chromosomes is reduced from the diploid number to the haploid number. Therefore, it is often referred to as a *reduction division.* The sequence of events in meiosis I is artificially divided into four phases: prophase I, metaphase

TABLE 25.1   A Review of the Stages of Meiosis

Stage		Description
Prophase I		During prophase I, the cell is preparing itself for division. The chromosomes are composed of two chromatids and coil so that they become shorter and thicker. Homologous chromosomes pair with each other. This pairing is called *synapsis*. While homologous chromosomes are paired, they can exchange equivalent sections of DNA. This process is called *crossing-over* and is important in mixing up the genes that will be divided into separate gametes. Long fibers known as *spindle fibers* become attached to the chromosomes.
Metaphase I		The synapsed homologous chromosomes move to the equator of the cell as single units. How they are arranged on the equator (which one is on the left and which one is on the right) is determined by chance. In the drawing, three green chromosomes from the father and one purple chromosome from the mother are lined up on the left. Similarly, one green chromosome from the father and three purple chromosomes from the mother are on the right. They could have aligned themselves in several other ways.
Anaphase I		During this stage, homologous chromosomes separate and their number is reduced from diploid to haploid. The two members of each pair of homologous chromosomes move away from each other toward opposite poles. The direction each takes is determined by how each pair was originally arranged on the spindle. Each chromosome is independently attached to a spindle fiber at its centromere. Each chromosome still consists of two chromatids. Because the homologous chromosomes and the genes they carry are being separated from one another, this process is called *segregation*. The way in which a single pair of homologous chromosomes segregates does not influence how other pairs of homologous chromosomes segregate; that is, each pair segregates independently of other pairs. This is known as *independent assortment* of chromosomes.
Telophase I		The two newly forming daughter cells are now haploid (*n*) since each contains only one of each pair of homologous chromosomes.
Prophase II		The chromosomes of each of the two haploid daughter cells shorten and become attached to spindle fibers by their centromeres.
Metaphase II		Chromosomes move to the equator of the cell.
Anaphase II		Centromeres divide, allowing the chromatids to separate toward the poles.
Telophase II		Four haploid (*n*) cells are formed from the division of the two meiosis I cells. Each of these sex cells is genetically different from all the others. These cells become the sex cells (egg or sperm) of a higher organism.

I, anaphase I, and telophase I. Meiosis II includes four phases: prophase II, metaphase II, anaphase II, and telophase II. The two daughter cells formed during meiosis I continue through meiosis II, so that four cells usually result from the two divisions. Table 25.1 is a review of the stages of meiosis as they would occur in a cell with diploid number of chromosomes equal to 8, (i.e., *n* = 4).

25.3 HUMAN SEXUALITY FROM DIFFERENT POINTS OF VIEW

Probably nothing interests people more than sex and sexuality. **Sex** is the nature of the biological differences between males and females. By **sexuality,** we mean all the factors that

A Closer Look

The Sexuality Spectrum

Although we tend to think of our species as being clearly divided into two genders, male and female, sexuality really is a spectrum that includes anatomical and behavioral components. Both anatomy and behavior are the result of a complex interplay between genetic and developmental processes that are influenced by environmental factors.

Anatomy

Hermaphrodites are organisms that have both ovaries and testes in the same body. This condition is extremely rare in humans. However, incidences of partial development of the genitalia (sex organs) of both sexes in one individual may be more frequent than most people realize. About 1 percent of births show some level of ambiguity related to sexual anatomy. These people are referred to as *intersexual* because their sexual anatomy is not clearly male or female. Sometimes, this abnormal development occurs because the hormone levels are out of balance at critical times in the development of the embryo. This hormonal imbalance may also be related to an abnormal number of sex-determining chromosomes, or it may be the result of abnormal functioning of the endocrine glands.

When children with abnormal combinations of sex organs are born, they are usually assessed by a physician in consultation with the parents to determine which sexual structures should be retained or surgically reconstructed. The physician might also decide that hormone therapy might be a more successful treatment. These decisions are not made easily because they involve children who have not fully developed their sexual nature. An increasingly vocal group advocates that children who are diagnosed with this condition not be surgically "corrected" as infants, recommending that, if the parents can cope with the unusual genitalia, they allow the child to grow older without having the surgery. They believe that people should choose for themselves once they are more mature. However, few long-term studies have examined whether delaying reconstructive surgery presents fewer social and psychological adjustment issues than performing reconstructive surgery in children.

Behavior

A person's gender is his or her sexual identity based on anatomy. However, a person also has a psychological gender. More and more frequently, we are becoming aware of individuals whose physical gender does not match their psychological gender. These individuals are often referred to as *transgender* persons. A male with normal male sex organs may "feel" like a female. The same situation may occur with structurally female individuals. Because some of these individuals might dress as a member of the opposite sex, they are sometimes called *cross-dressers* or *transvestites*. Some of these individuals may dress as the other sex in private but dress and behave in public appropriate to their anatomical sex. Others completely change their public and private behaviors to reflect their inner desire to function as the other sex. A male may dress as a female, work in a traditionally female occupation, and make social contacts as a female. Tremendous psychological and emotional pressures develop from this condition. Frequently, many of these individuals choose to undergo gender reassignment surgery—a sex-change operation.

Their goal is to interact with society without being detected as having been a different gender at one time. This surgery and the follow-up hormonal treatment can cost tens of thousands of dollars and take several years.

Homosexuality is a condition in which a person desires romantic and sexual relationships with members of their own sex. However, it is a complex behavioral pattern with many degrees of expression. Some individuals are exclusively homosexual, while others can be considered bisexual because they form sexual relationships with either males or females. Some are transgender individuals while others clearly accept their biological sex but prefer relationships with others of the same sex. However, it is becoming clear that sexual orientation in most cases is not a simple choice or a learned behavior. There appear to be differences in brain function and genetic makeup that are important. For example, certain studies suggest that genetic regions on chromosomes 7, 8, and 10 may influence homosexuality. Some regions on chromosome 7 and 8 have been linked with male sexual orientation, regardless of whether the male receives the chromosomes from his mother or father. The regions on chromosome 10 appear to be linked with male sexual orientation only if they were inherited from the mother. Sexuality ranges from strongly heterosexual to strongly homosexual. Human sexual orientation is a complex trait, and evidence suggests that there is no one gene that determines where a person falls on the sexuality spectrum. It is most likely a combination of various genes acting together and interacting with environmental factors.

contribute to one's female or male nature. These include the structure and function of the sex organs, the behaviors that involve these structures, psychological components, and the role culture plays in manipulating our sexual behavior. Males and females have different behavior patterns for a variety of reasons. Some behavioral differences are learned (patterns of dress, use of facial makeup), whereas others appear to be less dependent on culture (degree of aggressiveness, frequency of sexual thoughts). We have an intense interest in the facts about our own sexual nature and the sexual behavior of members of the opposite sex and that of peoples of other cultures.

There are several different ways to look at human sexuality. The behavioral sciences tend to focus on the behaviors associated with being male and female and what is considered appropriate and inappropriate sexual behavior. Sex is considered a strong drive, appetite, or urge by psychologists. They describe the sex drive as a basic impulse to satisfy a biological, social, or psychological need. Other social scientists (sociologists, cultural anthropologists) are interested in sexual behavior as it occurs in different cultures and subcultures. When a variety of cultures is examined, it becomes very difficult to classify various kinds of sexual behavior as normal or abnormal. What is considered abnormal in one culture may be normal in another. For example, public nudity is considered abnormal in many cultures but not in others.

Biologists have studied the sexual behavior of nonhuman animals for centuries. They have long considered the function of sex and sexuality in light of its value to the population or species. Sexual reproduction results in new combinations of genes that are important in the process of natural selection. Many biologists today are attempting to look at human sexual behavior from an evolutionary perspective and speculate on why certain sexual behaviors are common in humans. The behaviors of courtship, mating, rearing of the young, and the division of labor between the sexes are complex in all social animals, including humans. These are demonstrated in the elaborate social behaviors surrounding mate selection and the establishment of families. It is difficult to draw the line between the biological development of sexuality and the social establishment of customs related to the sexual aspects of human life (see "A Closer Look: The Sexuality Spectrum"). However, the biological mechanism that determines whether an individual will develop into a female or male has been well documented.

Sexuality ranges from strongly heterosexual to strongly homosexual. Human sexual orientation is a complex trait, and evidence suggests that there is no one gene that determines where a person falls on the sexuality spectrum. It is most likely a combination of various genes acting together and interacting with environmental factors (figure 25.4). The primary biological goal of sexual intercourse (coitus, mating) is the union of

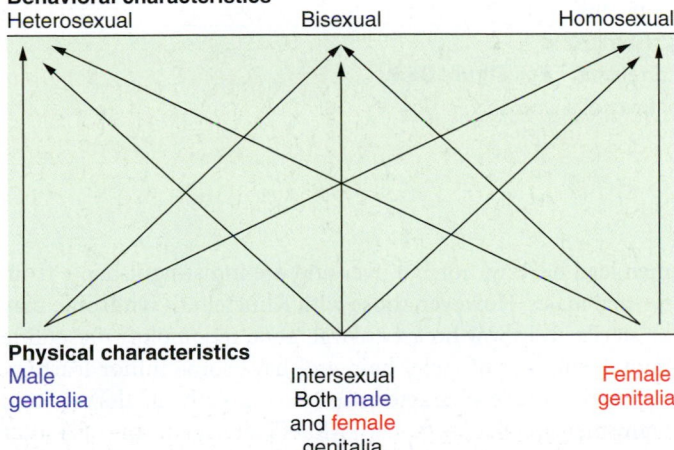

Behavioral characteristics

Heterosexual — Bisexual — Homosexual

Physical characteristics

Male genitalia — Intersexual Both male and female genitalia — Female genitalia

*Humans with both male and female sex characteristics are rare; most are not fertile, but that does not rule out sexual interest or potential for enjoyment of sexual activity.

FIGURE 25.4 A person's sexuality involves both anatomical and psychological components. This figure shows how behavioral and physical sexual characteristics interrelate. At the ends of the behavioral spectrum, individuals can be strongly heterosexual or strongly homosexual, or (in the center) they might be bisexual, attracted to both sexes. On the anatomical spectrum they may be clearly anatomically male or female or be transsexual.

sperm and egg to form offspring. However, this interaction can also improve the health of and bring pleasure to willing partners.

25.4 CHROMOSOMAL DETERMINATION OF SEX AND EARLY DEVELOPMENT

Two of the forty-six chromosomes in a human adult are the **sex-determining chromosomes.** The other twenty-two pairs that do not determine the sex of an individual are called **autosomes.** There are two kinds of sex-determining chromosomes: the **X chromosome** and the **Y chromosome** (see box figure 25.2 in "A Closer Look: Karyotyping and Down Syndrome"). The two sex-determining chromosomes, X and Y, do not carry equivalent amounts of information, nor do they have equal functions. The X chromosome carries typical genetic information about the production of specific proteins in addition to its function in determining sex. For example, the X chromosome carries information on blood clotting, color vision, and many other characteristics. The Y chromosome, however, appears to be primarily concerned with determining male sexual differentiation and has few other genes on it.

When a human sperm cell is produced, it carries twenty-two autosomes and a sex-determining chromosome. Unlike eggs, which always carry an X chromosome, half the sperm cells carry an X chromosome and the other half carry a Y chromosome. If an X-carrying sperm cell fertilizes an X-containing egg cell, the resultant embryo will develop into a female. If a Y-carrying sperm cell fertilizes the egg, a male embryo develops. It is the presence or absence of the *SRY* (*s*ex-determining *r*egion *Y*) gene typically located on the short arm of the Y chromosome that determines the sex of the developing individual. The SRY gene produces a chemical called the testes determining factor (TDF) and acts as a master switch that triggers the events that converts the embryo into a male. Without a functioning SRY gene, the embryo would become female.

Chromosomal Abnormalities and Sexual Development

Evidence that the Y chromosome and SRY gene controls male development comes as a result of many kinds of studies including individuals who have an abnormal number of chromosomes. An abnormal meiotic division that results in sex cells with too many or too few chromosomes is called *nondisjunction.* If nondisjunction affects the X and Y chromosomes, a gamete might be produced that has only twenty-two chromosomes and lacks a sex-determining chromosome, or it might have twenty-four, with two sex-determining chromosomes. If a cell with too few or too many sex chromosomes is fertilized, sexual development is usually affected. If a normal egg cell is fertilized by a sperm cell with no sex chromosome, the offspring will have only one X chromosome. These people are designated as XO. They develop a collection of characteristics known as

A Closer Look

Karyotyping and Down Syndrome

It is possible to examine cells and count chromosomes. Among the easiest cells to examine in this way are white blood cells. They are dropped onto a microscope slide so that the cells are broken open and the chromosomes are separated. Photographs are taken of chromosomes from cells in the metaphase stage of mitosis. The chromosomes in the photographs can then be cut and arranged for comparison to known samples (box figure 25.2). This picture of an individual's chromosomal makeup is referred to as that person's *karyotype*.

In the normal process of meiosis, diploid cells have their number of chromosomes reduced to haploid. This involves segregating homologous chromosomes into separate cells during the first meiotic division. Occasionally, a pair of homologous chromosomes does not segregate properly during gametogenesis and both chromosomes of a pair end up in the same gamete. This kind of division is known as *nondisjunction*. One example of the effects of nondisjunction is the condition known as Down syndrome. If an abnormal gamete with two number 21 chromosomes has been fertilized by another containing the typical one copy of chromosome number 21, the resulting zygote would have forty-seven chromosomes (e.g., twenty-four from the female plus twenty-three from the male parent) (box figure 25.3). Since the zygote divides by mitosis, the child who develops from this fertilization would have forty-seven chromosomes in every

cell of his or her body and would have the symptoms characteristic of Down syndrome. These may include thickened eyelids, some mental impairment, and faulty speech (box figure 25.4). Premature aging is probably the most significant impact of this genetic disease. On the other hand, a child born with only one chromosome 21 rarely survives.

It is known that the mother's age at childbirth plays an important part in the occurrence of *trisomies* (three of one of the kinds of chromosomes instead of the normal two), such as Down syndrome. In women, gametogenesis begins early in life, but cells destined to become eggs are put on hold during meiosis I. Beginning at puberty and ending at menopause, one of these cells completes meiosis I monthly. This means that cells released for fertilization later in life are older than those released earlier in life. Therefore, it was believed that the chances of abnormalities such as nondisjunction increase as the age of the egg increases. However, the evidence no longer supports this age-egg link. Currently, the increase in frequency of trisomies with age has been correlated with a decrease in the activity of a woman's

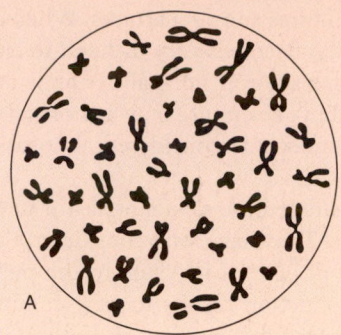

A

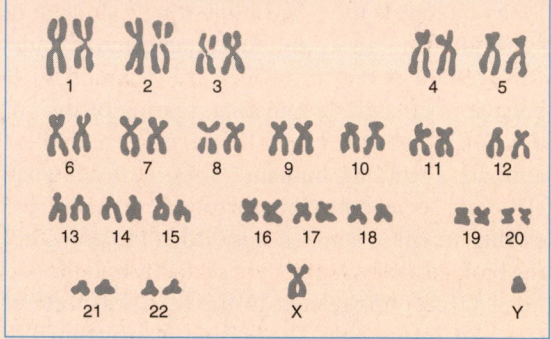

B

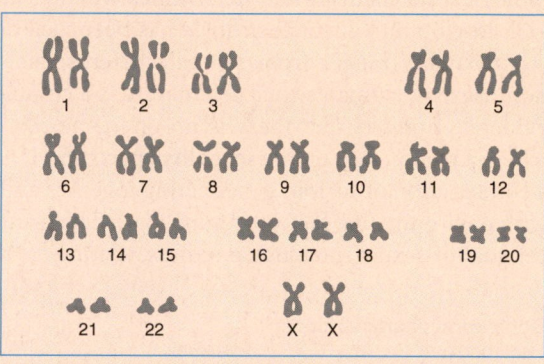

C

Box Figure 25.2

Turner's syndrome (figure 25.5). An individual with this condition is female, is short for her age, and fails to mature sexually, resulting in sterility. In addition, she may have a thickened neck (termed *webbing*), hearing impairment, and some abnormalities in the cardiovascular system. When the condition is diagnosed, some of the physical conditions can be modified with treatment. Treatment involves the use of growth-stimulating hormones to increase growth rate and female sex hormones to stimulate sexual development, although sterility is not corrected.

An individual who has XXY chromosomes is basically male (figure 25.6). This genetic condition is termed *Klinefelter's syndrome* and is probably the most common chromosomal variation found in humans. The largest percentage of these

men lead healthy, normal lives and are indistinguishable from normal males. However, those with Klinefelter's syndrome may be sterile and show breast enlargement, incomplete masculine body form, lack of facial hair, and have some minor learning problems. These characteristics vary greatly in degree, and many men are diagnosed with Klinefelter's syndrome only after they undergo testing to determine why they are infertile. This condition is present in about 1 in 500 to 1,000 men. Treatment may involve testosterone therapy and breast-reduction surgery in males who have significant breast development.

Because both conditions involve abnormal numbers of X or Y chromosomes, they provide strong evidence that these chromosomes are involved in determining sexual development.

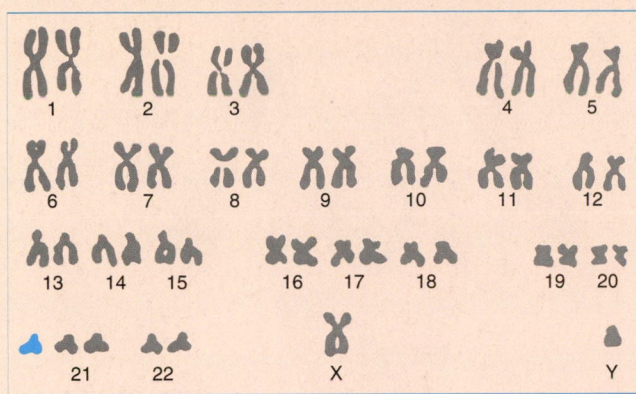

Box Figure 25.3

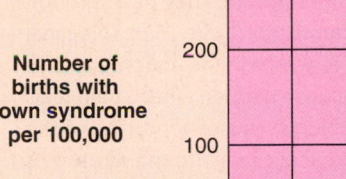

Number of births with Down syndrome per 100,000

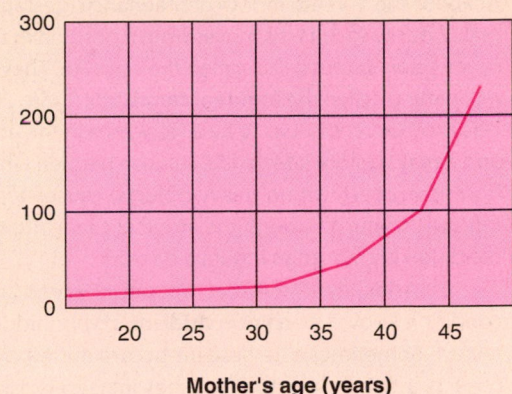

Mother's age (years)

Box Figure 25.5

Box Figure 25.4

immune system. As she ages, her immune system is less likely to recognize the difference between an abnormal and a normal embryo. This means that she is more likely to carry an abnormal fetus to full term.

Box figure 25.5 illustrates the frequency of occurrence of Down syndrome at different ages in women. Notice that the frequency increases very rapidly after age thirty-seven. For this reason, many physicians encourage couples to have their children in their early to mid-twenties and not in their late thirties or early forties. Physicians normally encourage older women who are pregnant to have the cells of their fetus checked to see if they have the normal chromosome number. It is important to know that the male parent can also contribute the extra chromosome 21. However, it appears that this occurs less than 30 percent of the time.

Sometimes a portion of chromosome 14 may be cut out and joined to chromosome 21. A person with this 14/21 transfer is monosomic and has only forty-five chromosomes; one 14 and one 21 are missing and replaced by the translocated 14/21. Statistically, about 15 percent of the children of carrier mothers inherit the 14/21 chromosome and have Down syndrome. Fewer of the children born to fathers with the 14/21 translocation inherit the abnormal chromosome and are Downic.

Whenever an individual is born with a chromosomal abnormality, it is recommended that both parents have a karyotype evaluation in an attempt to identify the possible source of the problem. This is not to fix blame but to provide information on the likelihood that another pregnancy would also result in a child with a chromosomal abnormality.

The early embryo resulting from fertilization and cell division is neither male nor female but becomes female or male later in development—based on the sex-determining chromosomes that control the specialization of the cells of the undeveloped, embryonic gonads into female ovaries or male testes. This specialization of embryonic cells is termed **differentiation.** The embryonic gonads begin to differentiate into testes about five to seven weeks after **conception** (fertilization) if the SRY gene is present and functioning. The Y chromosome seems to control this differentiation process in males because the gonads do not differentiate into female sex organs until later and then only if two X chromosomes are present. It is the absence of the Y chromosome that determines female sexual differentiation.

Fetal Sexual Development

Development of embryonic gonads begins very early during fetal growth. First, a group of cells begins to differentiate into primitive gonads at about week 5 (figure 25.7). By week 5 or 7 if a Y chromosome is present, a gene product (TDF) from the chromosome will begin the differentiation of these gonads into testes; they will develop into ovaries beginning about week 12 if two X chromosomes are present (Y chromosome is absent).

As soon as the gonad has differentiated into an embryonic testis at about week 8, it begins to produce testosterone. The presence of testosterone results in the differentiation of male

A Closer Look

Cryptorchidism—Hidden Testes

At about the seventh month of pregnancy (gestation), in normal male fetuses each testis moves from a position in the abdominal cavity to an external sac, called the scrotum. They pass through an opening called the **inguinal canal** (box figure 25.6). This canal closes off but continues to be a weak area in the abdominal wall, and it may rupture later in life. This can happen when strain (e.g., from improperly lifting heavy objects) causes a portion of the intestine to push through the inguinal canal into the scrotum, a condition known as an inguinal hernia.

Occasionally, the testes do not descend, resulting in a condition known as **cryptorchidism** (*crypt*=hidden; *orchidos*= testes). Sometimes, the descent occurs during puberty; if not, there is a twenty-five to fifty times increased risk for testicular cancer. Because of this increased risk, surgery can be done to allow the undescended testes to be moved into the scrotum. Sterility will result if the testes remain in the abdomen. This happens because normal sperm cell development cannot occur in a very warm environment. The temperature in the abdomen is higher than the temperature in the scrotum. Normally, the temperature of the testes is very carefully regulated by muscles that control their distance from the body. Physicians have even diagnosed cases of male infertility as being caused by tight-fitting pants that hold the testes so close to the body that the temperature increase interferes with normal sperm development. Recent evidence has also suggested that teenage boys and young men working with computers in the laptop position for extended periods may also be at risk for lowered sperm counts.

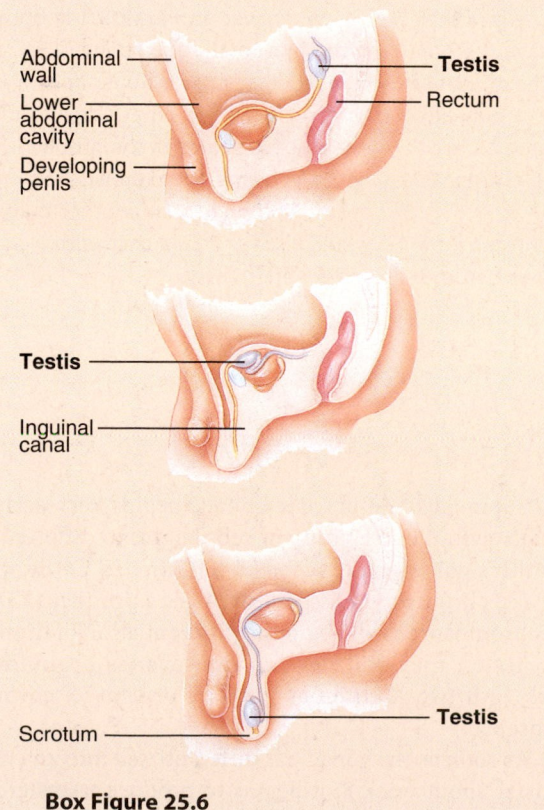

Box Figure 25.6

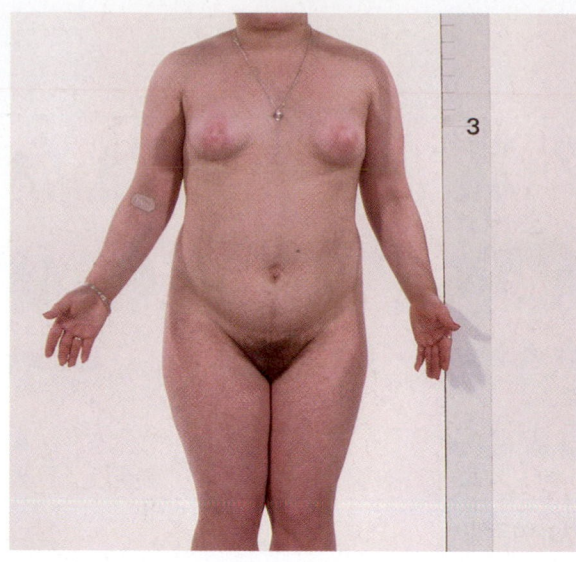

FIGURE 25.5 Turner's syndrome individuals have forty-five chromosomes. They have only one of the sex chromosomes and it is an X chromosome. Individuals with this condition are females, have delayed growth, and fail to develop sexually. This woman is less than 150 cm (5 ft) tall and lacks typical secondary sexual development for her age. She also has a "webbed neck" that is common among Turner's syndrome individuals.

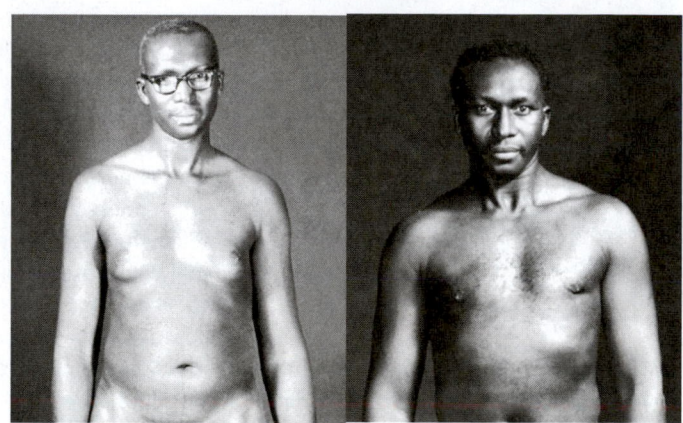

FIGURE 25.6 Klinefelter's syndrome individuals have two X chromosomes and a Y chromosome; they are male, are sterile, and often show some degree of breast development and female body form. They are typically tall. The two photos show a Klinefelter's individual before and after receiving testosterone hormone therapy.

sexual anatomy, and the absence of testosterone results in the differentiation into female sexual anatomy.

25.5 SEXUAL MATURATION OF YOUNG ADULTS

Following birth, sexuality plays only a small part in physical development for several years. Culture and environment shape the responses that the individual will come to recognize as normal behavior. During **puberty** (the developmental period when

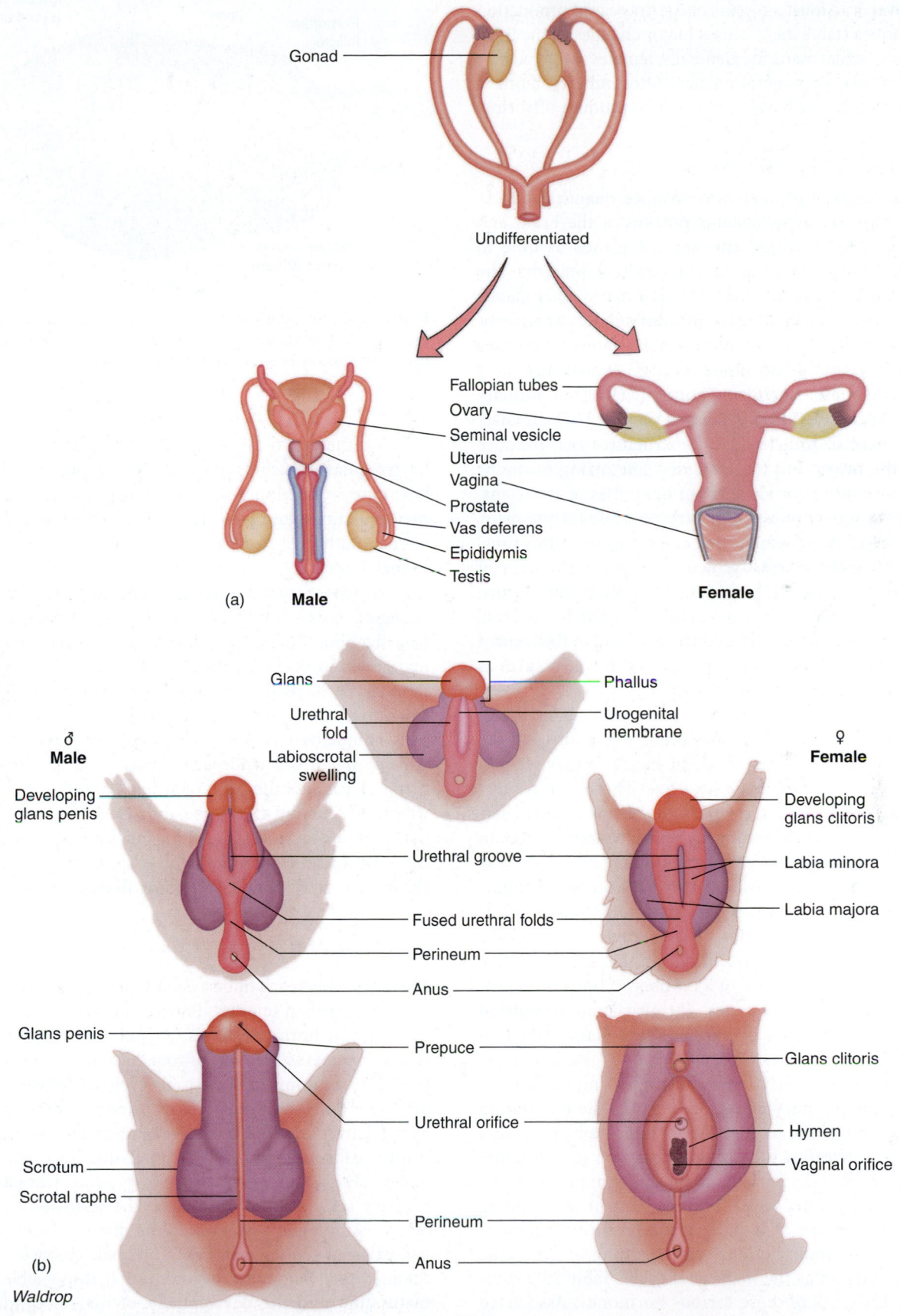

Gonad

Undifferentiated

Fallopian tubes
Ovary
Seminal vesicle
Uterus
Vagina
Prostate
Vas deferens
Epididymis
Testis

(a) **Male** **Female**

Glans — — Phallus
Urethral fold — — Urogenital membrane
Labioscrotal swelling

♂
Male

♀
Female

Developing glans penis
Developing glans clitoris

Urethral groove — Labia minora

Fused urethral folds — Labia majora

Perineum

Anus

Glans penis
Prepuce
Glans clitoris

Urethral orifice
Hymen

Scrotum
Scrotal raphe
Vaginal orifice

Perineum

(b)
Anus

Waldrop

FIGURE 25.7   The early embryo grows without showing any sexual characteristics. The male and female sexual organs eventually develop from common basic structures. (*A*) shows the development of the internal anatomy. (*B*) shows the development of external anatomy.

the body changes and becomes able to reproduce), normally between eleven and fourteen years of age, increased production of sex hormones (table 25.2) causes major changes as the individual reaches sexual maturity. Generally, females reach puberty six months to two years before males. After puberty, humans are sexually mature and have the capacity to produce offspring.

The Maturation of Females

Female children typically begin to produce quantities of sex hormones from the hypothalamus portion of the brain and the pituitary gland, ovaries, and adrenal glands at nine to twelve years of age. This marks the onset of puberty. The **hypothalamus** controls the functioning of many other glands throughout the body, including the **pituitary gland.** At puberty, the hypothalamus begins to release a hormone known as **gonadotropin-releasing hormone (GnRH),** which stimulates the pituitary to release luteinizing hormone (LH) and **follicle-stimulating hormone (FSH).** Increased levels of FSH stimulate the development of **follicles,** saclike structures that produce oocytes in the ovary, and the increased luteinizing hormone stimulates the ovary to produce larger quantities of **estrogens.** The increasing supply of estrogen is responsible for the many changes in sexual development that can be noted at this time. These changes include breast growth, changes in the walls of the uterus and vagina, increased blood supply to the clitoris, and changes in the pelvic bone structure. The **clitoris** is a small, elongated erectile structure located between and at the head of the labia; it develops from the same embryonic structures as the male penis.

Estrogen also stimulates the female adrenal gland to produce **androgens,** male sex hormones. The androgens are responsible for the production of pubic hair, and they seem to have an influence on the female sex drive. Those features that are not primarily involved in sexual reproduction but are characteristic of a sex are called **secondary sexual characteristics.** In women, breast development, the distribution of body hair, patterns of fat deposits, wider hips, and a higher voice are examples.

A major development during this time is the establishment of the **menstrual cycle.** This involves the periodic growth and shedding of the lining of the uterus. These changes are under the control of a number of hormones produced by the pituitary and ovaries. The ovaries are stimulated to release their hormones by the pituitary gland, which is in turn influenced by the ovarian hormones. Both follicle-stimulating hormone (FSH) and luteinizing hormone (LH) are produced by the pituitary gland. FSH causes the maturation and development of the ovaries. LH is important in causing ovulation and converting the ruptured follicle into a structure known as the *corpus luteum.* The corpus luteum produces the hormone **progesterone,** which is important in maintaining the lining of the uterus. Changes in the levels of progesterone result in a periodic buildup and shedding of the lining of the uterus known as the menstrual cycle. Table 25.2 summarizes the activities of these various hormones. Associated with the menstrual cycle is the periodic release of sex cells from the surface of the ovary, called **ovulation** (figure 25.8).

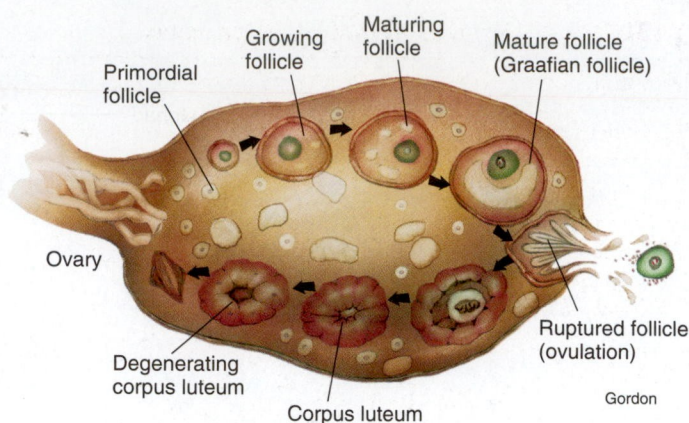

FIGURE 25.8 In the ovary, the egg begins development inside a sac of cells known as a follicle. Each month, one of these follicles develops and releases its product. This release through the wall of the ovary is known as ovulation.

Initially, these two cycles, menstruation and ovulation, may be irregular, which is normal during puberty. Eventually, hormone production becomes regulated so that ovulation and menstruation take place on a regular monthly basis in most women, although normal cycles may vary from twenty-one to forty-five days.

As girls progress through puberty, curiosity about the changing female body form and new feelings lead to self-investigation. Studies have shown that sexual activity such as manipulation of the clitoris, which causes pleasurable sensations, is performed by a large percentage of young women. Self-stimulation, frequently to orgasm, is a common result. This stimulation is termed *masturbation,* and it should be stressed that it is considered a normal part of sexual development. **Orgasm** is the peak of sexual arousal. During orgasm, whether occurring during masturbation or sexual intercourse, the heart rate increases, breathing quickens, and blood pressure rises; muscles throughout the body spasm but mainly those in the vagina, uterus, anus, and pelvic floor.

The Maturation of Males

Males typically reach puberty about two years later (ages eleven to fourteen) than females. Puberty in males also begins with a change in hormone levels. At puberty, the hypothalamus releases increased amounts of gonadotropin-releasing hormone (GnRH), resulting in increased levels of follicle-stimulating hormone (FSH) and luteinizing hormone. These are the same changes that occur in female development. Luteinizing hormone is often called interstitial cell-stimulating hormone (ICSH) in males. LH stimulates the testes to produce **testosterone,** the primary sex hormone in males. The testosterone produced by the embryonic testes caused the differentiation of internal and external genital anatomy in the male embryo. At puberty, the increased amount of testosterone is responsible for sexual maturation and the development of male secondary sexual characteristics, and is important in the maturation and production of sperm.

TABLE 25.2   Human Reproductive Hormones

Hormone	Production Site	Target Organ	Function
Gonadotropin-releasing hormone (GnRH)	Hypothalamus	Anterior pituitary	Stimulates the release of FSH and LH from anterior pituitary
Luteinizing hormone (LH) or interstitial cell-stimulating hormone (ICSH)	Anterior pituitary	Ovary, testes	Stimulates ovulation in females and sex-hormone (estrogens and testosterone) production in both males and females
Follicle-stimulating hormone (FSH)	Anterior pituitary	Ovary, testes	Stimulates ovary and testis development; stimulates egg production in females and sperm production in males
Estrogens	Ovary	Entire body	Stimulates development of female reproductive tract and secondary sexual characteristics
Testosterone	Testes	Entire body	Stimulates development of male reproductive tract and secondary sexual characteristics
Progesterone	Corpus luteum of ovary	Uterus, breasts	Causes uterine thickening and maturation; maintains pregnancy; contributes to milk production
Androgens	Testes, adrenal glands	Entire body	Stimulates development of male reproductive tract and secondary sexual characteristics in males and females
Oxytocin	Posterior pituitary	Uterus, breasts	Causes uterus to contract and breasts to release milk
Prolactin, lactogenic, or luteotropic hormone	Anterior pituitary	Breasts, ovary	Stimulates milk production; also helps maintain normal ovarian cycle
Human chorionic gonadotropin	Placenta	Corpus luteum	Maintains corpus luteum so that it continues to secrete progesterone and maintain pregnancy

The major changes during puberty include growth of the testes and scrotum, pubic-hair development, and increased size of the penis. Secondary sex characteristics also begin to become apparent. Facial hair, underarm hair, and chest hair are some of the most obvious. The male voice changes as the larynx (voice box) begins to change shape. Body contours also change, and a growth spurt increases height. In addition, the proportion of the body that is muscle increases and the proportion of body fat decreases. At this time, a boy's body begins to take on the characteristic adult male shape, with broader shoulders and heavier muscles.

In addition to these external changes, increased testosterone causes the production of seminal fluid by the accessory glands. FSH stimulates the production of sperm cells. The release of sperm cells and seminal fluid begins during puberty and is termed *ejaculation*. This release is generally accompanied by the pleasurable sensations of orgasm. The sensations associated with ejaculation may lead to self-stimulation, or masturbation. Masturbation is a common and normal activity as a boy goes through puberty. Studies of adult sexual behavior have shown that nearly all men masturbate at some time during their lives.

25.6 SPERMATOGENESIS

The biological reason for sexual activity is the production of offspring. The process of producing gametes includes meiosis and is called gametogenesis (gamete formation). The term **spermatogenesis** is used to describe gametogenesis that takes place in the testes of males. The two bean-shaped testes are composed of many small sperm-producing tubes, or **seminiferous tubules,** and collecting ducts that store sperm. These are held together by a thin covering membrane. The seminiferous tubules join together and eventually become the epididymis, a long, narrow convoluted tube in which sperm cells are stored and mature before ejaculation (figure 25.9).

Leading from the epididymis is the vas deferens, or sperm duct; this empties into the urethra, which conducts the sperm out of the body through the **penis** (figure 25.10). Before puberty, the seminiferous tubules are packed solid with diploid cells called spermatogonia. These cells, which are found just inside the tubule wall, undergo *mitosis* and produce more spermatogonia. Beginning about age eleven, some of the spermatogonia specialize and begin the process of *meiosis*, whereas others continue to divide by mitosis, ensuring a constant and continuous supply of spermatogonia. Spermatogenesis needs to occur below body temperature, which is why the testicles are in a sack, the scrotum, outside the body. Once spermatogenesis begins, the seminiferous tubules become hollow and can transport the mature sperm.

Spermatogenesis involves several steps. It begins when some of the spermatogonia in the walls of the seminiferous tubules differentiate and enlarge to become *primary spermatocytes*. These diploid cells undergo the first meiotic division, which produces two haploid *secondary spermatocytes*. The secondary spermatocytes go through the second meiotic division, resulting in four haploid **spermatids,** which lose much of their cytoplasm and develop long tails to mature into

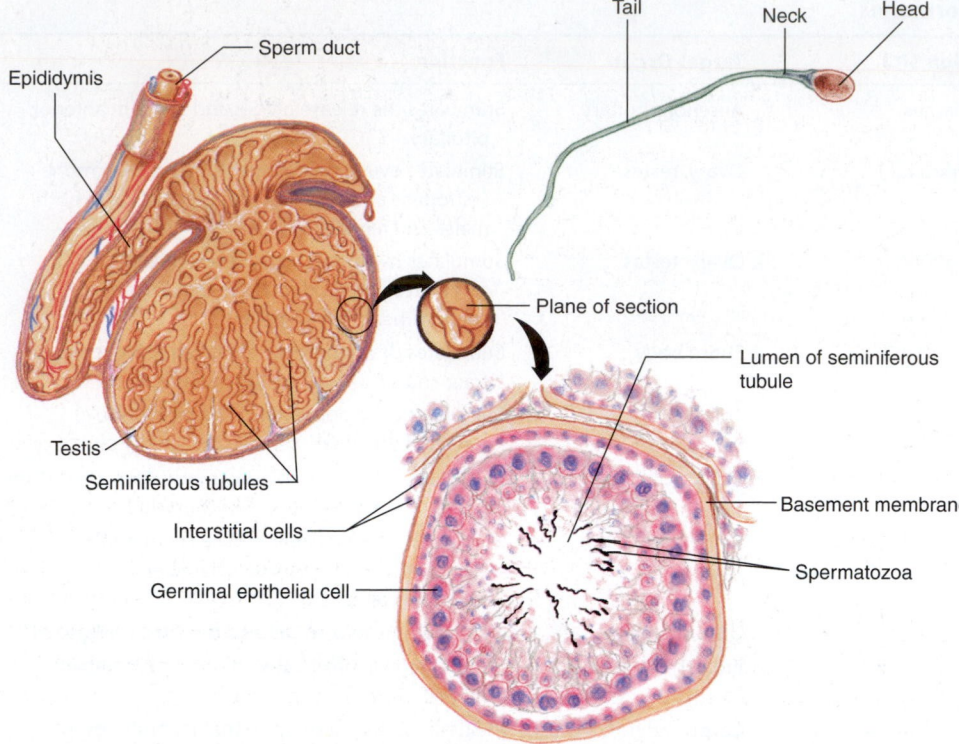

Tail   Neck   Head

Sperm duct

Epididymis

Plane of section

Lumen of seminiferous tubule

Testis

Seminiferous tubules

Basement membrane

Interstitial cells

Spermatozoa

Germinal epithelial cell

FIGURE 25.9 The testis consists of many tiny tubes called seminiferous tubules. The walls of the tubes consist of cells that continually divide, producing large numbers of sperm. The sperm leave the seminiferous tubules and enter the epididymis, where they are stored prior to ejaculation through the sperm duct. Sperm cells have a head region, with an enzyme-containing cap and the DNA. They also have a neck region, with ATP-generating mitochondria, and a tail flagellum, which propels the sperm.

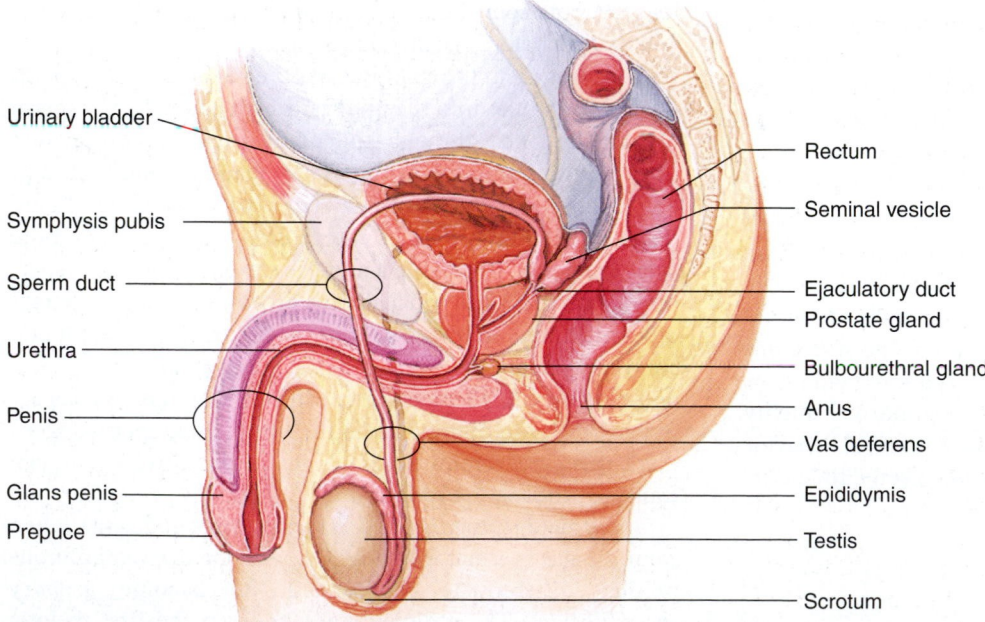

Urinary bladder

Symphysis pubis

Sperm duct

Urethra

Penis

Glans penis

Prepuce

Rectum

Seminal vesicle

Ejaculatory duct

Prostate gland

Bulbourethral gland

Anus

Vas deferens

Epididymis

Testis

Scrotum

FIGURE 25.10 The male reproductive system consists of two testes that produce sperm, ducts that carry the sperm, and various glands. Muscular contractions propel the sperm through the vas deferens past the seminal vesicles, prostate gland, and bulbourethral gland, where most of the liquid of the semen is added. The semen passes through the urethra of the penis to the outside of the body.

sperm. The sperm have only a small amount of food reserves. Therefore, once they are released and become active swimmers, they live no more than seventy-two hours. However, if the sperm are placed in a special protective solution, the temperature can be lowered drastically to −196°C. Under these conditions, the sperm become deactivated, freeze, and can survive for years outside the testes. This has led to the development of sperm banks. Artificial insemination (placing stored sperm into the reproductive tract of a female) or *in vivo* fertilization of cattle, horses, and other domesticated animals with sperm from sperm banks is common. The same techniques can be used in humans but are most often used when men produce insufficient sperm so that the sperm must be collected and stored to increase the number of sperm to enhance the likelihood of fertilization.

Semen, also known as *seminal fluid,* is a mixture of sperm and secretions from three *accessory glands.* These are the *seminal vesicles, prostate, and bulbourethral glands.* They produce secretions that nourish and activate the spermatozoa, and clear the urethral tract before ejaculation. They also serve to lubricate the tract and act as the vehicle to help propel the spermatozoa.

Seminal vesicles secrete an alkaline fluid that contains fructose and hormones. The alkaline nature of the fluid helps to neutralize the acid environment in the female reproductive tract. This improves the chances that sperm will make their way through the tract to the egg. The fructose provides energy for the sperm. The hormones stimulate contractions in the male's system that help to propel sperm. Seminal vesicle secretions make up about 60 percent of the total seminal fluid. Bulbourethral gland secretions are also alkaline and help neutralize the vaginal tract making fertilization more likely. The prostate gland (note spelling) produces a thin milky fluid with a characteristic odor that makes up about 25 percent of semen and contains sperm-activating enzymes.

Are Sperm Counts Falling?

In recent years, there have been several studies that suggest that sperm counts have been falling. One British study suggests that sperm counts have fallen by 29 percent in twelve years. Another study found that men with higher levels of certain pesticides in their blood had lower sperm counts. There are many critics of the various studies who question the way the data were collected or the conclusions reached. For example, some data were collected at fertility clinics, where men might be expected to have low sperm counts. Other studies had very small sample sizes or had other problems with the way the study was designed and carried out. Regardless of the quality of the data, people with particular views have used these studies to support their concerns about environmental chemicals (pesticides, chemicals in plastics), the use of birth control pills, consumption of high amounts of beef, and a variety of other environmental factors. Some have even suggested that tight pants or the heat from laptop computers could raise the temperature of the testes and reduce sperm counts.

Myths, Mistakes, and Misunderstandings

Mumps and Sterility

Misunderstanding: Both boys and girls might become sterile if they get the mumps.

In fact, mumps is a viral infection primarily of the salivary glands. It causes them to swell, and the swelling reaches its peak in about forty-eight hours and may last two weeks. Inflammation and swelling of the testes (called orchitis) can also occur in about 15 to 25 percent of the male cases. In rare cases, the swollen testes rupture their thin covering membrane, leading to sterility. In females, inflammation of the ovaries (called oophoritis) occurs in only 5 percent of the mumps cases and does not cause sterility, since the ovaries are not bound by a membrane and can swell without rupturing.

Prostate cancer can be a significant problem for men. Age appears to be the greatest risk factor of this disease. It has been estimated that 75 percent of men over sixty-five years of age are diagnosed with this form of cancer. This may be because prostate cancer is a slow, progressive disease. Prostate cancer is more common in African-American men in the United States than in any other group. Treatment of the disease may be watchful waiting, changes in diet, surgery, radiation therapy, hormonal therapy, or a combination of these. Surgery, if not done very carefully, can lead to problems with urination or an inability to obtain an erection. A recent study found that 30 percent of men who had their prostate removed and experienced erectile dysfunction regained their ability to have an erection after taking erectile dysfunction medications for nine months. The prostate also uses zinc when functioning properly. Low levels of dietary zinc can lead to enlargement, which may pinch the urethra, interfering with urination. Dietary supplements of zinc, selenium, and vitamin E are currently being researched as ways of combating this swelling.

Spermatogenesis in human males takes place continuously throughout a male's reproductive life, although the number of sperm produced decreases as a man ages. Sperm counts can be taken and used to determine the probability of successful fertilization. A healthy male probably releases about 150 million sperm with each ejaculation. A man must be able to release at least 100 million sperm per milliliter to be fertile. Men with sperm counts under 50 million/ml are often infertile, and those with sperm counts below 20 million/ml are clinically infertile. These vast numbers of sperm are necessary because so many die during their journey that large numbers are needed for the few survivors to reach the egg. In addition, each sperm contains enzymes in its head that are needed to digest through mucus and protein found in the female reproductive tract. Millions of sperm contribute in this way to the process of fertilization, but only one is involved in fertilizing the egg.

25.7 OOGENESIS

The term **oogenesis** refers to the production of egg cells. This process starts before a girl is born during prenatal development of the ovary. It occurs when diploid oogonia cease dividing by *mitosis* and enlarge to become *primary oocytes*. All of the primary oocytes that a woman will ever have are already formed before her birth. At this time, they number approximately two million, but that number is reduced by cell death to between three-hundred and four-hundred thousand cells by the time of puberty. Primary oocytes begin to undergo meiosis and pause in prophase I. Oogenesis pauses at this point, and all the primary oocytes remain just under the surface of the ovary.

At puberty and on a regular basis thereafter, the sex hormones stimulate a primary oocyte to continue its maturation process. It completes the first meiotic division of a single primary oocyte, which began before the woman's birth, about once a month. But in telophase I, the two cells that form receive unequal portions of cytoplasm. You might think of it as a lopsided division. The smaller of the two cells is called a *polar body,* and the larger haploid cell is the *secondary oocyte* (people commonly refer to this cell as an *egg,* or *ovum,* although technically it is not). The other primary oocytes remain in the ovary. Ovulation begins when the soon-to-be-released secondary oocyte, encased in a saclike structure known as a follicle, grows and moves near the surface of the ovary. When this maturation is complete, the follicle ruptures and the secondary oocyte is released. It is swept into the **oviduct** (fallopian tube) by ciliated cells and travels toward the **uterus** (figure 25.11). Because of the action of the luteinizing hormone, the follicle from which the oocyte ovulated develops into a glandlike structure, the corpus luteum, which produces hormones (progesterone and

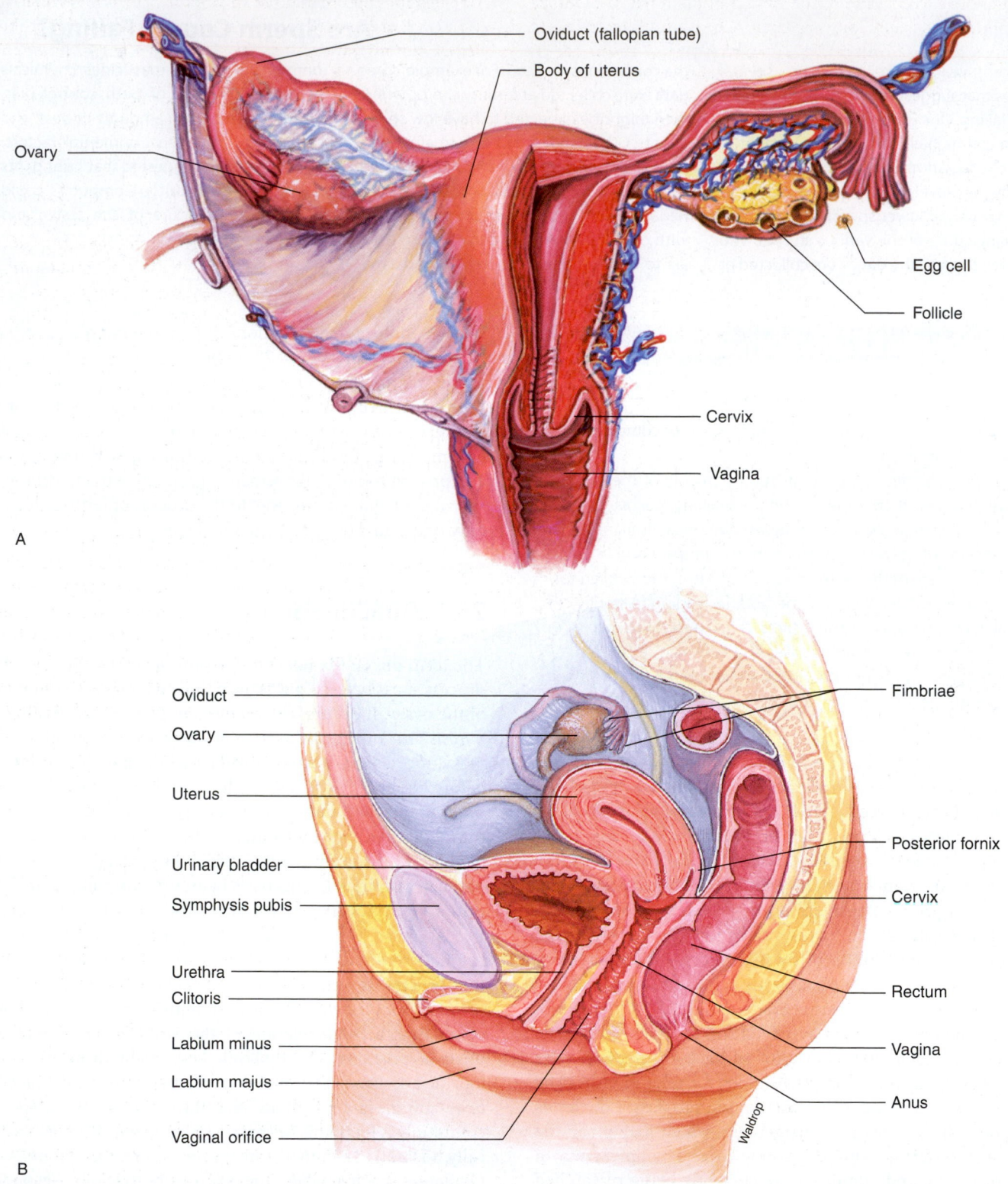

FIGURE 25.11 In the human female reproductive system (*A*), after ovulation the cell travels down the oviduct to the uterus. If it is not fertilized, it is shed when the uterine lining is lost during menstruation. (*B*) The human female reproductive system, side view.

estrogen) that prevent the release of other secondary oocytes. If the secondary oocyte is fertilized, it completes meiosis with the sperm DNA inside, and the haploid egg nucleus and sperm nucleus unite to form the zygote. If the secondary oocyte is not fertilized, it passes through the **vagina** to the outside during menstruation. During her lifetime, a female releases about three hundred to five hundred secondary oocytes. Obviously, few of these cells are fertilized.

One distinguishing characteristics is the relative age of male and female sex cells. In males, sperm production is continuous throughout life. Sperm do not remain in the tubes of the male reproductive system for very long. They are either released shortly after they form or die and are harmlessly absorbed.

In females, meiosis begins before birth, but the oogenesis process is not completed, and the cell is not released for many years. A secondary oocyte released when a woman is thirty-seven years old began meiosis thirty-seven years before! During that time, the cell was exposed to many influences, a number of which may have damaged the DNA or interfered with the meiotic process. The increased risk of abnormal births in older mothers may be related to the age of their eggs. Such alterations are less likely to occur in males because new gametes are being produced continuously. Also, defective sperm appear to be much less likely to be involved in fertilization.

Hormonal Control of Female Sexual Cycles

Hormones control the cycle of changes in breast tissue, in the ovaries, and in the uterus. In particular, estrogen and progesterone stimulate milk production by the breasts and cause the lining of the uterus to become thicker and filled with blood vessels before ovulation. This ensures that if fertilization occurs, the resultant embryo will be able to attach itself to the wall of the uterus and receive nourishment. If the cell is not fertilized, the lining of the uterus, *endometrium,* is shed. This is known as *menstruation, menstrual flow,* the *menses,* or a *period.*

The activities of the ovulatory cycle and the menstrual cycle are coordinated. During the first part of the menstrual cycle, increased amounts of FSH cause the follicle to increase in size. Simultaneously, the follicle secretes increased amounts of estrogen that cause the lining of the uterus to increase in thickness. When ovulation occurs, the remains of the follicle are converted into a corpus luteum by the action of LH. The corpus luteum begins to secrete progesterone, and the nature of the uterine lining changes by becoming more vascularized. This is choreographed so that if an embryo arrives in the uterus shortly after ovulation, it meets with a uterine lining prepared to accept it. If pregnancy does not occur, the corpus luteum degenerates, resulting in a reduction in the amount of progesterone needed to maintain the lining of the uterus, and the lining is shed. At the same time that hormones are regulating ovulation and the menstrual cycle, changes are taking place in the breasts. The same hormones that prepare the uterus to receive the embryo also prepare the breasts to produce milk. These changes in the breasts, however, are relatively minor unless pregnancy occurs.

25.8 HORMONAL CONTROL OF FERTILITY

An understanding of how various hormones manipulate the menstrual cycle, ovulation, milk production, and sexual behavior has led to the medical use of certain hormones. Some women are unable to have children because they do not release oocytes from their ovaries or they release them at the wrong time. Physicians can now regulate the release of oocytes from the ovary using certain hormones, commonly called *fertility drugs.* These hormones can be used to stimulate the release of

oocytes for capture and use in what is called *in vitro* fertilization (*IVF* or *test-tube* fertilization) or to increase the probability of natural conception; that is, *in vivo* fertilization (*in-life* fertilization).

Unfortunately, the use of these techniques often results in multiple embryos being implanted in the uterus. This is likely to occur because the drugs may cause too many secondary oocytes to be released at one time. In the case of *in vitro* fertilization, because there is a high rate of failure and the process is expensive, typically several early-stage embryos are inserted into the uterus to increase the likelihood that one will implant. If several are successful, multiple embryos implant. The implantation of multiple embryos makes it difficult for one embryo to develop properly and be carried through the entire nine-month gestation period. When we understand the action of hormones better, we may be able to control the effects of fertility drugs and eliminate the problem of multiple implantations.

A second medical use of hormones is in the control of conception by the use of birth-control pills—oral contraceptives. Birth-control pills have the opposite effect of fertility drugs. They raise the levels of estrogen and progesterone, which suppresses the production of FSH and LH, preventing the release of secondary oocytes from the ovary. Hormonal control of fertility is not as easy to achieve in men because there is no comparable cycle of gamete release. However, a new, reversible conception control method for males has been developed. It relies on using a combination of progestin, a hormone used in female contraceptive pills, and testosterone. The combination of the two hormones temporarily turns off the normal signals from the brain that stimulate sperm production. The use of drugs and laboratory procedures to help infertile couples have children has also raised the technical possibility of cloning in humans (table 25.3).

25.9 FERTILIZATION, PREGNANCY, AND BIRTH

In most women, a secondary oocyte is released from the ovary about fourteen days before the next menstrual period. The menstrual cycle is usually said to begin on the first day of menstruation. Therefore, if a woman has a regular twenty-eight-day cycle, the cell is released approximately on day 14 (figure 25.12). If a woman normally has a regular twenty-one-day menstrual cycle, ovulation would occur about day 7 in the cycle. If a woman has a regular forty-day cycle, ovulation would occur about day 26 of her menstrual cycle. Some women, however, have very irregular menstrual cycles, and it is difficult to determine just when the oocyte will be released to become available for fertilization. Once the cell is released, it is swept into the oviduct and moved toward the uterus. If sperm are present, they swarm around the secondary oocyte as it passes down the oviduct, but only one sperm penetrates the outer layer to fertilize it and cause it to complete meiosis II. The other sperm contribute enzymes that digest away the protein and mucus barrier between the egg and the successful sperm.

TABLE 25.3 Common Causes of Infertility

Lifestyle causes	Heavy use of alcohol and drugs. Low body fat or anorexia in women. Tight clothing in men may raise the temperature in the scrotum and affect sperm development. Stress may cause irregular ovulation in women or reduce sperm count in men.
Infections	Sexually transmitted diseases often result in scarring or blockage of reproductive tubes. Pelvic inflammatory disease (PID) is the most common cause of infertility in women.
Physical causes	Fibroids and endometriosis may cause blockage. Retrograde ejaculation—the semen is forced into the bladder rather than being ejaculated.
Developmental causes	Undescended testes. Swollen veins (varicocele) in scrotum. Undeveloped ovaries or testes (developmental defect, infection, etc.).
Hormonal causes	Any imbalance in the timing and quantity of the several sex hormones can result in lack of ovulation. The uterus may not be prepared to accept the embryo. Low progesterone levels may cause premature shedding of the uterine lining. Low testosterone levels result in low sperm counts.
Immune system causes	Female may develop antibodies against her partner's sperm. Male may develop an autoimmune response to his own sperm.
Illness and medication causes	Diseases such as diabetes, kidney disease, and high blood pressure contribute to infertility. Tranquilizers and blood pressure drugs may interfere with erection.

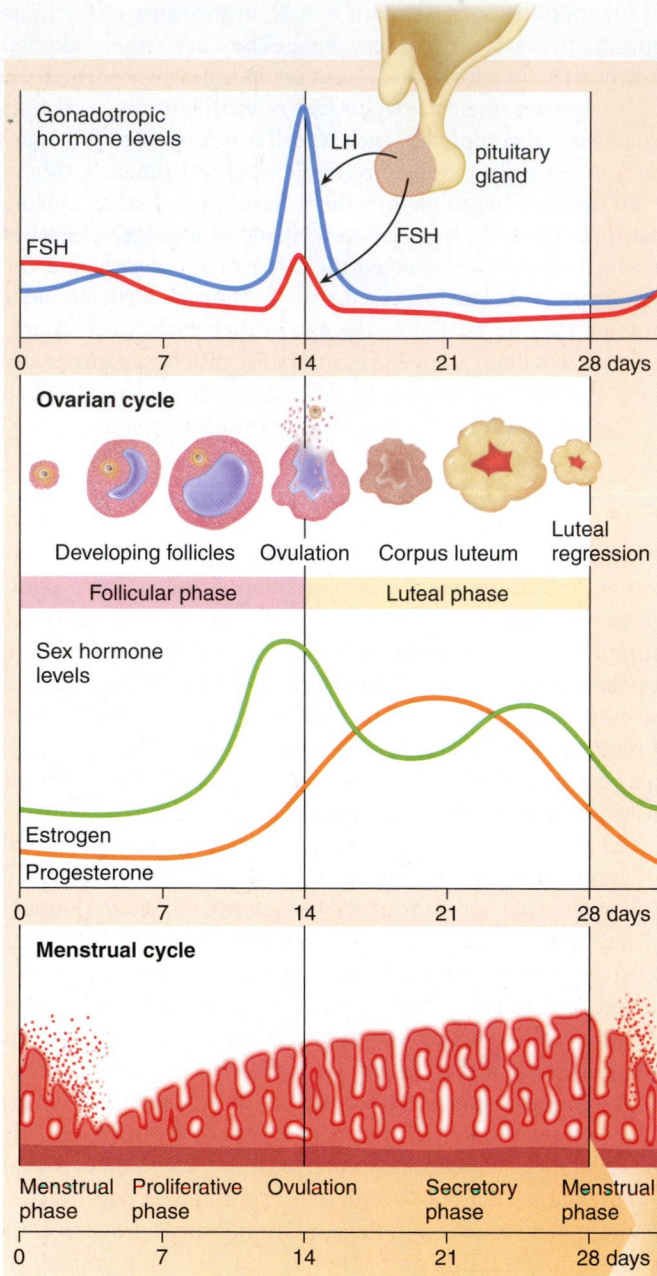

FIGURE 25.12 The release of an egg (ovulation) is timed to coincide with the thickening of the lining of the uterus. The uterine cycle in humans involves the preparation of the uterine wall to receive the embryo if fertilization occurs. Knowing how these two cycles compare, it is possible to determine when pregnancy is most likely to occur by noting when menstruation begins.

During this second meiotic division, the second polar body is pinched off and the *ovum* (egg) is formed. Because chromosomes from the sperm are already inside, they simply intermingle with those of the ovum, forming a diploid zygote or fertilized egg. As the zygote continues to travel down the oviduct, it begins to divide by mitosis into smaller and smaller cells without having the mass of cells increase in size (figure 25.13). This division process is called *cleavage*. Eventually, a solid ball of cells is produced, known as the morula stage of embryological development. The morula travels down the oviduct and continues to divide by mitosis. The result is called a *blastocyst*. The blastocyst becomes embedded, or implanted, when it reaches the lining of the uterus.

The next stage in the development is known as the gastrula stage because the gut is formed during this time (*gastro* = stomach). In many kinds of animals, the gastrula is formed by a complex folding of the blastocyst walls. The embryo eventually develops a tube that becomes the gut. The formation of the primitive gut is just one of a series of changes that result in an embryo that is recognizable as a miniature human being.

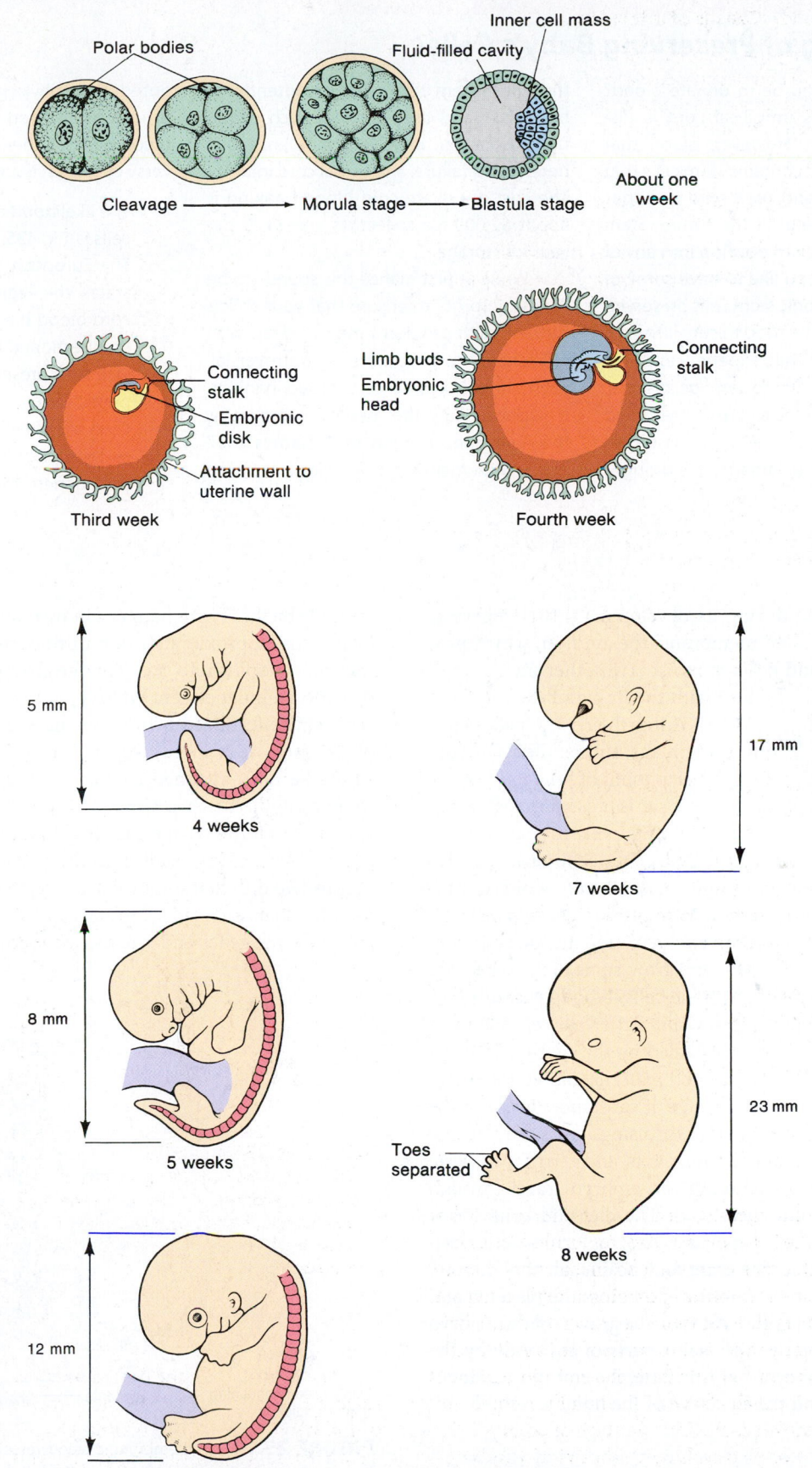

FIGURE 25.13   During the period of time between fertilization and birth, many changes take place in the embryo. Here we see some of the changes that take place during the first eight weeks.

Science and Society

Thinking of Preserving Baby's Cells?

How young can you be to donate blood? Normally, a baby's umbilical cord is discarded after birth. However, blood that remains in the cord contains stem cells that can be collected and preserved in hopes that it may be useful in the future. Stem cells have the ability to develop into any of your cells. Would you like to have some of your child's embryonic stem cells preserved so that they might be used to cure illness or repair injury? If the child experiences tissue or organ problems due to damage, disease, age, or genetic defects, these preserved cells might be used to generate tissues to repair or replace the damage. It is thought that these stem cells have the potential to be cloned and used to treat such conditions as cancer, brain injury, juvenile diabetes, renal failure, and spinal cord injuries. The cost of private cord blood banking is about $2,000 for collection and $125 per year for storage.

While at first glance this sounds to be "the way to go" to ensure that your child's future health problems may be dealt with efficiently, the procedure is controversial. Even though public cord blood banking is supported by the medical community, the American Academy of Pediatrics 2007 Policy Statement on Cord Blood Banking noted that physicians should be aware of unsubstantiated claims of private cord blood banks. Other aspects of this controversy center on issues and such facts as:

- The likelihood of using your own stem cells is 1 in 435.
- The European Union Group on Ethics states the legitimacy of commercial cord blood banks for such use should be questioned because they sell a service that presently has no real therapeutic value.
- Cord blood cells have the same genes as the donor and cannot be used to treat genetic diseases of the donor.

Most of the time during its development, the embryo is enclosed in a water-filled membrane, the amnion, which protects it from blows and keeps it moist. Two other membranes, the chorion and allantois, fuse with the lining of the uterus to form the **placenta** (figure 25.14). A fourth sac, the yolk sac, is well developed in birds, fish, amphibians, and reptiles. The yolk sac in these animals contains a large amount of food used by the developing embryo. Although a yolk sac is present in mammals, it is small and does not contain yolk. The nutritional needs of the embryo are met through the placenta. The placenta also produces the hormone chorionic gonadotropin, which stimulates the corpus luteum to continue producing progesterone and thus prevents menstruation and ovulation during gestation.

As the embryo's cells divide and grow, some of them become differentiated into nerve cells, bone cells, blood cells, or other specialized cells. To divide, grow, and differentiate, cells must receive nourishment. This is provided by the mother through the placenta, in which both fetal and maternal blood vessels are abundant, allowing for the exchange of substances between the mother and embryo. The materials diffusing across the placenta include oxygen, carbon dioxide, nutrients, and a variety of waste products. The materials entering the embryo travel through blood vessels in the umbilical cord. The diet and behavior of the mother are extremely important. Any molecules consumed by the mother can affect the embryo. Cocaine, alcohol, heroin, and chemicals in cigarette smoke can all cross the placenta and affect the development of the embryo. The growth of the embryo results in the development of major parts of the body by the tenth week of pregnancy. After this time, the embryo continues to increase in size, and the structure of the body is refined.

Twins

In the United States, women giving birth have a 1 in 40 chance of delivering twins and a 1 in 650 chance of triplets or other multiple births. Twins happens in two ways. In the case of identical twins (approximately one-third of twins), during cleavage the embryo splits into two separate groups of cells. Each develops into an independent embryo. Because they come from the same single fertilized ovum, they have the same genes and are of the same sex. Should separation be incomplete, the twins would be born attached to one another, a condition referred to as conjoined twins. Conjoined twins occur once in every seventy thousand to one-hundred thousand live births.

Fraternal twins result from the fertilization of two separate eggs by two different sperm. Therefore, they resemble each other no more than do regular brothers and sisters. They do not contain the same genes and are not necessarily the same sex.

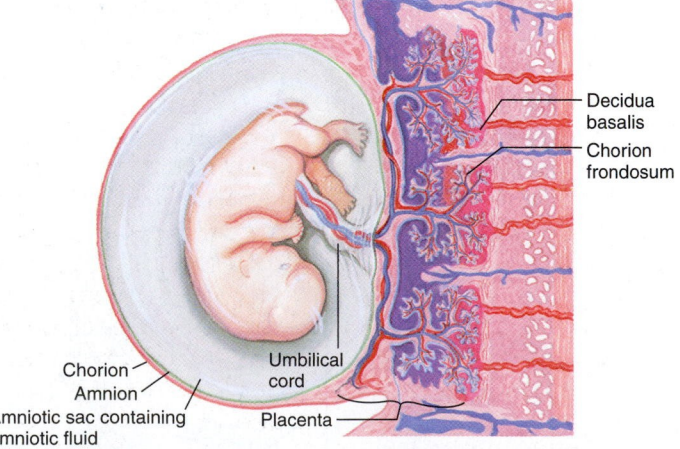

FIGURE 25.14 The embryonic blood vessels that supply the developing embryo with nutrients and remove the metabolic wastes are separate from the blood vessels of the mother. Because of this separation, the placenta can selectively filter many types of incoming materials and microorganisms.

Robert Geoffrey Edwards (1925–) and Patrick Christopher Steptoe (1913–1988)

Robert Geoffrey Edwards and Patrick Christopher Steptoe were British researchers—a physiologist and an obstetric surgeon, respectively—who devised a technique for fertilizing a human egg outside the body and transferring the fertilized embryo to the uterus of a woman. A child born following the use of this technique is popularly known as a "test-tube baby."

Robert Edwards was educated at the University of Wales and the University College of North Wales, Bangor, and from 1951 to 1957 at the University of Edinburgh. In 1957, Edwards went to the California Institute of Technology but returned to England the following year to the National Institute of Medical Research at Mill Hill. He remained there until 1962, when he took an appointment at Glasgow University. A year later, he moved again to the Department of Physiology at Cambridge University. From 1985 to 1989, he was professor of human reproduction at Cambridge.

During his research in Edinburgh, Edwards successfully replanted mouse embryos into the uterus of a mouse and wondered if the same process could be used to replant a human embryo into the uterus of a woman. Edwards was able to obtain human eggs from pieces of ovarian tissue removed during surgery. He found that the ripening process was very slow, the first division beginning only after twenty-four hours.

He studied the maturation of eggs of different species of mammals and in 1965 attempted the fertilization of human eggs. He left mature eggs with sperm overnight and found just one where a sperm had passed through the outer membrane, but it had failed to fertilize the egg. In 1967, Edwards read a paper by Patrick Steptoe describing the use of a new instrument, known as the *laparoscope,* to view the internal organs, which he saw had a possible application to his own research.

Patrick Christopher Steptoe was educated at King's College and St. George's Hospital Medical School, London. He was appointed chief assistant obstetrician and gynecologist at St. George's Hospital, London, in 1947, and senior registrar at the Whittington Hospital, London, in 1949. From 1951 to 1978, he was senior obstetrician and gynecologist at Oldham General Hospital, and from 1969, director of the Centre for Human Reproduction.

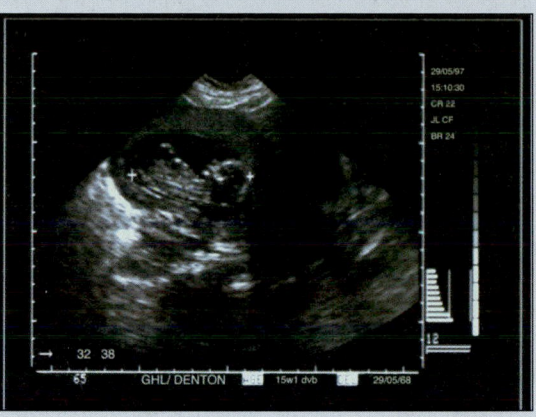

Box Figure 25.7

The paper describing laparoscopy that interested Edwards dealt with Steptoe's method of exploring the interior of the abdomen without a major operation. Steptoe inserted the laparoscope through a small incision near the navel, and by means of this telescopelike instrument, with its object lens inside the body and its eyepiece outside, he was able to examine the ovaries and other internal organs.

Early in 1968, Edwards and Steptoe repeated experiments on the fertilization of human eggs. Steptoe treated volunteer patients with a fertility drug to stimulate maturation of the eggs in the ovary. Edwards devised a simple piece of apparatus to be used with the laparoscope for collecting mature eggs from human ovaries. The mature eggs were removed and Edwards then prepared them for fertilization using sperm provided by the patient's husband. For a year, they continued experiments of this kind until they were sure that the fertilized eggs were developing normally. The next step was to see if an eight-celled embryo would develop.

In 1971, Edwards and Steptoe were ready to introduce an eight-celled embryo into the uterus of a volunteer patient. In 1975, an embryo did implant but in the stump of a fallopian tube, where it could not develop properly and was a danger to the mother. It was removed, but it did demonstrate that the basic technique was sound. In 1977, it was decided to abandon the use of the fertility drug and remove the egg at precisely the right natural stage of maturity; an egg was fertilized and then reimplanted (a process called *in vitro fertilization*) in the mother two days later. The patient became pregnant; twelve weeks later, the position of the baby was found to be satisfactory, and its heartbeat could be heard. During the last eight weeks of pregnancy, the mother was kept under close medical supervision, and a healthy girl—Louise Brown—was delivered by cesarean section on July 25, 1978.

In vitro fertilization has also been used to overcome the infertility in men that is due to a low sperm count. Edwards's research has further added to knowledge of the development of the human egg and young embryo, and Steptoe's laparoscope is a valuable instrument capable of wider application.

Source: Modified from the *Hutchinson Dictionary of Scientific Biography.* © RM, 2011. All rights reserved. Helicon Publishing is a division of RM.

Birth

The process of giving birth is also known as *parturition* or birthing. At the end of about nine months, hormone changes in the mother's body stimulate contractions of the muscles of the uterus during a period before birth called *labor*. These contractions are stimulated by the hormone oxytocin, which is released from the posterior pituitary. The contractions normally move the baby headfirst through the vagina, or birth canal. One of the first effects of these contractions may be bursting of the amnion (bag of water) surrounding the baby. Following this, the uterine contractions become stronger, and shortly thereafter, the baby is born. In some cases, the baby becomes turned in the uterus before labor. If this occurs, the feet or buttocks appear first.

Science and Society

Multiple Births, IVF, and Hormones

The worldwide incidence of multiple pregnancies (twins, triplets, quadruplets, and more) is increasing as advanced types of infertility treatments such as in vitro fertilization (IVF) have become more ordinary. *In vitro* fertilization is a method that uses hormones to stimulate egg production, removing the oocyte from the ovary and fertilizing it with donated sperm. The fertilized egg

is incubated to stimulate cell division in a laboratory dish and then placed in the uterus.

In the United States, the number of twin births has risen more than 50 percent in the past thirty years. The incidence of higher-order multiple pregnancies (triplets or greater) has increased about one hundred times. According to the National Center for Health Statistics, multiple pregnancies in the United States have seen the greatest increase among women in their thirties and forties. This trend is due in part to the fact that older women are less able to get pregnant naturally and are more likely to undergo infertility treatment.

One of the reasons why IVF is so successful is that it usually involves implanting more than one embryo, increasing the chance of a pregnancy. IVF babies are twenty times more likely to be born as multiple birth babies; one study shows that about 45 percent of all IVF newborns are born as multiple birth babies.

The use of hormones to stimulate super-ovulation, coupled with insemination, is a more common method of infertility treatment. Super-ovulation involves using the hormone gonadotropin to stimulate ovulation of more than one egg. The more eggs released, the greater the chance that a multiple pregnancy will result. Because it is a more commonly used method, and because it is so potent, super-ovulation accounts for as many if not more multiple births than IVF.

Such a birth is called a *breech birth.* This can be a dangerous situation because the baby's source of oxygen is being cut off as the placenta begins to separate from the mother's body.

If for any reason the baby does not begin to breathe on its own, it will not receive enough oxygen to prevent the death of nerve cells; thus, brain damage or death can result.

Occasionally, a baby may not be able to be born normally because of its position in the uterus, the location of the placenta on the uterine wall, the size of the birth canal, the number of babies in the uterus, or many other reasons. A common procedure to resolve this problem is the surgical removal of the baby through the mother's abdomen. This procedure is known as a cesarean, or C-section. The procedure was apparently named after the Roman emperor Julius Caesar, who was said to have been the first child to be delivered by this method. Currently, over 20 percent of births in the United States are by cesarean section. While C-sections are known to have been performed before Caesar, the name stuck. This rate reflects the fact that many women who are prone to problem pregnancies are having children rather than forgoing pregnancy. In addition, changes in surgical techniques have made the procedure much safer. Finally, many physicians who are faced with liability issues related to problem pregnancy may encourage cesarean section rather than normal birth.

Following the birth of the baby, the placenta, also called the *afterbirth,* is expelled. Once born, the baby begins to function on its own. The umbilical cord collapses, and the baby's lungs, kidneys, and digestive system must now support all its bodily needs. This change is quite a shock, but the baby's loud protests fill the lungs with air and stimulate breathing.

Over the next few weeks, the mother's body returns to normal, with one major exception. The breasts, which have undergone changes during the period of pregnancy, are ready to produce milk to feed the baby. Following birth, prolactin, a hormone from the pituitary gland, stimulates the production of milk, and oxytocin stimulates its release. If the baby is breast-fed, the stimulus of the baby's sucking will prolong the time during which milk is produced. This response involves both the nervous and endocrine systems. The sucking stimulates nerves in the nipple and breast, which results in the release of prolactin and oxytocin from the pituitary.

In some cultures, breast-feeding continues for two to three years, and the continued production of milk often delays the reestablishment of the normal cycles of ovulation and menstruation. Many people believe that a woman cannot become pregnant while she is nursing a baby. However, because there is so much variation among women, relying on this as a natural conception-control method is not a good choice. Many women have been surprised to find themselves pregnant again a few months after delivery.

25.10 CONTRACEPTION

Throughout history, people have tried various methods of conception control (figure 25.15). In ancient times, conception control was encouraged during times of food shortage or when tribes were on the move from one area to another in search of a new home. Writings as early as 1500 B.C. indicate that the Egyptians used a form of tampon medicated with the ground powder of a shrub to prevent fertilization. This may sound primitive, but we use the same basic principle today to destroy sperm in the vagina. As you read about the various methods of contraception, remember that no method described is 100 percent effective for avoiding pregnancy and preventing sexually transmitted diseases (STDs) except abstinence.

Chemical Methods

Contraceptive jellies and foams make the environment of the vagina more acidic, which diminishes the sperm's chances of survival. The spermicidal (sperm-killing) foam or jelly is placed in the vagina before intercourse. When the sperm make contact with the acidic environment, they stop swimming and soon die. Aerosol foams are an effective method of conception control, but interfering with the hormonal regulation of ovulation is more effective. Contraceptive foams and jellies provide very little protection from sexually transmitted diseases.

Hormonal Control Methods

The first successful method of hormonal control was "the pill," or "birth-control pill," which contained a combination of the hormones estrogen and progesterone. Today, there are two kinds of birth-control pills: those that contain estrogen and progesterone and those that contain progesterone only. The

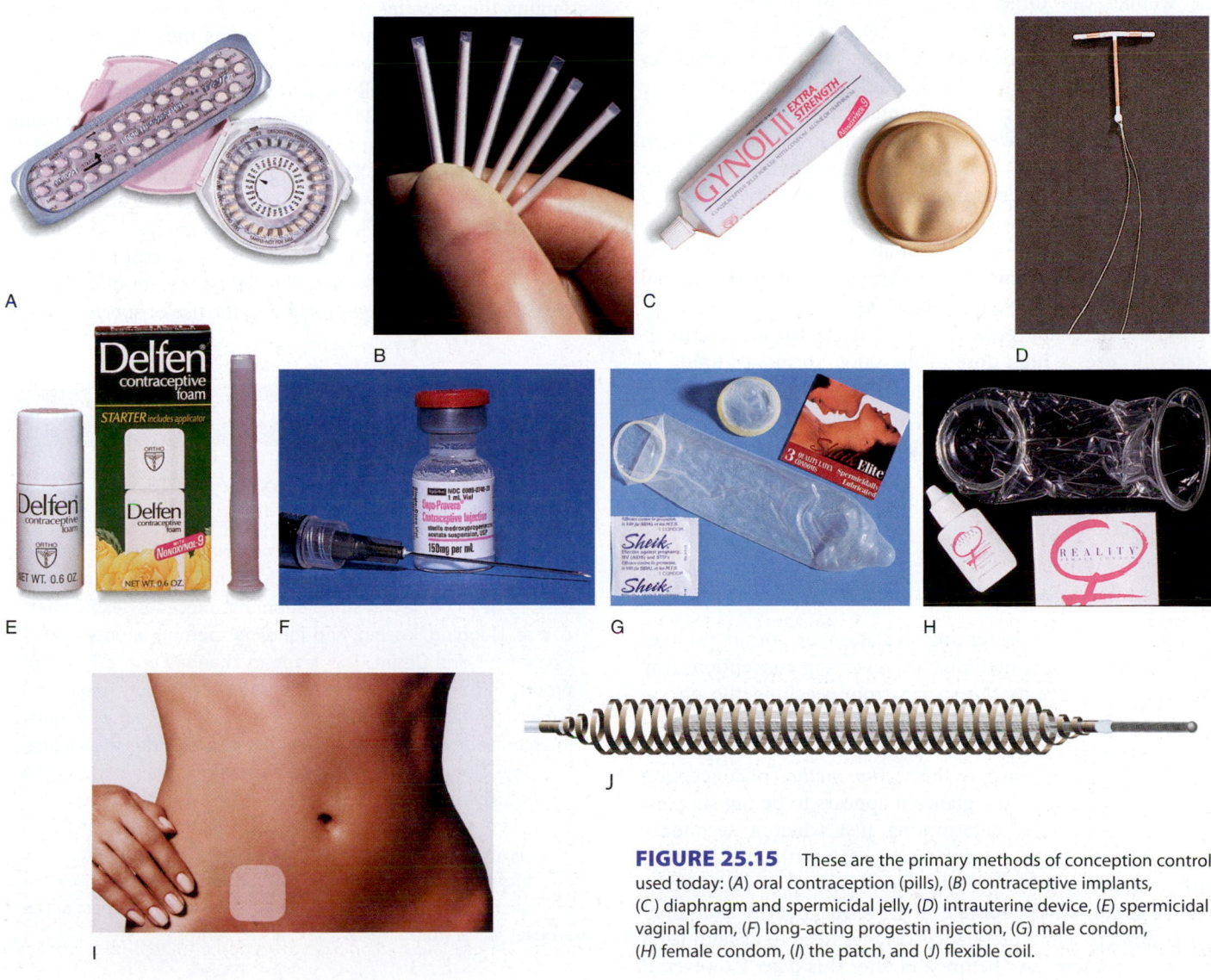

FIGURE 25.15 These are the primary methods of conception control used today: (*A*) oral contraception (pills), (*B*) contraceptive implants, (*C*) diaphragm and spermicidal jelly, (*D*) intrauterine device, (*E*) spermicidal vaginal foam, (*F*) long-acting progestin injection, (*G*) male condom, (*H*) female condom, (*I*) the patch, and (*J*) flexible coil.

combination birth-control pill works primarily by preventing ovulation and secondarily by interfering with implantation. The quantity and balance of hormones (estrogen and progesterone) in the pill fool the ovaries into functioning as if the woman were already pregnant. Therefore, ovulation does not occur, so conception is highly unlikely. The combination pills are taken daily for twenty-one days and then stopped (some brands have twenty-one pills with hormones and the last seven do not contain hormones). This allows for menstruation to occur.

The progesterone-only pill (POP, or mini pill) is taken continuously without a break. The progesterone in these pills has several effects that help to prevent conception. Progesterone thickens the mucus at the cervix, alters the uterine lining to make implantation unlikely, and appears to interfere with the movement of sperm and egg.

Other hormone methods involve delivering the hormones by way of implants, adhesive patches, or vaginal rings that slowly release hormones and prevent the maturation and release of eggs from the follicle. The major advantage of an implant is its convenience. Once the implant has been inserted, the woman can forget about contraceptive protection for several years. If she wants to become pregnant, the implants are removed, and her normal menstrual and ovulation cycles return over a period of weeks. Similarly, discontinuing the use of the patch allows normal ovulation to resume.

The vaginal ring releases a continuous low dose of estrogen and progestin for twenty-one days. At the end of the twenty-one days, the ring is removed for seven days to allow the menstrual period to occur. A new ring is then inserted monthly.

Contraceptive hormones can also be used to stop menstrual periods, preventing the symptoms of premenstrual syndrome and improving a woman's sex life.

The emergency contraceptive pill (ECP), or "morning-after pill," uses a high dose of the same hormones found in oral contraceptives, which prevents the woman from becoming pregnant in the first place. In fact, "the pill" in higher dosages can be used as an ECP. The common medication available in the United States is known as Plan B®. These pills work by inhibiting ovulation and thickening mucus, which makes it difficult for the sperm to get to the egg.

Timing Method

Killing sperm and preventing ovulation or sperm production are not the only methods of preventing conception. Any method that prevents the sperm from reaching the oocyte prevents conception. One method is to avoid intercourse during those times of the month when a secondary oocyte may be present. This is known as the *rhythm method* of conception control. Although at first glance it appears to be the simplest and least expensive, determining just when a secondary oocyte is likely to be present can be very difficult. A woman with a regular twenty-eight-day menstrual cycle will typically ovulate about fourteen days before the onset of the next menstrual flow. To avoid pregnancy, couples need to abstain from intercourse a few days before and after this date. However, if

a woman has an irregular menstrual cycle, there may be only a few days each month for intercourse without the possibility of pregnancy. In addition to calculating safe days based on the length of the menstrual cycle, a woman can better estimate the time of ovulation by keeping a record of changes in her body temperature and vaginal pH. Both changes are tied to the menstrual cycle and can therefore help a woman predict ovulation. In particular, at about the time of ovulation, a woman has a slight rise in body temperature—less than 1°C. Thus, one should use an extremely sensitive thermometer. A digital-readout thermometer on the market spells out the word "yes" or "no." However, all variations of the rhythm method have high failure rates—up to 25 percent per year.

Barrier Methods

Other methods of conception control that prevent the sperm from reaching the secondary oocyte include the diaphragm, cap, sponge, and condom. The diaphragm is a specially fitted membranous shield that is inserted into the vagina before intercourse and positioned so that it covers the cervix, which contains the opening of the uterus. Because of anatomical differences among females, diaphragms must be fitted by a physician. The effectiveness of the diaphragm is increased if spermicidal foam or jelly is also used. The vaginal cap functions in a similar way. The contraceptive sponge, as the name indicates, is a small amount of absorbent material that is soaked in a spermicide. The sponge is placed within the vagina and chemically and physically prevents the sperm cells from reaching the oocyte. The contraceptive sponge is no longer available for use in the United States but is still available in many other parts of the world. The failure rate for diaphragms, cervical caps, and sponges including the use of spermicides is

Myths, Mistakes, and Misunderstandings

Is it Sex?

Misunderstanding: Oral sex is not sex.

In fact, the phrase *sexual intercourse* involves several behaviors associated with sexual reproduction. Foreplay is the term used to describe sexual stimulation that precedes sexual intercourse. Hugging, kissing, and fondling (petting) arouse sexual excitement and desire. This leads to changes in the levels of production and an increase in blood flow in both male and female genitals in anticipation of vaginal intercourse. Vaginal intercourse involves inserting the erect penis into the vagina. The physiological reactions that occur affect many parts of the reproductive and nervous systems and are essential to successful sexual reproduction. They can also lead to the sensation known as orgasm. Two other forms of sexual intercourse practiced are anal (penis in anus) and oral (penis in mouth or oral stimulation of vagina or clitoris). These two variations may also be part of the arousal phase (foreplay) of a sexual encounter.

A Closer Look

Sexually Transmitted Diseases

Sexually transmitted diseases (STDs) were formerly called venereal diseases (VDs). Although these kinds of illnesses are most frequently transmitted by sexual activity, many can also be spread by other methods of direct contact, such as hypodermic needles, blood transfusions, and blood-contaminated materials. Currently, the Centers for Disease Control and Prevention (CDC) in Atlanta, Georgia, recognize more than 25 diseases as being sexually transmitted (box table 25.1).

The United States has the highest rate of sexually transmitted disease among the industrially developed countries. There are about 19 million new infections each year. About half of the infections occur in young adults aged 15–24 years of age. Box table 25.2 lists the most common STDs and estimates of the number of new cases each year. Some of the most important STDs are described here because of their high incidence in the population and the inability

to bring some of them under control. For example, there is no known cure for HIV, responsible for AIDS.

Despite efforts to educate the public and the availability of effective methods for treatment, several STDs are actually increasing in frequency. In particular, *Chlamydia* infections have increased to more than a million new infections per year. There has also been an increase in the number of cases of syphilis, especially among gay and bisexual men. In addition, a penicillin-resistant strain of *Neisseria gonorrhoeae* has led to an increase in the number of cases of gonorrhea.

The spread of STDs during sexual intercourse is significantly diminished by the use of condoms. Other types of sexual contact (e.g., hand, oral, anal) and transmission from a mother to the fetus during pregnancy help maintain some of these diseases in the population. Therefore, public health organizations—such as the U.S. Public Health

Service, the CDC, and state and local public health agencies—regularly keep an eye on the number of cases of STDs. All of these agencies are involved in attempts to raise the general public health to a higher level. Their investigations have resulted in the successful control of many diseases and the identification of special problems, such as those associated with the STDs. Because the United States has an incidence rate of STDs that is 50 to 100 times higher than that of other industrially developed countries, there is still much that needs to be done.

The high-risk behaviors associated with contracting STDs include sex with multiple partners and the failure to use condoms. Whereas some STDs are simply inconvenient or annoying, others severely affect health and can result in death. As one health official stated, "We should be knowledgeable enough about our own sexuality and the STDs to answer the question, 'Is what I'm about to do worth dying for?'"

Box Table 25.1

Sexually Transmitted Diseases

Disease	Agent
Genital herpes	Virus
Gonorrhea	Bacterium
Syphilis	Bacterium
Acquired immunodeficiency syndrome (AIDS)	Virus
Candidiasis	Yeast
Chancroid	Bacterium
Genital warts	Virus
Gardnerella vaginalis	Bacterium
Genital *Chlamydia* infection	Bacterium
Genital cytomegalovirus infection	Virus
Genital *Mycoplasma* infection	Bacterium
Group B *Streptococcus* infection	Bacterium
Nongonococcal urethritis	Bacterium
Pelvic inflammatory disease (PID)	Bacterium
Molluscum contagiosum	Virus
Crabs	Body lice
Scabies	Mite
Trichomoniasis	Protozoan
Hepatitis B	Virus
Gay bowel syndrome	Variety of agents

Box Table 25.2

Diagnosed Cases of Sexually Transmitted Diseases United States Reported in 2008

Reported cases by state health departments

Diseases	Cases Reported in 2008	Comments
Chancroid	25	
Hepatitis B	4,519	Up to 1.4 million people may have undiagnosed disease
HIV/AIDS	37,041	About 1.1 million people infected
Syphilis	46,277	Incidence is increasing
Vaginal trichomoniasis	204,000	Probably over 7 million cases/year
Genital herpes	292,000	About 16% of population infected (about 50 million)
Gonorrhea	336,742	700,000 cases/year estimated
Genital warts	385,000	About 1% of sexually active persons have genital warts
Chlamydia	1,213,523	Estimated 2.3 million infected
Other vaginal infections	3,571,000	Most common sexually transmitted disease

between 14 and 50 per 100 women per year; the methods are not an effective protection against sexually transmitted disease.

The male condom is probably the most popular contraceptive device. It is a thin sheath that is placed over the erect penis before intercourse. In addition to preventing sperm from reaching the secondary oocyte, this physical barrier also helps prevent the transmission of the microbes that cause STDs, such as syphilis, gonorrhea, and AIDS, from being passed from one person to another during sexual intercourse. The most desirable condoms are made of a thin layer of latex that does not reduce the sensitivity of the penis. Latex condoms have also been determined to be the most effective in preventing transmission of the AIDS virus. The condom is most effective if it is prelubricated with a spermicidal material such as nonoxynol-9. This lubricant also has the advantage of providing some protection against the spread of the HIV virus. The failure rate for latex condoms is 11 per 100 women per year. Except for abstinence, latex condoms are the best protection against sexually transmitted diseases.

Recently developed condoms for women are now available. One called the Femidom is a polyurethane sheath that, once inserted, lines the contours of the woman's vagina. It has an inner ring that sits over the cervix and an outer ring that lies flat against the labia. Research shows that this device protects against STDs and is as effective a contraceptive as the condom used by men. The failure rate for such barrier protection is 21 per 100 women per year; the method may give some protection against sexually transmitted diseases but is not as effective as latex condoms.

The intrauterine device (IUD) is not a physical barrier that prevents the gametes from uniting. How this mechanical device works is not completely known. It may in some way interfere with the implantation of the embryo. The IUD must be fitted and inserted into the uterus by a physician, who can also remove it if pregnancy is desired. IUDs are used successfully in many countries. Current research with new and different intrauterine implants indicates that they are able to prevent pregnancy, and one is currently available in the United States. The IUD can also be used for "emergency contraception"—in cases of unprotected sex (forced sex) or failure of a conception control method (a condom slips or breaks). The IUD must be inserted within seven days of unprotected sex. The failure rate for the IUD is less than 1 per 100 women per year; the device provides no protection from sexually transmitted disease.

Another barrier method is tubal implant sterilization. In this conception control method, a soft, flexible coil is inserted into both fallopian tubes (figure 25.15J). After being inserted, the coil stimulates a tissue growth resulting in blockage of the fallopian tube. This device is greater than 99 percent efficient and should be considered a form of sterilization. It is considered to be permanent and irreversible and does not protect against STDs.

Surgical Methods

Two contraceptive methods that require surgery are tubal ligation and vasectomy (figure 25.16). Tubal ligation is the cutting and tying off of the oviducts and can be done on an outpatient basis in most cases. An alternative to sterilization by tubal ligation involves the insertion of small, flexible devices called micro-inserts into each fallopian tube. Once inserted, tissue grows into the inserts, blocking the tubes. Ovulation continues as usual, but the sperm and egg cannot unite. Vasectomy is not the same as castration. Castration is the surgical removal of testes. Vasectomy can be performed in a physician's office and does not require hospitalization. A small opening is made

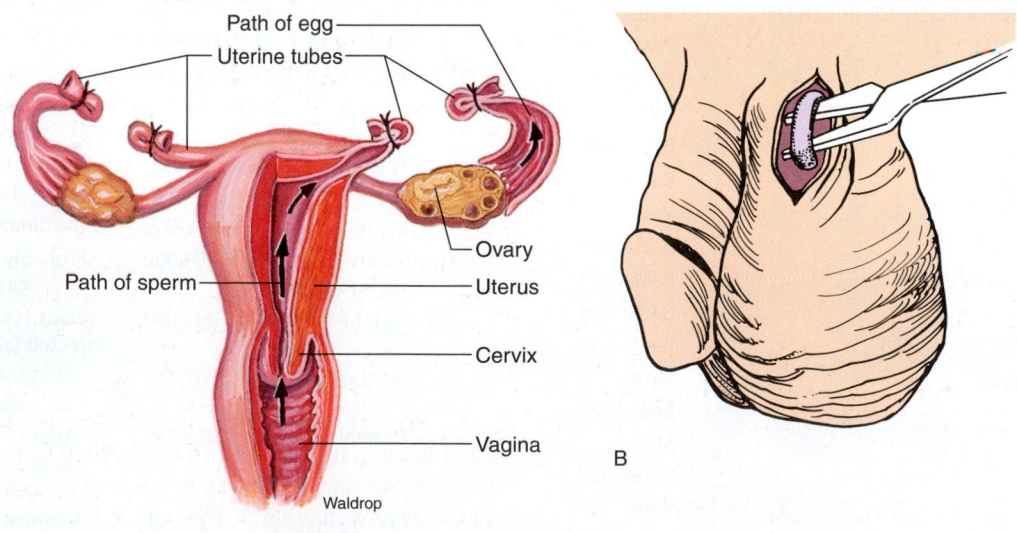

Path of egg
Uterine tubes
Ovary
Path of sperm
Uterus
Cervix
Vagina

Waldrop

A

B

FIGURE 25.16　Two very effective contraceptive methods require surgery. Tubal ligation (*A*) involves severing the oviducts and suturing or sealing the cut ends. This prevents the sperm cell and the secondary oocyte from meeting. This procedure is generally considered ambulatory surgery, or at most requires a short hospitalization period. Vasectomy (*B*) requires minor surgery, usually in a clinic under local anesthesia. Following the procedure, minor discomfort may be experienced for several days. The severing and sealing of the vas deferens prevents the release of sperm cells from the body by ejaculation.

above the scrotum, and the spermatic cord (vas deferens) is cut and tied. This prevents sperm from moving through the ducts to the outside. Because most of the sperm-carrying fluid (semen) is produced by the seminal vesicles, prostate gland, and bulbourethral glands, a vasectomy does not interfere with normal ejaculation. The sperm that are still being produced die and are reabsorbed in the testes. Neither tubal ligation nor vasectomy interferes with normal sex drives. However, these medical procedures are generally not reversible and should not be considered by those who may want to have children at a future date. The failure rate for sterilization is less than 1 per 100 women per year; the procedure provides no protection from sexually transmitted disease.

25.11 TERMINATION OF PREGNANCY

Another medical procedure often associated with birth control is abortion, which has been used throughout history. Abortion involves various medical procedures that cause the death and removal of the developing embryo. Abortion is obviously not a method of conception control; rather, it prevents the normal development of the embryo and causes its death. Abortion is a highly charged subject. Some people believe that abortion should be prohibited by law in all cases. Others think that abortion should be allowed in certain situations, such as in pregnancies that endanger the mother's life or that are the result of rape or incest. Still others think that abortion should be available to any woman under any circumstances. Regardless of the moral and ethical issues that surround abortion, it is still a common method of terminating unwanted pregnancies throughout the world.

The abortion techniques used in the United States today all involve the possibility of infections, particularly if done by poorly trained personnel. The three most common techniques are scraping the inside of the uterus with special instruments (called a *D and C* or *dilation and curettage*), injecting a saline solution into the uterine cavity, or using a suction device to remove the embryo from the uterus. In the future, abortion may be accomplished by a medication prescribed by a physician. One drug, RU-486, is currently used in about 15 percent or more of the elective abortions in France. It has received approval for use in the United States. RU-486 is an antiprogestin and works by blocking progesterone receptors on cells. The medication is administered orally under the direction of a physician, and several days later, a hormone is administered. This usually results in the onset of contractions that expel the fetus. A follow-up examination of the woman is made after several weeks to ensure that there are no serious side effects of the medication.

25.12 CHANGES IN SEXUAL FUNCTION WITH AGE

Although there is a great deal of variation, at about age fifty, a woman's hormonal balance begins to change because of changes in the ovaries' production of hormones. At this time, the menstrual cycle becomes less regular and ovulation is often unpredictable. Over several years, the changes in hormone levels cause many women to experience mood swings and physical symptoms, including cramps and hot flashes. **Menopause** is the period when a woman's body becomes nonreproductive, because reproductive hormones stop being produced. This causes the ovaries to stop producing eggs, and menstruation ends. Occasionally, the symptoms associated with menopause become so severe that they interfere with normal life and the enjoyment of sexual activity. A physician might recommend *hormone replacement therapy* (*HRT*), which involves administering either estrogen alone or estrogen and progestin together, to augment the natural production of estrogen and progesterone. Normally, the sexual enjoyment of a healthy woman continues during menopause and for many years thereafter.

Although human males do not experience a relatively abrupt change in their reproductive or sexual lives, recent evidence indicates that men also experience hormonal and emotional changes similar to those seen as women go through menopause. As men age, their production of sperm declines and they may experience a variety of problems related to their sexuality. The word *impotence* is used to describe problems that interfere with sexual intercourse and reproduction. These may include a lack of sexual desire, problems with ejaculation or orgasm, and erectile dysfunction (ED). Erectile dysfunction is the recurring inability to get or keep an erection firm enough for sexual intercourse. Most incidences of ED at any age are physical, not psychological. In older men, this is usually the result of injury, disease, or the side effects of medication. Damage to nerves, arteries, smooth muscles, or other tissue associated with the penis is the most common cause of ED. Diseases linked with ED include diabetes, kidney disease, chronic alcoholism, multiple sclerosis, atherosclerosis, vascular disease, and neurologic disease. Blood pressure drugs, antihistamines, antidepressants, tranquilizers, appetite suppressants, and certain ulcer drugs have been associated with ED. Other possible causes are smoking, which reduces blood flow in veins and arteries, and lowered amounts of testosterone. ED is frequently treated with psychotherapy, behavior modification, oral or locally injected drugs, vacuum devices, and surgically implanted devices. ED is not an inevitable part of aging. Rather, sexual desires tend to wane slowly as men age. They produce fewer sperm cells and less seminal fluid. Nevertheless, healthy individuals can experience a satisfying sex life during aging.

Human sexual behavior is quite variable. The same is true of older persons. The range of responses to sexual partners continues, but generally in a diminished form. People who were very active sexually when young continue to be active, but less so, as they reach middle age. Those who were less active tend to decrease their sexual activity also. It is reasonable to state that one's sexuality continues from before birth until death.

SUMMARY

The human sex drive is a powerful motivator for many activities in our lives. Although it provides for reproduction and improvement of the gene pool, it also has a nonbiological, sociocultural dimension. Sexuality begins before birth, as sexual anatomy is determined by the sex-determining chromosome complement that we receive at fertilization. Females receive two X chromosomes. Males receive one X and one Y *sex-determining chromosome*. It is the presence and activity of the *SRY gene* that causes male development and its absence or inactivity that allows female development.

At *puberty*, hormones influence the development of secondary sex characteristics and the functioning of gonads. As the *ovaries* and *testes* begin to produce gametes, fertilization becomes possible. Sexual reproduction involves the production of gametes by meiosis in the ovaries and testes. The production and release of these gametes is controlled by the interaction of hormones. In males, each cell that undergoes *spermatogenesis* results in four sperm; in females, each cell that undergoes *oogenesis* results in one oocyte and two polar bodies. Successful sexual reproduction depends on proper hormone balance, proper meiotic division, fertilization, placenta formation, proper diet of the mother, and birth. *Hormones* regulate ovulation and menstruation and may also be used to encourage or discourage ovulation. *Fertility drugs*, *the patch*, and *birth-control pills*, for example, involve hormonal control. A number of contraceptive methods have been developed, including the diaphragm, condom, IUD, spermicidal jellies and foams, contraceptive implants, the sponge, tubal ligation, and vasectomy.

Hormones continue to direct our sexuality throughout our lives. Even after menopause, when fertilization and pregnancy are no longer possible for a female, normal sexual activity can continue in both women and men.

KEY TERMS

androgens (p. **646**)
autosomes (p. **641**)
clitoris (p. **646**)
conception (p. **643**)
cryptorchidism (p. **644**)
differentiation (p. **643**)
diploid (p. **636**)
egg (p. **636**)
estrogens (p. **646**)
follicle (p. **646**)
follicle-stimulating hormone
   (FSH) (p. **646**)
gamete (p. **636**)
gametogenesis (p. **636**)

gonad (p. **638**)
gonadotropin-releasing hormone
   (GnRH) (p. **646**)
haploid (p. **636**)
hypothalamus (p. **646**)
inguinal canal (p. **644**)
meiosis (p. **637**)
menopause (p. **661**)
menstrual cycle (p. **646**)
oogenesis (p. **649**)
orgasm (p. **646**)
ovaries (p. **638**)
oviduct (p. **649**)
ovulation (p. **646**)

penis (p. **647**)
pituitary gland (p. **646**)
placenta (p. **654**)
progesterone (p. **646**)
puberty (p. **644**)
secondary sexual characteristics
   (p. **646**)
semen (p. **648**)
seminiferous tubules (p. **647**)
sex (p. **639**)
sex-determining chromosome
   (p. **641**)
sexual reproduction (p. **636**)
sexuality (p. **639**)

sperm (p. **636**)
spermatids (p. **647**)
spermatogenesis (p. **647**)
testes (p. **638**)
testosterone (p. **646**)
uterus (p. **649**)
vagina (p. **650**)
X chromosome (p. **641**)
Y chromosome (p. **641**)
zygote (p. **636**)

APPLYING THE CONCEPTS

Answers are located in appendix F.

1. Crossing-over between segments of homologous chromosomes results in
   a. new gene combinations.
   b. zygotes.
   c. diploid cells.
   d. segregation of genes.

2. An event unique to prophase I is
   a. segregation.
   b. synapsis.
   c. reduction division.
   d. independent assortment.

3. The fact that each homologous pair of chromosomes in humans separates and moves to the poles without being influenced by the other pairs is
   a. segregation.
   b. disintegration.
   c. independent assortment.
   d. fertilization.

4. An organism having a diploid number of 12 forms gametes having
   a. 6 chromosomes.
   b. 12 chromosomes.
   c. 18 chromosomes.
   d. 24 chromosomes.

5. Gametogenesis produces
   a. sex cells.
   b. gonads.
   c. zygotes.
   d. testes.

6. Human females typically are fertile during which point of their ovarian cycle?
   a. only after sexual intercourse
   b. immediately following their period
   c. about day 13
   d. continuously

7. Which of the following sexually transmitted diseases is caused by a virus?
   a. gonorrhea
   b. scabies
   c. candidiasis
   d. genital herpes

8. Which pituitary hormone stimulates the release of LH and FSH, stimulating the development of a follicle?
   a. estrogen
   b. gonadotropin-releasing hormone (GnRH)
   c. testosterone
   d. androgen

9. Without this, the embryo would become female.
   a. FSH gene   c. SRY gene
   b. estrogen   d. x-bodies

10. Which of these is not an accessory gland of males?
    a. follicle
    b. seminal vesicle
    c. bulbourethral gland
    d. prostate gland

11. Fertilization normally occurs in the
    a. uterus.   c. oviduct.
    b. ovary.   d. vagina.

12. Which is not a hormonal conception control method?
    a. the pill
    b. condom
    c. the patch
    d. morning-after pill

13. The diploid number of a typical human is
    a. 23
    b. 46
    c. 72
    d. 47

14. When this process occurs, pieces of genetic material are exchanged between the chromosomes you received from your mother and father.
    a. segregation
    b. mutation
    c. crossing-over
    d. assortment

15. The cell formed as a result of fertilization that divides by mitosis.
    a. SRY
    b. sperm
    c. zygote
    d. oocyte

QUESTIONS FOR THOUGHT

1. How do haploid cells differ from diploid cells?
2. What is unique about metaphase I?
3. How does mitosis differ from meiosis?
4. What are advantages and disadvantages of the rhythm method of conception control?
5. Diagram fertilization as it would occur between a sperm and an egg with the haploid number of 3.
6. What is HRT?
7. What structures are associated with the human female reproductive system? What are their functions?
8. What structures are associated with the human male reproductive system? What are their functions?
9. What are the differences between oogenesis and spermatogenesis in humans?
10. List the hormones associated with the functioning of the male and female reproductive systems. Describe their function.
11. What is the difference between the origin of fraternal and identical twins?
12. List three likely causes of infertility in males and three in females.
13. What advantage does a sexually reproducing, diploid organism have in comparison to an asexually reproducing, haploid organism?
14. Describe two general types of mutations and the role they play in generating genetic diversity.
15. Why is meiosis I referred to as reduction division?

FOR FURTHER ANALYSIS

Sexual orientation is a complex concept because there is so much confusion and a lack of information among a large part of society. Some very knowledgeable professionals lead into an explanation of this topic by describing intersex conditions and the complex intersections of what many refer to as "nature and nurture." The intersex concept is one that describes one's sexual orientation as being at some point on a continuum from strongly heterosexual at one end to strongly homosexual on the other. Each person's orientation is somewhere on this continuum. Research this issue by exploring the following:

1. The biological bases of sexual orientation
2. How various cultural biases influence sexual behavior

3. The influence various religious organizations have on an individual's thinking

Lately, there has been news on circumstances surrounding surgically changing one's sex. Thinking critically about this subject, answer the following questions:

1. What genetic reasons are associated with such a decision?
2. What cultural reasons support or rebuff such a decision?
3. What would be your reaction if a relative or close personal friend were to make the decision to change their sex?

How Might Hormones in the Environment Affect Human Reproduction?

The practice of medicine has increasingly moved in the direction of using medications containing steroids to control disease or regulate function, as in birth-control pills. To meet these demands, the pharmaceutical industry produces enormous amounts of these drugs. However, what happens to these drugs once they enter the body? Although some of the drug is destroyed in controlling disease or regulating a function, such as ovulation, a certain amount is not and is excreted. There is some concern about the amount of medi1cal steroids entering the environment in this way. What effects might they have on the public, who unintentionally ingest these as environmental contaminants? Consider the topics of sexual reproduction, the regulation of hormonal cycles, and fetal development, and explain (1) how you would determine "acceptable levels" of such contaminants, (2) what might happen if these levels were exceeded, and (3) what steps might be taken to control such environmental contamination.

26

Mendelian and Molecular Genetics

Charles Darwin proposed that human facial expressions are universal. Recent and continuing research is lending support to this hypothesis. Researchers found that, in fact, facial expressions are genetically determined.

CORE **CONCEPT**

All living things survive and reproduce only as a result of maintaining molecular genetic information that is passed from one generation to the next.

OUTLINE

665

OVERVIEW

Why do you have a particular blood type or hair color? Why do some people have the same skin color as their parents, while others have a skin color different from that of their parents? Why do flowers show such a wide variety of colors? Why is it that generation after generation of plants, animals, and microbes look so much like members of their own kind? What are genetically modified organisms, and how are they produced? These questions and many others can be better answered if you have an understanding of Mendelian and molecular genetics.

PART I
MENDELIAN GENETICS UPDATED

This chapter is divided into two parts. The first presents the fundamental concepts needed to understand Mendelian genetics, and Part II covers topics related to molecular genetics.

26.1 GENETICS, MEIOSIS, AND CELLS

A **gene** is a portion of DNA that determines a characteristic. Through meiosis and reproduction, genes can be transmitted from one generation to another. The study of genes, how genes produce characteristics, and how the characteristics are inherited is the field of biology called **genetics.** The first person to systematically study inheritance and formulate laws about how characteristics are passed from one generation to the next was an Augustinian monk named Gregor Mendel (1822–1884). Mendel's work was not generally accepted until 1900, when three men, working independently, rediscovered some of the ideas that Mendel had formulated more than thirty years earlier. Because of his early work, the study of the pattern of inheritance that follows the laws formulated by Gregor Mendel is often called **Mendelian genetics.**

To understand this chapter, you need to know some basic terminology. One term that you have already encountered is *gene.* Mendel thought of a gene as a particle that could be passed from the parents to the *offspring* (children, descendants, or progeny). Today, we know that genes are actually composed of specific sequences of DNA nucleotides. The particle concept is not entirely inaccurate, because a particular gene is located at a specific place on a chromosome called its *locus* (*loci* = pl.; location). When we study genetics, we study how these particles or genes are passed from parents to their offspring or children.

Another important idea to remember is that all sexually reproducing organisms have a diploid ($2n$) stage. Since gametes are haploid (n) and most organisms are diploid, the conversion of diploid to haploid cells during meiosis is an important process.

$$2(n) \rightarrow \text{meiosis } (n) \rightarrow \text{gametes}$$

The diploid cells have two sets of chromosomes—one set inherited from each parent.

$$n + n \text{ gametes} \rightarrow \text{fertilization} \rightarrow 2n$$

Therefore, each individual has two chromosomes of each kind and two genes for each characteristic (refer to figure 25.1), one from the mother and one from the father. When sex cells are produced by meiosis, reduction division occurs, and the diploid number is reduced to haploid. Therefore, the sex cells produced by meiosis have only one chromosome of each of the homologous pairs that were in the diploid cell that began meiosis. Diploid organisms usually result from the fertilization of a haploid egg by a haploid sperm. Therefore, they inherit one gene of each type from each parent. For example, each of us has two genes for earlobe shape: one came with our father's sperm, the other with our mother's egg (figure 26.1).

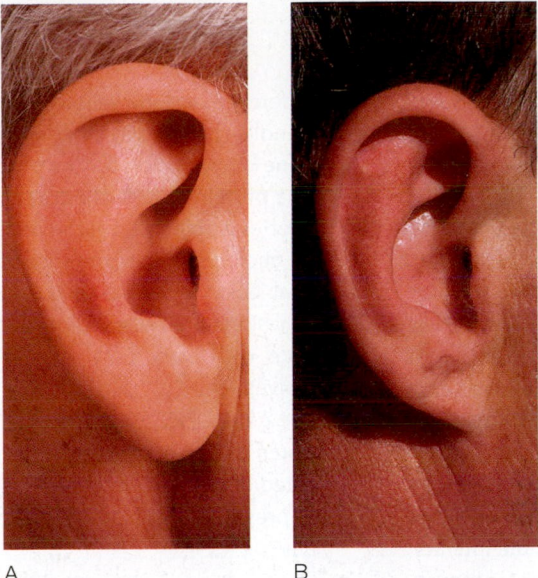

A   B

FIGURE 26.1 Whether your earlobe is free (*A*) or attached (*B*) depends on the genes you have inherited. As genes express themselves, their actions affect the development of various tissues and organs. Some people's earlobes do not separate from the sides of their heads in the same manner as do those of others. How genes control this complex growth pattern and why certain genes function differently from others is yet to be clarified. How much variation do you see among fellow class members?

26.2 SINGLE-GENE INHERITANCE PATTERNS

In diploid organisms, there may be two different forms of the gene. In fact, there may be *several* alternative forms or **alleles** of each gene within a population. The word *gene* is used to refer to the genetic material in general, whereas the term *allele* more specifically refers to the alternative forms of the genetic material for a certain characteristic. In people, for example, there are two alleles for earlobe shape. One allele produces an earlobe that is fleshy and hangs free, while the other allele produces a lobe that is attached to the side of the face and does not hang free. The type of earlobe that is present is determined by the type of allele (gene) received from each parent and the way in which these alleles interact with one another. Alleles are located on the same pair of homologous chromosomes—one allele on each chromosome. These alleles are also at the same specific location, or locus (figure 26.2).

The **genome** is a set of all the genes necessary to specify an organism's complete list of characteristics. The term *genome* is used in two ways. It may refer to the diploid (2*n*) or haploid (*n*) number of chromosomes in a cell. Be sure to clarify how your instructor uses this term. The **genotype** of an organism is a listing of the genes present in that organism. It consists of the cell's DNA code; therefore, you cannot see the genotype of an organism. It is not yet possible to know the complete genotype of most organisms, but it is often possible to figure out the genes present that determine a particular characteristic. For example, there are three possible genotypic combinations of the two alleles for earlobe shape. Genotypes are typically represented by upper- and lowercase letters. In the case of the earlobe trait, the allele for free earlobes is designated "*E*" while that for attached earlobes

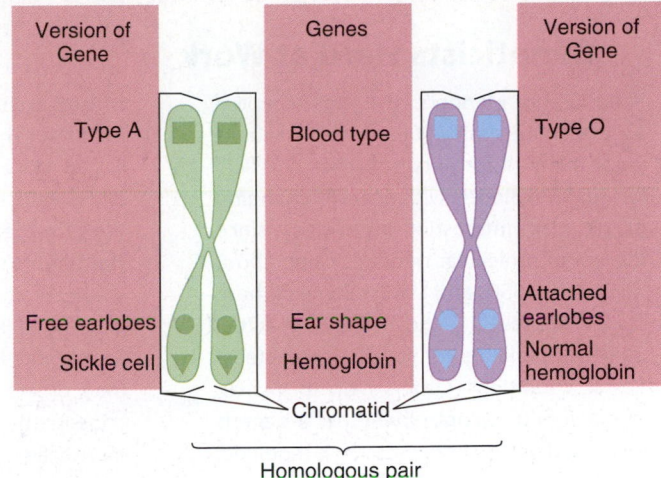

Homologous pair

FIGURE 26.2 Homologous chromosomes contain genes for the same characteristics at the same place. Note that the attached-earlobe allele is located at the ear-shape locus on one chromosome and the free-earlobe allele is located at the ear-shape locus on the other member of the homologous pair of chromosomes. The other two genes are for hemoglobin structure (alleles for normal and sickled cells) and blood type (alleles for blood types A and O).The examples presented here are for illustrative purposes only. We do not really know if these particular genes are on these chromosomes.

is "*e*." A person's genotype could be (1) two alleles for attached earlobes (*ee*), (2) one allele for attached earlobes and one allele for free earlobes (*Ee*), or (3) two alleles for free earlobes (*EE*).

How would individuals with each of these three genotypes look? The observable characteristics of an organism are known as its **phenotype** and are determined by the ways in which each combination of alleles expresses (shows) itself. A gene *expresses* itself by producing product molecules (proteins) that control and direct chemical activities or become structural components of the organism. The way combinations of alleles express themselves is also influenced by the environment of the organism. The phrase *gene expression* refers to the degree to which a gene goes through protein synthesis to show itself as a physical feature of the individual.

A person with two alleles for attached earlobes will have earlobes that do not hang free. A person with one allele for attached earlobes and one allele for free earlobes will have a phenotype that exhibits free earlobes. An individual with two alleles for free earlobes will also have free earlobes. Notice that there are three genotypes but only two phenotypes. The individuals with the free-earlobe phenotype can have different genotypes.

Alleles	Genotypes	Phenotypes
E = free earlobes	*EE*	free earlobes
e = attached earlobes	*Ee*	free earlobes
	ee	attached earlobes

The expression of some alleles is directly influenced by the presence of other alleles of the same gene. For any particular pair of alleles in an individual, the two alleles from the two parents are either identical or not identical. A person is **homozygous** for a trait when they have the combination of two identical alleles for that particular characteristic, for example, *EE* and *ee*.

A Closer Look

Geneticists Hard at Work

Since Gregor Mendel's work was accepted as "law" in the early 1900s, geneticists have made many important discoveries. This field has really exploded with new life-changing or just plain interesting information since the era of molecular genetics came about during the 1950s and 1960s. Some of these discoveries revealed the existence of actual genes responsible for specific characteristics or conditions. Others help to explain the factors that control whether a gene is expressed or how its expression is modified.

Here are just a few recent revelations from the scientists working in the field of genetics:

- Certain soil bacteria have been discovered that have genes that allow them to feed exclusively on antibiotics. This is of concern because these bacteria live in close association with human and livestock pathogens.

- While the genetic abnormality causing Huntington's disease causes neurons in the brain to be destroyed, it also plays a role in destroying cancer cells. People with Huntington's are less likely than others to suffer from cancer. It appears that the *huntingtin* gene has more than one effect.

- The inheritance of "dominant black" coat color in domestic dogs involves a gene that is distinct from, but interacts with, the genes responsible for conventional coat pigmentation. Variations in this gene are responsible for the color differences among yellow-, black-, and brindle-colored dog breeds. This same gene is responsible for the production of a protein (ß-defensin) that in other species is able to aid in the destruction of microbes. The presence of black coat color in wolves is the result of occasional interbreeding between dogs with black coat color and grey wolves.

- The gene DISC1 (Disrupted-in-Schizophrenia 1) has been strongly implicated in cases of schizophrenia, major depression, bipolar disorder, and autism.

A person with two alleles for freckles is said to be homozygous for that trait. A person with two alleles for no freckles is also homozygous. If an organism is homozygous, the characteristic expresses itself in a specific manner. A person homozygous for free earlobes has free earlobes, and a person homozygous for attached earlobes has attached earlobes.

An individual is designated as **heterozygous** when they have two different allelic forms of a particular gene, for example, *Ee.* The heterozygous individual received one allelic form of the gene from one parent and a different allele from the other parent. For instance, a person with one allele for freckles and one allele for no freckles is heterozygous. If an organism is heterozygous, these two different alleles interact to determine a characteristic. The phenotype produced by the alleles can vary, depending on how the two different alleles interact.

TABLE 26.1

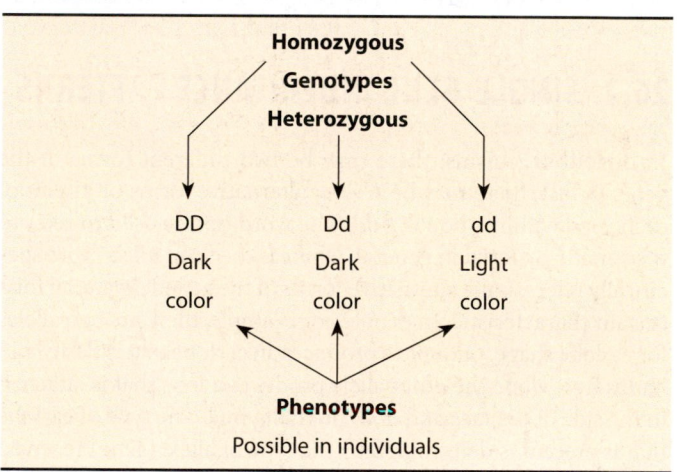

26.3 A SIMPLE MODEL OF INHERITANCE— DOMINANT AND RECESSIVE ALLELES

Phenotypes are determined by the type of allele (gene) received from each parent and the way in which these alleles interact with one another. The simplest rule to use to predict phenotype from genotype describes when one allele in the pair expresses itself and hides the other. This is an inheritance pattern called *dominance* and *recessiveness.* A **dominant allele** (denoted by a capital letter, such as *E* in the earlobe example) masks the effect of other alleles for the trait (table 26.1). For example, if a person has one allele for free earlobes and one allele for attached earlobes, that person has a phenotype of free earlobes. We say the allele for free earlobes is dominant. A **recessive allele** (denoted by a lowercase letter, here *e*) is one that, when present with another allele, has its actions overshadowed by the other; it is masked by the effect of the other allele. For example, if a person is heterozygous (has one allele for free earlobes and one allele for attached earlobes, *Ee*), the person will have a phenotype of free earlobes. The allele for free earlobes is dominant. While *EE* or *Ee* may show the free earlobe phenotype, recessive phenotypes are only observable when two recessive alleles (*ee*) are present.

Many people have some common misconceptions about recessive alleles. One is that recessive alleles are all bad or harmful. Don't think that recessive genes are necessarily bad. *The term recessive has nothing to do with the significance or value of the gene—it simply describes how it can be expressed. Recessive alleles are not less likely to be inherited but must be present in a homozygous condition to express themselves. This may make recessive alleles less likely to be observed. A second misconception is that recessive alleles occur less frequently in a population.*

This is not always the case. The number of fingers a person has is determined by a gene. The recessive allele for this gene is five fingers. The dominant allele is six. Clearly, the five-finger allele is more frequent in our population because we see many more five-fingered individuals than individuals with six fingers.

26.4 MENDEL'S LAWS OF HEREDITY

Heredity problems are concerned with determining which alleles are passed from the parents to the offspring and how likely it is that various types of offspring will be produced.

The Augustinian monk Gregor Mendel developed a method of predicting the outcome of inheritance. He performed experiments that showed the inheritance patterns of certain characteristics in common garden pea plants (*Pisum sativum*). From his work, Mendel concluded which traits were dominant and which were recessive. Some of his results are shown in table 26.2.

What made Mendel's work unique was that he studied only one trait at a time. Previous investigations had tried to follow numerous traits at the same time. When this was attempted, the total set of characteristics was so cumbersome to work with that no clear idea could be formed of how the offspring inherited traits. Mendel used traits with clear-cut alternatives, such as purple or white flower color, yellow or green seed pods, and tall or dwarf pea plants. He was very lucky to have chosen pea plants in his study because they naturally self-pollinate. When self-pollination occurs in pea plants over many generations, it is possible to develop a population of plants that is homozygous for a number of characteristics. Such a population is known as a *pure line.*

Mendel took a pure line of pea plants having purple flower color, removed the male parts (anthers), and discarded them so that the plants could not self-pollinate. He then took anthers from a pure-breeding white-flowered plant and pollinated the antherless purple flowers.

When the pollinated flowers produced seeds, Mendel collected, labeled, and planted them.

When these seeds germinated and grew, they eventually produced flowers.

You might be surprised to learn that all of the plants resulting from this cross had purple flowers. One of the prevailing hypotheses of Mendel's day would have predicted that the purple and white colors would have blended, resulting in flowers that were lighter than the parental purple flowers. Another hypothesis would have predicted that the offspring would have had a mixture of white and purple flowers. The unexpected result—all of the offspring produced flowers like those of one parent and no flowers like those of the other—caused Mendel to examine other traits as well and formed the basis for much of the rest of his work. He repeated his experiments using pure strains for other traits. Pure-breeding tall plants were crossed with pure-breeding dwarf plants. Pure-breeding plants with yellow pods were crossed with pure-breeding plants with green pods. The results were all the same: the offspring showed the characteristic of one parent and not the other.

Next, Mendel crossed the offspring of the white-purple cross (all of which had purple flowers) with each other to see what the third generation would be like. Had the characteristic of the original white-flowered parent been lost completely? This second-generation cross was made by pollinating these purple flowers that had one white parent among themselves. The seeds produced from this cross were collected and grown.

When these plants flowered, three-fourths of them produced purple flowers and one-fourth produced white flowers.

After analyzing his data, Mendel formulated several genetic laws to describe how characteristics are passed from one generation to the next and how they are expressed in an individual.

Mendel's **law of dominance**—When an organism has two different alleles for a given trait, the allele that is expressed, overshadowing the expression of the other allele, is said to be dominant. The gene whose expression is overshadowed is said to be recessive.

Mendel's **law of segregation**—When gametes are formed by a diploid organism, the alleles that control a trait separate from one another into different gametes, retaining their individuality. Parental alleles of a gene separate at meiosis into gametes:

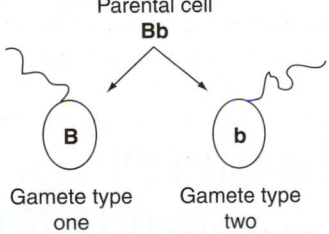

Parental cell
Bb

B   b

Gamete type   Gamete type
one   two

Mendel's **law of independent assortment**—Members of one *pair of alleles* separate from each other independently of the members of the other pair of alleles. In other words, if two different genes are separate chromosomes, where the one allele goes is not influenced by where the other goes.

At the time of Mendel's research, biologists knew nothing of chromosomes or DNA or of the processes of mitosis and meiosis. Mendel assumed that each gene was separate from other genes. It was fortunate for him that most of the characteristics he picked to study were found on separate chromosomes. If two or more of these genes had been located on the same chromosome (*linked genes*), he probably would not have been able to formulate his laws. The discovery of chromosomes and DNA has led to modifications in Mendel's laws, but it was Mendel's work that formed the foundation for the science of genetics.

TABLE 26.2 **Dominant and Recessive Traits in Pea Plants**

Characteristic	Dominant Allele	Recessive Allele
Plant height	Tall	Dwarf
Pod shape	Full	Constricted
Pod color	Green	Yellow
Seed surface	Round	Wrinkled
Seed color	Yellow	Green
Flower color	Purple	White

People Behind the Science

Gregor Johann Mendel (1822–1884)

Gregor Johann Mendel was an Austrian monk who discovered the basic laws of heredity, thereby laying the foundation of modern genetics—although the importance of his work was not recognized until after his death.

Mendel was born Johann Mendel on July 22, 1822, in Heinzendorf, Austria, the son of a peasant farmer. He studied for two years at the Philosophical Institute in Olomouc, after which, in 1843, he entered the Augustinian monastery in Brünn, Moravia, taking the name Gregor. In 1847, he was ordained a priest. During his religious training, Mendel taught himself a certain amount of science. In 1850, he tried to pass an examination to obtain a teaching license but failed, and in 1851, he was sent by his abbot to the University of Vienna to study physics, chemistry, mathematics, zoology, and botany. Mendel left the university in 1853 and returned to the monastery in Brünn in 1854. He then taught natural science in the local Technical High School until 1868, during which period he again tried, and failed, to gain a teaching certificate that would have enabled him to teach in more advanced institutions. It was also in the period 1854 to 1868 that Mendel performed most of his scientific work on heredity. He was elected abbot of his monastery at Brünn in 1868, and the administrative duties involved left him little time for further scientific investigations. Mendel remained abbot at Brünn until his death on January 6, 1884.

Mendel began the experiments that led to his discovery of the basic laws of heredity in 1856. Much of his work was performed on the edible pea (*Pisum* sp.), which he grew in the monastery garden. He carefully self-pollinated and wrapped (to prevent accidental pollination by insects) each individual plant, collected the seeds produced by the plants, and studied the offspring of these seeds. He found that dwarf plants produced only

Box Figure 26.1

dwarf offspring and that the seeds produced by this second generation also produced only dwarf offspring. With tall plants, however, he found that both tall and dwarf offspring were produced and that only about one-third of the tall plants bred true, from which he concluded that there were two types of tall plants, those that bred true and those that did not. Next, he cross-bred dwarf plants with true-breeding tall plants, planted the resulting seeds, and then self-pollinated each plant from this second generation. He found that all the offspring in the first generation were tall but that the offspring from the self-pollination of this first generation were a mixture of about one-quarter true-breeding dwarf plants, one-quarter true-breeding tall plants, and one-half nontrue-breeding tall plants. Mendel also studied other characteristics in pea plants, such as flower color, seed shape, and flower position, finding that, as with height, simple laws governed the inheritance of these traits. From his findings, Mendel concluded that each parent plant contributes a factor that determines a particular trait and that the pairs of factors in the offspring do not give rise to a merger of traits. These conclusions, in turn, led him to formulate his famous law of segregation and law of independent assortment of characters, which are now recognized as two of the fundamental laws of heredity.

Mendel reported his findings to the Brünn Society for the Study of Natural Science in 1865, and in the following year, he published *Experiments with Plant Hybrids,* a paper that summarized his results. But the importance of his work was not recognized at the time. It was not until 1900, when his work was rediscovered by Hugo De Vries, Carl Erich Correns, and Erich Tschermak von Seysenegg, that Mendel achieved fame—sixteen years after his death.

Source: Modified from the *Hutchinson Dictionary of Scientific Biography.* © RM, 2011. All rights reserved. Helicon Publishing is a division of RM.

26.5 STEPS IN SOLVING HEREDITY PROBLEMS: SINGLE-FACTOR CROSSES

People have long been interested in understanding why the offspring of plants and animals resemble or don't resemble their parents. Mendel's work helped formulate a way to answer such questions. The first type of problem we will consider is the easiest type, a single-factor cross. A **single-factor cross** (sometimes called a monohybrid cross: *mono* = one; *hybrid* = combination) is a genetic cross or mating in which a single characteristic is followed from one generation to the next.

For example, in humans, the allele for Tourette syndrome (TS) is inherited as a dominant allele. For centuries, people displaying this genetic disorder were thought to be possessed by the devil since they displayed such unusual behaviors. These motor and verbal behaviors or tics are involuntary and range from mild (e.g., leg tapping, eye blinking, face twitching) to the more violent forms such as the shouting of profanities, head

jerking, spitting, compulsive repetition of words, or even barking like a dog. The symptoms result from an excess production of the brain messenger, dopamine.

A Single-Factor Problem: If both parents are heterozygous (have one allele for Tourette syndrome and one allele for no Tourette syndrome), what is the probability that they can have a child without Tourette syndrome? With Tourette syndrome?

Step 1: Assign a symbol for each allele.

Usually a capital letter is used for a dominant allele and a small letter for a recessive allele. Use the symbol *T* for Tourette and *t* for no Tourette.

Allele	Genotype	Phenotype
T = Tourette	*TT*	Tourette syndrome
t = normal	*Tt*	Tourette syndrome
	tt	Normal

Muscular Dystrophy and Genetics

Because most genes are comprised of thousands of nucleotide base pairs, there are many different changes in these sequences that can result in the formation of multiple "bad" gene products and, therefore, an abnormal phenotype. The same gene can have many different "bad" forms. The fact that a phenotypic characteristic can be determined by many different alleles for a particular characteristic is called *genetic heterogeneity.* For example, two of the best-known forms of *muscular dystrophy* (*MD*) are Duchenne's and Becker's. Duchenne's (DMD) is characterized by a severe, progressive weakening of the muscles, the appearance of a false muscle atrophy (degeneration) in the calves of the legs, onset in early childhood, and a high likelihood of death in the thirties. DMD is caused by a mutation in the dystrophin gene located on the X chromosome and is a recessive trait. Becker's (BMD) is caused by a mutation in the same dystrophin gene but at a different location. BMD is a milder form of MD.

Step 2: Determine the genotype of each parent and indicate a mating.

Because both parents are heterozygous, the male genotype is Tt. The female genotype is also Tt. The × between them is used to indicate a mating.

$$Tt \times Tt$$

Step 3: Determine all the possible kinds of gametes each parent can produce.

Remember that gametes are haploid; therefore, they can have only one allele instead of the two present in the diploid cell. Because the male has both the Tourette syndrome allele and the normal allele, half of his gametes will contain the Tourette syndrome allele and the other half will contain the normal allele. Because the female has the same genotype, her gametes will be the same as his.

For genetic problems, a *Punnett square* is used. A **Punnett square** is a box figure that allows you to determine the probability of genotypes and phenotypes of the progeny of a particular cross. Remember, because of the process of meiosis, each gamete receives only one allele for each characteristic listed. Therefore, the male will produce sperm with either a T or a t; the female will produce ova with either a T or a t. The possible gametes produced by the male parent are listed on the left side of the square and the female gametes are listed on the top. In our example, the Punnett square would show a single dominant allele and a single recessive allele from the male on the left side. The alleles from the female would appear on the top.

Step 4: Determine all the gene combinations that can result when these gametes unite.

To determine the possible combinations of alleles that could occur as a result of this mating, simply fill in each of the empty squares with the alleles that can be donated from each parent. Determine all the gene combinations that can result when these gametes unite.

	T	t
T	TT	Tt
t	Tt	tt

Step 5: Determine the phenotype of each possible gene combination.

In this problem, three of the offspring, TT, Tt, and Tt, have Tourette syndrome. One progeny, tt, is normal. Therefore, the answer to the problem is that the probability of having offspring with Tourette syndrome is 3/4; for no Tourette syndrome, it is 1/4.

Take the time to learn these five steps. All single-factor problems can be solved using this method; the only variation in the problems will be the types of alleles and the number of possible types of gametes the parents can produce.

Female genotype
Tt

Male genotype
Tt

Possible female gametes
T and t

Possible male gametes
T and t

	T	t
T		
t		

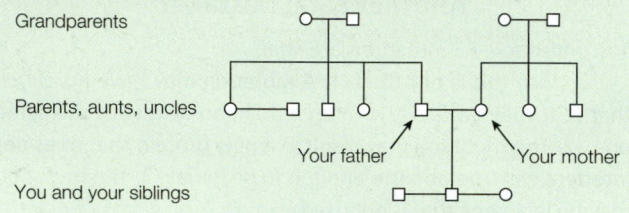
26.6 MORE COMPLEX MODELS OF INHERITANCE

So far, we have considered phenotypes that are determined by dominance and recessiveness between two alleles. In fact, dominant/recessive inheritance is probably the least common inheritance model since most inheritance patterns do not fit this simple pattern. In most situations, phenotypes are not only determined by the type of allele (gene) received but also by the alternative ways in which they interact with each other and how the environment influences their expression. There are six generally recognized patterns: X-linked characteristics, codominance, incomplete dominance, multiple alleles, polygenic inheritance, and pleiotropy.

X-Linked Genes

Alleles located on the same chromosome tend to be inherited together. They are said to be linked. The process of crossing-over, which occurs during prophase I of meiosis I, may split up these linkage groups. Crossing-over happens between homologous chromosomes donated by the mother and the father, and results in a mixing of genes. The closer two genes are to each other on a chromosome, the more probable it is that they will be inherited together.

People and many other organisms have two types of chromosomes. Autosomes (twenty-two pairs) are not involved in sex determination and have alleles for the same traits on both members of the homologous pair of chromosomes. *Sex chromosomes* are a pair of chromosomes that control the sex of an organism. In humans and some other animals, there are two types of sex chromosomes—the X chromosome and the Y chromosome. The Y chromosome is much shorter than the X chromosome and has few genes for traits found on the X chromosome in humans. One genetic trait that is located on the Y chromosome contains the testis-determining gene—SRY. Females are normally produced when two X chromosomes are present because they do not have the SRY gene. Males are usually produced when one X chromosome and one Y chromosome are present.

Genes found together on the X chromosomes are said to be **X-linked.** Because the Y chromosome is shorter than the X chromosome, it does not have many of the alleles that are found on the comparable portion of the X chromosome. Therefore, in a man, the presence of a single allele on his only X chromosome will be expressed, regardless of whether it is dominant or recessive because there is no corresponding allele on the Y chromosome. A Y-linked trait in humans is the SRY gene. This gene controls the differentiation of the embryonic gonad to a male testis. By contrast, more than one hundred genes are on the X chromosome. Some of these X-linked genes can result in abnormal traits such as *color deficiency* (formerly referred to as color blindness), *hemophilia, brown teeth,* and at least two forms of *muscular dystrophy* (Becker's and Duchenne's).

Codominance

In some inheritance situations, alleles are neither dominant nor recessive to one another, and both express themselves in the heterozygous individual. This inheritance pattern is called codominance. In a case of codominance, the phenotype of each allele is expressed in the heterozygous condition. Consequently, a person with the heterozygous genotype can have a phenotype very different from either of their homozygous parents. For codominant alleles, all uppercase symbols are used and superscripts are added to represent the different alleles. The uppercase letters call attention to the fact that each allele can be detected phenotypically to some degree even when in the presence of its alternative allele. For example, the coat colors (C) of horses, cattle, and other animals may be phenotypically roan. The roan gene adds white hairs among the other body hairs, whatever their color. The roan gene can be applied to any color. The most common base roan color is red. The color is the result of individual hairs that are two different colors. A red roan would have both red and white hairs intermingled. Animals that are true red would have the genotype $C^R C^R$, red roan would have $C^R C^W$, and true white would have $C^W C^W$ (figure 26.3).

Incomplete Dominance

In the inheritance pattern known as partial or **incomplete dominance**, alleles lack dominant relationships. The results of a cross are heterozygous offspring that have an intermediate phenotype that is not distinct from either homozygous parent.

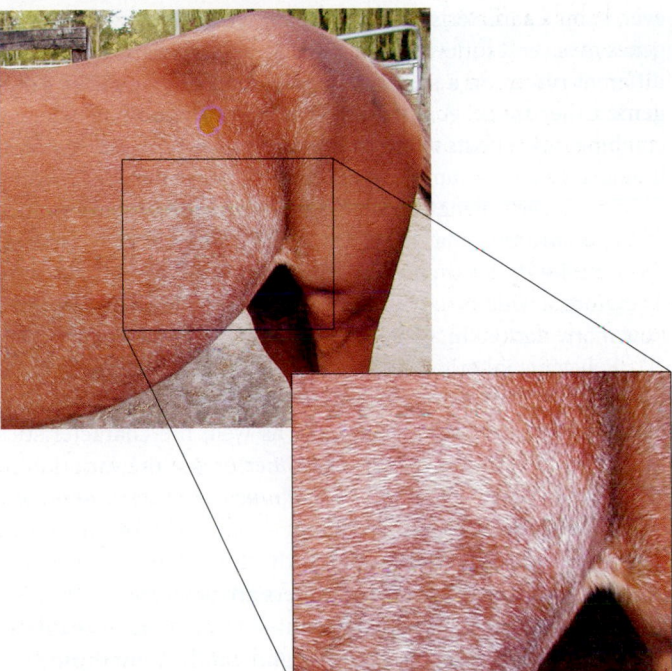

FIGURE 26.3 The color of this horse, an Arab, displays the codominant color pattern called roan. Notice that there are places on the body where both white and red hairs are found side by side.

Another example of incomplete dominance occurs in certain colors of horses. Palomino and buckskin horses show a cream-color allele for the basic color gene (C^{cr}). In these horses, the C^{cr} allele dilutes the hair pigment to yellow. In the homozygous condition ($C^{cr}C^{cr}$), the horse is called cremello; both black and red pigments are intermingled in a single hair to produce yellow.

A classic example of incomplete dominance in plants involves the color of the petals of snapdragons. The gene products interact to produce a blended result. There are two alleles for the color of these flowers. Because neither allele is recessive, we cannot use the traditional capital and small letters as symbols for these alleles. Instead, the allele for white petals is given the symbol F^w, and the one for red petals is given the symbol F^R (figure 26.4).

There are three possible combinations of these two alleles:

Genotype	Phenotype
F^wF^w	White flower
F^RF^R	Red flower
F^RF^w	Pink flower

Notice that there are only two different alleles, red and white, but there are three phenotypes, red, white, and pink. Both the red-flower allele and the white-flower allele partially express themselves when both are present, and this results in pink.

Multiple Alleles

So far, we have discussed only traits that are determined by two alleles, for example, *A, a*. However, there can be more than two different alleles for a single trait. All the various forms of the same gene (alleles) that control a particular trait are referred to as

Blame That Trait on Your Mother!

Within a eukaryotic cell, the bulk of DNA is located in the nucleus. It is this genetic material that controls the majority of biochemical processes of the cell. It has long been recognized that the mitochondria also contain DNA, mtDNA. This genetic material works in conjunction with that located in the nucleus. However, an interesting thing happens when a sperm is formed as a result of meiosis. All the mitochondria are packed into the "necks" and not the "heads" of the sperm. When a sperm penetrates a secondary oocyte, the head enters but the majority of the mitochondria in the neck remain outside. Therefore, fathers rarely, if ever, contribute mitochondrial genetic information to their children. Should a mutation occur in the mother's mitochondrial DNA, she will pass the abnormality on to her children. Should a mutation occur in the father's mitochondrial DNA, he will not transmit the abnormality. This unusual method of transmission is called *mitochondrial* (or *maternal*) *inheritance*. Most abnormalities transmitted through mitochondrial genes are associated with muscular weakness because the mitochondria are the major source of ATP in eukaryotic cells. People with one form of this rare type of genetic abnormality, *Leber's hereditary optic neuropathy,* show symptoms of sudden loss of central vision due to optic nerve death in young adults with onset at about age twenty. Both males and females may be affected.

A F^RF^R B F^WF^W

C F^WF^R

FIGURE 26.4 The colors of these snapdragons are determined by two alleles for petal color, F^W and F^R. There are three different phenotypes because of the way in which the alleles interact with one another. In the heterozygous condition, neither of the alleles dominates the other.

multiple alleles. However, one person can have only a maximum of two of the alleles for the characteristic. A good example of a characteristic that is determined by multiple alleles is the ABO blood type. There are three alleles for blood type:

Allele*

I^A = blood has type A antigens on red blood cell surface

I^B = blood has type B antigens on red blood cell surface

i = blood type O has neither type A nor type B antigens on red blood cell surface

*The symbols I^A and I^B stand for the technical term for the antigenic carbohydrates attached to red blood cells, the *Immunogens*. These alleles are located on human chromosome 9. The ABO system is not the only one used to type blood. Others include the Rh, MNS, and Xg systems.

Note in the midst of this example of multiple alleles that in the ABO system, A and B show *codominance* when they are together in the same individual. However, both A and B are dominant over the O allele. These three alleles can be combined as pairs in six different ways, resulting in four different phenotypes:

Genotype		Phenotype
$I^A I^A$	=	Blood type A
$I^A i$	=	Blood type A
$I^B I^B$	=	Blood type B
$I^B i$	=	Blood type B
$I^A I^B$	=	Blood type AB
ii	=	Blood type O

Multiple-allele problems are worked as single-factor problems. Some examples are in the practice problems at the end of this chapter.

Polygenic Inheritance

Thus far, we have considered phenotypic characteristics that are determined by alleles at a single place and chromosome. How-ever, some characteristics are determined by the interaction of genes at several different loci (on different chromosomes or at different places on a single chromosome). This is called **polygenic inheritance.** A number of different pairs of alleles may combine their efforts to determine a characteristic. Skin color in humans is a good example of this inheritance pattern. According to some experts, genes for skin color are located at a minimum of three chromosomal locations or loci. At each of these loci, the allele for dark skin is dominant over the allele for light skin. Therefore, a wide variety of skin colors is possible depending on how many dark-skin alleles are present (figure 26.5).

Polygenic inheritance is very common in determining characteristics that appear on a gradient in nature. In the skin-color example, and in many other traits as well, the characteristics cannot be categorized in terms of *either/or,* but the variation in phenotypes can be classified as *how much* or *by what amount.*

For instance, people show great variations in height. There are not just tall and short people—there is a wide range. Some people are as short as 1 m, and others are taller than 2 m. This quantitative trait is probably determined by a number of different genes. Intelligence also varies significantly, from those who are severely retarded to those who are geniuses. Many of these traits may be influenced by outside environmental factors such as diet, disease, accidents, and social factors. These are just a few examples of polygenic inheritance patterns.

Pleiotropy

Even though a single gene produces only one product, it often has a variety of effects on the phenotype of the person. This is called *pleiotropy.* The term **pleiotropy** (*pleio-* =changeable) describes the multiple effects a single gene has on a phenotype. A good example of pleiotropy is *Marfan syndrome* (figure 26.6), a disease suspected to have occurred in U.S. President Abraham Lincoln. Marfan is a disorder of the body's connective tissue but can also have effects in many other organs, includ-

Locus 1	d^1d^1	d^1D^1	d^1D^1	D^1D^1	D^1d^1	D^1d^1	D^1D^1
Locus 2	d^2d^2	d^2d^2	d^2D^2	D^2d^2	D^2d^2	D^2D^2	D^2D^2
Locus 3	d^3d^3	d^3d^3	d^3d^3	d^3d^3	D^3D^3	D^3D^3	D^3D^3
Total number of dark-skin genes	0	1	2	3	4	5	6

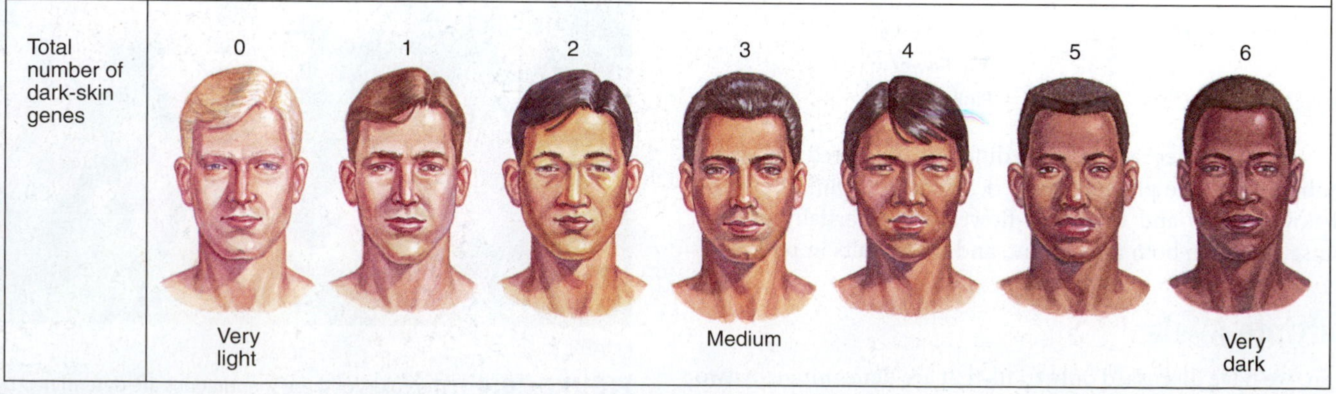

Very light — Medium — Very dark

FIGURE 26.5 Skin color in humans is an example of polygenic inheritance. There are several different genes for skin color located on different chromosomes, each with dark and light alleles. The total number of dark "D" alleles present have an additive effect on skin color. The top portion of the figure shows examples of genotypes that can produce the different skin colors. The number of dark "D" alleles is more important than how the "D" alleles are distributed in the different genes.

Connections . . .

Your Skin Color, Gene Frequency Changes, and Natural Selection

For centuries we have classified humans into "races" based on superficial traits. One of the most obvious is skin color. Fill out any survey and you will probably find a category asking you to identify yourself according to race. Skin color is almost always what comes to mind when you decide if you are Caucasian, African American, Hispanic, or mixed. But are there genes involved in this trait? When did humans develop different skin colors? What factors may have led to and stabilized the existence of different-colored groups of people?

Yes, skin color is regulated by your genes. In fact, several genes are involved in the polygenic inheritance of pigment production. Scientists had thought that beginning about 40,000 years ago, modern humans in Europe began to grow paler as they migrated farther north. They hypothesized that pale skin allows more sunlight to penetrate the skin. This allowed more UV light to stimulate the production of vitamin D, used in bone growth and many other essential pathways. One of the genes that apparently causes pale skin, known as SLC24A5, has been identified in many Europeans (but not in Asians). There are two forms of this gene that control the production of proteins used for skin pigmentation. The proteins produced by these two alleles differ by only one amino acid. Almost all Africans and East Asians carry the "dark" form of the gene, whereas 98 percent of Europeans have the other—the "pale" gene. Careful analysis of DNA now suggests that the source of the pale gene was a mutation and that it increased in frequency in European populations more recently than previously thought, most likely between 6,000 and 12,000 years ago.

---

ing the eyes, heart, blood, skeleton, and lungs. Symptoms generally appear as a tall, lanky body with long arms and spiderfingers, scoliosis (curvature of the spine), and depression of the chest wall. Many people with Marfan syndrome are nearsighted because they have a dislocation of the lens of the eye. The white of the eye (sclera) may appear bluish. Heart problems include dilation of the aorta and abnormalities of the heart valves.

Death may be caused by a dissection (tear) in the aorta from the rupture in a weakened and dilated area of the aorta called an aortic *aneurysm*. Thus, the alleles that produce Marfan syndrome have multiple effects on height, eyesight, and other traits.

Another example of pleiotropy is the genetic abnormality PKU (phenylketonuria). In this example, a single gene affects many different chemical reactions that depend on the way a cell

A

B

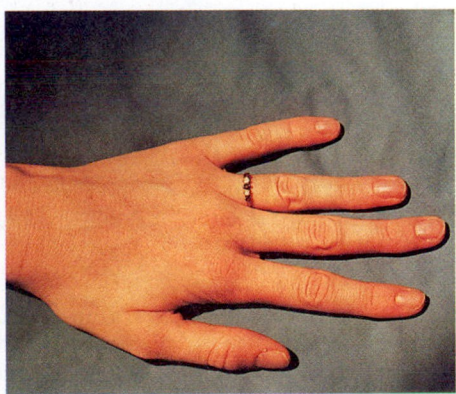

C

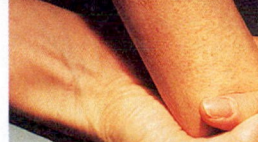
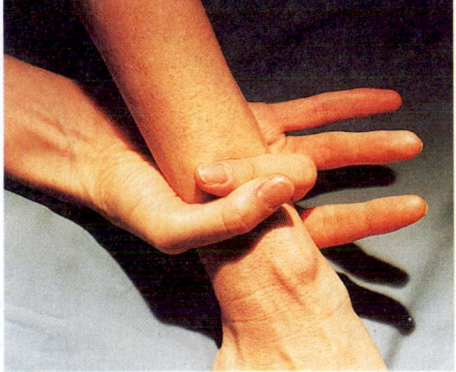

D

FIGURE 26.6 It is estimated that about 40,000 (1 out of 10,000) people in the United States have this autosomal dominant abnormality. Notice the common lanky appearance to the body and face of (A) this person with Marfan syndrome and (B) former U.S. President Abraham Lincoln. Photos (C) and (D) illustrate their unusually long fingers.

Connections . . .

Genes, Development, and Evolution

One of the important discoveries of modern molecular genetics is the remarkable similarity in the kinds of genes found in all organisms. This has important implications for understanding the evolution of organisms. It appears that once a new, valuable gene is created through the process of mutation, it is preserved in evolutionary descendants. One example is a group of genes known as *homeotic genes*. These genes regulate how an organism's body is formed by helping to define which end of the developing embryo is the head and which is the tail. As the embryo develops and regular body segments form, the homeotic genes also help define what each segment becomes. In insects, one segment might give rise to

antennae, while another gives rise to wings or legs. Homeotic genes were first discovered in the fruit fly (*Drosophila melanogaster*), which has been a favorite species for students of animal genetics for a hundred years. Fruit flies are ideal for genetic studies for several reasons: they are easy and inexpensive to raise in the lab, a new generation can be produced every ten days, and large numbers of offspring are produced.

It is now known that homeotic genes control the same developmental processes in all organisms that are bilaterally symmetrical (their left side mirrors their right side). This trend is so overwhelming that some scientists have suggested that the presence of one type of homeotic genes,

the *Hox* genes, should be used to define the Animal kingdom.

Essentially the same genes with the same functions can be found in widely different animals, such as fruit flies, earthworms, sea urchins, tapeworms, and humans. This means that the study of fruit flies can be used to discover how the same genes function in humans and other animals. Because homeotic genes are involved in regulating embryonic development and cellular differentiation, these studies can be used to help identify the causes of human embryonic developmental abnormalities and other diseases like cancer.

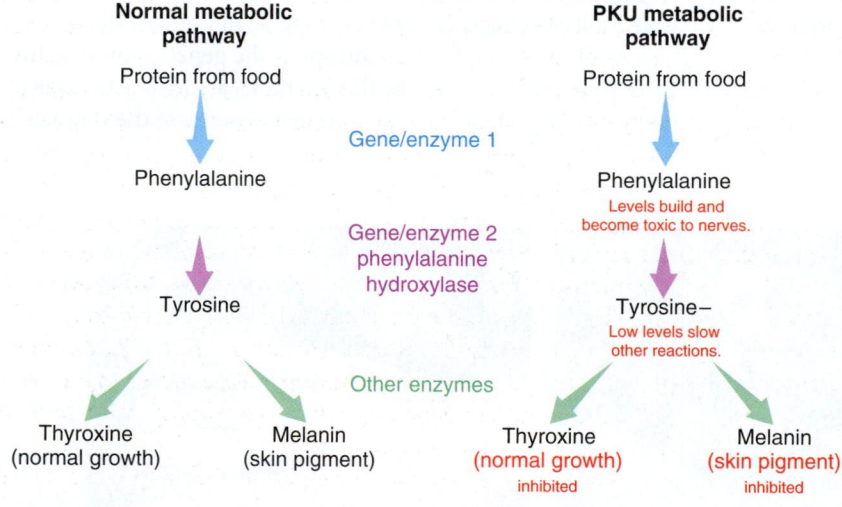

FIGURE 26.7 PKU is a recessive disorder located on chromosome 12. The diagram on the left shows how the normal pathway works. The diagram on the right shows an abnormal pathway. If the enzyme phenylalanine hydroxylase is not produced because of a mutated gene, the amino acid phenylalanine cannot be converted to tyrosine and is converted into phenylpyruvic acid, which accumulates in body fluids. The buildup of phenylpyruvic acid causes the death of nerve cells and ultimately results in mental retardation. Because phenylalanine is not converted to tyrosine, subsequent reactions in the pathway are also affected.

metabolizes the amino acid phenylalanine, commonly found in many foods (refer to figure 26.7).

In this genetic abnormality, people are unable to convert the amino acid phenylalanine into the amino acid tyrosine. The buildup of phenylalanine in the body prevents the normal development of the nervous system. Such individuals suffer

from PKU and may become mentally retarded. There is a concern in the medical profession that there may be many adults with PKU who do not know they have the condition. Children diagnosed with PKU are not tracked throughout their life, and many parents do not tell their children they have the condition. After these children become adults, they could easily be unaware of their medical history. The problem comes if a female PKU patient becomes pregnant and does not have proper prenatal care because neither she nor her physician is aware of her medical history.

Environmental Influences on Gene Expression

One of the complicating aspects of genetics is that expression of an allele's phenotype depends on so much more than just the genotype of the organism. Sometimes the physical environment determines whether or not dominant or recessive alleles are expressed. For example, in humans, genes for freckles do not express themselves fully unless the person's skin is exposed to sunlight (figure 26.8). Another example is the allele for six fingers (*polydactylism*), which is dominant over the allele for five fingers in humans. Some people who have received the allele for six fingers have a fairly complete sixth finger; in others, it may appear as a little stub. In another case, a dominant allele causes the formation of a little finger that cannot be bent as a normal little finger. However, not all people who are believed to have inherited that allele will have a stiff little finger. In some cases, this dominant characteristic is not expressed or perhaps only shows on one hand. Thus, there may be variation

A B C

FIGURE 26.8   The expression of many genes is influenced by the environment. The allele for dark hair in the cat (*A*) is sensitive to temperature and expresses itself only in the parts of the body that stay cool. (*B* and *C*) The allele for freckles expresses itself more fully when a person is exposed to sunlight.

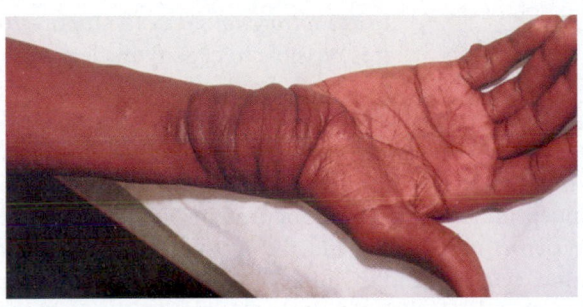

FIGURE 26.9   Neurofibromatosis I is seen in many forms, including benign fibromatous skin tumors, "café-au-lait" spots, nodules in the iris, and possible malignant tumors. It is extremely variable in its expressivity; that is, the traits may be almost unnoticeable or extensive. An autosomal dominant trait, it is the result of a mutation and the production of a protein (neurofibromin) that normally would suppress the activity of a gene that causes tumor formation.

in the degree to which an allele expresses itself in an *individual.* Geneticists refer to this as *variable expressivity.* A good example of this occurs in the genetic abnormality *neurofibromatosis type I* (NF1) (figure 26.9). In some cases, it may not be expressed in the *population* at all. This is referred to as a *lack of penetrance.* Other genes may be interacting with these dominant alleles, causing the variation in expression.

Both internal and external environmental factors can influence the expression of genes. For example, at conception, a male receives genes that will eventually determine the pitch of his voice. However, these genes are expressed differently after puberty. At puberty, male sex hormones are released. This internal environmental change results in the deeper male voice. A male who does not produce these hormones retains a higher-pitched voice in later life. Comparable changes can occur in females when an abnormally functioning adrenal gland causes the release of large amounts of male hormones.

Diet is an external environmental factor that can influence the phenotype of an individual. *Diabetes mellitus,* a metabolic disorder in which glucose in the blood is not properly metabolized and is passed out of the body in the urine, has a genetic basis. Some people who have a family history of diabetes are

thought to have inherited the trait for this disease. Evidence indicates that they can delay the onset of the disease by reducing the amount of sugar in their diet.

Epigenetics and Gene Expression

Epigenetics is the study of changes in gene expression caused by factors other than alterations in a cell's DNA. The term actually means "in addition to genetics," (i.e., non-genetic factors that cause the cell's genes to express themselves differently). When an epigenetic change occurs, it may last for the life of the cell and may even be passed on to the next generation. This is what happens when a cell (e.g., stem cell) undergoes the process of differentiation. Stem cells are called pluripotent because they have the potential to be any kind of cell found in the body (e.g., muscle, bone, or skin cell). However, once they become differentiated they lose the ability to become other kinds of cells and so do the cells they produce by cell division. For example, if a pluripotent cell were to express muscle protein genes and not insulin protein genes, it would differentiate into a muscle cell, not an insulin-producing cell.

The following are four examples of events that can cause an epigenetic effect:

1. Adding a methyl group to a cytosine in the gene changes it to methylcytosine. Since methylcytosine cannot be read during translation, the gene is turned off.
2. Altering the shape of the histones around the gene. Modifying histones ensures that a differentiated cell would stay differentiated and not convert back into being a pluripotent cell.
3. Having the protein that has already been transcribed return to the gene and keep it turned on.
4. Splicing RNA into sequences not originally determined by the gene.

Some compounds are considered epigenetic carcinogens; that is, they are able to cause cells to form tumors, but they do not change the nucleotide sequence of a gene. Examples include certain chlorinated hydrocarbons used as fungicides and some nickel-containing compounds.

While molecular genetics is only about fifty years old, this science has progressed rapidly. Our understanding of DNA structure and functions contributes greatly to our ability to control the genetics of organisms to our advantage.

26.7 THE CENTRAL DOGMA

As scientists began to understand the chemical makeup of the **nucleic acids,** an attempt was made to understand how DNA and RNA relate to inheritance, cell structure, and cell activities. The concept that resulted is known as the *central dogma.* The belief can be written in this form:

 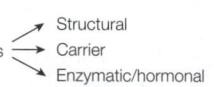

What this concept map says is that at the center of it all is DNA, the genetic material of the cell, and (going to the left) it is capable of reproducing itself, a process called **DNA replication**. Going to the right, DNA is capable of supervising the manufacture of RNA (a process known as **transcription**), which in turn is involved in the production of protein molecules, a process known as **translation.**

DNA replication occurs in cells in preparation for the cell division processes of mitosis and meiosis. Without replication, daughter cells would not receive the library of information required to sustain life. The transcription process results in the formation of a strand of RNA that is a copy of a segment of the DNA on which it is formed. Some of the RNA molecules become involved in various biochemical processes, while others are used in the translation of the RNA information into proteins. Structural proteins are used by the cell as building materials (feathers, collagen, hair), while others are used to direct and control chemical reactions (enzymes or hormones) or carry molecules from place to place (hemoglobin).

It is the processes of transcription and translation that result in the manufacture of all enzymes. Each unique enzyme molecule is made from a blueprint in the form of a DNA nucleotide sequence, or *gene.* Some of the thousands of enzymes manufactured in the cell are the tools required for transcription and translation to take place. *The enzymes made by the process carry out the process of making more enzymes!*

DNA ← (replication) ← DNA → (transcription) → RNA → (translation) → Proteins

Enzymes involved in

Tools are made to make more tools! The same is true for DNA replication. Enzymes made from the DNA blueprints by transcription and translation are used as tools to make exact copies of the genetic material! More blueprints are made so that future generations of cells will have the genetic materials necessary for them to manufacture their own regulatory and structural proteins. Without DNA, RNA, and enzymes functioning in the proper manner, life as we know it would not occur.

DNA has four properties that enable it to function as genetic material. It is able to (1) *replicate* by directing the manufacture of copies of itself; (2) *mutate,* or chemically change, and transmit these changes to future generations; (3) *store* information that determines the characteristics of cells and organisms; and (4) use this information to *direct* the synthesis of structural and regulatory proteins essential to the operation of the cell or organism.

The Structure of DNA and RNA

Nucleic acid molecules are enormous and complex polymers made up of monomers called **nucleotides.** Each nucleotide is composed of a sugar molecule (S) containing five carbon atoms, a phosphate group (P), and a molecule containing nitrogen that will be referred to as a *nitrogenous base* (B) (figure 26.10). It is possible to classify nucleic acids into two main groups based on the kinds of sugars and nitrogenous bases used in the nucleotides (i.e., DNA and RNA).

In cells, DNA is the nucleic acid that functions as the original blueprint for the synthesis of proteins. It contains the sugar *deoxyribose;* phosphates; and the nitrogenous bases adenine, guanine, cytosine, and thymine (A, G, C, T). RNA is a type of nucleic acid that is directly involved in the synthesis of protein. It contains the sugar *ribose;* phosphates; and adenine, guanine, cytosine, and uracil (A, G, C, U). There is no thymine in RNA and no uracil in DNA.

DNA and RNA differ in one other respect. DNA is actually a double molecule. It consists of two flexible strands held together between their protruding bases. The two strands are twisted about each other in a coil or double helix that resembles a twisted ladder (figure 26.11). The two strands of the molecule are held together because they "fit" each other like two jigsaw puzzle pieces that interlock with one another and are stabilized by weak chemical forces, hydrogen bonds. The four kinds of bases always pair in a definite way: adenine (A) with thymine (T), and guanine (G) with cytosine (C). Notice that the large molecules (A and G) pair with the small ones (T and C), thus keeping the two complementary (matched) strands parallel. The bases that pair are said to be **complementary bases.**

A : T

and

G : C

You can "write" a message in the form of a stable DNA molecule by combining the four different DNA nucleotides (A, T, G, C) in particular sequences. The four DNA nucleotides are being used as an alphabet to construct three-letter words. To make sense out of such a code, it is necessary to read in one direction. Reading the sequence in reverse does not always make sense, just as reading this paragraph in reverse would not make sense.

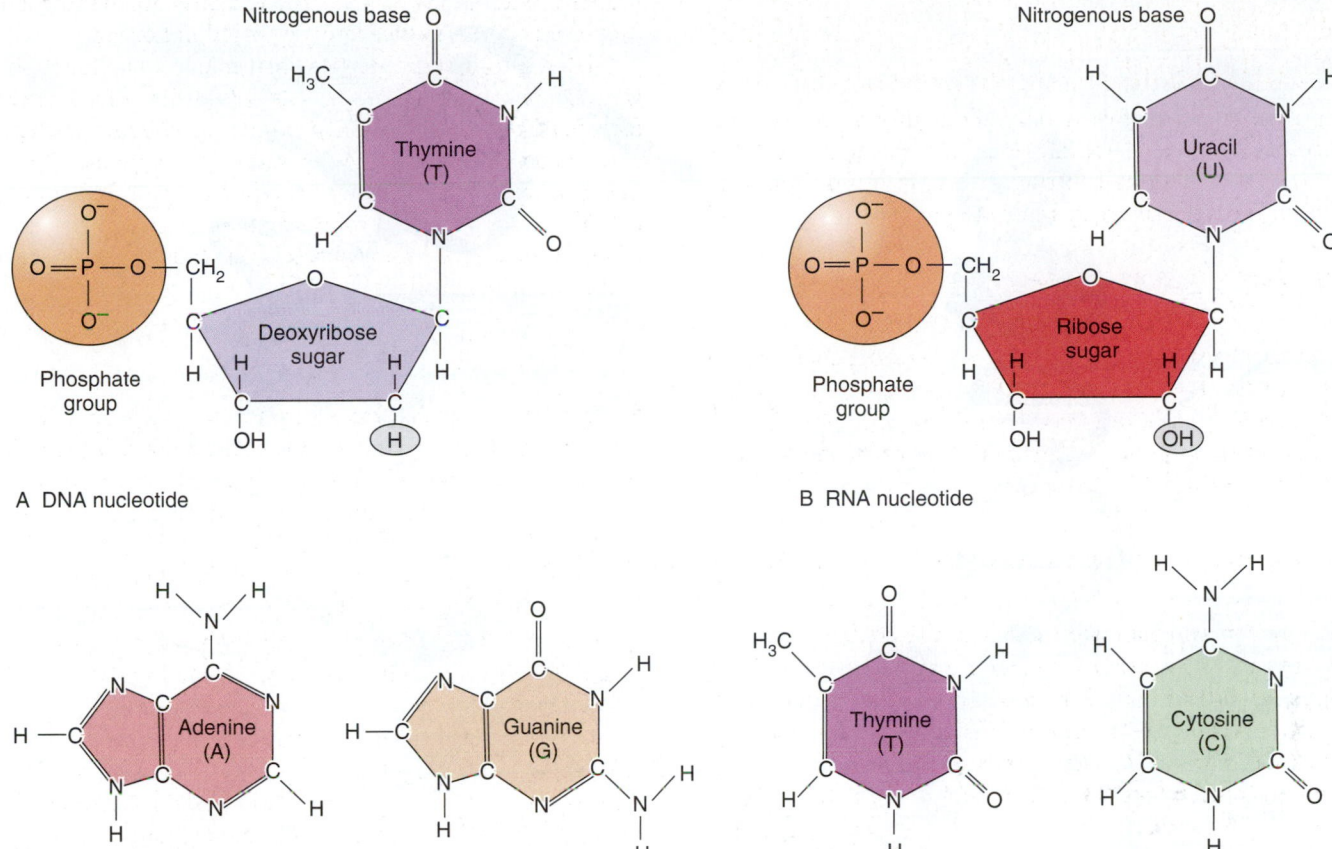

A DNA nucleotide

B RNA nucleotide

C The four nitrogenous bases that occur in DNA

FIGURE 26.10 (*A*) The nucleotide is the basic structural unit of all nucleic acid molecules. A thymine nucleotide of DNA is comprised of phosphate, deoxyribose sugar, and the nitrogenous base, thymine (T). Notice in the nucleotides that the phosphate group is written in "shorthand" form as a P inside a circle. (*B*) The RNA uracil nucleotide is comprised of a phosphate, ribose sugar, and the nitrogenous base, uracil (U). Notice the difference between the sugars and how the bases differ from one another. (*C*) Using these basic components (phosphate, sugars, and bases), the cell can construct eight common types of nucleotides. Can you describe all eight?

The genetic material of humans and other eukaryotic organisms is *strands* of coiled *double-stranded DNA*, which has histone proteins attached along its length. When eukaryotic chromatin fibers coil into condensed, highly knotted bodies, they are seen easily through a microscope after staining with dye. Condensed like this, a chromatin fiber is referred to as a **chromosome** (figure 26.12B). The genetic material in bacteria is also double-stranded DNA, but the ends of the molecule are connected to form a *loop* and they do not form condensed chromosomes (figure 26.13).

Each chromatin DNA strand is different because each strand has a different chemical code. Coded DNA serves as a central cell library. Tens of thousands of messages are in this storehouse of information. This information tells the cell such things as (1) how to produce enzymes required for the digestion of nutrients, (2) how to manufacture enzymes that will metabolize the nutrients and eliminate harmful wastes, (3) how to repair and assemble cell parts, (4) how to reproduce healthy offspring, (5) when and how to react to favorable and unfavorable changes in the environment, and (6) how to coordinate and regulate all of life's essential functions. If any of these functions is not performed properly, the cell may die. The importance of

maintaining essential DNA in a cell becomes clear when we consider cells that have lost their DNA. For example, human red blood cells lose their nuclei as they mature and become specialized for carrying oxygen and carbon dioxide throughout the body. Without DNA, they are unable to manufacture the essential cell components needed to sustain themselves. They continue to exist for about 120 days, functioning only on enzymes manufactured earlier in their lives. When these enzymes are gone, the cells die. Because these specialized cells begin to die the moment they lose their DNA, they are more accurately called *red blood corpuscles (RBCs)*: "little dying red bodies."

DNA Replication

Because all reproducing cells must maintain a complete set of genetic material, there must be a doubling of DNA in order to have enough to pass on to the offspring (refer to chapter 20 and "The Importance of Cell Division"). DNA replication is the process of duplicating the genetic material prior to its distribution to daughter cells. When a cell divides into two daughter cells, each new cell must receive a complete copy of the parent

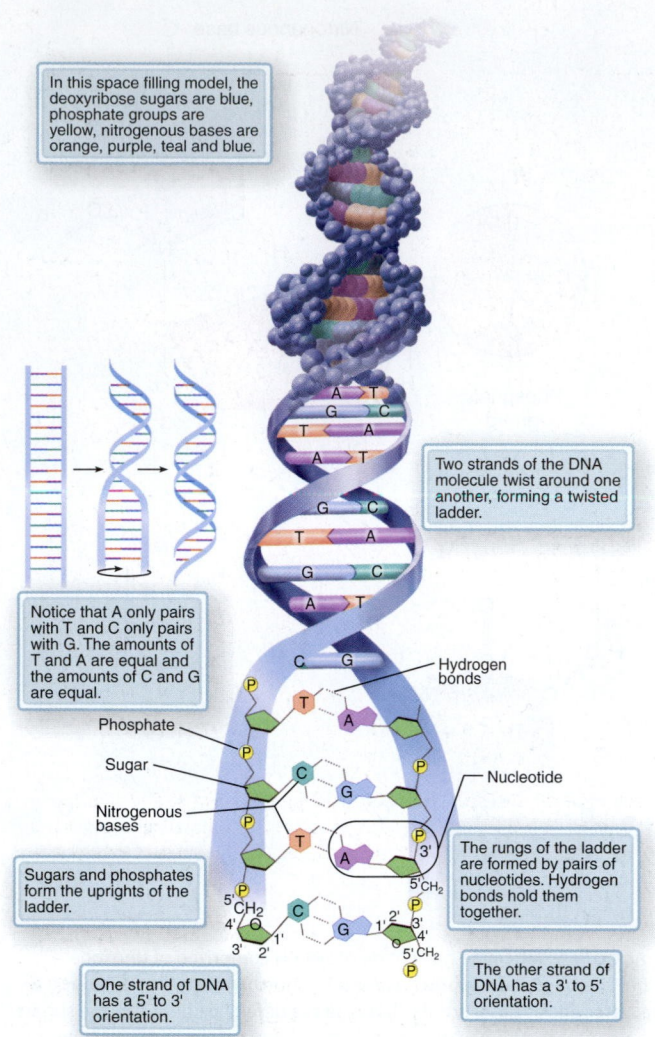

In this space filling model, the deoxyribose sugars are blue, phosphate groups are yellow, nitrogenous bases are orange, purple, teal and blue.

Two strands of the DNA molecule twist around one another, forming a twisted ladder.

Notice that A only pairs with T and C only pairs with G. The amounts of T and A are equal and the amounts of C and G are equal.

Hydrogen bonds

Phosphate

Sugar

Nitrogenous bases

Nucleotide

Sugars and phosphates form the uprights of the ladder.

The rungs of the ladder are formed by pairs of nucleotides. Hydrogen bonds hold them together.

One strand of DNA has a 5' to 3' orientation.

The other strand of DNA has a 3' to 5' orientation.

FIGURE 26.11 Polymerized deoxyribonucleic acid (DNA) is a helical molecule. The nucleotides within each strand are held together by covalent bonds. The two parallel strands are linked by hydrogen bonds between the base-paired nitrogen bases.

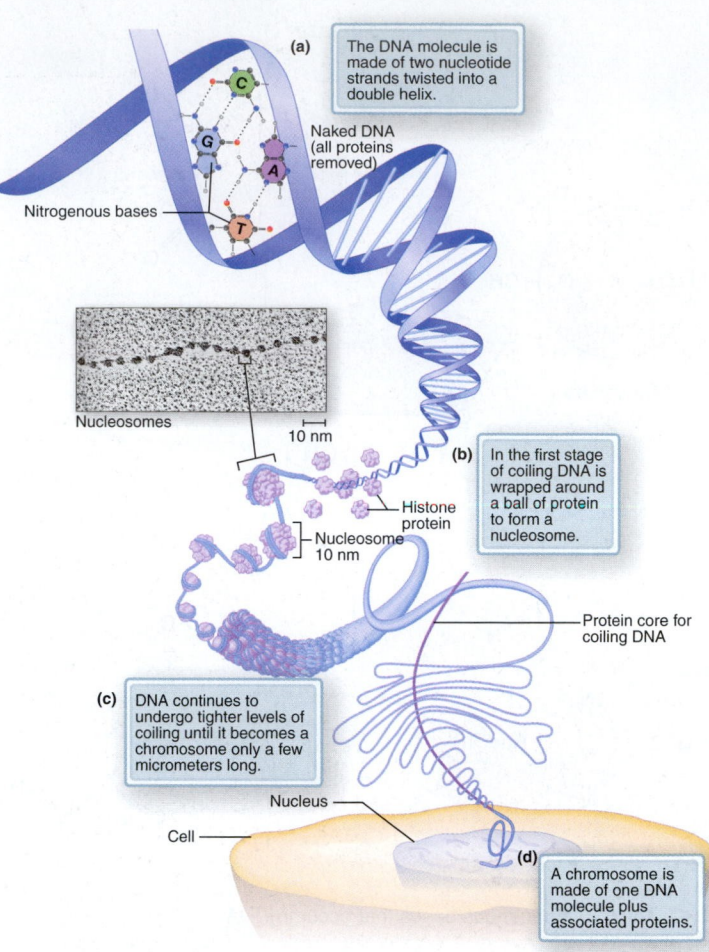

(a) The DNA molecule is made of two nucleotide strands twisted into a double helix.

Naked DNA (all proteins removed)

Nitrogenous bases

Nucleosomes

10 nm

(b) In the first stage of coiling DNA is wrapped around a ball of protein to form a nucleosome.

Histone protein

Nucleosome 10 nm

Protein core for coiling DNA

(c) DNA continues to undergo tighter levels of coiling until it becomes a chromosome only a few micrometers long.

Nucleus

Cell

(d) A chromosome is made of one DNA molecule plus associated proteins.

FIGURE 26.12 During certain stages in the life cycle of a eukaryotic cell, the DNA is tightly coiled to form a chromosome. To form a chromosome, the DNA molecule is wrapped around a group of several histone proteins. Together the histones and the DNA form a structure called the nucleosome. The nucleosomes are stacked together in coils to form a chromosome.

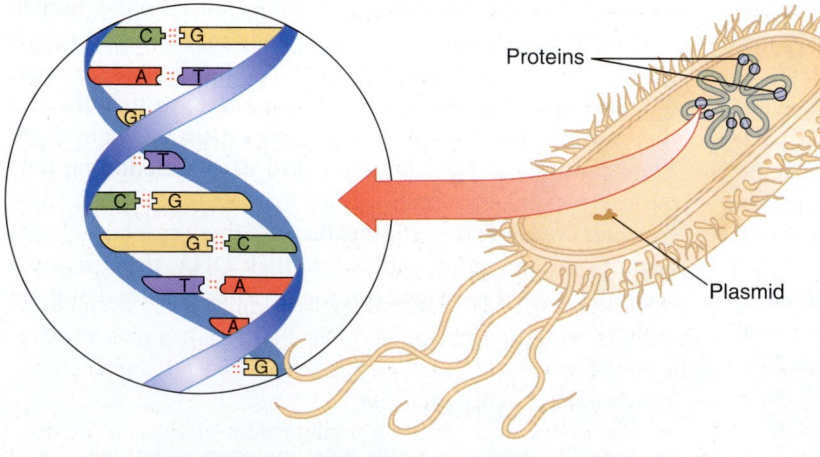

Proteins

Plasmid

Prokaryote

FIGURE 26.13 The nucleic acid of prokaryotic cells (the bacteria) has a different kind of protein compared to eukaryotic cells. In addition, the ends of the giant nucleoprotein molecule overlap and bind with one another to form a loop. The additional small loop of DNA is the plasmid, which contains genes that are not essential to the daily life of the cell.

cell's genetic information, or it will not be able to manufacture all the proteins vital to its existence. Accuracy of duplication is essential to guarantee the continued existence of that type of cell. Should the daughters not receive exact copies, they would most likely die.

1. The DNA replication process begins as an enzyme breaks the attachments between the two strands of DNA. In eukaryotic cells, this occurs in hundreds of different spots along the length of the DNA (figure 26.14).

2. Moving along the DNA, the enzyme "unzips" the halves of the DNA (figure 26.14A). Hydrogen bonds hold each in position (AT, GC), while the new nucleotide is covalently bonded between the sugar and phosphate of the new backbone.

3. Proceeding in opposite directions on each side, the enzyme *DNA polymerase* moves down the length of the DNA, attaching new DNA nucleotides into position. Rela-

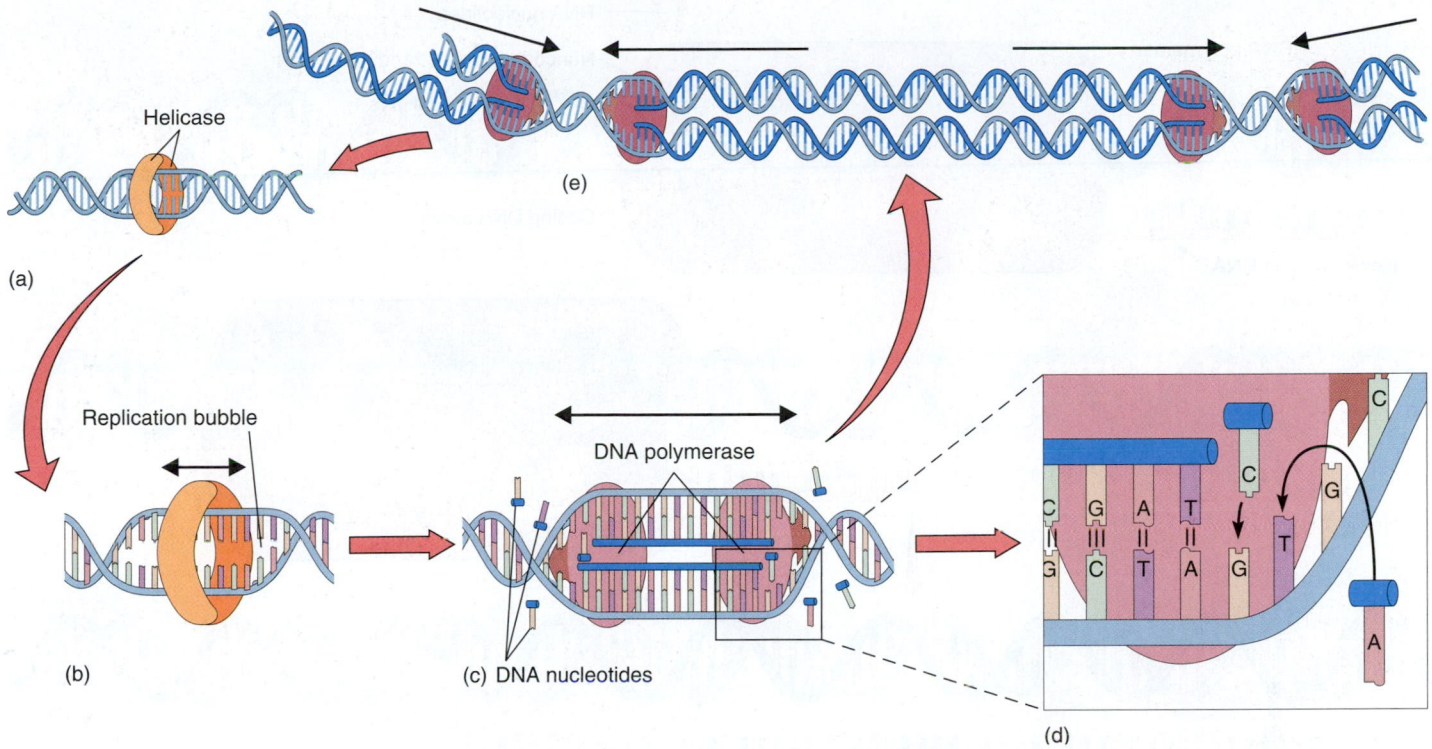

FIGURE 26.14 (*A*) *Helicase* enzymes bind to the DNA molecule. (*B*) The enzymes separate the two strands of DNA. (*C, D*) As the DNA strands are separated, new DNA nucleotides are added to the new strands by DNA polymerase. The new DNA strands are synthesized according to base-pairing rules for nucleic acids. (*E*) By working in two directions at once along the DNA strand, the cell is able to replicate the DNA more quickly. Each new daughter cell receives one of these copies.

tively weak hydrogen bonds hold each in position (AT, GC), while the new nucleotide is bonded to its neighboring nucleotide more strongly with covalent bonds between the sugar and phosphate of the new backbone (figure 26.14B).

4. The enzyme that speeds the addition of new nucleotides to the growing chain works along with another enzyme to make sure that no mistakes are made. If the wrong nucleotide appears to be headed for a match, the enzyme will reject it in favor of the correct nucleotide. If a mistake is made and a wrong nucleotide is paired into position, specific enzymes have the ability to replace it with the correct one (figure 26.14C).

5. Replication proceeds in both directions and at multiple sites along the chromatin, appearing as "bubbles" (figure 26.14D).

6. The complementary molecules (AT, GC) pair with the exposed nitrogenous bases of both DNA strands by forming new hydrogen bonds (figure 26.14C).

7. Once properly aligned, a bond is formed between the sugars and phosphates of the newly positioned nucleotides. A strong sugar and phosphate backbone is formed in the process (figure 26.14C).

8. This process continues until all the replication "bubbles" join (figure 26.14D).

A new complementary strand of DNA forms on each of the old DNA strands, resulting in the formation of two double-stranded DNA molecules. In this way, the exposed nitrogenous bases of the original DNA serve as a *template,* or pattern, for the formation of the new DNA. As the new DNA is completed, it twists into its double-helix shape.

The completion of the DNA replication process yields two double helices that are identical in their nucleotide sequences. Half of each is new; half is the original parent DNA molecule. The DNA replication process is highly accurate. It has been estimated that there is only one error made for every 2×10^9 nucleotides. A human cell contains forty-six chromosomes consisting of about 3,000,000,000 (3 billion) base pairs. This averages to about 1.5 errors per cell! Don't forget that this figure is an estimate. While some cells may have five errors per replication, others may have more, and some may have no errors at all. It is also important to note that some errors may be major and deadly, while others are insignificant. Because this error rate is so small, DNA replication is considered by most to be essentially error-free. Following DNA replication, the cell now contains twice the amount of genetic information and is ready to begin the process of distributing one set of genetic information to each of its two daughter cells.

The distribution of DNA involves splitting the cell and distributing a set of genetic information to the two new daughter cells. In this way, each new cell has the necessary information to control its activities. The mother cell ceases to exist when it divides its contents between the two smaller daughter cells. A cell does not really die when it reproduces itself; it merely starts over again.

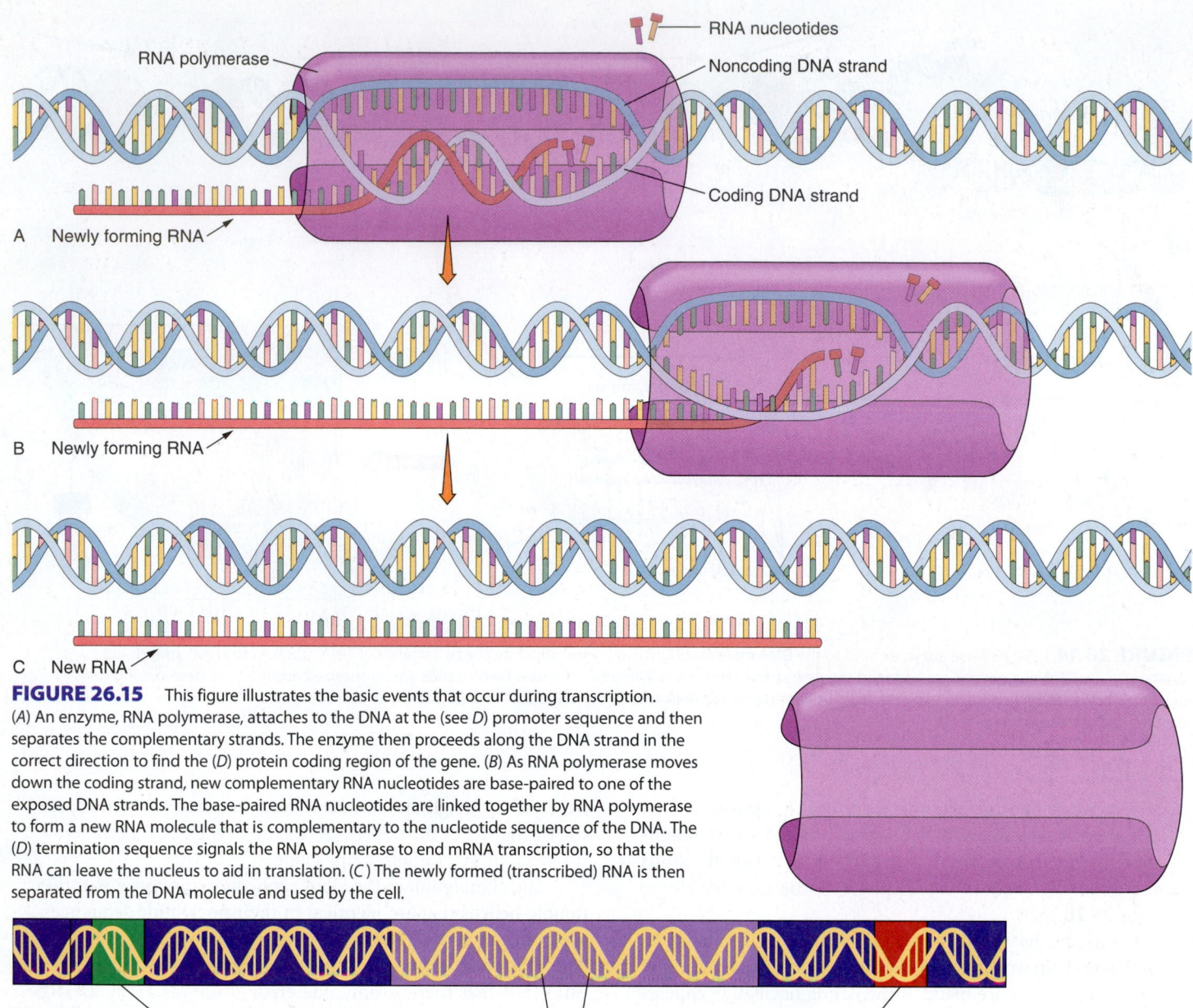

A Newly forming RNA

RNA polymerase

RNA nucleotides

Noncoding DNA strand

Coding DNA strand

B Newly forming RNA

C New RNA

FIGURE 26.15 This figure illustrates the basic events that occur during transcription. (A) An enzyme, RNA polymerase, attaches to the DNA at the (see D) promoter sequence and then separates the complementary strands. The enzyme then proceeds along the DNA strand in the correct direction to find the (D) protein coding region of the gene. (B) As RNA polymerase moves down the coding strand, new complementary RNA nucleotides are base-paired to one of the exposed DNA strands. The base-paired RNA nucleotides are linked together by RNA polymerase to form a new RNA molecule that is complementary to the nucleotide sequence of the DNA. The (D) termination sequence signals the RNA polymerase to end mRNA transcription, so that the RNA can leave the nucleus to aid in translation. (C) The newly formed (transcribed) RNA is then separated from the DNA molecule and used by the cell.

D Promoter sequence   Protein code   Termination sequence

DNA Transcription

DNA functions in the manner of a reference library that does not allow its books to circulate. Information from the originals must be copied for use outside the library. The second major function of DNA is the making of these single-stranded, complementary RNA copies of DNA. This operation is called transcription (*scribe* = to write), which means to transfer data from one form to another. In this case, the data are copied from DNA language to RNA language. The same base-pairing rules that control the accuracy of DNA replication apply to the process of transcription. Using this process, the genetic information stored as a DNA chemical code is carried in the form of an RNA copy to other parts of the cell. It is RNA that is used to guide the assembly of amino acids into structural and regulatory proteins. Without the process of transcription, genetic information would be useless in directing cell functions.

Although many types of RNA are synthesized from the genes, the three most important are *messenger RNA (mRNA), transfer RNA (tRNA),* and *ribosomal RNA (rRNA).*

Transcription begins in a way that is similar to DNA replication. The DNA strands are separated by an enzyme, exposing the nitrogenous-base sequences of the two strands. However, unlike DNA replication, transcription occurs only on one of the two DNA strands, which serves as a template, or pattern, for the synthesis of RNA (see figure 26.15). This side is also referred to as the genetic *coding strand* of the DNA. But which strand is copied? Where does it start and when does it stop? Where along the sequence of thousands of nitrogenous bases does the chemical code for the manufacture of a particular enzyme begin and where does it end? If transcription begins randomly, the resulting RNA may not be an accurate copy of the code, and the enzyme product may be useless or deadly to the cell. To

Basic Steps of Translation

1. An mRNA molecule is placed in the smaller of the two portions of a ribosome so that six nucleotides (two codons) are locked into position.

2. The larger ribosomal unit is added to the ribosome/mRNA combination.

3. A tRNA with bases that match the second mRNA codon attaches to the mRNA. The tRNA is carrying a specific amino acid. Once attached, a second tRNA carrying another specific amino acid moves in and attaches to its complementary mRNA codon right next to the first tRNA/amino acid complex.

4. The two tRNAs properly align their two amino acids so that the amino acids may be chemically attached to one another (box figure 26.2).

5. Once the two amino acids are connected to one another by a covalent peptide bond, the first tRNA detaches from its amino acid and mRNA codon.

6. The ribosome moves along the mRNA to the next codon (the first tRNA is set free to move through the cytoplasm to attach to and transfer another amino acid).

7. The next tRNA/amino acid unit enters the ribosome and attaches to its codon next to the first set of amino acids.

8. The tRNAs properly align their amino acids so that they may be chemically attached to one another, forming a chain of three amino acids.

9. Once three amino acids are connected to one another, the second tRNA is released from its amino acid and mRNA (this tRNA is set free to move through the cytoplasm to attach to and transfer another amino acid).

10. The ribosome moves along the mRNA to the next codon, and the fourth tRNA arrives (box figure 26.3).

11. This process repeats until all the amino acids needed to form the protein have attached to one another in the proper sequence. This amino acid sequence was encoded by the DNA gene.

12. Once the final amino acid is attached to the growing chain of amino acids, all the molecules (mRNA, tRNA, and newly formed protein) are released from the ribosome. The "stop" mRNA codon signals this action.

13. The ribosome is again free to become involved in another protein-synthesis operation.

14. The newly synthesized chain of amino acids (the new protein) leaves the ribosome to begin its work. However, the protein may need to be altered by the cell before it will be ready for use (box figure 26.4).

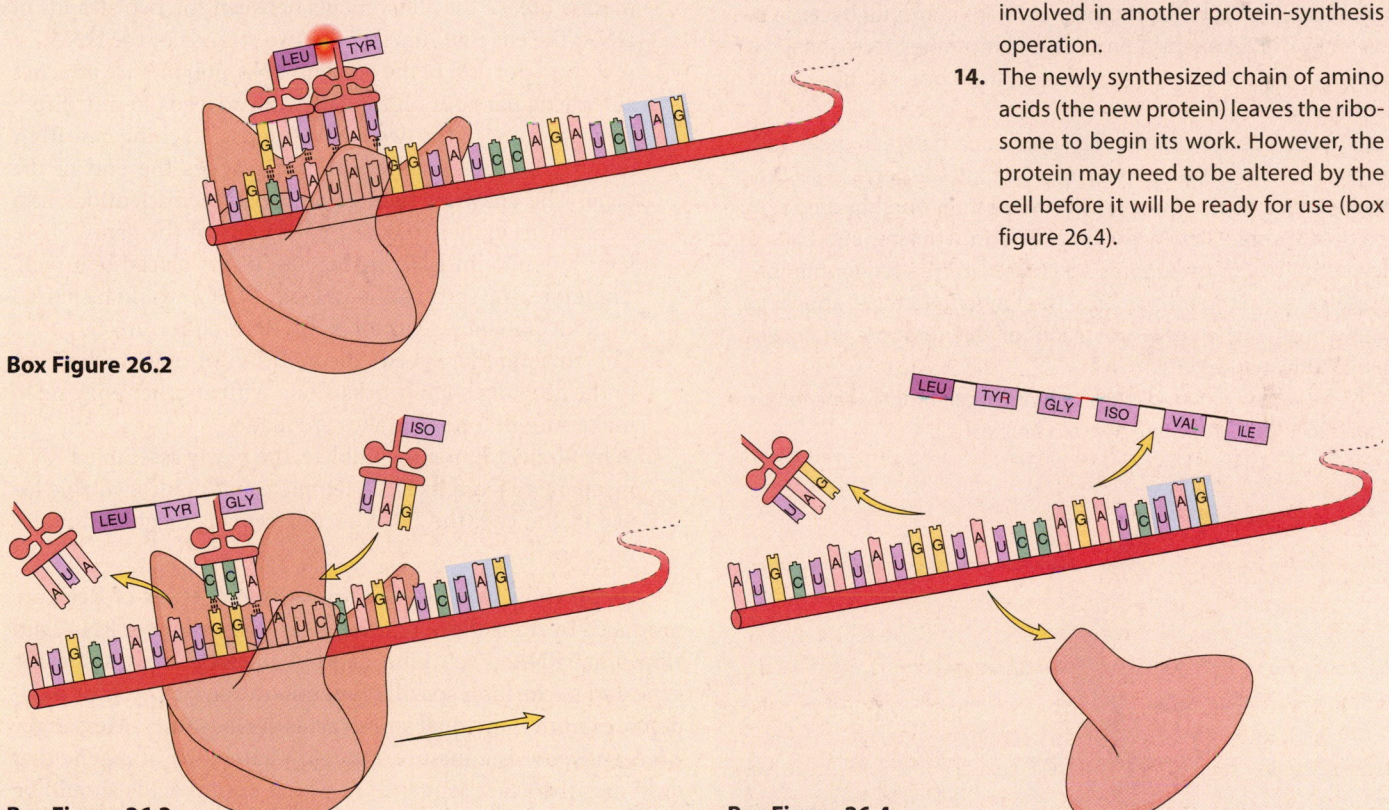

Box Figure 26.2

Box Figure 26.3

Box Figure 26.4

answer these questions, it is necessary to explore the nature of the genetic code itself.

We know that genetic information is in chemical-code form in the DNA molecule. When the coded information is used or *expressed,* it guides the assembly of particular amino acids into structural and regulatory polypeptides and proteins. If DNA is molecular language, then each nucleotide in this language can be thought of as a letter within a four-letter alphabet. Each word, or code, is always three letters (nucleotides) long, and only three-letter words can be written. A **DNA code** is a triplet nucleotide sequence that codes for one of the twenty common amino acids. The number of codes in this language is limited because there are only four different nucleotides, which are used only in groups of three. The order of these three letters is just as important

Connections . . .

Telomeres

Each end of a chromosome contains a sequence of nucleotides called a **telomere**. In humans, these chromosome "caps" contain many copies of the following nucleotide base-pair sequence:

TTAGGG

AATCCC

Telomeres are very important segments of the chromosome:

1. They are required for chromosome replication.
2. They protect the chromosome from being destroyed by dangerous DNAase enzymes (enzymes that destroy DNA).
3. They keep chromosomes from bonding to one another end to end.

Evidence shows that the loss of telomeres is associated with cell "aging," whereas not removing them has been linked to cancer. Every time a cell reproduces itself, it loses some of its telomeres. However, in cells that have the enzyme telomerase, new telomeres are added to the ends of the chromosome each time the cells divide. Therefore, cells that have telomerase do not age as other cells do, and cancer cells are immortal because of this enzyme. Telomerase enables chromosomes to maintain, if not increase, the length of telomeres from one cell generation to the next.

There is also an interesting connection between the length of telomeres and the number of skin moles a person has. Evidence indicates that people with more than one hundred skin moles age more slowly, show fewer skin wrinkles, and have a decreased risk of osteoporosis. People with the higher number of moles have longer telomeres. The longer telomeres appear to delay the age when cells are no longer able to divide and maintain healthy tissues.

The yellow regions on the drawing of a chromosome in box figure 26.5 indicate where the telomeres are.

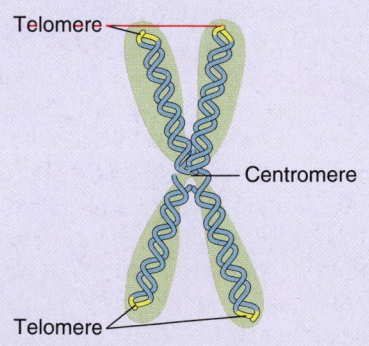

Box Figure 26.5

in DNA language as it is in our language. We recognize that CAT is not the same as TAC. If all the possible three-letter codes were written using only the four DNA nucleotides for letters, there would be a total of sixty-four combinations.

$$4^3 = 4 \times 4 \times 4 = 64$$

When codes are found at a particular place along a coding strand of DNA and the sequence has meaning, the sequence is a gene. "Meaning" in this case refers to the fact that the gene can be transcribed into an RNA molecule, which in turn may control the assembly of individual amino acids into a polypeptide.

When a gene is transcribed into RNA, the double-stranded DNA is "unzipped," and an enzyme known as *RNA polymerase* attaches to the DNA at the promoter region. It is from this region that the enzymes will begin to assemble RNA nucleotides into a complete, single-stranded copy of the gene, including initiation and termination codes. Triplet RNA nucleotide sequences complementary to DNA codes are called **codons**. Remember that there is no thymine in RNA molecules; it is replaced with uracil. Therefore, the initiation code in DNA (TAC) would be base-paired by RNA polymerase to form the RNA codon AUG. When transcription is complete, the newly assembled RNA is separated from its DNA template and made available for use in the cell; the DNA recoils into its original double-helix form.

In summary:

1. The process begins as one portion of the enzyme RNA polymerase breaks the attachments between the two strands of DNA; the enzyme "unzips" the two strands of the DNA.
2. A second portion of the enzyme RNA polymerase attaches at a particular spot on the DNA. It proceeds in one direction along one of the two DNA strands, attaching new RNA nucleotides into position until it reaches the end of the gene. The enzymes then assemble RNA nucleotides into a complete, single-stranded RNA copy of the gene. There is no thymine in RNA molecules; it is replaced by uracil. Therefore, the start codon in DNA (TAC) would be paired by RNA polymerase to form the RNA codon AUG.
3. The enzyme that speeds the addition of new nucleotides to the growing chain works along with another enzyme to make sure that no mistakes are made.
4. When transcription is complete, the newly assembled RNA is separated from its DNA template and made available for use in the cell; the DNA recoils into its original double-helix form.

As previously mentioned, three general types of RNA are produced by transcription: messenger RNA, transfer RNA, and ribosomal RNA. Each kind of RNA is made from a specific gene and performs a specific function in the synthesis of polypeptides from individual amino acids at ribosomes. **Messenger RNA (mRNA)** is a mature, straight-chain copy of a gene that describes the exact sequence in which amino acids should be bonded together to form a polypeptide.

Transfer RNA (tRNA) molecules are responsible for picking up particular amino acids and transferring them to the ribosome for assembly into the polypeptide (refer to "A Closer Look: Basic Steps of Translation"). All tRNA molecules are shaped like a cloverleaf. This shape is formed when they fold, and some of the bases form hydrogen bonds that hold the molecule together. One end of the tRNA is able to attach to a specific amino acid. Toward the midsection of the molecule, a triplet nucleotide sequence can base-pair with a codon on mRNA. This triplet nucleotide sequence on tRNA that is

TABLE 26.3   **Amino Acid–mRNA Nucleic Acid Dictionary**

Second letter

First letter	U	C	A	G	Third letter
U	UUU ⎫ Phe UUC ⎭ UUA ⎫ Leu UUG ⎭	UCU ⎫ UCC ⎬ Ser UCA ⎪ UCG ⎭	UAU ⎫ Tyr UAC ⎭ UAA Stop UAG Stop	UGU ⎫ Cys UGC ⎭ UGA Stop UGG Try	U C A G
C	CUU ⎫ CUC ⎬ Leu CUA ⎪ CUG ⎭	CCU ⎫ CCC ⎬ Pro CCA ⎪ CCG ⎭	CAU ⎫ His CAC ⎭ CAA ⎫ Gln CAG ⎭	CGU ⎫ CGC ⎬ Arg CGA ⎪ CGG ⎭	U C A G
A	AUU ⎫ AUC ⎬ Ile AUA ⎭ AUG Met or start	ACU ⎫ ACC ⎬ Thr ACA ⎪ ACG ⎭	AAU ⎫ ASN AAC ⎭ AAA ⎫ Lys AAG ⎭	AGU ⎫ Ser AGC ⎭ AGA ⎫ Arg AGG ⎭	U C A G
G	GUU ⎫ GUC ⎬ Val GUA ⎪ GUG ⎭	GCU ⎫ GCC ⎬ Ala GCA ⎪ GCG ⎭	GAU ⎫ Asp GAC ⎭ GAA ⎫ Glu GAG ⎭	GGU ⎫ GGC ⎬ Gly GGA ⎪ GGG ⎭	U C A G

complementary to a codon of mRNA is called an **anticodon. Ribosomal RNA (rRNA)** is a highly coiled molecule and is used, along with protein molecules, in the manufacture of all ribosomes, the cytoplasmic organelles where tRNA, mRNA, and rRNA come together to help in the synthesis of proteins.

Translation or Protein Synthesis

The mRNA molecule is a coded message written in the biological world's universal nucleic acid language. The code is read in one direction, starting at the initiator. The information is used to assemble amino acids into protein by a process called translation. The word *translation* refers to the fact that nucleic acid language is being changed to protein language. To translate mRNA language into protein language, a dictionary is necessary. Remember, the four letters in the nucleic acid alphabet yield sixty-four possible three-letter words. The protein language has twenty words in the form of twenty common amino acids. Thus, there are more than enough nucleotide words for the twenty amino acid molecules because each nucleotide triplet codes for an amino acid.

Table 26.3 is an amino acid–nucleic acid dictionary. Notice that more than one codon may code for the same amino acid. Some would contend that this is needless repetition, but such "synonyms" can have survival value. If, for example, the gene or the mRNA becomes damaged in a way that causes a particular nucleotide base to change to another type, the chances are still good that the proper amino acid will be read into its proper

position. But not all such changes can be compensated for by the codon system, and an altered protein may be produced. Changes can occur that cause great harm. Some damage is so extensive that the entire strand of DNA is broken, resulting in improper protein synthesis or a total lack of synthesis. Any change in DNA is called a **mutation.** Recall that some mutations may be so minor that they have no effect, while others may be harmful or beneficial. Although most mutations are likely neutral or harmful, beneficial mutations do occur.

The construction site of the protein molecules (i.e., the translation site) is on the ribosome, a cellular organelle that serves as the meeting place for mRNA and the tRNA that is carrying amino acid building blocks. Ribosomes can be found free in the cytoplasm or attached to the endoplasmic reticulum.

Thus, the mRNA moves through the ribosomes with the mRNA's codon sequence allowing for the chemical bonding of a specific sequence of amino acids. Remember that the sequence was originally determined by the DNA.

A protein's three-dimensional shape is determined by its amino acid sequence. This shape determines the activity of the protein molecule. The protein may be a structural component of a cell or a regulatory protein, such as an enzyme. Any change in amino acids or their order changes the action of the protein molecule. The protein insulin, for example, has a different amino acid sequence than the digestive enzyme trypsin. Both proteins are essential to human life and must be produced constantly and accurately. The amino acid sequence of each is determined by a different gene. Each gene is a particular sequence of DNA nucleotides. Any alteration of that sequence can directly alter the protein structure and, therefore, the survival of the organism.

Alterations of DNA

Several kinds of changes to DNA may result in mutations. Phenomena that are either known or suspected causes of DNA damage are called **mutagenic agents.** Agents known to cause damage to DNA are certain viruses (e.g., papillomavirus), weak or "fragile" spots in the DNA, X radiation (X rays), and chemicals found in foods and other products, such as chemicals from burning tobacco. All have been studied extensively, and there is little doubt that they cause mutations. *Chromosomal aberrations* is the term used to describe major changes in DNA. Four types of aberrations are inversions, translocations, duplications, and deletions. An *inversion* occurs when a chromosome is broken and this piece becomes reattached to its original chromosome but in reverse order. It has been cut out and flipped around. A *translocation* occurs when one broken segment of DNA becomes integrated into a different chromosome. *Duplication* occurs when a portion of a chromosome is replicated and attached to the original section in sequence. *Deletion* aberrations result when the broken piece becomes lost or is destroyed before it can be reattached.

In some individuals, a single nucleotide of the gene may be changed. This type of mutation is called a *point mutation.* An example of the effects of altered DNA may be seen in human red blood cells. Red blood cells contain the oxygen-transport molecule, hemoglobin. Normal hemoglobin molecules are composed of 150 amino acids in four chains—two alpha and two beta. The nucleotide sequence of the gene for the beta chain is known, as is the amino acid sequence for this chain. In normal individuals, the sequence begins like this:

Val-His-Leu-Thr-Pro-Glu-Glu-Lys . . . .

A single nucleotide change in the DNA sequence results in a mutation that produces a new amino acid sequence in all the red blood cells:

Val-His-Leu-Thr-Pro-Val-Glu-Lys . . . .

This single nucleotide change (known as a *missense point mutation*), which causes a single amino acid to change, may seem minor. However, it is the cause of **sickle-cell anemia,** a disease that affects the red blood cells by changing them from a circular to a sickle shape when oxygen levels are low (figure 26.16). When this sickling occurs, the red blood cells do not flow smoothly through capillaries. Their irregular shapes cause them to clump, clogging the blood vessels. This prevents them from delivering their oxygen load to the oxygen-demanding tissues. A number of physical disabilities may result, including physical weakness, brain damage, pain and stiffness of the joints, kidney damage, rheumatism, and, in severe cases, death.

Changes in the structure of DNA may have harmful effects on the next generation if they occur in the sex cells. Some damage to DNA is so extensive that the entire strand of DNA is broken, resulting in the synthesis of abnormal proteins or a total lack of protein synthesis. A number of experiments indicate that

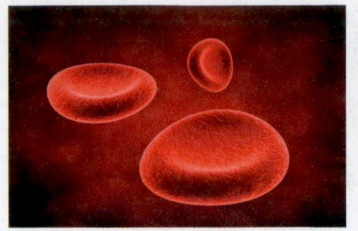
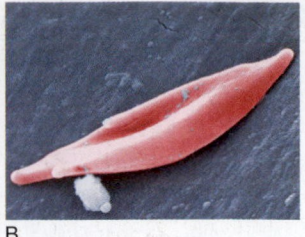

A   B

FIGURE 26.16   (A) A normal red blood cell is shown in comparison with (B) a cell having the sickle shape. This sickling is the result of a single amino acid change in the hemoglobin molecule.

TABLE 26.4   Types of Chromosomal Aberrations

Normal Sequence	THE ONE BIG FLY HAD ONE RED EYE*
KIND OF MUTATION	**SEQUENCE CHANGE**
Missense	THQ ONE BIG FLY HAD ONE RED EYE
Nonsense	THE ONE BIG
Frameshift	THE ONE QBI GFL YHA DON ERE DEY
Deletion	THE ONE BIG HAD ONE RED EYE
Duplication	THE ONE BIG FLY FLY HAD ONE RED EYE
Insertion	THE ONE BIG WET FLY HAD ONE RED EYE
Expanding mutation:	
Parents	THE ONE BIG FLY HAD ONE RED EYE
Children	THE ONE BIG FLY FLY FLY HAD ONE RED EYE
Grandchildren	THE ONE BIG FLY FLY FLY FLY FLY FLY HAD ONE RED EYE

Based on R. Lewis, *Human Genetics, Concepts and Applications,* 2d ed., Dubuque, IA: Wm. C. Brown Publishers, 1997.

*A sentence composed of three-letter words can provide an analogy to the effect of mutations on a gene's nucleotide sequence.

many street drugs, such as LSD (lysergic acid diethylamide), are mutagenic agents and cause DNA to break. Abnormalities have also been identified that are the result of changes in the number or sequence of bases. One way to illustrate these various kinds of mutations is seen in table 26.4.

26.8   USING DNA TO OUR ADVANTAGE

Advances in the field of biotechnology have allowed scientists to do many exciting things. To understand how these accomplishments are achieved by the scientists, we need to explore two key strategies: (1) genetic modification of an organism and (2) DNA sequencing.

Strategy One: Genetic Modification of Organisms

Since biblical times, civilizations have attempted to improve the quality of their livestock and agricultural stock. Cows that produced more milk or more tender meat were valued over those that were dry or tough. Initial attempts to develop

Connections . . .

Mutation Leads to Personal Energy Crisis

Ten-year-old Latisha Franklin has suffered from her own personal energy crisis since she was four. Latisha has been diagnosed with an uncommon illness, an abnormality called mitochondrial encephalopathy, or MELAS. MELAS is the acronym for *m*itochondrial *e*ncephalopathy, *l*actic *ac*idosis, and *s*troke-like episodes. It is caused by mutations in the DNA found in her mitochondria, mDNA. Mitochondria manufacture proteins using their own DNA. Enzymes help to produce useful chemical bond energy for cells, ATP. mDNA differs from the chromosomes found in the nucleus. They are much smaller and circular. Any changes (mutations) in mDNA can have far-reaching effects on the body's ability to control energy production.

Latisha has suffered encephalopathy in the form of epilepsy-like seizures and migraine-like headaches. She has also had severe muscle pain caused by excess lactic acid in her muscles, and stroke-like symptoms leading to paralysis and confusion. The mutations that cause MELAS and the chemical changes that occur in mitochondria have been identified; however, there is no cure. Medical professionals can only manage symptoms.

---

improved agricultural stock were limited to selective breeding programs where organisms with the desired characteristics were allowed to breed. With a greater understanding of the genetics, scientists irradiated cells to produce mutations that were then screened to determine if they were desirable. While this approach is a very informative way to learn about the genetics of an organism, it lacked the ability to create a specific change. Creating mutations is a very haphazard process. However, today the results are achieved in a much more directed manner that uses biotechnology's ability to transfer DNA from one organism to another. Techniques such as gene cloning and transformation (introducing new DNA sequences into an organism to alter their genetic makeup) allow scientists to introduce very specific characteristics to an organism. These techniques use enzymes that manipulate DNAs to form complementary base pairs. Once the new DNA sequences are stable and transferred into the host cell, the cell is genetically altered and begins to read the new DNA and produce new cell products such as enzymes and other important products. The resulting new form of DNA is called **recombinant DNA.** Organisms that contain these genetic changes are referred to as **genetically modified (GM)**. Genetic engineers identify and isolate sequences of nucleotides from a living or dead cell. Organisms such as viruses, bacteria, fungi, plants, and animals or their offspring have been engineered so that they contain genes from at least one unrelated organism.

As this highly sophisticated procedure has been refined, it has become possible to quickly and accurately splice genes from a variety of species into host bacteria or other host cells by a process called *gene cloning*. This makes possible the synthesis of large quantities of important products. For example, recombinant DNA procedures are responsible for the production of human insulin, used in the control of diabetes; nutritionally enriched "golden rice," capable of supplying underdeveloped nations with missing essential amino acids; interferon, used as an antiviral agent; human growth hormone, used to stimulate growth in children lacking this hormone; and somatostatin, a brain hormone also implicated in growth. Over two hundred such products have been manufactured using these methods.

Although some of these chemicals have been produced in small amounts from genetically engineered microorganisms, crops such as turnips, rice, soybeans, potatoes, cotton, corn, and tobacco can generate tens or hundreds of kilograms of specialty chemicals per year. Many of these GM crops also have increased nutritional value and yet can be cultivated using traditional methods. Such crops have the potential of supplying the essential amino acids, fatty acids, and other nutrients now lacking in the diets of people in underdeveloped or developing nations. Researchers have also shown, for example, that turnips can produce interferon (an antiviral agent), tobacco can create antibodies to fight human disease, oilseed rape plants can serve as a source of human brain hormones, and potatoes can synthesize human serum albumin that is indistinguishable from the genuine human blood protein.

Some of the likely rewards of biotechnology are (1) production of additional, medically useful proteins; (2) mapping of the locations of genes on human chromosomes; (3) a more complete understanding of how genes are regulated; (4) production of crop plants with increased yields; and (5) development of new species of garden plants (figure 26.17).

Gene Therapy

The field of biotechnology allows scientists and medical doctors to work together and potentially cure genetic disorders. Unlike contagious diseases, genetic diseases cannot be caught or transmitted because they are caused by a genetic predisposition for a particular disorder—not separate, disease-causing organisms such as a bacterium or virus. Gene therapy involves inserting genes, deleting genes, or manipulating the action of genes in order to cure or lessen the effect of genetic diseases. These activities are very new and experimental. While these lines of investigation create hope for many, there are many problems that must be addressed before gene therapy becomes a realistic treatment for many disorders.

The strategy for treating someone with gene therapy varies depending on the individual's disorder. When designing a gene therapy treatment, scientists have to ask themselves, "Exactly what is the problem?" Is the mutant gene not working at all? Is it working normally, but is there too little activity? Is there too much protein being made? Or possibly, is the problem that the gene is acting in a unique and new manner? If there is no gene activity or too little gene activity, the scientists need to somehow introduce a more active version of the gene. If there is too

FIGURE 26.17 Using our understanding of how DNA codes for proteins allows scientists to develop treatments for disease as well as develop crops with greater yield and nutritional value.

much activity or if the problem is caused by the gene having a new activity, this excess activity must first be stopped and then the normal activity restored.

To stop a mutant gene from working, scientists must change it. This typically involves inserting a mutation into the protein-coding region of the gene or the region that is necessary to activate the gene. Scientists have used some types of viruses to do this in organisms other than humans for some time now. The difficulty in this technique is to mutate only that one gene without disturbing the other genes of the cells and creating more mutations in other genes. Developing reliable methods to accomplish this is a major focus of gene therapy. Once the mutant gene is silenced, the scientists begin the work of introducing a "good" copy of the gene. Again, there are many difficulties in this process:

- Scientists must find a way of returning the corrected DNA to the cell.
- The corrected DNA must be made a part of the cell's DNA so that it is passed on with each cell division, it doesn't interfere with other genes, and it can be transcribed by the cell as needed.
- Finally, cells containing the corrected DNA must be reintroduced to the patient.

Many of these techniques are experimental, and the medical community is still evaluating their usefulness in treating many disorders, as well as the risks that these techniques pose to the patient.

Recently, the first efforts to determine the human genome, the entire human DNA sequence, were completed. Many

Connections...

Human Genome Project and Its Applications

The first draft of the human genome was completed early in 2003, when the complete nucleotide sequence of all twenty-three pairs of human chromosomes was determined. By sequencing the human genome, it is as if we have now identified all the words in the human gene "dictionary." Continued analysis will provide the definitions for these words—what these words tell the cell to do. The information provided by the human genome project is increasingly useful in diagnosing diseases and providing genetic counseling to those considering having children. This information can identify human genes and pro-

teins that can be targets for drugs and new gene therapies. Once it is known where an abnormal gene is located and how it differs in base sequence from the normal DNA sequence, steps could be taken to correct the abnormality. Further defining the human genome will also result in the discovery of new families of proteins and will help explain basic physiological and cell biological processes common to many organisms. All this information will increase the breadth and depth of the understanding of basic biology.

It was originally estimated that there were between 100,000 and 140,000 genes

in the human genome, because scientists were able to detect so many different proteins. DNA sequencing data indicate that there are only about 20,000 protein-coding genes—only about twice as many as in a worm or a fly. Our genes are able to generate several different proteins per gene because of alternative splicing. Alternative splicing occurs much more frequently than previously expected. Knowing this information provides insights into the evolution of humans and will make future efforts to work with the genome through bioengineering much easier.

---

scientists feel that advances in medical treatments will occur more quickly by having this information available. Three new fields of biology have grown out of these efforts: *genomics, transcriptomics,* and *proteomics.* Genomics is the study of the DNA sequence and looks at the significance of how different genes and DNA sequences are related to each other. When genes are identified from the DNA sequence, transcriptomics looks at when, where, and how much a gene is expressed. Finally, proteomics examines the proteins that are predicted from the DNA sequence. From these types of studies, we are able to identify gene families that can be used to determine how humans have evolved on a molecular level, how genes are used in an organism throughout its body and over its life span, and how to identify common themes from one protein to the next.

Strategy Two: Sequencing

The second strategy of biotechnology works with comparing of nucleotide sequences from different cells. The closer organisms are genetically, the more closely their DNA sequence will resemble one another. Base sequence analysis can be done two ways. Researchers can look at the DNA sequences directly for comparison. However, this is a very tedious and time-consuming process. A frequently used alternative is to use enzymes to cut the DNA in specific places to create DNA fragments of different lengths, separate out DNA fragments by size, and analyze these for similarities and differences. This second approach scans bigger stretches of DNA more quickly but not as closely. This method can be used by biologists to better understand evolutionary relationships between what at first glance might be unrelated species. This second category of research involves directly manipulating DNA using the more sophisticated techniques such as the polymerase chain reaction (PCR), genetic fingerprinting, and cloning.

The PCR Reaction

Both the cloning and sequencing strategies of biotechnology have been greatly aided by the development of a technique

called PCR. In 1989, the American Association for the Advancement of Science named DNA polymerase "Molecule of the Year." The value of this enzyme in the *polymerase chain reaction (PCR)* is so great that it could not be ignored. Just what is the PCR, how does it work, and what can you do with it?

PCR is a laboratory procedure for copying selected segments of DNA. A single cell can provide enough DNA for analysis and identification! Having a large number of copies of a "target sequence" of nucleotides enables biochemists to more easily work with DNA. This is like increasing the one "needle in the haystack" to such large numbers (one-hundred billion in only a matter of hours) that they are not hard to find, recognize, and work with. The types of specimens that can be used include semen, hair, blood, bacteria, protozoa, viruses, mummified tissue, and frozen cells. The process requires the DNA specimen, free DNA nucleotides, synthetic "primer" DNA, DNA polymerase, and simple lab equipment such as a test tube and a source of heat.

Having decided which target sequence of nucleotides (which "needle") is to be replicated, scientists heat the specimen of DNA to separate the coding and noncoding strands. Molecules of synthetic "primer" DNA are added to the specimen. These primer molecules are specifically designed to attach to the ends of the target sequence. Next, a mixture of nucleotides is added so that they can become the newly replicated DNA. The presence of the primer, attached to the DNA and added nucleotides, serves as the substrate for the DNA polymerase. Once added, the polymerase begins making its way down the length of the DNA from one attached primer end to the other. The enzyme bonds the new DNA nucleotides to the strand, replicating the molecule as it goes. It stops when it reaches the other end, having produced a new copy of the target sequence. Because the DNA polymerase will continue to operate as long as enzymes and substrates are available, the process continues, and in a short time, there are billions of small pieces of DNA, all replicas of the target sequence.

So what, you say? Well, consider the following. Using the PCR, scientists have been able to:

1. More accurately diagnose such diseases as sickle-cell anemia, cancer, Lyme disease, AIDS, and Legionnaires' disease;
2. Perform highly accurate tissue typing for matching organ transplant donors and recipients;
3. Help resolve criminal cases of rape, murder, assault, and robbery by matching a suspect's DNA to that found at the crime scene;
4. Detect specific bacteria in environmental samples;
5. Monitor the spread of genetically engineered microorganisms in the environment;
6. Check water quality by detecting bacterial contamination from feces;
7. Identify viruses in water samples;
8. Identify disease-causing protozoa in water;
9. Determine specific metabolic pathways and activities occurring in microorganisms;
10. Determine races, distribution patterns, kinships, migration patterns, evolutionary relationships, and rates of evolution of long-extinct species;
11. Accurately settle paternity suits;
12. Confirm identity in amnesia cases;
13. Identify a person as a relative for immigration purposes;
14. Provide the basis for making human antibodies in specific bacteria;
15. Possibly provide the basis for replicating genes that could be transplanted into individuals suffering from genetic diseases; and
16. Identify nucleotide sequences peculiar to the human genome (an application currently underway as part of the Human Genome Project).

Stem Cells

Stem cells are cells that are self-renewing and have not yet completed determination or differentiation, so they have the potential to develop into many different cell types. Scientists can generate stem cells by nuclear transfer techniques, but these cells occur naturally throughout the body. They are involved in many activities, including tissue regeneration, wound healing, and cancer.

If scientists had the ability to control determination and differentiation, that ability could allow the manipulation of an organism's cells or the insertion of cells into an organism to allow the regrowth of damaged tissues and organs in humans. This could aid in the cure or treatment of many medical problems, such as the repair of damaged knee cartilage, heart tissue from a heart attack, or damaged nerve tissue from spinal or head injuries. Some kinds of degenerative diseases occur because specific kinds of cells die or cease to function properly. Parkinson's disease results from malfunctioning brain cells, and many forms of diabetes are caused by malfunctioning cells in the pancreas. If stem cells could be used to replace these malfunctioning cells, normal function could be restored and the diseases cured.

A cloning experiment has great scientific importance because it represents an advance in scientists' understanding of the processes of *determination* and *differentiation*. Recall that determination is the process a cell goes through to select which genes it will express. A differentiated cell has become a particular cell type because of the proteins that it expresses. Differentiation is more or less a permanent condition. The techniques that produced Dolly and other cloned animals use a differentiated cell and reverse the determination process so that this cell is able to express all the genes necessary to create an entirely new organism. Until this point, scientists were not sure that this was possible.

Embryonic and Adult Stem Cells

Scientists have also explored other methods of obtaining stem cells (figure 26.18). Embryonic stem cells are cells obtained from embryos and reach an intermediary level of determination at which they are committed to becoming a particular *tissue* type, but not necessarily a particular *cell* type. An example of this intermediate determination occurs when stems cells

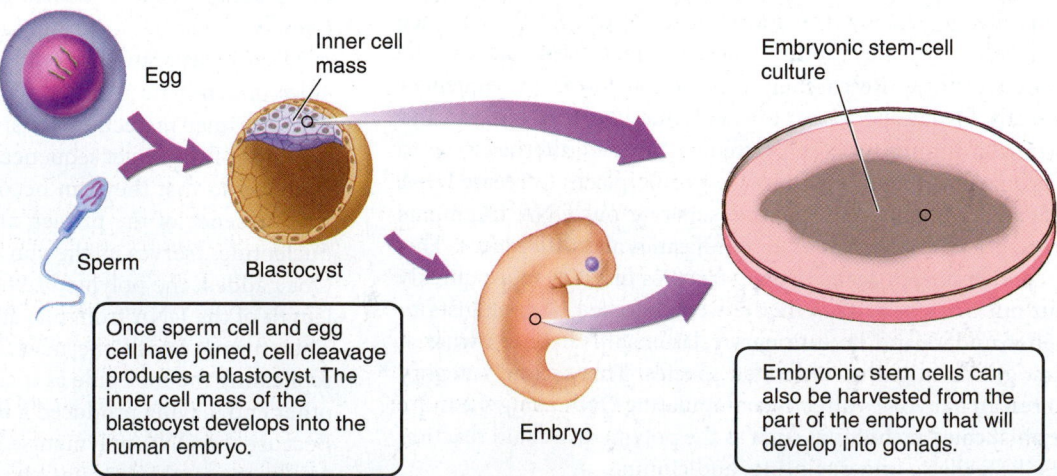

Egg

Sperm

Inner cell mass

Blastocyst

Embryo

Embryonic stem-cell culture

Once sperm cell and egg cell have joined, cell cleavage produces a blastocyst. The inner cell mass of the blastocyst develops into the human embryo.

Embryonic stem cells can also be harvested from the part of the embryo that will develop into gonads.

FIGURE 26.18 After fertilization of an egg with sperm, the cell begins to divide and form a mass of cells. Each of these cells has the potential to become any cell in the embryo. Embryonic stem cells may be harvested at this point or at other points in the determination process.

Science and Society

Cloning—What This Term Really Means

The term **clone** (Greek word *klön* for "twig") refers to exact copies of biological entities such as genes, cells, or organisms. The term refers to the outcome, not the way that the results are achieved. Many whole organisms "clone" themselves (e.g., bacteria undergo asexual binary fission, and many plants reproduce "clonally" through asexual means, sometimes referred to as *vegetative regeneration*). One process of making exact copies of the cells of a cat (box figure 26.6), sheep, or other organism in the laboratory is technically known as "*somatic cell nuclear transfer.*" The purpose of somatic cell nuclear transfer is not the "cloning" of a human or other adult but the establishment of a clonal cell line. In this technique, the nucleus of a somatic (body) cell of an adult donor is transferred into an oocyte, or "egg cell," whose own nucleus has been removed. Should such hybrid cells be cultured *in vitro* (*vitro* = glass; e.g., in test tubes) for a few weeks, the result is the production of a line of cloned cells (e.g., human embryonic stem cells). These can be used for a variety of medically valuable therapies or research (e.g., investigating certain genetic or other diseases). Using the other technique, *reproductive cloning*, the hybrid cell can be implanted into the uterus of a host or surrogate mother, where it may develop into a genetic copy or clone of the whole individual who donated the nucleus. A great variety of animals have already been cloned in this fashion, including cats, sheep, mice, monkeys, pigs, cattle, rabbits, mules, deer, horses, fish, and frogs.

In the film *Godsend*, starring Robert De Niro, Greg Kinnear, and Rebecca Romijn-Stamos, parents agree to clone their accidentally killed, eight-year-old son but wish they hadn't when the clone starts behaving erratically.

Questions to Discuss

1. Based on what you have learned, do you think cloning as depicted in *Godsend* would be possible?
2. What are the ethical and moral implications of taking such action?
3. Does it make a difference if the organism cloned is a human or a cat?

A

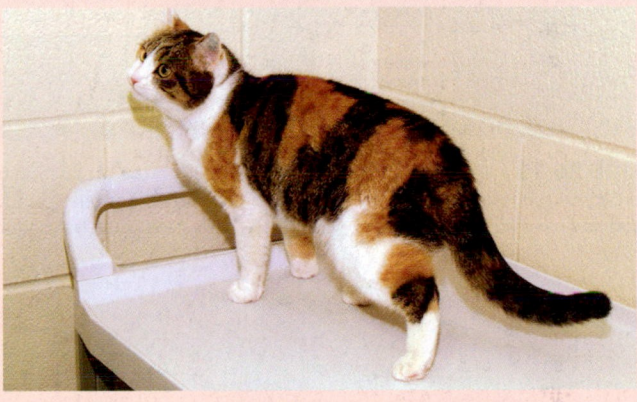

B

Box Figure 26.6   "CC," the cloned kitty, with surrogate mother and her genetic mother, "Rainbow." Out of 87 implanted cloned embryos, CC (*Copy Cat!*) is the only one to survive—comparable to the success rate in sheep, mice, cows, goats, and pigs. (*A*) Notice that she is completely unlike her tabby surrogate mother. (*B*) "Rainbow" is her genetic donor, and both are female calico domestic shorthair cats.
Source: Nature, AOP, published online: 14 February 2002; DOI: 10.1038/nature723.

become determined to be any one of several types of nerve cells but have not yet committed to becoming any one nerve cell. Scientists call these partially determined stem cells "tissue specific." These types of stem cells can be found in adults. One example is hematopoietic stem cells. These cells are able to become the many different types of cells found in blood—red blood cells, white blood cells, and platelets. The disadvantage of using these types of stem cells is that they have already become partially determined and do not have the potential to become every cell type.

Personalized Stem Cell Lines

Scientists hope that eventually it will be possible to produce embryonic stem cells from somatic cells by using somatic cell nuclear transfer techniques similar to those used for cloning a sheep. This technique would involve transferring a nucleus from the patient's cell to a human egg that has had its original nucleus removed. The human egg would be allowed to grow and develop to produce embryonic stem cells. If the process of determination and differentiation can be controlled, new tissues or even organs could be developed through what is termed regenerative medicine.

Under normal circumstances, organ transplant patients must always worry about rejecting their transplant and must take strong immunosuppressant drugs to avoid organ rejection. Tissue and organs grown from customized stem cells would have the benefit of being immunologically compatible with the patient; thus, organ rejection would not be a concern (figure 26.19).

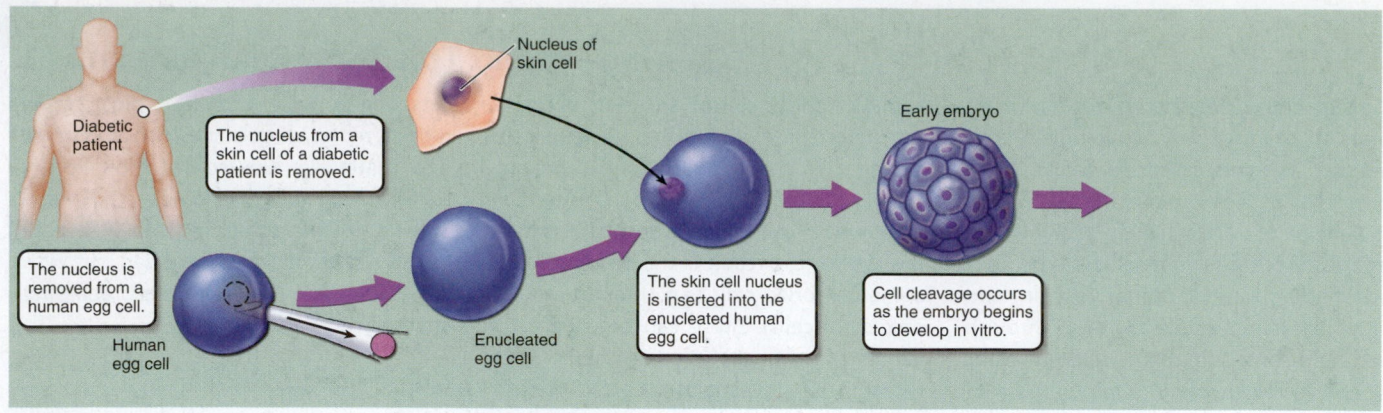

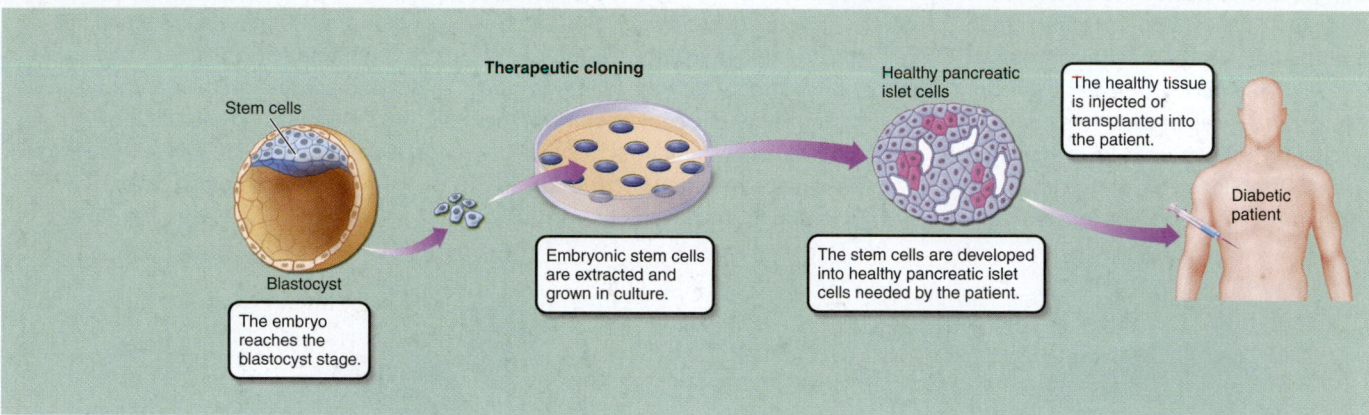

FIGURE 26.19 One potential use of biotechnology is the production of customized stem cell lines. In this application a somatic cell from a patient would be inserted into a human egg from which the nucleus has been removed. The egg would divide and generate stem cells. These cells could then be cultured and used for therapy. In this example the stem cells could be used to create pancreatic cells to treat a diabetic patient.

The potential therapeutic value of stem cells has resulted in the founding of many clinics around the world. These clinics offer stem cell–based therapies to patients with a variety of medical conditions. However, the benefits of these therapies are as yet unproven and, in fact, have the potential for serious harm. "Stem cell tourism" is a phrase being used to describe this industry because desperate patients travel to these clinics in hopes that such therapies will save their lives. This new industry is teeming with "medical tourist traps" offering unproven medical treatments to unsuspecting consumers. Unfortunately, the days of customized stem cells, stem cell therapies, and organ culture are still in the future.

Genetic Fingerprinting

With another genetic engineering accomplishment, *genetic fingerprinting*, it is possible to show the nucleotide sequence differences between individuals since no two people have the same nucleotide sequences. While this sounds like an easy task, the presence of many millions of base pairs in a person's chromosomes makes this a lengthy and sometimes impractical process. Therefore, scientists don't really do a complete fingerprint but focus only on certain shorter, repeating patterns in the DNA. By focusing on these shorter, repeating nucleotide sequences, it is possible to determine whether samples from two individuals have these same repeating segments. Scientists use a small number of sequences that are known to vary a great deal among individuals and compare those to get a certain probability of a match. The more similar the sequences, the more likely the two samples are from the same person. The less similar the sequences, the less likely the two samples are from the same person. In criminal cases, DNA samples from the crime site can be compared to those taken from suspects. If 100 percent of the short, repeating sequence matches, it is highly probable that the suspect was at the scene of the crime and may be the guilty party. This same procedure can also be used to confirm the identity of a person, as in cases of amnesia, murder, or accidental death (figure 26.20).

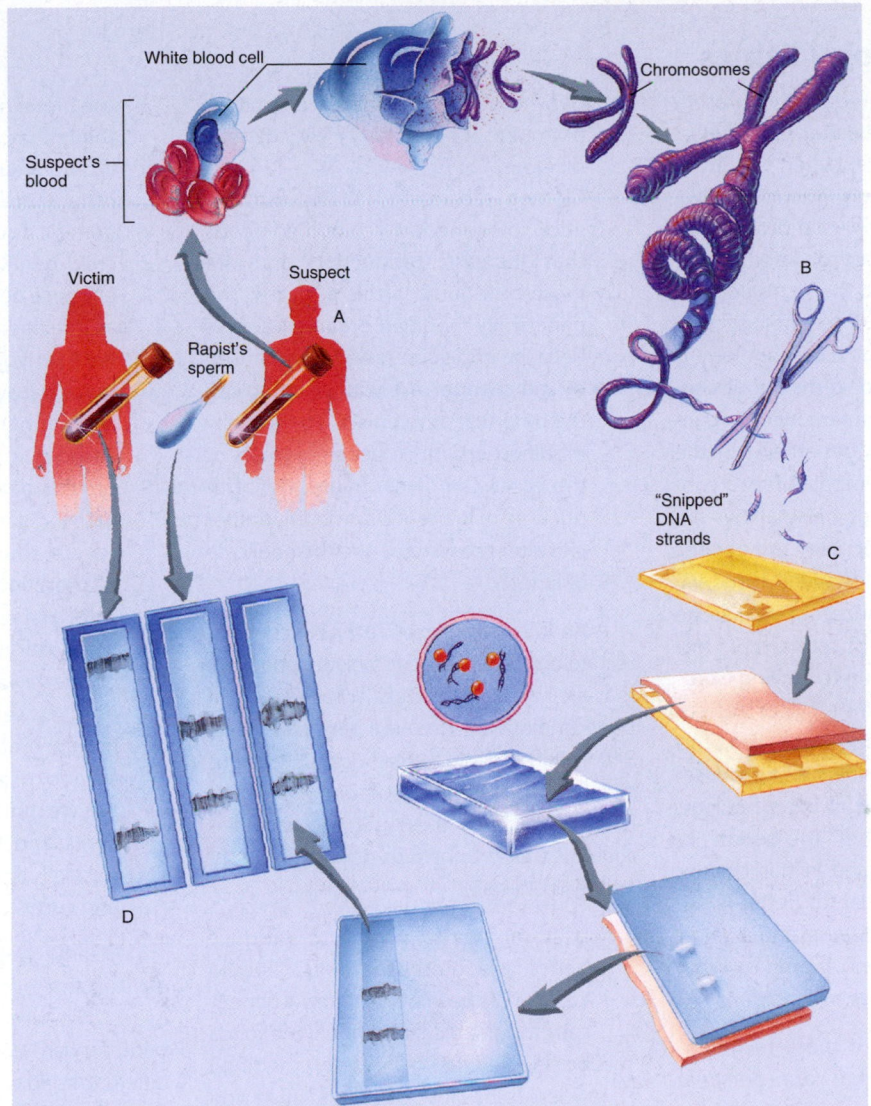

FIGURE 26.20 Because every person's DNA is unique (*A*), when samples of an individual's DNA are subjected to restriction enzymes, the cuts will occur in different places and DNA chunks of different sizes will result. (*B*) Restriction enzymes have the ability to cut DNA at places where specific sequences of nucleotides occur. When the chunks are caused to migrate across an electrophoresis gel (*C*), the smaller fragments migrate farther than the larger fragments, producing a pattern known as a "DNA fingerprint" (*D*). Because of individual differences in DNA sequences, these sites vary from person to person. As a result, the DNA fingerprint that separates DNA fragments on the basis of size can appear different from one person to another. Several controls are done in this type of experiment. One of the controls is the presence of the victim's DNA. Why is this important for the interpretation of this test?

SUMMARY

Genes are units of heredity composed of specific lengths of *DNA* that determine the characteristics an organism displays. Specific genes are at specific loci on specific chromosomes. The *phenotype* displayed by an organism is the result of the effect of the environment on the ability of the genes to *express* themselves.

Diploid organisms have two genes for each characteristic. The alternative forms of genes for a characteristic are called *alleles*. There may be many different alleles for a particular characteristic. Organisms with two identical alleles are *homozygous* for a characteristic; those with different alleles are *heterozygous*. Some alleles are *dominant* over other alleles that are said to be *recessive*.

Sometimes two alleles express themselves, and often a gene has more than one recognizable effect on the phenotype of the organism. Some characteristics may be determined by several different pairs of alleles. In humans and some other organisms, phenotypes are determined not only by the type of allele (gene) received but also by the alternative ways in which they interact with each other and how the environment influences their expression and epigenetic

Science and Society

Biotechnology Ethics

Scientific advances frequently present society with ethical questions that must be resolved. For example, when first introduced, immunization and in vitro fertilization were highly controversial procedures. How will new technology be used safely? Who will benefit? Should the technology be used to make a profit? Biotechnology is no different. Many feel that biotechnology is dangerous. There are concerns about contaminating the environment with organisms that are modified genetically in the lab. What would be the impact of such contamination? Biotechnology also allows scientists to examine molecularly the genetic characteristics of an individual. How will this ability to characterize individuals be used? How will it be misused? Others feel that biotechnology is akin to playing God.

One way to explore the ethics of biotechnology is to weigh its pros against its cons. This method of thinking considers all the consequences and implications of biotechnology. Which outweighs the other? The benefits of nearly everything discussed in this chapter include a greater potential for better medical treatment, cures for disease, and a better understanding of the world around us. What price must we pay for these advances?

- The development of these technologies may mean that our personal genetic information becomes public record. How might this information be misused? Insurance companies might deny coverage or charge exorbitant premiums for individuals with genetic diseases.
- Cloning technology allows the creation of genetically modified foods that increase production and are more nutritious. Is this ethical if the genetically modified organism suffers because of disorders and pain caused by the change? Are you willing to risk the potential problems of a genetically modified organism becoming part of the ecosystem? How might the introduction of genetically modified species alter ecosystems and their delicate balance?

Another way to explore the ethics of biotechnology is to ask if it violates principles that are valued by society. What aspects of biotechnology threaten the principles of the Bill of Rights? Basic human rights? Religious beliefs? Quality-of-life issues? Animal rights issues? Which of these sets of principles should be used to help us decide if biotechnology is ethical?

- Is it inherently wrong to produce genetically modified foods? Should companies be allowed to grow genetically modified crops? Should the foods be labeled as genetically modified when sold? How does this affect you as a consumer or simply as a person?
- Is it inherently wrong to manipulate genes? Are humans wise enough to use biotechnology safely? Is this something that only God should control? Would you have your child genetically altered as a fetus to prevent a genetic disease? Would you have your child genetically altered as a fetus to enhance desirable characteristics, such as intelligence, or even to control gender? Do you feel that one situation is morally justified but the other is not?
- Is it inherently wrong to harvest embryonic stem cells? Stem cells may provide new avenues of treatment for many disorders. Although there are several sources of stem cells, the cells of most interest are embryonic stem cells. Harvesting these cells destroys the embryo. Even if the embryo is not yet aware of its environment and does not sense pain, is it ethical to use human embryos to advance the treatment of disease?
- Are we morally obligated to search for cures and treatments? Can we stop research if people still need treatment and cures?

Clearly, these are issues that our society will debate for some time. Many of these issues have been debated for decades and bring forward very strong feelings and very different world views. As you continue to hear more about biotechnology in your day-to-day life, consider how that form of biotechnology may affect you.

---

factors. There are six generally recognized patterns: *multiple allelic, polygenic, pleiotropic, codominant, incomplete dominant,* and *X-linked characteristics.*

The successful operation of a living cell depends on its ability to accurately use the genetic information found in its DNA. The *enzymes* that can be synthesized using the information in DNA are responsible for the efficient control of a cell's metabolism. However, the production of protein molecules is under the control of the *nucleic acids,* the primary control molecules of the cell. The structure of the nucleic acids, *DNA* and *RNA,* determines the structure of the proteins, whereas the structure of the proteins determines their function in the cell's life cycle. *Protein synthesis* involves the decoding of the DNA into specific protein molecules and the use of the intermediate molecules, mRNA and tRNA, at the ribosome. Errors in any of the codons of these molecules may produce observable changes in the cell's functioning and can lead to cell death.

DNA replication results in an exact doubling of the genetic material. The process virtually guarantees that identical strands of DNA will be passed on to the next generation of cells.

Methods of manipulating DNA have led to the controlled transfer of genes from one kind of organism to another. This has made it possible for bacteria to produce a number of human gene products.

KEY TERMS

alleles (p. **667**)
anticodon (p. **685**)
chromosome (p. **679**)
clone (p. **691**)
codon (p. **684**)
complementary base (p. **678**)
DNA code (p. **683**)
DNA replication (p. **678**)
dominant allele (p. **668**)
epigenetics (p. **677**)
gene (p. **666**)
genetically modified (GM)
   organisms (p. **687**)

genetics (p. **666**)
genome (p. **667**)
genotype (p. **667**)
heterozygous (p. **668**)
homozygous (p. **667**)
incomplete dominance
   (p. **672**)
law of dominance (p. **669**)
law of independent assortment
   (p. **669**)
law of segregation (p. **669**)
Mendelian genetics (p. **666**)
messenger RNA (mRNA) (p. **684**)

multiple alleles (p. **674**)
mutagenic agent (p. **686**)
mutation (p. **685**)
nucleic acids (p. **678**)
nucleotides (p. **678**)
phenotype (p. **667**)
pleiotropy (p. **674**)
polygenic inheritance (p. **674**)
Punnett square (p. **671**)
recessive allele (p. **668**)
recombinant DNA (p. **687**)
ribosomal RNA (rRNA)
   (p. **685**)

sickle-cell anemia (p. **686**)
single-factor cross (p. **670**)
stem cell (p. **690**)
telomere (p. **684**)
transcription (p. **678**)
transfer RNA (tRNA)
   (p. **684**)
translation (p. **678**)
X-linked gene (p. **672**)

APPLYING THE CONCEPTS

Answers are located in appendix F.

Part I

1. An example of a phenotype is
   a. AB type blood.
   b. an allele for type A and B blood.
   c. a sperm with an allele for A blood.
   d. lack of iron in the diet causing anemia.

2. A person's height is determined by the interaction of numerous alleles. This is an example of
   a. autosomes.
   b. pleiotropy.
   c. single-factor crosses.
   d. polygenic inheritance.

3. In order for a recessive X-linked trait to appear in a female, she must inherit a recessive allele from
   a. neither parent.
   b. both parents.
   c. her father only.
   d. her mother only.

4. It was noticed that in certain flowers, if a flower had red petals, they were usually small petals. If the petals were white, pink, or orange, they were usually large petals. This could be the result of
   a. codominance.
   b. polygenic inheritance.
   c. incomplete dominance.
   d. None of these is correct.

5. A woman with blood type O and a man with blood type AB have a child together. What are the possible blood types of this child?
   a. AB or O
   b. A, B, or O
   c. A or B
   d. AB, A, B, or O

6. "She has really long fingers and toes and is exceptionally tall." This statement expresses the _____ of an individual.
   a. genotype
   b. phenotype
   c. monohybridization
   d. locus placement

Part II

7. One difference between mRNA and mature tRNA is the:
   a. kinds of polypeptide components.
   b. mRNA is straight while tRNA is cloverleaf-shaped.
   c. part of RNA from which they were coded.
   d. age of the molecules only.

8. While one strand of double-stranded DNA is being transcribed to mRNA:
   a. the complementary strand makes tRNA.
   b. the complementary strand is inactive.
   c. the complementary strand at this point is replicating.
   d. mutations are impossible during this short period.

9. A major difference between the genetic data of prokaryotic and eukaryotic cells is that in prokaryotes, the
   a. genes are RNA, not DNA.
   b. histones are arranged differently.
   c. double-stranded DNA is circular.
   d. double-stranded DNA is absent in bacteria.

10. If the DNA gene strand has the base sequence CCA, which of the following DNA nitrogenous bases would pair with them during DNA replication?
    a. CCA
    b. GGU
    c. GGU
    d. GGT

11. Using the Amino Acid–mRNA Nucleic Acid Dictionary (table 26.3), the codon ACG would control the placement of which amino acid in the synthesis of a new protein?
    a. proline
    b. isoleucine
    c. glycine
    d. threonine

12. This new field of biotechnology examines the proteins that are predicted from the DNA sequence.
    a. human genome
    b. codon technology
    c. proteomics
    d. hormone therapy

13. This term refers to exact copies of biological entities such as genes, cells, or organisms.
    a. clone
    b. PRC
    c. bioethics
    d. gene therapy
14. This is the process a cell goes through to select which genes it will express.
    a. differentiation
    b. determination
    c. activation
    d. actuation
15. Phenomena that are either known or suspected causes of DNA damage are called
    a. hallucinogenic agents.
    b. activation agents.
    c. GM agents.
    d. mutagenic agents.

QUESTIONS FOR THOUGHT

Part I

1. What is the probability of each of the following sets of parents producing the given genotypes in their offspring?

	Parents	Offspring	Genotype
a.	AA	3 aa	Aa
b.	Aa	3 Aa	Aa
c.	Aa	3 Aa	aa

2. If an offspring has the genotype Aa, what possible combinations of parental genotypes can exist?

3. In humans, the allele for albinism is recessive to the allele for normal skin pigmentation.
   a. What is the probability that a child of a heterozygous mother and father will be an albino?
   b. If a child is normal, what is the probability that it is a carrier of the recessive albino allele?

4. Dwarfism (achondroplasia) is the result of a mutation in the gene for the production of fibroblast growth factor. It is inherited as an autosomal dominant disorder and often occurs as a new mutation. What might be the explanation if two people of normal height and body stature have five children, one of whom displays achondroplasia?

5. A woman with color-deficient vision marries a man with normal vision. They have ten children—six boys and four girls.
   a. How many are expected to have normal vision?
   b. How many are expected to have color-deficient vision?

6. Hemophilia is a disease that prevents the blood from clotting normally. It is caused by a recessive allele located on the X chromosome. A boy has the disease; neither his parents nor his grandparents have the disease. What are the genotypes of his parents and grandparents?

7. The roan color of horses is an example of codominance. Why?

Part II

8. What are the differences between DNA and RNA?

9. List the sequence of events that takes place when a DNA message is translated into protein.

10. Chromosomal and point mutations both occur in DNA. In what ways do they differ? How is this related to recombinant DNA?

11. How does DNA replication differ from the manufacture of an RNA molecule?

12. If a DNA nucleotide sequence is CATAAAGCA, what is the mRNA nucleotide sequence that would base-pair with it?

13 What amino acids would occur in the protein chemically coded by the sequence of nucleotides in question 12?

14. Give an example of a potential use of biotechnology related to stem cells.

15. What role is played by DNA polymerase?

FOR FURTHER ANALYSIS

1. There are two approaches to gene therapy: somatic cell and germ cell line therapy. Compare and contrast these two approaches. Give examples of genetic abnormalities that might be controlled with each method.

2. It is important to remember that Mendel's work was introduced only a little over a hundred years ago. But in that time, science has progressed from his most basic hypothesis to a level of understanding that has already enabled the control of some of our most notorious genetic abnormalities. List five such diseases. List the kinds of genetic medicine that are proposed to help with each.

3. What is the genetic basis of the following diseases and what are their symptoms?
   a. Kearns-Sayre syndrome
   b. Leber's hereditary optic neuropathy (LHON)
   c. Myoclonic epilepsy and red ragged fibers (MERRF)

4. Polydactyly is the most common genetic abnormality of the human hand and is classified into three types.
   a. What are these types?
   b. Diagram how such phenotypes would arise beginning with a mutation in the DNA, following through the central dogma and ending with how this phenotype would be produced.

5. It is known that the difference between the genomes of humans and chimpanzees is about 1.5 percent. If there is only this small difference between what appears to be two strikingly different organisms, what percentage difference do you think exists among humans? Based on this information, comment on the validity of separating humans of the world into separate races.

INVITATION TO INQUIRY

Smoking and Mutations: What Do You Really Know?

For decades, cancer has been empirically (based on commonplace experiences) linked to the use of tobacco, but only recently have researchers showed a cause-and-effect relationship between the two. Studies of patients with lung cancer revealed that 60 percent exhibited mutations in a gene known as p53. Use the Internet to answer the following questions:

1. What is the p53 gene?
2. What is apoptosis or "programmed cell death"?
3. What does the compound in tobacco (benzopyrene) have to do with this gene?
4. What is known about the p53 gene and lung cancer?

Appendix A

Mathematical Review

A.1 WORKING WITH EQUATIONS

Many of the problems of science involve an equation, a short-hand way of describing patterns and relationships that are observed in nature. Equations are also used to identify properties and to define certain concepts, but all uses have well-established meanings, symbols that are used by convention, and allowed mathematical operations. This appendix will assist you in better understanding equations and the reasoning that goes with the manipulation of equations in problem-solving activities.

Background

In addition to a knowledge of rules for carrying out mathematical operations, an understanding of certain quantitative ideas and concepts can be very helpful when working with equations. Among these helpful concepts are (1) the meaning of inverse and reciprocal, (2) the concept of a ratio, and (3) fractions.

The term *inverse* means the opposite, or reverse, of something. For example, addition is the opposite, or inverse, of subtraction, and division is the inverse of multiplication. A *reciprocal* is defined as an inverse multiplication relationship between two numbers. For example, if the symbol n represents any number (except zero), then the reciprocal of n is $1/n$. The reciprocal of a number $(1/n)$ multiplied by that number (n) always gives a product of 1. Thus, the number multiplied by 5 to give 1 is $1/5$ $(5 \times 1/5 = 5/5 = 1)$. So $1/5$ is the reciprocal of 5, and 5 is the reciprocal of $1/5$. Each number is the *inverse* of the other.

The fraction $1/5$ means 1 divided by 5, and if you carry out the division, it gives the decimal 0.2. Calculators that have a $1/x$ key will do the operation automatically. If you enter 5, then press the $1/x$ key, the answer of 0.2 is given. If you press the $1/x$ key again, the answer of 5 is given. Each of these numbers is a reciprocal of the other.

A *ratio* is a comparison between two numbers. If the symbols m and n are used to represent any two numbers, then the ratio of the number m to the number n is the fraction m/n. This expression means to divide m by n. For example, if m is 10 and n is 5, the ratio of 10 to 5 is 10/5, or 2:1.

Working with *fractions* is sometimes necessary in problem-solving exercises, and an understanding of these operations is needed to carry out unit calculations. It is helpful in many of these operations to remember that a number (or a unit) divided by itself is equal to 1; for example,

$$\frac{5}{5} = 1 \qquad \frac{\text{inch}}{\text{inch}} = 1 \qquad \frac{5 \text{ inches}}{5 \text{ inches}} = 1$$

When one fraction is divided by another fraction, the operation commonly applied is to "invert the denominator and multiply." For example, 2/5 divided by 1/2 is

$$\frac{\dfrac{2}{5}}{\dfrac{1}{2}} = \frac{2}{5} \times \frac{2}{1} = \frac{4}{5}$$

What you are really doing when you invert the denominator of the larger fraction and multiply is making the denominator $(1/2)$ equal to 1. Both the numerator $(2/5)$ and the denominator $(1/2)$ are multiplied by 2/1, which does not change the value of the overall expression. The complete operation is

$$\frac{\dfrac{2}{5}}{\dfrac{1}{2}} \times \frac{\dfrac{2}{1}}{\dfrac{2}{1}} = \frac{\dfrac{2}{5} \times \dfrac{2}{1}}{\dfrac{1}{2} \times \dfrac{2}{1}} = \frac{\dfrac{4}{5}}{\dfrac{2}{2}} = \frac{\dfrac{4}{5}}{1} = \frac{4}{5}$$

Symbols and Operations

The use of symbols seems to cause confusion for some students because it seems different from their ordinary experiences with arithmetic. The rules are the same for symbols as they are for numbers, but you cannot do the operations with the symbols until you know what values they represent. The operation signs, such as $+$, $\div$, $\times$, and $-$, are used with symbols to indicate the operation that you *would* do if you knew the values. Some of the mathematical operations are indicated several ways. For example, $a \times b$, $a \cdot b$, and ab all indicate the same thing, that a is to be multiplied by b. Likewise, $a \div b$, a/b, and $a \times 1/b$ all indicate that a is to be divided by b. Since it is not possible to carry out the operations on symbols alone, they are called *indicated operations*.

Operations in Equations

An equation is a shorthand way of expressing a simple sentence with symbols. The equation has three parts: (1) a left side, (2) an equal sign $(=)$, which indicates the equivalence of the two sides, and (3) a right side. The left side has the same value and units as the right side, but the two sides may have a very different appearance. The two sides may also have the symbols that indicate mathematical operations ($+$, $-$, $\times$, and so forth) and may be in certain forms that indicate operations (a/b, ab, and so forth). In any case, the equation is a complete expression that states the left side has the same value and units as the right side.

Equations may contain different symbols, each representing some unknown quantity. In science, the expression "solve the equation" means to perform certain operations with one symbol (which represents some variable) by itself on one side of the equation. This single symbol is usually, but not necessarily, on the left side and is not present on the other side. For example, the equation $F = ma$ has the symbol F on the left side. In science, you would say that this equation is solved for F. It could also be solved for m or for a, which will be considered shortly. The equation $F = ma$ is solved for F, and the *indicated operation* is to multiply m by a because they are in the form ma, which means the same thing as $m \times a$. This is the only indicated operation in this equation.

A solved equation is a set of instructions that has an order of indicated operations. For example, the equation for the relationship between a Fahrenheit and Celsius temperature, solved for °C, is $C = 5/9(F - 32)$. A list of indicated operations in this equation is as follows:

1. Subtract 32° from the given Fahrenheit temperature.
2. Multiply the result of (1) by 5.
3. Divide the result of (2) by 9.

Why are the operations indicated in this order? Because the bracket means 5/9 of the *quantity* $(F - 32°)$. In its expanded form, you can see that $5/9(F - 32°)$ actually means $5/9(F) - 5/9(32°)$. Thus, you cannot multiply by 5 or divide by 9 until you have found the quantity of $(F - 32°)$. Once you have figured out the order of operations, finding the answer to a problem becomes almost routine as you complete the needed operations on both the numbers and the units.

Solving Equations

Sometimes it is necessary to rearrange an equation to move a different symbol to one side by itself. This is known as solving an equation for an unknown quantity. But you cannot simply move a symbol to one side of an equation. Since an equation is a statement of equivalence, the right side has the same value as the left side. If you move a symbol, you must perform the operation in a way that the two sides remain equivalent. This is accomplished by "canceling out" symbols until you have the unknown on one side by itself. One key to understanding the canceling operation is to remember that a fraction with the same number (or unit) over itself is equal to 1. For example, consider the equation $F = ma$, which is solved for F. Suppose you are considering a problem in which F and m are given, and the unknown is a. You need to solve the equation for a so it is on one side by itself. To eliminate the m, you do the *inverse* of the indicated operation on m, dividing both sides by m. Thus,

Solve an Equation

The equation for finding the kinetic energy of a moving body is $KE = 1/2mv^2$. You need to solve this equation for the velocity, v.

Answer

The order of indicated operations in the equation is as follows:

1. Square v.
2. Multiply v^2 by m.
3. Divide the result of (2) by 2.

To solve for v, this order is *reversed* as the "canceling operations" are used:

Step 1: Multiply both sides by 2

$$KE = \frac{1}{2}mv^2$$

$$2\,KE = \frac{2}{2}mv^2$$

$$2\,KE = mv^2$$

Step 2: Divide both sides by m

$$\frac{2\,KE}{m} = \frac{mv^2}{m}$$

$$\frac{2\,KE}{m} = v^2$$

Step 3: Take the square root of both sides

$$\sqrt{\frac{2\,KE}{m}} = \sqrt{v^2}$$

$$\sqrt{\frac{2\,KE}{m}} = v$$

or

$$v = \sqrt{\frac{2\,KE}{m}}$$

The equation has been solved for v, and you are now ready to substitute quantities and perform the needed operations.

$$F = ma$$

$$\frac{F}{m} = \frac{ma}{m}$$

$$\frac{F}{m} = a$$

Since m/m is equal to 1, the a remains by itself on the right side. For convenience, the whole equation may be flipped to move the unknown to the left side,

$$a = \frac{F}{m}$$

Thus, a quantity that indicated a multiplication (ma) was removed from one side by an inverse operation of dividing by m.

Consider the following inverse operations to "cancel" a quantity from one side of an equation, moving it to the other side:

If the Indicated Operation of the Symbol You Wish to Remove Is:	Perform This Inverse Operation on Both Sides of the Equation
multiplication	division
division	multiplication
addition	subtraction
subtraction	addition
squared	square root
square root	square

A.2 SIGNIFICANT FIGURES

The numerical value of any measurement will always contain some uncertainty. Suppose, for example, that you are measuring one side of a square piece of paper as shown in figure A.1. You could say that the paper is *about* 3.5 cm wide and you would be correct. This measurement, however, would be unsatisfactory for many purposes. It does not approach the true value of the length and contains too much uncertainty. It seems clear that the paper width is larger than 3.4 cm but shorter than 3.5 cm. But how much larger than 3.4 cm? You cannot be certain if the paper is 3.44, 3.45, or 3.46 cm wide. As your best estimate, you might say that the paper is 3.45 cm wide. Everyone would agree that you can be certain about the first two numbers (3.4) and they should be recorded. The last number (0.05) has been estimated and is not certain. The two certain numbers, together with one uncertain number, represent the greatest accuracy possible with the ruler being used. The paper is said to be 3.45 cm wide.

A *significant figure* is a number that is believed to be correct with some uncertainty only in the last digit. The value of the width of the paper, 3.45 cm, represents three significant figures. As you can see, the number of significant figures can be determined by the degree of accuracy of the measuring instrument being used. But suppose you need to calculate the area of the paper. You would multiply 3.45 cm × 3.45 cm and the product for the area

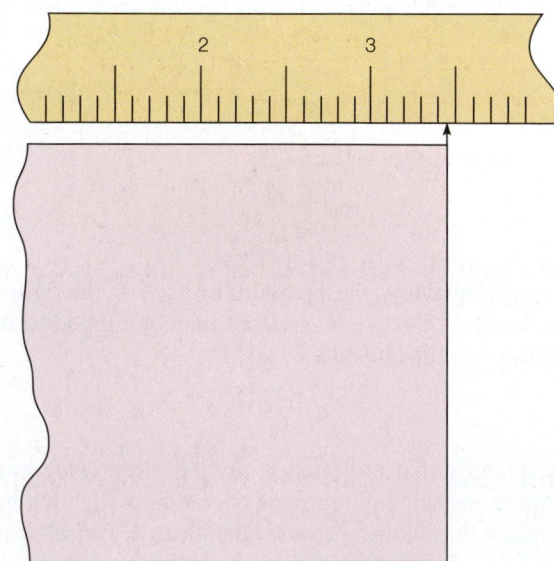

FIGURE A.1

would be 11.9025 cm². This is a greater accuracy than you were able to obtain with your measuring instrument. The result of a calculation can be no more accurate than the values being treated. Because the measurement had only three significant figures (two certain, one uncertain), then the answer can have only three significant figures. The area is correctly expressed as 11.9 cm².

There are a few simple rules that will help you determine how many significant figures are contained in a reported measurement:

1. All digits reported as a direct result of a measurement are significant.
2. Zero is significant when it occurs between nonzero digits. For example, 607 has three significant figures, and the zero is one of the significant figures.
3. In figures reported as *larger than the digit one,* the digit zero is not significant when it follows a nonzero digit to indicate place. For example, in a report that "23,000 people attended the rock concert," the digits 2 and 3 are significant but the zeros are not significant. In this situation, the 23 is the measured part of the figure, and the three zeros tell you an estimate of how many attended the concert, that is, 23 thousand. If the figure is a measurement rather than an estimate, then it is written *with a decimal point after the last zero* to indicate that the zeros *are* significant. Thus, 23,000 has *two* significant figures (2 and 3), but 23,000. has *five* significant figures. The figure 23,000 means "about 23 thousand," but 23,000. means 23,000. and not 22,999 or 23,001.
4. In figures reported as *smaller than the digit 1,* zeros after a decimal point that come before nonzero digits *are not* significant and serve only as place holders. For example, 0.0023 has two significant figures, 2 and 3. Zeros alone after a decimal point or zeros after a nonzero digit indicate a measurement, however, so these zeros *are* significant. The figure 0.00230, for example, has three significant figures since the 230 means 230 and not 229 or 231. Likewise, the figure 3.000 cm has four significant figures because the presence of the three zeros means that the measurement was actually 3.000 and not 2.999 or 3.001.

Multiplication and Division

When multiplying or dividing measurement figures, the answer may have no more significant figures than the *least* number of significant figures in the figures being multiplied or divided. This simply means that an answer can be no more accurate than the least accurate measurement entering into the calculation and that you cannot improve the accuracy of a measurement by doing a calculation. For example, in multiplying 54.2 mi/h × 4.0 h to find out the total distance traveled, the first figure (54.2) has three significant figures but the second (4.0) has only two significant figures. The answer can contain only two significant figures since this is the weakest number of those involved in the calculation. The correct answer is therefore 220 mi, not 216.8 mi. This may seem strange since multiplying the two numbers together gives the answer of 216.8 mi. This answer, however, means a greater accuracy than is possible, and the accuracy cannot be improved over the weakest number involved in the calculation. Since the weakest number (4.0) has only two significant figures the answer must also have only two significant figures, which is 220 mi.

The result of a calculation is *rounded* to have the same least number of significant figures as the least number of a measurement involved in the calculation. When rounding numbers, the last significant figure is increased by 1 if the number after it is 5 or larger. If the number after the last significant figure is 4 or less, the nonsignificant figures are simply dropped. Thus, if two significant figures are called for in the answer of the previous example, 216.8 is rounded up to 220 because the last number after the two significant figures is 6 (a number larger than 5). If the calculation result had been 214.8, the rounded number would be 210 miles.

Note that *measurement figures* are the only figures involved in the number of significant figures in the answer. Numbers that are counted or **defined** are not included in the determination of significant figures in an answer. For example, when dividing by 2 to find an average, the 2 is ignored when considering the number of significant figures. Defined numbers are defined exactly and are not used in significant figures. Since 1 kilogram is *defined* to be exactly 1,000 grams, such a conversion is not a measurement.

Addition and Subtraction

Addition and subtraction operations involving measurements, as with multiplication and division, cannot result in an answer that implies greater accuracy than the measurements had before the calculation. Recall that the last digit to the right in a measurement is uncertain; that is, it is the result of an estimate. The answer to an addition or subtraction calculation can have this uncertain number *no farther from the decimal place than it was in the weakest number involved in the calculation.* Thus, when 8.4 is added to 4.926, the weakest number is 8.4, and the uncertain number is .4, one place to the right of the decimal. The sum of 13.326 is therefore rounded to 13.3, reflecting the placement of this weakest doubtful figure.

The rules for counting zeros tell us that the numbers 203 and 0.200 both have three significant figures. Likewise, the numbers 230 and 0.23 only have two significant figures. Once you remember the rules, the counting of significant figures is straightforward. On the other hand, sometimes you find a

number that seems to make it impossible to follow the rules. For example, how would you write 3,000 with two significant figures? There are several special systems in use for taking care of problems such as this, including the placing of a little bar over the last significant digit. One of the convenient ways of showing significant figures for difficult numbers is to use scientific notation, which is also discussed in the "Scientific Notation" section in this appendix. The convention for writing significant figures is to display one digit to the left of the decimal. The exponents are not considered when showing the number of significant figures in scientific notation. Thus, if you want to write three thousand showing one significant figure, you would write 3×10^3. To show two significant figures, it is 3.0×10^3, and for three significant figures it becomes 3.00×10^3. As you can see, the correct use of scientific notation leaves little room for doubt about how many significant figures are intended.

A.3 CONVERSION OF UNITS

The measurement of most properties results in both a numerical value and a unit. The statement that a glass contains 50 cm^3 of a liquid conveys two important concepts: the numerical value of 50 and the referent unit of cubic centimeters. Both the numerical value and the unit are necessary to communicate correctly the volume of the liquid.

When working with calculations involving measurement units, *both* the numerical value and the units are treated mathematically. As in other mathematical operations, there are general rules to follow.

1. Only properties with *like units* may be added or subtracted. It should be obvious that adding quantities such as 5 dollars and 10 dimes is meaningless. You must first convert to like units before adding or subtracting.
2. Like or unlike units may be multiplied or divided and treated in the same manner as numbers. You have used this rule when dealing with area (length $\times$ length = length^2, for example, cm $\times$ cm = cm^2) and when dealing with volume (length $\times$ length $\times$ length = length^3, for example, cm $\times$ cm $\times$ cm = cm^3).

You can use these two rules to create a *conversion ratio* that will help you change one unit to another. Suppose you need to convert 2.3 kg to grams. First, write the relationship between kilograms and grams:

$$1,000 \text{ g} = 1 \text{ kg}$$

Next, divide both sides by what you wish to convert *from* (kilograms in this example):

$$\frac{1,000 \text{ g}}{1 \text{ kg}} = \frac{1 \text{ kg}}{1 \text{ kg}}$$

One kilogram divided by 1 kg equals 1, just as 10 divided by 10 equals 1. Therefore, the right side of the relationship becomes 1 and the equation is:

$$\frac{1,000 \text{ g}}{1 \text{ kg}} = 1$$

The 1 is usually understood, that is, not stated, and the operation is called *canceling*. Canceling leaves you with the fraction 1,000 g/1 kg, which is a conversion ratio that can be used to convert from kilograms to grams. You simply multiply the conversion ratio by the numerical value and unit you wish to convert:

$$= 2.3 \text{ kg} \times \dfrac{1{,}000 \text{ g}}{1 \text{ kg}}$$

$$= \dfrac{2.3 \times 1{,}000}{1} \dfrac{\text{kg} \times \text{g}}{\text{kg}}$$

$$= \boxed{2{,}300 \text{ g}}$$

The kilogram units cancel. Showing the whole operation with units only, you can see how you end up with the correct unit of grams:

$$\text{kg} \times \dfrac{\text{g}}{\text{kg}} = \dfrac{\text{kg} \cdot \text{g}}{\text{kg}} = \text{g}$$

Since you did obtain the correct unit, you know that you used the correct conversion ratio. If you had blundered and used an inverted conversion ratio, you would obtain

$$2.3 \text{ kg} \times \dfrac{1 \text{ kg}}{1{,}000 \text{ g}} = .0023 \dfrac{\text{kg}^2}{\text{g}}$$

which yields the meaningless, incorrect units of kg^2/g. Carrying out the mathematical operations on the numbers and the units will always tell you whether or not you used the correct conversion ratio.

A.4 SCIENTIFIC NOTATION

Most of the properties of things that you might measure in your everyday world can be expressed with a small range of numerical values together with some standard unit of measure. The range of numerical values for most everyday things can be dealt with by using units (1s), tens (10s), hundreds (100s), or perhaps thousands (1,000s). But the actual universe contains some objects of incredibly large size that require some very big numbers to describe. The Sun, for example, has a mass of about 1,970,000,000,000,000,000,000,000,000,000 kg. On the other hand, very small numbers are needed to measure the size and parts of an atom. The radius of a hydrogen atom, for example, is about 0.00000000005 m. Such extremely large and small numbers are cumbersome and awkward since there are so many zeros to keep track of, even if you are successful in carefully counting all the zeros. A method does exist to deal with extremely large or small numbers in a more condensed form. The method is called *scientific notation*, but it is also sometimes called *powers of ten* or *exponential notation*, since it is based on exponents of 10. Whatever it is called, the method is a compact way of dealing with numbers that not only helps you keep track of zeros but provides a simplified way to make calculations as well.

In algebra you save a lot of time (as well as paper) by writing $(a \times a \times a \times a \times a)$ as a^5. The small number written to the right and above a letter or number is a superscript called an *exponent*. The exponent means that the letter or number is to be multiplied by itself that many times; for example, a^5 means a multiplied by itself five times; or $a \times a \times a \times a \times a$. As you can see, it is much easier to write the exponential form of this operation than it is to write it out in the long form. Scientific notation uses an exponent to indicate the power of the base 10. The exponent tells how many times the base, 10, is multiplied by itself. For example,

$$10{,}000 = 10^4$$
$$1{,}000 = 10^3$$
$$100 = 10^2$$
$$10 = 10^1$$
$$1 = 10^0$$
$$0.1 = 10^{-1}$$
$$0.01 = 10^{-2}$$
$$0.001 = 10^{-3}$$
$$0.0001 = 10^{-4}$$

This table could be extended indefinitely, but this somewhat shorter version will give you an idea of how the method works. The symbol 10^4 is read as "ten to the fourth power" and means $10 \times 10 \times 10 \times 10$. Ten times itself four times is 10,000, so 10^4 is the scientific notation for 10,000. It is also equal to the number of zeros between the 1 and the decimal point; that is, to write the longer form of 10^4, you simply write 1, then move the decimal point four places to the *right*—10 to the fourth power is 10,000.

What is 26,000,000 in scientific notation?

Answer

Count how many times you must shift the decimal point until one digit remains to the left of the decimal point. For numbers larger than the digit 1, the number of shifts tells you how much the exponent is increased, so the answer is

$$2.6 \times 10^7$$

which means the coefficient 2.6 is multiplied by 10 seven times.

The power of ten table also shows that numbers smaller than 1 have negative exponents. A negative exponent means a reciprocal:

$$10^{-1} = \frac{1}{10} = 0.1$$

$$10^{-2} = \frac{1}{100} = 0.01$$

$$10^{-3} = \frac{1}{1000} = 0.001$$

To write the longer form of 10^{-4}, you simply write 1, then move the decimal point four places to the *left*; 10 to the negative fourth power is 0.0001.

Scientific notation usually, but not always, is expressed as the product of two numbers: (1) a number between 1 and 10 that is called the *coefficient* and (2) a power of ten that is called the *exponent*. For example, the mass of the Sun that was given in long form earlier is expressed in scientific notation as

$$1.97 \times 10^{30} \ \text{kg}$$

and the radius of a hydrogen atom is

$$5.0 \times 10^{-11} \ \text{m}$$

In these expressions, the coefficients are 1.97 and 5.0, and the power of ten notations are the exponents. Note that in both of these examples, the exponent tells you where to place the decimal point if you wish to write the number all the way out in the long form. Sometimes scientific notation is written without a coefficient, showing only the exponent. In these cases, the coefficient of 1.0 is understood, that is, not stated. If you try to enter a scientific notation in your calculator, however, you will need to enter the understood 1.0, or the calculator will not be able to function correctly. Note also that 1.97×10^{30} kg and the expressions 0.197×10^{31} kg and 19.7×10^{29} kg are all correct expressions of the mass of the Sun. By convention, however, you will use the form that has one digit to the left of the decimal.

It was stated earlier that scientific notation provides a compact way of dealing with very large or very small numbers, but it provides a simplified way to make calculations as well. There are a few mathematical rules that will describe how the use of scientific notation simplifies these calculations.

To *multiply* two scientific notation numbers, the coefficients are multiplied as usual, and the exponents are *added*

Solve the following problem concerning scientific notation:

$$\frac{(2 \times 10^4) \times (8 \times 10^{-6})}{8 \times 10^4}$$

Answer

First, separate the coefficients from the exponents,

$$\frac{2 \times 8}{8} \times \frac{10^4 \times 10^{-6}}{10^4}$$

then multiply and divide the coefficients and add and subtract the exponents as the problem requires,

$$2 \times 10^{\{[(4) + (-6)] = [4]\}}$$

Solving the remaining additions and subtractions of the coefficients gives

$$2 \times 10^{-6}$$

algebraically. For example, to multiply (2×10^2) by (3×10^3), first separate the coefficients from the exponents,

$$(2 \times 3) \times (10^2 \times 10^3),$$

then multiply the coefficients and add the exponents,

$$6 \times 10^{(2+3)} = 6 \times 10^5$$

Adding the exponents is possible because $10^2 \times 10^3$ means the same thing as $(10 \times 10) \times (10 \times 10 \times 10)$, which equals $(100) \times (1,000)$, or 100,000, which is expressed as 10^5 in scientific notation. Note that two negative exponents add algebraically, for example $10^{-2} \times 10^{-3} = 10^{[(-2) + (-3)]} = 10^{-5}$. A negative and a positive exponent also add algebraically, as in $10^5 \times 10^{-3} = 10^{[(+5) + (-3)]} = 10^2$.

If the result of a calculation involving two scientific notation numbers does not have the conventional one digit to the left of the decimal, move the decimal point so it does, changing the exponent according to which way and how much the decimal point is moved. Note that the exponent increases by one number for each decimal point moved to the left. Likewise, the exponent decreases by one number for each decimal point moved to the right. For example, 938. $\times 10^3$ becomes 9.38×10^5 when the decimal point is moved two places to the left.

To *divide* two scientific notation numbers, the coefficients are divided as usual and the exponents are *subtracted*. For example, to divide (6×10^6) by (3×10^2), first separate the coefficients from the exponents,

$$(6 \div 3) \times (10^6 \div 10^2)$$

then divide the coefficients and subtract the exponents,

$$2 \times 10^{(6-2)} = 2 \times 10^4$$

Note that when you subtract a negative exponent, for example, $10^{[(3) - (-2)]}$, you change the sign and add, $10^{(3 + 2)} = 10^5$.

Appendix B

Solubilities Chart

	Acetate	Bromide	Carbonate	Chloride	Fluoride	Hydroxide	Iodide	Nitrate	Oxide	Phosphate	Sulfate	Sulfide
Aluminum	S	S	—	S	s	i	S	S	i	i	S	d
Ammonium	S	S	S	S	S	S	S	S	—	S	S	S
Barium	S	S	i	S	s	S	S	S	S	i	i	d
Calcium	S	S	i	S	i	s	S	S	s	i	s	d
Copper(I)	—	s	i	s	i	—	i	—	i	—	d	i
Copper(II)	S	S	i	S	S	i	S	S	i	i	S	i
Iron(II)	S	S	i	S	S	i	S	S	i	i	S	i
Iron(III)	S	S	i	S	s	i	S	S	i	i	S	d
Lead	S	s	i	s	i	i	s	S	i	i	i	i
Magnesium	S	S	i	S	i	i	S	S	i	i	S	d
Mercury(I)	s	i	i	i	d	d	i	S	i	i	i	i
Mercury(II)	S	s	i	S	d	i	i	S	i	i	i	i
Potassium	S	S	S	S	S	S	S	S	S	S	S	i
Silver	s	i	i	i	S	—	i	S	i	i	i	i
Sodium	S	S	S	S	S	S	S	S	d	S	S	S
Strontium	S	S	s	S	i	s	S	S	—	i	i	i
Zinc	S	S	i	S	S	i	S	S	i	i	S	i

S = soluble
i = insoluble
s = slightly soluble
d = decomposes

Relative Humidity Table

Dry-Bulb Temperature (°C)	Difference Between Wet-Bulb and Dry-Bulb Temperatures (° C)																			
	1	2	3	4	5	6	7	8	9	10	11	12	13	14	15	16	17	18	19	20
0	81	64	46	29	13															
1	83	66	49	33	17															
2	84	68	52	37	22	7														
3	84	70	55	40	26	12														
4	86	71	57	43	29	16														
5	86	72	58	45	33	20	7													
6	86	73	60	48	35	24	11													
7	87	74	62	50	38	26	15													
8	87	75	63	51	40	29	19	8												
9	88	76	64	53	42	32	22	12												
10	88	77	66	55	44	34	24	15	6											
11	89	78	67	56	46	36	27	18	9											
12	89	78	68	58	48	39	29	21	12											
13	89	79	69	59	50	41	32	23	15	7										
14	90	79	70	60	51	42	34	26	18	10										
15	90	80	71	61	53	44	36	27	20	13	6									
16	90	81	71	63	54	46	38	30	23	15	8									
17	90	81	72	64	55	47	40	32	25	18	11									
18	91	82	73	65	57	49	41	34	27	20	14	7								
19	91	82	74	65	58	50	43	36	29	22	16	10								
20	91	83	74	66	59	51	44	37	31	24	18	12	6							
21	91	83	75	67	60	53	46	39	32	26	20	14	9							
22	92	83	76	68	61	54	47	40	34	28	22	17	11	6						
23	92	84	76	69	62	55	48	42	36	30	24	19	13	8						
24	92	84	77	69	62	56	49	43	37	31	26	20	15	10	5					
25	92	84	77	70	63	57	50	44	39	33	28	22	17	12	8					
26	92	85	78	71	64	58	51	46	40	34	29	24	19	14	10	5				
27	92	85	78	71	65	58	52	47	41	36	31	26	21	16	12	7				
28	93	85	78	72	65	59	53	48	42	37	32	27	22	18	13	9	5			
29	93	86	79	72	66	60	54	49	43	38	33	28	24	19	15	11	7			
30	93	86	79	73	67	61	55	50	44	39	35	30	25	21	17	13	9	5		
31	93	86	80	73	67	61	56	51	45	40	36	31	27	22	18	14	11	7		
32	93	86	80	74	68	62	57	51	46	41	37	32	28	24	20	16	12	9	5	
33	93	87	80	74	68	63	57	52	47	42	38	33	29	25	21	17	14	10	7	
34	93	87	81	75	69	63	58	53	48	43	39	35	30	28	23	19	15	12	8	5
35	94	87	81	75	69	64	59	54	49	44	40	36	32	28	24	20	17	13	10	7

Appendix D

Problem Solving

Students are sometimes apprehensive when assigned problem exercises. This apprehension comes from a lack of experience in problem solving and not knowing how to proceed. This reading is concerned with a basic approach and procedures that you can follow to simplify problem-solving exercises. Thinking in terms of quantitative ideas is a skill that you can learn. Actually, no mathematics beyond addition, subtraction, multiplication, and division is needed. What is needed is knowledge of certain techniques. If you follow the suggested formatting procedures and seek help from the mathematical review appendix as needed, you will find that problem solving is a simple, fun activity that helps you to learn to think in a new way. Here are some more considerations that will prove helpful.

1. Read the problem carefully, perhaps several times, to understand the problem situation. If possible, make a sketch to help you visualize and understand the problem in terms of the real world.

2. Be alert for information that is not stated directly. For example, if a moving object "comes to a stop," you know that the final velocity is zero, even though this was not stated outright. Likewise, questions about "how far?" are usually asking about distance, and questions about "how long?" are usually asking about time. Such information can be very important in procedure step 1 (see procedure steps in right column), the listing of quantities and their symbols. Overlooked or missing quantities and symbols can make it difficult to identify the appropriate equation.

3. Understand the meaning and concepts that an equation represents. An equation represents a relationship that exists between variables. Understanding the relationship helps you to identify the appropriate equation or equations by inspection of the list of known and unknown quantities (see procedure step 2). You will find a list of the equations being considered at the end of each chapter. Information about the meaning and the concepts that an equation represents is found within each chapter.

4. Solve the equation *before* substituting numbers and units for symbols (see procedure step 3). A helpful discussion of the mathematical procedures required, with examples, is in appendix A.

5. Note if the quantities are in the same units. A mathematical operation requires the units to be the same; for example, you cannot add nickels, dimes, and quarters until you first convert them all to the same unit of money. Likewise, you cannot correctly solve a problem if one time quantity is in seconds and another time quantity is in hours. The quantities must be converted to the same units before anything else is done (see procedure step 4). There is a helpful section on how to use conversion ratios in appendix A.

6. Perform the required mathematical operations on the numbers and the units as if they were two separate problems (see procedure step 6). You will find that following this step will facilitate problem-solving activities because the units you obtain will tell you if you have worked the problem correctly. If you just write the units that you *think* should appear in the answer, you have missed this valuable self-check.

7. Be aware that not all learning takes place in a given time frame and that solutions to problems are not necessarily arrived at "by the clock." If you have spent a half an hour or so unsuccessfully trying to solve a particular problem, move on to another problem or do something entirely different for a while. Problem solving often requires time for something to happen in your brain. If you move on to some other activity, you might find that the answer to a problem that you have been stuck on will come to you "out of the blue" when you are not even thinking about the problem. This unexpected revelation of solutions is common to many real-world professions and activities that involve thinking.

Until you are comfortable in a problem-solving situation, you should follow a formatting procedure that will help you organize your thinking. Here is an example of such a formatting procedure:

Step 1: List the quantities involved together with their symbols on the left side of the page, including the unknown quantity with a question mark.

Step 2: Inspect the given quantities and the unknown quantity as listed, and *identify* the equation that expresses a relationship between these quantities. A list of the equations that express the relationships discussed in each chapter is found at the end of that chapter. *Write* the identified equation on the right side of your paper, opposite the list of symbols and quantities.

Step 3: If necessary, *solve* the equation for the variable in question. This step must be done before substituting any numbers or units in the equation. This simplifies things and keeps down the confusion that may otherwise result.

If you need help solving an equation, see the section on this topic in appendix A.

Step 4: If necessary, *convert* any unlike units so they are all the same. If the equation involves a time in seconds, for example, and a speed in kilometers per hour, you should convert the km/h to m/s. Again, this step should be done at this point to avoid confusion and incorrect operations in a later step. If you need help converting units, see the section on this topic in appendix A.

Step 5: *Substitute* the known quantities in the equation, replacing each symbol with both the number value and units represented by the symbol.

Step 6: *Perform* the required mathematical operations on the numbers and on the units. This performance is less confusing if you first separate the numbers and units, as shown in the following example and in the examples throughout this text, and then perform the mathematical operations on the numbers and units as separate steps, showing all work.

Step 7: *Draw a box* around your answer (numbers and units) to communicate that you have found what you were looking for. The box is a signal that you have finished your work on this problem.

EXAMPLE PROBLEM

Mercury is a liquid metal with a mass density of 13.6 g/cm³. What is the mass of 10.0 cm³ of mercury?

SOLUTION

The problem gives two known quantities, the mass density (ρ) of mercury and a known volume (V), and identifies an unknown quantity, the mass (m) of that volume. Make a list of these quantities:

$$\rho = 13.6 \, \text{g/cm}^3$$
$$V = 10.0 \, \text{cm}^3$$
$$m = ?$$

The appropriate equation for this problem is the relationship between mass density (ρ), mass (m), and volume (V):

$$\rho = \frac{m}{V}$$

The unknown in this case is the mass, m. Solving the equation for m, by multiplying both sides by V, gives:

$$V\rho = \frac{m\cancel{V}}{\cancel{V}}$$
$$V\rho = m, \text{ or}$$
$$m = V\rho$$

Now you are ready to substitute the known quantities in the equation and perform the mathematical operations on the numbers and on the units:

$$m = \left(13.6 \, \frac{\text{g}}{\text{cm}^3}\right)(10.0 \, \text{cm}^3)$$

$$= (13.6)(10.0)\left(\frac{\text{g}}{\text{cm}^3}\right)(\text{cm}^3)$$

$$= 136 \, \frac{\text{g} \cdot \cancel{\text{cm}^3}}{\cancel{\text{cm}^3}}$$

$$= \boxed{136 \, \text{g}}$$

TIPS ON TAKING A MULTIPLE-CHOICE EXAM

Many multiple-choice exams are designed to determine whether or not you can recall or apply a particular piece of information. Therefore, the examiner is expecting you to search your memory for a definition, example, or other tidbit of information. It is factual information that is used in formulating answers to the more complex, critical thinking questions found in essay exams. Here are some guidelines that can be helpful:

1. Read the question in its entirety (question and all answers) before selecting an answer. You don't want to "jump to a conclusion" that might be wrong before knowing your options.

2. If you know the answer, select it and move on. You might want to place a check mark on the exam that notes your level of confidence in the answer. If you are highly confident in your response, the check mark will remind you not to waste time reading this question again.

3. If you are unsure of the answer, move on to the next question. Remember that this exam covers a rather limited block of material and there may be another question in the exam that, when you read it, reminds you of the answer to this question. Again, make a note on the exam that will remind you to return to this unanswered question. You don't want to forget to make another attempt at an answer.

4. If you find yourself stumped, try this process of elimination as you read each question.

   a. Think about the question and search out the most *incorrect* answer. Cross it out.

   b. Look for the next-most incorrect answer and cross it out.

   c. Now you are, in most cases, down to choosing between two answers, one of which is the correct response.

   d. Think about what you have read, heard in class, talked about in study groups, or know from other experiences.

   e. Select the answer you believe to be the correct response and move on.

5. Don't forget to go back and reattempt unanswered questions.

Solutions for Second Example Exercises

Note: Solutions that involve calculations of measurements are rounded up or down to conform to the rules for significant figures as described in appendix A.

Chapter 1

1.2

$m = 15.0$ g
$V = 4.50$ cm^3
$\rho = ?$

$$\rho = \frac{m}{V}$$

$$= \frac{15.0\,\text{g}}{4.50\,\text{cm}^3}$$

$$= \boxed{3.33 \frac{\text{g}}{\text{cm}^3}}$$

Chapter 2

2.2

$\bar{v} = 8.00$ km/h
$t = 10.0$ s
$d = ?$

The bicycle has a speed of 8.00 km/h, and the time factor is 10.0 s, so km/h must be converted to m/s:

$$\bar{v} = \frac{0.2778 \frac{\text{m}}{\text{s}}}{\frac{\text{km}}{\text{h}}} \times 8.00 \frac{\text{km}}{\text{h}}$$

$$= (0.2778)(8.00)\frac{\text{m}}{\text{s}} \times \frac{\text{h}}{\text{km}} \times \frac{\text{km}}{\text{h}}$$

$$= 2.22 \frac{\text{m}}{\text{s}}$$

$$\bar{v} = \frac{d}{t}$$

$$\bar{v}t = \frac{dt}{t}$$

$$d = \bar{v}t$$

$$= \left(2.22 \frac{\text{m}}{\text{s}}\right)(10.0\,\text{s})$$

$$= (2.22)(10.0)\frac{\text{m}}{\text{s}} \times \frac{\text{s}}{1}$$

$$= \boxed{22.2\,\text{m}}$$

2.4

$v_i = 0\frac{\text{m}}{\text{s}}$ $\qquad a = \frac{v_f - v_i}{t}$ $\quad \therefore \quad v_f = at + v_i$

$v_f = ?$ $\qquad\qquad\qquad\qquad = \left(5\frac{\text{m}}{\text{s}^2}\right)(6\,\text{s}) + 0$

$a = 5\frac{\text{m}}{\text{s}^2}$ $\qquad\qquad\qquad\quad = (5)(6)\frac{\text{m}}{\text{s}^2} \times \frac{\text{s}}{1}$

$t = 6\,\text{s}$ $\qquad\qquad\qquad\qquad = \boxed{30\frac{\text{m}}{\text{s}}}$

2.6

$m = 20$ kg
$F = 40$ N
$a = ?$

$$F = ma \quad \therefore \quad a = \frac{F}{m}$$

$$= \frac{40 \frac{\text{kg} \cdot \text{m}}{\text{s}^2}}{20\,\text{kg}}$$

$$= \frac{40}{20} \frac{\text{kg} \cdot \text{m}}{\text{s}^2} \times \frac{1}{\text{kg}}$$

$$= \boxed{2\frac{\text{m}}{\text{s}^2}}$$

2.8

$m = 60.0$ kg
$w = 100.0$ N
$g = ?$

$$w = mg \quad \therefore \quad g = \frac{w}{m}$$

$$= \frac{100.0 \frac{\text{kg} \cdot \text{m}}{\text{s}^2}}{60.0\,\text{kg}}$$

$$= \frac{100.0}{60.0} \frac{\text{kg} \cdot \text{m}}{\text{s}^2} \times \frac{1}{\text{kg}}$$

$$= \boxed{1.67\frac{\text{m}}{\text{s}^2}}$$

2.10

$m = 0.25$ kg
$r = 0.25$ m
$v = 2.0$ m/s
$F = ?$

$$F = \frac{mv^2}{r}$$

$$= \frac{(0.25\,\text{kg})\left(2.0\,\dfrac{\text{m}}{\text{s}}\right)^2}{0.25\,\text{m}}$$

$$= \frac{(0.25\,\text{kg})\left(4.0\,\dfrac{\text{m}^2}{\text{s}^2}\right)}{0.25\,\text{m}}$$

$$= \frac{(0.25)(4.0)}{0.25}\,\frac{\text{kg}\cdot\text{m}^2}{\text{s}^2}\times\frac{1}{\text{m}}$$

$$= 4.0\,\frac{\text{kg}\cdot\text{m}}{\text{s}^2}$$

$$= \boxed{4.0\,\text{N}}$$

2.12

$G = 6.67\times10^{-11}\text{N}\cdot\text{m}^2/\text{kg}^2$
$m_e = 6.0\times10^{24}$ kg
$d = 12.8\times10^6$ m
$g = ?$

$$g = \frac{Gm_e}{d^2}$$

$$= \frac{\left(6.67\times10^{-11}\,\dfrac{\text{N}\cdot\text{m}^2}{\text{kg}^2}\right)(6.0\times10^{24}\,\text{kg})}{(12.8\times10^6\,\text{m})^2}$$

$$= \frac{\left(6.67\times10^{-11}\,\dfrac{\text{N}\cdot\text{m}^2}{\text{kg}^2}\right)(6.0\times10^{24}\,\text{kg})}{1.64\times10^{14}\,\text{m}^2}$$

$$= \frac{(6.67\times10^{-11})(6.0\times10^{24})}{1.64\times10^{14}}\,\frac{\text{N}\cdot\text{m}^2}{\text{kg}^2}\times\frac{\text{kg}}{1}\times\frac{1}{\text{m}^2}$$

$$= 2.44\,\frac{\dfrac{\text{kg}\cdot\text{m}}{\text{s}^2}}{\text{kg}}$$

$$= \boxed{2.44\,\frac{\text{m}}{\text{s}^2}}$$

Chapter 3

3.2

$w = 50$ lb
$d = 2$ ft
$W = ?$

$$W = Fd$$

$$= (50\,\text{lb})\,(2\,\text{ft})$$

$$= (50)\,(2)\quad\text{ft}\times\text{lb}$$

$$= \boxed{100\,\text{ft}\cdot\text{lb}}$$

3.4

$w = 150$ lb
$h = 15$ ft
$t = 10.0$ s
$P = ?$

$$P = \frac{wh}{t}$$

$$= \frac{(150\,\text{lb})(15\,\text{ft})}{10.0\,\text{s}}$$

$$= \frac{(150)(15)}{10.0}\quad\frac{\text{ft}\cdot\text{lb}}{\text{s}}$$

$$= 225\,\frac{\text{ft}\cdot\text{lb}}{\text{s}}$$

$$= \frac{225\,\dfrac{\text{ft}\cdot\text{lb}}{\text{s}}}{550\,\dfrac{\dfrac{\text{ft}\cdot\text{lb}}{\text{s}}}{\text{hp}}}$$

$$= 0.41\,\frac{\text{ft}\cdot\text{lb}}{\text{s}}\times\frac{\text{s}}{\text{ft}\cdot\text{lb}}\times\text{hp}$$

$$= \boxed{0.41\,\text{hp}}$$

3.6

$m = 5.00 \text{ kg}$
$g = 9.8 \text{ m/s}^2$
$h = 5.00 \text{ m}$
$W = ?$

$$W = Fd$$
$$W = mgh$$
$$= (5.00 \text{ kg})\left(9.8\frac{\text{m}}{\text{s}^2}\right)(5.00 \text{ m})$$
$$= (5.00)(9.8)(5.00) \frac{\text{kg} \cdot \text{m}}{\text{s}^2} \times \text{m}$$
$$= 245 \text{ N} \cdot \text{m}$$
$$= \boxed{250 \text{ J}}$$

3.8

$m = 100.0 \text{ kg}$
$v = 6.0 \text{ m/s}$
$W = ?$

$$W = K.E.$$
$$K.E. = \frac{1}{2}mv^2$$
$$= \frac{1}{2}(100.0 \text{ kg})\left(6.0 \frac{\text{m}}{\text{s}}\right)^2$$
$$= \frac{1}{2}(100.0 \text{ kg})\left(36 \frac{\text{m}^2}{\text{s}^2}\right)$$
$$= \frac{1}{2}(100.0)(36) \frac{\text{kg} \cdot \text{m}}{\text{s}^2} \times \text{m}$$
$$= 1,800 \text{ N} \cdot \text{m}$$
$$= \boxed{1,800 \text{ J}}$$

Chapter 4

4.2

$T_C = 20°$
$T_F = ?$

$$T_F = \frac{9}{5}T_C + 32°$$
$$= \frac{9}{5}20° + 32°$$
$$= \frac{180°}{5} + 32°$$
$$= 36° + 32°$$
$$= \boxed{68°}$$

4.6

$m = 2 \text{ kg}$
$Q = 1.2 \text{ kcal}$
$\Delta T = 20 \text{ C}°$
$c = ?$

$$Q = mc\Delta T \quad \therefore \quad c = \frac{Q}{m\Delta T}$$
$$= \frac{1.2 \text{ kcal}}{(2 \text{ kg})(20.0 \text{ C}°)}$$
$$= \frac{1.2}{(2)(20.0)} \frac{\text{kcal}}{\text{kgC}°}$$
$$= \boxed{0.03 \frac{\text{kcal}}{\text{kgC}°}}$$

Chapter 5

5.3

$f = 2,500 \text{ Hz}$
$v = 330 \text{ m/s}$
$\lambda = ?$

$$v = \lambda f \quad \therefore \quad \lambda = \frac{v}{f}$$
$$= \frac{330 \frac{\text{m}}{\text{s}}}{2,500 \frac{1}{\text{s}}}$$
$$= \frac{330}{2,500} \frac{\text{m}}{\text{s}} \times \frac{\text{s}}{1}$$
$$= 0.13 \text{ m} \quad \text{or} \quad \boxed{13 \text{ cm}}$$

5.5

$t = 1.00 \text{ s}$
$v = 1,147 \text{ ft/s}$
$d = ?$

$$v = \frac{d}{t} \quad \therefore \quad d = vt$$
$$= \left(1147 \frac{\text{ft}}{\text{s}}\right)(1.00 \text{ s})$$
$$= (1147)(1.00) \quad \frac{\text{ft}}{\text{s}} \times \frac{\text{s}}{1}$$
$$= 1147 \text{ ft}$$

Sound traveled from the source, to a reflecting surface, then back to the source, so the distance is

$$1147 \text{ ft} \times \frac{1}{2} = \boxed{574 \text{ ft}}$$

Chapter 6

6.3

$V = 120$ V
$R = 30\ \Omega$
$I = ?$

$$V = IR \quad \therefore \quad I = \frac{V}{R}$$

$$= \frac{120\,V}{30\,\dfrac{V}{A}}$$

$$= \frac{120}{30}\ \frac{V}{1} \times \frac{A}{V}$$

$$= \boxed{4\,A}$$

6.5

$I = 0.5$ A
$V = 120$ V
$P = ?$

$$P = IV$$
$$= (0.5\,A)\,(120\,V)$$
$$= (0.5)\,(120)\ \frac{C}{s} \times \frac{J}{C}$$
$$= 60\frac{J}{s}$$
$$= \boxed{60\,W}$$

6.7

I = 0.5 A
V = 120 V
P = IV = 60 W
$Rate$ = \$0.10/kWh
$Cost$ = ?

$$cost = \frac{(watts)\,(time)\,(rate)}{1{,}000\ \dfrac{W}{kW}}$$

$$= \frac{(60W)\,(1.00h)\left(\$0.10\,\dfrac{W}{kW}\right)}{1{,}000\ \dfrac{W}{kW}}$$

$$= \frac{(60)\,(1.00)\,(0.10)}{1{,}000}\ \frac{W}{1} \times \frac{h}{1} \times \frac{\$}{kWh} \times \frac{kW}{W}$$

$$= \$0.006 \ \ or \ \boxed{0.6\ of\ a\ cent\ per\ hour}$$

Chapter 7

7.4

$f = 7.00 \times 10^{14}$ Hz
$h = 6.63 \times 10^{-34}$ J $\cdot$ s
$E = ?$

$$E = hf$$
$$= (6.63 \times 10^{-34}J \cdot s)\left(7.00 \times 10^{14}\frac{1}{s}\right)$$
$$= (6.63 \times 10^{-34})\,(7.00 \times 10^{14})\ J \cdot s \times \frac{1}{s}$$
$$= \boxed{4.64 \times 10^{-19}J}$$

Chapter 8

8.2

$f = 7.30 \times 10^{14}$ Hz
$h = 6.63 \times 10^{-34}$ J $\cdot$ s
$E = ?$

$$E = hf$$
$$= (6.63 \times 10^{-34}J \cdot s)\left(7.30 \times 10^{14}\frac{1}{s}\right)$$
$$= (6.63 \times 10^{-34})\,(7.30 \times 10^{14})\ J \cdot s \times \frac{1}{s}$$
$$= \boxed{4.84 \times 10^{-19}J}$$

Answers for Applying the Concepts

Chapter 1	1. b	2. b	3. a	4. d	5. c	6. c	7. b	8. b	9. c	10. d				
Chapter 2	1. b	2. d	3. b	4. a	5. c	6. a	7. b	8. a	9. c	10. b	11. b	12. c		
Chapter 3	1. d	2. d	3. d	4. c	5. b	6. d	7. b	8. c	9. c	10. c	11. b	12. b		
Chapter 4	1. c	2. c	3. d	4. b	5. a	6. c	7. b	8. b	9. b	10. a	11. c	12. a		
Chapter 5	1. c	2. a	3. b	4. a	5. a	6. d	7. c	8. a	9. d	10. c	11. a	12. d		
Chapter 6	1. b	2. b	3. a	4. b	5. c	6. a	7. d	8. b	9. b	10. c	11. a	12. c		
Chapter 7	1. a	2. c	3. a	4. d	5. d	6. d	7. a	8. c	9. b	10. a	11. d	12. c		
Chapter 8	1. d	2. a	3. c	4. d	5. d	6. b	7. a	8. b	9. c	10. c	11. c	12. c		
Chapter 9	1. c	2. a	3. a	4. b	5. b	6. c	7. b	8. c	9. b	10. a	11. a	12. c		
Chapter 10	1. c	2. a	3. c	4. b	5. a	6. d	7. a	8. c	9. a	10. c	11. a	12. b		
Chapter 11	1. b	2. c	3. c	4. a	5. c	6. a	7. b	8. c	9. b	10. a	11. c	12. a		
Chapter 11	1. b	2. c	3. c	4. a	5. c	6. a	7. b	8. c	9. b	10. a	11. c	12. a		
Chapter 12	1. c	2. a	3. a	4. d	5. c	6. b	7. d	8. c	9. c	10. a	11. c	12. d		
Chapter 13	1. d	2. b	3. a	4. d	5. d	6. b	7. c	8. d	9. b	10. c	11. a	12. d		
Chapter 14	1. c	2. c	3. a	4. c	5. d	6. a	7. c	8. b	9. a	10. b	11. a	12. c		
Chapter 15	1. c	2. b	3. a	4. b	5. a	6. a	7. a	8. c	9. b	10. a	11. d	12. b		
Chapter 16	1. b	2. a	3. d	4. c	5. b	6. d	7. c	8. b	9. a	10. d	11. d	12. b		
Chapter 17	1. b	2. b	3. d	4. a	5. b	6. d	7. b	8. a	9. c	10. b	11. a	12. d		
Chapter 18	1. c	2. d	3. a	4. d	5. b	6. a	7. a	8. c	9. a	10. c	11. a			
Chapter 19	1. c	2. d	3. c	4. a	5. b	6. d	7. d	8. d	9. a	10. d	11. c	12. c	13. b	14. b	15. c
Chapter 20	1. d	2. b	3. b	4. b	5. c	6. b	7. d	8. b	9. b	10. c	11. a	12. d	13. a	14. b	15. a
Chapter 21	1. d	2. b	3. d	4. b	5. a	6. b	7. a	8. a	9. d	10. a	11. d	12. b	13. b	14. a	15. b
Chapter 22	1. c	2. d	3. d	4. d	5. c	6. c	7. b	8. c	9. d	10. d	11. b	12. c	13. d	14. a	15. c
Chapter 23	1. d	2. a	3. d	4. c	5. b	6. c	7. d	8. c	9. d	10. a	11. c	12. d	13. a	14. b	15. d
Chapter 24	1. d	2. c	3. c	4. a	5. b	6. c	7. a	8. c	9. c	10. c	11. b	12. b	13. a	14. c	15. a
Chapter 25	1. a	2. b	3. a	4. a	5. a	6. c	7. d	8. b	9. c	10. a	11. c	12. b	13. b	14. c	15. c
Chapter 26	1. a	2. d	3. b	4. c	5. c	6. b	7. b	8. b	9. c	10. d	11. d	12. c	13. a	14. b	15. d

Solutions for Group A Parallel Exercises

Note: Solutions that involve calculations of measurements are rounded up or down to conform to the rules for significant figures described in appendix A.

Chapter 1

1.1 Answers will vary but should have the relationship of 100 cm in 1 m, for example, 178 cm = 1.78 m.

1.2 Since density is given by the relationship $\rho = m/V$, then

$$\rho = \frac{m}{V} = \frac{272\ g}{20.0\ cm^3}$$

$$= \frac{272}{20.0}\ \frac{g}{cm^3}$$

$$= \boxed{13.6\ \frac{g}{cm^3}}$$

1.3 The volume of a sample of lead is given and the problem asks for the mass. From the relationship of $\rho = m/V$, solving for the mass (m) tells you that the density (ρ) times the volume (V) equals the mass, or $m = \rho V$. The density of lead, 11.4 g/cm^3, can be obtained from table 1.3, so

$$\rho = \frac{m}{V}$$

$$V\rho = \frac{m\cancel{V}}{\cancel{V}}$$

$$m = \rho V$$

$$m = \left(11.4\ \frac{g}{cm^3}\right)(10.0\ cm^3)$$

$$= 11.4 \times 10.0\ \frac{g}{cm^3} \times cm^3$$

$$= 114\ \frac{g\cdot\cancel{cm^3}}{\cancel{cm^3}}$$

$$= \boxed{114\ g}$$

1.4 Solving the relationship $\rho = m/V$ for volume gives $V = m/\rho$, and

$$\rho = \frac{m}{V}$$

$$V\rho = \frac{m\cancel{V}}{\cancel{V}}$$

$$\frac{V\cancel{\rho}}{\cancel{\rho}} = \frac{m}{\rho}$$

$$V = \frac{m}{\rho}$$

$$V = \frac{600\ g}{3.00\ \dfrac{g}{cm^3}}$$

$$= \frac{600}{3.00}\ \frac{g}{1} \times \frac{cm^3}{g}$$

$$= 200\ \frac{\cancel{g}\cdot cm^3}{\cancel{g}}$$

$$= \boxed{200\ cm^3}$$

1.5 A 50.0 cm^3 sample with a mass of 34.0 grams has a density of

$$\rho = \frac{m}{V} = \frac{34.0\ g}{50.0\ cm^3}$$

$$= \frac{34.0}{50.0}\ \frac{g}{cm^3}$$

$$= \boxed{0.680\ \frac{g}{cm^3}}$$

According to table 1.3, 0.680 g/cm^3 is the density of gasoline, so the substance must be gasoline.

1.6 The problem asks for a mass and gives a volume, so you need a relationship between mass and volume. Table 1.3 gives the density of water as 1.00 g/cm^3, which is a density that is easily remembered. The volume is given in liters (L), which should first be converted to cm^3 because this is the unit in which density is expressed. The relationship of $\rho = m/V$ solved for mass is ρV, so the solution is

$$\rho = \frac{m}{V} \therefore m = \rho V$$

$$= \left(1.00 \frac{g}{cm^3}\right)(40{,}000 \text{ cm}^3)$$

$$= 1.00 \times 40{,}000 \frac{g}{cm^3} \times cm^3$$

$$= 40{,}000 \frac{g \cdot cm^3}{cm^3}$$

$$= 40{,}000 \text{ g}$$

$$= \boxed{40 \text{ kg}}$$

1.7 From table 1.3, the density of aluminum is given as 2.70 g/cm³. Converting 2.1 kg to the same units as the density gives 2,100 g. Solving $\rho = m/V$ for the volume gives

$$V = \frac{m}{\rho} = \frac{2{,}100 \text{ g}}{2.70 \dfrac{g}{cm^3}}$$

$$= \frac{2{,}100}{2.70} \frac{g}{1} \times \frac{cm^3}{g}$$

$$= 777.78 \frac{g \cdot cm^3}{g}$$

$$= \boxed{780 \text{ cm}^3}$$

1.8 The length of one side of the box is 0.1 m. Reasoning: Since the density of water is 1.00 g/cm³, then the volume of 1,000 g of water is 1,000 cm³. A cubic box with a volume of 1,000 cm³ is 10 cm (since $10 \times 10 \times 10 = 1{,}000$). Converting 10 cm to m units, the cube is 0.1 m on each edge.

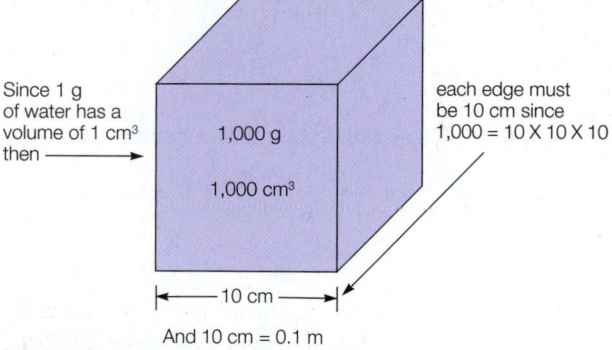

Since 1 g of water has a volume of 1 cm³ then → 1,000 g

1,000 cm³

each edge must be 10 cm since $1{,}000 = 10 \times 10 \times 10$

10 cm

And 10 cm = 0.1 m

FIGURE A.4

1.9 The relationship between mass, volume, and density is $\rho = m/V$. The problem gives a volume but not a mass. The mass, however, can be assumed to remain constant during the compression of the bread, so the mass can be obtained from the original volume and density, or

$$\rho = \frac{m}{V} \therefore m = \rho V$$

$$= \left(0.2 \frac{g}{cm^3}\right)(3{,}000 \text{ cm}^3)$$

$$= 0.2 \times 3{,}000 \frac{g}{cm^3} \times cm^3$$

$$= 600 \frac{g \cdot cm^3}{cm^3}$$

$$= 600 \text{ g}$$

A mass of 600 g and the new volume of 1,500 cm³ means that the new density of the crushed bread is

$$\rho = \frac{m}{V}$$

$$= \frac{600 \text{ g}}{1{,}500 \text{ cm}^3}$$

$$= \frac{600}{1{,}500} \frac{g}{cm^3}$$

$$= \boxed{0.4 \frac{g}{cm^3}}$$

1.10 According to table 1.3, lead has a density of 11.4 g/cm³. Therefore, a 1.00 cm³ sample of lead would have a mass of

$$\rho = \frac{m}{V} \therefore m = \rho V$$

$$= \left(11.4 \frac{g}{cm^3}\right)(1.00 \text{ cm}^3)$$

$$= 11.4 \times 1.00 \frac{g}{cm^3} \times cm^3$$

$$= 11.4 \frac{g \cdot cm^3}{cm^3}$$

$$= 11.4 \text{ g}$$

Also according to table 1.3, copper has a density of 8.96 g/cm³. To balance a mass of 11.4 g of lead, a volume of this much copper would be required:

$$\rho = \frac{m}{V} \therefore V = \frac{m}{\rho}$$

$$= \frac{11.4 \text{ g}}{8.96 \dfrac{g}{cm^3}}$$

$$= \frac{11.4}{8.96} \frac{g}{1} \times \frac{cm^3}{g}$$

$$= 1.27232 \frac{g \cdot cm^3}{g}$$

$$= \boxed{1.27 \text{ cm}^3}$$

Chapter 2

2.1 Listing the quantities with their symbols, we can see the problem involves the quantities found in the definition of average speed:

$$\bar{v} = 350.0 \text{ m/s}$$

$$t = 5.00 \text{ s}$$

$$d = ?$$

$$\bar{v} = \frac{d}{t} \therefore d = \bar{v}t$$

$$= \left(350.0 \frac{m}{s}\right)(5.00 \text{ s})$$

$$= (350.0)(5.00) \frac{m}{s} \times s$$

$$= \boxed{1{,}750 \text{ m}}$$

2.2 The initial velocity, final velocity, and time are known and the problem asked for the acceleration. Listing these quantities with their symbols, we have

$$v_i = 0 \text{ m/s}$$
$$v_f = 15.0 \text{ m/s}$$
$$t = 10.0 \text{ s}$$
$$a = ?$$

These are the quantities involved in the acceleration equation, which is already solved for the unknown:

$$a = \frac{v_f - v_i}{t}$$

$$a = \frac{15.0 \text{ m/s} - 0 \text{ m/s}}{10.0 \text{ s}}$$

$$= \frac{15.0}{10.0} \frac{\text{m}}{\text{s}} \times \frac{1}{\text{s}}$$

$$= \boxed{1.50 \frac{\text{m}}{\text{s}^2}}$$

2.3 The distance (d) and the time (t) quantities are given in the problem, and

$$\bar{v} = \frac{d}{t}$$

$$= \frac{160 \text{ km}}{2.0 \text{ h}}$$

$$= \frac{160}{2.0} \frac{\text{km}}{\text{h}}$$

$$= \boxed{80 \text{ km/h}}$$

The units cannot be simplified further. Note two significant figures in the answer, which is the least number of significant figures involved in the division operation.

2.4 Listing the known and unknown quantities:

$$m = 40.0 \text{ kg}$$
$$a = 2.4 \text{ m/s}^2$$
$$F = ?$$

These are the quantities found in Newton's second law of motion, $F = ma$, which is already solved for force (F). Thus,

$$F = (40.0 \text{ kg})\left(2.4 \frac{\text{m}}{\text{s}^2}\right)$$

$$= 40.0 \times 2.4 \frac{\text{kg} \cdot \text{m}}{\text{s}^2}$$

$$= \boxed{96 \text{ N}}$$

2.5 List the known and unknown quantities for the first situation, using an unbalanced force of 18.0 N to give the object an acceleration of 3 m/s²:

$$F_1 = 18 \text{ N}$$
$$a_1 = 3 \text{ m/s}^2$$

For the second situation, we are asked to find the force needed for an acceleration of 10 m/s²:

$$a_2 = 10 \text{ m/s}^2$$
$$F = ?$$

These are the quantities of Newton's second law of motion, $F = ma$, but the mass appears to be missing. The mass can be found from

$$F = ma \therefore m_1 = \frac{F_1}{a_1}$$

$$= \frac{18 \dfrac{\text{kg} \cdot \text{m}}{\text{s}^2}}{3 \dfrac{\text{m}}{\text{s}^2}}$$

$$= \frac{18}{3} \frac{\text{kg} \cdot \text{m}}{\text{s}^2} \times \frac{\text{s}^2}{\text{m}}$$

$$= 6 \text{ kg}$$

Now that we have the mass, we can easily find the force needed for an acceleration of 10 m/s²:

$$F_2 = m_2 a_2$$

$$= (6 \text{ kg})\left(10 \frac{\text{m}}{\text{s}^2}\right)$$

$$= 6 \times 10 \frac{\text{kg} \cdot \text{m}}{\text{s}^2}$$

$$= \boxed{60 \text{ N}}$$

2.6 Listing the known and unknown quantities:

$$m = 70.0 \text{ kg} \qquad w = mg$$
$$g = 9.8 \text{ m/s}^2 \qquad = (70.0 \text{ kg})\left(9.8 \frac{\text{m}}{\text{s}^2}\right)$$
$$w = ?$$

$$= 70.0 \times 9.8 \text{ kg} \times \frac{\text{m}}{\text{s}^2}$$

$$= 686 \frac{\text{kg} \cdot \text{m}}{\text{s}^2}$$

$$= \boxed{690 \text{ N}}$$

2.7 Listing the known and unknown quantities:

$$m = 100 \text{ kg}$$
$$v = 6 \text{ m/s}$$
$$p = ?$$

These are the quantities found in the equation for momentum, $p = mv$, which is already solved for momentum (p). Thus,

$$p = mv$$

$$= (100 \text{ kg})\left(6 \frac{\text{m}}{\text{s}}\right)$$

$$= \boxed{600 \frac{\text{kg} \cdot \text{m}}{\text{s}}}$$

(Note the lowercase p is the symbol used for momentum. This is one of the few cases where the English letter does not provide a clue about what it stands for. The units for momentum are also somewhat unusual for metric units since they do not have a name or single symbol to represent them.)

2.8 Listing the known and unknown quantities:

$$w = 13,720 \, \text{N}$$
$$v = 91 \, \text{km/h}$$
$$p = ?$$

The equation for momentum is $p = mv$, which is already solved for momentum (p). The weight unit must be first converted to a mass unit:

$$w = mg \therefore m = \frac{w}{g}$$

$$= \frac{13,720 \, \frac{\text{kg} \cdot \text{m}}{\text{s}^2}}{9.8 \, \frac{\text{m}}{\text{s}^2}}$$

$$= \frac{13,720}{9.8} \, \frac{\text{kg} \cdot \text{m}}{\text{s}^2} \times \frac{\text{s}^2}{\text{m}}$$

$$= 1,400 \, \text{kg}$$

The km/h unit should next be converted to m/s. Using the conversion factor from inside the front cover:

$$\frac{0.2778 \, \frac{\text{m}}{\text{s}}}{1 \, \frac{\text{km}}{\text{h}}} \times 91 \, \frac{\text{km}}{\text{h}}$$

$$0.2778 \times 91 \, \frac{\text{m}}{\text{s}} \times \frac{\text{h}}{\text{km}} \times \frac{\text{km}}{\text{h}}$$

$$25.2798 \, \frac{\text{m}}{\text{s}}$$

$$25 \, \frac{\text{m}}{\text{s}}$$

Now, listing the converted known and unknown quantities:

$$m = 1,400 \, \text{kg}$$
$$v = 25 \, \text{m/s}$$
$$p = ?$$

and solving for momentum (p),

$$p = mv$$

$$= (1,400 \, \text{kg}) \left(25 \, \frac{\text{m}}{\text{s}} \right)$$

$$= \boxed{35,000 \, \frac{\text{kg} \cdot \text{m}}{\text{s}}}$$

2.9 Listing the known and unknown quantities:

Bullet $\rightarrow m = 0.015 \, \text{kg}$ Rifle $\rightarrow m = 6 \, \text{kg}$
Bullet $\rightarrow v = 200 \, \text{m/s}$ Rifle $\rightarrow v = ? \, \text{m/s}$

Note the mass of the bullet was converted to kilograms. This is a conservation of momentum question, where the bullet and rifle can be considered as a system of interacting objects:

Bullet momentum = −rifle momentum

$$(mv)_b = -(mv)_r$$

$$(mv)_b - (mv)_r = 0$$

$$(0.015 \, \text{kg}) \left(200 \, \frac{\text{m}}{\text{s}} \right) - (6 \, \text{kg}) v_r = 0$$

$$\left(3 \, \text{kg} \cdot \frac{\text{m}}{\text{s}} \right) - (6 \, \text{kg} \cdot v_r) = 0$$

$$\left(3 \, \text{kg} \cdot \frac{\text{m}}{\text{s}} \right) = (6 \, \text{kg} \cdot v_r)$$

$$v_r = \frac{3 \, \text{kg} \cdot \frac{\text{m}}{\text{s}}}{6 \, \text{kg}}$$

$$= \frac{3}{6} \, \frac{\text{kg}}{1} \times \frac{1}{\text{kg}} \times \frac{\text{m}}{\text{s}}$$

$$= \boxed{0.5 \, \frac{\text{m}}{\text{s}}}$$

The rifle recoils with a velocity of 0.5 m/s.

2.10 A unit conversion is needed:

$$\left(90.0 \, \frac{\text{km}}{\text{h}} \right) \left(0.2778 \, \frac{\frac{\text{m}}{\text{s}}}{\frac{\text{km}}{\text{h}}} \right) = 25.0 \, \text{m/s}$$

a. $F = ma \therefore m = \dfrac{F}{a}$ and $a = \dfrac{v_f - v_i}{t}$, so

$$m = \frac{F}{\frac{v_f - v_i}{t}} = \frac{5,000.0 \, \frac{\text{kg} \cdot \text{m}}{\text{s}^2}}{\frac{25.0 \, \text{m/s} - 0}{5.0 \, \text{s}}}$$

$$= \frac{5,000.0 \, \frac{\text{kg} \cdot \text{m}}{\text{s}^2}}{5.0 \, \frac{\text{m}}{\text{s}^2}}$$

$$= \frac{5,000.0}{5.0} \, \frac{\text{kg} \cdot \text{m}}{\text{s}^2} \times \frac{\text{s}^2}{\text{m}}$$

$$= 1,000 \, \frac{\text{kg} \cdot \text{m} \cdot \text{s}^2}{\text{m} \cdot \text{s}^2}$$

$$= \boxed{1.0 \times 10^3 \, \text{kg}}$$

b. $w = mg$

$$= (1.0 \times 10^3 \, \text{kg}) \left(9.8 \, \frac{\text{m}}{\text{s}^2} \right)$$

$$= (1.0 \times 10^3)(9.8) \, \text{kg} \times \frac{\text{m}}{\text{s}^2}$$

$$= 9.8 \times 10^3 \, \frac{\text{kg} \cdot \text{m}}{\text{s}^2}$$

$$= \boxed{9.8 \times 10^3 \, \text{N}}$$

2.11

$$F = \frac{mv^2}{r}$$

$$= \frac{(0.20 \text{ kg})\left(3.0 \dfrac{\text{m}}{\text{s}}\right)^2}{1.5 \text{ m}}$$

$$= \frac{(0.20 \text{ kg})\left(9.0 \dfrac{\text{m}^2}{\text{s}^2}\right)}{1.5 \text{ m}}$$

$$= \frac{0.20 \times 9.0}{1.5} \frac{\text{kg} \cdot \text{m}^2}{\text{s}^2} \times \frac{1}{\text{m}}$$

$$= 1.2 \frac{\text{kg} \cdot \text{m} \cdot \cancel{\text{m}}}{\text{s}^2 \cdot \cancel{\text{m}}}$$

$$= \boxed{1.2 \text{ N}}$$

2.12 Newton's laws of motion consider the resistance to a change of motion, or mass, and not weight. The astronaut's mass is

$$w = mg \therefore m = \frac{w}{g}$$

$$= \frac{1{,}960.0 \dfrac{\text{kg} \cdot \text{m}}{\text{s}^2}}{9.8 \dfrac{\text{m}}{\text{s}^2}}$$

$$= \frac{1{,}960.0}{9.8} \frac{\text{kg} \cdot \text{m}}{\text{s}^2} \times \frac{\text{s}^2}{\text{m}}$$

$$= 200 \text{ kg}$$

a. From Newton's second law of motion, you can see that the 100 N rocket gives the 200 kg astronaut an acceleration of:

$$F = ma \therefore a = \frac{F}{m}$$

$$= \frac{100 \dfrac{\text{kg} \cdot \text{m}}{\text{s}^2}}{200 \text{ kg}}$$

$$= \frac{100 \text{ kg} \cdot \text{m}}{200 \text{ s}^2} \times \frac{1}{\text{kg}}$$

$$= 0.5 \text{ m/s}^2$$

b. An acceleration of 0.5 m/s² for 2.0 s will result in a final velocity of

$$a = \frac{v_f - v_i}{t} \therefore v_f = at + v_i$$

$$= (0.5 \text{ m/s}^2)(2.0 \text{ s}) + 0 \text{ m/s}$$

$$= \boxed{1 \text{ m/s}}$$

Chapter 3

3.1 Listing the known and unknown quantities:

$$F = 200 \text{ N}$$
$$d = 3 \text{ m}$$
$$W = ?$$

These are the quantities found in the equation for work, $W = Fd$, which is already solved for work (W). Thus,

$$W = Fd$$

$$= \left(200 \frac{\text{kg} \cdot \text{m}}{\text{s}^2}\right)(3 \text{ m})$$

$$= (200)(3) \text{ N} \cdot \text{m}$$

$$= \boxed{600 \text{ J}}$$

3.2 Listing the known and unknown quantities:

$$F = 440 \text{ N}$$
$$d = 5.0 \text{ m}$$
$$w = 880 \text{ N}$$
$$W = ?$$

These are the quantities found in the equation for work, $W = Fd$, which is already solved for work (W). As you can see in the equation, the force exerted and the distance the box was moved are the quantities used in determining the work accomplished. The weight of the box is a different variable and one that is not used in this equation. Thus,

$$W = Fd$$

$$= \left(440 \frac{\text{kg} \cdot \text{m}}{\text{s}^2}\right)(5.0 \text{ m})$$

$$= 2{,}200 \text{ N} \cdot \text{m}$$

$$= \boxed{2{,}200 \text{ J}}$$

3.3 Note that 10.0 kg is a mass quantity and not a weight quantity. Weight is found from $w = mg$, a form of Newton's second law of motion. Thus, the force that must be exerted to lift the backpack is its weight, or $(10.0 \text{ kg}) \times (9.8 \text{ m/s}^2)$, which is 98 N. Therefore, a force of 98 N was exerted on the backpack through a distance of 1.5 m, and

$$W = Fd$$

$$= \left(98 \frac{\text{kg} \cdot \text{m}}{\text{s}^2}\right)(1.5 \text{ m})$$

$$= 147 \text{ N} \cdot \text{m}$$

$$= \boxed{150 \text{ J}}$$

3.4 Weight is defined as the force of gravity acting on an object, and the greater the force of gravity, the harder it is to lift the object. The force is proportional to the mass of the object, as the equation $w = mg$ tells you. Thus, the force you exert when lifting is $F = w = mg$, so the work you do on an object you lift must be $W = mgh$.

You know the mass of the box and you know the work accomplished. You also know the value of the acceleration due to gravity, g, so the list of known and unknown quantities is:

$$m = 102 \text{ kg}$$
$$g = 9.8 \text{ m/s}^2$$
$$W = 5{,}000 \text{ J}$$
$$h = ?$$

The equation $W = mgh$ is solved for work, so the first thing to do is to solve it for h, the unknown height in this problem (note that height is also a distance):

$$W = mgh \therefore h = \frac{W}{mg}$$

$$= \frac{5,000 \frac{kg \cdot m}{s^2} \times m}{(102\,kg)\left(9.8\frac{m}{s^2}\right)}$$

$$= \frac{5,000.0}{102 \times 9.8} \frac{kg \cdot m}{s^2} \times \frac{m}{1} \times \frac{1}{kg} \times \frac{s^2}{m}$$

$$= \frac{5,000}{999.6} m$$

$$= \boxed{5\,m}$$

3.5 A student running up the stairs has to lift herself, so her weight is the required force needed. Thus, the force exerted is $F = w = mg$, and the work done is $W = mgh$. You know the mass of the student, the height, and the time. You also know the value of the acceleration due to gravity, g, so the list of known and unknown quantities is:

$$m = 60.0\,kg$$
$$g = 9.8\,m/s^2$$
$$h = 5.00\,m$$
$$t = 3.92\,s$$
$$P = ?$$

The equation $P = \dfrac{mgh}{t}$ is already solved for power, so:

a. $$P = \frac{mgh}{t}$$

$$= \frac{(60.0\,kg)\left(9.8\frac{m}{s^2}\right)(5.00\,m)}{3.92\,s}$$

$$= \frac{(60.0)(9.8)(5.00)\left(\frac{kg \cdot m}{s^2}\right) \times m}{(3.92)\quad s}$$

$$= \frac{2940}{3.92} \frac{N \cdot m}{s}$$

$$= 750\,\frac{J}{s}$$

$$= \boxed{750\,W}$$

b. A power of 750 watts is almost one horsepower.

3.6 Listing the known and unknown quantities:

$$m = 2,000\,kg$$
$$v = 72\,km/h$$
$$KE = ?$$

These are the quantities found in the equation for kinetic energy, $KE = 1/2mv^2$, which is already solved. However, note that the velocity is in units of km/h, which must be changed to m/s before doing anything else (it must be m/s because all energy and work units are in units of the joule [J]. A joule is a newton-meter, and a newton is a kg·m/s²).

Using the conversion factor from inside the front cover of your text,

$$\frac{0.2778\frac{m}{s}}{1.0\frac{km}{h}} \times 72\frac{km}{h}$$

$$(0.2778)(72)\frac{m}{s} \times \frac{h}{km} \times \frac{km}{h}$$

$$20\frac{m}{s}$$

and

$$KE = \frac{1}{2}mv^2$$

$$= \frac{1}{2}(2,000\,kg)\left(20\frac{m}{s}\right)^2$$

$$= \frac{1}{2}(2,000\,kg)\left(400\frac{m^2}{s^2}\right)$$

$$= \frac{1}{2} \times 2,000 \times 400\frac{kg \cdot m^2}{s^2}$$

$$= 400,000\frac{kg \cdot m}{s^2} \times m$$

$$= 400,000\,N \cdot m$$

$$= \boxed{4 \times 10^5\,J}$$

Scientific notation is used here to simplify a large number and to show one significant figure.

3.7 Recall the relationship between work and energy—that you do work on an object when you throw it, giving it kinetic energy, and the kinetic energy it has will do work on something else when stopping. Because of the relationship between work and energy, you can calculate (1) the work you do, (2) the kinetic energy a moving object has as a result of your work, and (3) the work it will do when coming to a stop, and all three answers should be the same. Thus, you do not have a force or a distance to calculate the work needed to stop a moving car, but you can simply calculate the kinetic energy of the car. Both answers should be the same.

Before you start, note that the velocity is in units of km/h, which must be changed to m/s before doing anything else (it must be m/s because all energy and work units are in units of the joule [J]. A joule is a newton-meter, and a newton is a kg·m/s²). Using the conversion factor from inside the front cover,

$$\frac{0.2778\frac{m}{s}}{1.0\frac{km}{h}} \times 54.0\frac{km}{h}$$

$$0.2778 \times 54.0\frac{m}{s} \times \frac{h}{km} \times \frac{km}{h}$$

$$15.0\frac{m}{s}$$

and

$$KE = \frac{1}{2}mv^2$$

$$= \frac{1}{2}(1{,}000.0 \text{ kg})\left(15.0\,\frac{\text{m}}{\text{s}}\right)^2$$

$$= \frac{1}{2}(1{,}000.0 \text{ kg})\left(225\,\frac{\text{m}^2}{\text{s}^2}\right)$$

$$= \frac{1}{2} \times 1{,}000.0 \times 225\,\frac{\text{kg}\cdot\text{m}^2}{\text{s}^2}$$

$$= 112{,}500\,\frac{\text{kg}\cdot\text{m}}{\text{s}^2} \times \text{m}$$

$$= 112{,}500 \text{ N}\cdot\text{m}$$

$$= \boxed{1.13 \times 10^5 \text{ J}}$$

Scientific notation is used here to simplify a large number and to easily show three significant figures. The answer could likewise be expressed as 113 kJ.

3.8 a. How much energy was used by a 1,000 kg car climbing a hill 51.02 m high is answered by how much work the car did. In this case, $W = Fd$ and the force exerted is the weight of the car, $w = mgh$. Thus,

$$w = mgh$$

$$= (1{,}000 \text{ kg})\left(9.8\,\frac{\text{m}}{\text{s}^2}\right)(51.02 \text{ m})$$

$$= 1{,}000 \times 9.8 \times 51.02 \text{ kg} \times \frac{\text{m}}{\text{s}^2} \times \text{m}$$

$$= 499{,}996\,\frac{\text{kg}\cdot\text{m}}{\text{s}^2} \times \text{m}$$

$$= 500{,}000 \text{ N}\cdot\text{m}$$

$$= 5 \times 10^5 \text{ J (or 500 kJ)}$$

b. How much potential energy the car has is found in the potential energy equation, $PE = mgh$. Note the potential energy is path independent, that is, depends only on the vertical height of the hill. Thus,

$$PE = mgh$$

$$= (1{,}000 \text{ kg})\left(9.8\,\frac{\text{m}}{\text{s}^2}\right)(51.02 \text{ m})$$

$$= 1{,}000 \times 9.8 \times 51.02 \text{ kg} \times \frac{\text{m}}{\text{s}^2} \times \text{m}$$

$$= 499{,}996\,\frac{\text{kg}\cdot\text{m}}{\text{s}^2} \times \text{m}$$

$$= 500{,}000 \text{ N}\cdot\text{m}$$

$$= \boxed{5 \times 10^5 \text{ J (or 500 kJ)}}$$

As you can see, the potential energy of the car is exactly the same as the amount of energy used to climb the hill.

3.9 a.
$$W = Fd$$
$$= (10 \text{ lb})(5 \text{ ft})$$
$$= (10)(5) \text{ ft} \times \text{lb}$$
$$= 50 \text{ ft}\cdot\text{lb}$$

b. The distance of the bookcase from some horizontal reference level did not change, so the gravitational potential energy does not change.

3.10 The force (F) needed to lift the book is equal to the weight (w) of the book, or $F = w$. Since $w = mg$, then $F = mg$. Work is defined as the product of a force moved through a distance, or $W = Fd$. The work done in lifting the book is therefore $W = mgh$, and:

a.
$$W = mgh$$

$$= (2.0 \text{ kg})(9.8 \text{ m/s}^2)(2.00 \text{ m})$$

$$= (2.0)(9.8)(2.00)\,\frac{\text{kg}\cdot\text{m}}{\text{s}^2} \times \text{m}$$

$$= 39.2\,\frac{\text{kg}\cdot\text{m}^2}{\text{s}^2}$$

$$= 39.2 \text{ J} = \boxed{39 \text{ J}}$$

b.
$$PE = mgh = \boxed{39 \text{ J}}$$

c.
$$PE_{\text{lost}} = KE_{\text{gained}} = mgh = \boxed{39 \text{ J}}$$

(or)

$$v = \sqrt{2gh} = \sqrt{(2)(9.8 \text{ m/s}^2)(2.00 \text{ m})}$$

$$= \sqrt{39.2 \text{ m}^2/\text{s}^2}$$

$$= 6.26 \text{ m/s}$$

$$KE = \frac{1}{2}mv^2 = \left(\frac{1}{2}\right)(2.0 \text{ kg})(6.26 \text{ m/s})^2$$

$$= \left(\frac{1}{2}\right)(2.0 \text{ kg})(39.2 \text{ m}^2/\text{s}^2)$$

$$= (1.0)(39.2)\,\frac{\text{kg}\cdot\text{m}^2}{\text{s}^2}$$

$$= \boxed{39 \text{ J}}$$

3.11
$$KE = \frac{1}{2}mv^2$$

$$= \frac{1}{2}(60.0 \text{ kg})\left(2.0\,\frac{\text{m}}{\text{s}}\right)^2$$

$$= \frac{1}{2}(60.0 \text{ kg})\left(4.0\,\frac{\text{m}^2}{\text{s}^2}\right)$$

$$= 30.0 \times 4.0 \text{ kg} \times \left(\frac{\text{m}^2}{\text{s}^2}\right)$$

$$= \boxed{120 \text{ J}}$$

$$KE = \frac{1}{2}mv^2$$

$$= \frac{1}{2}(60.0 \text{ kg})\left(4.0\,\frac{\text{m}}{\text{s}}\right)^2$$

$$= \frac{1}{2}(60.0 \text{ kg})\left(16\,\frac{\text{m}^2}{\text{s}^2}\right)$$

$$= 30.0 \times 16 \text{ kg} \times \left(\frac{\text{m}^2}{\text{s}^2}\right)$$

$$= \boxed{480 \text{ J}}$$

Thus, doubling the speed results in a four-fold increase in kinetic energy.

3.12 a. The force needed is equal to the weight of the student. The English unit of a pound is a force unit, so

$$W = Fd$$

$$= (170.0 \text{ lb})(25.0 \text{ ft})$$

$$= \boxed{4,250 \text{ ft·lb}}$$

b. Work (W) is defined as a force (F) moved through a distance (d), or $W = Fd$. Power (P) is defined as work (W) per unit of time (t), or $P = W/t$. Therefore,

$$P = \frac{Fd}{t}$$

$$= \frac{(170.0 \text{ lb})(25.0 \text{ ft})}{10.0 \text{ s}}$$

$$= \frac{(170.0)(25.0)}{10.0} \frac{\text{ft·lb}}{\text{s}}$$

$$= 425 \frac{\text{ft·lb}}{\text{s}}$$

One hp is defined as $550 \dfrac{\text{ft·lb}}{\text{s}}$ and

$$\frac{425 \text{ ft·lb/s}}{550 \dfrac{\text{ft·lb/s}}{\text{hp}}} = \boxed{0.77 \text{ hp}}$$

(Note that the student's power rating [425 ft·lb/s] is less than the power rating defined as 1 horsepower [550 ft·lb/s]. Thus, the student's horsepower must be *less* than 1 horsepower. A simple analysis such as this will let you know if you inverted the ratio or not.)

Chapter 4

4.1 Listing the known and unknown quantities:

body temperature $T_F = 98.6°$

$$T_C = ?$$

These are the quantities found in the equation for conversion of Fahrenheit to Celsius, $T_C = \dfrac{5}{9}(T_F - 32°)$, where T_F is the temperature in Fahrenheit and T_C is the temperature in Celsius. This equation describes a relationship between the two temperature scales and is used to convert a Fahrenheit temperature to Celsius. The equation is already solved for the Celsius temperature, T_C. Thus,

$$T_C = \frac{5}{9}(T_F - 32°)$$

$$= \frac{5}{9}(98.6° - 32°)$$

$$= \frac{333°}{9}$$

$$= \boxed{37°C}$$

4.2 $Q = mc\,\Delta T$

$$= (221 \text{ g})\left(0.093 \frac{\text{cal}}{\text{gC}°}\right)(38.0°C - 20.0°C)$$

$$= (221)(0.093)(18.0) \text{ g} \times \frac{\text{cal}}{\text{gC}°} \times C°$$

$$= 370 \frac{\text{g·cal·}C°}{\text{gC°}}$$

$$= \boxed{370 \text{ cal}}$$

4.3 First, you need to know the energy of the moving bike and rider. Since the speed is given as 36.0 km/h, convert to m/s by multiplying times 0.2778 m/s per km/h:

$$\left(36.0 \frac{\text{km}}{\text{h}}\right)\left(0.2778 \frac{\text{m/s}}{\text{km/h}}\right)$$

$$= (36.0)(0.2778) \frac{\text{km}}{\text{h}} \times \frac{\text{h}}{\text{km}} \times \frac{\text{m}}{\text{s}}$$

$$= 10.0 \text{ m/s}$$

Then,

$$KE = \frac{1}{2}mv^2$$

$$= \frac{1}{2}(100.0 \text{ kg})(10.0 \text{ m/s})^2$$

$$= \frac{1}{2}(100.0 \text{ kg})(100 \text{ m}^2/\text{s}^2)$$

$$= \frac{1}{2}(100.0)(100) \frac{\text{kg·m}^2}{\text{s}^2}$$

$$= 5,000 \text{ J}$$

Second, this energy is converted to the calorie heat unit through the mechanical equivalent of heat relationship, that 1.0 kcal = 4,184 J, or that 1.0 cal = 4.184 J. Thus,

$$\frac{5,000 \text{ J}}{4,184 \dfrac{\text{J}}{\text{kcal}}}$$

$$1.195 \frac{\text{J}}{1} \times \frac{\text{kcal}}{\text{J}}$$

$$\boxed{1.20 \text{ kcal}}$$

4.4 First, you need to find the energy of the falling bag. Since the potential energy lost equals the kinetic energy gained, the energy of the bag just as it hits the ground can be found from

$$PE = mgh$$

$$= (15.53 \text{ kg})(9.8 \text{ m/s}^2)(5.50 \text{ m})$$

$$= (15.53)(9.8)(5.50) \frac{\text{kg·m}}{\text{s}^2} \times \text{m}$$

$$= 837 \text{ J}$$

In calories, this energy is equivalent to

$$\frac{837 \text{ J}}{4,184 \dfrac{\text{J}}{\text{kcal}}} = 0.200 \text{ kcal}$$

Second, the temperature change can be calculated from the equation giving the relationship between a quantity of heat (Q), mass (m), specific heat of the substance (c), and the change of temperature:

$$Q = mc\,\Delta T \therefore \Delta T = \frac{Q}{mc}$$

$$= \frac{0.200 \text{ kcal}}{(15.53 \text{ kg})\left(0.200 \dfrac{\text{kcal}}{\text{kgC}°}\right)}$$

$$= \frac{0.200}{(15.53)(0.200)} \frac{\text{kcal}}{1} \times \frac{1}{\text{kg}} \times \frac{\text{kgC}°}{\text{kcal}}$$

$$= 0.064 \frac{\text{kcal·kg}C°}{\text{kcal·kg}}$$

$$= \boxed{6.4 \times 10^{-2} \, C°}$$

4.5 The Calorie used by dietitians is a kilocalorie; thus, 250.0 Cal is 250.0 kcal. The mechanical energy equivalent is 1 kcal = 4,184 J, so (250.0 kcal)(4,184 J/kcal) = 1,046,250 J.

Since $W = Fd$ and the force needed is equal to the weight (mg) of the person, $W = mgh = (75.0 \text{ kg})(9.8 \text{ m/s}^2)(10.0 \text{ m}) = 7,350$ J for each stairway climbed.

A total of 1,046,250 J of energy from the french fries would require 1,046,250 J/7,350 J per climb, or 142.3 trips up the stairs.

4.6 For unit consistency,

$$T_C = \frac{5}{9}(T_F - 32°) = \frac{5}{9}(68° - 32°) = \frac{5}{9}(36°) = 20°C$$

$$= \frac{5}{9}(32° - 32°) = \frac{5}{9}(0°) = 0°C$$

Glass bowl:

$$Q = mc\,\Delta T$$

$$= (0.5 \text{ kg})\left(0.2\,\frac{\text{kcal}}{\text{kg°C}}\right)(20 \text{C°})$$

$$= (0.5)(0.2)(20)\,\frac{\text{kg}}{1} \times \frac{\text{kcal}}{\text{kgC°}} \times \frac{\text{C°}}{1}$$

$$= \boxed{2 \text{ kcal}}$$

Iron pan:

$$Q = mc\,\Delta T$$

$$= (0.5 \text{ kg})\left(0.11\,\frac{\text{kcal}}{\text{kgC°}}\right)(20 \text{C°})$$

$$= (0.5)(0.11)(20) \text{ kg} \times \frac{\text{kcal}}{\text{kgC°}} \times \text{C°}$$

$$= \boxed{1 \text{ kcal}}$$

4.7 Note that a specific heat expressed in cal/gC° has the same numerical value as a specific heat expressed in kcal/kgC° because you can cancel the k units. You could convert 896 cal to 0.896 kcal, but one of the two conversion methods is needed for consistency with other units in the problem.

$$Q = mc\,\Delta T \therefore m = \frac{Q}{c\Delta T}$$

$$= \frac{896 \text{ cal}}{\left(0.056\,\dfrac{\text{cal}}{\text{gC°}}\right)(80.0 \text{C°})}$$

$$= \frac{896}{(0.056)(80.0)}\,\frac{\text{cal}}{1} \times \frac{\text{gC°}}{\text{cal}} \times \frac{1}{\text{C°}}$$

$$= 200 \text{ g}$$

$$= \boxed{0.20 \text{ kg}}$$

4.8 Since a watt is defined as a joule/s, finding the total energy in joules will tell the time:

$$Q = mc\,\Delta T$$

$$= (250.0 \text{ g})\left(1.00\,\frac{\text{cal}}{\text{gC°}}\right)(60.0 \text{ C°})$$

$$= (250.0)(1.00)(60.0) \text{ g} \times \frac{\text{cal}}{\text{gC°}} \times \text{C°}$$

$$= 1.50 \times 10^4 \text{ cal}$$

This energy in joules is $(1.50 \times 10^4 \text{ cal})\left(4.184\,\dfrac{\text{J}}{\text{cal}}\right) = 62,800 \text{ J}$

A 300 watt heater uses energy at a rate of $300\,\dfrac{\text{J}}{\text{s}}$, so

$$\frac{62,800 \text{ J}}{300 \text{ J/s}} = 209 \text{ s} \quad \text{is required, which is} \quad \frac{209 \text{ s}}{60\,\dfrac{\text{s}}{\text{min}}} = 3.48 \text{ min,}$$

or

$$\boxed{\text{about } 3\frac{1}{2} \text{ min}}$$

4.9
$$Q = mc\Delta \therefore c = \frac{Q}{m\Delta T}$$

$$= \frac{60.0 \text{ cal}}{(100.0 \text{ g})(20.0°C)}$$

$$= \frac{60.0}{(100.0)(20.0)}\,\frac{\text{cal}}{\text{gC°}}$$

$$= \boxed{0.0300\,\frac{\text{cal}}{\text{gC°}}}$$

4.10 To change water at 80.0°C to steam at 100.0°C requires two separate quantities of heat that can be called Q_1 and Q_2. The quantity Q_1 is the amount of heat needed to warm the water from 80.0°C to the boiling point, which is 100.0°C at sea level pressure ($\Delta T = 20.0$ C°). The relationship between the variables involved is $Q = mc\Delta T$. The quantity Q_2 is the amount of heat needed to take 100.0°C water through the phase change to steam (water vapor) at 100.0°C. The phase change from a liquid to a gas (or gas to liquid) is concerned with the latent heat of vaporization. For water, the latent heat of vaporization is given as 540.0 cal/g.

$m = 250.0$ g $\qquad\qquad Q_1 = mc\Delta T$

$L_{v(water)} = 540.0 \text{ cal/g} = (250.0 \text{ g})\left(1.00\,\dfrac{\text{cal}}{\text{gC°}}\right)(20.0 \text{ C°})$

$Q = ? \qquad\qquad\qquad = (250.0)(1.00)(20.0) \text{ g} \times \dfrac{\text{cal}}{\text{gC°}} \times \text{C°}$

$$= 5,000\,\frac{\text{g·cal·C°}}{\text{g·C°}}$$

$$= 5,000 \text{ cal}$$

$$= 5.00 \text{ kcal}$$

$$Q_2 = mL_v$$

$$= (250.0 \text{ g})\left(540.0\,\frac{\text{cal}}{\text{g}}\right)$$

$$= 250.0 \times 540.0\,\frac{\text{g·cal}}{\text{g}}$$

$$= 135,000 \text{ cal}$$

$$= 135.0 \text{ kcal}$$

$$Q_{Total} = Q_1 + Q_2$$

$$= 5.00 \text{ kcal} + 135.0 \text{ kcal}$$

$$= \boxed{140.0 \text{ kcal}}$$

4.11 To change 20.0°C water to steam at 125.0°C requires three separate quantities of heat. First, the quantity Q_1 is the amount of heat needed to warm the water from 20.0°C to 100.0°C ($\Delta T = 80.0$ C°). The quantity Q_2 is the amount of heat needed to take 100.0°C water to steam at 100.0°C. Finally, the quantity Q_3 is the amount of heat needed to warm the steam from 100.0° to 125.0°C. According to table 4.3, the c for steam is 0.480 cal/g°C.

$m = 100.0$ g
$\Delta T_{\text{water}} = 80.0$ C°
$\Delta T_{\text{steam}} = 25.0$ C°
$L_{v(\text{water})} = 540.0$ cal/g
$c_{\text{steam}} = 0.480$ cal/gC°

$$Q_1 = mc\,\Delta T$$
$$= (100.0\,\text{g})\left(1.00\,\frac{\text{cal}}{\text{gC}°}\right)(80.0\,\text{C}°)$$
$$= (100.0)(1.00)(80.0)\,\text{g} \times \frac{\text{cal}}{\text{gC}°} \times \text{C}°$$
$$= 8{,}000\,\frac{\text{g}\cdot\text{cal}\cdot\cancel{\text{C}°}}{\cancel{\text{g}}\cancel{\text{C}°}}$$
$$= 8{,}000\,\text{cal}$$
$$= 8.00\,\text{kcal}$$

$$Q_2 = mL_v$$
$$= (100.0\,\text{g})\left(540.0\,\frac{\text{cal}}{\text{g}}\right)$$
$$= 100.0 \times 540.0\,\frac{\text{g}\cdot\text{cal}}{\cancel{\text{g}}}$$
$$= 54{,}000\,\text{cal}$$
$$= 54.00\,\text{kcal}$$

$$Q_3 = mc\,\Delta T$$
$$= (100.0\,\text{g})\left(0.480\,\frac{\text{cal}}{\text{gC}°}\right)(25.0\,\text{C}°)$$
$$= (100.0)(0.480)(25.0)\,\text{g} \times \frac{\text{cal}}{\text{gC}°} \times \text{C}°$$
$$= 1{,}200\,\frac{\text{g}\cdot\text{cal}\cdot\cancel{\text{C}°}}{\cancel{\text{g}}\cdot\cancel{\text{C}°}}$$
$$= 1{,}200\,\text{cal}$$
$$= 1.20\,\text{kcal}$$

$$Q_{\text{total}} = Q_1 + Q_2 + Q_3$$
$$= 8.00\,\text{kcal} + 54.00\,\text{kcal} + 1.20\,\text{kcal}$$
$$= \boxed{63.20\,\text{kcal}}$$

4.12 a. Step 1: Cool the water from 18.0°C to 0°C.

$$Q_1 = mc\,\Delta T$$
$$= (400.0\,\text{g})\left(1.00\,\frac{\text{cal}}{\text{gC}°}\right)(18.0\,\text{C}°)$$
$$= (400.0)(1.00)(18.0)\,\text{g} \times \frac{\text{cal}}{\text{gC}°} \times \text{C}°$$
$$= 7{,}200\,\frac{\text{g}\cdot\text{cal}\cdot\cancel{\text{C}°}}{\cancel{\text{g}}\cancel{\text{C}°}}$$
$$= 7{,}200\,\text{cal}$$
$$= 7.20\,\text{kcal}$$

Step 2: Find the energy needed for the phase change of water at 0°C to ice at 0°C.

$$Q_2 = mL_f$$
$$= (400.0\,\text{g})\left(80.0\,\frac{\text{cal}}{\text{g}}\right)$$
$$= 400.0 \times 80.0\,\frac{\text{g}\cdot\text{cal}}{\cancel{\text{g}}}$$
$$= 32{,}000\,\text{cal}$$
$$= 32.0\,\text{kcal}$$

Step 3: Cool the ice from 0°C to ice at −5.00°C.

$$Q_3 = mc\,\Delta T$$
$$= (400.0\,\text{g})\left(0.50\,\frac{\text{cal}}{\text{gC}°}\right)(5.00\,\text{C}°)$$
$$= 400.0 \times 0.50 \times 5.00\,\text{g} \times \frac{\text{cal}}{\text{gC}°} \times \text{C}°$$
$$= 1{,}000\,\frac{\text{g}\cdot\text{cal}\cdot\cancel{\text{C}°}}{\cancel{\text{g}}\cdot\cancel{\text{C}°}}$$
$$= 1{,}000\,\text{cal}$$
$$= 1.0\,\text{kcal}$$

$$Q_{\text{total}} = Q_1 + Q_2 + Q_3$$
$$= 7.20\,\text{kcal} + 32.0\,\text{kcal} + 1.0\,\text{kcal}$$
$$= \boxed{40.2\,\text{kcal}}$$

Chapter 5

5.1
$$v = f\lambda$$
$$= \left(10\,\frac{1}{\text{s}}\right)(0.50\,\text{m})$$
$$= 5\,\frac{\text{m}}{\text{s}}$$

5.2 The distance between two *consecutive* condensations (or rarefactions) is one wavelength, so $\lambda = 3.00$ m and

$$v = f\lambda$$
$$= \left(112.0\,\frac{1}{\text{s}}\right)(3.00\,\text{m})$$
$$= 336\,\frac{\text{m}}{\text{s}}$$

5.3 a. One complete wave every 4 s means that $T = 4.00$ s. (Note that the symbol for the *time of a cycle* is T. Do not confuse this symbol with the symbol for temperature.)

b.
$$f = \frac{1}{T}$$
$$= \frac{1}{4.0\,\text{s}}$$
$$= \frac{1}{4.0}\,\frac{1}{\text{s}}$$
$$= 0.25\,\frac{1}{\text{s}}$$
$$= \boxed{0.25\,\text{Hz}}$$

5.4 The distance from one condensation to the next is one wavelength, so

$$v = f\lambda \therefore \lambda = \frac{v}{f}$$

$$= \frac{330 \frac{m}{s}}{260 \frac{1}{s}}$$

$$= \frac{330}{260} \frac{m}{s} \times \frac{s}{1}$$

$$= \boxed{1.3 \text{ m}}$$

5.5 a. $v = f\lambda = \left(256 \frac{1}{s}\right)(1.34 \text{ m}) = \boxed{343 \text{ m/s}}$

b. $= \left(440.0 \frac{1}{s}\right)(0.780 \text{ m}) = \boxed{343 \text{ m/s}}$

c. $= \left(750.0 \frac{1}{s}\right)(0.457 \text{ m}) = \boxed{343 \text{ m/s}}$

d. $= \left(2{,}500.0 \frac{1}{s}\right)(0.137 \text{ m}) = \boxed{343 \text{ m/s}}$

5.6 The speed of sound and time are given and you are looking for a distance.

$$v = 1{,}100 \text{ ft/s}$$
$$t = 4.80 \text{ s}$$
$$d = ?$$

Calculating the total distance the sound traveled,

$$v = \frac{d}{t} \therefore d = vt$$

$$= \left(1{,}100 \frac{ft}{s}\right)(4.80 \text{ s})$$

$$= (1{,}100)(4.80)\frac{ft}{s} \times s$$

$$= 4{,}800 \text{ ft}$$

Since the sound travels from you to the cliff, then back to you, the cliff must be half the total distance:

$$\frac{4{,}800 \text{ ft}}{2} = \boxed{2{,}400 \text{ ft}}$$

5.7 The speed of the sound and the time between the lightning and thunder are given:

$$v = 1{,}140 \text{ ft/s}$$
$$t = 4.63 \text{ s}$$
$$d = ?$$

The distance that a sound with this velocity travels in the given time is

$$v = \frac{d}{t} \therefore d = vt$$

$$= \left(1{,}140 \frac{ft}{s}\right)(4.63 \text{ s})$$

$$= (1{,}140)(4.63)\frac{ft}{s} \times s$$

$$= \boxed{5{,}280 \text{ ft or 1 mile}}$$

5.8 a. $v = f\lambda \therefore \lambda = \frac{v}{f}$

$$= \frac{1{,}125 \frac{ft}{s}}{440 \frac{1}{s}}$$

$$= \frac{1{,}125}{440} \frac{ft}{s} \times \frac{s}{1}$$

$$= 2.56 \frac{ft \cdot s}{s}$$

$$= \boxed{2.6 \text{ ft}}$$

b. $v = f\lambda \therefore \lambda = \frac{v}{f}$

$$= \frac{5{,}020}{440} \frac{ft}{s} \times \frac{s}{1}$$

$$= 11.4 \text{ ft} = \boxed{11 \text{ ft}}$$

Chapter 6

6.1 First, recall that a negative charge means an excess of electrons. Second, the relationship between the total charge (q), the number of electrons (n), and the charge of a single electron (e) is $q = ne$. The fundamental charge of a single ($n = 1$) electron (e) is 1.60×10^{-19} C. Thus

$$q = ne \therefore n = \frac{q}{e}$$

$$= \frac{1.00 \times 10^{-14} \text{ C}}{1.60 \times 10^{-19} \frac{C}{electron}}$$

$$= \frac{1.00 \times 10^{-14}}{1.60 \times 10^{-19}} \frac{C}{1} \times \frac{electron}{C}$$

$$= 6.25 \times 10^4 \frac{C \cdot electron}{C}$$

$$= \boxed{6.25 \times 10^4 \text{ electron}}$$

6.2

$$\frac{electric}{current} = \frac{charge}{time}$$

or

$$I = \frac{q}{t}$$

$$= \frac{6.00 \text{ C}}{2.00 \text{ s}}$$

$$= 3.00 \frac{C}{s}$$

$$= \boxed{3.00 \text{ A}}$$

6.3

$$R = \frac{V}{I}$$

$$= \frac{120.0 \text{ V}}{4.00 \text{ A}}$$

$$= 30.0 \frac{V}{A}$$

$$= \boxed{30.0 \ \Omega}$$

6.4

$$R = \frac{V}{I} \therefore I = \frac{V}{R}$$

$$= \frac{120.0 \text{ V}}{60.0 \, \frac{\text{V}}{\text{A}}}$$

$$= \frac{120.0}{60.0} \, \text{V} \times \frac{\text{A}}{\text{V}}$$

$$= \boxed{2.00 \text{ A}}$$

6.5 a.

$$R = \frac{V}{I} \therefore V = IR$$

$$= (1.20 \text{ A})\left(10.0 \, \frac{\text{V}}{\text{A}}\right)$$

$$= \boxed{12.0 \text{ V}}$$

b. Power = (current)(potential difference)

or

$$P = IV$$

$$= \left(1.20 \, \frac{\text{C}}{\text{s}}\right)\left(12.0 \, \frac{\text{J}}{\text{C}}\right)$$

$$= (1.20)(12.0) \frac{\text{C}}{\text{s}} \times \frac{\text{J}}{\text{C}}$$

$$= 14.4 \, \frac{\text{J}}{\text{s}}$$

$$= \boxed{14.4 \text{ W}}$$

6.6 Note that there are two separate electrical units that are rates: (1) the amp (coulomb/s) and (2) the watt (joule/s). The question asked for a rate of using energy. Energy is measured in joules, so you are looking for the power of the radio in watts. To find watts ($P = IV$), you will need to calculate the current (I) since it is not given. The current can be obtained from the relationship of Ohm's law:

$$I = \frac{V}{R}$$

$$= \frac{3.00 \text{ V}}{15.0 \, \frac{\text{V}}{\text{A}}}$$

$$= 0.200 \text{ A}$$

$$P = IV$$

$$= (0.200 \text{ C/s})(3.00 \text{ J/C})$$

$$= \boxed{0.600 \text{ W}}$$

6.7

$$\frac{(1,200 \text{ W})(0.25 \text{ h})(\$0.10/\text{kWh})}{1,000 \, \frac{\text{W}}{\text{kW}}}$$

$$\frac{(1,200)(0.25)(0.10)}{1,000} \, \frac{\text{W}}{1} \times \frac{\text{h}}{1} \times \frac{\$}{\text{kWh}} \times \frac{\text{kW}}{\text{W}}$$

$$\boxed{\$0.03} \text{ (3 cents)}$$

6.8 The relationship between power (P), current (I), and voltage (V) will provide a solution. Since the relationship considers

power in watts, the first step is to convert horsepower to watts. One horsepower is equivalent to 746 watts, so:

$$(746 \text{ W/hp})(2.00 \text{ hp}) = 1,492 \text{ W}$$

$$P = IV \therefore I = \frac{P}{V}$$

$$= \frac{1,492 \, \frac{\text{J}}{\text{s}}}{12.0 \, \frac{\text{J}}{\text{C}}}$$

$$= \frac{1,492}{12.0} \, \frac{\text{J}}{\text{s}} \times \frac{\text{C}}{\text{J}}$$

$$= 124.3 \, \frac{\text{C}}{\text{s}}$$

$$= \boxed{124 \text{ A}}$$

6.9 a. The rate of using energy is joule/s, or the watt. Since 1.00 hp = 746 W,

inside motor: (746 W/hp)(1/3 hp) = 249 W

outside motor: (746 W/hp)(1/3 hp) = 249 W

compressor motor: (746 W/hp)(3.70 hp) = 2,760 W

$$249 \text{ W} + 249 \text{ W} + 2,760 \text{ W} = \boxed{3,258 \text{ W}}$$

b. $$\frac{(3,258 \text{ W})(1.00 \text{ h})(\$0.10/\text{kWh})}{1,000 \text{ W/kW}} = \$0.33 \text{ per hour}$$

c. $(\$0.33/\text{h})(12 \text{ h/day})(30 \text{ day/mo}) = \boxed{\$118.80}$

6.10 The solution is to find how much current each device draws and then to see if the total current is less or greater than the breaker rating:

Toaster: $I = \dfrac{V}{R} = \dfrac{120 \text{ V}}{15 \text{ V/A}} = 8.0 \text{ A}$

Motor: $(0.20 \text{ hp})(746 \text{ W/hp}) = 150 \text{ W}$

$$I = \frac{P}{V} = \frac{15 \text{ J/s}}{120 \text{ J/C}} = 1.3 \text{ A}$$

Three 100 W bulbs: $3 \times 100 \text{ W} = 300 \text{ W}$

$$I = \frac{P}{V} = \frac{300 \text{ J/s}}{120 \text{ J/C}} = 2.5 \text{ A}$$

Iron: $I = \dfrac{P}{V} = \dfrac{600 \text{ J/s}}{120 \text{ J/C}} = 5.0 \text{ A}$

The sum of the currents is 8.0A + 1.3A + 2.5A + 5.0A = 16.8A, so the total current is greater than 15.0 amp and the circuit breaker will trip.

6.11 a.

$V_p = 1,200 \text{ V}$

$N_p = 1 \text{ loop}$

$N_s = 200 \text{ loops}$

$V_s = ?$

$$\frac{V_p}{N_p} = \frac{V_s}{N_s} \therefore V_s = \frac{V_p N_s}{N_p}$$

$$= \frac{(1,200 \text{ V})(200 \text{ loops})}{1 \text{ loop}}$$

$$= \boxed{240,000 \text{ V}}$$

b. $I_p = 40\ A$ $\quad V_p I_p = V_s I_s \ \therefore\ I_s = \dfrac{V_p I_p}{V_s}$

$I_s = ?$

$\qquad = \dfrac{1{,}200\ V \times 40\ A}{240{,}000\ V}$

$\qquad = \dfrac{1{,}200 \times 40}{240{,}000}\ \dfrac{\text{V}\cdot\text{A}}{\text{V}}$

$\qquad = \boxed{0.2\ A}$

6.12 a. $V_s = 12\ V$ $\qquad \dfrac{V_p}{N_p} = \dfrac{V_s}{N_s} \ \therefore\ \dfrac{N_p}{N_s} = \dfrac{V_p}{V_s}$

$\quad I_s = 0.5\ A$

$\quad V_p = 120\ V$

$\qquad\qquad\qquad\qquad = \dfrac{120\ \text{V}}{12\ \text{V}} = \dfrac{10}{1}$

$\quad \dfrac{N_p}{N_s} = ?$

or

$\boxed{10\ \text{primary to 1 secondary}}$

b. $I_p = ?$ $\quad V_p I_p = V_s I_s \ \therefore\ I_p = \dfrac{V_s I_s}{V_p}$

$\qquad\qquad I_p = \dfrac{(12\ V)(0.5\ A)}{120\ V}$

$\qquad\qquad\quad = \dfrac{12 \times 0.5}{120}\ \dfrac{\text{V}\cdot\text{A}}{\text{V}}$

$\qquad\qquad\quad = \boxed{0.05\ A}$

c. $\qquad P_s = ?$ $\quad P_s = I_s V_s$

$\qquad\qquad\quad = (0.5\ A)(12\ V)$

$\qquad\qquad\quad = 0.5 \times 12\ \dfrac{\text{C}}{\text{s}} \times \dfrac{\text{J}}{\text{C}}$

$\qquad\qquad\quad = 6\ \dfrac{\text{J}}{\text{s}}$

$\qquad\qquad\quad = \boxed{6\ W}$

Chapter 7

7.1 The relationship between the speed of light in a transparent material (v), the speed of light in a vacuum ($c = 3.00 \times 10^8$ m/s), and the index of refraction (n) is $n = c/v$. According to table 7.1, the index of refraction for water is $n = 1.33$ and for ice is $n = 1.31$.

a. $c = 3.00 \times 10^8$ m/s $\qquad n = \dfrac{c}{v} \ \therefore\ v = \dfrac{c}{n}$

$\quad n = 1.33$

$\quad v = ?$ $\qquad\qquad\qquad = \dfrac{3.00 \times 10^8\ \text{m/s}}{1.33}$

$\qquad\qquad\qquad\qquad = \boxed{2.26 \times 10^8\ \text{m/s}}$

b. $c = 3.00 \times 10^8$ m/s $\qquad v = \dfrac{3.00 \times 10^8\ \text{m/s}}{1.31}$

$\quad n = 1.31$

$\quad v = ?$ $\qquad\qquad\qquad = \boxed{2.29 \times 10^8\ \text{m/s}}$

7.2

$d = 1.50 \times 10^8$ km $\qquad v = \dfrac{d}{t} \ \therefore\ t = \dfrac{d}{v}$

$\ = 1.50 \times 10^{11}$ m

$c = 3.00 \times 10^8$ m/s $\qquad\qquad = \dfrac{1.50 \times 10^{11}\ \text{m}}{3.00 \times 10^8\ \dfrac{\text{m}}{\text{s}}}$

$t = ?$

$\qquad\qquad = \dfrac{1.50 \times 10^{11}}{3.00 \times 10^8}\ \text{m} \times \dfrac{\text{s}}{\text{m}}$

$\qquad\qquad = 5.00 \times 10^2\ \dfrac{\text{m}\cdot\text{s}}{\text{m}}$

$\qquad\qquad = \dfrac{5.00 \times 10^2\ \text{s}}{60.0\ \dfrac{\text{s}}{\text{min}}}$

$\qquad\qquad = \dfrac{5.00 \times 10^2}{60.0}\ \text{s} \times \dfrac{\text{min}}{\text{s}}$

$\qquad\qquad = \boxed{8.33\ \text{min}}$

7.3

$d = 6.00 \times 10^9$ km $\qquad v = \dfrac{d}{t} \ \therefore\ t = \dfrac{d}{v}$

$\ = 6.00 \times 10^{12}$ m

$c = 3.00 \times 10^8$ m/s $\qquad\quad = \dfrac{6.00 \times 10^{12}\ \text{m}}{3.00 \times 10^8\ \dfrac{\text{m}}{\text{s}}}$

$t = ?$

$\qquad\qquad = \dfrac{6.00 \times 10^{12}}{3.00 \times 10^8}\ \text{m} \times \dfrac{\text{s}}{\text{m}}$

$\qquad\qquad = 2.00 \times 10^4\ \text{s}$

$\qquad\qquad = \dfrac{2.00 \times 10^4\ \text{s}}{3{,}600\ \dfrac{\text{s}}{\text{h}}}$

$\qquad\qquad = \dfrac{2.00 \times 10^4}{3.600 \times 10^3}\ \text{s} \times \dfrac{\text{h}}{\text{s}}$

$\qquad\qquad = \boxed{5.56\ \text{h}}$

7.4 From equation 7.1, note that both angles are measured from the normal and that the angle of incidence (θ_i) equals the angle of reflection (θ_r), or

$$\theta_i = \theta_r \ \therefore\ \boxed{\theta_i = 10°}$$

7.5 $\quad v = 2.20 \times 10^8$ m/s $\qquad n = \dfrac{c}{v}$

$\quad c = 3.00 \times 10^8$ m/s

$\quad n = ?$ $\qquad\qquad\qquad = \dfrac{3.00 \times 10^8\ \dfrac{\text{m}}{\text{s}}}{2.20 \times 10^8\ \dfrac{\text{m}}{\text{s}}}$

$\qquad\qquad\qquad = 1.36$

According to table 7.1, the substance with an index of refraction of 1.36 is $\boxed{\text{ethyl alcohol.}}$

7.6 a. From equation 7.3:

$\lambda = 6.00 \times 10^{-7}$ m $c = \lambda f \therefore f = \dfrac{c}{\lambda}$

$c = 3.00 \times 10^8$ m/s

$f = ?$

$$= \dfrac{3.00 \times 10^8 \, \dfrac{m}{s}}{6.00 \times 10^{-7} \, m}$$

$$= \dfrac{3.00 \times 10^8}{6.00 \times 10^{-7}} \, \dfrac{\cancel{m}}{s} \times \dfrac{1}{\cancel{m}}$$

$$= 5.00 \times 10^{14} \, \dfrac{1}{s}$$

$$= \boxed{5.00 \times 10^{14} \, \text{Hz}}$$

b. From equation 7.4:

$f = 5.00 \times 10^{14}$ Hz $E = hf$

$h = 6.63 \times 10^{-34}$ J·s

$E = ?$

$$= (6.63 \times 10^{-34} \, \text{J·s})\left(5.00 \times 10^{14} \, \dfrac{1}{s}\right)$$

$$= (6.63 \times 10^{-34})(5.00 \times 10^{14}) \, \text{J·}\cancel{s} \times \dfrac{1}{\cancel{s}}$$

$$= \boxed{3.32 \times 10^{-19} \, \text{J}}$$

7.7 First, you can find the energy of one photon of the peak intensity wavelength (5.60×10^{-7}m) by using equation 7.3 to find the frequency, then equation 7.4 to find the energy:

Step 1:
$$c = \lambda f \therefore f = \dfrac{c}{\lambda}$$

$$= \dfrac{3.00 \times 10^8 \, \dfrac{m}{s}}{5.60 \times 10^{-7} \, m}$$

$$= 5.36 \times 10^{14} \, \text{Hz}$$

Step 2: $E = hf$

$$= (6.63 \times 10^{-34} \, \text{J·s})(5.36 \times 10^{14} \, \text{Hz})$$

$$= 3.55 \times 10^{-19} \, \text{J}$$

Step 3: Since one photon carries an energy of 3.55×10^{-19} J and the overall intensity is 1,000.0 W, each square meter must receive an average of

$$\dfrac{1,000.0 \, \dfrac{J}{s}}{3.55 \times 10^{-19} \, \dfrac{J}{\text{photon}}}$$

$$\dfrac{1.000 \times 10^3}{3.55 \times 10^{-19}} \, \dfrac{\cancel{J}}{s} \times \dfrac{\text{photon}}{\cancel{J}}$$

$$\boxed{2.82 \times 10^{21} \, \dfrac{\text{photon}}{s}}$$

7.8 a.

$f = 4.90 \times 10^{14}$ Hz $c = \lambda f \therefore \lambda = \dfrac{c}{f}$

$c = 3.00 \times 10^8$ m/s

$\lambda = ?$

$$= \dfrac{3.00 \times 10^8 \, \dfrac{m}{s}}{4.90 \times 10^{14} \, \dfrac{1}{s}}$$

$$= \dfrac{3.00 \times 10^8}{4.90 \times 10^{14}} \, \dfrac{m}{\cancel{s}} \times \dfrac{\cancel{s}}{1}$$

$$= \boxed{6.12 \times 10^{-7} \, m}$$

b. According to table 7.2, this is the frequency and wavelength of orange light.

7.9

$f = 5.00 \times 10^{20}$ Hz $E = hf$

$h = 6.63 \times 10^{-34}$ J·s

$E = ?$

$$= (6.63 \times 10^{-34} \, \text{J·s})\left(5.00 \times 10^{20} \, \dfrac{1}{s}\right)$$

$$= (6.63 \times 10^{-34})(5.00 \times 10^{20}) \, \text{J·}\cancel{s} \times \dfrac{1}{\cancel{s}}$$

$$= \boxed{3.32 \times 10^{-13} \, \text{J}}$$

7.10

$\lambda = 1.00$ mm

$\quad = 0.001$ m

$f = ?$

$c = 3.00 \times 10^8$ m/s

$h = 6.63 \times 10^{-34}$ J·s

$E = ?$

Step 1: $c = \lambda f \therefore f = \dfrac{c}{\lambda}$

$$= \dfrac{3.00 \times 10^8 \, \dfrac{m}{s}}{1.00 \times 10^{-3} \, m}$$

$$= \dfrac{3.00 \times 10^8}{1.00 \times 10^{-3}} \, \dfrac{\cancel{m}}{s} \times \dfrac{1}{\cancel{m}}$$

$$= 3.00 \times 10^{11} \, \text{Hz}$$

Step 2: $E = hf$

$$= (6.63 \times 10^{-34} \, \text{J·s})\left(3.00 \times 10^{11} \, \dfrac{1}{s}\right)$$

$$= (6.63 \times 10^{-34})(3.00 \times 10^{11}) \, \text{J·}\cancel{s} \times \dfrac{1}{\cancel{s}}$$

$$= \boxed{1.99 \times 10^{-22} \, \text{J}}$$

Chapter 8

8.1 Energy is related to the frequency and Planck's constant in equation 8.1, $E = hf$.

$$\text{For } n = 6, E_H = 6.05 \times 10^{-20} \text{ J}$$
$$\text{For } n = 2, E_L = 5.44 \times 10^{-19} \text{ J}$$
$$E = ? \text{ J}$$

$$E = E_H - E_L$$
$$= (-6.05 \times 10^{-20} \text{ J}) - (-5.44 \times 10^{-19} \text{ J})$$
$$= \boxed{4.84 \times 10^{-19} \text{ J}}$$

8.2 $\text{For } n = 6, E_H = -6.05 \times 10^{-20}$ J
$\text{For } n = 2, E_L = -5.44 \times 10^{-19}$ J
$h = 6.63 \times 10^{-34}$ J·s
$f = ?$

$$hf = E_H - E_L \therefore f = \frac{E_H - E_L}{h}$$
$$= \frac{(-6.05 \times 10^{-20} \text{ J}) - (-5.44 \times 10^{-19} \text{ J})}{6.63 \times 10^{-34} \text{ J·s}}$$
$$= \frac{4.84 \times 10^{-19}}{6.63 \times 10^{-34}} \frac{\text{J}}{\text{J·s}}$$
$$= 7.29 \times 10^{14} \frac{1}{\text{s}}$$
$$= \boxed{7.29 \times 10^{14} \text{ Hz (violet)}}$$

8.3 $(n = 1) = -13.6$ eV

Since the energy of the electron is -13.6 eV, it will require 13.6 eV (or 2.17×10^{-18} J) to remove the electron.

8.4

$q/m = -1.76 \times 10^{11}$ C/kg
$q = -1.60 \times 10^{-19}$ C
$m = ?$

$$\text{mass} = \frac{\text{charge}}{\text{charge / mass}}$$
$$= \frac{-1.60 \times 10^{-19} \text{ C}}{-1.76 \times 10^{11} \frac{\text{C}}{\text{kg}}}$$
$$= \frac{-1.60 \times 10^{-19}}{-1.76 \times 10^{11}} \cancel{C} \times \frac{\text{kg}}{\cancel{C}}$$
$$= \boxed{9.09 \times 10^{-31} \text{ kg}}$$

Chapter 9

9.1 Recall that the number of outer energy level electrons is the same as the family number for the representative elements:

a. Li: 1   d. Cl: 7
b. N: 5   e. Ra: 2
c. F: 7   f. Be: 2

9.2 The same information that was used in question 1 can be used to draw the dot notation:

(a) $\overset{\cdot}{\underset{\cdot}{B}} \cdot$ (c) Ca: (e) $\cdot \overset{\cdot\cdot}{\underset{\cdot\cdot}{O}} \cdot$

(b) $\cdot \overset{\cdot\cdot}{Br} \cdot$ (d) K·   (f) $\cdot \overset{\cdot\cdot}{\underset{\cdot\cdot}{S}} \cdot$

9.3 The charge is found by identifying how many electrons are lost or gained in achieving the noble gas structure:

a. Boron 3+
b. Bromine 1−
c. Calcium 2+
d. Potassium 1+
e. Oxygen 2−
f. Nitrogen 3−

9.4 The name of some common polyatomic ions are in box table 9.2. Using this table as a reference, the names are
a. hydroxide
b. sulfite
c. hypochlorite
d. nitrate
e. carbonate
f. perchlorate

9.5 The Roman numeral tells you the charge on the variable charge elements. The charges for the polyatomic ions are found in box table 9.2. The charges for metallic elements can be found in tables 9.2 and 9.3. Using these resources and the crossover technique, the formulas are as follows:
a. $Fe(OH)_3$
b. $Pb_3(PO_4)_2$
c. $ZnCO_3$
d. NH_4NO_3
e. $KHCO_3$
f. K_2SO_3

9.6 Box table 9.3 has information about the meaning of prefixes and stem names used in naming covalent compounds. (a), for example, asks for the formula of carbon tetrachloride. Carbon has no prefixes, so there is one carbon atom, and it comes first in the formula because it comes first in the name. The *tetra*-prefix means four, so there are four chlorine atoms. The name ends in *-ide*, so you know there are only two elements in the compound. The symbols can be obtained from the list of elements on the inside back cover of this text. Using all this information from the name, you can think out the formula for carbon tetrachloride. The same process is used for the other compounds and formulas:
a. CCl_4 d. SO_3
b. H_2O e. N_2O_5
c. MnO_2 f. As_2S_5

9.7 Again using information from box table 9.3, this question requires you to reverse the thinking procedure you learned in question 6.
  a. carbon monoxide
  b. carbon dioxide
  c. carbon disulfide
  d. dinitrogen monoxide
  e. tetraphosphorus trisulfide
  f. dinitrogen trioxide

9.8 **a.** $2 SO_2 + O_2 \rightarrow 2 SO_3$
  b. $4 P + 5 O_2 \rightarrow 2 P_2O_5$
  c. $2 Al + 6 HCl \rightarrow 2 AlCl_3 + 3 H_2$
  d. $2 NaOH + H_2SO_4 \rightarrow Na_2SO_4 + 2 H_2O$
  e. $Fe_2O_3 + 3 CO \rightarrow 2 Fe + 3 CO_2$
  f. $3 Mg(OH)_2 + 2 H_3PO_4 \rightarrow Mg_3(PO_4)_2 + 6 H_2O$

9.9 **a.** General form of $XY + AZ \rightarrow XZ + AY$ with precipitate formed: Ion exchange reaction.
  b. General form of $X + Y \rightarrow XY$: Combination reaction.
  c. General form of $XY \rightarrow X + Y + \ldots$: Decomposition reaction.
  d. General form of $X + Y \rightarrow XY$: Combination reaction.
  e. General form of $XY + A \rightarrow AY + X$: Replacement reaction.
  f. General form of $X + Y \rightarrow XY$: Combination reaction.

9.10 **a.** $C_5H_{12}(g) + 8 O_2(g) \rightarrow 5 CO_2(g) + 6 H_2O(g)$
  b. $HCl(aq) + NaOH(aq) \rightarrow NaCl(aq) + H_2O(l)$
  c. $2 Al(s) + Fe_2O_3(s) \rightarrow Al_2O_3(s) + 2 Fe(l)$
  d. $Fe(s) + CuSO_4(aq) \rightarrow FeSO_4(aq) + Cu(s)$
  e. $MgCl_2(aq) + Fe(NO_3)_2(aq) \rightarrow$ No reaction (all possible compounds are soluble and no gas or water was formed)
  f. $C_6H_{10}O_5(s) + 6 O_2(g) \rightarrow 6 CO_2(g) + 5 H_2O(g)$

Chapter 11

11.1 **a.** cobalt-60: 27 protons, 33 neutrons
  b. potassium-40: 19 protons, 21 neutrons
  c. neon-24: 10 protons, 14 neutrons
  d. lead-208: 82 protons, 126 neutrons

11.2 **a.** $^{60}_{27}Co$
  b. $^{40}_{19}K$
  c. $^{24}_{10}Ne$
  d. $^{204}_{82}Pb$

11.3 **a.** cobalt-60: Radioactive because odd numbers of protons (27) and odd numbers of neutrons (33) are usually unstable.
  b. potassium-40: Radioactive, again having an odd number of protons (19) and an odd number of neutrons (21).
  c. neon-24: Stable, because even numbers of protons and neutrons are usually stable.
  d. lead-208: Stable, because even numbers of protons and neutrons, *and* because 82 is a particularly stable number of nucleons.

11.4 **a.** $^{56}_{26}Fe \rightarrow \ ^{0}_{-1}e + \ ^{56}_{27}Co$
  b. $^{7}_{4}Be \rightarrow \ ^{0}_{-1}e + \ ^{7}_{5}B$
  c. $^{64}_{29}Cu \rightarrow \ ^{0}_{-1}e + \ ^{64}_{30}Zn$
  d. $^{24}_{11}Na \rightarrow \ ^{0}_{-1}e + \ ^{24}_{12}Mg$
  e. $^{214}_{82}Pb \rightarrow \ ^{0}_{-1}e + \ ^{214}_{83}Bi$
  f. $^{32}_{15}P \rightarrow \ ^{0}_{-1}e + \ ^{32}_{16}S$

11.5 **a.** $^{235}_{92}U \rightarrow \ ^{4}_{2}He + \ ^{231}_{90}Th$
  b. $^{226}_{88}Ra \rightarrow \ ^{4}_{2}He + \ ^{222}_{86}Rn$
  c. $^{239}_{94}Pu \rightarrow \ ^{4}_{2}He + \ ^{235}_{92}U$
  d. $^{214}_{83}Bi \rightarrow \ ^{4}_{2}He + \ ^{210}_{81}Tl$
  e. $^{230}_{90}Th \rightarrow \ ^{4}_{2}He + \ ^{226}_{88}Ra$
  f. $^{210}_{84}Po \rightarrow \ ^{4}_{2}He + \ ^{206}_{82}Pb$

11.6 Thirty-two days is four half-lives. After the first half-life (8 days), 1/2 oz will remain. After the second half-life (8 + 8, or 16 days), 1/4 oz will remain. After the third half-life (8 + 8 + 8, or 24 days), 1/8 oz will remain. After the fourth half-life (8 + 8 + 8 + 8, or 32 days), 1/16 oz will remain, or 6.3×10^{-2} oz.

11.7 The Fe-56 nucleus has a mass of 55.9206 u, but the individual masses of the nucleons are

$$26 \text{ protons} \times 1.00728 \text{ u} = 26.1893 \text{ u}$$
$$30 \text{ neutrons} \times 1.00867 \text{ u} = \frac{30.2601 \text{ u}}{56.4494 \text{ u}}$$

The mass defect is thus

$$56.4494 \text{ u}$$
$$\underline{-55.9206 \text{ u}}$$
$$0.5288 \text{ u}$$

The atomic mass unit (u) is equal to the mass of a mole (g), therefore 0.5288 u = 0.5288 g. The mass defect is equivalent to the binding energy according to $E = mc^2$. For a molar mass of Fe-56, the mass defect is

$$E = (5.29 \times 10^{-4} \text{ kg})\left(3.00 \times 10^8 \ \frac{m}{s}\right)^2$$
$$= (5.29 \times 10^{-4} \text{ kg})\left(9.00 \times 10^{16} \ \frac{m^2}{s^2}\right)$$
$$= 4.76 \times 10^{13} \ \frac{kg \cdot m^2}{s^2}$$
$$= 4.76 \times 10^{13} \text{ J}$$

For a single nucleus,

$$\frac{4.76 \times 10^{13} \text{ J}}{6.02 \times 10^{23} \text{ nuclei}} = 7.90 \times 10^{-11} \text{ J/nuclei}$$

Glossary

A

abiotic factors nonliving parts of an organism's environment

absolute humidity a measure of the actual amount of water vapor in the air at a given time—for example, in grams per cubic meter

absolute magnitude a classification scheme to compensate for the distance differences to stars; calculations of the brightness that stars would appear to have if they were all at a defined, standard distance

absolute scale temperature scale set so that zero is at the theoretical lowest temperature possible, which would occur when all random motion of molecules has ceased

absolute zero the theoretical lowest temperature possible, which occurs when all random motion of molecules has ceased

abyssal plain the practically level plain of the ocean floor

acceleration a change in velocity per change in time; by definition, this change in velocity can result from a change in speed, a change in direction, or a combination of changes in speed and direction

accretion disk fat bulging disk of gas and dust from the remains of the gas cloud that forms around a protostar

acetylcholine a neurotransmitter secreted into the synapse by many axons and received by dendrites

acetylcholinesterase an enzyme present in the synapse that destroys acetylcholine

achondrites homogeneously textured stony meteorites

acid any substance that is a proton donor when dissolved in water; generally considered a solution of hydronium ions in water that can neutralize a base, forming a salt and water

acid-base indicator a vegetable dye used to distinguish acid and base solutions by a color change

acquired characteristics characteristics an organism gains during its lifetime that are not genetically determined and therefore cannot be passed on to future generations

active transport use of a carrier molecule to move molecules through a cell membrane in a direction opposite that of the concentration gradient; the carrier requires an input of energy other than the kinetic energy of the molecules

adenine a double-ring nitrogenous-base molecule in DNA and RNA; the complementary base of thymine or uracil

adenosine triphosphate (ATP) a molecule formed from the building blocks of adenine, ribose, and phosphates; it functions as the primary energy carrier in the cell

aerobic cellular respiration the biochemical pathway that requires oxygen and converts food, such as carbohydrates, to carbon dioxide and water; during this conversion, it releases the chemical-bond energy as ATP molecules

air mass a large, more or less uniform body of air with nearly the same temperature and moisture conditions throughout

air mass weather the weather experienced within a given air mass; characterized by slow, gradual changes from day to day

alcohol an organic compound with a general formula of ROH, where R is one of the hydrocarbon groups; for example, methyl or ethyl

aldehyde an organic molecule with the general formula RCHO, where R is one of the hydrocarbon groups; for example, methyl or ethyl

alkali metals members of family IA of the periodic table, having common properties of shiny, low-density metals that can be cut with a knife and that react violently with water to form an alkaline solution

alkaline earth metals members of family IIA of the periodic table, having common properties of soft, reactive metals that are less reactive than alkali metals

alleles alternative forms of a gene for a particular characteristic (e.g., attached-earlobe and free-earlobe are alternative alleles for ear shape)

alpha particle the nucleus of a helium atom (two protons and two neutrons) emitted as radiation from a decaying heavy nucleus; also known as an alpha ray

alpine glaciers glaciers that form at high elevations in mountainous regions

alternating current an electric current that first moves one direction, then the opposite direction with a regular frequency

alternation of generations a term used to describe that aspect of the life cycle in which there are two distinctly different forms of an organism; each form is involved in the production of the other, and only one form is involved in producing gametes

alveoli tiny sacs that are part of the structure of the lungs where gas exchange takes place

amino acids organic molecules that join to form polypeptides and proteins

amp unit of electric current; equivalent to C/s

ampere full name of the unit amp

amplitude the extent of displacement from the equilibrium condition; the size of a wave from the rest (equilibrium) position

anaphase the third stage of mitosis, characterized by dividing of the centromeres and movement of the chromosomes to the poles

androgens male sex hormones produced by the testes that cause the differentiation of typical internal and external genital male anatomy

angle of incidence angle of an incident (arriving) ray or particle to a surface; measured from a line perpendicular to the surface (the normal)

angle of reflection angle of a reflected ray or particle from a surface; measured from a line perpendicular to the surface (the normal)

angular momentum quantum number in the quantum mechanics model of the atom, one of four descriptions of the energy state of an electron wave; this quantum number describes the energy sublevels of electrons within the main energy levels of an atom

annular eclipse occurs when the penumbra reaches the surface of Earth; as seen from Earth, the Sun forms a bright ring around the disk of the Moon

anorexia nervosa a nutritional deficiency disease characterized by severe, prolonged weight loss for fear of becoming obese

Antarctic Circle parallel identifying the limit toward the equator where the Sun appears above the horizon all day for six months during the summer; located at 66.5°S latitude

anther the sex organ in plants that produces the pollen that contains the sperm

antibody a globular protein molecule made by the body in response to the presence of a foreign or harmful molecule called an antigen; these molecules are capable of combining with the foreign molecules or microbes to inactivate or kill them

anticline an arch-shaped fold in layered bed rock

anticodon a sequence of three nitrogenous bases on a tRNA molecule capable of forming hydrogen bonds with three complementary bases on an mRNA codon during translation

anticyclone a high-pressure center with winds flowing away from the center; associated with clear, fair weather

antinode region of maximum amplitude between adjacent nodes in a standing wave

aorta the large blood vessel that carries blood from the left ventricle to the majority of the body

apogee the point at which the Moon's elliptical orbit takes the Moon farthest from Earth

apoptosis also known as "programmed cell death"; death of specific cells that has a genetic basis and is not the result of injury

apparent local noon the instant when the Sun crosses the celestial meridian at any particular longitude

apparent local solar time the time found from the position of the Sun in the sky; the shadow of the gnomon on a sundial

apparent magnitude a classification scheme for different levels of brightness of stars that you see; brightness values range from one to six with the number one (first magnitude) assigned to the brightest star and the number six (sixth magnitude) assigned to the faintest star that can be seen

apparent solar day the interval between two consecutive crossings of the celestial meridian by the Sun

aquifer a layer of sand, gravel, or other highly permeable material beneath the surface that is saturated with water and is capable of producing water in a well or spring

Archaea the domain in which are found prokaryotic organisms that live in extreme habitats

archaeology the study of past human life and culture from material evidence

Arctic Circle parallel identifying the limit toward the equator where the Sun appears above the horizon all day for one day up to six months during the summer; located at 66.5°N latitude

area the extent of a surface; the surface bounded by three or more lines

arid dry climate classification; receives less than 25 cm (10 in) precipitation per year

aromatic hydrocarbon organic compound with at least one benzene ring structure; cyclic hydrocarbons and their derivatives

arteries the blood vessels that carry blood away from the heart

arterioles small arteries located just before capillaries that can expand and contract to regulate the flow of blood to parts of the body

artesian term describing the condition where confining pressure forces groundwater from a well to rise above the aquifer

asbestos the common name for any one of several incombustible fibrous minerals that will not melt or ignite and can be woven into a fireproof cloth or used directly in fireproof insulation; about six different commercial varieties of asbestos are used, one of which has been linked to cancer under heavy exposure

assimilation the physiological process that takes place in a living cell as it converts nutrients in food into specific molecules required by the organism

asteroids small rocky bodies left over from the formation of the solar system; most are accumulated in a zone between the orbits of Mars and Jupiter

asthenosphere a plastic, mobile layer of Earth's structure that extends around Earth below the lithosphere

astronomical unit the radius of Earth's orbit is defined as one astronomical unit (AU)

atom the smallest unit of an element that can exist alone or in combination with other elements

atomic mass unit relative mass unit (u) of an isotope based on the standard of the carbon-12 isotope, which is defined as a mass of exactly 12.00 u; one atomic mass unit (1 u) is 1/12 the mass of a carbon-12 atom

atomic number the number of protons in the nucleus of an atom

atomic weight weighted average of the masses of stable isotopes of an element as they occur in nature, based on the abundance of each isotope of the element and the atomic mass of the isotope compared to carbon-12

atria thin-walled sacs of the heart that receive blood from the veins of the body and empty into the ventricles; singular, *atrium*

atrioventricular valves located between the atria and ventricles of the heart preventing blood from flowing backward from the ventricles into the atria

autosomes chromosomes that typically carry genetic information used by the organism for characteristics other than the primary determination of sex

autotroph an organism that is able to make its food molecules from inorganic raw materials by using basic energy sources such as sunlight

autumnal equinox one of two times a year that daylight and night are of equal length; occurs on or about September 23 and identifies the beginning of the fall season

axis the imaginary line about which a planet or other object rotates

axon a neuronal fiber that carries information away from the nerve cell body

B

background radiation ionizing radiation (alpha, beta, gamma, etc.) from natural sources; between 100 and 500 millirems/yr of exposure to natural radioactivity from the environment

Balmer series a set of four line spectra, narrow lines of color emitted by hydrogen atom electrons as they drop from excited states to the ground state

band of stability a region of a graph of the number of neutrons versus the number of protons in nuclei; nuclei that have the neutron to proton ratios located in this band do not undergo radioactive decay

barometer an instrument that measures atmospheric pressure, used in weather forecasting and in determining elevation above sea level

basal metabolic rate (BMR) the amount of energy required to maintain normal body activity while at rest

base any substance that is a proton acceptor when dissolved in water; generally considered a solution that forms hydroxide ions in water that can neutralize an acid, forming a salt and water

basilar membrane a membrane in the cochlea containing sensory cells that are stimulated by the vibrations caused by sound waves

basin a large, bowl-shaped fold in the land into which streams drain; also a small enclosed or partly enclosed body of water

beat rhythmic increases and decreases of volume from constructive and destructive interference between two sound waves of slightly different frequencies

benign tumor an abnormal cell mass that does not fragment and spread beyond its original area of growth

benthic aquatic organisms that live on the bottom

beta particle high-energy electron emitted as ionizing radiation from a decaying nucleus; also known as a beta ray

big bang theory current model of galactic evolution in which the universe was created from an intense and brilliant explosion from a primeval fireball

bile a product of the liver, stored in the gallbladder, which is responsible for the emulsification of fats

binding energy the energy required to break a nucleus into its constituent protons and neutrons; also the energy equivalent released when a nucleus is formed

binomial system of nomenclature a naming system that uses two Latin names, genus and specific epithet, for each type of organism

biodiversity the number of different kinds of organisms found in an area

biogenesis the concept that life originates only from preexisting life

biological species concept the concept that species can be distinguished from one another by their inability to interbreed

biology the science that deals with the study of living things and how living things interact with things around them

biomass (energy) any material formed by photosynthesis, including plants, trees, and crops, and any garbage, crop residue, or animal waste; the mass of living material in an area

biomes large regional communities primarily determined by climate

biotic factors living parts of an organism's environment

black hole the theoretical remaining core of a supernova with an intense gravitational field

blackbody radiation electromagnetic radiation emitted by an ideal material (the blackbody) that perfectly absorbs and perfectly emits radiation

blood the fluid medium consisting of cells and plasma that assists in the transport of materials and heat

Bohr model model of the structure of the atom that attempted to correct the deficiencies of the solar system model and account for the Balmer series

boiling point the temperature at which a phase change of liquid to gas takes place through boiling; the same temperature as the condensation point

boundary the division between two regions of differing physical properties

breaker a wave whose front has become so steep that the top part has broken forward of the wave, breaking into foam, especially against a shoreline

British thermal unit the amount of energy or heat needed to increase the temperature of 1 pound of water 1 degree Fahrenheit (abbreviated Btu)

bronchi major branches of the trachea that ultimately deliver air to bronchioles in the lungs

bronchioles small tubes that deliver air to the alveoli in the lung; they are capable of contracting

bulimia a nutritional deficiency disease characterized by a binge-and-purge cycle of eating

C

calorie the amount of energy (or heat) needed to increase the temperature of 1 gram of water 1 degree Celsius

Calorie the nutritionist's "calorie"; equivalent to 1 kilocalorie

cancer any abnormal growth of cells that has a malignant potential

capillaries tiny blood vessels through the walls of which exchange between cells and the blood takes place

carbohydrates organic compounds that include sugars, starches, and cellulose; carbohydrates are used by plants and animals for structure, protection, and food

carbon cycle the series of processes and organisms through which carbon atoms pass in ecological systems

carbon film a type of fossil formed when the volatile and gaseous constituents of a buried organic structure are distilled away, leaving a carbon film as a record

carbonation in chemical weathering, a reaction that occurs naturally between carbonic acid (H_2CO_3) and rock minerals

carcinogens agents responsible for causing cancer

carnivores animals that eat other animals

carrier protein proteins that pick up molecules from one place in a cell or multicellular organism and transport them to another; for example, certain blood proteins

carrying capacity (ecosystem) the *optimum* population size an area can support over an extended period of time

cast sediments deposited by groundwater in a mold, taking the shape and external features of the organism that was removed to form the mold, then gradually changing to sedimentary rock

cathode rays negatively charged particles (electrons) that are emitted from a negative terminal in an evacuated glass tube

cell the basic structural unit that makes up all living things

cell membrane the outer-boundary membrane of the cell; also known as the *plasma membrane*

cell nucleus the central part of a cell that contains the genetic material

cell plate a plant-cell structure that begins to form in the center of the cell and proceeds to the cell membrane, resulting in cytokinesis

cell wall an outer covering on some cells; may be composed of cellulose, chitin, or peptidoglycan depending on the kind of organism

cellulose a polysaccharide abundant in plants that forms the fibers in cell walls that provide structure for plant materials

Celsius scale referent scale that defines numerical values for measuring hotness or coldness, defined as degrees of temperature; based on the reference points of the freezing point of water and the boiling point of water at sea-level pressure, with 100 degrees between the two points

cementation process by which spaces between buried sediment particles under compaction are filled with binding chemical deposits, binding the particles into a rigid, cohesive mass of a sedimentary rock

Cenozoic the most recent geologic era; the time of recent life, meaning the fossils of this era are identical to the life found on Earth today

central nervous system the portion of the nervous system consisting of the brain and spinal cord

centrifugal force an apparent outward force on an object following a circular path that is a consequence of the third law of motion

centrioles organelles composed of microtubules located just outside the nucleus

centripetal force the force required to pull an object out of its natural straight-line path and into a circular path; *centripetal* means "center seeking"

centromere the unreplicated region where two chromatids are joined

Cepheid variables a bright variable star that can be used to measure distance

cerebellum a region of the brain connected to the medulla oblongata that receives many kinds of sensory stimuli and coordinates muscle movement

cerebrum a region of the brain that surrounds most of the other parts of the brain and is involved in consciousness and thought

chain reaction a self-sustaining reaction where some of the products are able to produce more reactions of the same kind; in a nuclear chain reaction, neutrons are the products that produce more nuclear reactions in a self-sustaining series

chemical bond an attractive force that holds atoms together in a compound

chemical change a change in which the identity of matter is altered and new substances are formed

chemical energy a form of energy involved in chemical reactions associated with changes in internal potential energy; a kind of potential energy that is stored and later released during a chemical reaction

chemical equation concise way of describing what happens in a chemical reaction

chemical reaction a change in matter where different chemical substances are created by forming or breaking chemical bonds

chemical weathering the breakdown of minerals in rocks by chemical reactions with water, gases of the atmosphere, or solutions

chemistry the science concerned with the study of the composition, structure, and properties of substances and the transformations they undergo

Chinook a warm wind that has been warmed by compression; also called Santa Ana

chloroplast an energy-converting, membranous, saclike organelle in plant cells containing the green pigment chlorophyll

chondrites subdivision of stony meteorites containing small spherical lumps of silicate minerals or glass

chondrules small spherical lumps of silicate minerals or glass found in some meteorites

chromatid one of two component parts of a chromosome formed by replication and attached at the centromere

chromatin areas or structures within the nucleus of a cell composed of long, loosely arranged molecules of deoxyribonucleic acid (DNA) in association with proteins

chromatin fibers see *nucleoproteins*

chromosomal aberrations changes in the gene arrangement in chromosomes; for example, translocation and duplication mutations

chromosomes complex tightly coiled structures within the nucleus composed of various kinds of histone proteins and DNA that contains the cell's genetic information

class a group of closely related orders found within a phylum

cleavage furrow an indentation of the cell membrane of an animal cell that pinches the cytoplasm into two parts

climate the general pattern of weather that occurs in a region over a number of years

climate change departure from the expected average pattern of climate for a given region over time

clitoris a small, elongated erectile structure located between and at the head of the labia

clones genetically identical individuals that were reproduced asexually

coalescence process (meteorology) the process by which large raindrops form from the merging and uniting of millions of tiny water droplets

cochlea the part of the ear that converts sound into nerve impulses

coding strand one of two DNA strands that serves as a template, or pattern, for the synthesis of RNA

codon a sequence of three nucleotides of an mRNA molecule that directs the placement of a particular amino acid during translation

cold front the front that is formed as a cold air mass moves into warmer air

combination chemical reaction a synthesis reaction in which two or more substances combine to form a single compound

comets celestial objects originating from the outer edges of the solar system that move about the Sun in highly elliptical orbits; solar heating and pressure from the solar wind form a tail on the comet that points away from the Sun

commensalism a relationship between two organisms in which one organism is helped and the other is not affected

community all of the kinds of interacting organisms within an ecosystem

compaction the process of pressure from a depth of overlying sediments squeezing together the deeper sediments and squeezing out water

competition a kind of interaction between organisms in which both organisms are harmed to some extent

complementary base a nitrogenous base in nucleic acids that can form hydrogen bonds with another base of a specific nucleotide

complex carbohydrates macromolecules formed as a result of the chemical combining of simple sugars (monomers) to form polysaccharides; for example, starch

compound a pure chemical substance that can be decomposed by a chemical change into simpler substances with a fixed mass ratio

compressive stress a force that tends to compress the surface as Earth's plates move into each other

concentration an arbitrary description of the relative amounts of solute and solvent in a solution;

a larger amount of solute makes a concentrated solution, and a small amount of solute makes a dilute concentration

conception fertilization

condensation (sound) a compression of gas molecules; a pulse of increased density and pressure that moves through the air at the speed of sound

condensation (water vapor) when more vapor or gas molecules are returning to the liquid state than are evaporating

condensation nuclei tiny particles such as tiny dust, smoke, soot, and salt crystals that are suspended in the air on which water condenses

condensation point the temperature at which a gas or vapor changes back to a liquid

conduction the transfer of heat from a region of higher temperature to a region of lower temperature by increased kinetic energy moving from molecule to molecule

cones light-sensitive cells in the retina of the eye that respond to different colors of light

consistent law principle the laws of physics are the same in all reference frames that are moving at a constant velocity with respect to each other

constancy of speed principle the speed of light in empty space has the same value for all observers regardless of their velocity

constructive interference the condition in which two waves arriving at the same place, at the same time and in phase, add amplitudes to create a new wave

consumers organisms that must obtain energy in the form of organic matter

continental air mass dry air masses that form over large land areas

continental climate a climate influenced by air masses from large land areas; hot summers and cold winters

continental drift a concept that continents shift positions on Earth's surface, moving across the surface rather than being fixed, stationary landmasses

continental glaciers glaciers that cover a large area of a continent; for example, Greenland and the Antarctic

continental shelf a feature of the ocean floor; the flooded margins of the continents that form a zone of relatively shallow water adjacent to the continents

continental slope a feature of the ocean floor; a steep slope forming the transition between the continental shelf and the deep ocean basin

control group the situation used as the basis for comparison in a controlled experiment

control processes mechanisms that ensure that an organism will carry out all metabolic activities in the proper sequence (coordination) and at the proper rate (regulation)

control rods rods inserted between fuel rods in a nuclear reactor to absorb neutrons and thus control the rate of the nuclear chain reaction

controlled experiment an experiment that allows for a comparison of two events that are identical in all but one respect

convection transfer of heat from a region of higher temperature to a region of lower temperature by the displacement of high-energy molecules; for example, the displacement of warmer, less dense air (higher kinetic energy) by cooler, more dense air (lower kinetic energy)

convection cell complete convective circulation pattern; also, slowly turning regions in the plastic asthenosphere that might drive the motion of plate tectonics

convection zone (of a star) part of the interior of a star according to a model; the region directly above the radiation zone where gases are heated by the radiation zone below and move upward by convection to the surface, where they emit energy in the form of visible light, ultraviolet radiation, and infrared radiation

convergent boundaries boundaries that occur between two plates moving toward each other

core (of Earth) the center part of Earth, which consists of a solid inner part and liquid outer part; makes up about 15 percent of Earth's total volume and about one-third of its mass

core (of a star) dense, very hot region of a star where nuclear fusion reactions release gamma and X-ray radiation

Coriolis effect the apparent deflection due to the rotation of Earth; it is to the right in the Northern Hemisphere

corpus luteum a glandlike structure produced in the ovary that produces hormones (progesterone and estrogen) that prevent the release of other eggs

correlation the determination of the equivalence in geologic age by comparing the rocks in two separate locations

coulomb unit used to measure quantity of electric charge; equivalent to the charge resulting from the transfer of 6.24 billion particles such as the electron

Coulomb's law relationship between charge, distance, and magnitude of the electrical force between two bodies

covalent bond a chemical bond formed by the sharing of a pair of electrons

covalent compound chemical compound held together by a covalent bond or bonds

creep the slow downhill movement of soil down a steep slope

crest the high mound of water that is part of a wave; also refers to the condensation, or high-pressure part, of a sound wave

critical angle limit to the angle of incidence when all light rays are reflected internally

critical mass mass of fissionable material needed to sustain a chain reaction

crossing-over the process that occurs when homologous chromosomes exchange equivalent sections of DNA during meiosis

crude oil petroleum pumped from the ground that has not yet been refined into usable products

crust the outermost part of Earth's interior structure; the thin, solid layer of rock that rests on top of the Mohorovicic discontinuity

cryptorchidism a developmental condition in which the testes do not migrate from the abdomen through the inguinal canal to the scrotum

curie unit of nuclear activity defined as 3.70×10^{10} nuclear disintegrations per second

cycle a complete vibration

cyclone a low-pressure center where the winds move into the low-pressure center and are forced upward; a low-pressure center with clouds, precipitation, and stormy conditions

cytokinesis division of the cytoplasm of one cell into two new cells

cytoplasm the more fluid portion of the protoplasm that surrounds the nucleus

cytosine a single-ring nitrogenous-base molecule in DNA and RNA; complementary to guanine

D

dark energy a recently discovered mystery force that is apparently accelerating the expanding universe

dark matter missing matter of the universe that is believed to exist but is invisible and unseen

data measurement information used to describe something

data points points that may be plotted on a graph to represent simultaneous measurements of two related variables

daughter cells two cells formed by cell division

daughter chromosomes chromosomes produced by DNA replication that contain identical genetic information; formed after chromosome division in anaphase

daughter nuclei two nuclei formed by mitosis

daylight saving time setting clocks ahead one hour during the summer to more effectively utilize the longer days of summer, then setting the clocks back in the fall

deceleration phase a period in a population growth curve following exponential growth in which the rate of population growth begins to slow

decibel scale a nonlinear scale of loudness based on the ratio of the intensity level of a sound to the intensity at the threshold of hearing

decomposers organisms that use dead organic matter as a source of energy

decomposition chemical reaction a chemical reaction in which a compound is broken down into the elements that make up the compound into simpler compounds or into elements and simpler compounds

deflation the widespread removal of base materials from the surface by the wind

delta a somewhat triangular deposit at the mouth of a river formed where a stream flowing into a body of water slowed and lost its sediment-carrying ability

denature a change in the chemical and physical properties of a molecule as a result of the breaking of chemical bonds within the molecule; for example, the change in egg white as a result of cooking

dendrites neuronal fibers that receive information from axons and carry it toward the nerve-cell body

denitrifying bacteria bacteria capable of converting nitrite (NO_2^-) to nitrogen gas (N_2), which is released into the atmosphere

density the compactness of matter described by a ratio of mass per unit volume

deoxyribonucleic acid (DNA) a polymer of nucleotides that serves as genetic information; in prokaryotic cells, it is a double-stranded DNA loop and contains attached HU proteins, and in eukaryotic cells, it is found in strands with attached histone proteins; when tightly coiled, it is known as a chromosome

deoxyribose a 5-carbon sugar molecule; a component of DNA

depolarized having lost the electrical difference existing between two points or objects

destructive interference the condition in which two waves arriving at the same point at the same time out of phase add amplitudes to create zero total disturbance

dew condensation of water vapor into droplets of liquid on surfaces

dew point temperature the temperature at which condensation begins

diaphragm a muscle separating the lung cavity from the abdominal cavity that is involved in exchanging the air in the lungs

diastolic blood pressure the blood pressure recorded while the heart is relaxing

diastrophism all-inclusive term that means any and all possible movements of Earth's plates, including drift, isostatic adjustment, and any other process that deforms or changes Earth's surface by movement

diet the food and drink consumed by a person from day to day

Dietary Reference Intakes current U.S. Department of Agriculture guidelines that provide information on the amounts of certain nutrients members of the public should receive

differentiation the process of forming specialized cells within a multicellular organism

diffuse reflection light rays reflected in many random directions, as opposed to the parallel rays reflected from a perfectly smooth surface such as a mirror

diffusion net movement of a kind of molecule from an area of higher concentration to an area of lesser concentration

diploid having two sets of chromosomes: one set from the maternal parent and one set from the paternal parent

direct current an electrical current that always moves in one direction

direct proportion when two variables increase or decrease together in the same ratio (at the same rate)

disaccharides two monosaccharides joined together with the loss of a water molecule; examples are sucrose (table sugar), lactose, and maltose

dispersion the effect of spreading colors of light into a spectrum with a material that has an index of refraction that varies with wavelength

divergent boundaries boundaries that occur between two plates moving away from each other

divide line separating two adjacent watersheds

DNA code a sequence of three nucleotides of a DNA molecule

DNA polymerase an enzyme that bonds DNA nucleotides together when they base pair with an existing DNA strand

DNA replication the process by which the genetic material (DNA) of the cell reproduces itself prior to its distribution to the next generation of cells

domain a classification group above that of kingdoms

dome a large, upwardly bulging, symmetrical fold that resembles a dome

dominant allele an allele that expresses itself and masks the effects of other alleles for the trait

Doppler effect an apparent shift in the frequency of sound or light due to relative motion between the source of the sound or light and the observer

double bond covalent bond formed when two pairs of electrons are shared by two atoms

dune a hill, low mound, or ridge of windblown sand or other sediments

duodenum the first part of the small intestine, which receives food from the stomach and secretions from the liver and pancreas

dwarf planet an object that is orbiting the Sun, is nearly spherical, but has not cleared matter from its orbital zone and is not a satellite

E

earthquake a quaking, shaking, vibrating, or upheaval of Earth's surface

earthquake epicenter point on Earth's surface directly above an earthquake focus

earthquake focus place where seismic waves originate beneath the surface of Earth

echo a reflected sound that can be distinguished from the original sound, which usually arrives 0.1 s or more after the original sound

eclipse when the shadow of a celestial body falls on the surface of another celestial body

ecology the branch of biology that studies the relationships between organisms and their environment

ecosystem an interacting unit consisting of a collection of organisms and abiotic factors

egg the haploid sex cell produced by sexually mature females

ejaculation the release of sperm cells and seminal fluid through the penis

El Niño changes in the atmospheric pressure systems, ocean currents, water temperatures, and wind patterns that seem to be linked to worldwide changes in the weather

elastic rebound the sudden snap of stressed rock into new positions; the recovery from elastic strain that results in an earthquake

elastic strain an adjustment to stress in which materials recover their original shape after a stress is released

electric charge a fundamental property of electrons and protons; electrons have a negative electric charge and protons have a positive electric charge

electric circuit consists of a voltage source that maintains an electrical potential, a continuous conducting path for a current to follow, and a device where work is done by the electrical potential; a switch in the circuit is used to complete or interrupt the conducting path

electric current the flow of electric charge

electric field force field produced by an electric charge

electric field lines a map of an electric field representing the direction of the force that a test charge would experience; the direction of an electric field shown by lines of force

electric generator a mechanical device that uses wire loops rotating in a magnetic field to produce electromagnetic induction in order to generate electricity

electrical energy a form of energy from electromagnetic interactions; one of five forms of energy—mechanical, chemical, radiant, electrical, and nuclear

electrical force a fundamental force that results from the interaction of electrical charge and is billions and billions of times stronger than the gravitational force; sometimes called the "electromagnetic force" because of the strong association between electricity and magnetism

electrical insulators electrical nonconductors, or materials that obstruct the flow of electric current

electrical nonconductors materials that have electrons that are not moved easily within the material—for example, rubber; electrical nonconductors are also called electrical insulators

electrolyte water solution of ionic substances that conducts an electric current

electromagnet a magnet formed by a solenoid that can be turned on and off by turning the current on and off

electromagnetic force one of four fundamental forces; the force of attraction or repulsion between two charged particles

electromagnetic induction process in which current is induced by moving a loop of wire in a magnetic field or by changing the magnetic field

electron subatomic particle that has the smallest negative charge possible and usually found in an orbital of an atom, but gained or lost when atoms become ions

electron configuration the arrangement of electrons in orbitals and suborbitals about the nucleus of an atom

electron dot notation notation made by writing the chemical symbol of an element with dots around the symbol to indicate the number of outer orbital electrons

electron pair a pair of electrons with different spin quantum numbers that may occupy an orbital

electrostatic charge an accumulated electric charge on an object from a surplus or deficiency of electrons; also called "static electricity"

element a pure chemical substance that cannot be broken down into anything simpler by chemical or physical means; there are over one hundred known elements, the fundamental materials of which all matter is made

endocrine glands secrete chemical messengers into the circulatory system

endocrine system a number of glands that communicate with one another and other tissues through chemical messengers transported throughout the body by the circulatory system

endoplasmic reticulum (ER) folded membranes and tubes throughout the eukaryotic cell that provide a large surface upon which chemical activities take place

endosymbiotic theory a theory suggesting that some organelles found in eukaryotic cells may have originated as free-living prokaryotes

energy the ability to do work

English system a system of measurement that originally used sizes of parts of the human body as referents

entropy the measure of disorder in thermodynamics

environment the surroundings; anything that affects an organism during its lifetime

environmental resistance the collective set of factors that limit population growth

enzymes protein molecules, produced by organisms, that are able to control the rate at which chemical reactions occur

eon the largest blocks of time in Earth's geologic history: the Hadean, Archean, Proterozoic, and Phanerozoic eons

epigenetics the study of changes in gene expression caused by factors other than alterations in a cell's DNA.

epinephrine a hormone produced by the adrenal medulla that increases heart rate, blood pressure, and breathing rate

epiphyte a plant that lives on the surface of another plant without doing harm

epochs subdivisions of geologic periods

equation a statement that describes a relationship in which quantities on one side of the equal sign are identical to quantities on the other side

equation of time the cumulative variation between the apparent local solar time and the mean solar time

equinoxes Latin meaning "equal nights"; time when daylight and night are of equal length, which occurs during the spring equinox and the autumnal equinox

eras the major blocks of time in Earth's geologic history: the Cenozoic, Mesozoic, Paleozoic, and Precambrian eras

erosion the process of physically removing weathered materials; for example, rock fragments are physically picked up by an erosion agent such as a stream or a glacier

essential amino acids those amino acids that cannot be synthesized by the human body and must be part of the diet (e.g., lysine, tryptophan, and valine)

esters class of organic compounds with the general structure of RCOOR', where R is one of the hydrocarbon groups—for example, methyl or ethyl; esters make up fats, oils, and waxes, and some give fruit and flowers their taste and odor

estrogen one of the female sex hormones that cause the differentiation of typical female internal and external genital anatomy; responsible for the changes in breasts, vagina, uterus, clitoris, and pelvic bone structure at puberty

estuary a special category of aquatic ecosystem that consists of shallow, partially enclosed areas where freshwater enters the ocean

ether class of organic compounds with the general formula ROR', where R is one of the hydrocarbon groups—for example, methyl

or ethyl; mostly used as industrial and laboratory solvents

Eubacteria the domain in which are found organisms commonly known as bacteria

Eucarya the domain in which are found all eukaryotic organisms

eukaryotic cells one of the two major types of cells; characterized by cells that have a true nucleus, as in plants, fungi, protists, and animals

euphotic zone the upper layer of the ocean, where the Sun's rays penetrate

evaporation process of more molecules leaving a liquid for the gaseous state than returning from the gas to the liquid; can occur at any given temperature from the surface of a liquid

evolution the continuous genetic adaptation of a population of organisms to its environment over time

excited state as applied to an atom, describes the energy state of an atom that has electrons in a state above the minimum energy state for that atom; as applied to a nucleus, describes the energy state of a nucleus that has particles in a state above the minimum energy state for that nuclear configuration

exfoliation the fracturing and breaking away of curved, sheetlike plates from bare rock surfaces via physical or chemical weathering, resulting in dome-shaped hills and rounded boulders

exocrine glands use ducts to secrete to the surface of the body or into hollow organs of the body; for example, sweat glands or pancreas

exosphere the outermost layer of the atmosphere, where gas molecules merge with the diffuse vacuum of space

experiment a re-creation of an event in a way that enables a scientist to gain valid and reliable empirical evidence

experimental group the group in a controlled experiment that is identical to the control group in all respects but one

exponential growth phase a period of time during population growth when the population increases at an accelerating rate

external energy the total potential and kinetic energy of an everyday-sized object

external parasite a parasite that lives on the outside of its host

F

facilitated diffusion diffusion assisted by carrier molecules

Fahrenheit scale referent scale that defines numerical values for measuring hotness or coldness, defined as degrees of temperature; based on the reference points of the freezing point of water and the boiling point of water at sea-level pressure, with 180 degrees between the two points

family (elements) vertical columns of the periodic table consisting of elements that have similar properties

family (organisms) a group of closely related genera within an order

fats organic compounds of esters formed from glycerol and three long-chain carboxylic acids

that are also called triglycerides; called fats in animals and oils in plants

fatty acid one of the building blocks of true fats and phospholipids; composed of a longchain carbon skeleton with a carboxylic acid functional group at one end; for example, linoleic acid

fault a break in the continuity of a rock formation along which relative movement has occurred between the rocks on either side

fault plane the surface along which relative movement has occurred between the rocks on either side; the surface of the break in continuity of a rock formation

ferromagnesian silicates silicates that contain iron and magnesium; examples include the dark-colored minerals olivine, augite, hornblende, and biotite

fertilization the joining of haploid nuclei, usually from an egg and a sperm cell, resulting in a diploid cell called the zygote

fiber natural (plant) or industrially produced polysaccharides that are resistant to hydrolysis by human digestive enzymes

first law of motion every object remains at rest or in a state of uniform straight-line motion unless acted on by an unbalanced force

first law of thermodynamics a statement of the law of conservation of energy in the relationship between internal energy, work, and heat

first quarter the moon phase between the new phase and the full phase when the Moon is perpendicular to a line drawn through Earth and the Sun; one-half of the lighted Moon can be seen from Earth, so this phase is called the first quarter

floodplain the wide, level floor of a valley built by a stream; the river valley where a stream floods when it spills out of its channel

fluids matter that has the ability to flow or be poured; the individual molecules of a fluid are able to move, rolling over or by one another

folds bends in layered bedrock as a result of stress or stresses that occurred when the rock layers were in a ductile condition, probably under considerable confining pressure from deep burial

foliation the alignment of flat crystal flakes of a rock into parallel sheets

follicle the saclike structure near the surface of the ovary that encases the soon-to-be released secondary oocyte

follicle-stimulating hormone (FSH) the pituitary secretion that causes the ovaries to begin to produce larger quantities of estrogen and to develop the follicle and prepare the egg for ovulation

food chain a sequence of organisms that feed on one another, resulting in a flow of energy from a producer through a series of consumers

Food Guide Pyramid a tool developed by the U.S. Department of Agriculture to help the general public plan for good nutrition; guidelines for required daily intake from each of five food groups. Replaced by the MyPlate Food Guide in 2011.

food web a system of interlocking food chains

force a push or pull capable of changing the state of motion of an object; since a force has magnitude (strength) as well as direction, it is a vector quantity

formula describes what elements are in a compound and in what proportions

formula weight the sum of the atomic weights of all the atoms in a chemical formula

fossil any evidence of former life

fossil fuels organic fuels that contain the stored radiant energy of the Sun converted to chemical energy by plants or animals that lived millions of years ago; coal, petroleum, and natural gas are the common fossil fuels

Foucault pendulum a heavy mass swinging from a long wire that can be used to provide evidence about the rotation of Earth

fovea centralis the area of sharpest vision on the retina, containing only cones, where light is sharply focused

fracture strain an adjustment to stress in which materials crack or break as a result of the stress

free fall when objects fall toward Earth with no forces acting upward; air resistance is neglected when considering an object to be in free fall

free-living nitrogen-fixing bacteria soil bacteria that provide nitrogen compounds that can be taken up by plants through their roots

freezing point the temperature at which a phase change of liquid to solid takes place; the same temperature as the melting point for a given substance

frequency the number of cycles of a vibration or of a wave occurring in one second, measured in units of cycles per second (hertz)

freshwater water that is not saline and is fit for human consumption

freshwater communities aquatic communities that have low salt concentrations

front the boundary, or thin transition zone, between air masses of different temperatures

frost ice crystals formed by water vapor condensing directly from the vapor phase; frozen water vapor that forms on objects

frost wedging the process of freezing and thawing water in small rock pores and cracks that become larger and larger, eventually forcing pieces of rock to break off

fuel rod long zirconium alloy tubes containing fissionable material for use in a nuclear reactor

full moon the moon phase when Earth is between the Sun and the Moon and the entire side of the Moon facing Earth is illuminated by sunlight

functional group the atom or group of atoms in an organic molecule that is responsible for the chemical properties of a particular class or group of organic chemicals

fundamental charge smallest common charge known; the magnitude of the charge of an electron and a proton

fundamental forces forces that cannot be explained in terms of any other force; gravitational, electromagnetic, weak, and strong nuclear force

fundamental frequency the lowest frequency (longest wavelength) that can set up standing waves in an air column or on a string

fundamental properties a property that cannot be defined in simpler terms other than to describe how it is measured; the fundamental properties are length, mass, time, and charge

fungus the common name for the kingdom Fungi; examples, yeast and mold

G

g symbol representing the acceleration of an object in free fall due to the force of gravity

galactic clusters gravitationally bound subgroups of as many as 1,000 stars that move together within the Milky Way galaxy

galaxy group of billions and billions of stars that form the basic unit of the universe; for example, Earth is part of the solar system, which is located in the Milky Way galaxy

gallbladder an organ attached to the liver that stores bile

galvanometer a device that measures the size of an electric current from the size of the magnetic field produced by the current

gamete a haploid sex cell

gametogenesis the generating of gametes; the meiotic cell-division process that produces sex cells; oogenesis and spermatogenesis

gametophyte stage a life-cycle stage in plants in which a haploid sex cell is produced by mitosis

gamma ray very short wavelength electromagnetic radiation emitted by decaying nuclei

gases a phase of matter composed of molecules that are relatively far apart moving freely in a constant, random motion and have weak cohesive forces acting between them, resulting in the characteristic indefinite shape and indefinite volume of a gas

gasohol solution of ethanol and gasoline

gastric juice the secretions of the stomach that contain enzymes and hydrochloric acid

Geiger counter a device that indirectly measures ionizing radiation (beta and/or gamma) by detecting "avalanches" of electrons that are able to move because of the ions produced by the passage of ionizing radiation

gene a unit of heredity located on a chromosome and composed of a sequence of DNA nucleotides

gene flow the movement of genes from one generation to another or from one place to another

gene frequency a measure of how often a specific gene for a characteristic shows up in the individuals of a population

gene pool all of the genes of a population of organisms

genetically modified (GM) organisms that contain genetic changes

general theory of relativity geometric theory of gravity from an analysis of space and time and how it interacts with a mass

generative processes actions that increase the size of an individual organism (growth) or increase the number of individuals in a population (reproduction)

genetic recombination the gene mixing that occurs as a result of sexual reproduction

genetics the study of genes, how genes produce characteristics, and how the characteristics are inherited

genome a set of all the genes necessary to specify an organism's complete list of characteristics

genotype the catalog of genes of an organism, whether or not these genes are expressed

genus (plural, *genera*) a group of closely related species within a family

geographic barriers geographic features that keep different portions of a species from exchanging genes

geographic isolation a condition in which part of the gene pool is separated by geographic barriers from the rest of the population

geologic time scale a "calendar" of geologic history based on the appearance and disappearance of particular fossils in the sedimentary rock record

geomagnetic time scale time scale established from the number and duration of magnetic field reversals during the past 6 million years

geosynchronous satellite a satellite with a period of one day, turning with Earth and appearing to move around a fixed point in the sky

geothermal energy energy from beneath Earth's surface

giant planets the large outer planets of Jupiter, Saturn, Uranus, and Neptune, which all have similar densities and compositions

glacier a large mass of ice on land that formed from compacted snow and slowly moves under its own weight

gland an organ that manufactures and secretes a material either through ducts or directly into the circulatory system

globular clusters symmetrical and tightly packed clusters of as many as a million stars that move together as subgroups within the Milky Way galaxy

glycerol an alcohol with three hydroxyl groups per molecule; for example, glycerin (1, 2, 3-propanetriol)

glycogen a highly branched polysaccharide synthesized by the human body and stored in the muscles and liver; serves as a direct reserve source of energy

glycol an alcohol with two hydroxyl groups per molecule; for example, ethylene glycol, which is used as an antifreeze

Golgi apparatus a stack of flattened, smooth, membranous sacs; the site of synthesis and packaging of certain molecules in eukaryotic cells

gonad a generalized term for organs in which meiosis occurs to produce gametes; ovary or testis

gonadotropin-releasing hormone (GnRH) a hormone released at puberty by the hypothalamus that stimulates the pituitary gland to release luteinizing hormone (LH) and follicle-stimulating hormone (FSH)

granite light-colored, coarse-grained igneous rock common on continents; igneous rocks formed by blends of quartz and feldspars, with small amounts of micas, hornblende, and other minerals

greenhouse effect the process of increasing the temperature of the lower parts of the atmosphere through redirecting energy back toward the surface; the absorption and reemission of infrared radiation by carbon dioxide, water vapor, and a few other gases in the atmosphere

ground state energy state of an atom with electrons at the lowest energy state possible for that atom

groundwater water from a saturated zone beneath the surface; water from beneath the surface that supplies wells and springs

growth-stimulating hormone (GSH) a hormone produced by the anterior pituitary gland that stimulates tissues to grow

guanine a double-ring nitrogenous-base molecule in DNA and RNA; the complementary base of cytosine

gyre the great circular systems of moving water in each ocean

H

habitat the place or part of an ecosystem occupied by an organism

hail a frozen form of precipitation, sometimes with alternating layers of clear and opaque, cloudy ice

hair hygrometer a device that measures relative humidity from changes in the length of hair

half-life the time required for one-half of the unstable nuclei in a radioactive substance to decay into a new element

halogen member of family VIIA of the periodic table, having common properties of very reactive nonmetallic elements common in salt compounds

haploid having a single set of chromosomes resulting from the reduction division of meiosis

hard water water that contains relatively high concentrations of dissolved salts of calcium and magnesium

Hardy-Weinberg concept the concept that population must be infinitely large, have random mating, no mutations, no migration, and no selection for specific characteristics in order to prevent evolution from taking place

heart the muscular pump that forces the blood through the blood vessels of the body

heat a measure of the internal energy that has been absorbed or transferred from one body to another

heat of formation energy released in a chemical reaction

Heisenberg uncertainty principle you cannot measure both the exact momentum and the exact position of a subatomic particle at the same time—when the more exact of the two is known, the less certain you are of the value of the other

hemoglobin an iron-containing molecule found in red blood cells, to which oxygen molecules bind

herbivores animals that feed directly on plants

hertz unit of frequency; equivalent to one cycle per second

Hertzsprung-Russell diagram diagram to classify stars with a temperature-luminosity graph

heterotroph an organism that requires a source of organic material from its environment; it cannot produce its own food

heterozygous describes a diploid organism that has two different alleles for a particular characteristic

high short for high-pressure center (anticyclone), which is associated with clear, fair weather

high latitudes latitudes close to the poles; those that for a period of time during the winter months receive no solar radiation at noon

high-pressure center another term for *anticyclone*

homeostasis the process of maintaining a constant internal environment as a result of constant monitoring and modification of the functioning of various systems

homologous chromosomes a pair of chromosomes in a diploid cell that contain similar genes at corresponding loci throughout their length

homozygous describes a diploid organism that has two identical alleles for a particular characteristic

hormones chemical messengers secreted by endocrine glands to regulate other parts of the body

horsepower measurement of power defined as a power rating of 550 ft·lb/s

host an organism that a parasite lives in or on

hot spots sites on Earth's surface where plumes of hot rock materials rise from deep within the mantle

humid moist climate classification; receives more than 50 cm (20 in) precipitation per year

humidity the amount of water vapor in the air; see *relative humidity*

hurricane a tropical cyclone with heavy rains and winds exceeding 120 km/h

hydration the attraction of water molecules for ions; a reaction that occurs between water and minerals that make up rocks

hydrocarbon an organic compound consisting of only the two elements hydrogen and carbon

hydrocarbon derivatives organic compounds that can be thought of as forming when one or more hydrogen atoms on a hydrocarbon have been replaced by an element or a group of elements other than hydrogen

hydrogen bonding a strong bond that occurs between the hydrogen end of one molecule and the fluorine, oxygen, or nitrogen end of another molecule

hydrologic cycle water vapor cycling into and out of the atmosphere through continuous evaporation of liquid water from the surface and precipitation of water back to the surface

hydronium ion a molecule of water with an attached hydrogen ion, H_3O

hypothalamus a region of the brain located in the floor of the thalamus and connected to the pituitary gland that is involved in sleep and arousal; emotions such as anger, fear, pleasure, hunger, sexual response, and pain; and automatic functions such as temperature, blood pressure, and water balance

hypothesis a tentative explanation of a phenomenon that is compatible with the data and provides a framework for understanding and describing that phenomenon

I

ice-crystal process a precipitation-forming process that brings water droplets of a cloud together through the formation of ice crystals

ice-forming nuclei small, solid particles suspended in air; ice can form on the suspended particles

igneous rocks rocks that formed from magma, which is a hot, molten mass of melted rock materials

immune system a system of white blood cells specialized to provide the body with resistance to disease; there are two types, antibody-mediated immunity and cell-mediated immunity

impulse a change of motion is brought about by an impulse; the product of the size of an applied force and the time the force is applied

incandescent matter emitting visible light as a result of high temperature; for example, a lightbulb, a flame from any burning source, and the Sun are all incandescent sources because of high temperature

incident ray line representing the direction of motion of incoming light approaching a boundary

incomplete dominance describes a situation in which the phenotype of a heterozygote is intermediate between the two homozygotes on a phenotypic gradient; that is, the phenotypes appear to be "blended" in heterozygotes

incus the ear bone that is located between the malleus and the stapes

independent assortment the segregation, or assortment, of one pair of homologous chromosomes independently of the segregation, or assortment, of any other pair of chromosomes

index fossils distinctive fossils of organisms that lived only a brief time; used to compare the age of rocks exposed in two different locations

index of refraction the ratio of the speed of light in a vacuum to the speed of light in a material

inertia a property of matter describing the tendency of an object to resist a change in its state of motion; an object will remain in unchanging motion or at rest in the absence of an unbalanced force

inferior vena cava a major vein that returns blood to the heart from the lower body

infrasonic sound waves having too low a frequency to be heard by the human ear; sound having a frequency of less than 20 Hz

inguinal canal opening in the floor of the abdominal cavity through which the testes in a human male fetus descend into the scrotum

initiation code the code on DNA with the base sequence TAC that begins the process of transcription

inorganic chemistry the study of all compounds and elements in which carbon is not the principal element

insulators materials that are poor conductors of heat—for example, heat flows slowly through materials with air pockets because the molecules making up air are far apart; also, materials that are poor conductors of electricity—for example, glass or wood

intensity a measure of the energy carried by a wave

interference phenomenon of light where the relative phase difference between two light waves produces light or dark spots, a result of light's wavelike nature

intermediate-focus earthquakes earthquakes that occur in the upper part of the mantle, between 70 to 350 km below the surface of Earth

intermolecular forces forces of interaction between molecules

internal energy sum of all the potential energy and all the kinetic energy of all the molecules of an object

internal parasite a parasite that lives inside its host

international date line the 180° meridian is arbitrarily called the international date line; used to compensate for cumulative time zone changes by adding or subtracting a day when the line is crossed

interphase the stage between cell divisions in which the cell is engaged in metabolic activities

interstitial cell-stimulating hormone (ICSH) the chemical messenger molecule released from the pituitary gland that causes the testes to produce testosterone, the primary male sex hormone; same as follicle-stimulating hormone in females

intertropical convergence zone a part of the lower troposphere in a belt from 10°N to 10°S of the equator where air is heated, expands, and becomes less dense and rises around the belt

intrusive igneous rocks coarse-grained igneous rocks formed as magma cools slowly deep below the surface

inverse proportion the relationship in which the value of one variable increases while the value of the second variable decreases at the same rate (in the same ratio)

inversion a condition of the troposphere when temperature increases with height rather than decreasing with height; a cap of cold air over warmer air that results in trapped air pollution

ion an atom or a particle that has a net charge because of the gain or loss of electrons; polyatomic ions are groups of bonded atoms that have a net charge

ion exchange reaction a reaction that takes place when the ions of one compound interact with the ions of another, forming a solid that comes out of solution, a gas, or water

ionic bond chemical bond of electrostatic attraction between negative and positive ions

ionic compounds chemical compounds that are held together by ionic bonds; that is, bonds of electrostatic attraction between negative and positive ions

ionization process of forming ions from molecules

ionization counter a device that measures ionizing radiation (alpha, beta, gamma, etc.) by indirectly counting the ions produced by the radiation

ionized an atom or a particle that has a net charge because it has gained or lost electrons

ionosphere refers to that part of the atmosphere—parts of the thermosphere and upper mesosphere—where free electrons and ions reflect radio waves around Earth and where the northern lights occur

iron meteorites meteorite classification group whose members are composed mainly of iron

island arcs curving chains of volcanic islands that occur over belts of deep-seated earthquakes; for example, the Japanese and Indonesian islands

isomers chemical compounds with the same molecular formula but different molecular structure; compounds that are made from the same numbers of the same elements but have different molecular arrangements

isotope atoms of an element with identical chemical properties but with different masses; isotopes are atoms of the same element with different numbers of neutrons

J

jet stream a powerful, winding belt of wind near the top of the troposphere that tends to extend all the way around Earth, moving generally from the west in both hemispheres at speeds of 160 km/h or more

joint a break in the continuity of a rock formation without a relative movement of the rock on either side of the break

joule metric unit used to measure work and energy; can also be used to measure heat; equivalent to newton-meter

K

Kelvin scale a temperature scale that does not have arbitrarily assigned referent points, and zero means nothing; the zero point on the Kelvin scale (also called absolute scale) is the lowest limit of temperature, where all random kinetic energy of molecules ceases

ketone an organic compound with the general formula $RCOR'$, where R is one of the hydrocarbon groups; for example, methyl or ethyl

keystone species species in an ecosystem that affects many aspects of the ecosystem and whose removal causes significant alteration of the ecosystem

kidneys the primary organs involved in regulating blood levels of water, hydrogen ions, salts, and urea

kilocalorie the amount of energy required to increase the temperature of 1 kilogram of water 1 degree Celsius; equivalent to 1,000 calories

kilogram the fundamental unit of mass in the metric system of measurement

kinetic energy the energy of motion; can be measured from the work done to put an object in motion, from the mass and velocity of the object while in motion, or from the amount of work the object can do because of its motion

kinetic molecular theory the collection of assumptions that all matter is made up of tiny atoms and molecules that interact physically, that explain the various states of matter, and that have an average kinetic energy that defines the temperature of a substance

kingdom a category within a domain used in the classification of organisms

Kuiper Belt a disk-shaped region of small icy bodies some 30 to 100 AU from the Sun; the source of short-period comets

L

lack of dominance the condition of two unlike alleles both expressing themselves, neither being dominant

lag phase a stage of a population growth curve during which both natality and mortality are low

lake a large inland body of standing water

large intestine the last portion of the food tube; primarily involved in reabsorbing water

last quarter the moon phase between the full phase and the new phase when the Moon is perpendicular to a line drawn through Earth and the Sun; one half of the lighted Moon can be seen from Earth, so this phase is called the last quarter

latent heat refers to the heat "hidden" in phase changes

latent heat of fusion the heat absorbed when 1 gram of a substance changes from the solid to the liquid phase, or the heat released by 1 gram of a substance when changing from the liquid phase to the solid phase

latent heat of vaporization the heat absorbed when 1 gram of a substance changes from the liquid phase to the gaseous phase, or the heat released when 1 gram of gas changes from the gaseous phase to the liquid phase

latitude the angular distance from the equator to a point on a parallel that tells you how far north or south of the equator the point is located

lava magma, or molten rock, that is forced to the surface from a volcano or a crack in Earth's surface

law of conservation of energy energy is never created or destroyed; it can only be converted from one form to another as the total energy remains constant

law of conservation of mass same as law of conservation of matter; mass, including single atoms, is neither created nor destroyed in a chemical reaction

law of conservation of momentum the total momentum of a group of interacting objects remains constant in the absence of external forces

law of dominance when an organism has two different alleles for a trait, the allele that is expressed and overshadows the expression of the other allele is said to be dominant; the allele whose expression is overshadowed is said to be recessive

law of independent assortment members of one allelic pair will separate from each other independently of the members of other allele pairs

law of segregation when gametes are formed by a diploid organism, the alleles that control a trait separate from one another into different gametes, retaining their individuality

light ray model model using lines to show the direction of motion of light to describe the travels of light

light-year the distance that light travels through empty space in one year, approximately 9.5×10^{12} km

limiting factors the identifiable factors that prevent unlimited population growth

limnetic zone the region in a lake that does not have rooted vegetation

line spectrum narrow lines of color in an otherwise dark spectrum; these lines can be used as "fingerprints" to identify gases

linear scale a scale, generally on a graph, where equal intervals represent equal changes in the value of a variable

lines of force lines drawn to make an electric field strength map, with each line originating on a positive charge and ending on a negative charge; each line represents a path on which a charge would experience a constant force and lines closer together mean a stronger electric field

linkage group genes located on the same chromosome that tend to be inherited together

lipid large, nonpolar, organic molecules that do not easily dissolve in polar solvents such as water; there are three different types of lipids: *true fats* (pork chop fat or olive oil), *phospholipids* (the primary component of cell membranes), and *steroids* (most hormones)

liquids a phase of matter composed of molecules that have interactions stronger than those found in a gas but not strong enough to keep the molecules near the equilibrium positions of a solid, resulting in the characteristic definite volume but indefinite shape of a liquid

liter a metric system unit of volume usually used for liquids

lithosphere solid layer of Earth's structure that is above the asthenosphere and includes the entire crust, the Moho, and the upper part of the mantle

littoral zone the region in a lake with rooted vegetation

liver an organ of the body responsible for secreting bile, filtering the blood, detoxifying molecules, and modifying molecules absorbed from the gut

locus the spot on a chromosome where an allele is located

loess very fine dust or silt that has been deposited by the wind over a large area

longitude angular distance of a point east or west from the prime meridian on a parallel

longitudinal wave a mechanical disturbance that causes particles to move closer together and farther apart in the same direction that the wave is traveling

longshore current a current that moves parallel to the shore, pushed along by waves that move accumulated water from breakers

loudness a subjective interpretation of a sound that is related to the energy of the vibrating source, related to the condition of the transmitting medium, and related to the distance involved

low latitudes latitudes close to the equator; those that receive vertical solar radiation at noon during some part of the year

luminous an object or objects that produce visible light; for example, the Sun, stars, lightbulbs, and burning materials are all luminous

lunar eclipse occurs when the Moon is full and the Sun, Moon, and Earth are lined up so the shadow of Earth falls on the Moon

lung a respiratory organ in which air and blood are brought close to one another and gas exchange occurs

L-wave seismic waves that move on the solid surface of Earth much as water waves move across the surface of a body of water

lymph liquid material that leaves the circulatory system to surround cells

lymphatic system a collection of thin-walled tubes that collects, filters, and returns lymph from the body to the circulatory system

lysosome a specialized, submicroscopic organelle that holds a mixture of hydrolytic enzymes

M

macromolecule very large molecule, with a molecular weight of thousands or millions of atomic mass units, that is made up of a combination of many smaller, similar molecules

magma a mass of molten rock material either below or on Earth's crust from which igneous rock is formed by cooling and hardening

magnetic field model used to describe how magnetic forces on moving charges act at a distance

magnetic poles the ends, or sides, of a magnet about which the force of magnetic attraction seems to be concentrated

magnetic quantum number from the quantum mechanics model of the atom, one of four descriptions of the energy state of an electron wave; this quantum number describes the energy of an electron orbital as the orbital is oriented in space by an external magnetic field, a kind of energy sub-sublevel

magnetic reversal the flipping of polarity of Earth's magnetic field as the north magnetic pole and the south magnetic pole exchange positions

main sequence stars normal, mature stars that use their nuclear fuel at a steady rate; stars on the Hertzsprung-Russell diagram in a narrow band that runs from the top left to the lower right

malignant tumor an abnormal, nonencapsulated growth of tumor cells that may spread to other parts of the body

malleus the ear bone that is attached to the tympanum

manipulated variable in an experiment, a quantity that can be controlled or manipulated; also known as the independent variable

mantle middle part of Earth's interior; a 2,870 km (about 1,780 mi) thick shell between the core and the crust

marine climate a climate influenced by air masses from over an ocean, with mild winters and cool summers compared to areas farther inland

marine communities aquatic communities that have a high salt content

maritime air mass a moist air mass that forms over the ocean

mass a measure of inertia, which means a resistance to a change of motion

mass defect the difference between the sum of the masses of the individual nucleons forming a nucleus and the actual mass of that nucleus

mass movement erosion caused by the direct action of gravity

mass number the sum of the number of protons and neutrons in a nucleus defines the mass number of an atom; used to identify isotopes; for example, uranium 238

masturbation stimulation of one's own sex organs

matter anything that occupies space and has mass

meanders winding, circuitous turns or bends of a stream

mean solar day a time period that is 24 hours long and is averaged from the mean solar time

mean solar time a uniform time averaged from the apparent solar time

measurement the process of comparing a property of an object to a well-defined and agreed-upon referent

mechanical energy the form of energy associated with machines, objects in motion, and objects having potential energy that results from gravity

mechanical weathering the physical breaking up of rocks without any changes in their chemical composition

medulla oblongata a region of the more primitive portion of the brain connected to the spinal cord that controls such automatic functions as blood pressure, breathing, and heart rate

meiosis the specialized pair of cell divisions that reduces the chromosome number from diploid ($2n$) to haploid (n)

melting point the temperature at which a phase change of solid to liquid takes place; the same temperature as the freezing point for a given substance

Mendelian genetics the pattern of inheriting characteristics that follows the laws formulated by Gregor Mendel

menopause the period when the ovaries stop producing viable secondary oocytes and the body becomes nonreproductive

menstrual cycle (menses, menstrual flow, period) the repeated building up and shedding of the lining of the uterus

meridians north-south running arcs that intersect at both poles and are perpendicular to the parallels

Mesozoic a geologic era; the time of middle life, meaning some of the fossils for this time period are similar to the life found on Earth today, but many are different from anything living today

messenger RNA (mRNA) a molecule composed of ribonucleotides that functions as a copy of the gene and is used in the cytoplasm of the cell during protein synthesis

metabolism the total of all the chemical reactions and energy changes that take place in an organism

metabolic processes the total of all chemical reactions within an organism; for example, nutrient uptake and processing, and waste elimination

metal matter having the physical properties of conductivity, malleability, ductility, and luster

metamorphic rocks previously existing rocks that have been changed into a distinctly different rock by heat, pressure, or hot solutions

metaphase the second stage in mitosis, characterized by alignment of the chromosomes at the equatorial plane

metastasize the process of cells of tumors moving from the original site and establishing new colonies in other regions of the body

meteor the streak of light and smoke that appears in the sky when a meteoroid is made incandescent by compression of Earth's atmosphere

meteor shower event when many meteorites fall in a short period of time

meteorite the solid iron or stony material of a meteoroid that survives passage through Earth's atmosphere and reaches the surface

meteoroids remnants of comets and asteroids in space

meteorology the science of understanding and predicting weather

meter the fundamental metric unit of length

metric system a system of referent units based on invariable referents of nature that have been defined as standards

microclimate a local, small-scale pattern of climate; for example, the north side of a house has a different microclimate than the south side

microvilli tiny projections from the surfaces of cells that line the intestine

middle latitudes latitudes equally far from the poles and equator; between the high and low latitudes

mineral (geology) a naturally occurring, inorganic solid element or chemical compound with a crystalline structure

minerals (biology) inorganic elements that cannot be manufactured by the body but are required in low concentrations; essential to metabolism

miscible fluids two or more kinds of fluids that can mix in any proportion

mitochondrion a membranous organelle resembling a small bag with a larger bag inside that is folded back on itself; serves as the site of aerobic cellular respiration

mitosis a process that results in equal and identical distribution of replicated chromosomes into two newly formed nuclei

mixture matter composed of two or more kinds of matter that has a variable composition and can be separated into its component parts by physical means

model a mental or physical representation of something that cannot be observed directly that is usually used as an aid to understanding

moderator a substance in a nuclear reactor that slows fast neutrons so the neutrons can participate in nuclear reactions

mold the preservation of the shape of an organism by the dissolution of the remains of a buried organism, leaving an empty space where the remains were

molecule from the chemical point of view, a particle composed of two or more atoms held together by an attractive force called a chemical bond; from the kinetic theory point of view, smallest particle of a compound or gaseous

element that can exist and still retain the characteristic properties of a substance

momentum the product of the mass of an object times its velocity

monomer individual, repeating units or segments of complex molecules that chemically combine to form long chainlike molecules called polymers; for example, monosaccharides link to form polysaccharides

monosaccharides simple sugars containing 3 to 6 carbon atoms; the most common kinds are 6-carbon molecules such as glucose and fructose

moraines deposits of rocks and other mounded materials bulldozed into position by a glacier and left behind when the glacier melted

morphological species concept the concept that species can be distinguished from one another by structural characteristics

mortality the number of individuals leaving the population by death per thousand individuals in the population

motor neurons those neurons that carry information from the central nervous system to muscles or glands

motor unit all of the muscle cells stimulated by a single neuron

mountain a natural elevation of Earth's crust that rises above the surrounding surface

multiple alleles a term used to refer to conditions in which there are several different alleles for a particular characteristic, not just two

mutagen any chemical, object, or energy that causes a change in the genetic information of an organism

mutagenic agent anything that causes permanent change in DNA

mutation any change in the genetic information of a cell

mutualism a relationship between two organisms in which both organisms benefit

MyPlate Food Guide a tool developed by the U.S. Department of Agriculture to help the general public plan for good nutrition; guidelines for required daily intake from each of five food groups

N

natality the number of individuals entering the population by reproduction per thousand individuals in the population

natural frequency the frequency of vibration of an elastic object that depends on the size, composition, and shape of the object

natural selection the processes that encourage the passage of beneficial genes and discourage the passage of harmful or unfavorable genes from one generation to the next

neap tide period of less-pronounced high and low tides; occurs when the Sun and Moon are at right angles to one another

nebulae a diffuse mass of interstellar clouds of hydrogen gas or dust that may develop into a star

negative electric charge one of the two types of electric charge; repels other negative charges and attracts positive charges

negative ion atom or particle that has a surplus, or imbalance, of electrons and, thus, a negative charge

nephrons millions of tiny tubular units that make up the kidneys, which are responsible for filtering the blood

nerve cell see *neuron*

nerve impulse a series of changes that take place in the neuron, resulting in a wave of depolarization that passes from one end of the neuron to the other

nerves bundles of neuronal fibers

nervous system a network of neurons that carry information from sense organs to the central nervous system and from the central nervous system to muscles and glands

net force the resulting force after all forces have been added; if a net force is zero, all the forces have canceled each other and there is not an unbalanced force

neuron the cellular unit consisting of a cell body and fibers that makes up the nervous system; also called nerve cell

neurotransmitter a molecule released by the axons of neurons that stimulates other cells

neutralized acid or base properties have been lost through a chemical reaction

neutron neutral subatomic particle usually found in the nucleus of an atom

neutron star very small superdense remains of a supernova with a center core of pure neutrons

new crust zone zone of a divergent boundary where new crust is formed by magma upwelling at the boundary

new moon the moon phase when the Moon is between Earth and the Sun and the entire side of the Moon facing Earth is dark

newton a unit of force defined as $kg \cdot m/s^2$; that is, a 1 newton force is needed to accelerate a 1 kg mass 1 m/s^2

niche the functional role of an organism

nitrifying bacteria bacteria that can convert ammonia (NH_3) into nitrite-containing (NO_2^-) compounds, which in turn can be converted into nitrate-containing (NO_3^-) compounds

nitrogen cycle nitrogen in the atmosphere is acted on and used by many bacteria and other organisms and ultimately returned to the atmosphere

nitrogenous base a category of organic molecules found as components of the nucleic acids; there are five common types: thymine, guanine, cytosine, adenine, and uracil

noble gas members of family VIII of the periodic table, having common properties of colorless, odorless, chemically inert gases; also known as rare gases or inert gases

node regions on a standing wave that do not oscillate

noise sounds made up of groups of waves of random frequency and intensity

nonelectrolytes water solutions that do not conduct an electric current; covalent compounds that form molecular solutions and cannot conduct an electric current

nonferromagnesian silicates silicates that do not contain iron or magnesium ions; examples include the minerals of muscovite (white mica), the feldspars, and quartz

nonmetal an element that is brittle (when a solid), does not have a metallic luster, is a poor conductor of heat and electricity, and is not malleable or ductile

nonsilicates minerals that do not have the silicon-oxygen tetrahedra in their crystal structure

noon the event of time when the Sun moves across the celestial meridian

norepinephrine a hormone produced by the adrenal medulla that increases heart rate, blood pressure, and breathing rate

normal a line perpendicular to the surface of a boundary

normal fault a fault where the hanging wall has moved downward with respect to the foot wall

north pole (of a magnet) short for "north seeking"; the pole of a magnet that points northward when it is free to turn

nova a star that explodes or suddenly erupts and increases in brightness

nuclear energy the form of energy from reactions involving the nucleus, the innermost part of an atom

nuclear fission nuclear reaction of splitting a massive nucleus into more stable, less massive nuclei with an accompanying release of energy

nuclear force one of four fundamental forces, a strong force of attraction that operates over very short distances between subatomic particles; this force overcomes the electric repulsion of protons in a nucleus and binds the nucleus together

nuclear fusion nuclear reaction of low-mass nuclei fusing together to form more stable and more massive nuclei with an accompanying release of energy

nuclear membrane the structure surrounding the nucleus that separates the nucleoplasm from the cytoplasm

nuclear reactor steel vessel in which a controlled chain reaction of fissionable materials releases energy

nucleic acids complex molecules that store and transfer genetic information within a cell; constructed of fundamental monomers known as nucleotides; the two common forms are DNA and RNA

nucleoli (singular, *nucleolus*) nuclear structures composed of completed or partially completed ribosomes and the specific parts of chromosomes that contain the information for their construction

nucleons name used to refer to both the protons and neutrons in the nucleus of an atom

nucleoproteins the double-stranded DNA with attached proteins; also called chromatin fibers

nucleosomes histone clusters with their encircling DNA

nucleotide the building block of the nucleic acids, composed of a 5-carbon sugar, a phosphate, and a nitrogenous base

nucleus (atom) tiny, relatively massive and positively charged center of an atom containing protons and neutrons; the small, dense center of an atom

numerical constant a constant without units; a number

nutrients molecules required by the body for growth, reproduction, or repair

nutrition collectively, the processes involved in taking in, assimilating, and utilizing nutrients

O

obese the condition of being 15 percent to 20 percent above the individual's ideal weight

obligate intracellular parasites infectious particles (viruses) that can function only when inside a living cell

observed lapse rate the rate of change in temperature compared to change in altitude

occluded front a front that has been lifted completely off the ground into the atmosphere, forming a cyclonic storm

ocean the single, continuous body of salt water on the surface of Earth

ocean basin the deep bottom of the ocean floor, which starts beyond the continental slope

ocean currents streams of water within the ocean that stay in about the same path as they move over large distances; steady and continuous onward movement of a channel of water in the ocean

ocean wave a moving disturbance that travels across the surface of the ocean

oceanic ridges long, high, continuous, suboceanic mountain chains; for example, the Mid-Atlantic Ridge in the center of the Atlantic Ocean Basin

oceanic trenches long, narrow, deep troughs with steep sides that run parallel to the edge of continents

octet rule a generalization that helps keep track of the valence electrons in most representative elements; atoms of the representative elements (A families) attempt to acquire an outer orbital with eight electrons through chemical reactions

offspring descendants of a set of parents

ohm unit of resistance; equivalent to volts/amps

Ohm's law the electric potential difference is directly proportional to the product of the current times the resistance

oil field petroleum accumulated and trapped in extensive porous rock structure or structures

oils organic compounds of esters formed from glycerol and three long-chain carboxylic acids that are also called triglycerides; called fats in animals and oils in plants

olfactory epithelium the cells of the nasal cavity that respond to chemicals

omnivores animals that are carnivores at some times and herbivores at others

oogenesis the specific name given to the gametogenesis process that leads to the formation of eggs

Oort cloud a spherical "cloud" of small, icy bodies from 30,000 AU out to a light-year from the Sun; the source of long-period comets

opaque materials that do not allow the transmission of any light

orbital the region of space around the nucleus of an atom where an electron is likely to be found

order a group of closely related families within a class

ore mineral mineral deposits with an economic value

organ a structure composed of two or more kinds of tissues that perform a particular function

organ system a group of organs that performs a particular function

organelles cellular structures that perform specific functions in the cell; the function of an organelle is directly related to its structure

organic acids organic compounds with a general formula of RCOOH, where R is one of the hydrocarbon groups; for example, methyl or ethyl

organic chemistry the study of compounds in which carbon is the principal element

organism an independent living unit

orgasm a complex series of responses to sexual stimulation that results in intense frenzy of sexual excitement

osmosis the net movement of water molecules through a selectively permeable membrane

osteoporosis a disease condition resulting from the demineralization of the bone, resulting in pain, deformities, and fractures; related to a loss of calcium

out-of-Africa hypothesis states that modern humans (*Homo sapiens*) originated in Africa, as had several other hominid species, and migrated from Africa to Asia and Europe and displaced species such as *H. erectus* and *H. heidelbergensis* that had migrated into these areas previously

oval window the membrane-covered opening of the cochlea, to which the stapes is attached

ovaries the female sex organs that produce haploid sex cells—the eggs or ova

overtones higher resonant frequencies that occur at the same time as the fundamental frequency, giving a musical instrument its characteristic sound quality

overweight person whose weight is outside his or her healthy range, by either gaining or losing weight, he or she will become healthier; a BMI of 25 or more

oviduct the tube (fallopian tube) that carries the oocyte to the uterus

ovulation the release of a secondary oocyte from the surface of the ovary

oxbow lake a small body of water, or lake, that formed when two bends of a stream came together and cut off a meander

oxidation the process of a substance losing electrons during a chemical reaction; a reaction between oxygen and the minerals making up rocks

oxidation-reduction reaction a chemical reaction in which electrons are transferred from one atom to another; sometimes called "redox" for short

oxidizing agents substances that take electrons from other substances

oxidizing atmosphere an atmosphere that contains molecular oxygen

oxytocin a hormone released from the posterior pituitary that causes contraction of the uterus

P

paleontology the science of discovering fossils, studying the fossil record, and deciphering the history of life

Paleozoic a geologic era; time of ancient life, meaning the fossils from this time period are very different from anything living on Earth today

pancreas an organ of the body that secretes many kinds of digestive enzymes into the duodenum

parallels reference lines on Earth used to identify where in the world you are northward or southward from the equator; east and west running circles that are parallel to the equator on a globe with the distance from the equator called the latitude

parasite an organism that lives in or on another organism and derives nourishment from it

parasitism a relationship which involves one organism living in or on another living organism from which it derives nourishment

parts per billion concentration ratio of parts of solute in every 1 billion parts of solution (ppb); could be expressed as ppb by volume or as ppb by weight

parts per million concentration ratio of parts of solute in every 1 million parts of solution (ppm); could be expressed as ppm by volume or as ppm by weight

Pauli exclusion principle no two electrons in an atom can have the same four quantum numbers; thus, a maximum of two electrons can occupy a given orbital

pelagic describes aquatic organisms that float or swim actively

penis the portion of the male reproductive system that deposits sperm in the female reproductive tract

penumbra the zone of partial darkness in a shadow

pepsin an enzyme produced by the stomach that is responsible for beginning the digestion of proteins

percent by volume the volume of solute in 100 volumes of solution

percent by weight the weight of solute in 100 weight units of solution

perception recognition by the brain that a stimulus has been received

perigee when the Moon's elliptical orbit brings the Moon closest to Earth

perihelion the point at which an orbit comes closest to the Sun

period (geologic time) subdivisions of geologic eras

period (periodic table) horizontal rows of elements with increasing atomic numbers; runs from left to right on the element table

period (wave) the time required for one complete cycle of a wave

periodic law similar physical and chemical properties recur periodically when the elements are listed in order of increasing atomic number

peripheral nervous system the fibers that communicate between the central nervous system and other parts of the body

permeability the ability to transmit fluids through openings, small passageways, or gaps

permineralization the process that forms a fossil by alteration of an organism's buried remains by circulating groundwater depositing calcium carbonate, silica, or pyrite

petroleum oil that comes from oil-bearing rock, a mixture of hydrocarbons that is believed to have formed from ancient accumulations of buried organic materials such as remains of algae

pH scale scale that measures the acidity of a solution with numbers below 7 representing acids, 7 representing neutral, and numbers above 7 representing bases

phagocytosis the process by which the cell wraps around a particle and engulfs it

pharynx the region at the back of the mouth cavity; the throat

phase change the action of a substance changing from one state of matter to another; a phase change always absorbs or releases internal potential energy without a temperature change

phases of matter the different physical forms that matter can take as a result of different molecular arrangements, resulting in characteristics of the common phases of a solid, liquid, or gas

phenotype the physical, chemical, and behavioral expression of the genes possessed by an organism

phospholipids a class of complex water-insoluble organic molecules that resemble fats but contain a phosphate group (PO_4) in their structure; a component of cellular membranes

photoelectric effect the movement of electrons in some materials as a result of energy acquired from absorbed light

photon a quantum of energy in a light wave; the particle associated with light

photosynthesis a series of reactions that takes place in chloroplasts and results in the storage of sunlight energy in the form of chemical-bond energy

phylogeny the science that explores the evolutionary relationships among organisms and seeks to reconstruct evolutionary history

phylum a subdivision of a kingdom

physical change a change of the state of a substance but not the identity of the substance

phytoplankton planktonic organisms that carry on photosynthesis

pistil the sex organ in plants that produces eggs or ova

pitch the frequency of a sound wave

pituitary gland the gland at the base of the brain that controls the functioning of other glands throughout the organism

placenta an organ made up of tissues from the embryo and the uterus of the mother that allows for the exchange of materials between the mother's bloodstream and the embryo's bloodstream; it also produces hormones

Planck's constant proportionality constant in the relationship between the energy of vibrating molecules and their frequency of vibration; a value of 6.63×10^{-34} Js

plane of the ecliptic the plane of Earth's orbit

planet an object that is orbiting the Sun, is nearly spherical, and is large enough to clear all matter from its orbital zone

plankton aquatic organisms that are so small and weakly swimming that they are simply carried by currents.

plasma (biology) the watery matrix that contains the molecules and cells of the blood

plasma (physics) a phase of matter; a very hot gas consisting of electrons and atoms that have been stripped of their electrons because of high kinetic energies

plasma membrane the outer-boundary membrane of the cell; see *cell membrane*

plastic strain an adjustment to stress in which materials become molded or bent out of shape under stress and do not return to their original shape after the stress is released

plate tectonics the theory that Earth's crust is made of rigid plates that float on the asthenosphere

pleiotropy the multiple effects that a gene may have on the phenotype of an organism

point mutation a change in the DNA of a cell as a result of a loss or change in a single nitrogenous-base

polar air mass cold air mass that forms in cold regions

polar body the smaller of two cells formed by unequal meiotic division during oogenesis

polar climate zone climate zone of the high latitudes; average monthly temperatures stay below 10°C (50°F), even during the warmest month of the year

polar molecule a molecule that has a negative charge on one part and a positive charge on another part

polarized light whose constituent transverse waves are all vibrating in the same plane; also known as plane-polarized light

Polaroid a film that transmits only polarized light

polyatomic ion ion made up of many atoms

polygenic inheritance the concept that a number of different pairs of alleles may combine their efforts to determine a characteristic

polymers huge, chainlike molecules made of hundreds or thousands of smaller repeating molecular units called monomers

polypeptide chain polymers of amino acids; sometimes called proteins

polysaccharides polymers consisting of monosaccharide units joined together in straight or branched chains; starches, glycogen, or cellulose

pond a small body of standing water, smaller than a lake

pons a region of the brain immediately anterior to the medulla oblongata that connects to the cerebellum and higher regions of the brain and controls several sensory and motor functions of the head and face

population a group of organisms of the same species located in an area

population growth curve a graph of the change in population size over time

porosity the ratio of pore space to the total volume of a rock or soil sample, expressed as a percentage; freely admitting the passage of fluids through pores or small spaces between parts of the rock or soil

positive electric charge one of the two types of electric charge; repels other positive charges and attracts negative charges

positive ion atom or particle that has a net positive charge due to an electron or electrons being torn away

potential energy energy due to position; energy associated with changes in position (e.g., gravitational potential energy) or changes in shape (e.g., compressed or stretched spring)

power the rate at which energy is transferred or the rate at which work is performed; defined as work per unit of time

Precambrian the time before the time of ancient life, meaning the rocks for this time period contain very few fossils

precession the slow wobble of the axis of Earth similar to the wobble of a spinning top

precipitation water that falls to the surface of Earth in the solid or liquid form

predation a relationship between two organisms that involves the capturing, killing, and eating of one by the other

predator an organism that captures, kills, and eats another animal

pressure defined as force per unit area; for example, pounds per square inch (lb/in^2)

prey an organism captured, killed, and eaten by a predator

primary carnivores carnivores that eat herbivores and are therefore on the third trophic level

primary coil part of a transformer; a coil of wire that is connected to a source of alternating current

primary consumers organisms that feed directly on plants—herbivores

primary loop part of the energy-converting system of a nuclear power plant; the closed pipe system that carries heated water from the nuclear reactor to a steam generator

primary oocyte the diploid cell of the ovary that begins to undergo the first meiotic division in the process of oogenesis

primary spermatocyte the diploid cell in the testes that undergoes the first meiotic division in the process of spermatogenesis

prime meridian the referent meridian (0°) that passes through the Greenwich Observatory in England

principal quantum number from the quantum mechanics model of the atom, one of four descriptions of the energy state of an electron wave; this quantum number describes the main energy level of an electron in terms of its most probable distance from the nucleus

principle of uniformity a frame of reference that the same processes that changed the landscape in the past are the same processes you see changing the landscape today

prions proteins that can be passed from one individual to another and cause disease

probability the chance that an event will happen, expressed as a percent or fraction

producers organisms that produce new organic material from inorganic material with the aid of sunlight

progesterone a hormone released by the corpus luteum that is important in maintaining the lining of the uterus

prokaryotic cells one of the two major types of cells; they do not have a typical nucleus bound by a nuclear membrane and lack many of the other membranous cellular organelles; for example, members of the Eubacteria and Archaea

promoter a region of DNA at the beginning of each gene, just ahead of an initiator code

proof a measure of ethanol concentration of an alcoholic beverage; proof is double the concentration by volume; for example, 50 percent by volume is 100 proof

properties qualities or attributes that, taken together, are usually unique to an object; for example, color, texture, and size

prophase the first phase of mitosis during which individual chromosomes become visible

proportionality constant a constant applied to a proportionality statement that transforms the statement into an equation

protein synthesis the process whereby the tRNA utilizes the mRNA as a guide to arrange the amino acids in their proper sequence according to the genetic information in the chemical code of DNA

proteins macromolecular polymers made of smaller molecules of amino acids, with molecular weight from about six thousand to 50 million; proteins are amino acid polymers with roles in biological structures or functions; without such a function, they are known as polypeptides

proton subatomic particle that has the smallest possible positive charge, usually found in the nucleus of an atom

protoplanet nebular model a model of the formation of the solar system that states that the planets formed from gas and dust left over from the formation of the Sun

protoplasm the living portion of a cell as distinguished from the nonliving cell wall

protostar an accumulation of gases that will become a star

proxy data data from natural sources used to infer temperature change, rainfall change, or some other climate condition

pseudoscience use of the appearance of science to mislead; the assertions made are not valid or reliable

psychrometer a two-thermometer device used to measure the relative humidity

puberty a time in the life of a developing individual characterized by the increasing production of sex hormones, which cause it to reach sexual maturity

pulmonary artery the major blood vessel that carries blood from the right ventricle to the lungs

pulmonary circulation the flow of blood through certain chambers of the heart and blood vessels to the lungs and back to the heart

pulsar the source of regular, equally spaced pulsating radio signals believed to be the result of the magnetic field of a rotating neutron star

Punnett square a method used to determine the probabilities of allele combinations in an offspring

pure substance materials that are the same throughout and have a fixed definite composition

pure tone sound made by very regular intensities and very regular frequencies from regular repeating vibrations

P-wave a pressure, or compressional, wave in which a disturbance vibrates materials back and forth in the same direction as the direction of wave movement

P-wave shadow zone a region on Earth between 103° and 142° of arc from an earthquake where no P-waves are received; believed to be explained by P-waves being refracted by the core

pyloric sphincter a valve located at the end of the stomach that regulates the flow of food from the stomach to the duodenum

Q

quad 1 quadrillion Btu (10^{15} Btu); used to describe very large amounts of energy

quanta fixed amounts; usually referring to fixed amounts of energy absorbed or emitted by matter (singular, *quantum*)

quantities measured properties; includes the numerical value of the measurement and the unit used in the measurement

quantum mechanics model of the atom based on the wave nature of subatomic particles, the mechanics of electron waves; also called wave mechanics

quantum numbers numbers that describe energy states of an electron; in the Bohr model of the atom, the orbit quantum numbers could be any whole number 1, 2, 3, and so on out from the nucleus; in the quantum mechanics model of the atom, four quantum numbers are used to describe the energy state of an electron wave

R

rad a measure of radiation received by a material (radiation absorbed dose)

radiant energy the form of energy that can travel through space; for example, visible light and other parts of the electromagnetic spectrum

radiation the transfer of heat from a region of higher temperature to a region of lower temperature by greater emission of radiant energy from the region of higher temperature

radiation zone part of the interior of a star according to a model; the region directly above the core where gamma and X rays from the core are absorbed and reemitted, with the radiation slowly working its way outward

radioactive decay the natural spontaneous disintegration or decomposition of a nucleus

radioactive decay constant a specific constant for a particular isotope that is the ratio of the rate of nuclear disintegration per unit of time to the total number of radioactive nuclei

radioactive decay series series of decay reactions that begins with one radioactive nucleus that decays to a second nucleus that decays to a third nucleus and so on until a stable nucleus is reached

radioactivity spontaneous emission of particles or energy from an atomic nucleus as it disintegrates

radiometric age age of rocks determined by measuring the radioactive decay of unstable elements within the crystals of certain minerals in the rocks

range the geographical distribution of a species

rarefaction a thinning or pulse of decreased density and pressure of gas molecules

ratio a relationship between two numbers, one divided by the other; the ratio of distance per time is speed

real image an image generated by a lens or mirror that can be projected onto a screen

recessive allele an allele that, when present with a dominant allele, does not express itself and is masked by the effect of the dominant allele

recombinant DNA DNA that has been constructed by inserting new pieces of DNA into the DNA of another organism, such as a bacterium

red giant stars one of two groups of stars on the Hertzsprung-Russell diagram that have a different set of properties than the main sequence stars; bright, low-temperature giant stars that are enormously bright for their temperature

redox (chemical) reaction short name for oxidation-reduction reaction

reducing agent supplies electrons to the substance being reduced in a chemical reaction

reduction division (also **meiosis**) a type of cell division in which daughter cells get only half the chromosomes from the parent cell

referent referring to or thinking of a property in terms of another, more familiar object

reflected ray a line representing direction of motion of light reflected from a boundary

reflection the change when light, sound, or other waves bounce backward off a boundary

refraction a change in the direction of travel of light, sound, or other waves crossing a boundary

regulator protein proteins that help determine the activities that will occur in a cell or multicellular organism; for example, enzymes and some hormones

relative dating dating the age of a rock unit or geological event relative to some other unit or event

relative humidity ratio (times 100 percent) of how much water vapor is in the air to the maximum amount of water vapor that could be in the air at a given temperature

reliable a term used to describe results that remain consistent over successive trials

rem measure of radiation that considers the biological effects of different kinds of ionizing radiation

replacement chemical reaction reaction in which an atom or polyatomic ion is replaced in a

compound by a different atom or polyatomic ion

replacement (fossil formation) process in which an organism's buried remains are altered by circulating groundwaters carrying elements in solution; the removal of original materials by dissolutions and the replacement of new materials an atom or molecule at a time

representative elements name given to the members of the A-group families of the periodic table; also called the main-group elements

reservoir a natural or artificial pond or lake used to store water, control floods, or generate electricity; a body of water stored for public use

resonance when the frequency of an external force matches the natural frequency of a material and standing waves are set up

response the reaction of an organism to a stimulus

responsive processes those abilities to react to external and internal changes in the environment; for example, irritability, individual adaptation, and evolution

retina the light-sensitive region of the eye

reverberation apparent increase in volume caused by reflections, usually arriving within 0.1 second after the original sound

reverse fault a fault where the hanging wall has moved upward with respect to the foot wall

revolution the motion of a planet as it orbits the Sun

rhodopsin a light-sensitive pigment found in the rods of the retina

ribonucleic acid (RNA) a polymer of nucleotides formed on the template surface of DNA by transcription; three forms that have been identified are mRNA, rRNA, and tRNA

ribose a 5-carbon sugar molecule that is a component of RNA

ribosomal RNA (rRNA) a globular form of RNA; a part of ribosomes

ribosomes small structures composed of two protein and ribonucleic acid subunits; involved in the assembly of proteins from amino acids

Richter scale expresses the intensity of an earthquake in terms of a scale with each higher number indicating ten times more ground movement and about thirty times more energy released than the preceding number

ridges long, rugged mountain chains rising thousands of meters above the abyssal plains of the ocean basin

rift a split or fracture in a rock formation, land formation, or in the crust of Earth

rip current strong, brief current that runs against the surf and out to sea

RNA polymerase an enzyme that attaches to the DNA at the promoter region of a gene and assists in combining RNA nucleotides when the genetic information is transcribed into RNA

rock a solid aggregation of minerals or mineral materials that have been brought together into a cohesive solid

rock cycle understanding of igneous, sedimentary, or metamorphic rock as a temporary state in an ongoing transformation of rocks into new types; the process of rocks continually changing from one type to another

rock flour rock pulverized by a glacier into powdery, silt-sized sediment

rods light-sensitive cells in the retina of the eye that respond to low-intensity light but do not respond to different colors of light

rotation the spinning of a planet on its axis

runoff water moving across the surface of Earth as opposed to soaking into the ground

S

salivary amylase an enzyme present in saliva that breaks starch molecules into smaller molecules

salivary glands glands that produce saliva

salt any ionic compound except one with hydroxide or oxide ions

San Andreas fault in California, the boundary between the North American Plate and the Pacific Plate that runs north-south for some 1,300 km (800 miles) with the Pacific Plate moving northwest and the North American Plate moving southeast

saturated air air in which an equilibrium exists between evaporation and condensation; the relative humidity is 100 percent

saturated molecule an organic molecule that has the maximum number of hydrogen atoms possible

saturated solution the apparent limit to dissolving a given solid in a specified amount of water at a given temperature; a state of equilibrium that exists between dissolving solute and solute coming out of solution

scientific law a relationship between quantities, usually described by an equation in the physical sciences; is more important and describes a wider range of phenomena than a scientific principle

scintillation counter a device that indirectly measures ionizing radiation (alpha, beta, gamma, etc.) by measuring the flashes of light produced when the radiation strikes a phosphor

sea a smaller part of the ocean with characteristics that distinguish it from the larger ocean

sea breeze cool, dense air from over water moving over land as part of convective circulation

seafloor spreading the process by which hot, molten rock moves up from the interior of Earth to emerge along mid-oceanic rifts, flowing out in both directions to create new rocks and spread apart the seafloor

seamounts steep, submerged volcanic peaks on the abyssal plain

second the standard unit of time in both the metric and English systems of measurement

second law of motion the acceleration of an object is directly proportional to the net force acting on that object and inversely proportional to the mass of the object

second law of thermodynamics a statement that the natural process proceeds from a state of higher order to a state of greater disorder

secondary carnivores carnivores that feed on primary carnivores and are therefore at the fourth trophic level

secondary coil part of a transformer, a coil of wire in which the voltage of the original alternating current in the primary coil is stepped up or down by way of electromagnetic induction

secondary consumers animals that eat other animals—carnivores

secondary loop part of a nuclear power plant; the closed pipe system that carries steam from a steam generator to the turbines, then back to the steam generator as feedwater

secondary oocyte the larger of the two cells resulting from the unequal cytoplasmic division of a primary oocyte in meiosis I of oogenesis

secondary sexual characteristics characteristics of the adult male or female, including the typical shape that develops at puberty: broader shoulders, heavier long-bone muscles, development of facial hair, axillary hair, and chest hair, and changes in the shape of the larynx in the male; rounding of the pelvis and breasts and changes in deposition of fat in the female

secondary spermatocyte cells in the seminiferous tubules that go through the second meiotic division, resulting in four haploid spermatids

sedimentary rocks rocks formed from particles or dissolved minerals from previously existing rocks that were deposited from air or water

sediments accumulations of silt, sand, or gravel that settled out of the atmosphere or out of water

segregation the separation and movement of homologous chromosomes to the opposite poles of the cell

seismic waves vibrations that move as waves through any part of Earth, usually associated with earthquakes, volcanoes, or large explosions

seismograph an instrument that measures and records seismic wave data

selecting agents specific environmental factors that favor the passage of certain characteristics from one generation to the next and discourage others

selectively permeable the property of a membrane that allows certain molecules to pass through it, but interferes with the passage of others

semen the sperm-carrying fluid produced by the seminal vesicles, prostate glands, and bulbourethral glands of males

semiarid climate classification between arid and humid; receives between 25 and 50 cm (10 and 20 in) precipitation per year

semicircular canals a set of tubular organs associated with the cochlea that sense changes in the movement or position of the head

semiconductors elements that have properties between those of a metal and those of a nonmetal, sometimes conducting an electric current and sometimes acting like an electrical insulator depending on the conditions and their purity; also called metalloids

semilunar valves pulmonary artery and aorta valves that prevent the flow of blood backward into the ventricles

seminal vesicle a part of the male reproductive system that produces a portion of the semen

seminiferous tubules sperm-producing tubes in the testes

sensory neurons those neurons that send information from sense organs to the central nervous system

sex the nature of biological differences between males and females

sex chromosomes a pair of chromosomes that determines the sex of an organism; X and Y chromosomes

sex-determining chromosome the chromosomes X and Y that are primarily responsible for determining if an individual will develop as a male or female

sexual reproduction the propagation of organisms involving the union of gametes from two parents

sexuality a term used in reference to the totality of the aspects—physical, psychological, and cultural—of our sexual nature

shallow-focus earthquakes earthquakes that occur from the surface down to 70 km deep

shear stress produced when two plates slide past one another or by one plate sliding past another plate that is not moving

shell model of the nucleus model of the nucleus that has protons and neutrons moving in energy levels or shells in the nucleus (similar to the shell structure of electrons in an atom)

shells the layers that electrons occupy around the nucleus

shield volcano a broad, gently sloping volcanic cone constructed of solidified lava flows

shock wave a large, intense wave disturbance of very high pressure; the pressure wave created by an explosion, for example

sickle-cell anemia a disease caused by a point mutation; this malfunction produces sickle-shaped red blood cells

sidereal day the interval between two consecutive crossings of the celestial meridian by a particular star

sidereal month the time interval between two consecutive crossings of the Moon across any star

sidereal year the time interval required for Earth to move around its orbit so that the Sun is again in the same position against the stars

silicates minerals that contain silicon-oxygen tetrahedra either isolated or joined together in a crystal structure

sill a tabular-shaped intrusive rock that formed when magma moved into the plane of contact between sedimentary rock layers

simple harmonic motion the vibratory motion that occurs when there is a restoring force opposite to and proportional to a displacement

single bond covalent bond in which a single pair of electrons is shared by two atoms

single-factor cross a genetic study in which one characteristic is followed from the parental generation to the offspring

small intestine the portion of the digestive system immediately following the stomach; responsible for digestion and absorption

small solar system bodies all objects orbiting the Sun that are not planets or dwarf planets

soil a mixture of unconsolidated weathered earth materials and humus, which is altered, decay-resistant organic matter

solenoid a cylindrical coil of wire that becomes electromagnetic when a current runs through it

solids a phase of matter with molecules that remain close to fixed equilibrium positions due to strong interactions between the molecules, resulting in the characteristic definite shape and definite volume of a solid

solstices times when the Sun is at its maximum or minimum altitude in the sky, known as the summer solstice and the winter solstice

solubility dissolving ability of a given solute in a specified amount of solvent, the concentration that is reached as a saturated solution is achieved at a particular temperature

solute the component of a solution that dissolves in the other component; the solvent

solution a homogeneous mixture of ions or molecules of two or more substances

solvent the component of a solution present in the larger amount; the solute dissolves in the solvent to make a solution

soma the cell body of a neuron, which contains the nucleus

sonic boom sound waves that pile up into a shock wave when a source is traveling at or faster than the speed of sound

sound quality characteristic of the sound produced by a musical instrument; determined by the presence and relative strengths of the overtones produced by the instrument

south pole (of a magnet) short for "south seeking"; the pole of a magnet that points southward when it is free to turn

special theory of relativity analysis of how space and time are affected by motion between an observer and what is being measured

speciation the process of generating new species

species a group of organisms that can interbreed naturally to produce fertile offspring

specific heat each substance has its own specific heat, which is defined as the amount of energy (or heat) needed to increase the temperature of 1 gram of a substance 1 degree Celsius

speed a measure of how fast an object is moving—the rate of change of position per change in time; speed has magnitude only and does not include the direction of change

sperm the haploid sex cells produced by sexually mature males

spermatids haploid cells produced by spermatogenesis that change into sperm

spermatogenesis the specific name given to the gametogenesis process that leads to the formation of sperm

spin quantum number from the quantum mechanics model of the atom, one of four descriptions of the energy state of an electron wave; this quantum number describes the spin orientation of an electron relative to an external magnetic field

spinal cord the portion of the central nervous system located within the vertebral column that carries both sensory and motor information between the brain and the periphery of the body

spindle fibers an array of microtubules extending from pole to pole; used in the movement of chromosomes

spontaneous generation the theory that living organisms arose from nonliving material

spontaneous mutation a change in the DNA of an organism for which there is no known cause

spring equinox one of two times a year that daylight and night are of equal length; occurs on or about March 21 and identifies the beginning of the spring season

spring tides unusually high and low tides that occur every two weeks because of the relative positions of Earth, Moon, and Sun

stable equilibrium phase a stage in a population growth curve following rapid growth in which there is both a decrease in natality and an increase in mortality so that the size of the population is stable

standard atmospheric pressure the average atmospheric pressure at sea level, which is also known as normal pressure; the standard pressure is 29.92 in or 760.0 mm of mercury (1,013.25 millibar)

standard time zones 15° wide zones defined to have the same time throughout the zone, with the time defined as the mean solar time at the middle of each zone

standard unit a measurement unit established as the standard upon which the value of the other referent units of the same type are based

standing waves condition where two waves of equal frequency traveling in opposite directions meet and form stationary regions of maximum displacement due to constructive interference and stationary regions of zero displacement due to destructive interference

stapes the ear bone that is attached to the oval window

starch a group of complex carbohydrates (polysaccharides) that plants use as a stored food source and that serves as an important source of food for animals

stationary front occurs when the edge of a weather front is not advancing

steam generator part of a nuclear power plant; the heat exchanger that heats feedwater from the secondary loop to steam with the very hot water from the primary loop

stem cells cells that have not yet completed determination or differentiation, so they have the potential to develop into many different cell types

step-down transformer a transformer that decreases the voltage

step-up transformer a transformer that increases the voltage

steroids a group of lipid molecules characterized by an arrangement of interlocking rings of carbon; they serve as hormones that aid in regulating body processes

stimulus any change in the internal or external environment of an organism that it can detect

stony meteorites meteorites composed mostly of silicate minerals that usually make up rocks on Earth

stony-iron meteorites meteorites composed of silicate minerals and metallic iron

storm a rapid and violent weather change with strong winds, heavy rain, snow, or hail

strain adjustment to stress; a rock unit might respond to stress by changes in volume or shape, or by breaking

stratosphere the layer of the atmosphere above the troposphere where temperature increases with height

stream body of running water

stress a force that tends to compress, pull apart, or deform rock; stress on rocks in Earth's solid outer crust results as Earth's plates move into, away from, or alongside each other

structural protein protein molecules whose function is to provide support and shape to a cell or multicellular organism; for example, muscle protein fibers

structural similarities one of the characteristics of living things; describes the fact that all living things are composed of cells, either prokaryotic or eukaryotic

subduction zone the region of a convergent boundary where the crust of one plate is forced under the crust of another plate into the interior of Earth

sublimation the phase change of a solid directly into a vapor or gas

submarine canyons a feature of the ocean basin; deep, steep-sided canyons that cut through the continental slopes

subspecies regional groups within a species that are significantly different structurally, physiologically, or behaviorally yet are capable of exchanging genes by interbreeding with other members of the species

summer solstice in the Northern Hemisphere, the time when the Sun reaches its maximum altitude in the sky, which occurs on or about June 22 and identifies the beginning of the summer season

superconductors some materials in which, under certain conditions, the electrical resistance approaches zero

supercooled water in the liquid phase when the temperature is below the freezing point

superior vena cava a major vein that returns blood to the heart from the head and upper body

supernova a rare catastrophic explosion of a star into an extremely bright but short-lived phenomenon

supersaturated containing more than the normal saturation amount of a solute at a given temperature

surf the zone where breakers occur; the water zone between the shoreline and the outermost boundary of the breakers

surface wave a seismic wave that moves across Earth's surface, spreading across the surface like water waves spread on the surface of a pond from a disturbance

S-wave a sideways or shear wave in which a disturbance vibrates materials from side to

side, perpendicular to the direction of wave movement

S-wave shadow zone a region of Earth more than 103° of arc away from the epicenter of an earthquake where S-waves are not recorded; believed to be the result of the core of Earth that is a liquid, or at least acts like a liquid

swell regular groups of low-profile, long-wavelength waves that move continuously

symbiosis a close physical relationship between two kinds of organisms

symbiotic nitrogen-fixing bacteria bacteria that live in the roots of certain kinds of plants, where they convert nitrogen gas molecules into compounds that the plants can use to make amino acids and nucleic acids

synapse the space between the axon of one neuron and the dendrite of the next, where chemicals are secreted to cause an impulse to be initiated in the second neuron

synapsis the condition in which the two members of a pair of homologous chromosomes come to lie close to one another

syncline a trough-shaped fold in layered bedrock

synodic month the interval of time from new moon to new moon (or any two consecutive identical phases)

systemic circulation the flow of blood through certain chambers of the heart and blood vessels to the general body and back to the heart

systolic blood pressure the blood pressure recorded in a large artery while the heart is contracting

T

talus steep, conical, or apronlike accumulations of rock fragments at the base of a slope

taxonomy the science of classifying and naming organisms

telophase the last phase in mitosis, characterized by the formation of daughter nuclei

temperate climate zone climate zone of the middle latitudes; average monthly temperatures stay between 10°C and 18°C (50°F and 64°F) throughout the year

temperature how hot or how cold something is; a measure of the average kinetic energy of the molecules making up a substance

tensional stress the opposite of compressional stress; occurs when one part of a plate moves away from another part that does not move

termination code the DNA nucleotide sequence at the end of a gene with the code ATT, ATC, or ACT that signals "stop here"

terrestrial planets planets Mercury, Venus, Earth, and Mars that have similar densities and compositions as compared to the outer giant planets

testes the male sex organs that produce haploid cells called sperm

testosterone the male sex hormone produced in the testes that controls male sexual development

thalamus a region of the brain that relays information between the cerebrum and lower portions of the brain, providing some level of awareness in that it determines pleasant and unpleasant stimuli and is involved in sleep and arousal

theory a broad, detailed explanation that guides the development of hypotheses and interpretations of experiments in a field of study

theory of natural selection the idea that some individuals within a population will have favorable combinations of genes that make it very likely that those individuals will survive, reproduce, and pass their genes to the next generation

thermohaline circulation the movement of the waters of the oceans caused by surface currents and changes in density due to differences in temperature or salinity

thermometer a device used to measure the hotness or coldness of a substance

third law of motion whenever two objects interact, the force exerted on one object is equal in size and opposite in direction to the force exerted on the other object; forces always occur in matched pairs that are equal and opposite

thrust fault a reverse fault with a low-angle fault plane

thunderstorm a brief, intense electrical storm with rain, lightning, thunder, strong winds, and sometimes hail

thymine a single-ring nitrogenous-base molecule in DNA but not in RNA; it is complementary to adenine

thyroid-stimulating hormone (TSH) a hormone secreted by the pituitary gland that stimulates the thyroid to secrete thyroxine

thyroxine a hormone produced by the thyroid gland that speeds up the metabolic rate

tidal bore a strong tidal current, sometimes resembling a wave, produced in very long, very narrow bays as the tide rises

tidal currents a steady and continuous onward movement of water produced in narrow bays by the tides

tides periodic rise and fall of the level of the sea from the gravitational attraction of the Moon and Sun

tissue a group of specialized cells that work together to perform a specific function

tornado a long, narrow, funnel-shaped column of violently whirling air from a thundercloud that moves destructively over a narrow path when it touches the ground

total internal reflection condition where all light is reflected back from a boundary between materials; occurs when light arrives at a boundary at the critical angle or beyond

total solar eclipse eclipse that occurs when Earth, the Moon, and the Sun are lined up so the Moon completely covers the disk of the Sun; the umbra of the Moon's shadow falls on the surface of Earth

trachea a major tube supported by cartilage that carries air to the bronchi; also known as the windpipe

transcription the process of manufacturing RNA from the template surface of DNA; three forms of RNA that may be produced are mRNA, rRNA, and tRNA

transfer RNA (tRNA) a molecule composed of ribonucleic acid; it is responsible for transporting a specific amino acid into a ribosome for assembly into a protein

transform boundaries in plate tectonics, boundaries that occur between two plates sliding horizontally by each other along a long, vertical fault; sudden jerks along the boundary result in the vibrations of earthquakes

transformer a device consisting of a primary coil of wire connected to a source of alternating current and a secondary coil of wire in which electromagnetic induction increases or decreases the voltage of the source

transition elements members of the B-group families of the periodic table

translation the assembly of individual amino acids into a polypeptide

transparent term describing materials that allow the transmission of light; for example, glass and clear water are transparent materials

transportation the movement of eroded materials by agents such as rivers, glaciers, wind, or waves

transverse wave a mechanical disturbance that causes particles to move perpendicular to the direction that the wave is traveling

triglyceride organic compound of esters formed from glycerol and three long-chain carboxylic acids; also called fats in animals and oils in plants

triple bond covalent bond formed when three pairs of electrons are shared by two atoms

trophic level a step in the flow of energy through an ecosystem

tropic of Cancer parallel identifying the northern limit where the Sun appears directly overhead; located at 23.5°N latitude

tropic of Capricorn parallel identifying the southern limit where the Sun appears directly overhead; located at 23.5°S latitutde

tropical air mass a warm air mass from warm regions

tropical climate zone climate zone of the low latitudes; average monthly temperatures stay above 18°C (64°F), even during the coldest month of the year

tropical cyclone a large, violent circular storm that is born over the warm, tropical ocean near the equator; also called hurricane (Atlantic and eastern Pacific) and typhoon (in western Pacific)

tropical year the time interval between two consecutive spring equinoxes; used as standard for the common calendar year

troposphere layer of the atmosphere from the surface to where the temperature stops decreasing with height

trough the low mound of water that is part of a wave; also refers to the rarefaction, or low-pressure part of a sound wave

true fats also known as *neutral fats;* a category of lipids composed of glycerol and fatty acids; for example, pork chop fat or olive oil

tsunami very large, fast, and destructive ocean wave created by an undersea earthquake, landslide, or volcanic explosion; a seismic sea wave

tumor a mass of undifferentiated cells not normally found in a certain portion of the body

turbidity current a muddy current produced by underwater landslides

tympanum the eardrum

typhoon the name for hurricanes in the western Pacific

U

ultrasonic sound waves too high in frequency to be heard by the human ear; frequencies above 20,000 Hz

umbra the inner core of a complete shadow

unconformity a time break in the rock record

undertow a current beneath the surface of the water produced by the return of water from the shore to the sea

underweight person whose weight is outside his or her healthy range; by either gaining or losing weight, they will become healthier; a BMI of 18.5 or less

unit in measurement, a well-defined and agreed-upon referent

universal law of gravitation every object in the universe is attracted to every other object with a force directly proportional to the product of their masses and inversely proportional to the square of the distance between the centers of the two masses

unpolarized light light consisting of transverse waves vibrating in all conceivable random directions

unsaturated molecule an organic molecule that does not contain the maximum number of hydrogen atoms; a molecule that can add more hydrogen atoms because of the presence of double or triple bonds

uracil a single-ring nitrogenous-base molecule in RNA but not in DNA; it is complementary to adenine

uterus the organ in female mammals in which the embryo develops

V

vacuole a large sac within the cytoplasm of a cell, composed of a single membrane

vagina the passageway between the uterus and outside of the body; the birth canal

valence the number of covalent bonds an atom can form

valence electrons electrons of the outermost shell; the electrons that determine the chemical properties of an atom and the electrons that participate in chemical bonding

valid a term used to describe meaningful data that fit into the framework of scientific knowledge

Van Allen belts belts of radiation caused by cosmic-ray particles becoming trapped and following Earth's magnetic field lines between the poles

vapor the gaseous state of a substance that is normally in the liquid state

variables changing quantities usually represented by a letter or symbol

veins the blood vessels that return blood to the heart

velocity describes both the speed and direction of a moving object; a change in velocity is a change in speed, in direction of travel, or both

ventifacts rocks sculpted by wind abrasion

ventricles the powerful muscular chambers of the heart whose contractions force blood to flow through the arteries to all parts of the body

vernal equinox another name for the spring equinox, which occurs on or about March 21 and marks the beginning of the spring season

vibration a back-and-forth motion that repeats itself

villi tiny fingerlike projections in the lining of the intestine that increase the surface area for absorption

viroids infectious particles that are composed solely of a small, single strand of RNA

virtual image an image where light rays appear to originate from a mirror or lens; this image cannot be projected on a screen

virus a nucleic acid particle coated with protein that functions as an obligate intracellular parasite

vitamins organic molecules that cannot be manufactured by the body but are required in very low concentrations for good health

volcanism volcanic activity; the movement of magma

volcano a hill or mountain formed by the extrusion of lava or rock fragments from a mass of magma below

volt the ratio of work done to move a quantity of charge

voltage source a device that does work in moving a quantity of charge

volume how much space something occupies

vulcanism volcanic activity; the movement of magma

W

warm front the front that forms when a warm air mass advances against a cool air mass

water table the boundary below which the ground is saturated with water

watershed the region or land area drained by a stream; a stream drainage basin

watt metric unit for power; equivalent to J/s

wave a disturbance or oscillation that moves through a medium

wave equation the relationship of the velocity of a wave to the product of the wavelength and frequency of the wave

wave front a region of maximum displacement in a wave; a condensation in a sound wave

wave height the vertical distance of an ocean wave between the top of the wave crest and the bottom of the next trough

wave mechanics alternate name for quantum mechanics derived from the wavelike properties of subatomic particles

wave period the time required for two successive crests or other successive parts of the wave to pass a given point

wavelength the horizontal distance between successive wave crests or other successive parts of the wave

weak acid acids that only partially ionize because of an equilibrium reaction with water

weak base a base only partially ionized because of an equilibrium reaction with water

weathering slow changes that result in the breaking up, crumbling, and destruction of any kind of solid rock

white dwarf stars one of two groups of stars on the Hertzsprung-Russell diagram that have a different set of properties than the main sequence stars; faint, white-hot stars that are very small and dense

wind a horizontal movement of air that moves along or parallel to the ground, sometimes in currents or streams

wind abrasion the natural sand-blasting process that occurs when wind-driven particles break off small particles of rock and polish the rock they strike

wind chill factor a factor that compares heat loss from bodies in still air with those in moving air; moving air removes heat more rapidly and causes a person to feel that the air is colder than its actual temperature; the cooling power of wind

winter solstice in the Northern Hemisphere, the time when the Sun reaches its minimum altitude, which occurs on or about December 22 and identifies the beginning of the winter season

work the magnitude of applied force times the distance through which the force acts; can be thought of as the process by which one form of energy is transformed to another

X

X chromosome the chromosome in a human female egg (and in one-half of sperm cells) that is associated with the determination of sexual characteristics

X-linked gene a gene located on one of the sex-determining X chromosomes

Y

Y chromosome the sex-determining chromosome in one-half of the sperm cells of human males responsible for determining maleness

Z

zone of saturation zone of sediments beneath the surface in which water has collected in all available spaces

zooplankton small, weakly swimming animals of many kinds

zygote a diploid cell that results from the union of an egg and a sperm; a fertilized egg

Credits

Index

Charge, 6
Charles, A. C., 15
Charles' law, 15
Chemical bonding, 217, 539
Chemical bonds, 201, 202, 203–207, 209
Chemical change, 199, 200–202
Chemical detection, 627
Chemical energy, 62–65, 201, 202, 450
Chemical equations, 202, 210, 212–214
Chemical fertilizers, 583
Chemical formula, 211–212
Chemical reactions
  atmosphere and, 378
  catalytic converter and, 217
  chemical bonds and, 203–207, 209
  chemical change and, 199, 200–202
  chemical equations and, 210, 212–214
  chemical formulas and, 211–212
  compounds and, 198–200, 208–210
  defined, 201
  elements and, 200–202, 205, 208–210, 215, 216
  ions and, 202–203
  living things and, 465, 486, 487, 488
  nuclear reactions and, 244
  Pauling and, 218
  proteins and, 453
  sedimentary rocks and, 335
  types of, 214–217
  valence electrons and, 202–203
Chemicals, cancer and, 492
Chemical symbols, 212
Chemical weathering, 364–366, 372
Chemoautotrophs, 501
Chernobyle, 257–258
Chesapeake Bay, 430
Chile saltpeter, 236
Chill bumps, 81
Chinook, 383
Chlorate, 208, 236
Chlorides, 236
Chlorine, 192, 200, 204, 205, 209, 215, 331, 380, 420, 423, 441, 442
Chlorine tablets, 236
Chlorite, 332
Chlorofluorocarbons, 441
Chlorophyll, 480, 484, 487, 530, 534
Chloroplasts, 480–481, 487, 493, 534

Cholesterol, 454, 458, 459, 460, 472, 613, 614
Chondrites, 303
Chondrules, 303
Chromate, 208
Chromatid, 490, 491
Chromatin, 482, 490, 493
Chromium, 206
Chromosomal aberrations, 638, 686
Chromosomal abnormalities, 641–643
Chromosomes, 455, 482, 490, 491, 492, 641–644, 645, 679
Chylomicrons, 460
Chyme, 609
Cigarette smoking, 606
Cilia, 482, 493
Ciliary muscle, 161
Circuit breaker, 136, 137
Circular motion, 44–45
Circular orbits, 186, 187
Circulatory system, 599–604, 610
Cirrostratus, 390
Cirrus, 389, 390, 391, 397
City weathering, 368, 401
Class, 528
Classification schemes, 31, 276, 335, 336, 356, 389, 526–533, 539
Clausius, Rudolf, 78
Cleavage, 652, 654
Climactic influence, 405–406
Climate, 401–403, 405–409
Climate change, 37, 406–409, 552, 553, 584, 594
Climate groups, 401–403, 405
Climate zones, 402, 403, 405, 406
Clitoris, 646
Cloning/clone, 651, 687, 691
Clouds, 378, 387, 388–391, 393
Coal, 66, 67–68, 69, 70, 330
Cobalt, 192, 194, 206, 421
Cobalt-58, 253
Cobalt-60, 247, 253
Cobalt chloride, 236
Cochlea, 628, 629
Coding strand, 455, 682
Codominance, 672, 674
Codons, 684
Coenzymes, 539
Coherent motion, 96
Cohesion, 77
Cold and quiet stage (Moon), 292
Cold front, 394–395, 397
Collagen, 451
College Chemistry (Pauling), 217
Colloids, 469

Color, 4, 157, 160, 163–164, 185, 268–269
Color deficiency, 672
Coma, 301
Combination reactions, 215, 216
Combustion, 214
Comets, 297, 300–301, 302, 303
Commensalism, 579
Commensal organisms, 534
Common names, 209–210
Community/communities, 468, 564, 569–575
Community interactions, 568–569
Compact disc (CD), 171
Compaction, 336
Comparative anatomy studies, 529–530
Competition, 580–581
Complementary bases, 678
Complete proteins, 619
Complex carbohydrates, 449, 450, 451
Compound microscopes, 471
Compound motion, 35–36
Compounds, 77, 190, 198–200, 201, 205, 208–210, 211–212, 215
Compression, 108, 361, 383, 391, 440
Compressive stress, 354, 356, 357
Concave lens, 161
Concave mirror, 158
Concentration, 234, 238, 409
Concentration gradient, 473
Conception, 643
Concepts, 2–3, 10, 11
Condensation, 92–93, 108, 109, 110, 112, 117, 199, 385–388, 391, 392, 397
Condensation nuclei, 388, 391
Condensation point, 90
Conduction, 86–87, 88, 89, 96, 99, 392
Conductivity, 191, 238
Conductor, 130–132, 137, 140, 173
Cones, 627–628
Conglomerate, 336, 337
Conjecture, 15
Conjoined twins, 654
Consistent law principle, 172
Constancy of speed principle, 172
Constant speed, 26, 27
Constructive interference, 114
Consumers, 565, 566
Continental air mass, 392
Continental crust, 338, 339
Continental divide, 416
Continental drift, 341, 348

Continental glaciers, 370
Continental shelf, 429
Continental slope, 430
Continent-continent plate convergence, 345
Continents, 338, 340–341, 343, 552
Continuous spectrum, 183
Continuous structure, 179
Contraception, 657–658, 660–661
Contractile vacuole, 493
Contrail, 387
Control group, 13
Controlled experiment, 13
Control processes, 465, 466, 467
Control rods, 254
Convection, 87, 88, 89, 96, 99, 163, 361, 379, 381, 389, 390, 392
Convection current, 87
Convection zone, 267, 268
Convective cells, 345, 347, 381, 391
Convergent boundaries, 343–344
Converging lens, 162
Conversion tables/factors, 40
Convex lens, 161
Convex mirror, 158
Coordination, defined, 467
Copper, 9, 192, 194, 200, 203, 206, 216, 332, 335, 421
Core, 267, 268, 271, 338, 339–340
Corey, Robert, 217
Coriolis effect, 314
Correlation, 548
Correns, Carl Erich, 670
Cosmic Background Explorer (COBE), 276
Coulomb, 128, 130, 132, 133–134
Coulomb's law, 129
Covalent bonds, 203, 205–207, 209, 226, 486
Covalent compounds, 208–209, 210, 211–212, 226, 229, 230
Crater, 362
Creep, 367
Crest, 110, 112, 114, 118, 385, 424–425
Cretaceous period, 552
Creutzfeldt-Jakob diseases, 452, 540
Crick, Francis, 14, 217
Critical angle, 159, 160
Critical density, 279
Critical mass, 253
Cro-Magnons, 556
Cross-dressers, 640
Crossing-over, 638
Crossover technique, 211, 213
Crude oil, 335, 436, 440

Human population growth, 591–592, 594
Human reproductive hormones, 647
Humans
classification of, 530
evolution of, 553–558
Human sexuality, 639–641
Humidity, 93, 386–387, 388, 389
Hurricane Andrew, 400
Hurricane Camille, 400
Hurricane Charley, 400
Hurricane Katrina, 400
Hurricanes, 396, 398–401
Hutton, James, 352, 371, 541
Huygens, Christian, 164, 165
Hydration, 228, 364, 366
Hydrocarbon derivatives, 441–442
Hydrocarbons, 214, 217, 299, 436–439
Hydrochloric acid, 230, 232, 233, 234, 362, 444
Hydroelectric plants, 68
Hydrogen, 111, 131, 179, 181, 182, 184–186, 190, 201, 205–207, 209, 211, 214–216, 226, 259, 261, 268, 271, 287, 295, 299, 304, 325, 449, 456
Hydrogen bonding, 226, 227
Hydrogen carbonate, 208
Hydrogen chloride, 229, 233
Hydrogen gas, 70, 71, 267
Hydrogen peroxide, 237
Hydrogen sulfate, 208
Hydrogen sulfide, 325, 362
Hydrologic cycle, 415, 422
Hydrolysis reaction, 449, 454
Hydronium ion, 230, 233, 234
Hydrophilic, 470
Hydrophobic, 470
Hydropower, 66, 68, 69
Hydroxide, 208, 216, 217, 233, 234, 333
Hydroxyl functional group, 442
Hyperopia, 162
Hypertonic, 475
Hypochlorite, 208
Hypothalamus, 646
Hypothesis, 13, 15
Hypotonic, 475

I

Ice, 160, 225, 226, 227, 231, 295, 299, 300, 364, 370, 397
Ice ages, 406–407, 408
Ice caps, 415
Ice-crystal process, 391
Ice-forming nuclei, 391
Identical twins, 654
Igneous rocks, 335, 341, 348, 356, 542, 543

Immiscible fluids, 227
Immune system, 599
Impotence, 661
Impulse, 43–44
Incandescent, 153, 154
Incident angle, 159
Incident ray, 157, 158, 159
Inclination of Earth's axis, 312
Inclined plane, 57
Incoherent light, 168
Incoherent motion, 96
Incomplete dominance, 672–673
Incomplete proteins, 619
Index fossils, 548
Index of refraction, 160
Indirect solar gain, 88, 89
Individual adaptation, 466
Individual organism, 468
Inert gases, 190
Inertia, 6, 32, 36, 37–38, 40, 45, 47, 111
Inertial confinement, 261
Infectious diseases, 515
Infertility, 652
Infrared radiation, 63, 65, 80, 353, 401
Infrasonic, 108
Ingestion, 610
Inguinal canal, 644
Inheritance model, 668–669, 672–677
Inorganic chemistry, 436
Inorganic compound, 436
In phase, 114
Instantaneous speed, 27
Insulator, 130–131
Intensity, 115–116, 117, 358
Interference, 114–115, 164, 170
Interglacial warming period, 406–407
Intermediate filaments, 481, 493
Intermittent stream, 415
Internal energy, 82–83, 94, 95, 96
Internal parasite, 579
Internal potential energy, 89, 201
Internal receptors, 629
International Astronomical Union (IAU), 286
International Bureau of Weights and Measures, 6
International date line, 319–321
International Red Cross, 260
International System of Units (SI), 5, 9, 250
Interphase, 490, 491
Intertropical convergence zone, 383, 385
Introduction to Quantum Mechanics, 217
Invasive species, 576
Inversely proportional, 39

Inverse proportion, 11
Inverse square proportion, 11
Inversion, 379, 380, 686
Invert sugar, 450
In vitro fertilization (IVF), 651, 655, 656, 691
In vivo fertilization, 651
Io, 296
Iodine, 192, 205, 209, 251
Iodine-129, 253
Iodine-131, 248, 253
Ion channels, 476, 477
Ion exchange reactions, 215, 216–217
Ionic bonds, 203, 204–205, 226
Ionic compounds, 205, 208, 210, 211, 228, 230, 231
Ionization, 229
Ionization counter, 250
Ionosphere, 381
Ion-polar molecule force, 228
Ions, 127, 131, 191–192, 194, 202–205, 208, 209, 213, 216, 229, 246, 331, 381, 419, 423
Iron, 9, 192, 194, 200, 203, 206, 208, 215, 272, 287, 302, 330, 331, 335, 339, 364, 366, 421, 422, 613
Iron-55, 253
Iron meteorites, 303, 339
Iron oxides, 292, 304, 365
Island arcs, 344, 348
Isobutane, 437, 438
Isolated atoms, 203
Isolated solar gain, 88, 89
Isolation, 519–520, 521
Isomers, 437
Isotonic, 474, 475
Isotopes, 182, 188, 244, 245, 246, 248, 249, 251, 252, 253, 259, 541–542

J

Jet stream, 384–385
Joule, 56, 68, 83, 95, 133–134
Joule, Jamaes Prescott, 68
Julian calendar, 321
Jupiter, 41, 274, 287, 288, 294–300, 302, 304, 305, 306, 310, 313

K

Kant, Immanuel, 303
Karyotyping, 642–643
Kelvin scale, 80, 81, 82
Kerosene, 440, 441
Ketones, 443–444, 450
Keystone species, 569
Kidney machines, 475
Kidneys, 475, 622
Kilo-, 6
Kilocalorie, 83–84, 611, 612

Kilogram, 6, 40, 611
Kinetic energy, defined, 61–62
Kinetic molecular theory, 76–78, 88–94, 107
Kinetic theory of gases, 173
Kingdom, 527–528, 530
Kinnear, Greg, 691
Klinefelter's syndrome, 642, 644
Krebs cycle, 487
Krypton, 190
Krypton-85, 253
Kuiper Belt, 300

L

Labor, 655
Laccolith, 363
Lack of penetrance, 677
Lacteal, 610
Lactic acid, 232, 446
Lactogenic hormone, 647
Lactose, 451
Lag phase, 590
Lake, 416, 418
Lamarck, Jean-Baptiste de, 506, 512, 513
Land, Edwin H., 166
Landslide, 367
La Niña, 404
Lanthanide series, 191
Laplace, Pierre, 303
Large intestine, 610
Laser Ranging Retro-Reflector Experiment, 346
Lasers, 168, 171
Last quarter, 322
Latent heat, 89
Latent heat of fusion, 90, 91
Latent heat of vaporization, 90–91
Latitude, 315, 316, 402, 405
Lava flows, 362
Law, defined, 14, 15
Law of conservation of energy, 65–66, 95, 567
Law of conservation of mass, 212
Law of conservation of momentum, 43, 44
Law of dominance, 669
Law of gravitation, 45–50
Law of independent assortment, 669
Law of inertia, 37–38
Law of reflection, 156, 157, 158
Law of refraction, 156
Law of segregation, 669
Law of thermodynamics, 94
Lead, 9, 111, 200, 206, 451, 627
Lead-206, 248
Lead-207, 247
Lead-208, 247
Lead-210, 253
Leap year, 321

Table of Atomic Weights (Based on Carbon-12)

Name	Symbol	Atomic Number	Atomic Weight	Name	Symbol	Atomic Number	Atomic Weight
Actinium	Ac	89	(227)	Mendelevium	Md	101	258.10
Aluminum	Al	13	26.9815	Mercury	Hg	80	200.59
Americium	Am	95	(243)	Molybdenum	Mo	42	95.94
Antimony	Sb	51	121.75	Neodymium	Nd	60	144.24
Argon	Ar	18	39.948	Neon	Ne	10	20.179
Arsenic	As	33	74.922	Neptunium	Np	93	(237)
Astatine	At	85	(210)	Nickel	Ni	28	58.71
Barium	Ba	56	137.34	Niobium	Nb	41	92.906
Berkelium	Bk	97	(247)	Nitrogen	N	7	14.0067
Beryllium	Be	4	9.0122	Nobelium	No	102	259.101
Bismuth	Bi	83	208.980	Osmium	Os	76	190.2
Bohrium	Bh	107	264	Oxygen	O	8	15.9994
Boron	B	5	10.811	Palladium	Pd	46	106.4
Bromine	Br	35	79.904	Phosphorus	P	15	30.9738
Cadmium	Cd	48	112.40	Platinum	Pt	78	195.09
Calcium	Ca	20	40.08	Plutonium	Pu	94	244.064
Californium	Cf	98	242.058	Polonium	Po	84	(209)
Carbon	C	6	12.0112	Potassium	K	19	39.098
Cerium	Ce	58	140.12	Praseodymium	Pr	59	140.907
Cesium	Cs	55	132.905	Promethium	Pm	61	144.913
Chlorine	Cl	17	35.453	Protactinium	Pa	91	(231)
Chromium	Cr	24	51.996	Radium	Ra	88	(226)
Cobalt	Co	27	58.933	Radon	Rn	86	(222)
Copper	Cu	29	63.546	Rhenium	Re	75	186.2
Curium	Cm	96	(247)	Rhodium	Rh	45	102.905
Dubnium	Db	105	(262)	Rubidium	Rb	37	85.468
Dysprosium	Dy	66	162.50	Ruthenium	Ru	44	101.07
Einsteinium	Es	99	(254)	Rutherfordium	Rf	104	(261)
Erbium	Er	68	167.26	Samarium	Sm	62	150.35
Europium	Eu	63	151.96	Scandium	Sc	21	44.956
Fermium	Fm	100	257.095	Seaborgium	Sg	106	(266)
Fluorine	F	9	18.9984	Selenium	Se	34	78.96
Francium	Fr	87	(223)	Silicon	Si	14	28.086
Gadolinium	Gd	64	157.25	Silver	Ag	47	107.868
Gallium	Ga	31	69.723	Sodium	Na	11	22.989
Germanium	Ge	32	72.59	Strontium	Sr	38	87.62
Gold	Au	79	196.967	Sulfur	S	16	32.064
Hafnium	Hf	72	178.49	Tantalum	Ta	73	180.948
Hassium	Hs	108	(269)	Technetium	Tc	43	(99)
Helium	He	2	4.0026	Tellurium	Te	52	127.60
Holmium	Ho	67	164.930	Terbium	Tb	65	158.925
Hydrogen	H	1	1.0079	Thallium	Tl	81	204.37
Indium	In	49	114.82	Thorium	Th	90	232.038
Iodine	I	53	126.904	Thulium	Tm	69	168.934
Iridium	Ir	77	192.2	Tin	Sn	50	118.69
Iron	Fe	26	55.847	Titanium	Ti	22	47.90
Krypton	Kr	36	83.80	Tungsten	W	74	183.85
Lanthanum	La	57	138.91	Uranium	U	92	238.03
Lawrencium	Lr	103	260.105	Vanadium	V	23	50.942
Lead	Pb	82	207.19	Xenon	Xe	54	131.30
Lithium	Li	3	6.941	Ytterbium	Yb	70	173.04
Lutetium	Lu	71	174.97	Yttrium	Y	39	88.905
Magnesium	Mg	12	24.305	Zinc	Zn	30	65.38
Manganese	Mn	25	54.938	Zirconium	Zr	40	91.22
Meitnerium	Mt	109	(268)				

Note: A value given in parentheses denotes the number of the longest-lived or best-known isotope.